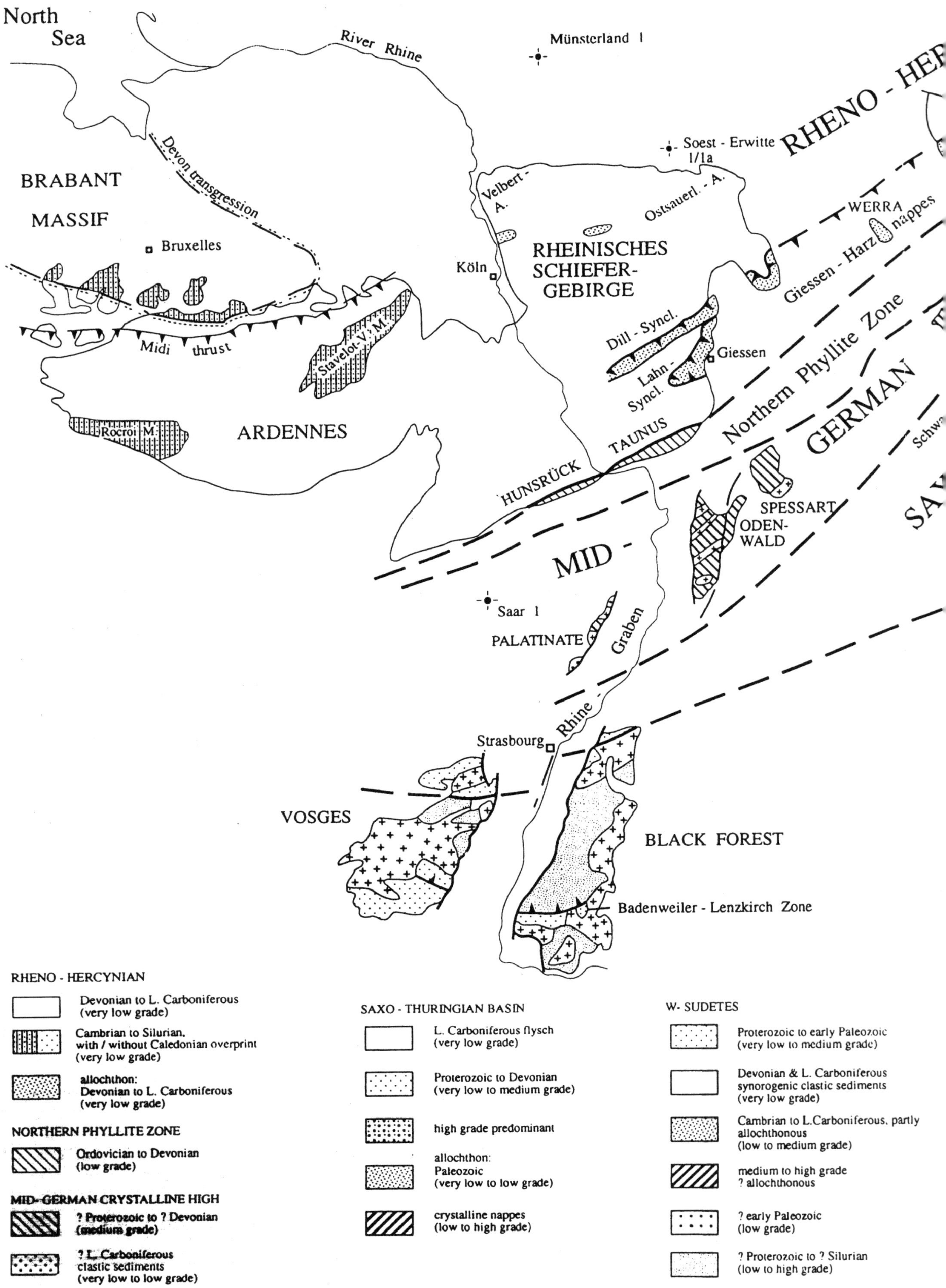
North Sea
River Rhine
Münsterland 1
RHENO - HE
Soest - Erwitte 1/1a
BRABANT MASSIF
Devon transgression
Bruxelles
Velbert - A.
Ostsauerl. - A.
WERRA
nappes
Giessen - Harz
RHEINISCHES SCHIEFER-GEBIRGE
Köln
Dill - Syncl.
Giessen
Lahn - Syncl.
Northern Phyllite Zone
Midi thrust
Stavelot V. M.
Rocroi M.
ARDENNES
TAUNUS
HUNSRÜCK
GERMAN
Schwa
SPESSART
ODEN-WALD
SA
MID -
Saar 1
PALATINATE
Graben
Rhine
Strasbourg
VOSGES
BLACK FOREST
Badenweiler - Lenzkirch Zone
RHENO - HERCYNIAN
Devonian to L. Carboniferous (very low grade)
Cambrian to Silurian, with / without Caledonian overprint (very low grade)
allochthon: Devonian to L. Carboniferous (very low grade)
NORTHERN PHYLLITE ZONE
Ordovician to Devonian (low grade)
MID-GERMAN CRYSTALLINE HIGH
? Proterozoic to ? Devonian (medium grade)
? L. Carboniferous clastic sediments (very low to low grade)
SAXO - THURINGIAN BASIN
L. Carboniferous flysch (very low grade)
Proterozoic to Devonian (very low to medium grade)
high grade predominant
allochthon: Paleozoic (very low to low grade)
crystalline nappes (low to high grade)
W- SUDETES
Proterozoic to early Paleozoic (very low to medium grade)
Devonian & L. Carboniferous synorogenic clastic sediments (very low grade)
Cambrian to L.Carboniferous, partly allochthonous (low to medium grade)
medium to high grade ? allochthonous
? early Paleozoic (low grade)
? Proterozoic to ? Silurian (low to high grade)

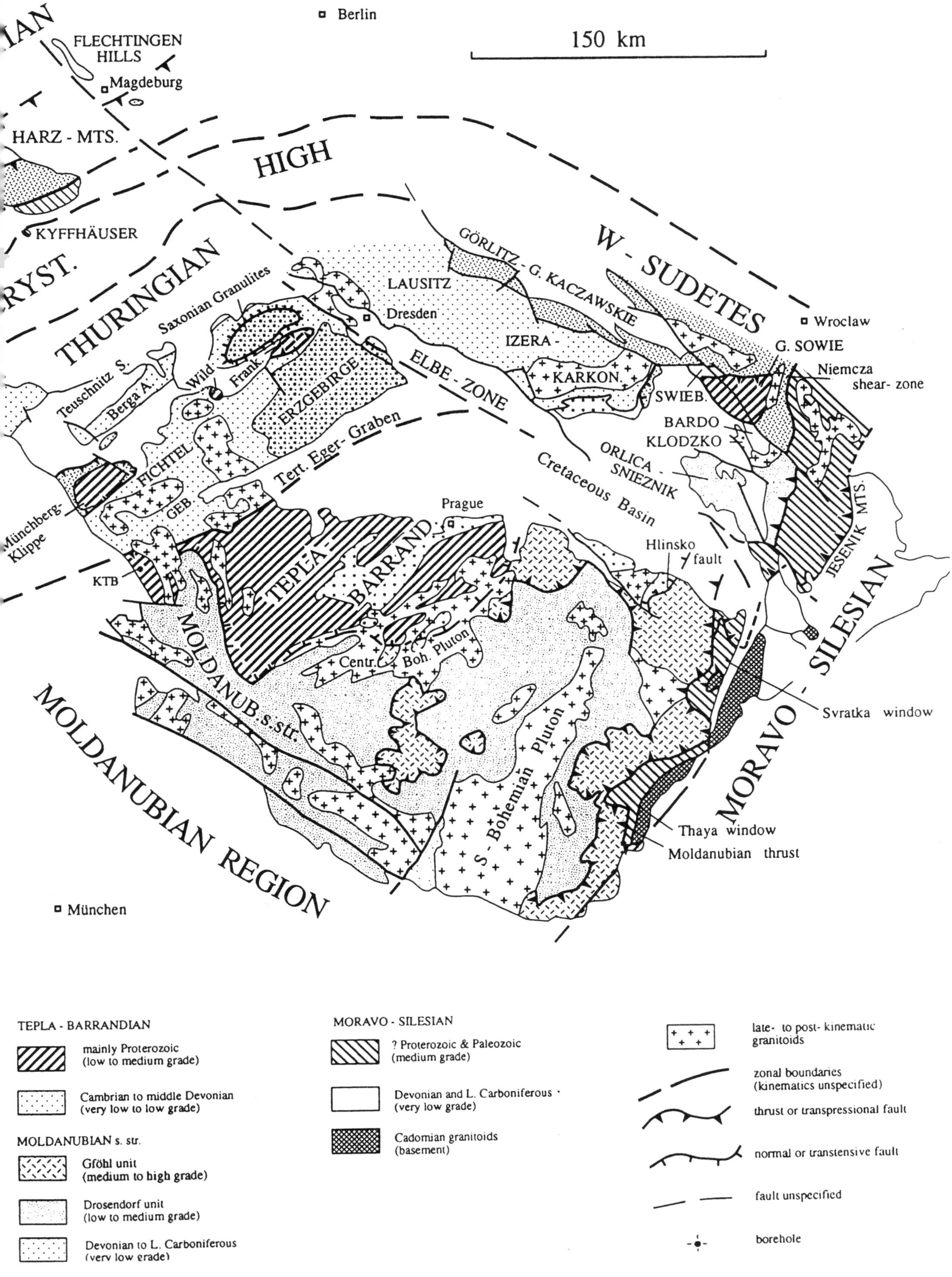
Berlin
150 km
FLECHTINGEN HILLS
Magdeburg
HARZ - MTS.
HIGH
KYFFHÄUSER
RYST.
THURINGIAN
W - SUDETES
GÖRLITZ - G. KACZAWSKIE
LAUSITZ
Dresden
Wroclaw
G. SOWIE
Saxonian Granulites
IZERA -
KARKON.
SWIEB.
Niemcza shear- zone
Teuschnitz S.
Berga A.
Wild
Frank
ERZGEBIRGE
ELBE - ZONE
BARDO
KLODZKO
ORLICA - SNIEZNIK
Tert. Eger - Graben
FICHTEL GEB
Cretaceous Basin
Münchberg-Klippe
Prague
Hlinsko fault
JESENIK MTS.
KTB
TEPLA - BARRAND
MOLDANUB
Centr. Boh. Pluton
MORAVO - SILESIAN
Svratka window
MOLDANUB.s.str.
S - Bohemian Pluton
MOLDANUBIAN REGION
Thaya window
Moldanubian thrust
München
TEPLA - BARRANDIAN
mainly Proterozoic (low to medium grade)
Cambrian to middle Devonian (very low to low grade)
MOLDANUBIAN s. str.
Gföhl unit (medium to high grade)
Drosendorf unit (low to medium grade)
Devonian to L. Carboniferous (very low grade)
MORAVO - SILESIAN
? Proterozoic & Paleozoic (medium grade)
Devonian and L. Carboniferous (very low grade)
Cadomian granitoids (basement)
late- to post- kinematic granitoids
zonal boundaries (kinematics unspecified)
thrust or transpressional fault
normal or transtensive fault
fault unspecified
borehole

Pre-Permian Geology of Central and Eastern Europe

Springer-Verlag Berlin Heidelberg GmbH

Project 233

IUGS UNESCO IGCP

Terranes In The Circum-Atlantic Paleozoic Orogens

R.D. Dallmeyer, Athens, USA
J.D. Keppie, Halifax, Canada
(Directors)

Volumes already published in the series:

R.D. Dallmeyer and E. Martínez García (Eds.):
Pre-Mesozoic Geology of Iberia (1990)
R.D. Dallmeyer and P.P. Lécorché (Eds.):
The West African Orogens and Circum-Atlantic Correlatives (1991)
J.D. Keppie (Ed.):
Pre-Mesozoic Geology in France and Related Areas (1994)
R.D. Dallmeyer, W. Franke, and K. Weber (Eds.):
Pre-Permian Geology of Central and Eastern Europe (1995)

R.D. Dallmeyer W. Franke
K. Weber (Eds.)

Pre-Permian Geology of Central and Eastern Europe

With 233 Figures and 30 Tables

Springer

R.D. DALLMEYER
Department of Geology
University of Georgia
Athens, Georgia 30602-2501
USA

W. FRANKE
Justus-Liebig-Universität
Senckenbergstrasse 3
35390 Giessen
Germany

K. WEBER
Institut für Geologie und Dynamik der Lithosphäre
Universität Göttingen
Goldschmidtstrasse 3
37077 Göttingen
Germany

Cataloging-in-Publication Data applied for

Die Deutsche Bibliothek – CIP-Einheitsaufnahme
Pre-permian geology of Central and Eastern Europe : with 30 tables / R. D. Dallmeyer ... (ed.). – Berlin ; Heidelberg ; New York ; Barcelona ; Budapest ; Hong Kong ; London ; Milan ; Paris ; Tokyo : Springer, 1995

NE: Dallmeyer, Robert D. [Hrsg.]

ISBN 978-3-642-77520-8 ISBN 978-3-642-77518-5 (eBook)
DOI 10.1007/978-3-642-77518-5

Originally published by Springer-Verlag Berlin Heidelberg New York in 1995
Softcover reprint of the hardcover 1st edition 1995

Typesetting: Best-set Typesetter Ltd., Hong Kong

SPIN: 10040688 32/3136/SPS – 5 4 3 2 1 0 – Printed on acid-free paper

Preface

In August 1990, Project 233 of the International Geological Correlation Program hosted an international conference in Göttingen/Giessen, Germany. Discussions were focused on the *Tectonothermal and Stratigraphic Evolution of the Central European Orogens*. The meeting marked the first opportunity for completely open scientific exchange following the recent political reformations in central Europe. This exciting new atmosphere of international cooperation resulted in presentation of a wealth of information which was new to scientists from both sides of former political boundaries. It was apparent that a unique opportunity was available to prepare a systematic overview in a volume dealing with the geology of Central Europe. The present book represents an outgrowth of this conference, but is not merely a compilation of the papers presented in Göttingen/Giessen. Instead, it represents a coordinated volume designed to present a balanced, comprehensive view of our present understanding of the tectonothermal and stratigraphic evolution of the Central European orogens.

We gratefully acknowledge the help of the national funding agencies, who have financed much of the research work summarized in this book, and of the International Geological Correlation Programme (IGCP, project no. 233) which provided and helped to finance the organisational framework.

We are indebted to Springer-Verlag for thorough copy-editing and production of this book, and we sincerely appreciate the efforts of all the reviewers whose comments have greatly helped to improve the quality of this volume. We also thank the various contributors for their diligence and perseverance in manuscript preparation. We are also grateful to be part of the rebirth of scientific collaboration now possible in a formerly divided world.

Athens, Giessen, Göttingen, 1994

R.D. Dallmeyer
W. Franke
K. Weber

Contents

List of Contributors

J. AICHLER
Czech Geological Survey
P.O. Box 65
79001 Jesenik
Czech Republic

H.-J. ANDERLE
Geological Survey of Hessen
Leberberg 9
65193 Wiesbaden
Germany

V. BACHTADSE
Institut für Allgemeine
und Angewandte Geophysik
Ludwig-Maximilians-Universität
Theresienstr. 41
80333 München
Germany

M. BANAŚ
University of Mining and Metallurgy
Al. Mickiewicza 30
30059 Krakow
Poland

P. BLÜMEL
Institut für Mineralogie
Technische Hochschule Darmstadt
Strukturgeologie/Petrographie
Fachbereich 11
Schnittspahnstr. 8
64287 Darmstadt
Germany

J. CHÁB
Czech Geological Survey
Malostranské nám. 19
11821 Prague 1
Czech Republic

J. CHALOUPSKÝ(†)
Department of Geology
Faculty of Science
Charles University
Albertov 6
12843 Prague 2
Czech Republic

I. CHLUPÁČ
Department of Geology
Faculty of Science
Charles University
Albertov 6
12843 Prague 2
Czech Republic

R.D. DALLMEYER
Department of Geology
University of Georgia
Athens, GA 30602-2501
USA

H. DILL
Bundesanstalt für Geowissenschaften
und Rohstoffe
Stilleweg 2
30655 Hannover
Germany

U. DITTMAR
Institut für Geologie
Universität Würzburg
Am Hubland
97074 Würzburg
Germany

A. DUDEK
Department of Geology
Charles University
Albertov 6
12843 Prague 2
Czech Republic

J. Dvořák
Czech Geological Survey
The BRNO Branch
Leitnerova 22
60200 Brno
Czech Republic

F. Falk
Institut für Geowissenschaften
Universität Jena
Burgweg 11
07749 Jena
Germany

J. Fiala
Academy of Sciences
of the Czech Republic
Geological Institute
Rozvojová 135
Suchdol
16500 Prague 6
Czech Republic

F. Finger
Institute of Mineralogy
Institute of Geology and Paläontology
University of Salzburg
Hellbrunnerstr. 34/III
5020 Salzburg
Austria

P.A. Floyd
Department of Geology
University of Keele
Staffordshire ST5 5BG
United Kingdom

B. Fojt
Masaryk University
Kotlàrská 2
61137 Brno
Czech Republic

H.-J. Förster
Zentralinstitut für Physik der Erde
Telegrafenberg
14473 Potsdam
Germany

D. Franke
Waidmannsweg 6
16548 Glienicke
Germany

W. Franke
Institut für Geowissenschaften
und Lithosphärenforschung
Justus-Liebig-Universität
Senckenbergstr. 3
35390 Giessen
Germany

G. Frasl
Institute of Mineralogy
Institute of Geology and Paläontology
University of Salzburg
Hellbrunnerstr. 34/III
5020 Salzburg
Austria

H.J. Franzke
Geologisches Institut
Leibnizstr. 10
38678 Clausthal-Zellerfeld
Germany

A. Frischbutter
GeoForschungs Zentrum Potsdam
Telegrafenberg
14473 Potsdam
Germany

H. Fritz
Institut für Geologie und Paläontologie
Karl Franzens Universität
Heinrichstr. 26
8010 Graz
Austria

G. Fuchs
Geologische Bundesanstalt Wien
Rasumowsky-Gasse 23–25
1030 Wien
Austria

P. Giese
Fachbereich Geowissenschaften
Fachrichtung Geophysik
Freie Universität Berlin
Rheinbabenallee 49
14199 Berlin
Germany

N. Hermsdorf
GeoForschungsZentrum
Telegrafenberg A 27
14473 Potsdam
Germany

G. Hirschmann
Niedersächsisches Landesamt
für Bodenforschung
Stilleweg 2
30655 Hannover
Germany

V. Höck
Institut für Geologie und Paläontologie
Universität Salzburg
Hellbrunnerstrasse 34/III
5020 Salzburg
Austria

F.V. Holub
Department of Petrology
Charles University
Albertov 6
12843 Prague 2
Czech Republic

J. Holubec
Geological Institute
of the Czechoslovak Academy
of Science
Rozvojová 135
16500 Prague 6
Czech Republic

E. Jelínek
Department of Geology
Charles University
Albertov 6
12843 Prague 2
Czech Republic

H. Kemnitz
GeoForschungsZentrum
Telegrafenberg A 27
14473 Potsdam
Germany

M. Klečka
Institute of the Rock
Structure and Mechanics
V Holesovickách 41
18209 Prague 8 – Liben
Czech Republic

Th. Klügel
Institut für Geologie
Universität Würzburg
Am Hubland
97074 Würzburg
Germany

W. Kramer
GeoForschungsZentrum
Telegrafenberg
14473 Potsdam
Germany

H. Kreuzer
Bundesanstalt für Geowissenschaften
und Rohstoffe
Stilleweg 2
30655 Hannover
Germany

A. Krohe
Institut für Mineralogie
Westfälische Wilhelms-Universität
Corrensstr. 24
48149 Münster
Germany

R. Kryza
Wrocław University
Department of Geological Sciences
Cybulskiego 30
50205 Wrocław
Poland

M. Kurze
Bergakademie Freiberg
Institut für Geologie
B.V. Cotta-Str. 2
09596 Freiberg
Germany

U. Linnemann
Institut für Geowissenschaften
Burgweg 11
07749 Jena
Germany

M. Lobkowicz
Geological Survey
Malostranské nám. 19
11821 Prague 1
Czech Republic

J. Mašek
Czech Geological Survey
Malostranské nám. 19
11821 Prague 1
Czech Republic

H.-J. Massonne
Institut für Mineralogie
Ruhr-Universität Bochum
Postfach 104863
44780 Bochum
Germany

D. Matějka
Department of Mineralogy
Geochemistry and Crystallography
Albertov 6
12843 Prague 2
Czech Republic

S. Meisl
Erbsenacker 38
65207 Wiesbaden
Germany

R. Melka
Karl Franzens Universität
Institut für Geologie und Paläontologie
Heinrichstr. 26
8010 Graz
Austria

Z. Mísař
Department of Geology
Charles University
Albertov 6
12843 Prague 2
Czech Republic

K. Mochnacka
University of Mining and Metallurgy
Al. Mickiewicza 30
30059 Krakow
Poland

P. Mrázek
Geological Institute
Czech Academy of Science
Rozvojová 135
Suchdol
16500 Prague 6
Czech Republic

S. Nasir
Department of Geology
UAE University
P.O. Box 17551
Al-Ain
United Arab Emirates

F. Neubauer
Institute of Geology
and Paleontology
University of Graz
Heinrichstr. 26
8010 Graz
Austria

M. Okrusch
Mineralogisches Institut
der Universität Würzburg
Am Hubland
97074 Würzburg
Germany

O. Oncken
GeoForschungsZentrum
Telegrafenberg
14473 Potsdam
Germany

K. Petrakakis
Institute of Petrology
University Vienna
Dr. Karl-Lueger-Ring 1
1010 Vienna
Austria

K. Pošmourný
Geological Survey Prague
Malostranské nám. 19
11821 Prague
Czech Republic

Z. Pouba
Department of Economic Geology
Charles University
Albertov 6
12843 Prague 2
Czech Republic

H. Rauche
Institut für Geologie
Ruhr-Universität
Universitätsstr. 150
44801 Bochum
Germany

W. Schmidt
Institute of Mineralogy
Freiberg University
of Mining and Technology
Brennhausgasse 14
09596 Freiberg
Germany

K. Schulmann
Department of Petrology
Charles University
Albertov 6
12843 Prague 2
Czech Republic

W. Schubert
Mineralogisches Institut
der Universität Würzburg
Am Hubland
97074 Würzburg
Germany

U. Schüssler
Mineralogisches Institut
Universität Würzburg
Am Hubland
97074 Würzburg
Germany

M. Schwab
Institut für Geowissenschaften
Universität Halle
06099 Halle
Germany

E. Seidel
Mineralogisch-Petrographisches Institut
Universität Köln
Zülpicher Str. 49
50674 Köln
Germany

H.C. Soffel
Institut für Allgemeine
und Angewandte Geophysik
Ludwig-Maximilians-Universität
Theresienstr. 41
80333 München
Germany

E. Stein
Mineralogisches Institut
Technische Hochschule Darmstadt
Schnittspahnstrasse 9
64287 Darmstadt
Germany

V. Suchý
Geological Institute
Czech Academy of Science
Rozvojová 135
Suchdol
16500 Prague 6
Czech Republic

J. Synek
Geological Institute
Czech Academy of Sciences
Rozvojová 135
Suchdol
16500 Prague 6
Czech Republic

J.A. Tait
Institut für Allgemeine
und Angewandte Geophysik
Ludwig-Maximilians-Universität
Theresienstr. 41
80333 München
Germany

M. Thöni
Institute of Geology
University of Vienna
Vienna
Austria

G. Tischendorf
Zentralinst. für Physik der Erde
Telegrafenberg
14473 Potsdam
Germany

T.H. Torsvik
Geological Survey of Norway
P.O. Box 3006 Lade
7002 Trondheim
Norway

M. Urban
Department of Geology
Charles University
Albertov 6
12843 Prague 2
Czech Republic

Z. Urbanek
Instytut Nauk Geologicznych PAN
Podwale 75
50449 Wroclaw
Poland

M. Vanecek
Mostecká 18
11800 Prague 1
Czech Republic

Z. VEJNAR
Geological Institute
Czech Academy of Science
Rozvojová 135
Suchdol
16500 Prague 6
Czech Republic

S. VRÁNA
Czech Geological Survey
Malostranské nám. 19
11821 Prague 1
Czech Republic

J. WALDHAUSROVÁ
Geological Institute
Czech Academy of Science
Rozvojová 135
Suchdol II
16500 Prague 6
Czech Republic

H.W. WALTHER
Saaleweg 17
30179 Hannover
Germany

K. WEBER
Institut für Geologie und
Dynamik der Lithosphäre
Universität Göttingen
Goldschmidtstr. 3
37077 Göttingen
Germany

J.I. WENDT
Max-Planck-Institut für Chemie
Saarstr. 23
55122 Mainz
Germany

C.D. WERNER
Institute of Mineralogy
Freiberg University of
Mining and Technology
Brennhausgasse 14
09596 Freiberg
Germany

W. WERNER
Geologisches Landesamt
Baden-Württemberg
Albertstr. 5
79104 Freiburg
Germany

A.P. WILLNER
Institut für Mineralogie
Ruhr Universität Bochum
Postfach 102148
44780 Bochum
Germany

A. ŻELAŹNIEWICZ
Polish Academy of Sciences
Institute of Geological Sciences
Laboratory of Geology of Sudetes
Podwale 75
50449 Worclaw
Poland

I Introduction

R.D. Dallmeyer, W. Franke, and K. Weber

The main exposures of Variscan basement and their internal subdivision are shown by the map on the lining.

Central and Eastern Europe have a polyorogenic history, which may locally be traced back into the early Proterozoic. Much more widespread, though still fragmentary, are remnants of a Cadomian (Panafrican) orogeny (approx. 620–540 Ma). This event probably affected much of Europe, which was then part of Gondwana. During the early Paleozoic, the northern part of Gondwana disintegrated into several microplates, which were separated by marine basins with at least minor amounts of oceanic crust. The northward drift of Gondwana and related northern outboard microplates, and their eventual connection with Laurentia and Baltica, have shaped the complex orogenic collage of the European Variscides.

A first step towards the final assembly was the collision, in Silurian time, of one Avalonian outboard terrane with Baltica. This Caledonian event may be traced in northern sectors of continental Europe, including the Brabant, Rocroi, and Stavelot-Venn Massifs of Belgium, probably continuous with areas along Tornquist's zone (hence: North-German/Polish Caledonides).

After a brief extensional episode in the early to middle Devonian (largely restricted to the Rhenohercynian belt), remaining basins to the south of Avalonia were closed during the late Devonian and earliest Carboniferous. From the early Carboniferous onwards, Variscan Europe underwent a protracted period of intracontinental deformation and metamorphism.

Each of the active margins created during the period of plate convergence developed into a separate orogenic belt. The forelands contain weakly metamorphosed synorogenic clastic sediments, underlain by deposits of the opposed passive margins; the areas of orogenic uplifts are now occupied by metamorphic and plutonic rocks. In spite of numerous heterogeneities, it is possible to trace most of these separate orogens across Europe (e.g., Lotze 1945; Franke and Engel 1982; Holder and Leveridge 1986; Matte 1986; Franke 1989).

In central Europe, the first regional tectonic subdivision was proposed by Kossmat (1927). He was not aware of what is now known as the North-German/Polish Caledonides (the greater part of which are only known from recent drilling activities), but clearly recognized the synorogenic clastic belt at the northern margin of the Rhenish Massif, overlying the deposits of a Rhenohercynian basin. The Saxothuringian zone adjacent to the south is made up of one crystalline belt, which was later termed Mitteldeutsche Kristallinschwelle (Mid-German Crystalline High) by Scholtz 1930 and Brinkmann 1948. This represents the orogenic uplift responsible for the synorogenic clastic sediments in the Rhenohercynian foreland basin to the north. The southern part of Kossmat's Saxothuringian zone is another sedimentary basin which is referred to in this volume as the Saxothuringian basin. Its primary relationships with the Mid-German Crystalline High are largely unknown. The development of the Saxothuringian basin was closely related, at least during the convergent stage, to a crystalline hinterland adjacent to the south (Kossmat's Moldanubian region). Northern parts of this hinterland formed an active margin during the closure of the Saxothuringian basin.

Unmetamorphosed sequences within the Moldanubian crystalline are difficult to interpret. Devonian to Carboniferous sedimentary and volcanic rocks in the Vosges and the Black Forest probably represent a post-collisional, intramontane basin. Paleozoic sequences within the Barrandian synform unconformably overlie Cadomian basement. They may represent: (1) the epicratonic, southern part of the Saxothuringian basin, (2) part of a Barrandian basin proper, or (3) part of a further, cryptic basin within southern sectors of the Moldanubian region. The existence

Dallmeyer et al. (Eds.)
Pre-Permian Geology of Central and Eastern Europe

of such a "Moldanubian basin" may be inferred from nappes, which include eclogites, granulites, and peridotites (Gföhl unit) over less metamorphosed, southern portions of the Moldanubian region (Drosendorf unit). Mafic and ultramafic rocks possibly represent Paleozoic ocean floor. Tollmann (1982), Matte (1986), and Matte et al. (1990) have correlated this Moldanubian basin with remnants of a "Massif Central Ocean" in France.

Pioneers of Variscan geology, like Franz Kossmat, Eduard Suess, and Franz-Eduard Suess not only laid the basis for a geological subdivision, but also published far-sighted tectonic concepts. F.E. Suess (1912) and F. Kossmat (1927) clearly identified long-distance nappe transport in the Rhenohercynian and Saxothuringian zones. These concepts were neglected or even radically discarded for some decades afterwards, but have been rehabilitated by numerous modern studies. Tectonic transport was NW-directed in the Rhenohercynian and Saxothuringian orogens, and appears to have a southeasterly component within the crystalline Moldanubian rocks. The same bilateral symmetry is also expressed in the systematic migrations of Variscan folding, metamorphism, flysch deposition, and partly also active margin vulcanicity. These features reflect bilateral subduction, at least during earlier parts of plate convergence. The Tepla-Barrandian (Bohemian) part of the Moldanubian region, which may be correlated with the Armorican Massif in France, constituted a median massif and forms the divide between the N- and the S-facing parts of the orogen.

During later parts of the collisional history, tectonic patterns changed to dextral strike slip along microplate boundaries, and also along faults which parallel the SW margin of the E-European Platform (Tornquist's Zone). Waning stages of convergence were characterized by widespread HT/LP metamorphism and extensive intrusions of granite, followed by Late Carboniferous to Permian volcanism and renewed extension.

Basement exposures in central Europe do not represent the eroded stumps of a Variscan mountain belt, but had been leveled down already in late Carboniferous/Permian times, and were subsequently covered by Permo-Mesozoic sediments. They owe their renewed exhumation to geodynamic events relating to the Alpine collision (e.g., Ziegler 1987). Late Cretaceous reverse faulting has exhumed the Harz Mts. and large parts of the Bohemian Massif; the central part of the Rhenish Massif straddles an active asthenospheric bulge, and the Black Forest and Vosges are shoulder uplifts of the Rhine Graben.

This volume has been organized to present a systematic description of the major Variscan tectonostratigraphic zones in Central Europe. Basement outcrops in Southwest Germany (Vosges and Black Forest) have not been included because they are separated from the Bohemian massif by a large area of Mesozoic cover, and are easier to correlate with the French Massif Central. Following an introductory chapter reviewing regional geophysical characteristics, each Variscan tectonic zone is discussed in a separate chapter which begins with an introductory statement outlining its overall characteristics, location, boundaries, history of deformation, and major points of controversy. Following are systematic descriptions of the stratigraphic, paleontological, structural, igneous, metamorphic, and metallogenetic characteristics. The Caledonian record in central Europe is reviewed in Chapter IX. Paleomagnetic constraints and the overall geodynamic evolution of central Europe are discussed in the last two chapters.

References

Brinkmann R (1948) Die Mitteldeutsche Schwelle. Geol Rundsch 36: 65–66

Franke W (1989) Tectonostratigraphic units in the Variscan belt of central Europe. Geol Soc Am Spec Pap 230: 67–90

Franke W, Engel W (1982) Variscan sedimentary basins on the Continent and relations with Southwest England. Proc Ussher Soc 5: 259–269

Holder MT, Leveridge BE (1986) Correlation of the Rhenohercynian Variscides. J Geol Soc Lond 143: 141–147

Kossmat F (1927) Gliederung des varistischen Gebirgsbaues. Abh Sächs Geol Landesamt 1: 39

Lotze F (1945) Zur Gliederung der Varisziden der Iberischen Meseta. Geotekton Forsch 6: 78–82

Matte Ph (1986) Tectonics and plate tectonics model for the Variscan belt of europe. Tectonophysics 126: 329–374

Matte Ph, Maluski H, Rajlich P, Franke W (1990) Terrane boundaries in the Bohemian Massif: result of large-scale Variscan shearing. Tectonophysics 177: 151–170

Scholtz H (1930) Das varistische Bewegungsbild, entwickelt aus der inneren Tektonik eines Profils von der Böhmischen Masse bis zum Massiv von Brabant. Fortschr Geol Paläontol 8 (25): 235–316

Suess FE (1912) Vorläufige Mitteilung über die Münchberger Deckscholle. Sitzungsber Akad Wiss Wien, mathematisch-naturwissenschaftliche Klasse, 121 (10): 253

Tollmann A (1982) Großräumiger variszischer Deckenbau im Moldanubikum und neue Gedanken zum Variszikum Europas. Geotekton Forsch 64: 1–91

Ziegler P (1987) Late Cretaceous and Cenozoic intraplate compressional deformations in the Alpine foreland – a geodynamic model. Tectonophysics 137: 389–420

II Geophysical Characteristics

II.A Main Features of Geophysical Structures in Central Europe

P. Giese

1 Introduction

Over the last four decades, the upper lithosphere of central Europe has been investigated intensively by geophysical studies applying a wide spectrum of different methods, including deep seismic refraction and reflection profiling, magnetotelluric experiments, and mapping of the anomalies of the magnetic and gravity field. A large number of detailed and review publications are available dealing with the crustal structure, its composition and evolution. The most recent reports resulting from the European Geotraverse (EGT) project have been published within the series of European Science Foundation Proceedings (Freeman and Mueller 1990; Freeman et al. 1990; Blundell et al. 1992).

The Phanerozoic crust of central and western Europe is clearly distinguished from the Precambrian crust of northern and eastern Europe by some prominent geophysical features. An average crustal thickness of only 28–32 km is typical for central and western Europe. This is in sharp contrast with an average crustal thickness of 40–50 km in northern and eastern Europe (Meissner et al. 1987). Further on in northern and eastern Europe, magnetic anomalies show a more irregular and larger scale pattern, which probably reflects subdivision into major crustal blocks. In western Europe the anomalies are smaller.

2 Structural Boundaries of Central Europe

The eastern margin of central Europe is formed by the Tornquist-Teisseyre Zone (Trans-European Suture Zone), which separates the mobile belts of western Europe from the stable platform of eastern and northern Europe (Figs. 1 and 2). This prominent zone starts between Denmark and SW Sweden, continues to Poland, passes east of the Carpathian Mountains, and extends up to the western Ukraine. This fault zone is characterized by an anomalously thick crust, as shown in the Moho contour map of Fig. 2.

The southern border of central Europe is defined by the mountain belts of the Northern Carpathians, the Eastern Alps and the Swiss Alps, which were created during the closure of the Tethys ocean. Near these mountain belts, crustal thickness increases from 30–35 km in the northern foreland to more than 50 km beneath the axes of these young mountain belts.

The western margin of central Europe is defined by a Cenozoic rift system, which extends from the Upper Rhine Graben via the Middle Rhine segment and the Lower Rhine embayment into the North Sea rift.

3 Crustal Thickness in Central Europe

Crustal thickness is one of the basic parameters for all discussions on crustal evolution. For about 40 years, the Earth's crust and upper mantle in central Europe have been investigated by deep seismic sounding studies. This activity is documented in a large number of publications and internal reports, describing the data and discussing the results obtained. In this chapter, only the main references can be cited, which may be used as further data sources.

The first data compilation of DSS data in central Europe was published by the German Research Group for Explosion Seismology (1964). A comprehensive data collection comprising all record sections recorded up to 1972 can be found in the monograph edited by Giese et al. (1976). Mostaanpour (1984) compiled typical crustal parameters for the whole western European area. The seismic refraction measurements culminated in the seismic long-range survey along the central segment of the EGT (EUGEMI working group 1990; Thybo 1990; Thybo et al. 1990; Aichroth et al. 1992; Blundell et al. 1992).

Dallmeyer et al. (Eds.)
Pre-Permian Geology of Central and Eastern Europe

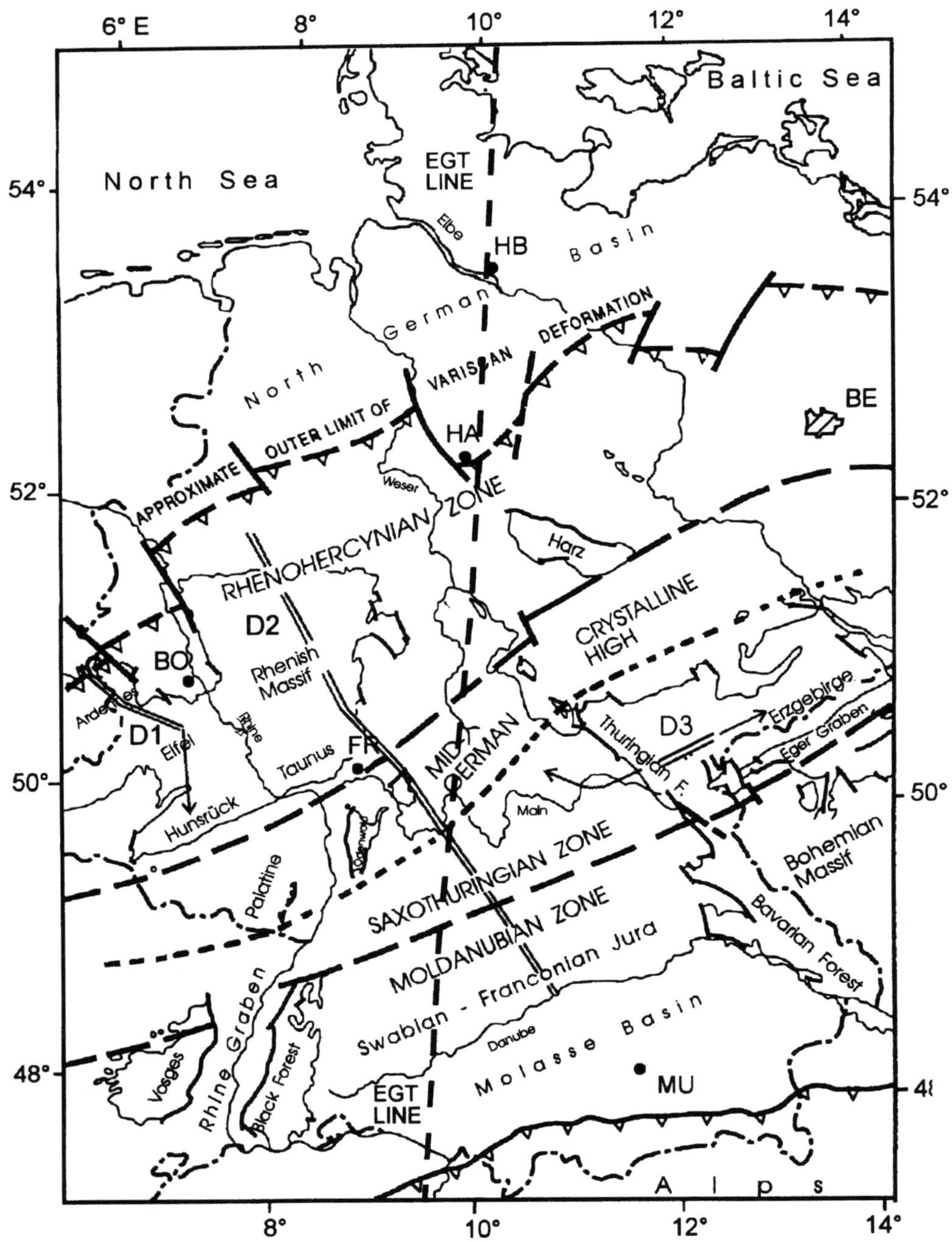

Fig. 1. Map showing the main morphotectonic regions and the Variscan zones in central Europe. (After Franke 1992.) *Dashed lines* indicate the position of the DEKORP-2 profile and the EGT section. *BE* Berlin; *BO* Bonn; *FR* Frankfurt/Main; *HA* Hannover; *HB* Hamburg; *MU* Munich

Further extensive seismic refraction studies, together with seismic reflection measurements, were carried out in the Black Forest and the Oberpfalz as a pre-site survey for the German Deep Continental Drilling Program (KTB) (Gajewski and Prodehl 1987; DEKORP Research Group 1988; Gebrande et al. 1989; Lüschen et al. 1989; Schmoll et al. 1989). Further detailed seismic investigations have covered the Rhenish Massif (Mooney and Prodehl 1978; Bartelsen et al. 1982; Mechie et al. 1983; Gajewski and Prodehl 1985; Gajewski et al. 1987) and southwest Germany (Lüschen et al. 1987; 1990; Holbrook et al. 1988; Gajewski 1989; Zeis et al. 1990). Crustal parameters for eastern Germany were presented by Bormann et al. (1987, 1989) and Schulze

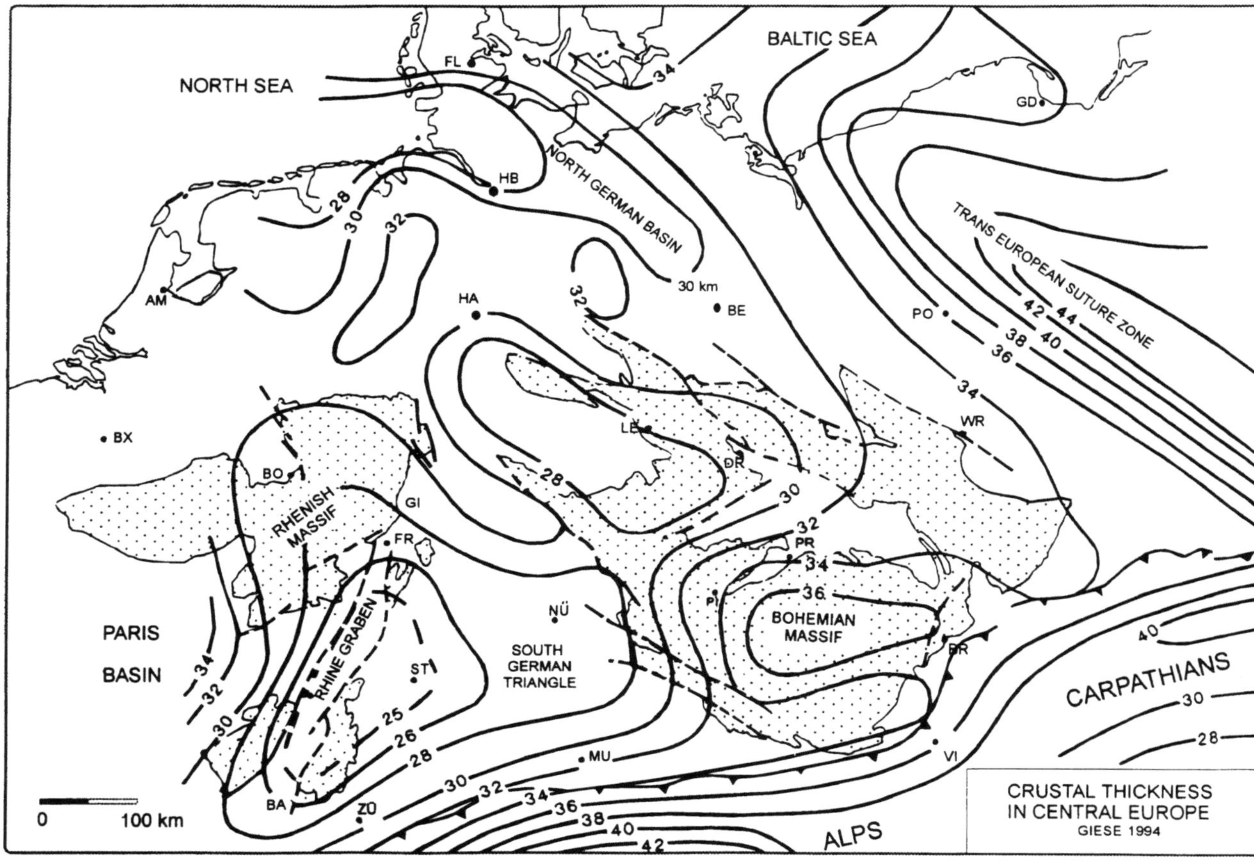

Fig. 2. Contour map of crustal thickness in central Europe. The accuracy is about ±1 km. The spacing of the contour lines is 2 km. The data sources are found in the text. *BA* Basel; *BE* Berlin; *BO* Bonn; *BR* Brno; *DR* Dresden; *FR* Frankfurt/Main; *HA* Hannover; *HB* Hamburg; *LE* Leipzig; *MU* Munich; *NÜ* Nürnberg; *PO* Poznan; *PR* Prague; *ST* Stuttgart; *WI* Vienna; *WR* Wroclaw; *ZÜ* Zürich

and Bormann (1990). Guterch et al. (1986) published data on crustal structure in Poland, while Ibrmayer et al. (1983) presented results of deep seismic sounding studies in the Bohemian Massif and adjacent regions. The first crustal data of the Netherlands were reported by Remmelts and Duin (1990). Further details and data sources which have been used for this compilation are available in Giese (1990), Giese and Prodehl (1986), Aichroth and Prodehl (1990), Prodehl and Giese (1990) and Prodehl and Aichroth (1992).

Whereas earlier seismic soundings were mainly based upon deep seismic refraction experiments, deep seismic reflection studies became more and more important from 1980 onwards, and revealed details of the internal structures which were hitherto unknown. The results of the German deep seismic reflection program (DEKORP) are documented in an atlas showing the seismic cross section of the DEKORP network (Meissner and Bortfeld 1990). Detailed interpretations of these lines are presented in a number of publications (DEKORP Research Group 1985, 1988, 1990, 1991, 1994; Reichert 1988; several papers in Meissner et al. 1991).

Figure 2 presents the most recent contour map of crustal thickness compiled from sources cited above. The main features of the morphology of the crust/mantle boundary are discussed in the following section.

3.1 North-Central Europe

The northern part of central Europe comprises mainly the North German Basin with its extension into northern Poland. Although in the North German Basin, the thickness of the sedimentary cover measures up to 10 km, the base of the crust has been detected at depths between 26 and 30 km. In the region of the Schleswig-Holstein gravity high, the crust/mantle boundary shows a slight elevation. A second prominent gravity high occurs in Mecklenburg, and it is debated whether the Moho discontinuity in this "Pritzwalk gravity high" has been uplifted or not. Although the gravity data strongly suggest crustal thinning in this area (Bachmann and Grosse 1989), this interpretation is not supported by seismic reflection data (Horst, pers. comm.). This gravity high is probably caused by a body of high density intruded into the lower crust. Since the Permo-Carboniferous North-German/Polish Basin records crustal extension in the area, a mafic body in the lower crust could be attributed to this process. Because crustal thickness across Germany is roughly uniform, it must be concluded that post-Variscan subsidence and sedimentation in the North-German Basin are compensated by a corresponding crustal thinning. A distinct thickening of the crust is observed eastwards across the Tornquist-Teisseyre suture zone.

3.2 Middle Part of Central Europe

In central Germany, the crustal thickness is rather uniform and ranges between 28 and 30 km. A thickening of the crust is observed on the western side towards the Paris Basin as well as on the eastern margin E of the Tornquist-Teisseyre suture zone. A slightly reduced crustal thickness is observed beneath the Thuringian Basin.

3.3 South-Central Europe

In the southern part of central Europe, the Moho depth contours allow differentiation between the following areas:

- The Rhine Graben is characterized by a distinctly reduced crustal thickness along the axis of the graben and its shoulders. The rise of the crust/mantle boundary appears to be asymmetric. Beneath the eastern flank, the Black Forest, the crust is thinner than beneath the western flank, the Vosges and Palatinate (Meier and Eisbacher 1991; Wenzel et al. 1991).
- The block of the Bohemian Massif shows, in its central part, a maximum crustal thickness of 36 km.
- Along the southern margin of central Europe, the crust/mantle boundary dips down beneath the Alps and the Northern Carpathians. It descends from 28 km north of the Molasse basin to 38 km beneath the Northern Calcareous Alps. This increase can be attributed mostly to the thickening of the sedimentary fill in the Molasse Basin: this behavior reflects flexural bending of the foreland crust caused by the tectonic edifice of the Alps.

Taken altogether, crustal thickness appears mainly to reflect post-Variscan processes such as Permo-Carboniferous extension and Cenozoic rifting; there is no distinct remnant of a Variscan orogenic "root" visible. The only indication of a plate boundary is the increase in crustal thickness across the Trans-European Suture Zone.

4 Bouguer Gravity Anomalies

A new map showing the Bouguer anomalies of central Europe (Fig. 3) compiled by Grosse et al. (1990) reveals very clearly a number of regionally important intracrustal inhomogeneities which are closely related to tectonic structures, but are partly hidden beneath young sedimentary cover. The following description of gravity anomalies is mainly based on Grosse et al. (1990).

4.1 North-Central Europe

In southern Denmark, the Ringkøbing-Fyn gravity high is caused by basement uplifts. A further anomaly is situated in the North German Basin (high of Schleswig-Holstein und Pritzwalk, Mecklenburg). It is puzzling, at first sight, that

Fig. 3. Contour map of Bouguer gravity compiled by Grosse and Conrad (1990). The spacing of the isolines is 5 mgal. *BE* Berlin; *BO* Bonn; *BR* Bremen; *DO* Dortmund; *DR* Dresden; *ER* Erfurt; *FB* Freiburg; *FL* Flensburg; *FR* Frankfurt/Main; *HA* Hannover; *HB* Hamburg; *GÖ* Göttingen; *LE* Leipzig; *LI* Linz; *MA* Magdeburg; *MU* Munich; *NÜ* Nürnberg; *PR* Prague; *RE* Regensburg; *ST* Stuttgart; *ZÜ* Zürich; *RF* Ringkøbing-Fyn high; *NGB* North German Basin; *STZ* Sorgenfrei-Tornquist zone; *TTZ* Tornquist-Teisseyre zone; *SHB* South Hercynian basin; *PW* Pritzwalk high; *BH* Bramsche high; *SS* Saverne-Sarrebourg high; *RM* Rhenish Massif; *KG TB* Kraichgau-Tauber high; *MOLD* Moldanubian zone; *SAX* Saxothuringian zone; *RHENO* Rhenohercynian zone; *BM* Bohemian massif; *EG* Eger Graben; *S-H* Schleswig-Holstein high; *MB* Magdeburg high; *VL* Vlotho high

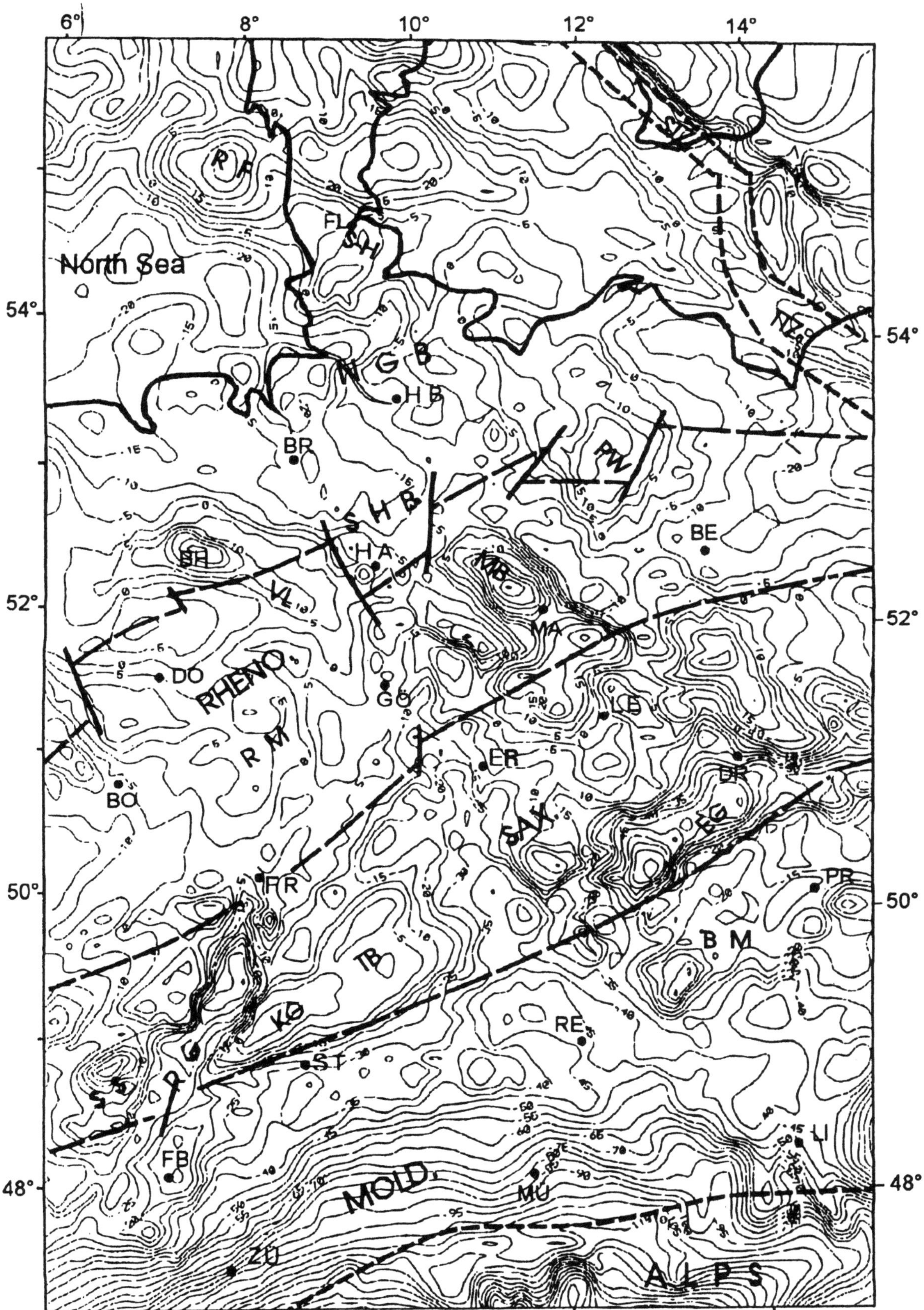
North Sea
RF
FL
SH
NGB
HB
BR
SHB
HA
BH
VL
PW
BE
KB
MA
DO
RHENO
GO
LE
RM
ER
DR
BO
SAX
EG
FR
PR
BM
TB
KO
RE
ST
FB
LI
MOLD.
MU
ZU
ALPS
NZ

the axis of these positive anomalies correlates with the depocenters of the North German Basin in Schleswig-Holstein and Mecklenburg. Models derived from seismic refraction and gravity data show a substantial relief on the Moho and suggest a large volume of mafic intrusions (Bachmann and Grosse 1989). These authors proposed that these features were developed in response to E-W-directed crustal extension which brought about the onset of subsidence and sedimentation in the North German Basin.

The southern margin of the North German Basin is defined by the North German Line, defined by Bachmann and Grosse (1989) in the stripped Bouguer map of the North German Basin (Hermes 1986). Along the southern margin of the gravimetrically "quiet" region in northern Germany, there are a number of gravity highs (the concealed "massifs" of Bramsche-Vlotho and Magdeburg), whose southern margin is used to define the "North German Line". This string of largely positive anomalies is oriented "en echelon" with the North German Line. Thermal aureoles permit identification of these "massifs" as early Cretaceous, mid-crustal intrusive bodies. Their emplacement, as well as Jurassic to early Cretaceous subsidence of the sub-Hercynian Basin are the expression of transcontinental shear belt. According to Conrad (1980) and Schubert and Conrad (1981), this structure extends from southern Poland to southern England.

4.2 Middle and Southern Part of Central Europe

The Rhenish massif shows only a weak positive anomaly. The most pronounced gravity anomaly in central Germany is the gravity minimum in the southeastern part of the Saxothuringian belt following the NE-trending axis of the Erzgebirge. This extraordinary gravity low with a minimum of − 76 mgal near Karlovy Vary does not correspond to any Moho depression (Figs. 2 and 3). Hence, it must be attributed to low density material, e.g., a belt of late-Variscan granites at deeper crustal levels (Grosse et al. 1961).

The southwestern region of central Europe is characterized by the well-defined gravity minimum of the Upper Rhine Graben. This minimum is largely attributable to the sedimentary infill of the Tertiary rift structure. In addition, positive axes along both flanks of the graben show up remarkably clearly in the stripped map by Rousset and Bayer (1990). Correlation of the Saverne-Sarrebourg (Vosges) gravity high with the southwestern part of the Kraichgau-Tauber anomaly, which is found on the eastern side of the Rhine Valley, suggests a 30-km left-lateral Permo-Carboniferous offset along a preexisting Upper Rhine lineament (Edel and Fluck 1989; Rousset and Bayer 1990).

The Kraichgau-Tauber anomaly, a broad, NE-trending, low-amplitude gravity high is the most prominent intracrustal feature in the South German Triangle. It occupies the southern part of the Saxothuringian zone, and its southern margin roughly coincides with the Saxothuringian-Moldanubian boundary. A possible equivalent of the Kraichgau high occurs near Hof. It corresponds to a dome-like structure at shallow depth revealed by seismic reflection (see below) and probably relates to the metamorphic core complex of the Saxonian granulites (Franke et al. 1994).

South of the Swabian-Franconian Jura, the increasingly negative gravity values are caused by the thickening of the Molasse basin and the downdip of the crust/mantle boundary towards the Alpine front.

Careful inspection of the Bouguer map of central Europe reveals a line running from Helgoland in the north via Göttingen down to Regensburg in the south. It separates two domains with different patterns of anomalies: the eastern block is dominated by NW-trending gravity axes, with only subordinate elements parallel to the Variscan structural trend (SW-NE); the Variscan trend is very apparent in the western block (Grosse et al. 1990). Vollbrecht et al. (1989) and Edel and Weber (1993) interpret this line as a late-Variscan dextral shear zone. In fact, important NW-trending faults, partly with co-related pull-apart basins, are known to have been formed in Permo-Carboniferous time; however, the Variscan structural zonation, as known from surface outcrops and boreholes, does not reveal any significant offset along the line in question.

5 Magnetic Anomalies in Central Europe

Figure 4 shows an aeromagnetic map for central Europe recently compiled by Wonik and Hahn (1989, 1990). These authors attempted a geo-

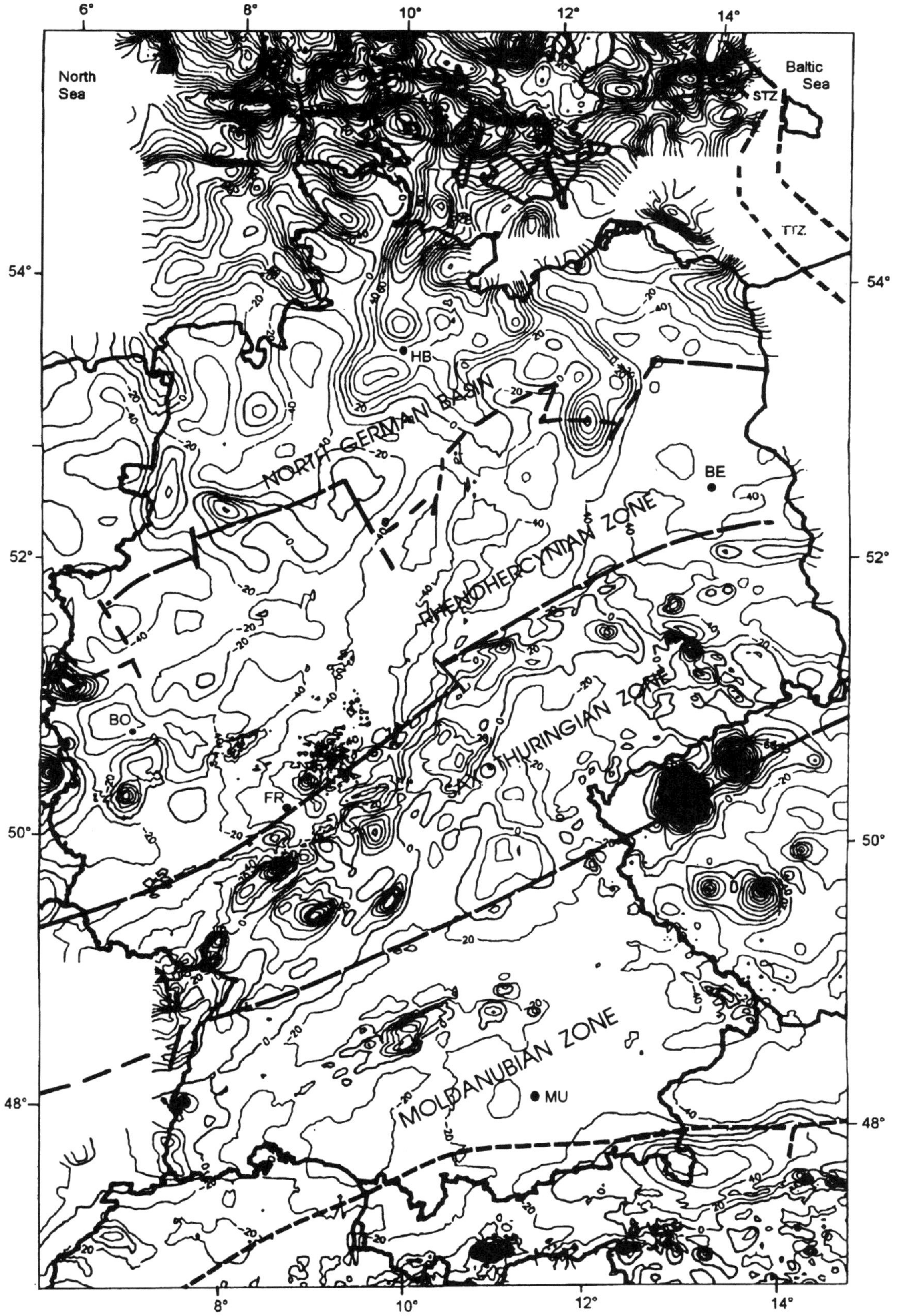

Fig. 4. Contour map of anomalies of total magnetic intensity. Altitude 3000 m above sea level, reference field DGRF 1980, Epoch 1980, 0; contour line spacing is 20 nT. (Wonik and Hahn 1990)

logical interpretation of the most important anomaly groups in central Europe. The present chapter focuses on the interpretation of longer wavelength anomalies.

Three types of sources may be responsible for the magnetic anomalies. Firstly, anomalies of unknown character can be seated in the concealed pre-Variscan basement. Secondly, anomalies of varying wavelength may be caused by Variscan and late-Variscan mafic intrusions and extrusions. Finally, a large number of anomalies, mostly concentric and high-amplitude, are related to Cenozoic volcanic activities, which took place in the central and southern parts of central Europe.

The map of aeromagnetic anomalies (Fig. 4) shows from north to south a number of zones with distinct patterns. North of the Tornquist-Teisseyre line, in southern Denmark and southern Sweden, the pattern consists of irregular, short-wavelength anomalies which can be attributed to basement rocks at outcrop or in the subsurface.

Southwards, the North German Basin is dominated by a series of anomalies extending from Schleswig-Holstein in the northwest to Pritzwalk (Mecklenburg) in the southwest. These anomalies coincide roughly with those of the gravity field (Fig. 3), suggesting that magmatic rocks of higher magnetization intruded mid- and/or deep-crustal levels during the late Variscan extensional processes. Only weak magnetic anomalies are recognizable in the southern part of the North German Basin.

The region east of the Tornquist-Teisseyre suture zone is again characterized by well expressed magnetic anomalies. A number of local anomalies are present in the Rhenish massif and its surrounding areas which are related to Devon and early Carboniferous as well as Cenozoic volcanic rocks. The most pronounced magnetic anomalies are concentrated along a SW-NE trending zone in the middle part of central Europe which coincides roughly with the Saxothuringian zone of the Variscan belt. This zone can be traced from the Paris basin in the west up to the Elbe line in Saxony, where it is probably displaced by a dextral offset of ca. 50 km (Franke et al., Chap. V.B.2, this Vol.). These anomalies probably relate to Variscan magmatic bodies, such as granodiorites at the SE flank of the Odenwald and Spessart Mountains.

In the southern part of central Europe, there are two prominent regions with magnetic anomalies. The young volcanism in the Eger Graben and its surroundings in northwestern Bohemia has generated very pronounced, more or less concentric magnetic anomalies. It should be noted that the Eger Graben reactivates the Saxothuringian-Moldanubian suture zone. SW-NE trending magnetic anomalies are recognizable in the Swabian Jura (NW of Munich), which are attributed to volcanic and subvolcanic rocks of Tertiary age as well.

A further zone of magnetic anomalies is found in the northern Alps. In the northern Calcareous Alps, the most prominent magnetic high occurs in the region of Berchtesgaden. Its westernmost end starts near the river Isar (south of Munich) and extends to the Vienna basin in the east where it continues along the Northern Carpathians. This anomaly is caused by magnetized bodies being situated with their upper surface at a depth of 6 to 10 km. It is still being debated whether these bodies form the base of the north-Alpine allochthon or already belong to the autochthonous basement.

6 A Crustal Cross Section Between the North German Basin and the Alps

In this section, the main features of crustal structure are described along a crustal section of the central part of the European Geotraverse (EGT) running between the North German Basin in the north and the northern Alps in the south (Fig. 5a,b).

The central European segment of the European Geotraverse between the northern margin

Fig. 5. *Upper section* Crustal section along the EGT central segment between the shot point K near Kiel and shot point D, located in the northern margin of the Swiss Alps (Aichroth and Prodehl 1990; Prodehl and Giese 1990; Prodehl and Aichroth 1992). The contour interval of the isovelocity lines is 0.1 km/s. Discontinuous velocity changes are indicated by *thick lines* with two velocity values *above* and *below* the boundary; *Lower section* Crustal section along the central segment of the EGT with a petrological interpretation which is based on seismic, gravimetric, magnetic, geological, and xenolith data. The *vertical columns* indicate locations (North Hessen, Heldburg, and Urach where xenolith data are available from young volcanic eruptions (Mengel and Kern 1990). The subdivision of the Variscan mountain system in central Europe as proposed by Kossmat from geological considerations is also confirmed by the block-like structure and composition of the Earth's crust

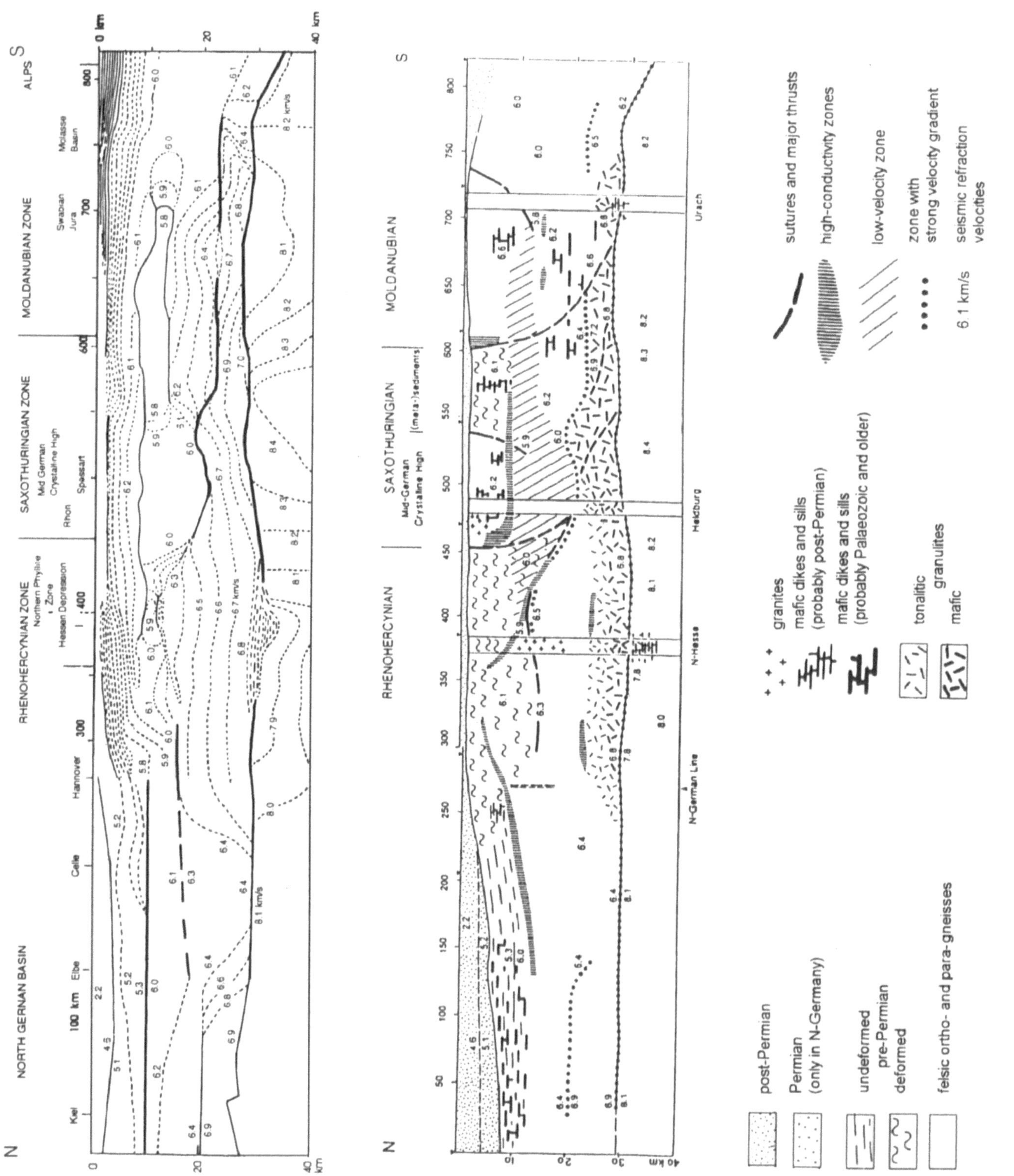
N
S
NORTH GERMAN BASIN
RHENOHERCYNIAN ZONE
SAXOTHURINGIAN ZONE
MOLDANUBIAN ZONE
ALPS
Kiel
Elbe
Celle
Hannover
Northern Phyllite Zone
Hessen Depression
Rhon
Spessart
Mid German Crystalline High
Swabian Jura
Molasse Basin
RHENOHERCYNIAN
SAXOTHURINGIAN
Mid-German Crystalline High
(meta-)sediments
MOLDANUBIAN
N-German Line
N-Hesse
Heldburg
Urach
post-Permian
Permian (only in N-Germany)
undeformed
pre-Permian
deformed
felsic ortho- and para-gneisses
granites
mafic dikes and sills (probably post-Permian)
mafic dikes and sills (probably Palaeozoic and older)
tonalitic
granulites
mafic
sutures and major thrusts
high-conductivity zones
low-velocity zone
zone with strong velocity gradient
6.1 km/s
seismic refraction velocities

of the Alps and the Baltic Sea near Kiel was the target of a detailed seismic refraction survey in 1986 (EUGEMI Working Group 1990; Aichroth and Prodehl 1990; Aichroth et al. 1992; Prodehl and Aichroth 1992). In addition, deep-seismic reflection studies between the Nördlinger Ries and the northern margin of the Rhenish Massif (DEKORP Research Group 1985, 1990, 1991, 1994) have revealed particular structural details within the crust. In some areas, the velocity distribution has been revealed by additional wide-angle measurements (Bittner et al. 1987; Giese 1990). The following description is mainly based on reports published by Prodehl and Giese (1990) and Prodehl and Aichroth (1992).

The crust/mantle boundary rises from 34 km depth under the northern margin of the Alps to 28 km under the Swabian-Franconian Jura. Further north, the crustal thickness varies only slightly around and average depth of about 29 km (with an error of ± 2 km). The Moho depth under the North German Basin is about 30–32 km. Between the Elbe line and the Ringkϕbing-Fyn high the Moho rises to a depth of about 26 km and the data indicate at least one fault-like step of 1–2 km before the crust thickens toward this basement high.

The P-wave data were interpreted in detail between the Molasse basin and the southern end of the North German Basin and a two-dimensional P-velocity model of the crust was derived (Fig. 5a) applying ray-tracing and ray-theoretical seismogram calculations for laterally inhomogeneous media. For the North German Basin the data quality justifies only the construction of a simplified model (Fig. 5a). Data and model are described in detail by Aichroth et al. (1992). Here, the results are only briefly summarized.

The internal crustal structure is quite complex. In the upper crust of the Moldanubian and Saxothuringian zones of the Variscan orogen, the velocities are lower than in the Rhenohercynian area. A velocity of 6 km/s is reached at a depth of 12 km under the northern margin of the Alps and at 7 km depth near the river Main. Under the Mid-German Crystalline Rise the 6.0 and 6.2 km/s isolines are observed to rise, in accordance with the presence of high-grade rocks at the surface. Northwards, i.e., towards the low-grade Rhenohercynian belt, the 6 km/s isoline descends again to a depth of 8 km under the Hesse depression and reaches 10 km depth under the North German Basin. The northward slope of the isolines around Hannover roughly corresponds to the North German Line derived from the gravity map and relates to Mesozoic wrenching, magmatism, and subsidence (see above).

The middle crust of the central segment is characterized by velocities between 5.8 and 6.1 km/s at a depth range between 7 and 20 km. In some parts, the middle crust contains a weak velocity reversal, e.g., between the shot points F and G from 7 to 12 km depth, while between the shot points G and H an average velocity of about 6 km/s is maintained from 8 to 20 km depth.

The structure of the lower crust varies considerably along the central segment. Towards the Alps, the average velocity of the consolidated crust (the crust between the top of the basement and the crust/mantle boundary) is as low as 6.2 km/s, in contrast to the area north of shot point E, where the velocities above the Moho vary between 6.8 and 7.2 km/s. Only between the shot points F and G is the lower crust subdivided by a boundary at 20–23 km depth, which disappears again between shot points H and J where the velocity increases gradually from 6.3 km/s at 13 km to 6.8 km/s at 29 km depth. The well-developed intracrustal discontinuity in the northern Saxothuringian zone may be seen as a relict of Variscan crustal doubling. Between Hannover and the river Elbe (Fig. 5a) the velocity of the lower crust is about 6.4 km/s. North of the Elbe line, a velocity of 6.9 km/s is observed. The velocities in the uppermost mantle show considerable variation. The highest velocity of 8.4 km/s is observed beneath the Saxothuringian zone and the lowest values of 7.8–7.9 km/s are recorded under the Hesse depression.

In summary, the velocity distribution reveals roughly some Variscan structures and sutures. Some elements, such as the northward velocity slope at the North German Line and the mid-crustal low-velocity zone, clearly post-date the Variscan orogeny. It cannot be excluded, however, that the higher velocities in the lower crust and mantle of the Saxothuringian belt represent eclogites derived from the southward subduction of Rhenohercynian ocean floor.

7 DEKORP Seismic Reflection Profiles Across the Variscan Orogen

The seismic reflection lines DEKORP 2-North and 2-South (D2) reveal an almost complete cross section through the Variscan orogen between the Northern German Basin and the Alpine Front (Fig. 6). The positions of this line and portions of the lines DEKORP 1 (D1) and DEKORP 3 (D3) are shown in Fig. 1. In the northern part of this line (Fig. 6, upper section), the pattern of seismic reflecting elements can be characterized as follows (DEKORP Research Group 1990):

- upper crust (0–3 s TWT): highly reflective, with numerous elements correlatable with bedding or faults at the surface;
- middle crust (3–6 s TWT): less reflective; reflections more or less straight and subhorizontal. At the top and at the base of this zone, there are levels with especially strong and laterally consistent bands of reflections;
- lower crust (below 6 s TWT): relatively transparent in the north, increasingly reflective towards the south.

This northern part of DEKORP-2-N reveals the Cretaceous of the Münsterland basin and the underlying folded Paleozoic rocks. The northward-decreasing intensity of folding is depicted in great detail by the highly reflective Late Carboniferous coal beds and deeper reflections down to the level of the Devonian carbonates. The lower crust in the northern part of the DEKORP line is characterized by a relatively transparent zone, which wedges out toward the south under the northern margin of the Siegen anticline. Comparisons with a similar feature in the ECORS profile "Nord de la France" suggest that the transparent zones in both sections correspond to a pre-Paleozoic basement, such as it underlies the Brabant Massif. Further south, the lower crust is increasingly reflective, probably because it has been affected by Devonian Extension.

In the Devonian and older rocks of the Rhenish Massif further south, the seismic profile is characterized by stronger, southward dipping reflections, which exhibit the listric curvature and flat-ramp geometry characteristic of faults. In the northern part of the section, the thrusts appear to be blind. From the Ebbe anticline southwards, prominent reflections can be correlated with important thrust faults known from the surface, such as the Ebbe, Siegen, Müsen and Sackpfeife thrusts, as well as further important thrust faults in the Lahn and Dill synclines.

At the base of the upper crust (about 10 km depth), the thrusts flatten out in a relatively transparent zone between 3–5 s TWT, partly with stronger reflections at its bottom and top. This zone apparently contains the basal dêcollement, in which the thrust faults are rooted (see also Chap. III.B.2). The supposed dêcollement with off-branching, SE-dipping reflections can be traced southwards under the Wetterau, down to

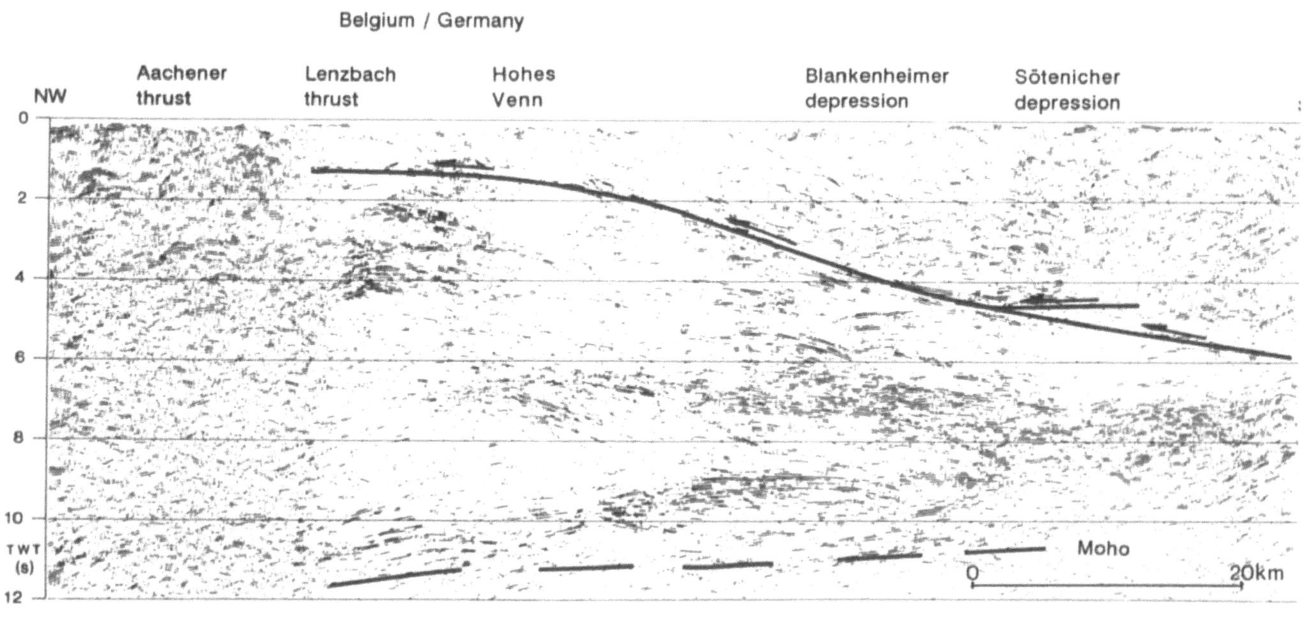

Fig. 6. Part of the DEKORP Line 1, showing a section of the northern part (D1 in Fig. 1). The most prominent feature is the southward-dipping Aachen thrust plane (DEKORP Research Group 1991). For details, see text

the northern margin of the Mid-German Crystalline High (Spessart). The transparent nature of the mid-crustal dêcollement zone possibly correlates with a high-conductivity layer detected in the magnetotelluric survey (see below).

A parallel profile DEKORP 1 (D1) has been recorded west of the river Rhine, a portion of this profile is shown in Fig. 7 (DEKORP Research Group 1991). In these sections, the fold and thrust structure is well recognizable, though less prominent than in the Rhenish Massif east of the Rhine. It is important to note that, at least in the northern part of DEKORP 1, the basal dêcollement occurs at relatively shallow depths (4–6 km), where it is obviously guided by the top of the Caledonian Brabant Massif and the rheological contrast with the overlying Devonian and Carboniferous sediments. This difference between "thin-skinned" deformation to the W and "thick-skinned" tectonics E of the Rhine river is probably due to a much stronger crustal extension in the eastern Rhenish Massif, which has brought about stronger subsidence and sedimentation, as well as extensive synsedimentary volcanism (largely absent from the areas W of the Rhine). Differential extension was probably accomodated by a transform fault grossly traced out by the Rhine valley (see Chap. III.B.1).

The southern part of DEKORP 3 beautifully depicts the northern boundary fault (a reactivation of the terrane boundary between the Northern Phyllite Zone and the Mid-German Crystalline Rise), and the internal structure of the Permo-Carboniferous Saar-Nahe Basin (DEKORP Research Group 1994). In the seismic sections through the southern, more internal parts of the Variscan belt (e.g., Fig. 6, lower section), intracrustal structures are much less clear than in the northern part. In general, the upper crust is less reflective than the lower crust, which appears strongly laminated (DEKORP Research Group 1985; Behr and Heinrichs 1987). A number of elements shows apparent dips to the NW as well as to the SE (Fig. 6, lower section). The paucity of reflections, as compared with the northern part of the section, is probably due to polyphase folding and metamorphism, as well as to the ubiquitous late Variscan granites. The boundaries between the main Variscan zones (Rhenohercynian-Saxothuringian and Saxothuringian-Moldanubian) are not very obvious. A typical feature of this southern part are clusters of diffracting elements in the lower crust, e.g., near Dinkelsbühl. Beneath the Ries impact crater, crustal reflectivity is poor, probably caused by the deep reaching destroying effect of the impact.

The crustal structure of the Saxothuringian and Moldanubian zones is also revealed by a set of profiles near the western margin of the Bohemian Massif (DEKORP 4-KTB and ISO 89), especially where the reflections are correlatable with surface outcrops. A most prominent example of Variscan thrust tectonics is depicted by profiles DEKORP 4-KTB, DEKORP 3 (D3), which clearly reveal the base as well as the internal structure of the Münchberg nappe pile, a tectonic outlier of the Moldanubian emplaced on its Saxothuringian foreland (Chap. V.B.2). This structure could be clearly revealed along the DEKORP 3 profile (Fig. 8) The northern part of DEKORP 4 also shows a prominent, dome-shaped set of reflections, which probably correlates with the outcrop of the Saxonian Granulites 150 km to the NE; the granulites are a prime example of a Variscan metamorphic core complex (see Chap. V.B.2).

The southern part of DEKORP 4 has revealed a strong band of SE-dipping reflections, partly with high seismic velocities, which descends southeastwards from approx. 4 s TWT under the KTB drilling site and reaches the crust/mantle boundary under the NW part of the Moldanubian. It is possible that this element represents mafic rocks emplaced at the Saxothuringian-Moldanubian suture (DEKORP Research Group 1988).

The seismic lines also reveal the topography of the crust/mantle boundary. In the DEKORP 2 lines, the Moho discontinuity is observed to rise southwards from about 11 s TWT under the northern part of the Rhenish Massif to 9 s under the Spessart and then descends again to 10 s in southern Germany. This slight elevation is probably due to Tertiary rifting in the Rhine Graben and its northern continuation into the Hesse Basin.

In summary, the DEKORP seismic reflection lines exhibit, in the north, a beautiful example of a foreland fold and thrust belt, but are less informative in their southern parts, especially in crystalline units. It is important to note that there are no indications, in the reflection lines, of Variscan orogenic roots, which must have been present under the Mid-German Crystalline High and the Moldanubian crystallines. Apparently, the Moho and the lower crust with its lamellar

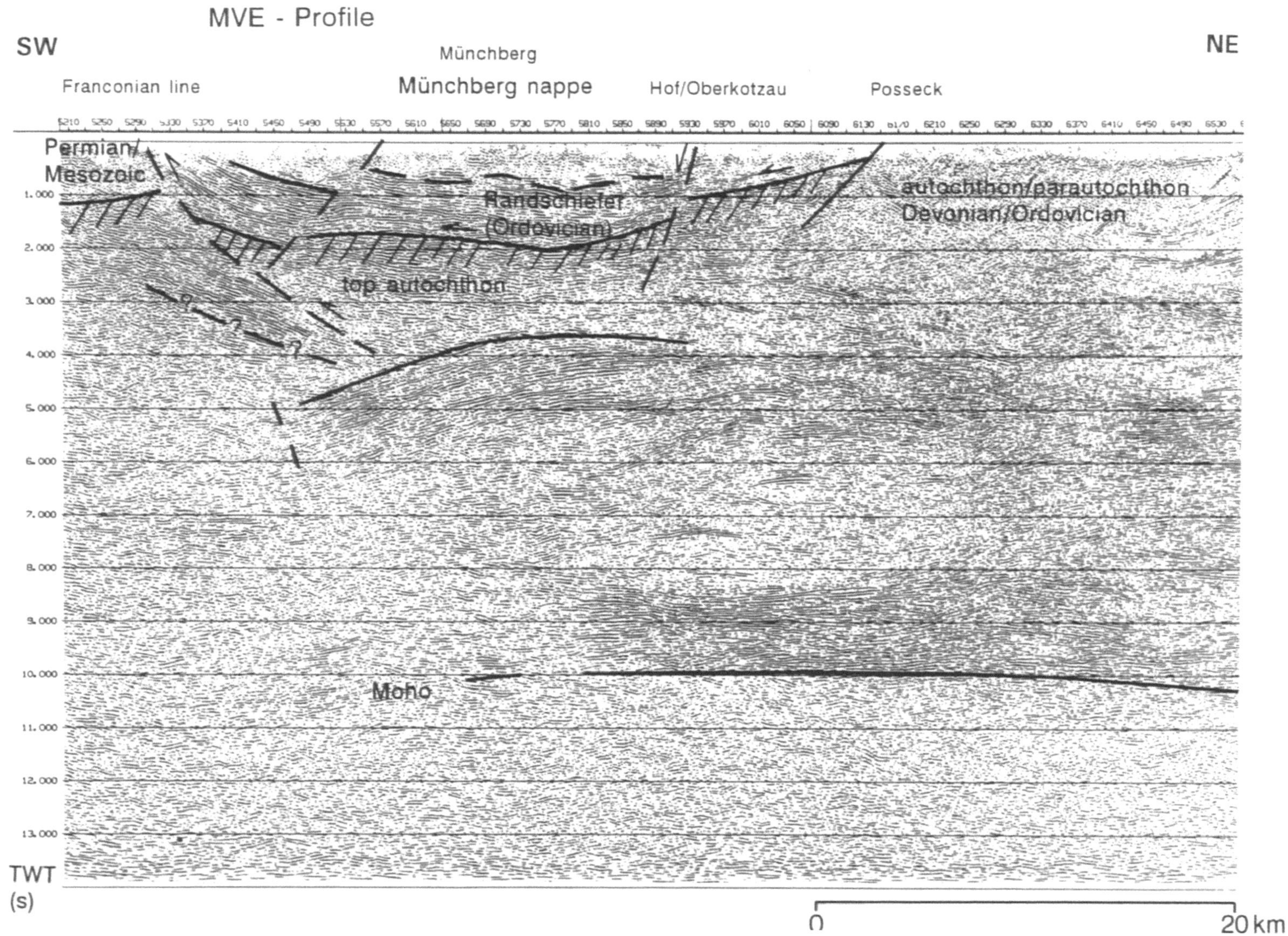

Fig. 7. Part of the DEKORP Line 3, showing that part which passes the Münchberg nappe (D3 in Fig. 1). The base of this crystalline nappe is clearly recognizable in the record section (DEKORP Research Group 1994). For details, see text

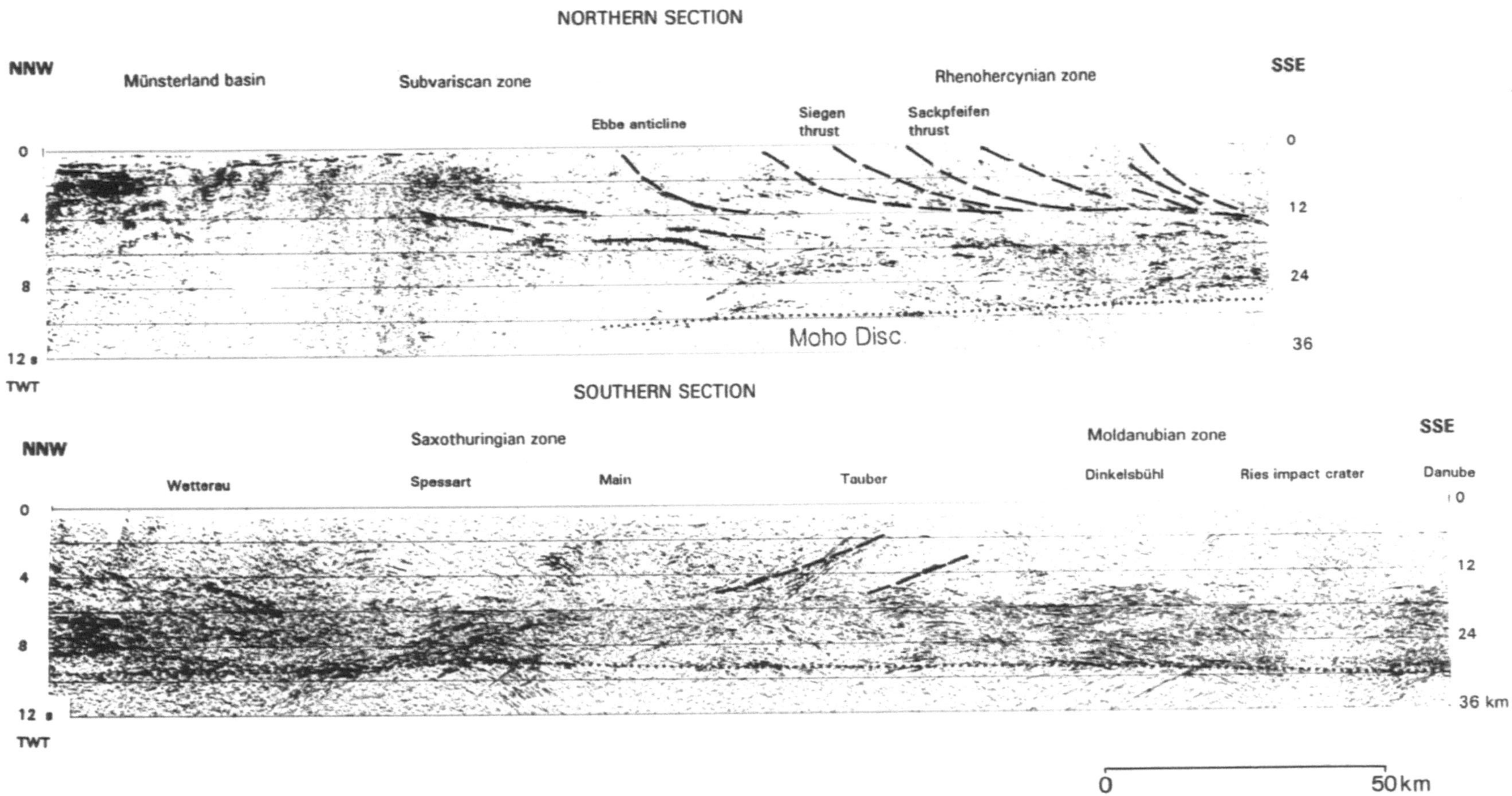

Fig. 8. Line drawing of the DEKORP-2 seismic reflection line, which runs from NNW to SSE between the Northern German Basin and the Danube in southern Germany (Meissner and Bortfeld 1990). The position of this line is shown in Fig. 1. The *upper section* covers the line between the Northern German Basin and the southern margin of the Rhenish Massif, the *lower section* is situated in southern Germany and passes the Saxothuringian and the Moldanubian zone

signature have been re-equilibrated by late and post-Variscan processes, such as Permo-Carboniferous and Cenozoic extension and magmatism.

8 Electrical Conductivity Anomalies in Central Europe

Intensive magnetotelluric studies have been carried out in the last decades in central Europe to reveal the electrical structure of the crust. A number of high conductivity anomalies, modeled as horizontally elongated zones as well as crustal layers, have been detected. The main results of these studies have been summarized by the EREGT group (1990), ERCEUGT Group (1994), Jödicke (1990) and Volbers et al. (1990). In Fig. 9, the main anomalies are displayed in a lithospheric N-S section. From south to north, the following conductivity anomalies have been detected:

Along the northern rim of the Alps the low-resistivity sediments of the Bavarian molasse basin with a thickness up to ca. 7000 m dominate the resistivity distribution (Dittus 1983).

A low-resistivity zone has been detected along the boundary between the Moldanubian and the Saxothuringian zone. This zone appears to extend along ca. 500 km between the Rhine valley in the southwest and the SW margin of the Bohemian Massif in the northeast. This low-resistivity structure of about 30 km width can be modeled by a continuous succession of medium-sized conductive bodies located at a depth of 8–9 km (Berktold, unpubl. report).

The most pronounced low-resistivity zone has been found south of the boundary between the Saxothuringian and the Rhenohercynian zone, along the transition zone between the Mid-German Crystalline Rise and the southern part of the Northern Phyllite Zone. The length of this transition zone is at least 300 km. Here, the upper boundary of the low-resisitivity body is located at approx. 7 km depth and dips southward to about 10 km depth. Both anomalies show a clearly expressed SW-NE trend, suggesting that their origin is related to Variscan structures and processes. The petrophysical nature of these anomalies (graphitic components or fluids) is still under discussion.

In the southern part of the Rhenish Massif there may exist a second low-resisitivity layer in the lower crust (Bahr 1985; Jödicke 1990). It seems to be restricted to that crustal segment which displays distinctive patterns of increased reflectivity (DEKORP Research Group 1990).

The Hesse Basin – Leine Graben anomaly, centered near Göttingen, trends SSW-NNW, i.e., parallel to the axis of the Permo-Mesozoic Hesse Basin and some Tertiary grabens of the same orientation. This suggests a post-Variscan origin of the anomaly.

The fourth elongated anomaly, known as the North-German-Polish conductivity anomaly, is located in northern Germany. Its broad shape indicates a wide zone of high-conductivity layers in the upper 10 km of the crust. The northern part of the section in Fig. 9 displays a very low-resistivity layer at a depth interval between 5 and 15 km (Losecke et al. 1986; Jödicke 1990; Volbers et al. 1990). While Losecke et al. (1986) concluded from 1-D and 2-D model calculations that its thickness is in the range of 5–10 km, Jödicke (1990) favors thicknesses of 0.1 km or even less which require extremely high conductivities. Jödicke (1990) presents arguments for an electronic conductivity caused by meta-anthracite/graphite components in widespread early Paleozoic black shales.

9 Conclusions

As usual in continental crustal regions, interpretation of geophysical signatures is greatly hampered by polyphase reworking during repeated compressional, extensional and magmatic events. In the case of the Variscides, the geological history can be traced back to 2 Ga and comprises at least two complete orogenic cycles (Cadomian and Variscan), with remote signs of the Alpine collision. This should be kept in mind as a note of caution.

The relatively uniform crustal thickness is probably due to Permo-Carboniferous extension and magmatism, with local overprints by the Cenozoic extensional and magmatic pulse. Reflection seismic profiles reveal the basement of the London/Brabant Massif, the fold and thrust architecture as well as the southern margin of the Rhenohercynian belt adjacent to the south, tectonic klippen and metamorphic dome structures in the Saxothurungian belt, and local indications of terrane boundaries. The Variscan terranes are also detectable in the Bouguer map, and partly also in the magnetic map, where it is

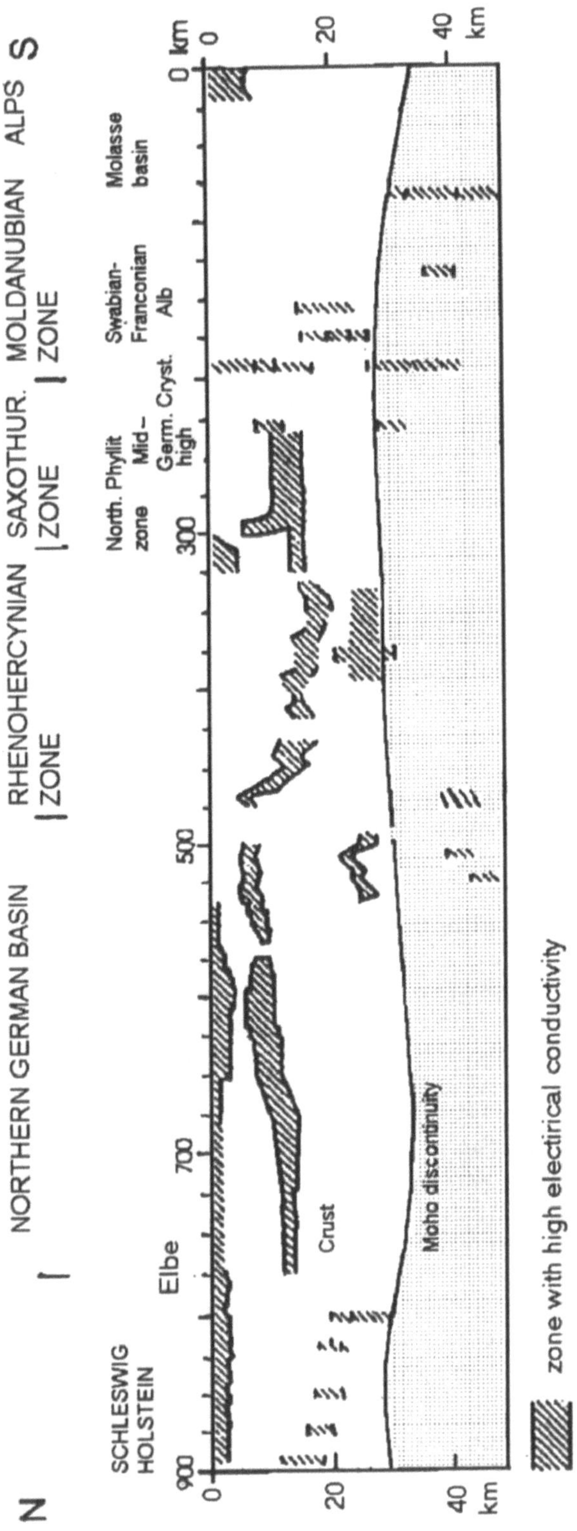

Fig. 9. A N-S section through central Europe displaying the main features of electrical conductivity distribution in the crust as published by the ERCEUGT group (1990). High-conductivity zones are indicated by *shading*. The central part of this section coincides with that of the DEKORP-2 line. The crustal thickness is shown by the Moho discontinuity

especially the Saxothuringian belt that is characterized by a multitude of positive anomalies. Lastly, terrane boundaries are especially well depicted in narrow anomalies of the electrical conductivity.

In summary, the geophysical imagery can be reconciled with the structural subdivision and geotectonic development deduced from surface outcrops.

Acknowledgments. The review of geophysical results presented in this chapter would not have been possible without the help and cooperation of a large number of colleagues and institutions. The author acknowledges with gratitude all these contributions. Sincere thanks for many stimulating discussions and suggestions are due to Wolfgang Franke, Gießen. He read this paper very critically and gave many helpful comments and suggestions. Special thanks to Elena Börner for her help with the text.

References

Aichroth B, Prodehl C (1990) EGT central segment refraction seismics. In: Freeman R, Mueller St (eds) Proc 6th EGT Worksh: Data compilation and synoptic interpretation. Eur Sci Found, Strasbourg, pp 105–127

Aichroth B, Prodehl C, Thybo H (1992) Crustal structure along the central segment of the EGT. Tectonophysics 176: 43–57

Bachmann GH, Grosse S (1989) Struktur und Entstehung des Norddeutschen Beckens – Geologische und geophysikalische Interpretation einer verbesserten Bouguer-Schwerekarte. Nds Akad Geowiss Veröff 2: 23–48

Bahr K (1985) Magnetotellurische Messung des elektrischen Widerstandes der Erdkruste und des oberen Mantels in Gebieten mit lokalen und regionalen Leitfähigkeitsanomalien. PhD Thesis, Geophys Inst, Göttingen

Bartelsen H, Lüschen E, Krey Th, Meissner R, Schmoll J, Walther Ch (1982) The combined seismic reflection-refraction investigation of the Urach geothermal anomaly. In: Hänel R (ed) The Urach geothermal project (Swabian Alb, Germany). Schweizerbarth, Stuttgart, pp 247–262

Behr HJ, Heinrichs T (1987) Geological interpretation of DEKORP 2-S: a deep seismic reflection profile across the Saxothuringian and possible implications for the late Variscan structural evolution of central Europe. Tectonophysics 142: 173–202

Berktold A (1994) Regional elektrisch leitfähige Strukturen in der Erdkruste Zentraleuropas. Münch Geophys Abh, unpubl report

Bittner R, Trappe H, Meissner R (1987) Piggyback seismic experiments during deep crustal reflection surveys. Ann Geophys 5 (B): 381–388

Blundell D, Freeman R, Mueller S (eds) (1992) A continent revealed in the European geotraverse. Cambridge University Press, Cambridge, 275 pp

Bormann P, Bankwitz P, Schulze A, Apitz E, Conrad W, Neunhöfer H, Porstendorfer G, Rugenstein B, Wolter J (1987) Methodical contributions to complex geophysical and geological investigations of the Earth's crust in the G.D.R. Part I. Geophysical methods and results. Gerlands Beitr Geophys 96 (3/4): 291–310

Bormann P, Bankwitz P, Conrad W, Oelsner Ch (1989) Crustal structure and development in central Europe. Gerlands Beitr Geophys 98 (5): 353–367

Conrad W (1980) Potentialfeldanomalien und geologischer Tiefenbau im Bereich des Südwestrandes der Osteuropäischen Tafel. Z Geol Wiss 8: 439–448

DEKORP Research Group (1985) First results and preliminary interpretation of deep-reflection seismic recordings along profile DEKORP 2-South. J Geophys 57: 137–163

DEKORP Research Group (1988) Results of the DEKORP 4/KTB Oberpfalz deep seismic reflection investigations. J Geophys 62: 69–101

DEKORP Research Group (1990) Crustal structure of the Rhenish Massif: results of the deep seismic reflection lines DEKORP 2-North and 2-North-Q. Geol Rundsch 79/3: 523–566

DEKORP Research Group (1991) Results of the DEKORP 1 (BELCORP-DEKORP) deep seismic reflection studies in the western part of the Rhenish Massif. Geophys J Int 106: 203–227

DEKORP Research Group (1994) MVE-West – preliminary geologic interpretation of a deep near-vertical reflection profile between the Rhenish and Bohemian Massifs, Germany. Z Geol Wiss (in press)

Dittus H (1983) Magnetotellurik in Südost-Bayern. Diplomarbeit, Inst Allg Angew Geophys, Ludwig-Maximilians-Univ, München

Edel JB, Fluck P (1989) The upper Rhenish shield basement (Vosges, Upper Rhinegraben, Schwarzwald): main structural features, deduced from magnetic, gravimetric and geological data. Tectonophysics 169: 303–316

Edel JB, Weber K (1993) The Cadonian terranes, wrench faulting and thrusting in the central European Variscides – geophysical and geological evidence. Unpubl report

ERCEUGT Group (1992) An electrical resistivity transect from the Alps to the Baltic Sea (central segment of the EGT). In: Freeman R, Giese P, Huch M, Mueller St (eds) The European geotraverse, part 8. Tectonophysics 207: 123–139

EREGT Group (1990) An electrical resistivity transect from the Alps to the Baltic Sea (centeral segment of the EGT). In: Freeman R, Mueller St (eds) Proc 6th EGT Worksh: Data compilation and synoptic interpretation. Eur Sci Found, Strasbourg, pp 299–313

EUGEMI Working Group (1990) The European geotraverse seismic refraction experiment of 1986 from Genova, Italy to Kiel, Germany. In: Freeman R, Mueller St (eds) The European geotraverse, part 6. Tectonophysics 176: 43–57

Franke W (1992) Phanerozoic structures and events in central Europe. In: Blundell D, Freeman R, Mueller St (eds) A continent revealed in the European geotraverse. Cambridge University Press, Cambridge, pp 164–179

Freeman R, Mueller St (eds) (1990) Proc 6th EGT Worksh: Data compilation and synoptic interpretation. Eur Sci Found, Strasbourg, 457 pp

Freeman R, Giese P, Mueller St (eds) (1990) Proc 5th EGT Study Centre. Eur Sci Found, Strasbourg, 404 pp

Gajewski D (1989) Compressional and shear-wave velocity models of the Schwarzwald derived from seismic refraction data. In: Emmermann R, Wohlenberg J (eds) The continental deep drilling program (KTB). Springer, Berlin Heidelberg New York, pp 363–383

Gajewski D, Prodehl C (1985) Crustal structure beneath the Swabian Jura, SW Germany, from seismic refraction investigations. J Geophys 56: 69–80

Gajewski D, Prodehl C (1987) Seismic refraction investigation of the Black Forest. Tectonophysics 142: 27–48

Gajewski D, Holbrook WS, Prodehl C (1987) A three-dimensional crustal model of southwest Germany derived from seismic refraction data. Tectonophysics 142: 49–70

Gebrande H, Bopp M, Neurieder P, Schmidt T (1989) Crustal structure in the surroundings of the KTB drill site as derived from refraction and wide-angle seismic observations. In: Emmermann R, Wohlenberg J (eds) The continental deep drilling program (KTB). Springer, Berlin Heidelberg New York, pp 151–176

German Research Group for Explosion Seismology (1964) Crustal structure in western Germany. Z Geophys 30: 209–234

Giese P (1990) Crustal thickness in central Europe. In: Freeman R, Giese P, Mueller St (eds) Proc 5th EGT Study Centre. Eur Sci Found, Strasbourg, pp 115–119

Giese P, Prodehl C (1986) Main features of crustal structure in the central section of the EGT based on seismic refraction data. In: Freeman R, Mueller St, Giese P (eds) Proc 3rd Worksh Eur Geotraverse (EGT) project in the central segment. Eur Sci Found, Strasbourg, pp 83–92

Giese P, Prodehl C, Stein A (eds) (1976) Explosion seismology in central Europe in data and results. Springer, Berlin Heidelberg New York, 429 pp

Grosse S, Kopf M, Sonntag K (1961) Ergebnisse der Gravimetermessungen im Westerzgebirge. Freiberger Forschungsh C 1 (10): 57–102

Grosse S, Conrad W, Behr HJ, Heinrichs T (1990) Major gravity axis and anomalies in central Europe. In: Freeman R, Giese P, Mueller St (eds) Proc 5th EGT Study Centre. Eur Sci Found, Strasbourg, pp 35–146

Guterch A, Grad M, Materzok J, Perchuc E (1986) Deep structure of the Earth's crust in the contact zone of the Palaeozoic and Precambrian platforms in Poland (Tornquist-Teisseyre-Zone). Tectonophysics 128: 251–279

Hermes HJ (1986) Calculation of pre-Zechstein Bouguer anomaly in northwest Germany. First Break 4 (I 1)

Holbrook WS, Gajewski D, Krammer A, Prodehl C (1988) An interpretation of wide-angle shear-wave data in southwest Germany: Poisson's ratio and petrological implications. J Geophys Res 93 (12) 81–106

Ibrmayer J, Blizkovsky M, Fejfar M, Ibrmayer I, Mayerova M, Novotny M, Tomek C (1983) Some new results of deep seismic sounding in Czechoslovakia. Proc 28th Int Geophys Symp, Balatonszemes 1983, Budapest, pp 819–835

Jödicke H (1990) Zone hoher elektrischer Krustenleitfähigkeit im Rhenohercynikum und seinem nördlichen Vorland. Diss Math Nat Fak Westfälische Wilhelms-Univ, Münster

Losecke W, Knödel K, Müller W (1986) Technologische Weiterentwicklung und Erprobung der Remote Reference Magnetotellurik. Abschlußber BMFT-Proj RG 8109 8, Bundesanst Geowiss Rohstoffe, Hannover

Lüschen E, Wenzel F, Sandmeier K-J, Menges D, Rühl Th, Stiller M, Janoth W, Keller F, Söllner W, Thomas R, Krohe A, Stenger R, Fuchs K, Wilhelm H, Eisbacher G (1987) Near-vertical and wide-angle seismic surveys in the Black Forest, SW Germany. J Geophys 62: 1–30

Lüschen E, Wenzel F, Sandmeier K-J, Menges D, Rühl Th, Stiller M, Janoth W, Keller F, Söllner W, Thomas R, Krohe A, Stenger R, Fuchs K, Wilhelm H, Eisbacher G (1989) Near-vertical and wide-angle seismic surveys in the Schwarzwald. In: Emmermann R, Wohlenberg J (eds) The German Continental Deep Drilling Program (KTB). Springer, Berlin Heidelberg New York, pp 297–362

Lüschen E, Nolte B, Fuchs K (1990) Shear-wave evidence for an anisotropic lower crust beneath the Black Forest, southwest Germany. In: Leven JH, Finlayson DM, Wright C, Dooley JC, Kennett BLN (eds) Seismic probing of continents and their margins. Tectonophysics 173: 483–493

Mechie J, Prodehl C, Fuchs K (1983) The long-range seismic refraction experiment in the Rhenish Massif. In: Fuchs K, von Gehlen K, Mälzer H, Murawski H, Semmel A (eds) Plateau uplift in the Rhenish Massif in a case history. Springer, Berlin Heidelberg New York, pp 260–275

Meier L, Eisbacher HG (1991) Tectonics 10 (3): 621–630

Meissner R, Bortfeld RK (eds) (1990) DEKORP-Atlas: Results of Deutsches Kontinentales Reflexionsseismisches Programm. Springer, Berlin Heidelberg New York, 19 pp and 80 sect

Meissner R, Bartelsen H, Murawski H (1980) Seismic reflection and refraction studies for investigating fault zones along the geotraverse Rhenoherzynikum. Tectonophysics 64: 59–84

Meissner R, Wever Th, Flüh EH (1987) The Moho in Europe in implication for crustal development. Ann Geophys 5B: 357–364

Meissner R, Brown L, Dürbaum HJ, Franke W, Fuchs K, Seiffert F (eds) (1991) Continental lithosphere: deep seismic reflections. Am Geophys Union Geodyn Ser 22

Mengel K, Kern H (1990) Petrologic versus seismic Moho and crustal root mysteries of the Hercynian orogen. Proc 5th EGT Study Centre. Eur Sci Found, Strasbourg, pp 169–176

Mengel K, Sachs PM, Stosch HG, Wörner G, Loock G (1991) Crustal xenolithes from Cenozoic volcanic fields of West Germany: implications for crustal structure and evolution. Tectonophysics 195: 271–289

Mooney WD, Prodehl C (1978) Crustal structure of the Rhenish Massif and adjacent areas; a reinterpretation of existing seismic-refraction data. J Geophys 44: 573–601

Mostaanpour M (1984) Einheitliche Auswertung krustenseismischer Daten in Westeuropa. Darstellung von Krustenparametern und Laufzeitanomalien. Berl Geowiss Abh B 10: 96

Prodehl C, Aichroth B (1992) Seismic investigations along the European Geotraverse and its surroundings in Central Europe. Terra Nova 4 (1): 14–24

Prodehl C, Giese P (1990) Seismic Investigations around the EGT in central Europe. In: Freeman R, Giese P, Mueller St (eds) Proc 5th EGT Study Centre. Eur Sci Found, Strasbourg, pp 77–97

Reichert C (1988) DEKORP in Deutsches Kontinentales Reflexionsseismisches Programm. Geol Jahrb A 109: 163–173

Remmelts G, Duin EJ Th (1990) Results of a regional deep seismic survey in the Netherlands. In: Pinet B, Bois C (eds) The potential of deep seismic profiling for hydrocarbon exploration. Editions Technip, Paris, pp 335–343

Rousset D, Bayer R (1990) Nouvelles cartes gravimetriques du sud du Fosse Rhenan. C R Acad Sci Paris 310 (11): 199–206

Schmoll J, Bittner R, Dürbaum HJ, Heinrichs T, Meissner R, Reichert C, Rühl T, Wiederhold H (1989) Oberpfalz deep seismic reflection survey and velocity studies. In: Emmermann R, Wohlenberg J (eds) The continental deep drilling program (KTB). Springer, Berlin Heidelberg New York, pp 99–149

Schubert H, Conrad W (1981) Beitrag zur Interpretation ausgewählter geophysikalischer Anomalien Mitteleuropas unter geodynamischen Aspekten. Z Geol Wiss 9 (10): 1179–1189

Schulze A, Bormann P (1990) Deep seismic sounding in eastern Germany. In: Freeman R, Giese P, Mueller St (eds) Proc 5th EGT Study Centre. Eur Sci Found, Strasbourg, pp 109–115

Thybo H (1990) A seismic velocity model along the EGT profile from the North-German Basin into the Baltic Shield. In: Freeman R, Giese P, Mueller St (eds) Proc 5th EGT Study Centre. Eur Sci Found, Strasbourg, pp 99–108

Thybo H, Kiorhoe LL, Moeller C, Schönharting G, Berthelsen A (1990) Geophysical and tectonic modelling of EUGENO-S profiles. In: Freeman R, Mueller St (eds) 6th EGT Worksh: Data compilations and synoptic interpretation. Eur Sci Found, Strasbourg, pp 93–104

Volbers R, Jödicke H, Untiedt J (1990) Magnetotelluric study of the earth's crust along the deep seismic reflection profile DEKORP 2-N. Geol Rundsch 79 (3): 581–601

Vollbrecht A, Weber K, Schmoll J (1989) Structural model for the Saxothuringian-Moldanubian suture in the Variscan basement of the Oberpfalz (northeastern Bavaria, F.R.G.) interpreted from geophysical data. Tectonophysics 157, 123–133

Wenzel F, Brun JB, ECORS-DEKORP Working Group (1991) A deep reflection seismic line across the Northern Rhine Graben. Earth Planet Sci Lett 104: 140–150

Wonik T, Hahn A (1989) Karte der Magnetfeldanomalien in der Bundesrepublik Deutschland, Luxemburg, Schweiz und Österreich (westlicher Teil) 1 : 1 000 000. Geol Jahrb E 43: 21

Wonik T, Hahn A (1990) Preliminary map of aeromagnetic anomalies for the EGT central segment. In: Freeman R, Giese P, Mueller St (eds) Proc 5th EGT Study Centre. Eur Sci Found, Strasbourg, pp 147–156

Zeis St, Gajewski D, Prodehl C (1990) Crustal structure of southern Germany from seismic refraction data. In: Freeman R, Mueller St (eds) The European geotraverse, part 6. Tectonophysics 176: 59–86

III Rhenohercynian Foldbelt

III.A Introduction

W. Franke

The Rhenohercynian Belt was defined by Kossmat (1927). This subdivision of the Variscan Belt is mainly based upon metamorphic grades, magmatic belts, and sedimentary facies, and can be translated, therefore, into modern plate-tectonic terms: very low grade sedimentary belts contain the records of pre-orogenic extensional basins and of foreland basins. They alternate with crystalline zones, which represent deeply eroded active margins. It is therefore still convenient to refer to Kossmat's zonation. However, modern findings require slight revisions of the classical definition.

To the NW of the Rhenohercynian Belt, Kossmat recognized a foreland (Westfälische Zone) with synorogenic, partly coal-bearing clastic sequences. We include this external part of the foreland basin into the Rhenohercynian Belt.

The main part of the Rhenohercynian Belt is occupied by Early to Middle Devonian shelf clastics and carbonates, which represent the passive southern margin of the Old Red Continent (Avalonia plus Laurentia plus Baltica, see Chaps. X, XI, this Vol.). They are overlain by hemipelagic shales and early Carboniferous flysch sediments (Chap. III.B.1).

This sequence is overlain by an extensive nappe, which consists of early Devonian MORB-type metabasalts (Chap. III.B.3), condensed pelites, and the earliest flysch clastics (Frasnian through to early Carboniferous). This association represents an oceanic part of the Rhenohercynian basin, which has been transported over the passive margin sequences of the foreland.

The southeastern margin of the Rhenohercynian Belt is formed by the Northern Phyllite Zone. It consists of sedimentary and volcanic rocks overprinted by Carboniferous deformation and greenschist-grade metamorphism. The paleogeographic derivation of the protoliths is heterogeneous and partly doubtful: the Northern Phyllite Zone is a tectonic collage which probably contains remnants of the distal, southeastern part of the Rhenohercynian passive margin, together with rocks indicative of a Silurian magmatic arc, and exotic sedimentary sequences probably derived from the Armorican microplate (Chap. III.C.1, this Vol.).

To the SE, the Rhenohercynian is bounded by the northwestern, crystalline part of Kossmat's Saxothuringian Zone. This crystalline belt was not named separately by Kossmat, but recognized later as a separate entity (Scholtz 1930) and named Mitteldeutsche Kristallinschwelle (Mid-German Crystalline High) by Brinkmann (1948). The Crystalline High (see Chap. IV, this Vol.) is now interpreted as the active, southern margin of the Rhenohercynian basin, i.e., as a site of synorogenic magmatism and as the source of synorogenic clastic sediments transported toward the NW into the Rhenohercynian foreland basin. Age and tectonic setting of the protoliths of the Crystalline Rise are largely enigmatic.

Those rocks which predominate at the present-day surface record crustal extension in Devonian time, i.e., clearly later than in all other Variscan basins, which represent Cambro-Ordovician rift zones. However, palaeomagnetic data, faunal affinities, geochemical data from the Northern Phyllite Zone and the Mid-German Crystalline High and some exotic rocks from southern parts of the Rhenohercynian belt strongly suggest that the Devonian basin had an oceanic precursor (Rheic ocean), which opened in the early Ordovician, was closed before the Devonian, and then re-opened in the Devonian to form the Rhenohercynian proper (Chaps. X, XI, this Vol.). Subsequent to this complex development, the Rhenohercynian belt was subject to plate convergence, largely at the same time as the neighbouring regions.

It is important to note that the Rhenohercynian Belt in central Europe can be correlated with regions along strike all over Europe (see the discussion in Engel et al. 1983; Franke and Engel 1982; Holder and Leveridge 1986). Towards the

Dallmeyer et al. (Eds.)
Pre-Permian Geology of Central and Eastern Europe

E, the Rhenohercynian of the Rhenish Massif can be traced via the Harz Mts., the Flechtingen Hills, and the subsurface E of the Bohemian Massif into the Moravo-Silesian Belt in Austria, Czechia and Slovakia (Chap. VIII.B.1, this Vol.). To the W of the Ardennes, the Rhenohercynian Belt is covered by the Mesozoic of the Paris Basin, but re-emerges in SW England and southern Ireland. A further extension probably occurs in S Portugal (see also Dallmeyer and Martinez-Garcia 1990). Remarkably, both the main extensional and the compressional events are grossly coeval, at least in the British Isles, in Central Europe, and in Moravia. These relationships justify a cylindroidal approach to the description and interpretation of the Rhenohercynian Belt.

References

Brinkmann R (1948) Die Mitteldeutsche Schwelle. Geol Rundsch 56: 56–66

Dallmeyer RD, Martinez-Garcia E (eds) (1990) Pre-Mesozoic geology of Iberia. Springer, Berlin Heidelberg New York, 416 pp

Engel W, Franke W (1983a) Flysch sedimentation: its relation to tectonism in the European Variscides. In: Martin H, Eder FW (eds) Intracontinental fold belts. Springer, Berlin Heidelberg New York, pp 290–321

Engel W, Franke W, Langenstrassen F (1983b) Paleozoic sedimentation in the Northern Branch of the Mid-European Variscides–essay of an interpretation. In: Martin H, Eder FW (eds) Intracontinental fold belts. Springer, Berlin Heidelberg New York, pp 9–41

Franke W, Engel W (1982) Variscan Sedimentary basins on the continent and relations with southwest England. Proc Ussher Soc 5: 259–269

Holder MT, Leveridge B (1986) Correlation of the Rhenohercynian Variscides. J Geol Soc Lond 143: 141–147

Kossmat F (1927) Gliederung des varistischen Gebirgsbaues. Abh Sächs Geol Landesamt 1: 39

Scholtz R (1930) Das varistische Bewegungsbild entwickelt aus der inneren Tektonik eines Profiles von der Böhmischen Masse bis zum Massif von Brabant. Fortschr Geol Paläontol 8 (25): 235–316

III.B Autochthon and Nonmetamorphic Nappe Units

III.B.1 Stratigraphy

W. Franke

1 Introduction

Within the unmetamorphosed part of the Rhenohercynian Belt, it is possible to distinguish one major, autochthonous part, and a number of allochthonous units. The **autochthon** is characterized by neritic and hemipelagic sequences, mostly of Devonian age, which represent the passive, southern margin of the Old Red Continent (see Chaps. I and III.A). The **allochthonous units** comprise ocean floor and active margin sequences, as well as rocks of uncertain affinities. The autochthon has been studied for more than 100 years, and is now relatively well understood in terms of biostratigraphy, sedimentology and paleogeography. Summaries have been published by Meischner (1971), Franke et al. (1978), Walliser (1981), Engel et al. (1983a), and Franke (1989). Therefore, the main characteristics of the autochthon will be summarized here only as far as is necessary for a geotectonic interpretation. The main stress is laid upon the allochthonous units, whose definition and interpretation are vital for plate tectonic considerations.

As mentioned in Chapter III.A, the Rhenohercynian Belt can be traced at least from SW England through to Eastern Germany, and its geological record is largely consistent along strike. It is sufficient, therefore, to describe the essential Rhenohercynian features in their main area of outcrop, the E part of the Rhenish Massif, with reference to neighboring areas. Short paragraphs deal with the main diverging features in the Rhenish Massif W of the Rhine river and in the Harz Mts. Discussion of the primary paleogeographic relationships requires some degree of speculation, especially in the possible and proven allochthons. To avoid repetition, interpretative scenarios are given in the end of the relative paragraphs.

2 Rhenish Massif (E Part) and Related Areas

2.1 Basement

All of the Rhenohercynian Belt is underlain by a crystalline basement. A raft of "Eckergneis" contained in the roof of a Late Carboniferous pluton in the Harz Mts (Fig. 4) has yielded zircons with a lower intercept of 560 Ma (Baumann et al. 1991). Together with the age record of detrital zircons (Von Hoegen et al. 1990), these data suggest a Cadomian (Panafrican) consolidation preceding the Paleozoic extension. Clear indications of a Caledonian tectonic event are only available in the Brabant Massif W of the Rhine River. However, it is possible that Caledonian basement uplifts are present also E of the Brabant Massif. This is suggested by the presence of coarse middle Devonian conglomerates in the Velbert Anticline (Fig. 1; Neumann-Mahlkau 1982), and slates of presumably Ordovician age in the Soest-Erwitte well (Fig. 1; Krebs 1978; Teichmüller 1978).

The sedimentary record of the Paleozoic outcrop E of the River Rhine spans the Late Ordovician through Upper Carboniferous. The Early and pre-Ordovician development is largely unknown. Ordovician through Early Devonian rocks are exposed in some anticlines in the northern part of the Rhenish Massif. The different stratigraphic units are separated by faults, but there is no indication of any Caledonian event between the Ordovician and Devonian. The Ordovician consists of shales and is overlain by Silurian marls; a complete record is available from the Earliest Devonian onwards (Gruppenbericht 1981).

A reflection seismic survey has failed to reveal the position of the pre-Ordovician basement; it probably lies below the brittle/ductile transition and has therefore not produced a major décolle-

Dallmeyer et al. (Eds.)
Pre-Permian Geology of Central and Eastern Europe

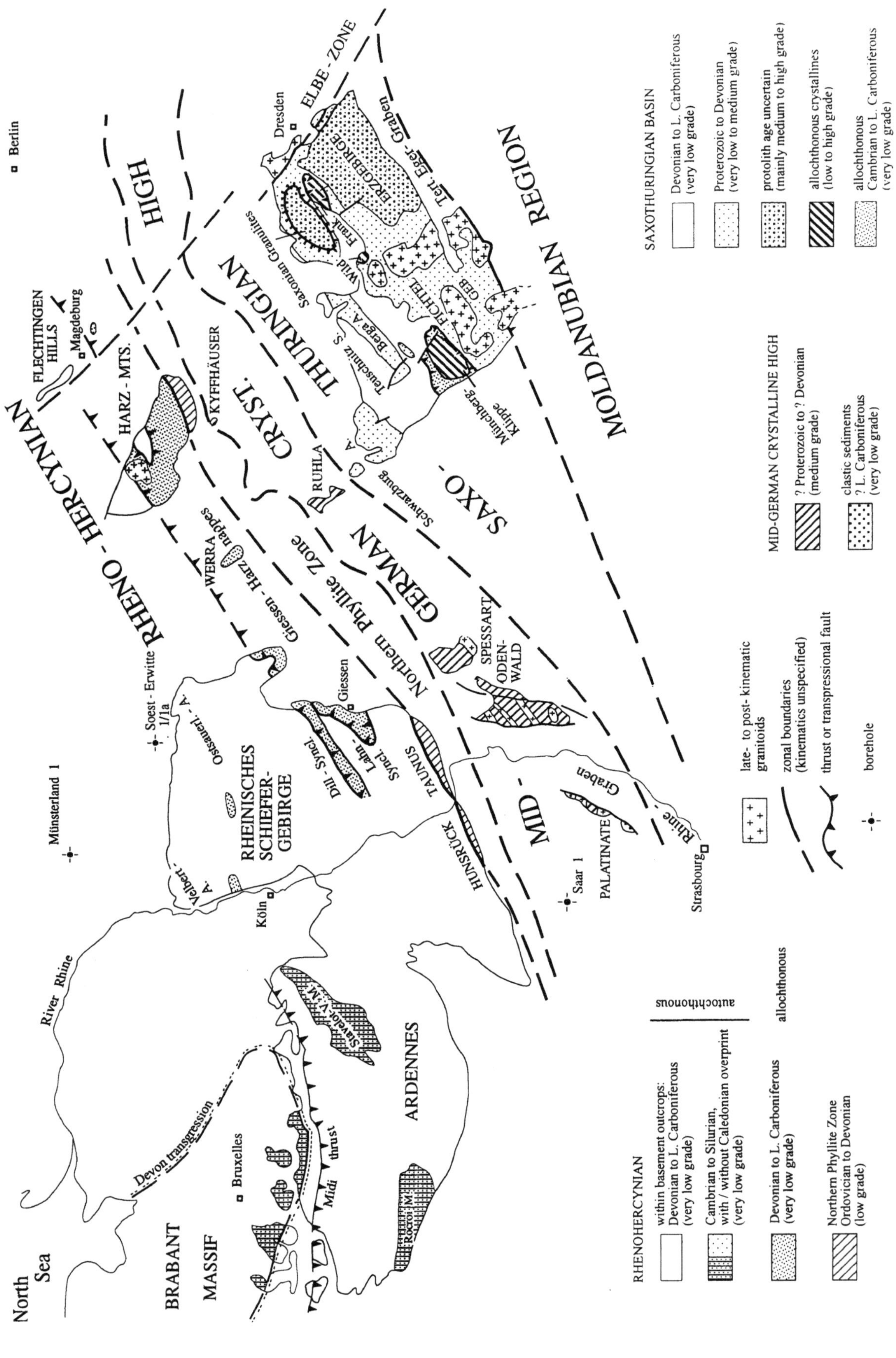

Fig. 1. Outcrops of Variscan basement in the Rhenohercynian Belt and in the NW part of the Saxothuringian Belt

ment (DEKORP Research Group, Franke et al. 1990).

2.2 Initial Rifting and Neritic Clastic Sedimentation

Notwithstanding the age and composition of the underlying basement, it can be stated that the Devonian record is controlled by a major phase of crustal extension. This is documented in thick sequences of near-shore and shelf sediments, as well as synsedimentary volcanic rocks. In the older literature, the neritic facies are referred to as "Rheinische Fazies". Summaries are available in Langenstrassen (1983) and Meyer and Stets (1980). Siliciclastic materials were derived from the Old Red Continent to the N, and transported far to the S. U/Pb systematics of detrital zircons in the Devonian clastics point to Laurentia- or Baltica-related sources (Haverkamp et al. 1991). Debris of ultramafic rocks (Schulz-Dobrick and Wedepohl 1983; Press 1986) was probably derived from Caledonian ophiolites somewhere to the N.

The Old-Red-derived clastic sediments make up most of the Rhenish Massif and are also present in the Harz Mts. (see the summaries in Walliser and Alberti 1983; Wachendorf 1986). Early Devonian (Gedinnian, Siegenian) clastic sediments in the southernmost part of the Rhenish Massif (Hunsrück, Taunus Mts.) are possibly derived from a southeasterly source, which is interpreted as an early expression of the Mid-German Crystalline High (Hahn 1990). This latter element, however, is an active margin, which becomes first apparent in the Early Upper Devonian flysch laid down on oceanic crust further S (Giessen-Harz Nappe). Besides, the Crystalline High has produced immature clastics (graywackes), and not quartzites. It is more likely, therefore, that the southern source was a preexisting, anorogenic high, possibly accentuated by uplift and tilting during the Early Devonian extension.

A prolonged break in the clastic supply from the north allowed the growth of platform carbonates and reef buildups on the shelf, during the Middle Devonian and early Upper Devonian (Frasnian) (see below). This applies both to the northern shelf and the Taunus area in the southern part of the Rhenish Massif.

Due to continued extension and subsidence, and possibly aided by the leveling-out of the source areas, the clastic shelf receded northwards with time (see Fig. 2). Therefore, in any given section, shelf clastics and carbonates are overlain by hemipelagic deposits.

Crustal extension is also documented in late Early Devonian (Emsian) felsic volcanic rocks (trachytes and rhyolites; see Kirnbauer 1991 and literature therein, and Chap. III.B.3 for petrogenetic considerations).

2.3 Pelagic Interlude

Continued crustal extension brought about a pelagic regime ("Hercynische Fazies"). It consists of shales with siliciclastic turbidites derived from the northwestern shelf. Pelagic sedimentation was highly sensitive to the submarine topography created by the tilting of blocks at extensional growth faults (Krebs 1975; Engel et al. 1983a; Werner 1988), and by strong submarine volcanism (see below). Rises are usually capped by condensed pelagic limestones (see the summary in Franke and Walliser 1983), and also served to pond siliciclastic or limestone turbidites (e.g., Franke et al. 1975).

From the early Middle Devonian onwards, volcanism was predominantly basaltic. It occurred in discrete belts at the northern margin of the Rhenish Massif (essentially Late Middle Devonian, see Fig. 2), as well as in the Dill and Lahn Synclines. These latter two are now separated by the narrow Hörre-Kellerwald belt, but could well have formed one coherent paleogeographic entity, if one interprets the Hörre-Kellerwald rocks as a nappe (see Chap. III.B.3).

Basaltic volcanism was especially active during the late Middle Devonian (Givetian) and Early Carboniferous, but occasional volcanic intercalations are also present between these peaks. The geochemistry of these rocks suggests an intraplate tectonic setting (see Sect.3).

Several volcanic mounds grew up close to the water table, and were colonized by coral/stromatoporoid reefs. Since the thickness of these volcanics does not exceed some hundreds of meters, this relationship testifies to the moderate depth of the pelagic realm. Summaries of the reef development, including those seated on volcanic mounds as well as the shelf reefs, have been published by Burchette (1981), Krebs (1974), and Mountjoy and Krebs (1983). Fuchs (1990) and

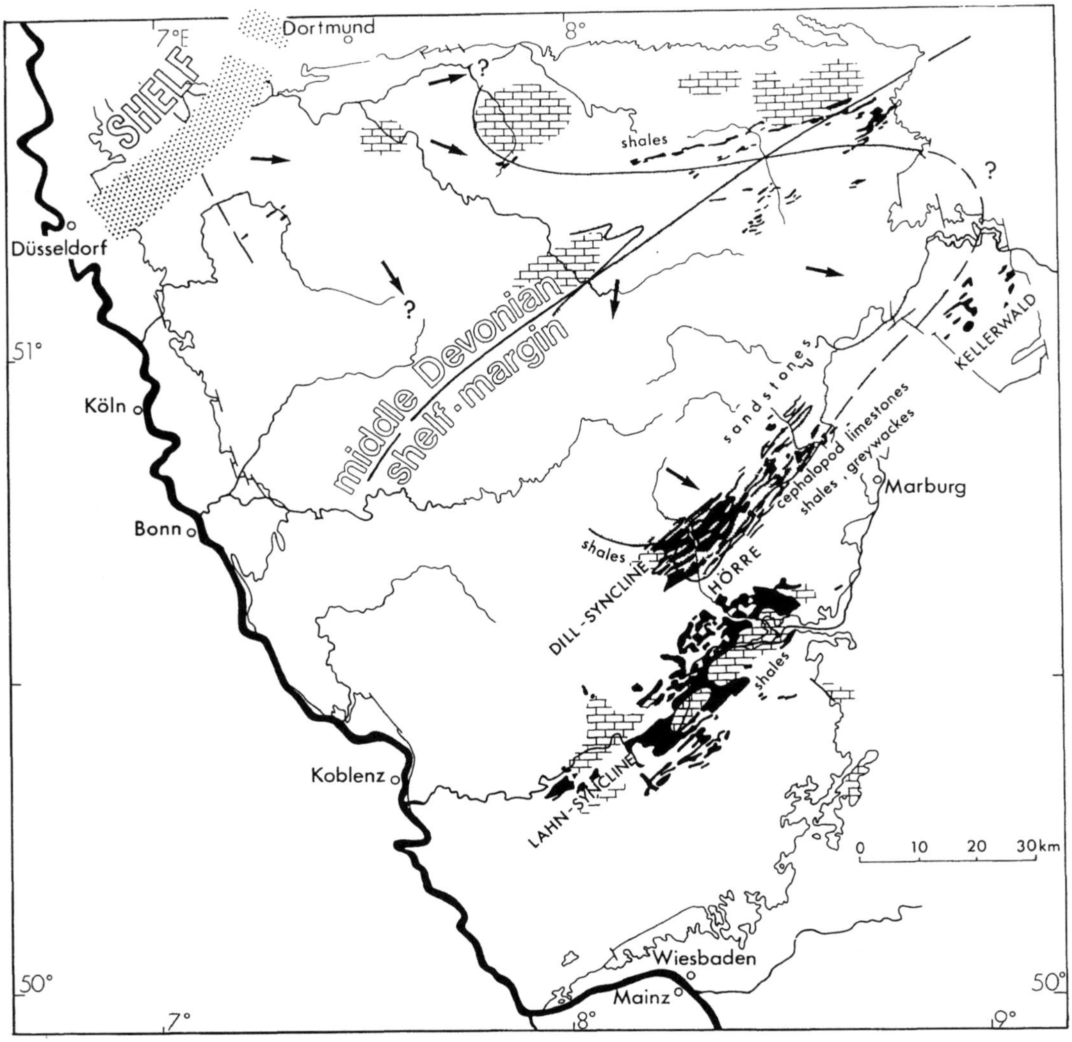

Fig. 2. Paleogeography of the Rhenish Massif E of the Rhine River during the Middle and Late Devonian. *Black* Devonian and Carboniferous synsedimentary volcanic rocks; *brick ornament* Givetian-Frasnian reef carbonates; *coarse stipple* margin of clastic shelf during the Famennian and basal Carboniferous; *fine stipple* limitation of shelf-derived Famennian sandstone turbidites. (Franke et al. 1978)

Franke (1973) have described reef complexes in the Harz Mts.

The Dill Syncline of the Rhenish Massif (pelagic realm with volcanics and reef growth) is clearly correlatable with the NW part of the Harz Mts. An equivalent of the Lahn Syncline exists in the Elbingerode window of the Harz Mts. The equivalent of the Hunsrück/Taunus shelf area is probably concealed, in the Harz Mts., by the unit interpreted below as the Hörre-Gommern (SE part) allochthon.

Pelagic sedimentation is punctuated by some global events, which have produced brief anoxic intervals, now documented as black shales. One of these intervals is probably connected with the worldwide extinction of the coral/stromatoporoid reefs in the Late Frasnian (Schindler 1990; Buggisch 1991). A further widespread shale horizon occurs at the Devonian Carboniferous boundary (see the contributions in Flajs and Feist 1988). The Tournaisian (Early Carboniferous II) black shales and cherts, and the overlying greenish gray cherts (Early to Middle Visean) are another conspicuous member of the German successions. They are correlatable at least all across Europe (Engel et al. 1983a). In the early Late Visean, there is a similarly widespread recurrence of anoxic conditions. Further examples of global events are treated in Walliser (1986).

Within the Early Carboniferous pelites and cherts, there are intercalations of limestone turbidites. Some of these are derived from the Early Carboniferous carbonate platform at the northwestern margin of the Rhenohercynian basin (Sect. 4), but many others appear to relate to intrabasinal rises which are either eroded or else covered by the Permian-Mesozoic east of the Rhenish Massif (Witten 1979; Eder et al. 1983). It is in these beds that Meischner (1962, 1964) first recognized and described the internal organization of limestone turbidites ("allodapische Kalke"). Only very few beds of mixed siliciclastic/carbonate composition appear to be derived from the southern shelf, i.e., the source area of the synorogenic greywacke turbidites (Meischner 1962; Eder et al. 1983).

2.4 Synorogenic Clastic Sedimentation

The pelagic sediments, volcanic rocks, and reefs are overlain by flysch sediments derived from an active margin to the south, the Mid-German Crystalline High (Chap. IV). Summaries of the synorogenic sedimentation have been published by Engel and Franke (1983) and Franke and Engel (1986).

With the exception of one Early Middle Devonian graywacke outcrop in the Kellerwald (Meischner 1968, 1991), the oldest synorogenic clastics occur in the Early Upper Devonian (Frasnian) of the Giessen-Harz allochthon (see Sect. 3.2). In the autochthon of the Rhenish Massif E of the Rhine, some first graywacke turbidites occur in the Early Carboniferous IIδ; graywackes are widespread from the Early Carboniferous IIIα onwards. Throughout the Early Carboniferous III and the earliest Late Carboniferous (Namurian), the graywacke front migrated northwestwards across the Rhenish Massif (Fig. 3). A change in mineral composition with time, from graywackes towards quartzitic sandstones, may be taken to indicate that an increasing amount of older (meta-)sediments in the Rhenohercynian foreland was accreted to the tectonic front and contributed to the flysch (Engel et al. 1983c). In Late Namurian time, subsidence was overcompensated by the clastic input, and sedimentation continued in the paralic environment of the Ruhr coal district at the NW margin of the Rhenish Massif (see the reviews in Geol. Landesamt Nordrhein-Westfalen 1984, 1985, and Jankowski 1990). The Rhenish Massif E of the Rhine, in Carboniferous time, exhibits all features typical of a foreland basin, generated by the tectonic load of the advancing active margin.

3 Allochthonous Units

3.1 General Remarks

Discussion of the Rhenohercynian allochthon is hampered by the low relief and resulting scarcity of outcrops. The tectonic nature of several units is therefore doubtful.

One unequivocally allochthonous element is the **Giessen-Werra-Harz nappe** (Fig. 1). As described below in more detail, it contains – besides MORB-type basalts and overlying condensed pelagic sediments – a thick sequence of Devonian flysch clastics. Two kinds of evidence testify to the allochthonous nature of this unit:

- the Giessen-Harz rocks are underlain and surrounded by time-equivalent rocks of completely different facies;
- the flysch clastics are derived from the Mid-German Crystalline High to the SE, but the belt of rocks now situated between the clastics and their source is completely devoid of time-equivalent coarser-grained clastic sediments. To explain this facies array, previous authors (e.g., Wittig 1974) have invoked a submarine canyon incised into the sand-free facies, through which the flysch clastics were funneled into the northwestern basin, where they fanned out laterally. However, modern submarine canyons cut down into thick sequences of time-equivalent neritic clastics, and not into thin hemipelagic shales. A tectonic emplacement of the Giessen-Harz graywackes is therefore much more likely.

Essentially the same considerations apply to a narrow belt of rocks which can be traced from the Rhenish Massif via the Harz Mts. into the Flechtingen Hills in E Germany: the **Hörre-Kellerwald-Acker-Gommern zone** (Figs. 1, 3, 4). Like the Giessen-Harz nappe, it contains (besides other lithologies) synorogenic clastic sediments of Late Devonian to early Carboniferous age, which are derived from the Mid-German Crystalline High. By analogy with the Giessen-Harz nappe, it is easier to envisage a tectonic mode of emplacement across the sand-free facies of the southeastern Rhenish Massif. Besides, there are close similarities between the areas on either side of the Hörre-Gommern zone, which include late Emsian quartzites, and widespread Givetian through Viséan bimodal volcanism (see Engel et al. 1983b). Also, the Famennian sandstone turbidites widespread in the northwestern Rhenish Massif (Fig. 2) have also been encountered at one locality south of the Hörre (Weck 1986), and Huckriede (1992) relates coarse-grained Givetian clastic sediments in the Dill syncline (north of the Hörre) to source rocks in the Lahn syncline south of the Hörre. Lastly, an allochthonous nature of the quartzites is suggested by the occurrence of a small slice of Hörre-Kellerwald quartzite at the base of the Giessen nappe (Bischoff and Stoppel 1957). The autochthonist view – mainly based upon "facies transitions" between the supposed allochthon and the autochthon – is laid out in Meischner (1991) and Schwan (1991).

Both the Giessen-Harz nappe and the Hörre-Gommern unit contain tectonic and/or sedimentary inserts of condensed carbonates and clastic sediments of Silurian up to Early Carboniferous age. These rocks are often summarized as the "**Hercynian limestones**" (after their dominantly "hercynian", i.e., pelagic character). The Hercynian limestones represent a third facies realm. It is clearly discernible from the parautochthon, which contains time-equivalent neritic clastic sediments, reef carbonates, and volcanics, and is also in marked contrast with the main portion of the Giessen-Harz and Hörre-Gommern units (pelagic shales and graywacke turbidites). A paleogeographic and tectonic model for the Hercynian limestones is presented below.

The proven and suspected allochthons form narrow belts, which are described below following their stacking order from the SE toward the NW. The widely scattered inserts of the Hercynian limestone facies are treated together in the last paragraph.

3.2 Giessen-Harz Nappe

This unit is exposed (Figs. 1, 3, 4) at the southeastern margin of the Rhenish Massif, possibly in the southern Kellerwald Mts., in the Werra Mts., and in the Harz Mts. (S Harz and Selke nappes).

At its base, there are tectonic slices of metabasalt with the geochemical fingerprints of MOR-type basalts (see Chap. III.B.3). It is relevant for the tectonic interpretation of these rocks, that the metabasalts in the parautochthon uniformly reveal the characteristics of intraplate volcanics. In a few cases, the basalts are overlain, without any tectonic disconformity, by condensed shales and bedded cherts with conodonts and radiolarians. In one section, the Late Lower Devonian (Emsian) through Early Frasnian is contained within only 4 m of pelitic sediments. In another locality, the cherts overlying the metabasalt have yielded Eifelian conodonts (Birkelbach et al. 1988). Hence, the metabasalts must be of pre-Eifelian or pre-Late Emsian age. Presumably, the metabasalts of the Giessen-Harz nappe were extruded during the Early Devonian.

The main part of the Giessen-Harz nappe consists of graywacke turbidites and intercalated shales. The oldest graywackes are of Early Frasnian age in the Giessen area, Frasnian in the southern Kellerwald (Jahnke and Paul 1968), Middle Frasnian in the Werra Mts. (Wittig 1968), and Early Famennian in the South Harz nappe (e.g., Walliser and Alberti 1983). In most of these areas, graywacke sedimentation can be shown to extend higher up into the Famennian; near Giessen, an Early Carboniferous age is possible for the southern part of the graywacke nappe (Dörr 1990).

Near Gießen, the graywackes occur as a thin tectonic klippe with a maximum thickness of 200 m only. However, drillings in the southern Harz Mts. have encountered approx. 2000 m of Devonian greywackes (Schust et al. 1991). Besides, the grade of metamorphism in the southern part of the Rhenish Massif (Hunsrück-Taunus) requires an overburden of 5–12 km (Oncken 1988), part of which is probably provided by the Gießen-Harz Nappe.

It is important to note that an almost identical sequence of lithologies exists in S Cornwall. Even the composition of the graywackes in both areas is more or less identical. However, the English sequences complement the German record, in that they also contain a unit with peridotites, gabbros and sheeted dikes of Devonian age (Lizard complex, see Floyd et al. 1990 for a review).

3.3 Hörre-Gommern Unit (SE part)

The main, southeastern part of the Hörre-Gommern unit is composed of clastic sediments intercalated with shales and bedded cherts, the latter with conodonts and radiolarians. Graywacke turbidites are most frequent, but there are also sandstones. As in the Giessen-Harz nappe, the grawackes contain a broad spectrum of metamorphic and magmatic rock fragments derived from the Mid-German Crystalline High (Homrighansen 1979).

In addition, there are packets of detrital limestones ("Flinzkalke"), which are composed of peloids, intraclasts, and sparse fragments of benthic organisms. The limestones have a minor admixture of terrigenous clasts: besides quartz and some feldspar, there are small fragments of felsic and mafic magmatic rocks. Debris of metamorphic rocks appears to be extremly rare. In the Hörre, these carbonates occur in the late Devonian and Tournaisian (see Fig. 5); in the Harz Mts., they are detectable at various levels from the Middle Devonian through to the Late Famennian (Lütke 1976; Walliser and Alberti 1983).

The detrital carbonates were probably derived from carbonate accumulations on near-by rises (Sect. 3.5), transported into the basin by turbidity currents, and partly redistributed by contour currents (Lütke 1976).

3.4 Hörre-Gommern Unit (NW Part)

The NW part of the Hörre-Gommern unit is composed of a conspicuous sequence of sandstones and quartzites, which comprise the Famennian through Early Carboniferous (possibly up into the Early Carb. IIIα). These rocks form a narrow belt of outcrop (Figs. 1, 3, 4), which can be traced from the Rhenish Massif (Hörre, Kellerwald) via the Harz Mts. into the Flechtingen Hills in eastern Germany, along a distance of approx. 300 km, thus providing the backbone for structural correlations in the Rhenohercynian belt of central Europe.

The quartzite beds partly show large-scale cross-bedding, heavy mineral laminae, and fragmentary benthic organisms (Schmidt 1933). These features are compatible with deposition in shallow water; however, graded bedding in sandstones and quartzites, as well as the frequent intercalation of bedded cherts with radiolarians and conodonts, rather suggest deeper-water origin, where stronger currents might have been produced by contour currents. Sedimentological and petrographic details are available in Homrighausen (1979).

The original site of deposition of the quartzites is entirely enigmatic. There is no indication anywhere of a suitable source within the Rhenohercynian autochthon. A derivation from the southern margin of the Rhenohercynian basin appears more likely, but is difficult to reconcile with the presence of SE-derived graywackes of the same age in the SE part of the Hörre-Gommern zone. Formation of quartzite by the reworking of graywacke, as suggested by Meischner (1968), is unlikely, since quartzites usually result from multiple reworking, and because the heavy mineral spectra of the quartzite and the graywackes are different (Trautnitz 1980). Franke and Engel (1988) have suggested reworking of older quartzites in an accretionary setting, and subsequent tectonic emplacement as the frontal unit of one major allochthon composed of the Giessen-Harz plus Hörre-Gommern units.

If the Hörre-Gommern unit is interpreted as allochthonous, one has to include Devonian and Early Carboniferous chert sequences to the NW of the quartzite belt, which still contain intercalations of quartzites (Meischner 1968, 1991: Kellerwald; Meyer 1968: Harz Mts.).

3.5 "Hercynian Limestones"

In close association with the proven and suspect allochthons, there are numerous small bodies of condensed hemipelagic carbonates. These clearly differ from both the parautochthon to the NW and the coeval clastic sediments to the SE. They contain conspicuous shelly faunas closely related with Bohemia. From their numerous occurrences in the Harz Mts., they have been assigned to a "Hercynian" facies realm.

The Hercynian limestones occur in a variety of tectonic settings (Fig. 4). In the eastern part of the Rhenish Massif, they are contained in tectonic lenses spread out along the NW margin of the Hörre and Kellerwald clastic belts (interpreted above as allochthonous). The "Bickener Schuppe" of the eastern Rhenish Massif and the "Ense Schuppe" of the Kellerwald horst span the Late Lower Devonian through Famennian.

In addition to these coherent sequences at the northwestern margin of the Hörre-Kellerwald belt, condensed pelagic carbonates also occur in a variety of positions, all associated in one or the other way with fault-bounded graywacke units (this applies likewise to the Harz Mts.):

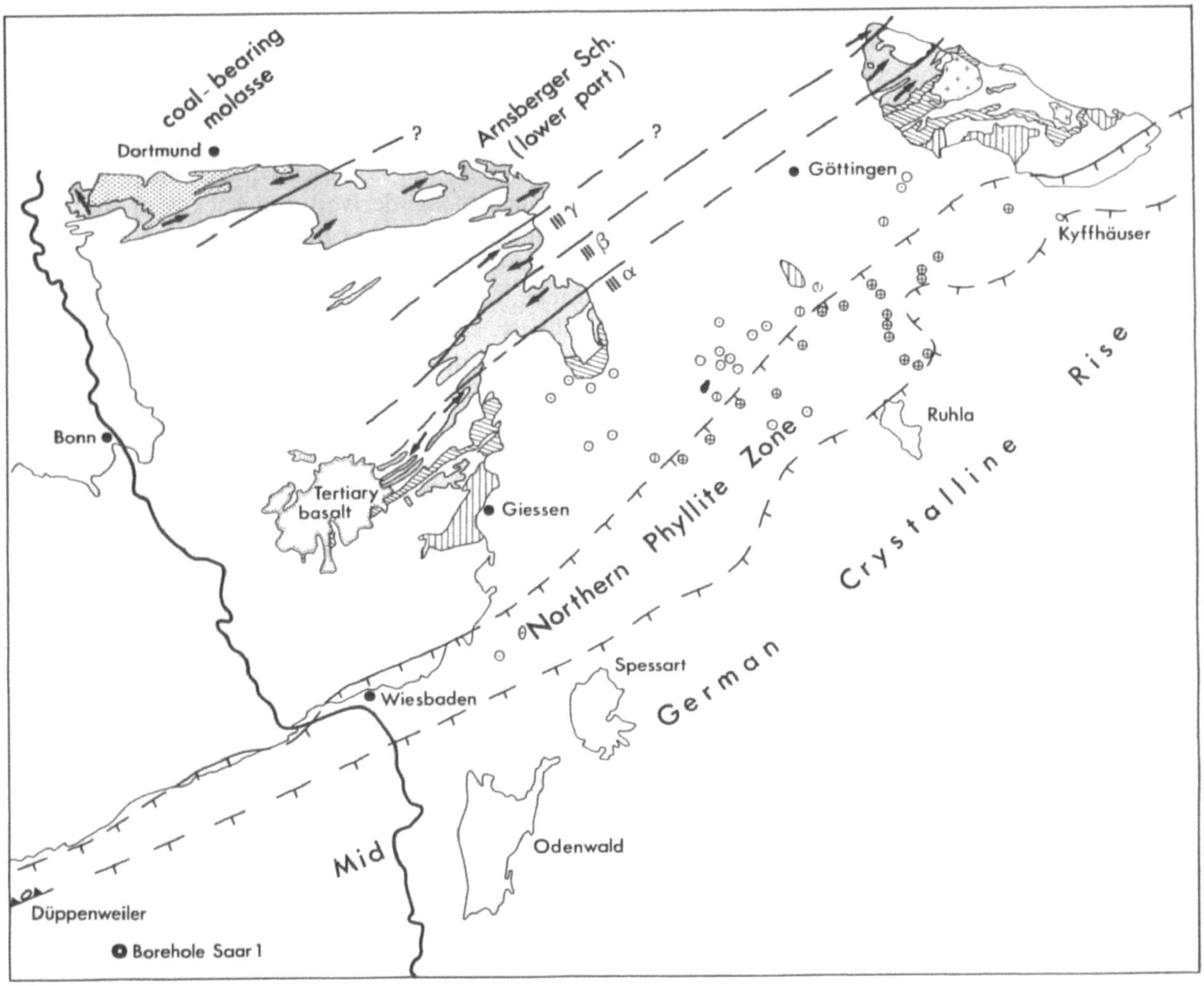

Fig. 3. Areal distribution of allochthonous and autochthonous synorogenic clastic sediments in the Rhenish Massif E of the Rhine River and Harz Mts., showing the NW-ward advance of the flysch front. *Vertical hatches* Gießen-Harz Nappe; *diagonal hatches* Hörre-Gommern ?allochthon; *fine stipple* autochthonous flysch; *coarse stipple* coal-bearing molasse. Boreholes between Harz Mts. and Rhenish Massif (*circles*): *crosses* phyllites; *vertical line* shales, carbonates; *points* graywackes. (Engel and Franke 1983)

- Devonian pelagic carbonates as clasts in graywackes (e.g., Giessen graywacke, Dörr 1990).
- Silurian graptolite shales, as well as Silurian, Devonian and – locally – earliest Carboniferous hemipelagic carbonates as olistoliths in debris flows of Early Carboniferous age of the Harz Mts. (Harzgerode Zone, see modern descriptions and bibliography in Stoppel 1977; Lutzens 1979; Walliser and Alberti 1983; Wachendorf 1986; Buchholz et al. 1990).
- Isolated, yet internally coherent sequences of Silurian and Devonian hemipelagic limestones and shales, situated within graywacke belts (e.g., "Silberstollen Rise" of the Kellerwald: Meischner 1968, 1991).
- Fault-bounded units situated at the margins (i.e., at the structural base) of greywacke belts (e.g., the Paleozoic sequence of the "Lindener Mark" near Gießen: Kegel 1953, or at the NW margin of the Kehna graywacke, SE of the Hörre).

The latter two types often contain not only condensed carbonates, but also **condensed terrigenous clastic sediments** of Late Lower Devonian (**early Emsian**) age, such as the "Erbsloch Grauwacke" of the Kellerwald, the "Rothäuser Grauwacke" of the Harz Mts. (for both see Jahnke 1971), the "Dalmaniten-Sandstein" of the "Lindener Mark" near Gießen (Kegel 1953), or

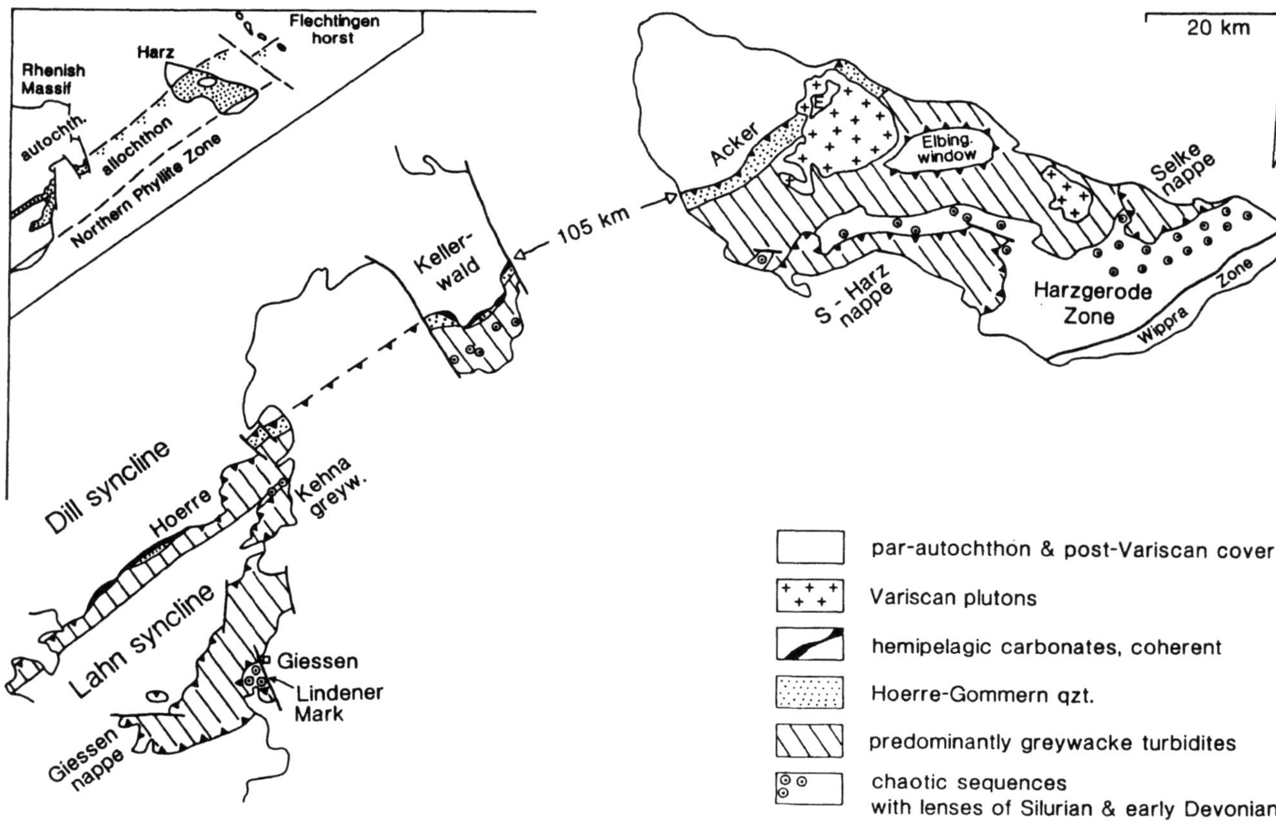

Fig. 4. Allochthonous units (proven and suspected) in the Rhenish Massif and Harz Mts

conglomerates at the base of the Kehna graywacke unit between the Hörre and the Giessen nappe (Bender et al. 1974). It is important to note that the Erbsloch graywacke and related rocks contain quartz, sedimentary and magmatic rock fragments, but next to no metamorphic debris. The Erbsloch-type graywackes contain a rich benthonic fauna (Jahnke 1971). These clastic sediments have apparently been laid down in a neritic environment, and must not be confounded with the late Devonian to Carboniferous synorogenic graywacke turbidites of the Rhenohercynian belt.

The **Lindener Mark** in and around Giessen is a small tectonic slice pinched between reef carbonates of the parautochthon and the base of the Giessen nappe. Its tectonic position and its composition provide clues for the interpretation of the enigmatic Hercynian limestones and related lithologies. A temporary outcrop in Giessen has exposed turbidites and a debris flow, which both contain a mix of siliciclastic components, benthic organisms, and clasts of Silurian and ?early Devonian limestones, up to 1 m in size (Birkelbach et al. 1988). Shaly interbeds with *Nowakia* cf. *acuaria* (det. G. Alberti, Hamburg) indicate a Pragian (= middle early Devonian) age of clastic sedimentation. It is important to state that redeposition occurred already in early Devonian time – in contrast to the early Carboniferous olistostromes of the Harz Mts. An association of Silurian and early Devonian fossils in a debris flow of the western Harz (Stoppel 1977) may be taken to indicate that early Devonian resedimentation phenomena are more widespread than hitherto suspected.

The ostracods and trilobites in the Silurian to Pragian carbonates of the Lindener Mark show affinities to Bohemia and Northern Africa (Alberti 1983; Bahlburg 1985; Schallreuther 1991); affinities with Thuringia and Bohemia have been pointed out for ostracods from early Devonian (late Lockkovian, Pragian) limestone pebbles in the early Emsian conglomerate at the base of the Kehna graywacke (Groos-Uffenorde and Jahnke 1973; Bender et al. 1974).

At Giessen, in the immediate vicinity of the debris flow, there occur isolated lenses of **Ordovician quartzites** ("Andreasteich-Quarzit") with trilobites of Arenig/Llanvirn and Caradoc age (Kegel 1953; Huckriede 1960; Struve 1962). These quartzites resemble the Gondwana-related "Armorican quartzites"). The quartzite lenses at Giessen either represent a separate tectonic slice at the base of the Gießen nappe, or else olistoliths (like the Silurian carbonates). It is important to note that olistoliths of Ordovician quartzites and of Silurian limestones also occur in the Meneage breccia at the base of the Lizard nappe in South Cornwall (Sadler 1974; Franke and Engel 1982; Floyd et al. 1990).

All of the carbonates and mixed siliciclastic/carbonate sediments of the Hercynian limestone association have been tectonically displaced, and some of the carbonates have also undergone resedimentation in early Devonian and/or early Carboniferous time. Under these circumstances, any **palinspastic restoration** must remain speculative.

In the classical view (e.g., Meischner 1991), the condensed pelagic sequences were deposited on narrow rises, separated by basins, which were later filled up, one by one, by the advancing front of flysch sediments. In this model, basin and rise sediments admittedly underwent strong tectonic imbrication, but have largely retained their original palinspastic relationships, and are part of the Rhenohercynian parautochthon.

However, the Bohemian and Thuringian affinities of the Silurian and early Devonian carbonates as well as the Ordovician quartzites locally associated with them strongly suggest derivation from the Armorican microplate, whereas the Rhenohercynian parautochthon represents Avalonia (see Chap. XI). Therefore, at least these older Hercynian limestones and the rocks associated with them should be interpreted as allochthonous, and derived from a position to the south of the present-day Rhenish Massif – this is demonstrable at least for the rocks of the Lindener Mark. In this instance, the limestones might have been deposited in a trench/slope-break environment at the southern, active margin of the Rhenohercynian basin, as has been proposed – without convincing evidence – by Franke and Engel (1988) and Oczlon (1992). This interpretation appears unlikely, since the carbonates have been dragged along at the base of the northwestward advancing Giessen nappe; hence, their original position should have been to the NW of the Rhenohercynian ocean. These considerations rather support the model of Lütke (1990), who referred the limestones to the passive, northern margin of the basin, i.e., to the distal southern margin of the Old Red Continent, out of reach of the Old-Red-derived clastic sediments dominant in the autochthon. In Lütke's interpretation, extensional faulting at the passive margin has created an area with highs and lows; condensed carbonates with a mix of pelagic and benthic organisms were deposited in elevated areas. The detrital limestones ("Flinzkalke") intercalated among the graywackes in the SE part of the Hörre-Gommern unit could well be derived from such rises (Lütke 1976). The passive margin model for the Hercynian limestones is supported by the tectonic relics of a condensed limestone facies in the southernmost part of the Rhenish Massif W of the Rhine River (see Sect. 4).

The very subordinate siliciclastic components which occur in the limestones, and the Late Lower Devonian clastic sediments ("Erbsloch" and other "graywackes") could be derived from older sedimentary, magmatic and subordinate metamorphic rocks exposed by extensional tilting.

The passive margin model combined with the Armorican faunal affinities of the Silurian/early Devonian sediments would suggest that the Ordovician to Pragian deposits of the Lindener Mark and related units were originally deposited on the Armorican microplate, were then docked to the southern margin of the Old Red continent, and became separated from their original setting during the Devonian opening of the Rhenohercynian basin. This scenario is discussed in Chapter XI.

4 Special Developments in the Rhenish Massif W of the Rhine River

The main trends of the sedimentary record are clearly represented also in the areas W of the River Rhine: an initial phase of neritic clastic and carbonate sedimentation, succeeded by hemipelagic environments, some flysch clastics, and – finally – coal-bearing molasse (see the summaries in Kasig and Wilder 1983; Walter et al. 1985). Several characteristics, however, deserve to be mentioned.

In the Ardennes and in the Brabant Massif further N, there are **extensive outcrops of Cambrian through Silurian** sediments. The Cambro-Ordovician is composed of several thousand m of quartzites and shales (Walter 1980; von Hoegen et al. 1985). These sequences are unconformably overlain by Devonian near-shore and fluvial conglomerates, which are of Gedinnian age in the southern Ardennes and become successively younger towards the NW (Frasnian at the southern margin of the Brabant Massif). This diachronous clastic facies records **Devonian transgression after some "Caledonian" uplift**. It is difficult to identify Caledonian tectonic fabrics; any Caledonian deformation and metamorphism can only have been very weak. Still, a "Caledonian" unconformity is unknown in the eastern part of the Rhenish Massif.

At the southern margin of the Rhenish Massif W of the Rhine (Hunsrück), a tectonic raft of paragneiss ("Wartenstein gneiss") exposes **pre-Variscan basement**. Isotopic studies (Rb/Sr mineral and whole rock, K/Ar, Ar/Ar on hornblende and mica: Meisl et al. 1989) indicate metamorphism at 550 ± 20 Ma. The Wartenstein gneiss apparently represents Cadomian basement exposed in a synsedimentary horst zone. Gedinnian redbeds (alluvial fan sediments) directly overlie the gneisses (Oncken, pers. comm. 1993). The Cambro-Ordovician sequences of the Ardennes were either eroded in pre-Gedinnian times, or else have never been deposited.

In the southernmost extremity of the Hunsrück, a baryte mine has encountered a **condensed sequence of hemipelagic carbonates**, which span the late Emsian to latest Famennian (Müller and Stoppel 1981). From an area nearby to the E, Dittmar and Oncken (1992) described middle- to late Devonian olistostromes with carbonate clasts (Oncken, pers. comm. 1993).

The synsedimentary horst of the Wartenstein gneiss and the carbonate-bearing units are referred, by Dittmar and Oncken (1992), to the distal southern margin of the Old Red Continent. These paleogeographic units have not been detected, so far, in the Rhenish Massif E of the Rhine.

A further peculiarity of the area W of the Rhine concerns the **cessation of neritic clastic sedimentation**, and the ensueing formation of shelf carbonates, which occurred already at about the **Lower/Middle Devonian boundary**; E of the Rhine, this change took place in the Late Middle Devonian.

There are only few and small intrusive bodies of felsic or mafic composition, which have been dated as Devonian (Goffette et al. 1991; Kramm and Buhl 1985), or else are tentatively assigned to the Devonian. Apart from these local occurrences, there is **no synsedimentary volcanism** in the areas W of the River Rhine. The volcanic belts at the N margin of the Rhenish Massif and in the Lahn- and Dill-area all terminate to the E of the Rhine (Fig. 2).

In the areas south of the Stavelot-Venn anticline, the youngest beds preserved belong into the Early Upper Devonian; it is unknown, therefore, if Early Carboniferous **flysch** has also been deposited in southern parts of the Rhenish Massif W of the Rhine river. The only flysch clastics known occur in the southern extremity of the area (Hunsrück Mts.); they have been dated as Frasnian and Famennian (Berger et al. 1991; Meyer 1970, 1975), and, therefore, rather compare with the allochthonous flysch of the Giessen-Harz nappe than with the younger clastics of the autochthon.

However, differences between the areas on both sides of the Rhine river become clearly apparent to the north of the Stavelot-Venn anticline, in the Dinant syncline. In this area, there is no transition from flysch into molasse clastics (as in the main part of the eastern Rhenish Massif), but installment of a **carbonate platform from the latest Devonian onwards**. These carbonates represent the southeasternmost part of the Carboniferous Limestone Platform, which covers large areas in Belgium, northern France, and the British Isles. These platform carbonates can be traced eastwards towards Düsseldorf, just E of the Rhine. In all other parts of the Rhenish Massif, and also in boreholes in northern Germany, the platform facies is laterally replaced by dark shales with limestone turbidites (Eder et al. 1983). Only far to the NE, on the south shore of the Baltic Sea, Early Carboniferous, platform carbonates have been encountered in wells (see D. Franke 1990 for a summary).

In the N part of the Rhenish Massif W of the Rhine, this carbonate platform is overlain by coal-bearing molasse sediments of Late Carboniferous age. These paralic sequences can be traced eastwards across the Rhine river, westwards into Belgium and Northern France, and, lastly, into South Wales.

A facies break along a roughly NS-trending line has also been detected in wells in the North Sea, where it has been referred to a synsedimentary "Loke Shear" by Berthelsen (1992). Judging from its effects in the Rhenish Massif, this important transverse element could have acted as a synsedimentary transform fault system, E of which Devonian extension was more pronounced. This interpretation could explain, in the areas E of the Rhine, the importance of synsedimentary volcanism (largely absent from the areas further W), as well as the predominance of thick clastic sequences during late middle Devonian and Carboniferous times (as opposed to time-equivalent carbonates in the W).

5 Special Developments in the Harz Mts. and Flechtingen Horst

The Harz Mts. are situated along-strike to the NE of the Rhenish Massif, and exhibit the same geological record. Some differences in the outcrop pattern are largely due to the more limited extent of the Paleozoic outcrop, and to a different level of erosion, which is apparent from the large extent of allochthonous units. The parautochthon is confined to the northwestern part and to a tectonic window (Elbingerode window) in the north-central part of the Harz Mts.

As already pointed out in Section 2.1, the Harz Mts. have a crystalline Cadomian basement documented in the raft of Ecker Gneis floating in the Late Carboniferous Brocken Pluton (SE of the Acker, Fig. 4). Cambrian and Ordovician rocks have not been detected so far, so that the presence or absence of a Caledonian event is uncertain. The oldest rocks exposed in the autochthon are of Late Early Devonian (Emsian) age, and belong to the clastic shelf at the southern margin of the Old Red Continent. The switch from neritic towards hemipelagic sedimentation, which is due to continued crustal extension, occurred in the Early Eifelian (Early Middle Devonian), i.e., at the same time as in along-strike parts of the Rhenish Massif. This parallelism indicates that the extensional faults that controlled subsidence and facies patterns were oriented more or less parallel with the later tectonic trend.

This is also suggested by the later development of the autochthon in the northwestern Harz, which is characterized by Late Middle Devonian volcanism, Givetian to Frasnian reef growth, Famennian to Early Carboniferous hemipelagic carbonates, pelites and cherts, as well as Early Carboniferous basaltic volcanism. This succession of lithologies and their stratigraphic ages correspond largely to that in along-strike parts of the eastern Rhenish Massif (Dill syncline); minor differences in the timing of vulcanicity have been pointed out by Meischner and Schneider (1970, p. 67).

Lastly, the development of the northwestward-prograding flysch basin exposed in the autochthon corresponds exactly to the development of the eastern Rhenish Massif (see Fig. 3 and the compilation in Ribbert 1975). This indicates that not only crustal extension but also convergence occurred along faults which parallel the present-day tectonic trend.

The well-known correlation between the Rhenish Massif and Harz Mts. is based not only upon the geological record of the autochthon in the northwestern Harz, but also on comparisons of units to the SE, which have been proven or at least suspected to be allochthonous: this applies especially to the Kellerwald quartzite and graywacke units to the SE (Meischner 1968; Schmidt 1931, 1933). The Hörre-Kellerwald quartzite of the Rhenish Massif is continued in the Acker-Bruchberg quartzite of the Harz Mts. Together with the Sieber and Tanner graywackes, the Acker quartzite represents the Harz segment of the Hörre-Gommern unit, which is probably allochthonous (Sects. 3.3, 3.4). The Südharz and Selke graywackes in the southern Harz Mts. are equivalents of the Werra and Gießen graywackes tö the SW; their allochthonous nature is generally accepted (Gießen-Harz nappe).

As can be deduced from the derivation of the term, the Hercynian limestones (Sect. 3.5) are widespread in the Harz Mts. Sheared mudstones with isolated blocks of Silurian through to Early Carboniferous sediments occur in several belts of

Fig. 5I–VII. Stratigraphic columns documenting facies and thickness for the main parts of the Rhenohercynian Belt E of the Rhine River. **I** Lahn Syncline. Less elevated parts of the mid-Devonian volcanic relief are capped by thin pelagic carbonates and Lahn-Dill-type hematite ores **(IA)**, higher volcanic rises serve as foundations for reef growth **(IB)**. Deeper parts of the basin with shales and limestone turbidites (not shown). **II** Taunus Mts. **III** Condensed (hemi-) pelagic carbonates, now found at the NW margin of the Hörre-Kellerwald-Gommern ?allochthon, and as olistoliths in Devonian and Carboniferous debris flow deposits. **IV** Devonian-Carboniferous bedded chert facies, adjacent to the Hörre-Kellerwald-Gommern quartzite. The column presented here is a diagrammatic representation of the Braunauer Kieselschiefer-Becken in the Kellerwald (Meischner 1968, 1991). **V** Hörre-Kellerwald-Gommern quartzite belt (here: Kellerwald quartzite and underlying sequences, after Meischner 1968, 1991, diagrammatic). **VI** Hörre-Kellerwald-Gommern Unit, SE-part (here: Hörre belt of the Rhenish Massif, after Bender 1978, Bender and Homrighausen 1979, diagrammatic). **VII** Gießen-Harz Nappe (here: Gießen nappe, after Birkelbach et al. 1988, diagrammatic)

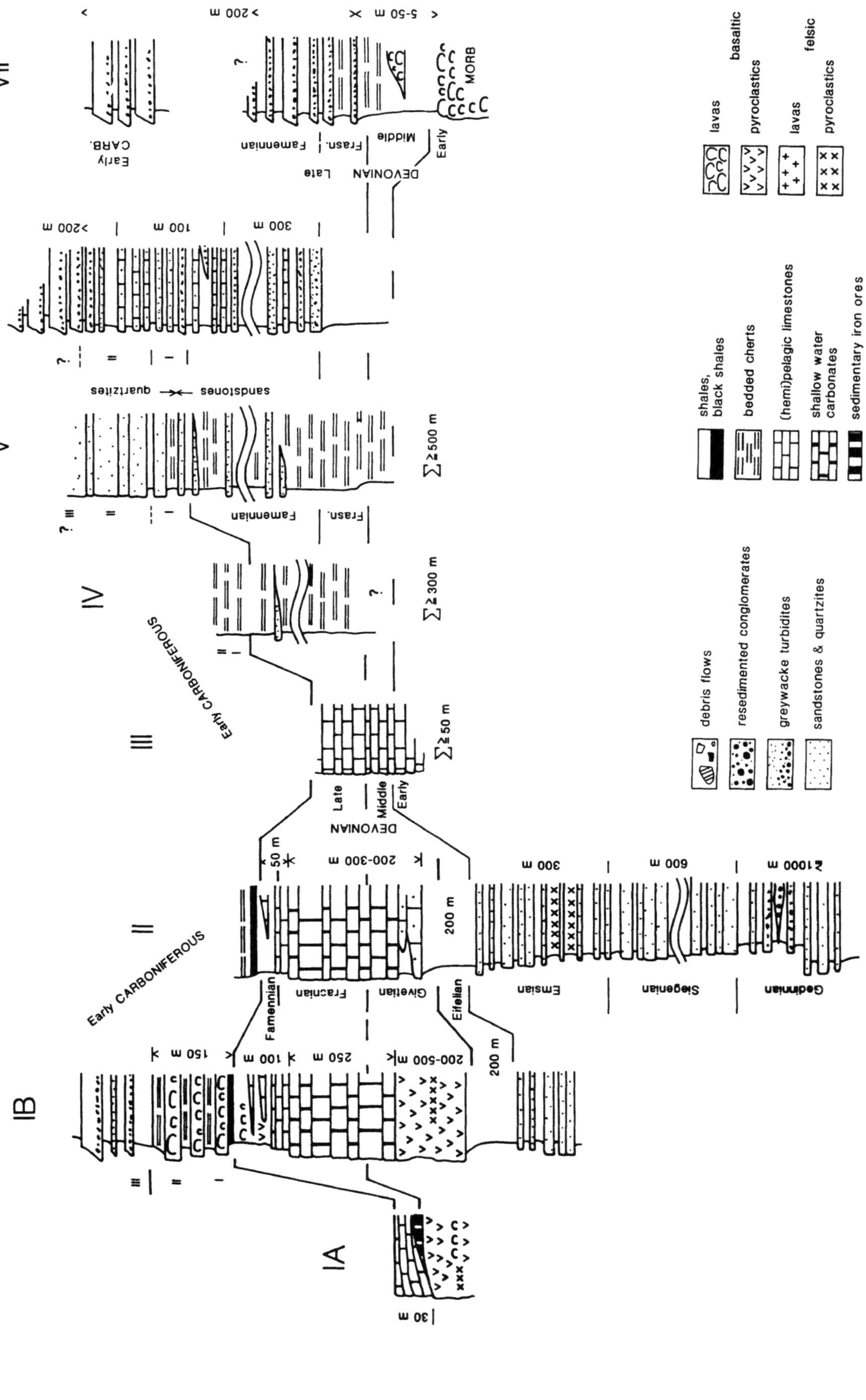

IA
IB
II
III
IV
V
VI
VII
Early CARBONIFEROUS
Early CARBONIFEROUS
DEVONIAN
Late
Middle
Early
Famennian
Frasnian
Givetian
Eifelian
Emsian
Siegenian
Gedinnian
Early CARB.
Frasn.
MORB
sandstones
quartzites
30 m
150 m
100 m
250 m
200-500 m
200 m
50 m
200-300 m
300 m
600 m
≥1000 m
Σ ≥ 50 m
Σ ≥ 300 m
Σ ≥ 500 m
>200 m
5-50 m
debris flows
resedimented conglomerates
greywacke turbidites
sandstones & quartzites
shales, black shales
bedded cherts
(hemi)pelagic limestones
shallow water carbonates
sedimentary iron ores
lavas
pyroclastics
basaltic
lavas
pyroclastics
felsic

outcrop which structurally underlie the various graywacke units, especially in the Harzgerode Zone. These chaotic formations are much more widespread than in the Rhenish Massif. The main constituents of the "Hercynian limestone" realm and its correlatives in the Rhenish Massif have already been discussed in Section 3.5. It should be stressed here again, that the occurrence of these rocks in the form of isolated lenses and the poor quality of the outcrops does, in many cases, not permit to identify the mode of emplacement with certainty: isolated blocks may represent sheared parautochthon, olistoliths, or tectonic shavings carried along at the base of nappes (for contrasting interpretations see, e.g., Lütke 1978; Walliser and Alberti 1983; Wachendorf 1986; Buchholz et al. 1990; Hahlbeck 1990).

Where the chaotic formations contain Famennian and even early Carboniferous limestone blocks, their sedimentary and/or tectonic displacement must have occurred in Carboniferous time. However, some of the blocks (Silurian and early Devonian limestones, Erbsloch-type grywackes) are identical with rock types of the Lindener Mark at Giessen, where redeposition has taken place already in the early Devonian. It appears quite possible that early Devonian passive-margin debris flows of the Lindener Mark type were incorporated, in early Carboniferous time, into a northwestward-driving accretionary wedge, resedimented a second time into the Harzgerode assemblage, and finally overthrust, displaced, and strongly sheared by the Südharz/Selke nappe. Any such odyssee of a limestone block is far from being actually proven, but may serve to highlight the problems met in the Rhenohercynian allochthons.

With regard to the significant areal extent of graywacke units, thrusts, and chaotic formations, it is not surprising that modern tectonic concepts have been conceived in the Harz Mts.: Reichstein (1965) revived earlier nappe tectonic models and suggested extensive olistostromes (with Hercynian limestone clasts) overthrust by the Südharz and Selke graywacke nappes. The extent of the allochthon is still a much-debated question (see, e.g., Wachendorf 1986 and Walliser and Alberti 1983 for contrasting interpretations).

Carboniferous flysch sediments are also present in the Flechtingen hills, a narrow horst zone 50 km NE of the Harz Mts. In the SE part of the Flechtingen horst, at Gommern SE of Magdeburg, small outcrops of early Carboniferous quartzite represent the easternmost known extension of the Hörre-Kellerwald-Acker quartzite slice (Pfeiffer 1967). With respect to the zonation of the Harz Mts., the Flechtingen horst shows a dextral offset of ca. 30 km, probably by late-Variscan dextral wrenching along a northwestern continuation of the Elbe zone.

Taken altogether, the Harz Mts. and Flechtingen horst – unlike the Rhenish Massif W of the Rhine river – do not show any significant differences with the eastern Rhenish Massif. This supports cylindroidal tectonic models for the German segment of the Rhenohercynian belt.

References

Alberti G (1983) Unterdevonische Nowakiidae (Dacryoconarida) aus dem Rheinischen Schiefergebirge, aus Oberfranken und aus N-Afrika (Algerien, Marokko). Senckenb Lethaea 64 (2/4): 295–313

Bahlburg H (1985) Zur faziellen Entwicklung des hercynischen Paläozoikums der Lindener Mark (Rheinisches Schiefergebirge) bei Gießen. Neues Jahrb Geol Paläontol Monatsh 11: 643–651

Baumann A, Grauert B, Mecklenburg S, Vinx R (1991) Isotopic age determinations of crystalline rocks of the Upper Harz Mountains, Germany. Geol Rundsch 80: 669–690

Bender P (1978) Die Entwicklung der Hörre-Zone im Devon und Unterkarbon. Zt Dtsch Geol Ges 129: 131–140

Bender P, Homrighausen R (1979) Die Hörre-Zone, eine Neudefinition auf lithostratigraphischer Grundlage. Geol Palaeontol 13: 257–260

Bender P, Jahnke H, Ziegler W (1974) Ein Unterdevon-Profil bei Marburg a.d. Lahn. Notizbl Hess Landesamtes Bodenforsch Wiesb 102: 25–45

Berger E, Torres P, Wefers J (1991) Zur Stratigraphie der Metamorphen Zone des Hunsrücksüdrands im Rheinischen Schiefergebirge. Neues Jahrb Geol Paläont Monatsh 12: 737–746

Berthelsen A (1992) From Precambrian to Variscan Europe. In: Blundell D, Freeman R, Mueller St (eds) A continent revealed – the European Geotraverse. Cambridge Univ Press, Cambridge

Birkelbach M, Dörr W, Franke W, Michel H, Stibane F, Weck R (1988) Die geologische Entwicklung der östlichen Lahnmulde (Exkursion C am 7. April 1988). Jahresber Mitt Oberrhein Geol Ver NF Stuttgart 70: 43–74

Bischoff G, Stoppel D (1957) Zum Alter des Wollenberg-Kellerwald-Quarzits (Rheinisches Schiefergebirge). Neues Jahrb Geol Paläont Monatsh 1957: 14–24

Buchholz P, Wachendorf H, Zweig M (1990) Resedimente der Prä-flysch- und der Flysch-Phase – Merkmale für Beginn und Ablauf orogener Sedimentation im Harz. Neues Jahrb Geol Paläontol Abh Stuttgart 179 (1): 1–40

Buggisch W (1991) The global Frasnian-Famennian "Kellwasser Event". Geol Rundsch Stuttgart 80 (1): 49–72

Burchette TP (1981) European Devonian Reefs: A review of current concepts and models. In: Toomey DF (ed) European fossil reef models. Soc Econ Paleontol Mineral Tulsa Spec Publ 30: 85–142

DEKORP Research Group: Franke W, Bortfeld RK, Brix M, Drozdzewski G, Dürbaum HJ, Giese P, Janoth W, Jödicke H, Reichert C, Scherp A, Schmoll J, Thomas R, Thünker M, Weber K, Wiesner MG, Wong HK (1990) Crustal structure of the Rhenish Massif: results of deep seismic reflection lines DEKORP 2-North and 2-North-Q. Geol Rundsch Stuttgart 79/3: 523–566

Dittmar U, Oncken O (1992) Anatomie und Kinematik eines passiven varistischen Kontinentalhanges – zum Strukturbau des südwestlichen Rheinischen Schiefergebirges. Frankf Geowiss Arb A 11: 34–37

Dörr W (1990) Stratigraphie, Stoffbestand und Fazies der Gießener Grauwacke (östliches Rheinisches Schiefergebirge). Geol Abh Hessen Wiesb 91: 94 pp

Eder FW, Engel W, Franke W, Sadler P (1983) Devonian and Carboniferous limestone-turbidites of the Rheinisches Schiefergebirge and their tectonic significance. In: Martin H, Eder FW (eds) Intracontinental fold belts. Springer, Berlin Heidelberg New York, pp 93–124

Engel W, Franke W (1983) Flysch-Sedimentation: Its relations to tectonism in the European Variscides. In: Martin H, Eder FW (eds) Intracontinental fold belts. Springer, Berlin Heidelberg New York, pp 289–321

Engel W, Franke W, Langenstrassen F (1983a) Paleozoic sedimentation in the northern branch of the Mid-European Variscides; Essay of an interpretation. In: Martin H, Eder FW (eds) Intracontinental fold belts. Springer, Berlin Heidelberg New York, pp 9–41

Engel W, Franke W, Grote C, Weber K, Ahrendt H, Eder FW (1983b) Nappe tectonics in the southeastern part of the Rheinisches Schiefergebirge. In: Martin H, Eder FW (eds) Intracontinental fold belts. Springer, Berlin Heidelberg New York, pp 267–287

Engel W, Flehmig W, Franke W (1983c) The mineral composition of Rhenohercynian flysch sediments and its tectonic significance. In: Martin H, Eder FW (eds) Intracontinental fold belts. Springer, Berlin Heidelberg New York, pp 171–184

Flajs G, Feist R (1988) Index conodonts, trilobites and environment of the Devonian-Carboniferous Boundary beds at La Serre (Montagne Noire, France). Cour Forsch-Inst Senckenberg 100: 53–107

Floyd PA, Leveridge BE, Franke W, Shail R, Dörr W (1990) Provenance and depositional environment of Rhenoherynian synorogenic greywackes from the Giessen Nappe, Germany. Geol Rundsch Stuttgart 79/3: 611–626

Franke D (1990) Der präpermische Untergrund der Mitteleuropäischen Senke – Fakten und Hypothesen. In: Geologie und Kohlenwasserstofferkundung in Präzechstein der DDR – Nordostdeutsche Senke. Nds Akad Geowiss Veröff Hannover 4: 19–75

Franke W (1973) Fazies, Bau und Entwicklungsgeschichte des Iberger Riffes (Mitteldevon bis Unterkarbon III, NW-Harz, W-Deutschland). Geol Jahrb Hannover A 11: 3–127

Franke W (1989) Tectonostratigraphic units in the Variscan belt of central Europe. Geol Soc Am Spec Pap 230: 67–90

Franke W, Engel W (1982) Variscan sedimentary basins on the Continent, and relations with southwest England. Proc Ussher Soc 5: 259–269

Franke W, Engel W (1986) Synorogenic sedimentation in the Variscan Belt of Europe. Bull Soc Géol Fr 1: 25–33

Franke W, Engel W (1988) Tectonic settings of synorogenic sedimentation in the Variscan Belt of Europe. In: Besly BM, Kelling G (eds) Sedimentation in a synorogenic basin complex. Blackie, Glasgow, pp 8–17

Franke W, Walliser OH (1983) Pelagic carbonates in the Variscan Belt: their tectonic and sedimentary environments. In: Martin H, Eder FW (eds) Intracontinental fold belts. Springer, Berlin Heidelberg New York

Franke W, Eder W, Engel W (1975) Sedimentology of a Lower Carboniferous shelf-margin (Velbert-Anticline, Rheinisches Schiefergebirge, W Germany). Neues Jahrb Geol Paläontol Abh 150 (3): 314–353 Stuttgart

Franke W, Eder W, Engel W, Langenstrassen F (1978) Main aspects of geosynclinal sedimentation in the Rhenohercynian Zone. Z Dtsch Geol Ges 129 (1): 210–216

Fuchs A (1990) Charakter und Ende der devonischen Riffentwicklung im Elbingeröder Komplex (Harz). Facies Erlangen 23: 97–108

Geologisches Landesamt Nordrhein-Westfalen (1984, 1985) Nordwestdeutsches Oberkarbon: Fortschritte in der Geologie von Rheinland und Westfalen. 32 (1984): 399, 33 (1985): 323

Goffette O, Liégeois J-P, André L (1991) Age U/Pb sur zircon dévonien moyen à supérieur du magmatisme bimodal du Massif de Rocroi (Ardenne, France): implications géodynamiques. CR Acad Sci Paris 312 II: 1155–1161

Groos-Uffenorde H, Jahnke H (1973) Die Fauna der Kalkgerölle aus dem unterdevonischen Konglomerat bei Marburg. Notizbl Hess Landesamt Bodenforsch 101: 80–98

Gruppenbericht (1981) Hamburger Arbeitskonferenz Ebbe-Antiklinorium. Mitt Geol Paläontol Inst Univ Hamburg 50: 5–16

Hahlbeck S (1990) Subaquatisch-gravitativ umgelagerte Sedimente des Devons und Karbons um Bad Lauterberg (SW-Harz). Mitt Geol Inst Univ Hannover 31: 105

Hahn H-D (1990) Fazies grobklastischer Gesteine des Unterdevons (Graue Phyllite bis Taunusquarzit) im Taunus (Rheinisches Schiefergebirge). Diss Fachbereich Geowiss Philipps-Univ Marburg/Lahn

Haverkamp J, Kramm U, Walter R (1991) U/Pb isotope variations of detrital zircons in sediments of the Rhenohercynian and their significance for the Paleozoic geotectonic development of NW central Europe. Terra Abstr 3: 207

Homrighausen R (1979) Petrographische Untersuchungen an sandigen Gesteinen der Hörre-Zone (Rheinisches Schiefergebirge, Oberdevon und Unterkarbon). Geol Abh Hessen Wiesb 79: 84 pp
Huckriede R (1960) Die Ordoviz-Gerölle des hessischen Tertiärs und ihre Verwendung in einer pleistozänen Geröllkultur. Neues Jahrb Geol Paläontol Abh Stuttgart 111 (2): 234–256
Huckriede H (1992) Das Barytlager von Günterod in der Dillmulde (Rheinisches Schiefergebirge, Deutschland): eine allochthone Scholle in givetischen Debris-Flow-Sedimenten. Geol Jahrb Hessen 120: 117–144
Jahnke H (1971) Fauna und Alter der Erbslochgrauwacke (Brachio-poden und Trilobiten, Unter-Devon, Rheinisches Schiefergebirge und Harz). Gött Arb Geol Paläontol Göttingen 9: 105
Jahnke H, Paul J (1968) Das Alter der Grauwacken im südlichen Kellerwald (Oberdevon und Unterkarbon, Rheinisches Schiefergebirge). Notizbl Hess Landesamt Bodenforsch Wiesb 96: 68–84
Jankowski B (1990) Sedimentologie des Oberkarbons in Nordwest-Deutschland. DGMK-Ber Hamburg 384-5: 317
Kasig W, Wilder H (1983) The sedimentary development of the western Rheinische Schiefergebirge and the Ardennes (Germany/Belgium). In: Martin H, Eder FW (eds) Intracontinental fold belts. Springer, Berlin Heidelberg New York, pp 185–211
Kegel W (1953) Das Paläozoikum der Lindener Mark bei Giessen. Abh Hess Landesamt Bodenforsch Wiesb 7: 1–55
Kirnbauer T (1991) Geologie, Petrographie und Geochemie der Pyroklastika des Unteren Ems/Unter-Devon (Porphyroide) im südlichen Rheinischen Schiefergebirge. Geol Abh Hessen Wiesb 92: 228
Kramm U, Buhl D (1985) U/Pb zircon dating of the Hill tonalite, Venn-Stavelot Massif, Ardennes. Neues Jahrb Geol Paläontol 171 (1/3): 329–337
Krebs W (1974) Devonian Carbonate Complexes of Central Europe. In: Laporte LF (ed) Reefs in time and space. Soc Econ Paleontol Mineral Spec Publ 18: 155–208
Krebs W (1975) Geologische Aspekte der Tiefenexploration im Paläozoikum Norddeutschlands und der südlichen Nordsee. Erdöl-Erdgas-Z 91: 277–284
Krebs W (1978) Die Kaledoniden im nördlichen Mitteleuropa. Z Dtsch Geol Ges A 27: 489–498
Langenstrassen F (1983) Neritic sedimentation of the Lower and Middle Devonian in the Rheinisches Schiefergebirge east of the River Rhine. In: Martin H, Eder FW (eds) Intracontinental fold belts. Springer, Berlin Heidelberg New York, pp 43–77
Lütke F (1976) Sedimentologische und geochemische Untersuchungen zur Genese der Flinzfazies im Harz (Givet und Oberdevon). Z Dtsch Geol Ges Hannover 127: 499–508
Lütke F (1978) Grundzüge der faziellen und paläogeographischen Entwicklung im südlichen Unter- und Mittelharz. Senckenb Lethaea 58 (6): 473–513
Lütke F (1990) The Rhenian shelf sea-ocean-volcanic arc domain with special reference to the structural development of the Harz Mountains (abs). In: Terranes in the Circum-Atlantic Paleozoic orogens. Int Conf on Paleozoic orogens in central Europe. Göttingen-Giessen, Aug–Sept 1990
Lutzens H (1979) Zur geotektonischen Entwicklung des Harzvariszikums mit besonderer Berücksichtigung synparoxysmaler Resedimentationsprozesse im Mittelharz. Veröff Zentralinst Physik Erde Potsdam 58: 1–19
Meischner KD (1962) Rhenaer Kalk und Posidonienkalk im Kulm des nordöstlichen Rheinischen Schiefergebirges und der Kohlenkalk von Schreufa (Eder). Abh Hess Landesamtes Bodenforsch 39: 47
Meischner D (1964) Allodapische Kalke, Turbidite in Riff-nahen Sedimentations-Becken. Dev Sedimentol 3: 156–191
Meischner D (1968) Stratigraphische Gliederung des Kellerwaldes. Notizbl Hess Landesamt Bodenforsch Wiesbaden 96: 18–30
Meischner D (1971) Clastic sedimentation in the Variscan Geosyncline east of the River Rhine. In: Müller G (ed) Sedimentation of parts of central Europe. 7th Int Sedimentol Congr Guidebook. Kramer, Frankfurt, pp 9–43
Meischner D (1991) Kleine Geologie des Kellerwaldes (Exkursion F am 5. April 1991). Jahresber Mitt Oberrhein Geol Ver NF 73: 115–142
Meischner D, Schneider J (1970) Ober-Devon und älteres Unter-Karbon zwischen Acker und Diabas-Zug im Oberharz. Neues Jahrb Geol Paläontol Abh 135 (1): 42–81
Meisl S, Kreuzer H, Höhndorf A (1989) Metamorphose-Bedingungen und -Alter des Kristallins am Wartenstein bei Kirn/Nahe. Vortrag, 5. Rundgespräch "Geodynamik des europäischen Variszikums", Braunschweig 16.–18.11.1989
Meyer DE (1970) Stratigraphie und Fazies des Paläozoikums im Guldenbachtal/SE-Hunsrück. Ph D Thesis, Univ Bonn, 307 pp
Meyer DE (1975) Geologischer Überblick über den südöstlichen Hunsrück und Beschreibung einer Exkursionsroute. Decheniana 128: 87–106
Meyer K-D (1968) Schwebende Probleme des Harz-Paläozoikums. Z Dtsch Geol Ges Hannover 117 (2/3): 346–367
Meyer W, Stets J (1980) Zur Paläogeographie von Unter- und Mitteldevon im westlichen und zentralen Rheinischen Schiefergebirge. Z Dtsch Geol Ges Hannover 131: 725–751
Mountjoy EW, Krebs W (1983) Diagenesis of Devonian Reefs and Buildups, Western Canada and Europe – a comparison. Z Dtsch Geol Ges Hannover 134: 5–60
Müller G, Stoppel D (1981) Zur Stratigraphie und Tektonik im Bereich der Schwerspatgrube "Korb" bei Eisen (N-Saarland). Z Dtsch Geol Ges Hannover 132: 325–352
Neumann-Mahlkau P (1982) Die Gerölle in den Schwarzbachtal-Konglomeraten bei Ratingen und ihre paläogeographische Aussage. Senckenb Lethaea 63 (1/4): 79–95
Oczlon MS (1992) Examples of Palaeozoic contourites, tempestites and turbidites – classification and palaeogeographic approach. Heidelb Geowiss Abh 53: 57–159

Oncken O (1988) Geometrie und Kinematik der Taunuskammüberschiebung – Beitrag zur Diskussion des Deckenproblems im südlichen Schiefergebirge. Geol Rundsch 77: 551–575

Pfeiffer H (1967) Der Magdeburg-Flechtinger Kulm und seine stratigraphische und regionale Stellung. Geologie 16: 781–790

Press S (1986) Detrital spinels from alpinotype source rocks in middle Devonian sediments of the Rhenish Massif. Geol Rundsch 75 (2): 333–340

Reichstein M (1965) Motive und Probleme erneuter Deckenbauvorstellungen für den Harz Geologie 14: 1039–1076

Ribbert KH (1975) Stratigraphische und sedimentologische Untersuchungen im Unterkarbon nördlich des Oberharzer Diabaszuges (NW-Harz). Gött Arb Geol Paläontol 18: 58

Sadler P (1974) Trilobites fro the Gorran Quartzites, Ordovician of South Cornwall. Palaeontology 17: 71–93

Schallreuther R (1991) Mikrofossilien aus dem Ostrakodenkalk (Silur) der Lindener Mark bei Gießen (Hessen). Neues Jahrb Geol Paläontol Monatsh 2: 105–118

Schindler E (1990) Die Kellwasser-Krise (hohe Frasne-Stufe, Ober-Devon). Gött Arb Geol Paläontol 46: 115

Schmidt H (1931) Die ursprünglichen Zusammenhänge zwischen Harz und Rheinischem Schiefergebirge Naturwissenschaften 19 (45): 911–916

Schmidt H (1933) Der Kellerwald-Quarzit – mit einer Beschreibung seiner Fauna und der aus der Tanner-Grauwacke. Geol Paläontol Abh Jena 19 (23) 5: 3–55 (= 297–349)

Schulz-Dobrick B, Wedepohl KH (1983) The chemical composition of sedimentary deposits in the Rhenohercynian Belt of Central Europe. In: Martin H, Eder FW (eds) Intracontinental fold belts. Springer, Berlin Heidelberg New York, pp 211–230

Schust F, Thomas U, Schwandtke E (1991) Erkenntnisfortschritte zur Geologie des Harzes (V). Z geol Wiss 19 (6): 625–635

Schwan W (1991) Geologie des Acker-Bruchberg-Ilsenburg-Zuges (Oberharz) – Derzeitiger Forschungsstand und Diskussion der Probleme. Zentralbl Geol Paläontol Stuttgart I 7: 787–850

Stoppel D (1977) Schlammstrom-Sedimente im Oberdevon des Südwestharzes und des südlichen Kellerwaldgebirges. Z Dtsch Geol Ges Hannover 128: 81–97

Struve W (1962) Einige Trilobiten aus dem Ordovizium von essen und Thüringen (Phacopina, Asaphina). Senckenb Lethaea 43 (2): 151–180

Teichmüller M (1978) Nachweis von Graptolithen-Periderm in geschieferten Gesteinen mit Hilfe kohlenpetrographischer Methoden. Neues Jahrb Geol Paläontol Monatsh 1978: 430–447

Trautnitz HM (1980) Zirkonstratigraphie nach vergleichender morphologischer Analyse und statistischen Rechenverfahren – dargestellt am Beispiel klastischer Gesteine im Harz. Diss Univ Erlangen, 159 pp

Von Hoegen J, Lemme B, Zielinski J, Walter R (1985) Cambrian and Lower Ordovician in the Stavelot-Venn Massif. A model for depositional history. Neues Jahrb Geol Paläontol Abh Stuttgart 171 (1/3): 217–235

Von Hoegen J, Kramm U, Walter R (1990) The Brabant Massif as a part of Armorica/Gondwana: U/Pb isotopic evidence from detrital zircons. Tectonophysics 185: 37–50

Wachendorf H (1986) Der Harz – variszischer Bau und geodynamische Entwicklung. Geol Jahrb Hannover A 91: 3–67

Walliser OH (1981) The geosynclinal development of the Rheinische Schiefergebirge (Rhenohercynian Zone of the Variscides; Germany). In: Zwart HJ, Dornsiepen UF (eds) The Variscan Orogen in Europe. Geol Mijnbouw 60: 89–96

Walliser OH (ed) (1986) Global bio-events – a critical approach. Lect Notes Earth Sci 8: 442, Springer, Berlin Heidelberg New York

Walliser OH, Alberti H (1983) Flysch, olistostromes and nappes in the Harz Mountains. In: Martin H, Eder FW (eds) Intracontinental fold belts. Springer, Berlin Heidelberg New York, pp 145–170

Walter R (1980) Lower Paleozoic paleogeography of the Brabant Massif and its southern adjoining areas. In: Bless MJM, Bouckaert J, Paproth E (eds) Pre-Permian around the Brabant Massif in Belgium, the Netherlands and Germany. Neth Rijks Geol Dienst Mededelingen 32 (2): 14–25

Walter R, Spaeth G, Kasig W (1985) An outline of the geological structure of the northeastern Hohes Venn area and its northern foreland. Neues Jahrb Geol Paläontol Abh 171 (1/3): 107–216

Weck R (1986) Geologische Untersuchungen am Westrand des Krofdorfer Forstes (Blatt 5317 Rodheim-Bieber). Diploma Thesis, Univ Gießen, 97 pp (unpubl)

Wedepohl K-H, Meyer K, Muecke GK (1983) Chemical composition and genetic relations of metavolcanic rocks from the Rhenohercynian Belt of Northwest Germany: In: Martin H, Eder FW (eds) Intracontinental fold belts. Springer, Berlin Heidelberg New York, pp 231–256

Werner W (1988) Synsedimentary faulting and sediment-hosted submarine hydrothermal mineralization – a case study in the Rhenish Massif, Germany. Gött Arb Geol Paläontol 36: 208

Witten W (1979) Stratigraphie, Sedimentologie und Paläogeographie der Kieselkalke im Unterkarbon II gamma/delta bis III alpha des nordöstlichen Rheinischen Schiefergebirges. Geol Abh Hessen 80: 132

Wittig R (1968) Stratigraphie und Tektonik des gefalteten Paläozoikums im Unterwerra-Sattel. Notizbl Hess Landesamt Bodenforsch Wiesb 96: 31–67

Wittig R (1974) Tiefoberdevonische Grauwacken unter der Hessischen Senke. Notizbl Hess Landesamt Bodenforsch Wiesb 102: 46–57

III.B.2 Structure

O. ONCKEN and K. WEBER

1 Introduction

The Rhenohercynian of central Europe represents a foreland fold-thrust belt. It developed from the sedimentary prism of a passive continental margin and is the lower plate with respect to the active northern margin of the Saxothuringian zone to the south (Mid-German Crystalline High). Its contractional structures are not uniform and depend on the early stretching history, the distance to the suture, the distribution of pre-Variscan basement blocks, and the thermal history.

Basin formation in the Rhenish Massif started on basement constituted of lower Paleozoic sediments and medium-grade gneisses of Cadomian age. The early (continental) rifting event is not reliably documented. Peak crustal stretching occured during the late Lower Devonian, recorded by strong subsidence, formation of growth faults, and widespread volcanism. During the Devonian and Lower Carboniferous, a thick sedimentary pile (3–12 km) of shallow marine clastics and carbonates accumulated. E of the Rhine, there are belts with important bimodal volcanic rocks. The southernmost imbricates of the Rhenohercynian zone probably represent the strongly telescoped slope and rise sequences bordering a minor southern oceanic basin indicated by MOR-type metabasalts (Meisl 1990; Dittmar and Oncken 1992). Preserved relics of a former supracrustal nappe system in the southeastern massif, the Giessen nappe, also contain MOR-type volcanics with Mid-Devonian cherts. These are superposed by flysch of Upper Devonian age (Engel et al. 1983; Grösser and Dörr 1986). The small former ocean basin, which was parent to these rocks, must be rooted between the Rhenohercynian and Saxothuringian zones. Subduction and closure of this basin began in the Upper Devonian. Collision is indicated by the deposition of the first autochthonous flysch during the Tournaisian of the Harz Mts. and the Upper Viséan of the eastern Rhenish Massif (Engel et al. 1983). The earlier extensional structures are stongly overprinted by collisional structures particularly in the southern part of the Rhenohercynian. They are better preserved in the middle and northern part, and are represented by growth faults in the Early through Late Devonian clastic shelf (Weber 1981; Werner 1989; Oncken 1989). Former listric normal faults are reactivated as reverse faults during collision (Oncken 1989; Franke et al. 1990).

East and west of the River Rhine, the DEKORP 1a–c and 2-N sections explore the Rhenohercynian crust from the Molasse Basin in the NW to the Taunus/Hunsrück Mts in the SE (Franke et al. 1990; DEKORP research group 1985, 1991). The Ardennes are crossed by an ECORS section (Cazes et al. 1985; Le Gall 1992).

2 Structures

The large-scale internal architecture of the belt is different in the arcuate part in the west, the Ardennes, from that in the linear part in the central and eastern parts, the Rhenish Massif, and the Harz Mts. Moreover, the Rhenohercynian shows several distinct changes in structural style in cross section from the northern foreland to the suture with the Saxothuringian microplate. Upright large-scale structures dominate the north, partly above a thin-skinned décollement (Ardennes). Towards the south the thick folded and cleaved Devonian basin filling progressively forms large imbricate fans separated by prominent SE-dipping thrusts (Siegen, Boppard, and Taunuskamm thrusts). Relics of an allochthonous nappe system are preserved in the eastern Rhenish Massif and the Harz Mts.

2.1 Ardennes

The structure of the Ardennes is characterized by thin-skinned tectonics in the roof of the London-

Dallmeyer et al. (Eds.)
Pre-Permian Geology of Central and Eastern Europe

Brabant Massif (Meissner et al. 1980, 1981; Cazes et al. 1985; DEKORP Research Group 1990; Le Gall 1992). The chain of major basement anticlines bordering the southern rim of the Ardennes are ramp anticlines and antiformal stacks formed above a major infracrustal ramp at the margin of the Brabant Massif. This margin originated during Lower Devonian crustal stretching and separates unrifted crust in the northwest from strongly rifted crust in the southeast (von Winterfeld and Walter 1993). The boundary coincides with the outer rim of the seismically transparent middle and lower crust in the DEKORP 1-A and DEKORP 2-N sections (DEKORP Research Group 1991). A frontal thrust is developed only west of the river Rhine and is represented by the Aachen Thrust (Meissner et al. 1981; DEKORP Research Group 1991) and its western prolongation, the Faille du Midi (Bois et al. 1986; Raoult and Meillez 1987; Bouckaert et al. 1988; Figs. 1, 3; see also Bertrand 1883 for early ideas on the nappe concept). The Aachen thrust crops out as an imbricate fan and grades southwards into a very gently (4°) SE-dipping detachment at a depth of 3 to 4 km. (Meissner et al. 1980, 1981; Durst 1985; DEKORP Research Group 1991; von Winterfeld and Walter 1993). To the north, the Aachen Thrust may die out as a blind thrust underneath the Wurm Syncline, which represents the eastern continuation of the Belgian Namur Syncline. Both form the transition into a narrow, weakly deformed molasse basin bordering the Brabant Massif.

The Stavelot-Venn Massif exposes the Caledonian Basement of the Brabant Massif. It has been displaced across the oblique east-dipping ramp onto the flat of the Aachen Thrust. The transition into the Eifel Syncline/thrust unit is characterized by asymmetric upright folds and backthrusting along the Malsbenden Fault. The structure reflects the flat/ramp-geometry of the detachment below the Eifel Syncline and the Stavelot-Venn Massif. Across the ramp the upper crustal detachment of the Aachen Thrust descends to the mid-crustal decollement below the Eifel North-South Zone the nature of which has been

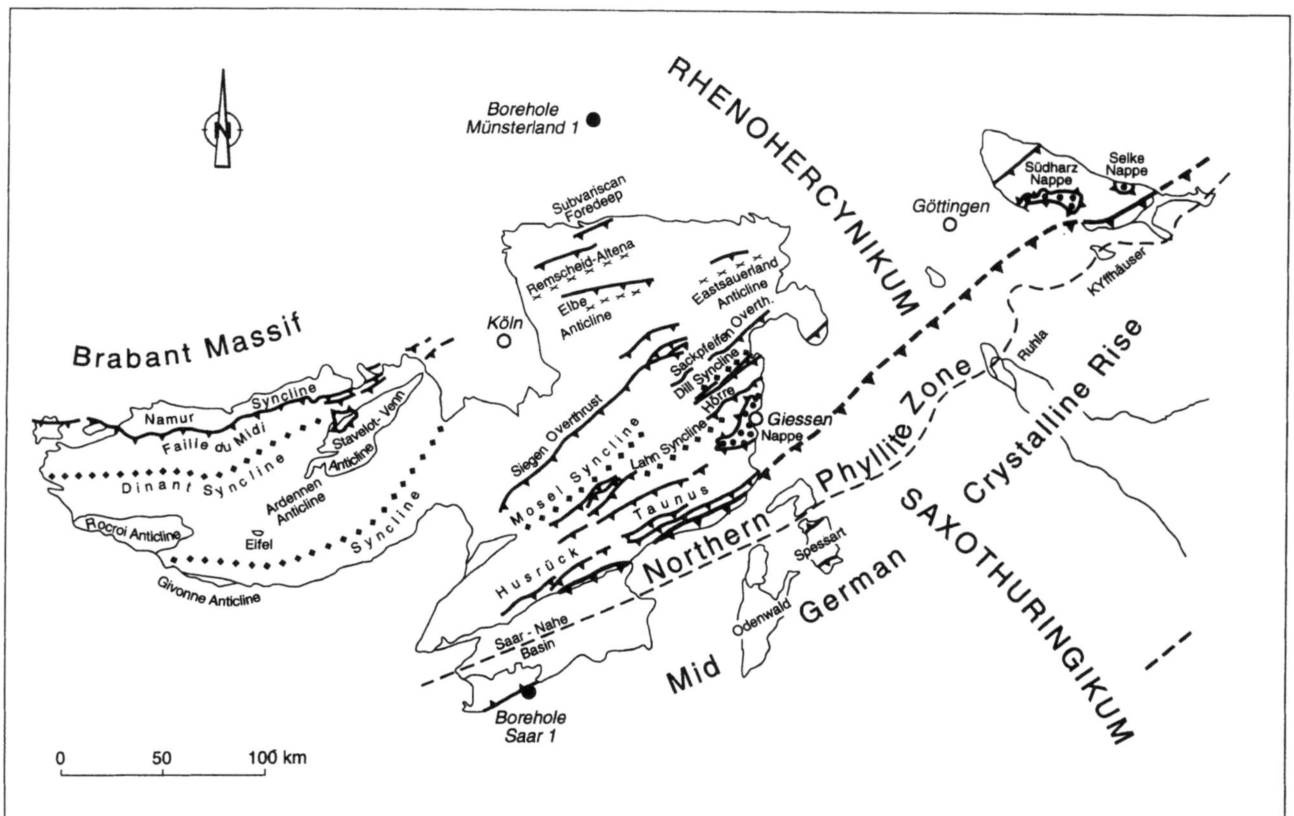

Fig. 1. Structural map of the Rhenohercynian belt

at the heart of a long-lasting debate. This zone forms a major structural depression of the Eifel Nappe trending N-S. Recent crustal scale balancing has shown that its western margin is controlled by the above ramp which is oblique to the general strike. Its eastern end is caused by another oblique burried ramp which is imaged by the western end of the large ramp anticline of the East Eifel (von Winterfeld and Walter 1993; Fig. 3).

2.2 Rhenish Massif

This intracrustal boundary marks the transition from thin-skinned into thick-skinned tectonics to the east and southeast which controls the different deformation style of the Rhenish Massif and its foreland east of the river Rhine. The structural style of the broad Molasse Basin is well known from drillholes and coal mines west of the DEKORP 2N-line (Drozdzewski et al. 1985). It is characterized by wide, open synclines and narrow, upright thrust-related anticlines. Some of the early thrusts, which show ramp and flat geometry, are folded. Faulting is most intensive inside narrower anticlinal structures. It results from incongruent folding and tends to compensate insufficient ductile strain inside fold structures. Thin-skinned tectonics and frontal thrusts, comparable to the Faille du Midi – Aachen Thrust, are not known east of the River Rhine (Brix et al. 1988; Franke et al. 1990).

The overall structural style of the central and eastern Rhenish Massif is dominated by small to large-scale NW-facing folds and thrusts. These contractional structures branch off from a mid-crustal decollement, which can be clearly seen in the DEKORP 2-N seismic section and which was predicted by Weber (1978, 1981) on the basis of the kinematics of folding and thrusting. This master décollement rises from about 6 s TWT in the south to about 4 s TWT below km 150 in the north (see Fig. 2). It forms an approximately 3 km wide zone, which is more or less identical with the transparent zone and becomes narrower to the north. The nature of the transparent zone itself is not known. Section balancing of the imbricate sheets of the foreland propagating schuppen and fold structures (Oncken 1989; Behrmann et al. 1991; Dittmar et al. 1994) suggest that the crustal scale décollement at its actual depth of roughly 13–18 km is located within the basement below the Devonian basin filling (see Fig. 3; Behrmann et al. 1991).

Although no seismic evidence is found west of the river Rhine, the surface structure indicates an important ramp structure which delineates the northwestern rim of the Mosel Syncline west of the River Rhine and its continuation, the Dill

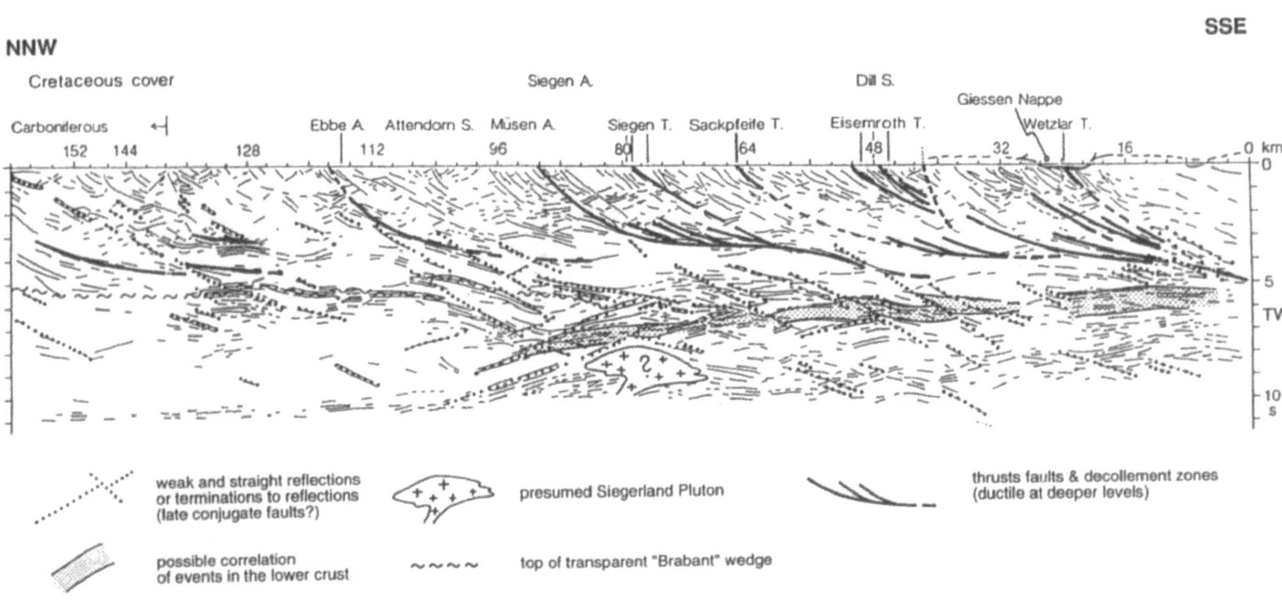

Fig. 2. The crustal structure of the eastern Rhenish Massif and its interpretation in the DEKORP 2 North reflection seismic section. (Line drawing from migrated section; from DEKORP research group, Franke et al. 1990)

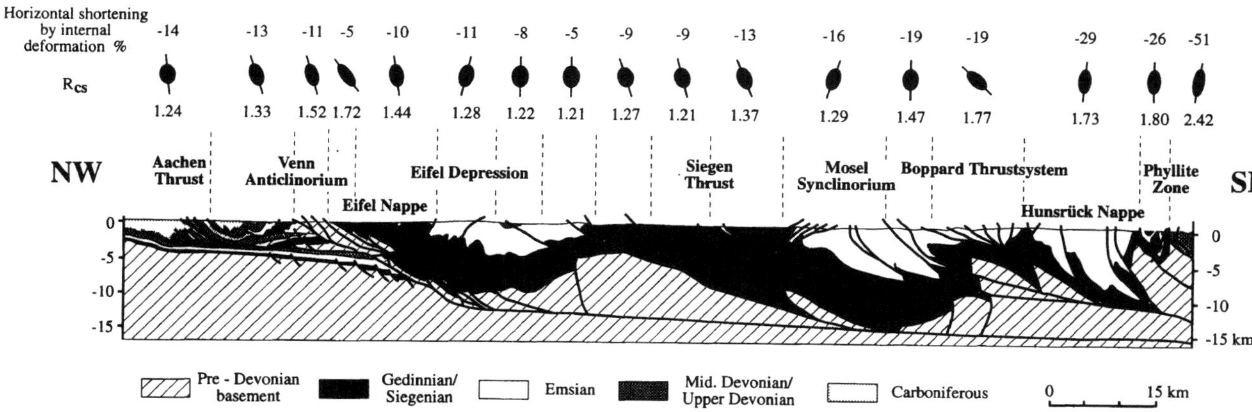

Fig. 3. Cross section through western Rhenish Massif parallel to DEKORP 1 seismic section showing transition from thin-skinned architecture of the Ardennes to the medium-skinned style of the central and southern Massif. *Ellipses above the section* show orientation and averaged shape of finite strain of homogeneously deformed units. (Section by Dittmar et al. 1994)

Syncline, east of the River Rhine. The structure of the Mosel Syncline has been a matter of discussion since Scholtz (1930) and Kienow (1934) interpreted the large fold- and cleavage fan south of the Siegen (Mayen) Thrust as primary structures of a large and very deep southeast-facing syncline (see also Meyer and Stets 1975). On the basis of very detailed structural analyses, Hoeppener (1957) drew the conclusion that the fan and meiler structure resulted from rotation of primarily NW-facing folds and their corresponding cleavage. Hoeppener (1957) explained the rotation by differential uplift and subsidence. In analogy to similar structures and deformation paths in the Taunus Mts, Weber (1978) applied his Taunus model (see below) to the Mosel Syncline and interpreted the fold and cleavage fans as resulting from backward rotation of thrust sheets during displacement along listric thrust planes. Detailed structural analyses and balanced cross sections carried out by Oncken (1989) revealed a complex flat and ramp geometry and confirmed the rotational deformation path related to flats and ramps of different size on top of the major detachments. Moreover, recent studies suggest that, similar to the southern rim of the Massif (see Oncken et al., Chap. III.C.2), large-scale backfolding of older imbricate fans is involved in the formation of these structural associations. The Siegen Thrust may branch off from a gently south-dipping master decollement in the middle crust. East of the River Rhine, the Siegen Thrust splits into several thrusts forming the Siegen Schuppen-Anticline with the Muesen leading thrust in the NW and the more steeply dipping trailing thrusts of the Siegen thrust zone in the SE.

The Boppard thrust (Oncken 1989) forms the southern boundary of the Mosel Syncline and, together with its continuation east of the Rhine (Weber 1978, 1981; Sauerland 1980), presumably forms the basal thrust of the strongly folded and highly imbricated Hunrück and Taunus Nappe (see also Dittmar and Oncken 1992). These thrusts and the Taunuskamm thrust (Oncken 1988a) of the southern Rhenish Massif are examples of such wedge-shaped foreland propagating fold and thrust systems in the roof of gently southeast-dipping master detachments. The kinematics of the fan-shaped structure show northwestward rotation and overprinting of the frontal horse by steeply dipping crenulation cleavage and associated D2-folds after translation across a ramp. To the south, the older horses are increasingly rotated backwards. The former NW-facing thrusts were transformed into apparent normal faults and the primarily SE-dipping second cleavage and associated D2 folds were rotated into a NW-dipping attitude. However, also here, antithetic rotation of imbricates is amplified by backfolding (see above). Extensional kink bands are the youngest fabrics. Where the pronounced anisotropy of the first cleavage is more or less vertical, a horizontal crenulation cleavage can be observed which is contemporaneous to the formation of the extensional kink bands. They are most frequent in the centre of the fan and are interpreted as collapse structures (Weber 1978, 1981).

The southernmost thrust sheet is the Metamorphic Zone of the southern Hunrück and

Taunus (Northern Phyllite Zone). It forms a multiply folded, imbricate belt of partly mylonitic metasediments and metavolcanics. The basal thrust has been rotated into a vertical or steeply NW-dipping position and is overprinted by sinistral as well as dextral strike-slip displacement. This thrust sheet is cut by the South Hunsrück Boundary Fault and continues probably below the Saar-Nahe Basin underneath the seismically transparent wedge of the Mid-German Crystalline Rise (see Chap. IV.C.3).

Two systems of mostly minor wrench faults are dominant in the Rhenohercynian. These are NS-trending sinistral and EW-trending dextral systems which correspond to the NW-SE shortening. The dextral system, however, is more pronounced than the sinistral increments of the deformation path (Weber 1978, 1980, 1981; Oncken 1988a,b). Analysis of incremental strains from pressure shadows, veins, and striated faults across the Massif (Fig. 4; Oncken 1988a,b; Dittmar et al. 1994) show a change from foreland-directed stretching/transport of the imbricates to late westward displacement which increases in importance towards the southern rim of the Massif. These features might be related to the bulk dextral transpression in the Variscan Belt of Western and Central Europe (Arthaud and Matte 1977).

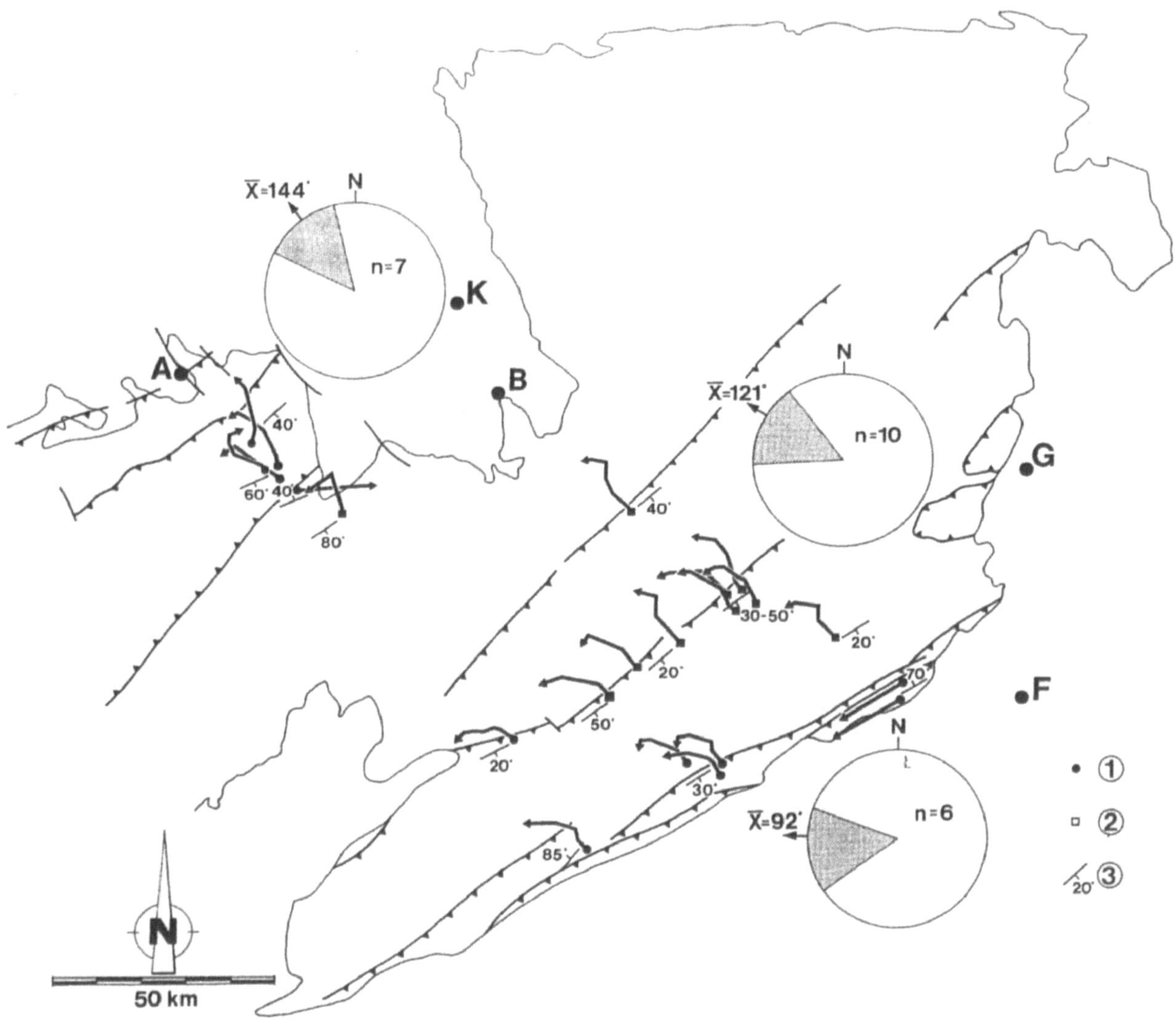

Fig. 4. Incremental strains measured from pyrite pressure shadows and vein generations in cleavage plane. *Rose diagrams* show azimuthal scatter of the orientation of the finite stretching vectors for northern Massif, southern, and southern rim of the Rhenish Massif. *Arrows* give arithmetic mean of measured directions (see Dittmar et al. 1994 for details). *Length of arrows* is normalized to a standard length

2.3 Harz

The Harz Mts. to the northeast owe their oblique exposure of Variscan rocks to an uplift along a transpressional wrench fault system at its northeastern border during Cretaceous times (Wrede 1988). The internal structural geometry of the Harz Mts. largely resembles that of the eastern Rhenish Massif. Large-scale features of the latter can be traced across the outcrop gap to the northeast along strike. The northern part (Upper Harz) is again dominated by large upright to NW-facing folds with minor imbrication. The middle and southeastern parts (Central and Lower Harz) are formed by equally NW-facing strongly imbricated and deformed Devonian and Lower Carboniferous sequences (Schwab 1976; Walliser and Alberti 1983; Wachendorf 1986) which have been assembled in several different fault-bounded zones. The southernmost unit (Harzgeröder Zone) contains a thick sequence of cleaved and sheared Lower Carboniferous wildflysch sediments which have been deposited as gravitational slide masses (Schwab 1979). They are overlain by further thrust units (see below). At its southern margin, the Harzgeröde Zone rests with tectonic contact against the steeply dipping Wippra Zone which forms part of the Northern Phyllite Zone bordering the southern rim of the Rhenohercynian belt. A long-lived controversy in the past has focused on whether these imbricates form more or less far traveled nappes (Kossmat 1927; Reichstein 1965; Engel et al. 1983; Walliser and Alberti 1983), or a foreland-propagating imbricate fan with inclusion of minor and major slide masses (Schwab 1979; Lutzens 1991), or else facies-controlled inversion of earlier extensional structures (Gallwitz 1958) in an otherwise autochthonous scenario (see below).

In contrast to the more western parts of the Rhenohercynian belt, the Harz Mts. have been intruded by two major postorogenic granite and gabbro plutons at the Carboniferous/Permian boundary (Vinx 1983).

2.4 Allochthonous Units

Nappe tectonic concepts for the southeastern Rheinisches Schiefergebirge were originally proposed by Kossmat (1927). The Upper Devonian and Lower Carboniferous Giessen Greywacke (Henningsen 1961; Engel et al. 1983; Stibane et al. 1984; Birkelbach et al. 1988) rests unconformably on Lower through Upper Devonian rocks. The base of the Giessen Nappe shows highly strained cataclasites and low temperature mylonites without quartz recrystallization. At the base of the nappe, and in upthrusted units within the nappe, there are tectonic slices of NMORB-type metabasalts (Wedepohl et al. 1983; Grösser and Dörr 1986) of probably Lower Devonian age (Birkelbach et al. 1988). A number of exotic rocks of Ordovician, Silurian, Lower Devonian, and Lower Carboniferous age at the margins of the Giessen Greywacke, including the phyllonitic Solmstal Beds (Emsian), are interpreted as allochthonous rafts at the base of the nappe (Engel et al. 1983). Similar units of Frasnian/Famennian graywackes and phyllitic to phyllonitic slates are found in the Werra-Grauwackengebirge between the Rhenish Massif and in the Harz Mts (Südharz and Selke Nappes) itself. They represent klippen of a formerly coherent nappe complex that probably rooted in the suture zone north of the Mid-German Crystalline rise.

The internal structure of the Giessen Nappe is characterized by recumbent NW-facing folds, that have been imported with the nappe. The basal contact is more or less horizontal and cuts the folds and thrusts in the autochthon. The Giessen Nappe represents a supracrustal nappe. Its mode of emplacement is still being debated. Oncken (1988a) discussed its origin as the roof thrust of the Taunuskamm Nappe. The Taunuskamm Nappe itself overlays the Taunus Nappe, which has been described above. The roof thrust of the Taunus Nappe represents the floor thrust of the Taunuskamm Nappe. Since the Giessen Nappe shows no internal flat and ramp geometry and must be derived from the active margin south of the Rhenohercynian Suture, it possibly forms the highest nappe complex of the southern Rhenish Massif.

3 Deformation of the Rhenohercynian Crust

The generalized cross section in Fig. 3 shows the finite strain pattern across the western Rhenish Massif, the included ellipses visualizing the average orientation and shape of the finite strain ellipse for homogeneously deformed units (see Dittmar et al. 1994). The cross section shows a general increase of strain to the southeast in discrete steps across major thrusts (especially at the

Boppard Thrust system). Net orogenic shortening of the upper crust of the fold and thrust belt is approximately 50% (also see Wunderlich 1964). It is mainly achieved by folding and tectonic stacking of the deformed basin filling. Overall shortening values vary between 10 and 30% at the northern front to 60–70% at the southern internal margin of the belt. The contribution of distributed deformation – as recorded in the finite strain data – to overall horizontal shortening reaches values between 0% (orogenic front) and 50% (southern rim of the Massif).

Distribution of strain in the orogenic wedge is controlled by different factors at different scales. Above the grain scale, heterogeneity is caused by changes in lithology and average grain size, by the structural position in folds and the distance from thrusts in the hangingwall (at deeper crustal levels), and by the regional distribution of the metamorphic grade, i.e., by the dominant deformation mechanism. Due to the latter, an exponential increase of strain is observed with approach to the ductile lower crust (Dittmar et al. 1994). At high crustal levels deformation is strongly localized in a network of discrete faults. The deformation is progressively delocalized with the onset of thermally activated, especially grain-size sensitive deformation mechanisms and forms shear strain gradients in thrust hangingwalls. The strain pattern controlled by these aspects is very complex in terms of both magnitude and orientation of the finite strain axes. The latter generally do not exhibit a simple relationship to tectonic structures, with exception of a close relationship to the dominant pervasive cleavage (Dittmar et al. 1994) which was largely formed at peak thermal conditions (Weber 1976; Ahrendt et al. 1983; Oncken 1991). Furthermore, the pattern is affected by late noncoaxial strain increments especially in the southern part of the thrust belt with its late W-ward motion (in shallow dipping imbricates) or strike-parallel motion (in steeply dipping imbricates).

Apart from the Northern Phyllite Zone and southern parts of the Taunus Mts, the bulk Rhenohercynian crust has never been thickened strongly. This is also apparent from the fact that no metamorphic basement is exposed. Shape and orientation of the bulk finite strain ellipse derived from cleavage fabrics (Weber 1976, 1981) and from finite strain measurements (Dittmar et al. 1994), however, show that the upper crust has been thickened considerably in the plane of the cross section (Fig. 3). Thickening of the upper crust has been compensated by postorogenic isostatically driven unroofing of the overlying nappe stack and possibly also by extensional flow in the lower crust particularly south of the seismically transparent wedge in the DEKORP 2-N section (Fig. 2). The horizontal or shallow dipping reflectors in the lower crust of the Rhenohercynian Belt south of the transparent wedge suggest an extensional flow regime by bulk Moho-parallel simple shear in the lower crust (cf. Weber 1981). The detachment of the middle crust (transparent zone) partitions crustal flow in contractional and extensional strain regimes of the upper and lower crust.

References

Ahrendt H, Clauer N, Hunziker JC, Weber K (1983) Migration of folding and metamorphism in the Rheinisches Schiefergebirge deduced from K-Ar and Rb-Sr age determinations. In: Martin H, Eder W (eds) Intracontinental fold belts. Springer, Berlin Heidelberg New York, pp 323–338

Arthaud F, Matte P (1977) Late Paleozoic strike-slip faulting in southern Europe and northern Africa: results of a right-lateral shear zone between the Appalachians and the Urals. Geol Soc Am Bull 88: 1302–1305

Behrmann J, Drozdzewski G, Heinrichs T, Huch M, Meyer W, Oncken O (1991) Crustal-scale balanced cross sections through the Variscan fold belt, Germany: the central EGT-segment. Tectonophysics 196: 1–21

Bertrand M (1883) Rapports de structure des Alpes de Glaris et du bassin houiller du Nord. Bull Soc Geol Fr 12: 318–330

Birkelbach M, Dörr W, Franke W, Michel H, Stibane F, Weck R (1988) Die geologische Entwicklung der östlichen Lahnmulde. Exkursion C am 7. April 1988, Jahresber Mitt Oberrhein Geol Vereinig 70: 43–74

Bois C, Cazes M, Damotte B, Galdeano A, Hirn A, Mascle A, Matte P, Raoult JF, Torreilles G (1986) Deep seismic reflection profiling of the crust in northern France: the project ECORS. In: Barazangi X, Brown LD (eds) Reflection seismology: a global perspective. Am Geophys Union, Washington DC, Geodyn Ser 13: 21–29

Bouckaert J, Fock W, Vandenberghe N (1988) First results of the Belgian geotraverse 1986 (BELCORP). Ann Geol Belg 111: 279–290

Brix MR, Drozdzewski G, Greiling RO, Wolf M, Wrede V (1988) The N Variscan margin of the Ruhr coal district (western Germany): structural style of a buried thrust front? Geol Rdsch 77: 115–126

Cazes M, Torreilles G, Bois C, Damotte B, Galdéano A, Hirn A, Mascle A, Matte Ph, Van Ngoc P, Raoult J-F (1985) Structure de la croûte hercynienne du

Nord de la France: premiers résultats du profil ECORS. Bull Soc Géol Fr 8 I (6): 925–941

DEKORP Research Group (1985) First results and preliminary interpretation of deep-reflection seismic recordings along profile DEKORP 2-South. J Geophys 57: 137–165

DEKORP Research Group (1991) Results of the DEKORP 1 (BELCORP-DEKORP) deep seismic reflection studies in the western part of the Rhenish Massif. Geophys J Int 106: 203–227

Dittmar U, Oncken O (1992) Anatomie und Kinematik eines passiven varistischen Kontinentalrandes – zum Strukturbau des südwestlichen rheinischen Schiefergebirges. Frankf Geowiss Arb A 11: 34–37

Dittmar U, Meyer W, Oncken O, Schievenbusch T, Walter R, Winterfeld C v (1994) Strain partitioning across a fold and thrust belt: the Rhenish Massif, Mid-European Variscides. J Struct Geol 16: 1335–1352

Drozdzewski G, Engel H, Wolf R, Wrede V (1985) Beiträge zur Tiefentektonik westdeutscher Steinkohlenlagerstätten. Geol Landesamt Nordrhein-Westfalen (Krefeld), 235 pp

Durst H (1985) Interpretation of a reflection seismic profile across the northeastern Stavelot-Venn Massif and its northern foreland. Neues Jahrb Geol Paläontol Abh 171: 441–446

Engel W, Franke W, Langenstrassen F (1983a) Paleozoic sedimentation in the northern branch of the Mid-European Variscides – essay of an Interpretation. In: Martin H, Eder W (eds) Intracontinental fold belts. Springer, Berlin Heidelberg New York, pp 9–41

Engel W, Franke W, Grote C, Weber K, Eder W, Ahrendt H (1983b) Nappe tectonics in the southeastern part of the Rheinisches Schiefergebirge. In: Martin H, Eder FW (eds) Intracontinental fold belts. Springer, Berlin Heidelberg New York, pp 267–287

Franke W, Bortfeld RK, Brix M, Drozdzewski G, Dürbaum HJ, Giese P, Janoth N, Jödicke H, Reichert C, Scherp A, Schmoll J, Thomas R, Thünker K, Weber K, Wiesner MG, Wong HK (1990) Crustal structure of the Rhenish Massif: results of deep seismic reflection lines DEKORP 2North and DEKORP 2-North-Q-. Geol Rdsch 79: 523–566

Gallwitz H (1958) Die Bedeutung der Stratigraphie für die tektonische Analyse des Harzes. Geologie 7: 448–464

Grösser J, Dörr W (1986) MOR-Typ-Basalte im östlichen Rheinischen Schiefergebirge. Neues Jahrb Geol Paläontol Monatsh 12: 705–712

Henningsen D (1961) Untersuchungen über Stoffbestand und Paläogeographie der Gießener Grauwacke. Geol Rdsch 52: 600–626

Hoeppener R (1957) Zur Tektonik des SW-Abschnittes der Moselmulde. Geol Rdsch 46: 318–348

Kienow S (1934) Die innere Tektonik des Unterdevons zwischen Mosel und Nahe. Jahrb Preuss Geol Landesamtes 54: 58–95

Kossmat F (1927) Gliederung des varistischen Gebirgsbaues. Abh Sächs Geol Landesamtes 1: 39

Le Gall B (1992) The deep structure of the Ardennes Variscan thrust belt from structural and ECORS seismic data. J Struct Geol Oxford 14 (5): 5531–5546

Lutzens H (1991) Flysch, Olisthostrome und Gleitdecken im Unter- und Mittelharz. Z Geol Wiss 19/6: 617–623

Meisl S (1990) Metavolcanic rocks in the "Northern Phyllite Zone" at the southern margin of the Rhenohercynian Belt. In: Field guide Mid German Crystalline Rise & Rheinisches Schiefergebirge. Int Conf on Paleozoic orogens in central Europe, Göttingen-Gießen 1990, pp 25–42

Meissner R, Bartelsen H, Murawski H (1980) Seismic reflection and refraction studies for investigating fault zones along the Geotraverse Rhenoherzynikum. Tectonophysics 64: 59–84

Meissner R, Bartelsen H, Murawski H (1981) Thin-skinned tectonics in the northern Rhenish Massif. Nature 290: 399–401

Meyer W, Stets J (1975) Das Rheinprofil zwischen Bonn und Bingen. Z Dtsch Geol Ges 126: 15–29

Murata H, Weber K (1983) Experimental Study of the Subfluence Tectonics in the Rheinisches Schiefergebirge. In: Martin H, Eder FW (eds) Intracontinental fold belts. Springer, Berlin Heidelberg New York, pp 339–354

Oncken O (1988a) Geometrie und Kinematik der Taunuskammüberschiebung, Beitrag zur Diskussion des Deckenproblems im südlichen Schiefergebirge. Geol Rdsch 77: 551–575

Oncken O (1988b) Aspects of the reconstruction of the paleostress history of a fold and thrust belt (Rhenish Massif, FRG). Tectonophysics 152: 1940

Oncken O (1989) Geometrie, Deformationsmechanismen und Kinematik großer Störungzonen der hohen Kruste (Beispiel Rheinische Schiefergebirge). Geotekton Forsch 73: 1–215

Oncken O (1991) Aspects of the structural and paleogeothermal evolution of the Rhenish Massif. Ann Soc Géol Belg 113: 139–159

Raoult JF, Meilliez F (1987) The Variscan Front and the Midi Fault between the Channel and the Meuse River. J Struct Geol Oxford 9 (4): 473–479

Reichstein M (1965) Motive und Probleme erneuter Deckenbauvorstellungen für den Harz. Geologie 14: 1039–1079

Sauerland V (1980) Tektonische Entwicklung des westlichen Taunus zwischen Taunuskamm überschiebung und Lahnmulde. Diss Univ Göttingen

Scholtz H (1930) Das varistische Bewegungsbild, entwickelt aus der inneren Tektonik eines Profils von der Böhmischen Masse bis zum Massiv von Brabant. Fortschr Geol Paläontol 8: 235–316

Schwab M (1976) Beiträge zur Tektonik der Rhenoherzynischen Zone im Unterharz. Jahrb Geol 5/6: 9–117

Schwab M (1979) Gravitational slide masses in the Harz. Veröff Zentralinst Physik Erde 58: 23–46

Stibane F, Dörr W, Michel H (1984) Zur stratigraphischen Stellung der Gießener Grauwacke (Rheinisches Schiefergebirge). Neues Jahrb Geol Paläontol Monatsh: 173–178

Vinx R (1983) Magmatische Gesteine des Westharzes. Fortschr Mineral 61 (Beih 2): 3–30
Von Winterfeld C, Walter R (1993) Die variszische Deformationsfront des nordwestlichen Rheinischen Schiefergebirges – ein bilanziertes geologisches Tiefenprofil über die Nordeifel. Neues Jahrb Geol Paläontol Monatsh 1993: 305–320
Wachendorf H (1986) Der Harz – variszischer Bau und geodynamische Entwicklung. Geol Jahrb A91: 3–67
Walliser OH, Alberti H (1983) Flysch, olisthostromes and nappes in the Harz Mountains. In: Martin H, Eder FW (eds) Intracontinental fold belts. Springer, Berlin Heidelberg New York, pp 145–169
Weber K (1976) Gefügeuntersuchungen an transveralgeschieferten Gesteinen aus dem östlichen Rheinischen Schiefergebirge. Neues Jahrb D 15: 1–99
Weber K (1981) Kinematic and metamorphic aspects of cleavage formation in very low-grad metamorphic slates. Tectonophysics 78: 291–306
Weber K (1978) Das Bewegungsbild im Rhenoherzynikum – Abbild einer variszischen Subfluenz. Z Dtsch Geol Ges 129: 249–281
Weber K (1980) Die Entwicklung tektonischer Gefüge in gefalteten und geschieferten Spilitkörpern aus dem Ostsauerländer Hauptsattel (Rheinisches Schiefergebirge). Berl Geowiss Abh A20: 223–235
Weber K (1981) The structural development of the Rheinisches Schiefergebirge. In: Zwart HJ, Dornsiepen UF (eds) The Variscan Orogen in Europe. Geol Mijnbouw 60: 149–159
Weber K, Behr HJ (1983) Geodynamic interpretation of the mid-European Variscides. In: Martin H, Eder FW (eds) Intracontinental fold belts. Springer, Berlin Heidelberg New York, pp 323–338
Wedepohl KH, Meyer K, Muecke GK (1983) Chemical composition and genetic relations of meta-volcanic rocks from the Rhenohercynian Belt of northwest Germany. In: Martin H, Eder FW (eds) Intracontinental fold belts. Springer, Berlin Heidelberg New York, pp 231–256
Werner W (1989) Synsedimentary Faulting and Sediment-Hosted Submarine-Hydrothermal Mineralization in the Late Palaeozoic Rhenish Basin (Germany). Geotekton Forsch 71: 1–305
Wrede V (1988) Der nördliche Harzrand – flache Abscherbahn oder wrench-fault-system? Geol Rdsch 77: 104–114
Wunderlich H (1964) Maß, Ablauf und Ursachen orogener Einengung am Beispiel des Rheinischen Schiefergebirges, Ruhrkarbons und Harzes. Geol Rdsch 54: 561–582

III.B.3 Igneous Activity

P.A. Floyd

1 Introduction

The external zone of the Variscan orogen in northern Europe can be traced from SW England via the Rhenish Massif to the Harz Mountains (a distance of ca. 1300 km) and then further into eastern Europe. Although Upper Paleozoic outcrops are discontinuous along the Rhenohercynian Zone, their common history is emphasized by similar sedimentary sequences, the nature of associated bimodal acid-basic volcanism and comparable tectonometamorphic events (Matthews 1977; Franke and Engel 1982; Holder and Leveridge 1986a; Franke 1989a). In broad terms, the sedimentary and volcanic history of the Rhenohercynia Zone records the formation of small late Silurian/early Devonian rift-generated basins that were progressively closed by plate convergence throughout the late Devonian to late Carboniferous (Behr et al. 1984; Franke 1989b). The presence of extensive submarine basaltic rocks testifies to periods of crustal extension, whereas the chemical characterization of their primary composition indicates that both mid-ocean ridge basalts (MORB) and within-plate basalts (WPB) are represented in the Rhenohercynian Zone (Herrmann and Wedepohl 1970; Floyd 1982a, 1984; Floyd et al. 1983; Wedephol et al. 1983; Grosser and Dorr 1986; Schmincke and Sunkel 1987; Platen et al. 1989; Platen 1991). These features imply that extension of the Paleozoic Mid-European continental plate was sufficient to generate some oceanic crust, of which the Lizard ophiolite is also representative (Bromely 1979; Kirby 1979; Holder and Leveridge 1986b). The Rhenohercynian Zone of northern Europe is thus envisaged as comprising ensialic basins (with WPB) that were sufficiently attentuated in their internal zones to produce a narrow segment of oceanic crust (with MORB).

This chapter reviews and compares the geochemical data for basalt lavas in the SW England and German sectors of the Rhenohercynian Zone to (1) determine the main chemical features and degree of variation, and (2) assess their eruptive setting on the basis of chemical discrimination and comparison with possible modern analogues. A review of the chemical data is now appropriate in the light of the wide compositional range and different comagmatic groups exhibited by Rhenohercynian basalts and also the variation of MORB types associated with spreading centres, in both large ocean basins and back-arc basins. The relevance of the basalts to plate tectonic reconstructions of the Variscan orogen is particularly important in the light of the apparent lack of destructive margin volcanism during basin closure.

2 Volcanic Environment

In simple terms, there are two modes of emplacement of basic magma in the Rhenohercynian Zone: (a) submarine basalt pillow lavas, sometimes with minor associated hyaloclastites, and (b) high-level dolerite intrusions (invariably sill-like). Some intrusions with pillowed or irregular tops were emplaced near the water-sediment interface and may be genetically related to spatially associated lavas. This paper is only concerned with the basaltic lava flows, which due to subsequent low-grade regional metamorphism are strictly metabasalts (spilites), although relict mineral assemblages and stable trace element geochemistry enable their original composition to be determined.

Volcanism in the Rhenohercynian Zone occurred intermittently throughout the Devonian and into the Lower Carboniferous with major activity culminating in the Upper Devonian-Dinantian. Submarine basalts are mainly associated with sedimentary basinal facies, although interpretation of their tectonic environment suggests two different eruptive regimes: (1) as the volcanic carapace of ocean crust, and (2) local volcanism associated with basin margin rifts. The

Dallmeyer et al. (Eds.)
Pre-Permian Geology of Central and Eastern Europe

evidence for the ocean crust environment is based on the associated pelagic sediments and the basalt chemistry, which has a MORB-like signature. Most of these basalts have been located within, or at the base of allochthonous units in both SW England and Germany. For example, basaltic clasts are common throughout the South Cornish melange (Barnes 1983, 1984) and variably tectonized pillow lava units within the Giessen Nappe (Grosser and Dorr 1986; Platen 1991). Both occurrences are associated with Middle and Upper Devonian sediments and represent the generation of limited ocean crust in the internal sector of the Rhenohercynian Zone. To the north of the oceanic area, rifting produced linear fractures in continental crust at the margins of ensialic basins, and up which magma penetrated to develop local pillow lava horizons. These stratigraphically limited, parautochthonous volcanic sequences are typical of the more external basins to the north of the southerly derived nappe sheets, and range in age from Upper Devonian through to Visean. The basalts are chemically distinct, being typically WPB, and exhibit a much greater range of composition and magma type. For example, the Upper Devonian pillow lavas of the north Cornish coast are alkali basalts (Floyd 1982a) extruded along the northern margin of the Trevose basin (Selwood and Thomas 1986), whereas the Lower Carboniferous alkaline Tintagel volcanics (Rice-Birchall and Floyd 1988) lie on the southern margin of the Bude basin (Matthews 1977). Similarly, in the Lahn-Dill region of the Rheinisches Schiefgebirge and the Harz Mountains to the north of the Giessen Nappe are Upper Devonian and Lower Carboniferous basinal sequences with tholeiitic and alkaline WPB (Herrmann and Wedepohl 1970; Wedepohl et al. 1983; Platen 1991).

Apart from the allochthonous lavas with a MORB-like chemical signature, the Lizard ophiolite is the only remnant of a predominantly plutonic section of oceanic crust in the Rhenohercynian Zone (Badham and Kirby 1975; Bromley 1979; Kirby 1979). Only the basaltic dykes from the Lizard Complex will be considered in this chapter, as they initially represented liquid compositions, relative to the metagabbros which are largely cumulates (Kirby 1978). However, as the dykes are not chemically related to the basalt clasts within the melange below the Lizard thrust (Barnes 1984; Floyd 1984), the dismembered ophiolite appears to be without a volcanic carapace, unless this is now represented by the highly foliated Landewednack hornblende schists.

3 Plate Tectonic Models

The Variscan orogen is a complex fold belt for which there are a number of plate tectonic models, although they will not be considered in detail here (for recent reviews and problems, see Franke 1989b). The chemical signature of the basalt lavas is important in this context and provides some evidence as to their eruptive setting. Most models recognize the presence of both ensialic basins, with intraplate basalts, as well as some oceanic crust. However, the actual plate tectonic setting of the Devonian oceanic basins in the Rhenohercynian Zone and the mechanism for their generation is in dispute (e.g., Behr et al. 1984).

To provide a general framework for the subsequent chemical data, current plate tectonic models for the Rhenohercynian Zone are outlined below:

1. Back-arc basin related to northerly directed subduction of an ocean to the south of Armorica (Anderton et al. 1979; Floyd 1982a; Leeder 1982; Ziegler 1984). In this model, back-arc oceanic crust is represented by the Lizard ophiolite and allochthonous Devonian basalts. If the Lizard ocean crust was both generated and subsequently obducted in the Devonian (Barnes and Andrews 1984; Davies 1984), the width of ocean crust produced over this time period was probably relatively narrow, assuming moderate spreading rates. This model has been extended by Isaac (1985) into allochthonous Upper Devonian-Visean sequences of central SW England, which are interpreted as a dismembered flysch-ophiolite complex (Chandler and Isaac 1982). However, Selwood and Thomas (1986) suggest that the evidence for a Carboniferous ocean in the allochthon is tenuous and unsupported by lithofacies analysis. Also, most models envisage closure of the oceanic area by the end of the Devonian via continent collision (Franke 1989a).

2. Intracratonic strike-slip pull-apart basins (Badham 1982; Barnes and Andrews 1986; Holdsworth 1989). Basin development was initiated in an intracontinental transform system with sediment-fill being derived from the adjacent cratonic area to the north. Progressive oblique attenuation of continental crust produced Lizard-

type ocean crust in (possibly) a number of small basins.

3. Small ocean basins related to southerly-directed subduction under an active arc, on, or to the north of, the Normannia High-Mid-German Crystalline Rise (Anderson 1975; Behr et al. 1984; Holder and Leveridge 1986a; Floyd and Leveridge 1987; Liew and Hofmann 1988; Franke 1989b). Although no arc basalts have been positively identified in the Rhenohercynian Zone, evidence for the previous existence of an arc coeval with the ocean basin was considered to be preserved within Devonian synorogenic sediments (Floyd and Leveridge 1987; Floyd et al. 1990). These sediments contain large quantities of arc-derived calc-alkali acidic clasts, although Mid-Proterozoic model ages indicate they were eroded from much older arc systems that were not tectonically connected to the Devonian ocean (Floyd et al. 1991a). If southerly directed subduction was involved in the subsequent closure of the Devonian Rhenohercynian ocean, then typical arc volcanism is missing or the arc was not located on the southern margin of the zone, but elsewhere. For example, early Carboniferous I-type granitoids in the northern Vosges, Black Forest, and Odenwald could represent the plutonic roots of arc magmatism related to such subduction (Liew and Hofmann 1988). Also, recent chemical data by Meisl (1990) on metavolcanic rocks from the Northern Phyllite Zone (at the southern margin of the Rhenohercynian Zone) has indicated the presence of both MORB and WPB, as well as an extensive calc-alkali meta-andesite to metarhyolite suite (South Taunus area) – the latter of which could represent the "missing" volcanic arc. However, the age of the acidic metavolcanics is uncertain as, on the basis of preserved microfossils, the associated sediments appear to range from late Silurian to Devonian (Reitz 1989).

What mechanism caused the initial (?)early Devonian rifting in the Rhenohercynian Zone is open to speculation, although it has been suggested that a heat source might have been provided by a relict Caledonian spreading axis over which continental crust migrated (Matthews 1978; Franke 1989a).

4 Primary Petrographic Features

Completely fresh basalt lavas are absent from the Rhenohercynian Zone, as the original mineralogy has been invariably replaced by low-grade secondary assemblages. However, relict mineral assemblages and textures, together with stable element geochemistry, provide a clue to the primary characteristics of many lavas. Petrographic features of Devonian and Carboniferous basaltic lavas from the Rhenohercynian Zone have been described by many workers, including Middleton (1960), Juteau and Rocci (1974), Floyd (1976, 1982b, 1983), Schmincke and Pritchard (1981), Wedepohl et al. (1983), Schmincke and Sunkel (1987), and Platen (1991); other and previous descriptions in the Variscan literature have tended to interpret the whole "spilitic" assemblage as primary, rather than secondary (e.g., Hentschel 1952; Lehmann 1974).

Most lavas are submarine pillow basalts, with thin flows, associated hyaloclastites, and pillow breccias being relatively minor. The stratigraphically limited, but often wide areal extent, of many pillow lava horizons suggests that they formed numerous, small, fissure-fed, interconnected lava mounds on the seafloor, and that few developed into typical seamounts as observed in modern oceans.

Many lavas are typically vesicular, with alkali basalts often showing several concentric zones of vesicles, whereas tholeiites with MORB features may be non-vesicular or exhibit very small vesicles in flow margins. Glassy selvages to pillows are rarely observed, being generally spalled off during cooling or now totally replaced by secondary chlorite. However, in a rare example, Schmincke and Pritchard (1981) have described spalled pillow margins of fresh brown sideromelane in fine-grained hyaloclastites from Carboniferous volcanics in the Rhenish massif. The rapidly chilled nature of the lavas is inferred from relict quench textures exhibiting curved plagioclase microlites (some with forked terminations), and plumose clinopyroxene variolites. Any original glass matrix is now completely replaced by secondary minerals, although the more crystalline interiors of large pillows are composed of primary, matrix clinopyroxene, plagioclase and opaques. Electron microprobe analysis shows the primary plagioclase to be nearly always replaced by albite (± other secondary phases), although preserved relict segments are labradorite (Wedepohl et al. 1983). Pyroxene relicts in tholeiites are typically calcic augite with variable Ti contents (Floyd and Rowbotham 1979; Schmincke and Sunkel 1987).

Very few basalts are completely aphyric and a number of phenocryst assemblages have been recognized. For example, plagioclase phyric alkali basalts are common in Devonian and Carboniferous sequences from north Cornwall (Floyd 1982b; Rice-Birchall and Floyd 1988; Floyd and Rice-Birchall 1995), whereas Cr spinel-olivine-clinopyroxene-plagioclase and clinopyroxene-plagioclase phyric basalts have been documented in the Lahn-Dill area (Schmincke and Sunkel 1987). In general, primary features are better preserved in the German sector of the Rhenohercynian Zone than in SW England. Overall, the interpretation of petrographic features indicate the presence of variably evolved (spinel-olivine to plagioclase-clinopyroxene phenocrystic assemblages), alkali and tholeiitic, aphyric and phyric lavas throughout most of the tectonic zone.

5 Low-Grade Alteration

The classic "spilite" volcanics of the Rhenohercynian Zone are now recognized as low-grade altered basalts, generally characterized by assemblages typical of the pumpellyite or, more rarely, greenschist facies of regional metamorphism (e.g., Herrmann and Wedepohl 1970; Hermann et al. 1974; Floyd 1976, 1983, 1984; Barnes and Andrews 1981; Robinson 1981; Robinson and Read 1981; Floyd and Rowbotham 1982; Wedepohl et al. 1983; Schmincke and Sunkel 1987).

The basaltic lavas were subjected to two phases of alteration: (a) initial submarine weathering soon after extrusion on the seafloor, and (b) subsequent regional metamorphism that accompanied Variscan deformation. The present mineralogical and chemical composition of the basalts largely reflects the effects of the second pervasive, higher-temperature, hydrothermal process. However, sideromelane, palagonite and highly oxidized hyaloclastites associated with pumpellyite facies seamount basalts from the Lahn-Dill area not only reflect extreme disequilibrium conditions, but represent relicts of an earlier seafloor alteration (Schmincke and Sunkel 1987). Common low-grade metamorphic assemblages within the pillow lavas and their vesicles include: chlorite-quartz-carbonate, chlorite-albite-carbonate-sphene-rutile-quartz (dominant assemblages), pumpellyite-albite-actinolite-chlorite-clinozoisite, pumpellyite-albite-prehnite-quartz, and pumpellyite-albite-quartz-chlorite (e.g., Herrmann and Wedepohl 1970; Floyd and Rowbotham 1979; Barnes and Andrews 1981; Schmincke and Sunkel 1987). Some of the above references quote electron microprobe analyses of secondary phases, and show that chlorites range from brunsvigite to pychnochlorite, pumpellyites include both Al- and Fe-rich varieties, epidotes are Fe-rich, and plagioclase is generally pure albite or Ab_{5-10} with variable Or content. Secondary rutiles are particularly rich in Nb and Zr (Williams and Floyd 1981). The presence of diagnostic prehnite-pumpellyite facies assemblages indicates maximum temperatures of 300–350 °C and pressures around 2 kb (Liou et al. 1985).

In high-temperature (300–450 °C) experimental studies on seawater-basalt reactions, Mottl (1983) suggested that the observed secondary mineralogy was a direct result of the water-rock ratio. This model indicates that the characteristic Rhenohercynian metabasalt assemblage of chlorite-albite-quartz ± epidote ± actinolite, is the result of water-rock ratios of between 30–50. However, these values must be treated with caution, as the nature of circulating fluids during regional metamorphism is probably very different to those generated in seafloor hydrothermal systems.

Higher grades of metamorphism are exhibited by metabasic rocks in localized areas of SW England.

1. The Tintagel Volcanic Formation schists of north Cornwall are characterized by low green-schist chlorite-albite-carbonate-epidote assemblages and superimposed, static growth, biotite-actinolite porphyroblasts developed at temperatures up to ca. 500 °C (Robinson and Read 1981; Primmer 1985). The local elevation of temperatures (above the surrounding pumpellyite facies terrain) was a consequence of local crustal thickening and enhanced heat flow from the rising Bodmin Moor granite pluton (Primmer 1985).

2. The quartz-muscovite-chlorite-albite-epidote metavolcanic greenschists of the Start Complex (south Devon) are also greenschist facies (Floyd et al. 1993). Comparison of retrogressive white mica in the Start schists indicates that the last stages of metamorphism were similar to that experienced by the adjacent Devonian phyllites, but at a slightly higher pressure (Robinson 1981).

3. The highest grades of regional metamorphism in SW England are represented by the

metavolcanic Landewednack hornblende-epidote-plagioclase-quartz schists of the Lizard Complex, and are characteristic of the amphibolite facies. These schists are considered to represent part of the Lizard ophiolite and developed their strong foliation and high-grade during suboceanic shearing as a prelude to obduction (Vearncombe 1979).

During low-grade mineralogical degradation the basalts have also been changed chemically, with large-ion-lithophile (LIL) elements, like K, Rb, Ba and Sr, being mobilized. Elemental exchange between basalt and seawater during submarine weathering, and hydrothermal alteration are well documented (e.g., Hart et al. 1974; Humphris and Thompson 1978; Thompson 1983); similar element mobilities have been recorded in Rhenohercynian Zone (meta)basalts (e.g., Herrmann and Wedepohl 1970; Floyd 1976, 1983; Wedepohl et al. 1983; Schmincke and Sunkel 1987). In summary, chemical alteration in the Rhenohercynian pillow lavas relative to fresh submarine basalts is indicated by:

1. high H_2O^+, CO_2 (or loss-on-ignition) and Fe_2O_3/FeO ratios,
2. highly variable contents, but with generally net gains, in K, Rb, Cs, Na, U, Mg, and net losses in Si, Ca, Sr, Ba, Fe (unless highly carbonated when Ca and Sr increase),
3. loss of characteristic magmatic relationships between pairs of coherent elements, e.g., very variable K/Rb ratios and data scatter in plots involving a LIL element and a stable element, such as Zr, and
4. high whole rock $\delta^{18}O$ values (10–16‰, relative to ca. 6‰ for fresh basalts; Wedepohl et al. 1983); the range of which probably reflects mixing between secondary, hydrous phyllosilicates and fresh basalt (cf. Muehlenbachs and Clayton 1972, for altered MORB).

6 Geochemical Characteristics

Because of the variably altered nature of the Rhenohercynian Zone basalts it is necessary to employ relatively stable trace elements to determine chemical groups, petrogenesis and tectonic environment. This includes (1) stable incompatible elements, such as Ti, P, Nb, Y, Zr, Hf, Ta, Th, Sr(?), rare earth elements (REE), and (2) partly stable transition elements, such Ni, Cr, Sc, V. As the chemical fractionation index, FeO*/MgO, is only reliable at low degrees of alteration, Zr has been employed as a standard indicator of magmatic differentiation within the basaltic range of compositions.

This section reviews and compares the geochemical data of basalt lavas from the SW England and German sectors of the Rhenohercynian Zone, in order to determine their main features and degree of variation prior to discussion of tectonic discrimination.

6.1 Major Chemical Distinctions

Current geochemical research has identified two broad basalt groups within the Rheonhercynian zone: (1) those with a MORB-like composition (tholeiites), and (2) those with a wider range of compositions and WPB affinities (tholeiites and alkali basalts). The chemical composition represented by the two groups is shown by representative analyses in Tables 1 and 2 respectively. Although this fundamental distinction has often been made on the basis of chemical diagrams that discriminate different tectonic environments (e.g., Floyd 1984; Grosser and Dorr 1986; Platen 1991), on the broad scale the two groups exhibit different signatures in terms of chondrite-normalized REE patterns and incompatible element ratios, such as Zr/Nb, Zr/Y, Zr/Ce, Nb/Y, Hf/Th, Hf/Ta and La/Th. In general terms, lavas described as exhibiting MORB-like compositions show the following chemical characteristics: (1) depleted to mildly enriched chondrite-normalized REE patterns with $(La/Yb)_N$ ranging from about 0.3–2.0 (although essentially flat patterns are dominant with $(La/Yb)_N$ ca. 1.0–1.2), and differ from the markedly enriched REE patterns of Rhenohercynian WPB (Fig. 1); (2) distinctive, stable incompatible element ratios, such as Zr/Nb >10 (range 10–35), Hf/Th >5 (range 5–12), Hf/Ta >7 (range 7–15), La/Th >10 (range 10–50) and Zr/Y <5 (range 2–5)–the range exhibited may also be greater than WPB (Fig. 2); and (3) transition trace elements, such as Cr and Ni, are generally higher and FeO*/MgO lower than in the WPB group, although there is marked compositional overlap and no clear distinction can be made using these elements.

Both basaltic groups, however, show considerable within group variation which will be examined below in terms of magmatic processes

Table 1. Representative examples of Rhenohercynian lavas with MORB affinities

Number Sample	1 HZ-7	2 HZ-1	3 RS-8	4 GMB2	5 GMB17	6 MZ-13	7 MI-4	8 NH-3	9 RV-31	10 CP-5
Major oxides (wt%)										
SiO_2	45.70	46.90	53.90	47.88	46.14	49.50	48.04	46.73	48.50	45.69
TiO_2	1.53	1.96	1.64	1.66	1.45	1.40	1.96	1.98	1.90	2.15
Al_2O_3	18.90	14.50	14.40	14.13	17.97	15.78	14.69	15.98	17.62	14.41
Fe_2O_3	1.85	3.29	1.75	8.73	8.54	1.77	3.03	11.11	1.36	1.39
FeO	7.61	7.70	5.32	–	–	8.31	8.10	–	7.52	10.29
MnO	0.20	0.19	0.14	0.12	0.33	0.17	0.16	0.17	0.10	0.32
MgO	5.92	5.60	5.93	7.38	5.98	7.70	6.25	7.61	7.21	7.17
CaO	5.57	9.77	7.71	7.44	7.06	5.91	9.95	9.13	6.01	14.47
Na_2O	4.36	3.62	5.06	4.95	4.13	3.00	3.35	3.75	4.00	1.55
K_2O	1.13	0.60	0.55	0.43	1.32	0.58	1.50	0.19	1.67	0.20
P_2O_5	0.17	0.20	0.18	0.19	0.16	0.11	0.16	0.20	0.19	0.27
LOI	–	–	–	6.99	7.15	–	–	–	–	–
Total	100.20	100.00	99.50	99.90	100.23	98.92	100.66	100.27	100.21	99.28
H_2O^+	4.43	3.26	2.45	–	–	3.54	2.35	3.42	4.13	1.37
CO_2	2.78	2.41	0.41	–	–	1.15	1.12	–	–	–
Trace elements (ppm)										
Ba	–	–	–	47	95	326	90	42	97	104
Cr	274	225	375	313	293	320	110	232	149	272
Cu	51	56	51	58	25	–	–	50	–	–
Ga	17	28	14	18	18	–	–	19	–	–
Nb	5	5	5	6	7	5	7	6	6	19
Ni	82	68	197	114	65	74	43	102	50	102
Pb	6	8	5	10	10	–	–	7	–	–
Rb	20	45	10	6	38	8	27	1	17	6
Sc	–	–	–	40	35	35	45	–	29	31
Sr	238	262	304	104	292	152	284	92	201	327
V	–	–	–	336	298	–	–	320	–	–
Y	21	36	32	36	28	27	31	42	22	30
Zn	133	59	78	79	188	–	–	91	–	–
Zr	114	138	128	113	111	132	160	141	135	185
Cs	–	–	–	2.48	8.00	0.63	4.20	–	1.40	1.90
Hf	–	–	–	2.90	2.70	2.83	3.46	–	3.00	4.30
Ta	–	–	–	0.32	0.31	0.34	0.34	–	0.53	1.60
Th	–	–	–	0.32	0.31	0.89	0.34	–	0.43	1.80
U	–	–	–	0.42	0.13	0.31	0.76	–	0.21	0.62
Rare earth elements (ppm)										
La	4.00	5.30	5.20	5.28	5.08	6.80	9.20	–	–	–
Ce	14.00	21.00	19.00	12.77	15.06	18.90	22.20	–	–	–
Pr	–	–	–	2.29	1.86	2.90	3.10	–	–	–
Nd	8.00	13.00	12.00	13.80	12.36	13.00	16.70	–	–	–
Sm	2.70	4.10	4.00	3.94	3.13	3.60	5.00	–	–	–
Eu	0.98	1.39	1.31	1.46	1.23	1.30	2.10	–	–	–
Gd	3.30	5.30	5.10	5.73	4.33	4.80	7.90	–	–	–
Dy	3.30	5.80	5.20	5.89	4.26	4.30	7.10	–	–	–
Ho	0.77	1.25	1.03	1.28	0.95	0.89	1.30	–	–	–
Er	2.20	3.80	3.30	3.40	2.00	2.90	4.20	–	–	–
Yb	2.10	3.40	3.00	3.23	2.30	2.30	3.10	–	–	–
Lu	0.32	0.53	0.48	0.50	0.38	0.34	0.49	–	–	–

Samples 1–5 and 6–10 from the German and SW England sectors respectively. Major oxides and trace elements Ba to Zr determined by XRF; Cs to U by INAA; REE by ICP.

Table 2. Representative examples of Rhenohercynian lavas with WPB affinities

Number Sample	1 CH-1	2 PP-7R	3 KH-3	4 TV-1	5 CC-5	6 RS-19	7 GMB11	8 GMB14	9 LD-67	10 HZ-28
Major oxides (wt%)										
SiO_2	42.72	36.80	41.82	43.78	43.59	50.30	35.53	42.44	48.10	49.10
TiO_2	3.06	2.34	3.80	3.32	2.37	1.70	1.01	1.36	1.23	1.92
Al_2O_3	13.98	15.68	14.59	21.41	20.20	15.30	10.76	14.76	13.24	16.30
Fe_2O_3	3.49	2.74	2.93	15.07	0.77	3.98	10.14	9.69	5.52	2.85
FeO	6.43	8.02	10.99	–	8.23	5.44	–	–	1.46	7.60
MnO	0.22	0.20	0.21	0.14	0.38	0.14	0.09	0.13	0.16	0.36
MgO	6.31	4.38	4.25	4.70	3.29	6.99	8.27	4.60	4.59	7.85
CaO	8.37	13.90	8.63	1.81	17.29	6.22	15.45	15.78	10.37	5.55
Na_2O	5.54	4.35	4.90	0.20	1.21	5.20	2.36	2.62	3.35	4.87
K_2O	0.14	0.15	0.03	4.16	0.51	0.39	0.75	0.17	0.51	0.38
P_2O_5	0.45	0.25	0.86	0.31	0.38	0.20	0.18	0.19	0.24	0.24
LOI	–	–	–	5.29	–	–	15.32	8.03	–	–
Total	99.88	99.59	100.52	100.19	99.79	100.30	99.86	99.78	97.84	101.10
H_2O^+	2.78	2.88	3.09	–	1.57	3.56	–	–	3.53	4.03
CO_2	6.39	7.90	4.42	–	–	0.90	–	–	5.54	0.09
Trace elements (ppm)										
Ba	124	70	154	31	70	–	70	11	87	–
Cr	19	32	19	173	103	268	390	228	383	199
Cu	–	–	–	12	–	55	74	33	124	63
Ga	–	–	–	27	–	17	17	24	–	17
Nb	40	23	51	42	27	13	11	14	16	14
Ni	16	41	16	73	53	74	440	131	332	110
Pb	–	–	–	31	–	5	12	10	–	5
Rb	8	4	2	132	14	6	17	5	16	8
Sc	20	22	22	29	28	–	17	22	–	–
Sr	298	249	318	85	431	214	206	81	263	588
V	–	–	–	197	–	–	192	195	157	–
Y	25	20	34	40	25	19	13	17	17	25
Zn	–	–	–	163	–	81	174	98	107	92
Zr	239	152	264	330	223	120	74	97	107	117
Cs	1.65	2.40	2.00	10.20	8.00	–	3.52	10.50	–	–
Hf	5.35	4.70	8.00	8.20	5.20	–	1.70	2.50	–	–
Ta	2.93	1.91	3.68	2.85	3.05	–	0.48	0.86	–	–
Th	3.16	1.86	3.72	2.90	3.53	–	0.64	1.39	–	–
U	1.10	0.89	1.30	0.79	1.60	–	0.60	0.35	–	–
Rare earth elements (ppm)										
La	28.90	20.80	34.10	24.69	–	11.00	5.98	10.86	–	12.00
Ce	46.80	44.00	59.00	66.05	–	30.00	13.90	24.00	–	33.00
Pr	6.50	4.50	9.10	8.20	–	–	1.56	2.77	–	–
Nd	24.10	20.00	36.00	38.61	–	14.00	8.87	13.59	–	16.00
Sm	7.10	5.80	11.30	8.49	–	3.70	1.97	3.10	–	4.40
Eu	2.60	1.90	3.30	2.63	–	1.21	0.77	1.14	–	1.49
Gd	5.50	5.00	11.80	8.68	–	4.00	2.57	3.67	–	4.40
Dy	4.24	5.20	6.90	7.42	–	3.40	2.16	3.01	–	4.40
Ho	0.46	0.64	1.10	1.47	–	0.67	0.47	0.61	–	0.80
Er	2.70	1.90	3.70	3.90	–	1.80	0.88	1.38	–	2.40
Yb	1.90	2.00	2.90	2.94	–	1.40	0.96	1.23	–	1.80
Lu	0.32	0.34	0.35	0.37	–	0.23	0.17	0.21	–	0.31

Samples 1–5 are alkalic basalts (SW England), 6–10 are tholeiites (Germany). Major oxides and trace elements Ba to Zr determined by XRF; Cs to U by INAA; REE by ICP.

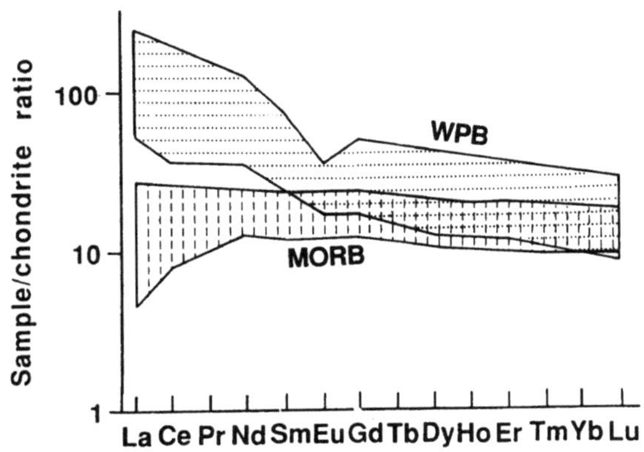

Fig. 1. Generalized (chondrite-normalized) rare earth element patterns illustrating the range of compositions within the two main basaltic types found in the Rhenohercynian Zone: depleted to mildly enriched mid-ocean ridge basalt (*MORB*) and variably enriched within-plate basalt *(WPB)*

(fractional crystallization, partial melting), mantle sources and also, relative stratigraphic location.

6.2 Basalts with MORB-Like Affinities

This group includes tholeiitic basalts from a number of geographical and tectonic locations:

1. SW England (Fig. 3)–dykes from the Lizard ophiolite; basalt clasts within the S Cornish melange (matrix is Upper Devonian, actual age of lavas unknown); pillow basalts in the Gramscatho Group at the base of the Veryan Nappe, S Cornwall (Middle-Upper Devonian); various pillow basalt horizons within the Mylor argillites of W Cornwall (Upper Devonian); metavolcanic greenschists of the Start Complex, S Devon (?undifferentiated Devonian).

2. Germany (Fig. 4)–pillow basalts at the base of the Giessen Nappe, Giessen (Lower Devonian); tectonized pillow basalts of the Albunger Palaozoikum unit, Lower Werra-Sattel (possibly Middle Devonian); pillow basalts at the base of the S Harz Nappe and within the Blankenburger Zone, Harz Mountains (Middle Devonian).

The chemical variation displayed by MORB-type lavas from SW England illustrates two features: (1) a geographical and/or tectonic distinction can be made between the Lizard ophiolite dykes and melange clasts (S Cornwall allochthon) on the one hand, and lavas within the Mylor Formation (W Cornwall parautochthon) on the other (Floyd 1984), and (2) the Start Complex metavolcanics can be chemically correlated with the allochthon basalts, and thus represent another segment of Rhenohercynian ocean floor that has only recently been recognized (Floyd et al. 1993). Representative analyses from the Start Complex metavolcanics are presented in Table 3. In the German sector most basalts are located within allochthonous slices and do not show similar distinctions.

Figures 5–7 illustrate some of the chemical variation shown by the two Cornish tectonic units and the German nappe basalts. As indicated by the broad decrease in Cr (similarly Ni) with increasing fractionation (using Zr abundance as a measure of fractionation), much of the internal variation is governed by mafic fractionation of mainly olivine, clinopyroxene and possibly spinel, although the last phase is only occasionally re-

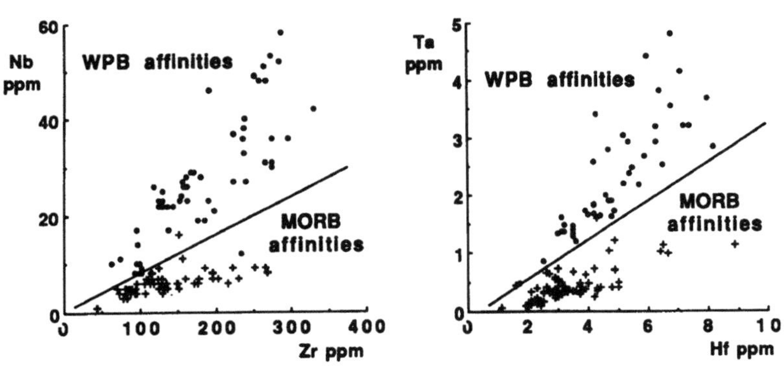

Fig. 2. Distinctive Zr/Nb and Hf/Ta ratios within Rhenohercynian Zone lavas with WPB and MORB affinities. Characterization of basalts initially based on chemical tectonomagmatic discrimination diagrams (e.g., in Herrmann et al. 1974; Floyd 1982a,b, 1984; Grosser and Dorr 1986; Platen 1991). Boundary line subsequently added to the diagrams on the basis of the distribution of the assigned tectonic environment

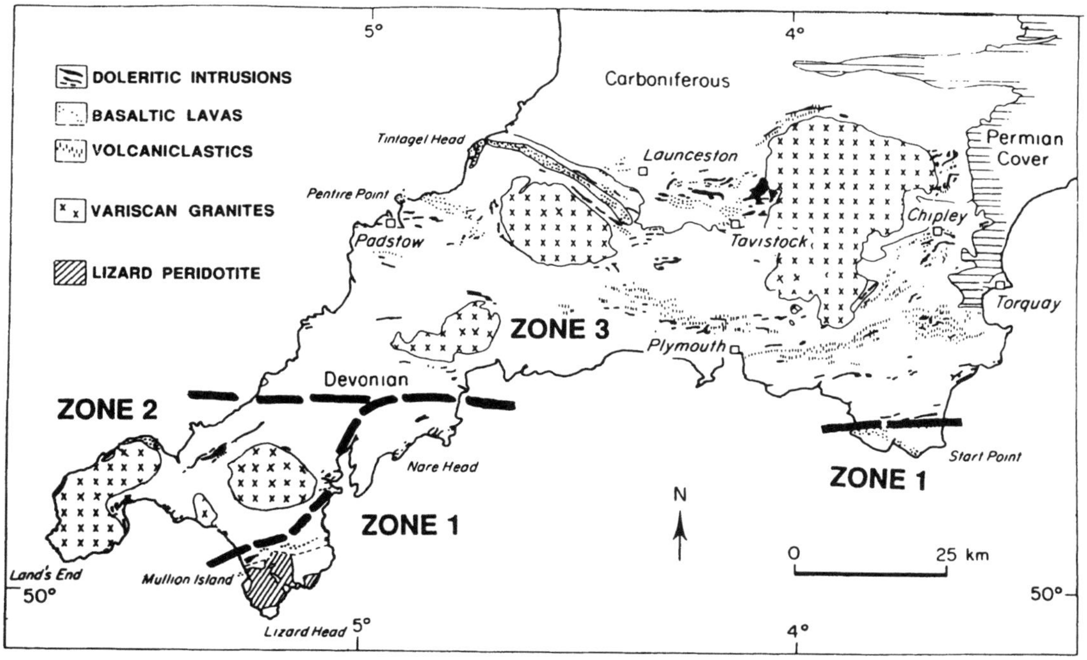

Fig. 3. Map of SW England showing the distribution of Devonian and Lower Carboniferous basaltic rocks (and associated volcaniclastics). Three broad tectonomagmatic zones can be recognized: *Zone 1* includes various Devonian allochthonous units in S Cornwall and the Start Complex (all lavas with MORB affinities); *Zone 2* includes the Devonian parautochthonous unit in W Cornwall (lavas in the Mylor Slate Formation with enriched MORB affinities); *Zone 3* represents mainly Devonian and Lower Carboniferous allochthonous units within N Cornwall and Devon (lavas with WPB affinities)

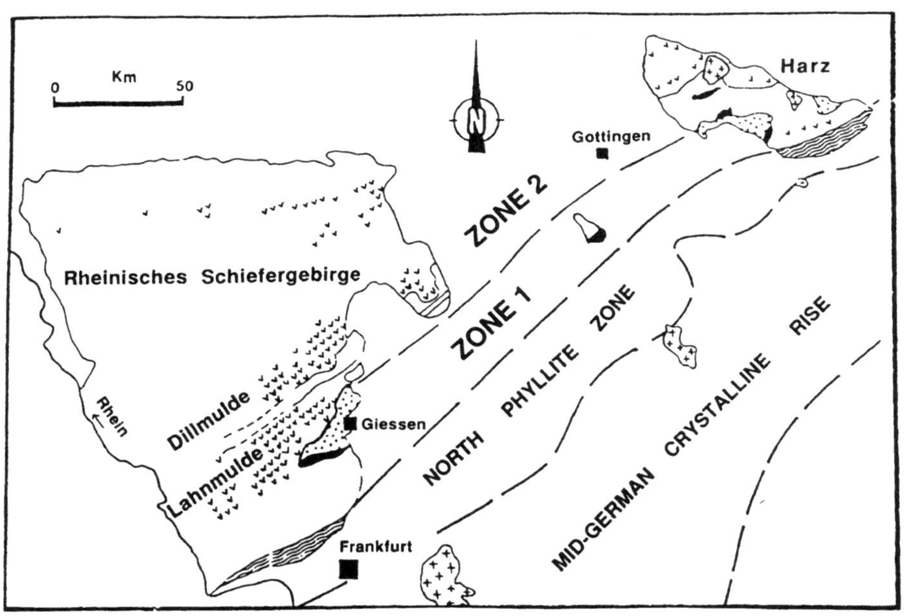

Fig. 4. Map of N Germany showing the distribution of Devonian and Lower Carboniferous basaltic rocks. (After Herrmann and Wedepohl 1970; Platen 1991.) *Zone 1* Includes Devonian MORB-like lavas at the base of the Giessen-S Harz Nappe (*black areas*), and *Zone 2* includes Devonian and Lower Carboniferous lavas with WPB affinities (*vvv ornament*)

Table 3. Basaltic metavolcanics from the Start Complex, S Devon, with N-MORB characteristics (Floyd et al. 1993)

Sample	SPV3	SPV8	SPV10	SPV15	SPV16	SPV23	SPV27	SPV37	SPV40	SPV43
Major oxides (wt%)										
SiO_2	49.09	48.63	52.59	47.07	47.03	49.77	48.19	47.68	45.51	49.05
TiO_2	1.24	1.39	1.35	1.36	1.34	1.14	1.36	1.42	1.70	1.24
Al_2O_3	15.04	14.35	15.76	15.31	16.43	18.80	16.47	16.63	17.44	14.66
Fe_2O_3	11.39	11.65	9.86	11.92	11.31	8.79	10.75	10.26	13.83	11.04
MnO	0.17	0.16	0.16	0.17	0.16	0.12	0.14	0.14	0.16	0.17
MgO	9.17	7.95	3.14	6.85	8.33	6.86	8.03	7.58	7.14	7.50
CaO	6.75	10.23	13.35	12.25	7.92	5.79	6.79	8.92	6.95	11.77
Na_2O	3.66	2.83	1.73	2.05	2.76	4.93	3.20	2.44	2.23	2.25
K_2O	0.13	0.12	0.20	0.05	0.59	0.25	0.94	1.14	1.17	0.09
P_2O_5	0.09	0.10	0.14	0.11	0.11	0.08	0.12	0.24	0.17	0.11
LOI	3.18	1.89	1.50	2.47	3.57	3.37	3.61	3.30	3.96	2.36
Total	99.91	99.31	99.77	99.61	99.56	99.90	99.61	99.77	100.25	100.23
Trace elements (ppm)										
Ba	14	15	15	5	24	16	24	16	83	2
Cr	322	350	316	286	454	248	383	387	341	350
Cu	64	74	31	45	184	33	111	76	83	71
Ga	14	12	21	19	13	13	16	21	19	12
Nb	4	5	4	5	4	3	4	6	7	3
Ni	87	86	100	82	161	139	134	109	99	88
Pb	6	3	6	6	6	6	6	12	8	6
Rb	7	8	8	5	15	8	25	17	41	3
S	114	121	116	113	116	116	118	123	186	119
Sc	39	43.8	38.7	43.8	38.7	29.5	40.1	39.9	47.9	45.7
Sr	54	119	311	131	147	155	163	204	155	135
V	278	305	214	312	262	163	267	273	318	270
Y	27	35	33	38	25	21	30	35	39	33
Zn	262	86	70	98	235	72	962	94	122	97
Zr	73	85	107	85	87	82	95	113	116	78
Cs	0.53	0.41	0.31	0.63	8.10	0.64	30.70	3.40	27.70	0.12
Hf	2.00	2.20	2.60	2.30	2.40	2.00	2.40	2.70	3.10	2.10
Ta	0.14	0.20	0.41	0.13	0.12	0.14	0.12	0.38	0.28	0.18
Th	0.04	0.06	0.04	0.06	0.07	0.11	0.11	0.37	0.36	0.13
U	0.18	0.09	0.16	0.12	0.09	0.11	0.17	0.22	0.31	0.27
Rare earth elements (ppm)										
La	1.53	2.61	2.87	2.33	2.88	2.75	5.26	5.78	5.69	2.56
Ce	7.32	7.96	10.18	8.57	7.93	8.80	10.80	14.88	13.55	7.77
Pr	1.22	1.45	1.79	1.41	1.69	1.40	2.33	2.42	2.30	1.41
Nd	7.89	8.99	11.09	9.29	9.59	8.59	12.08	12.09	11.77	8.69
Sm	2.44	3.01	3.31	3.12	3.17	2.44	3.94	3.52	3.67	2.92
Eu	0.93	1.11	1.23	1.17	1.19	0.96	1.47	1.27	1.34	1.09
Gd	3.79	4.58	4.70	4.82	4.56	3.50	5.61	4.73	5.19	4.50
Dy	4.40	5.21	5.07	5.61	5.11	3.62	6.20	4.96	5.85	5.16
Ho	0.99	1.19	1.12	1.27	1.14	0.81	1.40	1.11	1.31	1.17
Er	2.49	3.49	3.18	3.66	3.52	2.21	4.15	3.35	4.04	3.52
Yb	2.65	3.24	2.90	3.46	3.11	2.02	3.62	2.89	3.53	3.22
Lu	0.42	0.51	0.45	0.54	0.48	0.32	0.58	0.45	0.55	0.51

Major oxides and trace elements Ba to Zr determined by XRF; Cs to U by INAA; REE by ICP.

cognized. Note that hydrothermal alteration and (?)contact metamorphism by the nearby Land's End granite has disrupted this systematic trend in the W Cornish parautochthon basalts. The Lizard dykes, Start metavolcanics and Giessen Nappe basalts are all chemically primitive relative to the rest with high Ni and Cr contents similar to modern MORB. It is also within these three

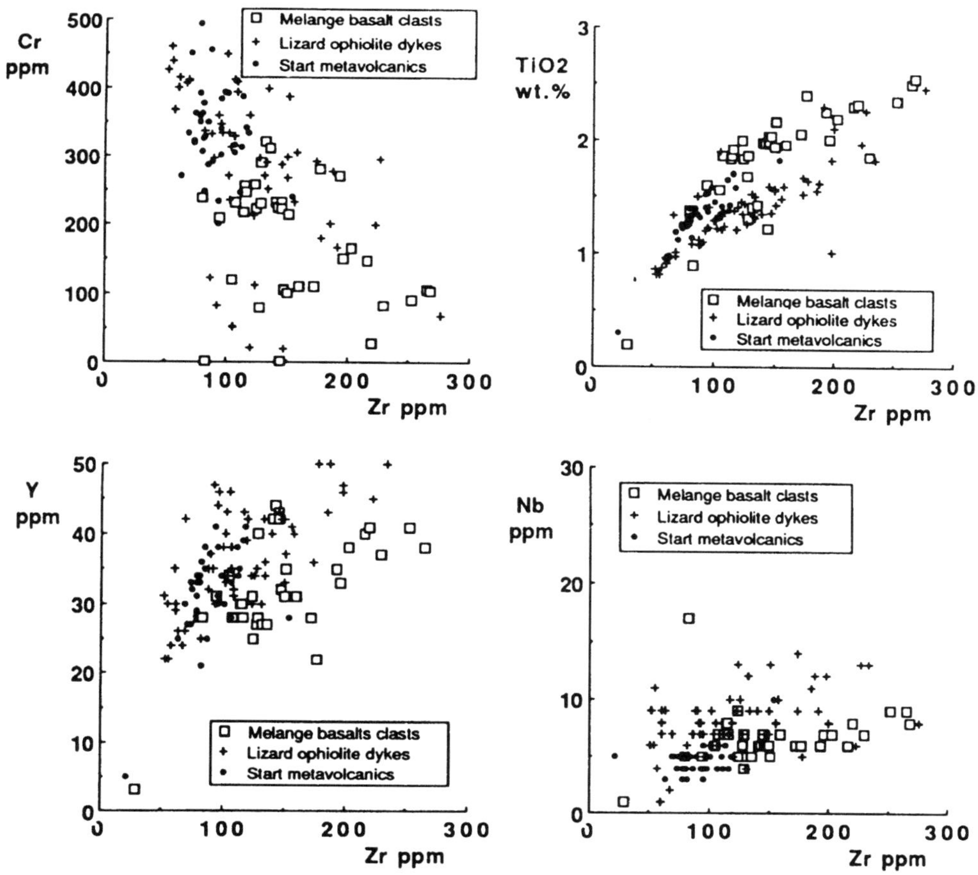

Fig. 5. Binary diagrams illustrating the compositional variation within basaltic lavas with MORB affinities from SW England. Zr is used as a stable fractionation index. The plots are indicative of the influence of mafic fractionation (see *Cr-Zr diagram*), differential partial melting (*Y-Zr diagram*) and variable sources (*TiO_2-Zr* and *Nb-Zr diagrams*) in the chemical suites displayed

groups that MORB with depleted incompatible element characters (normal-type MORB or N-MORB) is found, whereas most of the other groups have flat to mildly enriched normalized REE paterns and low Zr/Nb ratios, generally typical of transitional MORB (T-MORB). The W Cornish parautochthon basalts are distinct in having a greater incompatible element range, whereas the basalts at the base of the Veryan Nappe (Gramscatho Group) have affinities with the melange basalts, rather than those in the Mylor Formation. Zr/Y ratios vary both within and between suites and could imply variable partial melting of a common source or sources. For example, the variation shown by the Lizard dykes corresponds to the three dyke units recognized by Kirby (1984), and which exhibit a range of depleted to enriched normalized REE patterns. Two chemical suites within both the S Cornish melange and the Giessen Nappe basalts are also suggested by variable Zr/Y ratios, the parental melts of which were probably related via variable partial melting of their respective sources. Similarly, a number of chemical suites can be recognized in the Mylor Formation that correspond to specific lava horizons, each of which represent different partial melts derived from a common source. Variation in Ti/Zr and Zr/Nb ratios, which are often indicative of source characteristics, suggest that different mantle sources may have been available for the range of MORB-type lavas.

In summary, chemical variation is largely governed by variable crystal fractionation within different suites, the parental melts of which may be related via differential partial melting of a common source or possibly multiple sources. There are, however, chemical differences between (1) the W Cornish parautochthon basalts in the Mylor Formation and the allochthon basalts in

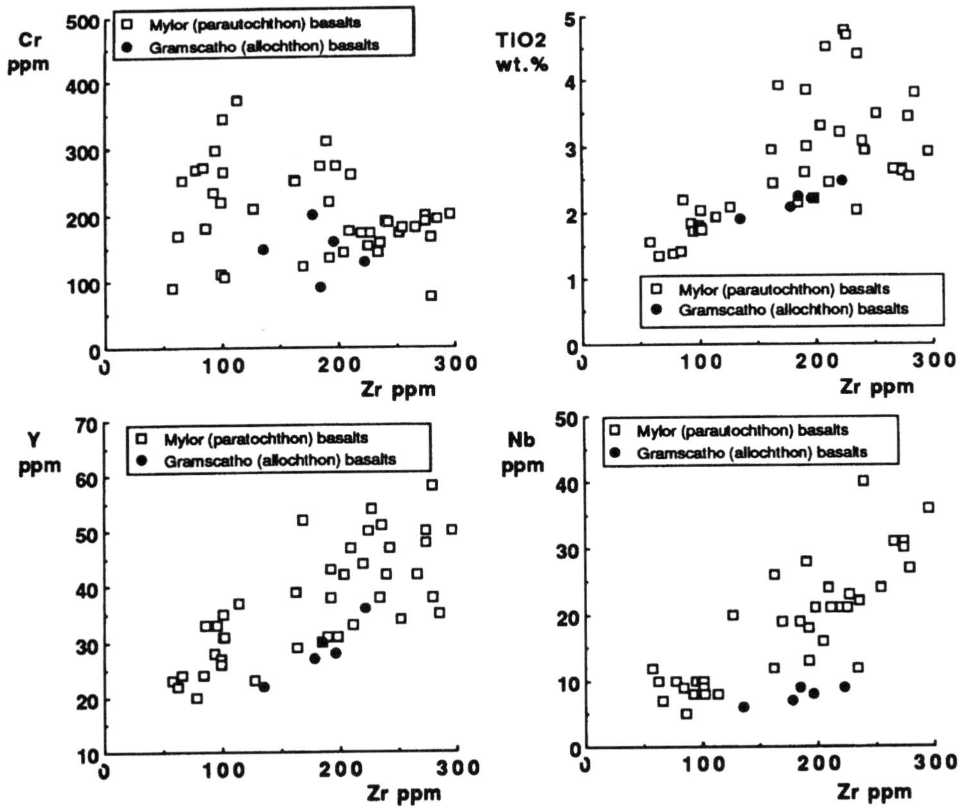

Fig. 6. Binary diagrams illustrating the compositional variation within basaltic lavas with enriched MORB affinities from the allochthonous and parautochthonous units of W Cornwall, SW England. The basalts in the Gramscatho Group (at the base of the Veryan Nappe) have chemical features similar to the MORB lavas of the S Cornish melange (cf. Fig. 5, but note change of scale for Y, Nb and TiO_2 in this figure)

both Cornwall and Germany, and (2) basalts with depleted N-MORB features and the more common type with mildly enriched T-MORB affinities; both of which can occur within all allochthonous units, including the Lizard ophiolite dykes. These fundamental distinctions between lavas with MORB-like features are probably a feature of their respective sources (see below).

6.3 Basalts with Intra-Plate Affinities

This group includes tholeiitic and alkalic basalts from a number of locations generally to the north (external Rhenohercynian Zone) of the MORB-bearing units, but again bounded by various northward-transported nappe slices:

a) SW England (Fig. 3)–various pillow lava horizons along the N Cornish coast near Padstow and inland towards the Bodmin Moor granite (Upper Devonian); pillow lavas in south and east Devon (some Middle, but largely Upper Devonian); Tintagel Volcanic Formation (Lower Carboniferous); scattered outcrops of lavas and pillowed basalts between Launceston and Tavistock), N Devon (Lower Carboniferous).

b) Germany (Fig. 4)–Dillmulde, Lahnmulde, Sauerland and Kellerwald, Rheinisches Schiefergebirge (Upper Devonian and Lower Carboniferous); Werra (Lower Carboniferous); Harzgeroder Zone, Harz Mountains (Middle Devonian).

Both the SW England and German sectors of the Rhenohercynian Zone containing WPB show a complete overlap in composition (Fig. 8) for both tholeiitic and alkalic basalts. In general, tholeiitic basalts tend to have slightly more primitive compositions (higher Cr) and lower Zr/Y ratios (ca. 4) than the alkalic basalts, which dominate the WPB group. On the basis of observed phenocryst phases in these lavas and the crude overall decrease in Cr (Fig. 8)

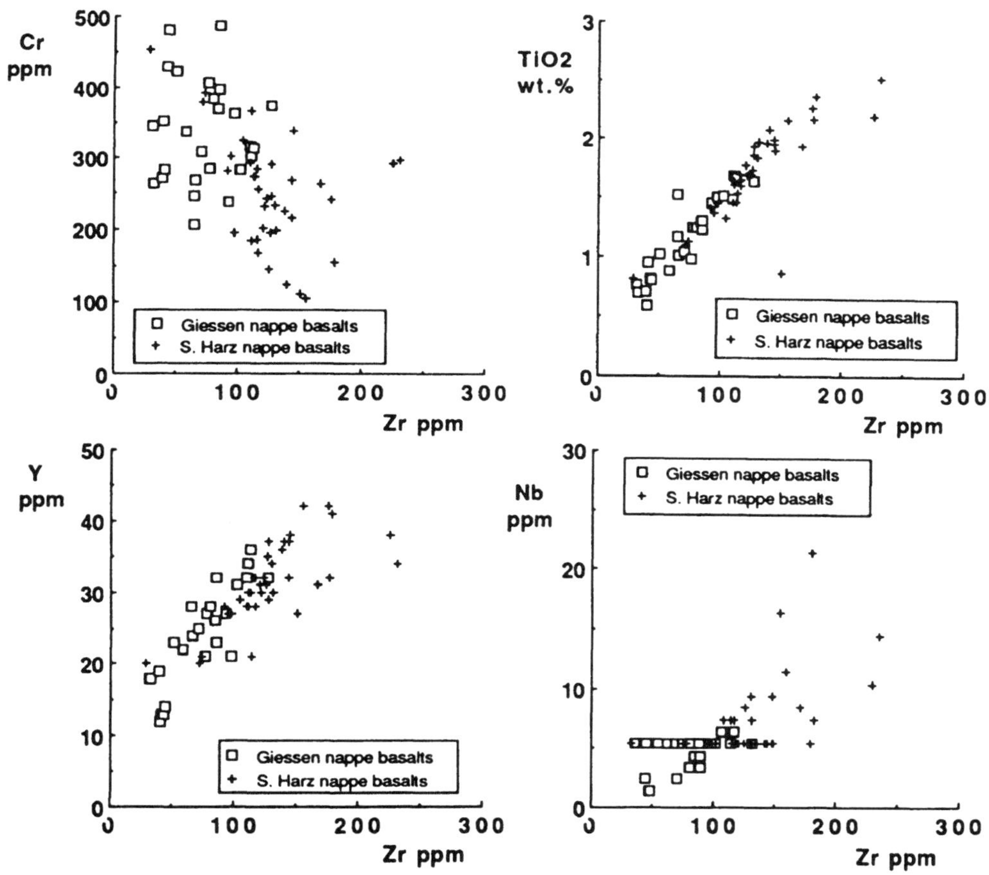

Fig. 7. Binary diagrams illustrating the compositional variation within basaltic lavas with MORB affinities from the base of the Giessen and S Harz Nappes, Germany. These plots can be compared directly with those of Fig. 5 from SW England

and other compatible transition trace elements (Ni, Sc) with increasing Zr, fractionation of spinel-olivine and olivine-clinopyroxene-plagioclase-titaniferous magnetite, has produced much of the variation observed. Although some of the scatter in the Cr-Zr plot may be due to secondary mobilization, many of the alkalic lavas are well evolved and exhibit Zr contents similar to hawaiites and mugearites. The range of Zr/Y ratios (from ca. 5–10) indicates that a number of different chemical suites are present that often characterize specific lava horizons. For example, both the Tintagel Volcanic Formation and Chipley pillow lavas are composed of single comagmatic alkaline suites, although their Zr/Y ratios vary, being 5.8 and 9.8 respectively. Different chemical suites may also be closely associated within the same volcanic formation; for example, the Herbornseelbach lava sequence, Dillmulde (Schmincke and Sunkel 1987). In both these cases, each suite of basalts could represent a different degree of partial melting of a common source.

Two other features are illustrated by Fig. 8: (1) there is complete overlap in composition between lavas of Upper Devonian (which includes a few Middle Devonian samples) and Lower Carboniferous age. There appears to be no systematic change in basalt composition during the period of maximum volcanic activity, with the exception of an increase in the dominance of alkalic basalts relative to tholeiites. (2) Similar to the MORB-like lavas, a number of chemical suites are present, although compositional overlap with the former group is only seen with the mildly enriched basalts within the parautochthon of W Cornwall (Fig. 6).

In summary, much of the variation is again due to differential fractional crystallization and partial melting that produced different comagmatic suites that often typify specific lava formations or horizons. One major distinction between the

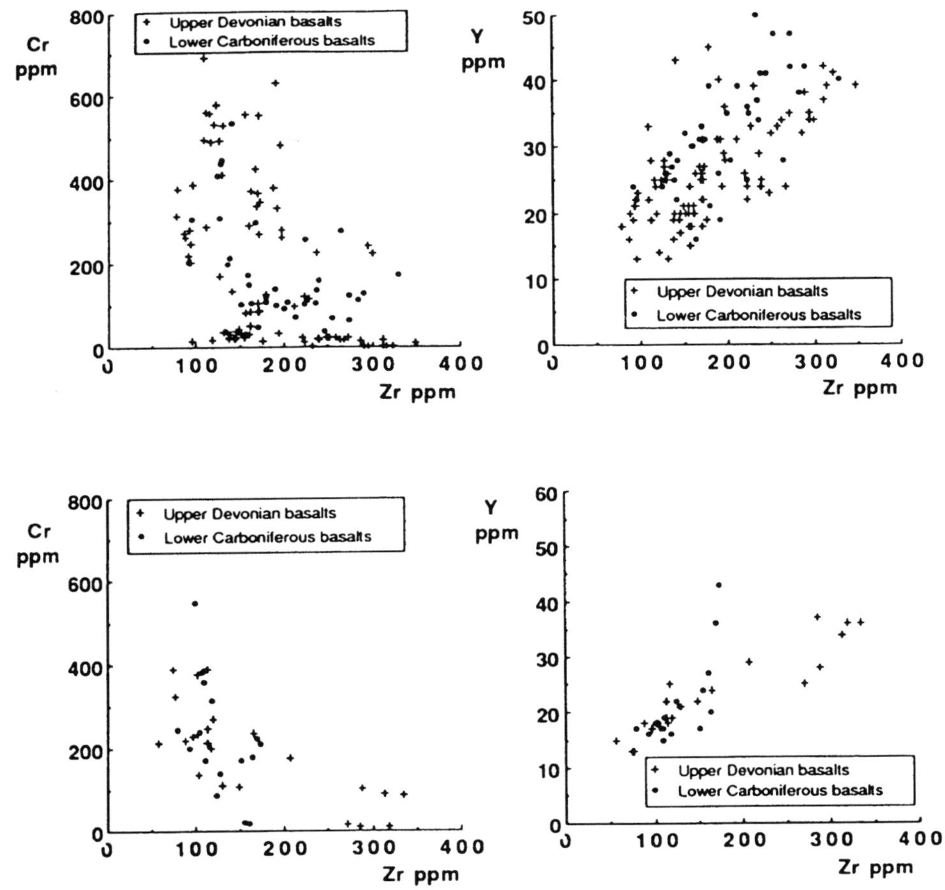

Fig. 8. Binary diagrams illustrating the compositional variation within basaltic lavas with WPB affinities from SW England (*top*) and Germany (*bottom*). Note that there is very little chemical difference between Upper Devonian and Lower Carboniferous basalts, and that the compositional range within each age group can be accounted for by variable fractionation (*Cr-Zr diagram*) and different chemical suites produced by variable partial melting (*Y-Zr diagram*)

WPB and the MORB groups is the additional variability caused by the presence of various alkalic basalt suites as well as tholeiites.

7 Mantle Sources

The above synthesis of Rhenohercynian Zone basalt compositions suggests that three fundamental incompatible multi-element patterns (chondrite-normalized) typify both the SW England and German sectors (illustrated in Fig. 9):

1. *Depleted patterns*–featured by some (late-stage) Lizard dykes, most basalts at the base of the Giessen Nappe, and all the Start Complex metavolcanics. Patterns are characteristic of N-MORB compositions.

2. *Flat to slightly enriched patterns*–featured by the majority of MORB-like Rhenohercynian Zone allochthonous volcanics, such as early Lizard dykes, S Cornish melange basic clasts and basalts at the base of nappe units (e.g., Veryan Nappe, Giessen Nappe, S Harz Nappe). This group also contains basalts from the parautochthonous Mylor Formation of W Cornwall, that are the most enriched of all the above units. Patterns are characteristic of T-MORB compositions.

3. *Variably enriched patterns*–featured by all Rhenohercynian WPB, irrespective of magma type or age, such as lavas from N Cornwall and Devon, Rheinisches Schiefergebirge, Werra and parts of the Harz Mountains. Patterns are characteristic of oceanic island basalt (OIB) compositions.

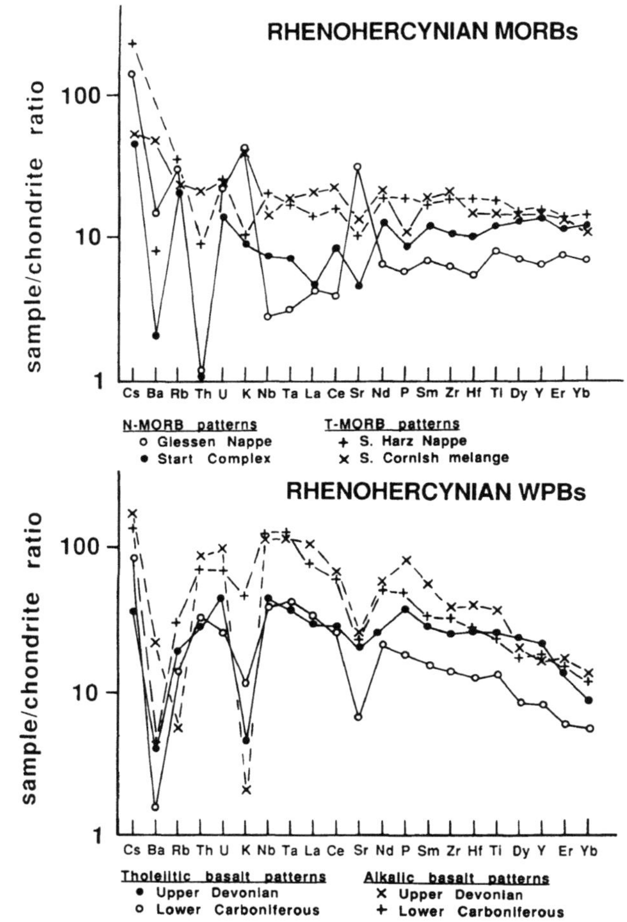

Fig. 9. Chondrite-normalized multi-element diagrams showing typical patterns for Rhenohercynian Zone lavas with MORB affinities (*top*) and WPB affinities (*bottom*). The MORB examples show both depleted (samples from the Start metavolcanics and basal Giessen Nappe) and mildly enriched (samples from basal S Harz Nappe and S Cornish melange) patterns. The WPB examples include both tholeiitic and alkalic basalts from the Upper Devonian and Lower Carboniferous

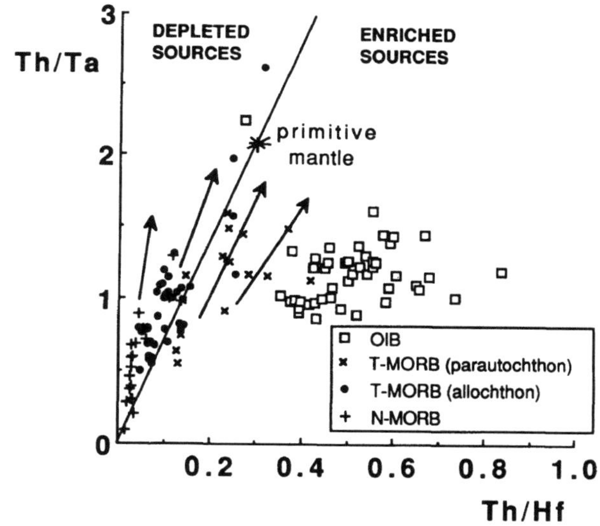

Fig. 10. Th/Ta-Th/Hf diagram showing the variably depleted and variably enriched sources for Rhenohercynian Zone basalts. Rhenohercynian MORBs define a number of separate subvertical trends, whereas the scattered WPB data probably reflects some degree of enriched source mixing. (Primitive mantle composition from Sun and McDonough 1989)

The incompatible element patterns indicate that Rhenohercynian basaltic melts were derived from at least two major mantle sources–variably depleted mantle (the MORB source) and variably enriched mantle (the OIB source). In the light of variations between ratios of highly incompatible elements or those with similar geochemical properties (see Ti/Zr and Zr/Nb relationships in Figs. 5–7), this can only be considered a broad generalization. For example, as shown by variations in Th/Ta and Th/Hf ratios (Fig. 10), MOR-type Rhenohercynian basalts define two main near-vertical trends (equivalent to melt batches) within the depleted mantle source field, whereas both OIB tholeiites and alkali basalts form more scattered patterns in the enriched mantle source field (cf. Joron and Treuil 1989). This implies that Rhenohercynian N-MORB was derived from a typical depleted MORB source, whereas those lavas with T-MORB features were derived from a slighly less depleted MORB mantle source.

The greater diversity of Rhenohercynian OIB is more of a problem in terms of the nature of the source–except that it is generally enriched. It is well known that OIB suites around the world are ubiquitously enriched (relative to MORB) and exhibit considerable variation in incompatible element ratios (e.g., Loubet et al. 1988; Saunders et al. 1988) and radiogenic isotopes (e.g., Zindler and Hart 1986). Sr, Nd and Pb isotopes have been particularly useful in identify mantle sources, such that four end-member compositions–depleted MORB mantle (DMM), enriched mantle I (EMI), enriched mantle II (EMII) and high μ (HIMU)–have been recognized to explain mantle isotopic heterogeneity (Hart 1988). Unfortunately, no radiogenic isotope data is available for Rhenohercynian basalts, although the systematic relationships between isotopes and ratios of highly incompatible trace elements (e.g., Palacz and Saunders 1986; Loubet et al. 1988; Sun and McDonough 1989; Weaver 1991) enables the latter to be used in assessing OIB sources. Particularly useful trace elements (and their respective ratios) include K, Rb, Ba, Th, Ta, Nb, Zr, La and Ce, although in this case the first three must be excluded, as their compositions reflect mobility during metamorphism of the Rhenohercynian basalts. Weaver (1991) has demonstrated that OIB mantle end-members (EMI, EMII, HIMU) have different trace element ratios and slightly different normalized patterns. In this context, and using Th/Nb and Zr/Nb ratios, the HIMU source (as represented by St. Helena) is distinct from both EMI and EMII sources (Fig. 11). The majority of the Devonian and Carboniferous Rhenohercynian alkali basalts define a crude trend between the HIMU and EMI sources, suggesting that they were derived from a mantle representing a mix between these two end-members, but strongly influenced by the former composition. Although the data plotted are from SW England, the few Lower Carboniferous samples available from Germany also plot within the same grouping. The few samples with enhanced EMII-type Th/Nb ratios are exclusively from E Devon and may have been influenced by the nearby Dartmoor granite which possibly increased the Th content by localized metasomatism. The designated mantle source for

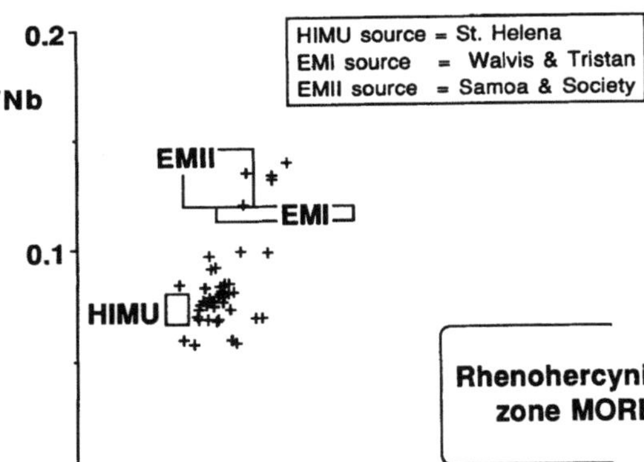

Fig. 11. Th/Nb-Zr/Nb diagram illustrating the OIB source characteristics of Rhenohercynian Zone alkalic WPB. The data distribution suggests that compositional variation within the alkalic basalt melts was due to mixing between a predominantly HIMU mantle component and an EMI mantle end-member. All Rhenohercynian Zone MORBs plot at low Th/Nb (<0.05) and high Zr/Nb (>15) ratios. Mantle end-member components (HIMU, EMI, EMII) based on averaged data in Weaver (1991)

Rhenohercynian Zone OIB is also supported by multi-element normalized patterns (Fig. 9) which exhibit a marked Nb-Ta-positive anomaly that is characteristic of the HIMU source (Weaver 1991).

Throughout the Rhenohercynian Zone, basaltic melts were derived from both DMM and HIMU (with variable EMI) sources. In comparison with petrogenetic models for modern MORB and OIB, melts from the former were derived by high degrees of partial melting at a shallow mantle level, whereas the latter were probably generated as numerous lower degree partial melts at greater depths.

8 Chemical Discrimination of Eruptive Environment

The application of stable incompatible trace elements to determining the tectonic environment of Rhenohercynian basalts has indicated the presence of both MORB and WPB. The simplest interpretation of the occurrence of the two types involves the generation of early Devonian basins floored by ocean crust, and Devonian to Lower Carboniferous ensialic basins with locally developed mounds and possibly small seamounts of pillow lavas.

A variety of discrimination diagrams (e.g., plotting the following groups of elements: Ti-Y-Zr; Zr/Y-Zr; Hf-Ta-Th; Ti-Cr; Ti-Zr; REE patterns) have been used to demonstrate the presence of the two major eruptive environments mentioned above (e.g., Floyd 1982a, 1984; Grosser and Dorr 1986; Platen 1991) and the apparent lack of any arc-related setting for the Rhenohercynian Zone. Even with the advent of new data (from the Start metavolcanics and Germany) there is little change in the general discrimination of the basalts, as shown by the Hf-Ta-Th diagram (Fig. 12). Rhenohercynian N-MORB types plot towards the top of the appropriate field (relatively high Hf), whereas the T-MORB types (including some of the Lizard dykes) plot further down in the enriched MORB field. The rest of the basalts plot in either the tholeiitic or alkali basalt WPB fields, although mildly enriched basalts from the Mylor Formation parauthochthon of W Cornwall fall almost exclusively in the enriched MORB/WPB tholeiitic field. Island arc basalts are absent.

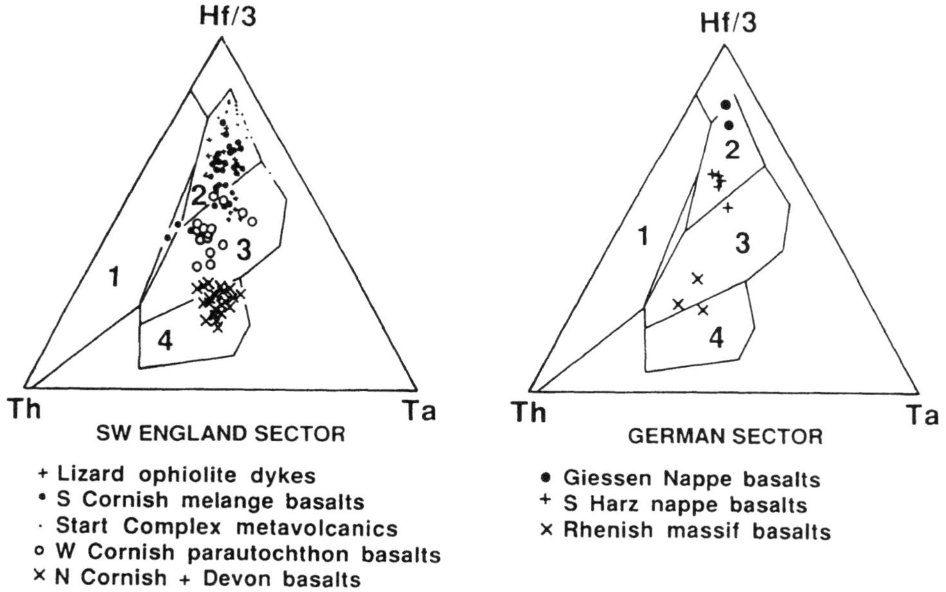

Fig. 12. Hf-Ta-Th tectonic environment discrimination diagram (Wood 1980) for Rhenohercynian Zone basalts from SW England (*left plot*) and German (*right plot*) sectors. Note the lack of volcanic arc basalts. *Field 1* Volcanic arc; *field 2* N-MORB; *field 3* E-MORB and tholeiitic WPB; *field 4* alkalic WPB

8.1 The WPB Environment

The intraplate Rhenohercynian Zone basalts have a chemical signature similar to tholeiites and alkali basalts from ocean islands and seamounts that are often fed via ascending mantle plumes. Whether a deep mantle plume origin for these particular Rhenohercynian Zone basalts is open to question, although a number of factors suggest it is unlikely: (1) their limited stratigraphic occurrence and small volume, (2) chemically diverse with each separate locality often having its own identity, (3) volcanism related to linear fractures along the margins of small basins, and (4) no age progression within linear volcanic belts. In general, the nature and character of Rhenohercynian Zone volcanism is very minor compared with major plume-fed activity in the oceans, such as Hawaii or the Azores. However, the mantle source for intraplate Rhenohercynian Zone basalts was relatively uniform in gross terms (HIMU-type OIB) and might have been represented by small enriched domains (mantle "plums", Sun 1985), that were variably melted and mixed with other mantle constituents on the local scale (cf. Zindler et al. 1984).

8.2 The MORB Environment

One of the main features of Rhenohercynian Zone basalts with MORB affinities is that they are dominated by lavas with flat to mildly enriched incompatible trace element patterns (see above). These basalts are generally referred to as T-MORB, although back-arc basin basalts (BABB) are also characterized by this feature (Saunders and Tarney 1984). This observation has a bearing on the tectonic environment of Rhenohercynian MORB–being formed in either a major ocean basin or a back-arc situation–that contains both N-MORB and T-MORB compositions.

The Mid-Atlantic Ridge exhibits axial segments with T-MORB chemistry (e.g., FAMOUS and Reykjanes Ridge; Tarney et al. 1979; Wood et al. 1979; Bougault et al. 1980), whose enrichment may be due to the proximity of a plume (e.g., White and Schilling 1978) or mantle metasomatic veining (e.g., Tarney et al. 1980). However, the T-MORB feature (with minor N-MORB segments) probably tells us little about the tectonic nature of the Rhenohercynian "ocean"; only that the composition of the mantle underlying the zone at the time of its formation was similar to that found in some slow-spreading modern oceans; but unlike modern ocean crust, the proportion of

T-MORB is far higher in the Rhenohercynian Zone crust and represents the dominant basaltic type.

Back-arc basins, on the other hand, are under the influence of nearby subduction zones and BABB may exhibit some of the chemical features of subduction-related magmas, but to a lesser extent. One particular feature of subduction-generated island arc tholeiites (IAT) is the depletion of high field strength (HFS) incompatible elements (Nb, Ta, Y, Yb, Zr, Hf, Ti) relative to large-ion-lithophile (LIL) elements (K, Rb, Ba, Sr, Th, La, Ce), such that LIL/HFS ratios are markedly higher than MORB (Pearce 1983; Saunders and Tarney 1984), for example, La/Nb in IAT >5, BABB 1–5, and MORB <1 (Floyd et al. 1991b). However, only a few Rhenohercynian Zone basalts have enhanced La/Nb (ca. 1.5, with the majority between 0.5 and 1.0) or Th/Yb (ca. 0.1, with the majority between 0.05 and 0.07) ratios indicative of a back-arc environment. At best the chemical evidence is equivocable and does not clearly support a typical back-arc environment relative to a major ocean. Also, to complicate matters further, the few enriched values could equally be caused by localized contamination with continental crust during the early rifting stage.

8.3 Gulf of California Model

However, it must be remembered that there are many different forms of back-arc basin and it is considered that the Gulf of California, in particular, may provide a reasonable geological model for the Rhenohercynian Zone overall. The Gulf of California is an immature ensialic basin which is developing oceanic crust via spreading from a series of central en echelon axial zones that are connected to the East Pacific Rise at its mouth. Continental extension in the Gulf area has taken place at short, spreading centres separated by major transform faults that partially isolate, small, deep, sediment-filled basins (e.g., Larson et al. 1968). The Gulf developed in response to the oblique collision of the Farallon plate, and subsequent subduction of the precursor to the East Pacific Rise, which effectively terminated the subduction process. The volcanic products produced during and after sessation of subduction are markedly different within the Gulf region. Results from DSDP Legs 64 and 65 (Saunders et al. 1982a; Saunders 1983) show that the nature and chemical composition of the basalts changes from the Gulf mouth to the Guaymas basin (situated half-way up the Gulf). Gulf mouth basalts are typical N-MORB flows with depleted (chondrite-normalized) light REE patterns and very low contents of highly incompatible elements, whereas those from the Gulf interior are thick flows and/or sills with mildly enriched REE patterns and generally higher incompatible element ratios (e.g., La/Yb, Th/Hf, Zr/Ti; Saunders et al. 1982b). Basalts from the Guaymas Basin have similar chemical characteristics to the T-MORB of the Rhenohercynian Zone, in terms of both REE patterns and highly incompatible element ratios. For example, utilizing Th and Hf relationships (Fig. 13), distinctions between the three main Rhenohercynian basaltic types are clearly illustrated–the MORB types of which can be matched with the depleted and enriched basalts from the Gulf mouth and Guaymas Basin, respectively. Volcanic islands within the Gulf (e.g., Isla Tortuga) are tholeiitic with nearly flat REE patterns (Batiza et al. 1979) and are, thus, dissimilar to the enriched Rhenohercynian WPB "seamounts". However, after ridge subduction in

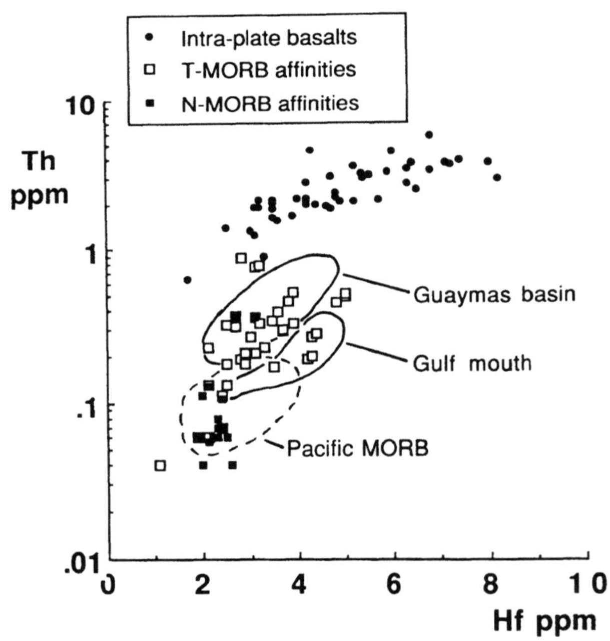

Fig. 13. Th-Hf diagram for all Rhenohercynian Zone basalts compared with variation exhibited by basalts in the Gulf of California. (Data from Saunders et al. 1982a,b; Saunders 1983) and N-MORB from the Pacific. (Data from Floyd et al. 1991c)

the Gulf region, the younger volcanics in Baja California are dominated by alkalic basalts with compositions indistinguishable from OIB (Rogers et al. 1985; Storey et al. 1989), and similar to Rhenohercynian Zone WPB. In the modern model, the post-ridge subduction volcanics are subaerial, more varied (with high-Mg andesites), and are developed along the continental margins of the basin, rather than the submarine rift-related volcanics seen in the Rhenohercynian Zone.

If the Gulf of California model has general applicability, it implies that we can recognize within the Rhenohercynian Zone the initial stages of ocean crust formation via the extension of a small immature ensialic basin. Also, by analogy, rifting of the continental crust was probably initiated by collision and subsequent subduction of an active ridge. Certainly the nature of the volcanic products have a close correspondence with those in the Rhenohercynian Zone. The development of the Rhenohercynian Zone goes further than our modern-day analogy–with the final closure of the basin via stacking of nappe slices. The Gulf of California is not a typical back-arc basin (as seen in the western Pacific) as there is no active arc and probably no subduction zone present now (Dickinson and Snyder 1979). The apparent lack of an active arc is also considered a feature of the Rhenohercynian Zone (see above).

In summary, it is considered that the Rhenohercynian Zone thus provides a fragmented picture of the initiation, development and closure of a narrow Gulf of California-type basin. The volcanic history documents: (1) basin inception (localized, early, calc-alkali acidic volcanism, Durrance 1985), (2) limited ocean crust formation (N- and T-MORB), (3) attenuation of a continental crust margin with development of separate depocentres margined by fractures (associated alkalic and tholeiitic WPB initially, followed by alkalic WPB prior to termination of volcanism) and (4) closure of basinal areas via thrusting from the south and cessation of volcanic activity.

9 Conclusions

One of the fundamental features of the Rhenohercynian Zone is the similarity between the SW England and German sectors in terms of stratigraphy, deformational history and volcanism. On the basis of basalt geochemistry, this chapter reinforces the similarities between the two regions. The following general conclusions thus refer to basalts found throughout the Rhenohercynian Zone discussed in this chapter.

1. Basaltic lavas within the Rhenohercynian Zone are strictly meta-basalts ("spilites") that exhibit varying degrees of low-grade secondary alteration superimposed on a magmatic mineralogy and texture. Relict features indicate the original presence of glass and quench textures typical of submarine pillow lavas, and fine- to medium-grained hypocrystalline interiors in thicker cooling units. Both aphyric and phyric tholeiitic and alkalic basalts are represented. Phenocryst assemblages range from relatively primitive spinel-olivine-bearing tholeiites to more evolved plagioclase-clinopyroxene-bearing basalts, although many porphyritic alkalic basalts are characterized by abundant plagioclase phenocrysts.

2. Apart from early seafloor weathering, most secondary assemblages in the basalts are characteristic of pervasive regional metamorphism that accompanied Variscan deformation. Critical assemblages indicate pumpellyite facies (with local areas of greenschist facies) of regional metamorphism with temperatures of 300–350 °C and pressures around 2 kb. The dominant secondary assemblages suggest water-rock ratios of 30–50.

3. Geochemical discrimination has identified the presence of both MORB and WPB within the Rhenohercynian Zone. In general terms, lavas with MORB affinities are located in allochthonous units derived from the interior of the Zone, whereas WPB are characteristic of the more external allochthonous and parautochthonous units. Within each major chemical grouping a number of comagmatic suites are recognized with internal variation largely due to dominant mafic, and felsic, fractionation. Each stratigraphically limited volcanic location representative of the WPB group may have its own characteristic incompatible element identity, although in any one region parental melts may be linked via partial melting of a common source.

4. In terms of chondrite-normalized incompatible element patterns, Rhenohercynian Zone basalts define three main groups: (1) those with depleted patterns (comparable to N-MORB), (2) those with flat to slightly enriched patterns (comparable to T-MORB), and (3) those with variably enriched patterns (the tectonically designated

WPB, with features typical of OIB). The MORB types were originally derived from a differentially depleted mantle. However, the dominance of T-MORB compositions in the lavas indicates that much of the Variscan mantle contained an enriched factor relative to modern DMM. Variation within Rhenohercynian Zone OIB reflects the presence of both tholeiitic and alkalic basalts derived from a predominantly HIMU mantle source mixed with a minor EMI component.

5. The coexistence of MORB and WPB reinforces the notion that the Rhenohercynian Zone is representative of an ensialic basinal area that via early crustal attenuation developed a narrow zone of oceanic crust. The WPB were located along the fractured margins of basins developed on a continental margin to the north of the oceanic area. On the basis of the basalt geochemistry, it is suggested that the atypical back-arc basin, Gulf of California, which contains both depleted and enriched MORB-types, is a reasonable modern analogue. This model could also account for the apparent lack of coeval subduction-related arc volcanism within the Rhenohercynian Zone.

Acknowledgments. I would like to acknowledge the detailed stratigraphic and structural work of many British and German colleagues, which have provided a framework within which to understand the geochemical variability of the basalt lavas. Special thanks are due to Prof. W. Franke and colleagues for stimulating discussions on the Variscan orogen, and the hospitality shown to me and British workers by Prof. R. Emmermann (Giessen) and Prof. K.H. Wedepohl (Gottingen). Finally, this chapter could not have been completed without the utilization of largely unpublished geochemical data from theses by Dr. K.M. Platen (Justus-Liebig University) and Dr. G.A. Kirby (Southampton University); to both, sincere thanks.

References

Anderson TA (1975) Carboniferous subduction complex in the Harz Mountains, Germany. Bull Geol Soc Am 86: 77–82

Anderton R, Bridges PH, Leeder MR, Selwood BW (1979) A dynamic stratigraphy of the British Isles. Allen & Unwin, London

Badham JPN (1982) Strike-slip orogens–an explanation for the Hercynides. J Geol Soc Lond 139: 495–506

Badham JPN, Kirby GA (1975) Ophiolites and the generation of ocean crust: data from the Lizard Complex. Bull Geol Soc Fr 4: 855–858

Barnes RP (1983) The stratigraphy of a sedimentary melange and associated deposits in South Cornwall. Proc Geol Assoc Lond 94: 217–229

Barnes RP (1984) Possible Lizard-derived material in the underlying Meneage Formation. J Geol Soc Lond 141: 79–85

Barnes RP, Andrews JR (1981) Pumpellyite-actinolite grade regional metamorphism in S Cornwall. Proc Ussher Soc 5: 139–146

Barnes RP, Andrews JR (1984) Hot or cold emplacement of the Lizard Complex. J Geol Soc Lond 141: 37–40

Barnes RP, Andrews JR (1986) Upper Palaeozoic ophiolite generation and obduction in south Cornwall. J Geol Soc Lond 143: 117–124

Batiza R, Futa K, Hedge CE (1979) Trace element and strontium isotope characteristics of volcanic rocks from Isla Tortuga: a young seamount in the Gulf of California. Earh Planet Sci Lett 43: 269–278

Behr HJ, Engel W, Franke W, Giese P, Weber K (1984) The Variscan belt in central Europe: main structures, geodynamic implications, open questions. Tectonophysics 109: 15–40

Bougault H, Joron JL, Treuil (1980) The primordial chondritic nature and large-scale heterogeneities in the mantle: evidence from high and low partition coefficient elements in oceanic basalts. Philos Trans R Soc Lond 297: 203–213

Bromley AV (1979) Ophiolitic origin of the Lizard Complex. J Camborne School Mines 79: 25–38

Chandler P, Isaac KP (1982) The geological setting geochemistry and significance of Lower Carboniferous basic volcanic rocks in central southwest England. Proc Ussher Soc 5: 279–288

Davies GR (1984) Isotope evolution of the Lizard Complex. J Geol Soc Lond 141: 3–14

Dickinson WR, Snyder WS (1979) Geometry of subducted slabs related to San Andreas transform. Geology 87: 609–621

Durrance EM (1985) Lower Devonian acid igneous rocks of south Devon: implications for Variscan plate tectonics. Proc Ussher Soc 6: 205–210

Floyd PA (1976) Geochemical variation in the greenstones of SW England. J Petrol 17: 522–545

Floyd PA (1982a) Chemical variation in Hercynian basalts relative to plate tectonics. J Geol Soc Lond 139: 505–520

Floyd PA (1982b) The Hercynian trough: Devonian and Carboniferous volcanism in southwestern Britain. In: Sutherland DS (ed) Igneous rocks of the British Isles. Wiley, London, pp 227–242

Floyd PA (1983) Composition and petrogenesis of the Lizard Complex and pre-orogenic basaltic rocks in southwest England. In: Hancock PL (ed) The Variscan foldbelt in the British Isles. Hilger, Bristol, pp 130–152

Floyd PA (1984) Geochemical characteristics and comparison of the basic rocks of the Lizard Complex and the basaltic lavas within the Hercynian troughs of SW England. J Geol Soc Lond 141: 61–70

Floyd PA, Leveridge BE (1987) Tectonic environment of the Devonian Gramscatho basin, South Cornwall: framework mode and geochemical evidence from

turbiditic sandstones. J Geol Soc Lond 144: 531–542
Floyd PA, Rice-Birchall B (1995) Basic igneous rocks and associated metamorphism. In: Williams B (ed) Camelford Memoir, British Geological Survey. (in press)
Floyd PA, Rowbotham G (1979) Chemical composition of relict clinopyroxenes from the Mullion Island lavas, Cornwall. Proc Ussher Soc 4: 419–429
Floyd PA, Rowbotham G (1982) Chemistry of primary and secondary minerals in titaniferous brown amphibole-bearing greenstones from north Cornwall. Proc Ussher Soc 5: 296–303
Floyd PA, Exley CS, Stone (1983) Variscan magmatism in southwest England: discussion and synthesis. In: Hancock PL (ed) The Variscan foldbelt in the British Isles. Hilger, Bristol, pp 178–185
Floyd PA, Leveridge BE, Franke W, Shail R, Dorr W (1990) Provenance and depositional environment of Rhenohercynian synorogenic greywackes from the Giessen Nappe, Germany. Geol Rdsch 79: 611–626
Floyd PA, Shail R, Leveridge BE, Franke W (1991a) Geochemistry and provenance of Rhenohercynian synorogenic sandstones: implications for tectonic environment discrimination. In: Morton AC, Todd SP, Haighton PDW (eds) Developments in sedimentary provenance studies. Geol Soc Lond Spec Publ 57: 173–188
Floyd PA, Kelling G, Gokcen SL, Gokcen N (1991b) Geochemistry and tectonic environment of basaltic rocks from the Misis ophiolite melange, south Turkey. Chem Geol 89: 263–280
Floyd PA, Castillo PR, Pringle M (1991c) Tholeiitic and alkalic basalts of the oldest Pacific Ocean crust. Terra Nova 3: 257–265
Floyd PA, Holdsworth RE, Steele SA (1993) Geochemistry of the Start Complex greenschists: Rhenohercynian MORB? Geol Mag 130: 345–352
Franke W (1989a) Tectonostratigraphic units in the Variscan Belt of central Europe. Geol Soc Am Spec Pap 230: 67–90
Franke W (1989b) Variscan plate tectonics in Central Europe – current ideas and open questions. Tectonophysics 169: 221–228
Franke W, Engel W (1982) Variscan sedimentary basins on the continent and relations with southwest England. Proc Ussher Soc 5: 259–269
Grosser J, Dorr W (1986) MOR-typ-basalte im ostlichen Rheinischen Schiefergebirge. Neues Jahrb Geol Palaontol Monatsh 12: 705–722
Hart SR (1988) Heterogeneous mantle domains: signatures, genesis and mixing chronologies. Earth Planet Sci Lett 90: 273–296
Hart SR, Erlank AJ, Kable EJD (1974) Sea floor basalt alteration: some chemical and Sr isotopic effects. Contrib Mineral Petrol 44: 219–230
Hentschel H (1952) Zur petrographie des Diabas-Magmatismus im Lahn-Dill-gebiet. Z Dtsch Geol Ges 104: 238–246
Herrmann AG, Wedepohl KH (1970) Untersuchungen an spilitischen Gesteinen der variskischen geosynkline in Nordwestdeutschland. Contrib Mineral Petrol 29: 255–274
Herrmann AG, Potts MJ, Knake D (1974) Geochemistry of the rare earth elements in spilites from the oceanic and continental cust. Contrib Mineral Petrol 44: 1–16
Holder MT, Leveridge BE (1986a) Correlation of the Rhenohercynian Variscides. J Geol Soc Lond 143: 141–147
Holder MT, Leveridge BE (1986b) A model for the tectonic evolution of south Cornwall. J Geol Soc Lond 143: 125–134
Holdsworth RE (1989) The Start-Perranporth line: a Devonian terrane boundary in the Variscan orogen of SW England. J Geol Soc Lond 143: 146: 419–422
Humphris SE, Thompson G (1978) Hydrothermal alteration of oceanic basalts by seawater. Geochim Cosmochim Acta 42: 107–126
Isaac KP (1985) Discussion of papers on the Hercynian back-arc marginal basin of SW England. J Geol Soc Lond 142: 927–929
Joron JL, Treuil M (1989) Hygromagmaphile element distributions in ocean basalts as fingerprints of partial melting and mantle heterogeneities: a specific approach and proposal of an identification and modelling method. In: Saunders AD, Norry MJ (eds) Magmatism in the ocean basins. Geol Soc Lond Spec Publ 42: 277–299
Juteau T, Rocci G (1974) Vers une meilleure connaissance du problème des spilites à partir de données nouvelles sur le cortège spilito-keratophyrique hercynotype. In: Amstutz GC (ed) Spilites and spilitic rocks. Springer, Berlin, Heidelberg New York, pp 253–329
Kirby GA (1978) Layered gabbros in the eastern Lizard, Cornwall, and their significance. Geol Mag 115: 199–204
Kirby GA (1979) The Lizard Complex as an ophiolite. Nature 282: 58–60
Kirby GA (1984) The petrology and geochemistry of dykes of the Lizard ophiolite complex, Cornwall. J Geol Soc Lond 141: 53–60
Larson RL, Menard HW, Smith SM (1968) Gulf of California: a result of ocean floor spreading and transform faulting. Science 161: 781–784
Leeder MR (1982) Upper Palaeozoic basins of the British Isles-Caledonide inheritance versus Hercynian plate margin processes. J Geol Soc Lond 139: 479–491
Lehmann E (1974) Spilitic magma: characteristics and mode of formation. In: Amstutz GC (ed) Spilites and spilitic rocks. Springer, Berlin, Heidelberg New York, pp 23–38
Liew TC, Hofmann AW (1988) Precambrian crustal components, plutonic associations, plate environment of the Hercynian fold belt of central Europe: indications from a Nd and Sr isotopic study. Contrib Mineral Petrol 98: 129–138
Liou JG, Maruyama S, Cho M (1985) Phase equilibria and mineral paragenesis of metabasites in low-grade metamorphism. Mineral Mag 49: 321–333
Loubet M, Sassi R, Donato GD (1988) Mantle heterogeneities: a combined isotope and trace element approach and evidence for recycling continental crust

materials in some OIB sources. Earth Planet Sci Lett 89: 299–315
Matthews SC (1977) The Variscan foldbelt in southwest England. Neues Jahrb Geol Palaontol Abh 154: 94–127
Matthews SC (1978) Caledonian connections of Variscan tectonism. Z Dtsch Geol Ges 129: 423–428
Meisl S (1990) Metavolcanic rocks in the "Northern Phyllite Zone" at the southern margin of the Rhenohercynian belt. In: Franke W (ed) Mid-German Crystalline Rise and Rheinisches Schiefergebirge (field guide), Conf Palaeozoic orogens in central Europe–geology and geophysics, Göttingen-Giessen, Justus-Liebig Univ Giessen, pp 25–42
Middleton GV (1960) Spilitic rocks in southeast Devonshire. Geol Mag 97: 192–207
Mottl MJ (1983) Metabasalts, axial hot springs and the structure of hydrothermal systems at mid-ocean ridges. Bull Geol Soc Am 94: 161–180
Muehlenbachs K, Clayton RN (1972) Oxygen isotope studies of fresh and weathered submarine basalts. Can J Earth Sci 9: 172–184
Palacz ZA, Saunders AD (1986) Coupled trace element and isotope enrichment in the Cook-Austral-Samoa Islands, southwest Pacific. Earth Planet Sci Lett 79: 270–280
Pearce JA (1983) Role of the subcontinental lithosphere in magma genesis at active continental margins. In: Hawkesworth CJ, Norry MJ (eds) Continental basalts and mantle xenoliths. Shiva, Cheshire, pp 230–249
Platen KM (1991) Geochemische untersuchungen an metabasalten aus dem Rhenohercynikum. Diplomarbeit Thesis, Justus-Liebig Univ Giessen, 134 pp
Platen KM, Emmermann R, Franke W (1989) Devonian MOR-type metabasalts from the Rhenohercynian zone in Germany. EUG V, Strasbourg, Terra Abstr 1/1: 365
Primmer TJ (1985) The pressure-temperature history of the Tintagel district, Cornwall: metamorphic evidence on the tectonic evolution of the area. Proc Ussher Soc 6: 218–223
Reitz E (1989) Devonische sporen aus phylliten vom Sudrand des Rheinischen Schiefergebirges. Geol Jahrb Hessen 117: 23–35
Rice-Birchall B, Floyd PA (1988) Geochemical and source characteristics of the Tintagel Volcanic Formation. Proc Ussher Soc 7: 52–55
Robinson D (1981) Metamorphic rocks of an intermediate facies series juxtaposed at the Start boundary, southwest England. Geol Mag 118: 297–301
Robinson D, Read D (1981) Metamorphism and mineral chemistry of greenschists from Trebarwith Strand, Cornwall. Proc Ussher Soc 5: 132–138
Rogers G, Saunders AD, Terrell DJ, Verma SP, Marriner GF (1985) Geochemistry of Holocene volcanic rocks associated with ridge subduction in Baja California, Mexico. Nature 315: 389–392
Saunders AD (1983) Geochemistry of basalts recovered from the Gulf of California during Leg 65 of the Deep Sea Drilling Project. In: Lewis BTR, Robinson P et al. (eds) Initial Reports of the DSDP 65: 591–621
Saunders AD, Tarney J (1984) Geochemical characteristics of basaltic volcanism within back-arc basins. In: Kokelaar BP, Howells MF (eds) Marginal basin geology. Geol Soc Lond Spec Publ 16: 59–76
Saunders AD, Fornari DJ, Joron JL, Tarney J, Treuil M (1982a) Geochemistry of basic igneous rocks, Gulf of California, Deep Sea Drilling Project Leg 64. In: Curray JR, Moore DG et al. (eds) Initial Reports of the DSDP 64: 595–642
Saunders AD, Fornari DJ, Morrison MA (1982b) The composition and emplacement of basaltic magmas produced during the development of continental margin basins: the Gulf of California, Mexico. J Geol Soc Lond 139: 335–346
Saunders AD, Norry MJ, Tarney (1988) Origin of MORB and chemically depleted mantle reservoirs: trace element constraints. J Petrol 29: 415–445
Schmincke HU, Pritchard G (1981) Carboniferous volcanic glass from submarine hyaloclastite, Lahn-Dill area, Germany. Naturwissenschaften 67: 615–616
Schmincke HU, Sunkel G (1987) Carboniferous submarine volcanism at Herbornseelbach, Lahn-Dill area, Germany. Geol Rdsch 76: 709–734
Selwood EB, Thomas JM (1986) Upper Palaeozoic successions and nappe structures in north Cornwall. J Geol Soc Lond 143: 75–82
Storey M, Rogers G, Saunders AD, Terrell DT (1989) San Quintin volcanic field, Baja California, Mexico: "within-plate" magmatism following ridge subduction. Terra Nova 1: 195–202
Sun SS (1985) Ocean islands–plums or plumes? Nature 316: 103–104
Sun SS, McDonough WF (1989) Chemical and isotopic systematics of ocean basalts: implications for mantle composition and processes. In: Saunders AD, Norry MJ (eds) Magmatism in the ocean basins. Geol Soc Lond Spec Publ 42: 313–345
Tarney J, Saunders AD, Weaver SD, Donnellan NCB, Hendry GL (1979) Minor element geochemistry of basalts from Leg 49, North Atlantic ocean. In: Luyendyk BP, Cann JR et al. (eds) Initial Reports of the DSDP 49: 657–691
Tarney J, Wood DA, Saunders AD, Cann JR, Varet J (1980) Nature of mantle heterogeneity in the North Atlantic: evidence from deep sea drilling. Philos Trans R Soc Lond 297: 179–202
Thompson G (1983) Basalt-seawater interaction. In: Rona PA, Bostrom K, Laubier L, Smith KL (eds) Hydrothermal processes at seafloor spreading centers. Plenum Press, New York, pp 225–278
Vearncombe JR (1979) The Lizard ophiolite and two phases of suboceanic deformation. In: Panayiotou A (ed) Proc Int Ophiolite Symp, Cyprus Geological Survey, Cyprus, pp 527–537
Weaver BL (1991) The origin of ocean island basalt end-member compositions: trace element and isotopic constraints. Earth Planet Sci Lett 104: 381–397
Wedepohl KH, Meyer K, Muecke GK (1983) Chemical composition and genetic relations of metavolcanic rocks from the Rhenohercynian belts of northwest Germany. In: Martin H, Eder FW (eds) Intra-

continental fold belts. Springer, Berlin Heidelberg New York, pp 231–256
White WM, Schilling JG (1978) The nature and origin of geochemical variation in Mid-Atlantic Ridge basalts from the central North Atlantic. Geochem Cosmochim Acta 42: 1501–1516
Williams CT, Floyd PA (1981) The localized distribution of U and other incompatible elements in spilitic pillow lavas. Contrib Mineral Petrol 78: 111–117
Wood DA (1980) The application of a Th-Hf-Ta diagram to problems of tectonomagmatic classification and to establishing the nature of crustal contamination of basaltic lavas of the British Tertiary Volcanic Province. Earth Planet Sci Lett 50: 11–30
Wood DA, Tarney J, Varet J, Saunders AD, Bougault H, Joron JL, Treuil M, Cann JR (1979) Geochemistry of basalts drilled in the North Atlantic by IPOD Leg 49: implications for mantle heterogeneity. Earth Planet Sci Lett 42: 77–97
Ziegler PA (1984) Caledonian and Hercynian crustal consolidation of western and central Europe–a working hypothesis. Geol Mijinbouw 63: 93–108
Zindler A, Hart S (1986) Chemical geodynamics. Annu Rev Earth Planet Sci 14: 493–571
Zindler A, Staudigal H, Batiza R (1984) Isotope and trace element geochemistry of young Pacific seamounts: implications for the scale of upper mantle heterogeneity. Earth Planet Sci Lett 70: 175–195

III.B.4 Metamorphic Evolution

O. ONCKEN, H.J. MASSONNE, and M. SCHWAB

1 Introduction

The Rhenohercynian Zone – corresponding to the external foreland fold and thrust belt of the European Variscides (e.g., Franke 1989) – developed from a passive continental margin sedimentary prism and is largely built up by deformed sedimentary rocks which underwent weak metamorphism during orogenic contraction. Due to this weak metamorphism, a specific set of different criteria to determine the metamorphic grade has been used in the past. Although all these indicators depend on somewhat differing conditions, the results form a rather coherent picture of the metamorphic evolution of the Rhenohercynian belt.

2 Data on Metamorphism

Data on the metamorphic grade within the Rhenohercynian Zone are mainly obtained from seven different methods:

Petrologic data from index minerals and mineral assemblages are mainly obtained from mafic volcanics contained in the sequence of the eastern Rhenish Massif (Meisl 1970; Weber 1972) and in parts of the Harz mountains (Müller and Strauss 1985). The rocks show the assemblage prehnite + pumpellyite ± actinolite ± epidote. Paragonite and pyrophyllite are typical for deeper buried sediments (Weber 1972; Oncken 1991). Lower Paleozoic rocks of the southern Venn anticline in the western Rhenish Massif may have the assemblage with andalusite, spessartine, and Mn-rich chloritoid (Kramm 1982; Kramm et al. 1985a). The Phyllite Zone at the southern margin of the belt may show pumpellyite + actinolite + chlorite + quartz in the west (Oncken 1991), quartz + albite + K-feldspar + phengite + chlorite or stilpnomelane + epidote + titanite + amphibole + hematite, and albite + chlorite + amphibole + epidote + phengite + titanite + hematite + quartz + stilpnomelane + K-feldspar in metarhyolites and metaandesites respectively, in the southern Taunus (Anderle et al. 1990). In the southern Harz, chloritoid, paragonite, and Mn-carpholite appear in metasediments (Kramm 1978; Löffler and Schwab 1981; Siedel 1989). Higher stratigraphic levels in the northern Massif and the Harz mountains are devoid of critical assemblages.

Thermobarometric data based essentially on phengite barometry yield synorogenic pressures in the southern Venn anticline and in the nappe units at the southern border of the Massif (Massonne 1989; Siedel 1989; Anderle et al. 1990).

Studies on the rank of organic matter dispersed in sediments have received considerable attention in the Rhenish Massif and the Harz. Some 1500 data have been published so far (e.g., Wolf 1972, 1975, 1978; Paproth and Wolf 1973; Jordan and Koch 1975; Teichmüller and Teichmüller 1979; Teichmüller et al. 1979; Kalckreuth 1979; Lütke and Koch 1983; Ecke et al. 1985; Kramm et al. 1985a). Values of vitrinite reflectance (%R_{max}) vary between <2% (northern border, Upper Devonian/Carboniferous in the north and the Eifel NS Zone, Northwest Harz) and 7% and thus mostly lie in the subanthracite to metaanthracite field. The 4% value has been established to approximately mark the 200 °C isotherm (boundary diagenesis anchimetamorphism; Frey 1986) and is used to plot the regional pattern of metamorphism (Fig. 1). Further analysis of rank data yields paleotemperatures, paleogeothermal gradients, and the thickness of eroded sediments in the northern Massif (Oncken 1984, 1987, 1991).

Conodont alteration: Only few studies using the Conodont alteration index (CAI) exist so far (e.g., Lütke and Koch 1983; Buggisch 1986). They do not indicate a simple relationship to other criteria.

Microthermometric studies on fluid inclusions: Koschinski (1979) has studied different genera-

Dallmeyer et al. (Eds.)
Pre-Permian Geology of Central and Eastern Europe

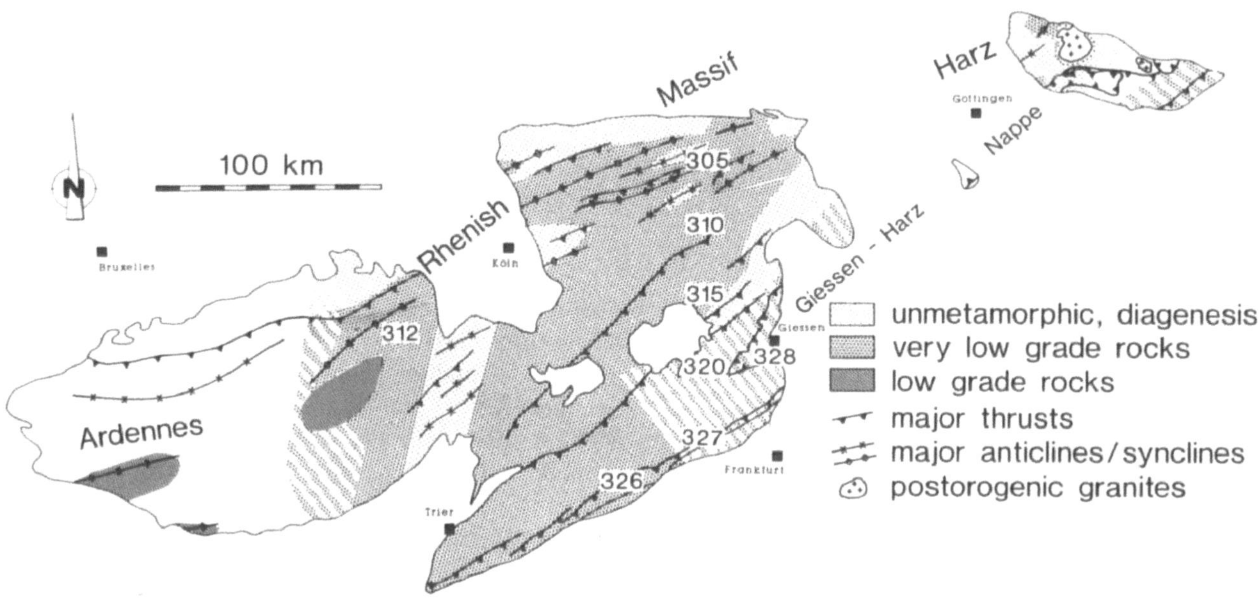

Fig. 1. Regional pattern of synorogenic temperatures in the Rhenohercynian Zone. Subdivision of metamorphic zones in the Rhenohercynian is based on organic rank (boundary unmetamorphic-very low grade $\approx$ 4% R_{max}) and on petrologic data for the low grade areas. *Interrupted pattern*. Temperatures inferred from petrologic, microfabric, and illite crystallinity data. Synorogenic pressures increase from <1 kb in the north to 3–6 kb at the southern rim. *Numbers* on the Rhenish Massif give K/Ar ages from Ahrendt et al. (1983) and Kramm et al. (1985b)

tions of quartz veins and their respective fluid inclusions in the eastern Rhenish Massif. The method yields a higher resolution of the thermal evolution during deformation than most other criteria. Cooling through very low grade conditions and declining deformation can be observed. Data for the Harz (Lüders et al. 1990), however, do not yield unequivocal P-T conditions.

Illite crystallinity analyses have equally led to a dense set of data (e.g., Weber 1972, 1976; Kasig and Spaeth 1975; Wolf 1975; Teichmüller et al. 1979; Buggisch 1986; Siedel 1989; Frank and Spaeth 1991). Since preparation techniques and applied crystallinity indexes vary greatly (see discussion by Buggisch 1986), a straightforward equivalence of the results is not achieved. On the whole, similar trends with relation to stratigraphic age and structure are evident, as found by rank analysis.

Fabrics controlled by thermally activated deformation mechanisms yield characteristic deformation textures (pressure solution, subgrain formation, polygonization, dynamic recrystallization, secondary grain growth, etc. in quartz, calcite, mica; see Oncken 1991) which allow the rough subdivision of the anchimetamorphic zone (200–300 °C) and resolution of the thermal evolution during deformation.

In summary, at least parts of the continental Rhenohercynian basin start with high paleogeothermal gradients at the outset – probably due to early crustal attenuation. Peak metamorphism at the out-crop level in the Rhenish Massif and the Harz mountains reaches very low-grade conditions with a general increase of synorogenic temperatures and pressures from north to south (Fig. 2): <150 °C to ca. 300 °C in the Rhenish Massif with a concomitant increase of burial from 1–5 km in the north (depending on stratigraphy) to ca. 10 km in the south. The general trend is disturbed by major thrusts, and by major synclinoria with younger strata as well as by the Eifel NS zone: these latter structures were never deeply buried and subsequently exhumed. The southern Venn anticline, and part of the Lower Paleozoic Ardennes, have reached greenschist facies conditions during deformation (300–420 °C at 2–3 kb). The Phyllite Zone in the south has straddled the boundary to greenschist facies conditions (300–350 °C, 3–6 kb). Criteria with a higher resolution power show late- to post-kinematic cooling and decompression. Some postorogenic thermal overprint

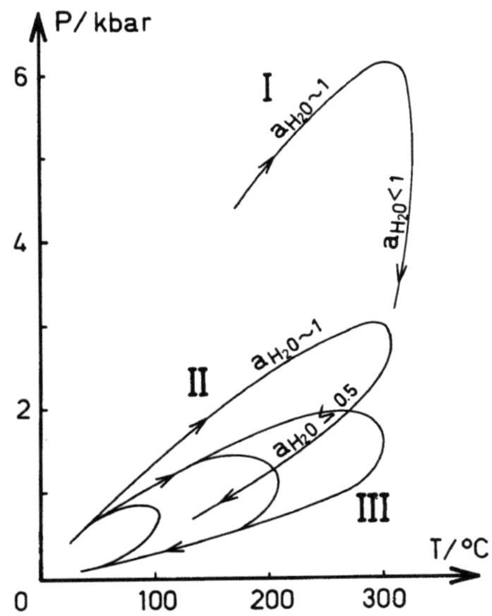

Fig. 2. Schematic P-T paths for several areas in the Rhenohercynian belt: *I* Northern Phyllite Zone at southern margin of Rhenish Massif; *II* Venn anticline (Ardennes), southern Rhenish Massif, and southeastern Harz; *III* northern and central Rhenish Massif as well as northwestern Harz (paths showing T-range depending on stratigraphic age of rocks involved, i.e., Ordovician through Lower Carboniferous)

has occurred in the contact aureoles of the Harz granites and in parts of the northeastern Rhenish Massif.

3 Timing of Metamorphism

K/Ar determinations on mica by Ahrendt et al. (1983) for the eastern Rhenish Massif yield decreasing ages from south to north (from 325 to 300 Ma; see also Kramm et al. 1985b; Fig. 1). The analyzed micas grew or recrystallized during cleavage formation (Weber 1976). The metamorphic temperatures (see above) did not reach the closing temperature of the K/Ar system in muscovite (exception: the southern Venn anticline). Thus, the above ages are interpreted as dating the peak of deformation and metamorphism as well as the northwards migration of the orogenic front through the Rhenohercynian basin. Indications of an earlier, Caledonian, deformation and weak metamorphism in Cambro-Ordovician beds of the Venn anticline are still under debate (Kramm et al. 1985b; Spaeth et al. 1985). The Ecker gneiss of the Western Harz, as well as tectonic gneiss-amphibolite lenses in the southwestern Rhenish Massif with Proterozoic/Cambrian ages (Meisl et al. 1989; Vinx and Schlüter 1989; Baumann et al. 1991), are thought to be part of the Rhenohercynian basement formed during the Cadomian cycle.

4 Metamorphism and Deformation

The combined evaluation of rank data, microfabrics, fluid inclusion data, and illite crystallinity permits a higher temporal resolution of the thermal history with respect to deformation. Metamorphic isograds (i.e., rank) follow stratigraphic boundaries in the northern Massif, Ardennes, and Harz with few exceptions (see Kalckreuth 1979; Teichmüller et al. 1979). Across a broad transitional zone this relationship changes towards the south: here, isograds cut across stratigraphic and tectonic boundaries with, sometimes, minor offsets at major thrusts (Oncken 1991). In the north, peak temperatures are thus reached during deformation near the point of maximum burial of the sedimentary pile. Towards the south, heating increasingly accompanies and outlasts deformation and thrusting with a concomittant reduction of former thermal differences between different strata.

Although metamorphic differences between thrust units and their respective footwalls are generally small in the south (Oncken 1991), the offset of isograds rules out a complete thermal equilibration across the thrusts during the peak thermal conditions. Microfabric studies reveal higher syn-s_1 temperatures in the hangingwalls compared to syn-s_1 conditions in the footwalls which moreover show a weak inverse metamorphic gradient. Syn-s_2 metamorphic differences between hangingwalls and footwalls cannot be resolved. Fluid inclusion and microfabric studies (Koschinski 1979; Oncken 1991) also point to considerable cooling through this stage. These aspects indicate that the main stage of tectonic stacking of the presently outcropping units followed peak metamorphic conditions. The apparent inverse metamorphic gradients below the thrusts probably reflect heating of the footwall units by the overriding warmer units, the latter showing evidence of concomitant cooling. Peak metamorphic conditions in hangingwall and footwall units are therefore, strictly speaking, not contemporaneous (cf. difference in K/Ar-age between for example, the Giessen Nappe body and the underlying autochthonous; Ahrendt et al. 1983).

5 Conclusions

The overall very low-grade metamorphism of the Rhenohercynian shows a general increase in temperatures and pressures in discrete steps from near-surface conditions in the foredeep to greenschist conditions at the southern border. The presently exposed surface thus constitutes an oblique section through the former orogenic wedge.

This section was affected by increasing synkinematic heating towards the south, which cannot be attributed to an eroded sedimentary cover. This aspect is thus interpreted as evidence of an increasing influence of formerly superincumbent nappe systems – now mostly eroded – from the Mosel-Lahn/Dill synclines in the central Massif towards the south (i.e., the Giessen-Harz Nappe; see Engel et al. 1983; Oncken 1991).

Moreover, exhumation and cooling of the presently exposed part of the wedge was associated with continuous shortening and thrusting, as evidenced by the metamorphic hiatus across thrusts and by the lack of a complete thermal equilibration during or after deformation. Obviously, the deformed Rhenohercynian basin filling, while eroding, went through some out-of-sequence thickening of its southern and central parts during the northward propagation of the deformation front and the growth of the orogenic wedge. At the same time, no evidence of major crustal thickening involving the lower crust is apparent, which may be concluded from the absence of mid-crustal rocks at the present surface. Accordingly, the present metamorphic pattern supports the view of a detachment of the upper crust (i.e., the Devonian basin filling) from the lower crust (i.e., Cadomian crystalline basement; cf. Behrmann et al. 1991).

The above-mentioned metamorphic zonation is responsible for a gradual change of the dominant deformation mechanism from frictional sliding on localized brittle fractures in the north through pressure solution-dominated fabric formation to a zone of incipient crystal-plastic deformation at the southern border of the Rhenohercynian (Oncken 1991). Along this transect through the brittle upper crust into the brittle-ductile transition, a concomitant change of the large-scale structural style can be established: from large-scale folding in the north to pervasive cleavage and imbrication in the south.

The metamorphic history of the Rhenohercynian may be summarized to start on hot crust during basin formation – probably related to previous crustal stretching (Oncken 1987). During shortening and imbrication of the shelf sequence, upon continental collision with the Saxothuringian, this setting is overprinted by high-P synorogenic conditions in the south followed by high-T low-P conditions during exhumation and postkinematic extension (see Anderle et al. 1990).

References

Ahrendt H, Clauer N, Hunziker JC, Weber K (1983) Migration of folding and metamorphism in the Rheinisches Schiefergebirge deduced from K-Ar and Rb-Sr age determinations. In: Martin H, Eder FW (eds) Intracontinental fold belts. Case Studies in the Variscan Belt of Europe and the Damara Belt in Namibia. Springer, Berlin Heidelberg New York, pp 323–338

Anderle HJ, Massonne HJ, Meisl S, Oncken O, Weber K (1990) Southern Taunus Mountains. In: Field guide Mid-German Crystalline Rise and Rheinisches Schiefergebirge, Int Conf Paleozoic Orogens in central Europe, Göttingen-Giessen 1990, pp 125–148

Baumann A, Grauert B, Mecklenburg S, Vinx R (1991) Isotopic age determinations of crystalline rocks of the Upper Harz Mountains. Geol Rdsch 80: 669–690

Behrmann J, Drozdzewski G, Huch M, Meyer W, Oncken O (1991) Synthetic crustal scale balanced cross sections for parts of the Central EGT segment. Tectonophysics 196: 1–21

Buggisch W (1986) Diagenese und Anchimetamorphose aufgrund von Conodontenfarbe (CAI) und "Illitkristallinität" (IC). Geol Jahrb Hessen 114: 181–200

Ecke HH, Hoffmann M, Ludewig B, Riegel W (1985) Ein Inkohlungsprofil durch den südlichen Hunsrück (südwestliches Rheinisches Schiefergebirge). Neues Jahrb Geol Paläontol Monatsh 7: 395–410

Engel W, Franke W, Grote C, Weber K, Ahrendt H, Eder FW (1983) Nappe tectonics in the south-eastern Part of the Rheinisches Schiefergebirge. In: Martin H, Eder FW (eds) Intracontinental fold belts. Case studies in the Variscan Belt of Europe and the Damara Belt in Namibia. Springer, Berlin Heidelberg New York, pp 267–287

Frank B, Spaeth G (1991) Die Illit-Kristallinitätsverhältnisse im nordöstlichen Teil des Stavelot-Venn-Massivs, Linksrheinisches Schiefergebirge. Neues Jahrb Geol Paläontol Monatsh 5: 255–283

Franke W (1989) Tectonostratigraphic units in the Variscan belt of central Europe. Geol Soc Am Spec Pap 230: 67–90

Frey M (1986) Very low-grade metamorphism of the Alps – an introduction. Schweiz Mineral Petrogr Mitt 66: 13–27

Jordan H, Koch J (1975) Inkohlungsuntersuchungen im Unterkarbon des Nordwestharzes. Geol Jahrb A 29: 33–43
Kalckreuth W (1979) Das Inkohlungsbild des Ostsauerländer Hauptsattels im Rhenoherzynikum mit besonderer Berücksichtigung der Trendflächenanalyse. Fortschr Geol Rheinland Westfalen 27: 277–321
Kasig W, Spaeth G (1975) Neue Ergebnisse über die Geologie der Kern- und Mantelschichten des Hohen Venns auf Grund von Profilaufnahmen bei der Verlegung der Erdgasleitung Aachen-Rheinfelden. Z Dtsch Geol Ges 126: 1–14
Koschinski G (1979) Mikrostrukturelle und mikrothermometrische Untersuchungen an Quarzmineralisationen aus dem östlichen Rheinischen Schiefergebirge. Thesis, Univ Göttingen, 156 pp
Kramm U (1978) Die Metamorphose Mn-reicher Pelite der Wippraer Zone/Unterharz. Fortschr Mineral 56/1: 67–68
Kramm U (1982) Die Metamorphose des Venn-Stavelot-Massivs, nordwestliches Rheinisches Schiefergebirge: Grad, Alter und Ursache. Decheniana 135: 121–178
Kramm U, Spaeth G, Wolf M (1985a) Variscan Metamorphism in the NE Stavelot-Venn Massif, Ardennes: a new approach to the question of regional dynamothermal or contact metamorphism. Neues Jahrb Geol Paläontol Abh 171: 311–327
Kramm U, Buhl D, Chernyshev IV (1985b) Caledonian or Variscan metamorphism in the Venn-Stavelot Massif, Ardennes? Arguments from a K-Ar and Rb-Sr study. Neues Jahrb Geol Paläontol Abh 171: 339–349
Löffler HK, Schwab M (1981) Über das Auftreten von Karpholith im Harz und seine Bedeutung für die mineralfazielle Entwicklung der Wippraer Einheit. Z Geol Wiss schaften 9: 1309–1313
Lüders V, Franzke HJ, Rauche H (1990) Mikrothermometrische Untersuchungen an Flüssigkeitseinschlüssen in Quarzen der Metamorphen Zone von Wippra (Harz) und angrenzender Sb-betonter Gangstrukturen. Beih Eur J Mineral 2: 166
Lütke F, Koch J (1983) Das Inkohlungsbild des Paläozoikums im Westharz und seine Interpretation. Geol Jahrb A 69: 1–42
Massonne HJ (1989) Metamorphoseentwicklung des Rhenoherzynikums am Beispiel eines Metagranodiorites aus dem Venn-Stavelot-Massiv. Kurzfassungen zum 5. Rundgespräch Geodyn Eur Varistikums, Braunschweig, 16.–18.11.1989
Meisl S (1970) Petrologische Studien im Grenzbereich Diagenese-Metamorphose. Abh Hess Landesamts Bodenforsch 57: 1–93
Meisl S, Kreuzer H, Höhndorf A (1989) Metamorphose-Bedingungen und -Alter des Kristallins am Wartenstein bei Kirn/Nahe. Kurzfassungen zum 5. Rundgespräch Geodyn Eur Varistikums, Braunschweig, 38 pp
Müller G, Strauss KW (1985) Beitrag zur Regionalmetamorphose des Harzes. Geol Rdsch 74: 87–94
Oncken O (1984) Zusammenhänge in der Strukturgenese des Rheinischen Schiefergebirges. Geol Rdsch 73: 619–649
Oncken O (1987) Heat flow and kinematics of the Rhenish Basin. In: Vogel A, Miller H, Greiling R (eds) The Rhenish Massif. Vieweg (Earth Evol Sci), Braunschweig, pp 63–78
Oncken O (1991) Aspects of the structural and paleogeothermal evolution of the Rhenish Massif. Ann Soc Géol Belg 113: 139–159
Paproth E, Wolf M (1973) Zur paläogeographischen Deutung der Inkohlung im Devon und Karbon des nördlichen Rheinischen Schiefergebirges. Neues Jahrb Geol Paläontol Monatsh 1973: 469–493
Siedel H (1989) Beiträge zur Petrographie und Geochemie epizonal metamorpher Pelite in der Wippraer Einheit (Harz). Thesis, Univ Halle, 142 pp
Spaeth G, Fielitz W, Frank B (1985) Caledonian deformation and very-low-grade metamorphism in the northeastern part of the Venn-Stavelot-Massif. Neues Jahrb Geol Paläontol Abh 171: 297–310
Teichmüller M, Teichmüller R (1979) Ein Inkohlungsprofil entlang der linksrheinischen Geotraverse von Schleiden nach Aachen und die Inkohlung in der Nord-Süd-Zone der Eifel. Fortschr Geol Rheinland Westfalen 27: 323–355
Teichmüller M, Teichmüller R, Weber K (1979) Inkohlung und Illitkristallinität. Vergleichende Untersuchungen im Mesozoikum und Paläozoikum von Westfalen. Fortschr Geol Rheinland Westfalen 27: 201–276
Vinx R, Schlüter J (1989) Der Eckergneis: Cadomisches Basement im Rhenoherzynikum ? Kurzfassungen zum 5. Rundgespräch Geodyn Eur Variszikums, Braunschweig, 52 pp
Weber K (1972) Kristallinität des Illits und andere Kriterien schwacher Metamorphose im nordöstlichen Rhenischen Schiefergebirge. Neues Jahrb Geol Paläontol Abh 141: 333–363
Weber K (1976) Gefügeuntersuchungen an transversalgeschieferten Gesteinen aus dem östlichen Rheinischen Schiefergebirge. Geol Jahrb 94/D15: 3–98
Wolf M (1972) Beziehungen zwischen Inkohlung und Geotektonik im nördlichen Rheinischen Schiefergebirge. Neues Jahrb Geol Paläontol Abh 141: 222–257
Wolf M (1975) Über die Beziehung zwischen Illit-Kristallinität und Inkohlung. Neues Jahrb Geol Paläontol Monatsh 1975: 437–447
Wolf M (1978) Inkohlungsuntersuchungen im Hunsrück (Rheinisches Schiefergebirge). Z Dtsch Geol Ges 129: 217–227

III.B.5 Metallogenesis

W. WERNER and H.W. WALTHER

1 Introduction

The Rhenohercynian Fold Belt (RFB) in Germany (Fig. 1) is one of the most important mineral belts in Europe; its main mineral districts are located in the Upper Harz Mts., and in the Siegerland, Sauerland, Lahn-Dill, and Bleiberg-Moresnet-Aachen-Stolberg regions of the Rhenish Massif (Fig. 2). The belt contains a great number of syn- and epigenetic ore deposits and occurrences in which the metals Fe, Zn, Pb, and Ba predominate; Cu, Sb, Co, Ni, Ag, Au, and F are locally enriched. In striking contrast to the Saxothuringian and Moldanubian zones, mineralization of Sn, W, and U are very rare.

Mining activities began during Celtic and Roman times. Main mining periods were from 1200 to 1350, 1500 to 1630, 1675 to 1775, and from 1850 to 1970. At the most famous orebody, i.e., at Rammelsberg/Harz Mts., operations were taken up during the 3rd century A.D. and terminated in 1988. Due to low metal prices, mining of lead and zinc ores was stopped at Meggen and Bad Grund at the end of March 1992. At present, mining activities are restricted to barite orebodies at Dreislar (Rhenish Massif) and Lauterberg (Harz Mts.).

2 Classification

The mineral deposits and occurrences of the RFB can be classified into (A) Devonian to Lower Carboniferous **synsedimentary to syndiagenetic** and (B) Late Lower Carboniferous to Tertiary **epigenetic mineralization**; the latter formed either in the course of the Variscan orogeny or during post-Variscan times, especially between the Triassic and the Tertiary.

Three fundamental processes led to synsedimentary to syndiagenetic mineralization: sedimentation, basin dewatering, and volcanism. The oolitic and detrital Fe ores of the Eifel (A1) are typical representatives of the sedimentary type. The pyrite-barite orebodies with Zn, Pb, Cu (Ag) such as Meggen and Rammelsberg (A2), are considered to be products of a focused ascent of metalliferous basin-dewatering brines, thus belonging to the class of SEDEX-type mineralization (Werner 1989a,b). Bound to Devonian submarine volcanism are barite and pyrite occurrences hosted in pelagic shales and carbonates and associated with keratophyric volcanics (A3) as well as the Lahn-Dill-type Fe ores (A4), the formation of which is connected with the spilitization of basalts. Contemporaneous sediment dewatering focused at syndepositional faults might be significant for both types (A3, A4) (cf. Flick et al. 1990). The low-grade Cu protores of the Marsberg type hosted in Lower Carboniferous black shales (A5) probably were formed by the interaction of all mechanisms listed above.

Group B comprises synorogenic mineral veins generally following cleavage or thrust planes (B1) and postorogenic (post-Variscan) veiniform, irregular, bedded, or disseminated mineralization (B2). Zn, Pb, Cu, Sb, Ag, and Co occur in both epigenetic subgroups, whereby Ba and F are restricted to post-Variscan formations.

3 Age of Mineralization

(A1) The sedimentary Fe ores of the Eifel region are of Emsian-Eifelian age. (A2) SEDEX-type ores formed during the Emsian to Frasnian period of accelerated crustal stretching; the most prominent examples are the Rammelsberg (Eifelian) and Meggen deposits (uppermost Givetian). (A3) The sediment-hosted lenticular bodies of massive pyrite and barite related to keratophyre volcanism formed during Eifelian to upper Givetian times (Eisen, Lohrheim, Elbingerode). (A4) The Lahn-Dill-type Fe mineralization had a climax in the upper Givetian-lower Frasnian period of basaltic volcanism. (A5) The stratiform and disseminated

Dallmeyer et al. (Eds.)
Pre-Permian Geology of Central and Eastern Europe

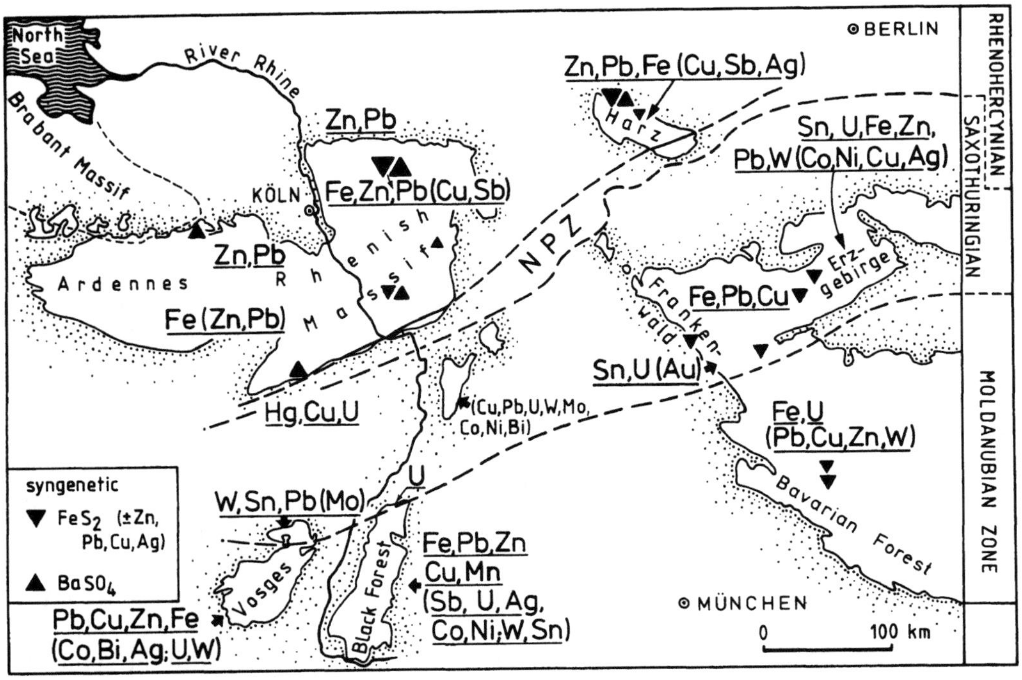

Fig. 1. Metallotectonic map of the Central European Variscides illustrating the characteristic metal associations in Precambrian to Late Paleozoic syngenetic, Variscan, and post-Variscan epigenetic mineral deposits. Barite-fluorite veins (occurring in all Variscan units) and mineral deposits related to transgressive post-Rotliegend strata are not shown. Subdivision of Variscan belts after Franke (1989)

pyritic Cu protores in black shales and lydites (Marsberg type) are of Lower Carboniferous age. (B1) The syntectonic (Variscan) veins are of Carboniferous age; a number of geological criteria allow classification of the veins of the Siegerland-Wied district (Fig. 2) as upper Visean-lower Namurian formations (Walther in: Fenchel et al. 1985). (B2) The widespread mineral occurrences of the post-Variscan cycle developed during Permian to Tertiary extension and uplift of the Rhenish Massif.

4 Summarizing Description of the Mineralization Types, Metallogenetic Aspects

4.1 Synsedimentary (Syngenetic) Mineralization

(A1) In the western Rhenish Massif (Fig. 2), there occur sedimentary ores with 14–21% Fe hosted in upper Emsian to Eifelian shallow-water carbonates. They are mainly composed of hematite-chamosite ooides and mineralized fossil debris. The sedimentary Fe ores of the Eifel area are products of Devonian lateritic weathering on the Old Red Continent and Fe precipitation in an intertidal environment (Neumann-Redlin et al. 1977; Simon 1979). Similar ores are developed in the Ardennes (Namur, Liège).

(A2) In contrast to the Saxothuringian and Moldanubian zones the RFB contains SEDEX-type barite and pyrite deposits and occurrences with ± Zn, Pb, Cu, Ag. Listed from W to E: Belgium: Chaudfontaine (ba); Germany: Plettenberg-Affeln (py, ba), Meggen (py, ba, sph, ga), Oberhundem (py), Günterod (ba, ga), Rammelsberg (py, sph, ga, cpy, ba, Au, Ag). The common host rocks are hemipelagic pelites with variable amounts of silt and sand and with little tuffaceous material from distant volcanism. Reef debris from nearby bioherms of the shelf represents the host rock at Chaudfontaine (Dejonghe 1985), but is also associated at Meggen (Krebs 1981) documenting shallow water conditions during sulfide precipitation (Fig. 3). The Meggen orebody, composed of one large massive-sulfide lens of 59.2 Mt with 36% S, 8% Zn, 1% Pb, 0.04% Cu on an average and a surrounding barite seam (9.5 Mt with 97% $BaSO_4$), was under exploitation from 1861 to 1992; barite mining ceased in 1977. The orebody is hosted by upper

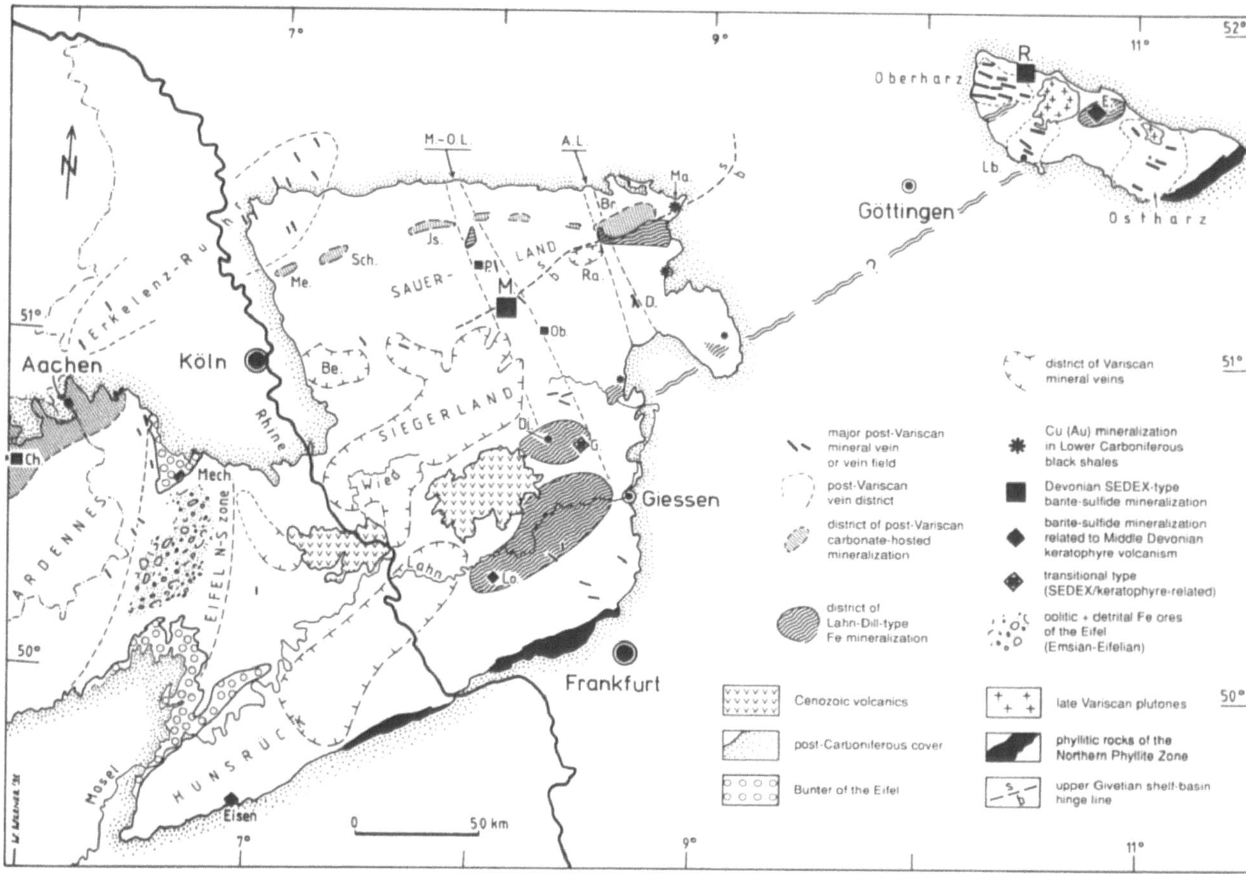

Fig. 2. The generalized pattern of mineralization in the Rhenohercynian Fold Belt. The figure depicts the metallogenetic classification of main mineral districts, vein fields, and of some selected occurrences as well as regionally extending structures which are important for mineralization. *Ch* Chaudfontaine; *Be* Bensberg district; *Br* Brilon district; *D* Dreislar; *Di* Dillenburg; *E* Elbingerode; *G* Günterod; *Is* Iserlohn district; *Lb* Lauterberg; *Lo* Lohrheim; *M* Meggen; *Ma* Marsberg; *Me* Mettmann; *Mech* Mechernich; *Ob* Oberhundem; *P* Plettenberg; *R* Rammelsberg; *Ra* Ramsbeck district; *Sch* Schwelm; *A.L.* Altenbüren Lineament; *M.-O.L.* Menden-Oberscheld Lineament; *wavy line* likely continuation of the volcanic rise Lahn-Dill-Kellerwald-Central Harz Mts.

Givetian dark gray, silt-laminated shales; stratigraphic equivalents are platform micrites of the shelf and the shelf-basin transition, the upper part of which overlie the orebody (Clausen 1978). The footwall beds consist of interbedded pyritic dark gray shales and siltites. Over a period of about 13 Ma, episodic submarine-hydrothermal activity occurred within the Meggen basin, resulting in syngenetic pyrite and siderite within the Eifelian to Frasnian beds (Werner 1989a, p. 205).

The Rammelsberg deposit previously containing 27 Mt with 13.7% Zn, 5.9% Pb, 1.1% Cu (Sperling and Walcher 1990), 100 g/t Ag, 22% $BaSO_4$ and the surrounding 5 Mt of Banderz (a sulfide/shale intercalation with 9% base metals) made mining possible over a period of more than 1000 years. Submarine mineralization took place at the hinge between the Goslar trough (with >1000 m of dark gray to black Wissenbach shales) and the Westharz rise (with <200 m of contemporaneous pelitic sediments).

SEDEX deposits are considered products of long-term basin dewatering focused at syndepositional fault zones; metals are preserved when anoxic conditions make sulfide precipitation possible (Badham 1981; Morganti 1981; Carne and Cathro 1982; Goodfellow and Jonasson 1986). The conditions leading to the escape of metalliferous formation waters are (1) differential subsidence of basin elements, (2) elevated heat flow focused at deeply penetrating fracture systems, and (3) increased permeability at syn-

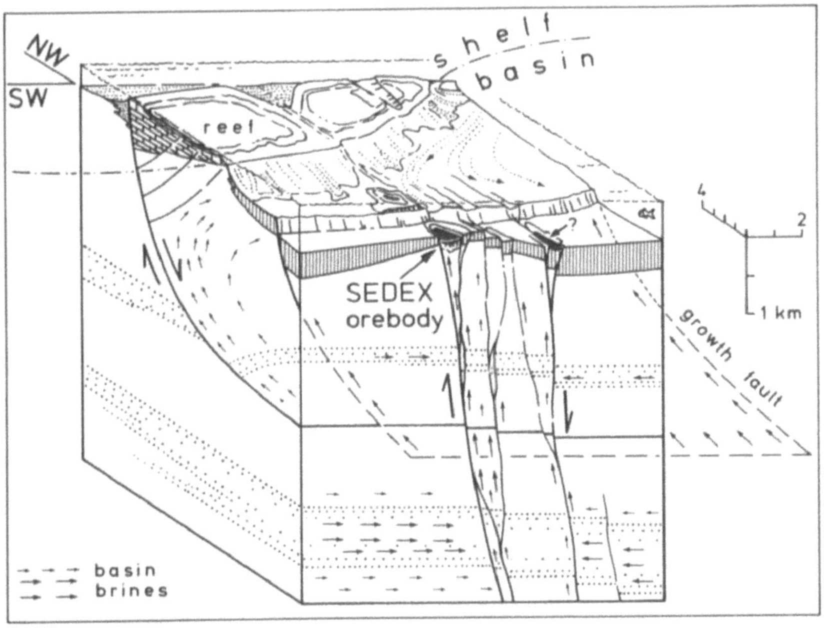

Fig. 3. Schematic block diagram illustrating the likely paleogeology of SEDEX-type massive sulfide orebodies in the RFB (example Meggen). Basin-dewatering brines had maximum discharge at the seafloor, where syndepositional fault systems intersected. The NW-SE-oriented system (the later M.-O.L.) also led to the formation of restricted pools with anoxic bottom waters, where metalliferous brines could be trapped

depositional faults. Favorable tectonic settings are intracontinental rifts and stretched passive continental margins with psammopelitic marine sedimentation, moderate heat flow (little to no volcanism), and slow but long-lasting syndepositional faulting (Werner 1989a,b, 1990a).

(A3) Lens-shaped bodies of pyrite and/or barite with low base-metal contents are situated on an ENE-WSW-oriented volcanic belt trending from the Hunsrück to the central Harz Mts. (Fig. 2). The occurrences at Eisen (Hunsrück) and Lohrheim (Lahn area), both with less than 0.5 Mt pyrite or barite, and the Elbingerode deposit with 15 Mt of 15–20% pyrite (Scheffler, pers. comm.) are hosted in Eifelian black shales or Givetian/Frasnian limestones, contain tuffs and/or are underlain by keratophyre volcanics. The sediment-hosted submarine hydrothermal pyrite-barite deposits of Eisen and Lohrheim are considered to represent transitional types between Rio-Tinto massive sulfide and Meggen-type (SEDEX) mineralizations (Werner 1989a; Emmermann et al. 1993). The pyrite orebody of the Einheit mine (Elbingerode complex), which is considered to be a product of synsedimentary to syndiagenetic hydrothermal leaching of the footwall sediments and volcanics (Scheffler 1975), formed at the intersection of ENE-WSW- and NW-SE-trending syndepositional fault zones (H. Weller/Greifswald, pers. commun.).

(A4) The numerous volcano-sedimentary Fe deposits of the Lahn-Dill type are confined to the upper Givetian to lowermost Frasnian beds of the RFB east of the River Rhine, i.e., to the Lahn, Dill, Ostsauerland and Balve areas in the Rhenish Massif as well as to the Elbingerode Complex and Oberharz Diabaszug in the Harz Mts. Of these orebodies (for description see Lippert in: Walther et al. 1986), several hundred mines produced about 118 Mt of iron ore until 1973. The largest single orebody with 6 Mt was exploited in the Fortuna mine/Lahn area. The orebodies formed at the flanks of volcanic seamounts composed mainly of pyroclastic material. They are generally overlain by hemipelagic pelites or cephalopod-limestone banks. There are two groups of genetic models applied to the Lahn-Dill-type mineralization. (1) Fe and Si were derived from cooling basaltic magmas (e.g., Lehmann 1972; Quade 1976). (2) Fe, Si, Ca were leached from the volcanics by saline waters – "hydrous metamorphic basalt alteration" (Hentschel 1960; Wedepohl et al. 1983: p. 239). Flick and Nesbor (1988) conclude that the Devonian Lahn-Dill iron ores and the Lower Carboniferous iron cherts both are products of submarine low-temperature (about

120 °C) syndiagenetic alteration of footwall volcanoclastics. There is a controversy as to whether the volcanics were leached by circulating heated seawater (Wedepohl et al. 1983) or by sediment-dewatering fluids (Flick et al. 1990). A combination of both mechanisms seems most probable (Fig. 4), since the ores are confined to volcanoes (main Fe source rocks, local heat-flow maximum), which are lined up upon major NE-SW- and NNW-SSE-trending fracture systems (favoring heat flow and the migration of basin brines).

(A5) In the Lower Carboniferous black shales and lydites of the eastern Rhenish Massif disseminated to stratiform cupriferous pyrite mineralization formed under euxinic conditions (Marsberg, Korbach, Dexback, Bergfreiheit; Fig. 2). Basin brines leaching the underlying black shales and mafic volcanics were probably important for the formation of syndiagenetic Cu protores; along rejuvenated faults multiple metamorphic to supergene remobilization and precipitation led to ore-grade enrichment (Stribrny et al. 1988; Werner 1989a, p. 104ff).

4.2 Epigenetic Mineralizations

(B1) The Variscan mineral veins contain siderite, sphalerite, and silver-bearing galena as the main ore minerals. In striking contrast to the post-Variscan formations, the ores typically contain Fe-rich sphalerite and Ag-rich tetrahedrite and are barren in barite (Schaeffer 1984; Krahn 1988). The veins opened in already cleaved psammopelitic sequences of large anticlines as a consequence of variable rock competencies and changing deformation patterns (cf. Hannak 1964; Weber 1977; Podufal in: Walther et al. 1986). Typical structural features of veins which formed during the Variscan orogeny are, for instance, mineralized breccia fillings cut by Variscan thrusts (Hesemann 1978), mylonitization of sulfide ore in Variscan thrusts (Weber 1977), drag and distortion of siderite veinlets into cleavage planes (Fenchel and Lusznat, in Fenchel et al. 1985). According to Hein (1993), the siderite veins originate from CO_2-undersaturated fluids of low salinity and precipitated at temperatures between 180 and 320 °C "immediately after peak metamorphism predating the postkinematic magmatism of the Rhenohercynian Belt".

The Siegerland veins, out of which 175 Mt of siderite (with 35% Fe and 6–7% Mn) were exploited, are up to 12 km long, up to 10 m wide, and are proven to extend down to a depth of 1000 m (Walther and Dill 1995). The veins comprise an early quartz-danaite phase, main siderite-quartz phase, and a late phase with quartz, chalcopyrite, silver-bearing galena, and antimonite (Fenchel et al. 1985). The main ore minerals of the veins in Bensberg district (production: 1.6 Mt of Zn and Pb) are sphalerite, galena, and siderite accompanied by a variety of accessory Pb, Cu, Fe, As, and Sb minerals (e.g., Lehmann and Pietzner 1970; Hesemann and Lehmann in: Walther et al. 1986). At Ramsbeck, where 16.7 Mt ore with 4.4% Zn and 2.1% Pb were

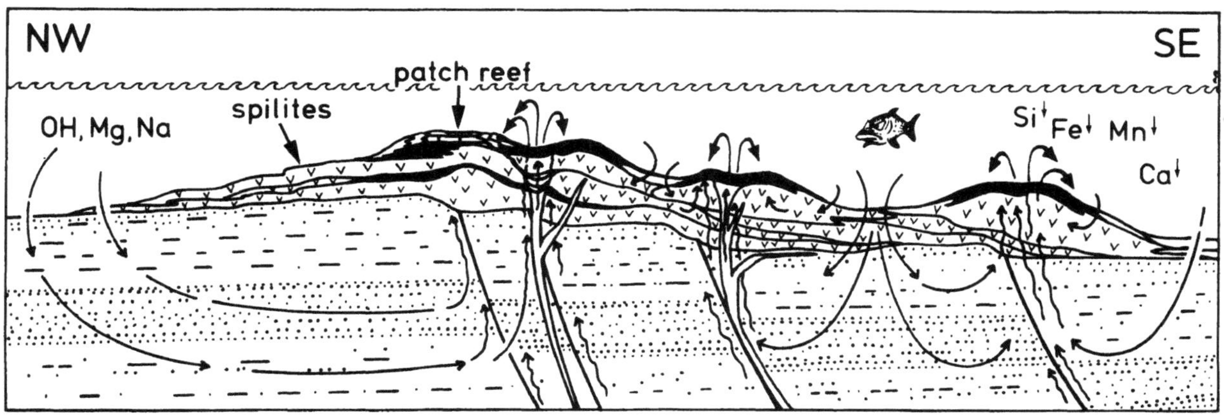

Fig. 4. Numerous small iron orebodies of the Lahn-Dill type (*black*) formed on top of volcanic seamounts as a result of low-temperature hydrous alteration of submarine basalts (spilitization) and sediments. Convecting heated seawater as well as sediment-dewatering fluids could rapidly ascend at growth faults and volcanic vents

mined from 1840 to 1974, the veins are built up by quartz, calcite, ankerite, and siderite, with the main ore minerals sphalerite, pyrite, and galena (Bauer et al. 1979; Podufal 1983). In the southern Rhenish Massif the Siegerland-vein-type mineralization of the Ems district and the Mühlenbach mine delivered 12.6 Mt ore with 0.75 Mt Zn and 0.4 Mt Pb (with 900 ppm Ag), and the so-called schistosity veins of the Hunsrück-Lower Lahn district (containing the Holzappel, Werlau, Tellig and Altlay mines) produced 7.5 Mt of Zn-Pb-Cu-Ag ore with about 0.5 Mt Zn and 0.27 Mt Pb. In the Harz Mts. the only hydrothermal mineralizations of Variscan age are tungsten- and tourmaline-bearing quartz veins which, however, never had economic significance (Kuschka and Franzke 1974).

(B2) The post-Variscan or "Saxonian" deposits and occurrences are veiniform, irregular, disseminated, or stratiform. In contrast to the synorogenic mineral veins these ores are generally coarse-grained and barite is an abundant gangue mineral; fluorite occurs locally. The numerous post-Variscan occurrences of the eastern Rhenish Massif are composed of about 50% calcite, 20% quartz, 10% barite, 6% hematite, and 5% silver-poor Pb, Fe, Zn sulfides (Schaeffer 1984, 1986). Post-Variscan carbonate-hosted enrichments of (colloform) sphalerite, calamine, galena and Fe sulfides reached some economic significance in the Aachen-Stolberg and Iserlohn-Schwelm districts (Fig. 2). Up to 90-m-thick beds of Lower Triassic (Middle Bunter) sandstones and conglomerates of the Mechernich and Maubach areas are Pb- and Zn-mineralized; about 3 Mt of Pb were mined in the Mechernich district until 1969 (Henneke 1977), some 0.27 Mt Pb and 0.1 Mt Zn near Maubach (Siemes and Breuer 1992). Probably of early post-Variscan age (Permian?) is the recently explored Altenbüren mineralization at the western flank of the Brilon reef complex; marcasite, barite and sphalerite are hosted in thick collapse breccias of reef and back-reef sediments (Werner 1990c). The deposits and occurrences at Maubach, Mechernich, Dreislar, and Altenbüren are – like many smaller ones in the Eifel and Sauerland areas – bound to transverse lineaments, which developed from synsedimentarily active fault systems. Similar relations between "ancient lineaments" and post-Variscan mineralization are observed in the Harz Mts. (Mohr 1978).

The hydrothermal mineral veins of the Oberharz district/Harz Mts. (Fig. 2) are mainly composed of quartz, calcite, Fe-Mg carbonates, barite, galena (rich in Ag and Sb) and sphalerite (rich in Cd) (Sperling et al. 1979). Ore precipitation might have begun in the Permian, but the bulk of the veins seems to be of Mesozoic age (e.g., Mertz et al. 1989; Hagedorn and Lippolt 1993). Some 37 Mt of ore containing 1.4 Mt Zn, 1.9 Mt Pb, and about 5000 t Ag have been produced since 1540; from the large Grund vein deposit 19.1 Mt of ore with 10% Pb + Zn were mined from 1900 to 1992 (Sperling, in Walther et al. 1986; Stedingk and Stoppel 1993). The veins of the SW-Harz district contain >7 Mt of barite (with little fluorite), thus representing one of the largest accumulations of this mineral in Europe (Stoppel et al. 1983). The mineral veins of the Ostharz (Fig. 2) contain >8.5 Mt of fluorite ore. The last operating mines (Strassberg, Rottleberode) were closed in 1990.

5 Regional Metallogenetic Relations

With the exception of the sedimentary Fe ores of the Eifel and Ardennes regions, all other types of mineralization mentioned above were formed by hydrothermal processes, which were related either to basin extension (A2, A3, A4), synorogenic NW-directed compression, folding, and thrusting (B1), or to continent-wide extension (B2). The syngenetic ores formed upon a thinned and intensely block-faulted continental crust, which during the Variscan orogeny suffered NW-migrating compression, folding and thrusting – accompanied by syntectonic veiniform mineralization. The bulk of the post-Variscan mineralization formed penecontemporaneously with the Alpidic orogenic cycle and is probably connected with the opening of the Atlantic ocean (Walther 1982b).

Devonian and Lower Carboniferous sedimentation and volcanism in the Rhenohercynian Basin were affected by fault zones trending NE-SW and NNW-SSE to N-S (Langenstrassen 1983; Engel et al. 1983; Bottke 1978; Werner 1989a). The SEDEX-type deposits and occurrences of the Chaudfontaine, Sauerland (Meggen, Plettenberg), and Rammelsberg areas all string on an ENE-WSW-trending belt oriented subparallel to a line with sediment-hosted pyrite and barite occurrences associated with bimodal volcanics (Fig. 2). In the eastern Rhenish Massif all the SEDEX-type mineralizations, two Lahn-Dill Fe districts (Oberscheld, Wetzlar as well as the small ore field of Balve) and several areas with siderite/Fe silicate concretions in Emsian beds are situated on a Devonian NNW-SSE-trending main fracture system, later converted to the Menden-Oberscheld Lineament (Werner 1989a: Figs. II.-42 and II.-43). Most post-Variscan veins of the central RFB occur on the three N-S/NNW-SSE lineaments illustrated in Fig. 2.

The Variscan vein districts, however, are bound to large anticlines, which formed during the orogenetic uplift of Devonian basins of a size of some tens of kilometers ("second-order depression", see Krebs 1978). Several second-order depressions terminated at the synsedimentary precursor of the Menden-Oberscheld fault system, which may explain the lack of Variscan veins along this lineament. It is characteristic of the RFB that main paleogeographical elements (basins, paleo-uplifts) and major fault systems strongly affected the synsedimentary to metamorphic heat flow (Weber 1972; Hoyer et al. 1974; Kalkreuth 1979; Teichmüller et al. 1979; Oncken 1987; Werner 1989a).

Figure 1 illustrates the typical metal associations of the syn- and epigenetic ore deposits of the Central European Variscides. The Rhenohercynian and sub-Variscan mineral belts are characterized by the predominance of Fe, Zn and Pb, whereas the Saxothuringian and Moldanubian basement contains widespread mineralization of Sn, U, and W (± Bi-Co-Ni). About 97% of the Zn and Pb ores of the Federal Republic (with >32 Mt of Pb and Zn) occur within the western Harz Mts., the Rhenish Massif, and within the Ruhr district (sub-Variscan foredeep), from which about 40% are concentrated in syngenetic and 60% in epigenetic orebodies (Walther 1981; Walther and Dill 1995). Geological and geochemical studies document that the synsedimentary, synorogenic (Variscan) and post-Variscan sediment-hosted hydrothermal mineralization of Fe, Zn, Pb, Cu, barite, and fluorite led to the derivation of the metals from the Paleozoic psammopelitic successions and the intercalated volcanics, which were leached by heated brines during periods of crustal fracturing and elevated heat flow (cf. compilation in: Werner 1990a). Isotope data clearly show that the metals of the syn- and epigenetic mineral deposits of the RFB were periodically leached out of epicontinental clastic material from Devonian to Tertiary times (Wedepohl et al. 1978; Krahn 1988; Lüders et al. 1993).

Considering that the RFB, the Saxothuringian, and the Moldanubian zones all underwent similar Variscan and post-Variscan deformation and heating, the striking contrast between the three belts concerning the abundance of metals (Fig. 1) indicates that the southern Variscan units are composed of different crustal material compared with the RFB. Since it seems likely that RFB was also part of the northern margin of Gondwana until the Ordovician-Silurian period of crustal extension (Franke et al. 1990), one may assume that Gondwana was likewise built up by heterogeneous crustal blocks accreted during Cadomian events.

References

Badham JPN (1981) Shale-hosted Pb-Zn deposits: products of exhalation of formation waters? Trans Inst Min Metall B 90: 70–76

Bauer G, Ebert A, Hesemann J, Kamp H von, Müller D, Pietzner H, Podufal P, Scherp A, Wellmer FH (1979) Die Blei-Zink-Erzlagerstätten von Ramsbeck und Umgebung. Geol Johrb D 33: 1–377

Bottke H (1978) Zur faziesgebundenen Tektonik der Briloner Scholle (Ostsauerland, Rheinisches Schiefergebirge). Z Dtsch Geol Ges 129: 141–151

Carne RC, Cathro RJ (1982) Sedimentary exhalative (sedex) zinc-lead-silver deposits, northern Canadian Cordillera. CIM Bull 75 (840): 66–78

Clausen CD (1978) Geologische Karte von Nordrhein-Westfalen 1:25000. Erläuterungen zu Bl 4814, Lennestadt, Geologisches Laudesamt Nordrhein-Westfalen, pp 1–474

Dejonghe L (1985) Contribution à l'étude métallogénique du synclinorium de Verviers (Belgique). Thesis, Univ Pierre et Marie Curie, Paris

Emmermann KH, Hindel R, Krimmel M, Zinner HJ (1993) Sulfid-Baryt-Mineralisationen in der südwestlichen Lahnmulde. Ein Beitrag zur Genese und Prospektion synsedimentär-exnalativer Vorkommen im Rheinischen Schiefergebirge. Mainzer Geowiss Mitt 22: 7–38

Engel W, Franke W, Langenstrassen F (1983) Palaeozoic sedimentation in the northern branch of the Mid-European Variscides – essay of an interpretation. In: Martin H, Eder FW (eds) Intracontinental fold belts. Springer, Berlin Heidelberg New York, pp 9–41

Fenchel W, Gies H, Gleichmann HD, Hellmund W, Hentschel H, Heyl KE, Hüttenhain H, Langenbach U, Lippert HJ, Lusznat M, Meyer W, Pahl A, Rao

MS, Reichenbach R, Stadler G, Vogler H, Walther HW (1985) Die Sideriterzgänge im Siegerland-Wied-Distrikt. Geol Jahrb D77: 3–517 (Sammelwerk Deutsche Eisenerzlagerstätten I Eisenerze im Grundgebirge (Varistikum) 1.)
Flick H, Nesbor D (1988) Der Vulkanismus in der Lahnmulde. Jauresber Mitt Oberrhein Geol Ver 70: 411–475
Flick H, Nesbor H, Behnisch R (1990) Iron ore of the Lahn-Dill type formed by diagenetic seeping of pyroclastic sequences – a case study on the Schalstein section at Gänsberg (Weilburg). Geol Rdsch 79: 401–415
Franke W (1989) Tectonostratigraphic units in the Variscan belt of Central Europe. Geol Soc Am Spec Pap 230: 67–90
Franke W, Bortfeld RK, Brix M, Drozdzewski G, Dürbaum HJ, Giese P, Janoth W, Jödicke H, Reichert C, Scherp A, Schmoll J, Thomas R, Thünker M, Weber K, Wiesner MG, Wong HK (1990) Crustal structure of the Rhenish Massif: results of deep seismic reflection lines DEKORP 2-North and 2-North-Q. Geol Rdsch 79: 523–566
Goodfellow WD, Jonasson IR (1986) Environment of formation of the Howards Pass (XY) Zn-Pb deposit, Selwyn Basin, Yukon. CIM Bulle Spec 37: 19–50
Hagedorn B, Lippolt HJ (1993) Isotopic age constrains for epigentic mineralizations in the Harz Mountaines (Germany) from K-Ar, $^{40}Ar/^{39}Ar$ and Rb-Sr data of authigenic K-feldspars. Monogr Ser Miner Deposits 30: 87–102
Hannak W (1964) Ergebnisse der Untersuchungen im Blei-Zink-Erzbezirk des südlichen Rheinischen Schiefbergebirges. Erzmetall 17: 291–298
Hein UF (1993) Synmetamorphic Variscan siderite mineralisation of the Rhenish Massif, central Europe. Mineral Mag 57: 451–467
Henneke J (1977) Die bergwirtschaftliche Bedeutung der Blei-Zink-Erzlagerstätte Mechernich. Glückauf Forsch-H 38: 9–18
Hentschel H (1960) Zur Frage der Bildung der Eisenerze vom Lahn-Dill Typ. Freiberg Forsch-H C79: 82–105
Hesemann J (1978) Der Blei-Zink-Erzbezirk des Bergischen Landes (Rheinisches Schiefergebirge) als Prototyp einer frühorogenen und palingenen Vererzung. Decheniana 131: 292–299
Hoyer P, Clausen CD, Leuteritz K, Teichmüller R, Thome KN (1974) Ein Inkohlungsprofil zwischen dem Gelsenkirchener Sattel des Ruhrkohlenbeckens und dem Ostsauerländer Hauptsattel des Rheinischen Schiefergebirges. Fortschr Geol Rheinl Westfalen 24: 161–172
Kalkreuth W (1979) Das Inkohlungsbild des Ostsauerländer Hauptsattels im Rhenoherzynikum mit besonderer Berücksichtigung der Trendflächenanalysen. Fortschr Geol Rheinl Westfalen 27: 277–321
Krahn L (1988) Buntmetall-Vererzung und Blei-Isotopie im linksrheinischen Schiefergebirge und in angrenzenden Gebieten. Thesis, Rheinisch-Westfälische Tech Hochschule, Aachen, 199 pp
Krebs W (1978) Moderne Suchmethoden auf Buntmetalle im Mitteleuropäischen Grundgebirge. Zum geologischen Rahmen schichtgebundener Sulfid-Baryt-Lagerstätten. Erdöl Kohle 31: 125–139
Krebs W (1981) The geology of the Meggen ore deposit. In: Wolf KH (ed) Handbook of strata-bound and stratiform ore deposits. Elsevier, Amsterdam 9: 509–549
Kuschka E, Franzke HJ (1974) Zur Kenntniss der Hydrothermalite des Harzes. Z Geol Wiss 2: 1417–1436
Langenstrassen F (1983) Neritic sedimentation of the Lower and Middle Devonian in the Rheinische Schiefergebirge east of the River Rhine. In: Martin H, Eder FW (eds) Intracontinental fold belts. Springer, Berlin Heidelberg New York, pp 43–76
Lehmann E (1972) On the source of the iron in the Lahn ore deposits. Miner Deposita 7: 247–270
Lehmann H, Pietzner H (1970) Der Lüderich-Gangzug und das Gangvorkommen Nikolaus-Phönix im Bergischen Land – ein Beitrag zur Geologie und Geochemie. Fortschr Geol Rheinl Westfalen 17: 589–664
Lüders V, Möller P, Dulski P (1993) REE Fractionation in carbonates and fluorite. Monogr Ser Miner Deposits 30: 133–150
Mertz DF, Lippolt HJ, Schnorrer-Köhler G (1989) Early Cretaceous mineralizing activity in the St. Andreasberg ore district (Southwest Harz, FRG). Miner Deposita 24: 9–13
Mohr K (1978) Geologie und Minerallagerstätten des Harzes. Schweizerbart, Stuttgart, 387 pp
Morganti JM (1981) The geology and ore deposits of the Howards Pass stratiform ore deposits: some models and a new classification. Geosci Can 8: 65–75
Neumann-Redlin C, Walther HW, Zitzmann A (1977) The iron ore deposits of the Federal Republic of Germany. In: Walther HW, Zitzmann A (eds) The iron ore deposits of Europe and adjuacent areas I. Schweizerbart, Stuttgart, pp 165–186
Oncken O (1987) Heat flow and kinematics of the Rhenish Basin. In: Vogel A, Miller H, Greiling R (eds) The Rhenish Massif. Structure, evolution, mineral deposits, and present geodynamics. Vieweg, Braunschweig, pp 63–78
Podufal P (1983) Die Blei-Zink-Erzlagerstätte von Ramsbeck. GDMB Jahrestagung 1983 (Warstein) "Alter, Genese und wirtschaftliche Bedeutung der postvariszischen Gangmineralisation in Mitteleuropa (Oberperm bis Tertiär)". In: Walther HW (ed) Field guide to GDMB Meet 1983 (Warstein), Clausthal-Zellerfeld, pp 128–136
Quade H (1976) Genetic problems and environmental features of volcano-sedimentary iron-ore deposits of the Lahn-Dill type. In: Wolf KH (ed) Handbook of strata-bound and stratiform ore deposits. Elsevier, Amsterdam 7: 255–294
Schaeffer R (1984) Die postvariszischen Mineralisationen im nordöstlichen Rheinischen Schiefergebirge. Brswg Geol Paläontol Diss 3: 1–206
Schaeffer R (1986) Geochemische Charakteristik und Genese der jungmesozoisch-tertiären Vererzung im Sauerland (Rheinisches Schiefergebirge). Fortschr Geol Rheinl Westfalen 34: 337–381

Scheffler H (1975) Schwefelisotopenverhältnisse und Spurenelementgehalte von Sulfiden aus der Schwefelkieslagerstätte "Einheit" bei Elbingerode im Harz. Z Geol Wiss 3: 313–326

Siemes H, Breuer Ch (1992) Geostatistische Vorratsberechnung einer sedimentären Pb-Zn-Lagerstätte. Erzmetall 45: 48–56

Simon P (1979) Die marin-sedimentären Eisenerze in der Eifel. Geol Jahrb D31: 133–151

Sperling H, Walcher E (1990) Die Blei-Zink-Erzlagerstätte Rammelsberg. Geol Jahrb D91: 3–153

Sperling H, Stoppel D, Berthold G, Dennert H (1979) Beschreibung der Oberharzer Erzgänge. Geol Jahrb D34: 5–345

Stedingk K, Stoppel D (1993) History of mining operations and economic significance of the Harz vein deposits. Monogr Ser Miner Deposits 30: 1–3

Stoppel D, Gundlach H, Heberling E, Heinrich G, Hüser M, Kallies HB, Schaeffer R (1983) Schwer- und Flußspatlagerstätten des Südwestharzes. Geol Jahrb D54: 3–269

Stribrny B, Urban H, Weber H (1988) The Lower Carboniferous black shale formation, a possible source for noble and base metal deposition the NE Rhenish Massif, Federal Republic of Germany. Mineral Petrol 39: 129–143

Teichmüller M, Teichmüller R, Weber K (1979) Inkohlung und Illit-Kristallinität. Vergleichende Untersuchungen im Mesozoikum und Paläozoikum von Westfalen. Fortschr Geol Rheinl Westfalen 27: 201–276

Walther HW (1981) Quantitative regionale Metallogenese. Erzmetall 34: 432–438

Walther HW (1982a) Die varistische Lagerstättenbildung im westlichen Mitteleuropa. Z Dtsch Geol Ges 133: 667–698

Walther HW (1982b) On the Alpidic mineralization in the western Central Europe outside the Alps. Soc Geol Appl Miner Deposits Spec Publ 2: 598–606 (Ramdohr vol)

Walther HW, Dill HG (1995) Bodenschätze Mitteleuropas – ein Überblick. In: Walter R (ed) Geologie von Mitteleuropa, 6th edn. Schweizerbart, Stuttgart, pp 410–466

Walther HW, Emmermann KH, Fenchel W, Fuchs W and others (1986) Federal Republic of Germany. In: Dunning FW, Evans AM (eds) Mineral deposits of Europe 3. Central Europe. IMM & Miner Soc, London, pp 175–301

Weber K (1972) Kristallinität des Illits in Tonschiefern und andere Kriterien schwacher Metamorphose im nordöstlichen Rheinischen Schiefergebirge. Neues Jahrb Geol Paläontol Abh 141: 333–363

Weber K (1977) Bau und tektonische Entwicklung des Ostsauerländer Hauptsattels und der variszischen Ramsbecker Blei-Zinkerzlagerstätte. Fortschr Miner 55 (2): 48–63

Wedepohl KH, Delevaux MH, Doe BR (1978) The potential source of lead in the Permian Kupferschiefer bed of Europe and some selected Paleozoic mineral deposits in the Federal Republic of Germany. Contrib Miner Petrol 65: 273–281

Wedepohl KH, Meyer K, Mücke GK (1983) Chemical composition and genetic relation of metavolcanic rocks from the Rhenohercynian Belt of Northwest Germany. In: Martin H, Eder FW (eds) Intracontinental fold belts. Springer, Berlin Heidelberg New York, pp 231–256

Werner W (1989a) Synsedimentary faulting and sediment-hosted submarine-hydrothermal mineralization in the Late Palaeozoic Rhenish Basin (Germany). Geotekton Forsch 71: 1–305

Werner W (1989b) Contribution to the genesis of SEDEX-type mineralizations of the Rhenish Massif (Germany) – implications for future Pb-Zn exploration. Geol Rdsch 78: 571–598

Werner W (1990a) Examples of structural control of hydrothermal mineralization: fault zones in epicontinental sedimentary basins – a review. Geol Rdsch 79: 279–290

Werner W (1990b) The mineral deposits of the Rhenohercynian Zone in Germany: field guide IGCP 233 pre-conference excursion "Mid-German crystalline rise & Rheinisches Schiefergebirge", 55–79, August 1990, Göttingen. IGL, Giessen

Werner W (1990c) Die epigenetische Markasit-Schwerspat-Zinkblende-Vererzung "Altenbüren" (nordöstliches Rheinisches Schiefergebirge). Geol Jahrb D95: 139–176

III.C Metamorphic Units (Northern Phyllite Zone)

III.C.1 Stratigraphy

H.-J. ANDERLE, W. FRANKE, and M. SCHWAB

1 Introduction

The Northern Phyllite Zone was defined by Scholtz (1934) as a narrow metamorphic zone separating the Rhenohercynian belt in the north from the Mid-German Crystalline High in the south. The Southern Phyllite Zone, at the SE margin of the Crystalline High, represents a metamorphosed part of the Saxothuringian basin. The Northern Phyllite Zone is exposed in a number of outcrop areas (listed in order from the SW to the NE): an isolated outcrop occurs at Düppenweiler, south of the Hunsrück, within the Permian Saar-Nahe Basin (see Fig. 1); more extensive outcrops are available at the southern margin of the Rhenish Massif, on both sides of the Rhine River (S Hunsrück to the W, and S Taunus to the E). It can then be traced in boreholes NE of the Rhenish Massif (under the Permian and Mesozoic cover of the Hesse Basin), and reemerges at the southern margin of the Harz Mts (Wippra Zone). A northeastern continuation of the Wippra Zone is known from drillings in the NW-trending Flechtingen-Roßlau horst. This Pakendorf Zone is briefly treated together with the Wippra Zone.

The Northern Phyllite zone is characterized by its greenschist-grade metamorphism (Chap. III.C.4), which is intermediate between the very-low-grade rocks in the main part of the Rhenohercynian belt to the NW, and the medium- to high-grade rocks of the Mid-German Crystalline High to the SE. Its NW margin is a thrust fault. The SE margin of the Northern Phyllite Zone is nowhere exposed. Between the Northern Phyllite Zone and the Crystalline High, there is a narrow belt of unmetamorphosed graywackes, encountered in one minor outcrop at Erbstadt NNE of Frankfurt and in boreholes further NE (see Fig. 1). These graywackes probably represent part of the Giessen nappe, i.e., part of the Rhenohercynian ocean floor and its sedimentary cover (Chap. III.B.1).

Until recently, there were but few reliable age data for the Northern Phyllite Zone, and its speculative assignments to the Devonian or pre-Devonian need not be discussed here. The new findings and the consequences arising are briefly summarized below.

2 Düppenweiler

At the NE margin of the village of Düppenweiler, NW-dipping mylonitic greenschist-grade metapelites and quarzites are exposed in a small area (1500 × 400 m), which emerges from under the post-Variscan cover (Late Carboniferous to the SE and Early Permian to the NW). The contacts are faults. At its southern margin, the phyllites overlie, with a tectonic contact, a clastic sequence of uncertain age, which was earlier interpreted as a Devonian transgressive sequence, now in an overturned position under the phyllites of the basement. These clastic sediments have yielded spores of Early Devonian, Middle Devonian and early Carboniferous age (Streel 1976, 1981; Doubinger and Rauscher 1984; all cited in Schneider 1991), and limestone pebbles with middle- or late-Devonian corals (Firtion 1983; Mistiaen and Rohard 1983; both cited in Schneider 1991). The age of this clastic sequence can only be constrained as Early Carboniferous or younger. According to Wehrens (1985), the phyllites have been emplaced as a sedimentary klippe. The sedimentation age of the greenschist-grade metasediments is unknown.

3 S Hunsrück (Soonwald)

In the S Hunsrück, the Phyllite Zone is exposed along the southern margin of the Soonwald hills. It is mainly composed of dark to greenish metapelites and metabasalts with some quartzites, sandstones, and limestone lenses intercalated. To

Dallmeyer et al. (Eds.)
Pre-Permian Geology of Central and Eastern Europe

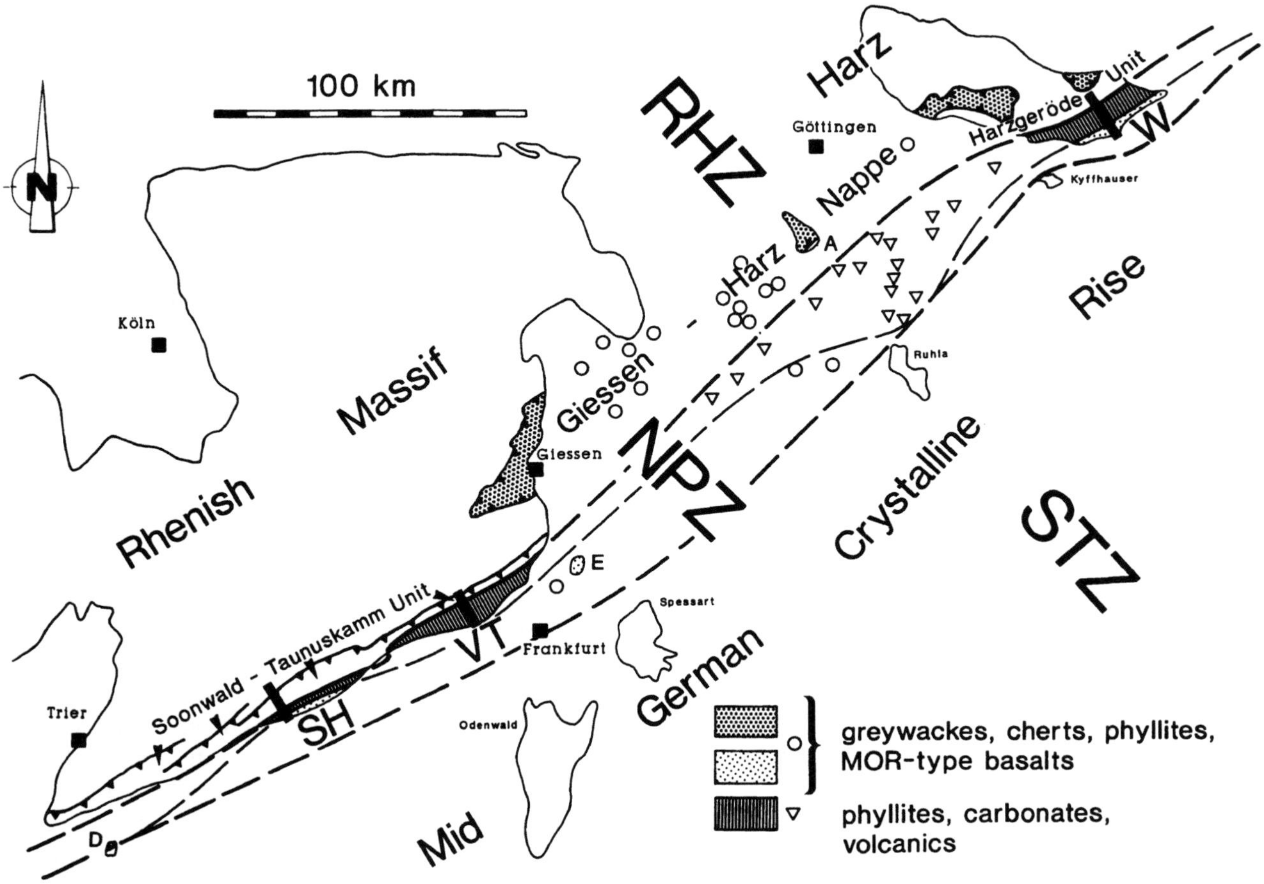

Fig. 1. Geological setting of the Northern Phyllite Zone in the Rhenish Massif and Harz Mts. (Oncken et al., this Vol.). RHZ = Rhenchercynian Zone, NPZ = Northern Phyllite Zone, STZ = Saxothuringian Zone, A = Albungen-Werna, D = Düppenweiler, E = Erbstadt, SH = South Hunstück (Soonwald), VT = Vordertaunus (South Taunus)

the NW, the Phyllite Zone is bounded against the Rhenohercynian belt proper by a thrust fault. To the NW of this fault, there occur small tectonic lenses of gneiss, one near the southwestern end of the Phyllite Zone (**Wartenstein gneiss**), the other near the northeastern end of the belt (**Schweppenhausen gneiss**). In the SE, the Phyllite Zone is bordered by terrestrial sedimentary and volcanic rocks of Early Permian age (Rotliegend of the Saar-Nahe Syncline). In its southwestern part, the Phyllite Zone of the S Hunsrück is divided in half by a thrust fault (**Wiesbachtal mylonite** of Schafft 1985; Fig. 2).

The Wartenstein gneiss represents the basement to the Devonian sequences of the Soonwald unit in the sense of Oncken et al. (Chap. III.C.2) and Dittmar and Oncken (1992), which is the southeasternmost part of the Rhenohercynian belt proper.

The lithostratigraphic sequence between Wartenstein gneiss and Wiesbachtal mylonite has recently been dated with conodonts and acritarchs. Berger et al. (1991) have described conodont faunas, which make it possible to establish a stratigraphic sequence. They have proven the existence of sediments dated into various stages of the Late Devonian (do I, II, and VI, with the probable intercalation of rocks of the do III–V). According to these authors, the Late Devonian is largely composed of metapelites (phyllites) with occasional, conodont-bearing limestone nodules (Late Devonian, do I–VI), with intercalations of metabasalt (do I), of graywackes (do I and II), and some quartzite (do II). Dittmar and Oncken (1992) have reported olistostromes from the Middle and early Upper Devonian. Red coloration in the pelites of the do II remind the Late Devonian red shales in the unmetamorphosed parts of the Rhenish Massif (see Chap. III.B.1). Late Devonian (do I) phyllites and dolomitic limestone nodules with conodonts have also been detected by Meyer (1966, 1970:126) SW of Schweppen-

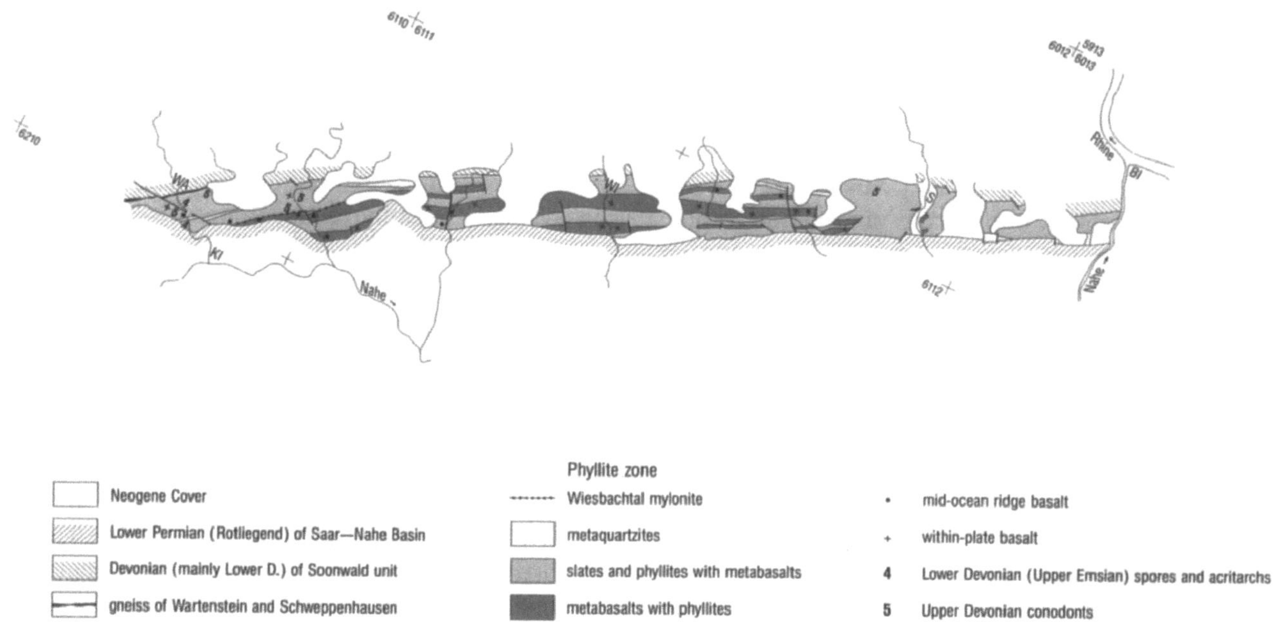

Fig. 2. Simplified geological map of the S Hunsrück (Soonwald). Phyllite zone compiled from Berger et al. (1991), Bierther (1941), Geib (1973), Meyer (1970), Wagner and Michels (1930), and Werner (1952). Wiesbachtal mylonite after Schafft (1985); fossil localities from Berger et al. (1991), Ecke et al. (1985), Meyer (1970), and Reitz (1989). Geochemistry of metabasalts after Meisl (Chap. III.C.3). *BI* Bingen; *KI* Kirn; *S* Schweppenhausen; *WA* Wartenstein castle; *WI* Winterbach. Sheet numbers of ordnance map 1:25 000 are added

hausen, along-strike to the NE. Late Early Devonian (probably Early Upper Emsian) acritarchs and spores have been detected in phyllites of the Hahnenbach-Serie (Bierther 1941) about 500 m south of Wartenstein castle by Ecke et al. (1985:402). Late Emsian spores and acritarchs have been described from two localities near Cramers Mühle NW of Kirn in the Vorsoonwald Serie by Ecke et al. (1985:404) and Reitz (1989:27). These localities are situated close to the Kallenfels quartzite, which had already been suspected as Emsian by Werner (1952), despite some lithological differences with the unmetamorphosed Ems-Quarzit in the Rhenish Massif further N. Berger et al. (1991) interpreted the K. quartzite as late Devonian (do II) because of conodont findings nearby. This problem highlights the need for more biostratigraphic data.

The southeasternmost part of the Northern Phyllite Zone in the S Hunsrück – SE of the Wiesbachtal mylonite – consists of greenish gray to black metapelites and metacherts together with metabasalts (**Vorsoonwald Serie** of Werner 1952). Some variegated phyllites, certain albitized rocks, and iron ores are intercalated. A metamorphosed iron ore near Winterbach (Franzke and Anderle, Chap. III.C.5) might be an equivalent of unmetamorphosed, Late Emsian oolitic ironstones at Wald-Erbach in the unmetamorphosed Rhenohercynian Soonwald unit.

The metapelites, -cherts and -basalts were interpreted, by Berger et al. (1991), as an equivalent of the black shales and cherts with metabasalts, which form a conspicuous part of the Early Carboniferous sequences in the Lahn and Dill synclines of the unmetamorphosed Rhenish Massif (Chap III.B.1). Since part of the metabasalts show the trace-element spectrum of MOR-basalts (Meisl 1990; Chap.III.C.3), Oncken et al. (Chap. III.C.2) and Dittmar and Oncken (1992) have suggested, instead, that these rocks represent a metamorphosed part of the Giessen nappe at the E margin of the Rhenish Massif, which contains MOR-type metabasalts overlain by late Early Devonian to Frasnian pelagic shales and bedded cherts (often black), and Frasnian flysch (Chap. III.B.1). Their opinion has lately been confirmed by Sommermann et al. (1994).

Two thirds of the metabasalts analyzed so far (Chap. III.C.3) turned out to be within-plate basalts, which occur mainly between Winterbach and Schweppenhausen (Fig. 2). This might suggest tectonic mixing of oceanic crust with rocks of the distended continental margin.

Eastwards from Schweppenhausen, the Vorsoonwald unit appears to continue as far as the Nahe River (Fig. 2). These rocks have been interpreted, by Wagner and Michels (1930), as Lower Devonian, but there is no biostratigraphic evidence.

Earlier opinions about the age and structural relationships of the S Hunsrück are summarized by Meyer (1975) and Meisl et al. (1982).

4 S Taunus

The S Taunus segment of the Northern Phyllite Zone consists, in its NW part, mainly of acid and intermediate metavolcanics with intercalations of metasediments, whereas the SE part is made up of green, greenish gray and dark gray metasediments with very scarce intercalations of metabasalts. According to Oncken et al. (Chap. III.C.2), the Phyllite Zone of the Taunus Mts. is bounded against the Taunuskamm unit to the NW by a thrust fault, but no major metamorphic break. To the SE, the phyllites are juxtaposed against Tertiary and Permian sequences (Fig. 3).

The oldest rocks of the Northern Phyllite Zone in the S Taunus are metavolcanites: Grünschiefer (greenschist), Quarzkeratophyr, Keratophyr, Serizitgneis and Felsokeratophyr. The greenschists and keratophyres developed from andesites, tranchyandesites and rhyodacites/dacites, the other rocks from rhyolites and rhyodacites/dacites (see Chap.III.C.3). Clear ignimbrite textures have been identified at one locality only, and the majority of the metavolcanics are interpreted to represent lavas. U-Pb and Pb-Pb studies on zircons gave late Ordovician to Silurian crystallization ages (Grünschiefer: 442 ± 22 Ma, Felsokeratophyr 433 + 9/−7, Serizitgneis 426 + 14/−15 Ma; Sommermann et al. 1990, 1992). The associated metasediments have not been dated with fossils).

The bulk of the metasediments in the Phyllite Zone of the S-Taunus appear SE of the metavolcanics. These metasediments are subdivided into a lower, northwestern part (**Eppstein Fm.**), and the overlying **Lorsbach Fm.** to the SE (see Stenger 1961 for details).

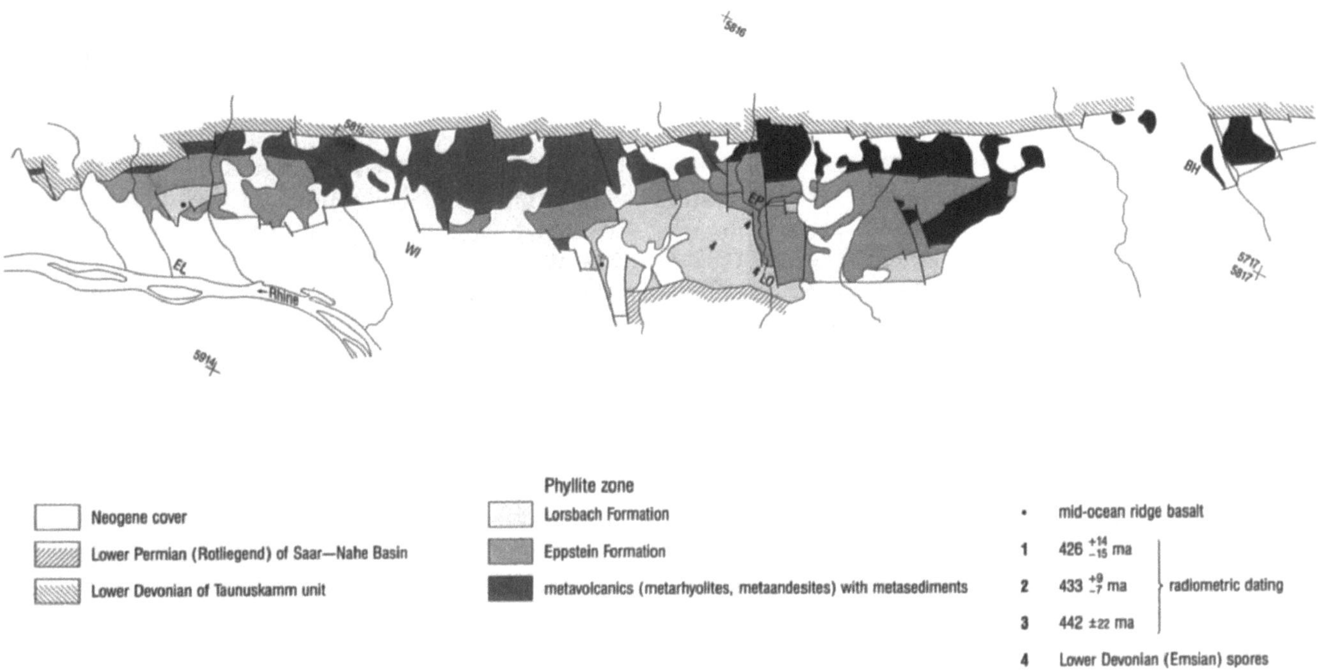

Fig. 3. Simplified geological map of the S Taunus. Phyllite Zone compiled from the geological maps 1:25000, with additional information from Anderle (1987), Nickel (1958), and Stenger (1961). Fossil and geochronology localities from Reitz (1989) and Sommermann et al. (1990, 1992). Geochemistry of metabasalts after Meisl (Chap. III.C.3). *BH* Bad Homburg; *EL* Eltville; *EP* Eppstein; *LO* Lorsbach; *WI* Wiesbaden. Sheet numbers of ordnance map 1:25000 are added

The Eppstein Fm. consists of greenish or purple sericite-chlorite phyllites with arenaceous laminae in some of the green portions. In places they contain some graywacke and arkose (Stenger 1961; Meisl in: Anderle et al. 1972). Until now they have not yielded any fossils. Their lithology resembles that of the "Bunte Schiefer" Fm. of the Taunuskamm unit in the NW.

The Lorsbach Fm. is more variable. It is composed of bluish gray to black slates with quartzites, with some intercalations of greenish gray quartzitic sericite slates and slates with sandy ribbons. Towards its southern margin, this sequence also contains graphitic slates, siliceous sericite-chlorite slates (meta-cherts?) and calcareous slates (meta-marls). There are also two localities with MOR-type metabasalts (Chap. III.C.3). Reitz (1989:24) has described spores of Emsian age from three localities in the Lorsbach Fm. The uppermost part of the formation with dark slates, cherts, and calcareous phyllites might be of Middle Devonian (as supposed by Bierther 1951) or even younger age. Earlier opinions about the age and structural relationships are summarized in Meisl et al. (1982).

It is vital for the paleogeographic interpretation that Quarzporphyr of Ordovician /Silurian age (U-Pb zircon, 434 + 34/−22, Sommermann et al. 1994) has recently been dated at the islet of Krausaue, in the bend of the Rhine river. Here, the metavolcanic rocks are clearly overlain by early Devonian rocks of the Taunuskamm unit. Similar relationships occur SW of Eltville (mostly outside Fig. 3) and at Bad Homburg. This confirms the interpretation of the clastic sedimentary rocks in the Eppstein and Lorsbach formations (above the volcanic rocks of the Phyllite Zone) as early Devonian.

5 S Harz (Wippra Zone)

The belt of low-grade rocks at the SE margin of the Harz Mts., the Wippra Zone, has been subdivided into seven lithological units (Fischer 1929), which are briefly described below in order from NW to SE (Fig. 4).

Zone 1 consists of graywackes, bedded cherts and shales, and occasional limestone lenses. This association compares well with that of the Harzgeröder Zone to the NW, and is therefore interpreted, in all recent publications, as part of this unit, i.e., excluded from the Northern Phyllite Zone proper.

Zone 2 (Phyllitische Tonschieferserie) is essentially composed of bluish gray to black slates with local intercalations of limestone bands and - lenses, siderite nodules, and graywacke beds. There are also some volcanic inserts (alkalibasalts, quartz-

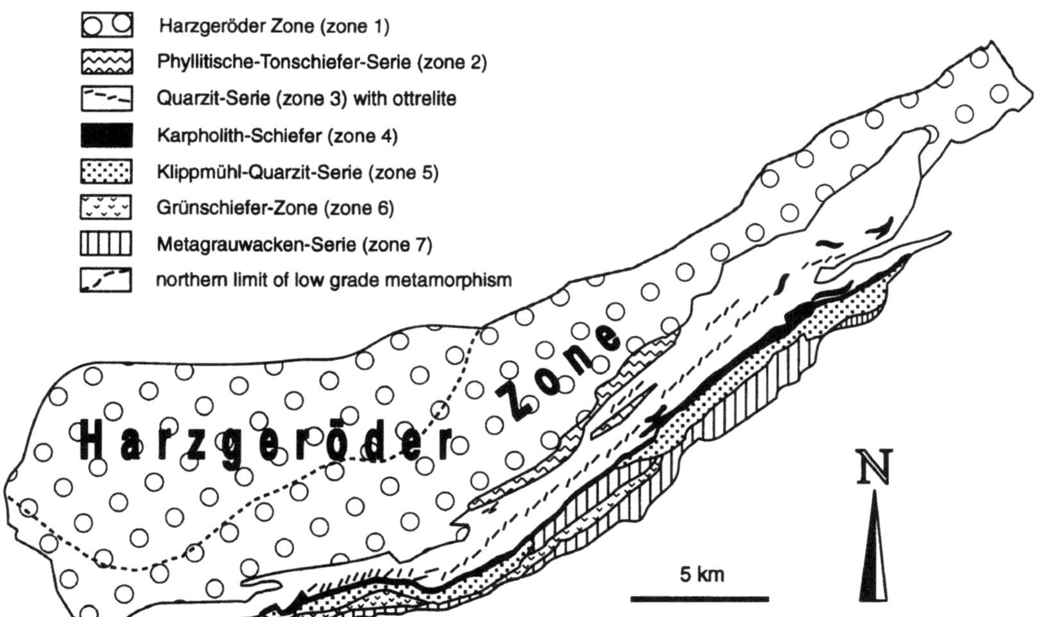

Fig. 4. Simplified geological map of the Wippra Zone (Harz Mts.). (After Wachendorf 1986)

keratophyres, and trachyte-/trachyandesite tuffs; Borsdorf and Estrada 1991).

Reichstein (pers. comm. in Franzke 1969) was the first to report Silurian fossils (Early Ludlow). This was recently confirmed by Sehnert (1991), who described eight localities with acritarch floras dating into the Late Llandovery to Early Wenlock. The fossil assemblage is taken to indicate an offshore shelf facies, and the predominance of *Domasia* reflects, after Cramer and Diez (1972), deposition under low subtropical latitudes.

Zone 3 (Quarzit-Serie) consists of metapelites, partly with quartzitic lamina, and lenses of quartzite. The metapelites often contain ottrelite (Mn-chloritoid). In addition, there are intercalations of metabasalts and comagmatic tuffs (with and without pyroxene), and ultramafic dikes (Pikrite). The Pikrite and part of the metabasalts are derived from tholeiites, whereas other metabasalts and all pyroclastics relate to alkali-basalts (Borsdorf and Estrada 1991). At nine localities, the metapelites have yielded acritarch assemblages of Arenig to Llanvirn age (Burmann 1973; Sehnert 1991); the elements of the microfloras are taken to indicate open marine environments, and a boreal admixture to otherwise Mediterranean affinities (Sehnert 1991). Poorly preserved conodonts are compatible with the palynological evidence, but do not permit further specification (Ackermann 1985). In some of the metapelites associated with the quarzites, there are reworked microfloras of ? Tremadoc, Arenig, and Llanvirn age. Reworking and ultimate deposition probably took place in Ashgill or Early Llandovery time.

Zone 4 (Karpholith-Schiefer) is a conspicuous sequence of variegated (purple and greenish) metapelites. Karpholith occurs in quartz segregates. There are no metavolcanic rocks. The age of the sequence is unknown; it is seen, by most authors, as stratigraphically coherent with zone 3. Besides muscovite, paragonite and chlorite, the metapelites contain chloritoid or sudoite, and partly also pyrophyllite or kaolinite (the former possibly a product of Paleozoic, and the latter, of recent weathering). However, the high content of Fe_2O_3 (often more than 10%) and other geochemical characteristics are taken to indicate submarine volcanic exhalations rather than terrestrial input (Siedel 1991).

Zone 5 (Klippmühl-Quarzit-Serie) consists of grey, greenish-grey or reddish quartzitic sandstones, alternating with variegated metapelites. There are but minor inserts of metabasalt. The arenites are less mature than those in zone 3. From two localities, Sehnert (1991) has described acritarchs of Arenig to Llanvirn age. A find of the conodont *Acodus* sp. (Ackermann 1985) is compatible with the palynomorph datings, yet does not permit any specification.

Zone 6 (Grünschiefer-Zone) is characterized by metavolcanic rocks such as metabasalts and pyroclastic rocks, intercalated with metapelites, arenaceous metapelites, and rare limestone lenses. The metavolcanic rocks show local impregnations with and thin beds of hematite. The geochemistry of the volcanics corresponds to primitive continental low-K-tholeiites. Some intrusive rocks compare well with MOR-type basalts (Werner 1991), but the general abundance of pyroclastics precludes the interpretation of zone 6 as oceanic. Franzke (1969) has described poorly preserved tentaculitids from a limestone lens. The stratigraphic range of tentaculitids spans the Early Ordovician to Early Late Devonian, with the greatest diversity and abundance in the Devonian. The association of mafic volcanic rocks with tentaculite-bearing limestones recalls the volcano-sedimentary sequences of the Rhenohercynian belt to the NW, but this attempt at a correlation is by no means conclusive.

Zone 7 (Meta-Grauwacken-Serie) consists of metagraywackes alternating with metapelites and -cherts. Some metabasalts show geochemical affinities with MOR-type basalts (Borsdorf and Estrada 1991). A poorly preserved conodont from a metachert has alternatively been interpreted as belonging to a Devonian (Reichstein 1964) or ?Arenigian species (Ackermann 1985). Sehnert (1991) described, from two metapelite localities, Devonian acritarchs (the material did not permit a more precise dating). At map scale, the metagraywackes and -pelites interfinger with the metavolcanics of zone 6; transitions without recognizable tectonic breaks have also been observed in good outcrops in the Dinsterbach valley (Franzke 1969). It appears possible, therefore, that zones 6 and 7 represent, in general terms, a stratigraphic succession, though most of the lithological contacts have been reworked by tectonic deformation.

The Pakendorf Zone, known from boreholes 40 km NE of the Harz Mts., contains quartzites, variegated pelites, and few metabasalts. These are correlated, by lithological comparisons, with the Ordovician sequences of the Wippra Zone (Borsdorf and Estrada 1991).

6 Paleogeographic Affinities

Because of the scarcity of age data and the polyphase deformation, any paleogeographic interpretation of the Northern Phyllite Zone must be regarded as preliminary and highly speculative. However, the following conclusions are worth consideration:

6.1 Ordovician/Silurian Volcanic Rocks

Franke and Oncken (1990) have interpreted the Ordovician/Silurian metavolcanic rocks of the S Taunus as the earliest expressions of Rhenothercynian rifting. In the light of geochemical, biogeographic, and paleomagnetic evidence discussed in Chapter XI, it is more probable that the volcanic rocks of the S Taunus represent a late Ordovician/Silurian island arc. Formation of the arc probably related to closure of a "Rheic" ocean, which had opened, in Ordovician time, between Avalonia and Armorica.

The composition of the Ordovician/Silurian metavolcanic rocks in the Wippra Zone of the Harz Mts. is different (felsic calc-alkaline rocks in the S Taunus, intraplate tholeiites and alkalibasalts in the Wippra Zone), and the Wippra Zone appears to contain quartzites coeval with the metabasalts; the quartzites indicate Armorican affinities of the Wippra Zone.

Ordovician/Silurian postcollisional granitoids and Silurian clastic sediments are also present in the Spessart segment of the Mid-German Crystalline High (Chap. IV), which suggests a magmatic and sedimentary development similar to that of parts of the Phyllite Zone. It is possible that, during this time, the Phyllite Zone and parts of the Mid-German Crystalline High were coherent, before they were separated by the Devonian rift process that created the Rhenohercynian ocean.

6.2 Early Devonian Setting

Early Devonian rocks in the facies of the southeasternmost Rhenish Massif (Taunuskamm unit) overlie both the Cadomian basement of the Wartenstein gneiss and the Silurian metavolcanic rocks of the S Taunus. Hence, the metavolcanic rocks of the S Taunus must have been accreted to the southern margin of the Rhenohercynian in late Silurian/earliest Devonian time.

6.3 Mid-/Late Devonian Setting

Successions dated or interpreted as Middle or Late Devonian, in the S Hunsrück, are mainly pelagic metapelites with subordinate metabasalts, and predominantly volcanic (intraplate metabasalts and related pyroclastics) in zone 6 of the Wippra belt. In spite of these differences in bulk lithology, both areas might have been coherent with one another and represent part of the extended, northern margin of the Rhenohercynian basin.

All along its southeastern margin, the Northern Phyllite Zone contains associations of metabasalts and dark metapelites (S Hunsrück, S Taunus), partly also with metagraywackes (Wippra). Part of the metabasalts have geochemical affinities with MOR basalts. All along its length, the Northern Phyllite Zone is accompanied, in the S, by a belt of graywackes, which separates the Northern Phyllite Zone from the Mid-German Crystalline High (Fig. 1). As proposed by several authors (e.g., Engel et al. 1983; Wachendorf 1986), all these rocks may be regarded as part of the Giessen-Harz nappe, preserved in a position close to the root zone. The southernmost part of the phyllite zone and the graywacke belt adjacent to the S would represent a more or less complete tectonic profile. Since the position of the sequence is subvertical (as observed in the surface outcrops of the Northern Phyllite Zone), one may expect, towards the S, the least metamorphosed parts of the tectonic sequence. This interpretation would serve to explain the very low metamorphic grade of some of these graywackes (e.g., at Erbstadt NE of Frankfurt, see Fig. 1).

If one accepts this concept, all units to the NW of the graywacke/-MORB-belt must form part of the passive, northern margin of the Rhenohercynian basin, as it existed in Devonian time.

In summary, the Northern Phyllite Zone can be characterized as a heterogenous tectonic assemblage of rocks. During the Ordovician and Silurian, its sedimentary and magmatic record is different from that of the Soonwald-Taunus unit adjacent to the NW, but shows affinities with the Spessart segment of the Mid-German Crystalline High. These earliest magmatic constituents of the Phyllite Zone have been explained as the incipient stage of the Rhenohercynian rift, but probably represent an Ordovician/Silurian magmatic arc. The original site of this arc is unknown, but from the Early Devonian onwards, it was overlain by clastic sequences which also occur in the Rhenohercynian parautochthon further norther. In early Devonian time, the Phyllite Zone was situated to the NW of the opening Rhenohercynian ocean, proximal to its passive northern margin.

Acknowledgments. We are indebted to H.J. Franzke/Clausthal-Zellerfeld, D.E. Meyer/Essen, J. Nagel/Bonn, and W. Riegel/Göttingen for helpful comments.

References

Ackermann S (1985) Beiträge zur Stratigraphie und Lithologie der Wippraer Formation. Diss, Univ Halle, 136 pp

Anderle H-J (1987) Entwicklung und Stand der Unterdevon-Stratigraphie im südlichen Taunus. Geol Jahrb Hessen 115: 81–98

Anderle H-J, Ehrenberg K-H, Meisl S (1972) Metamorphe Zone und Unterdevon im Taunus. Jahresber Mitt Oberrhein Geol Ver NF 54: 123–139

Berger E, Torres P, Wefers J (1991) Zur Stratigraphie der Metamorphen Zone des Hunsrücksüdrandes im Rheinischen Schiefergebirge (vorläufige Mitteilung). Neues Jahrb Geol Paläontol Monatsh 1991/Heft 12: 737–746

Bierther W (1941) Geologie des unteren Hahnenbachtales bei Kirn a. d. Nahe. Jahresber Reichsst Bodenforsch 61: 109–156

Bierther W (1951) Devon? am Südrand des Rheinischen Schiefergebirges bei Lorsbach im Taunus. Notizbl Hess Landesamt Bodenforsch (VI) 2: 15–21

Borsdorf K-H, Estrada S (1991) Ordovizischer Magmatismus im östlichen Rhenoherzynikum. Zentralbl Geol Paläontol (Stuttgart) Teil I 5: 1289–1296

Burmann G (1973) Das Ordovizium der nördlichen Phyllitzone. Teil II. Wippraer Zone. Z Geol Wiss, Themenheft (Berl) 1: 9–43

Cramer FH, Diez de Cramer MDCR (1972) North American Silurian palynofacies and their spatial arrangement: Acritarchs. Palaeontographica B (Stuttgart) 138: 107–180

Dittmar U, Oncken O (1992) Anatomie und Kinematik eines passiven varistischen Kontinentalrandes – zum Strukturbau des südwestlichen Rheinischen Schiefergebirges. Frankf Geowiss Arb Ser A 11: 34–37

Doubinger J, Rauscher R (1984) Palynologische Untersuchungen an Bohrkernen aus den Bohrungen Düppenweiler/Saar. Unpubl Rep, Strasbourg

Ecke H-H, Hoffman M, Ludewig B, Riegel W (1985) Ein Inkohlungsprofil durch den südlichen Hunsrück (südwestliches Rheinisches Schiefergebirge). Neues Jahrb Geol Paläontol Monatsh 1985: 395–410

Engel W, Franke W, Grote C, Weber K, Ahrendt H, Eder FW (1983) Nappe tectonics in the southeastern part of the Rheinisches Schiefergebirge. In: Martin H, Eder FW (eds) Intracontinental Fold Belts. Springer, Berlin Heidelberg New York, pp 267–287

Firtion F (1983) Korallen aus einem Geröll der Bohrung FB la/Düppenweiler/Saar. Unpubl MS

Fischer G (1929) Die Gesteine der Metamorphen Zone von Wippra mit besonderer Berücksichtigung der Grünschiefer. Abh Preuß Geol Landesanst NF (Berl) 121: 64

Franke W, Oncken O (1990) Geodynamic evolution of the North-Central Variscides–a comic strip. In: Freeman R, Giese P, Mueller St (eds) The European geotraverse: integrative studies. Eur Sci Found Strasbourg, pp 187–194

Franzke H-J (1969) Ergebnisse einer geologischen Neuaufnahme der Grünschiefer von Wippra (Harz). Hercynia NF (Leipzig) 6: 187–206

Geib KW (1973) Geologische Karte von Rheinland-Pfalz 1:25000. Blatt 6112 Waldböckelheim. Geol LA Rheinland-Pfalz

Hahn H-D (1990) Fazies grobklastischer Gesteine des Unterdevons (Graue Phyllite bis Taunusquarzit) im Taunus (Rheinisches Schiefergebirge). PhD Thesis, Univ Marburg, 173 pp

Meisl S (1990) Metavolcanic rocks in the "Northern Phyllite Zone" at the southern margin of the Rhenohercynian Belt. In: Franke W (ed) Field guide: Mid-German Crystalline Rise and Rheinisches Schiefergebirge. Int Conf Paleozoic orogens in central Eur, Göttingen-Giessen, Aug–Sept 1990, pp 25–42

Meisl S, Anderle H-J, Strecker G (1982) Niedrigtemperierte Metamorphose im Taunus und im Soonwald. Fortschr Miner 60 (2): 43–69

Meyer DE (1966) Mittel- und Oberdevon im Guldenbachtal (SE-Hunsrück, Rheinisches Schiefergebirge). Vorläufige Mitteilung über neue Ergebnisse zur Stratigraphie. Notizbl Hess Landesamt Bodenforsch 94: 302–309

Meyer DE (1970) Stratigraphie und Fazies des Paläozoikums im Guldenbachtal/SE-Hunsrück am Südrand des Rheinischen Schiefergebirges. Thesis, Univ Bonn, 307 pp

Meyer DE (1975) Geologischer Überblick über den südöstlichen Hunsrück und Beschreibung einer Exkursionsroute. Decheniana 128: 87–106

Mistiaen B, Rohard JC (1983) Bestimmung von Korallen in Karbonatgeröllen aus den Bohrungen Düppenweiler/Saar. Unpubl Rep, Lille

Nickel W (1958) Silurisch-devonische Grenzschichten im Taunus (Graue Phyllite zwischen Wiesbaden und Ehlhalten). Diploma Thesis, Univ Frankfurt, 49 pp (unpubl)

Reichstein M (1964) Stratigraphische Konzeptionen zur metamorphen Zone des Harzes. Geologie Berl 13, 1: 5–25

Reitz E (1989) Devonische Sporen aus Phylliten vom Südrand des Rheinischen Schiefergebirges. Geol Jahrb Hessen 117: 23–35

Schafft R (1985) Makro- und mikrostrukturelle Untersuchungen der Metamorphen Zone im südlichen Hunsrück. Thesis Univ Göttingen, 88 pp

Schmidt K (1976) Das "kaledonische Ereignis" in Mittel- und Südwesteuropa. Nova Acta Leopold NF 45: 381–401

Schneider H (1991) Sammlung Geologischer Führer, 84: Saarland. Borntraeger, Berlin, 271 pp

Scholtz H (1934) Die Tektonik des Steinkohlenbeckens im Saar-Nahe-Gebiet und die Entstehungsweise der Saar-Saale-Senke. Z Dtsch Geol Ges 85: 316–382

Sehnert M (1991) Beiträge zur Biostratigraphie epizonaler Metamorphite in der Wippraer Einheit (Harz). Diss, Univ Halle, 119 pp

Siedel H (1991) Ordovizischer basischer Vulkanismus als Ursache für die Entstehung Al-Fe-Mn-reicher Pelite in der Wippraer Zone, Unterharz Zentralbl Geol Paläontol (Stuttgart) Teil I 5: 1367–1368

Sommermann A-E, Meisl S, Todt W (1990) U/Pb-Alter von Zirkonen aus Metavulkaniten des Südtaunus. Ber Dtsch Mineral Ges Beih Eur J Mineral 2 (1): 244

Sommermann A-E, Meisl S, Todt W (1992) Zirkonalter von drei verschiedenen Metavulkaniten aus dem Südtaunus. Geol Jahrb Hessen 120: 67–76

Sommermann A-E, Anderle H-J, Todt W (1994) Das Alter des Quarzkeratophyrs der Krausaue bei Rüdesheim am Rhein (Blatt 6013 Bingen, Rheinisches Schiefergebirge). Geol Jahrb Hessen 122: 143–157

Stenger B (1961) Stratigraphische und gefügetektonische Untersuchungen in der metamorphen Taunus-Südrand-Zone (Rheinisches Schiefergebirge). Abh Hess Landesamt Bodenforsch 36: 1–68

Streel M (1976) Palynologische Untersuchungen an Kernen der Bohrungen Düppenweiler/Saar. Unpubl Rep, Liège

Streel M (1981) Weitere palynologische Untersuchungen der sedimentären Serien von Düppenweiler/Saar. Unpubl Rep, Liège

Wachendorf H (1986) Der Harz – variszischer Bau und geodynamische Entwicklung. Geol Jahrb Hannover A 91: 3–67

Wagner W, Michels F (1930) Geologische Karte von Hessen 1:25000. Blatt Bingen-Rüdesheim, Hess Staatsverlag, Darmstadt

Wehrens K (1985) Sedimentologische Untersuchungen im karbonischen Alluvialfächer von Düppenweiler/Saar. Beih Geol Landesaufn Saarl 5: 304 pp

Werner C-D (1991) Initialer Magmatismus im Unter- und Mittelharz. Zentralbl Geol Paläontol (Stuttgart) Teil I 5: 1389–1390

Werner H-H (1952) Geologie der Winterbacher Mulde im südöstlichen Hunsrück. Geol Jahrb 66: 627–660

Note Added in Proof. Reitz, Anderle and Winkelmann (1995) have dated metapelites with intercalated intermediate metavolcanics from a drillhole NW of Wiesbaden (Fig. 3) as Early Ordovician (Arenig) by acritarchs (Geol Jb Hessen 123, in press).

III.C.2 Structure

O. Oncken, H.J. Franzke, U. Dittmar, and Th. Klügel

1 Introduction

The Northern Phyllite Zone – named by Scholtz (1930) – marks the boundary between the Saxothuringian Zone in the southeast and the Rhenohercynian Zone in the northwest. This position, taken by several authors to represent the site of a former suture between the two Variscan zones (see Weber and Behr 1983; Matte 1986; Weber 1986; Ziegler 1986; Franke 1989; Franke and Oncken 1990), has attracted considerable attention in recent years. Towards the north, the Phyllite Zone is confined by the equally narrow Taunuskamm-Soonwald Unit in the Rhenish Massif, and by the Harzgeröde unit in the Harz mountains (see Fig. 1). Both units form thrust bodies evolving from the sedimentary prism of the Rhenohercynian shelf proper (Schwab 1979; Oncken 1988b), partly having the internal geometry of an elongate horst system at the southern shelf margin (Taunuskamm-Soonwald Unit). Towards the south, the Phyllite Zone is confined by the Taunus-Hunsrück borderzone fault. This fault – partly buried beneath Permian and Tertiary sediments – forms a steeply dipping fault system crossing in part the entire crust with an apparently complex movement history (Ahorner and Murawski 1975; Anderle 1987; Oncken 1988a; DEKORP Research Group 1991).

The nature and the evolution of the contact to the medium-grade rocks of the presumably former active continental margin of the northern Saxothuringian Zone (or Mid-German Crystalline Rise) is unclear because of a lack of exposures. The results of the DEKORP 1 seismic line (see DEKORP Research Group 1991) suggest large-scale thrusting of the Saxothuringian Zone over the Phyllite Zone (cf. also Behr and Heinrichs 1987; Weber 1990; and balanced sections by Behrmann et al. 1991).

Alongstrike, the Phyllite Zone is subdivided into three areas of outcrop from west to east: the South Hunsrück Unit, the Vordertaunus Unit, and the Wippra Unit (Fig. 1). Stratigraphy and lithology differ significantly among these units as well as from the adjoining Rhenohercynian shelf sequence (see chapters on stratigraphy, this Vol.). The internal structure, as well as the kinematic and metamorphic evolution of all three units, however, is largely interchangeable and also shows only minor differences to the neighboring Rhenohercynian nappe units. Thus, the question of the former relationship to the continental Rhenohercynian basin, and of the paleogeographic setting of the three Phyllite Zone units is difficult to assess in all cases at present. The actual debate centers around two interpretations of the Phyllite Zone, or parts of, either representing the southern prolongation, i.e., the slope of the Rhenohercynian passive continental margin (e.g., Franke and Oncken 1990; Dittmar and Oncken 1992), or the Phyllite Zone being an exotic terrane – of Saxothuringian provenance? – within the accretionary wedge formed in front of the approaching Saxothuringian active margin (e.g., Weber 1990; Jacob et al. 1990).

2 Internal Structure

The general structure of all three units of the Phyllite Zone is characterized by a rather uniform strike of folds, thrusts, and cleavages of ca. 60° parallel to the boundary between the Variscan zones (cf. Figs. 1 and 2). The dominating cleavage, generally parallel or at an acute angle to bedding, and related fold-axial planes dip moderately to steeply to the SE (Wippra and southeastern Vodertaunus Units) or show an overturned, NW-dipping orientation (South Hunsrück and northwestern part of Vordertaunus Unit; Kubella 1951; Bierther 1953; Stenger 1961; Anderle 1976; Schafft 1985; Doutsos and Prüfert 1986; Jacob et al. 1990). This structure is identical to fabric orientations in the immediately adjoining Taunuskamm-Soonwald and Harzgeröde Units. On a larger scale, across the southern Rhenohercynian belt the mentioned fabrics form a major vergency

Dallmeyer et al. (Eds.)
Pre-Permian Geology of Central and Eastern Europe

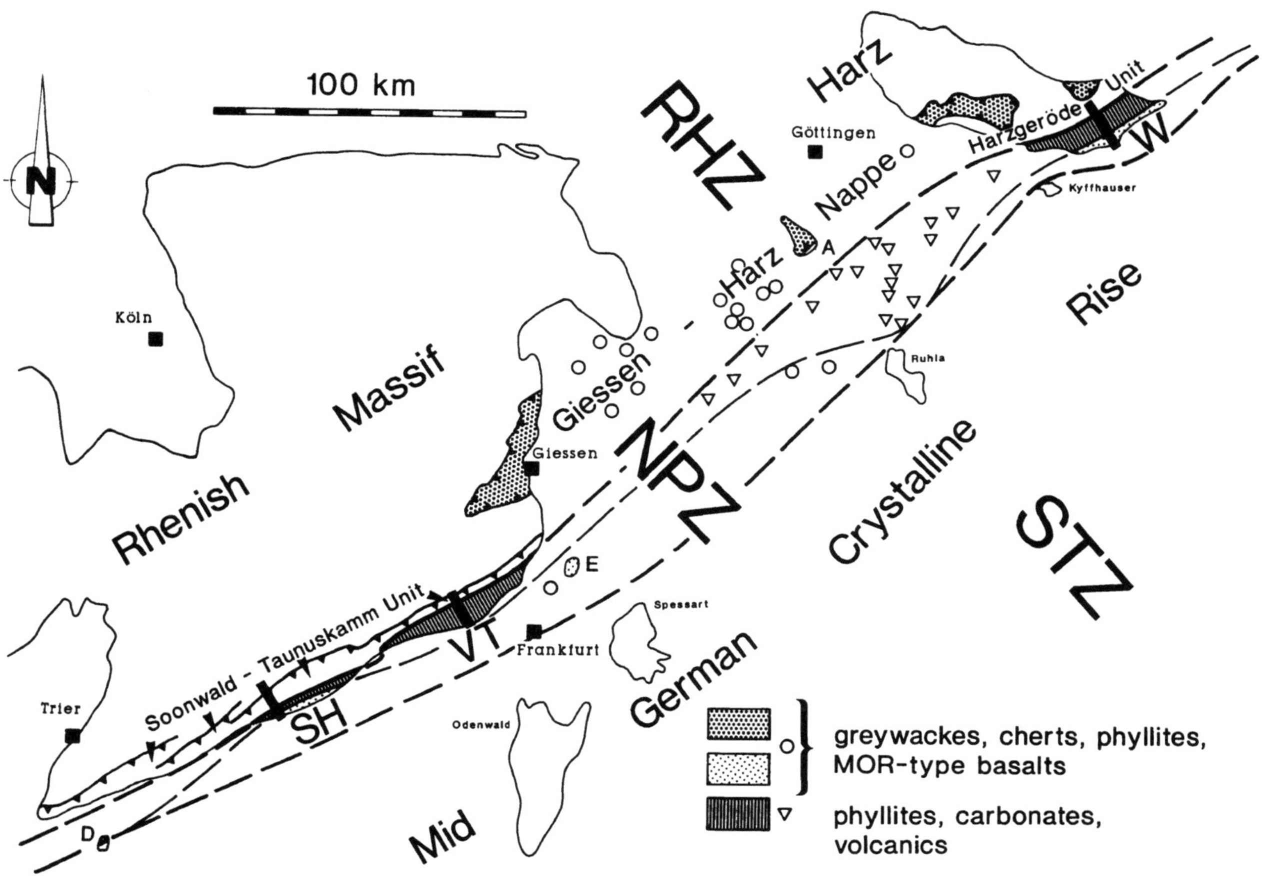

Fig. 1. Schematic tectonic map of the Rhenohercynian Zone (*RHZ*) and Saxothuringian Zone (*STZ*) showing the three units of the Northern Phyllite Zone (*NPZ*) and remains of the Giessen-Harz Nappe; *SH* South Hunsrück Unit; *VT* Vordertaunus Unit; *W* Wippra Unit; *D* Düppenweiler; *E* Erbstadt; *A* Albungen. The Phyllite Zone is subdivided into a Lower Paleozoic metasediments and -volcanics unit and the Devonian metabasalt-graywacke association to the south; all boundaries are major fault contacts. *Heavy black bars* show trace of cross sections in Fig. 4. The map includes drilled units. (Compiled from Fig. 1 in Weber and Behr 1983)

fan which flattens out towards the NW. The general steep dip of structures to greater depth is also supported by the largely transparent nature of the upper crust (down to 3 s TWT) in recent reflection seismic surveys across the Rhenohercynian-Saxothuringian boundary (DEKORP Research Group 1985, 1991; Behr and Heinrichs 1987).

The axial plunge of the first, mostly nearly isoclinal folds is subhorizontal; near the contact to larger shear zones steeply dipping fold axes may occur. The related cleavage may show an inconspicuous mineral elongation lineation in the vicinity of shear zones oriented either subhorizontal or dipping to the east – in rare cases to the west (Fig. 2). Microscopic evidence (see below) in several cases shows relics of an earlier cleavage, an observation which is in contrast to the fabric evolution of the Taunuskamm-Soonwald and the Harzgeröde Units.

A number of vein generations (quartz, quartz-albite, epidote) crosscuts the sequence. Most veins are deformed during formation of the principal cleavage. Many veins, however, intrude along the dominant foliation. Some of these are boudinaged and sheared along this foliation, indicating a discontinuous movement history during the principal stage of deformation. A later vein generation is unaffected by the dominant foliation, but in most cases folded about a late crenulation cleavage along with the earlier veins.

The entire Phyllite Zone as well as the adjoining Rhenohercynian is overprinted by a later crenulation cleavage with associated folds (folding the

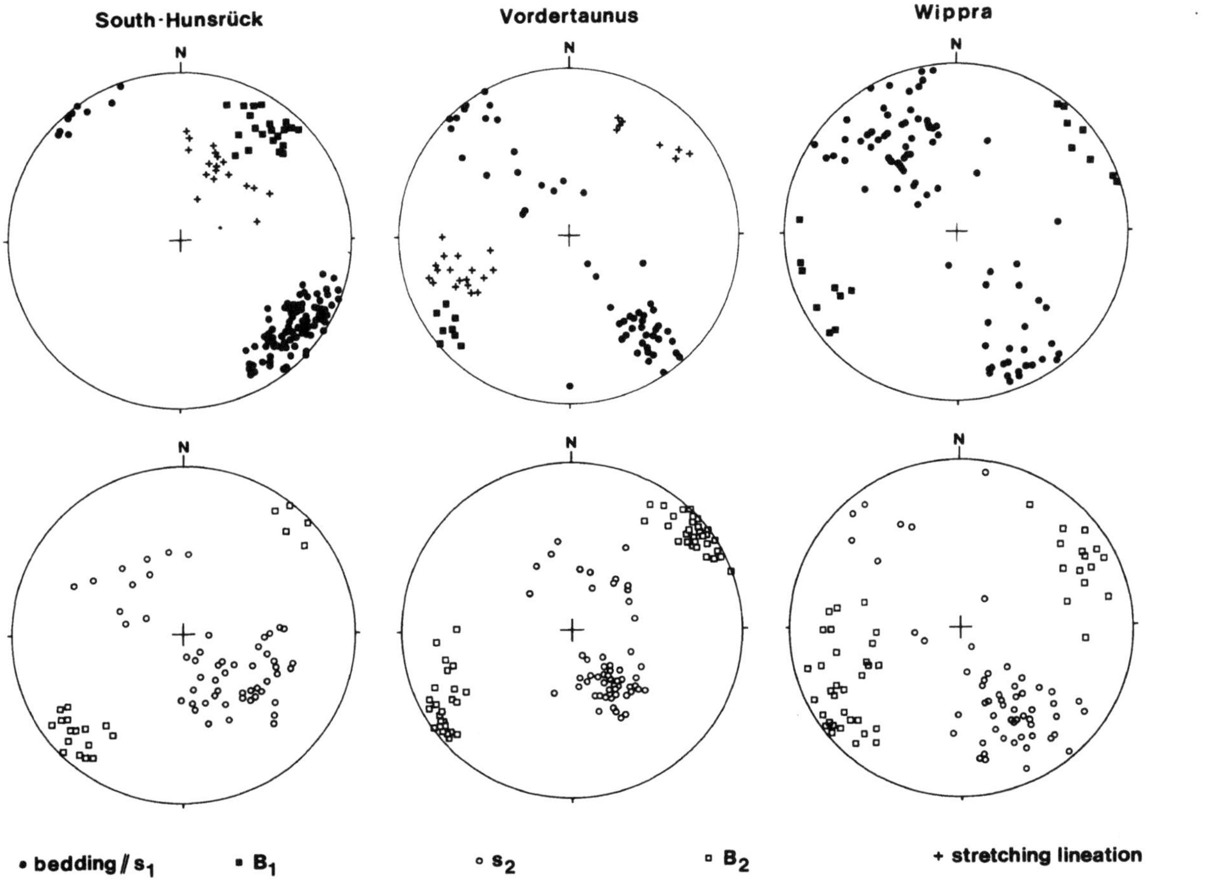

Fig. 2. Fabric orientation in the Phyllite Zone units in the Schmidt net (equal area projection of the lower hemisphere)

principal cleavage, bedding, and veins) and a mostly subhorizontal intersection lineation (Figs. 2, 3). This crenulation cleavage is subhorizontal to moderately NW-dipping in the west but may dip steeply NW in the Wippra Unit, forming again a major fan across the southern Rhenohercynian (see Fig. 4). The kinematics of most of these folds are related to large-scale backfolding and backthrusting in all units. A late extensional kink band generation of variable strike indicates late-orogenic extension of the belt (Weber 1978; Weber and Behr 1983; Anderle et al. 1990).

The internal stratigraphic coherence of the Phyllite Zone Units is strongly disrupted by shear zones (Fig. 4). However, in spite of internal repetitions, the direction of stratigraphic younging within the imbricates of the three units is to the southeast (cf. geochronologic data by Reitz 1989; Sehnert 1990; Sommermann et al. 1990; Berger et al. 1991). The imbricate stack of the Wippra Unit – with the oldest rocks in its central part – is interpreted as resulting from a major anticlinal structure (Jacob et al. 1990; Sehnert 1990). In the South Hunsrück Unit, the southernmost rock association is formed by MOR-type basaltic rocks (Meisl 1990) and phyllitic slates of Emsian and probably also younger age (Reitz 1989). The occurrence of a similar lithologic association of inferred Devonian age including graywackes (Sehnert 1990) in the equivalent position in the Wippra unit (zones 6 + 7; Fig. 4) suggests a close relationship with the first. This thrust slice rests with tectonic contact against the adjacent imbricates with Lower Devonian clastics to Upper Devonian shales and carbonates (Dittmar and Oncken 1992). The Early Upper Devonian includes large olisthostrome units containing material from the adjoining Taunuskamm Unit. The latter sequence is interpreted to reflect the stacked former continental slope environment. The above sediment type and a related abrupt decrease in sedimentary thickness to the north

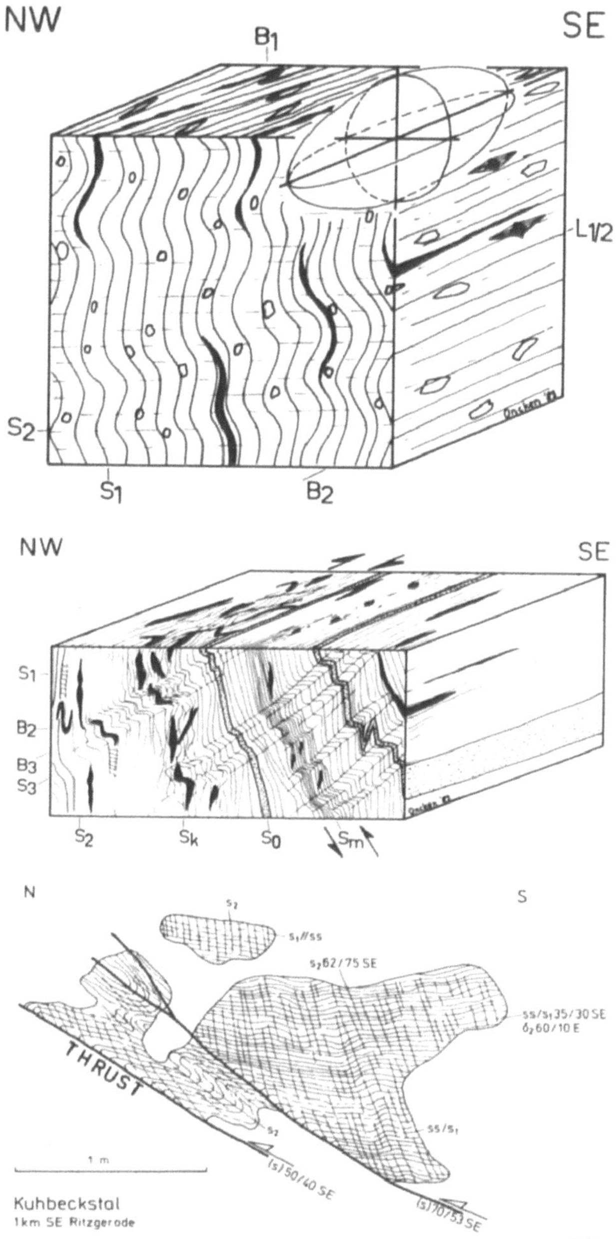

Fig. 3. Characteristics fabric associations in the Phyllite Zone. *Top* Schematic block diagram of fabric orientation of Lower Silurian metarhyolite at Fischbacher Kopf (Vordertaunus Unit) with respect to finite strain axes; diagram shows cleavage and fold generations, feldspar phenocrysts (*white*), pyrite grains (*black squares*) with quartz-filled pressure finges, and quartzalbite veins (*black*). *Center* Schematic block diagram of fabric orientation in ?Lower Devonian Eppstein phyllites (Vordertaunus unit); diagram shows fold and cleavage generations (S_{1-3}, B_{2-3}), mylonitic and foliated cataclastic shear zones (S_m and S_k), and synkinematic generations of quartz-albite veins (*black*). *Below* Fabric association and kinematics of phyllites from the Wippra Unit showing cleavage generations and shear zone pattern

moreover points to the formation of growth faults at the inferred shelf margin (Dittmar and Oncken 1992).

In contrast to the above younging relationship, stratigraphic younging in the Taunuskamm-Soonwald Unit (Rhenish Massif) immediately north of the contact to the Phyllite Zone is to the northwest. This apparent anticlinal juxtaposition of Phyllite Zone and southernmost Taunuskamm-Soonwald Unit at the sheared contact underlines the peculiar nature of their mutual contact (Fig. 4). Moreover, the age relationships across this contact are complex:

- In the South Hunsrück Unit, Upper Devonian metasediments of the Phyllite Zone (Berger et al. 1991) are in contact with Gedinnian to Emsian sediments and their metamorphic basement (cf. "Gneis von Wartenstein", Meisl et al. 1989) of the Taunuskamm-Soonwald Unit.
- In the Vordertaunus Unit, Late Ordovician/ Silurian metavolcanics (Sommermann et al. 1990) are in contact with Gedinnian metasediments from the Taunuskamm-Soonwald Unit. These metavolcanics also appear in two isolated inliers within the Taunuskamm-Soonwald Gedinnian near the western end of the Vordertaunus Unit, which suggests a common volcanic basement of the metasediments of both the Vordertaunus and the Taunuskamm Units.
- In the Wippra Unit, imbricated and sheared bodies of Ordovician and Silurian metasediments (units 2–4) are overthrust on a Lower Carboniferous olisthostrome unit (Harzgeröde Unit).

These contact relationships preclude a simple overthrust relationship of the Phyllite Zone over the Rhenohercynian units to the north for all three units and rather point to differences in structural relationships along the strike of the belt or suggest a complex kinematic history. The appearance of this boundary, moreover, is highly variable alongstrike: associated shear zones in the southern Hunsrück and Taunus along the contact to the Taunuskamm-Soonwald Unit are formed by thin to several-m-thick mylonites (with differentiated quartz-rich and phyllosilicate-rich layers) which form an anastomosing network dipping steeply NW (South Hunsrück, Vordertaunus) and SE (Wippra). The width of the sheared zone may reach some 300 m and may include tectonic slices of metavolcanics and

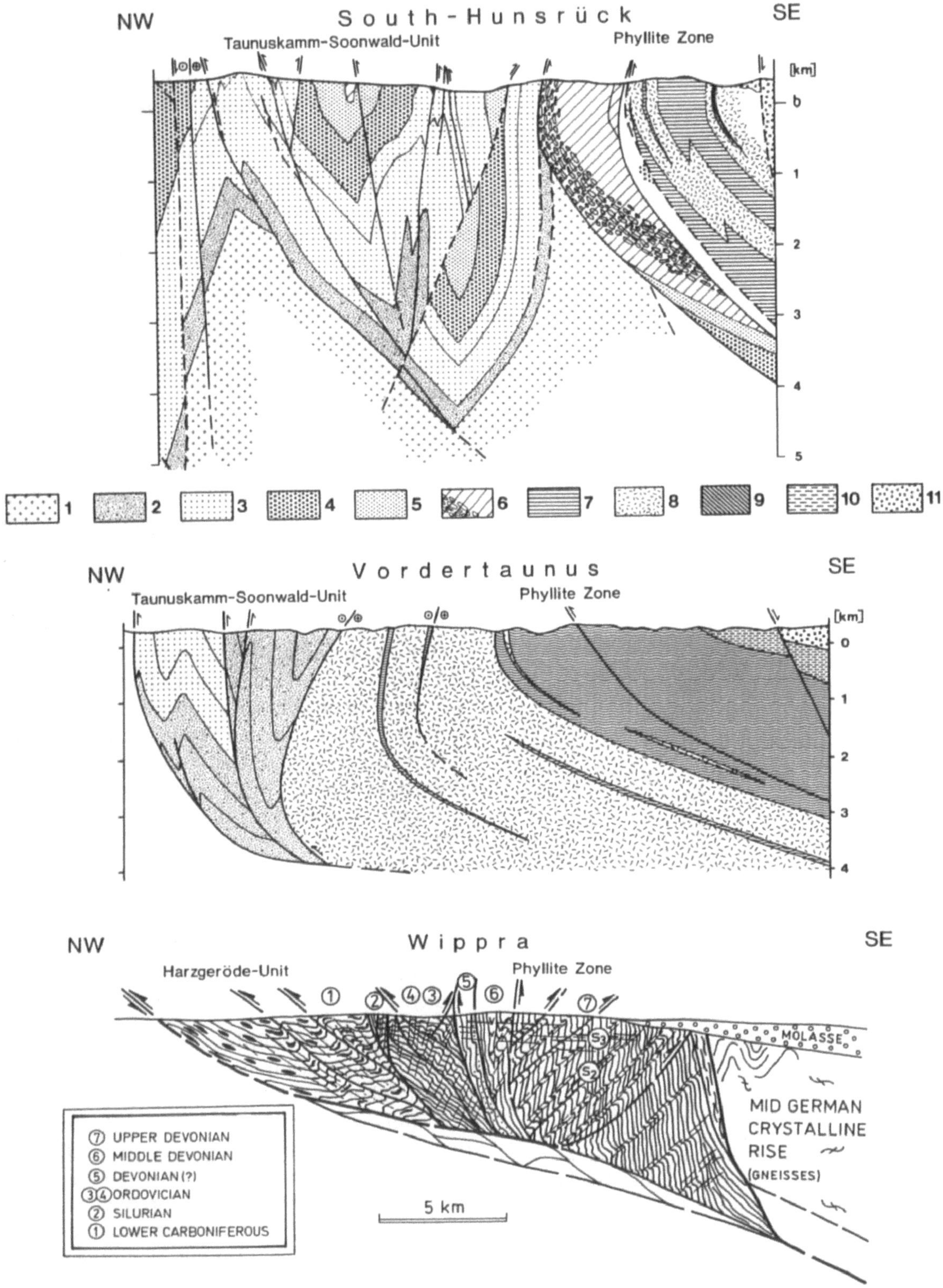

Fig. 4. Schematic cross sections across Phyllite Zone units (see Fig. 1 for location of sections). Legend for South Hunsrück and Vordertaunus: *1* pre-Variscan crystalline basement; *2* Gedinnian; *3* Siegenian; *4* Emsian; *5* Mid-Devonian; *6* Upper Devonian and ?Lower Carboniferous (with olisthostrome intercalations): *7* phyllites (Upper Emsian to ?Upper Devonian); *8* metavolcanics (MOR type in South Hunsrück and Late Ordovician/Silurian felsic to intermediate flows in the Vordertaunus); *9* Eppstein phyllites (?Silurian to Siegenian); *10* Lorsbach phyllites (Lower Emsian to ?Mid Devonian); *11* Permo-Carboniferous molasse. Sense of shear shown usually depicts only early increments

metasediments quite similar to nearby sediments (Taunus).

The mylonites – with a ductile fabric – may in rare cases be overprinted by cataclastic shear zones. Lineations and indicators of sense of shear (asymmetric porphyroclasts, shear bands) in mylonites and cataclasites show several shear increments. An early, not always equally well-preserved generation in all units points to uplift of the respective southern units suggesting early overthrusting to the northwest. Motion probably predated the rotation of the thrusts into a steep dip. The pervasive mylonitic fabric, however, is related to subsequent dextral shearing – at steep contacts – or east-directed oblique dextral backthrusting of the Taunuskamm-Soonwald Unit over the Phyllite Zone (oblique backthrusting with a sinistral component is also observed). The sequence of motions is not unequivocally resolved at present. This includes the possibility of earlier, now obliterated shear increments which may relate to the early stacking history during formation of the rarely preserved first cleavage.

Each of the three units of the Phyllite Zone is subdivided by a number of imbricates. The associated shear zones show the same kinematic pattern as along the boundary to the northern units. The formation of the shear zones was partly accompanied by intense mobilization of quartz. Generally, the early shear zone fabrics are strongly overprinted by the late crenulation cleavage and are folded around the related fold axes (Fig. 3), while the later oblique to strike-slip mylonites are only weakly overprinted by the late crenulation cleavage in most cases.

3 Fabrics

Corresponding to the very-low-grade to low-grade metamorphic conditions (Meisl et al. 1982; Anderle et al. 1990; Lüders et al. 1990), the syncrystalline deformation of all three units is generally driven by thermally activated deformation mechanisms: crystal plasticity (in largely monomineralic quartzites or pelites) associated with a variable degree of dynamic recovery and recrystallization, or diffusion-controlled mechanisms (e.g., pressure solution, cf. Schafft 1985; Oncken 1988b, 1991; see Fig. 5). The resulting fabric may show relics of an earlier equally penetrative cleavage which seems to have formed under prograde conditions (no recrystallization and only minor growth of layer silicates; Fig. 5). The principal cleavage (s_1), on the other hand, appears to have formed near the peak thermal conditions (mica grain growth, quartz and albite recrystallization) which controlled a prolonged period during isothermal decompression (cf. Chap. III.c.4).

White micas yield K/Ar-ages (see Ahrendt et al. 1983) which, since the micas crystallized synkinematically to s_1 below the white mica K-Ar closing temperatures, times deformation at ca. 325 Ma for the southern Rhenohercynian Zone. However, the Vordertaunus Unit also yields ages of about 310 Ma, the reason for this age group remaining obscure at present.

Preliminary observations on finite strain suggest that deformation during this stage of principal fabric formation shows a complex pattern: The stretching axis (from preferred orientations of stretched phenocrysts in the metavolcanics and detrital grains in metasediments) mostly dips at shallow angles. Aspect ratioes of the finite strain ellipsoids ranges between $R_{XZ} = 2$ to > 6. Pressure shadows around pyrite grains (cf. Stenger 1961) support the above relationship: incremental strains thus also seem to largely show an alongstrike stretching of the Vordertaunus Unit, the orientation partly paralleling the rare macroscopic stretching lineation. Also, the stretching of the rock association in the Vordertaunus Unit parallel to their strike is not coaxial throughout – in many instances stretched grains show asymmetric tails (σ-type, Passchier and Simpson 1986) with mostly a dextral sense of shear like in neighboring shear zones (see above; Fig. 5).

The above-mentioned late nonpenetrative crenulation cleavage is only associated with pressure solution, cold-working, and weak recovery of quartz.

4 Discussion and Conclusions

The rock association of the Phyllite Zone forms steeply dipping to overturned units. The succession of deformation fabrics and associated metamorphic conditions reflects the development from a compressional high P/moderate T to an extensional low P/low T regime (Anderle et al. 1990; Lüders et al. 1990). Rare relics of an earlier fabric hint at an early starting deformation during prograde conditions. Peak compressional

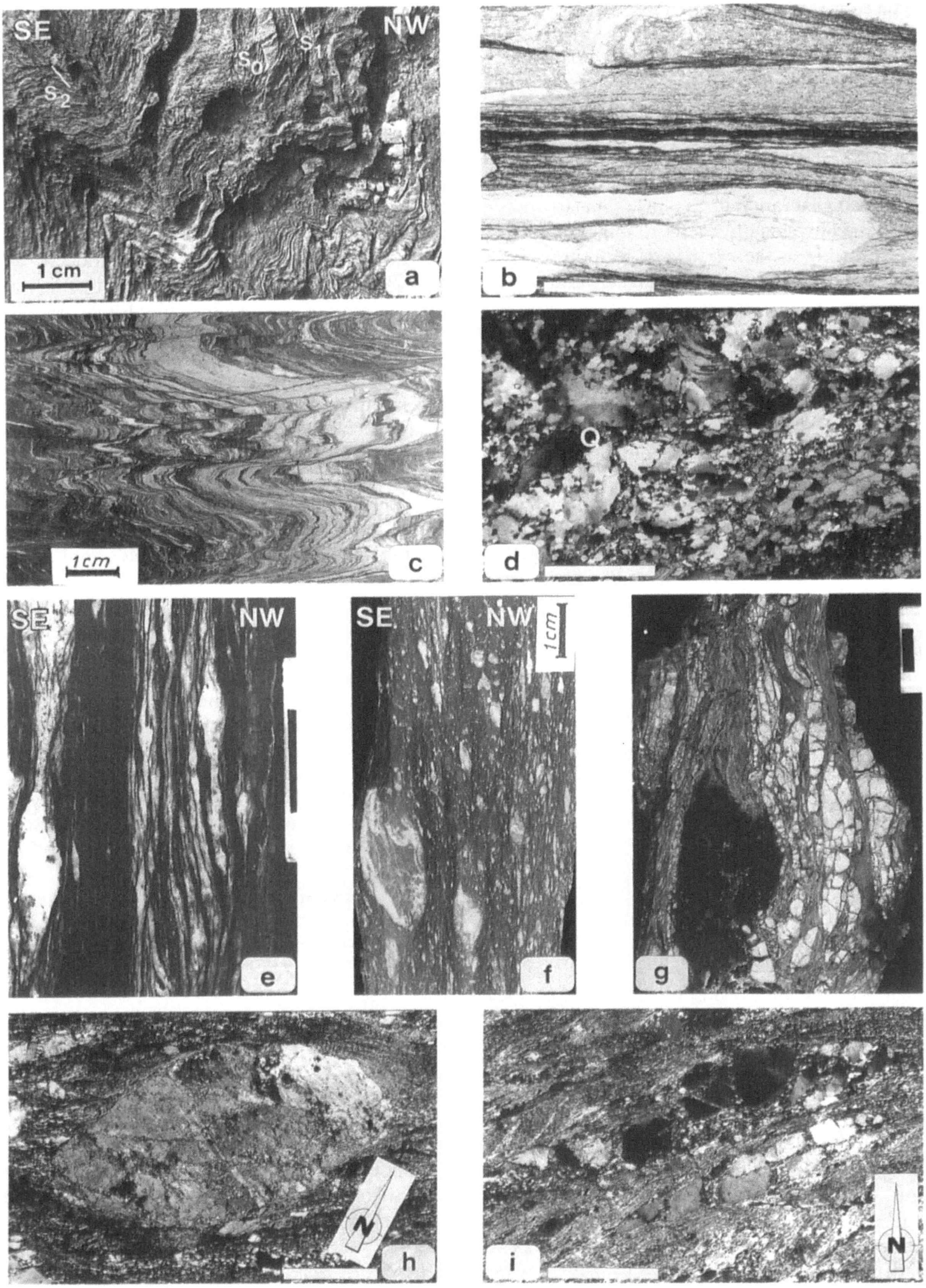
SE
NW
S_0
S_1
S_2
1 cm
a
b
1cm
c
Q
d
SE
NW
e
SE
NW
1cm
f
g
N
h
i
N

deformation and metamorphism are related to the formation of a pervasive mostly bedding-parallel foliation with minor folding. The metamorphic conditions of this stage must have been strongly controlled by a – now eroded – thick pile of nappe units, the remains of which are probably retained in the Giessen-Ostharz nappe (cf. Engel et al. 1983; Oncken 1988a). Although unsatisfactorily resolved at present, this stage probably involves the NW-directed overthrusting of the Phyllite Zone over the Taunuskamm-Soonwald and Harzgeröde Units – which both also show the same displacement direction (Oncken 1988a; Jacob et al. 1990).

Continuing fabric formation during isothermal decompression is mostly associated with further thickening of the thrust wedge by backfolding and backthrusting. The mentioned PTD path, moreover, suggests that removal of the above nappe system and exhumation of the Phyllite Zone was only partly controlled by erosion: supported by the established switch of compressional to late extensional fabrics, a large portion of the exhumation must have been driven by rapid extensional removal of the tectonic cover due probably to gravitational collapse of the thickened crust along the suture.

Although the stratigraphic sequences of the Phyllite Zone units are internally strongly deformed and stratigraphically disrupted, internal stacking and stratigraphic repetition do not seem to play a dominant role with exception of the contact to the adjoining nappe units to the north. Moreover, the amount of stratigraphic separation of the Phyllite Zone to the northern units appears to decrease systematically from the Harz to the western Rhenish Massif or even to reverse at the South Hunsrück Unit (Upper Devonian in the south against Siegenian in the north). This points to either a decrease of vertical displacement at the northern boundary of the Phyllite Zone from east to west, or to an increase of late backthrusting in the same direction – or else the classical northern boundary of the Phyllite Zone is not the same structural element throughout. In addition, inversion of earlier growth faults (see above) at the Phyllite Zone boundary (South Hunsrück) contributes to the complex picture.

These complexities of inconsistent vertical displacement can also be seen to agree with the so far known sense of shear, which suggests some dextral displacement of the Vordertaunus and South Hunsrück Units with respect to the Taunuskamm-Soonwald Unit. Besides the lack of major stratigraphic separation, this pattern of ductile wrench faulting may possibly also account for the anticlinal-like juxtaposition of the Phyllite Zone rocks with the southernmost Taunuskamm-Soonwald Unit sequence at the contact: the units might be seen to form a large-scale flower structure with a vergency fan, resulting possibly from oblique collision of the Saxothuringian and Rhenohercynian Zones in the sense of a dextral transpression (cf. Oncken 1988a,b). The Taunuskamm-Soonwald Unit, which shows evidence of late W-directed thrusting at the front (Oncken 1988b), moreover appears to evolve to a major pop-up structure with bidirectional oblique thrusting during this late stage (Dittmar and Oncken 1992).

Currently investigated open questions referring to the paleogeographic position of the Phyllite Zone can only be answered in part at present. The lithologic and stratigraphic content of the Phyllite Zone resembles that of the Rhenohercynian Basin to the north rather than that of the condensed pelagic sequences contained in the Giessen-Harz Nappe. The appearance of melange-type sediments containing the nearby shelf material in the South Hunsrück also suggests a continental slope environment (Dittmar and Oncken 1992). The latter may grade into the southerly adjoining imbricates with a MOR-basalt-phyllite association possibly pertaining to the continental rise and marking the transition

Fig. 5a–i. Micrographs of fabrics and shear zones. **a** Cleavage generations in Eppstein slates (Vordertaunus) with early cleavage (*S_0*), dominant differentiated crenulation cleavage (*S_1*), and final crenulation cleavage (*S_2*; compare same outcrop in Fig. 3 center). **b** Intrafolial B_1-folds in pervasively cleaved phyllite with development of differentiated layering (South Hunsrück). **c** Pervasive cleavage (parallel to layering) folded by late orogenic crenulation cleavage in metabasalts (South Hunsrück). **d** Recrystallised quartz (*Q*) and calcite (*C*) in metabasalt from South Hunsrück Unit. **e, f** Mylonites from boundary of Phyllite Zone to Taunuskamm-Soonwald Unit with shear bands and asymmetric porphyroclasts. **g** Cataclastically overprinted mylonite in retrogressive shear zone. **h** Asymmetric σ-type porphyroclast of feldspar phenocryst in felsic metavolcanics from Vordertaunus Unit in oriented section (= map view). **i** Stretched grain fabric from same outcrop as **h** in map view. **h, i** From outcrop shown in Fig. 3 top. Micrographs in **d, h, i** with crossed nicols. *Bar* 1 cm in **c, e, f, g**, 1 mm in all other cases to the small oceanic basin which was parent to the above nappe stack. This nappe system also contains MOR-type basalts. Meisl (1990) therefore concludes a possible continuation of the Giessen-Harz Nappe pile in these southernmost imbricates, which might thus be seen to form its root zone as already proposed by Engel et al. (1983). Further indications for this position of the root zone or suture of the Rhenohercynian-Saxothuringian Zones are minor outcropping and drilled occurences of unmetamorphic graywackes (see Fig. 1).

Including the above-described contact relationships and the lack of a clearcut metamorphic hiatus across the boundary to the north (cf. Meisl et al. 1982; Siedel 1989; Anderle et al. 1990), most of the Phyllite Zone thus most probably simply represents the slope of the Rhenohercynian passive margin (cf. Franke and Oncken 1990; Dittmar and Oncken 1992). As such, it may be seen to bridge part of the gap to the small Rhenohercynian ocean to the south. During Lower Carboniferous continental collision upon approach of the Saxothuringian terrane, the Phyllite Zone and its northern foreland were incorporated into the accretionary wedge and successively deformed as described above.

References

Ahorner L, Murawski H (1975) Erdbebentätigkeit und geologischer Werdegang der Hunsrück-Südrand-Störung. Z Dtsch Geol Ges 126: 63–82

Ahrendt H, Clauer N, Hunziker JC, Weber K (1983) Migration of folding and metamorphism in the Rheinisches Schiefergebirge deduced from K-Ar and Rb-Sr age determinations. In: Martin H, Eder FW (eds) Intracontinental fold belts. Case studies in the Variscan Belt of Europe and the Damara Belt in Namibia. Springer, Berlin Heidelberg New York, pp 323–338

Anderle HJ (1976) Der Südrand des Rhenoherzynikums im Taunus. Geol Jahrb Hessen 104: 279–284

Anderle HJ (1987) The evolution of the South Hunsrück and Taunus borderzone. Tectonophysics 137: 101–114

Anderle HJ, Massonne HJ, Meisl S, Oncken O, Weber K (1990) Southern Taunus Mountains. In: Field guide: Mid-German Crystalline Rise and Rheinisches Schiefergebirge. Int Conf on Paleozoic orogens in central Europe, Göttingen-Giessen, pp 125–148

Behr HJ, Heinrichs T (1987) Geological interpretation of DEKORP 2-S: A deep seismic reflection profile across the Saxothuringian and possible implications for the Late Variscan structural evolution of Central Europe. Tectonophysics 142: 173–202

Behrmann J, Drozdzewski G, Heinrichs T, Huch M, Meyer W, Oncken O (1991) Crustal scale balanced cross sections through the Variscan Fold Belt, Germany – the central EGT segment. Tectonophysics 196: 1–21

Berger E, Torres P, Wefers J (1991) Zur Stratigraphie der Metamorphen Zone des Hunsrücksüdrandes im Rheinischen Schiefergebirge. Neues Jahrb Geol Paläontol Monatsh 1991: 737–746

Bierther W (1953) Zur Stratigraphie und Tektonik der Metamorphen Zone im südlichen Rheinischen Schiefergebirge. Geol Rdsch 41: 173–181

DEKORP Research Group (1985) First results and preliminary interpretation of deep reflection seismic recordings along profile DEKORP 2-South. J Geophys 57: 137–163

DEKORP Research Group (1991) First results and preliminary interpretation of deep reflection seismic recordings along profile DEKORP 1. Geophys J Int 106: 203–227

Dittmar U, Oncken O (1992) Anatomie und Kinematik eines passiven varistischen Kontinentalrandes – zum Strukturbau des südwestlichen Rheinischen Schiefergebirges. Frankf Geowiss Arb A12: 34–37

Doutsos T, Prüfert J (1986) Bau und Entwicklung der Metamorphen Zone am Taunussüdrand (Rheinisches Schiefergebirge). Geol Jahrb Hessen 114: 125–149

Engel W, Franke W, Grote C, Weber K, Ahrendt H, Eder FW (1983) Nappe tectonics in the southeastern part of the Rheinisches Schiefergebirge. In: Martin H, Eder FW (eds) Intracontinental fold belts. Case studies in the Variscan Belt of Europe and the Damara Belt in Namibia. Springer, Berlin Heidelberg New York, pp 267–287

Franke W (1989) Tectonostratigraphic units in the Variscan belt of central Europe. Geol Soc Am Spec Pap 230: 67–90

Franke W, Oncken O (1990) Geodynamic evolution of the north-central Variscides: a comic strip. In: Freeman R, Giese P, Mueller S (eds) The European geotraverse: integrative studies. Eur Sci Found, Strasbourg, pp 187–194

Jacob G, Siedel H, Rauche H, Franzke J, Hoth F (1990) Transverse section crossing the metamorphic zone of Wippra/Harz mountains. In: Abstr Int Conf Terranes in the circum-Atlantic Paleozoic orogens, Göttingen-Giessen

Kubella K (1951) Zum tektonischen Werdegang des südlichen Taunus. Abh Hess Landesamtes Bodenforsch 3: 1–81

Lüders V, Franzke HJ, Rauche H (1990) Mikrothermometrische Untersuchungen an Flüssigkeitseinschlüssen in Quarzen aus Deformationsstrukturen der Zone von Wippra und Sb-betonten Gangstrukturen bei Wolfsberg (Unterharz). Beih Eur J Mineral 2: 166

Matte P (1986) Tectonics and plate tectonics model for the Variscan Belt of Europe. Tectonophysics 126: 329–374

Meisl S (1990) Metavolcanic rocks in the Northern Phyllite Zone at the southern margin of the Rhenohercynian belt. In: Field guide: Mid-German Crystalline Rise and Rheinisches Schiefergebirge. Int Conf

Paleozoic orogens in central Europe, Göttingen-Giessen, pp 25–42

Meisl S, Anderle HJ, Strecker G (1982) Niedrigtemperierte Metamorphose im Taunus und im Soonwald. DMG-Tagung 1982, Exkursion E3. Fort-schr Mineral 60/2: 43–69

Meisl S, Kreuzer H, Höhndorf A (1989) Metamorphose-Bedingungen und -Alter des Kristallins am Wartenstein bei Kirn/Nahe. Abstr 5. Rundgespräch Geodyn Eur Varistikums, Braunschweig, p 38

Oncken O (1988a) Aspects of the reconstruction of the paleo-stress history of a fold and thrust belt (Rhenish Massif, Federal Republic of Germany). Tectonophysics 152: 19–40

Oncken O (1988b) Geometrie und Kinematik der Taunuskammüberschiebung – Beitrag zur Diskussion des Deckenproblems im südlichen Schiefergebirge. Geol Rdsch 77: 551–575

Oncken O (1991) Aspects of the structural and paleogeothermal evolution of the Rhenish Massif. Ann Soc Géol Belg 113: 139–159

Passchier CW, Simpson C (1986) Porphyroclast systems as kinematic indicators. J Struct Geol 8: 831–843

Reitz E (1989) Devonische Sporen aus Phylliten vom Südrand des Rheinischen Schiefergebirges. Geol Jahrb Hessen 117: 23–35

Schafft R (1985) Makro- und mikrostrukturelle Untersuchungen der metamorphen Zone im südlichen Hunsrück. Thesis, Univ Göttingen, Göttingen, 88 pp

Scholtz H (1930) Das varistische Bewegungsbild entwickelt aus der inneren Tektonik eines Profils von der Böhmischen Masse bis zum Massif von Brabant. Fortschr Geol Paläontol 8 (25): IX + 316 pp

Schwab M (1979) Gravitational slide masses in the Harz. Veröff Zentralinst Physik Erde Potsdam 58: 23–46

Sehnert M (1990) Zur Biostratigraphie epizonaler Metamorphite der Wippraer Einheit/Unterharz. Abstr Congr Zeitmarken im Eur Variszikum, Clausthal-Zellerfeld, 15–18.11.1990

Siedel H (1989) Beiträge zur Petrographie und Geochemie epizonal metamorpher Pelite in der Wippraer Einheit (Harz). Thesis, Univ Halle, Halle, 142 pp

Sommermann AE, Meisl S, Todt W (1990) U-Pb-Alter von Zirkonen aus Metavulkaniten des Südtaunus. Beih Eur J Mineral 2: 244

Stenger B (1961) Stratigraphische und gefügetektonische Untersuchungen in der metamorphen Taunus-Südrand-Zone (Rheinisches Schiefergebirge). Abh Hess Landesamtes Bodenforsch 36: 1–68

Weber K (1978) Das Bewegungsbild im Rhenoherzynikum – Abbild einer varistischen Subfluenz. Z Dtsch Geol Ges 129: 249–281

Weber K (1986) The Mid-European Variscides in terms of allochthonous terrains. In: Freeman R, Mueller S, Giese P (eds) Proc 3rd Worksh EGT: the central segment. 14.–16.4 1986, Eur Sci Found, Strasbourg, pp 73–82

Weber K (1990) Crustal structure and the geological development of the Rhenohercynian/Saxothuringian terrane boundary in central Europe. In: Abstr Int Conf Terranes in the circum-Atlantic Paleozoic orogens, Göttingen-Giessen

Weber K, Behr HJ (1983) Geodynamic interpretation of the Mid-European Variscides. In: Martin H, Eder FW (eds) Intracontinental fold belts. Case Studies in the Variscan Belt of Europe and the Damara Belt in Namibia. Springer, Berlin Heidelberg New York, pp 427–469

Ziegler PA (1986) Geodyamic model for the Paleozoic Crustal Consolidation of Western and Central Europe. Tectonophysics 126: 303–328

III.C.3 Igneous Activity

S. Meisl

1 Introduction

The Northern Phyllite Zone is set between the unmetamorphosed sequences of the Rheinisches Schiefergebirge (to the north) and the high-grade metamorphic rocks of the Mid-German Crystalline Rise (to the south). The Phyllite Zone is exposed in the South Hunsrück (Soonwald) Mts., in the South Taunus Mts. and in the Wippra Zone of the Harz Mts., 150 km to the NE (see maps in Anderle et al., chap. III.C.1). The South Hunsrück and the Wippra Zone contain only metabasaltic rocks; the South Taunus has only very few metabasalts and is dominated by metavolcanic rocks of felsic (Serizitgneis, Felsokeratophyr) and intermediate composition (greenschist, Keratophyr, see Sect. 5.2). The relative proportion of the metavolcanic rocks is approx. 30–40% in the South Hunsrück and South Taunus, and approx. 10% in the Wippra Zone of the Harz Mts.

The age of the metavolcanic rocks is, in most cases, unknown. In the South Hunsrück, the metavolcanics are supposed to be of Devonian age, but biostratigraphic data are available only for the metasediments, whose relationship with the volcanics is uncertain. In the South Taunus, the accompanying metasediments have, until recently, been interpreted as pre-Devonian, but Reitz (1989) has detected Emsian spores in the Lorsbach Formation, which contains two intercalations of metabasalt. The intermediate to felsic metavolcanic rocks occur within the Eppstein Formation, which is supposed to be older than the Lorsbach Formation (Stenger 1961), but has until now not yielded fossils. W. Ackermann (unpubl.) has obtained, from a Serizitgneis at Ehlhalten, a Rb-Sr whole rock age of 394 Ma (cf. Meisl 1970), which would match the Lower Devonian age of the Lorsbach Formation, however, another Serizitgneis (Wiesbaden-Sonnenberg) yielded 304 Ma by the same method. A study on zircons is presently being carried out. The first measurements give 426 + 14/−15 Ma for a Serizitgneis (Fischbacher Kopf), and 433 + 9/−7 Ma for a Felsokeratophyr (Dachsbau). These data are taken to reflect magmatic crystallization in the late Ordovician or Silurian time (Sommermann et al. 1990).

In the meantime, also a greenschist sample with dacitic composition was dated (Rompf quarry, Ruppertshain). The age of 442 ± 22 Ma for this rock is slightly older than the more acidic rock types mentioned above. It is likely that at least the beginning of volcanic activity was in the Ordovician because volcanic rocks frequently occur in other Ordovician areas. The slight difference in age between Serizitgneis and greenschist corresponds to the normal differentiation trend, where volcanism starts with more basic members and becomes more acidic with time (Sommermann et al. 1992).

Dating of the metavolcanic rocks is as difficult in the Wippra Zone of the Harz Mts. as in the Rheinisches Schiefergebirge; there are few fossils, and the stratigraphic relationships between the dated sediments and the metavolcanic rocks are poorly understood. The metavolcanic rocks are supposed to be of Ordovician, Silurian, and Devonian age (Schwab 1984, 1989).

Petrological interpretation of magmatic rocks requires, besides the determination of the major elements, mineral composition and fabrics, analysis of the immobile trace elements (including REE). These data are available for approx. 300 samples from the South Taunus. The determinations comprise, among others, the trace elements P, Ti, Ga, Y, Zr, and Nb, and the REE La, Ce, Nd, Sm, Eu, Gd, Dy, Er, and Yb. In the South Hunsrück (Soonwald), these elements (plus Cr) have been determined in approx. 160 samples. The relevant data for the Wippra Zone in the Harz Mts. are taken from the literature (e.g., Werner 1984, 1986). Tectonic deformation in a ductile regime has wiped out most of the primary fabrics, and greenschist-grade metamorphism has brought about changes in the concentration of

Dallmeyer et al. (Eds.)
Pre-Permian Geology of Central and Eastern Europe

major and also some minor elements (see chaps. III.C.1, III.C.2, and III.C.4, this Vol.). Therefore, the present chapter is based mainly upon the relatively immobile trace elements.

Since most of the magmatic fabrics have been destroyed by tectono-metamorphic alteration, thin section petrography is only given in a few cases, where magmatic fabrics have survived. For petrographic details, the reader is referred to Meisl (1970).

2 Methods

The trace elements Ga, Y, Zr and Nb were determined by RFA after Leoni and Saitta (1976, modified by F. Rosenberg). Determination of REE was carried out by ICP from a peroxide dissolution after Schnetger (1990), combined with enrichment by ion-exchange after Edge and Ahrens (1962). Ti and Cr were analyzed by ICP from a Li-Tetraborate dissolution, whereas P was determined by colorometry. The author is greatly obliged to the analysts (Dr. F. Rosenberg and G. Wenzel, Wiesbaden). The analytical data have not been published so far, but are available from the author on request.

3 South Hunsrück (Soonwald)

The metavolcanics of the Soonwald are metabasalts, as can be deduced from relict minerals. Relict clinopyroxenes are widespread, yet ophitic fabrics have survived in only a few cases. The basaltic nature of the rocks is also indicated by the immobile trace elements: in the Zr/TiO2-Nb/Y diagram (Winchester Floyd 1977) they plot partly in the field of alkali-basalt with approx. 60 samples plotting in the subalkaline basalt field (Fig. 1). In the Zr/TiO2-Ce diagram (not figured), and in the Zr/TiO2-Ga diagram (Winchester and Floyd 1977), all these samples plot in the basalt field (not figured).

After the Ti-Cr discrimination diagram (Fig. 2) of Pearce and Gale (1977), these metabasalts correspond mainly to MOR-type basalts. Only four samples plot in the field of island arc basalt.

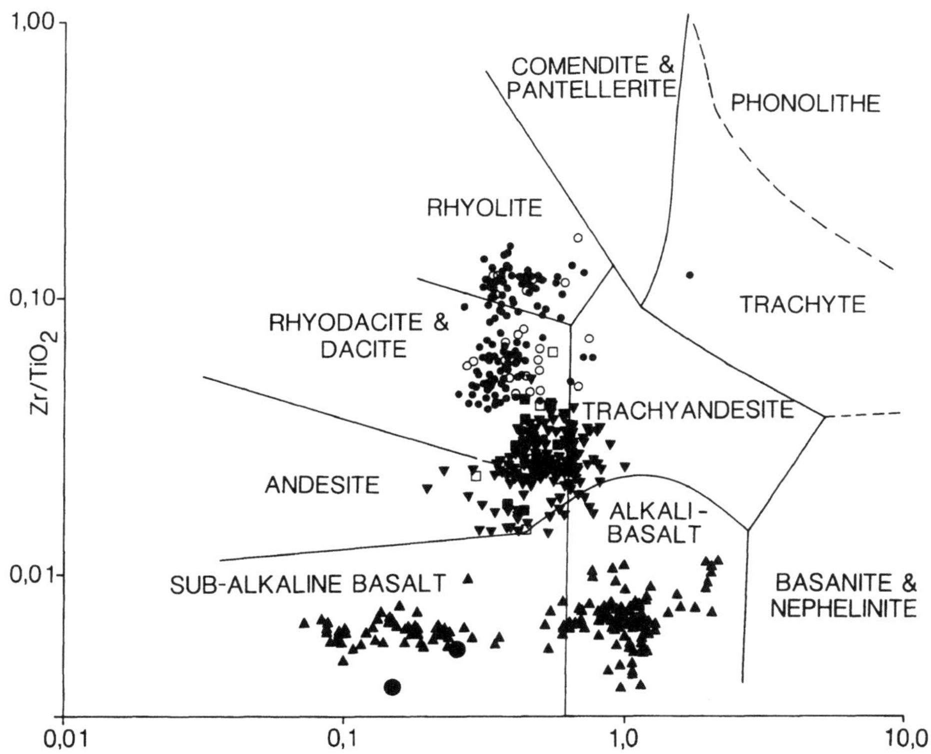

Fig. 1. Metavolcanic rocks of the South Hunsrück (Soonwald) and of the South Taunus plotted in the Zr/TiO2-Nb/Y diagram of Winchester and Floyd (1977). Symbols: ▲ metabasalts of the Soonwald; ● metabasalts of the South Taunus; ▼ greenschists; ■ Keratophyr; □ silicified Keratophyr; • Serizitgneis, ○ Felsokeratophyr of the South Taunus

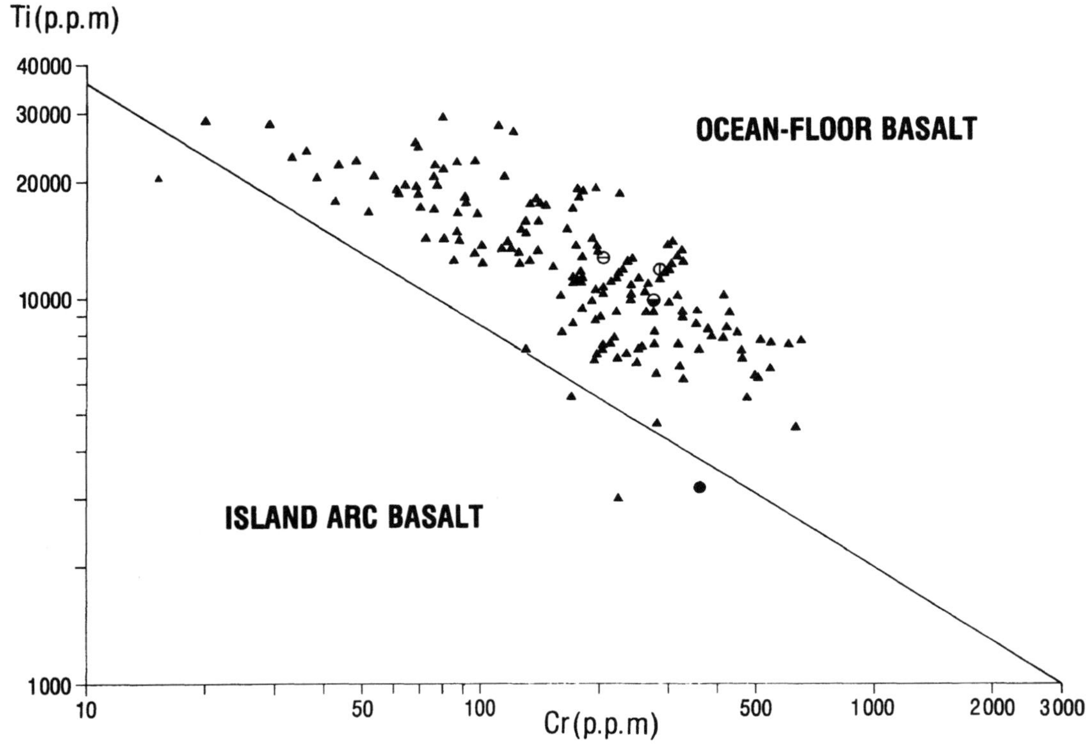

Fig. 2. Metabasalts of the Soonwald (▲) and of the South Taunus (●) plotted in The Ti-Cr diagram after Pearce and Gale (1977). The other plots represent mean compositions of metabasalts from the Wippra Zone (Harz Mts.): ◒ Ordovician; ⊖ Middle Devonian; ⦶ Upper Devonian

In the Ti-Zr diagram (Fig. 3, after Pearce and Gale 1977), the Soonwald metabasalts plot in the within-plate and ocean-floor fields. Two samples tending toward island-arc compositions are certainly not sufficient to suggest such a tectonic setting; the bulk of the data clearly indicates generation of the lavas at an oceanic ridge.

In the Zr/Y-Ti/Y diagram (Fig. 4, after Pearce and Gale 1977) which is taken to discriminate between intra-plate and plate-margin basalts, the samples show the same partitioning as in Fig. 1: the ones plotting in the intraplate field occur in the alkali-basalt field of Fig. 1, and the samples plotting in the plate-margin field occur in the subalkaline basalt field of Fig. 1. The combination of ocean-floor and divergent plate-boundary characteristics can be taken to indicate a MORB-type character for these samples. This is also apparent from the Ti/100-Zr-3Y triangle (Fig. 5, after Pearce and Cann 1973): the samples interpreted as MORB-type plot in field B, the intraplate basalts in D.

The Zr/4-2Nb-Y diagram (Fig. 6, after Meschede 1986) leads to the same interpretation. In this diagram, three clusters of data are recognizable: two of them are positioned in field A (intraplate basalts), the third lies in field D and corresponds to N-type MOR basalts. Three samples plot in field B (P-type MORB). Some very few plots in marginal parts of fields B and C are regarded as insignificant.

The P_2O_5-TiO_2 diagram after Ridley et al. 1974 (Fig. 7) can prove the occurrence of oceanic island tholeiites (OIT) beside alkalibasalts (ALK) and ocean-ridge basalts (ORB), although the OIT field partially overlaps the ALK- as well as the ORB field. Some samples plot outside the defined fields.

In the diagram for discrimination of alkalibasalts from tholeiites after Winchester and Floyd (1976), all ORB- and nearly all of the OIT type samples from Fig. 7 occur in the tholeiite field, whereas those from the overlap OIT/ALK and the samples outside the defined fields fall in the range of alkali-basalts (Fig. 8).

In the Ti-Zr diagram (Fig. 13) of Pearce (1983), the Soonwald metabasalts plot in the MORB field, or adjacent parts of the field for within-plate lavas. Though the MORB field, in this type of diagram, overlaps the island arc and within plate

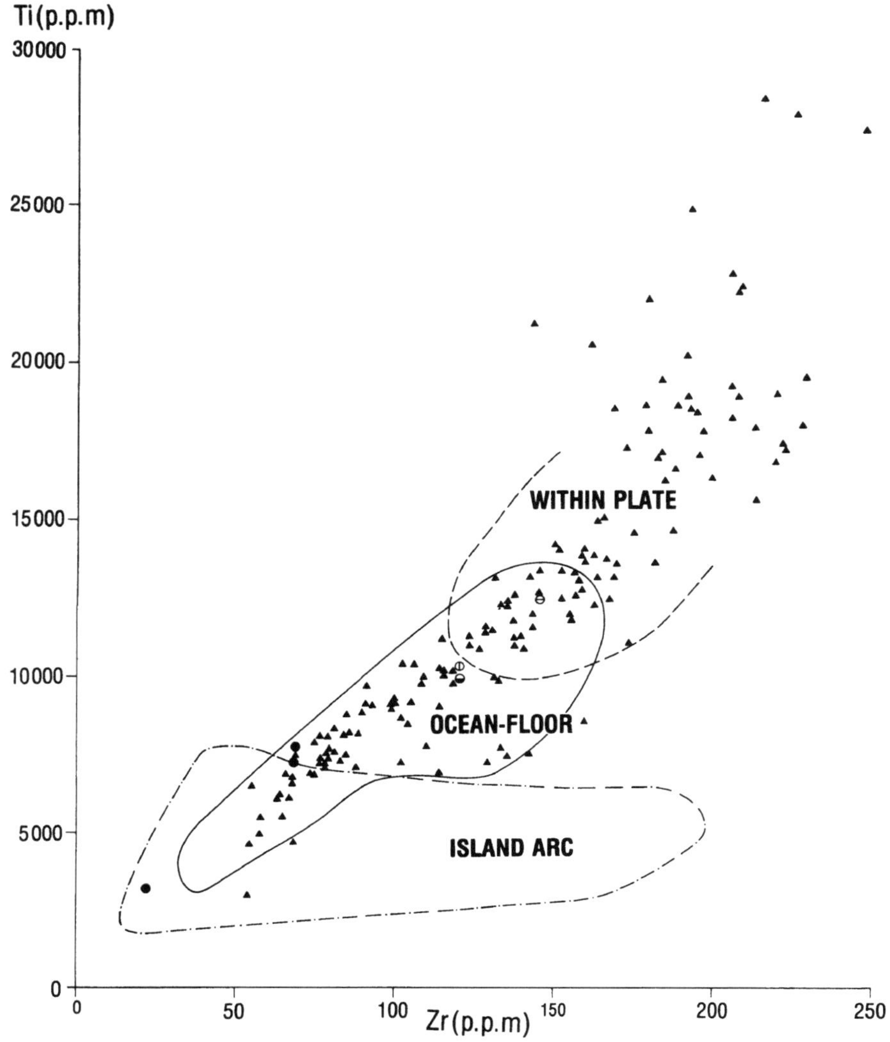

Fig. 3. Metabasalts of the Soonwald (▲) and metabasalts from the South Taunus (●) plotted in the Ti-Zr diagram after Pearce and Gale (1977). The other plots represent mean compositions of metabasalts from the Wippra Zone (Harz Mts.): ◒ Ordovician; ⊖ Middle Devonian; ⦶ Upper Devonian

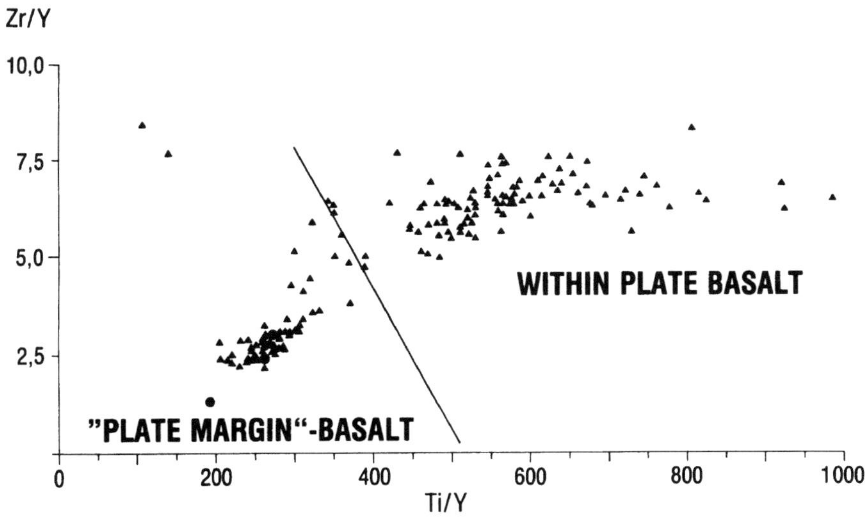

Fig. 4. Metabasalts of the Soonwald (▲) and of the South Taunus (●) in the Zr/Y-Ti/Y diagram after Pearce and Gale (1977)

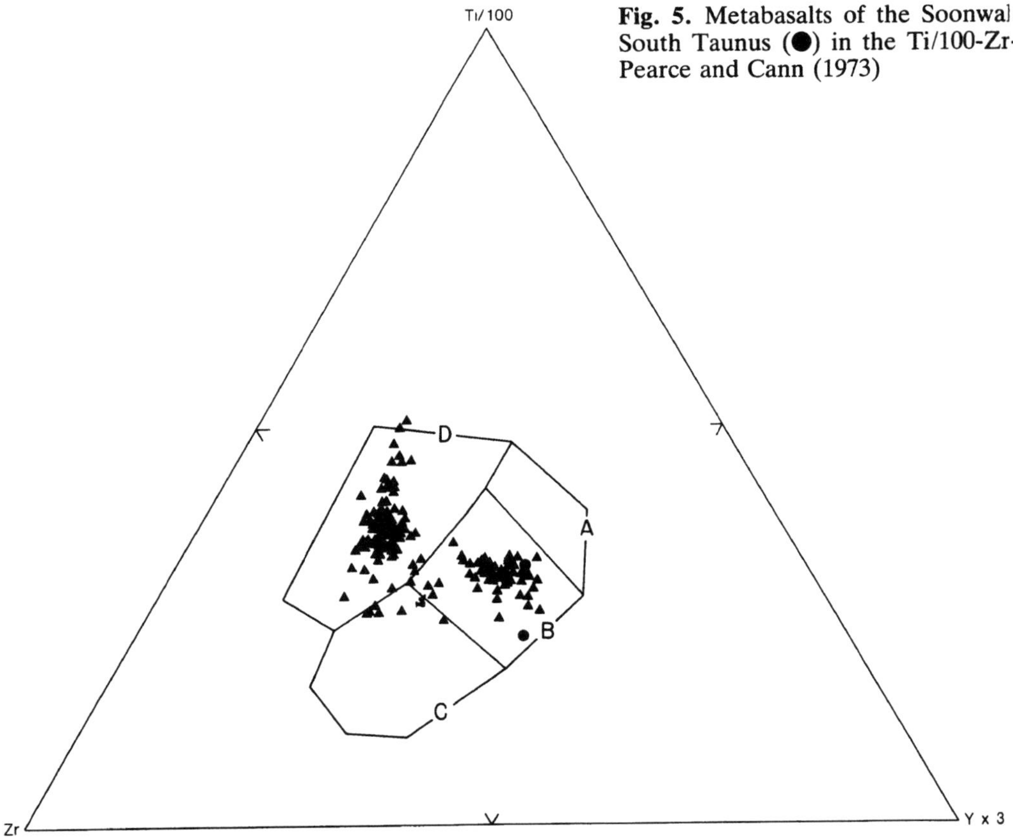

Fig. 5. Metabasalts of the Soonwald (▲) and of the South Taunus (●) in the Ti/100-Zr-3Y diagram after Pearce and Cann (1973)

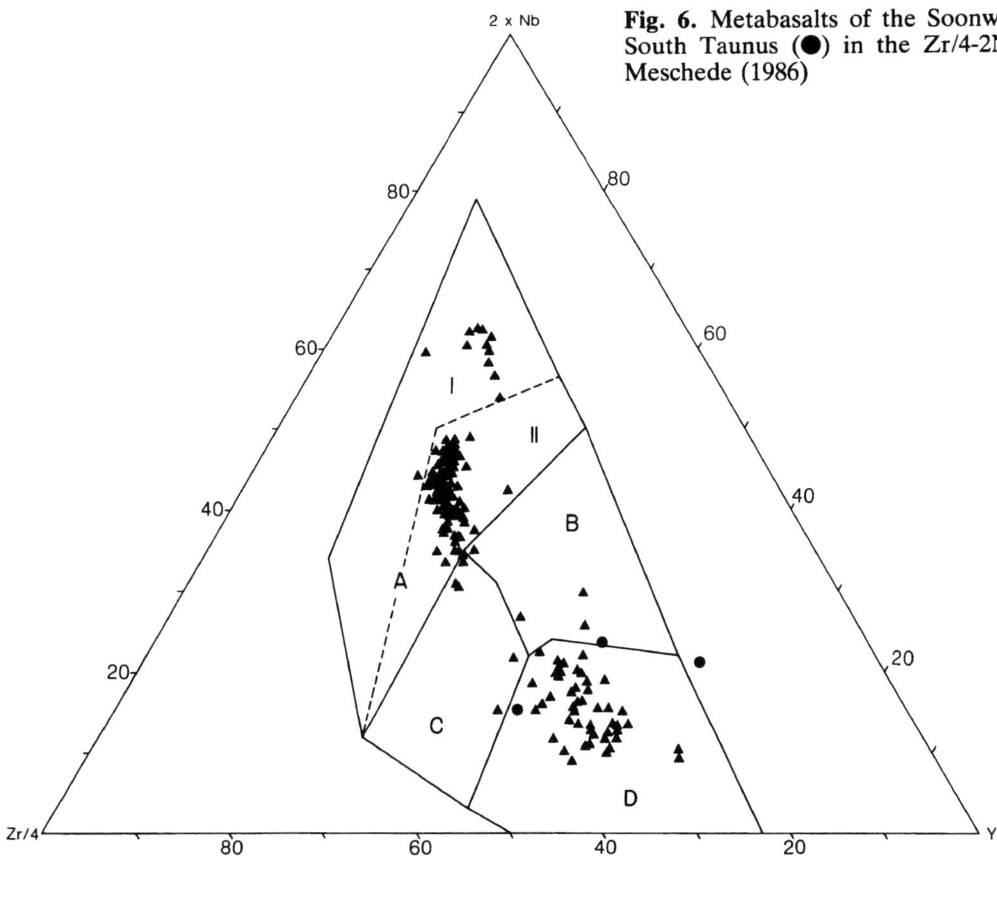

Fig. 6. Metabasalts of the Soonwald (▲) and of the South Taunus (●) in the Zr/4-2Nb-Y diagram after Meschede (1986)

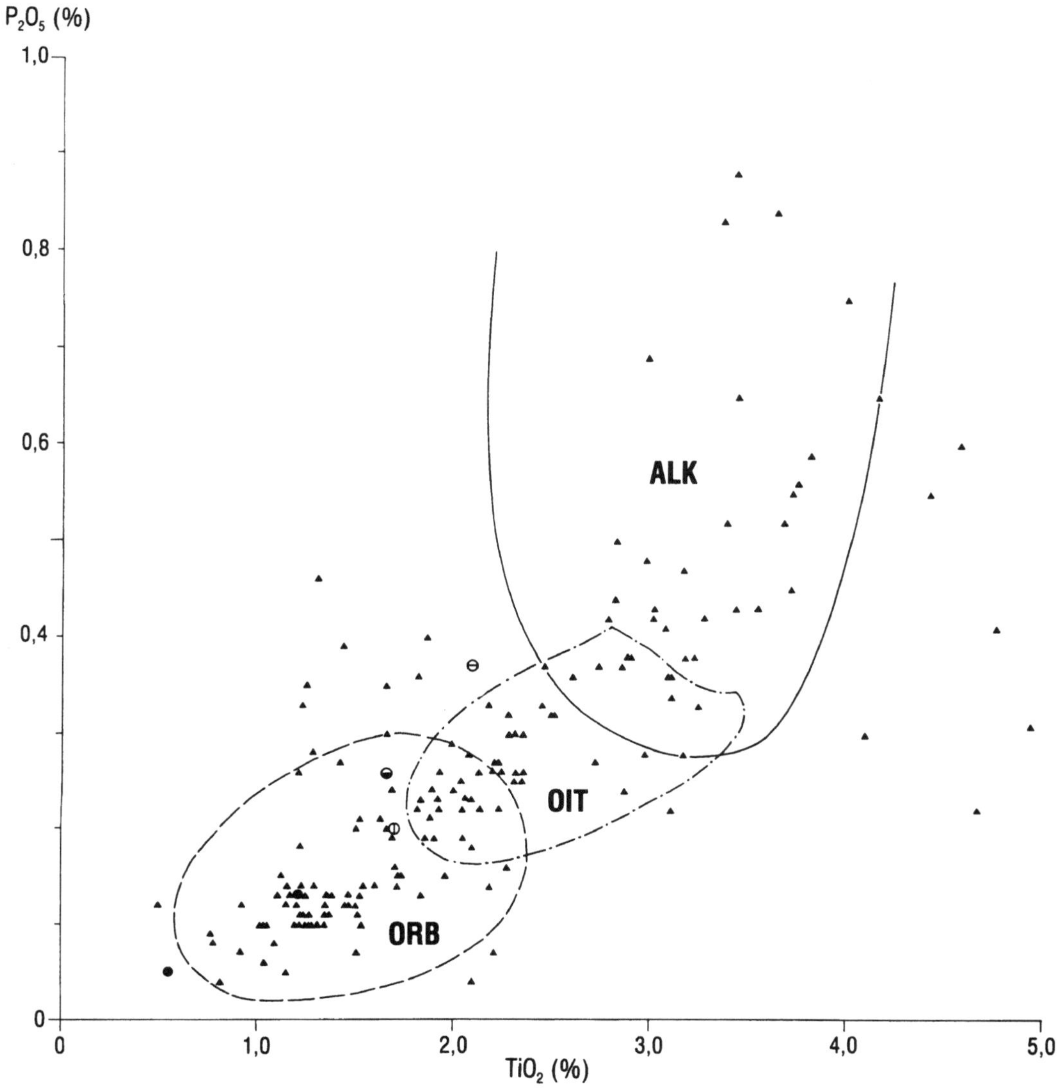

Fig. 7. Metabasalts of the Soonwald (▲) and of the South Taunus (●) in the P_2O5-TiO_2 diagram after Ridley et al. (1974). The other plots represent mean compositions of metabasalts from the Wippra Zone (Harz Mts.): ◒ Ordovician; ⊖ Middle Devonian; ⦶ Upper Devonian

fields, and is, therefore, not sufficient for the identification of MORB type basalts, this plot is well compatible with the MORB affinities basalts, this plot is well compatible with the MORB affinities apparent from the other diagrams.

The two groups apparent from Figs. 1, 4, 5, and 6 are also different in their REE spectra (Fig. 9a,b). The samples resembling intraplate basalts (ocean-island tholeiites and alkali-basalts) show strong enrichment of LREE and, in most cases, positive Eu-anomalies, which suggest accumulation of plagioclase in the rock. Although their position within field D of Fig. 6 identifies them as normal ocean-ridge basalts, the REE spectra of the samples resembling MORB-type rocks are not typical for normal ocean-ridge basalts. They show both negative or positive Eu anomalies, but otherwise flat patterns. Hence, they resemble basalts from "transitional ridge segments".

The relatively high number of samples makes it possible to recognize, in the Soonwald, patterns in the areal distribution of the different types of metabasalts. It appears that narrow bands of MOR-type metabasalts, oriented parallel with the northeast-trending tectonic structures, alternate with zones containing intraplatetypes. This alternation might have been produced by tectonic mixing of volcanic rocks derived from a

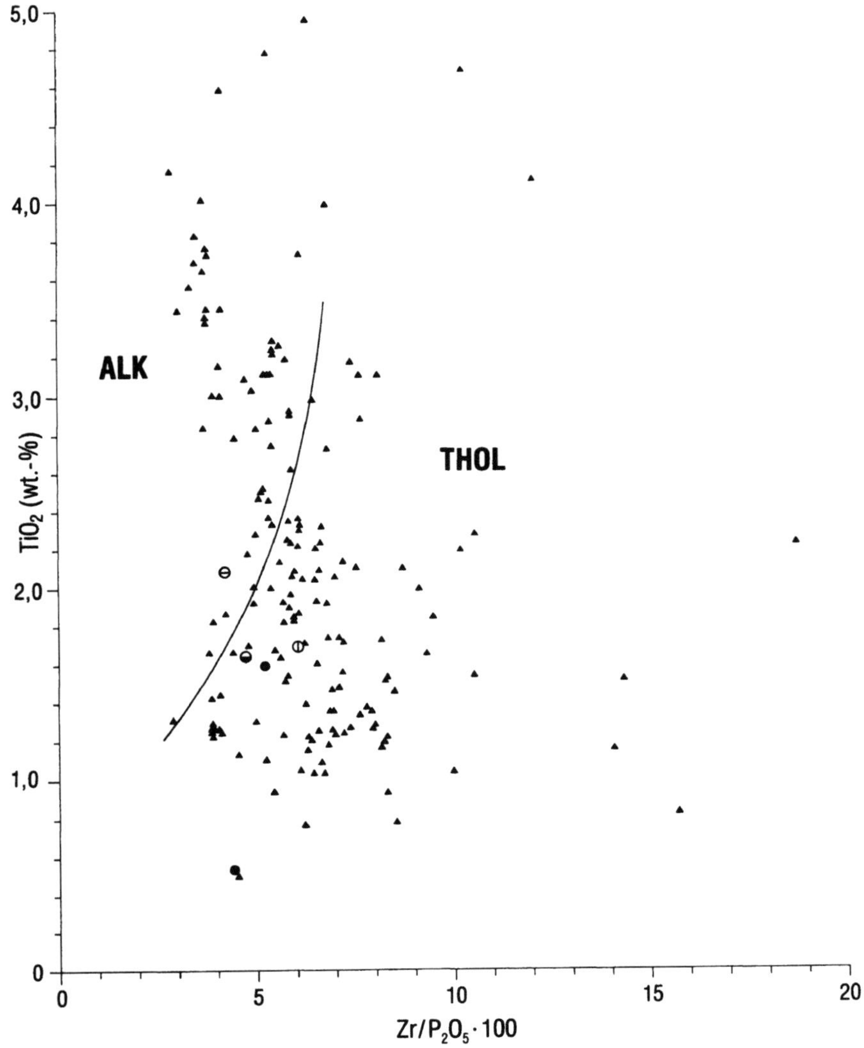

Fig. 8. Metabasalts of the Soonwald (▲) and the South Taunus (●) in The TiO_2-Zr/P2O_5 × 100 diagram after Winchester and Floyd (1976). The other plots represent mean compositions of metabasalts from the Wippra Zone (Harz Mts.): ◒ Ordovician; ⊖ Middle Devonian; ⦶ Upper Devonian

thinned continental margin and the adjacent ocean floor, originally situated at the southern margin of the Old Red Continent (see Franke, Chap. III.B.1).

The occurrence of MOR-type metabasalts in the Phyllite Zone of the Soonwald supports plate tectonic models for the Rhenohercynian Belt. It is possible that the Soonwald metabasalts are close to the root zone, and represent part of, the Gießen-Werra-Harz nappe, which likewise contains MORB-type rocks (Platen et al. 1989; Franke 1989a, 1990a,b; Wedepohl 1990; and the contributions by Franke, Chap. III.B.1, Anderle et al., Chap. III.C.1, this Vol.). Unfortunately, the age of the Soonwald metavolcanic rocks is still a matter of speculation; correlation with the Gießen-Werra-Harz nappe would imply an early Devonian age.

4 Harz Mts. (Wippra Zone)

The metabasalts supposed to be of **Ordovician** age correspond to relatively primitive lavas with a marked tendency toward oceanfloor basalts (Werner 1984). The chondrite-normalized REE patterns clearly show MORB trends (Werner 1986). On the basis of general geological considerations, whose discussion would go beyond the frame of this chapter, Werner has suggested generation of these lavas in an extended ensialic basin. However, the geochemical data rather

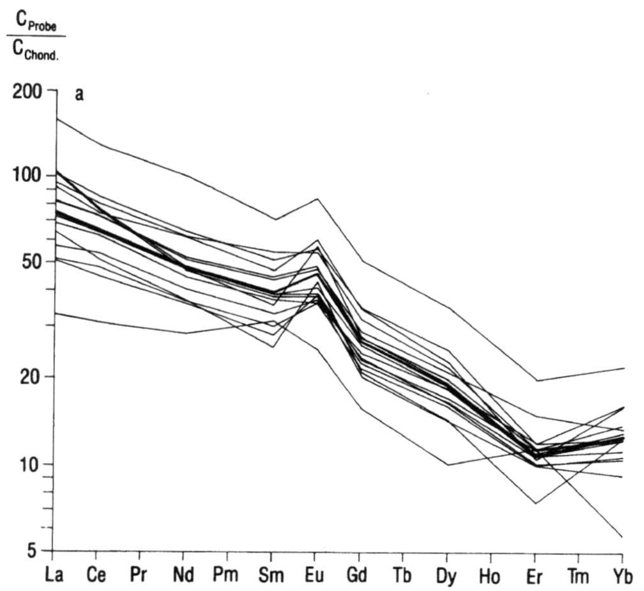

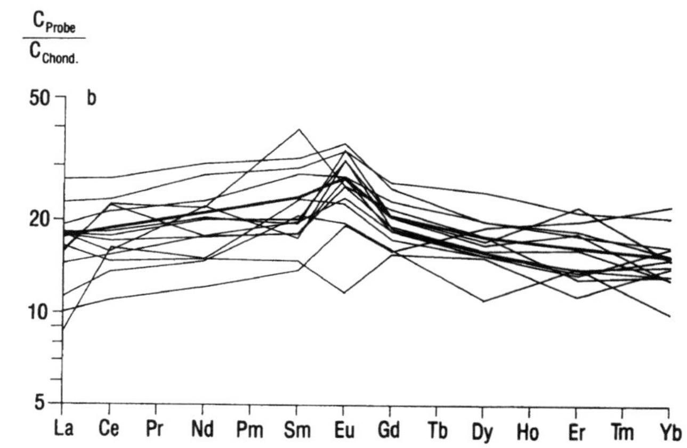

Fig. 9a,b. Chondrite-normalized REE patterns of metabasalts from the Soonwald showing averages for several samples. **a** Within-plate basalts, *thick line* representing the total average of about 100 samples. **b** MOR-type basalts, *thick line* representing the total average of about 60 samples. (Normalized against average composition of C1-chondrite according to Evensen et al. 1978)

indicate a truly oceanic environment. If the stratigraphic interpretation is correct, these rocks might represent part of the Ordovician basin set between E Avalonia (exposed, e.g., in the Rhenohercynian autochthon) and Armorica (?Mid-German Crystalline High, Saxothuringian, Tepla-Barrandian). The existence of such a basin is deduced from paleomagnetic data (see Bachtadse et al., Chap. X, and Franke et al., Chap. XI, this Vol.).

The spectrum of metavolcanic rocks assigned to the **Silurian** comprises tholeiites as well as quartz-normative basalts and are close to the root zone, and represent part of andesites (Werner 1984).

The belt of greenschists in the Wippra Zone, which is supposed to be of **Middle Devonian** age, consists of metabasalts and cogenetic pyroclastic rocks which resemble continental, low-K tholeiites of rather primitive character.

Primitive melts also occur in sequences interpreted as **Upper Devonian**. Their setting is interpreted, by Werner (1984), as continental, but the chemistry of these rocks suggests derivation from ocean-floor basalts.

The analytical data of Werner (1986) have been plotted in the discrimination diagrams of Figs. 2, 3, 7, and 8; but they could not be incorporated into Figs. 4, 5, and 6, because the samples have not been analyzed for Y. The plots represent means of 11 (Ordovician), 10 (Middle Devonian), and 3 (Upper Devonian) metabasalt samples. In Fig. 2, all these mean values plot in the ocean-floor basalt field. In Fig. 3, the three mean plots are contained within the same field, though the mean of the Middle Devonian samples lies in the overlap of the ocean-floor and the within-plate basalt fields. In Fig. 7, the Ordovician and Upper Devonian rocks plot in the ocean-ridge field, whereas the Middle Devonian ones lie outside, yet close to, the ocean-island tholeiite domain. In Fig. 8, the Middle Devonian metabasalts, strangely, occur among the alkali-basalts, whereas the others plot in the tholeiite field.

Like in the South Hunsrück (Soonwald), the affinities with MOR-type basalts observed in the metabasalts supposed to be of Devonian age suggest a probably allochthonous position for these rocks.

5 South Taunus

5.1 Metabasalts

In the South Taunus, metabasalts have been encountered at only two localities: Bubenhäuser Höhe south of Rauenthal (ordnance map sheet no. 5914 Eltville) and Heimlicher Berg 1.5 km NNE of Wiesbaden-Igstadt (5916 Hochheim). Both are intercalated among the metapelites of the Lorsbach Formation, which has yielded Emsian spores (see above).

Except for one of the two samples from the Bubenhäuser Höhe, which plots in Figs. 2, 3, and 13 in the field of island arc basalts, these metabasalts likewise resemble MOR-type basalts, as demonstrated by Figs. 3, 4, 5, 6, 7, and 13. The sample mentioned above plots on Fig. 6 outside the defined fields, but within field B on Fig. 5. In Fig. 1, the metabasalts from both localities appear in the field of subalkaline basalts and were likewise characterized as tholeiites in Fig. 8.

Figure 10 shows the chondrite-normalized REE-spectra of the three samples. They show 8- to 20-fold relative enrichment of REE. The sample from Heimlicher Berg shows the depletion of LREE, typical of N-type MOR basalts. The REE spectra from Bubenhäuser Höhe are less typical. Their Ce/Yb ratios of 0.53 and 0.75 are comparable with the value for the sample from Heimlicher Berg 0.70).

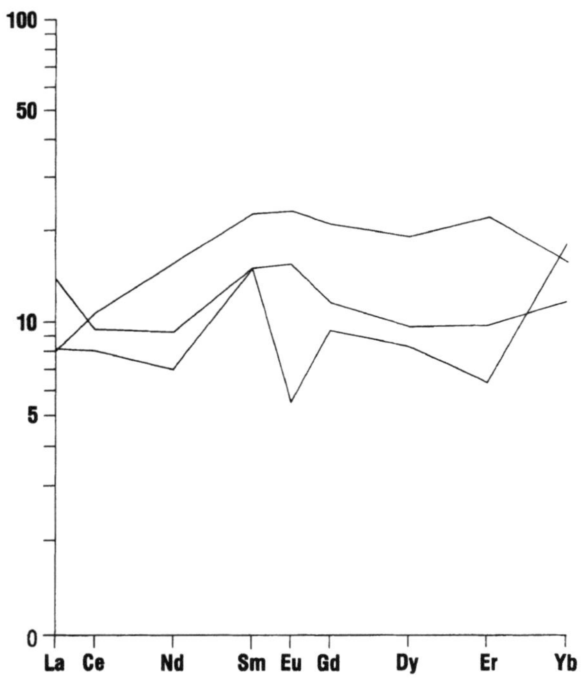

Fig. 10. Chrondrite-normalized REE patterns of metabasalts from the South Taunus. (Normalization as in Fig. 9)

5.2 Intermediate and Felsic Metavolcanic Rocks

Most of the metavolcanic rocks of the South Taunus are felsic in character (Serizitgneis, Felsokeratophyr) or intermediate (Grünschiefer = greenschist, Keratophyr). In order to allow cross references with the German literature, the present chapter uses the terms coined by early field geologists. Both Serizitgneis and Felsokeratophyr are metarhyolites or metarhyodacites, but Serizitgneis has a tectonic foliation, whereas Felsokeratophyr is massive. The same distinction by the tectonic fabric is used for the intermediate volcanics (Grünschiefer is foliated, Keratophy massive).

Porphyric fabrics, which are frequently found in the felsic types, are proof of the volcanic nature of the protoliths. The phenocrysts are albitized plagioclase (Ab98.5 Or1.0 An0.5), K-feldspar (Or97 Ab3), quartz, and pseudomorphs after biotite. In the intermediate members, the porphyric fabric is less pronounced, and the phenocrysts are smaller. In many cases, the radial or trachytic texture of the matrix feldspars has been preserved. Occasional vesicles are filled with albite, quartz, or epidote. Phenocrysts in the intermediate metavolcanics consist of albitized plagioclase, and subordinate K-feldspar, quartz, or pseudomorphs after clinopyroxene. One single relict clinopyroxene (among 148 samples) has been determined as augite (Wo 39.8, En 48.8, Fs 12.1).

According to their fabrics, the bulk of these rocks have been extruded as lavas. In a felsic to intermediate volcanic suite, one would expect a high proportion of pyroclastic rocks. However, indications of pyroclastic protoliths are extremely rare, and unequivocal evidence of ignimbrite texture has been found in only one single case (Meisl 1970, Fig. 56). It cannot be excluded that the relevant textures have been destroyed during deformation and metamorphism.

Taking into account some changes in the bulk chemistry during low-grade metamorphism, the relict magmatic minerals indicate derivation from rhyolitic or rhyodacitic (Serizitgneis, Felsokeratophyr) and andesitic to trachytic (greenschist, Keratophyr) protoliths (Hentschel and Meisl 1966). This is grossly corroborated by the immobile trace element data.

In Fig. 1, greenschists and Keratophyrs plot in the lower part of the rhyodacite + dacite field, the andesite field, and the trachyandesite field.

Hence, they can be identified as metaandesites, -latite-andesites, and -dacites. Only five stray samples plot in the basalt fields (though they are clearly not basalts). The Serizitgneis and Felsokeratophyr samples mainly plot in the upper part of the rhyodacite + dacite and in the rhyolite field. They represent metarhyolites and -rhyodacites.

The REE spectra of the meta-andesites and - dacites show strong enrichment of the LREE (Fig. 11a,b), and, in a few cases, negative Eu-anomalies. The metarhyolites and -rhyodacites are characterized by 50- to 200-fold enrichment of the LREE, by pronounced negative Eu-anomalies (plagioclase fractionation), and a moderate (range 10–60, average 40-fold) enrichment of the HREE relative to chondrite (Fig. 12a–f). The REE data are also compatible with derivation from high-K lavas; however, this hypothesis cannot be put to the test, since potassium has demonstrably been mobile during metamorphism.

Following Pearce (1983), the immobile trace elements Ti and Zr suggest a plate-tectonic setting for the intermediate and felsic volcanic rocks. With the exception of the metabasalt from Heimlicher Berg, which plots in the MORB field, all metavolcanic rocks of the South Taunus are within the normal range of arc lavas (Fig. 13).

The genetic relationships between the different types of metavolcanic rocks in the South Taunus are still not clearly understood. Since they are more or less of the same age, they could well represent members of a differentiated sequence.

Associations of andesites, dacites, and rhyolites with some basalt are usually interpreted as representing an island arc or active continental margin. Until recently, this interpretation has been questioned, because the rocks of the Northern Phyllite Zone were interpreted as metamorphosed equivalents of the very-low-grade Rhenohercynian belt to the north, which only records early to middle Devonian rifting and late Devonian to early Carboniferous convergence. The late Ordovician/Silurian magmatic rocks contained in the Phyllite Zone and also in the Mid-German Crystalline High (see Okrusch and Schubert, Chap. IV.D, this Vol.) have, therefore, been interpreted as an early part of the extensional history of the Rhenohercynian basin, which, by the early Devonian, attained an oceanic

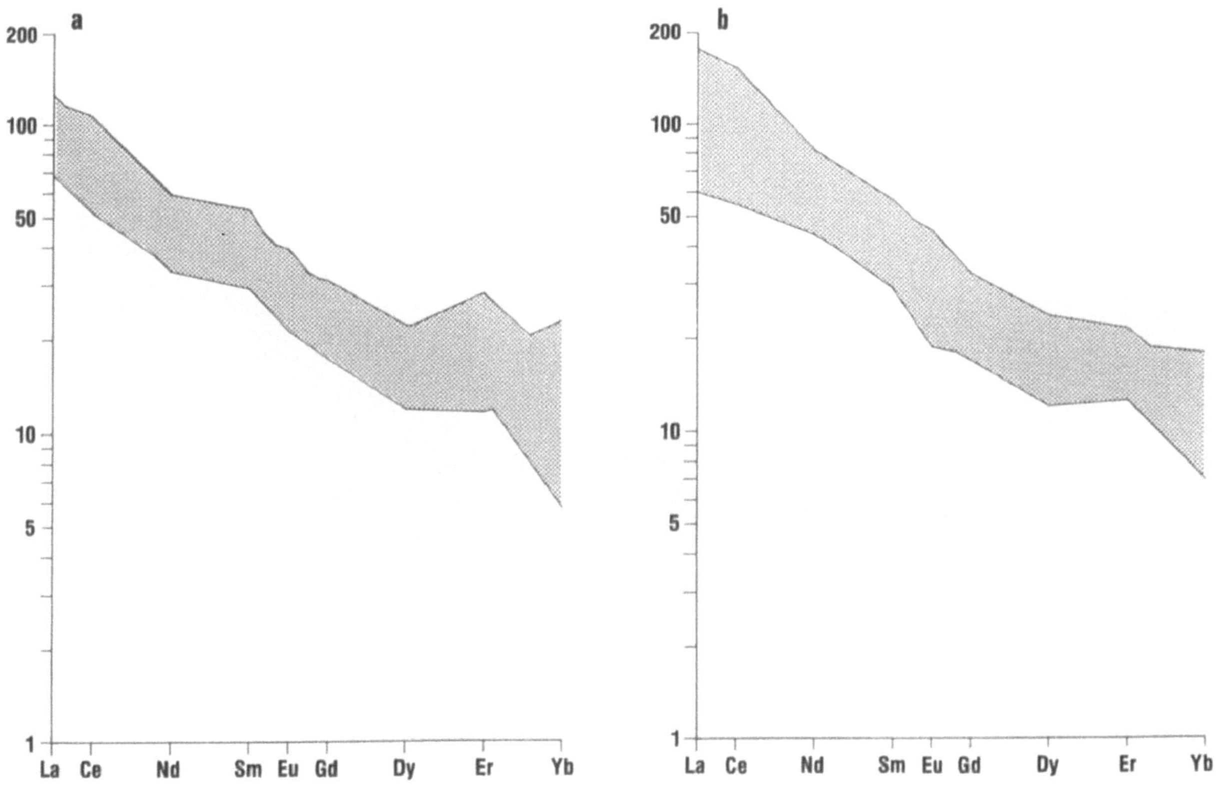

Fig. 11a,b. Ranges of chrondrite-normalized REE patterns of **a** greenschists, 25 selected samples, and **b** Keratophyr samples. (Normalization as for Fig. 9)

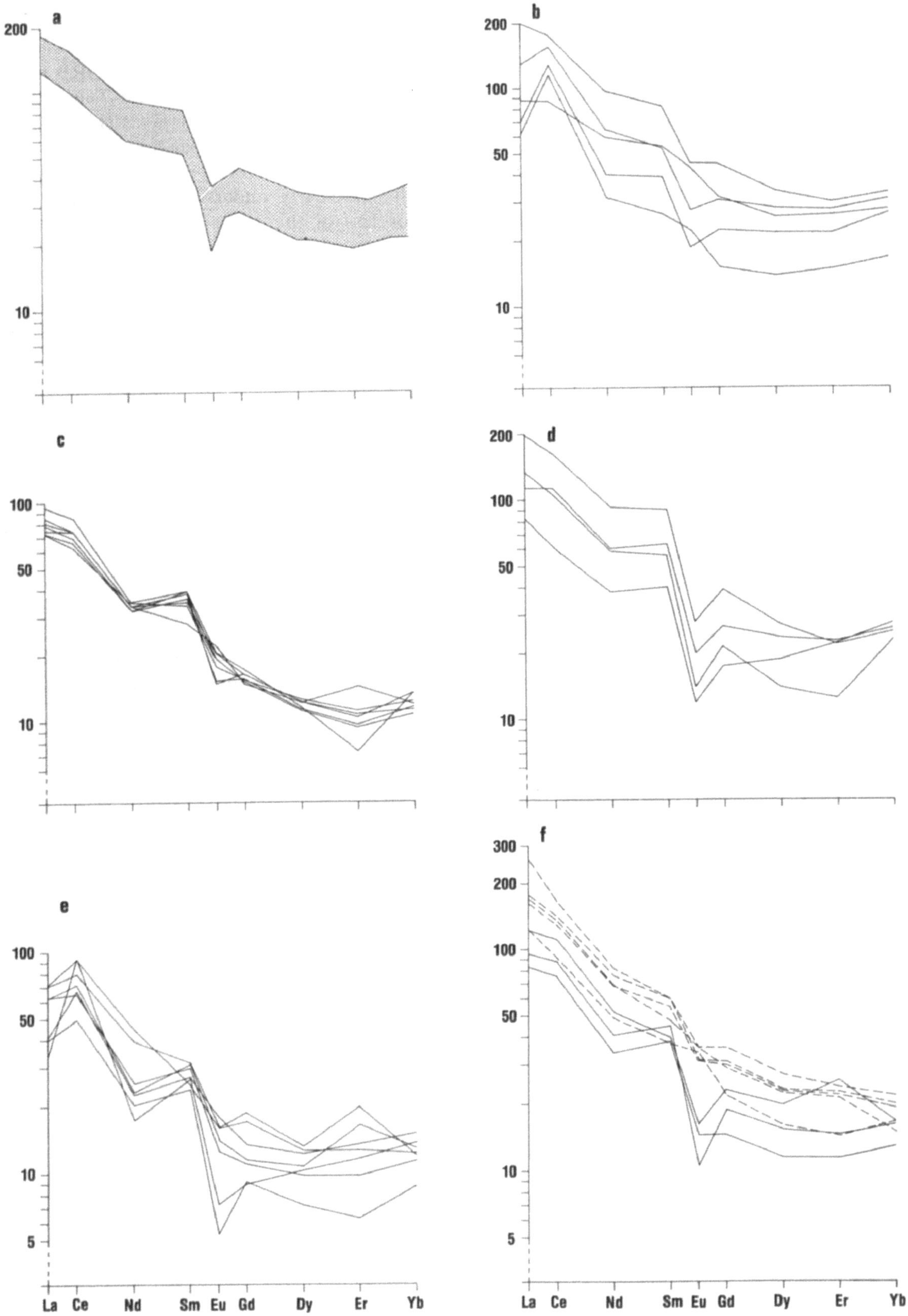

Fig. 12a–f. Chrondrite-normalized REE patterns of Serizitgneis (**a**, **b** 25 selected samples) and Felsokeratophy (**c**, **d**, **e**, **f**. **c** Type Holstein-Klippe. **d** Type Kitzelberg. **e** Type Dachsbau. **f** Type Ochsenhaag; *broken line*: Nonnenbuchwald), all in the South Taunus. (Normalization as in Fig. 9)

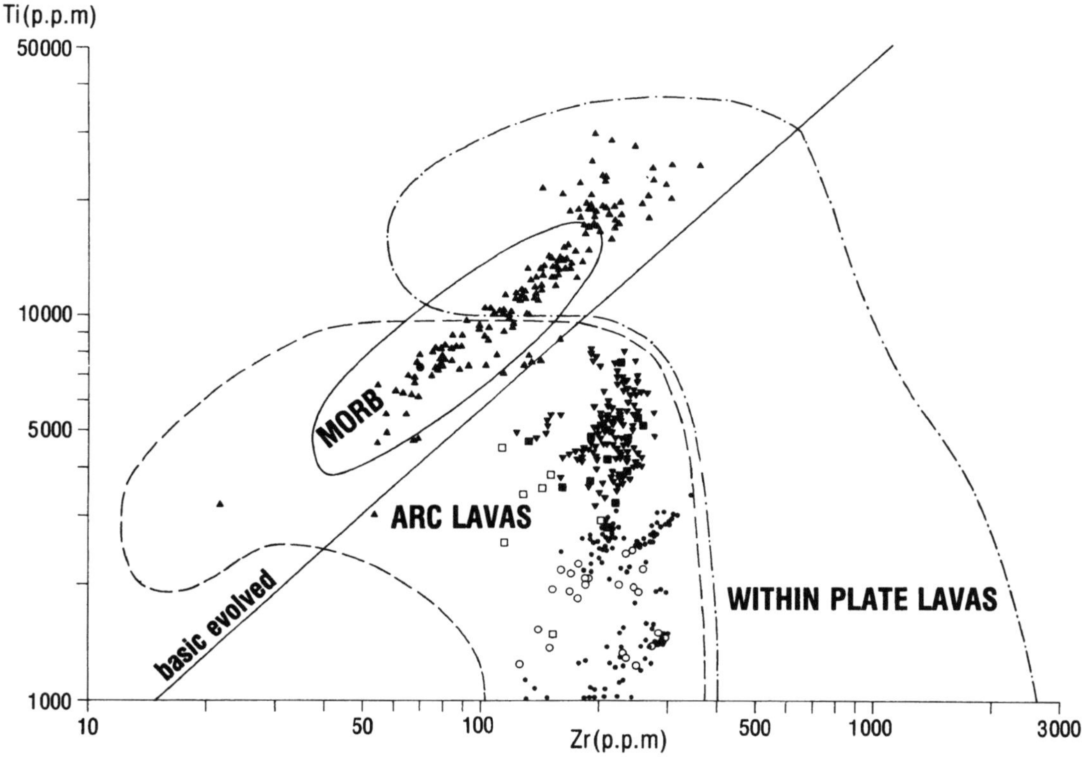

Fig. 13. Metavolcanic rocks of the South Taunus and of the Soonwald in the Ti-Zr diagram of Pearce (1983). Symbols South Taunus: ● metabasalt; ▼ greenschist; ■ Keratophyr; □ silicified Keratophyr; • Serizitgneis; ○ Felsokeratophyr. Symbol Soonwald: ▲ metabasalt

stage (Lizard-Gießen-Harz ocean, see Franke 1989a,b; Franke and Oncken 1990). However, the paleomagnetic scenario proposed in the contribution by Bachtadse et al. (Chap. X, this Vol.) implies opening and closure, within the Ordovician, of an oceanic basin in this area. The felsic to intermediate rocks encountered in the South Taunus might well represent the late Ordovician/Silurian closure of this basin.

6 Conclusions

The metavolcanic rocks of the Soonwald Mts. (South Hunsrück) and of the Wippra Zone in the Harz Mts. are metabasalts. In both regions, there are frequent occurrences of ocean-floor basalts with the geochemical characteristics of MORB.

In the Soonwald Mts., bands of MORB alternating with bands of intraplate basalts (oceanic island tholeiites and alkali-basalts) probably represent a tectonic mix. We are possibly dealing here with thinned continental and oceanic crust of Devonian age, at the southern, passive margin of the Old Red Continent.

In the South Taunus Mts. there are only two small intercalations of metabasalts in the Lower Devonian Lorsbach Formation, both showing MORB trends. As in the South Hunsrück, these metabasalts possibly represent parts of the early Devonian ocean floor contained in the Gießen-Werra-Harz nappe.

The metabasalts of the Wippra Zone are supposed to be partly of Ordovician and Silurian, partly of Middle and Upper Devonian age. Ordovician and Upper Devonian metabasalts clearly show MORB-trends. The stratigraphic age of all these rocks is somehow doubtful, but it can be stated that Ordovician MORB-type metabasalts are well compatible with the paleomagnetic evidence of an Ordovician ocean in the area (see Bachtadse et al., Chap. X, this Vol.).

Devonian MORB-type rocks in the Wippra Zone could likewise be interpreted as part of the Gießen-Werra-Harz nappe, but their assumed late Devonian age does not fit this concept, because the oceanic crust in the nappe is known to be pre-middle Devonian, and the late Devonian is already represented by flysch turbidites which indicate plate convergence (see Franke, Chap. III.B.1, this Vol.).

The South Taunus is dominated by meta-volcanic rocks of felsic and intermediate compositions. According to zircon-ages of 426 + 14/−15, 433 + 9/−7 and 442 ± 22 Ma, this volcanism took place from the Middle Ordovician until the early Silurian. The trace element data indicate an island arc or active margin setting, compatible with the opening and closure of an oceanic realm deduced from paleomagnetic data (Bachtadse et al., Chap. X, this Vol.).

References

Edge RA, Ahrens LH (1962) The determination of Sc, Y, Nd, Ce, and La in silicate rocks by a combined cation exchange-spectrochemical method. Anal Chim Acta: 355–362

Evensen NM, Hamilton PJ, O'Nions RK (1978) Rare earth abundances in chondritic meteorites. Geochim Cosmochim Acta 42: 1199–1212

Franke W (1989a) Tectonostratigraphic units in the Variscan Belt of Europe. Geol Soc Am Spec Pap 230: 67–90

Franke W (1989b) Variscan plate tectonics in Central Europe – current ideas and open questions. Tectonophysics 169: 221–228

Franke W (1990a) Rhenohercynian Basin and Mid-German Crystalline Rise – an introduction. Field guide to pre-conf excursion. Int Conf Paleozoic orogens in central Europe. Geol Geophys, Mid-German Crystalline Rise and Rheinisches Schiefergebirge, pp 1–16

Franke W (1990b) Lahn Syncline and Giessen Nappe. Field guide to pre-conf. excursion. Conf Pal Orog Centr Eur Geol Geophys, Mid-German Crystalline Rise and Rheinisches Schiefergebirge, pp 149–156

Franke W, Oncken O (1990) Geodynamic evolution of the north-central Variscides – a comic-strip. In: Freeman R, Giese P, Mueller St (eds) The European geotraverse: integrative studies. Results 5th Study Centre, Rauischholzhausen, Eur Sci Found, Strasbourg, pp 187–194

Hentschel H, Meisl S (1966) Exkursion in das magmatogne Vordevon des Taunus. Fortschr Mineral 42: 321–333

Leoni L, Saitta M (1976) X-ray fluorescene analysis of 29 trace elements in rocks and mineral standards. Redinconti Soc Ital Mineral Petrol 32: 497–510

Meisl S (1970) Petrologische Studien im Grenzbereich Diagenese Metamorphose. Abh Hess Landesamtes Bodenforsch 57: 1–93

Meisl S (1990) Metavolcanic rocks in the "Northern Phyllite Zone" at the southern margin of the Rhenohercynian Belt – Field guide to pre conf excursion. Int Conf Paleozoic orogens in central Europe. Geol Geophys, Mid-German Crystalline Rise and Rheinisches Schiefergebirge, pp 25–42

Meschede M (1986) A method of discriminating between different types of mid-ocean ridge basalts and continental tholeiites with the Nb-Zr-Y diagram. Chem Geol 56: 207–218

Pearce JA (1983) A users guide to basalt discrimination diagrams. Open Univ, Milton Keynes, pp 1–39 (Unpubl Rep)

Pearce JA, Cann JR (1973) Tectonic setting of basic volcanic rocks determined using trace element analysis. Earth Planet Sci Lett 19: 290–300

Pearce JA, Gale GH (1977) Identification of ore-deposition environment from trace-element geochemistry of associated igneous host rocks. Geol Soc Lond Spec Publ 7: 14–24

Platen K, Emmermann R, Franke W (1989) Devonische Mortyp Metabasalte im Rhenoherzynikum (abs.). 5. Rundgespräch Geodyn Eur Variszikums, Braunschweig, 42 pp

Reitz E (1989) Devonische Sporen aus Phylliten vom Südrand des rheinischen Schiefergebirges. Geol Jahrb Hessen 117: 23–35

Ridley WI, Rhodes JM, Reid AM, Jakes P, Shih C, Bass MN (1974) Basalt from Leg 6 of the Deep-Sea Drilling Project. J Petrol 15: 140–159

Schnetger B (1990) Bestimmung der Seltenen Erden mit ICPAES. In: Heinrichs H, Herrmann AG (eds) Prakt Anal Geochem: 498–504

Schwab M (1984) The Harz Mountains. In: Guidebook of excursion. PCIX, SC 3: 30–70

Schwab M (1989) Kenntnisstand und Probleme zur Geologie der Wippraer Zone im Harz (abs.). 5. Rundgespräch Geodyn Eur Variszikums, Braunschweig, pp 46–47

Sommermann AE, Meisl S, Todt W (1990) U-Pb-Alter von Zirkonen aus Metavulkaniten des Südtaunus. Fortschr Mineral 68: 244

Sommermann AE, Meisl S, Todt W (1992) Zirkonalter von drei verschiedenen Metavulkaniten aus dem Südtaunus. Geol Jahrb Hessen 119

Stenger B (1961) Stratigraphische und gefügetektonische Untersuchungen in der metamorphen Taunus-Südrand-Zone (Rheinisches Schiefergebirge). Abh Hess Landesamtes Bodenforsch 36: 1–68

Wedepohl KH (1990) Geochemical and petrological indicators for plate tectonic processes during the formation of the Rhenoherzynian Belt. Field guide to preconf excursion. Int Conf Paleozoic orogens in central Europe. Geol Geophys, Mid-German Crystalline Rise and Rheinisches Schiefergebirge, pp 17–24

Werner CD (1984) Diabases in the Lower Harz. In: Guidebook of excursion. PCIX, SC 3: 71–77

Werner CD (1986) Präkinematische variszische Magmatite. In: Autorenkollektiv (eds) Vergleichende Geochemie basischer Magmatite auf dem Gebiet der Deutschen Demokratischen Republik, Teil 1. Freiberger Forschungsh C 406: 47–148

Winchester JA, Floyd PA (1976) Geochemical magma type discrimination: application to altered and metamorphosed basic igneous rocks. Earth Planet Sci Lett 28: 459–469

Winchester JA, Floyd PA (1977) Geochemical discrimination of different magma series and their differentiations products using immobile elements. Chem Geol 20: 325–343

III.C.4 Metamorphic Evolution

H.-J. Massonne

1 Introduction

Rocks of the Northern Phyllite Zone comprise low-grade metasediments and metavolcanics. They are exposed in the southern Hunsrück (Soonwald), the southern Taunus Mountains and the Wippra Zone of the Harz Mountains as relatively small patches, with an area of some square kilometers, at the southern boundary of the Rhenohercynian Zone. In addition, some boreholes between the Taunus and Harz Mountains delivered metamorphic rocks of this zone.

Particularly, numerous metavolcanics with basaltic to rhyolitic protoliths (Meisl 1990) show low-variance mineral assemblages. Among them are: quartz-albite-K-feldspar-phengite-chlorite-amphibole-epidote-titanite-hematite-calcite in metaandesites of the southern Taunus Mountains (Anderle et al. 1990) and quartz-albite-phengite-chlorite-amphibole-epidote-pumpellyite-titanite-calcite in metabasalts of the Soonwald (Meisl 1986). Some rocks contain phengite + stilpnomelane (Anderle and Meisl 1974; Meisl 1986) as a critical mineral assemblage, which is not known from rocks of the Rhenohercynian Zone. Amphiboles of crossite and Mg-riebeckite composition, besides Ca-amphiboles, were reported by Meisl (1970) and Peters (1982) in metavolcanics of the southern Taunus Mountains. In contrast, amphiboles in metabasalts of the Rhenohercynian Zone are, in general, actinolites (e.g., Schreyer and Abraham 1978; Tanumihardja et al. 1986). Furthermore, phengites in some rocks of the Northern Phyllite Zone show typically Si contents above 3.4 per formula unit (Massonne and Schreyer 1983; Meisl 1986). Metamorphic biotite is absent in the entire Northern Phyllite Zone.

The metamorphic P-T conditions resulting from the observed mineral assemblages are as follows: Meisl et al. (1982) assumed temperatures of 400 to 450 °C for the southern Taunus Mountains and <400 °C for the Soonwald. However, due to the absence of biotite, the maximum metamorphic temperatures should be less than 350 °C (see Massonne 1991). Because of the widespread greenschist assemblage chlorite-epidote-actinolite, minimum temperatures are above 250 °C. The temperature range of 250 to 350 °C is compatible with the occurrence of sudoite, pyrophyllite, Mn-rich chloritoid, and carpholite in metapelites of the Wippra Zone (Kramm 1978; Siedel and Theye 1993). Early estimates of metamorphic pressures were between 2 and 3 kb (e.g., Meisl et al. 1982). This is supported by the study of Siedel and Theye (1993), at least for Al-rich metapelites of the Wippra Zone. However, the assemblage phengite + stilpnomelane is stable at metamorphic pressures exceeding 3 to 4 kb (Massonne 1991). Rocks with phengites having Si contents above 3.4 per formula unit (p.f.u.), in the presence of K-feldspar, quartz and chlorite or stilpnomelane, experienced minimum pressures as high as 5 kb (see Massonne 1991). Pressures above 8 kb are unlikely, because of the absence of the high-pressure indicator minerals: glaucophane, omphacite, and aragonite.

2 P-T Evolution of Metavolcanics from the Southern Taunus Mountains

Several metavolcanics of the Northern Phyllite Zone with low-variance mineral assemblage were the object of a detailed electron microprobe study by the present author. A brief description of the results obtained from rocks of the southern Taunus Mountains follows.

Representative compositions of minerals from two metaandesites and one metarhyolite are given in Table 1. Amphiboles, particularly in the metaandesites, are chemically zoned. Cores are rich in Mg-riebeckite component. In amphibole rims, actinolite dominates. An intermediate zone is composed of winchite. Phengites show a compositional range as well. The Si contents of phengite in K-feldspar + quartz + chlorite or

Dallmeyer et al. (Eds.)
Pre-Permian Geology of Central and Eastern Europe

Table 1. Structure formulas of minerals in metavolcanics of the Southern Taunus Mountains

	Phengite			Stilpnomelane		Feldspars	
	860909-6-53	14444-16′	-15′	Ex87-17-5	14444-26	860909-6-25	-23
Si	7.023	7.053	6.796	8.051	7.942	2.981	2.995
Al	3.635	3.589	3.673	1.118	1.310	0.999	1.004
Ti	0.004	0.004	0.007	0.001	0.001	–	–
Fe^{3+}	0.112	0.097	0.520	0.120	0.174	0.023	0.013
Fe^{2+}	0.450	0.728	0.437	2.747	4.039	–	–
Mn	0.004	0.013	0.019	0.331	0.206	–	–
Mg	0.872	0.616	0.648	2.632	1.327	–	–
Ca	–	–	–	0.020	0.019	0.003	–
Na	0.010	0.010	0.037	–	0.055	0.999	0.011
K	1.633	1.770	1.883	0.028	0.356	0.003	0.951
Ba	0.007	–	0.010	0.038	0.011	–	0.004

	Amphibole					
	860909-6-45	-41	-49	14444-11′	Ex87-17-19	-24
Si	7.885	7.768	7.868	7.901	7.932	7.940
Al	0.390	0.336	0.121	0.178	0.340	0.095
Fe^{3+}	1.196	0.934	0.225	0.059	1.387	0.150
Fe^{2+}	1.112	0.919	1.196	3.095	1.073	1.116
Mn	0.026	0.037	0.052	0.092	0.071	0.110
Mg	2.390	3.027	3.537	1.674	2.194	3.584
Ti	–	–	0.001	–	0.003	0.004
Ca	0.561	1.170	1.843	1.897	0.388	1.806
Na	1.510	0.809	0.202	0.120	1.594	0.218
K	0.007	0.004	0.011	0.041	0.011	0.013
Ba	0.002	–	0.008	0.002	0.011	0.012

	Chlorite		Hematite	Epidote		Titanite	
	860909-6-16	Ex87-17-40	Ex87-17-17	860909-6-18	14444-7	860909-6-19	14444-4
Si	5.829	5.695	–	3.000	3.000	1.000	1.000
Al	4.536	4.517	–	1.923	2.231	0.124	0.206
Ti	–	0.002	0.016	–	0.005	0.777	0.769
Fe^{3+}	–	0.089	1.969	1.033	0.747	0.097	0.024
Fe^{2+}	3.597	3.352	0.012				
Mn^{3+}	–	–	–	0.014	0.007	0.003	0.001
Mn^{2+}	0.086	0.201	0.001				
Mg	5.855	6.145	–				
Ca				1.984	1.952	0.959	1.004
Na				0.003	0.009	0.001	0.005

(860909-6: metaandesite, moat of the castle of Falkenstein; Ex87-17: metaandesite, road cut 1 km south of Schlangenbad; 14444: metarhyolite, quarry near hill top Fischbacher Kopf) calculated from representative electron microprobe analyses as follows: 42 val. + Na + K + Ba/2 and 12.1 cations for double formula unit of phengites; 47.375 val. + Na + K + (Ba + Ca)/2 and 15 cations for stilpnomelane; 16 val. for feldspars; 46 val. and 15 cations without Ca + Na + K + Ba for amphibole; 56 val. and 20 cations for chlorite; 8 val. and 3 cations for hematite; 3 Si p.f.u. for epidote; 1 Si p.f.u. for titanite.

stilpnomelane containing metavolcanics lie between 3.35 and 3.55 p.f.u. The decrease of the Si contents in phengites is accompanied by increasing contents of ferri-muscovite. Phengites with the highest Si contents are related to early metamorphic stages, whereas low-Si phengites are probably late metamorphic products.

The chemical trends of amphibole (Fig. 1) and phengite were used to derive the metamorphic P-T path I of Fig. 2. For this purpose, thermochemical calculations were undertaken applying the GE0-CALC software package of Brown et al. (1988) and the thermodynamic data set of Berman (1988) augmented by data for mica and amphibole components (Table 2). The $\Delta G_R = 0$ relation was solved for equilibria relevant to the metaandesites investigated here, with quartz, albite (activity = 0.97), K-feldspar

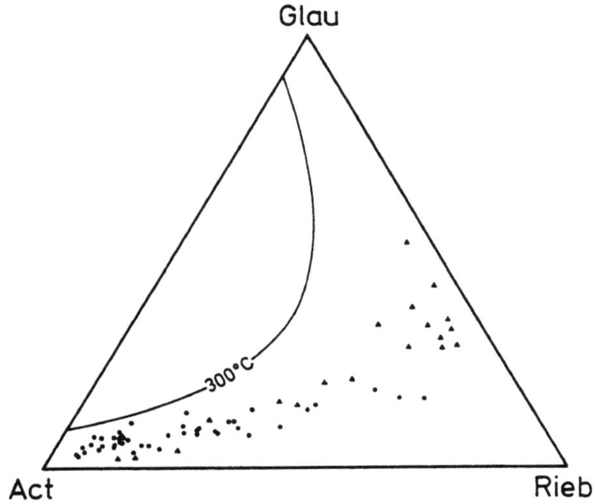

Fig. 1. Electron microprobe analyses of amphiboles from three metaandesites of the Southern Taunus Mountains (*closed circles* samples 860909-6 and Ex87-17: see Table 1; 14453: sample locality as for Ex87-17) plotted in a ternary diagram with respect to the Ca (Act) and Na (Glau and Rieb) content of the M4 site as well as the Fe^{3+} (Rieb) and Al (Glau) content. *Closed triangles* mark compositions of amphiboles from high-P/T metabasites of northeastern Oman reported by El-Shazly et al. (1990) for their metamorphic zone A. The 300 °C line is the solvus at this temperature marking the miscibility gap for amphiboles of the ideal system tremolite – glaucophane – Mg-riebeckite as can be calculated from activity formulations of Massonne (1992)

(a = 0.97), clinochlore component in chlorite (a ≈ 0.08 according to the activity formulation $X_{Si}{}^4 \cdot X_{Mg}{}^5 \cdot 3.1605$), hematite (a = 0.984), muscovite and Mg-Al-celadonite components in phengite as well as glaucophane and Mg-riebeckite components in amphibole. Formulations for the molar fractions and the activities of mica and amphibole components were given by Massonne (1992), e.g., $a_{Glau} = \{Na^{M4} \cdot [Al^{VI}/(Al^{VI} + Fe^{3+})]/2\}^2 \cdot \gamma_1 \cdot OH/(OH + F) \cdot \square \cdot \gamma_2 \cdot Mg/(Mg + Fe^{2+} + Mn)$. The resulting activities of the relevant amphibole and mica components are displayed in Fig. 2. Additional equilibria with H_2O were calculated to determine the water activities (see Massonne 1992) for various stages of the P-T path.

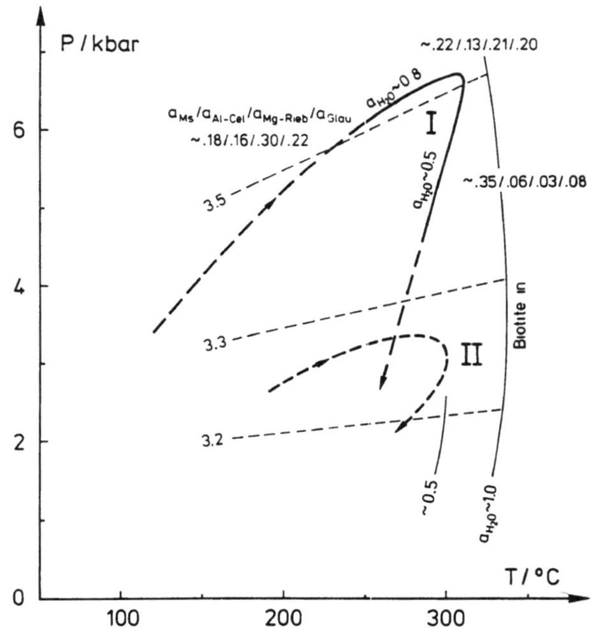

Fig. 2. Metamorphic P-T paths (*solid* corroborated by thermodynamic calculations with activities for amphibole and phengite components as given in this figure; *broken* estimated as discussed in the text) for the metavolcanic unit (*I*) and the metasedimentary unit (*II*) of the Northern Phyllite Zone of the Southern Taunus Mountains. The curves for the formation of biotite (+ potassic white mica, quartz and H_2O) from chlorite + K-feldspar considering different H_2O activities are shown by the *thin solid lines* according to Massonne (1991). *Thin broken lines* indicate calculated Si isopleths for phengite coexisting with chlorite, K-feldspar, quartz, and H_2O (see Massonne 1991). The *thin lines* are generally related to intermediate Mg to Fe^{2+} ratios in the corresponding minerals considering $K_D(Fe^{2+}/Mg) = 2$ for the chlorite-phengite pair

Table 2. Thermodynamic standard state properties of mica and amphibole components

		H_f J/mol	S J/K/mol	V J/b/mol	c_P	a_P, β_T
Muscovite (Ms)	$KAl_2[AlSi_3O_{10}](OH)_2$	−5976740(1)	293.157(1)	14.080(2)	(1)	(1)
Mg-Al-celadonite (Al-Cel)	$KMgAl[Si_4O_{10}](OH)_2$	−5832415(2)	288.527(2)	13.870(2)	(3)	As Ms
Glaucophane (Glau)	$■Na_2Mg_3Al_2[Si_8O_{22}](OH)_2$	−11960500(4)	535.000(4)	26.050(4)	(4)	(4)
Mg-riebeckite (Mg-Rieb)	$■Na_2Mg_3Fe^{3+}{}_2[Si_8O_{22}](OH)_2$	−11065000(5)	602.000(6)	27.130(6)	(3)	As Glau

Sources: (1) Berman 1988; (2) Massonne 1992; (3) Berman and Brown 1985; (4) Evans 1990; (5) Massonne unpubl.; (6) Holland 1989.

The prograde metamorphic stages (see Fig. 2) are characterized by high P/T ratios and relatively high water activities, which point to the presence of a H_2O-rich fluid phase. Peak P-T conditions are close to 6.5 kb and 300 °C. At this stage, amphiboles of winchite compositions and phengites with Si contents close to 3.5 p.f.u. appear in the investigated metaandesites. Such phengites occur in the metarhyolites, as well. A penetrative foliation probably formed at the peak P-T conditions. Na-rich amphiboles predate the corresponding deformation event. Exhumation of the metavolcanics is accompanied by a slight temperature decrease, a further deformation event leading to crenulation cleavage, and a decrease of the H_2O activity. The latter is probably caused by CO_2-bearing fluids leading to the formation of calcite + quartz + TiO_2 from titanite, which is frequently decomposed along grain boundaries.

3 Two Baric Rock Types Resulting from Phengite Barometry

X-ray powder diffractograms of metavolcanics frequently show a moderate to strong basal reflection (hkl = 0 0 10) of $2M_1$-phengite in the range d = 1.984 – 1.989. These d-values were reported by Massonne and Schreyer (1986) from (0 0 10)-reflections of synthetic $2M_1$-micas of the muscovite – Mg–Al–celadonite series with Si contents between 3.43 and 3.58 p.f.u. This coincidence of synthetic and natural phengites with similar Si contents is due to low contents of paragonite component in the natural phengites (see Table 1).

In contrast to the metavolcanics, many metapelites of the southern Taunus Mountains exhibit (0 0 10)-reflections of phengite, frequently coexisting with K-feldspar + chlorite + quartz, in the range d = 1.993 – 1.997. According to Massonne and Schreyer (1986), this points to phengite Si contents of 3.2 to 3.3 p.f.u. only, which were corroborated by electron microprobe studies. Thus, assuming the presence of an H_2O-rich fluid phase, maximum metamorphic pressures of these rocks were around 3 kb (see Si-isopleths of phengite in Fig. 2) and, hence, clearly lower than those of the metavolcanics. Metamorphic temperatures were estimated to be as high as 300 °C, because the quartz fabric is the result of recrystallization (Anderle et al. 1990). The inferred P-T path could be the one marked II in Fig. 2, which is similar to the path for a metagranodiorite of the Rhenohercynian Zone derived by Massonne (1989).

The result of the application of the above XRD method to many rock samples of the southern Taunus Mountains (see also Sachtleben 1988) is as follows: rock slices of the medium-pressure metavolcanic unit, outcropping up to a few kilometers in length and some 100 meters in width, occur mainly along the boundary between the low-pressure metasedimentary rocks of the Rhenohercynian Zone and the Northern Phyllite Zone. In addition, medium-pressure rock slices are imbedded in metasedimentary rocks of the low-pressure unit of the Northern Phyllite Zone. A contact of both baric rock units is, for instance, exposed in the quarry close to the hill top of the Fischbacher Kopf west of Fischbach (Anderle et al. 1990).

Units of two baric rock types could also be identified by the present author, using the above XRD method, for the Northern Phyllite Zone of the Hunsrück Mountains, the Wippra Zone (see also Siedel and Theye 1993), and in the area made accessible by boreholes. In these regions, the interrelation of metavolcanics and medium-pressure metamorphism is obvious, as well as in the southern Taunus Mountains. In addition, in the southern Hunsrück Mountains, slices composed of former medium-grade metamorphic rocks (e.g., Wartenstein Gneis) were affected by the medium-pressure, low-temperature metamorphism according to Si-rich phengites reported by Meisl (1986).

4 Geodynamic Consequences

The protoliths of the contrasting baric rock units of the Northern Phyllite Zone are of different origin. In the medium-pressure unit, volcanics predominate that erupted during early Silurian times (Sommermann et al. 1990). The dominant rock type in the low-pressure unit is sediments deposited from Ordovician to Devonian (Reitz 1989; Sehnert 1991). In addition, different geodynamic environments can be related to the contrasting baric rock types. Low-pressure metamorphism of the metasedimentary unit is effected by crustal thickening that may only be due to thin-skinned tectonics. However, the prograde metamorphic path I of Fig. 2 is probably the result of the burial of rocks in a subduction zone and their subsequent assimilation within an accretionary wedge overlying this zone. For example, El-Shazly et al. (1990) gave an analogous interpretation of rocks from northeastern Oman (Saih Hatat tectonic window, zone A), which show mineralogical features similar to those of the metavolcanics of the Northern Phyllite Zone, but are clearly younger and part of a much more coherent high-P/T metamorphic terrane. In contrast to the rocks from Oman, the position of the Variscan accretionary wedge is not clear, because the medium-pressure rocks in the Northern Phyllite Zone are allochthonous as a result of major thrusting and dextral strike-slip within the Northern Phyllite Zone (Anderle et al. 1990). Because the subduction event must be, at least, somewhat older than the K/Ar-age of about 325 Ma reported by Ahrendt et al. (1983), the accretionary prism formed can be related to a Variscan magmatic arc that is, for example, exposed in the Odenwald (Willner et al. 1991) lying south of the Taunus Mountains.

Acknowledgments. The author thanks T. Theye, K. Weber, and A. Willner for helpful comments to this chapter.

References

Ahrendt H, Hunziker J, Weber K (1983) K/Ar-Altersbestimmungen an schwach-metamorphen Gesteinen des Rheinischen Schiefergebirges. Z Dtsch Geol Ges 129: 229–247

Anderle H-J, Meisl S (1974) Geologischmineralogische Exkursion in den Südtaunus. Fortschr Miner 51: 137–156

Anderle H-J, Massonne H-J, Meisl S, Oncken O, Weber K (1990) Southern Taunus Mountains. Field guide to pre-conference excursion: Mid-German Crystalline Rise and Rheinisches Schiefergebirge; Conf Paleozoic Orogens in Central Europe – Geology and Geophysics, Göttingen-Giessen, Justus-Liebig Univ Giessen, pp 125–148

Berman RG (1988) Internally-consistent thermodynamic data for minerals in the system Na_2O-K_2O-CaO-MgO-FeO-Fe_2O_3-Al_2O_3-SiO_2-TiO_2-H_2O-CO_2. J Petrol 29: 445–522

Berman RG, Brown TH (1985) Heat capacity in the system Na_2O-K_2O-CaO-MgO-FeO-Fe_2O_3-Al_2O_3-SiO_2-TiO_2-H_2O-CO_2: representation, estimation, and high temperature extrapolation. Contrib Mineral Petrol 89: 168–183

Brown TH, Berman RG, Perkins EH (1988) Ge0-Calc: software package for calculation and display of pressure-temperature-composition phase diagrams using an IBM or compatible Personal Computer. Comput Geosci 14: 279–289

El-Shazly A, Coleman RG, Liou JG (1990) Eclogites and blueschists from northeastern Oman: petrology and P-T evolution. J Petrol 31: 629–666

Evans BW (1990) Phase relations of epidote-blueschists. Lithos 25: 3–23

Holland TJB (1989) Dependence of entropy on volume for silicate and oxide minerals: a review and predictive model. Am Mineral 74: 5–13

Kramm U (1978) Die Metamorphose Mn-reicher Pelite der Wippraer Zone/Unterharz. Fortschr Miner 56 (Beih 1): 67–68

Massonne H-J (1989) Metamorphoseentwicklung des Rhenoherzynikums am Beispiel eines Metagranodiorits aus dem Venn-Stavelot-Massiv. 5. Rundgespräch Geodynamik des europäischen Variszikums, Kurzfassungen, Braunschweig, 36

Massonne H-J (1991) High-pressure, low-temperature metamorphism of pelitic and other protoliths based on experiments in the system K_2O-MgO-Al_2O_3-SiO_2-H_2O. Habilitation Thesis, Ruhr-Univ Bochum, 172 pp

Massonne H-J (1992) Thermochemical determination of water activities relevant to eclogitic rocks. In: Kharaka YK, Maest AS (eds) Water-rock interaction WRI-7. Proc 7th Int Symp, Park City, Utah, pp 1523–1526

Massonne H-J, Schreyer W (1983) A new experimental phengite barometer and its application to a Variscan subduction zone at the southern margin of the Rhenohercynicum. Terra Abstr 3: 187

Massonne H-J, Schreyer W (1986) High-pressure syntheses and X-ray properties of white micas in the system K_2O-MgO-Al_2O_3-SiO_2-H_2O. Neues Jahrb Miner Abh 153: 177–215

Meisl S (1970) Petrologische Studien im Grenzbereich Diagenese-Metamorphose. Abh Hess Landesamtes Bodenforsch 57: 1–93

Meisl S (1986) Mineralogisch-petrographische Exkursion in den Soonwald. Fortschr Miner 64 (Beih 2): 35–95

Meisl S (1990) Metavolcanic rocks in the Northern Phyllite Zone at the Southern Margin of the Rhenohercynian belt. Field guide to pre-conference excursion: Mid-German Crystalline Rise and Rheinisches Schiefergebirge; Conf Paleozoic Orogens in Central Europe – Geology and Geophysics, Göttingen-Giessen, Justus-Liebig Univ Giessen, pp 25–42

Meisl S, Anderle H-J, Strecker G (1982) Niedrigtemperierte Metamorphose im Taunus und im Soonwald. Fortschr Miner 60 (Beih 2): 43–69

Peters M (1982) Petrographische, tektonische und mikrothermometrische Untersuchungen im Südtaunus zwischen Lorsbach, Eppstein, Ehlhalten und Fischbach (Blatt Königstein 5816). Diplom Thesis, Geol Paläontol Inst Univ Göttingen, 124 pp

Reitz E (1989) Devonische Sporen aus Phylliten vom Südrand des Rheinischen Schiefergebirges. Geol Jahrb Hessen 117: 23–35

Sachtleben V (1988) Das "Vordevon" des Falkensteiner Hains bei Königstein/Ts. Diplom Thesis, Geowiss Inst Univ Mainz, 135 pp

Schreyer W, Abraham K (1978) Prehnite/chlorite and actinolite/epidote bearing mineral assemblages in the metamorphic igneous rocks of La Helle and Challes, Venn-Stavelot Massif, Belgium. Ann Soc Geol Belg 101: 227–241

Sehnert M (1991) Zur Biostratigraphie epizonaler Metamorphite der Wippraer Einheit/Unterharz. Zentralbl Geol Paläontol 1991, Teil I: 1366

Siedel H, Theye T (1993) Very low grade metamorphism of pelites in the Wippra metamorphic Zone, Harz Mountains, Germany. Miner Monatsh 1993: 115–132

Sommermann A-E, Meisl S, Todt W (1990) U-Pb-Alter von Zirkonen aus Metavulkaniten des Südtaunus. Ber Dtsch Mineral Ges Beih Eur J Miner 2 (1): 244

Tanumihardja S, Strauss KW, Müller G (1986) Petrologie unterkarbonischer Spilite der NE-Lahnmulde / Rheinisches Schiefergebirge. Z Dtsch Geol Ges 137: 465–487

Willner AP, Massonne H-J, Krohe A (1991) Tectonothermal evolution of a part of a Variscan magmatic arc: the Odenwald in the Mid-German Crystalline Rise. Geol Rdsch 80: 369–389

III.C.5 Metallogenesis

H.J. Franzke and H.-J. Anderle

1 Introduction

The Northern Phyllite Zone of the Mid-German Variscides forms a narrow strip of less than 10 km in width between the Mid-German Crystalline Rise and the Rhenohercynian Zone (see Chap. III.B.5).

Mineral concentrations or ore deposits exploited formerly or up to present time are sparsely developed in the Phyllite Zone including the areas which are in discussion as Metamorphic Nappes. Therefore it will not be necessary to discriminate between both units. On the other hand the discussion of metallogenetic processes and mineral concentrations in the Wippra Unit (Harz Mountains) should be separated from that one in the Taunus and Hunsrück (Rhenish Massif, see Fig. 1).

2 Classification

In the sense of metallogenetic processes, the Northern Phyllite Zone can be described as the southern rim of the Rhenohercynian Zone containing nearly the same types of mineralizations resulting from the same minerogenetic processes (see Werner and Walther, Chap. III.B.5, this Vol.). Therefore we can follow the classification given by these autors and concentrate on the detailed description of the deposits.

The mineralizations can be classified principally into two groups: those formed during Devonian and Lower Carboniferous sedimentation before the Variscan orogenesis (group A) and the second group formed in Late and Post-Variscan Time (group B). There may be one exception (see p. 141 and no. 5 in Fig. 1). Further subdivisions (A1–A4 and B1–B2, see below and Werner and Walther, Chap. III.B.5) are based on genetic aspects.

Three processes, acting during the sedimentation and diagenesis of the Paleozoic rock sequence, governed the formation of "Pre-Sudetic" mineral concentrations: primary concentrations in sediments (A1), basin dewatering connected with keratophyric volcanism (A2/A3), and submarine basic volcanism (A4).

Group B starts with mineral veins following cleavage and thrust planes or contrasts of rock competency. They can be slightly deformed by the final impulses of the Variscan orogenesis (B1).

Group B2 comprises a sequence of hydrothermal veins formed between the beginning of the Late Variscan inversion tectonics in the Uppermost Carboniferous/Lower Permian and the "Saxonian" mineralizations of Mesozoic, probably lasting up to the Tertiary (pseudomorphic quartz veins in the Rhenish Massif).

The authors wish to emphasize that the classification of the described deposits and occurrences in several cases is tentative because of uncomplete and partly very old information.

3 Mineral Concentrations in the Hunsrück and Taunus

In Fig. 1 the geological environment of the Taunus and Hunsrück mineralizations is shown and the positions of the mineral occurrences mentioned are marked. The southern margin of the Rhenohercynian Zone in Hunsrück and Taunus is characterized by upturned stacked rock sequences with a fanning syncrystalline main foliation (Anderle 1987). SE of the line of divergence (Figs. 1 and 2), NNW-dipping ss/s1 fabrics prevail. A generally southward-increasing strain can be observed. A later subhorizontal to NNW-dipping crenulation cleavage together with almost homoaxial kink bands (especially in the Taunus area) correlates with late orogenic extension (Weber 1978). Metamorphic peak conditions are assessed as 300 °C and 6 kbar for the Taunus metavolcanites, 300 °C and 3 kb for the metasediments and should be of a similar range in the

Dallmeyer et al. (Eds.)
Pre-Permian Geology of Central and Eastern Europe

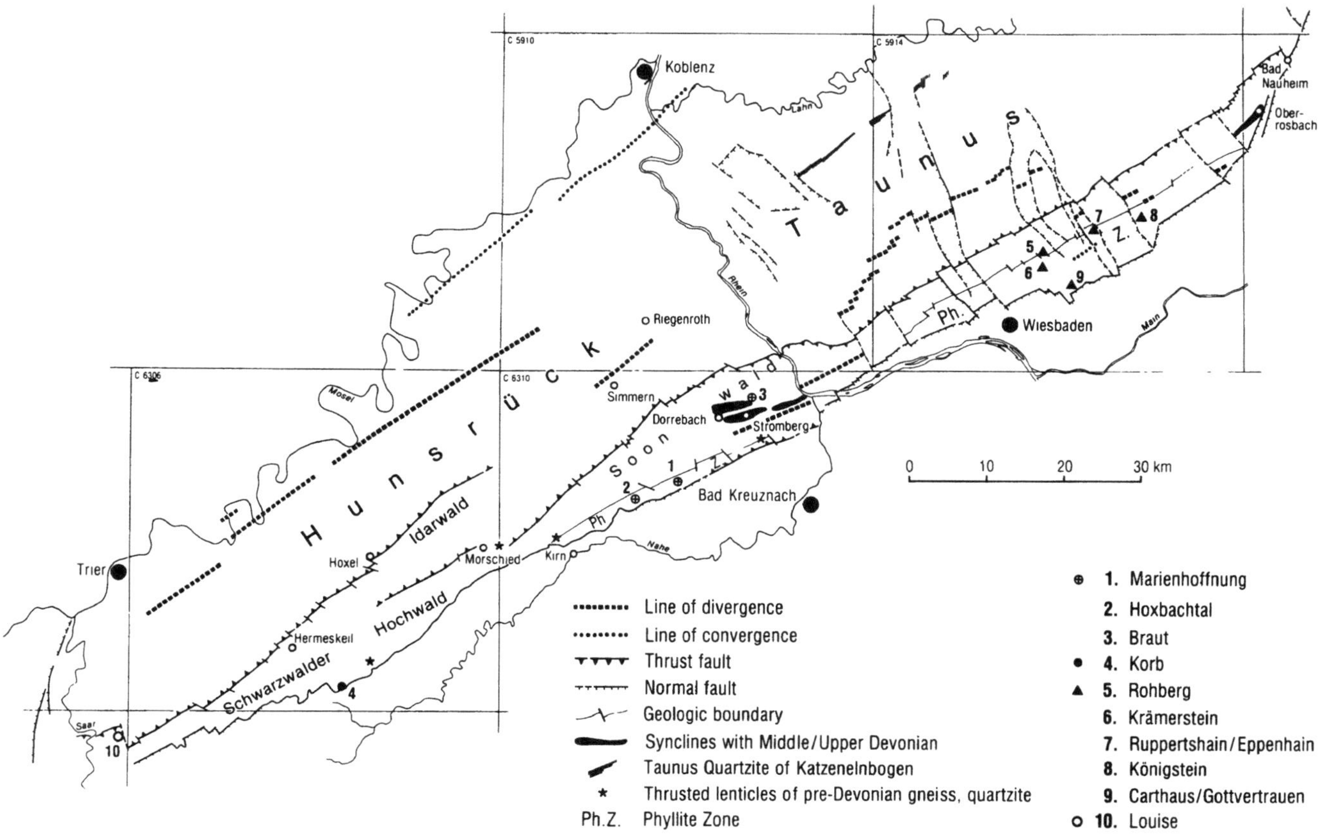

Fig. 1. Geological sketch map comprising the Northern Phyllite Zone in the Taunus and Hunsrück. The *bold numbers* correspond to those cited in the text. *C-numbers* are sheets of the topographic maps 1:100000

Soonwald area (see Meisl et al. 1982; Anderle et al. 1990).

The phyllite zones of Soonwald and Taunus, as bounded in Fig. 1, are two different tectonostratigraphic units, which are reflected in their rock sequences and mineralizations. The Soonwald Phyllite Zone shows basic metavolcanics and metapelites of probable Devonian age, both partly albitized, with intercalations of gray quartzites, and iron ores of type A1. The Taunus phyllite zone is made up of acid to intermediate metavolcanics of Ordovician/Silurian age (Sommermann et al. 1990) and metapelites with some graywacke intercalations, which are at least partly of Lower Devonian (Emsian) age, as proved by spores (Reitz 1989). The older mineralizations are tentatively described as type A3 and B1 here (as opposed to Anderle 1984). Almost no modern geological or minerogenetic investigations have been carried out on the Taunus occurrences.

Further geological information is given in Part III.C.

The numbers of the described deposits and occurrences of the Taunus and Hunsrück, located in Fig. 1, have been put in parentheses at the ends of the headlines.

3.1 Synsedimentary Mineralizations (Type A1)

3.1.1 Marienhoffnung Mine (see Fig. 1, No. 1)

At the southern margin of the Soonwald (SE part of the Hunsrück) several seam-like beds of ferrous schist appear in very low grade chlorite-sericite phyllites together with metabasalts, both of which are predominantly albitized. In the years 1907–1908, 1915–1919, and 1924 several magnetite lodes were mined.

Their thickness was 0.1–1.2 m, locally reaching up to 2 m. The crude ore contained 58–63% Fe, 4–7% SiO_2, 0.41–0.47% P, and 0.01–0.06% S (Rée 1979a).

The ore lenses are of changing extent and massive, as well as intensely crenulated (Cissarz 1927). The mineral paragenesis consists of magnetite, hematite, and pyrite. Pyrite is the youngest mineral (Cissarz 1927; Meisl et al. 1982).

The ores were formed either from iron-rich residual solutions (type A4) belonging to the basaltic volcanism (Chudoba and Obenauer 1932) or by the metamorphosis of synsedimentary iron ores (Bierther 1953) of Lower Devonian (type A1) like the Braut mine (No. 3).

3.1.2 Magnetite Quartzites of the Hoxbachtal (Soonwald, No. 2)

At other places in the imbricate zone of the southern Soonwald, magnetite/hematite-bearing quartzites form interbeds in the metamorphic series of albitized metadiabases and metapelites. The ore bands are included in long striking mylonitic quartzite interbeds in the western prolongation of the Marienhoffnung mine and form seams up to 0.1 m thick, containing up to 80% magnetite in zones of several meters.

Hematite is the product of an early martitization of the magnetite before the deformation of the metamorphic layering by mylonitization/recrystallization began (Wotzlaw 1988).

This means that the formation of the ore beds is explained by sedimentation of a sequence of Fe-rich sandstones and their metamorphic conversion to magnetite ores, as first described by Vierschilling (1910).

3.1.3 Braut Mine Near Wald-Erbach (No. 3)

Sandy shales of uppermost Lower Emsian of the Stromberg syncline (Soonwald) include four intercalations of red iron ores partly made up of true ooides with cores of calcite surrounded by hematite or sometimes silicate of iron. The seams have been worked from 1839 to 1904.

The ore seams and the surrounding host rocks strike SW-NE and dip steeply toward NW at the surface and SE at depth. The main lode forms a folded S-shaped band along the strike with an average thickness of 1–3 m, in the fold crests up to 6 m. The iron content was 50–55% at the surface, and decreased below 30% with depth (Vierschilling 1910; Rée 1979c).

The iron-bearing solutions are derived from an unknown neighboring mainland (Wolf 1930). They were concentrated during uppermost Lower Emsian age (Upper Vallender substage) according to a marine macrofauna of mainly bivalves and brachiopodes (Mittmeyer and Geib 1967; Meyer 1970).

The mine is situated outside the Phyllite Zone in the Stromberg syncline of the northerly adjoining duplex system (see Oncken 1988).

3.2 Deposits from Metalliferous Basin-Dewatering Brines (Type A3)

These mineral concentrations belong to the SEDEX-type mineralizations (Werner 1989). In the area under investigation, the Pb-Zn-Cu-(Ag) bearing pyrite-barite orebodies (type A2, type Rammelsberg/Meggen) are missing. Only two mineral concentrations of the type A3 have been mined (Korb and Rohberg mines).

3.2.1 Korb Mine Near Eisen (Hunsrück, No. 4)

The korb mine is located within the Rhenohercynian Zone close to the Northern Phyllite Zone.

During open pit mining of iron and manganese oxides, barite was discovered possibly before World War I. Subsurface mining of barite started about 1958. The mine was abandoned a few years ago. The barite lode is concordantly interbedded in a steeply NW-dipping and thrusted series of shales, carbonates, and phyllites (Schömer 1952; Hofmann 1965; Mihm 1968), which was first thought to be of Lower Devonian age and later turned out to comprise uppermost Lower Devonian through Lower Carboniferous strata (Krebs 1970; Gwosdz et al. 1974; Müller and Stoppel 1981) by conodont datings.

The orebody is overturned toward the SE (Fig. 2). Black shales in the footwall contain layers of pyrite and lenses of barite which intercalate with gray limestones of the Cephalopodenkalk type. Both are of syngenetic origin and interpreted as precipitations from heavy metal bearing brines (Gwosdz et al. 1974). The main body of barite forms a distorted lense of 110 m in length and maximally 22 m in thickness along the strike of the host rocks. This lense was investigated down to 113 m depth.

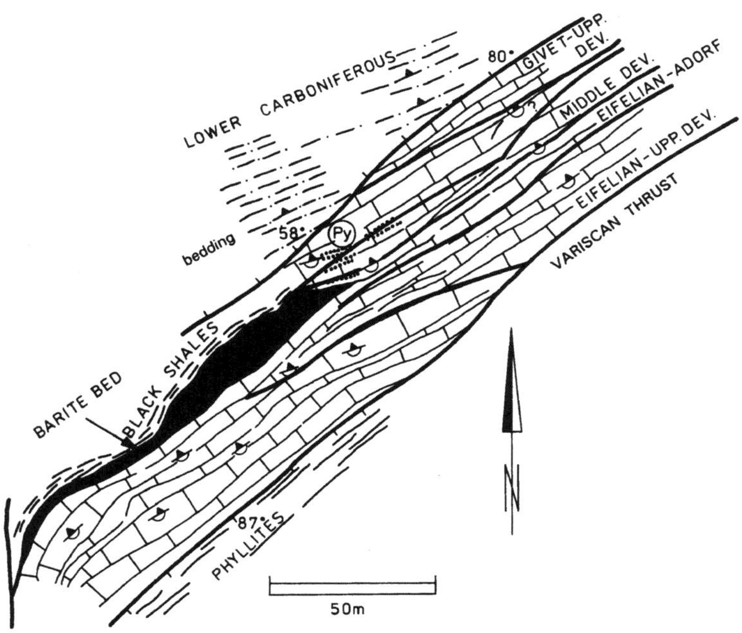

Fig. 2. Third level of the Korb mine near Eisen/Hunsrück (No. 4 in Fig. 1). The barite bed forms an intercalation in the overturned rock sequence of Devonian age. (After Müller and Stoppel 1981)

The gray to dark bluish gray barite is massive crystalline and contains pyrite, limonite, dolomite, and subordinately quartz. Intercalations of greenish sericite shales are assumed to be tuffs. Important was the observation of Hofmann (1966) and Mihm (1968) that pyritic bands are occasionally folded. This shows that the deposit suffered Variscan deformation. Similarities with the deposits Meggen and Rammelsberg point towards a syngenetic-synsedimentary origin (Hofmann 1965, 1966; Krebs 1970). Müller and Stoppel (1981) and Werner (1989) accept this genesis only for the gray variety of the barite. In their opinion, the variegated barite with higher percentages of calcite and dolomite is of metasomatic origin, having replaced carbonates of Middle Devonian age. Krahn (1988) reports distinct lead isotopes dissimilarities between Korb on one hand and Meggen and Rammelsberg on the other.

3.2.2 Rohberg Mine at Naurod Near Wiesbaden (No. 5)

Early in the 19th century, barite was discovered at Naurod. One principal vein has an acute-angled direction to and several smaller ones mainly follow the WSW-ENE striking main foliation of the host rock – so-called sericite gneiss (early Silurian meta-rhyodacite) – near its northern border to phyllitic slates with sandstone intercalations of lowermost Devonian. The dip of the veins is 50 to 80° NW to NNW. The indistinctly tectonically banded barite is mainly pure white or light bluish gray and granular. Some galena is interspersed. Quartz is partly intermingled or divides the barite lodes. Their thickness ranged from a few millimetres to 1.5 m (maximum 4–5 m). The main lode was followed for about 100 m by galleries. The known mineralized zone at the northern border of the Phyllite Zone at Naurod covers an area of 200 m across strike and at least 800 m along strike.

Strong deformation (undulosity, twinning), recovery (formation of subgrains, sutured grain boundaries) and dynamic recrystallization of barite and quartz prove a pre-Variscan origin of the mineralization. It is part of a sinistral Variscan shear zone along the main foliation. Criteria for synsedimentary formation are lacking.

As the barite forms veins in early Silurian volcanics (Sommermann et al. 1990, 1992), which were deformed during the Variscan orogeny, it can have formed during a time span of about 100 million years. The Rohberg mineralization only then belongs to type A3 of Werner and Walther (Chap. III.B.5., this Vol.) if it formed as feeder dikes for ascending brines during Devonian or Lower Carboniferous (for details see Anderle and Kirnbauer 1993).

3.3 Synorogenic Mineral Veins in Cleavage and Thrust Planes (Type B1)

This type of synorogenetic vein follows cleavage and thrust planes and rigidity contrasts in the host rocks. Shape and contents of the highly irregular veins are influenced by the final impulses of the Variscan deformations but never affected by metamorphic processes.

The mineralizations (sulfides of Sb, Pb, Zn, Cu, Ag, and Co) are the product of leaching processes from metal-enriched host rocks caused by hydrothermal circulation. The circulation cells were driven by syn- to late orogenetic heating due to crustal stacking and acid melt generation.

As already pointed out, information on the mineralizations in the Taunus Phyllitic Zone is obsolete and incomplete. All former mines are inaccessible, so that no modern data are available.

3.3.1 Krämerstein Mine at Naurod (No. 6)

Quartz veins following the main foliation of the surrounding meta-rhyodacites bear copper ore nodes with chalcopyrite, azurite, cuprite, bornite, malachite, chalcocite, pyrrhotite, hematite, calcits, fahlore, and granular barite (Sandberger 1850; Koch 1880d; Michels 1932). They were tested without success in the 18th century (Becht 1979). Pieces of the lode were found as xenoliths in a neighbouring basalt neck, which has been K-Ar dated as Eocene (57 Ma) by Lippolt et al. (1975).

3.3.2 Copper Mineralizations Near Ruppertshain (No. 7)

Chalcopyrite, bornite, and covellite have been observed in a meta-andesite quarry near Ruppertshain by Hentschel and Meisl (1966). The copper minerals replace elder pyrite, covellite replaces all these mineral phases. The question of genetic connections to the metamorphosis of the host rocks is not answered yet.

3.3.3 Abandoned Copper Mine Near Königstein (No. 8)

In the second hald of the 19th century, a deep gallery of an old copper mine was still partly open. The ores, chalcopyrite with galena, occurred in veins in phyllitic metasediments (Koch 1880c; Leppla 1924) here and at a few other places around Königstein.

3.3.4 Former Mineral Workings in the Surroundings of Carthaus/Gottvertrauen Igstadt (No. 9)

At the southern margin of the phyllitic metasediments, several test shafts and galleries for gold mining are reported from the Klingelbach valley north of Breckenheim near Wiesbaden.

They encountered pyritic and graphitic slates with small lenses of greasy quartz which contain traces of gold-bearing silver, quartzites with thin layers of white mica, veinlets of calcite, and layers of granular light gray limestone. Singular loadings of mm-thick, shining graphite in quartz segregations are also reported.

The latest mining tests were carried out in 1899–1901. The described rock series is covered by younger sediments (Koch 1880b; von Reinach 1905; Leppla and Steuer 1923; Leppla 1924).

The mineral occurrences 5–9 in the southern Taunus all gather in an area where the Northern Phyllite Zone is cross-cut by an old NW-SE-trending block border which neotectonically is marked by the Idstein depression. The minerogenetic importance of such intersection areas is described by Werner (1990) from the Rhenohercynian farther north.

3.4 Postorogenic Mineralizations (Type B2)

This group contains so-called Saxonian mineralizations, widespread epithermal vein mineralizations bound to the continent-wide stretching processes lasting from the Mesozoic to the Tertiary. These mineral veins at different places penetrate the Mesozoic cover or form seam-like lodes and infiltrations of sulfidic ores and carbonate/barite/fluorite/quartz masses in the cover rocks synchronous to the veins (Dill 1988; Merz et al. 1990).

3.4.1 Quartz-Hematite Veins of the Louise Mine (No. 10)

SE of Saarburg, several hydrothermal quartz-hematite veins occur in the Lower Devonian

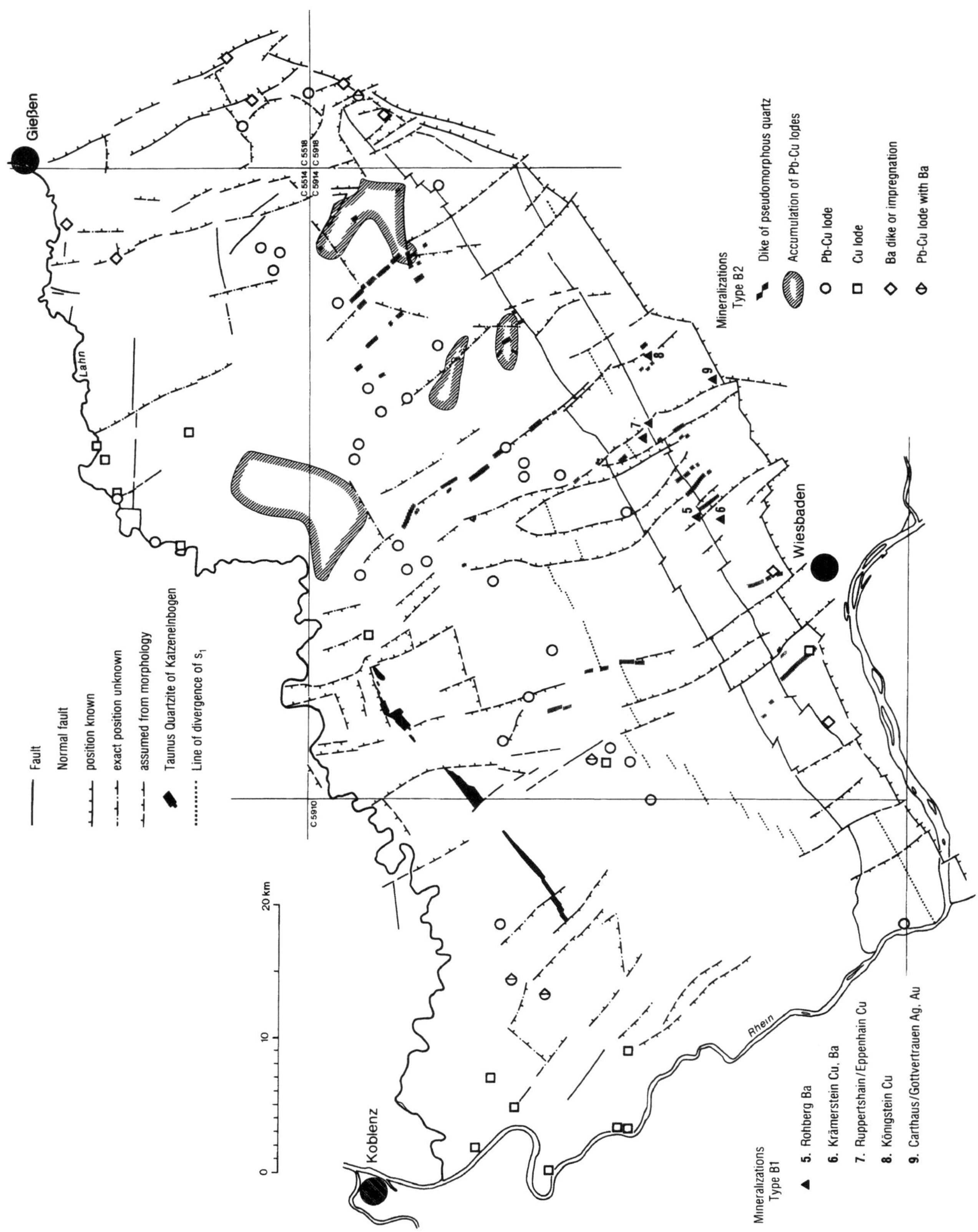

Fig. 3. Distribution of post-Variscan hydrothermal vein minerlizations in the Taunus area together with pre-Variscan mineralizations 5–9 from Fig. 1. (Completed after Anderle 1984)

Taunus Quartzite (Fig. 1). Eleven lodes were found in the Louise mine. The main lodes strike NNE-SSW and dip 60–90° toward SE and NW. Two cross-veins strike NW-SE and dip 65–87° NE. The veins were 1.5–5.0 m thick, the lodes about 0.5 m (max. 3.4 m). The ores comprise hematite and some Mn oxides. The mine was operated 1841–1880 and 1938–1943 (Rée 1979b). These mineralizations are probably of Permian age, followed by adularia-bearing sulfidic ores (Merz et al. 1990).

3.4.2 Mesozoic/Cenozoic Mineralizations

The so-called Saxonian mineralizations for the whole Taunus area have been compiled by Anderle (1984). Their distribution is shown in Fig. 3. They comprise large dikes with quartz pseudomorphous after barite and younger quartz generations like the typical cap quartz, furthermore quartz veins containing galena and chalcopyrite, sometimes fahlores and some, barite in veins and in Tertiary sands as at Bad Kreuznach, Wiesbaden, and near Bad Nauheim.

4 Mineral Concentrations in the Wippra Unit (Harz Mountains)

4.1 Geochemical Specialization and Variscan Mineral Concentrations (Types A1–A4)

Figure 4 shows the stratigraphic/lithological sequence of the Wippra Unit using the current information on the stratigraphic positions of

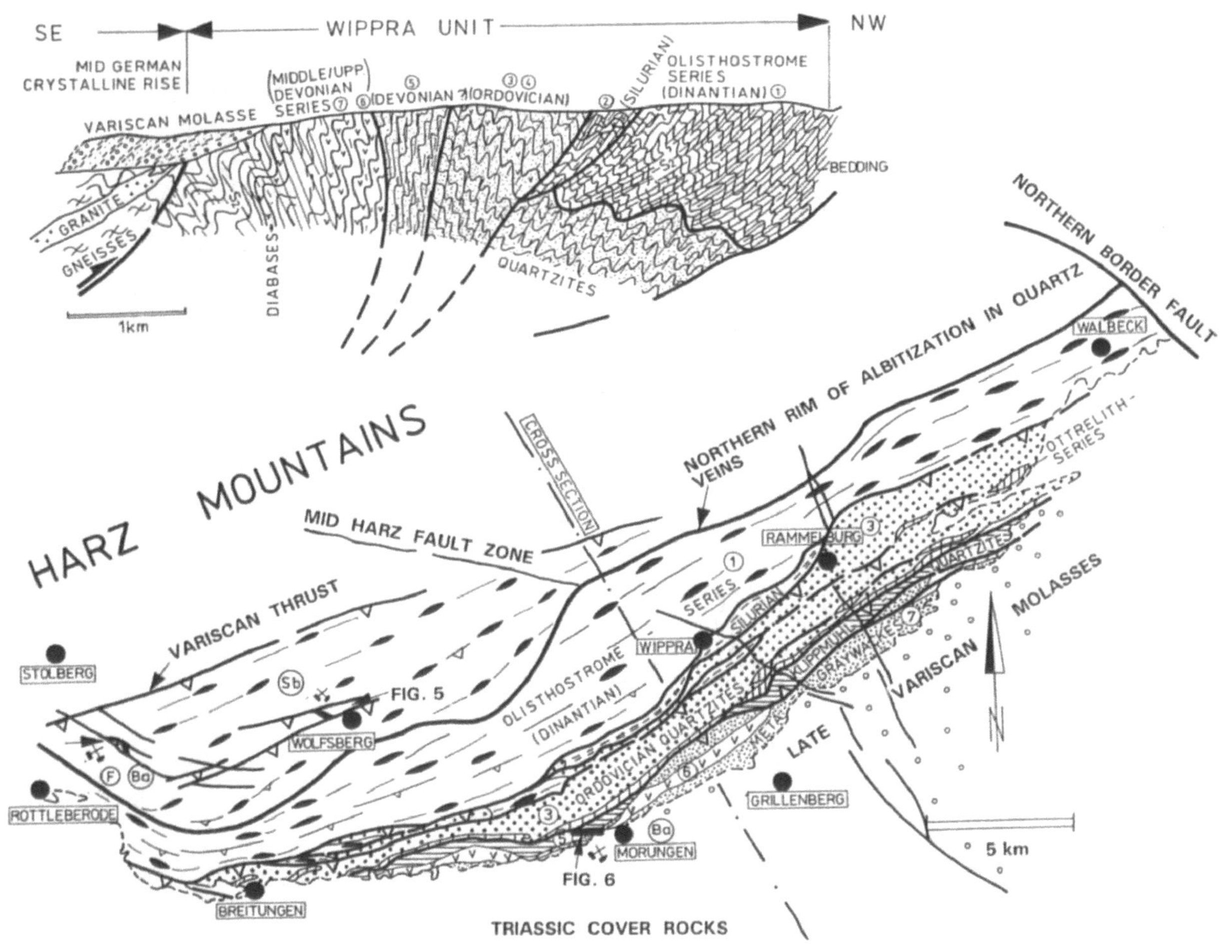

Fig. 4. Geological overview of the Wippra Unit/Harz Mountains and cross section showing the dominating steeply dipping rock sequences in front of the Crystalline Rise. (After Schwab and Franzke, unpubl.) See the position of the descriobed deposits Wolfsberg and Rottleberode

the seven rock series established since Fischer's (1929) investigations and improved by Reichstein (1964), Franzke (1969), Burmann (1973), Ackermann (1985), Siedel (1989), and Sehnert (1990). This very-low-grade metamorphic rock sequence (Fischer 1929; Franzke 1969; Siedel 1989) of about 4 km in thickness comprises phyllitic shales, quartzites (series 4 and 5), graywackes and siliceous shales (series 7) and diabases and their tuffs in the series 1, 2, 3, 6, and 7 (see also Oncken et al., Chap. III.C.2, this Vol.).

Their stratigraphic positions (see Schwab et al., Chap. III.C.1, this Vol.) could be clarified, except for the series 5 and 7, by means of conodonts (Reichstein 1964) and acritarchs (Burmann 1973 and Sehnert 1990). The cross section may give an impression about the mutual positions of the distinguished rock series and their actual tectonical position in the northern front area of the Mid-German Crystalline Rise.

From the primary geochemical specialization of these rock units (Ackermann 1985; Siedel 1989), we can recognize a reducing environment in the Silurian shales (series 2) expressed by an interbedding of clastic and organopelitic shales. This gave rise to the formation of organometallic complexes with increased contents of V, Cu, and Se, and the formation of thin ironstone bands and concretions, which were never of interest for greater mining activity in the past.

Another complex of metallogenetic interest is the series 4 (reddish and grayish shales with carpholite, Ordovician) where increased contents of Fe, Mn, Al, Co, Ba, Pb, Sr could be recognized. They are restricted to two beds of reddish shales (Estrada, Berlin, pers. comm.) which hint at volcanic-exhalative processes during the Lower Ordovician (uppermost part of the series 4) in an oxidizing environment, which should be responsible for the increased quantity of metals. After the metamorphic peak, Fe and Mn were redeposited on secondary joints and in small veinlets, striking parallel to the main cleavage and shear zones. These veinlets are filled with spherosiderite/rhodocrosite and carpholite/quartz. All these mineralizations never gave rise to any attempt at mining.

The described concentrations belong to synsedimentary concentration processes under reducing conditions in black shales (type A1, shales of Silurian, series 2) or by submarine synsedimentary hydrothermal exhalations (types A3 and A4).

The widespread spilitization of basaltic volcanics is a common feature in the Wippra Unit and forms a small iron source without any economical importance (Lahn-Dill type, A4 mineralizations).

4.2 Synorogenetic Veins in Cleavage and Thrust Planes (Type B1)

In the southeastern Harz Mountains, several occurrences of Sb-emphasized sulfidic hydrothermal veins are outcropping, which have been intensely mined in the last three centuries (Hesemann 1930).

Their special characteristics are the spacing of the vein structures parallel to the Variscan fold belt, especially that they follow cleavage and thrust planes of the Variscan deformation.

The development of the vein structures started with the formation of a breccia healed by quartz, which consists of angular fragments of the host rocks. It was followed by different sequences of sulfidic minerals embedded in quartz and partly siderite, which were of mining interest because of the Ag content of the fahlores. Younger "Saxonian" mineralizations are missing, except for small quantities of barite.

The Jost-Christian mine near Wolfsberg (position see Fig. 4) is directly located at the margin of the phyllitic zone mapped by means of the occurrence of albite in metamorphic segregation quartzes.

Typical for this type of hydrothermal veins in the Lower harz Mountains is their control (as to location of the deposits and their vein-space filling) by Variscan thrust elements activated up to the latest Variscan impulses.

Figure 5 explains the direction of the Variscan compression responsible for the formation of the complete pattern of Variscan structural elements, starting with the bedding-parallel first cleavage formed synchronously with the metamorphic peak, followed by the folding of these elements by the F2 deformation, finally overprinted by thrust tectonics along distinct bedding/first cleavage parallel shear zones.

The formation of the vein structures belonged to the interaction of compression/decompression, which allowed for the generation of hydrothermal circulation cells of local importance controlled by the thrust pattern.

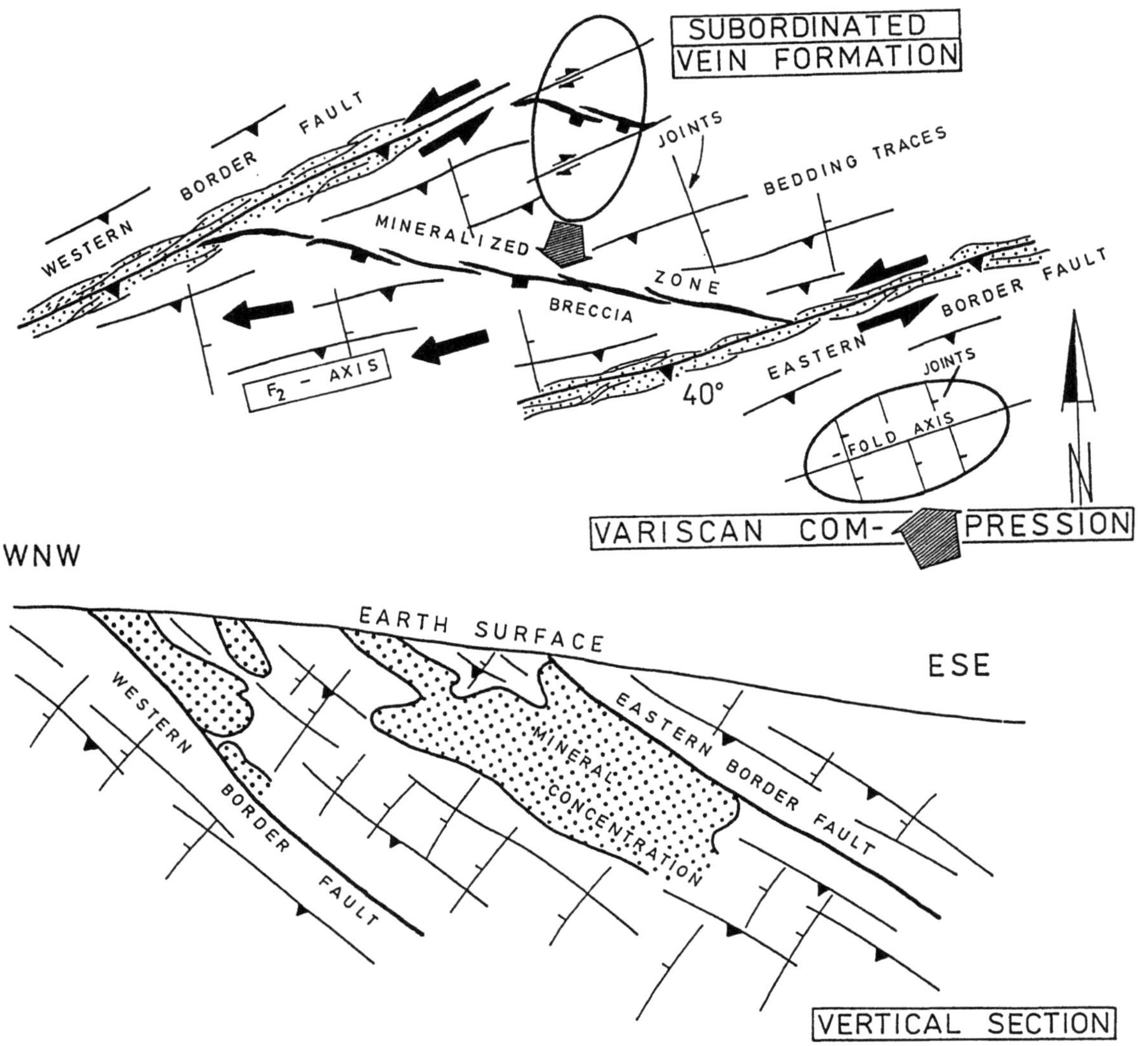

Fig. 5. Layout and parallel section describing the structural situation in the vein field of the Jost-Christian mine near Wolfsberg. For further explanation see text

The veins form diagonal shear elements opened and filled in decompressional phases of the final stages of the thrust tectonics probably after the principle of seismic pumping under partly high inner pressures of the hydrothermal system (Sibson et al. 1975). The veins are still affected by the final impulses of the Variscan thrust tectonics.

4.3 Postorogenic Mineralizations (Type B2)

Postorogenic hydrothermal mineralizations are widespread in the Harz Mountains. For the distribution and mineral paragenesis of these mineralizations in the Lower Harz Mountains see Kuschka and Franzke (1974) and Klaus (1976).

New radiometric datings yield Triassic/Jurassic ages for the galena-bearing vein mineralizations (Bielecki and Tischendorf 1991; Hagedorn and Lippolt 1990; Leveque and Haack 1990), so that the younger carbonate/barite/fluorite vein mineralizations should be upper Jurassic/Cretaceous in age.

In the phyllitic region only a few occurrences of postorogenic vein mineralizations are known. Their paragenetic sequences are comparable with the neighboring occurrences in the Harz Mountains and also in the lowermost part of the platform sediments ("Mansfelder Rücken").

At the Schlossberg near Morungen (location see Fig. 4) several E-W-trending and steeply dipping barite veins without any fluorite content were precipitated after the initial quartz-haematite mineralization of probably Permian age (Fig. 6).

The filling took place under normal fault conditions which formed a group of feather-like veins mined till the beginnung of our century. Older tectonical patterns (bedding/foliation 1) were converted to shear-and feather planes in course of the applied stress-field with minimal principal stresses in S-N.

The mineralization of the well-known fluorite vein deposit of Rottleberode (location see Fig. 4) is bound to a complicated set of NW-SE-trending and mainly SW-dipping normal faults, but also conjugate NE-dipping fault elements (Fig. 7).

The formation of the vein structures in Rottleberode started with the same quartz-healed breccia as described in Wolfsberg, which trace the central masterfault of the vein system. This was followed by joint-bound haematite mineralizations, which are restricted to the Variscan fundament and the Variscan molasse stage. This indicates Permian age for this mineralization. The pure intense sulfidic ores are followed by carbonates, mostly siderite and two phases of fluorite, interrupted by the precipitation of barite in higher levels of the vein structures.

The older fluorites in the deposit Rottleberode show a REE pattern with continuously decreasing contents towards the heavy RE elements. They were formed outside of granitic environments and may represent the primary fluorite source for later replacements (Lüders 1990).

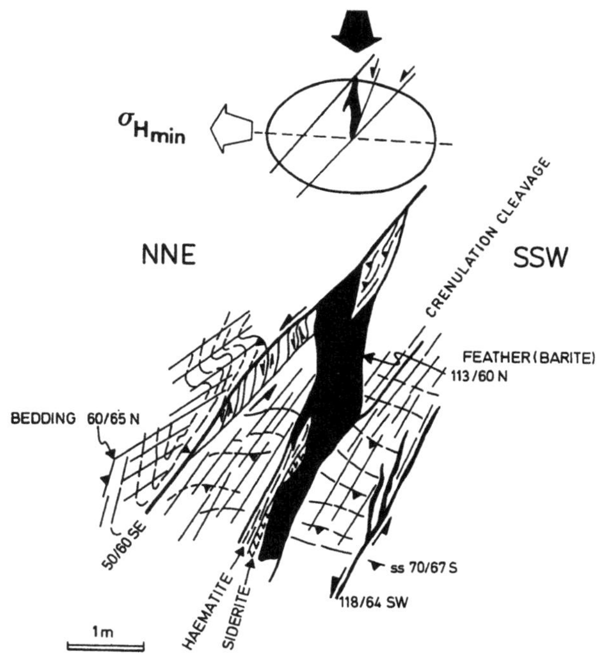

Fig. 6. Cross section of a barite vein near Morungen. A Variscan reverse fault and the transverse – second – or crenulation cleavage were converted to to barite-filled normal faults/feathers

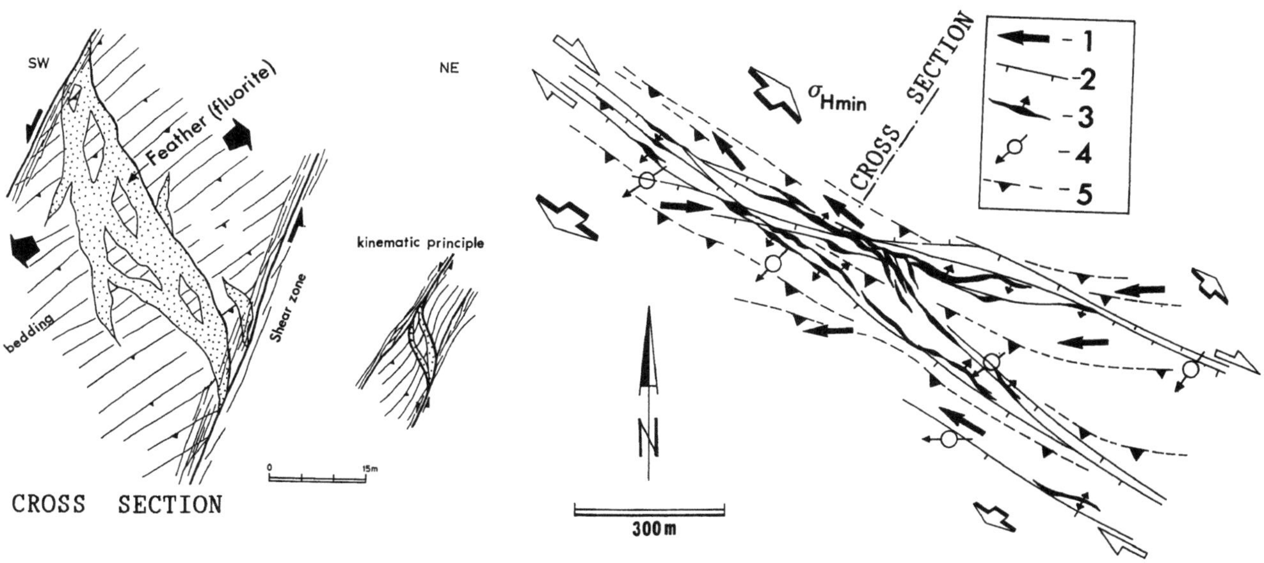

Fig. 7. Geological map and typical cross section from the 9th level (+180 m) of the fluorite deposit Rottleberode representing the structural pattern, the slip direction of the dominant normal faulting and kinematic principle for the opening of the vein elements. *1* Fold axis; *2* normal fault; *3* fluorite concentrations; *4* slickenside stria; *5* bedding

The REE pattern of the economically important younger fluorite, its homogenization temperatures between 100 and 150 °C and the high chlorinity of the fluid inclusions (4–5 M) together with structural arguments support the hypothesis, that the ore-forming fluids are derived from the older fluorite of probably Variscan age, disseminated in the underground, as well as from additionally supplied solutions (Lüders 1988; Lüders et al. 1990).

The REE pattern of a third type of fluorite, concentrated around the Ramberg granite, forms a typical negative Eu-anomaly, which hints at its genesis from granitic alteration processes.

The widespread barite mineralizations in the southern Harz Mountains are strictly bound to the footwall of the platform cover, so that different authors favor the hypothesis of solution mixing (Behr and Gerler 1987; Möller et al. 1990) of hot ascending metalliferous brines and cool descent of SO_4-rich brines for their genesis.

5 Regional Minerogenetic Remarks

Some sedimentary Fe-ores of the type A1 exist in the Soonwald area (occurrences 1, 2, and 3). The majority of all other mineralizations was, however, formed by hydrothermal circulations, which were bound to Early Paleozoic basin extension (deposit type A2–A4), the Variscan convergence tectonics (B1) and the Late/post-Variscan inversion tectonics (B2) lasting till the collision tectonics in the Alpine orogen in Upper Cretaceous/Lower Tertiary.

Regional structural studies (Rauche and Franzke 1990) carried out in the hydrothermal vein deposits of the Thuringian Forest and the eastern Harz Mountains presented arguments, that compressive impulses resulting from the alpine collisional front affected and displaced the barite/fluorite veins during Upper Cretaceous/Lower Terriary in this area by reverse faulting or thrusting.

The minerogenetic role of the Variscan terrane boundary between the Mid-German Crystalline Rise and the Northern Phyllite zone is hard to estimate as a result of the widespread covering by Mesozoic and Cenozoic strata.

All the described types and intensities of mineral concentrations in the Taunus/Hunsrück area are to be found in the transition zone between the Northern Phyllite Zone and the Rhenohercynian Zone of the European Variscides and at intersections with NW-SE-trending fault zones.

In the eastern part of the Northern Phyllite Zone (Harz Mountains) only very small and rare ore/mineral concentrations belonging to the Variscan metallogenetic cycle are known and we can never advocate a concentration of Variscan mineralization processes in this area.

In correspondence with the results of Werner (1989, 1990) and other authors we found that the emplacement of mineral concentrations was mainly influenced by the distribution of the regional fault patterns, especially their deep-reaching conjugating and crossing branches, Convergent movements were bound to the Variscan convergence tectonics; on the other hand, vertical block tectonics indicate extensional tectonical regimes in pre- and postorogenetic times combined with increased magmatic/volcanic activities and heat flow up to the Permian.

Since the Upper Permian, we recognize heat flows in the same range as in present time and mineralization processes in the Variscan consolidated Mid-European crust now took place beneath a widely impermeable barrier of cover rocks. The intense vertical block tectonics was now the generator for the movement of the hydrothermal circulation cells beneath the platform sediments.

A very important role played the availability of pre-existing primary mineral/metal concentrations in framing or better in underlying source rocks (group A), which could be replaced by means of tectonically controlled leaching processes.

Acknowledgments. Critical remarks by Thomas Kirnbauer, Wiesbaden, helped to improve the text on the Taunus mineralizations.

References

Ackermann S (1985) Beiträge zur Stratigraphie und Lithologie der Wippraer Formation. Dissertation, Sektion Geogr Univ Halle, Halle, p 134

Anderle H-J (1984) Postvaristische Bruchtektonik und Mineralisation im Taunus. Schriftenr Ges Dtsch Metallh Bergleute 41: 201–217

Anderle H-J (1987) The evolution of the South Hunsrück and Taunus border zone. In: Ziegler PA (ed) Compressional intra-plate deformations in the Alpine Foreland. Tectonophysics 137: 101–114

Anderle H-J, Massone H-J, Meisl S, Oncken O, Weber K (1990) Southern Taunus Mountains. In: Conf Paleozoic orogens in central Europe, Göttingen-Gießen, Aug-Sept 1990, Field guide: Mid-German Crystalline Rise and Rheinisches Schiefergebirge, pp 125–148

Anderle H-J, Kirnbauer Th (1993) Das Schwerspatvorkommen von Naurod im Taunus (Bl. 5815 Wehen) – eine prävaristische Gangmineralisation. Geol Jahrb Hessen, Wiesbaden 121 (1): 91–123

Becht A (1979) Ein Kupfer- und Silberbergwerk bei Naurod. Nauroder Hefte 3 (5): 1–8

Behr HJ, Gerler J (1987) Inclusions of sedimentary brines in post-Variscan mineralizations in the Federal Republic of Germany – a study by neutron activation analysis. Chem Geol Amsterdam 61: 45–67

Beschreibung (1893) Beschreibung der Bergreviere Wiesbaden und Diez. Marcus, Bonn, 252 pp

Bielicki K-H, Tischendorf G (1991) Isotopic composition of ore leads from central Europe. Lead isotope and Pb-Pb model age determination and their metallogenetic interpreation. Contrib Mineral Petrol, Heidelberg 106 (3): 440–462

Bierther W (1953) Zur Stratigraphie und Tektonik der metamorphen Zone im südlichen Rhein. Schiefergebirge. Geol Rdsch 41: 173–181

Burmann G (1973) Das Ordovizium der nördlichen Phyllitzone. Teil II: Wippraer Zone. Z Geol Wiss Berlin Themenheft 1: 9–43

Chudoba K, Obenauer K (1932) Über die metamorphen Gesteine bei Winterburg im Hunsrück. Neues Jahrb Min Geol Paläontol Beil A: 59–81

Cissarz A (1927) Ueber einige metamorphe Gesteine bei Winterburg im Hunsrück und die mit ihnen verknüpften Eisenerzlagerstätten. Z Prakt Geol 35: 86–91

Dill H (1988) Lagerstätten-Typisierung und metallogenetische Entwicklung im Umfeld der Grenzregion Saxothuringikum-Moldanubikum (Nordbayern)-ein Überblick. Geol Bavarica München 92: 133–150

Fischer G (1929) Die Gesteine der metamorphen Zone von Wippra mit besonderer Berücksichtigung der Grünschiefer. Abh Preuß Geol Landesanst NF Berlin 121: 64

Franzke HJ (1969) Ergebnisse einer geologischen Neuaufnahme der Grünschiefer von Wippra. Hercynia Leipzig 6: 187–206

Gwosdz W, Krüger H, Paul D, Baumann A (1974) Die Liegendschichten der devonischen Pyrit- und Schwerspat-Lager von Eisen (Saarland), Meggen und des Rammelsberges. Geol Rdsch 63: 74–93

Hagedorn B, Lippolt HJ (1990) Untersuchungen zur zeitlichen Einstufung epigenetischer Vorgänge im Harz. In: Friedrich G (ed) Metallogenetische Entwicklung im Harz. DFG- Schwerpunktprogramm "Intraformationale Lagerstättenbildung", Report 1: 169–183

Hentschel H, Meisl St (1966) Exkursion in das magmatogene Vordevon des Taunus am 9. September 1964. Fortschr Miner 42: 321–333.

Hesemann J (1930) Die Erzbezirke des Ramberges und von Tilkerode im Harz. Arch Lagerstättenfarsch Berlin 46: 92

Hofmann R (1965) Der Schwerspat von Eisen (Saarland). Aufschluß 1672: 302–303

Hofmann R (1966) Lagerstättenkundliche Untersuchungen im Bereich der Schwerspatgrube Eisen (südwestlicher Hunsrück). Neues Jahrb Geol Paläontol Mouatsh: 22–35

Klaus D (1976) Ergebnisse paragenetischer und tektonischer Untersuchungen der Gangsysteme des Unterharzes. Dissertation Bergakademie Freiberg, Sektion Geowiss, Freiberg/Sa, 136 pp

Koch C (1880a) Erläuterungen zur geologischen Spezialkarte von Preussen und den Thüringischen Staaten. Lfg 15, Bl Eltville, p 59

Koch C (1880b) Erläuterungen zur geologischen Spezialkarte von Preussen und den Thüringischen Staaten. Lfg 15, Bl Hochheim, p 40

Koch C (1880c) Erläuterungen zur geologischen Spezialkarte von Preussen und den Thüringischen Staaten. Lfg 15, Bl Königstein, p 46

Koch C (1880d) Erläuterungen zur geologischen Spezialkarte von Preussen und den Thüringischen Staaten. Lfg 15, Bl Platte, p 37

Krahn L (1988) Buntmetall-Vererzung und Blei-Isotopie im Linksrheinischen Schiefergebirge und in angrenzenden Gebieten. Thesis, Univ Aachen, 72 figs, 24 tabl, 199 pp

Krebs W (1970) Nachweis von Oberdevon in der Schwerspat-Grube Eisen (Saargebiet) und die Folgerungen für die Paläogeographie und Lagerstättenkunde des linksrheinischen Schiefergebirges. Neues Jahrb Geol Paläontol Monatsh 8: 465–480

Kuschka E, Franzke HJ (1974) Zur Kenntnis der Hydrothermalite des Harzes. Z Geol Wiss Berlin 2: 1417–1436

Leppla A (1924) Erläuterungen zur geologischen Karte von Preußen und benachbarten Bundesstaaten, 2. Aufl. Lfg 15, Bl Königstein, p 56

Leppla A, Steuer A (1923) Erläuterungen zur geologischen Karte von Preußen und benachbarten Bundesstaaten, 2. Aufl. Lfg 15, Bl Hochheim-Raunheim, p 64

Leveque J, Haack U (1990) Blei-Isotope Harzer Lagerstätten. In: Friedrich G (ed) DFG-Schwerpunktprogramm "Intraformationale Lagerstättenbildung", Aachen, Rep 2, pp 405–407

Lippolt HJ, Baranyi I, Todt W (1975) Die Kalium-Argon-Alter der postpermischen Vulkanite des nordöstlichen Oberrheingrabens. Aufschluß, Sonderband 27 (Odenwald): 205–212

Lüders V (1988) Geochemische Untersuchungen an Erz-und Gangartmineralen des Harzes. Berl Geowiss Abh D 188: 74

Lüders V (1990) Granitalteration im Harz und ihr möglicher Einfluß auf die hydrothermale Lagerstättenbildung. J Eur Mineral Beih 1: 116

Meisl S, Anderle H-J, Strecker G (1982) Niedrigtemperierte Metamorphose im Taunus und im Soonwald. DMG-Tagung 1982, Exkursion E3. Fortschr Miner 60 (Beih 2): 43–69

Merz DF, Lippolt HJ, Müller G (1990) Isotopic dating of adularia-bearing mineralizations: I. Saar-Nahe region/Southwest Germany. Eur J Mineral Stuttgart 2: 89–101

Meyer DE (1970) Stratigraphie und Fazies im Guldenbachtal/SE-Hunsrück am Südrand des Rheinischen Schiefergebirges. Thesis, Univ Bonn, 307 pp

Michels F (1932) Erläuterungen zur geologischen Kart von Preußen und benachbarten deutschen Ländern, 2. Aufl. Lfg 288, Bl Wehen, p 56

Mihm A (1968) Die Schwerspatgrube Korb bei Eisen (nördliches Saarland). Ann Univ Sarav R Math Nat Fak 6: 1–42

Mittmeyer H-G, Geib KW (1967) Gliederung des Unterdevons im Gebiet Warmsroth – Wald-Erbach (Stromberger Mulde). Notizbl Hess Landesamt Bodenforsch 95: 24–44

Möller P, Lüders V, Franzke HJ (1990) Zur Genese der hydrothermalen Sufid-, Baryt- und Fluorit-Ganglagerstätten des Harzes. In: Friedrich G (ed) Post-orogene intraformationale Bildung der hydrothermalen Pb-Zn-und F-Ba-Ganglagerstätten im Harz. DFG-Schwerpunktprogramm "Intraformationale Lagerstättenforschung", Aachen, Final Rep, pp 2–18

Müller G (1988) Gedanken zu den Mineralisationen im Saarland und in seiner Umgebung. Aufschluß 39: 257–268

Müller G, Stoppel D (1981) Zur Stratigraphie und Tektonik im Bereich der Schwerspatgrube "Korb" bei Eisen (N-Saarland). Z Dtsch Geol Ges 132: 325–352

Oncken O (1988) Geometrie und Kinematik der Taunuskammüberschiebung – Beitrag zur Disskussion des Deckenproblems im südlichen Schiefergebirge. Geol Rdsch 77: 551–575

Rauche H, Franzke HJ (1990) Stress field evolution at the northern part of the South German Block on the territory of the GDR. Gerlands Beitr Geophys Leipzig 99: 441–461

Rée Chr (1979a) Die metamorphen Eisenerzlager von Winterbach – Winterburg am Soonwald, Geol Jahrb D31: 111–114

Rée Chr (1979b) Die hydrothermalen Eisenerzgänge der Grube Louise bei Saarburg. Geol Jahrb D31: 115–117

Rée Chr (1979c) Die marin-sedimentären Eisenerzlager von Wald-Erbach bei Bingen am Rhein. Geol Jahrb D31: 119–121

Reichstein M (1964) Stratigraphische Konzeption zur metamorphen Zone des Harzes. Geologie Berlin 13 (1): 25–135

Reitz E (1989) Devonische Sporen aus Phylliten vom Südrand des Rheinischen Schiefergebirges. Geol Jahrb Hessen 117: 23–35

Sandberger F (1850) Ueber die geognostische Zusammensetzung der Umgegend von Wiesbaden. Jahrb Ver Naturkd Herzogthum Nassau 6: 1–27

Schömer R (1952) Die Grube Korb bei Sötern. Vorläufige Mitteilung. Ann Univ Sarav 1: 144–150

Sehnert M (1990) Zur Biostratigraphie epizonaler Metamorphite der Wippraer Einheit/Unterharz. In: Quade H (ed) Zeitmarken im europäischen Varizikum. Clausthal-Zellerfeld, p 60

Sibson RH, Moore J Mc, Rankin AH (1975) Seismic pumping- a hydrothermal fluid transport mechanism. J Geol Soc Lond 131: 653–659

Siedel H (1989) Beiträge zur Petrographie und Geochemie epizonal metamorpher Pelite in der Wippraer Zone (Harz). Dissertation, Sektion Geogr Univ Halle, 142 pp

Sommermann A-E, Meisl S, Todt W (1990) U/Pb-Alter von Zirkonen aus Metavulkaniten des Südtaunus. Ber Dtsch Mineral Ges Beih Eur J Mineral 2 (1): 244

Sommermann A-E, Meisl S, Todt W (1992) Zirkonalter von drei verschiedenen Metavulkaniten aus dem Südtaunus. Geol Jahrb Hessen 120: 67–76

Vierschilling A (1910) Die Eisen- und Manganerzlagerstätten im Hunsrüch und Soonwald. Z Prakt Geol 18: 393–431

Von Reinach A (1905) Gebirgsbau und Stratigraphie des Taunus. Jahrb Kgl Preuss Geol Landesanstalt Bergakad 1902, 23: 596–608

Weber K (1978) Das Bewegungsbild im Rhenoherzynikum – Abbild einer varistischen Subfluenz. Z Dtsch Geol Ges 129: 249–281

Werner W (1989) Synsedimentary faulting and sediment-hosted submarine-hydrothermal mineralizations in the Late Palaeozoic Rhenish Basin (Germany). Geotekt Forsch 71: 305

Werner W (1990) Examples of structural control of hydrothermal mineralization: fault zones in epicontinental sedimentary basins – a review. Geol Rdsch 79: 279–290

Wolf M (1930) Alter und Entstehung des Wald-Erbacher Roteisensteins (Grube Braut im Hunsrück) mit einer stratigraphischen Untersuchung der Umgebung. Abh Preuß geol Landesanstalt NF 123: 105

Wotzlaw A (1988) Kartierung der Nördlichen Phyllitzone im Hoxbachtal (SE-Hunsrück) und Analyse magnetischer Gefüge am Magnetitquarzit aus dem Hoxbachtal im Soonwald. Diplom arbeit, IGDL, Univ Göttingen, 73 pp

IV Mid-German Crystalline High

IV.A Introduction

K. Weber

The Mid-German Crystalline Rise (MGCR) forms part of the Saxothuringian Zone. The term was first introduced into the German literature by Scholtz (1930: "mitteldeutsche kristalline Schwelle") and later worked out in detail by Brinkmann (1984), Kneuper (1966), Behr (1966), Weber (1978, 1981, 1984), Weber and Behr (1983), and Hirschmann and Okrusch (1988).

The MGCR can be traced as a curvilinear structure from the Saar-Nahe Basin in the west, where it is known from the borehole Saar 1 and outcrops along the western Rhine Graben escarpment, via the Odenwald and the Spessart Mts., the Ruhla Crystalline Complex of the northern Thuringian Forest to the Kyffhäuser SE of the Harz Mts. Further east, in SW Poland, the MGCR is known only from boreholes and geophysical data. Along its extent in Germany, the Mid-German Crystalline Rise is rather inhomogeneous and may be composed of different allochthonous units or even terranes. The largest exposures are the Odenwald and the Spessart Crystalline Complexes. The western part of the Odenwald (Bergsträßer Odenwald) is geologically different from that of the eastern part (Böllstein Odenwald). Both are separated by an important fault (see Chap. IV.C.2). The Böllstein Odenwald belongs to the same structural unit as the Spessart Crystalline Complex. Eastern continuations of the Spessart are known from the Rhön Mts., where crystalline rocks with a similar composition and NW-SW zonation have been found in numerous drill holes under the Triassic cover (Weinelt et al. 1985; Schmidt et al. 1986). Correlation of the Spessart Crystalline Complex with the Ruhla Crystalline Complex of the Thuringian Forest has been attempted by several authors (see Hirschmann; Chap. IV.B), but the results are not convincing. There are more lithological similarities to the Saxothuringian Basin to the SE of the MGCR.

In the Saar-Nahe Basin, the borehole Saar 1 has encountered, at a depth of 5662 m, an undeformed granite (Lenz and Müller 1976), interpreted as part of the crystalline basement, which is overlain by unmetamorphosed Middle Devonian through Carboniferous and Permian sediments.

To the north, the Mid-German Crystalline Rise is bordered by the "Northern Phyllite Zone", which can be traced from the southern Hunsrück and Taunus Mts. via the northern Harz Mts. at least up to the Elbe Lineament (Chap. IV.B). East of the Thuringian Forest, the Mid-German Crystalline Rise is accompanied on the SE by a relatively narrow synclinal zone of Cambrian to Ordovician phyllitic rocks. This zone already belongs to the Saxothuringian basin.

References

Behr H (1966) Das metamorphe Grundgebirge im Thüringer Becken. Ber Dtsch Ges Geol Wiss A Geol Paläontol 11: 39–56

Brinkmann R (1948) Die Mitteldeutsche Schwelle. Geol Rdsch 36: 65–66

Hirschmann G, Okrusch M (1988) Spessart-Kristallin und Ruhlaer Kristallin als Bestandteile der Mitteldeutschen Kristallinzone – ein Vergleich. Neues Jahrb Geol Paläontol Abh 177: 1–39

Kneuper G (1966) Zur Entstehung der Saar-Nahe Senke. Z Dtsch Geol Ges 117: 312–322

Lenz H, Müller P (1976) Radiometrische Altersbestimmungen am Kristallin der Bohrung Saar 1. Geol Jahrb A27: 429–432

Schmidt FP, Gebreyohannes Y, Schliestedt M (1986) Das Grundgebirge der Rhön. Z Dtsch Geol Ges Hannover 137: 287–300

Scholtz H (1930) Das varistische Bewegungsbild, entwickelt aus der inneren Tektonik eines Profils von der Böhmischen Masse bis zum Massiv von Brabant. Fortschr Geol Paläontol 8 (25): 235–316

Weber K (1978) Das Bewegungsbild im Rhenoherzynikum – Abbild einer varistischen Subfluenz. Z Dtsch Geol Ges 129: 249–281

Weber K (1981) The structural development of the Rheinische Schiefergebirge. Geol Mijnbouw 60: 149–159

Weber K (1984) Variscan events: early Palaeozoic rift metamorphism and late Palaeozoic crustal shortening. Geol Soc Lond Spec Publ 14: 3–22

Dallmeyer et al. (Eds.)
Pre-Permian Geology of Central and Eastern Europe

Weber K, Behr H-J (1983) Geodynamic interpretation of the mid European Variscides. In: Martin H, Eder FW (eds) Intracontinental fold belts. Springer, Berlin Heidelberg New York, pp 427–469

Weinelt W, Okrusch M, Richter P (1985) Das kristalline Grundgebirge im nördlichen Hochspessart aufgrund der Ergebnisse neuer Tiefbohrungen. Geol Bavarica 87: 39–60

IV.B Lithological Characteristics

G. Hirschmann

1 Introduction

The Mid-German Crystalline Rise (MGCR) comprises metamorphic and magmatic complexes. The former are composed of metasedimentary and migmatic as well as metamagmatic rocks (orthogneisses) of different metamorphic type and degree (Fig. 1).

The magmatic complexes are represented by plutonic bodies of variable composition (mafic, intermediate, acid), age of intrusion (pre-Variscan, Variscan) and degree of deformation (late kinematic, postkinematic). In many cases the genetic origin is a matter of controversial discussion. Plutonic bodies are concentrated in some areas of the MGCR (Pfälzer Wald-Odenwald-southern Spessart, Thüringer Wald and Central Thuringia, Elbe Zone-South Brandenburg, parts of the Foresudetic Block). Most of them are restricted to the southern parts of the zone without going beyond its boundaries. A detailed description of the igneous activity is given in Chapter IV.D.

2 Lithology of Metamorphic Complexes

Hitherto, there has been no general and unified lithological subdivision of the whole MGCR, since the knowledge is very different and incomplete. The best information concerning the distinguishable units and their relative succession comes from the Spessart and Ruhla crystalline complexes.

2.1 Spessart

The principal lithological subdivision of the Spessart crystalline complex (Figs. 2, 4) has been known since Bücking (1892) and Thürach (1893), and was defined later by Bederke (1957), Matthes and Okrusch (1977), Okrusch (1983), and Hirschmann and Okrusch (1988) on the basis of numerous detailed investigations. The following units may be distinguished (from south to north):

Quartz Diorite-Granodiorite Complex

Elterhof formation: striated paragneisses with intercalations of amphibolites, marbles, calc-silicate rocks, hornblende gneisses, and graphitic quartzites. The amphibolites are concentrated in the northern (lower?) part of the formation.

Schweinheim-(?Haibach) formation: Haibach biotite gneiss with interlayers of Schweinheim muscovite-biotite schists and some intercalations of more quartzitic rocks. According to unpublished geochemical (A. Dombrowski) and geochronological data (F. Henjes-Kunst and A. Höhndorf) the Haibach gneiss is possibly derived from an igneous protolith similar to the Rotgneiss complex (M. Okrusch, pers. comm.). The homogeneousness of this formation is, therefore, unclear.

Rotgneiss Complex: metagranitic muscovite-biotite gneiss with some intercalated metabasites and ultramafics.

Mömbris formation: metapelitic staurolite-bearing paragneisses with minor intercalations of metapsammitic paragneisses and calc-silicate rocks. The upper part of the Mömbris formation and the transition to the following Geiselbach formation is characterized by several horizons of quartzite and metabasites (metabasite belt of Hörstein-Huckelheim).

Geiselbach formation: garnet-bearing mica-schists and quartz-micaschists with concordant intercalations of sericite quartzites in the lower part of the formation.

Alzenau formation: alternating amphibolites, biotite gneisses, granitic gneisses and minor quartzites and calc-silicate rocks.

The formations described above group into at least two tectonic units. The *lower unit* (1 in Fig. 2) comprises the Mömbris and Geiselbach for-

Dallmeyer et al. (Eds.)
Pre-Permian Geology of Central and Eastern Europe

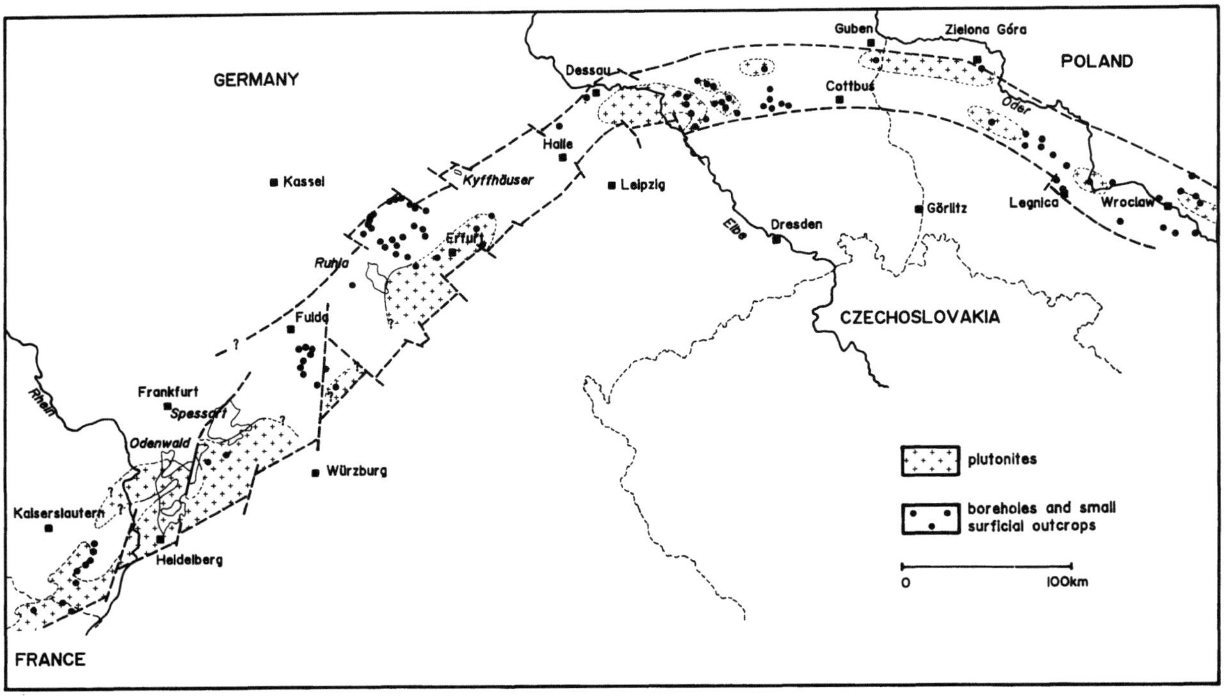

Fig. 1. Outline map of the Mid-German Crystalline Rise in Germany and SW Poland

mations and is separated from the Alzenau formation in the north by the Michelbach fault. According to palynological evidence, the Geiselbach formation or parts of it are of Upper Silurian age (Reitz 1987). The Mömbris formation is mostly considered to be in a normal stratigraphic succession with the Geiselbach formation and may be of Cambrian to Ordovician age (Matthes 1954; Bederke 1957). Possibly parts of it interfinger to the lower parts of the Geiselbach formation (Juckenack 1990).

Probably the Rotgneis complex is part of the lower tectonic unit. It is folded together with the adjacing Mömbris formation. The orthogneisses are derived from shallow granitic to granodioritic intrusions which occurred, according to Rb-Sr isochron ages, in Silurian or even in Ordovician times (see Chap. IV.D). This means that either the intrusions in a very shallow level are more or less synchronous with the sedimentation of the Mömbris and Geiselbach formations or that the orthogneisses represent a separate tectonic subunit.

Elterhof and Alzenau formations are partly of comparable character (Thürach 1893) and may be correlated as parts of an *upper tectonic unit* (Weber and Juckenack 1990; 2 in Fig. 2). The stratigraphic interpretations of the Alzenau and Elterhof formations vary from Precambrian (Bederke 1957; Neumann 1973; Fluck et al. 1980) to Cambrian (Braitsch 1957; Hirschmann and Okrusch 1988).

The position of the Schweinheim-(?Haibach) formation in relation to the lower or upper tectonic unit is not sufficiently clear. Possibly, Schweinheim micaschists and Elterhof formation represent a normal stratigraphic sequence, the Elterhof formation overlying the Schweinheim-(?Haibach) formation (Braitsch 1957; Hirschmann and Okrusch 1988 – as proposed in Fig. 4). In this case, a Precambrian age would be probable. On the other hand, the similarity between the Haibach gneisses and the Rotgneiss complex hints at a near connection with the lower unit. Possibly, mylonitic gneisses of the Elterhof formation support the assumption of a tectonic contact between Schweinheim micaschists and Elterhof formation (see Chap. IV.C).

2.2 Odenwald

The domelike metamorphic complex of the *Böllstein Odenwald* comprises a core of granitic and granodioritic gneisses with intercalated metabasites and some ultramafics and an envelope

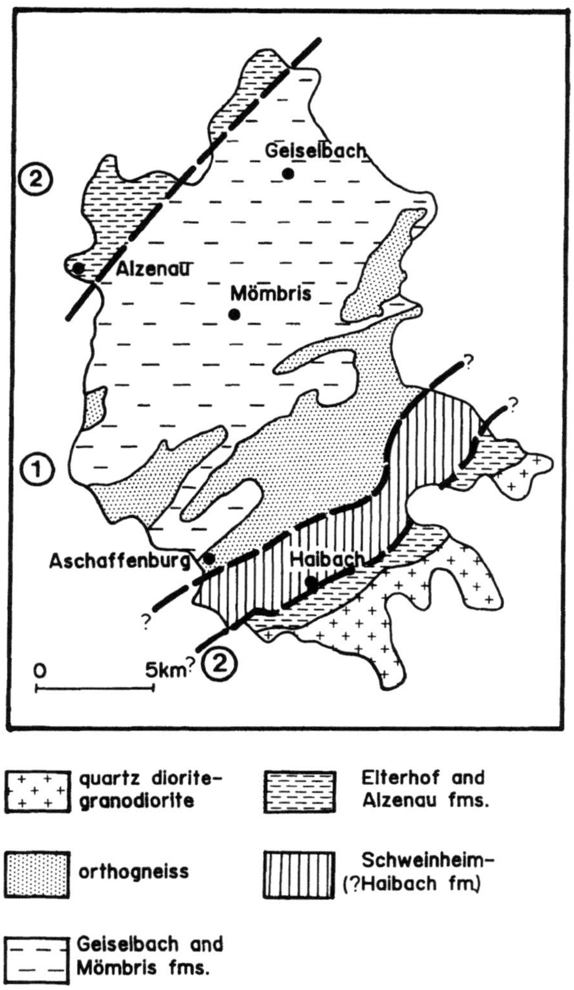

Fig. 2. Sketch map of the crystalline basement of the Spessart. *1* Lower tectonic unit; *2* upper tectonic unit

of micaschists and metagraywacke gneisses with small intercalations of calc-silicate rocks, amphibolites, and graphitic quartzites (Chatterjee 1960; Okrusch et al. 1975; Fluck et al. 1980).

In the *Bergsträsser Odenwald*, the metamorphic complexes form several narrow NE-SW-trending belts which are separated by huge masses of plutonites. Therefore it is very difficult to establish a lithostratigraphic sequence of the metamorphic units. All these belts are characterized by a variegated unit of metapelitic schists and metagraywacke gneisses with intercalations of quartzites (± graphite ± apatite), calc-silicate rocks, marbles and metabasites. Amphibole gneisses represent former marly graywackes or, perhaps, volcanoclastics. Alumina-rich metamorphics may be deduced from an old weathering horizon (Okrusch et al. 1975). Metabasaltic amphibolites with subordinate metaultramafics are abundant within the belts of Eberstadt-Rossdorf, Nieder-Beerbach-Ober-Ramstadt, Auerbach-Groß-Bieberau, Kolmbach-Laudenau, and Flockenbach-Schimmelberg (Schubert 1968; Okrusch et al. 1975). Possibly they represent a separate lithostratigraphic unit (metabasic unit) which may be distinguished from the variegated unit (Fig. 4), though an interfingering of metabasites and marbles has been observed in some places.

Since the Böllstein gneisses are derived from early Paleozoic intrusives, they cannot be regarded as the oldest (Precambrian) lithostratigraphic unit, as has been repeatedly discussed (Neumann 1966; Hoth et al. 1970; Nickel 1975). The variegated units of the Böllstein envelope and the Bergsträsser Odenwald display strong similarities with the Elterhof formation of the Spessart (Chatterjee 1960; Okrusch et al. 1975; Fluck et al. 1980). Perhaps the metabasic unit of the Bergsträsser Odenwald can be correlated with the Alzenau formation or the metabasite-rich (lower?) part of the Elterhof formation. The possibilities of stratigraphic interpretation (from Precambrian through Silurian or Devonian) have been discussed in detail by Nickel (1975) and Okrusch et al. (1975). At present, a lower Paleozoic age seems to be most probable, at least for the variegated units.

2.3 Pfälzer Wald

West of the Rhine valley, gneisses of migmatic or metagranitic character with dm- to m-thick intercalations of amphibolites are exposed near Albersweiler (Zaminer 1957; Flöttmann and Oncken 1990). The correlation with the metabasic unit of the Bergsträsser Odenwald (Fig. 4) is highly speculative. These basement rocks contrast strongly with nearby low grade to unmetamorphic Devonian and Carboniferous sediments (Burrweiler, Neustadt).

2.4 Rhön

In the northeast of the Spessart crystalline complex, the basement of the MGCR has been exposed by several boreholes in the Rhön area (Trusheim 1964; Schmidt et al. 1986). The rocks are interpreted as equivalents of the Spessart units (Schweinheim-?Haibach formation, Elterhof formation, Geiselbach formation, "Rotgneiss" complex) and represent apparently a very similar succession.

2.5 Ruhla

The lithostratigraphic subdivision of the Ruhla crystalline complex (Thüringer Wald) has been worked out by von Gaertner (1950), Neumann (1964, 1974), and Wunderlich (1989). The basement may be subdivided into a western, a central, and an eastern section. These sections are separated from each other by NNE-SSW-trending tectonic zones with intrusions of Variscan age (Ruhla granite, granites, and diorites of Trusetal and Brotterode).

The *western section* (Thal-Ruhla area = 1 in Fig. 3) is occupied by the Ruhla group which can be subdivided into a lower part and an upper part (Fig. 4). The lower part of the group (RI; Neumann 1974) consists of micaschists with intercalations of metabasites, albite gneisses, quartzites, quartzitic micaschists, graphitic micaschists, and rare calc-silicate rocks. The upper part (RII–IV) is represented by micaschists

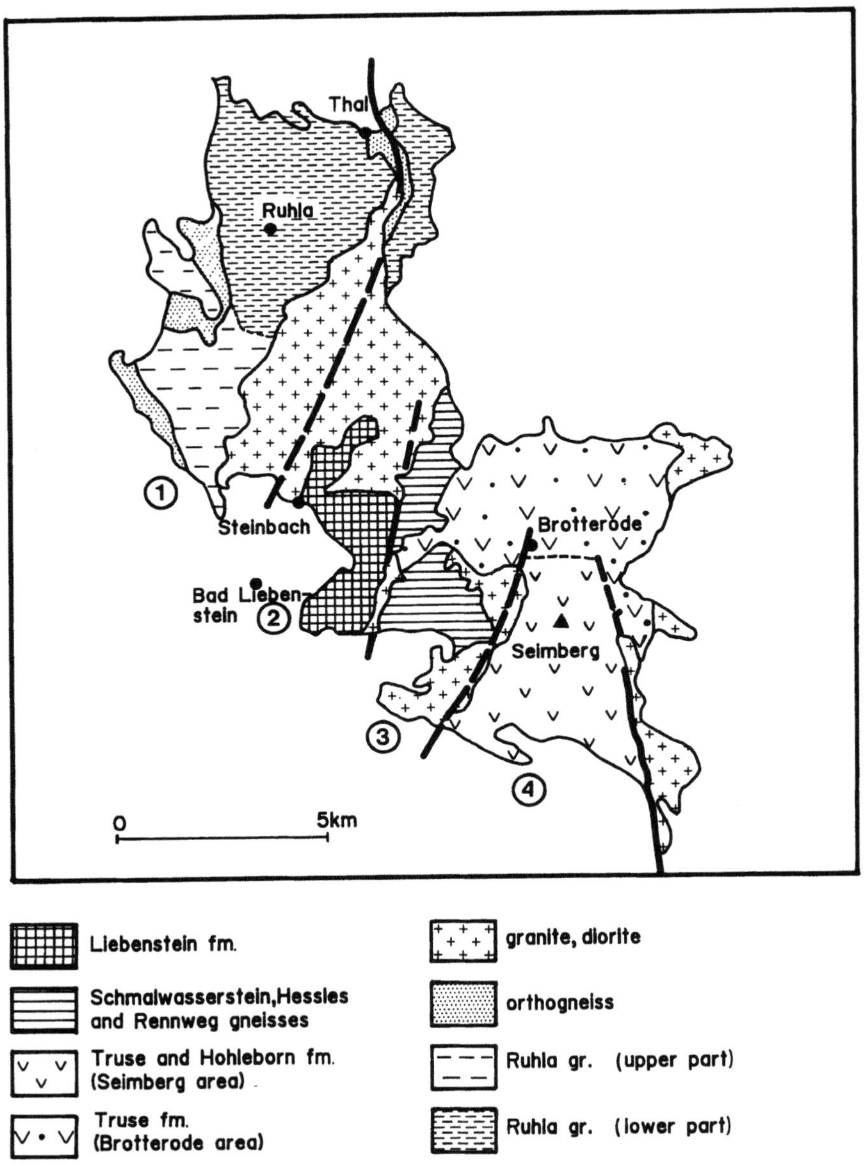

Fig. 3. Sketch map of the Ruhla crystalline complex. *1* Thal-Ruhla area; *2* Liebenstein area; *3* Schmalwasserstein-Rennweg area; *4* Seimberg and Brotterode areas

and metapelitic gneisses, quartzitic micaschists, and horizons of quartzites. Orthogneisses of granitic composition have been interpreted by Neumann (1974) to be parts of the normal lithostratigraphic succession. With respect to their regional distribution, it seems possible, however, to regard them as intrusions in fault or shear zones.

From comparison with Saxothuringian profiles and with the Spessart crystalline complex, the Ruhla group is, probably, of Cambrian (lower part) and Ordovician to Silurian age (upper part). The connection between the upper part of the Ruhla group and the Geiselbach formation (Spessart) is supported by boreholes in the Rhön area (RN 1, SR 12, Hettenhausen, Kleiner Hundskopf).

The *central section* comprises the Liebenstein area (2 in Fig. 3) and the Schmalwasserstein-Rennweg area (3 in Fig. 3). The Liebenstein area is occupied by the Liebenstein formation (Fig. 4). The lower part of it consists of biotite gneisses (Steinbach augen gneiss) with intercalations of quartz-feldspar schists and biotite schists. The upper part is composed of migmatic biotite gneisses, hornblende gneisses and metatectic amphibolites (Liebenstein migmatites). Pre- to synkinematic granites (Dorngehege gneiss) are intimately connected with the Liebenstein formation.

The boundary between the Liebenstein and the Schmalwasserstein-Rennweg areas is a N-S fault zone which is partly marked by dioritic and granitic intrusions (Wunderlich 1989). Characteristic rocks of the Schmalwasserstein-Rennweg area are strongly foliated, metablastic, or phyllonitic gneisses (Schmalwasserstein gneiss, Hessles gneis, Rennweg gneiss). Neumann (1964) regarded these different types as deformed upper parts of the Liebenstein formation ("Haubengneise"), but there are no clear transitions between the Liebenstein gneisses and the assumed derivates.

The *eastern section* of the Ruhla crystalline complex (4 in Fig. 3) is composed of the Truse and the Hohleborn formations (Fig. 4). The Truse formation in its typical development (Seimberg area) is characterized by biotite and muscovite-biotite gneisses with intercalations of amphibolites, amphibole-biotite schists, quartzites, graphitic quartzites, garnetiferous layers, and small calc-silicate or marble occurrences. At least a partial correlation with the migmatic complexes of the Brotterode area is likely. The Hohleborn formation is represented by an alternation of feldspar gneisses and muscovite-biotite gneisses (metagraywackes and metapelites) with intercalated metabasalts.

The age relations between the Truse formation and the Liebenstein formation are unclear. Indications of an angular unconformity (Neumann 1964) seem to be ambiguous (Hoth 1968; Wunderlich 1989). Quartzites of the Truse formation overlie the Rennweg gneiss rather with a tectonic than with a normal transgressive contact (Mittlerer Beerberg). It is, therefore, hard to judge whether the Liebenstein formation is older than the Truse formation and represents a pre-Brioverian (Moldanubian? Dalslandian?) basement (Neumann 1964, 1988) or not. Notwithstanding these uncertainties, the Liebenstein and Schmalwasserstein-Hessles gneisses are most probably derived from Proterozoic protoliths. Truse and Hohleborn formations may be Upper Proterozoic units with strong similarities to the Brioverian Proterozoic of Central Bohemia or Bohemicum (Hoth 1968; Neumann 1974; Wunderlich 1989); but, according to original stratigraphic interpretations (von Gaertner 1950; Neumann 1964) and lithological-geochemical comparison, the possibility of a (partial) correlation with the Paleozoic Ruhla group and the Elterhof/Alzenau and Mömbris formations of the Spessart cannot be ruled out (Hirschmann and Okrusch 1988; Estrada 1991).

2.6 Central Thuringia

In the area between the Thüringer Wald and the city of Halle, the basement is known from the surface outcrops of the Kyffhäuser (Hesemann 1930; Seim 1967; Neumann 1968, 1974) and from a number of boreholes (Erzberger et al. 1962; Gottesmann and Steinike 1962; Behr 1966; Hoth 1968; Lorenz 1972; Neumann 1974; Röllig et al. 1989).

A complex of gneisses with intercalations of metabasites and calc-silicate rocks is known from the Kyffhäuser and some boreholes near Mühlhausen (Obermehler 1), north of Weimar (Roldisleben 1) , and north of Halle (Plötz 472). It represents, possibly, equivalents of the Truse formation. Micaschists, sericitic gneisses, and garnetiferous phyllites in western Thuringia may be equivalents of upper parts of the Ruhla group or somewhat younger. However, in this area, the boundary between the MGCR and the Northern Phyllite Zone is not sufficiently clear. In western and central Thuringia, several boreholes exposed deformed granitoids, but the distinction between prekinematic orthogneisses and syn- to late kinematic granitoids is not always clear.

2.7 South Brandenburg

The lithological units of the MGCR in South Brandenburg and the adjacent area west of the Elbe river (Möbus 1963; Erzberger et al. 1964; Hoth 1968; Brause 1969; Röllig et al. 1989; Franzke et al. 1990) differ greatly from the units farther west. The paragneisses of Luckau are regarded to be the oldest unit (Fig. 4). The Prettin-Drehna formation of variegated character and low to medium grade metamorphism (phyllites, quartzitic phyllites, micaschists with intercalations of metabasites and carbonate rocks) is the most typical unit. It seems to be older than an unmetamorphic unit of graywacke composition (Calau, Herzberg) which can be compared with the Leipzig and Kamenz formations (North Saxon Group) of Upper Brioverien age. Metamorphic granitoids have long been known from the area of Dessau (Schüller 1951). Large plutonic bodies (Pretzsch, Prettin, Schönewalde) varying from gabbroic to granitic composition are, according to geochronological data, of early Paleozoic age (Röllig et al. 1990).

3 General Aspects and Correlations

As a whole, the MGCR shows a great compositional heterogeneity. Due to the fact that the basement is exposed only in isolated outcrops and boreholes, it is very difficult to recognize the mutual relations between lithological units. Paleontological and geochronological data are not sufficient to establish a general stratigraphic scheme for the entire zone. Hitherto existing attempts at lithostratigraphic interpretation and correlation rely greatly on the comparison with Paleozoic and Proterozoic units of the Saxothuringicum and Moldanubicum in the south of the MGCR, but the possibility of connections with units of the Rhenohercynicum in the north should also be taken into consideration. Probably, the MGCR is composed of several segments of different lithological character and tectonometamorphic development. According to the descriptions of Section 2, a subdivision into at least four segments or tectonostratigraphic units is proposed (Figs. 4, 5):

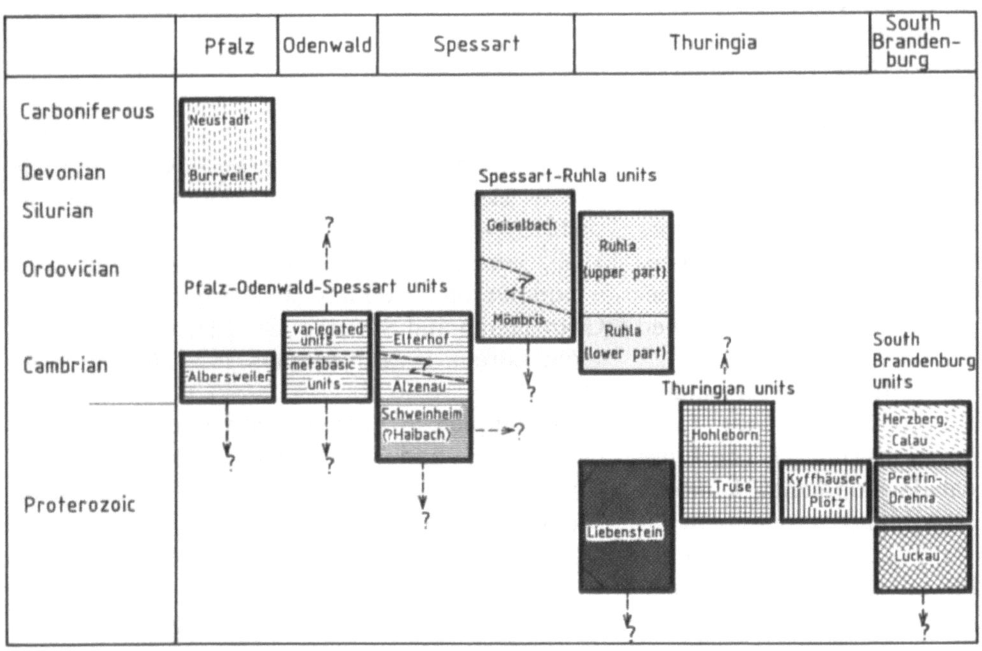

Fig. 4. Proposed stratigraphic scheme

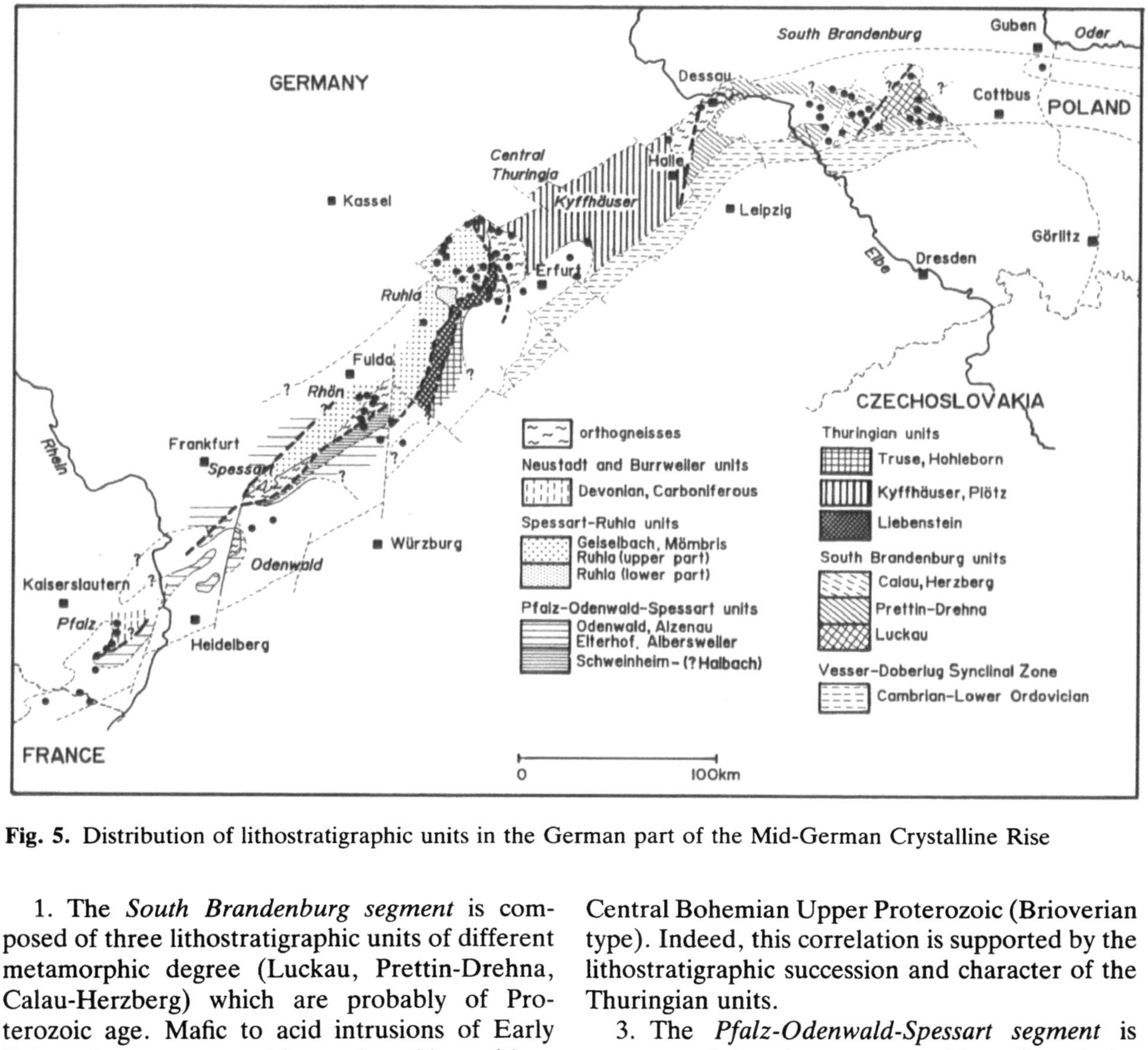

Fig. 5. Distribution of lithostratigraphic units in the German part of the Mid-German Crystalline Rise

1. The *South Brandenburg segment* is composed of three lithostratigraphic units of different metamorphic degree (Luckau, Prettin-Drehna, Calau-Herzberg) which are probably of Proterozoic age. Mafic to acid intrusions of Early Paleozoic age support this stratigraphic position. The South Brandenburg units exhibit strong similarities with units of the Lausitz and the Elbe zone (eastern Saxothuringicum-western Lugicum).

2. The *Thuringian segment* comprises the central and eastern parts of the Ruhla crystalline complex and large parts of central Thuringia. Among the Thuringian units, the Liebenstein formation is generally regarded as Proterozoic, but the assumption of a pre-Brioverian (Moldanubian? Dalslandian?) age has not been substantiated until now. The metabasite-rich units of the Kyffhäuser and some boreholes are probably equivalents of the Truse formation. Truse and Hohleborn formations are considered by some authors to be of early Paleozoic age, but most authors regard them as equivalents of the Central Bohemian Upper Proterozoic (Brioverian type). Indeed, this correlation is supported by the lithostratigraphic succession and character of the Thuringian units.

3. The *Pfalz-Odenwald-Spessart segment* is represented by the metamorphic units of the Odenwald, the upper tectonic unit of the Spessart (Elterhof and Alzenau formations) and, with some reservation, the rocks of Albersweiler/Pfälzer Wald. The variegated units of the Spessart (especially the Elterhof formation) and of the Odenwald are, according to correlations made by most authors, of Paleozoic age (Cambrian? Silurian/Devonian cannot be excluded). The same may apply to the metabasic units including the Alzenau formation (and Albersweiler?), since they are closely connected with the variegated units. Only the Schweinheim(-?Haibach) formation of the Spessart might be of Proterozoic age, but the correlation with the Liebenstein formation (Ruhla) and the interpretation as part of a pre-Brioverian basement are doubtful. As a whole, the Pfalz-Odenwald-Spessart units

display some similarities with Late Proterozoic/ Early Paleozoic profiles in some parts of the Saxothuringicum. Possibly the variegated and metabasic units are equivalents of the volcano-sedimentary formations of the Vesser area (Cambrian? of the Vesser-Doberlug Synclinal Zone; Bankwitz and Bankwitz 1990; Estrada 1991).

4. To the *Spessart-Ruhla segment* are ascribed the lower tectonic unit of the Spessart (Mömbris and Geiselbach formations) and their equivalents in the Rhön area as well as the Ruhla group (western segment of the Ruhla crystalline complex) and metapelitic rocks known from boreholes in western Thuringia. According to lithostratigraphic correlations and the biostratigraphic evidence for the Geiselbach formation the Spessart-Ruhla units probably represent the time interval from Cambrian through Silurian. Whereas the lower parts of the profiles may, according to earlier correlations, correspond to Cambrian and Ordovician units of the Saxothuringicum, the upper parts (Geiselbach formation – Silurian?) seem to differ greatly from the Saxothuringian development.

References

Bankwitz E, Bankwitz P (1990) Lower Palaeozoic series at the southern margin of the Central German Crystalline Zone. In: Franke W, Weber K (eds) IGCP 233, Int Conf Paleozoic orogens in central Europe, Göttingen-Giessen, Aug–Sept 1990, Abstr

Bederke E (1957) Alter und Metamorphose des kristallinen Grundgebirges im Spessart. Abh Hess Landesamt Bodenforsch Wiesbaden 18: 7–19

Behr H-J (1966) Das metamorphe Grundgebirge im Thüringer Becken. Ber Dtsch Ges Geol Wiss Berlin A11: 39–56

Braitsch O (1957) Beitrag zur Kenntnis der kristallinen Gesteine des südlichen Spessarts und ihrer geologisch-tektonischen Geschichte. Abh Hess Landesamt Bodenforsch 18: 21–72

Brause H (1969) Das verdeckte Altpaläozoikum der Lausitz und seine regionale Stellung. Abh Dtsch Akad Wiss, Klasse Bergbau Hüttenwes Montangeol Berlin 1968 (1): 143 pp

Bücking H (1892) Der nordwestliche Spessart. Abh Preuss Geol Landesanst NF Berlin 12: 274 pp

Chatterjee ND (1960) Geologische Untersuchungen im Kristallin des Böllsteiner Odenwaldes. Neues Jahrb Geol Paläontol Abh Stuttgart 111: 137–180

Erzberger R, Gottesmann W, Kaemmel T, Kampe A, Nöldeke W, Steinike K (1962) Untersuchungsergebnisse an Neuaufschlüssen in kristallinen Gesteinen zwischen Halle und Cottbus. Teil 1: Regionalgeologische Situation und Bedeutung der Bohraufschlüsse Plötz 472, Doberlug T 52 und Colochau 3/60. Geologie Berlin 11: 633–639

Erzberger R, Kaemmel T, Meissner B, Nöldeke W, Siegert C (1964) Zur Stellung des südbrandenburgischen Grundgebirges innerhalb der saxothuringischen Zone im Lichte von Neuaufschlüssen. Krystalinikum Prag 2: 23–38

Estrada S (1991) Zur Lithologie und Geochemie des Kambroordoviziums des Thüringer Waldes. 7. Rundgespräch Geodynamik des Variszikums, Freiberg/Sachsen, 8. bis 10. November 1991, Kurzfassungen: 13

Flöttmann Th, Oncken O (1990) Constraints on the tectonothermal evolution of the Mid-German Crystalline Rise west of the river Rhine. In: Franke W, Weber K (eds) IGCP 233, Int Conf Paleozoic orogens in central Europe, Göttingen-Giessen, Aug–Sept 1990, Abstr

Fluck P, Maass R, von Raumer JF (1980) The Variscan Units east and west of the Rhine Graben. Mém BRGM Orléans 108: 112–131

Franzke HJ, Röllig G, Kampe A (1990) Magmatic and structural evolution of the area of the Central German Crystalline Rise (CGCR) between Kyffhäuser mountains and Lusatia. In: Franke W, Weber K (eds) IGCP 233, Int Conf Paleozoic orogens in central Europe, Göttingen-Giessen, Aug–Sept 1990, Abstr

Gaertner H-R von (1950) Probleme des Saxothuringikums. Geol Jahrb Hannover/Celle 65: 409–450

Gottesmann W, Steinike K (1962) Untersuchungsergebnisse an Neuaufschlüssen in kristallinen Gesteinen zwischen Halle und Cottbus. Teil 2: Zur Petrographie der kristallinen Gesteine der Bohrung Plötz 472. Geologie Berlin 11: 640–662

Hesemann J (1930) Das Kyffhäuser-Kristallin. Jahrb Preuss Geol Landesanst Berlin 50 (II): 607–651

Hirschmann G, Okrusch M (1988) Spessart-Kristallin und Ruhlaer Kristallin als Bestandteile der Mitteldeutschen Kristallinzone – ein Vergleich. Neues Jahrb Geol Paläontol Abh Stuttgart 177: 1–39

Hoth K (1968) Präkambrium oder Kryptozoikum. Grundriss Geol DDR Berlin 1: 75–94

Hoth K, Hirschmann G, Lorenz W (1970) Das Jungpräkambrium im Bereich der Varisziden und Kaledoniden West- und Nord-Europas und seine Beziehungen zu den paläozoischen Entwicklungsetappen. Ber Dtsch Ges Geol Wiss Berlin A 15: 379–424

Juckenack C (1990) Beitrag der Anisotropie der magnetischen Suszeptibilität (AMS) für Struktur- und Gefügeuntersuchungen von Metamorphiten: Einzelbeispiele und regionale Anwendung im Spessart-Kristallin. Dissertation, Georg-August-Univ Göttingen, 176 pp

Lorenz W (1972) Alterskriterien für das Präkambrium am Nordrand der Böhmischen Masse. Teil I: Zur Abgrenzung und absoluten Altersdatierung. Geologie Berlin 21: 405–417

Matthes S (1954) Die Paragneise im mittleren kristallinen Vor-Spessart und ihre Metamorphose. Abh Hess Landesamt Bodenforsch Wiesb 8: 1–86

Matthes S, Okrusch M (1977) The Spessart crystalline complex, north-west Bavaria: rock series, metamorphism, and position within the Central German Crystalline Rise. Coll Int Cent Natl Rech Sci (Rennes 1974) Paris 243: 375–390

Möbus G (1963) Der Kristallin-Anteil der Tiefbohrung Luckau II, südlich von Berlin. Geologie Berlin 12: 264–282

Neumann W (1964) Zum Stand der Erforschung des kristallinen Grundgebirges von Ruhla-Brotterode. Abh Dtsch Akad Wiss Berlin, Klasse Bergbau Hüttenwes Montangeol Berlin 2: 327–348

Neumann W (1966) Versuch eines lithostratigraphischen Vergleiches von Grundgebirgsanschnitten im Bereich der Mitteldeutschen Schwelle. Geologie Berlin 15: 942–962

Neumann W (1968) Das Grundgebirge am Kyffhäuser (Nord-Thüringen) – Lithostratigraphie, Tektonik, Regionalmetamorphose und Migmatisation. Geologie Berlin 17 (Beih 63): 1–79

Neumann W (1973) Zum Stockwerkbau im Bereich der "Mitteldeutschen Kristallinzone" (speziell im Ruhlaer Kristallin). Veröff Zentralinst Physik Erde Potsdam 14: 391–409

Neumann W (1974) Mitteldeutsche Kristallinzone. In: Hoppe W, Seidel G (eds) Geologie von Thüringen. Gotha-Leipzig, pp 77–118

Neumann W (1988) Mitteldeutsche Kristallinzone. In: Klassische geologische Gebiete in Mitteleuropa. Exkursionsführer. Zentralinst Physik Erde, Potsdam, pp 180–197

Nickel E (1975) Geologische Position und Petrogenese des kristallinen Odenwaldes. Aufschluss Heidelberg Sonderbd 27: 1–25

Okrusch M (1983) The Spessart Crystalline Complex, Northwest Bavaria. DMG SFMC Joint Meet, Excursion E4. Fortschr Miner Stuttgart 61 (Beih 2): 135–169

Okrusch M, von Raumer J, Matthes S, Schubert W (1975) Mineralfazies und Stellung der Metamorphite im kristallinen Odenwald. Aufschluss Heidelberg Sonderbd 27: 109–134

Reitz E (1987) Palynologie in metamorphen Serien: I. Silurische Sporen aus einem granatführenden Glimmerschiefer des Vor-Spessart. Neues Jahrb Geol Paläontol Monatsh Stuttgart 11: 699–704

Röllig G, Viehweg M, Kampe A (1989) Zur Wirksamkeit von meridionalen und äquatorialen Strukturen im Gebiet der Mitteldeutschen Schwelle zwischen Halle und Schönewalde. Z Angew Geol Berlin 35: 297–300

Röllig G, Bräuer H, Viehweg M, Gruner B, Kabardin B, Wünsch K, Haase G (1990) Altersstellung und petrogenetische Charakteristik der Plutonite im Gebiet des Zentralteiles der Mitteldeutschen Schwelle. Z Angew Geol 36: 208–212

Schmidt F-P, Gebreyohannes Y, Schliestedt M (1986) Das Grundgebirge der Rhön. Z Dtsch Geol Ges Hannover 137: 287–300

Schubert W (1968) Die Amphibolite des prävaristischen Schieferrahmens im Bergsträsser Odenwald. Neues Jahrb Miner Abh Stuttgart 108: 69–110

Schüller A (1951) Petrographie und tektonische Stellung des metamorphen Granits von Dessau. Neues Jahrb Miner Abh 82: 121–142

Seim R (1967) Der Para-Biotit-Plagioklasgneis des Kyffhäuserkristallins, seine metatektischen und metablastischen Derivate und seine Einlagerungen. Geologie Berlin 16 (Beih 56): 1–85

Thürach H (1893) Über die Gliederung des Urgebirges im Spessart. Geognost Jh Kassel 5: 160 pp

Trusheim F (1964) Über den Untergrund Frankens. Geol Bavarica München 54: 92 pp

Weber K, Juckenack C (1990) The structure of the Spessart Mts. crystalline basement and its position in the frame of the European Variscides. In: Franke W, Weber K (eds) IGCP 233, Int Conf Paleozoic orogens in central Europe, Göttingen-Giessen, Aug–Sept 1990, Field Guide: Mid-German Crystalline Rise & Rheinisches Schiefergebirge, pp 101–114

Wunderlich J (1989) Neue Ergebnisse zur Geologie des südöstlichen Ruhlaer Kristallins. Freiberger Forschungsh Leipzig C 429: 7–32

Zaminer Ch (1957) Geologisch-petrographische Untersuchungen im Grundgebirge der Pfalz. Mitt Pollichia Bad Dürkheim III (4): 7–33

IV.C Structure

IV.C.1 The Spessart Crystalline Complex

K. Weber

The Spessart Crystalline Complex (SCC) is an approx. 25-km-wide, NE-trending asymmetric antiform with a steeply to vertically dipping southeastern and a more gently dipping northwestern limb (Figs. 1, 2 and 3). Lineations and fold axes plunge toward ENE and WSW. The core of the antiform is occupied by orthogneisses ("Rotgneis") which are homotactically deformed together with their hanging wall paragneisses of the Mömbris Fm. The Michelbach fault in the northwestern part of the SCC separates the variegated amphibolite facies Alzenau Fm. in the NW from the upper greenschist facies, quartzitic-pelitic Geiselbach Fm. in the SE. On the base of lithological and geochemical data (Okrusch, Chap. IV.E) the variegated Elterhof Fm. of the southeastern part is correlated with the variegated Alzenau Fm. in the northwestern part of the Spessart Mts. The Alzenau Fm. forms a tectonic klippe, whereas the Schweinheim-Haibach Fm., the Orthogneiss Complex as well as the Mömbris and Geiselbach Fms. are exposed in a tectonic window. HT-mylonitic gneisses ("Plattengneise") of the Elterhof Fm. suggest a tectonic contact between the Elternhof and Schweinheim-Haibach Fms. The Schweinheim-Haibach Fm. contains highly strained, fine-grained gneisses (i.e., the Haibach Gneiss) and might also be allochthonous. The whole structure represents a tectonic ramp structure with an out-of-sequence thrust of the Alzenau-Elterhof nappe (Figs. 1 and 2).

The only time markers are:

1. Late Silurian age of the Geiselbach Fm. according to palynological data (Reitz 1987).
2. Late Ordovician intrusion age (440 Ma Rb/Sr whole rock isochrone age; Nasir et al. 1991) of the granitic to granodioritic protoliths of the red gneisses, which indicates, that the Mömbris Fm. is not younger than Late Ordovician.
3. Upper Carboniferous (324 to 318 Ma; Lippolt 1986; Nasir et al. 1991) hornblende, muscovite, and biotite K/Ar ages. Since the maximum metamorphic temperature was in the range of 620 °C, and the Hbl, Ms, and Bio cooling ages are not significantly different from each other, Nasir et al. (1991) assume rapid uplift and cooling soon after culmination of Variscan deformation and amphibolite facies metamorphism.

D1-folds in the dm- to m-scale are most frequent in the staurolite bearing gneisses of the Mömbris Fm. and in quartzites and micaschists of the Geiselbach Fm. D1-folds are asymmetric, disharmonic internal folds (Figs. 2 and 3). In the quartzites of the Geiselbach Fm., these folds are cylindrical, but tend to be noncylindrical in some places of higher grade rocks, particularly in the staurolite gneisses of the Mömbris Fm. Two stretching lineations are observed. The younger lineation, L1,2, is parallel to the macroscopic fold axes, whereas the older lineation, L1,1, forms an angle to the axes of cylindrical folds (Figs. 4 and 5a). The angle between the two lineations is less than 30°, and may increase inside highly strained overturned limbs and towards the Michelbach fault. Where the angle is small, there are transitions between both orientations, so that the two lineations are not very well defined as separate directions. On the normal limbs the older lineation deviates dextrally and on the overturned limbs sinistrally from the fold axis. In cross sections perpendicular to the fold axes, both lineations have the characteristics of an intersection lineation between bedding and foliation. The younger lineation, which is parallel to the fold axes, is more widely spaced than the older and accompanied by crenulation and polygonization of biotite and muscovite.

Folding and stretching are interpreted as continuous processes in a sinistral transpressive strain field. Folds and L1,1 lineations, which are formed in the course of transpression, are initiated perpendicular to the direction of first increments of shortening and rotated in the sence of superim-

Dallmeyer et al. (Eds.)
Pre-Permian Geology of Central and Eastern Europe

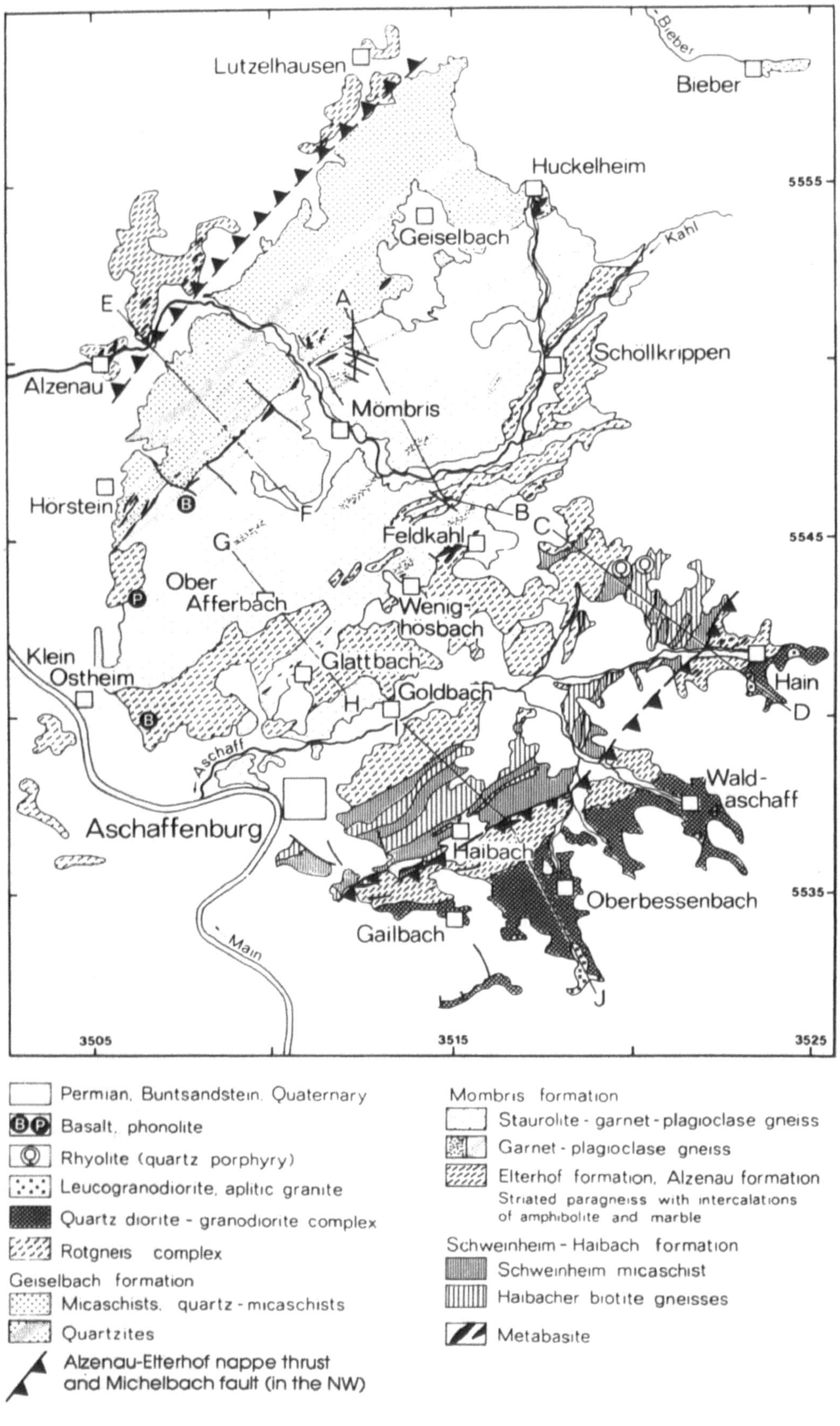

Fig. 1. Simplified geological map of the Spessart Crystalline Complex. (After Hirschmann and Okrusch 1988)

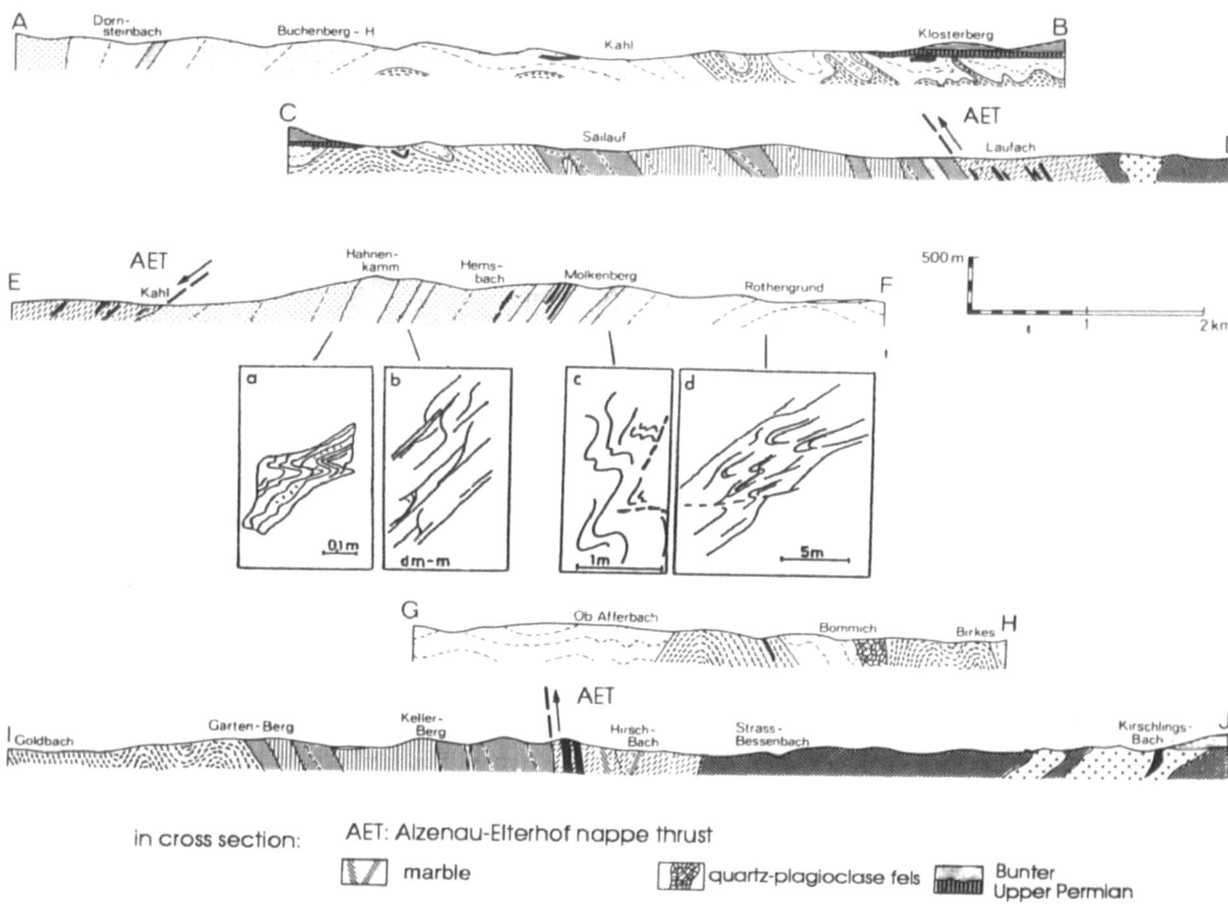

Fig. 2. Geological cross sections through the Spessart Crystalline Complex. Legend see Fig. 1 (After Hirschmann and Okrusch 1988). Insets *a–d* show NW-facing recumbent fold structures on the NW part of the Spessart Antiform

posed shear. During progressive transpression, further increments of folding deviate dextrally from earlier increments and younger lineations deviate dextrally in the normal limbs and sinistrally in the overturned limbs. This indicates refolding of foliation and lineation during progressive folding by rolling of hinges, i.e., without generation of new macroscopic folds (Fig. 4). In the allochthonous Alzenau Fm. nearly homoaxially refolded folds were observed, which are strongly stretched parallel to the fold axes. Younger increments of folding and stretching deviate slightly dextrally from earlier increments. In some cases fold structures of the Geiselbach Fm. are overprinted by the younger foliation L1,2. In that case the intersection lineation deviates dextrally from the fold axis in both limbs.

F1 folds and L1 lineations are refolded by the asymmetric large scale Spessart antiform around its NE-trending axis. Therefore, the Spessart antiform is younger than D1 folding and nappe transport. Due to updoming the NW-facing D1 folds are rotated into an upright position on its southeastern limb and into recumbent up to inverted attitude on the northwestern limb. A down-dip-stretching lineation in the HT-mylonitic Plattengneis of the Elterhof Fm. on the steeply dipping SE limb of the Spessart Antiform is related to the formation of the ramp structure. A horizontal or gently dipping crenulation cleavage is found in rocks with a subvertical metamorphic foliation. It is interpreted as postmetamorphic collapse structure, produced during uplift and crustal extension.

There are no indications for frequent SE-facing folds which are related to an asymmetric, SE-facing Spessart anticline as postulated by Doutsos (1979). Although there is no stratigraphic proof for the single case, the dominant NW displacement and tectonic reduction of the hanging wall

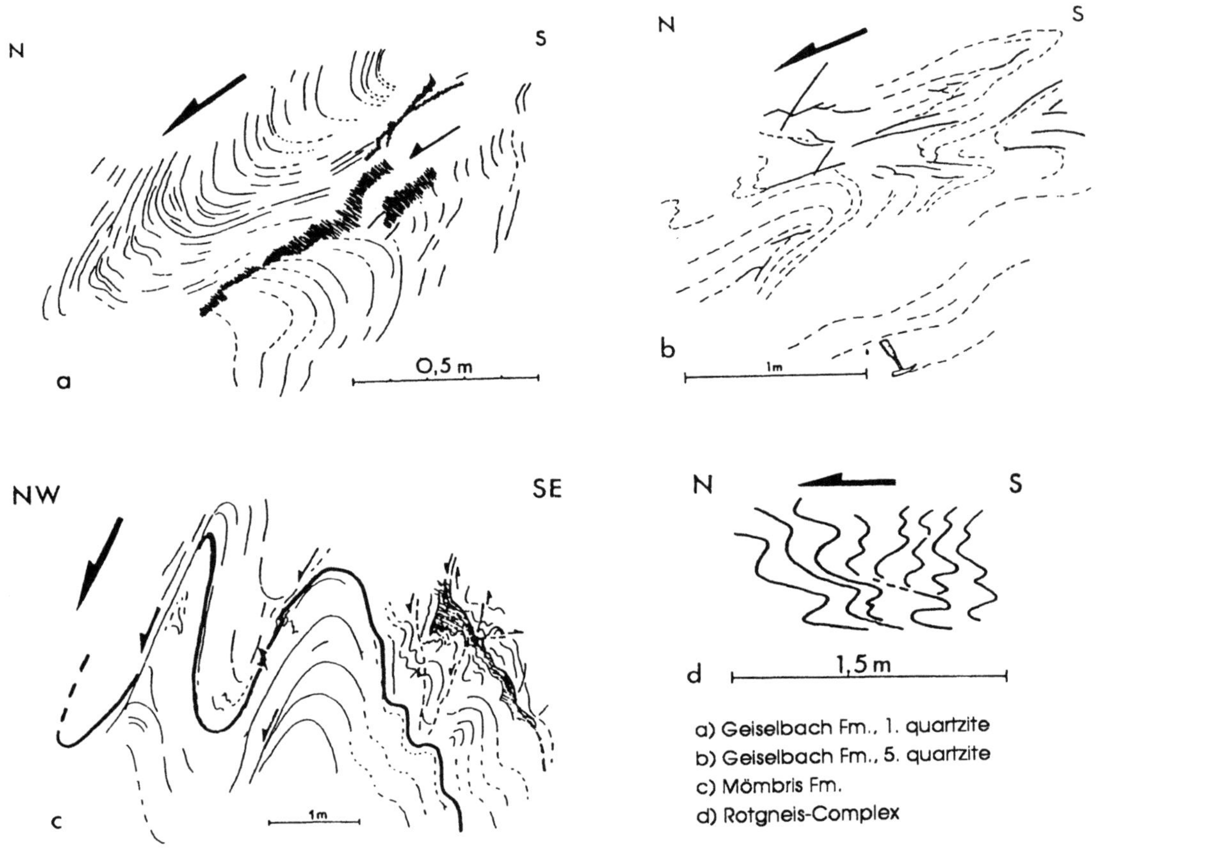

Fig. 3a–d. Fold structures from the northern and central part of the Spessart Antiform. The dominant NW displacement of the hanging wall limb of the folds point to NW-facing folds which are rotated into recumbent attitude by updoming of the Spessart Antiform (see also Fig. 2)

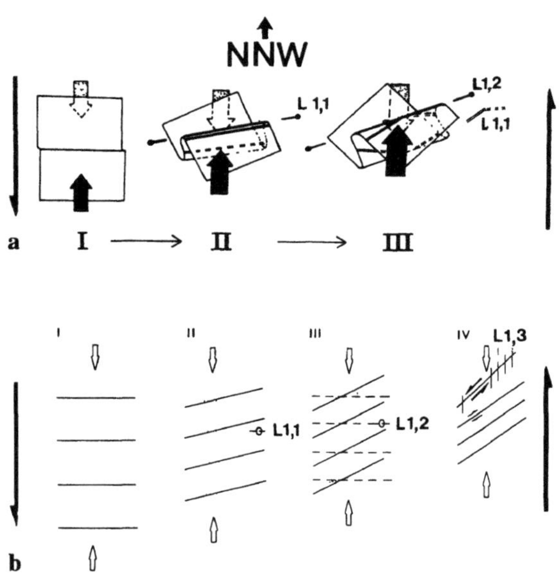

Fig. 4. a Progressive folding by rolling of hinges and refolding of lineation in the course of sinistral transpression. **b** More schematic presentation of the dextral deviation of the younger lineations from the fold axis during sinistral transpression. Stage *IV* shows the L1,3 lineation, which is related to the NW-dipping extensional crenulation cleavage in the mylonites and ultramylonites along the Michelbach fault zone

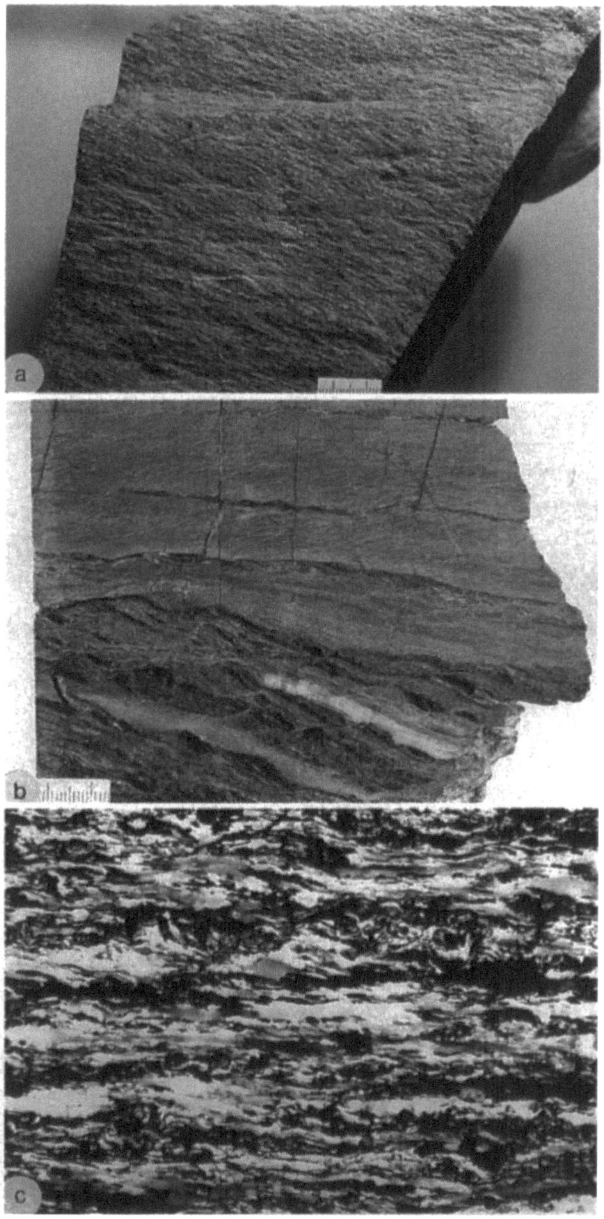

Fig. 5. a Quartzite of the Geiselbach Fm. (4. Quartzite) showing L1,1 and L1,2 lineation. L1,2 is parallel to the fold axis. Both lineations are intersection lineations and reveal some stretching parallel L. L1,2 deviates sinistrally from L1,1 due to the upside down orientation of the sample. **b** Sinistral extensional crenulation cleavage in ultramylonitic quartzite and mica schist of the Geiselbach Fm. near the Michelbach fault. **c** Mylonitic fabric and extensional crenulation cleavage of the sinistrally sheared quartzite of **b**

limb of the folds, the absence of a major fold structure, to which the seemingly overturned limbs could be assigned, the general stratigraphic younging, the presence of a normal metamorphic sequence (see Chap. IV.E), the decreasing grade of quartz recrystallization to the NW, as well as the NW-ward increasing dominance of basel glide over prism glide (Juckenack 1990) point to NW-facing folds which are rotated into recumbent attitude (Figs. 2 and 3).

The Michelbach fault (Figs. 1 and 2) has been interpreted as a SE-directed thrust fault (Thürach 1893; Bederke 1957; Plessmann 1957; Murawski 1958). The Michelbach fault is not exposed, but clearly documented by a metamorphic break between the Alzenau and Geiselbach Fms. It is interpreted as a NW-facing thrust fault by Weber (1981), Behr and Heinrichs (1987), and Hirschmann and Okrusch (1988), which was rotated into NW-dipping attitude during updoming of the Spessart antiform.

In the area of the inferred Michelbach fault, mylonites and ultramylonites are observed (Schneider 1962), which are strongly cataclastically overprinted in some places. Various shear criteria (Weber and Juckenack 1990) document sinistral displacement along the fault zone (Fig. 5). A N-plunging intersection lineation in mylonitic rocks along the Michelbach fault zone is related to a NW-dipping extensional crenulation cleavage (Fig. 5b), which results from sinistral transpression around the fault zone (Fig. 5c). It seems possible that the sinistral shear zone has reactivated the former thrust plane. New processing data of the reflection seismic DEKORP 2S line, however, point to a crustal scale fault. This sinsitral fault probably continues toward SSW in the Otzberg Fault, which separates the eastern (Böllstein) from the western (Bergstraße) Odenwald (see Chap. IV.C.4).

In the middle and upper crust. The offset of the Moho possibly is younger and related to the Tertiary Wetterau rift structure. The solid lines do not represent real fault zone, except at the surface. They represent tentative interpretations of displacements inside the crust based on the overall structural development; − and + indicate the sinistral displacement along the infered transform fault. This sinistral fault is interpreted as the northeastern prolongation of the Otzberg-fault, which separates the western part of the Odenwald from the Böllstein Odenwald in the east (see Chaps. IV.C.2 and IV.C.4). NW of the Spessart Mts this fault turns into the main thrust between the Northern Phyllite Zone and the Mid-German Crystalline Rise (see Chap. IV.C.4, Fig. 1).

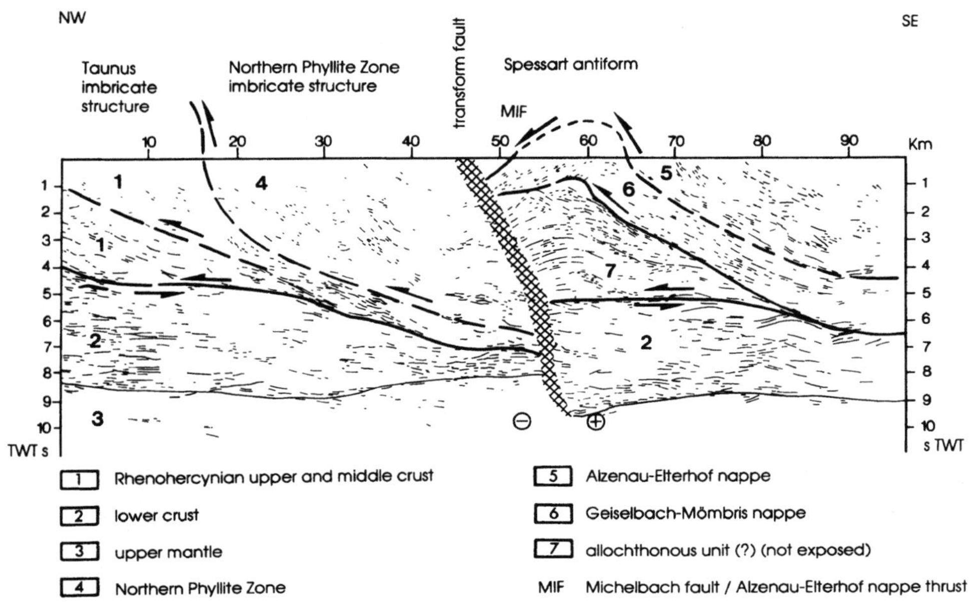

Fig. 6. Tentative interpretation of the seismic sections DEKORP 2N and 2S across the Rhenohercynican/Saxothuringian boundary. The upper crust of the Taunus Mts. and the Northern Phyllite Zone is rather transparent due to the steeply dipping attitude of bedding and foliation. The fault zone between the Northern Phyllite Zone and Spessart antiform, inferred from the surface geology, might be seen in the middle and upper crust. The offset of the Moho is possibly younger and related to the Tertiary Wetterau rift structure. The *solid lines* do not represent real fault zones, except at the surface. They represent tentative interpretations of displacements inside the crust based on the overall structural development. – and + indicate the sinistral displacement along the inferred transform fault. This sinistral fault is interpreted as the north-eastern prolongation of the Otzberg fault, which separates the western part of the Odenwald from the Böllstein Odenwald in the east (see Chaps. IV.C.2 and IV.C.4). NW of the Spessart Mts. this fault turns into the main thrust between the Northern Phyllite Zone and the Mid-German Crystalline Rise (see Chap. IV.C.4, Fig. 1)

References

Bederke E (1957) Alter und Metamorphose des kristallinen Grundgebirges im Spessart. Abh Hess Geol Landesamt Bodenforsch 18: 7–20

Behr HJ, Heinrichs T (1987) Geological interpretation of DEKORP 2-S. A deep seismic reflection profile across Saxothuringian. Tectonophysics 142: 173–202

Doutsos T (1979) Tektonische Analyse des nördlichen kristallinen Spessarts. Geol Bavarica 79: 127–176

Hirschmann G, Okrusch M (1988) Spessart-Kristallin und Ruhlaer Kristallin als Bestandteile der Mitteldeutschen Kristallinzone – ein Vergleich. Neues Jahrb Geol Paläontol Abh 177: 1–39

Juckenack C (1990) Beitrag der Anisotropie der magnetischen Suszeptibilität (AMS) für Struktur- und Gefügeuntersuchungen von Metamorphiten: Einzelbeispiele und regionale Anwendung im Spessart-Kristallin. Dissertation, Georg-August-Univ Göttingen, 176 pp

Lippolt HJ (1986) Nachweis altpaläozoischer Primäralter (Rb/Sr) und karbonischer Abkühlungsalter (K/Ar) der Muskovit-Biotit-Gneise des Spessarts und der Biotit-Gneise des Böllsteiner Odenwaldes. Geol Rdsch Hanover 75/3: 569–583

Murawaski H (1958) Der geologische Bau des zentralen Vorspessarts. Z Dtsch Geol Ges 110: 360–388

Nasir S, Okrusch M, Kreuzer H, Lenz H, Höhndorf A (1991) Geochronology of the Spessart crystalline complex, Mid-German Crystalline Rise. Mineral Petrol 43: 39–55

Plessman W (1957) Zur Baugeschichte des nordwestlichen kristallinen Spessart. Abh Hess Geol Landesamt Bodenforsch 18: 149–166

Reitz E (1987) Palynologie in metamorphen Serien: I. Silurische Sporen aus den granatführenden Glimmerschiefern des Vorspessart. Neues Jahrb Geol Paläontol Monatsh 11: 699–704

Schneider A (1962) Zur Tektonik und Metamorphose im nördlichen Spessartkristallin. Notizbl Hess Landesamt Bodenforsch 90: 246–272

Thürach H (1893) Über die Gliederung des Urgebirges im Spessart. Geognost Jh Kassel 5: 160

Weber K (1981) The structural development of the Rheinische Schiefergebirge. Geol Mijnbuow 60: 149–159

Weber K, Juckenack Chr (1990) The structure of the Spessart Mts. crystalline basement and its position in the frame of the European Variscides. Int Conf on Paleozoic orogens in central Europe, field guide Mid-German Crystalline Rise and Rheinisches Schiefergebirge, pp 101–114

IV.C.2 The Odenwald Crystalline Complex

A. KROHE and A.P. WILLNER

1 Subdivision of the Complex

The Odenwald Crystalline Complex is subdivided into four structural units (Fig. 1; Krohe 1992). These units also show strong variations in P-T histories (see Willner et al. 1991 for detailed P-T data and discussion). They can be considered as tectonic slivers of mid-crustal rocks that were juxtaposed during the Variscan orogeny along a complex system of shear zones characterized by divergent sinistral strike shear and normal shear. Calc-alkaline plutonic suites of each unit differ in composition and structural style. They intruded synkinematically into the shear zones (Krohe 1991) at different stages of the crustal evolution. Uplift and exhumation of the crystalline rocks were also governed by movements along these shear zones (Krohe 1992, 1994). From the northwest to the southeast the units are characterized as follows (Fig. 1):

The metamorphic assemblages of **unit I** consist of predominant amphibolites and metapelites and a plutonic suite of gabbros containing ultramafic cumulates, diorites and subordinate granitoids. P-T data for the thermal peak of metamorphism in the amphibolites infer a high initial geothermal gradient of about 80 °C/km (Willner et al. 1991). Typical for early gabbro intrusions are shallow intrusion depths (only about 1–2 kb; Trochim 1960; Matthes and Schubert 1971; Willner et al. 1991) and $Ar^{39/40}$ cooling ages of hornblende in the range of 360 Ma (Kirsch et al. 1988). Increasing intrusion depths of younger granodioritic to granitic intrusions (2–3 kb) imply an anticlockwise P-T path (Willner et al. 1991) (Fig. 2).

Metamorphic assemblages of **unit II** consist (1) of orthogneisses and amphibolites (Neutsch Complex after Nickel 1985; N in Fig. 1), (2) of orthogneisses, amphibolites, and few marbles (Felsberg/Auerbach Complex; F in Fig. 1. cf. Barth 1973) and (3) of paragneisses and amphibolites (Heppenheim Complex; H in Fig. 1, cf. von Raumer 1973). However, unit II is dominated by a voluminous complete calc-alkaline magmatic suite ranging from gabbro to granite (Neunkirchen Suite; n in Fig. 1, cf. Maggetti 1975; Nickel and Zurbriggen 1975; Melibokus Suite; m in Fig. 1). The southeastern part of unit II (Heppenheim Complex) experienced a complex P-T history. Medium pressure relics (8–9 kb/580 °C; Willner et al. 1991) are restricted to garnet-bearing kinzigites interleaved within quartzites (see Büsch et al. 1980 for description). Subsequently, these kinzigites were metamorphosed under amphibolite facies conditions with relatively low pressures (see also Büsch et al. 1980). This later metamorphism also affected the para- and orthogneisses of the Heppenheim Complex (Fig. 2). The typical mineral assemblages formed during that stage are presented in Chapter IV.E (Metamorphic evolution). P-T estimates using various methods yielded conditions of about 650 °C at 4–5 kb (Okrusch et al. 1975; Willner et al. 1991). Locally, the stability of muscovite + quartz was exceeded (von Raumer 1973). The retrograde P-T path intersects the muscovite + quartz breakdown curve at about 3 kb and 630 °C (Willner et al. 1991). This suggests an increase in the geothermal gradient from about 40 °C/km to 70 °C/km during retrograde metamorphism probably due to isothermal (i.e., rapid) uplift of the southern Unit II (Fig. 2). Uplift of the central and northern unit II toward a level corresponding to 3 kb during retrograde metamorphism occurs during activity of D_3-sinistral strike shear zones (see below): emplacement pressures of granitoids that intruded early into these shear zones (between the Felsberg-Auerbach Complex and the Neutsch Complex) range between 4 and 5 kb; the granitoids have been mylonitisized at depths corresponding to about 4 kb assuming synkinematic temperatures above 500 °C (see Willner et al. 1991; Krohe 1992). Granitoid suites that intruded late into these shear zones (Neunkirchen Suite) have been emplaced and mylonitisized at depths corresponding to 3 kb. Hence, emplacement pressures

Dallmeyer et al. (Eds.)
Pre-Permian Geology of Central and Eastern Europe

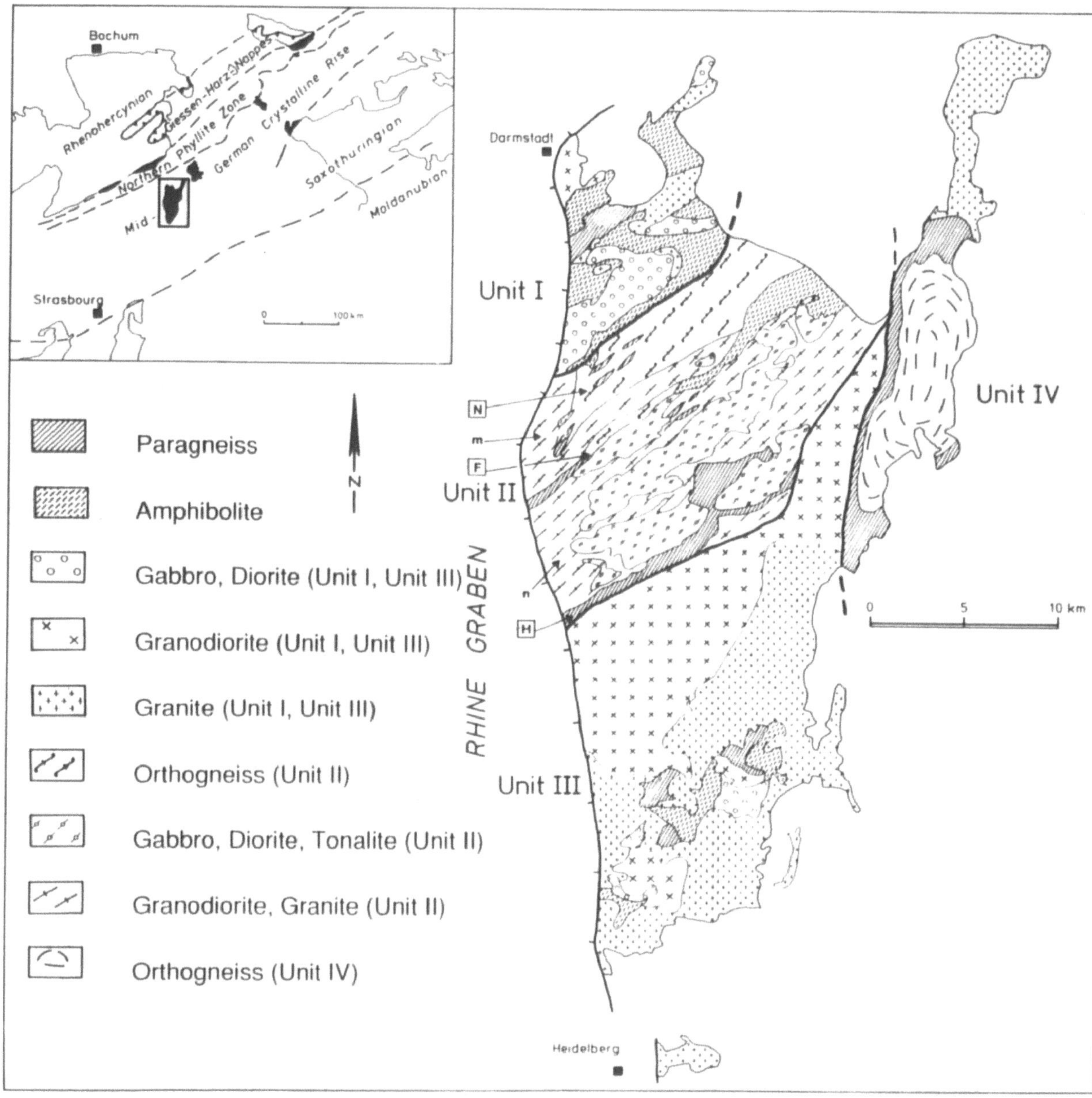

Fig. 1. Simplified structural map of the Odenwald Crystalline Complex and the location of the study area in the Variscan belt. (After Nickel 1985.) Metamorphic complexes and magmatic suites of unit II: N Neutsch Complex; F Felsberg/Auerbach Complex; H Heppenheim Complex; n Neutsch Magmatic Suite; m Melibokus Magmatic Suite

of granitoids of unit II are higher than those of the gabbros and diorites in unit I (see also Henes-Klaiber and Altherr 1991). K/Ar cooling ages of hornblende of both metamorphic and plutonic rocks, which are in the range of about 342 Ma (Kreuzer and Harre 1975; Rittmann 1984) are younger than in unit I.

In **unit III** the volume portion of metamorphic rocks is small. Partly migmatized metasedimentary rocks, amphibolites, and orthogneisses are intercalated in the center of a suite of huge granodioritic to granitic plutons. So far, no peak P-T data exist of this unit. Retrograde P-T data indicate an increase of the geothermal gradient during uplift (Willner et al. 1991; Fig. 2). Concordant zircon ages of 335 Ma in migmatic rocks (Todt 1979) probably date the age of the thermal peak of metamorphism. The granitoids of unit III

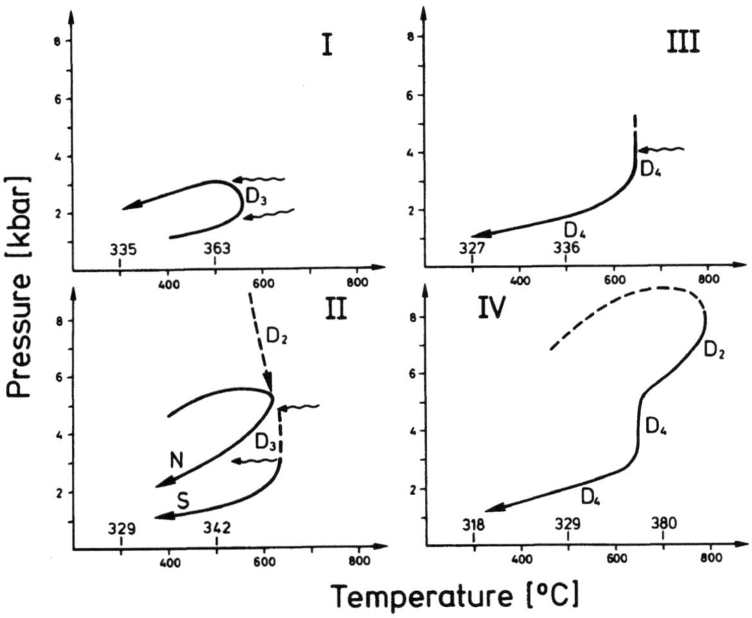

Fig. 2. Schematic P-T paths for *units I–IV* including mean cooling ages (for details see Willner et al. 1991). *Curved arrows* indicate intrusions

show younger K/Ar ages of hornblende than the late granitoids of unit II (about 335 Ma; Kreuzer and Harre 1975; Rittmann 1984), but higher emplacement pressures (around 4 kb; Willner et al. 1991). Consequently, in the time between 335 and 342 Ma, unit III was situated at a greater depth than unit II.

Unit IV is mainly composed of granitic and granodioritic gneisses containing lenses of paragneisses and amphibolites. Voluminous late intrusive bodies are lacking. P-T calculations with relic inclusion parageneses including garnet within granodiorite gneisses and garnet amphibolites yielded 8–9 kb and 550–800 °C (Willner et al. 1991). After this medium pressure event, unit IV was also metamorphosed under amphibolite facies conditions in the range of about 650 °C at 4.5–5 kb, followed by an isothermal decompression of about 1.5 kb associated with a retrograde increase of the geothermal gradient (Willner et al. 1991; Fig. 2). Uplift caused an NNW-SSW-trending antiform (Krohe 1994), which is the dominant structure of unit IV (Chatterjee 1960; Fig. 1). In unit IV the youngest K/Ar ages of hornblende (about 325 Ma; Kreuzer and Harre 1975; Lippolt 1986) were found. Lower U-Pb intercept ages of zircons of around 380 Ma (Todt 1979) probably date earlier increments of metamorphism (associated with the medium pressure event?), suggesting that a metamorphic evolution of unit IV continued for a long time.

2 Structural Evolution

Each unit underwent polyphase deformation. Fabrics which developed before and during the thermal peak of metamorphism are summarized as D_1/D_2 (Krohe 1992). They are not necessarily contemporaneous and do not show uniform kinematics in the four structural units, but all developed before juxtaposition of the metamorphic units.

D_1/D_2 structures are characterized as follows: progressive deformation caused polyphase folding, whereby at least two transposition foliations were formed (units I and II). D_1-planar fabrics are relic structures, defined by the long axes of quartz segregations and biotites, which were tightly folded or boudinaged during D_2. The D_1 planes are composite fabrics. The main D_2 structures are pervasive foliations and stretching lineations, defined by preferred dimensional orientation of amphibole, mica, and recrystallized aggregates of quartz and feldspar. Transposition of the D_1 surfaces generated a compositional banding that is subparallel to the D_2 foliation.

The orientation of the D_2 structures differs in the various metamorphic units. In unit I the trend of the D_2 foliation is roughly NE-SW. In the Neutsch Complex, it varies between NE-SW and NNW-SSE, in the Felsberg/Auerbach Complex and in the Heppenheim Complex the D_2-foliation trends uniformly NE-SW and contains a subhorizontal lineation (Hustiak 1990; Krohe 1992). In unit IV the D_2 foliation outlines the dominating NNE-SSW-trending antiformal structure (Chatterjee 1960; Fig. 1). An either NE- or SE-plunging stretching lineation developed. Since in unit I gabbros yielding cooling ages of about 360 Ma (Kirsch et al. 1988) intruded into the amphibolites deformed by D_2, D_2 must be older than 360 Ma. A maximum age of D_1/D_2 of around 420 Ma is given by Ludlowian spores found by Reitz (1987) in metasediments of the nearby Spessart Crystalline Complex, which mark the only biostratigraphically controlled sedimentation age in the entire Mid-German Crystalline Rise.

D_1/D_2-structures may have accommodated crustal shortening and thickening (Krohe 1992): after D_2 and before uplift and exhumation of rocks, in units II-IV an excess in crustal thickness must have been present (1) since the rocks exposed at present in that area experienced metamorphism at 4–5 kb corresponding to a depth of about 15–19 km, and (2) since the present thickness of the crust is still about 28–30 km (revealed by seismic reflection data; Dekorp Research Group 1985; Korsch and Schäfer 1991). Increasing crustal thicknesses from unit I to III have also been inferred by Henes-Klaiber and Altherr (1991) from compositions of basic magmas. The inferred excess in crustal thickness in units II–IV could have been generated by an early thrusting event. However, individual metamorphic complexes may have experienced different D_1/D_2 histories. In some units thrusting may also be responsible for burial of rocks to achieve the 8–9 kb pressure.

The younger episodes of deformation (Fig. 3) occurred at a retrograde stage and affected both the metamorphic and plutonic rocks. Juxtaposition of the units and intrusion of the plutonic suites occurred during these episodes of deformation. In the northwestern and southeastern units different structures developed (Krohe 1991, 1992).

D_3-structures (Fig. 3) are steeply dipping NE-trending sinistral shear zones, developed in intrusives and in their metamorphic host rocks. They show horizontal or gently dipping stretching lineations and have been observed only in unit I and particularly in unit II (Hustiak 1990; Krohe 1991, 1992). In the plutonic rocks NE-SW-trending foliations were generated either by magmatic or by solid state flow. In both metamorphic and plutonic rocks, sinistral shearing is indicated by various shear sense indicators. The plutonic rocks of unit II are supposed to have intruded synkinematically into dilatant regions within this zone of strike shear (see Krohe 1991). D_3-related shear zones and magmatism occupy an area about 10 km in width within unit II.

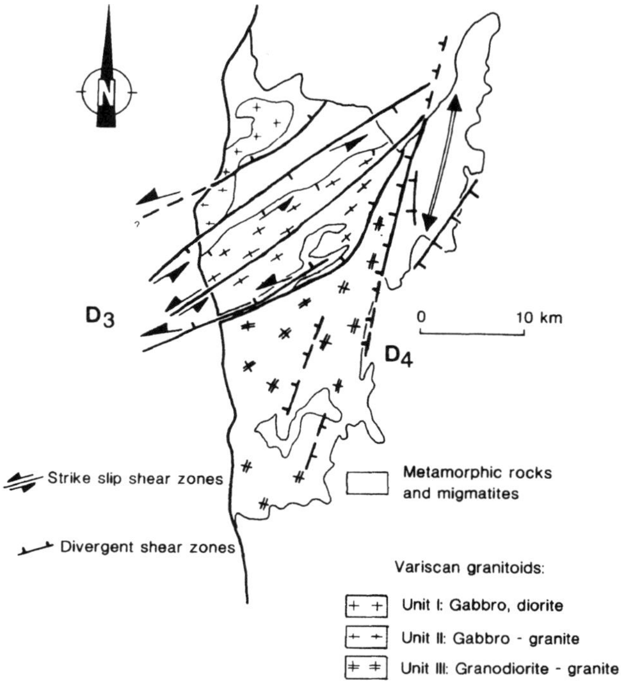

Fig. 3. Orientation of the D_3 and D_4 shear zones

D_4-structures (Fig. 3) are ductile normal fault zones. They predominate in the southeast, in units III and IV (Krohe 1992). They indicate WSW-ESE extension. In the metamorphic rocks of unit IV, NNE-SSW-trending and WNW- or SE-dipping D_4 shear zones with down dip lineations were formed. Probably synchronous with the formation of the D_4 normal shear zones, the preexisting D_2 foliation was warped around an antiformal structure forming the NNE-trending gneiss dome of unit IV. This dome is considered as a typical Variscan metamorphic core complex, as they are widely known in Western Europe (e.g., Martinez et al. 1988; Echtler and Malavieille 1990). Solid state foliations, formed in the plutonic rocks of unit III follow roughly the NNE-SSW trend of the D_4 shear zones of unit IV and also show down dip stretching lineations. Northwest-dipping shear zones (northwestern unit III) are characterized by WNW-directed normal shearing, and SW-dipping shear zones (southeastern unit III) by southeast directed displacement (Krohe 1992). A NNE-SSW-trending zone at the border between units III and IV was called the Otzberg Fault by Schälicke (1975).

Both D_3 and D_4 shearing started above 500 °C, i.e., above the blocking temperature of the K/Ar hornblende system, and continued through cooling into the (semi)-brittle field (i.e., below 500 °C). Thus, K/Ar and $^{39}Ar/^{40}Ar$ ages of hornblende roughly date the age of D_3 and D_4 (Krohe 1992). The decrease of these ages from unit to IV indicates the temporal change from divergent strike slip (between 360 and 335 Ma) into extension (between 335 and 320 Ma) in this area. Within each unit, average biotite ages are up to 8 Ma younger than hornblende ages (see data in Kreuzer and Harre 1975; Rittmann 1984; Lippolt 1986; Kirsch et al. 1988) and the mineral ages of plutonic and metamorphic rocks appear to show no marked differences. Since magma emplacement was synchronous with D_3 (unit II) and D_4 (unit III; cf. Krohe 1991, 1992), the mineral ages also date the intrusions.

Extension associated with D_3-strike shearing is supposed to have governed (Krohe 1991, 1992):

1. Juxtaposition of units I and II and subsidence in unit I (Fig. 4) that shows shallower intrusion depths and lower pressures of the thermal peak of metamorphism than unit II;

2. Subsidence of the Saar-Nahe Basin to the W of unit I (Figs. 4 and 5; Hering et al. 1976): from the Middle Devonian onward in this area, a marine basin developed which is probably confined to unit I of the Odenwald Crystalline Complex by the southwestern prolongation of the D_3 zone of strike shearing.

3. Juxtaposition of the metamorphic assemblages within unit II: the Neutsch Complex borders the Felsberg/Auerbach Complex by a steep D_3 shear zone which is more than 300 m in width, and the Felsberg/Auerbach Complex borders the Heppenheim Complex by the broad syn-D_3 Neunkirchen Magmatic Suite (Krohe 1991, 1992). Slip vectors of D_3-shear zones indicate uplift of the metamorphic wall rocks of the Heppenheim and Neutsch Complexes relative to

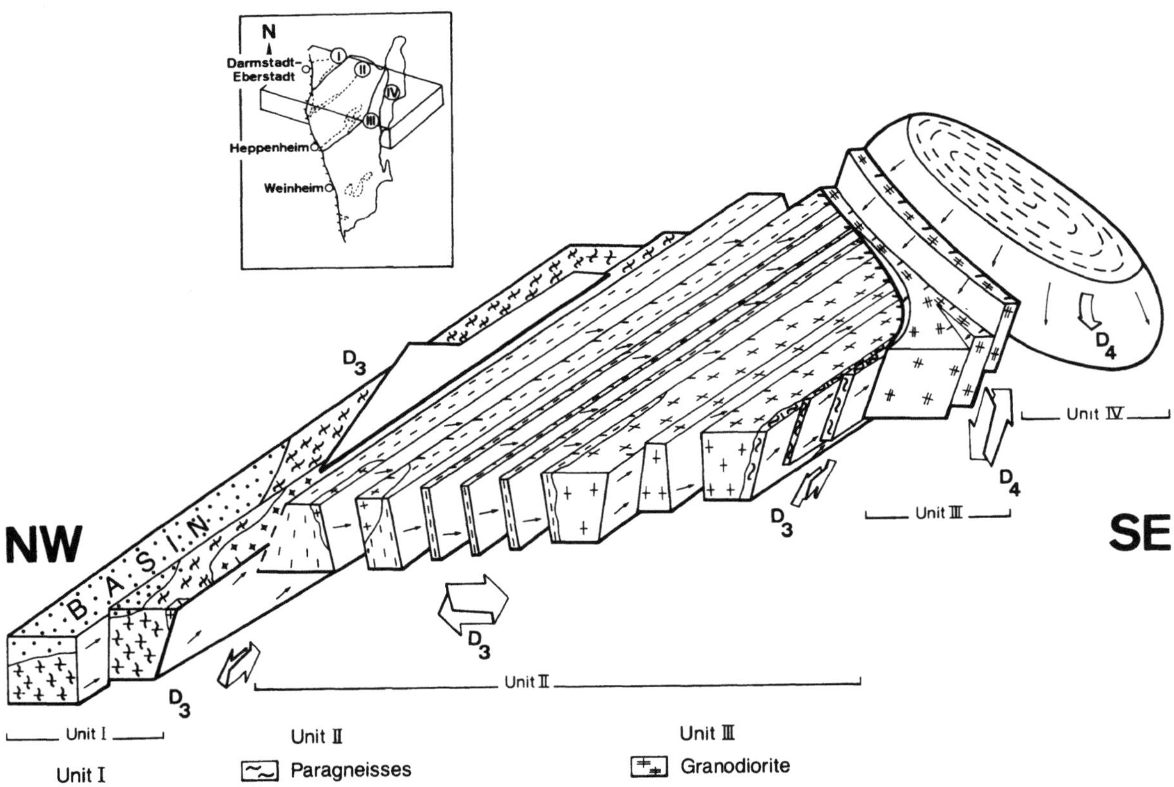

Fig. 4. Block diagram illustrating the relative movements between the different units during D_3 and D_3

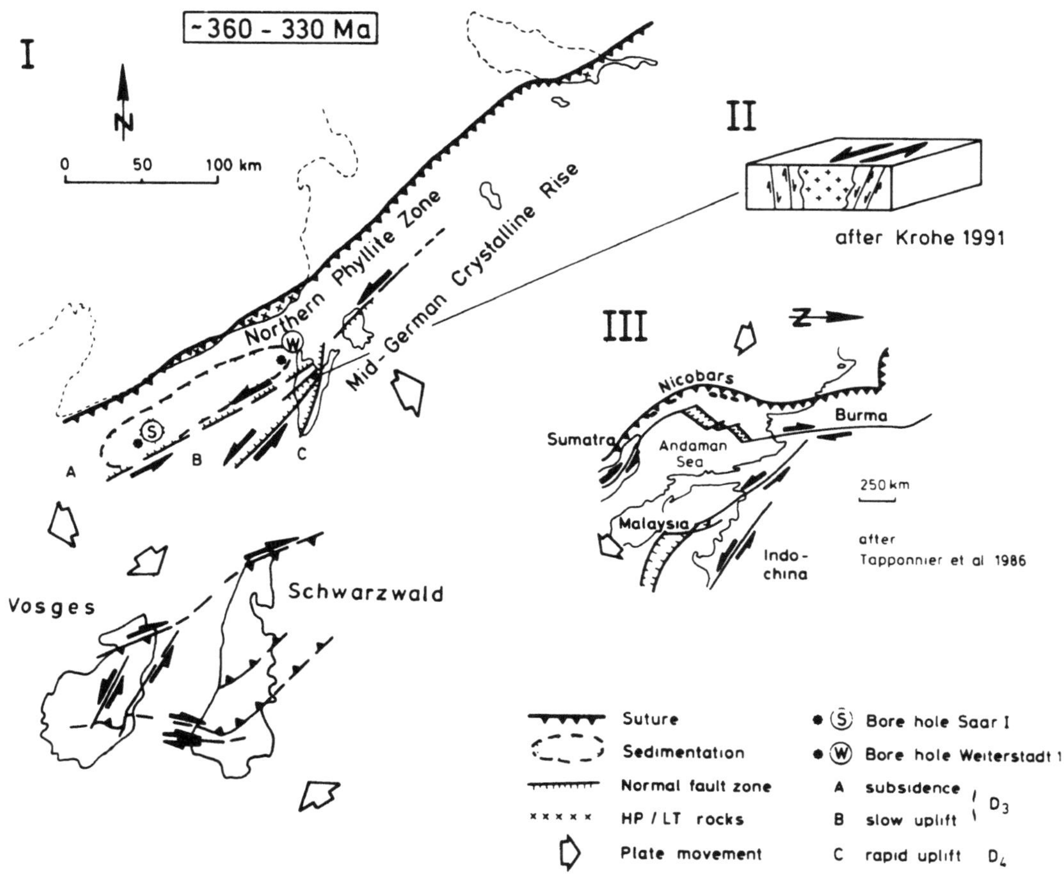

Fig. 5. **I** Structural sketch showing the main shear zones in the Mid-German Crystalline Rise and the Schwarzwald/Vosges area during the period 360–330 Ma. **II** Sketch of the transtensional setting of Unit II. **III** Structural sketch of the Burma/Sumatra subduction Zone (rotated 90° clockwise) as a recent model for the Central European Variscides

the plutons of the Neunkirchen Suite (Fig. 4). Divergent sinistral shear has also been observed in the adjacent Spessart Crystalline Complex (Weber and Juckenack 1990).

Extension associated with D_4-normal faulting is supposed to have governed uplift of unit III relative to unit II and of unit IV relative to III (Fig. 4; Krohe 1992). This is supported by the inferred shallower intrusion depths of the older granitoids of unit II relative to those of unit III (Willner et al. 1991) and by structural data (Krohe 1992). Unit IV is confined to unit III by a broad penetrative D_4 shear zone with extensional geometry (about 4 km in width), plunging beneath unit III (Otzberg Fault Zone). Nearly isothermal uplift derived from P-T data of the southern unit II, units III, and IV is considered to result from rapid uplift due to D_4 normal shearing.

D_3/D_4-related magmatism and unroofing of metamorphic rocks occurred roughly concomitant with the subduction and closure of the Rhenohercynian ocean. Thus, D_3/D_4 transtension/extension are considered to have occurred within the framework of Variscan oblique subduction of oceanic crust and SW-NE extrusion (Willner et al. 1991; Krohe 1992). Major parts of the Mid-German Crystalline Rise can be regarded as a trench-linked strike slip zone (in the sense of Woodcock 1986), which is located within a continental magmatic arc. Neogene equivalents of those strike-slip zones occurring, e.g., in the Sunda magmatic arc/Indonesia (Fig. 5; Tapponnier et al. 1986) accommodate lateral movements within the dimension of some hundreds of kilometers.

In the authors' opinion, the major tectonic imprint of the Odenwald Complex and probably of most of the Mid-German Crystalline Rise was caused by sinistral strike slip. It is concluded that in the Odenwald Crystalline Complex this episode (1) occurred at about 360–340 Ma, (2) juxtaposed unit I and II at a level corresponding to 3 kb and (3) postdated a period of early thrusting in some units. In the Rhenohercynian thrust belt to the N (Fig. 1), however, thrusting did not start before about 325 Ma (Ahrendt et al. 1983), this is 20–30 Ma after the termination of strike slip movements in units I–III of the Odenwald Crystalline Complex, which were nearly exhumed before 330 Ma. Thus, the tectonic history of the Mid-German Crystaline Rise is profoundly different from that of the Rhenohercynian zone. Probably the Rhenohercynian zone and the Mid-German Crystalline Rise were juxtaposed along late (i.e., post D_3 and D_4) NE-SW trending zones of dextral strike shear located within the Northern Phyllite Zone (Fig. 1).

Acknowledgments. The manuscript benefited from a careful review by W. Frisch. Financial support by the Deutsche Forschungsgemeinschaft to one of us (AK; grant Kr 266/2-2) is gratefully acknowledged.

References

Ahrendt H, Clauer N, Hunzinger JC, Weber K (1983) Migration of folding and metamorphism in the Rheinische Schiefergebirge deduced from K-Ar and Rb-Sr age dterminations. In: Martin H, Eder FW (eds) Intracontinental fold belts. Springer Berlin Heidelberg New York, pp 323–338

Barth H (1972) Petrologische Untersuchungen im Felsberg-Zug (Bergsträßer Odenwald). Abh Hess Landesamt Bodenforsch Wiesb 66: 1–85

Büsch W, Matthes S, Mehnert KH, Schubert W (1980) Zur genetischen Deutung der Kinzigite im Schwarzwald und Odenwald. Neues Jahrb Mineral Monatsh 137: 223–256

Chatterjee ND (1960) Geologische Untersuchungen im Gebiet des Böllsteiner Odenwaldes. Neues Jahrb Geol Abh 111: 137–180

Decorp Research Group (1985) First results and preliminary interpretation of deep reflection seismic recordings along the profile DEKORP 2S. J Geophys 57: 137–163

Echtler H, Malavieille J (1990) Extensional tectonics, basement uplift and Stephano-Permian collapse basin in a late Variscan metamorphic core complex (Montagne Noire, Southern Massif Central). Tectonophysics 177: 125–138

Henes-Klaiber U, Altherr R (1991) Die I-Typ Granitoide des Odenwaldes. Intrusionstiefen und Krustenmächtigkeiten. Eur J Min 3 (Beih 1): 163

Hering O, Zimmerle W, Ghazanfari A, Weber G, Krebs C, Tan CL, Nickel E, Reible P (1976) Die Tiefbohrung Saar 1. – Petrographische Beschreibung und Deutung der erbohrten Schichten. Geol Jahrb A 27: 91–305

Hustiak D (1990) Verformungsgeschichte einer sinistralen Blattverschiebung im Odenwald. Diploma Thesis, Univ Bochum, 132 pp (unpubl)

Kirsch H, Kober B, Lippolt HJ (1988) Age of intrusion and rapid cooling of the Frankenstein gabbro (Odenwald, SW-Germany) evidenced by 40Ar/39Ar and single-zircon 207Pb/206Pb measurements. Geol Rdsch 77: 693–711

Korsch RJ, Schäfer A (1991) Geological interpretation of DEKORP deep seismic reflection line profiles 1C and 9N across the Variscan Saar-Nahe basin, southwest Germany. Tectonophysics 191: 127–146

Kreuzer H, Harre W (1975) K/Ar Altersbestimmungen an Hornblenden und Biotiten des Kristallinen Odenwaldes. Aufschluß Sonderbd 27: 71–77

Krohe A (1991) Emplacement of synkinematic plutons controlled by transtensional tectonics in the Variscan Odenwald (SW-Germany). Geol Rdsch 80: 391–409

Krohe A (1992) Structural evolution of intermediate-crustal rocks in a strike-slip and extensional setting (Variscan Odenwald, SW Germany): differential upward transport of metamorphic complexes and changing deformation mechanisms. Tectonophysics 206: 357–386

Krohe A (1994) Verformungsgeschichte in der mittleren Kruste eines magmatischen Bogens – der variszische Odenwald als Modellregion. Geotekt Forsch 80: 1–147

Liew TC, Hofmann AW (1988) Precambrian crustal components, plutonic associations, plate environment of the Hercynian Fold Belt of Central Europe: indications from a Nd and Sr isotopic study. Contrib Mineral Petrol 98: 129–138

Lippolt HJ (1986) Nachweis altpaläozoischer Primäralter (Rb/Sr) und karbonischer Abkühlungsalter (K/Ar) der Muskovit-Biotit-Gneise des Spessarts und der Biotit-Gneise des Böllsteiner Odenwaldes. Geol Rdsch 75: 569–583

Maggetti M (1971) Die basischen Intrusiva im Heppenheim – Lindenfelser Zug (Mittlerer Bergsträßer Odenwald). Neues Jahrb Mineral Abh 115: 192–228

Maggetti M (1975) Die Tiefengesteine des Bergsträßer Odenwaldes. Aufschluß Sonderbd 27: 87–107

Maggetti M, Nickel E (1973) Hornblende-Diorite und Biotit-Diorite im kristallinen Odenwald. Neues Jahrb Mineral Abh 119: 232–265

Martinez FJ, Julivert M, Sebastian A, Arboleya ML, Gil Ibarguchi JI (1988) Structural and thermal evolution of high-grade areas in the northwestern parts of the Iberian Massif. Am J Sci 288: 969–996

Matthes S, Schubert W (1971) Der Original-Beerbachit im Odenwald, ein Amphibolit-Hornfels in Pyroxen-Hornfelsfazies. Contib Mineral Petrol 33: 62–86

Matthes S, Okrusch M, Richter P (1972) Zur Migmatit-Bildung im Odenwald. Neues Jahrb Mineral Abh 116: 225–267

Nickel E (1985) Odenwald, Vorderer Odenwald zwischen Darmstadt und Heidelberg, 2nd edn. Sammlung Geol Führer 65: 232, Borntraeger, Berlin

Nickel E, Maggetti M (1974) Magmenentwicklung und Dioritbildung in synorogen konsolidierten Grundgebirge des Bergsträßer Odenwaldes. Geol Rdsch 63: 618–654

Nickel E, Zurbriggen B (1975) Die Entwicklung der Neunkirchner Flasergranitoide (Gesichtspunkte zu Gefügestudien in Odenwaldgneisen VI). Aufschluß Sonderbd 27: 159–174

Okrusch M, Raumer JF von, Matthes S, Schubert W (1975) Mineralfazies und Stellung der Metamorphite im kristallinen Odenwald. Aufschluß Sonderbd 27: 109–134

Reitz E (1987) Palynologie in metamorphen Serien: I. Silurische Sporen aus einem granatführenden Glimmerschiefer des Vor-Spessart. Neues Jahrb Geol Palaentol Monatsh 1987: 699–704

Rittmann K (1984) Argon in Hornblende, Biotit, und Muskovit bei der geologischen Abkühlung. – 40Ar/39Ar Untersuchungen. Doctoral Thesis, Univ Heidelberg, 278 pp (unpubl)

Schälicke W (1975) Die Otzberg-Zone. Aufschluß Sonderbd 27: 47–57

Schubert W (1979) Metamorphe Gesteinsserien im Bergsträsser Odenwaldkristallin. Fortschr Miner 57 (Beiheft 2): 52–61

Tapponnier P, Peltzer G, Armijo R (1986) On the mechanics of the collision between India and Asia. In: Coward MP, Ries A (eds) Collision tectonics. Geol Soc Spec Publ 19: 115–157

Todt W (1979) U-Pb-Datierungen an Zirkonen des Kristallinen Odenwaldes. Fortschr Miner 57 (Beih 1): 153–154

Trochim HD (1960) Zur Petrogenese des Gabbro-Plutons vom Frankenstein (Odenwald). Doctoral Thesis, Univ Freiburg, 150 pp

Von Raumer JF (1973) Die mineralfazielle Stellung der Metapelite und Metagrauwacken zwischen Heppenheim und Reichelsheim (Odenwald). Neues Jahrb Mineral Abh 118: 313–336

Weber K, Juckenack Chr (1990) The structure of the Spessart Mts. crystalline basement and its position in the frame of the European Variscides. In: Franke W, Weber K (eds) IGCP 233 Int Conf on Paleozoic orogens in Central Europe. Field guide: Mid-German Crystalline Rise Rheinisches Schiefergebirge Göttingen/Giessen, Aug–Sept 1990, pp 101–114

Willner AP, Massonne H-J, Krohe A (1991) Tectonothermal evolution of a part of Variscan magmatic arc: the Odenwald in the Mid-German Crystalline Rise. Geol Rdsch 80: 369–389

Woodcock NH (1986) The role of strike slip fault systems at plate boundaries. R Soc Lond Phil Trans A-317: 13–29

IV.C.3 The Saar-Nahe Basin

K. Weber

In the drill hole Saar 1 (Fig. 1), the crystalline basement of the Saar-Nahe basin, represented by a weakly deformed aplitic albite granite is covered by a nonmetamorphic shelf sequence of Middle of Upper Devonian marine limestones with intercalations of clastic sedimentary rocks (Hering et al. 1976). Upper Devonian feldspathic sandstones are composed dominantly of volcanic material, indicating an island arc as the source area. The recalculated Rb/Sr age of the basement granite is 403 ± 24 Ma (Lenz and Müller 1976). The relatively wide spread of the isochron date indicates that the granite could also be of Silurian age and be part of the pre-Hercynian rift magmatism (Weber 1984), which intruded a Cadomian basement. Evidence of the existence of a Cadomian basement is given by the 525-Ma hornblende cooling ages (Meisl at al. 1989) in the lower amphibolite facies gneiss of Wartenstein, which forms a basement slice inside the allochthonous unit of the southern Hunsrück (Fig. 2).

Marine sedimentation continued into the Lower Carboniferous and is represented by uppermost Devonian and Lower Carboniferous black alum shales with intercalations of cherts and thin layers of middle to coarse-grained graywackes which are dominantly composed of volcanic fragments (Kelch and Reible 1976; Hering et al. 1976). The weakly deformed, low grade metamorphic Lower Carboniferous graywackes of Neustadt, Hambach, and Weiler at the western Rhine graben escarpment east of the Saar 1 drill hole are interpreted by Flöttmann and Oncken (1992) as mid-fan sediments of a fore-arc basin. The layer thickness/grain size relationship of the graywackes of the Saar 1 drill hole indicate proximal graywackes. The former depocenter of the flysch graywacke was probably located further to the NW. It formed presumably the southwestern prolongation of the former oceanic Giessen Graywacke trough and must have been eroded completely after nappe displacement across the western Rheinische Schiefergebirge. The Middle Devonian subsidence results from thermal contraction of the upper mantle, and indicates the beginning of subduction. The subsidence increases through Upper Devonian and Lower Carboniferous and leads to the formation of a fore-arc basin.

The black shales are covered by an approx. 20-m-thick conglomerate, which is dominantly composed of undeformed magmatic rocks (60% volcanic and 15% plutonic rocks), 20% diaphthoritic metamorphic rocks without garnet, staurolite, sillimanite, or kyanite (Hering et al. 1976). The conglomerate documents the beginning of the development of the Permo-Carboniferous molasse basin. The preceding uplift and erosion is interpreted by Krebs (1976) and Kneuper (1976) as representing the Sudetian Orogeny at the end of the Lower Carboniferous, which dominates in the Saxothuringian Zone. Thrusting and crustal stacking, which can be observed in the DEKORP 1C section below the wedge of crystalline rocks (Fig. 2) must be related to this presumably Upper Viséan collisional event. This timing correlates with the switch from subduction to collision in the Eastern Rheinische Schiefergebirge, where the deposition of the allochthonous flysch ends in the Middle Visean and the deposition of the authochthonous flysch starts in the Upper Visean (Engel at al. 1983; Engel and Franke 1983).

The DEKORP 1C section (Fig. 2) shows the Saar-Nahe Basin as a classical halfgraben structure, which is bounded to the NW by the South Hunsrück Boundary Fault (SHBF). As shown by Korsch and Schäfer (1991) the geometry of the basin cannot be related to a simple listric fault. The SHBF has behaved as a dextral strike-slip fault in Rotliegend time (Anderle 1987; Schwab 1987). SW-dipping slickensides point to dextral transtensional movement, which is superimposed on the post-Visean listric faulting. Both authors assume that some of these displacements could be related to Cenozoic extension, associated with the formation of the Rhine Graben. Pre-Namurian

Dallmeyer et al. (Eds.)
Pre-Permian Geology of Central and Eastern Europe

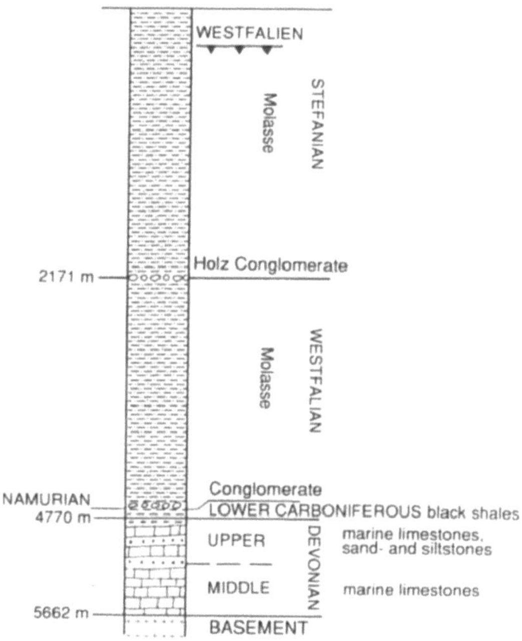

Fig. 1. Stratigraphic section of the borehole Saar 1

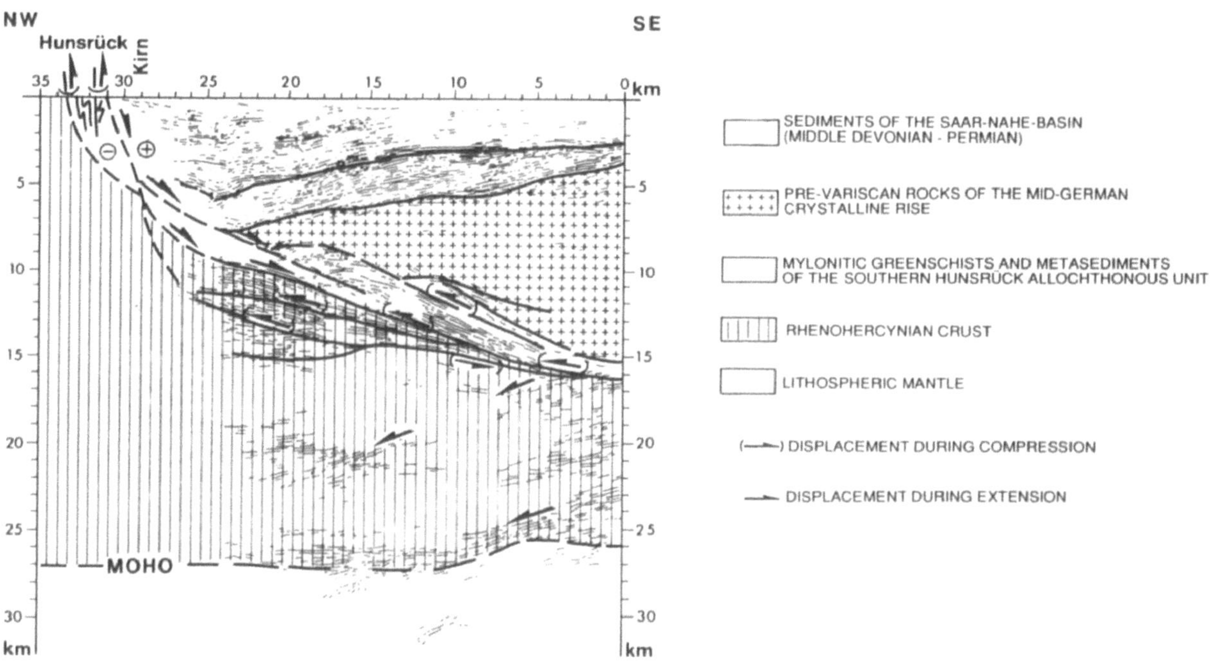

Fig. 2. Interpretation of the seismic section DEKORP 88-1C across the South Hunsrück Boundary Fault and the the Saar-Nahe Basin; − and + indicate dextral displacement along the South Hunsrück Boundary Fault

dextral transpression preceded dextral transtension, as shown by Oncken (1988), and is documented in the metamorphic rocks of the Southern Hunsrück by asymmetric dextral schlingenfolds. These folds represent a late strain increment, since all earlier compressional fabrics are refolded. Dextral transpression ends with the formation of the imbricated ramp structure of the southern Hunsrück, which traces out the later listric transtensional normal fault with its more or less vertical upper part and with several gently dipping branches in its lower part. Therefore, the former thrust planes below the basement slice of the Mid-German Crystalline Rise are reactivated as normal faults during transtension (Fig. 2). The seismic transparency of the crust below the Hunsrück area may result from steeply dipping reflectors formed during dextral transpression which has affected the whole Hunsrück area (Weber, unpubl.). An additional reason could be a data misfit during data processing resulting from the pull-down effect in the sedimentary rocks of the adjacent Saar-Nahe basin. There is no indication in the DECORP C1 reflection seismic record for a South Hunsrück Boundary Fault that can be traced down to the base of the crust. Rather, it can be seen that stretching in the upper crust was accommodated in the lower crust by NW-dipping ductile normal faulting (Figs. 2 and 3). From the geometry of these dipping reflectors there is no reason to assume that they represent pre-Hercynian thrusts.

Due to the unmetamorphic and very weakly deformed character of the Devonian and Lower Carboniferous sedimentary rocks of the Saar 1 drill hole, the displacement must have taken place passively on top of the seismically transparent crystalline basement slice of the Mid-German Crystalline Rise. The SE-dipping, duplex-shaped strong seismic reflectors below the transparent wedge of crystalline rocks in Fig. 2 might be correlated with the mylonitic greenschists and metasediments of the schuppen zone of the southern Hunrück. Due to the MORB affinity of

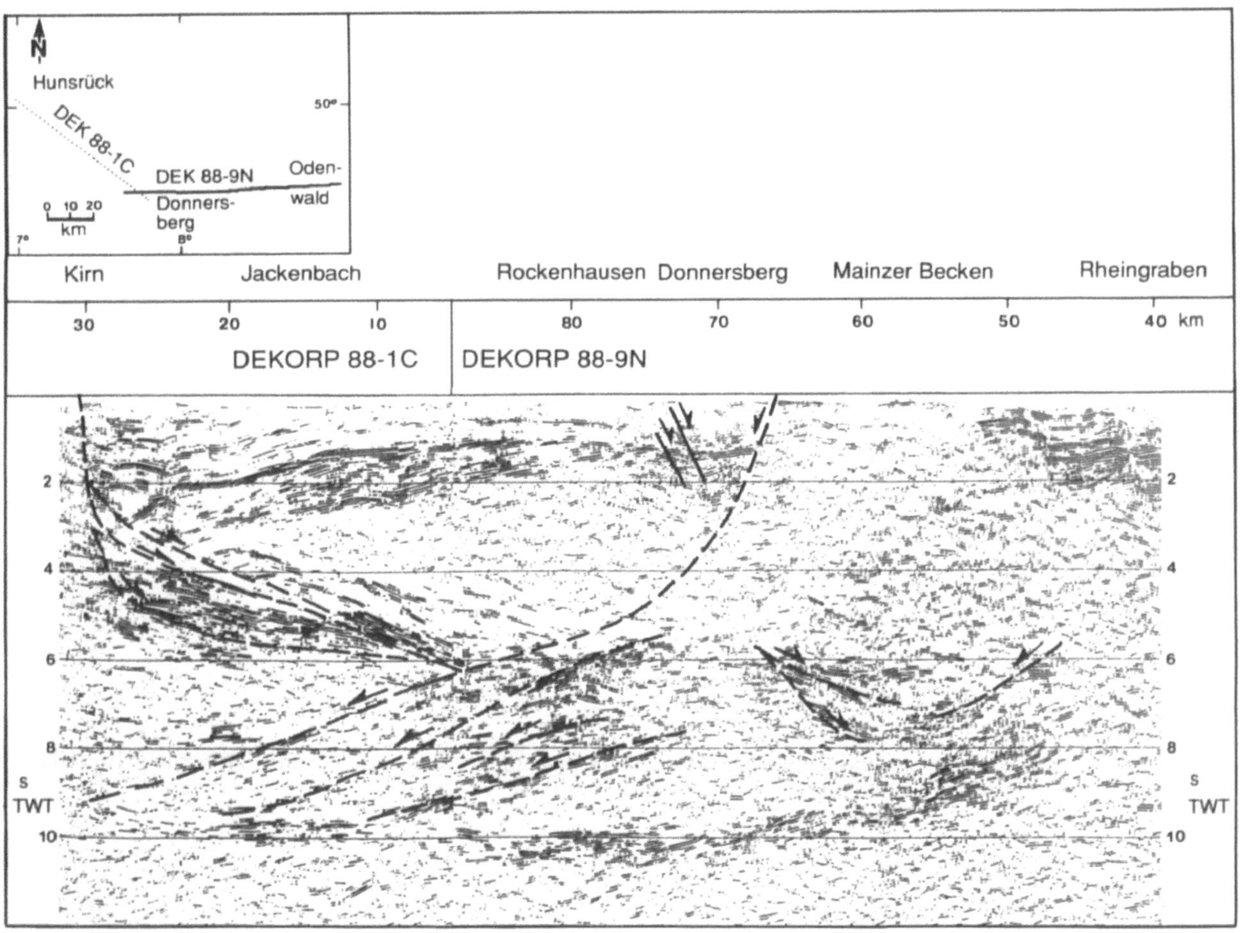

Fig. 3. Seismic sections DEKORP 88-1C and DEKORP 88-9C between the South Hunsrück Boundary Fault in the NW and the Rhine Graben in the E interpreted in terms of late Hercynian crustal stretching

the southern Hunsrück greenschists (Meisl 1990), these structural patterns might be interpreted as the basal part of an accretionary wedge which has been overridden by the transparent basement slice, which forms part of the Mid-German Crystalline Rise. The mylonitic greenschists and metasediments of the so-called Metamorphic Zone of the southern Hunsrück represents a relic of this accretionary wedge. Its eroded higher parts may be related to the Gießen Nappe of the eastern Rheinische Schiefergebirge.

Near the SHBF, the reflectors are complexly displaced by secondary normal faults; steep NW-dipping seimic events may represent pronounced rollover structures similar to that shown by Blundell (1990) from the Gulf of Mexico and by McClay et al. (1991) from analogue models. It might be possible that SE of the Donnersberg (Fig. 3) a less important pre-Saxonian boundary fault in connection with a graben structure exists, which can be related to NW-dipping normal faults in the lower crust. Characteristically, this graben lies on top of a horst structure which can be seen below 6s TWT (Fig. 3).

Uplift and erosion of up to 1500 m of sediments took place at the end of the Westfalian. The Holz Conglomerate at the base of the Stefanian was deposited with a low angle unconformity all over the Saar-Nahe Basin. Maximum erosion took place along the Saarbrücken-Pfalz Anticline, which represents the crest of the large scale rollover structure of the listric faulted Saar-Nahe Basin. This large-scale bend has been misinterpreted in the past as gentle folding at the end of the Westfalian. Compression took place no earlier than at the beginning of the Saxonian, forming the thrust bounded Saarbrücken Anticline (Kneuper 1976) and the gentle folds of the Saar-Nahe Basin (Schäfer 1989).

References

Anderle H-J (1987) The evolution of the South Hunsrück and Taunus borderzone. Tectonophysics 137: 101–114

Blundell DJ (1990) Some observations on basin evolution and dynamics. J Geol Soc Lond 148: 789–800

Engel W, Franke W (1983) Flysch sedimentation: its relations to tectonism in the European Variscides. In: Martin H, Eder W (eds) Intracontinental fold belts. Springer, Berlin Heidelberg New York, pp 289–321

Engel W, Franke W, Grote C, Weber K, Ahrendt H, Eder W (1983) Nappe tectonics in the southeastern part of the Rheinisches Schiefergebirge. In: Martin H, Eder W (eds) Intracontinental fold belts. Springer, Berlin Heidelberg New York, pp 289–321

Flöttmann Th, Oncken O (1992) Constraints on the evolution of the Mid-German Crystalline Rise – a study of outcrops west of the river Rhine. Geol Rundsch 81 (2): 515–543

Hering O, Zimmerle W, Ghazanfari A, Weber G, Krebs C, Tan CL, Nickel E, Reible P (1976) Die Tiefbohrung Saar 1. – Petrographische Beschreibung und Deutung der erbohrten Schichten. Geol Jahrb A 27: 91–305

Kelch H-J, Reible P (1976) Beschreibung der Spülproben und Kern der Bohrung Saar 1. in Die Tiefbohrung Saar 1. Geol Jahrb A27: 29–89

Kneuper G (1976) Regionalgeologische Folgerungen aus der Bohrung Saar 1. Geol Jahrb A27: 499–510

Korsch RJ, Schäfer A (1991) Geological interpretation of DEKORP deep seismic reflection line profiles 1C and 9N across the Variscan Saar-Nahe basin, southwest Germany. Tectonophysics 191: 127–146

Krebs W (1976) Zur geotektonischen Position der Bohrung Saar 1. Geol Jahrb A27: 489–498

Lenz H, Müller P (1976) Radiometrische Altersbestimmungen am Kristallin der Bohrung Saar 1. Geol Jahrb A27: 429–432

McClay KR, Waltham DA, Scott AD, Abousetta A (1991) Physical and seismic modelling of listric normal fault geometries. In: Roberts AH, Yielding G, Freeman B (eds) The geometry of normal faults. Geol Soc Lond Spec Publ 56: 231–239

Meisl S (1990) Metavolcanic rocks in the Northern Phyllite Zone at the southern margin of the Rhenohercynian Belt. Int Conf on Paleozoic orogens in central Europe, Göttingen-Giessen, Aug–Sept 1990, Field Guide: Mid German Crystalline Rise and Rheinisches Schiefergebirge, pp 25–42

Meisl S, Kreuzer H, Höhndorf A (1989) Metamorphosebedingungen und-Alter des Kristallins am Wartenstein bei Kirn/Nahe. 5. Rundgespräch "Geodynamik des europäischen Variszikums", Braunschweig, pp 38–39

Oncken O (1988) Geometrie und Kinematik der Taunuskammüberschiebung – Beitrag zur Diskussion des Deckenproblems im südlichen Rheinischen Schiefergebirge. Geol Rundsch 77: 551–575

Schäfer A (1989) Variscan molasse in the Saar-Nahe Basin (W Germany), Upper Carboniferous and Lower Permian. Geol Rdsch 78: 499–524

Schwab K (1987) Compression and right-lateral strike-slip movement at the Southern Hunsrücke Borderfault (south-west Germany). Tectonophysics 137: 115–126

Weber K (1984) Variscan events: early Palaeozoic rift metamorphism and late Palaeozoic crustal shortening. Geol Soc Lond Spec Publ 14: 3–22

IV.C.4 Structural Relationship Between Saar-Nahe Basin, Odenwald, and Spessart Mts.

K. WEBER

There is an obvious contrast in the structural and metamorphic development between the Saar-Nahe Basin, the Odenwald, and the Spessart Mts.

In the Spessart Mts. an Ordovician/Silurian magmatic event is documented by the S-type granitic and granodioritic intrusions, which were interpreted by Weber (1984) as early Variscan rift-related and by Okrusch and Richter (1986) as related to a postcollision extensional tectonic event. These granites were transformed into the Red Gneisses during Hercynian deformation and amphibolite facies metamorphism. The exact age of this metamorphism is not known. Weber (1984) supposed a metamorphic age of 380 Ma based on U/Pb zircon ages (Todt 1979) from the Böllstein Odenwald and 370 Ma Hbl cooling ages (Kreuzer and Harre 1975) from the northern Odenwald. The data presented in Chapter IV show that the geological histories of the Spessart and the Bergstrasse-Odenwald crystalline complexes are completely different. The Bergstrasse Odenwald represents a parautochthonous Variscan Island Arc, whereas the Böllstein Odenwald has more similarities with the Spessart Crystalline Complex, but the detailed relationship is not yet clear. Nasir et al. (1991) and Lippolt (1986) have shown that in the Spessart Cryrstalline Complex no significant difference in the K/Ar age data of Hbl, Ms, and Bio exists. All data scatter around 320 Ma. The temperature of Barrovian-type metamorphism is in the range of 600 °C and, considering the closing temperatures of these minerals, Nasir et al. (1991) came to the conclusion that the cooling ages cannot be far from peak metamorphism and the rocks suffered rapid uplift and cooling at about 320 Ma. I interpret this fast cooling history as resulting from upthrusting and supracrustal nappe displacement of the whole Spessart Crystalline Complex over the Middle Devonian through Lower Carboniferous island arc which may be buried now below the Spessart Mts. The arrival of the Spessart nappe complex at ≈320 Ma postdates the deposition of the Upper Visean (≈340 Ma) through Lower Namurian (≈330 Ma) autochthonous flysch of the Rheinische Schiefergebirge and the Harz Mts. This explains the high content of I-type magmatic pebbles and weakly metamorphosed sedimentary rocks that cannot be derived from the present Mid-German Crystalline Rise (Sommermann 1990). Pebbles showing Lower to Middle Paleozoic fossils and 550-Ma-old magmatic pebbles are proof of a Cadomian source area with a Paleozoic sedimentary cover (Sommermann 1990). The extent of this source area is not known, but has been possibly much wider than the present Mid-German Crystalline Rise.

The deposition of the allochthonous flysch outlines the subduction stage. The deposition of the autochthonous Upper Visean through Namurian flysch documents a downwarping of the Rhenohercynian crust and the collision of the Saxothuringian Island arc with the Rhenohercynian passive margin. A new tectonic pulse occurred in the Namurian when the Spessart Nappe Complex overthrust the island arc. At the same time, the deformation formed at the present southern Rheinische Schiefergebirge (325 Ma, Ahrendt et al. 1978, 1983) and peak metamorphic conditions were reached because of the tectonic overburden due to folding, thrusting, and nappe emplacement (Weber 1981; Weber and Behr 1983; Oncken 1988).

The crystalline basement and the pre-Namurian sedimentary cover of the Saar-Nahe Basin form the western prolongation of a fore-arc region NW of the Spessart Mts. It was displaced onto the southern Rheinische Schiefergebirge during the Namurian and its southern part was overridden by the Spessart crystalline nappe complex. The pre-Namurian of the Saar-Nahe Basin has not been metamorphosed due to the lack of a tectonic overburden. Only in its southeastern part can some tectonic overburden be assumed, because of the biotite bearing, weakly deformed, Lower Carboniferous graywackes of Neustadt, Hambach, and Weiler.

Dallmeyer et al. (Eds.)
Pre-Permian Geology of Central and Eastern Europe

The reason for the reduced crustal covergence west of the Odenwald Mts. was the system of sinistral strike-slip faults (see Krohe and Willner, Chap. IV.C.2) which belong to a transform fault system between the Rhenohercynian and the Moldanubian suture (Fig. 1; Weber 1987, 1990; Weber and Vollbrecht 1989). Inside that area, the transform fault reduced the overall crustal convergence in the west and prevented the Middle Devonian through Lower Carboniferous sediments from metamorphism. East of the transform fault, the crustal convergence was much stronger and led to sinistral transpression in the Spessart Mts. From the Namurian formation of the Saar-Nahe Molasse Basin onward, crustal extension dominates in the west. Since the strike of the Saar-Nahe Basin is SW-NE, sinistral strike-slip displacements along the NNE-SSW trending transform fault leads to sinistral transtensional strain in the basin and dextral strike-slip with a SW-dipping displacement vector along the Southern Hunsrück Boundary fault.

The Bergstrasse Odenwald forms the transition between the region of reduced crustal convergence in the west and strong crustal convergence and nappe tectonics in the east. As shown by Krohe (1991), the northern Odenwald must be replaced to the NE relative to its southern part. Therefore, it seems possible that during collision the northern parts escaped to the SW, i.e., from the zone of high compression into the region of less compressive strain or extension from the Namurian onward. This kind of "escape tectonics" is known from the Himalayas and explains quite well the transtensional strain path and the magmatic history of units I to III of the Odenwald. The northernmost unit escaped first and shows with its mantle-derived gabbroic high level intrusion the strongest affinity on an oceanic environment. The U/Pb zircon age of 362 ± 7 Ma (Kirsch et al. 1988) dates the intrusion and the contact metamorphism of the wall rocks of amphibolites and pelitic strata of unknown age. The geotectonic environment where this mantle-derived intrusion

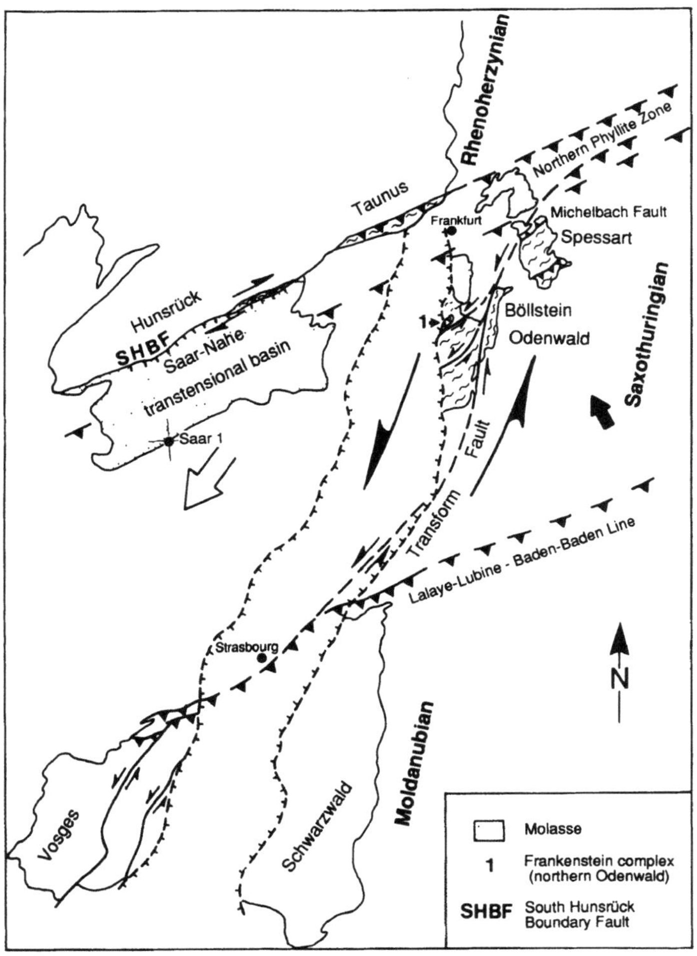

Fig. 1. Fault patterns and displacements in the area around the Otzberg Transform Fault system. For further explanation, see text

took place is unknown. The anticlockwise PT path and the high geothermal gradient (70–80 °C/km, Willner et al. 1991) point to an early oceanic rift environment and later shallow subsidence in a subduction zone. The terrane, however, escaped subduction by sinistral strike-slip displacement.

Units II and III are dominated by calc-alkaline intrusions of I-type granitoids (Hennes-Klaiber 1989; Hennes-Kleiber and Altherr 1991). They represent the deeper parts of an island arc, where the magmatites intruded into a metamorphic basement of unknown, possibly Cadomian age. Due to the great intrusion depth (Willner et al. 1991; see also Chap. IV.C.2), the Hbl-cooling ages of 360–340 Ma in unit II and of 335 Ma in unit III (Kreuzer and Harre 1975; Rittmann 1984) are not identical with the intrusion age in unit I. Nevertheless, they outline a minimum age of the island arc activity which cannot be far from peak metamorphic conditions, since Todt (1979) found concordant zircon ages of 335 Ma in migmatites of both units.

The nappe displacement of the Spessart Crystalline Complex postdates the Hbl-cooling ages of the Bergstrasse Odenwald, whereas the Bio ages are identical. This documents a different early cooling history. It is not very likely, but it cannot be excluded, that the Bergsträsser Odenwald or parts of it were overlain by the Spessart nappe complex. Further to the west, in the Saar-Nahe Basin, such an overburden must be excluded because of the very weak deformation and absent metamorphism. Therefore, an important fault must exist west of the Spessart Mts. Since the Böllstein Odenwald (unit IV) does not belong to the island arc of units II and III and reveals a cooling history similar to the Spessart Mts, it is most likely that the Otzberg Fault between Bergstrasse and Böllstein Odenwald represents this transform fault or forms part of a system of parallel faults which outline the Tertiary Rhine Graben. Later uplift and cooling below 300 °C took place at the same time (≈320 Ma) in both areas and also in places further to the east (Schmidt et al. 1986). The K/Ar biotite cooling age (or intrusion age) of 330 Ma obtained by Hess and Schmidt (1989) from an undeformed lamprophyre which cuts through the mylonites and cataclasites of th Otzberg Fault seems problematic. If this age is not too high (excess argon?), it remains unclear why the numerous biotite cooling ages from the Odenwald and Spessart Mts. are all clearly younger.

References

Ahrendt H, Hunzicker JC, Weber K (1978) K/Ar-Altersbestimmungen an schwach-metamorphen Gesteinen des Rheinischen Schiefergebirges. Z Dtsch Geol Ges 129: 229–247

Ahrendt H, Clauer N, Hunzicker JC, Weber K (1983) Migration of folding and metamorphism in the Rheinische Schiefergebirge deduced from K-Ar and Rb-Sr age determinations. In: Martin H, Eder FW (eds) Intracontinental fold belts. Springer, Berlin Heidelberg New York, pp 323–338

Hennes-Klaiber U (1989) Der Odenwald: Hinweise auf einen variszischen Arc-Magmatismus? Beih Eur J Miner 1 (1): 66

Hennes-Klaiber U, Altherr R (1991) Die I-Typ Granitoide des Odenwaldes. Intrusionstiefen und Krustenmächtigkeiten. Eur J Miner 3, Beih 1: 163

Hess JC, Schmidt G (1989) Zur Altersstellung der Kataklasite im Bereich der Otzberg-Zone, Odenwald. Geol Jahrb Hessen (Wiesbaden) 117: 69–77

Kirsch H, Kober B, Lippolt HJ (1988) Age of intrusion and rapid cooling of the Frankenstein gabbro (Odenwald, SW-Germany) evidenced by 40Ar/39Ar and single-zircon 207Pb/206Pb measurements. Geol Rundsch 77: 693–711

Kreuzer H, Harre W (1975) K/Ar Altersbestimmungen an Hornblenden und Biotiten des Kristallinen Odenwaldes. Aufschluβ Sdb (Heidelberg) 27: 71–77

Krohe A (1991) Emplacement of synkinematic plutons controlled by transtensional tectonics in the Variscan Odenwald (SW-Germany). Geol Rundsch 80: 391–409

Lippolt HJ (1986) Nachweis altpaläozoischer Primäralter (Rb-Sr) und karbonischer Abkühlungsalter (K-Ar) der Muskovit-Biotit-Gneise des Spessarts und der Biotit-Gneise des Böllsteiner Odenwaldes. Geol Rundsch 73 (3): 569–583

Nasir S, Okrusch M, Kreuzer H, Lenz H, Höhndorf A (1991) Geochronology of the Spessart Crystalline Complex, Mid-German Crystalline Rise. Miner Petrol 44: 39–55

Okrusch M, Richter P (1986) Orthogneisses of the Spessart crystalline complex, northwest Bavaria: indicators of the geotectonic environment? Geol Rundsch 75 (3): 555–568

Oncken O (1988) Geometrie und Kinematik der Taunuskammüberschiebung – Beitrag zur Diskussion des Deckenproblems im südlichen Rheinischen Schiefergebirge. Geol Rundsch 77: 551–575

Rittmann KL (1984) Argon in Hornblende, Biotit und Muskovit bei der geologischen Abkühlung – $^{40}Ar/^{39}Ar$-Untersuchungen. PhD Thesis, Univ Heidelberg, 278 pp

Schmidt FP, Gebreyohannes Y, Schliestedt M (1986) Das Grundgebirge der Rhön. Z Dtsch Geol Ges 137: 287–300

Sommermann A (1990) Petrographie und Geochemie der magmatogenen Gerölle in Konglomeraten des Kulms in Hinblick auf ihre Herkunft von der Mitteldeutschen Schwelle. Geol Jahrb Hessen (Wiesbaden) 118: 167–197

Todt W (1979) U-Pb-Datierungen an Zirkonen des kristallinen Odenwaldes. Fortschr Miner 57 (Beih 1): 153–154

Weber K (1981) The structural development of the Rheinische Schiefergebirge. Geol Mijnbuow 60: 149–159

Weber K (1984) Variscan events: early Palaeozoic rift metamorphism and late Palaeozoic crustal shortening. Geol Soc Lond Spec Publ 14: 3–22

Weber K (1987) The mid-European Variscides in terms of allochthonous terranes. In: Freeman R, Mueller St, Giese P (eds) Proc 3rd Worksh on the European Geotraverse (EGT) Project, pp 73–81

Weber K (1990) Crustal structure and the geological development of the Rhenohercynian/Saxothuringian terrane boundary in central Europe. Int Conf Paleozoic orogens in central Europe (Abstr)

Weber K, Behr H-J (1983) Geodynamic interpretation of the mid European Variscides. In: Martin H, Eder FW (eds) Intracontinental fold belts. Springer, Berlin Heidelberg New York, pp 427–469

Weber K, Vollbrecht A (1989) The crustal structure at the KTB drill site, Oberpfalz. In: Emmermam R, Wohlenberg J (eds) The German Continental Deep Drilling Program (KTB) Springer, Berlin Heidelberg New York, pp 5–36

Willner AP, Massonne H-J, Krohe A (1991) Tectonothermal evolution of a part of a Variscan magmatic arc: the Odenwald in the Mid-German Crystalline Rise. Geol Rundsch 80: 369–389

IV.D Igneous Activity (Pre- to Early Variscan Magmatism)

M. Okrusch, W. Schubert, and S. Nasir

The metamorphic sequences constituting the MGCR frequently contain intercalations of metabasites. Judging from structural and textural relics, from rarely preserved magmatic minerals and from geochemical constraints, most of these metabasites are derived from igneous protoliths, especially from mafic volcanites, including volcanoclastites, in places also from gabbros.

Orthogneisses of granitic to granodioritic composition constitute the gneiss dome of the Böllstein Odenwald; they form parts of the crystalline complexes in the Pfälzer Wald, the Spessart, the Ruhla, and the Kyffhäuser area, and were penetrated by various drillholes in the Rhön, the Dessau area, and in southern Brandenburg.

1 Pfälzer Wald

The crystalline basement of the Pfälzer Wald, forming the westernmost part of the MGCR, is very limited in exposure. The most extensive outcrop is a big quarry near Albersweiler (Zaminer 1957; Stellrecht 1971; Laue et al. 1990; Flöttmann and Oncken 1992). The main rock type is an **orthogneiss** of granitic to granodioritic composition and I-type geochemical character, containing about 10 vol% of biotite and subordinate hornblende in some layers. Intercalated within the orthogneisses are lenses and pods of biotite- and epidote-**amphibolites**; the contacts between both rock types are obliquely cut by the schistosity. The association is intruded by late- to postkinematic aplites and pegmatites.

2 Bergsträsser Odenwald

The western part of the Odenwald crystalline complex, the so-called Bergsträsser Odenwald, is predominantly formed by magmatic intrusions of Variscan age, covering a wide compositional range. The hosting metamorphic sequences, by contrast, are subordinate in regional extent and exposure. They consist of metapelitic to metapsammitic schists with intercalations of amphibolite as well as subordinate garnetcordierite-plagioclase gneiss (kinzigite), calcsilicate rocks, marble, graphite schist, graphite or apatite-bearing quartzites, and corundumhercynite fels (e.g., Okrusch et al. 1975).

The **metabasites** are arranged in five, SW-NE-trending belts. In places, the amphibolites display relics of primary ophitic or porphyric textures and relics of An-rich plagioclase, testifying to an igneous protolith. Based on their major and minor element chemistry as well as their Cr and Ni contents, Schubert (1968) deduced a tholeiitic composition for the amphibolites. Trace element data of Klemm and Weber-Diefenbach (1971), Nasir (unpubl.) and Poller (1991) point to a subalkaline character similar to modern island arc tholeiites. This is clearly demonstrated in relevant discrimination diagrams, the Ti vs. Zr plot of Pearce (1982) being given as an example (Fig. 1a). Conformably, the trace element patterns – normalized to the average of mid-ocean ridge basalts (MORB) – show the typical relative enrichment of Ce, compared to the neighboring Nb and P, a slight depletion in Ti and a marked depletion in Cr (Fig. 2a). According to Poller (1991), most of the amphibolites show a slight enrichment of the light rare-earth elements (LREE) with La/Lu_n ratios between 1.1 and 5.8, in one case even 20.3; the total REE content ranges from 37 to 166 ppm; the sample with the strongest LREE enrichment contains 288 ppm. In summary, the geochemical character of the Bergsträsser metabasites may indicate that the volcanosedimentary sequences were deposited in a back-arc situation, which conforms to the occurrence of metamorphosed limestones in the metabasite belt Bensheim-Reinheim. It should be stressed that the metabasites display very similar petrographical and geochemical characteristics throughout the Bergsträsser Odenwald (Poller 1991).

Dallmeyer et al. (Eds.)
Pre-Permian Geology of Central and Eastern Europe

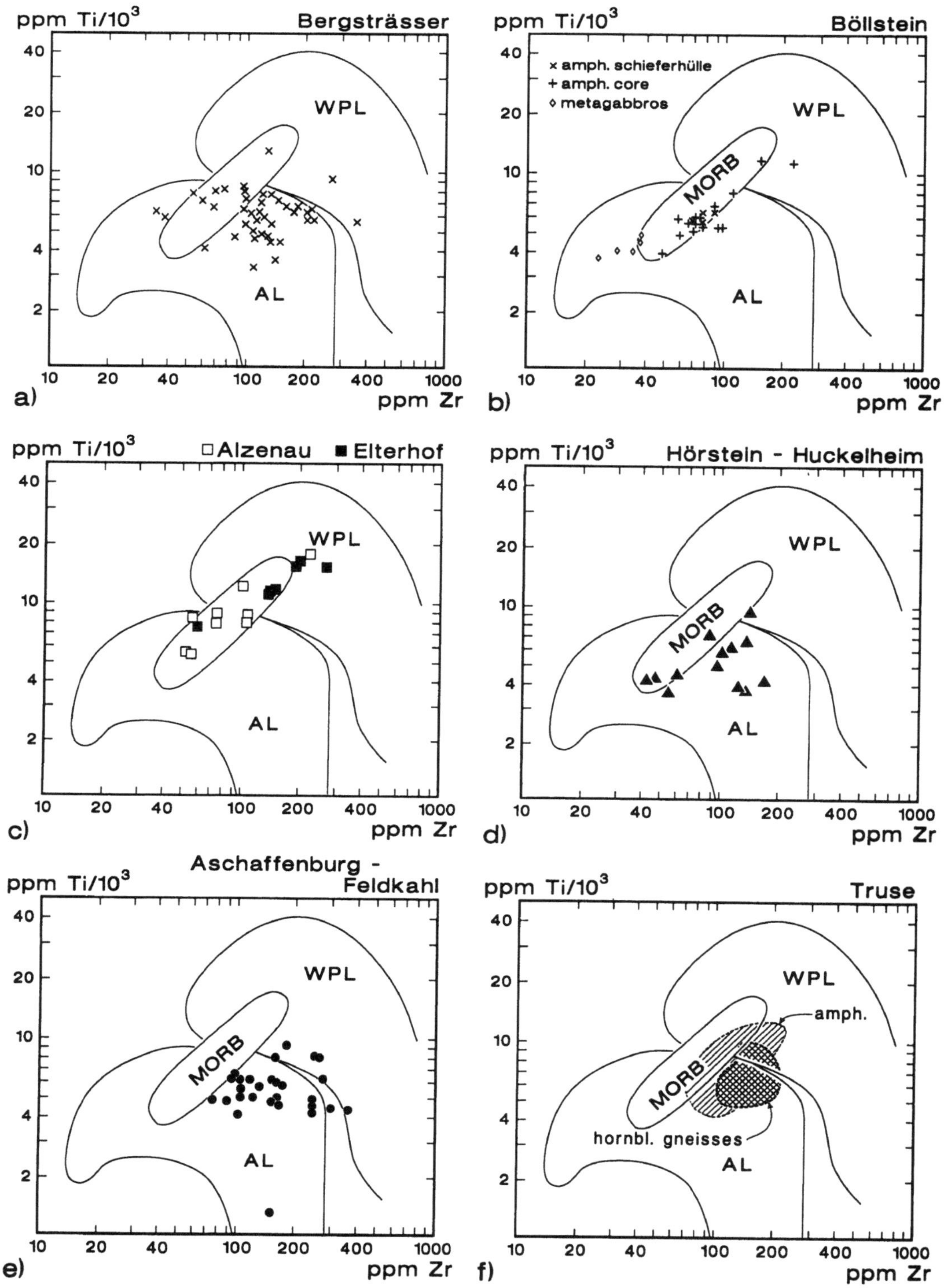

Fig. 1a–f. Discrimination diagram Ti vs. Zr after Pearce (1982) for metabasites from the crystalline complexes of **a** the Bergsträsser, **b** the Böllstein Odenwald, **c**, **d**, **e** the Spessart, and **f** the Ruhla area. Data sources see text

A characteristic companion of the amphibolites in the Bergsträsser Odenwald are **talc-bearing chlorite-amphibole felses**, shortly designated as "hösbachites" after an occurrence in the Spessart crystalline complex. They can be derived from ultramafics which presumably consisted of olivine, orthopyroxene, clinopyroxene, and Ca-amphibole in variable proportions (Schubert 1969). Since the

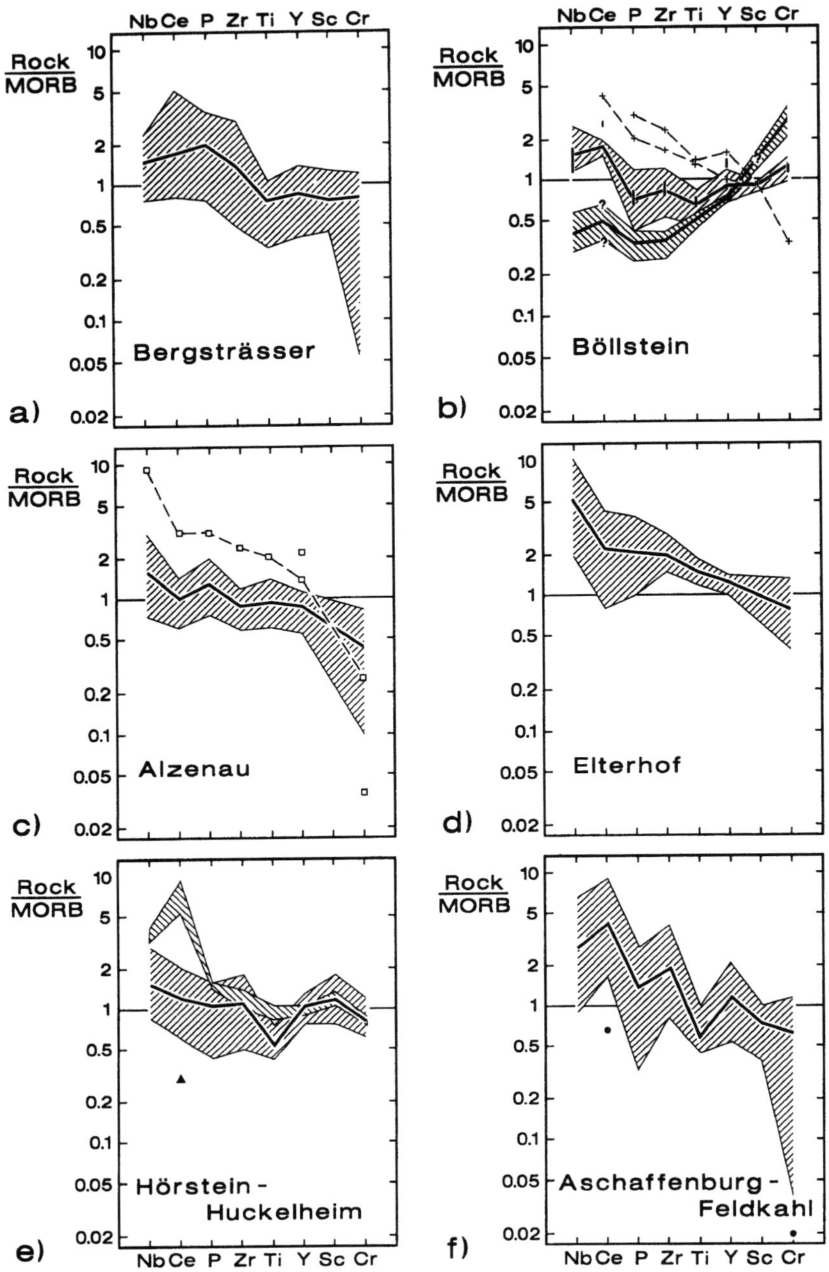

Fig. 2a–f. MORB-normalized trace element patterns for metabasites from the crystalline complexes of **a** the Bergsträsser Odenwald (n = 29; Ce: n = 17; Nb: n = 9); **b** the Böllstein Odenwald (metagabbros: n = 5; Ce: n = 1; amphibolites of the core: n = 11 + 2; Ce: n = 3; amphibolites of the Schieferhülle n = 4, Ce: n = 1); and the Spessart, **c** Alzenau formation (n = 8 + 1); **d** Elterhof formation (n = 6); **e** metabasite belt Hörstein Huckelheim (n = 10 + 3); **f** metabasite belt Aschaffenburg – Feldkahl – Rottenberg (n = 28). Compositional ranges are shown as *shaded areas*, average values as *solid lines*, exceptional values by *single symbols*. Normalizing values are after Pearce (1982). Data sources see text

hösbachites are associated with amphibolites showing ophitic relic textures, they may have formed as cumulates within thick basic sills. A similar interpretation is discussed by Matthes et al. (1995) for the hösbachites penetrated by the KTB pilot hole in the Oberpfalz area.

3 Böllstein Odenwald

The eastern part of the Odenwald crystalline complex, the Böllstein Odenwald, is formed by a dome-like structure of granodioritic and granitic orthogneisses with intercalations of metagabbro and amphibolite. This gneiss dome is surrounded by an envelope of metapelitic to metapsammitic schists, the so-called Schieferhülle, which contains intercalations of amphibolite, calc-silicate rocks, and quartzite. The age relationships between these individual rock types has been a matter of discussion for nearly a century. According to the most recent concept (Altenberger et al. 1990), the volcanosedimentary protoliths of the Schieferhülle were intruded by granodiorite and subsequently by granite magmas. These intrusives were affected, together with their country rocks, by a first deformation D_1 leading to the s_1 schistosity. In contrast to Chatterjee (1960), Altenberger et al. (1990) assume that the basic magmas intruded the granodioritic and granitic orthogneisses late- to posttectonically relative to D_1. The whole association underwent a second penetrative deformation, D_2, leading to isoclinal folds and the schistosity, s_2, whereas the dome-like structure is due to a third deformational event, D_3.

Geochemical investigation of Altenberger et al. (1990) show that the **granodioritic** and **granitic orthogneisses** of the Böllstein are derived from S-type granitoids, presumably formed at a convergent plate margin: In the discrimination diagrams Rb vs. Nb + Y (Fig. 3a) of Pearce et al. (1984) the data points plot predominantly in the fields of of volcanic-arc and syn-collision granites. Minimum melting compositions, typical for synorogenic, anatectic twomica granites (e.g., Batchelor and Bowden 1985) are displayed in the simplified granite system Q-Ab-Or (Fig. 4a), where the data points of the granodioritic and granitic Böllstein gneisses plot in the low temperature trough at low water vapor pressures, close to the cotectic lines at 1 and 0.5 kb, respectively (Altenberger et al. 1990). Rb-Sr total rock analyses of seven granodiorite gneisses did not yield an isochron, while seven granite gneisses gave a best-fit line corresponding to an imprecise age of 360 ± 56 Ma. Combining the results of all Böllstein gneiss samples, an apparent age of 413 ± 26 Ma is obtained which may date the intrusion of two closely related magma pulses (Lippolt 1986).

The **metagabbros** form stock-like bodies which display a primary gabbroic structure, texture and mineralogy in their cores. Textural evidence shows a magmatic crystallization sequence olivine → plagioclase → orthopyroxene and clinopyroxene, indicating a formation under relatively low pressures (Knauer et al. 1974). Towards the margins, there are all transitions to common

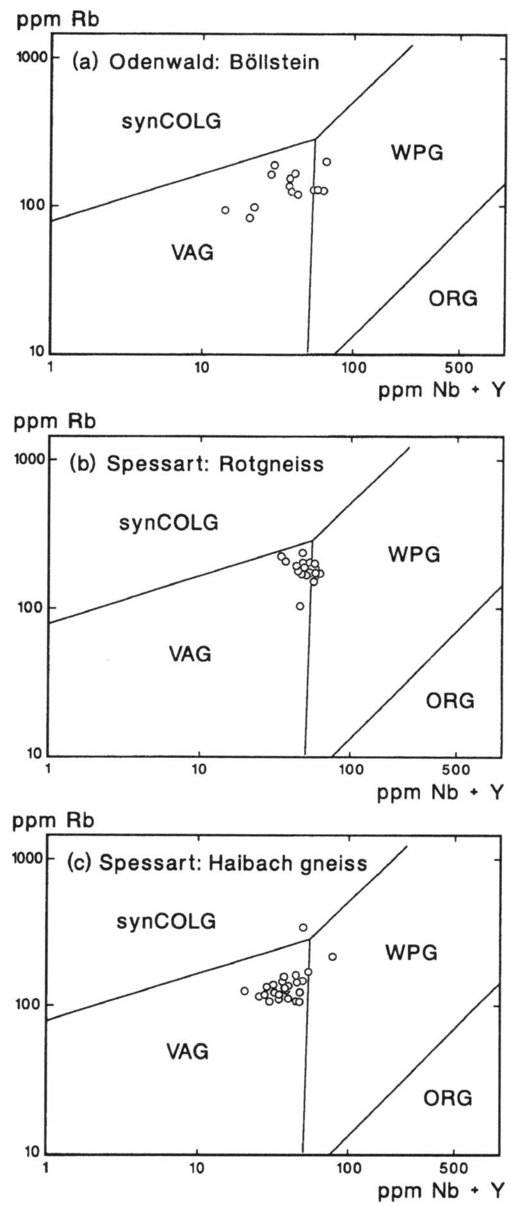

Fig. 3a–c. Variation diagram Rb vs. Y + Nb to discriminate between syn-collision granitoides (*synCOLG*), within-plate granitoids (*WPG*), volcanic-arc graniotoids (*VAG*) and ocean-ridge granitoids (*ORG*), after Pearce et al. (1984). Orthogneisses from **a** the Böllstein Odenwald; **b** the Rotgneiss complex; **c** the Haibach gneiss of the Spessart. Data sources see text

amphibolites. Conforming to the occurrence of pseudomorphs after olivine, the metagabbros display an Ol-normative chemical composition. They are depleted, relative to average MORB, in the high-field strength (HFS) elements Nb, P, Zr, Ti and, to a lesser extent, Y, whereas Cr is enriched (Figs. 1b, 2b); the LREE are slightly depleted with $La/Lu_n = 0.47$; the total amount of REE is at 16 ppm (Altenberger et al. 1990).

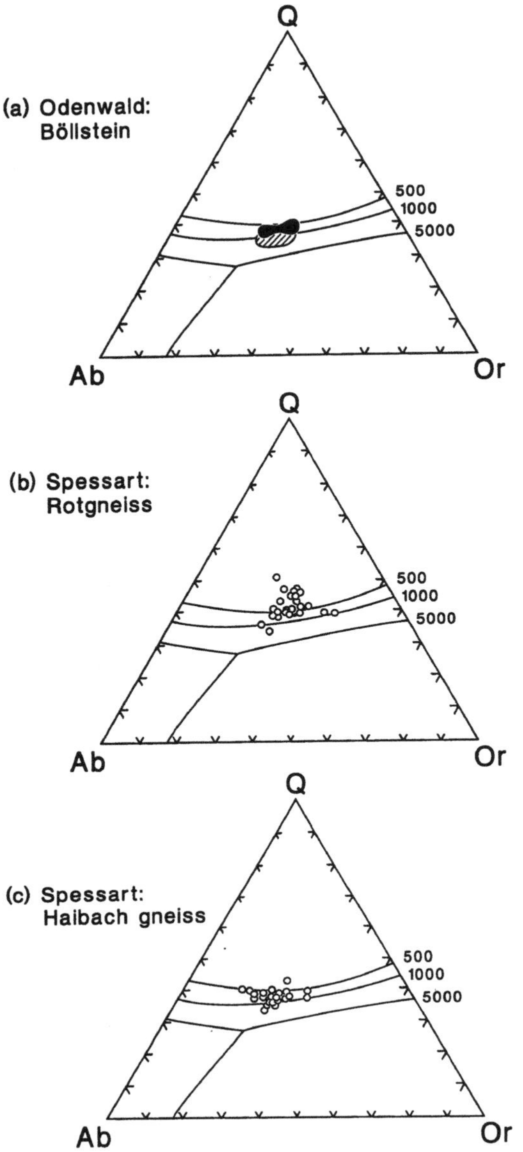

Fig. 4a–c. Chemical variation of orthogneisses from **a** the Böllstein Odenwald (granite gneisses *black field*; granodiorite gneisses *hatched field*); **b** the Rotgneiss complex; **c** the Haibach gneiss of the Spessart in the normative *Q-Ab-Or* triangle. Cotectic lines, ternary minima, or the ternary eutectic are given for H_2O pressures for 500, 1000, and 5000 bars (Tuttle and Bowen 1958; Luth et al. 1964). Data sources see text

Independent of the metagabbros are sill-like intercalations of **amphibolite**, which, in places, exhibit ophitic textures and relics of primary An-rich plagioclase. Judging from their major element composition, these amphibolites are derived from olivine to quartz tholeiites. Trace element data of Nasir (unpubl.), Altenberger et al. (1990), and Poller (1991) indicate MORB to IAB characters of the amphibolites of the Böllstein gneiss core (Figs. 1b, 2b), although some samples show a stronger enrichment of the HFS elements and a depletion in Cr. The amphibolites of the gneiss core are slightly LREE enriched with La/Lu_n ratios of 1.9–2.8 and total REE contents of 32–61 ppm (Altenberger et al. 1990; Poller 1991).

Interestingly, the amphibolites of the Schieferhülle strongly resemble those of the gneiss core in their chemical characteristics (Figs. 1b, 2b), except for somewhat higher Ce and Y values. The total amount of REE of one analyzed sample is at 69 ppm, the La/Lu_n ratio being at 2.5 (Altenberger et al. 1990).

Metamorphosed **ultramafics**, closely resembling the "hösbachites" of the Bergsträsser Odenwald, were recognized in only one occurrence of the gneiss core. They may contain relics of olivine, orthopyroxene, and primary amphibole, but are frequently transformed into the typical amphibolite-facies assemblage amphibole + chlorite + talc (Knauer et al. 1974).

4 Spessart

The Spessart crystalline complex, situated northeast of the Böllstein Odenwald, consists predominantly of gneisses and micaschists with intercalations of quartzite, marble, and amphibolite, whereas – in contrast to the Bergsträsser Odenwald – unmetamorphosed igneous rocks are rare. The SW-NE striking units are summarized by Hirschmann in Chapter IV.B.

4.1 Metabasites

The metabasites of the Spessart crystalline complex form constituents of three different lithostratigraphic units. Consequently, their respective protoliths should be of different age.

1. The variegated sequences of the **Alzenau** and the **Elterhof formations** – assumed to be of upper Proterozoic (Bederke 1957; Braitsch 1957b) or lower Cambrian age (Braitsch 1957b; Hirschmann and Okrusch 1988) – contain ample intercalations of amphibolite and hornblende gneiss. Together with the associated acid gneisses (Braitsch 1957a; Okrusch et al. 1967), these metabasites represent an early phase of igneous activity, a bimodal basalt-rhyolite volcanism.

According to the major and trace element data of El Shazly (1983) and Okrusch et al. (1985), the protoliths of the metabasites are subalkaline tholeiites or their volcanoclastic equivalents displaying a relatively high degree of chemical variation, e.g., with respect to their contents in Ti, P, Zr and their FeO^{tot}/MgO ratios. In relevant discrimination diagrams, e.g., the Ti vs. Zr plot of Pearce (1982) shown in Fig. 1c, the compositional fields of the metabasites from the two formations show a considerable overlap, but are not identical. Nearly all samples from the Alzenau formation plot within the MORB field; conformably, their MORB normalized trace element patterns compare well with modern ocean-floor tholeiites (Fig. 2c). On the other hand, most of the Elterhof samples and the Alzenau amphibolite Sp72-135 are more enriched in incompatible trace elements like Ce, P, Zr and Ti and, consequently, plot in the fields of modern within-plate or ocean-island basalts (Fig. 1c). Accordingly, their MORB normalized trace element patterns are similar to those of E-MORB or intraplate basalts (Fig. 2c,d). However, compared to MORB, the Cr and Ni contents of the Alzenau and Elterhof metabasites are generally depleted pointing to an island-arc or interarc-basin environment. This would be in accord with the occurrence of metamorphosed limestones in the Alzenau and especially in the Elterhof formation.

2. The **metabasite belt of Hörstein-Huckelheim**, in the border zone of the Geiselbach and the Mömbris formation, is formed by amphibolites and hornblende gneisses which are assumed to represent constitutents of the normal lithostratigraphic succession. According to their geochemical characteristics, these metabasites are derived from subalkaline, tholeiitic basalts which, except for the low Ti values, conform to a MORB-like trace-element pattern (Nasir 1986; Nasir and Okrusch 1991; see Fig. 2e). In the Ti vs. Zr diagram of Pearce (1982) the metabasites of the Hörstein-Huckelheim belt plot in the transitional region of modern mid-ocean ridge and island-arc basalts (Fig. 1d). The IAT signature is most pronounced for the massive biotite amphibolite at Hörstein which exhibits markedly higher contents of Nb and Ce, and the typical Ce enrichment relative to the neighboring elements Nb and P (Fig. 2e).

Judging from geochemical evidence, the hosting, predominantly pelitic to psammitic metasediments were deposited in a continental-margin environment, not far from the source area (Smoler 1987). Metasediments, stratigraphically 500–1000 m above the metabasite belt, yielded psilophytes of Silurian, possibly Ludlowan age (Reitz 1987). Combining all the available evidence, one may assume that the volcanic activity took place, in the late Ordovician to early Silurian, in a small oceanic basin. According to the paleogeographic reconstructions of Neugebauer (1988, Fig. 12), the Tornquist Sea, a gulf-like extension of the Iapetus Ocean, did exist during this time.

3. The **metabasite belt of Aschaffenburg-Feldkahl-Rottenberg** consists of amphibolites and hornblende gneisses which are intercalated within metasediments of the Mömbris formation – assumed to be of Cambrian to Ordovician age (Matthes 1954; Bederke 1957; Hirschmann and Okrusch 1988) – and within orthogneisses of the Rotgneiss complex. Amphibolites near Feldkahl display a well-preserved ophitic texture, testifying to a doleritic protolith (Matthes and Krämer 1955). In their chemical composition, the metabasites of the Aschaffenburg-Rottenberg belt exhibit a wide variation ranging from subalkaline basalts to andesites and even dacites (Hesselmann 1982; El Shazly 1983; Nasir 1986; Nasir and Okrusch 1991). Conformably, the Ti vs. Zr plot (Fig. 1e) and the MORB-normalized trace element patterns (Fig. 2f) are similar to those of modern island-arc volcanites. Field relationships, although poorly exposed, seem to indicate that this calc-alkaline volcanism was contemporaneous with, or somewhat younger than the intrusion of the Rotgneiss granitoids about 430 Ma ago. The geotectonic significance of these igneous activities will be discussed below.

4.2 Ultramafics

In the metabasite belt Aschaffenburg-Feldkahl-Rottenberg, talc-bearing chlorite-hornblende felses, the so-called hösbachites occur close to the village of Wenighösbach. They are similar, in their petrographical and geochemical character, to the hösbachites in the Bergsträsser and the Böllstein Odenwald, and are derived from ultramafic lenses (Matthes and Krämer 1955; Matthes and Schubert 1967; Nasir 1986, 1990), possibly formed as cumulates in mafic sills (Matthes et al. 1995). It is interesting to note that hösbachites from the Spessart crystalline complex were used to manufacture moulds for casting bronze axes and knives by the prehistoric Urnfield people in Franconia and Thuringia (Okrusch and Schubert 1986; Schubert et al. in prep.).

4.3 Orthogneisses

Two types of granitic to granodioritic orthogneisses occur in the Spessart crystalline complex: Muscovite-biotite gneisses of the so-called **Rotgneiss unit** are interfolded with staurolite-bearing metapelites of the Mömbris formation. Judging from geological, petrographical, geochemical, and geochronological evidence, the Rotgneiss is derived from a pre-Variscan granitoid intrusion which was deformed and metamorphosed, together with its sedimentary country rocks, during the Variscan orogenesis (Bederke 1957; Matthes and Okrusch 1965; Okrusch and Richter 1986). The **Haibach gneiss**, an equigranular biotite gneiss, is interlayered with metapelites of the Schweinheim formation. Although – in contrast to the Rotgneiss – textural relics of an igneous protolith were not recorded, recent geochemical investigations indicate that the Haibach gneiss was derived from a similar magmatic protolith (Dombrowski et al. 1994).

Both the Rotgneiss and the Haibach gneiss display peraluminous compositions with geochemical characteristics transitional between the S-type and I-type granitoids of Chappell and White (1974). In the discrimination diagrams Rb vs. SiO_2, Nb vs. Y, and Rb vs. Nb + Y of Pearce et al. (1984), the Rotgneiss and Haibach gneiss plot in the fields of volcanic arc and/or syncollision granites (Fig. 3b,c). This would indicate that the granitic magmas from which the protoliths of the Spessart orthogneisses were derived formed at a convergent plate margin (Dombrowski et al. 1995). Conformably, compositions typical of synorogenic, anatectic two-mica granites (e.g., Batchelor and Bowden 1985) are displayed in the ternary system Q-Ab-Or: the data points of the Rotgneiss and the Haibach gneiss concentrate in the low-temperature trough, close to the cotectic lines and ternary melting minima at low water vapor pressures (<1 kb; see Fig. 4b,c).

Rb-Sr isotopic analyses of eight Haibach gneiss, whole-rock samples from a single outcrop, the Wendelberg quarry, yielded an isochron age of 407 ± 7 Ma (MSWD = 2.2) with an initial $^{86}Sr/^{87}Sr$ ratio of 0.7077 ± 0.0007. This date is sustained by single zircon dating on three clear, inclusion-free, idiomorphic grains from the Schmerlenbach quarry, which yielded virtually identical $^{207}Pb/^{206}Pb$ ratios combining to a mean age of 410 ± 18 Ma (Dombrowski et al. 1995). An identical $^{207}Pb/^{206}Pb$ age of 418 ± 18 Ma was obtained by these authors on three clear, euhedral zircon crystals from two Rotgneiss occurrences. This result also confirms earlier Rb-Sr whole-rock dates on the Rotgneiss: Nasir et al. (1991) reanalyzed a suite of Rotgneiss samples from widely separated localities (previously investigated by Kreuzer et al. 1973, and Lippolt 1986, respectively) and obtained a best-fit line corresponding to an age of 439 ± 15 Ma (MSWD = 4.9) with a $^{87}Sr/^{86}Sr$ initial ratio of 0.7048 ± 0.0026. In summary, the Rb-Sr whole rock and single zircon dates can be interpreted as Silurian intrusion ages for the granitoid precursors of the Haibach gneiss and the Rotgneiss. It should be noted that the sedimentation of the Ordovician to Silurian Geiselbach formation was still going on when the granitoids intruded in the deeper, Cambro-Ordovician levels of the Mömbris-Geiselbach sediment pile.

Although clear, euhedral zircon grains are predominant in the Spessart orthogneisses, a second variety of stubby, rounded zircon grains was recognized in the investigated Rotgneiss samples. These zircons yielded highly variable and surprisingly old $^{207}Pb/^{206}Pb$ ages of 2278 ± 12, 2490 ± 12 and 2734 ± 10 Ma, testifying to an early Proterozoic to late Archean crustal component in the Rotgneiss protolith (Dombrowski et al. 1995). On the other hand, the low initial $^{87}Sr/^{86}Sr$ ratios recorded for the Haibach gneiss and the Rotgneiss may indicate that a significant juvenile element was involved in the magma generation. Dombrowski et al. (1995) therefore propose that the protoliths of the Spessart orthogneisses are formed, in mid- to late Silurian times, by mixing of anatectic continental crust with mantle-derived, possibly underplated juvenile material, presumably in an island-arc or Andean-type tectonic setting. Such a scenario would fit within recent paleomagnetic data of Trench and Torsvik (1992), which indicate a closure of the Tornquist Sea at about 430 Ma. The calc-alkaline volcanism documented in the metabasite belt Aschaffenburg-Feldkahl-Rottenberg may be related to this event as well. Alternatively, the emission of the calc-alkaline magmas may reflect a period of compressional tectonics at the advent of the Variscan orogenesis, i.e., crustal shortening and eventual closure of the Saxothuringian basin during the early Devonian (e.g., Weber and Behr 1983; Weber 1984; Neugebauer 1988; Franke 1989a,b, 1990; Franke and Oncken 1990).

5 Ruhla Area, Thüringer Wald

Although the similarities between the Spessart and the Ruhla crystalline complexes were early recognized (Matthes 1954; Bederke 1957), the lithostratigraphic correlation is still open to discussion (e.g., Neumann 1966, 1973, 1985; Hirschmann and Okrusch 1988; Wunderlich 1989; Estrada et al. 1992). Consequently, the stratigraphic position and age of the metamorphosed, pre- to early Variscan magmatites in the Ruhla area are less well established than in the Spessart.

5.1 Basic to Acid Metavolcanites

The lowermost unit of the Ruhla crystalline complex, the (?) Upper Proterozoic **Liebenstein formation** consists predominantly of graywackes, metamorphosed and migmatized under P-T conditions of the higher amphibolite facies. Frequent intercalations of amphibolites and hornblende gneisses are widely changed in their initial composition due to the strong migmatization (e.g., Koch 1940; Werner 1971a, 1974).

The **Truse formation**, a possible equivalent to the (?) Upper Proterozoic to Lower Cambrian Elterhof and Alzenau formations of the Spessart, is formed by muscovite-biotite-gneiss, garnet-biotite-plagioclase gneiss and acid quartz-plagioclase gneiss with intercalations of quartzite, garnet-quartzite and calc-silicate gneiss. The majority of these rocks is regarded as metagraywackes (e.g., Werner 1971b), whereas Estrada et al. (1992) claim that a high amount of acid to intermediate metavolcanites are incorporated in the sequence, similar to the Cambrian Vesser formation in the southern Thuringian synclinorium (e.g., Bankwitz et al. 1989, 1994).

According to Werner (1974) and Estrada et al. (1992), the metabasites of the Truse formation are derived from tholeiitic basalts, more rarely andesites, and the respective volcanoclastites; their trace element characteristics (see also Kramer et al. 1986) indicate an ocean-floor to island-arc affinity (Fig. 1f). It should be noted that some metabasites of the Truse formation underwent strong metablastic recrystallization leading to the formation of dioritic rocks (Koch 1940; Werner 1970, 1974).

The **Ruhla formation**, presumably ranging in age from the Lower Cambrian to the Silurian and forming possible equivalents of the Mömbris and Geiselbach formations of the Spessart, is composed predominantly of pelitic and quartzitic metasediments. Metabasalts and metaquartz keratophyres are intercalated especially in the lower parts, the R 1 member (e.g., Werner 1989). A comprehensive geochemical investigation is lacking so far.

5.2 Orthogneisses

The upper Proterozoic Liebenstein formation contains an orthogneiss of leucogranitic composition, the Dorngehege gneiss of uncertain intrusion age. The Katzenstein granite is presumably younger and intruded late- to posttectonically.

Whereas no unequivocal orthogneisses can be identified in the Truse formation, the Windsberg and the Silbergrund gneisses of the Ruhla formation must be regarded as intrusions of lower Paleozoic age, although they are intercalated in a stratiform way (e.g., Hirschmann and Okrusch 1988).

6 Kyffhäuser

According to Seim (1967), the Kyffhäuser crystalline complex is formed by a variegated metamorphic sequence, resembling the Truse formation in the Ruhla, as well as the Elterhof and Alzenau formations in the Spessart complex. It consists predominantly of biotite- and biotite-hornblende gneiss with intercalations of amphibolite, silicate marble, and calc-silicate fels. In addition, granitic to granodioritic orthogneisses and meta-aplites are present (e.g., Schüller 1952; Seim 1967; Zeh and Katzung 1994). The geochemical characteristics of these acid meta-igneous rocks were not investigated so far; the major and trace element geochemistry of the Kyffhäuser metabasites was determined by Schweder (1966).

The amphibolites and acid orthogneisses recorded in various drill holes in the eastern extension of the MGCR were not investigated geochemically (e.g., Erzberger et al. 1962; Gottesmann and Steinike 1962; Möbus 1963).

Acknowledgments. The manuscript materially benefited from discussions with W. Franke, O. Oncken, U. Schüssler and K. Weber, and from the critical reviews of G. Hirschmann and P.M. Sachs. We thank K.-P. Kelber for the line drawings.

References

Altenberger U, Besch T, Mocek B, Zaipeng Y, Yong S (1990) Geochemie und Geodynamik des Böllsteiner Odenwaldes. Mainzer Geowiss Mitt Mainz 19: 183–200

Bankwitz P, Bankwitz E, Kramer W (1989) Tectonic development and magmatism of Cambro-Ordovician sequence, Vesser area/Thuringian forest. In: Kramer W, Werner C-D (eds) Prevariscan mafic rocks in the Saxothuringian zone. Guidebook of Excursions in the German Democratic Republic May/June 1989: 73–89, Acad Sci GDR Potsdam

Bankwitz P, Bankwitz E, Kramer W, Pin C (1994) Early Paleozoic bimodal volcanism in the Vesser area, Thuringian Forest, eastern Germany. Zentralbl Geol Paläont Teil I 1992: 1113–1132

Batchelor RA, Bowden P (1985) Petrogenetic interpretation of granitoid rock series using multicationic parameters. Chem Geol 48: 43–55

Bederke E (1957) Alter und Metamorphose des kristallinen Grundgebirges im Spessart. Abh Hess Landesamt Bodenforsch Wiesb 18: 7–19

Braitsch O (1957a) Beitrag zur Kenntnis der kristallinen Gesteine des südlichen Spessarts und ihrer geologisch-tektonischen Geschichte. Abh Hess Landesamt Bodenforsch Wiesbaden 18: 21–72

Braitsch O (1957b) Zur Petrographie und Tektonik des Biotitgneises im südlichen Vorspessart. Abh Hess Landesamt Bodenforsch Wiesbaden 18: 73–99

Chappell BW, White AJR (1974) Two contrasting granite types. Pacific Geol Tokyo 8: 173–174

Chatterjee ND (1960) Geologische Untersuchungen im Kristallin des Böllsteiner Odenwaldes. Neues Jahrb Geol Paläontol Abh 111: 137–180

Dombrowski A, Henjes-Kunst F, Höhndorf A, Kröner A, Okrusch M, Richter P (1995) Orthogneisses in the Spessart Crystalline Complex, Northwest Bavaria: Silurian granitoid magmatism at an active continental margin. Geol Rdsch 84 (in press)

El Shazly S (1983) Petrography and geochemistry of the striated paragneiss series in the southern part of the Spessart crystalline complex. Dr rer nat Thesis, Tech Univ Braunschweig, 153 pp

Estrada S, Grunewald V, Wunderlich J (1992) Zur Eduktcharakteristik und Lithostratigraphie von Truse- und Hohleborn-Serie (Ruhlaer Kristallin). Neues Jahrb Geol Paläontol Abh 184: 389–419

Erzberger R, Gottesmann W, Kaemmel Th, Kampe A, Nöldeke W, Steinike K (1962) Untersuchungsergebnisse an Neuaufschlüssen in kristallinen Gesteinen zwischen Halle und Cottbus. Teil 1: Regionalgeologische Situation und Bedeutung der Bohraufschlüsse Plötz 472, Doberlug T52 und Collochau 3/60. Geologie Berlin 11: 633–639

Flöttmann TH, Oncken O (1992) Constraints on the evolution of the Mid German Crystalline Rise – a study of outcrops west of the river Rhine. Geol Rdsch 81: 515–543

Franke W (1989a) Tectonostratigraphic units in the Variscan belt of central Europe. Geol Soc Am Spec Pap 230: 693–708

Franke W (1989b) Variscan plate tectonics in Central Europe – current ideas and open questions. Tectonophysics 169: 221–228

Franke W (1990) Rhenohercynian Basin and Mid-German Crystalline Rise – an introduction. In: Franke W (ed) Mid-German Crystalline Rise and Rheinisches Schiefergebirge. Field guide to pre-conference excursion. IGCP 233 Terranes in the Circum-Atlantic Paleozoic Orogens. Conf on Paleozoic Orogens in Central Europe – Geology and Geophysics. Göttingen-Giessen, 169 pp

Franke W, Oncken O (1990) Geodynamic evolution of the North-Central Variscides – a comic strip. In: Freeman R, Giese P, Mueller St (eds) The European geotraverse: integrative studies. Results from the Fifth Study Centre, Rauischholzhausen, 26 March–7 April 1990, Eur Sci Found, Strasbourg, pp 187–194

Gottesmann W, Steinike K (1962) Untersuchungsergebnisse an Neuaufschlüssen in kristallinen Gesteinen zwischen Halle und Cottbus. Teil 2: Zur Petrographie der kristallinen Gesteine der Bohrung Plötz. Geologie Berlin 11: 640–662

Hesselmann H (1982) Petrographische und geochemische Untersuchungen an Gesteinen aus dem mittleren Vorspessart. Diploma Thesis, Tech Univ Braunschweig, 97 pp (unpubl)

Hirschmann G, Okrusch M (1988) Spessart-Kristallin und Ruhlaer Kristallin als Bestandteile der Mitteldeutschen Kristallinzone – ein Vergleich. Neues Jahrb Geol Paläontol Abh 177: 1–39

Klemm DD, Weber-Diefenbach K (1971) Geochemische Untersuchungen an Amphiboliten und Dioriten des nördlichen Odenwaldes. Neues Jahrb Mineral Abh 116: 80–111

Knauer E, Okrusch M, Richter P, Schmidt K, Schubert W (1974) Die metamorphe Basit-Ultrabasit-Assoziation in der Böllsteiner Gneiskuppel, Odenwald. Neues Jahrb Mineral Abh 122: 186–228

Koch W (1940) Metatexis und Metablastesis in Migmatiten des nordwestlichen Thüringer Waldes. Mineral Petrogr Mitt 51: 1–101

Kramer W, Pfeiffer L, Rösler H-J, Pilot J, Werner CD (1986) Vergleichende Geochemie basischer Magmatite auf dem Gebiet der Deutschen Demokratischen Republik. Teil II. Freiberger Forschungsh Leipzig C 407: 1–117

Kreuzer H, Lenz H, Harre W, Matthes S, Okrusch M, Richter P (1973) Zur Altersstellung der Rotgneise im Spessart, Rb/Sr-Gesamtgesteinsdatierungen. Geol Jahrb A9: 69–88

Laue S, Reischmann T, Emmermann KH (1990) Geochemical variation of granitoid rocks from the NW margin of the Rhinegraben. Ber Dsch Mineral Ges Eur J Mineral 2 (Beih 1): 156

Liew TC, Hofmann A (1988) Precambrian crustal components, plutonic associations, plate environments

of the Hercynian Fold Belt of central Europe; Indications from a Nd and Sr isotopic study. Contrib Mineral Petrol 98: 129–138
Lippolt HJ (1986) Nachweis altpaläozoischer Primäralter (Rb-Sr) und karbonischer Abkühlungsalter (K-Ar) der Muskovit-Biotit-Gneise des Spessart und der Biotit-Gneise des Böllsteiner Odenwald. Geol Rdsch 75: 569–583
Luth WC, Jahns RJ, Tuttle OF (1964) The granite system at pressures of 4–10 kilobars. J Geophys Res 69: 759–773
Matthes S (1954) Die Paragneise im mittleren kristallinen Vor-Spessart und ihre Metamorphose. Abh Hess Landesamt Bodenforsch Wiesb 8: 1–86
Matthes S, Krämer H (1955) Die Amphibolite und Hornblendegneise im mittleren kristallinen Vor-Spessart und ihre petrogenetische Stellung. Neues Jahrb Mineral Abh 88: 225–272
Matthes S, Okrusch M (1965) Petrographische Untersuchung der Rotgneise im Spessart. Geologie Berlin 14: 1148–1200
Matthes S, Okrusch M (1977) The Spessart crystalline complex, North-West Bavaria: rock series, metamorphism and position within the Central German crystalline rise. In: La chaine varisque d'Europe moyenne et occidentale. Coll Int Cent Natl Rech Sci (Rennes 1974) Paris 243: 375–390
Matthes S, Schubert W (1967) Der Chlorit-Hornblende-Fels von Wenighösbach im mittleren Vorspessart und seine Beziehungen zu spätkinematischen Chlorit-Amphiboliten. Veröff Geschichts Kunstverein Aschaffenburg 10: 15–46
Matthes S, Okrusch M, Röhr C, Schüssler U, Richter P, von Gehlen K (1995) Talc-chlorite-amphibole felses of the KTB pilot hole, Oberpfalz, Bavaria: Protolith characteristics and phase relationships. Mineral Petrol 52: 25–59
Möbus G (1963) Der Kristallin-Anteil der Tiefbohrung Luckau II, südlich von Berlin. Geologie Berlin 12: 264–282
Nasir S (1986) Die Metabasite im mittleren kristallinen Vorspessart: Petrographie – Geochemie – Phasenpetrologie. Dr rer nat Thesis, Univ Würzburg, 191 pp
Nasir S (1990) Coexisting cummingtonite-hornblende pairs in metamafic rocks from the central Spessart crystalline complex, NW-Bavaria, FRG. Chem Erde 50: 181–188
Nasir S, Okrusch M (1991) Metabasites from the Central Vor-Spessart, North-West Bavaria. Part 1: Geochemistry. Neues Jahrb Mineral Monatsh 1991: 500–522
Nasir S, Okrusch M, Kreuzer H, Lenz H, Höhndorf A (1991) Geochronology of the Spessart crystalline complex, mid-German crystalline rise. Mineral Petrol 43: 39–55
Neugebauer J (1988) The Variscan plate tectonic evolution: an improved "Iapetus model". Schweiz Mineral Petrogr Mitt 68: 313–333
Neumann W (1966) Versuch eines lithostratigraphischen Vergleiches von Grundgebirgsanschnitten im Bereich der Mitteldeutschen Schwelle. Geologie Berlin 15: 942–962
Neumann W (1973) Zum Stockwerkbau im Bereich der "Mitteldeutschen Kristallinzone" (speziell im Ruhlaer Kristallin). Veröff Zentralinst Phys Erde Potsdam 14: 391–409
Neumann W (1985) Ruhlaer Kristallin. Variszikum und Saxonikum. Exkursionsführer Akad Wiss DDR, Potsdam, pp 84–92
Okrusch M (1983) The Spessart crystalline complex, Northwest Bavaria. – DMG SFMC Joint Meeting. Excursion E 4. Fortschr Mineral 61 (Beih 2): 135–169
Okrusch M, Richter P (1986) Orthogneisses of the Spessart crystalline complex, Northwest Bavaria: indicators of the geotectonic environment? Geol Rdsch 75: 555–568
Okrusch M, Schubert W (1986) Das Gestein Hösbachit als Material für prähistorische Bronze-Gussformen aus dem fränkischen Raum. Mainfränk Stud Würzburg 37: 31–36
Okrusch M, Streit R, Weinelt W (1967) Erläuterungen zur geologischen Karte von Bayern 1:25000 Blatt Nr. 5920, Alzenau i. Ufr. Bayer Geol Landesamt München, 366 pp
Okrusch M, Raumer J von, Matthes S, Schubert W (1975) Mineralfazies und Stellung der Metamorphite im kristallinen Odenwald. Aufschluss Heidelberg Sonderbd 27 (Odenwald): 109–234
Okrusch M, Müller R, El Shazly S (1985) Die Amphibolite, Kalksilikatgesteine und Hornblendegneise der Alzenauer Gneisserie am Nordwest-Spessart. Geol Bavarica München 87: 5–37
Pearce JA (1982) Trace element characteristics of lavas from destructive plate boundaries. In: Thorpe RS (ed) Andesites. Elsevier, New York, pp 525–548
Pearce JA, Harris NBW, Tindle AG (1984) Trace element discrimination for the tectonic interpretation of granitic rocks. J Petrol 25: 956–983
Poller U (1991) Geochemie der Odenwälder Amphibolite und Ableitung ihrer geotektonischen Position. Diploma Thesis, Univ Würzburg, 130 pp (unpubl)
Reitz E (1987) Palynologie in metamorphen Serien: I. Silurische Sporen aus einem granatführenden Glimmerschiefer des Vor-Spessart. Neues Jahrb Geol Paläontol Monatsh 1987: 699–704
Schubert W (1968) Die Amphibolite des prävaristischen Schieferrahmens im Bergsträsser Odenwald. Neues Jahrb Mineral Abh 108: 69–110
Schubert W (1969) Chlorit-Hornblende-Felse des Bergsträsser Odenwaldes und ihre Phasenpetrologie. Contrib Mineral Petrol 21: 295–310
Schüller A (1952) Zur Grundgebirgsgeologie des Kyffhäusers. Geologica Berlin 11: 107–114
Schweder P (1966) Geochemische Untersuchungen an Gesteinen und gesteinsbildenden Mineralen des Kyffhäuserkristallins. Dr rer nat Thesis, Univ Rostock (cit. Seim 1967) (unpubl)
Seim R (1967) Der Para-Biotit-Plagioklasgneis des Kyffhäuserkristallins, seine metatektischen Derivate und seine Einlagerungen. Geol Beih Berlin 56: 1–85
Smoler M (1987) Petrographische, geochemische und phasenpetrologische Untersuchungen an Metasedimenten des NW-Spessart/Bayern. Dr rer nat Thesis, Univ Würzburg, 256 pp

Stellrecht R (1971) Geologisch-tektonische Entwicklung im Raum Albersweiler/Pfalz. Jahresber Mitt Oberrhein Geol Ver NF 53: 239–262

Trench A, Torsvik TH (1992) The closure of the Iapetus Ocean and Tornquist Sea: new paleomagnetic constraints. J Geol Soc Lond 149: 867–870

Tuttle OF, Bowen NL (1958) Origin of granite in the light of experimental studies in the system $NaAlSi_3O_8$-$KAlSi_3O_8$-SiO_2-H_2O. Geol Soc Am Mem 74: 153

Weber K (1984) Variscan events: Early Paleozoic continental rift metamorphism and late Paleozoic crustal shortening. In: Hutton DH, Sanderson DJ (eds) Variscan tectonics in the North Atlantic region. Blackwell, London, pp 3–22

Weber K, Behr H-J (1983) Geodynamic interpretation of the mid European Variscides. In: Martin H, Eder FW (eds) Intracontinental fold belts. Springer, Berlin Heidelberg New York, pp 427–469

Werner CD (1970) Geochemie der Brotteröder Serie im Ruhlaer Kristallin. Freiberger Forschungsh Leipzig C 259: 7–82

Werner CD (1971a) Geochemische Untersuchungen an Gesteinen der alten Baustufe des Ruhlaer Kristallins (Thüringer Wald). Ber Dtsch Ges Geol Wiss B Berlin 16: 41–91

Werner CD (1971b) Geochemische Untersuchungen an Metamorphiten der Seimberg-Serie im Ruhlaer Kristallin (Thüringer Wald). Ber Dtsch Ges Geol Wiss B Berlin 16: 491–526

Werner CD (1974) Metamorphose und Migmatisation im Ruhlaer Kristallin (Thüringer Wald). Freiberger Forschungsh Leipzig C 284: 1–134

Werner CD (1989) Central German crystalline zone. In: Kramer W, Werner C-D (eds) Prevariscan mafic rocks in the Saxothurigian zone. Guidebook of Excursions in the German Democratic Republic, May/June 1989, Acad Sci GDR, Centr Inst Phys Earth, Potsdam, pp 90–101

Wunderlich J (1989) Neue Ergebnisse zur Geologie des südöstlichen Ruhlaer Kristallins. Freiberger Forschungsh Leipzig C 429: 7–32

Zaminer C (1957) Geologisch-petrographische Untersuchungen im Grundgebirge der Pfalz. Mitt Polichia III (Reihe 4): 7–33

Zeh A, Katzung G (1994) Zur Geologie des Kyffhäuser-Kristallins. Neues Jahrb Geol Paläontol Monatsh 1994: 368–384

IV.E Metamorphic Evolution

M. Okrusch

1 Introduction

During the Variscan orogenesis, most of the complexes forming constituents of the MGCR, were subjected to regional metamorphism under conditions generally ranging from the lower to the upper amphibolite facies. Except for the Ruhla crystalline complex, no consistent isograd pattern can be mapped in these metamorphic complexes, indicating that the present erosion surfaces are virtually parallel to the metamorphic isograd surfaces. The constituents of the MGCR form individual blocks which experienced different peak metamorphic conditions (Fig. 1), underwent different P-T paths (Fig. 2) and were uplifted to different levels of erosion at different times (e.g., Ahrendt 1990; Okrusch 1990; Nasir et al. 1991; Willner et al. 1991; Krohe 1991; Flöttmann and Oncken 1992).

Unmetamorphosed graywackes of upper Proterozoic age have been recorded in boreholes in South Brandenburg as constituents of the MK (Hirschmann, this Vol., Chap. IV.B).

2 Pfälzer Wald

In the **Albersweiler** quarry mineral assemblages typical of the amphibolite facies are present in metabasites

plagioclase + common hornblende + biotite + quartz,
plagioclase + common hornblende + epidote + titanite,

and in granitoids

quartz + K-feldspar + plagioclase + biotite ± hornblende.

The widespread occurrence of late to postkinematic aplites and pegmatites of cotectic composition is taken as evidence that, at the peak of regional metamorphism, conditions of partial melting were nearly reached in the exposed metamorphic complex (Laue et al. 1990; Flöttmann and Oncken 1992). On the other hand, the hornblende-plagioclase geothermobarometer of Plyusnina (1982) yielded only temperatures of around 540 °C and pressures of 3–4 kb, and postkinematically recrystallized plagioclase rims and hornblende overgrowths gave even lower pressures of <2 kb and 520 °C (Flöttmann and Oncken 1992). The amphibolite-facies assemblages were lateron retrograded under greenschist-facies P-T conditions as demonstrated by the static crystallization of chlorite in the metabasites.

According to Flöttmann and Oncken (1992), the pressure-temperature-deformation (P-T-D) path of the Albersweiler complex is further constrained by two phases of mylonization.

Phase I started still within the stability field of biotite (>420 °C), but was later accompanied by the formation of synkinematic chlorite + sericite. Phase-II mylonites were formed in N-S trending strike slip faults at temperatures of around 300 °C as indicated by the unequilibrated grain boundaries of the recrystallized quartz. A K-Ar date on biotite of about 340 Ma (Horn et al. in Frenzel 1971, recalculated) is assumed to record cooling through the phase-I mylonitization.

In metapelites exposed at **Burrweiler**, a few km NW of Albersweiler, Flöttmann and Oncken (1992) recognized relics of the critical assemblage

quartz + biotite + cordierite + K-feldspar,

testifying to a first, pre- to early tectonic high-T/low-P stage, possibly in a thermal aureole. During a later, medium-pressure stage, the assemblags

quartz + biotite + muscovite
(Si 3.2–3.3, per formula unit = p.f.u.)
quartz + biotite + garnet (Alm_{80}),

were formed syntectonically with respect to the first phase of deformation D_1. The compositions of garnet (X_{Fe} = 0.87; Spear and Cheney 1989)

Dallmeyer et al. (Eds.)
Pre-Permian Geology of Central and Eastern Europe

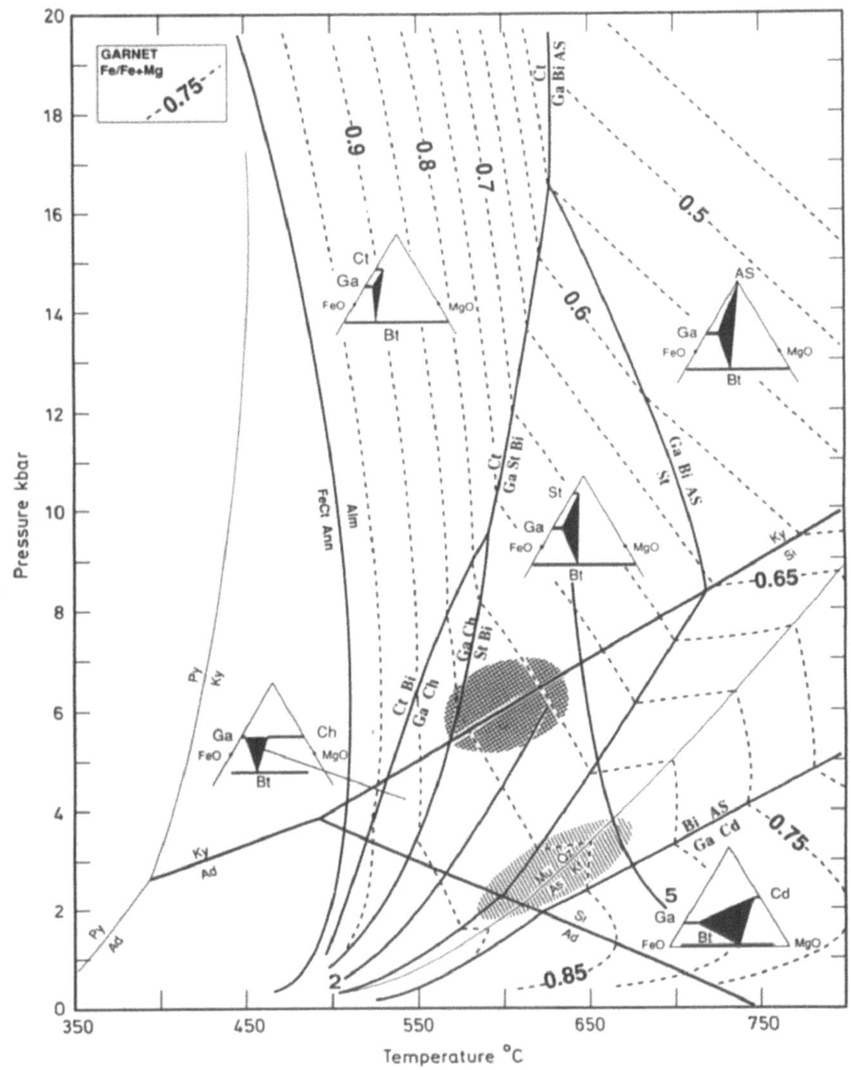

Fig. 1. Petrogenetic grid for the KFMASH system K_2O-FeO-MgO-Al_2O_3-SiO_2-H_2O after Spear and Cheney (1989) showing, among others, the reactions mentioned in the text. It should be noted that quartz, muscovite, K-feldspar and water vapor may be present as additional reaction partners (see text). Reaction (2) is only indicated by the *number*. Isopleths of X_{Alm} for garnet in assemblage that contain garnet + biotite are shown by *dashed lines*. The granite solidus (*curve 5*) is also given (Johannes 1984). P-T ranges for peak metamorphic conditions are given for the Bergsträsser Odenwald (*hatched field*) and the Spessart (*cross-hatched field*)

and muscovite (3.2–3.3 p.f.u.; Massonne and Schreyer 1987), lead to a P-T combination of about 550 °C and 6–8 kb. Breakdown of biotite to muscovite + hematite ± chlorite and the growth of late, undeformed porphyroblasts of muscovite (Si 3.1–3.2 p.f.u.) ± chlorite took place under lower-pressure, greenschist-facies conditions during a posttectonic retrograde event (Flöttmann and Oncken 1992). The resulting complicated P-T path is shown in Fig. 2a.

Lower Carboniferous graywackes and shales at **Neustadt** (Weinstrasse) and **Hambach** were weakly deformed and metamorphosed under greenschist-facies conditions leading to the posttectonic formation of muscovite and biotite. A similar sequence at **Weiler**, west of Wissembourg, experienced a weak, bedding-parallel pressure-solution cleavage formed by the alignment of muscovite, biotite, and chlorite flakes and of elongate, pressure-solved quartz grains. Un-

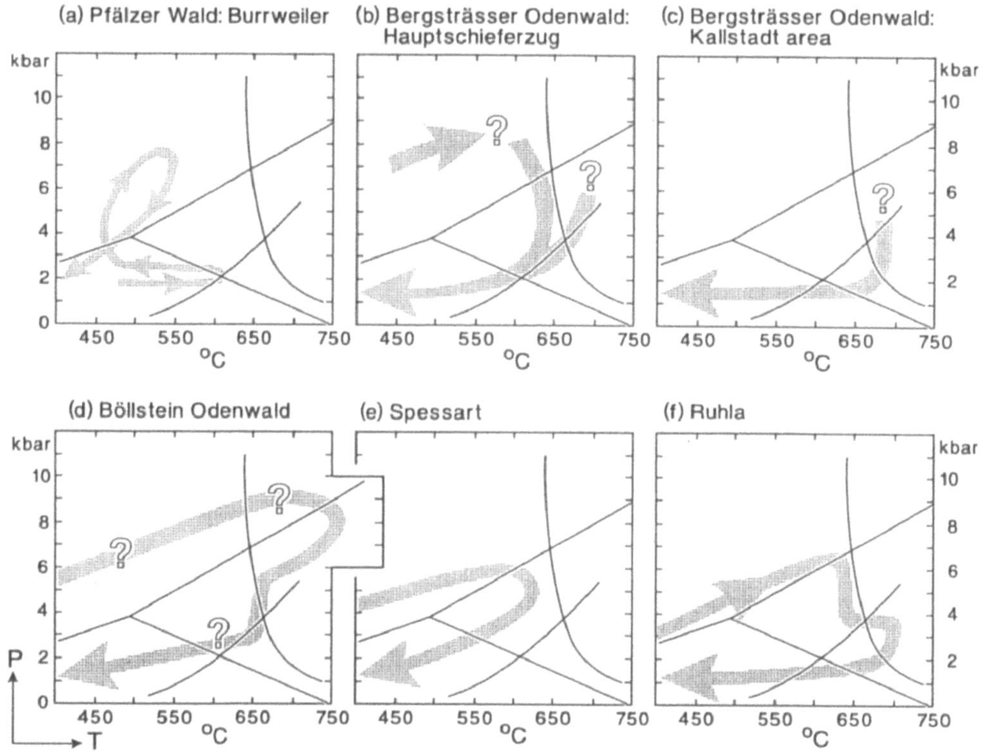

Fig. 2a–f. Synopsis of P-T paths in various parts of the Central German Crystalline Rise, in part highly tentative. Data sources are given in the text. For reference, the equilibrium curves for the Al_2SiO_5 polymorphs and the breakdown of muscovite + quartz as well as the granite solidus are shown

oriented, post-s_1 growth of biotite took place at the peak of metamorphism, still within the greenschist facies (Flöttmann and Oncken 1992).

3 Bergsträsser Odenwald

Before or near the thermal peak of metamorphism, the metamorphic sequences of the Bergsträsser Odenwald experienced two phases of penetrative deformation D_1 and D_2 (Okrusch et al. 1975; Krohe 1991, 1992; Willner et al. 1991). During a late episode of deformation, D_3, which postdates the thermal peak of metamorphism, steep NE-SW trending ductile shear zones developed, followed by a horizontal extension, D_4 (Krohe 1991, 1992).

The widespread occurrence of the assemblage

plagioclase + common hornblende ± diopside
± epidote ± biotite ± quartz + titanite

in metabasites indicates that the volcanosedimentary sequence constituting parts of the Bergsträsser Odenwald was metamorphosed under amphibolite-facies conditions (Schubert 1968; Okrusch et al. 1975). This is supported by the mineral assemblages in the famous Auerbach marble and related calc-silicate rocks

calcite + plagioclase + diopside ± grossular
+ epidote ± quartz,
calcite + plagioclase + wollastonite
+ grossular,
calcite + diopside + grossular + vesuvian,
scapolite + diopside ± plagioclase ± garnet
± quartz + titanite,

as summarized by Meisl (1972).

Index minerals in the metapelites of the Bergsträsser Odenwald – although not very frequent – allow a more precise estimate of the P-T conditions and testify to regional variations in the P-T evolution. This led Krohe (1991, 1992) to divide the Bergsträsser Odenwald into three different tectonometamorphic units.

In the metapelites of Krohe's (1992) **unit II** the critical assemblage (always + quartz + plagioclase)

muscovite + sillimanite + biotite
± almandine-rich garnet ± cordierite

dominates in the so-called Hauptschieferzug Heppenheim – Lindenfels – Reichelsheim whereas the assemblage

muscovite + andalusite + biotite + cordierite
+ garnet

is restricted to the area Kolmbach – Glattbach – Gadernheim and the vicinity of Hambach, northeast of Heppenheim (Matthes et al. 1972; von Raumer 1973).

Judging from these assemblages and assuming $P(H_2O) \approx P_{tot}$, as a first approximation, the P-T conditions reached at the peak of metamorphism were estimated by Okrusch (1990) using the available petrogenetic grids in the KFMASH system (e.g., Spear and Cheney 1989; see Fig. 1). The common occurrence of andalusite and sillimanite suggests that the possible P-T range for the Hauptschieferzug was close to the equilibrium curve

andalusite = sillimanite (1a)

(Holdaway 1971). In course of prograde metamorphism, i.e., at increasing temperatures and/or decreasing pressures, the equilibrium curves of the reactions

chlorite + muscovite + quartz = cordierite
+ biotite + andalusite/sillimanite + H_2O (2)

and

staurolite + muscovite + quartz = garnet
+ biotite + andalusite/sillimanite + H_2O (3)

(Spear and Cheney 1989) must have been crossed in Mg- and Fe-rich bulk compositions, respectively. On the other hand, the equilibrium curve of the reaction

muscovite + quartz = K-feldspar
+ andalusite/sillimanite + H_2O (4)

(Chatterjee and Johannes 1974) was not crossed in wide parts of the Bergsträsser Odenwald. The combination of the equilibrium curves (1a), (3), and (4) brackets the P-T conditions in a narrow range of 600–610 °C and 2.5 kb (Figs. 1, 2b). The $Fe^{2+}/(Fe^{2+} + Mg)$ ratios of garnets coexisting with biotite in the metapelites of the Hauptschieferzug vary between 0.92 and 0.87 and may show a slight increase from core to rim (Matthes et al. 1972; von Raumer 1973; Willner et al. 1991). Using the isopleths constructed by Spear and Cheney (1989), this would lead to a temperature range of about 550–600 °C at 2.5–3 kb. It should be noted, however, that the spessartine contents of these garnets cover a wide range between 5 and 25 mol%. For a garnet-biotite-muscovite gneiss of the Hauptschieferzug, Willner et al. (1991) calculated a conforming temperature of 610 °C, but a distinctly higher pressure of 4.9 kb by using the thermodynamic data set of Berman (1988).

The prograde transition from the low- to the high-grade amphibolite facies is locally recognized in the Hauptschieferzug of the western Odenwald. This is indicated by the breakdown of muscovite + quartz according to reaction (4) leading to the assemblages

K-feldspar + andalusite + biotite ± cordierite
± garnet,
K-feldspar + sillimanite + biotite ± cordierite
± garnet,

and by the rare formation of vein migmatites (Matthes et al. 1972; von Raumer 1973). If we interpret the migmatites as products of incipient in-situ melting, temperatures of 670 to 680 °C at 2.5–3 kb should have been locally reached judging from the equilibrium curve of the melting reaction

plagioclase (An_{20}) + K-feldspar + quartz
+ H_2O = melt (5)

and assuming $P(H_2O) \approx P_{tot}$ (Johannes 1984; see Fig. 1).

The famous kinzigites of Gadernheim and Laudenau in the Hauptschieferzug (Büsch et al. 1980) contain the assemblage

biotite + garnet + cordierite + sillimanite
+ andalusite,

the associated metabauxites (Adusumalli and Schubert 1991, 1992)

corundum + cordierite + plagioclase
+ hercynite + biotite ± sillimanite
± gedrite.

Both prograde muscovite and K-feldspar are absent, owing to a bulk rock composition in the K-poor volume of the KFMASH system.

At the peak of metamorphism, there is a clear coexistence of garnet with cordierite, but on the other hand biotite is still compatible with sillimanite and andalusite. This indicates that the equilibrium curve of reaction (6) was not crossed. Moreover, both Al-silicates are in mutual contact with each other. These facts and the rim composition of the kinzigite garnet with a $Fe^{2+}/(Fe^{2+} + Mg)$ ratio of 0.88 and 9 mol% spessartine indicate P-T conditions of about 600 °C/2.5 kb, broadly conforming to those estimated for the adjacent, nonmigmatitic metapelites (Okrusch 1990; Willner et al. 1991).

However, the Gadernheim kinzigites reveal evidence for a polymetamorphic history with a clockwise P-T path: The garnet cores show $Fe^{2+}/(Fe^{2+} + Mg)$ ratios as low as 0.75 and 5 mol% spessartine; they contain inclusions of biotite, plagioclase, and quartz, but not of cordierite which, however, is rarely included in marginal parts of the garnets (Büsch et al. 1980; Willner et al. 1991). The latter authors, applying the method of Spear and Selverstone (1983) to the assemblage of the garnet core with the biotite and plagioclase inclusions, calculated a pressure of 8.9 kb and a temperature of 575 °C. Such a P-T combination would be within the stability field of kyanite, which is not observed in the Gadernheim kinzigites. However, Willner et al. (1991) observed inclusions of staurolite in quartz and ilmenite, indicating pressures above the equilibrium curve of reaction (3), i.e., of at least 6 kb for the early stage of metamorphism (Figs. 1, 2b).

Retrograde muscovite and chlorite are frequently formed in the metasediments of the Hauptschieferzug. A strong postdeformative crystallization of muscovite porphyroblasts is possibly due to the thermal overprint by the intrusion of diorite and hornblende-gabbros, in the Heppenheim-Lindenfels area (Matthes et al. 1972).

Conspicuous migmatites occur in the Kallstadt area forming part of Krohe's (1992) **unit III** in the southern Bergsträsser Odenwald. Andalusite is absent in these rocks, the stable assemblage being

K-feldspar + sillimanite + biotite + garnet

(Matthes et al. 1972). This may indicate that the P-T regime was somewhat higher than in the Hauptschieferzug, but still at the low-temperature side of the equilibrium

$$\text{biotite} + \text{sillimanite} + \text{quartz} = \text{K-feldspar} + \text{garnet} + \text{cordierite} + H_2O. \qquad (6)$$

(Figs. 1, 2c). It is interesting to note that garnets, coexisting with biotite in melanosomes of the Kallstadt migmatites, show the same range of $Fe^{2+}/(Fe^{2+} + Mg)$ ratios of 0.87–0.93 as in the metapelites of the Hauptschieferzug thus indicating temperatures below 700 °C; their spessartine contents vary between 5 and 9 mol%.

The minimum age of regional metamorphism in units II and III of the Bergsträsser Odenwald is constrained by Early Carboniferous K-Ar and Ar-Ar dates around 342 Ma obtained on hornblendes, muscovites, and biotites from different metamorphic and igneous rocks (Kreuzer and Harre 1975, recalculated; Hellmann et al. 1982; Rittmann 1984). This fits well to a concordant U-Pb date of 335 Ma obtained on zircon fractions $<25\,\mu m$ from various migmatite samples of the Bergsträsser Odenwald (Todt 1979).

By contrast, a distinctly older age and a different P-T path was recorded for the shallow intrusion of the Frankenstein gabbro in the northern part of the Bergsträsser Odenwald (Kreher 1992), forming part of Krohe's **unit I**. Confirming earlier K-Ar dates of Kreuzer and Harre (1975, recalculated), Kirsch et al. (1988) obtained $^{40}Ar/^{39}Ar$ dates of 363 ± 7 Ma for hornblendes and 359 ± 3 Ma for plagioclases as well as $^{207}Pb/^{206}Pb$ single-zircon dates of 362 ± 9 Ma. As demonstrated by Matthes and Schubert (1971), the gabbro magma transformed rafts of country-rock amphibolite into pyroxene hornfelses ("beerbachites") with the assemblage

$$\text{plagioclase } (An_{50\text{-}70}) + \text{diopside} + \text{hypersthene} + \text{ilmenite} + \text{magnetite} \pm \text{quartz}.$$

By using the two-pyroxene geothermometer of Lindsley (1983), temperatures of about 900 °C were calculated by Willner et al. (1991) for the hornfels rafts in the gabbros, while the plagioclase-hornblende geothermobarometer of Plyusnina (1982) yielded 545 °C and 1.5 kb for a thermally not overprinted epidote amphibolite in the vicinity of the Frankenstein gabbro. Willner et al. (1991) observed discrete D_3 shear planes with recrystallized amphibole and plagioclase in both the gabbro and the country-rock amphibolites and gabbros indicating that temperatures >500 °C persisted after the emplacement of the Frankenstein pluton. Muscovite geobarometry and hornblende-plagioclase geothermometry applied to granites and tonalites intruding the Frankenstein gabbro indicate an anticlockwise P-T path with a pressure maximum of 2.5–3 kb at temperatures of 550–460 °C.

Regarding the age of the Bergsträsser Odenwald, it is interesting to note, that in the Weiterstadt 1 borehole, 6 km west of Darmstadt, amphibolites and granites were hit, which are overlain by fossiliferous black shales of Westfal C age (Marell 1989). An overlying basalt flow was radiometrically dated, with the ^{40}Ar-^{39}Ar method on biotite, at 332.2 ± 2.7 Ma (Kirsch 1984).

4 Böllstein Odenwald

Mineral assemblages in the metasediments forming the schist envelope of the Böllstein gneiss dome are

biotite + muscovite ± garnet + quartz (± plagioclase ± K-feldspar),
plagioclase + K-feldspar + quartz + biotite,

while the intercalated metabasites and meta-ultrabasites display

plagioclase + common hornblende ± diopside ± garnet ± biotite ± epidote ± quartz,
tschermakitic hornblende + spinel + garnet + plagioclase,
magnesio-hornblende + chlorite + talc + spinel

(Chatterjee 1960; Knauer et al. 1974; Willner et al. 1991). These parageneses demonstrate that the last important metamorphic event, which occurred after the D_1/D_2 deformation during the Variscan orogenesis, took place under conditions of the amphibolite facies. By applying the data set of Berman (1988) and the muscovite geobarometer of Massonne and Schreyer (1987), temperatures of 620–650 °C and a pressure of about 4.5 kb are deduced by Willner et al. (1991). These authors assume that the migmatite structures, which are frequently observed in the Böllstein Odenwald, were formed during this stage by incipient melting.

An earlier, perhaps pre-Variscan phase of amphibolite-facies metamorphism and anatexis was assumed by Chatterjee (1960), judging from textural and structural observations. Mineralogical evidence for an older metamorphic event, which took place under higher pressure conditions before and possibly during the D_1/D_2 deformation, was recognized by Knauer et al. (1974) and Willner et al. (1991) in garnet amphibolites of the Böllstein: garnet ($Gross_{25-28}$-Alm_{58-54}) coexisting with plagioclase (An_{38-44} or An_{94}) + hornblende I ± clinozoisite ± chlorite ± quartz breaks down to form symplectitic aggregates of plagioclase (An_{26-28}) + clinopyroxene + hornblende II. For the early assemblage, Willner et al. (1991) calculated 8 kb and 545 °C using Berman's (1988) data set. In a metapelite, the same authors also detected an earlier assemblage kyanite + biotite + garnet included in plagioclase; the $Fe^{2+}/(Fe^{2+} + Mg)$ ratio of garnet is 0.78 in the core, 0.84 in the rim, the spessartine contents are 20 and 27 mol%), respectively. This assemblage should have formed at the high-P side of the equilibrium curve

$$\text{kyanite} = \text{sillimanite} \tag{1b}$$

and the high-T side of the equilibrium curve

$$\text{staurolite} + \text{muscovite} = \text{kyanite} + \text{biotite} + \text{garnet} + H_2O, \tag{7}$$

which meet at an invariant point at about 8 kb/700 °C (Spear and Cheney 1989; Fig. 1). However, the relatively high almandine content of garnet would not fit these P-T conditions, suggesting that garnet, biotite, and kyanite are not in mutual equilibrium. Therefore, the P-T combination of 8.6 kb/805 °C, calculated by Willner et al. (1991) using Berman's (1988) data set (Fig. 2d) must be taken with caution, all the more as it drops out of the kyanite stability field!

The age of this older metamorphic event is presumably constrained by U-Pb zircon dating of Todt (1979). He recorded a concordant U-Pb age of 380 Ma for a zircon fraction <25 μm from the Böllstein granodiorite gneiss and virtually the same lower intercept age for a discordia defined by three zircon fractions from metasediments of the Schieferhülle. These data testify to a thermal event in early Devonian times. A lower age limit for the latest metamorphic event is provided by K-Ar dates on hornblendes, muscovites and biotites of the Böllstein gneiss dome ranging between 304 and 325 Ma (Kreuzer and Harre 1975; Lippolt 1986).

5 Spessart

As in the Bergsträsser and Böllstein Odenwald, metabasites with the assemblage

plagioclase + common hornblende ± quartz ± epidote ± diopside ± biotite ± garnet, testifying to amphibolite-facies metamorphism, are frequently intercalated within pelitic to psammitic meta-sediments consisting of

quartz ± plagioclase ± muscovite + biotite ± garnet

and orthogneisses with the assemblage

quartz + plagioclase + K-feldspar + biotite ± muscovite ± garnet.

Metamorphic temperatures and/or pressures display only small regional variations. Except for a minor retrograde overprint, there are no indications of a polymetamorphic evolution.

In the **central part** of the Spessart crystalline complex, the index mineral staurolite is widely distributed in metapelites of suitable bulk rock composition. Moreover kyanite and/or sillimanite are present at several places. Matthes (pers. comm.) observed a replacement of earlier kyanite by sillimanite. The typical metapelitic assemblage is

quartz + plagioclase + muscovite + biotite + garnet + staurolite ± kyanite/sillimanite,

whereas subordinate meta-psammitic intercalations display

quartz + plagioclase + muscovite + biotite + garnet.

P-T conditions attained at the peak of regional metamorphism are defined by the stability field of staurolite + biotite ± garnet. In course of the prograde metamorphic evolution, the equilibrium curve of the reaction

kyanite = sillimanite (1b)

(Holdaway 1971) was crossed. The lower temperature limit is given by the equilibrium curves of the reactions

chlorite + muscovite = staurolite + biotite + quartz + H_2O, (8)

experimentally determined by Hoschek (1967) and

garnet + chlorite + muscovite = staurolite + biotite + quaraz + H_2O (9)

calculated by Spear and Cheney (1989). A combination of the equilibrium curves (8) and (9) with (1b) leads to a minimum temperature of 570 °C at pressures of around 5 kb (Fig. 1).

According to microprobe analyses of Smoler (1987), $Fe^{2+}/(Fe^{2+} + Mg)$ ratios of garnets coexisting with biotite and staurolite range between 0.8 and 0.9, in average 0.85 (6–21 mol% spessartine). Using the isopleths calculated by Spear and Cheney (1989), temperatures should not have exceeded about 620 °C at about 6.5 kb (Fig. 1). Conformably, Nasir (1986) recorded 605 °C/6.5 kb and 620 °C/6 kb in two calc-silicate gneiss samples from the Feldkahl area using the garnet-clinopyroxene thermometer of Ellis and Green (1979) and the barometer of Newton and Perkins (1982).

This P-T range is well in accord with the absence of migmatite structures in the central part of the Spessart crystalline complex indicating that the minimum melting curve of granite according to the reaction

plagioclase (An_{20}) + K-feldspar + quartz + H_2O = melt (5)

(Johannes 1984) was not reached (Fig. 1).

The estimated pressure range conforms to the chemical composition of muscovites coexisting with biotite, staurolite and garnet. Microprobe analyses of Smoler (1987) yielded maximum Si values of 3.1 p.f.u. which leads to **minimum** pressures of about 4 kb, using the data of Massonne and Schreyer (1987).

In the associated orthogneisses of the Rotgneiss complex, muscovite coexists with K-feldspar and biotite (+ garnet) and thus provide a direct pressure estimate. Dombrowski et al. (1994) recorded that these muscovites are zoned. High Si values up to 3.3 p.f.u. in the cores testify to an early stage at lower T/P conditions, whereas the rim compositions of 3.2–3.1 Si p.f.u. indicate the metamorphic peak with higher T and lower P. Combination of the muscovite-biotite geothermometer of Hoisch (1991) with the garnet-biotite thermometer of Perchuk and Lavrent'eva yielded pressures of 4–7 kb and temperatures of 605–645 °C, whereas the garnet-biotite calibration of Ferry and Spear (1978) leads to temperatures 30–60 °C higher (Dombrowski et al. 1994).

Mainly based on textural evidence, Matthes (1954) recognized an increase in metamorphic grade towards the southern part of the Mömbris formation. On the other hand, the metapelites of the Geiselbach formation in the **northwestern part** of the Spessart crystalline complex, are totally devoid of staurolite which may indicate that the metamorphic grade decreases towards the northwest. This assumption was confirmed by detailed microprobe investigation of amphibolites and hornblende gneisses intercalated within the different stratigraphic units (Nasir 1986): Metabasites of the northwestern belt Hörstein-Huckelheim, concentrated in the border zone of the Geiselbach and the Mömbris formation display systematically lower hornblende-plagioclase temperatures than the metabasites of the southeastern belt Aschaffenburg-Feldkahl-Rottenberg intercalated within the central part of the Mömbris formation and the Rotgneiss complex, although the obtained temperatures are generally somewhat lower than the above estimates.

Application of the biotite-muscovite thermometer of Hoisch (1991) and the garnet-biotite thermometer of Perchuk and Lavrent'eva (1983) to the Schweinheim micaschists in the **southwestern part** of the Spessart crystalline complex leads to temperatures between 590 and 620 °C and pressures of 6–8 kb, which may indicate a stage with slightly higher P/T ratios on the prograde P-T path. Muscovites coexisting with biotite, K-feldspar and quartz (but not with garnet) in the intercalated Haibach gneiss contain 3.2–3.05 Si p.f.u. reflecting pressures between 6 and 3.5 kb (at 600–650 °C) based on the experimental data of Massonne and Schreyer (1987); see Dombrowski et al. (1994).

Sillimanite is the only Al_2SiO_5 polymorph detected in metarhyolites of the Elterhof formation in the southern Vorspessart. It may occur even together with K-feldspar, but not in direct contact (Braitsch 1957), indicating that the equilibrium curve of reaction (4) was not crossed. Garnets in metasediments of the Elterhof formation have $Fe^{2+}/(Fe^{2+} + Mg)$ ratios between 0.84 and 0.92.

Marbles of the Elterhof formation often consist of calcite + dolomite and may contain phlogopite, tremolite/actinolite, anthophyllite, olivine, chondrodite and humite as additional phases (Mosebach 1934). The occurrence of the assemblages calcite + olivine and calcite + tremolite puts narrow constraints on the temperatures reached in the southern part of the Spessart crystalline complex. At a total pressure of 5 kb, about 650 °C at $X(CO_2) = 0.15$ and about 700 °C at $X(CO_2) = 0.75$ can be inferred from the data of Metz (1976) and Kerrick and Slaughter (1976).

Calc-silicate gneisses of the Alzenau and Elterhof formation display the assemblage

plagioclase ± K-feldspar
+ common hornblende + (ferro-)salite
± almandine-grossular + titanite
± clinozoisite ± scapolite (Me_{81})

(Braitsch 1957; El Shazly 1983; Okrusch et al. 1985). Garnet-clinopyroxene geothermometry (Ellis and Green 1979) on two samples from the Alzenau and Elterhof formation, respectively, yields 630 °C and 635/680 °C (garnet core/rim), assuming a pressure of 5 kb.

In summary, the P-T conditions attained in the Spessart crystalline complex conform to a typical Barrovian, medium pressure type of amphibolite-facies metamorphism. This is at variance to the low-pressure, amphibolite-facies conditions realized during the main phase of regional metamorphism in the Bergsträsser Odenwald (Figs. 1, 2).

In restricted areas, mineral assemblages of the greenschist facies were approached or locally reached, by a younger retrograde metamorphism and phyllonitization. Biotite and garnet were partly or totally replaced by chlorite, staurolite by chlorite + sericite (Matthes 1954).

K-Ar dates on hornblendes, muscovites, and biotites of different rock types (including pegmatites) from various lithostratigraphic units and independent of the dated minerals range between 318 and 324 Ma (±4–8 Ma for hornblendes, ±2–3 Ma for micas), with only few exceptions (Lippolt 1986; Nasir et al. 1991; Dombrowski et al. 1994). A conformable Ar-Ar plateau age of 324 ± 3 Ma was obtained for one muscovite from the Haibach gneiss, and Rb-Sr dating on two whole rock – muscovite pairs from the Haibach gneiss yielded 323 and 326 (±7) Ma (Dombrowski et al. 1994). The nearly concordant radiometric dates on minerals with different blocking temperatures indicate a rapid uplift and cooling. In this case, the closing temperatures of the specific isotopic system is of minor importance and, consequently, the data should provide a direct information on the age of metamorphism (Schmädicke et al. 1995).

The radiometric dates for the Spessart crystalline complex conform well to the results obtained for the Böllstein Odenwald, for the Fichtelgebirge crystalline complex (autochthonous Saxothuringian) and for the southerly bordering autochthonous Moldanubian (e.g., Kreuzer et al. 1989; Hansen et al. 1989), but are distinctly younger than in the Bergsträsser Odenwald (Kreuzer and Harre 1975).

6 Ruhla Area, Thüringer Wald

The Ruhla crystalline complex can be divided into four metamorphic units

a) the Ruhla formation constituting its western part,
b) the Truse formation predominantly exposed in the southeastern part,
c) the Brotterode formation in the central-eastern part, and
d) the central gneiss unit.

These metamorphic units broadly conform to the lithostratigraphic and tectonic units formerly proposed in the literature (e.g., Neumann 1966; Werner 1974; Hirschmann and Okrusch 1988; Wunderlich 1989).

According to Neumann (e.g., 1964), the Liebenstein formation of the central gneiss unit was already metamorphosed under amphibolite-facies conditions during the Dalslandian orogenesis, in the Late Proterozoic. This concept was adopted by Werner (1974), but was recently disputed by Hirschmann and Okrusch (1988).

During the Variscan orogenesis, the Ruhla crystalline complex underwent polymetamorphic evolution (Koch 1940; Von Gaertner 1950; Neumann 1964; Werner 1974; Wunderlich 1989; Estrada et al. 1992; Zeh et al. 1994, 1995): A first, regional metamorphic event took place under medium pressures and peak conditions of the lower to upper amphibolite facies. Based on regional distribution, textural relationships and compositional differences of critical metamorphic minerals, Zeh et al. (1994, 1995) concluded that the four units experienced different tectonometamorphic evolutions and reached different P-T conditions at the peak of regional metamorphism (Zeh et al. 1994, 1995), detailed petrological analysis of which is under progress (Zeh in prep.). The regional metamorphic assemblages were subsequently overprinted under low-pressure, static conditions. In contrast to former investigators, Zeh et al. (1994, 1995) attribute this thermal overprint to the late Variscan intrusions.

a) During the **first, regional metamorphic event** the assemblage

common hornblende + plagioclase ± quartz
± garnet ± diopside

was formed in metabasites, frequently intercalated in metasedimentary schists of the Ruhla crystalline complex. The amphibolites were partly retrograded to greenschist assemblages during uplift, especially in the Ruhla formation. Marble intercalations contain hornblende, epidote, phlogopite, chlorite or olivine as additional phases (Koch 1940).

In metapelites of the **Ruhla formation**, the assemblage

biotite + garnet ± staurolite ± muscovite
+ plagioclase + quartz

was formed during the prograde regional metamorphism, contemporaneously with a first deformational event (D_1). The retrograde evolution during D_2 and D_3 led to a widespread chloritization of biotite, garnet and staurolite.

Staurolite-bearing assemblages with kyanite and/or corundum as additional critical phases, are recognized in metapelites of the **Truse formation**:

staurolite + garnet + biotite + muscovite
± kyanite + plagioclase + quartz,
kyanite + biotite + plagioclase + quartz,
kyanite + corundum + plagioclase.

These crystallized during the D_2 deformation, while D_1 is only documented by trails of mineral inclusions in cores of zoned garnets (Zeh et al. 1994b). The assemblages in the Truse metapelites indicate peak metamorphic temperatures of around 630 °C and minimum pressures of about 7 kb (Fig. 1). A later retrograde overprint was accompanied by the D_2 deformation.

In metapelites of the **Brotterode** formation the first event to be recognized is the crystallization, under virtually static conditions, of the regional assemblage

garnet + biotite + sillimanite + K-feldspar
+ plagioclase + quartz + graphite,

indicating higher temperatures than in the Truse formation. A subsequent deformational event took place under slightly decreasing temperatures reflected by the garnet zonation (Zeh et al. 1994; Zeh in prep.).

In ortho- and paragneisses of the **central gneiss unit**, one prograde and two retrograde deformational events can be distinguished, from which the first one is presumably related to D_2 in the Truse formation. Regional metamorphic assemblages conform to the amphibolite facies, but are strongly affected by the low pressure overprint.

b) The subsequent **thermal overprint under low-pressure conditions** is, at least in part, spatially related to late Variscan intrusions (Zeh et al. 1994, 1995). At the eastern margin of the **Ruhla formation**, the thermal influence of the Ruhla granite yielded the contact metamorphic assemblage

andalusite/fibrolite + biotite ± cordierite
± muscovite + plagioclase + quartz.

The western part of the **Truse formation** was thermally overprinted by the intrusions of the Trusetal granite and the Brotterode diorite leading to a zoned contact aureole with the following assemblages (always + plagioclase + quartz):

outer zone: andalusite + muscovite + biotite
+ garnet,
middle zone: fibrolite (+ andalusite) +
muscovite + biotite + garnet,
inner zone: sillimanite + biotite + cordierite
(+ muscovite).

The formation of the assemblage

plagioclase + sillimanite + cordierite + biotite
+ spinel + magnetite + corundum

in silica-deficient, Fe-Mg-Al rich lenses enclosed in acid gneisses of the *central gneiss unit* is also attributed to a thermal overprint under low-P/high-T conditions (Zeh et al. 1994; Zeh in prep.). The highest metamorphic grade is documented by intensive migmatisation of various lithotypes, including amphibolites, of the Brotterode formation and the central gneiss unit (Koch 1940). Peak conditions during this stage were estimated, by Estrada et al. (1992), at about 3.5 kb and 650–670 °C, but even higher temperatures are not unlikely (Figs. 1, 2f), especially if the H_2O activity was <1.

Pilot (in Werner 1974, recalculated) obtained K-Ar dates of 375 and 369 Ma on two hornblende, and of 353 and 347 Ma on two biotite separates from gneisses of the Liebenstein formation, providing a minimum age constraint for the latest metamorphic event. These K-Ar dates, which await further confirmation, are distinctly older than in the Spessart and the Böllstein Odenwald, and slightly older than in Krohe's (1991, 1992) units II and III of the Bergsträsser Odenwald. A biotite from the post-tectonic Ruhla granite yielded a (recalculated) K-Ar age of 322 Ma (Afanassjew in Neumann 1966).

7 Kyffhäuser

According to the detailed description of Seim (1967), the amphibolites and hornblende gneisses of the Kyffhäuser crystalline complex contain the assemblages

plagioclase ($An_{29\text{-}41}$) + common hornblende
+ biotite ± garnet ± quartz

and

plagioclase ($An_{36\text{-}38}$) + quartz + K-feldspar
+ common hornblende + biotite ± epidote,

typical of the amphibolite facies. Biotite-plagioclase gneisses rarely contain some garnet, biotite-microcline-plagioclase gneisses may bear some muscovite.

Calc-silicate gneisses display the assemblage

plagioclase (An_{26-32}) + microcline + quartz
+ diopside + hornblende,

while the assemblages

calcite + diopside + actinolite/tremolite
± phlogopite ± epidote ± plagioclase
± K-feldspar ± grossular ± scapolite,

and

calcite + dolomite + chlorite + plagioclase
+ microcline

are present in silicate marbles and subordinate dolomite marbles, respectively.

These mineral assemblages closely resemble those in the Elterhof formation of the Spessart crystalline complex and indicate similar P-T conditions at the peak of metamorphism (Seim 1967). Judging from the data of Metz (1976) and Kerrick and Slaughter (1976), the paragenesis tremolite + calcite has an upper stability limit of about 650 °C at $X(CO_2) = 0.15$ and about 700 °C at $X(CO_2) = 0.75$, at a total pressure of 5 kb, which would indicate the maximum temperatures reached in the Kyffhäuser. On the other hand, anatectic phenomena are widespread thus testifying to a minimum temperature of 650 °C at $P(H_2O) = 5$ kb (Fig. 1).

During a retrograde event in the P-T evolution of the Kyffhäuser crystalline complex, conditions of the greenschist facies were attained leading to the formation of epidote and chlorite in metabasites (Seim 1967).

K-Ar dating on one biotite and one biotite + muscovite concentrate yielded 354 and 346 Ma, similar to those in the Ruhla complex (Valny and Barvitiusova in Seim 1967). Further geochronological work in the Kyffhäuser crystalline complex is clearly demanded.

Acknowledgments. Thanks are due to C. Adusumalli, W. Franke, O. Oncken, W. Schubert, K. Weber and A. Zeh for stimulating discussions, to H. Murawski for a helpful suggestion, and to P. Blümel, G. Hirschmann and an unknown reviewer for critical reading of the manuscript.

References

Adusumalli C, Schubert W (1991) Petrologie der Al-übersättigten Gesteine von Laudenau & Klein Gumpen (Bergsträsser Odenwald). Ber Dtsch Mineral Ges Eur J Mineral 3 (Beih 1): 3

Adusumalli C, Schubert W (1992) Paläobauxite im Saxothuringikum der Varisciden. Ber Dtsch Mineral Ges Eur J Mineral 4 (Beih 1): 1

Ahrendt H (1990) An outline of the geological history of the Odenwald, the Spessart and the Rheinisches Schiefergebirge based on geochronological data. In: Franke W (ed) Mid-German Crystalline Rise & Rheinisches Schiefergebirge. Field guide to pre-conference excursion. Conf on Paleozoic orogens in central Europe – geology and geophysics, IGCP 233 Terranes in the Circum-Atlantic Paleozoic Orogens, Göttingen-Giessen, pp 93–97

Berman RG (1988) Internally consistent thermodynamic data for minerals in the system Na_2O-K_2O-CaO-MgO-FeO-Fe_2O_3-Al_2O_3-SiO_2-TiO_2-H_2O-CO_2. Petrol 29: 445–522

Braitsch O (1957) Beitrag zur Kenntnis der kristallinen Gesteine des südlichen Spessarts und ihrer geologisch-tektonischen Geschichte. Abh Hess Landesamt Bodenforsch Wiesbaden 18: 21–72

Büsch W, Matthes S, Mehnert KR, Schubert W (1980) Zur genetischen Deutung der Kinzigite im Schwarzwald und Odenwald. Neues Jahrb Mineral Abh 137: 223–256

Chatterjee ND (1960) Geologische Untersuchungen im Kristallin des Böllsteiner Odenwaldes. Neues Jahrb Geol Paläontol Abh 111: 137–180

Chatterjee ND, Johannes W (1974) Thermal stability and standard thermodynamic properties of synthetic $2M_1$-muscovite $KAl_2(AlSi_3O_{10}(OH)_2)$. Contrib Mineral Petrol 48: 89–114

Dombrowski A (1992) Der Haibacher Biotitgneis im südöstlichen Spessartkristallin. Diploma Thesis, Univ Würzburg, 113 pp

Dombrowski A, Okrusch M, Henjes-Kunst F (1994) Geothermobarometry and geochronology on mineral assemblages of orthogneisses and related metapelites of the Spessart Crytalline Complex, NW Bavaria, Germany. Chemie Erde 54: 85–101

Ellis DJ, Green DH (1979) An experimental study of the effect of Ca upon garnet – clinopyroxene Fe-Mg exchange equilibria. Contrib Mineral Petrol 71: 13–22

El Shazly S (1983) Petrography and geochemistry of the striated paragneiss series in the southern part of the Spessart crystalline complex. Dr rer nat Thesis, Tech Univ Braunschweig, 153 pp

Estrada S, Grunewald V, Wunderlich J (1992) Zur Eduktcharakteristik und Lithostratigraphie von Truse- und Hohleborn-Serie (Ruhlaer Kristallin). Neues Jahrb Geol Paläont Abh 184: 389–419

Flöttmann Th, Oncken O (1992) Constraints on the evolution of the Mid German Crystalline rise – a study of outcrops west of the river Rhine. Geol Rdsch 81: 515–543

Frenzel G (1971) Die Mineralparagenese der Albersweiler Lamprophyre (Südpfalz). Neues Jahrb Mineral Abh 115: 164–191

Hansen BT, Teufel S, Ahrendt H (1989) Geochronology of the Moldanubian-Saxothuringian transition zone, northeast Bavaria. In: Emmermann R, Wohlenberg J (eds) The German Continental Deep Drilling Program (KTB). Site-selection studies in the Oberpfalz and Schwarzwald. Springer, Berlin Heidelberg New York, pp 55–65

Hellmann KN, Lippolt HJ, Todt W (1982) Interpretation der Kalium-Argon-Alter eines Odenwälder Granitporphyrganges und seiner Nebengesteine. Aufschluss 33: 155–164

Hirschmann G, Okrusch M (1988) Spessart-Kristallin und Ruhlaer Kristallin als Bestandteile der Mitteldeutschen Kristallinzone – ein Vergleich. Neues Jahrb Geol Paläont Abh 177: 1–39

Holdaway MJ (1971) Stability of andalusite and the aluminium silicate phase diagram. Am J Sci 271: 97–131

Hoschek G (1967) Untersuchungen zum Stabilitätsbereich von Chloritoid und Staurolith. Contrib Mineral Petrol 14: 97–131

Johannes W (1984) Beginning of melting in the granite system Qz-Or-Ab-An-H_2O. Contrib Mineral Petrol 86: 264–273

Kerrick DM, Slaughter J (1976) Comparison of methods for calculating and extrapolating equilibria in the P-T-XCO_2 space. Am J Sci 276: 883–916

Kirsch H (1984) Die Alter einiger variscischer Plutonite und Vulkanite des Odenwaldes und des nördlichen Oberrheingrabens ($^{40}Ar/^{39}Ar$-Untersuchungen). Diploma thesis, Univ Heidelberg 122 pp (unpubl)

Kirsch H, Kober B, Lippolt HJ (1988) Age of intrusion and rapid cooling of the Frankenstein gabbro (Odenwald, SW-Germany) evidenced by $^{40}Ar/^{39}Ar$ and single-zircon $^{207}Pb/^{206}Pb$ measurements. Geol Rdsch 77: 693–711

Knauer E, Okrusch M, Richter P, Schmidt K, Schubert W (1974) Die metamorphe Basit-Ultrabasit-Assoziation in der Böllsteiner Gneiskuppel, Odenwald. Neues Jahrb Mineral Abh 122: 186–228

Koch W (1940) Metatexis und Metablastesis in Migmatiten des nordwestlichen Thüringer Waldes. Mineral Petrogr Mitt 51: 1–101

Kreher B (1992) Petrologie und Geochemie der Gabbrointrusion des Frankenstein (Odenwald). Dr rer nat Thesis, Univ Würzburg, 229 pp

Kreuzer H, Harre W (1975) K/Ar-Altersbestimmungen an Hornblenden und Biotiten des Kristallinen Odenwaldes. Aufschluss Sonderb 27 (Odenwald): 71–77

Kreuzer H, Seidel E, Schüssler U, Okrusch M, Lenz K-L, Raschka H (1989) K-Ar geochronology of different tectonic units at the northwestern margin of the Bohemian Massif. Tectonophysics 157: 149–178
Krohe A (1991) Emplacement of synkinematic plutons in the Variscan Odenwald (Germany) controlled by transtensional tectonics. Geol Rdsch 80: 391–409
Krohe A (1992) Structural evolution of intermediate-crustal rocks in a strike-slip and extensional setting (Variscan Odenwald, SW Germany): differential upward transport of metamorphic complexes and changing deformation mechanisms. Tectonophysics 205: 357–386
Laue S, Reischmann T, Emmermann KH (1990) Geochemical variation of granitoid rocks from the NW margin of the Rhinegraben. Ber Dtsch Mineral Ges Beih Eur J Mineral 2 (1): 156
Lindsley DH (1983) Pyroxene thermometry. Am Mineral 68: 477–493
Lippolt HJ (1986) Nachweis altpaläozoischer Primäralter (Rb-Sr) und karbonischer Abkühlalter (K-Ar) der Muskovit-Biotit-Gneise des Spessart und der Biotit-Gneise des Böllsteiner Odenwald. Geol Rdsch 75: 569–583
Marell D (1989) Das Rotliegende zwischen Odenwald und Taunus. Geol Abh Hessen Wiesbaden 89: 128
Massonne HJ, Schreyer W (1987) Phengite geobarometry based on the limiting assemblage with K-feldspar, phlogopite, and quartz. Contrib Mineral Petrol 96: 218–224
Matthes S (1954) Die Paragneise im mittleren kristallinen Vorspessart und ihre Metamorphose. Abh Hess Landesamt Bodenforsch Wiesb 8: 1–86
Matthes S, Okrusch M (1977) The Spessart crystalline complex, North-West Bavaria: rock series, metamorphism and position within the central German crystalline rise. In: La chaîne varisque d'Europe moyenne et occidentale. Coll Int Cent Natl Rech Sci (Rennes 1974) Paris 243: 375–390
Matthes S, Schubert W (1971) Der Original-Beerbachit im Odenwald, ein Amphibolit-Hornfels in Pyroxen-Hornfelsfazies. Contrib Mineral Petrol 33: 62–86
Matthes S, Okrusch M, Richter P (1972) Zur Migmatit-Bildung im Odenwald. Neues Jahrb Mineral Abh 116: 225–267
Meisl S (1972) Das kristalline Grundgebirge. In: Kupfahl H-G, Meisl S, Kümmerle E (eds) Erläuterungen zur Geologischen Karte von Hessen 1:25000, Blatt Nr. 6217 Zwingenberg a. d. Bergstrasse, 2. Aufl. Hess Landesamt Bodenforsch, Wiesbaden
Metz P (1976) Experimental investigation of the metamorphism of siliceous dolomites. III. Equilibrium data for the reaction: 1 tremolite + 11 dolomite = 8 forsterite + 13 calcite + 9 CO_2 + 1 H_2O for the total pressures of 3000 and 5000 bars. Contrib Mineral Petrol 58: 137–148
Mosebach R (1934) Die Körnigen Kalke von Auerbach-Hochstädten a. d. Bergstr. und der Umgebung von Aschaffenburg. Senckenbergiana Frankfurt a M 16: 175–188
Nasir S (1986) Die Metabasite im mittleren kristallinen Vorspessart: Petrographie – Geochemie – Phasenpetrologie. Dr rer nat Thesis, Univ Würzburg, 191 pp
Nasir S, Okrusch M, Kreuzer H, Lenz H, Höhndorf A (1991) Geochronology of the Spessart crystalline complex, mid-German crystalline rise. Mineral Petrol 43: 39–55
Neumann W (1964) Grundgebirge von Ruhla-Brotterode. Exkursionsführer 11. Jahrestagung 1964 Geol Ges DDR: 101–130, Berlin
Neumann W (1966) Versuch eines lithostratigraphischen Vergleiches von Grundgebirgsanschnitten im Bereich der Mitteldeutschen Schwelle. Geologie Berlin 15: 942–962
Newton RC, Perkins D (1982) Thermodynamic calibration of geobarometers based on the assemblages garnet – plagioclase – orthopyroxene (clinopyroxene) – quartz. Am Mineral 67: 203–222
Okrusch M (1983) The Spessart crystalline complex, Northwest Bavaria. DMG SFMC Joint Meeting 1983, Excursion E 4. Fortschr Mineral 61 (Beih 2): 135–169
Okrusch M (1990) Metamorphism in the Odenwald and Spessart crystalline mountains (Mid-German Crystalline Rise). In: Franke W (ed) Mid-German Crystalline Rise and Rheinisches Schiefergebirge. Field guide to pre-conference excursion. Conf on Paleozoic orogens in central Europe – geology and geophysics, IGCP 233 Terranes in the Circum-Atlantic Paleozoic Orogens, Göttingen-Giessen, pp 81–91
Okrusch M, Müller R, El Shazly S (1985) Die Amphibolite, Kalksilikatgesteine und Hornblendegneise der Alzenauer Gneisserie am Nordwest-Spessart. Geol Bavarica 87: 5–37
Okrusch M, Raumer J von, Matthes S, Schubert W (1975) Mineralfazies und Stellung der Metamorphite im kristallinen Odenwald. Aufschluss Sonderbd 27 (Odenwald): 109–134
Rittmann KL (1984) Argon in Hornblende, Biotit und Muskovit bei der geologischen Abkühlung – $^{40}Ar/^{39}Ar$-Untersuchungen. Dr rer nat Thesis, Univ Heidelberg, 278 pp
Schmädicke E, Mezger K, Cosca MA, Okrusch M (1995) Variscan Sm-Nd and Ar-Ar ages for eclogite-facies rocks from lthe Erzgebirge, Bohemian Massif. J metam Geol 13 (in press)
Schubert W (1968) Die Amphibolite des prävaristischen Schieferrahmens im Bergsträsser Odenwald. Neues Jahrb Mineral Abh 108: 69–110
Seim R (1967) Der Para-Plagioklasgneis des Kyffhäuserkristallins, seine metatektischen und metablastischen Derivate und seine Einlagerungen. Geologie Berlin (Beih 56): 1–85
Smoler M (1987) Petrographische, geochemische und phasenpetrologische Untersuchungen an Metasedimenten des NW-Spessart/Bayern. Dr rer nat Thesis, Univ Würzburg, 256 pp
Spear FS, Cheney JT (1989) A petrogenetic grid for pelitic schists in the system SiO_2-Al_2O_3-FeO-MgO-K_2O-H_2O. Contrib Mineral Petrol 101: 149–164
Spear FS, Selverstone J (1983) Quantitative P-T paths from zoned minerals: theory and tectonic application. Contrib Mineral Petrol 83: 348–357
Todt W (1979) U-Pb-Datierungen an Zirkonen des kristallinen Odenwaldes. Fortschr Mineral 57 (Beih 1): 153–154

Von Gaertner HR (1950) Probleme des Saxothuringikums. Geol Jahrb Hannover 65: 409–465
Von Raumer JF (1973) Die mineralfazielle Stellung der Metapelite und Metagrauwacken zwischen Heppenheim und Reichelsheim (Odenwald). Neues Jahrb Mineral Abh 118: 313–336
Werner CD (1974) Metamorphose und Migmatisation im Ruhlaer Kristallin (Thüringer Wald). Freiberger Forschungsh Leipzig C 284: 134
Willner AP, Massonne H-J, Krohe A (1991) Tectonothermal evolution of a part of a Variscan magmatic arc: the Odenwald in the Mid-German Crystalline Rise. Geol Rdsch 80: 369–389
Wunderlich J (1989) Neue Ergebnisse zur Geologie des südlichen Ruhlaer Kristallins. Freiberger Forschungsh Leipzig C 429: 7–32
Zeh A, Okrusch M, Schubert W, Werner CD (1994) Druck-Temperatur-Zeitpfade der variscischen Metamorphose in der Mitteldeutschen Kristallinschwelle: Eine Fallstudie zur tektonometamorphen Entwicklung des Ruhlaer Kristallins. Koll DFG-Schwerpunktprogramm "Orogene Prozesse – ihre Quantifizierung und Simulation am Beispiel der Varisciden" 30/31 March 1994, Würzburg, Abstr 3
Zeh A, Wunderlich J, Okrusch M, Werner CD, Mädler J (1995) Distribution of index minerals in the Ruhla Crystalline Complex, Mid-German Crystalline Rise: a first interpretation. Zentralbl Geol Paläontol Teil I (in press)

V Saxothuringian Basin

V.A Introduction

W. FRANKE

The Saxothuringian Zone of Kossmat (1927) is one of the most complex parts of the Variscan Belt in central Europe. It consists of a northwestern crystalline belt, the Mid-German Crystalline High (Chap. IV), and a southeastern part dominated by relatively low-grade sedimentary and volcanic rocks. These sequences have been laid down in a Cambrian through to Devonian rift basin, which was closed in Devonian to Early Carboniferous time – we refer to this as the Saxothuringian Basin.

The primary relationships between the Saxothuringian Basin and the Crystalline High are entirely unknown. Today, the crystalline rocks are juxtaposed against the relatively low grade material by a south-facing reverse fault, but it is possible that the boundary between the two parts of Kossmat's Saxothuringian Zone is a major strike slip zone (Chap. XI).

Though the Crystalline Rise was a tectonic uplift at least from the Frasnian onwards, there are no clastic sediments in the Saxothuringian Basin which could unequivocally be referred to sources in the Crystalline High. Since the age and paleogeographic provenance of the protoliths of the Crystalline High are largely unknown, correlations with the less metamorphosed sequences in the Saxothuringian basin are – at best – hypothetical (see Chap. IV).

To the southeast, the Saxothuringian basin was bounded, at least from the early Famennian onwards, by the Cadomian Tepla-Barrandian terrane, which represents the upper plate and active margin during the Variscan convergence.

The internal organization of the Saxothuringian Basin is complex, since it contains high-grade metamorphic rocks and fossiliferous sequences in close juxtaposition.

The main, (par-)autochthonous part of the Saxothuringian basin consists of Cambro-Ordovician clastic sediments and volcanic rocks, which represent the passive margin of a continental block to the NW. This continental crust has so far not been identified with certainty in the tectonostratigraphic units further N; it might be contained in the Mid-German Crystalline High. The early Paleozoic shelf deposits are overlain by late Ordovician to Early Carboniferous hemipelagic sediments and intraplate volcanic rocks, followed by distal flysch sediments. The parautochthonous sequence is usually referred to as the Thuringian facies realm.

Deeper water environments of the Saxothuringian basin are contained in the tectonic klippen of Münchberg, Wildenfels, and Frankenberg, which are derived from areas to the SE: these are Cambrian to earliest Ordovician neritic clastic sediments, overlain by an Ordovician sequence with bimodal intra-plate volcanic rocks and some sandstone turbidites, Silurian to Early Carboniferous bedded cherts, and a proximal Early Carboniferous flysch sequence. The latter contains near-source stratigraphic equivalents of the Thuringian flysch. The flysch basin was fed by a crystalline hinterland in the SE, and by a frontal accretionary wedge containing the same lithologies which are now seen to override the flysch in the Münchberg and other nappes. The sequences in the allochthons are referred to as the Bavarian facies realm.

An oceanic part of the Saxothuringian basin is represented by the eclogites contained within the crystalline nappes of the Münchberg klippe (see Chap. V.C). Some of the crystalline rocks above the eclogite probably form part of the continental terrane adjacent to the S (Tepla-Barrandian or Bohemian, Chap. VII.B).

It should be noted that large-scale allochthonism in the Saxothuringian basin has been identified already 80 years ago by F.E. Suess (1912), and was recognized, a little later, as the governing structural principle in this area as well as in other parts of the Variscan belt (Kossmat 1927).

While part of the crystalline units occur as tectonic klippen in synformal positions, other metamorphic units in the Saxothuringian

Dallmeyer et al. (Eds.)
Pre-Permian Geology of Central and Eastern Europe

basin are situated in the core of the extensive Fichtelgebirge-Erzgebirge antiform. With regard to the overlying Münchberg and related nappes, these rocks represent a "relative autochthon". Those members identified as Paleozoic belong to the Thuringian facies realm.

The core of the antiform, in the eastern Erzgebirge, contains kyanite-bearing sequences and eclogites. Between the core and the mantle of the antiform, there is an important metamorphic break, which juxtaposes the eclogites against rocks with maximum metamorphic pressures of no more than 8–10 kb. The extreme reduction of the metamorphic profile requires extension. A similar history can be inferred for the Saxonian Granulites, a metamorphic dome structure set within low- to very-low-grade Thuringian facies rocks to the NW of the Erzgebirge. The overall setting and especially the extreme reduction of the metamorphic profile suggest that the granulites represent a metamorphic core complex.

Irritatingly, the available isotopic ages from the eclogite- and granulite-bearing domes appear to suggest high-grade metamorphism around 360 and 340 Ma. This contradicts the sedimentary record of the cover, which shows uninterrupted sedimentation, both in the Thuringian autochthon and the Bavarian allochthon, from the Cambrian through to the late lower Carboniferous (ca. 330 Ma), and the angular unconformity at the base of an intramontane molasse sequence, which cuts the nappe thrusts, constrains tectonic deformation in the area at about the Visean/Namurian boundary (ca. 325 Ma). It is possible, therefore, that the high-grade rocks in the cores of the Erzgebirge and the Saxonian Granulite antiforms represent pre-Variscan basement, which was largely exhumed already before the onset of Paleozoic sedimentation, and then went through a second (Variscan) cycle of sedimentary plus tectonic burial, before they were exhumed again after the early Carboniferous convergent episode. Some post-Variscan extension is documented in tectonic fabrics and in the formation of intramontane basins (latest Visean to Permian). If the high-grade rocks are of pre-Variscan origin, the isotopic data would reflect reheating and uplift, instead of earlier peak metamorphism. We are dealing here with a severe problem, which requires high-resolution studies of the P-T paths and the isotopic systems.

Deeper structural levels were invaded by voluminous late- to post-tectonic granitoids (Late-early Carboniferous to Permian, Chap. V.B.3), which also stitch the Saxothuringian/Tepla-Barrandian terrane boundary.

References

Kossmat F (1927) Gliederung des varistischen Gebirgsbaues. Abh Sächs Geol Landesamt NF 1: 39 pp

Suess F-E (1912) Vorläufige Mitteilung über die Münchberger Deckscholle. Sitzungsber Akad Wiss Wien, Math-Naturwiss Kl 121 Abt IIa (10): 253

V.B Autochthon and Nonmetamorphic Nappe Units

V.B.1 Stratigraphy

F. Falk, W. Franke, and M. Kurze

1 Paleogeographic Subdivision and General Trends

In the Saxothuringian Basin, it is possible to distinguish two facies realms, which have been termed the Thuringian and the Bavarian facies by Wurm (1925). Modern studies have confirmed models already proposed early in this century, in which the Bavarian facies rocks represent nappes emplaced on (para)autochthonous sequences developed in Thuringian facies (Suess 1912; Wurm 1925; Behr et al. 1982; Franke 1984a,b, 1989a,b). This concept is well established for the Münchberg Klippe, which is the most extensive, best-preserved and most differentiated of the notional nappes. For the sister massifs of Wildenfels and Frankenberg (Fig. 1), this interpretation is still being questioned, mainly by East German colleagues (one of the authors, M.K., included). The easternmost outcrops of the Bavarian facies, in the Elbe valley (Nossen-Wilsdruff and Elbtal Schiefergebirge), are contained in a major NW-trending fault zone.

These latter rocks could represent thrust sheets, which were later transposed and displaced in a dextral shear zone, or – in a less mobilistic scenario – belong to a relatively autochthonous embayment of the Bavarian facies within the adjacent Thuringian realm.

Different Paleozoic facies, partly with the characteristic features of the Bavarian or Thuringian facies realm, are also detectable E of the Elbe lineament, e.g., in the Lausitz (Lusatia) or in the Sudetes (Brause 1969; Jaeger 1977; Chap. VI.B).

In the main, northwestern part of the Saxothuringian basin, the Paleozoic sequences are of very-low metamorphic grade. Greenschist and higher-grade alteration is only present in parts of the Frankenwald Transverse Zone (see also Chap. V.B.2), and – laterally extensive – in the core regions of the Fichtelgebirge-Erzgebirge Antiform.

Since the Saxothuringian sequences have been studied for more than 100 years, it is impossible to discuss the literature in detail. We therefore have mainly cited the most recent publications (with references) and some of the earlier key papers. More detailed stratigraphic reviews are available in Franke (1984b), Gandl (1981), and Wurm (1961) (all concerning the W German part of the ST), Grundriß der Geologie der DDR (1968), Hoppe and Seidel (1974 Thuringia), Kurze et al. (1994) and Pietzsch (1962) (both for Saxony).

2 Thuringian Facies

2.1 Proterozoic, Cambrian, Ordovician

2.1.1 Proterozoic/Cambrian Boundary

It is notoriously difficult to define the base of the Cambrian. In the Saxothuringian basin, the boundary is usually placed, by convention, between graywacke/shale sequences in the lower parts of the sections and the more mature clastic sediments above. In this way, the boundary is drawn, e.g., in the Schwarzburg antiform and in the Erzgebirge. This boundary – notwithstanding its exact biostratigraphic age – marks a change in the tectonic regime: the notional pre-Cambrian graywackes, which are widespread throughout the Bohemian Massif (cf. Chap. VII.B.1), probably reflect a Cadomian orogenic event, and the early Paleozoic quartzites are taken to represent the new marine transgression.

It must be noted, however, that there is no unequivocal evidence of a Cadomian unconformity in the Saxothuringian basin; unconformable relationships between the notional Cambrian and Proterozoic sequences in the Erzgebirge (Walther 1972; Lorenz and Hoth 1990) are difficult to ascertain because of the intense tectonometamorphic overprint during the Variscan orogeny. In any case, the existence

Dallmeyer et al. (Eds.)
Pre-Permian Geology of Central and Eastern Europe

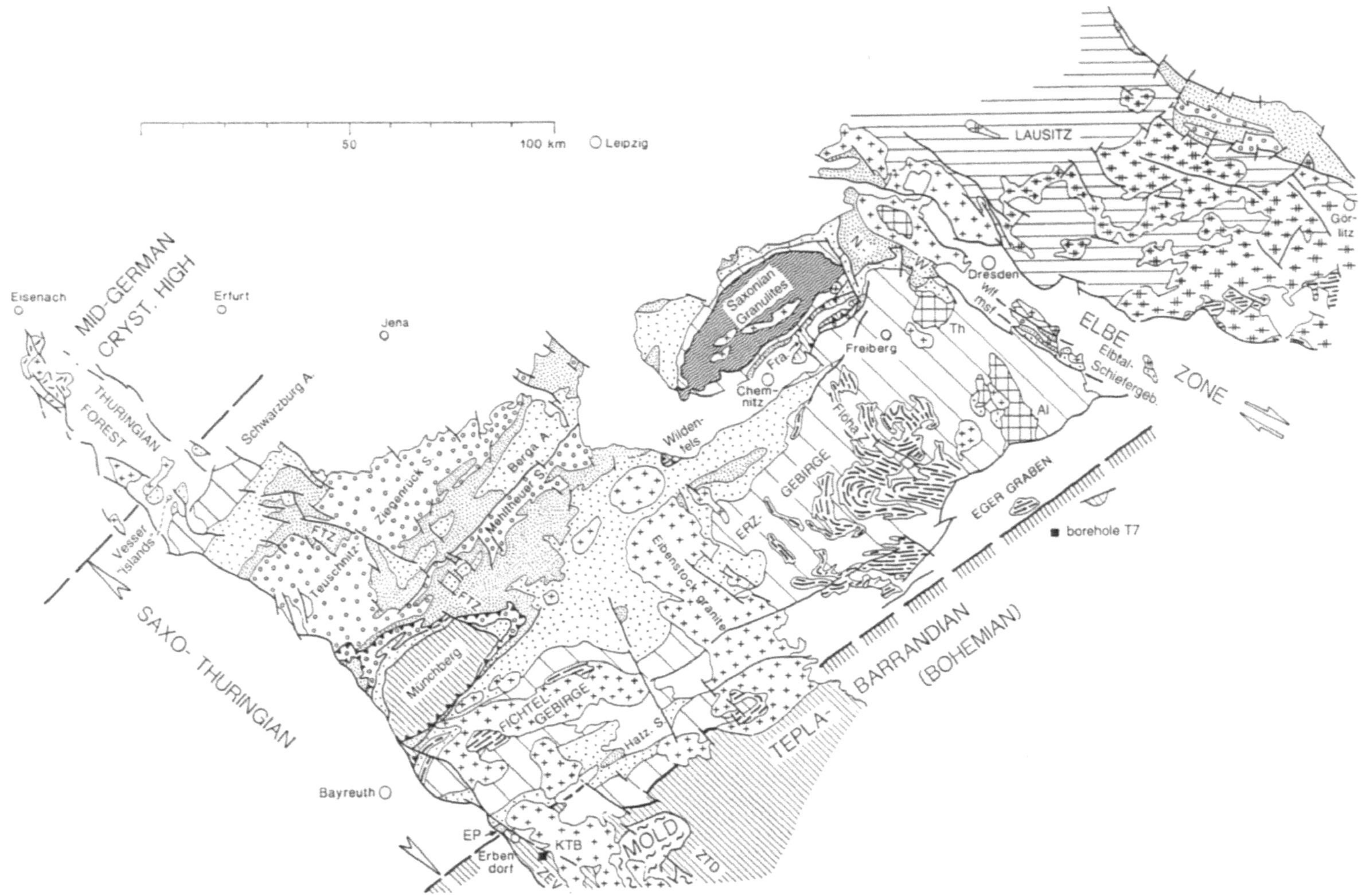

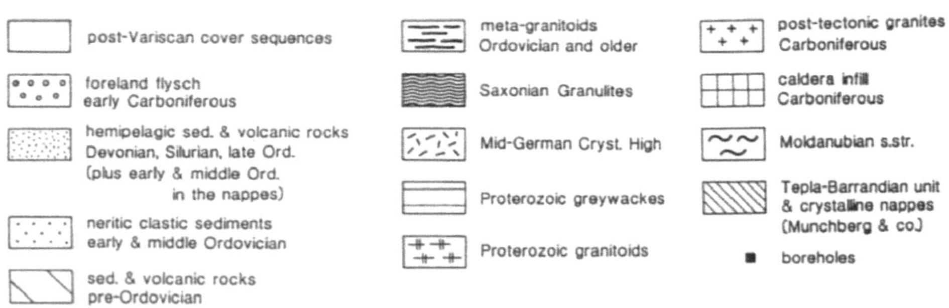

Fig. 1. Geological sketch map of the Saxothuringian Belt. Abbreviations: *A.* Antiform; *Al* Altenberg caldera; *EP* Erbendorf Paleozoic rocks; *Fra* Frankenberg Klippe; *FTZ* Frankenwald Transverse Zone; *KTB* German Continental Deep Drilling Site; *msf* Mid-Saxonian fault; *N.-W.* Nossen-Wilsdruff Schiefergebirge; *S.* Synform; *Th* Tharandt caldera; *wlf* West Lausitz fault; *ZEV* Zone of Erbendorf/Vohenstrauß; *ZTD* Zone of Tepla-Domazlice. Note that the base of the Ordovician is arbitrary

of a pre-Cambrian basement can be inferred from the shallow-water character of the Cambro-Ordovician sequences.

The presence of Proterozoic rocks in the eastern Erzgebirge has recently been proven by U-Pb work on zircons from granitoids (Kröner et al. 1991). Proterozoic ages in Saxony have also been detected in the Laas granodiorite (Wenzel et al. 1989) and in the Dohna granodiorite (Schmidt, in Autorenkollektiv 1991). The Cadomian unconformity is unequivocally represented in the Tepla-Barrandian (Bohemian) terrane to the SE of the Saxothuringian basin (see Chap. VII.B.1).

2.1.2 Cambrian

In the Schwarzburg Anticline, the transition from the graywacke-dominated Proterozoic into the Cambrian is well exposed. The Cambrian Goldisthal formation consists essentially of silty shales, near the base with intercalations of quartzites and (locally) felsic volcanic rocks (Falk 1964; Bankwitz 1970, 1977; Falk and Biewald 1990). These sediments were probably formed in a neritic realm. Their thickness ranges between 250 m at the SE-, and 700 m at the NW- flank of the antiform (Estrada et al. 1994).

Toward the NW, in the Vesser area, a thick, bimodal volcanic suite (Bankwitz and Bankwitz 1990) is attributed to the Cambrian, because it is overlain by quartzites which are correlated with the basal Ordovician quartzites in neighboring regions.

In the Frankenwald Transverse Zone, to the SE, a drilling has revealed pre-Ordovician carbonates, whose microfaunas have been attributed to the Lower or Middle Cambrian (Wucher 1967; Blumenstengel 1980).

Cambrian carbonates have also been encountered at the SE flank of the Saxonian Granulite Antiform (Freyer et al. 1982).

In the core of the Fichtelgebirge, carbonate-bearing sequences underlying the basal Ordovician quartzites probably belong to the Cambrian (von Gaertner 1942a). Their interpretation as Precambrian by Stettner (1980) is based upon the notion of a separate, pre-Variscan tectonic deformation, which has since been refuted by the tectonic studies of Stein (1988).

Lastly, Cambrian sequences are inferred to exist in the metamorphic terrain of the Western Erzgebirge (see Lorenz and Hoth 1990 for a latest review). It is now known, however, that these sequences comprise a mix of contrasting metamorphic grades including eclogites (e.g., Schmädicke 1991), so that the succession of rocks encountered represents (in parts) a tectonic rather than a stratigraphic sequence (see also Schmidt et al. 1994).

2.1.3 Ordovician

In all of the Saxothuringian basin, the basal part of the Ordovician is composed of quartzites and shales, locally with conglomerates (Leinhos 1956; Falk 1966). This sequence, the Frauenbach and Phycodes formations, may attain more than 2000 m in thickness (Wiefel 1974). Hummocky cross stratification, trilobate crawling trails, and the foraging trail *Phycodes circinatum* indicate shallow-water environments (Lützner and Mann 1988). Compositional grading in some quartzites probably reflects deposition from tempešites, and suggests that storm-controlled transport was one of the main factors in the dispersal of clastic material over the early Ordovician shelf (see Falk and Lützner 1991 for sedimentological details). Graptolite findings in the upper part of the Phycodes formation suggest a Tremadoc through to basal Arenig age (Erdtmann 1991; Schmidt 1939; Volk 1966). The position of the Cambrian/Ordovician boundary is paleontologically not defined, but conventionally placed at the base of the Frauenbach quartzites.

In many regions, there are intercalations of felsic and (subordinate) mafic volcanic rocks, which occur as effusives or sills. In the Schwarzburg Anticline, there are Tremadocian quartz-porphyries and related tuffs (Meinel 1974). Coarse-grained varieties with a Variscan metamorphic foliation have been described as gneiss (e.g., Hirschberger Gneis in the Frankenwald Transverse Zone, Schröder 1958; "Epigneis" of the Fichtelgebirge; probably also the youngest parts of "Rotgneis" (red gneiss) of the Eastern Erzgebirge – a fairly heterogeneous and possibly also heterochronous suite of felsic rocks, see Frischbutter 1990). Sparse geochronological data grossly confirm the stratigraphic position: zircons from the "epigneiss" have yielded a U/Pb age of 449 ± 4 Ma (Teufel 1988), and zircons from a pebble of "Hirschberg Gneiss" in Frasnian conglomerates (see below) have a U/Pb age of 504 ± 7 Ma (Dörr, unpubl.).

The remainder of the Ordovician is relatively thin (a few 100 m only). The Gräfenthal formation consists of shales with conspicuous intercalations of quartzites and oolitic ironstone seams, as well as occasional limestones associated with the latter. The lower part of this sequence spans most of the Arenig through Caradoc and is less than 250 m thick, the major portion being composed of an Arenigian shale member (Griffelschiefer = pencil slates). Part of the Llanvirn and the Llandeilo is probably contained in a depositional break between the ironstone seams (Fuchs 1990).

The Ashgill is represented by the "Lederschiefer", a shaly sequence dominated by diamictites of glacio-marine origin. Striated stones (Katzung 1961; Steiner and Falk 1981) record the influence of the Saharan glaciation at the Ordovician/Silurian boundary. This is in accord with the cold-water character of the conodonts from the Gräfenthal formation (Fuchs 1990) and the steep paleoinclinations derived from Ordovician volcanics in the Bavarian facies (Böhm et al. 1989).

Notwithstanding glacio-eustatic sea-level changes, the sedimentary record reflects a general deepening trend from the lower Ordovician neritic clastics toward the Silurian graptolite shales (see below). This is probably also expressed in trace fossils of the *Nereites*-Ichnofacies (sensu Seilacher 1967) in the "Hauptquarzit" of the Caradoc (Pfeiffer 1972; Benton 1982). Along with this trend, the rates of accumulation are seen to increase from the Cambro-Ordovician towards the Devonian (Lützner et al. 1986).

2.1.4 Silurian

The Llandovery and Wenlock are represented by black shales and cherts with graptolites, which are ubiquitous in the Variscan Belt and probably are related with the post-glacial rise of the sea level (see Leggett et al. 1981 and Wignall 1991 for a review of possible relationships). The anoxic period is briefly interrupted during part of the Ludlow by the deposition of a pelagic limestone (the "Ockerkalk"), which is likewise widespread in Europe (see Jaeger 1976, 1977, 1988), and must be related to large-scale hydrographic fluctuations. Locally, there occur intercalations of basic lavas and tuffs (see the compilation in Franke 1984b, p. 24).

2.2 Devonian and Basal Carboniferous

The older part of the Devonian (Gedinnian through Givetian) is composed of black shales with pelagic faunas such as graptolites and dacryoconarids. An intercalation of condensed pelagic limestone ("Tentakuliten-Knollenkalk" = nodular tentaculite lst.) comprises the Late Gedinnian through Early Emsian (Al-Rawi 1977; locally up to the Late Emsian, Alberti 1983).

Above the limestone member, tentaculite-bearing shales contain thin-bedded intercalations of quartzite with trace fossils ("*Nereites* Quartzite"). The quartzites are turbidites derived from an unknown source-region, probably situated to the NW (Steinbach and Blumenstengel 1974). The black shale facies may – at least locally – extend into the lowermost Frasnian (Gandl 1989).

Higher up in the Frasnian, there is a very conspicuous sequence of few hundred metres of sedimentary and volcanic rocks, which records a brief yet prominent interval of renewed rifting. Along the axis of the presentday outcrop of the Saxothuringian basin, there is a NE-trending belt of volcanic rocks, which can be traced towards the NE as far as the Elbe Valley Transverse Zone. The volcanic sequence comprises metabasalts and related pyroclastic rocks, partly associated with minor amounts of felsic volcanic materials. The volcanic rocks can be assigned to an intraplate setting (Wirth 1978; Werner 1981; Klenke and Poll 1985). Locally, the volcanic rocks are overlain by volcano-sedimentary hematite ores of the Lahn-Dill type (von Gaertner 1942b; Gräbe 1956; Wiefel 1976). Conodont faunas allow to constrain the volcanic period into a brief interval within the Early Frasnian (Gandl 1989).

Intercalated among the volcanic rocks, as well as immediately above and/or below, there are coarse-grained clastic sediments (conglomerates, graywackes, and silty shales), with clasts derived from the parautochthonous Ordovician, Silurian, and Devonian rocks. Their composition is often dominated by debris derived from the more or less coeval basaltic volcanics. Except for very few and doubtful cases discussed in Franke (1984b), there are no extrabasinal clasts or metamorphic rocks. The most conspicuous pebbles are derived from the Lower Ordovician granitoids (Eigenfeld 1938). The coarsest materials (pebbles of more than 1 m!) occur in a linear, NE-trending belt, which can be traced from the autochthon at the N margin of the Münchberg Klippe toward the N flank of the Saxonian Granulite Antiform (see, e.g., Gräbe et al. 1968; Leiteritz 1957). As laid out in Chapter V.B.2, the granulite antiform probably represents a metamorphic core complex, and the conspicuous association of Frasnian volcanic and sedimentary rocks can be taken to indicate uplift and extension.

Reef-building organisms and some few embryonic reefs on Frasnian volcanic mounds, together with frequent pyroclastic deposits, indicate limited water depths at least for the volcanic belt of the Thuringian facies realm.

The remainder of the Frasnian and the Famennian are composed of condensed red, green, or gray shales with nodular limestones (the latter especially on rises). The faunas are almost exclusively pelagic (ammonoids, conodonts, pelagic ostracodes, some trilobites). At the S flank of the Schwarzburg Anticline, few and thin beds of quartzite turbidites are intercalated among the variegated shales which comprise the Devonian/Carboniferous boundary (including the Dinantian I, Bartzsch and Weyer 1985, 1986). The same type of shales is widespread at least in Europe (Hangenberg-Schiefer), and contains some coarser-grade clastic sediments also in other regions (e.g., Rhenish Massif). Both the shales and the clastic intercalations probably relate to a global regression (Flajs and Feist 1988). This effect may also help to explain local stratigraphic breaks straddling the Devonian/Carboniferous boundary, which otherwise may be satisfactorily explained by condensation or nondeposition on submarine rises (see the discussion in Franke 1984b, p. 104). Numerous sections expose a continuous stratigraphic sequence; hence, an orogenic event (Bretonic Phase) at the Devonian/Carboniferous boundary, which has been postulated by previous authors (e.g., Gandl 1981) can be ruled out.

Some global hydrographic change must also held responsible for the installment of a black-shale environment in the middle Tournaisian (Dinantian II alpha), whose representation in the Thuringian realm is called Rußschiefer (sooty slate).

The typical Thuringian sequences can be traced eastwards into the northern mantle of the **Saxonian Granulite** and into the Elbe Valley Zone (though the latter also contains Bavarian elements). To the N of the granulites, there occur the early and middle Devonian tentaculite limestones and shales, mafic volcanic rocks and some metarhyolites (Quarzkeratophyr), some carbonates, and the Frasnian clastic sequences described above. At the southern flank of the granulite antiform, the metabasalt/limestone association is missing; instead, there are mafic tuffs, cherts, and reddish shales. Early and middle Devonian have not been detected so far. The absence of some conspicuous Thuringian elements, as well as the higher metamorphic grade (as compared with the northern flank of the granulites), might be due to the northward overthrusting of the Frankenberg complex.

S of the **Frankenberg metamorphic complex** (?klippe), the black cherty shales of the Silurian (with graptolites) are overlain by thick basaltic lavas and tuffs, associated with red shales and one very small occurrence of limestones. The volcanics correspond to modern intraplate rocks (Werner 1981). Some rhyolites (Quarzkeratophyr) are not dated, but probably belong to the same sequence. Big pebbles of Devonian shallow-water limestones (middle to early upper Devonian, see Kurze 1966; Sebastian 1990) occur in Early Carboniferous conglomerates, which either represent proximal flysch (Paech 1975) or else belong to the post-orogenic intramontane molasse.

The Devonian of the **Wildenfels area** is similar to that of the region S of Frankenberg, but contains much more limestones.

In the **Elbtal-Schiefergebirge**, the Thuringian-type Devonian contains (from bottom to top) dark shales with acritarchs of Early Devonian age (Linnemann et al. 1988), overlain by the Diabas-Kalkstein-Serie (metabasalt/limestone formation), which spans the Middle/Late Devonian boundary through to the Early Famennian (Schützenmeister 1963). Some quartzites and graywackes (Engert 1956) could represent equivalents of the Frasnian clastic belt.

In the **Nossen-Wilsdruff Schiefergebirge**, the Early Devonian consists of black shales, Tentakulitenkalk, and Tentakulitenschiefer (Tröger 1990), the latter with some intercalations of quartzite. The overlying Diabas-Kalkstein-Serie has yielded Frasnian corals and brachiopods (Gotte 1953, 1956; Werneburg and Schüppel in Kupetz 1987; Tröger, unpubl.).

2.3 Flysch Sedimentation (essentially Early Carboniferous)

In most of the Thuringian realm, synorogenic clastic sedimentation (flysch) started not before the basal Viséan (see below); only at the southern margin of the Fichtelgebirge, in the very low-grade Erbendorf unit, early Famennian graywackes and shales have been detected by Adam and Reuter (1981). These contain chromite (Ludwig 1968) and probably reflect obduction of ophiolite material. Metamorphosed graywackes of probably Early Devonian age have also been detected among the kyanite-grade gneisses of the Erbendorf-Vohenstrauß Zone (ZEV), in the northwestern extremity of the Tepla-Barrandian (Bohemian) terrane (Pflug and Prössl 1989). The occurrence of the earliest flysch sediments at the southern margin of the Saxothuringian basin is consistent with the concept of a northward prograding orogenic front (see below).

With the exception of the Devonian at Erbendorf, flysch sediments have only been preserved in the extensive areas of outcrop of the Teuschnitz-Ziegenrück and Blintendorf-Mehltheuer synforms (see Fig. 1). Graywacke sedimentation starts at about the Tournaisian-Visean boundary and extends into the Late Visean. Still within the Latest Visean, a coal-bearing, intramontane molasse oversteps unconformably the thrusted and folded older rocks (Borna-Hainichen formation, overlying the Frankenberg Massif in Saxony). This permits restriction of the age of tectonic deformation in the unmetamorphosed part of the Saxothuringian Basin into the Late Visean (approx. 330 Ma). The relevant biostratigraphic literature is compiled and evaluated in Franke (1984b).

Most of the Thuringian flysch (altogether ca. 3000 m) is made up of classical turbidites, with some conglomerate levels, which probably represent suprafan lobes. The sequence is devoid of body fossils, but contains abundant trace fossils of the *Nereites* ichnofacies, which is typical of turbidite basins (Stepanek and Geyer 1989). There are very few intercalations of felsic tuffs.

The Thuringian flysch shows the same spectrum of clasts as the Bavarian, which is partly derived from the pre-Carboniferous rocks now contained in the nappes overriding the flysch. There are also some beds with calcareous fossil debris – these latter are correlatable with more voluminous beds of the same type in the more proximal Bavarian facies (see below).

The highest part of the sequence (around the Teuschnitz conglomerate, see Fig. 2) contains also clasts which are probably derived from the Thuringian facies, such as *Phycodes*-bearing quartzites and undeformed Ordovician granitoids (Dörr et al. 1989) which may be taken to indicate that the Thuringian foreland was also incorporated into the sediment-producing wedge at the orogenic front.

3 Bavarian Facies

The Bavarian facies is restricted to certain tectonic units, at least part of which represent nappes. If not marked otherwise, the following description refers to the Münchberg area, which is the most complete and best preserved, and whose allochthonous nature is clearly established (see Chaps. V.B.2, C.1).

In the Münchberg area, the **normal Bavarian type** Paleozoic sequences are contained in nappe sheets inserted between the higher-grade crystalline nappes above and the very-low-grade parautochthon below. In accord with the overall inversion of the Münchberg tectonic sequence (see Chap. V.C.1), the oldest rocks (Ordovician) occur in the highest of the less metamorphosed nappes, and are underlain, in their turn, by tectonic units mainly composed of Devonian and – in the lowermost nappe – Early Carboniferous sediments. The thickness of the very low grade nappes is 2–3 km, with the Ordovician making up the upper half.

Many lithologies encountered in the nappes also occur as clasts in the underlying Early Carboniferous flysch, which has apparently been fed by the nascent tectonic pile.

The flysch clastics also contain blocks and pebbles of Silurian and Devonian rocks of a third facies, which occurs neither in the Thuringian parautochthon nor in the normal Bavarian type sequences of the Münchberg nappes. This peculiar facies is mainly composed of condensed limestones. It must have been present, in Early Carboniferous time, in the tectonic sequence of the source region, and has only survived in the form of clasts. The original paleogeographic context of this limestone facies is unknown; similarities with some carbonate members of the Thuringian facies realm could be taken to suggest, that the limestones are derived

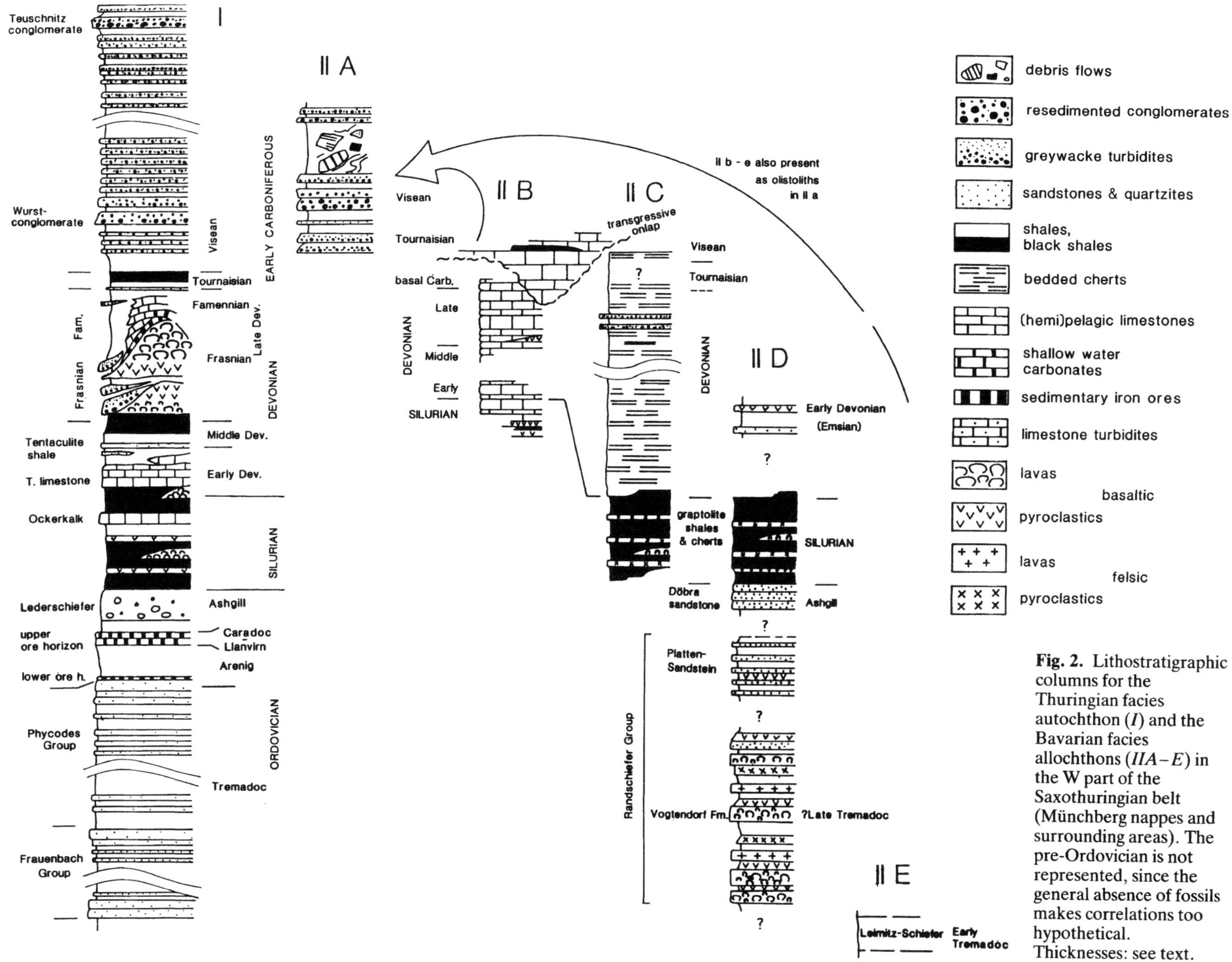

Fig. 2. Lithostratigraphic columns for the Thuringian facies autochthon (*I*) and the Bavarian facies allochthons (*IIA–E*) in the W part of the Saxothuringian belt (Münchberg nappes and surrounding areas). The pre-Ordovician is not represented, since the general absence of fossils makes correlations too hypothetical. Thicknesses: see text.

from a southeastern part of the Thuringian realm, and were incorporated into the advancing orogenic front as a nappe unit which has not survived. Since this erratic limestone facies is only known from clasts in the proximal Bavarian flysch, it is included in this chapter and referred to as the **Bavarian limestone facies** (IIB in Fig. 2).

3.1 Cambrian, Ordovician, and Silurian

3.1.1 Cambrian

The Cambrian of the Bavarian facies has survived only in the form of large olistoliths or gravitational nappes of up to 2 km^2 in extent. According to trilobite faunas, all these rocks appear to be of Middle Cambrian age. The oldest rocks are graywackes and shales; the younger parts of the sequence consist of sandstones and quartzites with intercalated shales, and a few occurrences of felsic tuffs. A detailed lithological description of the Cambrian sequences is available in Ludwig (1969). With the possible exception of the basal graywackes, the faunas and sedimentary structures consistently suggest deposition in shallow water.

3.1.2 Ordovician

Some other olistoliths consist of shales with a rich Tremadocian trilobite fauna (Sdzuy 1955, IIE in Fig. 2). Most of the Bavarian-type Ordovician is contained in a nappe, the bulk of which consists of a bimodal volcanic suite, with alkali basalts and their felsic differentiates (Wirth 1978; Wahnig 1986). Trilobites and brachiopods indicate a Late Tremadocian age for at least part of this sequence (Vogtendorf formation).

The volcanic sequence is overlain by shales and sandstone turbidites of the Randschiefer formation, which probably comprises the Middle and Late Ordovician. Sparse conodonts and sponge spicules do not permit a precise age dating (Sannemann 1955). The youngest member of the Ordovician is the Döbra sandstone (Ashgill), which probably reflects the glacio-eustatic regression.

3.1.3 Silurian

The Silurian is entirely represented by black shales and bedded cherts with some felsic and mafic tuffs. The pelites contain graptolites (see, e.g., Greiling 1957). The Bavarian limestone facies is represented by big blocks of the Orthocerenkalk (with orthocone nautilids) of the Late Wenlock and Ludlow.

3.2 Devonian

In the Münchberg nappes, most of the Devonian is represented by light-colored bedded cherts with radiolarians, which occur as a composite, often dismembered thrust sheet as well as in the form of olistoliths within the flysch. The bedded cherts have been found to conformably overlie the black Silurian cherts, and have also yielded numerous conodont faunas of Early, Middle and Late Devonian age, the Late Famennian included (see the literature compiled in Franke 1984b, and Schmidt-Effing 1988 for radiolarian evidence). This facies may extend into the Early Carboniferous (earliest Visean by radiolarians, Braun and Schmidt-Effing 1988). Locally, there are intercalations of greywacke turbidites, which have, at one locality, been dated as latest Frasnian (Zitzmann 1968, p. 600).

A different facies of the Early Devonian has been encountered in the upper part of the Randschiefer nappe – shales and tuffs have yielded microfloras of Early Emsian age, with reworked Arenigian taxa (Hammann et al. 1989).

Olistoliths of the Bavarian limestone facies comprise the Tentakulitenkalk of the Early Devonian (Pragian and Late Lochkovian, Alberti 1983), and pelagic limestones of the Frasnian and Famennian (hundreds of clasts; conodont literature compiled in Franke 1984b).

In none of the different Bavarian sequences has it been possible to identify a complete transition from the Devonian into the Carboniferous. In the cherts and shales, the relevant levels are probably present, but remain to be proven; in several blocks of the Bavarian limestone facies, there are depositional breaks (see the compilation in Mansourian 1979, p. 20, but also the discussion in Franke 1984b, p. 73 ff, p. 102 ff). In these cases, Famennian nodular limestones are overlain by Tournaisian or Early Visean shallow-water carbonates (see below).

The conspicuous Devonian cherts are also present in the **Elbtal-Schiefergebirge**, where they contain conodont faunas spanning at least the Givetian through Famennian (Freyer and Uhlig 1963; Kurze et al. 1994). Similar rocks occur in the **Nossen-Wilsdruff-Schiefergebirge** (Frasnian and Famennian bedded cherts, Kupetz 1987), and in the **Frankenberg area** (Kurze 1964, 1966, 1969). In an Early Carboniferous(?) debris flow in the Nossen-Wilsdruff area (Helbigsdorf), limestone clasts with Frasnian and Famennian conodonts (Reichstein and Schwab 1962) remind the Bavarian limestone facies of the Münchberg nappes.

An allochthonous position of all these units is possible, yet unproven. Kurze et al. (1994) concede that the Bavarian facies rocks are separated from the Thuringian sequences by important reverse faults or thrusts, but suggest derivation from deeper parts of the Saxothuringian basin which were not far distant from the Thuringian facies areas.

3.3 Early Carboniferous

Apart from the one occurrence of Early Visean bedded chert (see above), the Early Carboniferous of the Bavarian facies consists of coarse-grained synorogenic clastic sediments (flysch): resedimented conglomerates, coarse-grained graywacke turbidites, slumps and debris flows, and intercalated mudstones. This association can be referred to slope and inner fan environments in a depositional system, whose more distal parts are represented in the Thuringian facies (Behr et al. 1982; Franke 1984b). Numerous blocks of pre-Carboniferous rocks in Bavarian facies are included in the debris flows.

Another group of clasts is made up of benthic organisms (corals, forams, algae, shell-hash, crinoids), and also lithified carbonates (Carboniferous Limestone), which represent various shallow-water depositional environments (Franke 1984b), and have been dated, mainly by foraminifera, as Tournaisian and Visean (Mansourian 1979, with amendments in Franke 1984b). These resedimented carbonates contain siliciclastic grains and pebbles, which correspond to those encountered in the surrounding flysch. According to Behr et al. (1982) and Franke (1984b), the carbonates were deposited on a growing accretionary wedge, from which they received their siliciclastic components, and were episodically transported, together with pre-Carboniferous materials, into the adjacent flysch-basin. This setting would also serve to explain the unconformable onlap of early Carboniferous on Famennian carbonates preserved in some olistoliths.

Distal equivalents of the mixed (siliciclastic + carbonate) mass-flow deposits also occur in the lower part of the Thuringian Carboniferous to the NW, and in the Mehltheuer Syncline to the NE of the Münchberg Klippe (Gräbe in: Hoppe and Seidel 1974, p. 273).

The Early Carboniferous mass-flow deposits with clasts typical of the Bavarian facies (especially the light-colored Devonian and black Silurian cherts) are also detectable in the **Elbtal-Schie-fergebirge** (Kieselschiefer-Hornstein-Konglomerat, Pietzsch 1962; Hübschmann 1989). Kieselschiefer-Hornstein-Konglomerat also occurs at Prüfern in the **Nossen-Wilsdruff Schiefergebirge**. Several occurrences of olistostromes in this area have been described by Kittler et al. (1977). A survey of olistostromal deposits in the Saxothuringian belt is available in Kurze (1983).

It remains to point out that – with the exception of singular goniatite finds in Tournaisian shales at a few localities (Poll and Rossner 1981; Gandl 1972, p. 262) – the Early Carboniferous rocks are devoid of pelagic fossils, whereas the largely time-equivalent flysch sequences of the Rhenohercynian basin (see Chap. III.B.1) contain abundant goniatite faunas of Viséan and Namurian age. The Saxothuringian flysch has only been dated by means of resedimented benthic organisms and conodonts, which give maximum ages. The correlation of the Bavarian with some older part of the Thuringian flysch is therefore mainly based upon mineral composition and clast spectra.

4 Paleogeographic Interpretation

In both the Thuringian and the Bavarian facies realm, the oldest (Cambrian) rocks have been deposited in shallow water, probably on Cadomian crust.

Both facies realms show a marked tendency toward deeper water environments in the younger beds: neritic sequences are overlain by (hemi-)pelagic sediments. Together with the Cambro-Ordovician (and partly Devonian) volcanic rocks, this trend reflects an important Early Paleozoic phase of crustal extension.

The Thuringian realm occupies what is now a northwestern part of the Saxothuringian belt, and apparently represents the passive margin of a continental unit to the NW. This unit – the source region of the Cambro-Ordovician clastics – has not been identified so far; it might be contained in the metamorphic rocks of the Mid-German Crystalline High (see Chap. IV). It cannot be excluded, however, that strike-slip movements have displaced the Saxothuringian rocks, and that their present juxtaposition with the Crystalline High is incidental. This interpretation is backed up by the observation, that the Crystalline High, in early Carboniferous time, has not shed clastic detritus into the Saxothuringian basin, though it was an important source area for the Rhenohercynian flysch to the north (see Chap. III.B.1).

As a generalization, it can be said that the Bavarian-type sequences, throughout the Early Paleozoic, represent deeper water conditions than their Thuringian stratigraphic equivalents, and that a (hemi)pelagic regime was installed earlier than in the Thuringian facies. The Thuringian rocks can be attributed to neritic and hemipelagic realms at the northwestern margin of the Saxothuringian Basin, whereas the Bavarian facies largely consists of hemipelagic and pelagic sediments laid down in deeper parts of the basin, which – following the nappe concept – were originally situated more to the SE.

Since the Bavarian type rocks occur in nappes, it cannot be excluded that some of the pelagic sequences were originally underlain by oceanic crust. The existence of a Saxothuringian ocean floor is indicated by the MORB-type chemistry of eclogites overriding the Bavarian facies nappes (Chap. V.C.1).

The passive margin at the SE margin of the Saxothuringian basin has not been clearly identified so far. As in many orogenic belts, it may have been incorporated into the later active margin – in our case, it may be hidden among the metamorphic rocks in the Tepla-Barrandian (Bohemian) terrane (Chap. VII.B.1), where poorly preserved acritarchs from paragneisses appear to indicate an early Paleozoic age (Pflug and Prössl 1991).

The Cambrian through to middle Devonian sequences, which overlie the Cadomian basement of the Tepla-Barrandian terrane in the Prague synform, likewise represent an early Palaeozoic rift sequence. In contrast to the Saxothuringian belt, these sequences attained a deep-water stage not before the upper Givetian (the youngest beds preserved). Devonian limestones with reef faunas or crinoids and trilobites and only moderate thickness (a few hundred meters) were apparently laid down on stable continental crust. In Cambrian through middle Devonian time, this area could well have belonged to the SE part of the Saxothuringian basin; however, relationships with a "Moldanubian" basin to the S are equally possible.

In both the Bavarian and Thuringian realm, the Cambrian to Devonian, rift-related sequences are overlain by synorogenic clastic sediments derived from the southern (active) margin of the Saxothuringian basin. The oldest flysch clastics found in contact with the Thuringian facies sequence (continental foreland) are of Famennian age. By that time, any oceanic crust must have been completely subducted. Hence, the early Carboniferous flysch represents the infill of a foreland basin.

The apparent absence of all pelagic organisms (such as goniatites) at least from the younger clastic sequences may be taken to indicate that the Saxothuringian foreland basin, in Visean time, had become secluded from the open marine realm by the advanced plate convergence: the basin was bounded by the Mid-German Crystalline High to the NW, the active Bohemian margin to the SE, and (probably) the East-European platform to the NE. If one follows less mobilistic tectonic interpretations, the Early Carboniferous flysch basin in the eastern part of the Saxothuringian belt may even have been subdivided and barred by rises. Shortly after the deposition of the youngest flysch, still within the latest Visean, coal-bearing intramontane molasse was deposited on the Frankenberg crystalline (?)klippe and the surrounding rocks.

The tectonic development of the sedimentary nappes and their relationships with the overlying crystalline materials are discussed in Chapters V.B.2 and V.C.1.

References

Adam J-F, Reuter A (1981) Das Paläozoikum von Erbendorf (NE-Bayern) – Ergebnisse einer Neukartierung. Diplomarbeit, Georg-August-Univ. Göttingen (unpubl)

Alberti G (1983) Zum Auftreten des Devons der Thüringischen Faziesreihe auf Blatt Lehesten und Umgebung (Frankenwälder Querzone, Saxothuringicum; Bayern). Mitt Geol-Paläontol Inst Univ Hamburg 54: 1–8

Alberti G (1983) Zur Biostratigraphie und Paläontologie des kalkig entwickelten Unterdevons im Frankenwald. Geol Bl NO-Bayern 33 (3/4): 91–133

Alberti G (1985) Nachweis des "RICHTER'schen Konglomerates" im Devon des Gräfenthaler Horstes von Oberfranken. Senckenb Lethaea 66 (3/5): 383–392

Al-Rawi D (1975) Zur Petrographie der Nereitenquarzite (Devon) Thüringens und des Vogtlandes. Z Geol Wiss 3 (1): 71–78

Al-Rawi D (1977) Biostratigraphische Gliederung der Tentakuliten-Schichten des Frankenwaldes mit Conodonten und Tentakuliten (Unter- und Mittel-Devon; Bayern, Deutschland). Senckenb Lethaea 58 (1/3): 25–79

Autorenkollektiv (1991) Neue Ergebnisse stratigraphischer und petrographischer Untersuchungen im Paläozoikum und Mesozoikum des SE-Teils der Elbe-Zone. Ges Geol Wiss Exkursionsführer, Berlin, 59 pp

Bankwitz P (1970) Über die Basisfolge des Kambriums (Goldisthaler Schichten) und ihre Grenze zum liegenden Proterozoikum (Katzhütter Schichten) an der SE-Flanke des Schwarzburger Sattels. Geologie 19: 1023–1047

Bankwitz P (1977) Überblick über den Magmatismus und die Metamorphose im Altpaläozoikum Mittel- und Westeuropas. Veröff ZIPE 44 (1): 7–30

Bankwitz P, Bankwitz E (1990) Lower Paleozoic series at the southern margin of the Central German Crystalline Zone. In: Terranes in the circum-Atlantic Paleozoic orogens. (Abstr vol, unpubl)

Bartzsch K, Weyer D (1985) Zur Stratigraphie der Oberdevon-Quarzite von Saalfeld im Thüringischen Schiefergebirge. Freib Forschungsh C 400: 5–36

Bartzsch K, Weyer D (1985) Zur Stratigraphie der Oberdevon-Quarzite von Saalfeld im Thüringischen Schiefergebirge. Freib Forschungsh C 400: 5–36

Behr H-J, Engel W, Franke W (1982) Variscan wildflysch and nappe tectonics in the Saxothuringian Zone (Northeast Bavaria, West Germany). Am J Sci 282: 1438–1470

Benton MJ (1982) Dictyodora and associated trace fossils from the Paleozoic of Thuringia. Lethaia 15: 115–132

Blumenstengel H (1980) Zur Paläontologie und Biostratigraphie der Heinersdorfer Serie (Kambrium) im Thüringer Schiefergebirge. Freib Forschungsh C 348: 61–68

Böhm V, Franke W, Soffel H (1989) New paleomagnetic constraints on Ordovician paleogeography in Europe. Terra Abstr 1/1: 82

Braun A, Schmidt-Effing R (1988) Radiolarienfaunen aus dem tiefen Visé (Unter-Karbon) des Frankenwaldes (Bayern). Neues Jahrb Geol Paläontol Monatsh 11: 645–660

Brause H (1969) Das verdeckte Altpaläozoikum der Lausitz und seine regionale Stellung. Abh Dtsch Akad Wiss Berlin Klasse Bergbau Hüttenwes Montangeol 1968 (1): 143 pp

Dörr W, Franke W, Kramm U (1989) U/Pb-Daten von Zirkonen klastischer Sedimente – Beiträge zur Entwicklung der Böhmischen Masse. KTB-Rep 89-3: 351

Eigenfeld R (1938) Die granitführenden Konglomerate des Oberdevons und Kulms im Gebiet altkristalliner Sattelanlagen in Ostthüringen, Frankenwald und Vogtland. Abh Math Phys Kl Sächs Akad Wiss 42: 1–149

Engert P (1956) Beiträge zur Stratigraphie und Tektonik des Elbtal-Schiefergebirges. Geologie Beih 16: 72

Erdtmann B-D (1991) The post-Cadomian early Palaeozoic tectonostratigraphy of Germany. Ann Soc Géol Belgique 114: 19–43

Estrada S, Heuse T, Schulz E (1994) Zur regionalgeologischen Interpretation kambro-ordovizischer Schichten an der Nordwest-Flanke des Schwarzburger Antiklinoriums, Thüringen. Z geol Wiss 22: 537–553

Falk F (1964) Die konglomeratische Arkose der Goldisthaler Schichten. Abh DAW 2: 3–26

Falk F (1966) Die gröberklastischen Bildungen in der algonkisch-tiefordovizischen Schichtenfolge des Schwarzburger Sattels in Thüringen und ihr paläogeographischer Aussagewert. Geologie 15: 396–418

Falk F, Biewald W (1990) Kambrium und tiefstes Ordovizium an der SE-Flanke des Schwarzburger Antiklinoriums. Ges Geol Wiss Exkursionsführer 1990: 4–8

Falk F, Lützner H (1991) Schichtungsgefüge der Goldisthal- und Frauenbach-Folge (Kambro-Ordovizium) im Thüringischen Schiefergebirge. Wiss Beitr Ernst-Moritz-Arndt-Univ Greifswald, Math-Nat R, Fortschr Geol Greifswald, pp 27–31

Flajs G, Feist R (1988) Index conodonts, trilobites and environment of the Devonian-Carboniferous Boundary beds at La Serre (Montagne Noire, France). Cour Forsch-Inst Senckenberg 100: 53–107

Franke W (1984a) Late events in the tectonic history of the Saxothuringian Zone. In: Hutton DHW, Sanderson DJ (eds) Variscan tectonics of the North Atlantic Region. Geol Soc Spec Publ 14: 33–45

Franke W (1984b) Variszischer Deckenbau im Raume der Münchberger Gneismasse – abgeleitet aus der Fazies, Deformation und Metamorphose im umgebenden Paläozoikum. Geotekt on Forsch 68: 1–253

Franke W (1989a) Tectonostratigraphic units in the Variscan Belt of Europe. Geol Soc Am Spec Pap 230: 67–90

Franke W (1989b) The geological framwork of the KTB drilling site. In: Emmermann R, Wohlenberg J (eds) Exploration of the continental crust through drilling – the German continental deep drilling program. KTB: 37–45, Springer, Berlin Heidelberg New York

Freyer G (1990) Zur Ausbildung und Fossilführung des nordwest-sächsischen Kambriums. Kurzfassungen, 6. Rundgespr Geodynamik des europäischen Variszikums, Clausthal-Zellerfeld

Freyer G, Schreiber A (1978) Zur Biostratigraphie des Wildenfelser Oberdevons und ihre Bedeutung für die geologische Erkundung. Wiss Beitr Martin-Luther-Univ Halle 30: 129–139

Freyer G, Uhlig HI (1963) Conodontenfunde aus den Hornsteinen des Gebietes von Nenntmannsdorf (Elbtalschiefergebirge). Geologie 12/5: 621–622

Freyer G, Geissler E, Hoth K, Tran Thin Nh (1982) Die Altersstellung des Karbonatgesteinszuges

Rabenstein-Auerswalde und ihre Bedeutung für die Geologie des südöstlichen Schiefermantels des sächsischen Granulitgebirges. Z Geol Wiss Berlin 11: 1403–1424

Frischbutter A (1990) Prävariszische Granitoide der Fichtelgebirgisch-Erzgebirgischen Antiklinalzone und ihre Bedeutung für die Krustenentwicklung am Nordrand des Böhmischen Massivs. Veröff Zentralinst Physik Erde 69: 153

Fuchs A (1990) Bemerkungen zu Stratigraphie und Charakter der ordovizischen Conodontenfauna Thüringens. Neues Jahrb Geol Paläontol Monatsh 4: 193–214

Gandl J (1972) Zur gegenwärtigen Kenntnis der Unterkarbon-Stratigraphie im Frankenwald (NE-Bayern). Z Dtsch Geol Ges 123: 259–271

Gandl J (1981) Exkursion in das Paläozoikum des Frankenwaldes (Gebiet nordwestlich der Münchberger Gneismasse). Jahresber Mitt Oberrhein Geol Ver NF 63: 63–73

Gandl J (1989) Zum Alter der Schwärzschiefer und des hangenden Diabas-Vulkanismus im Devon des Frankenwaldes. Z Dtsch Geol Ges 140: 117–135

Gotte W (1953) Über Fossilfunde in der Diabas-Kalksteinserie der Elbtalzone. Geologie 2: 266–267

Gotte W (1956) Neue stratigraphische und tektonische Untersuchungen im Wilsdruff-Nossener Schiefergebirge. Ber Geol Ges DDR 1 (2): 199–201

Gräbe R (1956) Ausbildung und Entstehung der oberdevonischen Roteisenerze und ihre Nebengesteine im Schleizer Trog. Ber Geol Ges DDR 1: 155–197

Gräbe R, Schlegel G, Wucher K (1968) Environment and palaeogeography of the Devonian in the area of the Berga Anticline, Thuringia, Germany. In: Int Symp Devonian Syst, Calgary: 1283–1296

Grundriß der Geologie der DDR (1968) Band 1: Geologische Entwicklung des Gesamtgebietes. Akademie Verlag, Berlin, 454 pp

Greiling L (1957) Das Gotlandium des Frankenwaldes (Bayerische Entwicklung). Geol Jahrb 73: 301–356

Hamman W, Heunisch C, Schüssler U (1989) Organische Mikrofossilien (Chlorophyta, Acritarcha, Sporae dispersae, Scolecodonten) aus den Schichten des Streichengrundes, Unterdevon, im Raum Guttenberg-Kupferberg des Frankenwaldes. Beringeria 1: 57–113

Hoppe W, Seidel G (1974) Geologie von Thüringen. Haack, Gotha, 1000 pp

Hofmann J, Mathé G, Pilot J, Ullrich B, Wienholz R (1979) Fazies und zeitliche Stellung der Regionalmetamorphose im Erzgebirgskristallin. Z Geol Wiss 7: 1091–1106

Hübschmann M (1989) Das Kieselschiefer-Hornstein-Konglomerat des Elbtal-Schiefergebirges. Diplom arbeit, Bergakademie Freiberg, 40 pp

Jaeger H (1976) Das Silur und Unterdevon vom thüringischen Typus in Sardinien und seine regionalgeologische Bedeutung. Franz-Kossmat-Symp, Nova Acta Leopold NF 224: 263–299

Jaeger H (1977) Das Silur/Lochkov-Profil im Frankenberger Zwischengebirge (Sachsen). Freiberger Forschungsh C 326: 45–59

Jaeger H (1988) The Ordovician-Silurian boundary in the Saxothuringian Zone of the Variscan Orogen. Bull Br Mus Nat Hist Geol 43: 101–106

Katzung G (1961) Die Geröllführung des Lederschiefers (Ordovizium) an der SE-Flanke des Schwarzburger Sattels (Thüringen). Geologie 10: 778–802

Kittler W, Kurze M, Wienholz R (1977) Olisthostrome im Nossen-Wilsdruffer Schiefergebirge. Freiberger Forschungsh C 323: 27–35

Klenke T, Poll G (1985) Petrographie und Geochemie von Magmatiten des südlichen Frankenwaldes. Geol Bl NO-Bayern 34/35: 137–180

Kröner A, Frischbutter A, Bergner R, Hofmann J (1991) Zirkon-Evaporationsalter von granitoiden Gesteinen aus dem Erzgebirge und dem Rande der Lausitz und ihre geodynamische Bedeutung. Kurzfassungen 7. Rundgespr Geodynamik des europäischen Variscikums, Freiberg

Kupetz M (1987) Stratigraphische und regionale Stellung des variszisch gefalteten Paläozoikums im Nossen-Wilsdruffer Schiefergebirge. Freiberger Forschungsh C 415: 5–40

Kurze M (1964) Conodontenfunde in der Umrahmung des Zwischengebirgskristallins von Frankenberg-Hainichen und ihre Bedeutung für die stratigraphische Einstufung der Schichtenfolgen von Biensdorf, vom Treppenhauer und von Hausdorf ("Niederwiesaer Serie"). Monatsber Dtsch Akad Wiss 6/12: 930–936

Kurze M (1966) Die tektonisch-fazielle Entwicklung im Nordostteil des Zentralsächsischen Lineaments. Freiberger Forschungsh C 201: 11–89

Kurze M (1969) Ergebnisse einer Spezialkartierung im Altpaläozoikum südlich Frankenberg/Sachsen. Freiberger Forschungsh C 241: 45–55

Kurze M (1983) Erscheinungsformen und Probleme olisthostromatischer Sedimentation im Saxothuringicum und angrenzenden Bereichen. Z Geol Wiss 11 (1): 5–15

Kurze M, Linnemann U, Tröger K-A (1992) Weesensteiner Gruppe und Altpaläozoikums in der Elbtalzone. Geotekton Forsch 77: 101–167

Kurze M, Freyer G, Tröger K-A (1994) Das mittelsächsische Devon. In: Weddige K (ed) Das Devon Mitteleuropas. Senckenbergiana (in press)

Leggett JK, McKerrow WS, Cocks LRM, Rickards RB (1981) Periodicity in the early Paleozoic marine realm. J Geol Soc Lond 138: 167–176

Leinhos J (1956) Über einen geröllführenden Quarzit an der NW-Flanke des Schwarzburger Sattels. Ber Geol Ges DDR 2: 3–37

Leiteritz H (1957) Oberdevonkonglomerate am Nordwestrand des sächsischen Granulitgebirges. Abh Dtsch Akad Wiss Berlin 5: 1–95

Linnemann U, Kurze M, Sehnert M (1988) Nachweis von altpaläozoischen Acritarchen und einige neue Ergebnisse von lithostratigraphischen Untersuchungen im Elbtalschiefergebirge. Veröff Mus Westlausitz 12: 44–51

Lorenz W, Hoth K (1990) Lithostratigraphie im Erzgebirge – Konzeption, Entwicklung, Probleme und Perspektiven. Abh Staatl Mus Mineral Geol Dresden 37: 7–35

Ludwig V (1986) Zur Lithologie des "Kulms" bei Erbendorf/Oberpfalz (Bayern). Neues Jahrb Geol Paläontol Monatsh 7: 407–417
Ludwig V (1969) Lithologische Untersuchung des Kambriums im Frankenwald (Bayern). Geol Jahrb 87: 89–156
Lützner H, Mann M (1988) Arthropodenfährten aus der Phycoden-Folge (Ordovizium) des Schwarzburger Antiklinoriums. Z Geol Wiss 16: 493–501
Lützner H, Ellenberg J, Falk F (1986) Entwicklung der Sedimentationsrate und der Ablagerungsprozesse im Altpaläozoikum Thüringens. Z Geol Wiss 14: 83–93
Mansourian E (1979) Biostratigraphische und mikrofazielle Untersuchungen der Fränkischen Kohlenkalke unter besonderer Berücksichtigung der Foraminiferen. Geol Bavarica 79: 1–176
Meinel G (1974) Magmatismus und Metamorphose. In: Hoppe W, Seidel G (eds) Geologie von Thüringen. Maack, Gotha, pp 335–355
Paech (1975) Frühmolassen des variszischen Tektogens. Exkursionsführer 22. Jahrestag Ges Geol Wiss, Berlin, pp 33–40
Pfeiffer H (1972) Beiträge zur Tektonik der Frankenwälder Querzone zwischen Saalfeld und Lehesten. Geologie 11: 416–427
Pflug HD, Prössl KF (1989) Palynology in gneiss – results from the Continental Deep Drilling Program. Naturwissenschaften 76: 565–567
Pflug HD, Prössl KF (1991) Palynostratigraphical and paleobotanical studies in the pilot hole of the German continental deep drilling program: results and implications. Sci Drilling 2: 13–33
Pietzsch K (1962) Geologie von Sachsen. Dtsch Verlag Wiss, Berlin, 870 pp
Poll K, Rossner R (1981) Tektonische Probleme im Bereich des Poppengrüner Bahneinschnittes (Blatt Nr. 5735 Schwarzenbach/Wald und Nr. 5736 Helmbrechts). Geol Bl NO-Bayern 31 (1/4): 322–342
Reichstein M, Schwab M (1962) Nachweis hochoberdevonischer Schichten und bretonischer Bewegungen im Nossen-Wilsdruffer Schiefergebirge. Geologie 11: 17–22
Sannemann D (1955) Ordovicium und Oberdevon der bayerischen Fazies des Frankenwaldes nach Conodontenfunden. Neues Jahrb Geol Paläontol Abh 102 (1): 1–36
Schmädicke E (1991) Quartz pseudomorphs after coesite in eclogites from the Saxonian Erzgebirge. Eur J Mineral 3: 231–238
Schmidt W (1939) Die Grenzschichten Silur-Devon in Thüringen mit besonderer Berücksichtigung des Downton-Problems. Abh Preuß Geol Laudesamt NF 195: 1–99
Schmidt W, Werner C-D, Pilot J unter Mitarbeit von Alder F, Schilka W (1994) Eklogite und Granulite im Saxothuringikum. Ber Deutsch Min Ges, Eur J Mineral 6(2): 1–19
Schmidt-Effing R (1988) Eine Radiolarien-Fauna des Famenne (Ober-Devon) aus dem Frankenwald (Bayern). Geol Palaeontol 22: 33–41
Schreiber A (1967) Zur geologischen Stellung des Wildenfelser Zwischengebirges. Jahrb Geol 1 (1965): 325–359
Schreiber A (1992) Das Paläozoikum in der Zone Münchberg-Frankenberg und die Herkunft der dortigen Kristallinschollen – ein Beitrag zur Deckenfrage im Saxothuringikum. Geol Jahrb A (135): 135 pp
Schröder E (1958) Schiefergebirgstektonik und Grundgebirgstektonik in der Hirschberg-Greizer Zone (Ostthüringen). Geologie 7: 465–483
Schützenmeister W (1963) Über zwei Conodontenfunde aus der Diabas-Kalkstein-Serie bei Mühlbach-Häselich (Elbtalschiefergebirge). Geologie 12 (5): 615–620
Sdzuy K (1955) Die Fauna der Leimitz-Schiefer (Tremadoc). Abh Senckenb Naturforsch Ges 492: 1–74
Seilacher A (1967) Bathymetry of trace fossils. Marine Geol 5: 413–428
Sebastian U (1990) Zur Stratigraphie, Lithologie und Tektonik in der Umgebung des Sächsischen Granulitgebirges. Diplomarbeit, Bergakad Freiberg, 61 pp
Stein E (1988) Die strukturgeologische Entwicklung im Übergangsbereich Saxothuringikum/Moldanubikum in NE-Bayern. Geol Bavarica 92: 5–131
Steinbach W, Blumenstengel H (1974) Devon. In: Hoppe W, Seidel G (eds) Geologie von Thüringen. Haack, Gotha, pp 209–256
Steiner J, Falk F (1981) The Ordovician Lederschiefer of Thuringia. In: Hambrey MJ, Harland WB (eds) Earth's pre-Pleistocene glacial record. Cambridge, pp 579–581
Stepanek J, Geyer G (1989) Spurenfossilien aus dem Kulm (Unterkarbon) des Frankenwaldes. Beringeria 1: 1–55
Stettner G (1980) Zum geologischen Aufbau des Fichtelgebirges. Aufschluß 31: 391–403
Suess FE (1912) Vorläufige Mitteilung über die Münchberger Deckscholle. Sitzungsber Akad Wiss Wien Math-Naturwiss Kl 121 IIa (10): 1–253
Tröger K-A (1990) Nachweis von unterdevonischen Tentakulitenschiefern im Nossen-Wildruffer Schiefergebirge – tektonische und regionalgeologische Konsequenzen. Z Geol Wiss 18 (7): 587–594
Teufel S (1988) Vergleichende U-Pb- und Rb-Sr-Altersbestimmungen an Gesteinen des Übergangsbereiches Saxothuringikum/Moldanubikum, NE-Bayern. Gött Arb Geol Paläontol 35: 87
Volk M (1966) *Dictyonema* cf. *uralensis* OBUT aus den Phycodenschiefern bei Steinach. Geol Bl NO-Bayern 16: 162–163
Von Gaertner HR (1942a) Die Schichtgliederung der Phyllitgebiete in Thüringen und Nordbayern und ihre Einordnung in das stratigraphische Schema. Jb Reichsanst Bodenforsch 62: 54–80
Von Gaertner HR (1942b) Die geologische Stellung der Eisenerzlager in Thüringen und Oberfranken. Jahrb Reichsanst Bodenforsch 62: 81–108
Von Gaertner HR (1951) Probleme des Saxothuringikums. Geol Jahrb 65: 409–450
Wahnig M (1986) Geochemische Untersuchungen an Keratophyren des Frankenwaldes. Geol Bl NO-Bayern 36 (3/4): 203–230

Walther K (1972) Die mineralfazielle und tektonische Entwicklung der Annaberger und Marienberger Gneise. Freiberger Forschungsh C 269: 1–101

Wenzel Th, Schmidt W, Hengst M, Wolf D, Pilot J (1989) Geochemical study of some types of granitoids from the Massif of Meissen. Oral presentation, 5th Working Meet Isotopes in nature, Leipzig, 25.–29.9.1989

Wenzel Th, Hengst M, Pilot J (1991) Die plutonischen Gesteine des Meißener Massivs (Elbtal-Zone) – Zeitliche Rekonstruktion der magmatischen Entwicklung (Erstarrungs- und Reliktalter) mit Hilfe von Einzel-Zirkon und K-Ar-Altersbestimmungen. Ber Dtsch Mineral Ges Beih Eur J Mineral 3 (1): 292

Werner C-D (1981) Outline of the evolution of the magmatism in the GDR. In: Ophiolites and initialites of the northern border of the Bohemian Massif, Acad Sci GDR and Pol Acad Sci, Potsdam-Freiberg, Guide-book of excursions I: 17–68

Wiefel H (1974) Ordovizium. In: Hoppe W, Seidel G (eds) Geologie von Thüringen. Haack, Gotha, pp 165–194

Wiefel H (1976) Die geologische Entwicklung der Lahn-Dill-Erzlagerstätte Görkwitz bei Schleiz. Jahrb Geol 5/6: 451–588

Wignall PB (1991) Model for transgressive blackshales? Geology 19: 167–170

Wirth R (1978) Geochemie und Petrographie der paläozoischen Magmatite des Frankenwaldes. Diabase – Keratophyre – Pikrite. Dissertation, Univ Würzburg, 130 pp

Wozniak S (1969) Stratigraphie und Tektonik der Lößnitz-Zwönitzer Zwischenmulde. Dissertation, Bergakad Freiberg, 146 pp

Wucher K (1967) Ergebnisse der Kartierungsbohrungen Heinersdorf 1/60 und 2/62 (Thüringisches Schiefergebirge). Jahrb Geol 1: 297–323

Wurm A (1925) Geologie von Bayern. 1. Nordbayern, Fichtelgebirge, Frankenwald. In: Krenkel (ed) Hdb Bodenschätze Deutschlands (ed. Krenkel), II. Abt. Borntraeger, Berlin, 2: 373

Wurm A (1961) Geologie von Bayern, 2. Aufl. Borntraeger, Berlin, 555 pp

Zitzmann A (1968) Das Paläozoikum im Grenzbereich zwischen Bayerischer und Thüringischer Faziesreihe des Frankenwaldes. Geol Jahrb 86: 579–654

V.B.2 Structure

W. Franke, H. Rauche, and E. Stein

1 Introduction

The map pattern of the Saxothuringian Belt is largely controlled by a number of large, **NE-trending syn- and antiforms**, which were formed at a late stage of tectonic deformation and therefore affect both the stratigraphic sequence and the metamorphic isograds. These fold structures are transsected by **NW-trending fault zones**, and by **WNW- to NNW-trending magmatic bodies** reflecting SW-NE-directed extension.

In the following chapter we briefly discuss the tectonic boundaries of the Saxothuringian belt, the deformational history and array of the main structures along SE/NW-trending profiles, and – lastly – the transverse elements.

2 Structural Boundaries

The Saxothuringian belt is bounded to the NW and to the SE by two major polyphase fault zones, which were formed in the Permo-Carboniferous and reactivated as reverse faults in late- to post-Cretaceous times. To the SW, the Bohemian Massif is bounded by the fault system of the **Franconian Line**. As it is known from few surface outcrops (e.g., Schröder 1987; Franzke and Rauche 1991) and from seismic data (Schmoll et al. 1989), the fault zone dips toward the NE at 30 to 70°. The post-Variscan tectonic inventory of the crystalline basement close to the Franconian Line has been extensively described by Zulauf (1992; Zulauf et al. 1990). Late Cretaceous reverse faulting is also known from the **Elbe Zone**, which truncates the Saxothuringian belt to the NE. The Lusatian block adjacent to the NE already forms part of the Western Sudetes (see Chap. VI). The Cretaceous basin of the Elbe valley has first been fed and later overthrust over approx. 1 km by the Lusatian granodiorites (Müller and Wächter 1970).

Reverse faulting along the Franconian Line and the Elbe Zone represents foreland deformation of the Alpine collision (Ziegler 1987; Stackebrandt and Franzke 1988). The origin of NW-trending fault zones dates back to the Permo-Carboniferous: several Permo-Carboniferous depocenters set between the Franconian Line and the Eichenberg-Saalfeld fault further W (Schröder 1988) probably represent pull-apart basins (Andreas 1987; Lützner 1987). Similar relations are probable for the Permo-Carboniferous molasse in the Elbe Valley (Döhlen basin).

The **NW margin of the Saxothuringian outcrop** is formed by the onlap of Carboniferous and Permian sedimentary and volcanic rocks. Late Variscan extension is partly documented in NW-dipping normal faults with several 100 meters of downthrow, e.g., at the northwestern margin of the Schwarzburg anticline. The Permo-Carboniferous sequences conceal the boundary between the Saxothuringian basin and the Mid-German Crystalline High.

At its **SE margin**, the ST abuts against the high-grade rocks of the Moldanubian region (Moldanubian s.str. and Tepla-Barrandian units, see Chaps. V.C.1 and VII). This boundary is only exposed in its southwestern part, near the margin of the Bohemian Massif. The structural relationships in this area are difficult to interpret. The SW segment, contained between the NW-trending Berneck-Fichtelnaab fault and the Franconian Line, exposes a relatively high structural level. Here, it is possible to observe what probably represents the primary tectonic boundary of the ST belt: early Paleozoic rocks of the Fichtelgebirge are structurally overlain by a greenschist-grade unit (Erbendorf greenschist), which, in turn, is overlain by gneisses, amphibolites and eclogites of the Zone of Erbendorf-Vohenstrauß (ZEV). Both the greenschists and the ZEV rocks have much in common with the crystalline nappes of the Münchberg klippe, which is therefore regarded as a tectonic outlier of the ZEV and

Dallmeyer et al. (Eds.)
Pre-Permian Geology of Central and Eastern Europe

related rocks in the Moldanubian region (Chap. V.C.1). In this model, the SE margin of the Saxothuringian belt represents a zone of dextral transpression with a NW-ward driving thrust component developed at the northwestern, active margin of the Moldanubian region.

In the eastern segment of the exposed boundary zone, early Paleozoic metasediments of the Fichtelgebirge are juxtaposed against metasediments of the Moldanubian s.str., and both these units have undergone the same low-pressure/high-temperature metamorphism in early Carboniferous (Visean) time (see Chap. V.B.4), and dextral wrench-faulting (Zulauf 1992). Since Paleozoic protoliths have been identified also in the Moldanubian s.str., it is not possible at present to exactly delineate a boundary here, and it has even been argued that some NW part of the Moldanubian s.str. and the ST sequences of the Fichtelgebirge belong to one and the same unit (e.g., Stein 1988; Matte et al. 1990).

Further E, the boundary is contained in the Marianske Lazne mafic complex (northwestern most part of ZTD, Fig. 1). Relatively low grade rocks in its NW part are structurally overlain by serpentinites, eclogites and amphibolites (Kachlík 1992) which can be interpreted as equivalents of similar rocks in the ZEV and the Münchberg klippe (Chap. V.C.1).

Still further E, the SE boundary of the Saxothuringian belt is largely masked by late Variscan granite and by the fill of the Tertiary Eger Graben, which apparently has reactivated the Variscan Saxothuringian/Tepla-Barrandian plate boundary. At the SE margin of the Erzgebirge, Tertiary downfaulting amounts to ca. 1500 m. Boreholes through the Cenozoic cover of the boundary region have encountered eclogites (Fiala and Padera 1984; T 7 on Fig. 1). U-Pb data indicate two high-pressure events at 342 ± 4 and 424 ± 4 Ma (Gebauer 1991). The latter recalls the situation in the Marianske Lazne and ZEV areas.

It appears that the dose juxtaposition of Moldanubian (s.str.) and Saxothuringian rocks, as observed in the segment between the ZEV and the ZTD (Fig. 1), is not representative of the boundary zone.

3 Southwestern Transsect

3.1 Description

The Fichtelgebirge Antiform and the Hatzenreuth Synform at its southeastern margin are composed of Cambrian through to Silurian sequences which belong to the Thuringian facies realm (von Gaertner 1951, see also Chap. V.B.1). As proven by Stein (1988, 1992), there is no indication of pre-Variscan basement rocks in the core of the antiform: the deepest structural levels exhibit the same deformational history as the overlying Paleozoic sequences (see also Nöhring 1949). Metamorphism is of the low-pressure type and attains, in some central parts of the antiform, staurolite-andalusite grade; at the SW flank of the Hatzenreuth synform (i.e., near the Saxothuringian/Moldanubian boundary), Saxothuringian rocks were metamorphosed at 580–600 °C/ca. 3 kb (sillimanite-biotitemuscovite, see Chap. V.B.4).

As described by Stein (1988), the Fichtelgebirge has undergone a polyphase deformation. A first deformation (D1) created a prominent, synmetamorphic foliation and few NE-trending isoclinal folds. D2 brought about another penetrative foliation. In incompetent rocks, s1 and s2 are indistinguishable from each other and form a composite foliation. D2 folds are open to south-facing, with a tendency towards formation of sheath folds in domains with high simple shear. During D2, the peak metamorphic conditions still prevailed. D3 is characterized by NE-trending, upright, mostly open folds and compressional kinkbands formed at a late stage. Pelitic lithologies show subvertical crenulation cleavage. D3 postdates the low-pressure metamorphism and refolds the isograds. D3 has also created the Fichtelgebirge antiform and Hatzenreuth-Waldsassen synform. A late stage of the Variscan compressive deformation (D4) is documented in NW-trending, compressional kinkbands, which are either conjugate or else record relative uplift of the northeastern segment (Stein 1988, 1992; Stein and Kleemann 1990). This latter effect is responsible for the exposure of increasingly deeper structural levels towards the NE. D4 also brought about NE-trending extensional kinkbands. Where the attitude of the foliation is upright, extension is also recorded in recumbent folds with a subhorizontal crenulation cleavage. NW/SE-directed extension is also reflected in steeply NW-dipping normal faults, which reduce the tectonometamorphic profiles at the NW flank of the Fichtelgebirge (including the SE margin of the Münchberg klippe), and at the NW flank of the Erzgebirge.

A largely similar sequence of events has been described by Franke (1984a,b) from the less

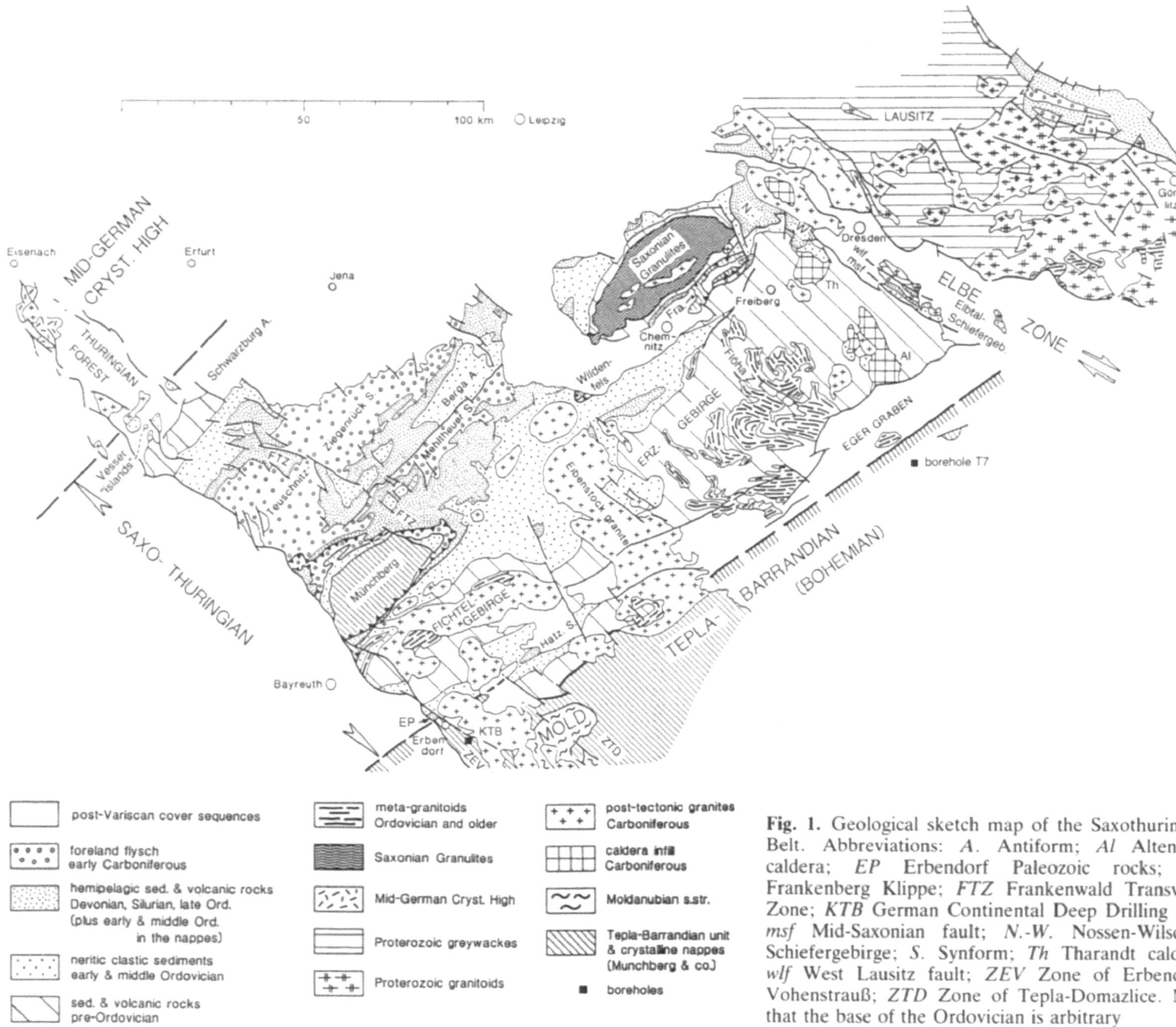

Fig. 1. Geological sketch map of the Saxothuringian Belt. Abbreviations: *A.* Antiform; *Al* Altenberg caldera; *EP* Erbendorf Paleozoic rocks; *Fra* Frankenberg Klippe; *FTZ* Frankenwald Transverse Zone; *KTB* German Continental Deep Drilling Site; *msf* Mid-Saxonian fault; *N.-W.* Nossen-Wilsdruff Schiefergebirge; *S.* Synform; *Th* Tharandt caldera; *wlf* West Lausitz fault; *ZEV* Zone of Erbendorf/ Vohenstrauß; *ZTD* Zone of Tepla-Domazlice. Note that the base of the Ordovician is arbitrary

metamorphosed, NW part of the Saxothuringian belt (Frankenwald = Franconian forest), which extends from the NW margin of the Fichtelgebirge into the Teuschnitz Synform. In the very-low-grade rocks of the Thuringian facies parautochthon, as well as in the very-low-grade nappes of the Münchberg Klippe, the oldest phase of deformation has produced open to tight, W- to NW-facing folds with an axial-plane fracture cleavage or slaty cleavage at the dm- to m-scale. These elements have been refolded by open, upright, or SE-facing D2 folds. D1 dies out toward the NW, so that, in the Berga Antiform and in the Teuschnitz Synform, the SE-facing "D2" deformation is actually the first (and the main) one. An equivalent of D3 has not been identified. It appears that, in the Frankenwald, the map-scale fold structures relate to D2. Late NW-trending compressional kink zones and local occurrences of gravity-induced folding of sub-vertical anisotropies (extension in NW/SE) recall D4 or D5 of the Fichtelgebirge.

The northwesternmost part of the SW transsect is occupied by the Schwarzburg Anticline, which brings to the surface pre-Devonian rocks, with greenschist grade in the core. Detailed descriptions of deformational fabrics are available in Bankwitz (1968) and Bankwitz and Bankwitz (1989). The oldest element is a slaty cleavage relating to the metamorphism up to greenschist grade. There are no indications of tectonic vergence. s1 and bedding have been deformed by upright to S-facing folds and an s2 crenulation cleavage. This D2 deformation has also produced the main antiformal structure and probably correlates with the deformation of the Berga A. and the Teuschnitz S. A steeply NW-dipping reverse fault on the southern flank of the antiform is probably cogenetic with D2. The D1 deformation of the Schwarzburg A. cannot be correlated with that of the areas further SE, since the D1 deformation in the Paleozoic rocks surrounding the Münchberg klippe is seen to die out toward the NW. The D1 structures encountered in the Schwarzburg area must therefore be treated as a separate development, which possibly has to be seen in context with that of the Mid-German Crystalline High rather than with that of the areas further SE. Isolated areas of greenschist-grade Early Paleozoic rocks ("Vesser Islands") to the NW of the Schwarzburg A., SE of the Suhl-Ilmtal granite, probably correspond to central parts of the Schwarzburg A. They show essentially the same deformational history (Bankwitz and Bankwitz 1990), but an additional thermal overprint by the granites. The Ilmtal-Suhl and Schleuse granites are polyphase granitic to dioritic intrusions (Meinel 1974) and were emplaced in the boundary region between the Saxothuringian basin and the Mid-German Crystalline High (see Behr 1966; Trusheim 1964, for borehole data). They postdate the ductile Variscan deformation, but the exact age of their emplacement is unknown.

The SE-facing deformation in Thuringia can be studied in more detail in the Berga Anticline. The main structures are upright to slightly SE-vergent folds and a correlated cleavage, which often shows pronounced fanning. A later increment of deformation has produced crenulation of the main cleavage (Schroeder et al. 1965; Bankwitz 1968).

In all parts of the Berga A., there is a younger fracture cleavage dipping gently towards the N, and correlated thrust surfaces with S- to SW-directed transport, with displacements up to more than 1000 m (Motz 1991; Rauche et al. 1991). In the northeasternmost part of the Berga A., there are some younger, NW- to W-trending open folds of large wavelength, which rotate the earlier fabrics. All these elements probably correlate with the NW-trending compressional features in the Frankenwald to the W.

NW-trending shear zones in the Berga A. show rotation of earlier fabrics in the sense of dextral shear (Benek 1958). Together with sinistral, NNE- to NE-trending faults, they form a conjugate system which indicates NNW/SSE-directed compression. The SW-directed thrusts are possibly correlated with this stress field.

3.2 Interpretation

According to Franke (1984a,b) and Stein (1988), the NW-facing D1 in the Fichtelgebirge and Frankenwald relates to the emplacement of the Münchberg nappe pile, whereas D2 represents either backthrusting (Stein 1988), or else a general reversal of tectonic polarity (Franke 1989a).

Apart from the vergence of folds, which shows considerable directional variation, the direction of tectonic transport during the emplacement of the nappes (D1) is also documented in large-scale relationships. The sole markings in the Early Carboniferous flysch basin are mostly oriented

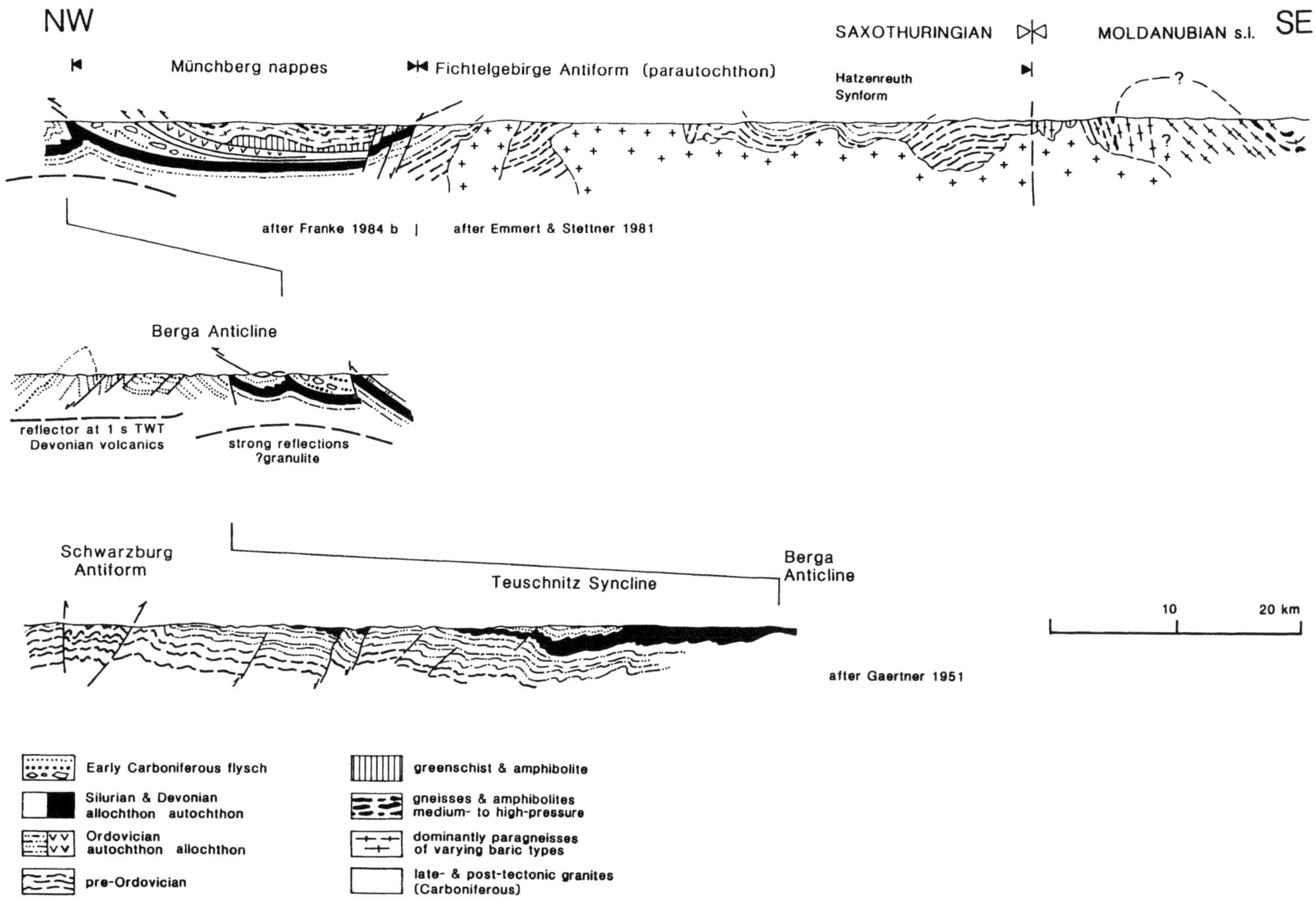

Fig. 2. Tectonic profile across the SW part of the Saxothuringian belt. (Largely after Franke 1984b)

SW-NE (Franke 1984b; Fig. 2), which constrains the orientation of the basin axis and, hence, of the tectonic front. Besides, the tectonic profiles in the klippen of Münchberg, Wildenfels and Frankenberg are so similar (see Chap. V.C.1), that their present-day alignment (SW-NE) probably correponds to the structural trend of the orogenic wedge in Early Carboniferous time. Pre-Carboniferous increments of deformation in the Saxothuringian/Moldanubian boundary region are controlled by dextral transpression (see Chap. V.C.1).

The higher grade of metamorphism in the Fichtelgebirge (max. pressures of 3 kb in early Paleozoic sediments, see Chap. V.B.4) must at least partially relate to tectonic overburden. Since Paleozoic sediments of the parautochthonous Thuringian facies underlying the Münchberg klippe and also at the southeastern margin of the Fichtelgebirge (Erbendorf Paleozoics) exhibit only very low metamorphic grade, this overburden can only to a part be explained with the Münchberg nappes, which – in the type area – are only ca. 5 km thick (Chap. V.C.1). It is therefore necessary to invoke tectonic thickening within the Thuringian facies parautochthon. In fact, metamorphic contrasts between the early Paleozoic of the central Fichtelgebirge and time-equivalent rocks in the Erbendorf area (at the SE margin of the Fichtelgebirge) suggest that the latter unit represents a thrust sheet which originally occupied a structurally higher position and is, therefore, less metamorphosed (Franke 1989a; Franke et al. 1992; Stein 1988).

The NW-ward fading of the D1 deformation in the flysch of the Teuschnitz Synform is in accord with the relatively external tectonic position of this area in the foreland of the tectonic front, which has imported the Münchberg and related klippen.

The renewed increase in the intensity of deformation toward the NW (Schwarzburg Antiform and beyond) is due to the proximity of the Mid-German Crystalline High acting as a backstop. The Crystalline High, in Early Carboniferous time, formed the NW-driving active margin of the Rhenohercynian belt (see Chap. III.B.1), and, at the same time, the northwestern limit to the Saxothuringian belt. The southeastern boundary of the Crystalline High is not exposed, but steeply NW-dipping reverse faults in the northwestern part of the Saxothuringian belt suggest that the Crystalline High overthrusts the Saxothuringian. Hence, the Mid-German Crystalline High appears as a crystalline "pop-up" with bivergent upthrusting, formed during a late stage of collision.

4 Northeastern Transsect

The northeastern transsect (Fig. 3) traverses the Erzgebirge antiform, the synformal structure containing the Frankenberg klippe (Central Saxonian Lineament), and the antiform of the Saxonian Granulites. Earlier accounts of the tectonometamorphic development are available in Bankwitz and Bankwitz (1982) and Frischbut-

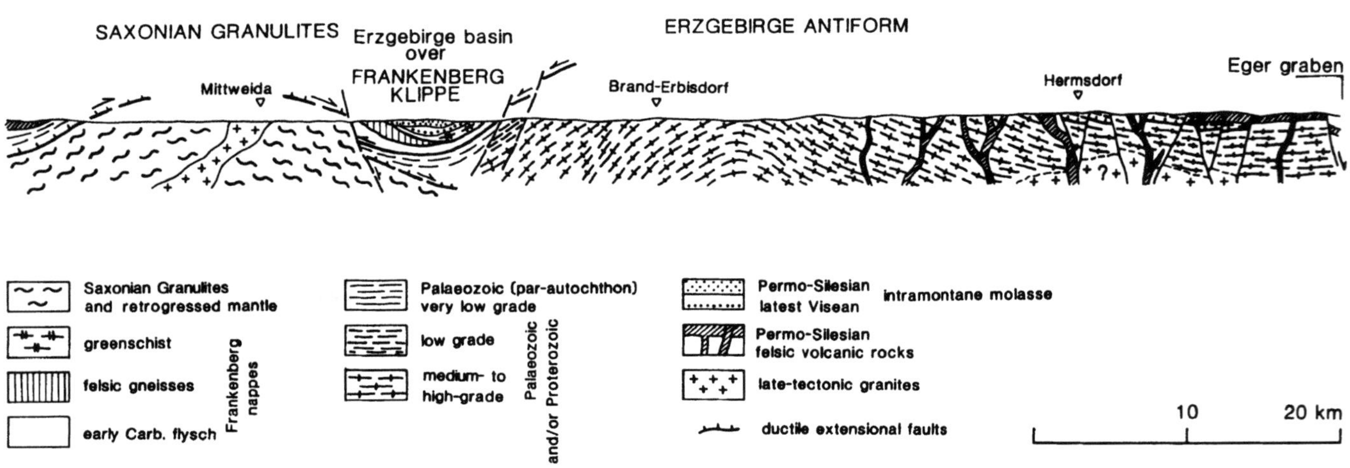

Fig. 3. Tectonic profile across the NE part of the Saxothuringian belt

ter (1990) for the Erzgebirge region, and in Kurze (1966) and Neumann (1988) for the Saxonian Lineament and the Granulites. In spite of these earlier publications and detailed mapping, the eastern part of the Saxothuringian belt is much less well understood than the western part. This was made clear by recent investigations, which have revealed an unexpectedly high proportion of high temperature and/or high pressure rocks. In addition, it is now clear that the sequence contains evidence of inverted isogrades (i.e., stacking), as well as of extreme reduction of the metamorphic profile (i.e., extension). Therefore, the classical "stratigraphic" subdivision of the Erzgebirge (e.g., Lorenz and Hoth 1990) can no longer be maintained. However, P-T paths, tectonic history, and isotopic ages are, at present, so poorly constrained, that it is only possible to give a preliminary account of the data and point out the problems of interpretation.

4.1 Description

The **Erzgebirge** has a core region of relatively high grade metamorphic rocks developed from a large variety of sedimentary and magmatic protoliths, the latter often summarized as Rotgneise. Rotgneis-metagranitoids in the eastern Erzgebirge have yielded 550 ± 14 to 555 ± 14 Ma, interpreted as magmatic crystallization ages (Pb/Pb evaporation technique, Kröner et al. 1991). Though these metagranitoids are bounded by ductile shear zones, it appears that at least parts of the core region must represent pre-Variscan basement. The high-grade core is structurally overlain, at the NW and SW flanks, by micaschists and phyllites. In the lower-grade rocks of the mantle, it is possible to identify Thuringian facies Paleozoic protoliths. As in the Fichtelgebirge, the updoming of the antiform postdates the main metamorphism. These rocks have been intruded by postkinematic (Carboniferous) granites.

Lenses of eclogites record formation at 850 °C and pressures in excess of 29 kb for the central Erzgebirge, and a westward decrease to 650 °C/20–24 kb (Schmädicke et al. 1992). The surrounding metasediments have also yielded indications of high (though not ultra-high) pressures: micaschists in the western Erzgebirge record 550 °C/10 kb, and garnet phyllites adjacent to the W experienced 500 °C/8 kbar (Roetzler et al. 1992). The westernmost report of pressure-dominated metamorphism comes from a locality immediately W of the Eibenstock granite, where early Paleozoic metasediments record 350–400 °C and pressures in excess of 6–7 kb, but probably below 9 kb (Soucek and Holub 1992).

The age of the metamorphism is unknown. Apart from the indications of pre-Variscan basement, there is a Rb-Sr whole rock isochron from metapelites of 485 Ma, which is interpreted, by Krentz (1985), as dating the peak metamorphic temperature. However, this method is not reliable, and the involvement of Paleozoic rocks into the deformation and metamorphism indicates at least some Variscan (Carboniferous) increment. Unpublished U-Pb and Ar/Ar data reveal equilibration of the isotopic systems at about 340 Ma (pers. comm. by A. von Quadt, A. Kröner, and E. Schmädicke).

The tectonic inventory of the Erzgebirge has been described by Bankwitz and Bankwitz (1982). Within the NE-trending major antiform, there is a marked EW-directed structural grain defined by fold axes and stretching lineations. The major fold structures mapped out in the Erzgebirge are not cylindroidal, but rather dome-shaped. This is due to the interference of up to four generations of folds. These refold a penetrative, metamorphic foliation which has also affected the eclogites. A younger foliation cuts the earlier planar fabrics at angles between 10 and 35°. Kinematic markers reveal that this foliation is extensional (Krohe 1991; Kroner and Voigt 1992; Willner et al. 1992). Extension is also apparent from the reduction of the metamorphic profile at the NW flank of the Erzgebirge (Lorenz 1989). This NW/SE-directed extension is also responsible for the formation of the sedimentary basin set between the Saxonian Granulites and the Erzgebirge (see below). In the Altenberg caldera (see Sect. 5.2), there are some isolated occurrences of greenschist grade rocks (Hermsdorf-Rehefeld phyllites with some marbles), which are in marked contrast with the surrounding high grade rocks. Duplex structures indicate transport towards the S. Pietzsch (1914) and Rauche (1991) have proposed that the phyllites were derived from the Elbe Zone and emplaced by thrusting. However, the reduction of the metamorphic sequence clearly requires crustal extension.

The space/time relationships between the compressive and extensional tectonic features and the increments of the metamorphic evolution remain to be analyzed in detail.

The less metamorphosed northern flank of the Erzgebirge Antiform is structurally overlain by the **Frankenberg complex** – a sister massif to the Münchberg, composed of a variety of metamorphic rocks and an underlying very low grade Paleozoic sequence with a special, "Bavarian-type" facies (see Chap. V.B.1). The Frankenberg complex escaped from erosion in the core of the Vogtland synform. Still in latest Early Carboniferous time, this synform was transformed into an intra-montane basin (Borna-Hainichen), in which approx. 1000 m of coarse-grained clastic sediments with some coal seams were deposited. This early molasse was still affected by some compressional movements, and is unconformably overlain by unfolded clastic sediments and volcanics of late Westphalian (B/C, D) and early Permian (Rotliegend) age. Normal faults reducing the metamorphic profile at the northwestern flank of the Erzgebirge and at the southeastern flank of the Saxonian Granulites (see below) probably relate to the tectonic subsidence of the Erzgebirge intramontane molasse basin.

The northwesternmost part of the transsect is occupied by the **Saxonian Granulites**, one of the most disputed elements of the Saxothuringian belt. Unlike the tectonic klippen of Münchberg, Wildenfels, and Frankenberg, the Saxonian Granulites do not occur in a synformal position, but represent an open, brachyanticlinal structure. A further peculiarity is the extreme reduction of the metamorphic profile, which juxtaposes, within less than 1 km of tectonic thickness, high-grade metamorphic rocks and fossiliferous Paleozoic sediments. The granulite outcrop is essentially composed of leucocratic granulites together with some eclogites and ultramafic rocks, including garnet-pyroxenites, and metagabbros. These high-grade rocks are overlain and (on the map) surrounded by a mantle of amphibolite to very low-grade Paleozoic sequences, which belong to the par-autochthonous ("Thuringian") facies realm (cf. Chap. V.B.1). The contact between core and mantle is a prominent ductile shear zone, in which the granulites are retrogressed (Kroner 1991).

The granulite facies overprints the high-pressure stage. For the felsic granulites, Rötzler (1992) has reconstructed peak conditions of 780–850 °C/11.5 ± 1 kb, followed by near-isothermal decompression. While the high-grade rocks in the core of the antiform were retrogressed, the schist mantle underwent prograde high-temperature metamorphism up to 660 °C/3.4 ± 0.5 kb.

According to Sm/Nd and U-Pb data of von Quadt and Gebauer (1988, 1990) and von Quadt (1993), eclogite metamorphism occurred in the Devonian, followed by granulite metamorphism in late Devonian to early Carboniferous time (U-Pb zircon ages ranging between 338 ± 2 and 352 ± 2 Ma).

The core and locally also the mantle of the granulite antiform have been intruded by late tectonic granitoids (G. of Mittweida and Berbersdorf). The Berbersdorf granite has been dated by zircons (338 ± 5 Ma; von Quadt 1993), and the Mittweida granite probably intruded at about 350 Ma (Pb-Pb feldspar, Bielicki and Hiller 1987; Rb/Sr-whole rock, Haase et al. 1987).

There is, as yet, no modern and comprehensive description of the structural inventory of the metamorphic rocks. The dominant fabric is a penetrative, closely spaced foliation, formed under granulite and amphibolite facies conditions. As already stated by Weber and Behr (1983), these synmetamorphic elements were formed in an extensional regime. Quartz fabrics record deformation in the flattening field (core granulites), giving way to plane strain and noncoaxial deformation at the margins (Lister and Dornsiepen 1982, re-interpreting quartz fabric data of Behr 1961, 1980). Mylonites in the roof of the granulites show top to the SE ductile displacement (Kroner 1991; Kroner and Voigt 1992). The emplacement of the granites post-dates the main foliation. According to Vulpius and Kroner (1992), Rötzler (1992) and Rötzler (pers. comm. 1992), at least part of these granites were intruded in an extensional regime. The latest event affecting the granulites is S-vergent folding and minor thrusting, observed in the schist mantle and outer granulites (Kroner et al. 1991).

4.2 Interpretation

The tectonometamorphic development of the **Erzgebirge** is difficult to interpret, especially because of its high-grade rocks. Since, in the W part of the Saxothuringian belt, high-pressure rocks are restricted to the Münchberg klippe, earlier authors like Scheumann (1935) and Behr (1983) have suggested that the high-pressure metamorphic rocks represent nappes. However, this interpretation is problematic, since the relevant rocks, in the Erzgebirge, are contained in the core of an antiform, i.e., represent a deep structural level, whereas the Münchberg klippe is preserved in a synformal position, i.e., occupies the highest structural level of all the Saxothuringian belt.

Still, the widespread occurrence of eclogites and other high-grade rocks implies tectonic stacking and subsequent exhumation. Since extensional features are well-developed and widespread, Kroner and Voigt (1992) and Willner et al. (1992) have suggested that the Erzgebirge dome represents a metamorphic core complex, in which the high-grade core was exhumed by multistage extension in an upper plate, subsequent to plate collision. These processes are attributed to the Variscan (Carboniferous) orogeny. However, early Carboniferous tectonic stacking and high-pressure metamorphism should have occasioned areally extensive isostatic uplift and exhumation, which is contradicted by the preservation of Cambrian to early Carboniferous rocks on the flanks of the antiform. Besides, the Fichtelgebirge/Erzgebirge antiform does not belong to an upper plate, but represents, instead, the foreland and lower plate to the Münchberg klippe and related allochthons. As an alternative to the Carboniferous stacking and collapse model, at least the highest grade rocks in the core of the antiform could be interpreted as part of a pre-Cambrian crystalline basement, which was later overprinted by the Variscan orogeny. Some Carboniferous extension is obvious, but those extensional features which are demonstrably younger than the Variscan metamorphism (such as tectonic subsidence of the intramontane basins) are probably not important enough to exhume eclogites.

Distinction between the Variscan and the multi-orogenic exhumation models requires more detailed assessment of the P-T-d paths and refined isotopic age dating. Anyhow, the clear indications of tectonic stacking and extension demonstrate that the lithological column described, e.g., by Lorenz and Hoth (1990) is a tectonic, and not a stratigraphic, sequence.

The **Frankenberg complex** to the NW of the Erzgebirge antiform is less problematic. Because of the overwhelming number of features common to the Frankenberg, Wildenfels, and Münchberg complexes, most authors have agreed that all these units should have the same tectonometamorphic history and mode of tectonic emplacement, either as tectonic klippen derived from a root zone to the SE of the Fichtelgebirge/Erzgebirge Antiform, or else as diapir-like protrusions of crystalline basement (see Suess 1912; Kossmat 1927; and the discussions in Schwan 1974; Behr et al. 1982; Franke 1984b). Since, in the case of the Münchberg complex, there is strong evidence for the nappe concept (including high-quality reflection seismic data), the Frankenberg rocks should likewise represent a klippe derived from the SE (see also Chap. V.C.1). It should be pointed out, however, that this analogy leaves at least one open question: widespread SE-vergent folds in the Paleozoic rocks underlying the Frankenberg crystallines appear to contradict NW-ward emplacement (see Schwan 1974). It remains to be tested if these folds represent a second stage of folding which postdates nappe emplacement, as has been demonstrated for the Münchberg area.

The **Saxonian Granulites** to the NW of the Frankenberg unit are characterized by an antiformal structure and the extreme reduction of the metamorphic profile. As pointed out by Franke (1989a,b, 1991, 1993), these features recall the metamorphic core complexes of the western US (Basin and Range Province). An extensional setting is also suggested by certain elements of the Devonian paleogeography. As discussed in Chapter V.B.1, the Saxonian Granulite is in line with an intrabasinal, conglomerate-producing uplift, set within a NE-trending belt with basaltic intraplate volcanic rocks. These latter elements were active in Frasnian time and probably record uplift and extension of the granulite axis.

The presence of eclogite-grade rocks implies a preceding stage of important crustal thickening prior to the formation and exhumation of the granulites. The age of these high-grade metamorphic increments is doubtful. U/Pb and Sm/Nd age data (von Quadt and Gebauer 1988, 1990; von Quadt 1993) suggest eclogite facies metamorphism in the Devonian, and granulite facies metamorphism in Late Devonian to Early Carboniferous time. However, a Devonian eclogite facies metamorphism would imply important crustal thickening and uplift, which is incompatible with the presence of a complete Cambrian to Carboniferous sequence in the mantle of the granulite antiform., Hence, it appears possible that the eclogites, and possibly also the granulites, relate to a pre-Variscan cycle, and only underwent re-heating, extension and uplift in the Devonian/early Carboniferous, whereby the U/Pb and Sm/Nd systems were reset (Franke 1993).

5 Transverse Elements

The inventory of ductile deformation described in the preceding paragraphs is transsected by a variety of N- to WNW-trending tectonic elements, most of which were formed in a brittle regime. These elements include compressional as well as extensional features, partly combined with dextral strike-slip.

5.1 Transverse Extensional Structures

SW/NE- to NNE-directed fractures are clearly documented in the **orientation of major granite bodies**, which cut across the older structural grain. This is true for the Leuchtenberg-Flossenbürg-Falkenberg intrusive complex which stitches the Saxothuringian/Moldanubian boundary near the SW margin of the Bohemian Massif, for the Bor granite in the W part of the Zone of Tepl-Taus (ZTT), and the Eibenstock-Karlovy Vary granite. All these granites intruded into rocks, which – during the time of intrusion – were situated at upper crustal levels (see Thomas and Tischendorf 1987, for a differentiated account).

A similar element is the **Frankenwald Transverse Zone**, a WNW-trending horst zone (Schwan 1956), which can be traced from the SE flank of the Schwarzburg A. toward the NE of the Münchberg Klippe, where it brings to the surface the lower nappes and the parautochthon. It is characterized by a relatively strong metamorphism (up to greenschist grade), already during the D1 and D2 deformations. This is indicated by the development of slaty cleavage (the neighboring areas show only fracture cleavage), by enhanced illite crystallinity (Franke 1984b) and metamorphic index minerals (Brand 1980). Higher temperatures have brought about a complex tectonic development which cannot be discussed here in detail (Schroeder 1966; Franke 1984b). Besides, the transverse zone is characterized by Permian remagnetization in the Paleozoic spilites (Kim and Soffel 1982), Permian resetting of K/Ar ages in shales (Ahrendt et al., in prep.), and the intrusion of posttectonic granites (Thüringische Granitlinie, Meinel in Hoppe and Seidel 1974), which is also reflected in a marked negative gravity anomaly (Plaumann 1982). Since, in the narrow transverse zone, there is no indication of a higher overburden, the metamorphic effects must be due to enhanced heat-flow, probably brought about by fluid advection in a deep-reaching fracture zone during late Visean through Permian time.

In the eastern Erzgebirge, around **Altenberg**, there is a broad, NNW-trending group of rocks which likewise cut across the older structures of the Erzgebirge. The most conspicuous feature is the occurrence of Late Carboniferous to Permian dikes and effusive sheets of rhyolites and some minor granite bodies. The dikes are not represented in the map (Fig. 1); they trend partly NNW, and partly WSW. The dikes are related to the formation of a caldera-like volcano-tectonic depression, which was filled with Permo-Carboniferous molasse sediments. Subsidence of the caldera has also preserved the "phyllites" of Hermsdorf-Rehefeld, which represent a higher structural level.

A similar caldera further to the NNW, near **Tharandt**, is probably situated in the same fracture zone as the Altenberg structure.

Some SW-NE-directed extension is also documented by narrow Permo-Carboniferous intramontane basins of the **Flöha Zone** and in the Elbe Zone (see below).

5.2 Transverse Compressional and Strike-Slip Elements

Apart from the late Cretaceous reverse faults, the western part of the Saxothuringian belt is largely devoid of major NW-trending faults. However, the NW-trending kinkbands (D4) of the Fichtelgebirge record some late Variscan NW/SE-directed compression.

The map pattern of the Erzgebirge reveals two NW-trending zones, which bring about the uplift of deeper structural levels towards the NE. One of these zones occurs E of the Eibenstock granite, where the axes of late, large-scale folds plunge to the SW and form a lateral ramp. Immediately to the E of this ramp, the first high-pressure rocks appear at the surface.

This effect is also expressed in the **Flöha zone** of the Erzgebirge – a narrow, NW-trending synform, which appears to overprint all earlier structures. It is bounded to the E by a NW-trending reverse fault, which brings to the surface the deeper structural levels of the eastern Erzgebirge. In a recent reflection seismic profile (DEKORP-MVE, unpubl.), it appears as a prominent reflection in the upper 2 s TWT dipping ca. 40° to the NE.

A similar set of reflections is apparent in the **Elbe Zone**, which is the boundary between the Saxothuringian of the Erzgebirge to the SW and the Sudetes to the NE (see the summary in Pietzsch 1956). Large parts of the Elbe Zone are covered by Cretaceous sediments. The most important Variscan tectonic element is situated at the SW margin of the Elbe Zone: the **Mid-Saxonian fault** ("Mittelsächsische Störung" of Pietzsch 1914, 1956) separates the high-grade rocks of the Eastern Erzgebirge from greenschist and very low-grade sequences of the "Elbtal Schiefergebirge" (see Chap. V.B.1 for the stratigraphy). E of the Mid-Saxonian fault, there are greenschist-grade rocks of uncertain protolith age, overlain by fossiliferous Devonian sequences of the Thuringian facies, which is typical of the Saxothuringian autochthon. Northeastwards, beyond another NW-trending fault, there follow Bavarian facies Paleozoic sequences, which are characteristic of the allochthons further W (Chap. V.B.1). These Bavarian rocks, in their turn, are separated by the **Westlausitz fault** from Proterozoic greywackes (Wesenstein g.), which already belong to the Proterozoic block of the Lausitz (Lusatia), i.e., to the Sudetes. Towards the NW, the continuations of the Mid-Saxonian fault zone produces a similar contrast between the Saxonian granulites and Paleozoic sequences to the E (Nossen-Wilsdruff "Schiefergebirge", see Chap. V.B.1).

The fault zone dips to the NE and is several hundreds of metres wide. It consists of mylonites and ultra-mylonites, derived from Erzgebirge gneisses and quartzites, and orthogneisses of the Elbe Zone (Wiedemann 1958; Rauche 1989, 1992). The fault zone also contains lenses of a strongly deformed syenogranite (Tourmaline Granite).

The kinematics of the mid-Saxonian fault has recently been studied by Rauche (1992). A subhorizontal stretching lineation with asymmetric feldspar porphyroclasts and s-c fabrics records ductile strike-slip with a dextral sense of shear. These features have been overprinted by semi-brittle and brittle reverse faulting (top to the SW). K-Ar cooling ages of white mica, biotite, and feldspar, separated from the fault rocks, indicate cooling from more than 420 °C down to less than 250 °C between 370 and 320 Ma. These data also constrain the transition from ductile to brittle behaviour, which in quartzofeldspatic rocks occurs at temperatures around 300 °C. The fault rocks were cooled through this temperature level between 350 and 320 Ma, which indicates a minimum age range for the ductile deformation.

The fault zone is covered by early Permian sediments of the Döhlen basin, which show an average dip of 25° towards NW. If one rotates bedding into the horizontal, the mineral lineation is restored into its original position with a SE-ward dip of 10–20°, and, hence, indicates dextral transtension. The extensional component of this displacement helps to explain the metamorphic contrast between the high-grade rocks of the eastern Erzgebirge and the greenschist to very low grade sequences to the E. A later increment of transtension is suggested by the NW-trending orientation of the latest Westphalian/early Permian Döhlen basin, which probably was formed as a pull-apart basin. This mechanism may also have controlled emplacement of the Meissen granite (NW of Dresden)

A complementary explanation for the metamorphic contrasts at the mid-Saxonian and related faults arises from the strike-slip component. Since the strike-slip displacements postdate the formation of the Fichtelgebirge/Erzgebirge antiform, dextral translation brings rocks from the less metamorphosed NW flank of the antiform in juxtaposition with its high-grade core. Such a displacement is consistent with the correlation of the Erzgebirge Antiform with the very similar Izera-Krkonosce Dome of the Sudetes, which would imply a dextral offset of ca. 50 km (see Rajlich 1987; Matte et al. 1990).

5.3 Interpretation

With the exception of the transtensional faulting in the Elbe zone, the NW-trending tectonic elements in the Saxothuringian belt were formed under brittle conditions, in a strain-hardened and cooled crust, late in the collisional stage. The orientation of compressional and extensional features records changes in the direction of the main horizontal stress. The patterns of magmatic dikes in the Erzgebirge have been used to constrain the changing stress field. It appears that the NW/SE-directed compression, which prevailed during most of the collisional stage, was interrupted by periods of SW/NE-directed compression in the Westfalian B, Westfalian C, and early Stefanian. (Lobin 1985; Wetzel 1989). This is largely consistent with the record of brittle deformation in the Saxothuringian/Moldanubian boundary region around the KTB drilling site, where Zulauf (1992) has reported NW/SE-compression succeeded by E/W and NE/SW compression of Westphalian to Stephanian age. NE/SW compression might also be responsible for the Westphalian part of the subsidence history in the NE-trending intramontane basin NW of the Erzgebirge. However, a detailed and reliable correlation of magmatic events, basin subsidence and fault activity is not possible at present, because the resolution of stratigraphic and isotopic age data is not sufficient to record the rapid changes of the late-Variscan stress field.

References

Andreas D (1987) Structural dual character of the Rotliegendes in the Thuringian Forest and its surroundings. Z Geol Wiss 16: 979–992

Bankwitz E, Bankwitz P (1990) Lower Paleozoic series at the southern margin of the Central German Crystalline Rise (Abstr). Terranes in the Circum-Atlantic Paleozoic orogens, Int Conf Paleozoic orogens in central Europe, Göttingen-Giessen

Bankwitz P (1968) Phyllittektonite verschiedenen Alters in einem Querprofil durch den Schwarzburger Sattel. Geologie 17: 720–752

Bankwitz P, Bankwitz E (1982) Zur Entwicklung der Erzgebirgischen und der Lausitzer Antiklinalzone. Z Angew Geol 28: 511–524

Bankwitz P, Bankwitz E (1989) Fabric elements and strain in the Schwarzburg Antiklinorium, Thuringian Slate Mountains. Z Geol Wiss 17 (7): 653–668

Behr HJ (1961) Beiträge für petrographischen und tektonischen Analyse des Sächsischen Granulitgebirger. Freib Forschungsh Leipfig C 119: 1–118

Behr HJ (1966) Das metmorphe Grundgebirge im Thüringer Becken. Ber Dtsch Ges Gol Wiss A 11: 39–56

Behr HJ (1980) Polyphase shear zones in the granulite belts along the margins of the Bohemian Massif. J Struct Geol 2 (1/2): 249–254

Behr HJ (1983) Intracrustal and subcrustal thrust-tectonis at the northern margin of the Bohemian Massif. In: Martin H (ed) Intracontinental fold belts. Case studies in the Variscan Belt of Europe and the Damara Belt in Namibia. Springer, Berlin Heidelberg New York

Behr HJ, Engel W, Franke W (1982) Variscan wild-flysch and nappe tectonics in the Saxothuringian Zone (Northeast Bavaria, West Germany). Am J Sci 282: 1438–1470

Benek R (1958) Tektonische Untersuchungen aus dem Bereich der Pohlener Störung südlich Gera, Thüringen. Biologie 7: 495–518

Bielicki KH, Hiller H (1987) Pb-Pb model ages of some dike granite samples and a Jamesonite sample of the Sächsisches Granulitgebirge, G.D.R. In: Gerstenberger H (ed) Contrib Geol Saxonian Granulite Massif. ZFI-Mitt, Ber Zentralinst Isotoper-u. Strahlenforsch, Leipzig 133: 99–114

Blümel P, Stein E (1990) Metamorphic units within the Central European Variscides. Abstr Int Conf Paleozoic orogens in central Europe, Göttingen

Brand R (1980) Die niedriggradige Metamorphose einer Diabas-Assoziation im Gebiet von Berg/Frankenwald. Neues Jahrb Miner Abh 139: 82–101

Fiala J, Padera K (1984) Eclogitic rocks from the basement of the Ceske Stredohori Mts. Krystalinikum 17: 33–45

Franke W (1984a) Late events in the tectonic history of the Saxothuringian Zone. In: Hutton DHW, Sanderson DJ (eds) Variscan tectonics of the North Atlantic Region. Geol Soc Spec Publ 14: 33–45

Franke W (1984b) Variszischer Deckenbau im Raume der Münch berger Gneismasse – abgeleitet aus der Fazies, Deformation und Metamorphose im umgebenden Paläozoikum. Geotekton Forsch 68 I–II: 1–253

Franke W (1989a) The geological framework of the KTB drill site, Oberpfalz. In: Emmermann R, Wohlenberg J (eds) The German Continental Deep Drilling Program (KTB). Springer, Berlin Heidelberg New York pp 37–54

Franke W (1989b) Tectonostratigraphic units in the Varican belt of central Europe. Geol Soc Am Spec Pap 230: 67–90

Franke W (1991) The Saxonian granulite – a metamorphic core complex? Terra Abstr 3 (1): 205

Franke W (1993) The Saxonian granulites – a metamorphic core complex? Geol Rdsch 82/3: 505–515

Franke W, Prössl KF, Schwarz J (1992) Devonische Grauwacken im Erbendorfer Paläozoikum – Alter, tektonische Stellung und geotektonische Bedeutung. KTB-Rep 92-4: 225–240

Franzke HJ, Rauche H (1991) Ablauf der rupturellen Deformationen an der fränkischen Linie des nordwestlichen Thüringer Waldes und Südthüringens. KTB Rep 91-1 A, Nachträge zum KTB-Koll 1990: 21–39

Franzke HJ, Röllig G, Kampe A (1990) Magmatic and structural evolution of the area of the Central German Crystalline Rise (CGCR) between Kyffhäuser Mountains and Lusatia (Abstr). Terranes in the Circum-Atlantic Paleozoic Orogens, Int Conf on Paleozoic orogens in central Europe, Göttingen-Giessen

Frischbutter A (1990) Prävariszische Granitoide der Fichtelgebirgisch-Erzgebirgischen Antiklinalzone und ihre Bedeutung für die Krustenentwicklung am Nordrand des Böhmischen Massivs. Akad Wiss DDR, Zentralinst Phys Erde Potsdam 69: 153 pp

Gebauer D (1991) Two Palaeozoic high-pressure events in a garnet-peridotite of northern Bohemia, Czechoslovakia. Terra Abstr Terra Nova 3 (suppl 6): 5

Haase G, Gerstenberger H, Werner CD (1987) Rb/Sr geochronology of granites from the Sächsisches Granulitgebirge. In: Gerstenberger H (ed) Contrib to the geology of the Saxonian Granulite Massif. ZFI-Mitt, Ber Zentralinst Isotopen- u. Strahlenforsch Leipzig 133: 89–98

Hoppe W, Seidel G (eds) (1974) Geologie von Thüringen. VEB Hermann Haack, Gotha Leipzig 1000 pp

Kachlík V (1992) Metabasites in the area o L. Kynzvart, Lazy and Prameny W. from the Marianské Lazne metaophiolite complex and problem of Saxothuringian-Tepla-Barrandian terrane boundary. Abstr 7th Geol Worksh, Styles of superposed Variscan nappe tectonics, Kutná Hora, 47 pp

Kim I-S, Soffel HC (1982) Paleomagnetic results from Lower Paleozoic diabases and pillow lavas from the Frankenwald area (Northwestern edge of the bohemian Massif). J Geophys 51: 24–28

Kossmat F (1927) Gliederung des varistischen Gebirgsbaues. Abh Sächs Geol Landesamt 1: 39

Krentz O (1985) Temperaturregime und Altersstellung der regionalen Metamorphose im mittleren Erzgebirge. Freiberger Forschungsh C 390: 12–28

Kröner A, Frischbutter A, Bergner R, Hofmann J (1991) Zirkon-Evaporationsalter von granitoiden Gesteinen aus dem Erzgebirge und dem Rande der Lausitz und ihre geodynamische Bedeutung. Kurzfassungen 7. Rundgespr Geodyn Eur Variszikums, Freiberg

Krohe A (1991) Großräumige Krustendehnung im NW-Erzgebirge – Ansätze zu einem neuen Konzept. Kurzfassungen 7. Rundgespr Geodyn Eur Variszikum, Freiberg, 28 pp

Kroner U (1991) Struktureller Vergleich einzelner Einheiten des Granulitkomplexes – erste Ergebnisse. Kurzfassungen 7. Rundgespr Geodyn Eur Variszikums, Freiberg, 29 pp

Kroner U, Voigt T (1992) Postkollisionale Extension – ein Modell für die Entwicklung am Nordrand der Böhmischen Masse. Frankf Geowiss Arb A 11: 104–107

Kroner U, Rauche H, Bergner R (1991) Südpolare Überschiebungstektonik am Nordrand des Sächsischen Granulitmassivs – möglicher Ausdruck einer hochkrustalen varistischen Indentation. Zentralb Geol Paläontol I (5): 1339–1343

Kurze M (1966) Die tektonisch-fazielle Entwicklung im Nordostteil des Zentralsächsischen Lineaments. Freib Forschungsh C 201: 89 pp

Lister GS, Dornsiepen UF (1982) Fabric transitions in the Saxony granulite terrain. J Struct Geol 4: 81–92

Lobin M (1985) Pflanzenfunde aus den Tuffiten des Teplicer Quarzporphyrs. Exkursionsführer Ges Geol Wiss DDR, Berlin 1983, 31 pp

Lorenz W (1989) Generalized map of the Erzgebirge Anticlinorium. Enclosure to: Tischendorf G (compiler), Silicic Magmatism and Metallogenesis of the Erzgebirge. Veröff Zentralinst Phys Erde 107

Lorenz W, Hoth K (1990) Lithostratigraphie im Erzgebirge – Konzeption Entwicklung, Probleme und Perspektiven. Abh Staatl Mus Mineral Geol Dresden 37: 7–35

Lützner H (1987) Sedimentology and basin development of intramontane Rotliegend basins in Central Europe. Z Geol Wiss 16:845–864

Matte Ph, Maluski H, Rajlich P, Franke W (1990) Terrane boundaries in the Bohemian Massif: result of large-scale Variscan shearing. Tectonophysics 177: 151–170

Meinel G (1974) Magmatismus und Metamorphose. In: Hoppe W, Seidel G (eds) Geologie von Thüringen. VEB Hermann Haack, Gotha Leipzig pp 335–355

Motz H (1991) Überschiebungstektonik und Schichtlücken im Ronneburger Erzfeld. Abstr 7. Rundgespr Geodyn Eur Variszikums, Freiberg, 31 pp

Müller B, Wächter K (1970) Beiträge zur Tektonik der Elbtalzone unter besonderer Berücksichtigung der Lausitzer Störung. Geodät Geophys Veröff Berlin III 18: 52

Neumann W (1988) Neuinterpretation des Gefüges der sächsischen Granulite unter dem Aspekt der Querfaltung. Z Geol Wiss Berlin 16 (8): 753–764

Nöhring F (1949) Kleintektonische Untersuchungen im Fichtelgebirge südlich Wunsiedel. Geol Rdsch 37: 78–79

Pietzsch K (1914) Tektonische Probleme in Sachsen. Geol Rdsch 5: 161–174

Pietzsch K (1956) Die Elbtalzone. Ber Geol Ges DDR 1: 117–135

Plaumann S (1982) Die Schwereanomalie im Bereich der Münchberger Gneismasse und ihre Interpretation. Z Dtsch Geol Ges 133: 649–665

Rajlich P (1987) Variszische duktile Tektonik im Böhmischen Massiv. Geol Rdsch 76: 755–786

Rauche H (1989) Aspekte der Deformation des Südwestrand der Mittelsächsischen Störung. Freib Forschungsh C 429: 60–71

Rauche H (1991) Spätvariszische Kompressionstektonik am Südwestrand der Elbezone. Z Geol Wiss Berlin 19 (4): 463–475

Rauche H (1992) Spätvariszische Spannungs- und Verformungsgeschichte der Gesteine am Südwestrand der Elbezone (östliches Saxothuringikum, Varisziden). Disserfation Univ Bochum, 65 pp

Rauche H, Benek R, Fengler HJ, Franzke HJ, Märtens P, Motz H, Reichardt C, Schuster D (1991) Zeitliche Stellung und regionale Einordnung variszischer und postvariszischer Deformationsprozesse am Geraer Vorsprung. Posterbeitr Tagung Ges Geol Wiss "Bau und Lagerstättenführung der Gera-Joachimsthaler Störungszone", Gera, 4.–7. Aug. 1991

Rötzler J (1992) Zur Petrogenese im Sächsischen Granulitgebirge – die pyroxenfreien Granulit und die Metapelite. Geotekton Forsch 77: 100

Rötzler K, Schumacher R, Maresch WV (1992) The metamorphic evolution in the Eastern Erzgebirge. Abstr 8. Rundgespr Geodyn Eur Variszikums, Cheb pp 34–35

Scheumann KH (1935) Die Rotgneise der Glimmerschieferdecke des sächsischen Granulitgebirges.

Ber Math Phys Kl Sächs Akad Wiss 87: 251–286
Schmädicke E, Okrusch M, Schmidt W (1992) Eclogite-facies rocks in the Saxonian Erzgebirge, Germany: high-pressure metamorphism under contrasting P-T conditions. Contrib Mineral Petrol 110: 226–241
Schmoll J, Bittner R, Dürbaum H-J, Heinrichs T, Meissner R, Reichert C, Rühl T, Wiederhold H (1989) Oberpfalz deep seismic reflection survey and velocity studies. In: Emmermann R, Wohlenberg J (eds) The German Continental Deep Drilling Program (KTB). Springer, Berlin Heidelberg New York, pp 99–149
Schröder B (1987) Inversion tectonics along the western margin of the Bohemian Massif. Tectonophysics 137: 93–100
Schröder B (1988) Outline of the Permo-Carboniferous basins at the western margin of the Bohemian Massif. Z Geol Wiss 16 (10): 993–1001
Schroeder E (1966) Zur tektonischen Zonen- und Stockwerksgliederung des Thüringisch-Vogtländischen Schiefergebirges. Geologic 15: 973–980
Schroeder E, Bankwitz P, Paech HJ (1965) Zur Tektonik des Thüringisch-Vogtländischen Schiefergebirges. Ber Geol Ges DDR 10 (3): 265–278
Schwan W (1956) Die Frankenwälder Querzone. Abh Dtsch Akad Wiss Kl Math Allg Naturwiss 1954 (6): 80
Schwan W (1974) Die Sächsischen Zwischengebirge und Vergleiche mit der Münchberger Gneismasse und anderen analogen Kristallinvorkommen im Saxothuringikum. Erl Geol Abh 99: 189
Soucek J, Holub F (1992) Blueschist-greenschist metamorphism of metabasites in the Western Krusnehory (Erzgebirge) Mts. Abstr 8. Rundgespr Geodyn Eur Variszikums, Cheb, 40 pp
Stackebrandt W, Franzke HJ (1988) Alpidic reactivation of the Variscan consolidated lithosphere – the activity of some fracture zones in Central Europe. Z Geol Wiss 17 (7): 699–712
Stein E (1988) Die strukturgeologische Entwicklung im Übergangsbereich Saxothuringicum/Moldanubikum in NE-Bayern. Geol Bavarica 92: 5–131
Stein E (1992) Die Altersstellung der Arzberger Serie abgeleitet aus der tektonischen Entwicklung des S' Saxothuringikums. KTB-Rep 92-4: 251–260
Stein E, Kleemann U (1990) Evidence for late Variscan nappe emplacement in the northwestern part of the Bohemian Massif, NE-Bavaria. Abstr Geol Vereinig 80th Annu Meet, 59 pp
Suess FE (1912) Vorläufige Mitteilung über die Münchberger Deckscholle. Sitzungsber Akad Wiss Wien, Math Naturwiss Kl 121, Abt IIa (10): 253
Tomas R, Tischendorf G (1987) Evolution of Variscan magmatic-metallogenetic processes in the Erzgebirge according to thermometric investigations. Z Geol Wiss 15 (1): 25–42
Trusheim F (1964) Über den Untergrund Frankens. Ergebnisse von Tiefbohrungen in Franken und Nachbargebieten 1953–1960. Geol Bavarica 54: 1–92
Vollbrecht A, Weber K, Schmoll J (1989) Structural model for the Saxothuringian/Moldanubian suture in the Variscan basement of the Oberpfalz (northeastern Bavaria, F.R.G.) interpreted from geophysical data. Tectonophysics 157: 123–133
Von Gaertner HR (1951) Probleme des Saxothuringikums. Geol Jahrb 65: 409–450
Von Quadt A (1993) The Saxonian Granulite Massif: new aspects from geochronological studies. Geol Rdsch 82: 516–530
Von Quadt A, Gebauer D (1988) Sm/Nd-, U/Pb-, Rb/Sr- dating of high pressure ultramafic to felsic rocks from the Moldanubian area of NE Bavaria (FRG) and the Saxonian Granulite Massif (GDR) (Abstr). Chem Geol 70: 15
Von Quadt A, Gebauer D (1990) Isotopic ages from the Saxonian Granulite Massif (GDR) and their geodynamic significance (Abstr). Terranes in the Circum-Atlantic Paleozoic Orogens, Int Conf on Paleozoic orogens in central Europe, Göttingen-Giessen
Vulpius B, Kroner U (1992) Die Intrusion des Berbersdorfer Granites (Sächs. Granulitgebirge) – lokaler Ausdruck einer regionalen Extension? Abstr 8. Rundgespr Geodyn Eur Variszikums, Cheb 46 pp
Weber K, Behr HJ (1983) Geodynamic interpretation of the mid-European Variscides. In: Martin H, Eder FW (eds) Intracontinental Fold Belts. Springer, Berlin Heidelberg New York, pp 427–469
Weber K, Vollbrecht A (1989) The crustal structure at the KTB Drill Site, Oberpfalz. In: Emmermann R, Wohlenberg J (eds) The German Continental Deep Drilling Program (KTB). Springer; Berlin Heidelberg New York, pp 5–36
Wetzel HU (1989) Further comments on relationships between plutonic and volcanic formations and ore deposition, structural aspects. In: Tischendorf G (ed) Silicic magmatism and metallogenesis of the Erzgebirge. Veröff Zentralinst Phys Erde 107: 200–209
Wiedemann F (1958) Geologische und petrographische Situation der Serizit- und Chloritgneise der Elbezone. Freib Forschungsh C 55: 89
Willner AP, Krohe A, Maresch WV (1992) Mehrphasige Exhumierungsgeschichte des Erzgebirgskristallins/Sachsen – petrologische und strukturgeologische Belege aus Orthogneisen des mittleren Erzgebirges. Abstr 8. Rundgespr Geodyn Eur Variszikums, Cheb 48 pp
Ziegler P (1987) Late Cretaceous and Cenozoic intraplate compressional deformations in the Alpine foreland – a geodynamic model. Tectonophysics 137: 389–420
Zulauf G (1992) Late to post-Variscan deformation phases and palaeostresses in the KTB pilot research well (Bohemian Massif, Germany). Tectonophysics 202: 1–21
Zulauf G, Kleinschmidt G, Oncken O (1990) Brittle deformation and graphitic cataclasites in the pilot research well KTB-VB (Oberpfalz, FRG). In: Knipe RJ, Rutter EH (eds) Deformation mechanisms, rheology and tectonies. Geol Soc Spec Publ 54: 97–103

V.B.3 Igneous Activity

G. TISCHENDORF, H.-J. FÖRSTER, A. FRISCHBUTTER, W. KRAMER,
W. SCHMIDT, and C.D. WERNER

1 Introduction

Repeatedly changing tectonic regimes during Late Proterozoic-Phanerozoic evolution resulted in the formation of numerous distinct plutonic-volcanic suites (Benek et al. 1973; Collective authorship 1986; Tischendorf et al. 1987b). In times of extensional tectonics, ultramafic and mafic rocks (tholeiitic basalt, ophiolite series, etc.) predominated. In stages soon after compressional movements, however, abundant intermediate to acidic magmas, in addition to subordinate mafic melts, were generated (diorite, granodiorite, monzonite, trachybasalt, lamprophyre, and particularly granitoids). Both mantle and crustal material contributed to the variety of magmatic associations (Fig. 1). Small granitoid intrusions are usually less differentiated; however, the more voluminous plutons exhibit, in most cases, a remarkable evolution from more primitive to strongly evolved and specialized magmatic members.

2 Precambrian Mafic Magmatic Series

Gabbroic rocks overlain by Cambrian sediments, diabases, and ultramafics of an ophiolite series (Werner 1985), pyroxene granulites (Granulitgebirge) and eclogites (Erzgebirge) form the earliest stage of the magmatic development in western Saxony. These rocks are mostly depleted and show geochemical features of oceanic-type rocks (cf. Fig. 2a,b). The pyroxene granulites are part of the volcanosedimentary units, whereas the gabbros, associated with harzburgitic-dunitic serpentinites (sometimes also with pyroxenite), are relicts of nappes at the border of the granulite complex and can be interpreted as obducted ophiolites. All the eclogites have MORB characteristics (Fig. 2b) and underwent granulite facies metamorphism in the lower crust (Klapova 1990).

3 Late Proterozoic-Early Paleozoic Metagranitoids

Multiply deformed gneisses of granitic composition are a characteristic feature of the crust at the northern margin of the Bohemian Massif. They are in part orthogneisses, primarily representing laccolithic and weakly differentiated calc-alkaline granites (Reitzenhain, Sayda). Their magmatic origin is proved by relict features of contact metamorphism in their metasedimentary country rocks of Late Proterozoic to Ordovician age. Other granitic gneisses, mostly augengneisses (Schwarzenberg), were formed in connection with processes of deformation and metamorphism and could have been derived from sediments (blastomylonites). Especially from the Fichtelgebirge, rocks are known which can be interpreted with some certainty as metavolcanics (epigneisses).

Based on geologic evidence, this granitic magmatism started in Late Proterozoic times. The peak thermal reactivation of the crust (without tectogenesis) is thought to be of Early Cambrian age, but it may reach to the Ordovician. Different types of orthogneisses in the central (Reitzenhain) and eastern Erzgebirge (Freiberg, Glashütte/Lauenstein) were found to have single-zircon $^{207}Pb/^{206}Pb$ ages between 555 and 550 Ma (Kröner et al. 1991). This age is thought to reflect the intrusion of all orthogneiss precursors, an interpretation which, based on field observations, must be doubted for the Reitzenhain metagranitoids. Other isotopic age determinations, mainly on so-called relict granites, yielded dates between 420 and 490 Ma (Hofmann et al. 1979). Presently, there are good reasons to interpret those rocks formerly believed to be nearly unaltered protoliths of orthogneisses as having formed later than the gneissification (Frischbutter 1990).

The Erzgebirgian Late Proterozoic-Early Paleozoic acid magmatism decreases progressively in age towards the north. This temperature pulse

Dallmeyer et al. (Eds.)
Pre-Permian Geology of Central and Eastern Europe

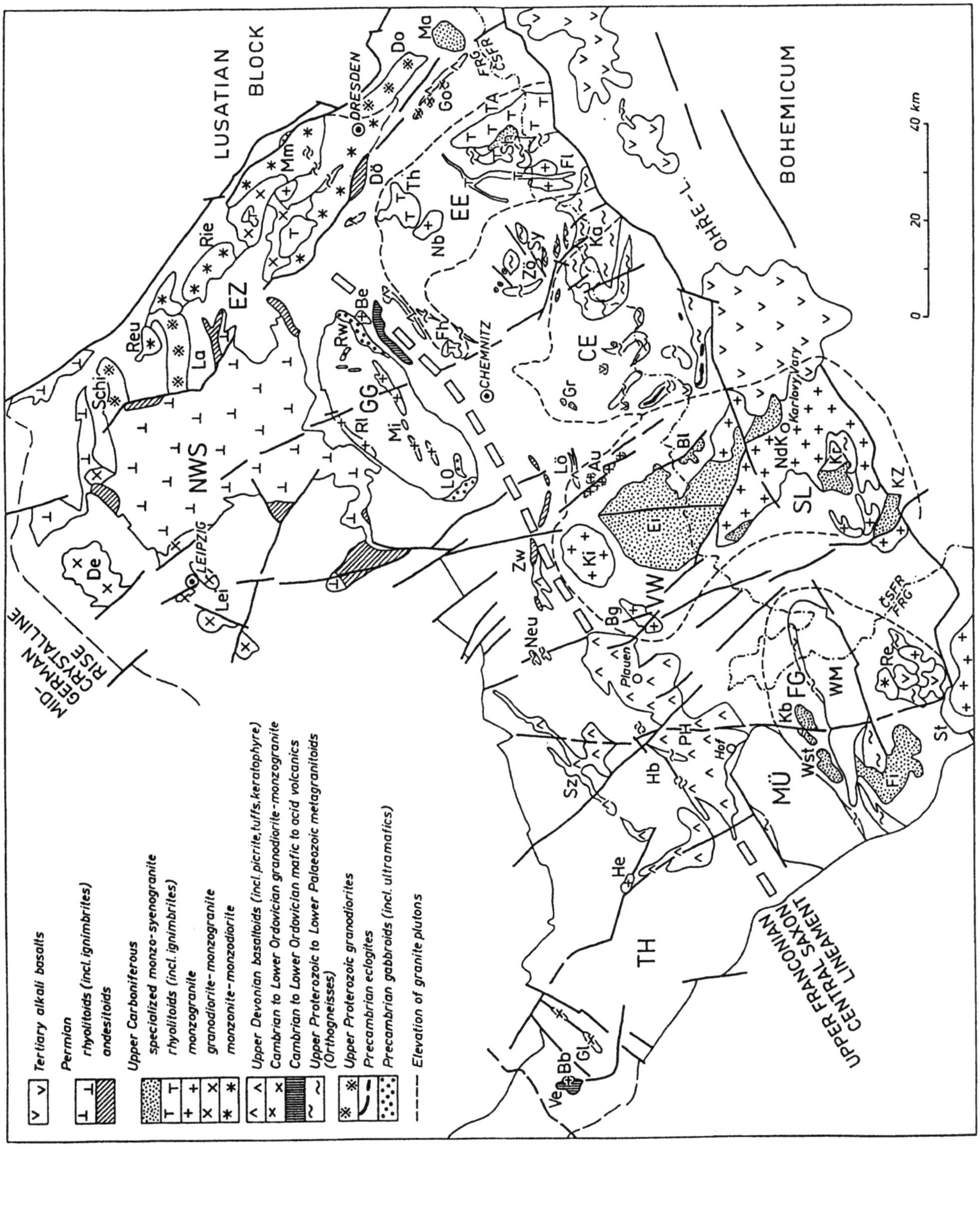
LUSATIAN BLOCK
BOHEMICUM
MID-GERMAN CRYSTALLINE RISE
UPPER FRANCONIAN CENTRAL SAXONIAN LINEAMENT
OHŘE-L.
DRESDEN
CHEMNITZ
LEIPZIG
Plauen
Karlovy Vary
Hof
FRG ČSFR
NWS
EZ
GG
EE
CE
SL
VW
FG
WM
MÜ
TH
0 20 40 km
Tertiary alkali basalts
Permian
rhyolitoids (incl. ignimbrites)
andesitoids
Upper Carboniferous
specialized monzo-syenogranite
rhyolitoids (incl. ignimbrites)
monzogranite
granodiorite-monzogranite
monzonite-monzodiorite
Upper Devonian basaltoids (incl. picrite, tuffs, keratophyre)
Cambrian to Lower Ordovician granodiorite-monzogranite
Cambrian to Lower Ordovician mafic to acid volcanics
Upper Proterozoic to Lower Palaeozoic metagranitoids (Orthogneisses)
Upper Proterozoic granodiorites
Precambrian eclogites
Precambrian gabbroids (incl. ultramafics)
Elevation of granite plutons

followed a pulse of deformation with a similar sense of progression. Both events mark a crustal stacking already in the Late Proterozoic. During this period granitoidic magmas were generated under conditions of compressional tectonics and a relatively low crustal thickness. They intruded during the phase of crustal relaxation which is connected with an Early Paleozoic tensional regime.

4 Cambrian – Early Ordovician Granodiorites-Monzogranites

These granitoids are small, high-level, epizonally altered S-type granites, sometimes sheet-like, within the Berga anticlinal zone (Hirschberg, Greiz) and at the south-western border of the Elbe Lineament (tourmaline granites near Gottleuba) which can be regarded as equivalents of the red gneiss complexes, but are positioned in a marginal position relative to the Erzgebirge anticlinorium. These less evolved and nonspecialized rocks, comparable with the red gneisses in the Erzgebirge, do not possess any metallogenetic potential (Tischendorf et al. 1987b).

Fig. 1. Igneous rocks of the Saxothuringian Basin. *Thüringer Schiefergebirge* (*TH*): Schmiedefeld-Vesser (*Ve*) Burgberg (*Bb*), Glasebach (*Gl*), Henneberg (*He*), Schleiz (*Sz*), Hirschberg (*Hb*), Neumühle (*Neu*); *Fichtelgebirge* (*FG*): Fichtelberg (*Fi*) (= Zentralstock), Weißenstadt-Marktleuthen (*WM*), Waldstein (*Wst*), Kornberg (*Kb*), Steinwald (*St*), Marktredwitz (*Re*), including Münchberger Gneismasse (*MÜ*); *western Erzgebirge-Vogtland* (*VW*) and *Slavkovsky les* (*SL*): Bergen (*Bg*), Kirchberg (*Ki*), Aue (*Au*), Lößnitz (*Lö*), Eibenstock (*Ei*), Horni Blatna (*Bl*), Nejdek-Karlovy Vary (*NdK*), Krasno (*Kr*), Kynzvart-Zandov (*KZ*), including Plauen-Hof (*PH*) and Zwickau (*Zw*); *central Erzgebirge* (*CE*): Greifensteine (*Gr*), Sv. Katheriny (*Ka*), Zöblitz (*Zö*), Sayda (*Sy*); eastern Erzgebirge (*EE*): Niederbobritzsch (*Nb*), Tharandt (*Th*), Flaje (*Fl*), Schellerhau (*Sh*), Teplice-Altenberg (*TA*), Flöha (*Fh*); Granulitgebirge (*GG*): Mittweida (*Mi*), Berbersdorf (*Be*), Rochlitz (*Rl*), Roßwein (*Rw*), Limbach-Oberfrohna (*LO*); north-west Saxony (*NWS*): Delitzsch (*De*), Leipzig (*Lei*); Elbe zone (*EZ*): Schildau (*Schi*), Reudnitz (*Reu*), Laas (*La*), Riesa (*Rie*), Meissen Massif (*Mm*), Döhlen (*Dö*), Dohna (*Do*) including Gottleuba (*Go*) and Markersbach (*Ma*)

5 Cambrian to Devonian Mafic Volcanic Associations

Cambro-Ordovician metabasites were metamorphosed in the greenschist facies (southern flank of the Mid-German Crystalline Rise) and metagraywackes of the western Erzgebirge gneiss-mica schist-zone reached the amphibolite facies. The "Prasinite" (greenschist) facies is featured in the 20-km-long metabasic zone neighboring the southern schist mantle of the Granulitgebirge within the Central Saxon Lineament. Mafic members of the more than 800-m-thick bimodal volcano-sedimentary sequence of Schmiedefeld-Vesser show MORB-features (cf. Fig. 2c) and seem to be connected with dilatational tectonics, crustal thinning, and rifting of back arc-affinity.

The Early Hercynian (initial) **Late Devonian mafic volcanic series** (Adorf stage) was progressively metamorphosed from low to upper greenschist facies (hornblende-mica-rocks), coming from the Central Saxon Lineament, the Vogtland syncline, and into the western Erzgebirge phyllite range. The basalts, often pillow lavas (spilites), diabases and picrites in sills, extend up to the Elbe zone (Werner et al. 1987). The main occurrences of this Ti- and Na-rich series (cf. Fig. 2d) with subordinate trachytes and rhyolites (keratophyres) are described in detail by Rösler and Werner (1979) and Kramer (1988).

6 Late Carboniferous – Early Permian Silicic Plutonic Rocks

The most variegated igneous rocks within the Saxothuringian zone undoubtedly were formed during the Hercynian orogenesis in response to continent-continent collision with crustal stacking and thrusting (cf. Fig. 1). These include:

- well developed, voluminous monzo-syenogranitic, partially highly specialized intrusive complexes within the Erzgebirge-Fichtelgebirge anticlinorium,
- several small, less fractionated monzogranitic bodies around and within the Granulitgebirge anticlinorium, and
- a mostly immature and heterogeneous monzonitic-monzogranitic intrusive complex as part of a long-living lineament (Elbe zone).

The petrogenesis, mineralogy (Table 1), geochemistry (Table 2) and ore-potential of the

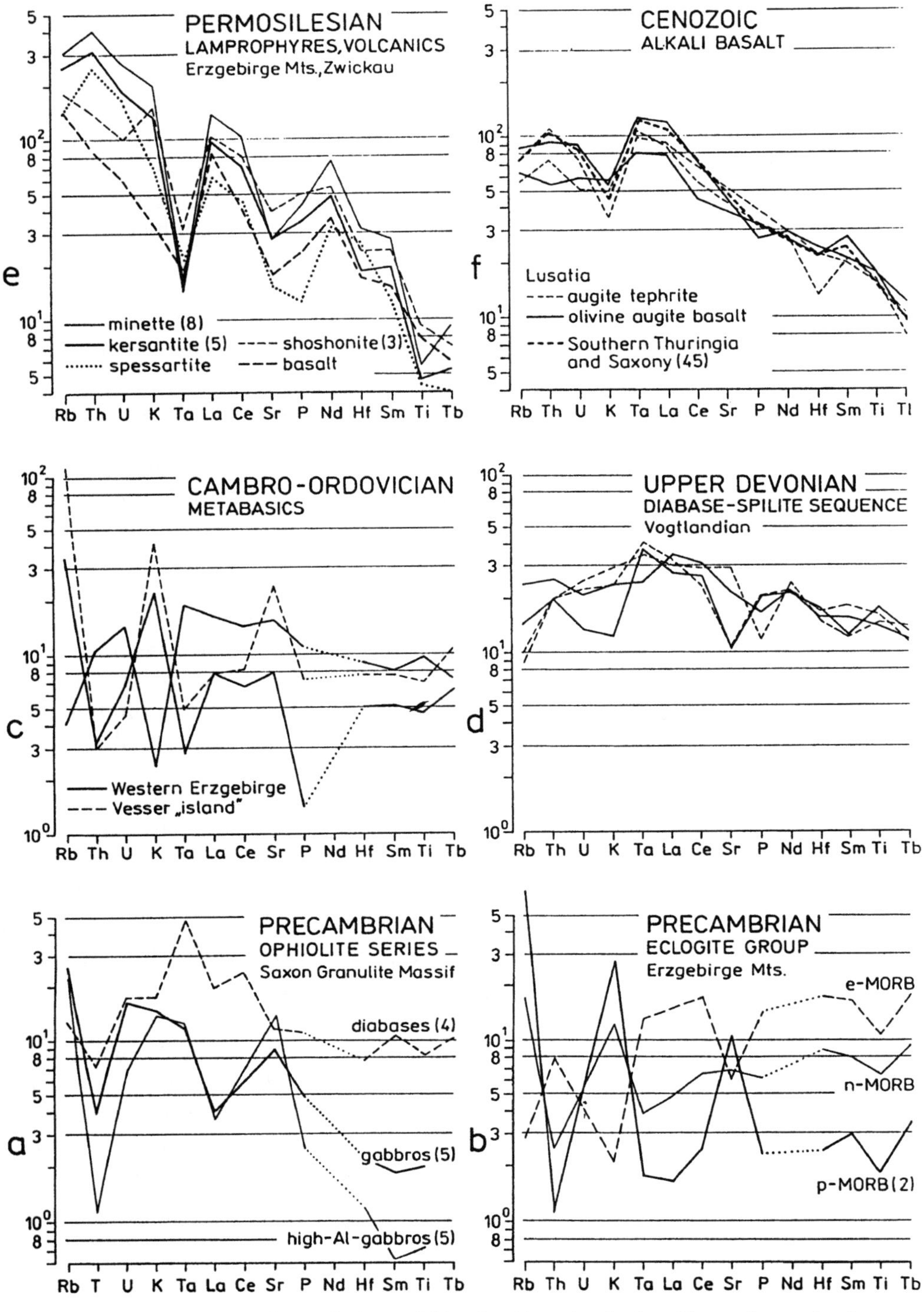

Fig. 2a–f. Mafic magmatic evolution within the Saxothuringian Zone shown by hygromagmatophile-element spider diagrams normalized to primordial mantle composition

granitoids exposed in these three areas are quite different.

The Hercynian granites of the **Erzgebirge-Fichtelgebirge** have been traditionally regarded as belonging to two major plutonic series, the Older (OIC) and Younger Intrusive Complexes, composed of biotite monzogranites and topaz-albite bearing Li mica monzo-syenogranites, respectively. This subdivision is substantiated by various petrochemical, petrographical, mineralogical,

Table 1. Mineral content of granitic-monzonitic rocks of the Erzgebirge, Granulitgebirge, and the Elbe zone

	Erzgebirge				Granulitgebirge				Elbe zone						
	OIC		YIC						Reudnitz		Riesa	Meissen massif			
	1	3	1	3	LG	BD	MG	GG	Encl.	MP	Mp	Encl.	MM	MG	RG
Quartz	27	36	40	34	27	27	30	28	–	–	2	–	8	23	30
Plagioclase	36	31	30[a]	33[a]	32	28	26	30	43	5	55	43	44	42	34
K-feldspar	28	30	21	22	35	37	40	38	2	52	12	–	29	28	33
Pyroxene	–	–	–	–	–	–	–	–	14	13	15	29	–	–	–
Hornblende	–	–	–	–	–	–	–	–	3	8	1	10	11	–	–
Biotite	8	2	5[b]	4[b]	4	7	3	3	34	18	12	17	6	5	2
Muscovite	<1	<1	2	3	1	<1	<1	<1	–	–	–	–	–	–	–
Accessories	<1	<1	<1	<1	1	<1	1	<1	4	4	3	11	2	2	1
Topaz	–	–	1	3	–	–	–	–	–	–	–	–	–	–	–
Rock type	Monzogranite		Syenogranite		Monzogranite				Diorite	Syenite	Monzodiorite	Diorite	Qtzmonzonite	Monzogranite	

[a] Ab-rich. [b] Li siderophyllite-protolithionite.
OIC – Older Intrusive Complex; YIC – Younger Intrusive Complex; LG – Lagergranite; BD – monzonite; MG – Main granite; RG – Riesenstein granite; rock type after the Streckeisen classification; Encl. – enclave.

Table 2. Minor and trace element contents of granitic – monzonitic rocks of the Erzgebirge

	Erzgebirge				Granulitgebirge				Elbe Zone						
	OIC		YIC						Reudnitz		Riesa	Meissen massif			
	1	3	1	3	LG	BD	MG	GG	Encl.	MP	MP	Encl.	MM	MG	RG
TiO_2	0.53	0.14	0.11	0.06	0.24	0.35	0.25	0.15	1.12	0.63	1.05	1.07	0.59	0.26	0.26
MgO	1.07	0.28	0.17	0.11	0.30	0.41	0.30	0.19	6.52	2.77	3.31	6.50	2.20	0.69	0.22
P_2O_5	0.21	0.05	0.30	0.38	0.13	0.11	0.06	0.05	0.88	0.41	0.50	0.77	0.44	0.11	0.10
F	0.10	0.05	0.59	0.70	0.06	0.12	0.04	nd	0.31	0.09	0.11	0.26	0.08	0.04	0.05
Ba	733	65	70	26	833	970	1380	1135	2049	1596	1713	1475	2013	1627	360
Co	7	2	3	2	4	5	4	4	41	45	32	36	19	9	7
Cr	14	8	9	6	4	6	7	6	143	71	21	90	54	25	11
La	63	24	9	2	37	33	36	29	94	115	100	195	113	56	33
Li	84	116	490	632	70	63	44	38	48	58	51	65	43	64	40
Ni	8	3	2	1	3	6	4	3	51	40	17	68	15	15	12
Pb	38	40	14	10	70	70	81	78	20	21	29	32	40	50	32
Rb	247	360	801	1050	287	318	276	242	119	156	90	190	177	179	234
Sn	8	8	44	50	14	6	7	4	nd	nd	8	nd	7	5	6
Sr	215	27	22	11	189	353	352	280	1313	1425	1061	2556	1532	1077	112
Th	29	32	12	5	61	60	71	54	nd	19	6	54	27	33	39
U	7	20	17	10	10	17	20	15	nd	5	4	10	8	12	4
V	43	6	3	2	13	22	11	8	189	101	186	155	85	29	12
Zn	55	39	51	47	20	29	20	12	86	60	100	90	71	48	32
Zr	239	90	62	27	282	164	227	191	402	329	318	370	316	162	65

OIC: Kirchberg massif, subintrusions 1 and 3, YIC: Eibenstock massif, subintrusions 1 and 3. For abbreviations and further explanations see Table 1.

and metallogenic features which in part show remarkable differences between the suites (Lange et al. 1972; Richter and Stettner 1979; Stemprok 1986; Tischendorf et al. 1987a; Tischendorf and Förster 1990; Breiter et al. 1991; Förster and Tischendorf 1994). In many cases, the granites form composite bodies built up by a succession of texturally and geochemically distinct subintrusions (early coarse-grained porphyritic to final fine-grained aplitic).

The typical geochemical patterns of OIC and YIC granites are shown in Fig. 3A and B. Among the strongly differentiated YIC rocks two geochemically distinct types can be discriminated. One type, exemplified by the Zinnwald granite pattern, displays a marked enrichment in Nb, Ta, Y, and HREE in addition to relatively high concentrations of U, Th, Zn, Zr, Hf, and Sc and low P content (Breiter et al. 1991; Förster and Tischendorf 1994). The more common and widely distributed type (e.g., Ei3 subintrusion) comprises peraluminous leucogranites which are extremely low in Y, REE, Th, Zr, Hf, and Sc, comparatively lower in Nb, Ta, and Zn but enriched in P. These

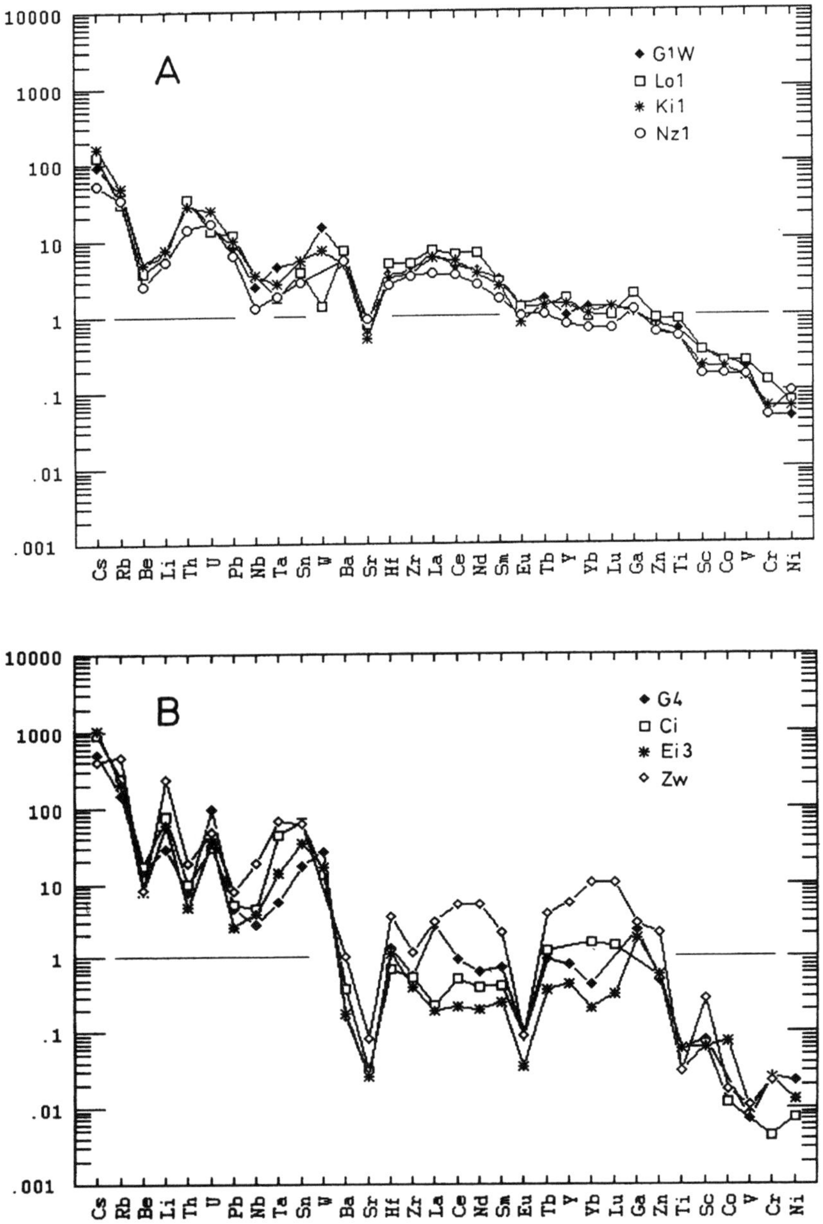

Fig. 3A,B. Lower-crust normalized spider diagrams for selected, less-fractionated OIC. (**A** G1W – Weißenstadt-Marktleuthen, Lo1 – Loket, Ki1 – Kirchberg, Nz1 – Niederbobritzsch) and most-evolved YIC granite subintrusions (**B** G4 – Zinngranit, Zentralmassif/Fichtelgebirge, Ci – Cista, Ei3 – Eibenstock, Zw – Zinnwald, eastern Erzgebirge/German part)

two types, however, do have common features, namely exceptionally high contents of Cs, Rb, Li, Sn, W, and marked depletion in Sr, Eu, and Ti. The progressive fractionation of OIC and YIC magmas is well displayed by their evolution paths in the Sr-Ba-Rb triangle (Fig. 4).

Currently available Rb-Sr and Sm-Nd isotopic data are limited (Lenz 1986; Gerstenberger 1989; Carl and Wendt 1993) and imply multiple intrusion of OIC granites between about 330 and 315 Ma and of YIC granites between approximately 310 and 290 Ma.

The exact nature of the granite protoliths remains unresolved. Sr_i ratios for OIC rocks in the range of 0.7062 to 0.7086 are discussed in terms of an extensive contribution of felsic, lower

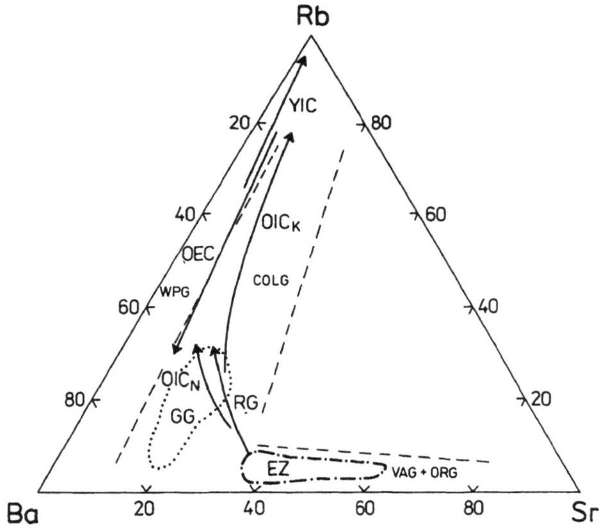

Fig. 4. Position and evolution paths of monzonitic and granitic rocks of the Erzgebirge, Granulitgebirge, and the Elbe zone in the Rb-Sr-Ba triangle (supplemented after Tischendorf and Förster 1990). *EZ* Elbe zone, additional *RG* Riesenstein granite; *GG* Granulitgebirge; *OIC$_N$* Older Intrusive Complex, Niederbobritzsch massif; *OIC$_K$* Kirchberg massif; *YIC* Younger Intrusive Complex; *OEC* Older Extrusive Complex

crustal, Cadomian metagranitoids in their petrogenesis (Tischendorf et al. 1992). Comparatively more scattered initial Sr isotopic ratios obtained for YIC granites suggest the incorporation of isotopically (and geochemically) variable sources. The interpretation of the Fichtelgebirge redwitzites (Sr_i = 0.7041–0.7058) as restites or cumulates of the Hercynian granites or as older products of manifold partial melting processes is still unsolved (Köhler et al. 1989).

The majority of granites match the geochemical characteristics of collision granites (COLG, Tischendorf and Förster 1992; Förster and Tischendorf 1994). The geochemical patterns of OIC samples resemble those of granitoids having a syncollisional genesis and late-orogenic position. Most of the YIC granites appear to represent late-collisional, post-orogenic types, except for those from the eastern Erzgebirge, which plot in the field of within-plate types on commonly used discrimination diagrams and exhibit compositional and geologic features of A-type granites. However, an anorogenic setting of these plutonic rocks is unlikely.

Several indications exist supporting the idea that the Erzgebirge-Fichtelgebirge granites originated from repeated melting events encompassing a time span of about 40 Ma. More or less pronounced differences with respect to geochemical patterns, fractionation regularities, isotopic signatures, and geochronology between the individual plutons imply multiple and genetically separate episodes of melting which led to the formation of independently evolving granite magmas (Förster and Tischendorf 1994).

Li and especially F enhanced liquid-liquid fractionation in evolved YIC granites and thus have contributed to the extraordinarily productivity of these granites with respect to Sn (Tischendorf 1986; Tischendorf and Förster 1990). Differences in the tin potential of the OIC and YIC evolution may also be related to distinct oxygen fugacities (OIC: lg $fO_2 = -13 \pm 1$; YIC: $= -17 \pm 1$) and relative fluorine fugacities (OIC: lg $(fHF/fH_2O) = -3.9 \pm 0.2$; YIC: $= -3 \pm 0.3$) (Förster and Tischendorf 1992).

The small monzogranitic bodies of the **Granulitgebirge** anticlinorium exhibit different textures (strongly deformed Lagergranite-LG in marginal position, undeformed Mittweida-Mühlau bodies-MM in central one), but similar petrographic (Table 1) and geochemical conditions (Table 2) (Gottesmann et al. 1995). Geochemically, they are similar to the Erzgebirge OIC granites (see also Fig. 4); however, they are depleted in Cs, Be, Li, Rb, Sn, and are significantly enriched in Th, U, Pb relative to these. Such differences may be caused by regionally specific partial melting and the style of differentiation. The epsilon Nd-T value of −5.6 for the Berbersdorf granite (BD; Quadt and Gebauer 1990), together with the Sr_i ratio (0.7090) for the Mittweida granite and the Ganggranite (GG) reported by Haase et al. (1987), hints at a significant contribution of crustal material to granite production. U-Pb zircon ages of 342 ± 15 (BD; Quadt and Gebauer 1990) correspond with Rb-Sr whole-rock isochron ages of 350 ± 5 Ma (MM, GG; Haase et al. 1987), indicating magma emplacement during the early stage of the Hercynian plutonic activity. Thus, the granites were probably related to a single event but underwent different strain conditions in the course of crystallization according to their more marginal or central position in the intrusion.

The oldest magmatic rocks of the **Elbe zone** (Cadomian, 650–600 Ma; Wenzel et al. 1993) are represented by small and usually deformed massifs of granodioritic composition (Schildau, Laas,

Dohna; cf. Fig. 1). The magmatic suite which characterizes the Elbe zone sensu strictu consists of several bodies along the line Riesa – Dresden, represented, among others, by the stocks of Reudnitz (Gottesmann 1966) and Riesa-Gröba, and the more voluminous massif of Meissen (Pfeiffer 1964; Schmidt 1995). The latter is composed of at least three individual sequences:

1. several varieties of quartz monzonitic rocks (Marginal Monzonites) with huge dioritic enclaves,
2. more homogenous biotite monzogranite (Main Granite) in the center, and finally,
3. a most fractionated stock of leucomonzogranite (Riesenstein granite).

Application of the single-zircon evaporation technique yielded a $^{207}Pb/^{206}Pb$ age of 355 ± 12 Ma for the pyroxene monzodiorite from Gröba (Gröbaite) which predates the emplacement of the Meissen massif (Wenzel et al. 1993).

Mineral and minor/trace element contents are compiled in Tables 1 and 2. The monzonites are characterized by high proportions of pyroxene (mainly diopsidic clinopyroxene and subordinate orthopyroxene) and hornblende. Titanite, epidote/orthite and magnetite (or martite) are the most frequent accessories. In Fig. 5, a general scheme of the evolution of felsic and mafic rock-forming minerals is given. Depending on varying conditions of magma mobilization and contamination, different varieties are generated. The main types are those of Reudnitz (syenitic) and of Plauenscher Grund (quartz monzonitic). High values of Ig (fHF/fH_2O) determined on biotites from the mafic members indicate a fluorine rich, but water poor source. The Marginal Monzonites are distinctly enriched in some compatible elements (Co, Cr, Ni, V, Zn but also Ba and Sr). Both geochemically and petrographically, the Main Granite reveals a certain kinship to the Marginal Monzonites which is not observed in the Riesenstein granite. This granite is more strongly enriched in incompatible elements (Rb) and depleted in compatible ones (Co, Cr, Ni, V, Sr, Ba). Its fractionation degree is similar to that of the Erzgebirge OIC and the Granulitgebirge granites (Fig. 4).

The monzonitic plutonism of the Elbe zone is seen as a direct consequence of the overthrusting of the Lusatian block onto the Saxothuringian zone, giving rise to melting of both mantle and crustal sources. The main rock-forming process of the Marginal Monzonites is magma mingling (dioritic/syenitic) and mixing (monzonites). An

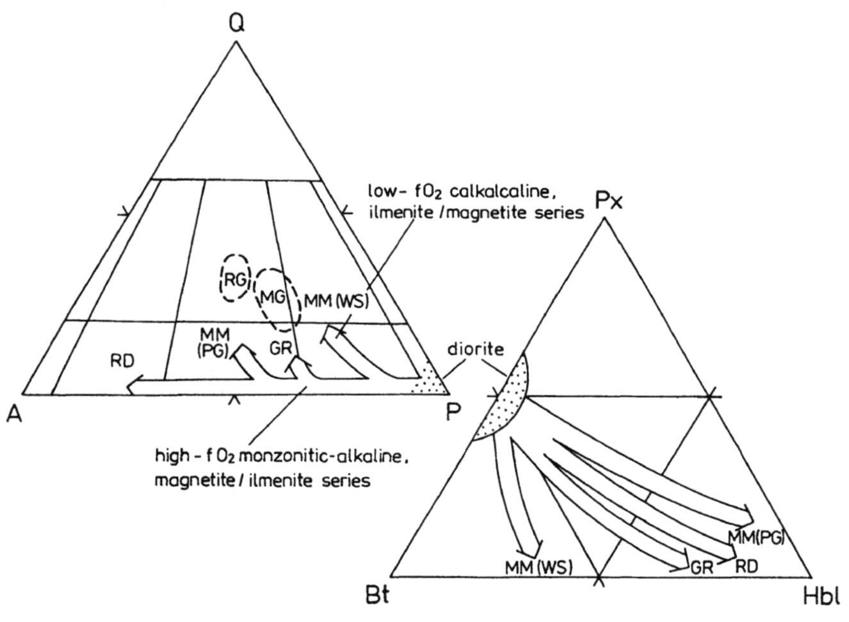

Fig. 5. Position of and relations between evolution paths of magmatic rock series of the Elbe zone in terms of felsic and mafic rock-forming minerals: *RD* syenite of Reudnitz; *MM* (*PG*) (*WS*) marginal monzonites, type Plauenscher Grund and Wesenstein, resp.; *GR* Riesa-Gröba massif, main phase (Gröbaite); *MG* Main granite; *RG* Riesenstein granite

comprehensive petrogenetic model for the Meissen massif, involving both open and closed system fractionation processes, has been proposed recently by Wenzel et al. (1991).

The hidden granodiorite (diorite-) complex of Delitzsch (cf. Fig. 1) at the southern flank of the **Mid-German Crystalline Rise** represents a moderately fractioned igneous suite intruded during the Late Carboniferous. Sr_i values of 0.7053 (diorites) and 0.7079 (granodiorites) are interpreted to reflect different magma sources (lower and upper crustal rocks, respectively) (Röllig et al. 1990b). The granitoids are classified as transitional S-I types and syn-collison intrusions.

The age of the Pretzsch-Prettin-Schönewalde plutonic complex which is situated in the NW of Delitzsch and composed of several dioritic, granitic, and presumably older gabrroic rocks is a matter of debate. Whereas some workers favor a Hercynian age, Röllig et al. (1990b) propose its intrusion during the Late Cambrian and Early Ordovician, based on Rb/Sr whole-rock isochron ages between 495 ± 3 and 536 ± 11 Ma.

7 Late Carboniferous – Early Permian Volcanism

Rhyolites, rhyodacites, andesites, and locally also, basalts and lamprophyres occur in spatial association with the granitoid sequences, but always in "subsequent" stages and in some places with unknown genetic connection. Such volcanics are widespread in NW Saxony (Gläser 1983), the Erzgebirge anticlinorium (Le Thi Lai 1978; Wetzel 1984; Rank and Pälchen 1989; Benek 1991), and in the Elbe zone (see Fig. 1). Rhyolitic rocks predominate by far. The individual volcanic series often start with a minor basaltic phase and finish with voluminous rhyolitic rocks. However, there is generally an increase of more mafic (andesitic-basaltic) volcanics in later series (Röllig and Schirmer 1978). In the eastern Erzgebirge the temporal sequence: phenocryst-free Teplice rhyolite (I) – phenocryst-bearing Teplice rhyolite (II) – Altenberg granite porphyry demonstrates a decrease of acidity and increase of basicity in a local structure and, therefore, extrusion of magma from vertically-fractionated magma chambers (cf. Tischendorf comp, 1989). This process is proved, for example, by the Ba-Sr-Rb distribution of the respective rocks (Fig. 4). Lamprophyric dykes (kersantite, minette, spessartite) are common in the Erzgebirge-Fichtelgebirge anticlinorium as well as in eastern Thuringia, and trachyandesitic (to shoshonitic) volcanics of adjacent intramontane basins, interpreted as comagmatic, bear chemical/mineralogical features of deep-seated, crustal-contaminated melts at "destructive margins" (Kramer 1976, 1988). The relation of part of the lamprophyres to the granite evolution is still in discussion (Tischendorf et al. 1992).

8 Late Cretaceous Ultramafic Lamprophyres and Carbonatites

In the region of Delitzsch, a series of mantle-derived (Sr_i = 0.7038 to 0.7041; Wand et al. 1989, cited in Röllig et al. 1990a) lamprophyric rocks (alnöites, aillikites, monchiquites, alkali picrites) and carbonatites (beforsites, rauhaugites, alvikites) has been discovered by drilling. This suite is locally associated with subeconomic LREE mineralization, especially bastnaesite.

Current genetic models for these rocks involve a polyphase formation which is spatially related to deep-seated fault zones. Integration of petrographical, geochemical, and isotopic data suggests magma generation from metasomatized portions of the mantle which were strongly enriched in certain incompatible elements (Röllig et al. 1990a).

9 Tertiary Alkali Basalts

Rift-bounded intra-continental alkali basalts connected with the northern graben shoulder of the Ohre rift are formed in the cratonal stage in the Erzgebirge and in the Elbe zone. This volcanism had its climax during the Miocene (Pfeiffer 1978). The trace element geochemistry of tephrites and nephelinites from southwestern Saxony shows that the overwhelming majority of these rocks is remarkably uniform (cf. Fig. 2f from Niese et al. 1988). The most prominent aberrations are the ultramafic alkaline "polcenitic" dyke rocks of the Elbe zone and carbonatitic rocks from the Roztoky volcanic rift centre (Kopecky 1978).

References

Benek R (1991) Aspekte einer Volumenbilanz paläovulkanischer Förderprodukte-Beispiel Teplice-Rhyolith (Ostdeutschland). Z Geol Wiss 19: 395–405

Benek R, Röllig G, Eigenfeld F, Schwab M (1973) Zur strukturellen Stellung des Magmatismus der Subsequenzperiode im DDR-Anteil der mitteleuropäischen Varisziden. Veröff Zentralinst Phys Erde Potsdam 14 (1): 203–244
Breiter K, Sokolova M, Sokol A (1991) Geochemical specialization of the tin-bearing granitoid massifs of NW Bohemia. Mineral Deposit 26: 298–306
Carl C, Wendt I (1993) Radiometrische Datierung der Fichtelgebirgsgranite. Z Geol Wiss 21: 49–72
Collective authorship (1986) Vergleichende Geochemie basischer Magmatite auf dem Gebiet der Deutschen Demokratischen Republik. Freiberger Forschungsh C 406: 159 and C 407: 117
Förster HJ, Tischendorf G (1992) Volatile signatures of the Hercynian postkinematic granites of the Erzgebirge: implications to related tin-tungsten-molybdenum metallogenesis. Chemie Erde 52: 49–61
Förster HJ, Tischendorf G (1994) The western Erzgebirge-Vogtland granites: Implications to the Hercynian magmatism in the Erzgebirge-Fichtelgebirge anticlinorium. In: Seltmann R, Kämpf H, Möller P (eds) Metallogeny of collisional orogens. Czech Geol Surv Prague: 35–48
Frischbutter A (1990) Prävariszische Granitoide der Fichtelgebirgisch-Erzgebirgischen Antiklinalzone und ihre Bedeutung für die Krustenentwicklung am Nordrand des Böhmischen Massivs. Veröff Zentralinst Phys Erde Potsdam 69: 1–153
Gerstenberger H (1989) Autometasomatic Rb enrichment in highly evolved granites causing lowered Rb-Sr isochron intercepts. Earth Planet Sci Lett 93: 65–75
Gläser W (1983) Beitrag zur Petrologie und Vulkanologie der andesitoiden Vulkanite Nordwestsachsens. Hallesches Jahrb Geowiss 8: 1–30
Gottesmann W (1966) Der Syenit von Reudnitz (Meißener Massiv). Jahrb Geol 2: 537–603
Gottesmann B, Tischendorf G, Wand U, Bielicki KH, Förster HJ, Haase G, Thomas R (1995) Die granitoiden Gesteine des Sächsischen Granulitmassivs: Petrographie, Geochemie und Altersstellung. Hallesches Jahrb Geowiss 19 (in press)
Haase G, Gerstenberger H, Werner CD (1987) Rb/Sr geochronology of granites from the Sächsisches Granulitgebirge. Mitt Zentralinst Isotopen Strahlenforsch Leipzig 133: 79–87
Hofmann J, Mathe G, Pilot J et al. (1979) Fazies und zeitliche Stellung der Regionalmetamorphose im Erzgebirgskristallin. Z Geol Wiss 7: 1091–1106
Klapova (1990) Eclogites of the Bohemian part of the Saxothuringicum. Rozpr Cesk Akad Ved, Rada Mat Prir Ved Roc 100: 1–86
Köhler H, Propach G, Troll G (1989) Die Redwitzite Nordbayerns. Ber Dtsch Mineral Ges 2: 2–17
Kopecky L (1978) Neoidic taphrogenic evolution and young alkaline volcanism of the Bohemian Massif. Sb Geol Ved Geol 31: 91–104
Kramer W (1976) Genese der Lamprophyre im Bereich der Fichtelgebirgisch-Erzgebirgischen Antiklinalzone – ein geochemisch-petrologischer Beitrag zum Problem der Kruste-Mantel-Beziehungen. Chemie Erde 35: 1–49
Kramer W (1988) Magmengenetische Aspekte der Lithosphärenentwicklung. Schriftenr Geol Wiss 26: 135
Kröner A, Frischbutter A, Berger R, Hofmann J (1991) Zirkon-Evaporationsalter von granitoiden Gesteinen aus dem Erzgebirge und am Rande der Lausitz und ihre geodynamische Bedeutung. Abstr 7. Rundgespr Geodyn Eur Variszikums, Freiberg, p 27
Lange H, Tischendorf G, Pälchen W, Klemm J, Ossenkopf W (1972) Zur Petrographie und Geochemie der Granite des Erzgebirges. Geologie 21: 457–493
Lenz H (1986) Rb/Sr-Gesamtgesteins-Altersbestimmung am Weißenstadt-Marktleuthener Porphyrgranit des Fichtelgebirges. Geol Jahrb E 34: 67–76
Le Thi Lai (1978) Petrographische und geochemische Bearbeitung der sauren variszischen gang- und deckenförmigen Magmatite des Erzgebirges als Beitrag zur metallogenetischen Untersuchung dieses Raumes. Thesis, EMAU, Greifswald
Niese S, Gleisberg B, Kramer W, Pfeiffer L, Werner CD (1988) Application of neutron activation analysis for investigating the genesis of Tertiary magmatites from the south of the GDR. J Radioanal Nucl Chem Articles 122: 253–259
Pfeiffer L (1964) Beiträge zur Petrologie des Meißener Massivs. Freiberger Forschungsh C 179: 222
Pfeiffer L (1978) Beitrag zur Petrochemie der sächsischen Tertiärvulkanite. Freiberger Forschungsh C 333: 163
Quadt AV, Gebauer D (1990) Isotopic ages from the Saxonian Granulite Massif (GDR) and their geodynamic significance. Ext Abstr Int Conf on Paleozoic orogens in central Europe, Göttingen
Rank G, Pälchen W (1989) Zur Geochemie der sauren postvariszischen Vulkanite im Raum Flöha – Karl-Marx-Stadt. Z Geol Wiss 17: 1087–1097
Richter P, Stettner G (1979) Geochemische und petrographische Untersuchungen der Fichtelgebirgsgranite. Geol Bavarica 78: 143
Röllig G, Schirmer B (1978) Zum Stoffbestand der subsequenten variszischen Vulkanite im Südteil der DDR. Z Geol Wiss 6: 1101–1118
Röllig G, Viehweg M, Reuter N (1990a) The ultramafic lamprophyres and carbonatites of Delitzsch/GDR. Z Angew Geol 36: 59–64
Röllig G, Bräuer H, Viehweg M, Gruner B, Kabardin B, Wünsch K, Haase G (1990b) Altersstellung und petrogenetische Charakteristik der Plutonite im Gebiet des Zentralteiles der Mitteldeutschen Schwelle. Z Angew Geol 36: 208–212
Rösler HJ, Werner CD (1979) Petrologie und Geochemie der variszischen Geosynklinalmagmatite Mitteleuropas. Freiberger Forschungsh C 336: 265 and C 344: 295
Schmidt W (1995) Plutonism in the Elbe Zone – petrographical contraints. Lithos (submitted)
Stemprok M (1986) Petrology and geochemistry of the Czechoslovak part of the Krusnehory Mts granite pluton. Sb Geol Ved Loz Geol Min 27: 111–156

Tischendorf G (1986) Variscan ensialic magmatism and metallogenesis in the Ore Mountains – modelling of the process. Chemie Erde 45: 75–104

Tischendorf G, Förster HJ (1990) Acid magmatism and related metallogenesis in the Erzgebirge. Geol J 25: 443–454

Tischendorf G, Förster HJ (1992) Central European granitoids and their tectonic setting: Implications from a new discrimination diagram. Zentralbl Geol Paläontol 1: 791–801

Tischendorf G, Geisler M, Gerstenberger H, Budzinski H, Vogler P (1987a) Geochemistry of Variscan granites of the Westerzgebirge-Vogtland Region – an example of tin deposit-generating granites. Chemie Erde 46: 213–235

Tischendorf G, Pälchen W, Röllig G, Lange H (1987b) Formationelle Gliederung, petrographisch-geochemische Charakteristik und Genese der Granitoide der Deutschen Demokratischen Republik. Chemie Erde 46: 7–23

Tischendorf G comp. (1989) Silicic Magmatism and Metallogenesis of the Erzgebirge. Veröff Zentralinst Physik Erde Potsdam 107: 316

Tischendorf G, Förster HJ, Bielicki KH, Haase G, Bankwitz P, Kramer W (1992) On the origin of Hercynian magmatites and ore deposits in the Erzgebirge: crustal signatures. Zentralbl Geol Paläontol 1: 131–146

Wenzel Th, Mercolli I, Oberhänsli R (1991) The plutonic rocks of the Meißen massif (Germany): evidence for open and closed system fractionation processes. Schweiz Mineral Petrogr Mitt 71: 371–390

Wenzel Th, Hengst M, Pilot J (1993) The plutonic rocks of the Elbe valley-zone (Germany): evidence for the magmatic development from single-zircon evaporation and K-Ar age determinations. Chem Geol 104: 75–92

Werner CD (1985) Upper Riphean ophiolites in Saxony (GDR). In: Debretsov NL (ed) Riphean-Paleozoic ophiolites of North Eurasia. Nauka, Novosibirsk, pp 106–119 (in Russian)

Werner CD, Loos G, Niese S (1987) Seltene Elemente in Initialiten der DDR. Chemie Erde 47: 129–166

Wetzel HU (1984) Spätvariszische Bruchtektonik und subsequente Gangmagmatite als Ausdruck der Krustenentwicklung im Osterzgebirge (Altenberger Scholle). Thesis, Zentralinst Phys Erde, Potsdam

V.B.4 Metamorphic Evolution

P. Blümel

1 Introduction

The Saxothuringian Basin, comprising Upper Proterozoic to Paleozoic series of 4–5 km thickness, is delimited by (citation clockwise from N): the Mid-German Crystalline Rise, the Permo-Mesozoic platform of Sachsen, the Lausitz basement, the core of the Erzgebirge Zone, the Teplá-Barrandean, the Moldanubian, and the Permo-Mesozoic of Franken and Hessen (Fig. 1). The sediments were deformed during the Upper Visean into six large SW-NE-trending fold structures (from N to S; Fig. 1):

1. the Schwarzburg-Nordsachsen Anticline,
2. the Teuschnitz-Ziegenrück Syncline,
3. the Berga Anticline,
4. the Vogtland-Mittelsachsen Syncline,
5. the Fichtelgebirge-Erzgebirge Anticline,
6. the Waldsassen Syncline.

These megastructures are cut by four transverse zones (Hempel 1974; from SW to NE):

1. the Frankenwälder Querzone (horst structure; Fig. 5),
2. the Südvogtländische Querzone (graben structure),
3. the Nordostvogtländische Querzone (horst structure),
4. the Westsächsische Querzone (graben structure).

Posttectonic granitoid magmatism extends northwards from the Moldanubian into the Saxothuringian as far as the Vogtland-Mittelsachsen Syncline (granites of Sparnberg, Eibenstock etc.).

The Saxothuringian slates are metamorphosed into the greenschist and even the amphibolite facies in two subareas (Fig. 1):

1. the core of the Schwarzburg Anticline (subarea 1),
2. the Fichtelgebirge-Waldsassen-Tirschenreuth area (subarea 2).

Subarea 1: the metapsammopelites known as the Katzhütter Schichten of Upper Proterozoic age show the mineral assemblage biotite + chlorite + muscovite + albite + quartz + ilmenite + Fe-sulfides + graphitoid (Bankwitz 1988; Meinel 1974). Maximum temperatures may be estimated at ca. 450 °C. The pressure of metamorphism is unknown. The metasediments (Cadomian? Variscan?) of the anticline are in tectonic contact with anchimetamorphic Cambro-Ordovician slates.

Subarea 2: south of the Vogtland-Mittelsachsen Syncline (and the Fichtelgebirgsnordrand Fault) the Saxothuringian slates are metamorphosed at conditions of low-pressure type (Fig. 2). In the Fichtelgebirge, Anticline fragments of a biotite zone and of an andalusite-staurolite zone are found: their isolation being referred to late-Variscan block faulting (Mielke and Schreyer 1969; Mielke et al. 1979). Continuous prograde regional metamorphism starts from the chlorite zone within the core of the Waldsassen Syncline (B3, Stein 1988) and increases in grade towards

Fig. 2. Regional low-pressure metamorphism in the Fichtelgebirge-Waldsassen-Tirschenreuth area (subarea II). In the Fichtelgebirge the metamorphic zonation is blurred by block faulting leaving fragments of medium-grade zones within a low-grade frame. It should be noted that prograde regional low-pressure metamorphism continues into the adjacent Moldanubian (further observations see text). *1* Chlorite zone; *2* biotite zone; *3* staurolite zone; *4* andalusite-staurolite zone; *5* andalusite zone; *6* sillimanite-muscovite zone; *7* sillimanite-potash feldspar zone; *8* cordierite-potash feldspar zone; *A* localities with andalusite and muscovite; *A-St* localities with andalusite-staurolite-muscovite. *FA* Fichtelgebirge Anticline; *WS* Waldsassen Syncline; *MM* Münchberg Mass; *ZTT* Zone of Teplá-Domažlice (Taus); *ZEV* Zone of Erbendorf-Vohenstrauß; *SAX* Saxothuringian; *MOLD* Moldanubian; *Bt.* Bayreuth; *Ch.* Chéb (Eger); *E.* Erbendorf; *Ta.* Tachov; *Ti.* Tirschenreuth; *Wa.* Waldsassen; O KTB well; dash-dotted lines, SAX/MOLD boundary in 2 versions

Dallmeyer et al. (Eds.)
Pre-Permian Geology of Central and Eastern Europe

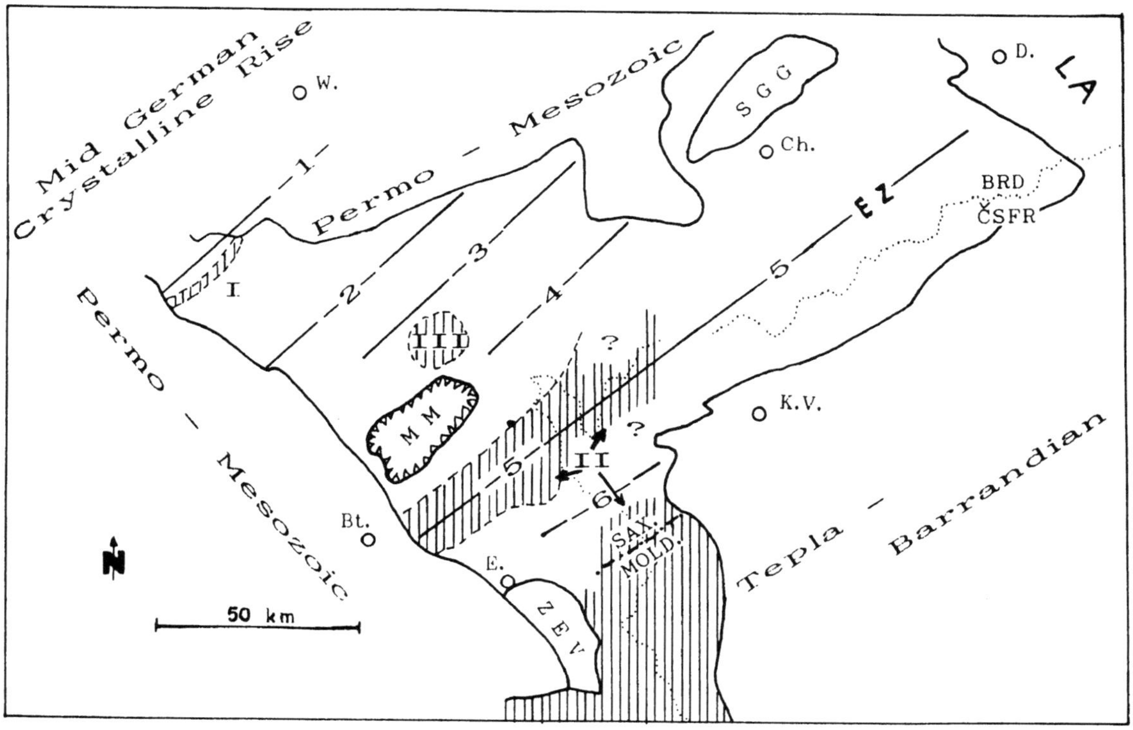

Fig. 1. Saxothuringian Basin with areas of metamorphism as discussed in the text and main fold structures. *I* Metamorphic Zone in the Schwarzburg Anticline (subarea I); *II* Metamorphic Zone in the Fichtelgebirge-Waldsassen-Tirschenreuth area (subarea II); *III* Metamorphic Zone of Hirschberg (subarea III). For Münchberg Mass (*MM*), Sächsisches Granulitgebirge (SGG) and Erzgebirge Zone (EZ) which are allochthonous see Chapter V.C.2. *LA* Lausitz Basement; *SAX* Saxothuringian; *MOLD* Moldanubian; *Bt.* Bayreuth; *Ch.* Chemnitz; *D.* Dresden; *E.* Erbendorf; *H.* Hof; *K.V.* Karlovy Vary (Karlsbad); *W.* Weimar; *Zw.* Zwickau

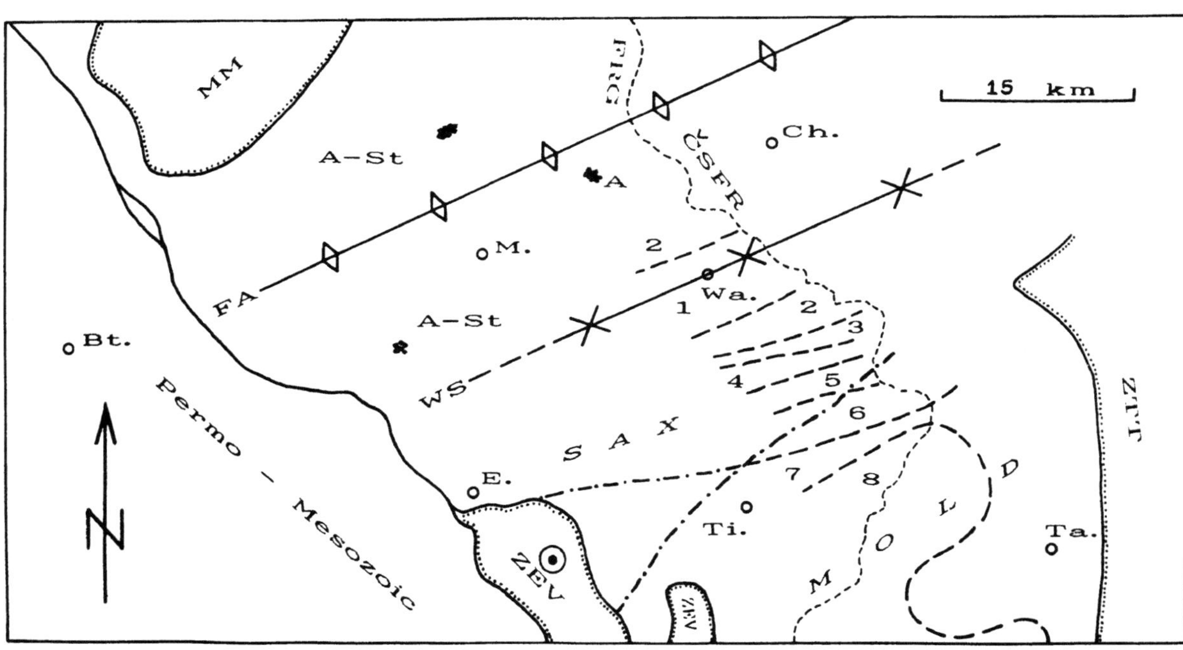

the Moldanubian with mineral zones as follows (Blümel 1982; Wagener-Lohse and Blümel 1984; Wagener-Lohse, unpubl.; Fig. 2):

- Biotite Zone.
- Garnet Zone.
- Staurolite Zone.
- Andalusite-Staurolite Zone.
- Andalusite Zone.
- Sillimanite-Muscovite Zone.

The composition of some index minerals is given in Fig. 3.

The mineral zones are narrowly spaced and probably have a steep northwesterly dip (as extrapolated from the situation in the adjacent Moldanubian). The isograds run obliquely to the B3 fold structure. They seem to continue across the border to the Czech Republic, yet their western extension is unknown due to a Miocene/Pliocene cover.

Biotite Zone (loc. Querenbach):

$alm_{76.3/62.1}$ $gro_{4.2/5.0}$
$pyp_{6.3/4.1}$ $and_{0.2/0.3}$
$spe_{13.0/28.5}$ (rim/core)
assemblage: garnet+biotite+chlorite+muscovite+albite+quartz+ilmenite+Fe-sulfide

Staurolite Zone (loc. Kalmreuth):

$Fe\text{-}staur_{67.7}Mg\text{-}staur_{8.0}Mn\text{-}staur_{1.8}Zn\text{-}staur_{22.5}$
assemblage: staurolite+biotite+muscovite+albite+quartz+ilmenite+Fe-sulfide

Staurolite Zone (loc. Poxdorf):

$Fe\text{-}staur_{86.1}Mg\text{-}staur_{11.4}Mn\text{-}staur_{1.6}Zn\text{-}staur_{0.9}$
assemblage: staurolite+biotite+chlorite+muscovite+albite+quartz+ilmenite+Fe-sulfide

Andalusite-Staurolite Zone (loc. Stollenhölzer):

$alm_{69.8/55.9}$ $gro_{3.6/7.5}$
$pyp_{6.9/4.9}$ $and_{1.5/0.8}$
$spe_{18.2/30.9}$ (rim/core)
assemblage: garnet+biotite+muscovite+plagioclase+quartz+ilmenite+Fe-sulfide

First Sillimanite Isograd (loc. Hendlmühle):

$alm_{66.8}pyp_{7.9}spe_{23.7}gro_{1.6}$(rim)
assemblage: sillimanite+andalusite+garnet+biotite+muscovite+plagioclase+quartz+ilmenite+Fe-sulfides

Fig. 3. Selected compositional data for garnet and staurolite in metapelites of the Waldsassen-Tirschenreuth area (subarea II; Wagener-Lohse, unpubl.)

The P-T conditions at the first sillimanite isograd (Fig. 4) correspond to that of the interval (on the andalusite/sillimanite phase boundary) between:

3.7 kb/500 °C and 2.3 kb/600 °C

using the P-T location of $muscovite_{ss}$ + $albite_{ss}$ + quartz/Al_sSiO_5 + k-$feldspar_{ss}$ + H_2O by Chatterjee and Flux (1986) and the Al_2SiO_5 triple point of Holdaway (1971). The P-T evolution of the Saxothuringian micaschists was obviously restricted to the low-pressure range, since sillimanite is found in pseudomorphs after andalusite.

The age of regional metamorphism is given by a concordant monazite age of 320 ± 3 MA (from sillimanite mica schist) and by a Rb/Sr isochron of 315 ± 5 MA (andalusite micaschist) (Teufel 1988). At that time, the first posttectonic granites are emplaced in the subarea (Weissenstadt-Marktleuthen; Wendt et al. 1986).

The granites leave narrow contact aureoles (width less than 100–150 m) mainly with cordierite + sillimanite. Near Marktredwitz, a complex gabbrodiorite-diorite-granodiorite pluton produced a pyroxene hornfels facies aureole in regional metapelites of low-pressure greenschist facies (Schulte 1991).

In the Frankenwälder Querzone, syndeformative anchizonal metamorphism evolved, as demonstrated by structural and illite crystallinity studies (Franke 1984). Anchimetamorphism here may be due to stratigraphic repetition or to somewhat increased heat flow (extension?) giving

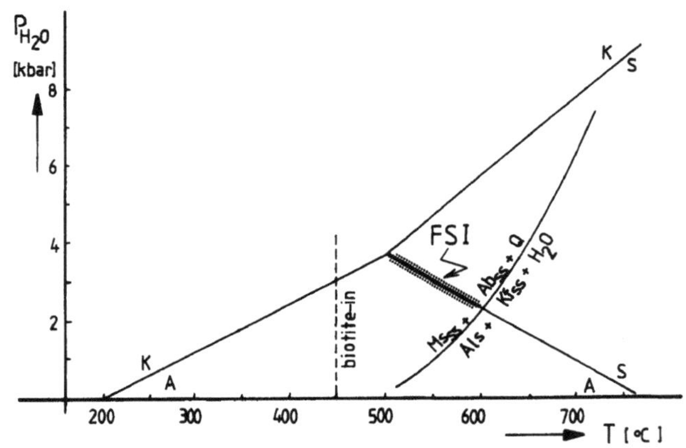

Fig. 4. P-T estimate for first sillimanite isograd (FSI) of subarea II (*shaded line*). $Muscovite_{ss}$ + $albite_{ss}$ + quartz/Al_2SiO_5 + k-$feldspar_{ss}$ + H_2O (Chatterjee and Flux 1986); Al_2SiO_5-diagram (Holdaway 1971)

way to subsequent transverse tectonics. Such sites are also favored by intrusions. Indeed, in the southeastern section of the Frankenwälder Querzone, near Hirschberg, an isolated metamorphic zone of low grades is found with a psammopelite/diabase association (subarea III of Fig. 1). Mineral zones are mapped in the Bavarian part of the metamorphic zone exclusively on the basis of the metadiabase assemblages (Brand 1980; Fig. 5). The critical mineral assemblages of the zones are given in Fig. 6. Since metamorphism in the associated Ordovician Griffelschiefer culminates in the assemblage andalusite + biotite + muscovite + albite + quartz, we are dealing with low-pressure metamorphism, the P-T conditions being less than 3.7 kb and somewhat more than 450 °C (Fig. 4). The contact nature of metamorphism is demonstrated by the broadly concentric pattern of mineral zones and the existence of a subsurface pluton (reached in drill holes)

1. Prehnite-pumpellyite zone:
   albite+prehnite+chlorite+pumpellyite+sphene (+quartz)
2. Pumpellyite-actinolite-chlorite zone:
   albite+epidote+chlorite+pumpellyite+actinolite+sphene
3. Albite-chlorite-actinolite zone:
   albite-chlorite+actinolite+epidote/clinozoisite (+quartz) (+stilpnomelane)
4. Zone of albite-epidote hornfels facies:
   albite/oligoclase+chlorite+actinolite+epidote (+quartz) (+diopside)

   In siliceous marble: grossular-andradite, wollastonite, white mica, graphite (Theobald 1915)

   In metapsammopelites: andalusite+biotite

Fig. 6. Mineral zones in metadiabase of the Hirschberg contact aureole. (Brand 1980): critical assemblages

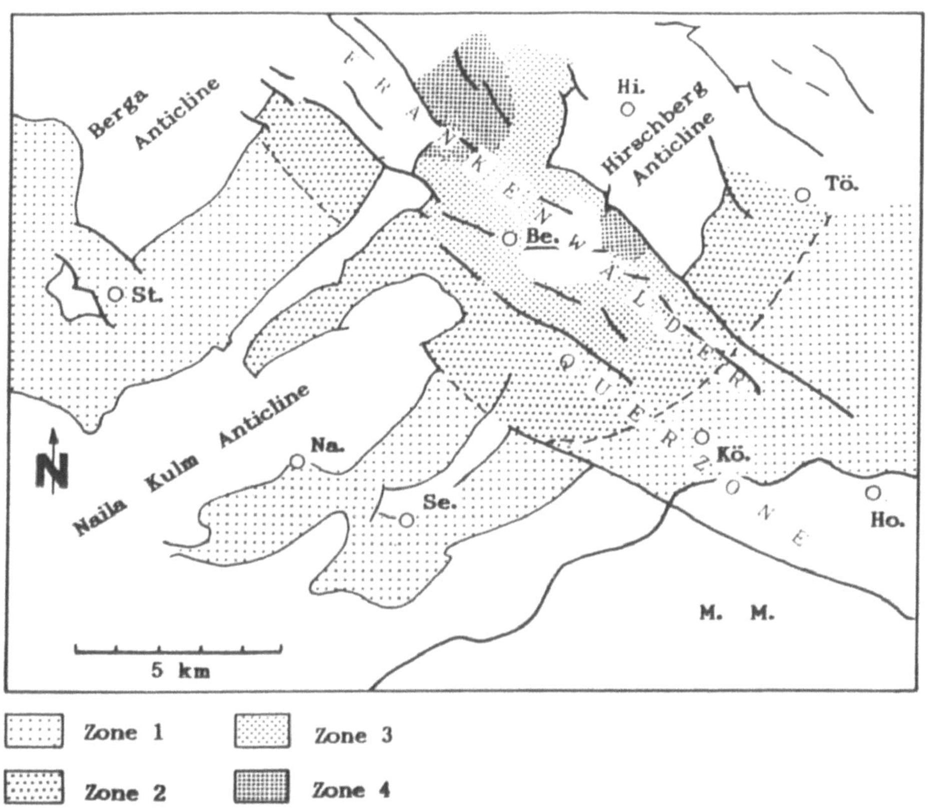

Fig. 5. Contact-aureole in Devonian metadiabase above subsurface pluton near Hirschberg overprinting anchizonal syndeformative metamorphism (not shown), with subsequent Frankenwälder Querzone. *Zone 1* Prehnite-pumpellyite; *Zone 2* pumpellyite-actinolite-chlorite; *Zone 3* albitechlorite-actinolite; *Zone 4* albite-epidote hornfels facies with andalusite + biotite in metapsammopelites. *M.M.* Münchberg Mass; *Be.* *Berga*; *Hi.* Hirschberg; *Ho.* Hof; *Kö.* Köditz; *Na.* Naila; *Se.* Selbitz; *St.* Steben; *Tö.* Töpen

with some outcropping apophyses of granite (Sparnberger Granit, etc.).

Regional low-pressure metamorphism extends along the Fichtelgebirge Anticline toward the NE as far as the Erzgebirgsgranite, which overprint the Ordovician phyllites and micaschists with andalusite-bearing aureoles. However, for the northeasterly adjoining Erzgebirge Zone, on the basis of a regional synthesis by Wienholz et al. (1979), metamorphism of the medium-pressure type was inferred and by consequence an allochthonous nature of the Erzgebirge Zone anticipated (Blümel 1984, 1986). Further work will have to concentrate on the relations of the U. Proterozoic gneiss core with the Paleozoic micaschists and phyllites.

2 Saxothuringian/Moldanubian Relations

As already suggested by Schreyer (1966), prograde regional low-pressure metamorphism continues from the Saxothuringian into the southeasterly adjoining Moldanubian of E Bavaria. The Saxothuringian mineral zones are followed without hiatus by the Moldanubian sillimanite-potash feldspar zone and the cordierite potash feldspar zone, such that both are welded together into a low-pressure zone (extending widely within the Mid-European Variscan; Blümel 1982, 1983, 1984; Wagener-Lohse and Blümel 1984; Wagener-Lohse, unpubl.). Similarly, the Variscan structures extend from the Saxothuringian into the Moldanubian over at least some 50 km (Stein 1988). Thus we may conclude that the Saxothuringian is the low-grade equivalent of the Moldanubian. Consequently, we may infer a low-pressure Moldanubian below the Saxothuringian Basin, which agrees well with the low-pressure character of the Schiefermantel of the Sächsische Granulitgebirge and of the Ruhla basement (Blümel 1984, 1986). However, this may be at variance with the low-grade Upper Proterozoic core of the Schwarzburg Anticline, thus calling for more complex models.

Prior to regional low-pressure metamorphism, the Saxothuringian Paleozoic (Thüringische Fazies) and the high-grade Moldanubian were at different tectonic positions as indicated by the mono- and low-pressure metamorphic imprint of the Saxothuringian and the poly-metamorphic evolution of the Moldanubian (see Chap. V.C.4). In the border region between both units in NE Bavaria late-metamorphic displacements concordant with the isograd pattern are indicated by diaphthorites and mylonitic quartz-feldspar rocks. This situation strongly contrasts with the Lalaye-Lubine Zone in the northern Vosges some 300 km further west where garnetiferous phyll(on)ites are in tectonic contact with cordierite-sillimanite gneiss. Other portions on this "two-point" tectonic line actually serve as thrust-zones between medium-pressure and low-pressure rocks of the Variscan and thus cannot help to corroborate a late-metamorphic first-order boundary between both Kossmat units (see Chap. V.C.4).

3 Relations Between Saxothuringian B Paleozoic and Exotic Nappes

Saxothuringian slates (Thuringian Fazies) and Saxonian granulites reequilibrated to low-pressure amphibolite facies are related by prograde metamorphism which, however, is tectonically reduced and telescoped. Intermediate grades are represented by schists with andalusite, garnet + staurolite, sillimanite + biotite and by gneisses and anatexites with cordierite + sillimanite (Rötzler 1992).

Prograde metamorphism is reported to continue into the Erzgebirge Basement (biotite zone, garnet zone, staurolite zone; Wienholz et al. 1979). The pressure conditions of the low-grade portion being unsettled, it may well be that both ends of the Saxothuringian Paleozoic show prograde low-pressure metamorphism. The geologic relations are furthermore complicated by the high-grade metamorphic Frankenberger Zwischengebirge overriding the Saxothuringian slates and by the Hermsdorf-Rehfeld phyllites resting in a horizontal (?tectonic) position onto the Erzgebirge Basement (for geologic maps, see Chap. V.C.2; for metamorphism see Chap. V.C.2).

References

Bankwitz P (1988) Schwarzburger Antiklinorium. In: Exkursionsführer Akademie der Wissenschaften DDR. ZIPE, Potsdam, pp 136–153

Blümel P (1982) Aufbau, Metamorphose und geodynamische Deutung des Variszischen Grundgebirges im Bereich der Bundesrepublik. Ruhr-Univ Bochum, Jahrb 1982: 169–201

Blümel P (1983) Intermediate and low-pressure facies series in the Variscan Basement of Southern Germany. Terra Cognita 3: 195

Blümel P (1984) Mitteldruck- und Niederdruckmetamorphose in den außeralpinen Varisziden Mitteleuropas. Fortschr Mineral 62 (1): 28–29

Blümel P (1986) Metamorphic processes in the Variscan crust of the Central Segment. In: Freeman R, Mueller St, Giese P (eds) Proc 3rd EGT Worksh – the Central Segment. Eur Sci Found, Strasbourg

Brand R (1980) Die niedriggradige Metamorphose einer Diabas-Assoziation im Gebiet von Berg/Frankenwald. Neues Jahrb Mineral Abh 139: 82–101

Chatterjee N, Flux S (1986) Thermodynamic mixing properties of muscovite-paragonite crystalline solutions at high temperatures and pressures, and their geologic applications. J Petrol 27: 677–693

Franke W (1984) Variszischer Deckenbau im Raume der Münchberger Gneismasse – abgeleitet aus der Fazies, Deformation und Metamorphose in umgebenden Paläozoikum. Geotekton Forsch Stuttgart 68: 1–253

Hempel G (1974) Variszische Tektogenese. In: Hoppe W, Seidel G (eds) Geologie von Thüringen. Hermann Haack, Gotha-Leipzig, pp 289–334

Holdaway MH (1971) Stability of andalusite and the aluminiumsilicate phase diagram. Am J Sci 271: 97–131

Meinel G (1974) Magmatismus und Metamorphose. In: Hoppe W, Seidel G (eds) Geologie von Thüringen. Hermann Haack, Gotha-Leipzig, pp 335–355

Mielke N, Schreyer W (1969) Mineralparagenesen in Metasedimenten des Fichtelgebirges. Geol Bavarica 60: 29–44

Mielke H, Blümel P, Langer K (1979) Regional low-pressure metamorphism of low and medium grade in metapelites and psammites of the Fichtelgebirge area. Neues Jahrb Mineral Abh 137: 83–112

Rötzler J (1992) Zur Petrogenese im Sächsischen Granulitgebirge – die pyroxenfreien Granulite und die Metapelite. Geotekton Forsch 77: 1–100

Schulte B (1991) Geologisch-petrographisch-tektonische Untersuchungen am Nordostrand des Redwitzitplutons südlich von Arzberg. Diploma Thesis, TH Darmstadt

Schreyer W (1966) Metamorpher Übergang Saxothuringikum/Moldanubikum östl. Tirschenreuth/Opf. nachgewiesen durch phasenpetrologische Analyse. Geol Rdsch 55: 491–509

Stein E (1988) Die strukturgeologische Entwicklung im Übergangsbereich Saxothuringikum/Moldanubikum in NE-Bayer. Geol Bavarica 92: 5–131

Teufel St (1988) Vergleichende U-Pb- und Rb-Sr-Altersbestimmungen an Gesteinen des Übergangsbereiches Saxothuringikum/Moldanubikum, NE-Bayern. Gött Arb Geol Paläontol 35, 87 pp

Theobald H (1915) Beitrag zur Kenntnis metamorpher Gesteine aus der Umgebung von Pottiga-Sparnberg an der oberen Saale. Chemie Erde 1: 1–42

Wagener-Lohse C, Blümel P (1984) Prograde Regionalmetamorphose vom Niederdruck-Typ in der Grenzzone Saxothuringikum/Moldanubikum E Tirschenreuth/NE-Bayern. Fortschr Mineral 62 (Beih 1): 254–255

Wendt I, Kreuzer H, Müller P, Schmid H (1986) Gesamtgesteins- und Mineraldatierungen des Falkenberger Granits. Geol Jahrb Hannover E 34: 5–66

Wienholz R, Hofmann J, Mathé G (1979) Über Metamorphose, Tiefbau und regionale Position des Erzgebirgskristallins. Z Geol Wiss Berlin 7: 385–395

V.B.5 Metallogenesis

G. Tischendorf, H. Dill, and H.-J. Förster

1 Introduction

The Saxothuringian Zone sensu strictu comprises a large part of the NW edge of the Bohemian Massif and a very small one of the northernmost Black Forest in the environs of Baden-Baden as well as of the Vosges (Fig. 1). Not surprisingly, the Saxothuringian Zone of the Black Forest and the Vosges is of less importance regarding metallogenesis than the area belonging to the Bohemian Massif, that bears a great many sites of mineralization, formerly operated for base metals, as well as some first-class Sn, W, Ag, and U deposits, that have only recently been shut down for economic reasons. These sites of mineralization have been the target of many studies, with a wealth of publications during the last decades, out of which only some comprehensive papers are mentioned: Chrt et al. (1966); Reh and Schröder (1974); Bernard (1980); Štemprok (1980); Dill (1985a); Legler (1985); Baumann et al. (1986); Pouba and llavsky (1986); Tischendorf (1986); Walther et al. (1986); Tischendorf comp. 1989; Tischendorf and Förster (1994).

The ore mineralization in the Saxothuringian Zone lasted from Upper Proterozoic until Quarternary and may be grouped into six classes (Fig. 2):

1. Early to Middle Paleozoic stratabound deposits,
2. Early Hercynian, metamorphogene thrust-bound deposits,
3. Hercynian granite-related deposits,
4. Late-Hercynian stratabound deposits,
5. Post-Hercynian unconformity-related deposits,
6. Supergene deposits.

This metallogenic subdivision corresponds with pre- and main stages of the Hercynian orogeny, evolution of post-Hercynian basin-and-range topography, and geomorphological/geological processes operative during Tertiary-Quarternary.

2 Early to Middle Paleozoic Stratabound Deposits

During the early Paleozoic, pyrite- and pyrrhotite-dominated base metal deposits evolved in a rift environment along the Erzgebirge-Fichtelgebirge Anticline. These basins are prevalently filled with arenaceous and argillaceous sediments embedded in volcanoclastic and some volcanic strata. The most prominent representatives of these massive syngenetic sulfide mineralizations are Elterlein/FRG (pyrite), Jahnsbach/FRG (Zn), Waldsassen/FRG, and tisova/Czech Republic, containing pyrite, pyrrhotite, Cu-, Pb-, and Zn-sulfides. Submarine exhalative processes during incipient stages of rifting were responsible for these polymetallic deposits. Ore-bearing fluids are likely to have emanated from the sedimentary basins. This idea is corroborated by S isotope variations in sulfides, that display a very wide spread of isotopically heavy sulfur and a trend covarying with marine sulfate age curve (Nielsen 1978; Legler et al. 1984; Dill 1989). In addition to these clastibound deposits, there are also some small carbonate-hosted sulfidic Fe-Zn-Cu-Pb mineralizations (e.g., Lengefeld, Hermsdorf) in the Cambrian Klinovec Group and Thum Group, respectively (Legler 1985). The early Paleozoic age of formation of these base metal deposits is proved by Pb-isotope data (Bielicki and Tischendorf 1991).

It is still a matter of debate whether the siderite-bearing Wundsiedel Marble is of syngenetic origin or the result of redeposition of a Fe low metal preconcentration. It has much in common with other occurrences of the iron skarn association (Tischendorf 1986), the reference of which is situated at Medenec/Czech Republic. According to Nemec (1979), this metal accumulation resulted from Cambro-Ordovician contact metasomatism of granitic rocks that through subsequent metamorphism were altered to the so-called red gneisses.

Dallmeyer et al. (Eds.)
Pre-Permian Geology of Central and Eastern Europe

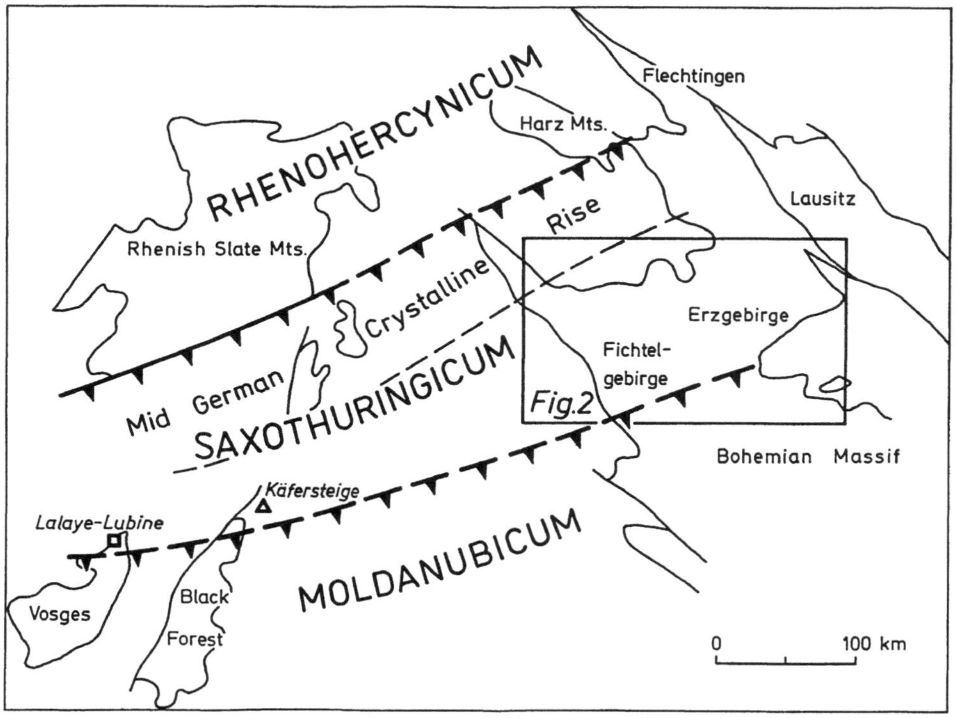

Fig. 1. Sketch map of mid-European Saxothuringian with the only two deposit regions being located in the Saxothuringian of the Black Forest and the Vosges. The southern margin of the Saxothuringian Zone is according to Franke (1989). The framed area denotes the position of the map shown in Fig. 2

The worldwide deposition of Ordovician ironstones may also be traced in the early Paleozoic sedimentary record of the mid-European Saxothuringian. Three seams containing chamosite, thuringite, magnetite, and siderite were exploited near Wittmansgereuth and Gebersreuth (Reh and Schröder 1974). Global eustatic sea-level rises accompanied by lateritic weathering on a barely vegetated land throughout Middle and Upper Ordovician were prerequisites for these sedimentary iron ores (Young 1989). Although of vast lateral extension, ore traps, where these alum shales have been enriched in metals like Mo, V, U, Au, and REEs by factors greater than 10, are rare in the Silurian and Lower Devonian Graptolite Shales. Mostly, they attain ore grade only in those sites where fracture-controlled redeposition has led to the formation of distinct ore shoots (e.g., Ronneburg, Gera; Lange et al. 1991). Their S isotope patterns resemble that of syn-(dia)-genetic Ordovician ore seams, with a very wide δ^{34}S range.

There is a striking difference with respect to S isotopes (narrow spread, close to zero), host rocks (bimodal volcanic and volcanoclastic rocks of calc-alkaline affinity) and metal composition (Fe-Zn-Cu, no PbS) between the formerly mentioned Ordovician mineralizations and stratabound deposits of Kupferberg-Neufang-Sparneck. They may tentatively be described as Besshi type.

In terms of Fe concentration, another peak may be observed in the Saxothuringian Zone during Middle to Upper Devonian with the formation of the Lahn-Dill exhalative hematite-magnetite ores, especially near Schleiz-Pörmitz (Rösler 1964). These deposits are bound to keratophyric volcanic and volcanoclastic rocks. The adoption of a model based on present-day submarine metalliferous brines is attractive, yet can hardly convincingly account for the preponderance of Fe minerals over base metals. Preconcentration of Fe in a relatively thick crust might be a plausible explanation for this iron preference in Upper Devonian emanations.

3 Early Hercynian, Metamorphogene Thrustbound Deposits

There is a group of elements – Sb, Au, Hg, (As) – that was prevalently accumulated in shear zones

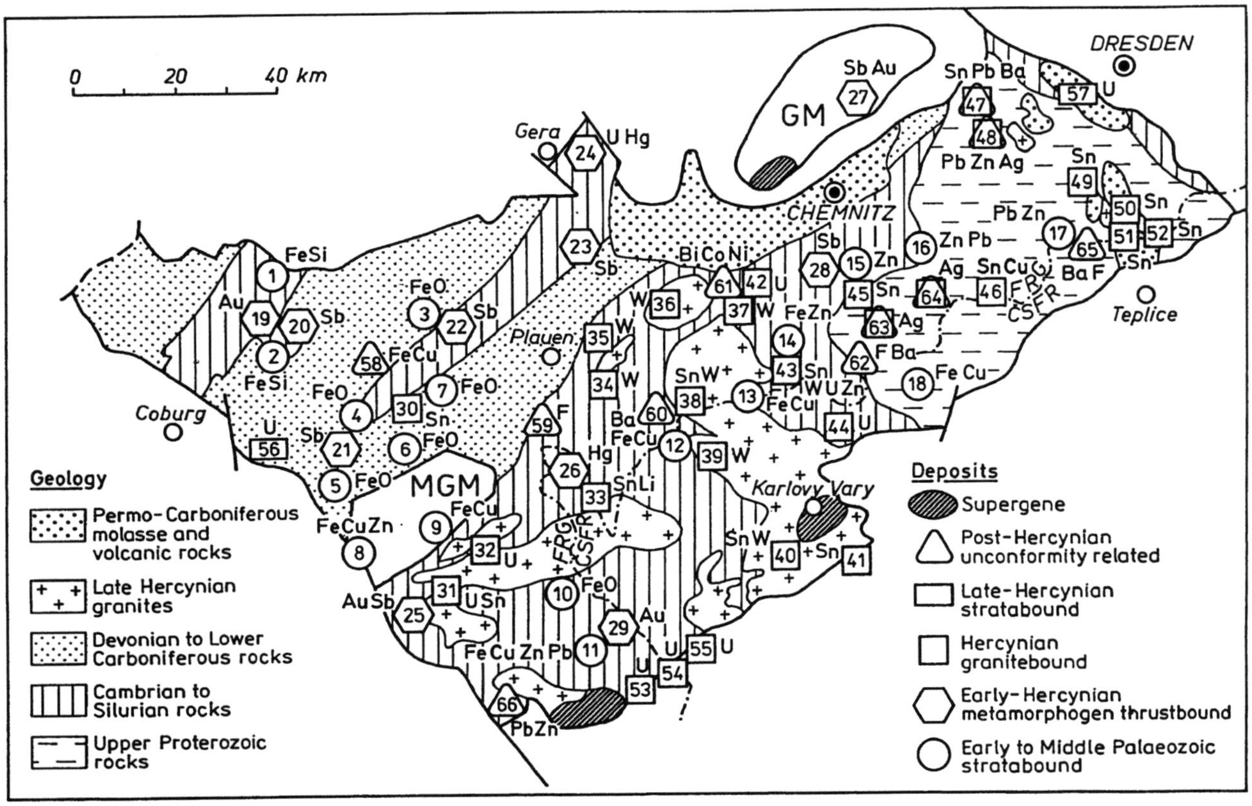

Fig. 2. Mineral deposits in the Saxothuringian of the Bohemian Massif. *Early to Middle Paleozoic stratabound deposits*: *1* Wittmansgereuth; *2* Ludwigstadt; *3* Schleiz; *4* Langenbach-Steinbach; *5* Stadtsteinbach-Presseck; *6* Weidesgrün; *7* Gebersreuth; *8* Kupferberg-Neufang; *9* Sparneck; *10* Arzberg; *11* Waldsassen; *12* Tisova; *13* Johanngeorgenstadt; *14* Elterlein; *15* Jahnsbach; *16* Lengenfeld; *17* Hermsdorf; *18* Medenec; *Early Hercynian, metamorphogene thrustbound deposits*: *19* Reichmannsdorf; *20* Gräfenthal; *21* Wolfersgrün; *22* Schleiz; *23* Greiz; *24* Gera; *25* Brandholz-Goldkronach; *26* Horni Luby; *27* Hartmannsdorf; *28* Hormersdorf; *29* Neualtenreuth; *Hercynian granite-related deposits*: *30* Sparnberg-Pottiga; *31* Weißenstadt; *32* Großschloppen-Hebanz; *33* Veřnerov; *34* Tirpersdorf; *35* Zobes; *36* Pechtelsgrün; *37* Zschorlau; *38* Gottesberg-Mühlleiten; *39* Rotava; *40* Horny Slavkov; *41* Krasno; *42* Hartenstein; *43* Breitenbrunn-Pöhla; *44* Jachimov; *45* Ehrenfriedersdorf; *46* Seiffen; *47* Halsbrücke; *48* Freiberg; *49* Sadisdorf; *50* Altenberg; *51* Zinnwald-Cinovec; *52* Krupka; *53* Poppenreuth; *54* Mähring; *55* Dylen; *Late Hercynian stratabound deposits*: *56* Stockheim; *57* Freital *Post-Hercynian unconformity-related deposits*: *58* Lichtenberg-Lobenstein; *59* Schönbrunn-Bösenbrunn; *60* Brundöbra; *61* Schneeberg; *62* Niederschlag; *63* Annaberg; *64* Moldava; *65* Erbendorf; *GM* Granulite Massif; *MGM* Münchberg Gneiss Massif

during Upper Devonian through Lower Carboniferous (Dill 1985b; Tischendorf comp 1989; Bielicki and Tischendorf 1991). Its mineralization-merits particular attention for its two-step concentration. Following an early Paleozoic protore stage (e.g., Hormersdorf, Gräfenthal Horst), Sb, Au, and Hg were eventually redeposited during the late Paleozoic along thrust zones of the Schwarzburg, Berga, and Fichtelgebirge-Erzgebirge anticlines. Two distinct mineral assemblages may be recognized: (1) monotonous/distal facies: stibnite, arsenopyrite, cinnabar – no gold, (2) polymetallic/proximal facies: stibnite, arsenopyrite, Sb sulfosalts – locally with native gold.

4 Granite-Related Deposits

The Upper Carboniferous to Lower Permian granitoids (330–290 Ma; Besang et al. 1976; Gerstenberger 1989) gave rise to the formation of a great variety of Sn-W, Pb-Zn, and U deposits in and around the Fichtelgebirge-Erzgebirge Anticline. We distinguish two main intrusive complexes exhibiting distinct metallogenic signatures (Tischendorf and Förster 1990): an older one (OIC, 330–310 Ma), mixed I/S-type, monzogranitic, moderately specialized and lacking Li-F accumulation, and a younger one (YIC, 305–290 Ma), mostly I-type with some A-type affinities of representatives located in the eastern Erzgebirge, monzo- to syenogranitic, highly specialized with strong Li-F accumulation, partially autometasomatically altered. To the OIC granites, wolframite-molybdenite-muscovite greisen and vein-type associations (Pechtelsgrün) are related. The cassiterite-wolframite-topaz-Li mica associations (Altenberg, Zinnwald, Ehrenfriedersdorf, Weißenstadt), however, show a close regional relationship to granites of the younger complex. During a subsequent stage, cassiterite-topaz-muscovite mineralizations poor in Li, and enriched in Cu-Bi (Gottesberg, Mühlleiten) were formed. For tin production, the first-order targets were the metagranitic metasomatic greisen bodies or the structurally controlled ore shoots within massifs and vein fillings above granite stocks and ridges. Metal accumulations are often connected with fluid-explosive brecciation processes (Seltmann and Schilka 1991). Greisenization in all stages is linked with huge element supply (Si, F, Sn) and removal (Na, K) (Gottesmann et al. 1994).

The concept of genetical relationships between W-Mo associations and OIC granites, as well as Sn-W-F (Li-bearing or -free) associations and YIC granites is supported not only by an identical chemical specialization, but also by differences in the physicochemical signatures of the magmatic fluid system (OIC: intermediate oxygen and low fluorine fugacities, YIC: low fO_2 and high fHF), which prevented (OIC) or stimulated (YIC) tin accumulation (Förster and Tischendorf 1989, 1992).

Younger than YIC tin associations are vein-type base metal mineralizations with arsenopyrite, pyrite, chalcopyrite, sphalerite, galena, quartz, and subordinate amounts of tetrahedrite and other sulfosalts, argentite, native silver, and carbonates (Freiberg, Schneeberg, Annaberg). Mineral assemblages of more than 1000 veins, cutting through Upper Proterozoic gneisses in the Freiberg ore district, show strong regional and temporal zonation (Baumann 1958). Geological and isotopic data point to a Lower Permian age of vein mineralization, probably contemporaneous with volcanic sequences of the YIC. Contrary to W-Mo (OIC) and Sn-W-F (YIC) associations, that are likely to have been orthomagmatically controlled, the younger base metal mineralization may have been brought about by residual solutions from deeper-seated magma chambers increasingly mixed with intraformational and meteoric solutions.

Within the Vosges ore province, the polymetallic vein-bound mineralizations of Lalaye-Lubine district belong to the Saxothuringian part of that metallogenic group (see Fluck et al. 1975). Sb-rich associations in the Schists de Ville show similarities to the metamorphogene thrust-bound type.

Uranium-bearing associations in the Erzgebirge (Aue, Schneeberg; Lange et al. 1991) seem to be oxidized equivalents of the base metal mineralizations described above. Harlass and Schützel (1965) distinguished three stages: (1) pitchblende with quartz, fluorite, hematite, coffinite, calcite, anhydrite; (2) pitchblende with fluorite, dolomite, hematite, selenides, and (3) remobilized pitchblende within an Bi-Co-Ni-As association. The oldest pitchblende yielded an upper intercept in the discordia plot at 295 ± 4 Ma (Carl et al. 1983) and shows a genetic relation to the YIC and related subsequent volcanics. The source of uranium is as yet not well defined. However, oxydized and halogen-rich solutions are thought to have mobilized U, especially from the low-Ca peraluminous YIC granites, where it is mainly located in uraninite (high U/Th whole rock ratio) representing an easily leachable metal preconcentration (Tischendorf and Förster 1994).

5 Late-Hercynian Stratabound Deposits

In the aftermaths of Sudetian movements and intrusion of granites into the Saxothuringian basement, narrow basins subsided in the environs of Stockheim, Erbendorf, Chemnitz-Zwickau, and Döhlen. There is no doubt that a great portion of detritus derived from reworking of volcanic material during Permo-Carboniferous. Its abundance in labile constituents such as glass shards or pumice particles, susceptible to leaching by intrastratal solutions, the alternating red and gray beds attesting to alternating redox conditions and permeability contrasts in these fluvio-lacustrine depositions (Lützner 1987), favored the precipitation of sooty pitchblende and carburan in terrigenous clastic rocks of Döhlen basin (Freital), Stockheim basin, Baden-Baden basin

(Müllenbach-Black Forest), and near St. Hippolyte (Vosges).

6 Post-Hercynian Unconformity-Related Deposits

The Saxothuringian Zone is not so abundant in fluorite-barite vein-type deposits as the adjacent parts of the Hercynian orogen. Nevertheless, these F-Ba mineralizations sparked off a discussion on their origin. Baumann (1979) advocated a simatic derivation, whereas Bernard and Škvor (1980) stressed a granite-related genesis. It cannot be denied that impregnative fluorite in Upper Carboniferous through Lower Permian volcanoclastic and volcanic rocks was formed by remobilization processes affecting anomalously F-rich late Hercynian granites, but any direct genetic link between these potential source rocks and F-Ba mineralization does not exist. F-Ba veins occur in the crystalline basement (Halsbrücke, Brunndöbra, Schönbrunn) and its overburden up to the Lower Triassic Bunter (e.g., Käfersteige, N Black Forest; see Möller et al. 1982). $^{87}Sr/^{86}Sr$ ratios, $\delta^{34}S$ values, and REE patterns in context with the position of F-Ba deposits relative to the late Paleozoic paleotopography point to the late Hercynian unconformity or peneplain as a reference plain for these veintype deposits and for a crustal origin. It is a first-order geohydraulic plain in close vicinity of which fluids were able to percolate through the crushed country rocks and to trigger barite-fluorite mineralization (Dill 1988).

Most of these mineralizations are governed by block-faulting. The huge Käfersteige fluorite deposit belongs to a dextral strike-slip system. Rhythmic tectonic movements often initiate rhythmic precipitations with cyclic variations of trace and rare earth elements in fluorite and of main elements in fluid inclusions (Kämpf et al. 1985). Convective systems in uplifting blocks may have caused leaching of pre-enriched rocks (evolved granites, their cumulates, or other suitable rocks) and initiated mineralizing processes over a long period. Rösler and Pilot (1967) already determined Mesozoic ages (220–150 Ma) of some of the fluorite-barite bearing mineralizations. According to Pb/Pb data interpreted in the sense of the AMOVs dynamic model, mineralizing stages can be assumed in the intervals 255-245 (Schönbrunn; Thomas et al. 1989), 215-205 (Gersdorf), 196-188 as well as 183–159 Ma (Halsbrücke; Bielicki and Tischendorf 1991).

Age and origin of the Bi-Co-Ni association (Schneeberg) with Co-Ni arsenides, native bismuth, and arsenic, as well as silver arsenides and sulfides is only poorly understood to date, but according to some new thorogenic Pb/Pb model ages in the interval 184–177 Ma (K.-H. Bielicki 1991, unpubl.), we assume a affilation at least to Mesozoic events. Up to now, source rocks of this assemblage have remained unspecified.

7 Supergene Deposits

Even throughout Tertiary and Quarternary, the Saxothuringian country rocks and Paleozoic to Mesozoic ore deposits have been on impact on supergene ore deposition. It encompasses uranium yellow ores, Fe-Mn-bearing duricrusts, Ni-hydrosilicate enrichments, and china clay deposits. Large areas underlain by feldspathic crystalline rocks are concealed by a kaolinized saprolite such as the Naab-Wondreb-Depression/FRG and the West-Bohemian china clay triangle Karlovy Vary-Podbořany-Kadan (Czech Republic). It has to be noted that kaolinization in western Bohemia is in part of pre-Tertiary age, beginning with the first exposure of crystalline rock to pervasive weathering during the Carboniferous (Kuzvart 1969). Peneconcordant redox-controlled uranium accumulations of the roll front type occur in Upper Cretaceous sandstones in the Elbe Graben south of Dresden near Königstein (Lange et al. 1991). Erosion of cassiterite enrichments in zones near the contact of the Hercynian younger granites led to the formation of alluvial placers which already in the Middle Ages were the objective of mining activities.

8 Summary and Conclusions

Based on the detailed investigation of each type of ore mineralization six metallogenic stages may be delineated for the Saxothuringian Zone.

Stage I comprehends Upper Proterozoic (?) or Cambrian to Silurian mineralizations commenced with rift-related iron sulphide deposits and faded out with volcanic-hosted Fe-Cu-Zn deposits and Fe skarns. Ore types and country rocks attest an extensional regime succeeded by a compressional one, probably of B-subduction style, at the end of

Silurian, supporting the view that the Cambrian to Silurian mineralizations are of Besshi type. During Middle to Upper Devonian, riftogene processes and deep-seated faulting again provoked the ascent of basic magma. Unlike the Rhenohercynian Zone abundant in Devonian stratabound base metal (e.g., Meggen, Rammelsberg) or barite deposits (e.g., Günterod, Lohrheim), the Saxothuringian is poor in rift-related base metal deposits, but enriched in hematite-magnetite deposits such as the Rhenohercynian. Subsequent thrustbound Sb-Au-Hg mineralization was the product of leaching processes during Hercynian crustal stacking and representative of stage II.

Mineralizations of stage III are related in time and space to Hercynian granites and volcanics (older magmatic complex: WMo, PbU?; younger magmatic complex: SnLi, SnCu, ZnPbU). They are mainly controlled by orthomagmatic processes, the fluids of which were increasingly influenced by intraformational and meteoric solutions. Moreover, some Permian Cu-U accumulations in molasse basins may also fall into this category.

Peneplaination, block tectonics, uplifting, and initiation of convection systems by granites of the HHP type were responsible for widespread leaching to form fluorite-barite-base metal as well as BiCoNi associations during Permian to Jurassic (stage IV).

A model which may explain the whole magmatic-metallogenic process from the Carboniferous to Triassic-Jurassic is illustrated in Fig. 3. It reflects the complicated relations between plutonic and subsequent volcanic formations, the various mineralizations related with them in the Erzgebirge ore province, and the position of Mesozoic ore associations which often occur

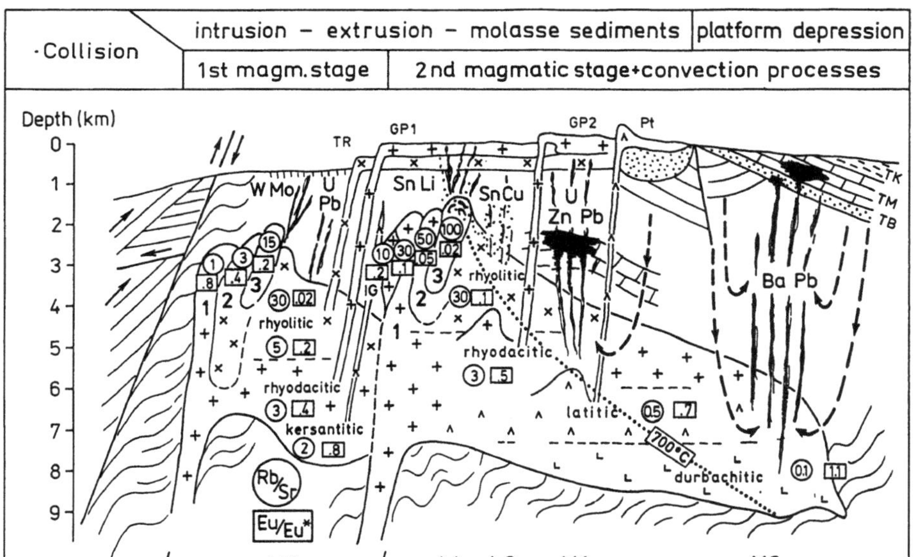

Fig. 3. Model of the Older (*OMC*) and the Younger Magmatic Complex (*YMC*) in the Erzgebirge, including in each case three intrusion phases after in situ differentiation and several extrusion phases after vertical compositional zoning as well as the corrdinated ore associations (*L* Leucrocratic metallogenesis; *M* melanocratic metallogenesis) in a depth-time diagram. *Rb/Sr* and *Eu/Eu* ratios* reflect the degree of fractionation. *Arrows at the right* express increasing influence of convection and leaching proceses during M1-M2 mineralization. 700 °C isotherm approximates temperature decrease from 290 to 230 Ma. *IG* Intermediate granites, marginal granites of YIC; *TR* Teplice rhyolite; *GP* 1,2-granite porphyries in different positions; *Pt* feldspar and micaceous porphyrites. (Tischendorf and Förster 1990)

in uplifted blocks above segments with highly fractionated Hercynian granites, but also reach and mineralize overlying Triassic sediments.

Recurrent vertical movements played a significant part also for stage V mineralization during Tertiary and Quarternary. Uplifting exposed basement rocks to weathering and were conducive to the generation of U, Fe-Mn, and Ni deposits, as well as kaolinized saprolite. Surrounding the Late Hercynian granite tin-bearing, alluvial placers were formed.

Lead isotope data, especially the $^{207}Pb/^{204}Pb$ ratio (Cambrian-Ordovician associations: 15.52–15.59 to Triassic-Jurassic ones: 15.59–15.65) constrain in the main crustal sources for the mineralizations (Höhndorf and Dill 1986; Bielicki and Tischendorf 1991).

References

Baumann L (1958) Tektonik und Genesis der Erzlagerstätte von Freiberg (Zentralteil). Freiberger Forschungsh C 46: 208

Baumann L (1979) Some aspects of mineral deposit formation and the metallogeny of Central Europe. Verh Geol Bundesanst Wien 1978: 205–220

Baumann L, Kölbel B, Kraft M et al. (1986) German Democratic Republic. In: Dunning SWF, Evans AM (eds) Mineral deposits of Europe, vol 3. Central Europe. Inst Mining and Metallurgy, London, pp 303–329

Bernard JH (1980) Paragenetic units of the European Variscan megazone. Freiberger Forschungsh C 354: 55–117

Bernard HJ, Škvor V (1980) The reactivation of ancient massif and metallogeny: the example of the Bohemian Massif. Econ Geol 75: 251–259

Besang C, Harre W, Kreuzer H, Lenz H, Müller P, Wendt I (1976) Radiometrische Datierung, geochemische und petrographische Untersuchungen der Fichtelgebirgsgranite. Geol Jahrb E 8: 3–71

Bielicki KH, Tischendorf G (1991) Isotopic composition of ore leads from Central Europe. Lead isotope and Pb-Pb model age determination and their metallogenetic interpretation. Contrib Mineral Petrol 106: 440–461

Carl C, Dill H, Kreuzer H, Wendt I (1983) U-Pb dating of ores in NE Bavaria. Terra Cognita: 195–196

Chrt J, Bolduan H, Bernstein KH et al. (1966) Die postmagmatische Mineralisation des Westteils der Böhmischen Masse. Sb Geol Věd Lg Praha 8: 113–192

Dill H (1985a) Die Vererzung am Westrand der Böhmischen Masse. Metallogenese in einer ensialischen Orogenzone. Geol Jahrb D 73: 3–461

Dill H (1985b) Antimoniferous mineralization from the mid-European Saxothuringian Zone: mineralogy/geology, geochemistry and ensialic origin. Geol Rdsch 74: 447–466

Dill H (1988) Geologic setting and age relationship of fluorite-barite mineralization in southern Germany with special reference to the late Paleozoic unconformity. Mineral Deposit 23: 16–23

Dill H (1989) Metallogenetic and geodynamic evolution in the Central European Variscides – a pre-well site study for the German Continental Deep Drilling Programme. Ore Geol Rev 4: 279–304

Fluck P, Weil R, Wimmenauer W et al. (1975) Géologie des gîtes minéraux des Vosges. Mém BRGM 87: 189

Förster HJ, Tischendorf G (1989) Reconstruction of the volatile characteristics of granitoidic magmas and hydrothermal solutions on the basis of dark micas: the Hercynian postkinematic granites and associated high-temperature mineralizations of the Erzgebirge (GDR). 1. Communication: calculation procedure and results. Chemie Erde 49: 7–20

Förster HJ, Tischendorf G (1992) Volatile signatures of the Hercynian postkinematic granites of the Erzgebirge: implications to related tin-tungsten-molybdenum metallogenesis. Chemie Erde 52: 49–61

Franke W (1989) Tectonostratigraphic units in the Variscan belt of central Europe. Geol Soc Am Spec Pap 230: 67–90

Gerstenberger H (1989) Autometasomatic Rb enrichments in highly evolved granites causing lowered Rb-Sr isochron intercepts. Earth Planet Sci Lett 93: 65–75

Gottesmann B, Wasternack J, Märtens S (1994) The Gottesberg tin deposit (Saxony): geological and metallogenic characteristics. In: Seltmann R, Kampf H, Möller P (eds) Metallogeny of collisional orogens. Czech Geol Surv, Prague, pp 110–115

Harlass E, Schützel H (1965) Zur paragenetischen Stellung der Gangpechblende in den hydrothermalen Lagerstätten des westlichen Erzgebirges. Z Angew Geol 11: 569–582

Höhndorf A, Dill H (1986) Lead isotope studies of stratabound, vein-type and unconformity-related Pb, Sb and Bi ore mineralizations from the western edge of the Bohemian Massif (F.R. Germany). Mineral Deposit 21: 329–336

Kämpf H, Bankwitz P, Strauch G et al. (1985) Local and regional processes and zoning in a hydrothermal Late Variscan vein mineralization from the southern part of the G.D.R. Gerlands Beitr Geophys 94: 426–434

Kuzvart M (1969) Kaolin deposits of Czechoslovakia In: Malkovský M, Vachtl J (eds) 23th Int Geol Congr, Academica Prague 15: 47–73

Lange G, Mühlstedt P, Freyhoff G, Schröder D (1991) Der Uranerzbergbau in Thüringen und Sachsen – ein geologisch-bergmännischer Überblick. Erzmetall 44: 162–171

Legler C (1985) Die schichtgebundenen Mineralisationen des Erzgebirges. Freiberger Forschungsh C 401: 93

Legler C, Pilot J, Schlichting M (1984) Blei- und Schwefelisotopenbestimmungen an schichtgebundenen Mineralisationen der Fichtelgebirgisch-

Erzgebirgischen Antiklinalzone. Freiberger Forschungsh C 389: 122–150
Lützner H (1987) Sedimentary and volcanic Rotliegendes of the Saale Depression. Excursion Guidebook. Zentralinst Phys Erde Potsdam, 197 pp
Möller P, Maus H, Gundlach H (1982) Die Entwicklung von Flußspatmineralisationen im Bereich des Schwarzwaldes. Jahrb Geol Landesamt Baden-Württemberg Freiburg 24: 35–70
Nemec D (1979) Genese der regional-metamorphen Skarne der Böhmischen Masse. Chemie Erde 38: 194–209
Nielsen H (1978) Isotopes in nature – sulfur. In: Wedepohl KH (ed) Handb geochem 16 B: 1–40
Pouba Z, Ilavsky J (1986) Czechoslovakia. In: Mineral deposits of Europe, vol 3. Central Europe. Inst Mining and Metallurgy London, pp 117–173
Reh H, Schröder N (1974) Erze. In: Hoppe W, Seidel G (eds) Geologie von Thüringen. Hermann Haack, Gotha, pp 867–997
Rösler HJ (1964) Genetische Probleme der Erze des sogenannten erweiterten Lahn-Dill-Typs. Ber Geol Ges DDR 9: 445–454
Rösler HJ, Pilot J (1967) Die zeitliche Einstufung der sächsisch-thüringischen Ganglagerstätten mit Hilfe der K-Ar-Methode. Freiberger Forschungsh C 209: 87–98
Seltmann R, Schilka W (1991) Metallogenetic aspects of breccia-related tin granites in the Eastern Erzgebirge. Z Geol Wiss 19: 485–490
Štemprok M (1980) Tin and tungsten deposits of the West-Central European Variscides. In: Ridge JD (ed) Proc 5th IAGOD-Symp, Stuttgart, pp 496–512
Thomas R, Strauch G, Bielicki KH, Haase G, Kämpf H, Klemm W (1989) Temperature-time-matter-related convection model of a granitic intrusion. In: Wand U, Strauch G (eds) Proc 5th Work Meet Isotopes in nature. Zentralinst Isotopen- und Strahlenforsch Leipzig, pp 397–409
Tischendorf G (1986) Variscan ensialic magmatism and metallogenesis in the Ore Mountains – modelling of the process. Chemie Erde 45: 75–104
Tischendorf G, Förster HJ (1990) Acid magmatism and related metallogenesis in the Erzgebirge. Geol J 25: 443–454
Tischendorf G, Förster HJ (1994) Hercynian granitic magmatism and related metallogenesis in the Erzgebirge: a status report. In: Gehlen Kv, Klemm DD (eds) Mineral deposits of the Engebirge/Krušné hory (Germany/Czech Republic). Monogr Ser Miner Deposit 31: 5–23
Tischendorf G comp. (1989) Silicic magmatism and metallogenesis of the Erzgebirge. Veröff Zentralinst Phys Erde Potsdam 107: 316
Walther HW, Emmermann KH, Fenchel W et al. (1986) Federal Republic of Germany. In: Dunning SWF, Evans AM (eds) Mineral deposits of Europe, vol 3. Central Europe. Inst Mining and Metallurgy London, pp 175–301
Young TP (1989) Eustatically controlled ooidal ironstone deposition: facies relationships of the Ordovician open-shelf ironstones of western Europe. In: Young TP, Taylor WEG (eds) Phanerozoic ironstones. Geol Soc Spec Publ 46: 51–63

V.C Exotic Metamorphic Nappes

V.C.1 Stratigraphy, Structure, and Igneous Activity

W. Franke, H. Kreuzer, M. Okrusch, U. Schüssler, and E. Seidel

1 Introduction

As laid out in Chapters V.B.1 and V.B.2, the parautochthonous Thuringian facies of the Saxothuringian Belt has been overthrust from the SE by nappes, which have survived in the core of the Vogtland Synform. These are the tectonic klippen of Münchberg, Wildenfels, and Frankenberg (in order from W to E; see Fig. 1). The klippen represent, in fact, erosional remnants of one or several nappe piles, which were originally more extensive, and possibly laterally continuous with each other. The tectonic sequence in each of these klippen is characterized by inversion of the stratigraphy and of the metamorphic grades. The lower nappes comprise (in order from bottom to top): proximal Early Carboniferous flysch, Silurian/Devonian bedded chert, and a volcano-sedimentary sequence of largely Ordovician age. These very-low-grade, fossiliferous sequences are overlain by a number of metamorphic thrust sheets. This chapter describes the lithology, tectonic sequence, and internal structure of the metamorphic nappes in the klippen, and of their probable equivalents at the northwestern margin of the Moldanubian Region, from which they are derived. We set out from the Münchberg klippe, which is the largest, most differentiated and best studied example of its kind, and then briefly discuss the closely related units. Details on the metamorphic evolution are available in the contribution by Blümel (Chap. V.C.2)

It should be noted that the Wildenfels and Frankenberg klippen are still held to be parautochthonous, especially by some East German workers. However, the lithological composition and tectonic structure of these units are very similar to that of the Münchberg klippe, whose allochthonous nature is fairly well established, so that a different tectonic origin is highly improbable. A detailed discussion of the pros and cons is available in Franke (1984a,b). A conclusive treatment is not possible because of the lack of modern studies (reflection seismics, geochronology, structural geology) in the Wildenfels and Frankenberg sequences.

2 Münchberg Nappe Pile

The Münchberg klippe consists of four metamorphic nappes, which are described in order from top to bottom (Fig. 2).

2.1 Hangend-Serie (HS)

The HS (= upper sequence) consists mainly of banded hornblende gneisses and amphibolites, as well as paragneisses with some calc-silicate rocks and marbles. The mafic rocks are characterized by the assemblage hornblende + plagioclase (oligoclase-andesine) + quartz + rutile (partly replaced by titanite) ± almandine-rich garnet ± zoisite. According to Pommerenke (1985), the chemical composition of the mafic rocks is similar to that of calc-alkaline basalts. At or near the base of the HS, there are a number of lenses of serpentinite, eclogite, and eclogite-amphibolite. Dark eclogites have a composition comparable with N-MORB, whereas light, kyanite-bearing eclogites are high-Al-basaltic in character (Matthes et al. 1975; Puchelt et al. 1978) and may be derived from plagioclase-rich gabbroic cumulates (Stosch and Lugmair 1987, 1990).

The eclogites have been metamorphosed at pressures of 15 to more than 20 kb and temperatures of 620 ± 30 °C (Franz et al. 1986; Okrusch et al. 1990a, 1991). They were overprinted under amphibolite-facies conditions at about 10 kb and 600–700 °C (Franz et al. 1986; Bosbach et al. 1991). Mylonitic metasediments associated with the eclogites have a very similar tectonometamorphic history (Klemd et al. 1991); however, it is still an open question whether the P-T-t path of the high-pressure rocks is representative for all

Dallmeyer et al. (Eds.)
Pre-Permian Geology of Central and Eastern Europe

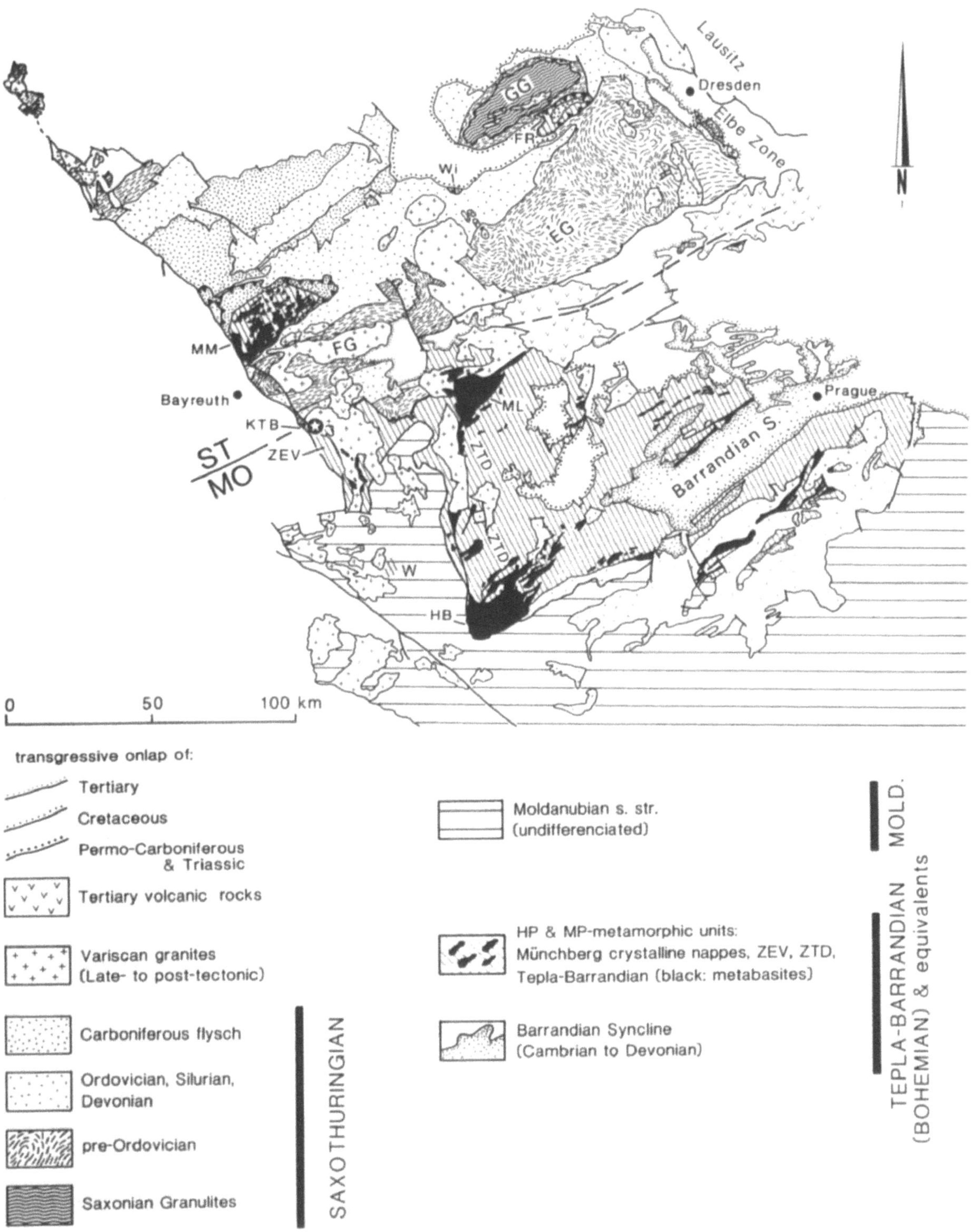

Fig. 1. Structural map of the northwestern part of the Bohemian Massif. Abbreviations: *ST* Saxothuringian; *MO* Moldanubian region; *GG* Saxonian Granulite Complex; *MM*, *Wi*, *FR* metamorphic klippen of Münchberg, Wildenfels, and Frankenberg, and underlying low-grade allochthons; *FG*, *EG* antiform of the Fichtelgebirge and Erzgebirge; *KTB* Continental Deep Drilling Site; *ZEV* zone of Erbendorf/Vohenstrauß (includi the Erbendorf greenschists at the northwesternm part of the ZEV); *ZTD* zone of Teplá/Domazlice; high-grade metabasites near Winklarn; *ML* Marians Lazne ophiolite; *HB* Hoher Bogen (Neukirchen-Kdyn igneous complex

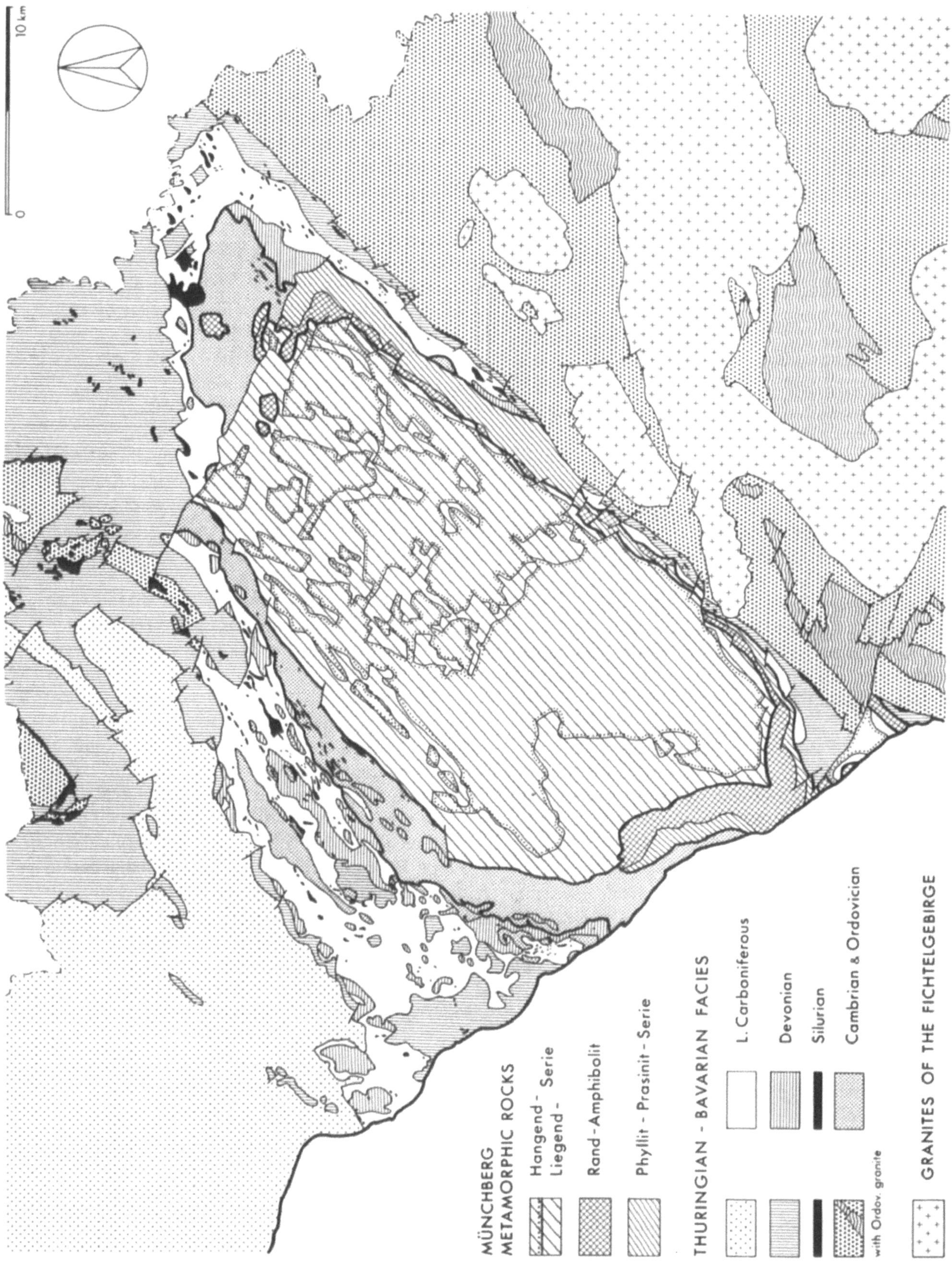

Fig. 2. Geological map of the Münchberg nappe pile and the surrounding parautochthon. (After Behr et al. 1982)

of the HS, as has been suggested by Grüner and Morteani (1990).

According to a Sr evolution diagram[1] of Söllner et al. (1981a), the maximum sedimentation age of metasediments in the HS is approx. 600 Ma, i.e., latest Proterozoic (assuming that the adopted or averaged Rb/Sr ratio is representative and has not been changed by later processes). The protolith of one type of eclogites crystallized in early Paleozoic time (480 ± 23 Ma by Sm-Nd whole-rock; Stosch and Lugmair 1990).

The age of the high-pressure metamorphism of the HS is still being debated. Müller-Sohnius et al. (1987) argue for a late Ordovician age around 435 Ma. Their data base comprises the paragneiss Rb-Sr isochron date of 435 ± 14 Ma (Söllner et al. 1981a, recalculated), a recalculation of zircon data from two eclogites analyzed by Gebauer and Grünenfelder (1979) treating them as being concordant (425 ± 5 and 411 ± 11 Ma), and – less convincingly – the similarity with eclogite ages in France (Ducrot et al. 1983). On the other hand, Stosch and Lugmair (1990) argue for an eclogitization around 395 Ma, followed by an amphibolite facies metamorphism within less than 15 Ma. For one sample with the highly retentive Sm-Nd system they derived an isochron age of 395 ± 4 Ma from whole-rock, eclogite minerals and a garnet core. If the one garnet core is neglected, all other data points define an isochron age of 384 ± 4 Ma. Kreuzer and Seidel (1989) report Ar-Ar plateau ages of 390 Ma for coarse 3T phengites in eclogite, but 377 Ma for muscovite in cross-cutting pegmatites and 380 Ma for hornblendes in amphibolites. This confirms earlier K-Ar data on hornblendes with a mean value of 380 Ma (Kreuzer et al. 1989b). Younger dates even from the same outcrop (Hammerschmidt and Franz 1988, 1991) are interpreted as local rejuvenations – a feature seldom observed in the Münchberg Massif, but ubiquitous in related rock units in NE-Bavaria (ZEV, ZTD, see below).

The narrow spread of zircon, hornblende, and mica ages is also observed in the underlying metamorphic nappes and indicates rapid cooling and uplift in middle Devonian time.

[1] Where necessary, Rb-Sr age estimates were recalculated with the IUGS-recommended constants (Steiger and Jäger 1977). Error estimates are at the 95% level of analytical confidence (2 sigma).

2.2 Liegend-Serie (LS)

The LS (= lower sequence) consists of pelitic and quartzofeldspathic paragneisses with intercalations of an extensive sheet of felsic orthogneiss and some smaller bodies of metagabbros and metagranodiorites. The gneisses show the assemblage quartz + albite + K-feldspar + muscovite ± biotite ± garnet.

Former contact aureoles are marked by metahornfelses. Isochemical conditions provided, the leucocratic to mesocratic intrusive rocks with a sodium predominance can be classified by their geochemical composition as tonalites, granodiorites, granites, and quartz-rich granites. Following the chemical classification of Chappell and White (1974) the granitoids compare well to S-type granites (Okrusch et al. 1990b).

Metagabbros and metagabbronorites still exhibit igneous structures and minerals. Regional metamorphism yielded the assemblage oligoclase/andesine + hornblende ± garnet ± zoisite ± white mica + quartz + rutile/titanite, typical of the amphibolite facies.

The peak of metamorphism in a metagabbro occurred at 600 °C and approx. 11 kb (Bosbach et al. 1991). Mineral relics of a high-pressure event were not recorded in the metagabbros or in any other rock type of the LS. However, geochemical considerations of Bosbach et al. (1991) would be compatible with a common magmatic origin of the light eclogite type in the HS and the metagabbros of the LS, which also display a high-Al basaltic geochemical character. According to these authors, the different, but converging P-T-t paths of the LS and the HS reflect a different tectonometamorphic evolution of these two units after magmatism and before amphibolite-facies metamorphism.

The age of the protoliths of the metasediments is poorly constrained; from a Sr evolution diagram, Söllner et al. (1981a) suggest a maximum age of sedimentation of 765 Ma, though older ages up to 1100 Ma cannot be excluded. Anyhow, deposition of the sediments must predate the crystallization age of the intrusions, which have yielded ages of approx. 500 Ma (499 ± 20 Ma, Rb-Sr whole-rock isochron in meta-granite, Söllner et al. 1981a; 500 Ma, U-Pb on zircon and monazite from meta-gabbro and meta-granite, Gebauer and Grünenfelder 1979).

The age of metamorphism in the LS is debated. Müller-Sohnius et al. (1987) have argued in favor of an Ordovician metamorphism, which is indicated by a Rb-Sr whole-rock isochron from a paragneiss (490 ± 22 Ma), corroborated by Rb-Sr muscovite dates averaging 484 Ma (Söllner et al. 1981a,b). According to Müller-Sohnius et al., the LS and the HS were brought in contact with each other not before approx. 385 Ma, and cooling recorded by hornblende and muscovite K-Ar and Ar-Ar dates, and muscovite Rb-Sr dates around 380 Ma (Söllner et al. 1981b; Kreuzer and Seidel 1989; Kreuzer et al. 1989b) was the first event common to both these units.

However, the metamorphic foliation and mineral lineation of the LS are not discernible form those in the HS, which strongly argues for a common medium-pressure metamorphic overprint. Besides, it is important to note that the Ordovician data interpreted as compression-related metamorphism conform with the intrusion age of the associated magmatic rocks, which appear to belong to the Cambro-Ordovician rifting event documented in all of the Saxothuringian Belt (see Chap. V.B.1). Likewise, none of the unmetamorphosed sequences in the Saxothuringian or Tepla-Barrandian (Bohemian) realms shows any indication of an orogenic event between 450 and 500 Ma. Therefore, several authors (e.g., Weber 1984) have considered a thermal event relating to the early Paleozoic rifting to explain the relevant geochronological data. At present, the interpretation of the Ordovician mineral ages is an unsolved problem.

2.3 Rand-Amphibolit-Serie (RA)

The RA (= marginal amphibolites) consists almost exclusively of massive, banded, and schistose amphibolites. The upper and lower parts of the RA have undergone pervasive mylonitization.

The typical mineral assemblage of the amphibolites is hornblende + andesine ± grossular-rich almandine ± epidote + titanite + opaques. A retrograde overprint, combined with penetrative deformation, is documented by the formation of actinolitic hornblende, albite, chlorite, and epidote, especially in the lower mylonites.

The chemical composition of these rocks conforms to subalkaline, tholeiitic basalts which exhibit a pronounced differentiation trend from Mg-rich, Ti- and Zr-poor to Fe-, Ti- and Zr-rich compositions (Oppermann 1985; Okrusch et al. 1989). The intercalation of few carbonate lenses suggests derivation from effusive or shallow-level intrusive rocks.

No radiometric data are available to constrain the age of the protoliths of the RA, and the peak of metamorphism. Half of the K-Ar hornblende data cluster closely around 384 ± 3 Ma, which conforms well with the respective data from the overlying units. However, data around 400 Ma in the other samples should be taken as a warning against the possibility of extraneous or inherited argon (Kreuzer et al. 1989b; Schüssler et al. 1986).

2.4 Prasinit-Phyllit-Serie (PPS)

The PPS (= prasinite-phyllite- sequence) consists of metamorphosed basic to intermediate lavas and pyroclastics, with the greenschist-facies assemblage actinolite + epidote + chlorite + albite + quartz, and of quartz-rich phyllites with quartz + albite + muscovite + chlorite ± garnet + graphite. The metavolcanics are subalkaline, with calc-alkaline tendency (Oppermann 1985). They are compatible with an island arc or back-arc environment. The phyllites are derived from silty shales. At the top of the sequence, i.e., along the boundary with the RA, there are numerous lenses of serpentinite, partly transformed into talc schists.

Acritarchs from the phyllites indicate a Proterozoic age of the protoliths (Pflug and Reitz 1987; Reitz and Höll 1988); the PPS might therefore represent an island arc belonging to the Cadomian cycle.

Metamorphism occurred under greenschist-facies conditions. Only at its basal contact with the underlying fossiliferous sequences does the PPS show some brittle deformation. The peak of metamorphism must have occurred prior to 370 Ma, which is the oldest K-Ar age of muscovites from the phyllites (Kreuzer et al. 1989b, backed up by unpublished Ar-Ar plateaus).

2.5 Tectonic Development

In the absence of conclusive evidence, we do not consider, in the following scenario, the possibility of an Early Ordovician compressive deformation (see Sect. 2.2). The polyphase tectonic history which we observe may easily have occurred within the complex Variscan orogeny, which possibly began in the Late Ordovician and spans at least the interval between approx. 390 and 320 Ma. Apart from fabric relics in helicitic mineral grains and mafic bodies, the oldest penetrative fabric (D1) is a strong metamorphic foliation and a correlated mineral lineation oriented SW-NE. This lineation occurs in the HS, LS, RA and PPS units. It has been interpreted, by Vollbrecht (1981) and Behr et al. (1982) in terms of the "rolling the dough" mechanism proposed by Lister and Price (1978), which would imply a NW-ward tectonic transport. However, recent studies of asymmetric clasts have only revealed indications of transport parallel with the lineation, with a sense of shear toward SW (Franke et al. 1992).

The strong foliation and NE-trending mineral lineation have been overprinted by several later generations of folds, which are increasingly more open and upright. In the HS and LS, the outcrop pattern is controlled by large, SW/NE-trending D4-folds facing WNW, except for some SE part of the Münchberg complex, where they are partly upright or ESE-facing. Polyphase folding is also well-documented in the lower units (RA, PPS; Franke et al. 1992). At least the last two generations are cylindroidal, and the orientation of the fold axes records a clockwise rotation of the maximum contraction from WSW/ENE toward WNW/ESE (Fig. 3). This development fills the gap between the top to the SW movements recorded in the earlier mineral lineation, and the NW-ward emplacement of the crystalline pile over the unmetamorphosed metasediments (Franke 1984a,b). The WSW-facing D3 folds in the PPS were formed still under greenschist facies conditions. The D4 folds formed under WNW/ESE-directed compression show cold deformation of quartz and greenschist facies minerals. Therefore, the K-Ar muscovite dates between 360 and 370 Ma (see above), which indicate cooling through 350 ± 50 °C (or neoformation at lower temperatures), tie in somewhere between the greenschist-facies D3 and the brittle D4 deformation. Hence, the rotation of the shortening direction must have been well under way around 360 Ma, i.e., around the Devonian/Carboniferous boundary.

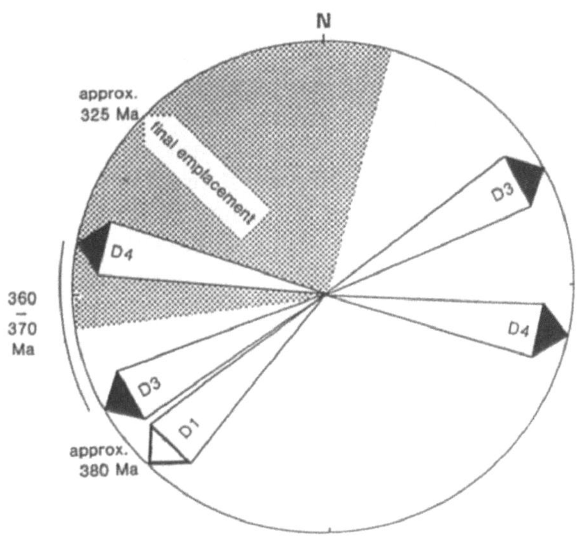

Fig. 3. Succession of tectonic events recorded in the metamorphic rocks of the western part of the Münchberg nappe pile, with tectonic transport directions and age data. (After Franke et al. 1992)

3 Wildenfels Complex

The Wildenfels complex is largely overlain by the Permo-Carboniferous, so that only a very small, southwestern part is accessible. The petrography of the rocks has been described by Schüller (1954).

As in the Münchberg klippe, the upper part consists of banded hornblende gneisses and garnet-amphibolites. There are also some felsic intrusive bodies (now orthogneiss). The mafic materials have partly been retrogressed into greenschist facies. The geochemistry of the mafic rocks corresponds to low-Ti MORB and high-Ti-low-K-tholeiites (Werner 1989).

This unit is underlain by a sequence of mica-schists and metaquartzites.

The underlying, unmetamorphosed Paleozoic sequences belong essentially to the Thuringian facies; there is, however, a sheet of Bavarian-type Early Carboniferous directly underlying the metamorphic units.

The main difference with the Münchberg klippe lies in the absence of separate greenschist (PPS) and amphibolite (RA) units, but the general sequence (mafics over metasediments) is similar.

The tectonic deformation of the Wildenfels units has been described in detail by Schwan (1974). The crystalline rocks show a strong foliation and a SW/NE-trending mineral lineation. Later folds face mostly towards the SE.

4 Frankenberg Complex

The Frankenberg complex consists of three main units, which are very similar to the Münchberg sequence. These are (from top to bottom):

4.1 HM-Series (HM)

The HM (hornblende-muscovite) series is composed of hornblende-bearing schists, metapelites, some orthogneisses, and frequent intercalations of mafic rocks (see, e.g., Schwan 1974). According to the geochemical studies of Werner (1989), the more massive, coarse-grained amphibolites correspond to MORB; thinner, finer-grained amphibolites resemble low-K-tholeiites and continental arc basalts. Some thicker metavolcanic units and correlated pyroclastic rocks are andesitic.

4.2 Frankenberger Gneis (FG)

This unit consists essentially of orthogneisses (often augengneiss), with some intercalations of metasediments. The petrography has been studied especially by Scheumann (1924a–c, 1932, 1938, 1957).

4.3 Prasinit-Serie (PS)

The PS is composed of various mafic lavas, few rhyolites, pyroclastic rocks, metapelites, cherts, marbles, and some quartzitic graywackes (see the description by Weg 1931). They are metamorphosed in greenschist and epidote-amphibolite facies. The geochemistry of the magmatic rocks allows identification of a complete island-arc suite, plus some MORB-type lavas (Werner 1989).

4.4 Tectonic Development

The fabrics and structures of the Frankenberg klippe have been described by Schwan (1974). Deformation of the crystalline rocks is characterized by a strong foliation and a SW-trending mineral lineation (as in the Münchberg and Wildenfels klippen). The later folds show a fan-shaped asymmetry which has encouraged earlier authors to suggest vertical uplift and bilateral extrusion of the crystalline rocks – a model which is incompatible with the brittle behavior of metamorphic rocks in near-surface levels.

5 NW Margin of the Moldanubian Region

As described in Chapter VII, the Moldanubian Region to the SE of the Saxothuringian allows definition of at least two major tectonometamorphic units: the Moldanubian sensu stricto (s.str.) and the Tepla-Barrandian (or Bohemian). The Moldanubian s.str. consists of monotonous paragneiss sequences with few intercalations of graphite schists, marbles, and mafic or felsic intrusives. The last (and dominant) metamorphism occurred around 330–320 Ma (Grauert et al. 1974; Teufel 1988; Kreuzer et al. 1989b), in a low-pressure/high-temperature regime. The local occurrence of relict kyanite indicates a preceding medium-pressure stage (see Chap. VII.4). Relics of high-pressure rocks near Oberviechtach-Winklarn are possibly allochthonous (see Sect. 5.2). Though the bulk lithology of the Moldanubian s.str. is grossly similar to that of the Münchberg Liegend-Serie (and its equivalents in the Frankenberg and Wildenfeld klippen), the different age pattern and metamorphic evolution preclude a correlation with these units. The Moldanubian s.str. is therefore not discussed any further. It should be noted, however, that the palinspastic interpretation of the Moldanubian s.str. is a crucial point for any geotectonic model. It cannot, for example, be ruled out that part of the Moldanubian s.str. is a higher-grade continuation of the Saxothuringian (Matte et al. 1990).

The Moldanubian s.str. is structurally overlain by the Tepla-Barrandian rocks (e.g., Franke 1989a; Weber and Vollbrecht 1989; Stein and Kleemann 1992) which are characterized by abundant mafic intercalations and by high- to medium-pressure metamorphic gradients. Important constituents of the Tepla-Barrandian are the Erbendorf Greenschist Zone (EGZ), the Zone of Erbendorf-Vohenstrauß (ZEV), and the Zone of Tepla-Domazlice (ZTD) with the metabasite complexes of Marianske Lazne, Cerna Hora, and Domazlice.

The contact between the two units is a ductile shear zone, which is poorly exposed. An increasing number of observations suggest tectonic transport of the hanging wall toward the SW (Heinicke 1987; Masch and Koch 1988; Kleemann et al. 1989; Seibert et al. 1990; Stein and Kleemann

1992; Tanner, unpubl.). The early Devonian medium- and high-pressure metamorphism in basal parts of the overlying Tepla-Barrandian unit has partly been rejuvenated to mid-Carboniferous values (Schüssler et al. 1986; Kreuzer et al. 1989b, 1992; Grauert et al. 1990). This applies also to areas where influence from neighboring granites may be excluded, e.g., in the southern part of the Zone of Tepla-Domazlice (ZTD). Monazite from a mylonite near the eastern margin of the ZEV has yielded a concordant U-Pb age of 333 Ma (Teufel 1988); the shear-zone analyzed by Kleemann et al. (1989) has been dated by the Rb/Sr thin-slab method at 305 ± 9 Ma (Grauert et al. 1990). These relationships, together with the ductile nature of the contact, suggest that thrusting was closely connected with the low-pressure metamorphism, which has been dated at 330–320 Ma by various methods (see the compilation in Hansen et al. 1989). Thrusting is postdated by the intrusion of the earliest post-tectonic granite (Leuchtenberg g.) at 325 Ma (Siebel 1991, pers. comm.).

5.1 Greenschist Unit of Erbendorf (EGZ)

Situated between the autochthonous Fichtelgebirge in the north and the Zone of Erbendorf-Vohenstrauß in the south, the EGZ appears to be in a tectonic position similar to the Prasinite-Phyllite Series in the Münchberg klippe. Interestingly, the metabasites in both tectonic units exhibit striking similarities in their chemical composition. The amphibolites and metagabbros which are the predominant constituents of the EGZ conform to Mg-rich tholeiites and calc-alkaline basalts and thus indicate an island arc or back-arc environment (Schüssler et al. 1989; Okrusch et al. 1989; Schüssler 1990). Another similarity is the widespread occurrence of serpentinites, whereas metasediments are subordinate.

The only difference lies in the metamorphic path, which is prograde at Münchberg, and retrograde at Erbendorf, where the main phase of regional metamorphism took place under conditions of the lower amphibolite facies with the assemblage hornblende + oligoclase + epidote + opaques in metabasites. However, due to a strong retrograde overprint, the greenschist-facies minerals albite, epidote, chlorite, and calcite were formed. By the intrusion of late-Variscan granites, a contact metamorphic overprint led to the formation of a thermal aureole with well-developed prograde zonation (Matthes 1990).

Ar-Ar spectra of hornblendes and muscovites are inclined between 380 and 390 Ma and 370–380 Ma, respectively. This indicates either slow cooling, or, more probably, moderate rejuvenation of rocks older than 390 Ma at 380 and 370 Ma, respectively (Kreuzer, unpubl.).

5.2 Zone of Erbendorf-Vohenstrauß (ZEV)

The ZEV consists of pelitic to psammitic metasediments with intercalations of calc-silicate rocks, graphite-quartzites and graphitic schists. Predominant meta-igneous rocks are garnet-bearing flaser amphibolites, schistose and banded amphibolites with calc-silicate interlayers, metagabbros and meta-ultramafics. Minor granitic orthogneisses occur, but have not been studied in detail.

For the metasediments in the KTB drillhole, which is situated in the northern part of the ZEV (Fig. 1), Hofmann et al. (1993) have presented Sm-Nd model ages and Rb-Sr dates which are taken to indicate mean crustal residence times of 1–1.2 Ga, which is younger than the ones in the gneisses of the near-by Moldanubian s.str. They state derivation from at least two sources, one older than 1.5 Ga, and one younger than 1 Ga. For the metabasic rocks, they conclude crustal formation around 0.7 Ga. They further argue for a partial Sr homogeneization in paragneisses and mafic rocks around 400 Ma.

Lower intercept U-Pb dates of discordant zircons indicate a weak metamorphism around 500 Ma in gneisses of the ZEV (Teufel 1988; Miller et al. 1990), i.e., in the early Ordovician. At about the same time (495–490 Ma), the precursors of metagabbroic rocks and amphibolites of the KTB pilot hole were emplaced, as it is recorded by upper intercept U-Pb ages of euhedral zircons (von Quadt 1990, pers. comm. 1991). The lower intercept of zircons from an orthogneiss in the ZEV is interpreted to record magmatic activity at 460 Ma (Ordovician; Teufel 1988).

The KTB pilot hole near Windisch-Eschenbach, which was drilled between September 1987 and April 1989 at a final depth of 4000 m, penetrated four characteristic rock series of the ZEV (e.g., Röhr et al. 1990):

1. 0 to 460 m: a variegated sequence of metabasites and metasediments consisting of garnet-bearing

flaser amphibolites with calc-silicate and minor marble layers, diopside-garnet bearing biotite-hornblende gneisses, muscovite-biotite gneisses and subordinate kyanite/sillimanite-garnet-biotite ± perthite gneisses.
2. 460 to 1160 m, 1610 to 2470 m and 2690 to 3575 m: monotonous garnet-bearing biotite gneisses with kyanite and/or sillimanite, predominantly metagraywackes, rare metapelites.
3. 1160 to 1610 m and 3575 to 4000 m: amphibolite-metagabbro sequence with minor lenses of ultramafic talc-chlorite-amphibole felses.
4. 2470 to 2690 m: a variegated sequence of biotite-hornblende gneisses with frequent intercalations of amphibolites, calc-silicate rocks, diopside-garnet-biotite-hornblende gneisses and sillimanite-biotite gneisses.

Relics of an ophitic texture and of igneous minerals like Ti-bearing augite, Ti-bearing brown hornblende, plagioclase, and ilmenite clearly indicate that the metagabbros of the ZEV have an igneous protolith. Judging from contact relationships observed in drill cores, the amphibolite-metagabbro sequences are derived either from basaltic flows, subsequently intruded by gabbro/dolerite dikes or sills or, alternatively, from a thick, structurally differentiated basaltic sill with doleritic portions and ultramafic cumulate layers (von Gehlen et al. 1991; Matthes et al. 1995; O'Brien et al. 1992). The age of the gabbroic/doleritic intrusion has been determined by a U-Pb upper intercept of 494 ± 3 Ma (Early Ordovician) of euhedral zircons (von Quadt 1990). In this respect, it is interesting to note that Pflug and Prössl (1990) recorded tracheides, miospores, and acritarchs of Early Devonian age in gneisses of the monotonous sequence (pilot hole: 800 to 1000 m) overlying the upper amphibolite-metagabbro sequence. This could be explained in two alternative ways: (a) The graywacke protoliths of the gneisses were deposited on top of an old (?meta-) basaltic/gabbroic sequence, e.g., in a continental margin environment, or (b) the two units were tectonically superimposed prior to the last, amphibolite facies metamorphism.

In their major, trace and rare earth element geochemistry, the flaser amphibolites from surface outcrops in the northern part of the ZEV and from the KTB pilot hole exhibit a tholeiitic composition with marked enrichment of the incompatible trace elements, conforming to an E-MORB or within-plate character. In contrast, schistose and striped amphibolites mainly concentrated in the southern part of the ZEV display a chemical composition similar to N-MORB. In their geochemical character, the metagabbros are transitional between both amphibolite types; however, with stronger variation which is attributed to differentiation processes in the igneous protolith (Schüssler et al. 1989; Schüssler 1990; Patzak 1991; Patzak et al. 1991).

Rare occurrences of retrograded eclogite in the ZEV at Kaimling (Busch 1970), in the VSP-1 hole at the KTB drill site and in the KTB pilot hole (Schalkwijk 1991; O'Brien et al. 1992), as well as mineral relics and reaction textures in metagabbros of the KTB pilot hole demonstrate that at least parts of the ZEV were subjected to a polyphase metamorphic evolution. According to O'Brien et al. (1992) a first eclogite stage with minimum pressures of 13 kb and temperatures of 580–650 °C is documented by the assemblage garnet + omphacite (at least Jd_{26}) + zoisite ± kyanite + rutile + quartz. A subsequent high-pressure granulite stage leads to the breakdown of omphacite forming symplectites of plagioclase and clinopyroxene (Jd_{9-14}) at 10.5–12.5 kb and 670–730 °C. A pervasive medium-pressure overprint under amphibolitefacies conditions led to the stable assemblage hornblende + plagioclase ± garnet ± salite ± titanite + ilmenite + sulfides in metagabbros and amphibolites. Symplectitic breakdown of garnet in the flaser amphibolites indicates a multistage development of this amphibolite-facies metamorphism. According to Reinhard (1990), the P-T path in the adjacent metasediments – which bear no high-pressure relics – is composed of a prograde branch, leading from the sillimanite to the kyanite field, and a retrograde branch, leading back to the sillimanite field. P-T conditions at the peak of amphibolite-facies metamorphism are estimated at 660–720 °C and 6–8 kb for the metasediments (Reinhard 1990), whereas various geobarometers applied to the metabasites led to pressures around 10 kb (Patzak et al. 1991). It is still open to discussion whether the metabasites and the metasediments experienced the same metamorphic evolution. The following questions arise:

1. Did the metasediments suffer a high-pressure evolution together with the metabasites?
2. Does the amphibolite-facies, medium-pressure overprint of the metabasites represent a renewed, prograde development, or is this

stage of purely retrograde character subsequent to the high-pressure events?

Due to the intrusion of post-tectonic granites, parts of the ZEV underwent thermal metamorphism leading to the formation of contact aureoles near Steinach and Windisch-Eschenbach (Voll 1960; Okrusch 1969, 1971; Schüssler 1990).

Preliminary results of Sm-Nd dating on eclogitic relics of the KTB pilot hole indicate an age of the eclogite facies metamorphism similar to that concluded by Stosch and Lugmair (1990) for that of the Münchberg klippe (von Quadt, pers. comm.).

For the western part of the ZEV, early Devonian dates around 380 Ma are reported for various minerals and methods: a lower U-Pb intercept of zircon from a garnet-kyanite gneiss yields 391 + 4/−2 Ma, and concordant monazite from the same sample 378 ± 3 Ma (Teufel 1988). Half of the K-Ar dates on hornblendes are closely around 380 Ma (Kreuzer et al. 1989b; Schüssler et al. 1990). However, the discrepancy between the concordant monazite and the discordant zircon, as well as a wide range of dates from apparently slightly older hornblendes may be taken to indicate different grades of the final amphibolite facies overprint at different locations. Different kinds of Ar-Ar spectra suggest a prograde character of this event, with some of the investigated locations clearly having been in a marginal position to the 380 Ma event (Kreuzer et al. 1989a, 1990, unpubl.). Various methods record low-temperature influence and impregnation with fluids around 360 Ma (Teufel 1988; Albat et al. 1989, 1990; Kreuzer et al. 1989b, 1990a).

In the eastern part of the ZEV, the hornblende dates give values around 325 Ma. The younger age with respect to the western ZEV is either due to the low-pressure metamorphic event which has effected the neighboring units (Saxothuringian and Moldanubian s.str.), or, alternatively could have been caused by the intrusion of the post-tectonic Leuchtenberg Granite and later pulses of intrusions (Köhler et al. 1974).

To the SE of the ZEV, near Oberviechtach and Winklarn (W in Fig. 1), there are isolated occurrences of amphibolitized eclogites, garnet-clinopyroxenite, and serpentinite, surrounded by the typical cordierite-sillimanite-gneisses of the Moldanubian s.str. (O'Brien 1989, 1991; O'Brien et al. 1990). According to von Quadt and Gebauer (1988), the eclogites and associated felsic rocks appear to be derived from an island arc setting and underwent high-grade metamorphism in Late Ordovician or Early Silurian time (Sm-Nd mineral and whole rock on eclogite: 427 ± 5 Ma, on felsic rock: 423 ± 8 Ma); a U-Pb monazite age of 320 Ma reflects the low-pressure metamorphic overprint which characterizes the surrounding Moldanubian rocks. The serpentinites represent oceanic mantle (von Gehlen and Schmitt 1989). It is still debated whether these rocks represent a tectonic outlier of the ZEV, or else document an early high-pressure development of the Moldanubian s.str., which, in the surrounding metasediments has been wiped out by later events.

5.3 Zone of Tepl-Taus (Teplá-Domazlice) (ZTD)

Because of the variegated lithology, a comparable metamorphic evolution and rather similar age pattern, this zone has been tentatively parallelized with the Münchberg klippe and the Zone of Erbendorf-Vohenstrauß. As in the other two zones, the last pervasive metamorphic event in the ZTD was of medium-pressure type, with an increase of metamorphic conditions from greenschist facies to amphibolite facies from NE to SW (Vejnar 1972, 1977; Blümel 1983). The age of this metamorphism is given by K-Ar and Ar-Ar dates around 380 Ma (Kreuzer et al. 1992). However, a strong influence of the younger, low-pressure metamorphic overprint in the neighboring Moldanubian s.str. is indicated by a disturbance of the K-Ar isotopic system or even a complete rejuvenation with a reset to midCarboniferous ages, especially in the southern part of the ZTD (Kreuzer et al., 1992).

Eclogite occurrences in the Marianske Lazne metabasite complex reveal an earlier high-pressure metamorphic evolution for at least parts of the ZTD under eclogite-facies and subsequent granulite-facies conditions (O'Brien 1991), similar to the P-T path of metagabbros and eclogites from the KTB pilot hole (see above). According to U-Pb dates of nearly concordant zircons, the magmatic age of the mafics is Ordovician (U-Pb upper intercept 496 ± 1 Ma, Pb-Pb isochron 495 ± 2 Ma). In a near-by gneissic sheet of intermediate igneous composition, discordant zircons with Pb-Pb dates between 510 and 450 Ma suggest an early Palaeozoic protolith. The U-Pb lower intercept date of 362 ± 12 Ma is well in accord with the early Devonian hornblende and mica dates cited above (zircon data from Bowes and Aftalion 1991). Silurian K-Ar mica dates reported by Gottstein (1970) from the Tepla highlands could not be confirmed (Kreuzer, unpubl.).

Metabasites of the Marianske Lazne complex and the Černa Hora area in the northern and middle part of the ZTD compare well, in their chemical composition, to normal midocean-ridge tholeiites, whereas most of the greenschists and amphibolites of the Domazlice crystalline basement in the southern part show an alkaline-basaltic tendency (Schüssler et al., 1992; Lohoff 1990).

The Domazlice crystalline basement contains large intrusions of basic to felsic composition, the Neukirchen-Kdyne igneous complex, which was interpreted by Vejnar (1986) as a Skaergaard-type layered intrusion of late Cadomian/Cambrian age. Marginal parts of this igneous complex were metamorphosed, together with their country rocks, presumably around 380 Ma; however, with an obvious later thermal influence (Kreuzer et al. 1992). The intrusion age of this complex is constrained by two independent geochronological data: zircon from the Stod granitoid, intruding the Neukirchen-Kdyne complex, yielded a concordant U-Pb age of 482 ± 3 Ma (Dörr unpubl.). A concordant U-Pb age of 511 ± 3 Ma was recorded for an idiomorphic zircon fraction from the gabbroamphibolite of the Blätterberg, interpreted as age of intrusion (Gebauer, in press), confirming an earlier date of around 530 Ma (Gebauer and Grünenfelder 1982).

5.4 Tectonic Development

The gneisses and metabasites of the ZEV show a polyphase ductile deformation starting with the HT mylonitization and ending with a greenschist-facies retrogression. The foliation and stretching lineations probably correspond to the metamorphism around 380 Ma. Shear sense and orientation of lineations are variable and difficult to interpret because of later deformations. These earlier elements were overprinted by NW- to NNW-trending folds, which have also affected the underlying rocks of the Moldanubian s.str. (D4 of Stein 1988). This deformation has produced the halfwindow-like, northward protrusion of the Moldanubian visible on the map (Fig. 1), as well as the SW-facing synformal structure of the ZEV (see, e.g., Hirschmann et al. 1990).

6 Metamorphic Nappes in the Erzgebirge?

If one adopts the view that the Wildenfels and Frankenberg metamorphics, like the Münchberg, represent nappes which have traveled across the area now occupied by the Fichtelgebirge-Erzgebirge Antiform, one may look out for remnants of these nappes in the roof of the antiform.

Although the metamorphic sequences of the Erzgebirge were pervasively overprinted under medium-pressure, amphibolite-facies conditions, eclogitic relics testifying to an older high-pressure evolution are widespread (Schmidt et al. 1990; Schmädicke 1991; Schmädicke et al. 1992). According to these authors, mineral assemblages in the eclogites indicate a considerable metamorphic contrast with the surrounding rocks: eclogites in the central Erzgebirge, associated with garnet-bearing ultramafics, display frequently relics of coesite and were formed at minimum pressures of 29 kb and temperatures around 850 °C; intercalated acid granulites experienced the same P-T conditions. The coesite-free eclogites of the western Erzgebirge were metamorphosed under lower pressures around 20 kbar and temperatures ranging between ca. 750 °C in the eastern, and 650 °C in the westernmost part (Schmädicke et al. 1992). It is interesting to note that the eclogites of the western Erzgebirge closely resemble the light eclogite type of the Münchberg klippe.

Besides, there are extensive sheets of orthogneiss (Rotgneiss) in the Eastern Erzgebirge, which contain abundant kyanite and therefore contrast with the low-pressure metamorphism of the surrounding areas (see Chap. VII.4). As suggested earlier by Scheumann (1935) and, in later time, e.g., by Behr (1983), the Rotgneiss sheets could represent nappes originally coherent with the more felsic members of the Münchberg, Wildenfels, and Frankenberg klippen.

However, field relationships suggest that these rocks do not occupy the high tectonic level which would permit their interpretation as klippen, but are contained in the core of the Erzgebirge antiform. This points to their setting in a metamorphic core complex rather than a tectonic klippe. The age of the eclogites is uncertain – they might have formed during the Variscan collision, or else date back to a pre-Variscan orogenic cycle, during which they were emplaced in position at or close to the surface, covered by Early Paleozoic sediments, and ultimately exhumed during the Variscan orogeny.

The widespread presence of medium-pressure mineral associations is in marked contrast with areas further southwest (Fichtelgebirge), which are characterized by low-pressure assemblages (see the contribution by Blümel, Chap. V.B.4). These contrasting Variscan geothermal gradients remain to be explained.

7 Tectonic Model

7.1 Palinspastic Reconstruction

The sedimentary and volcanic rocks in the unmetamorphosed part of the Saxothuringian Belt (see Chap. V.B.1), as well as the lithology and geochronology of the metamorphic rocks, clearly indicate an important phase of rifting during the Cambro-Ordovician. Metamorphosed volcanic and plutonic rocks correlatable with this event are present in the ZTD, the ZEV, in the HS and LS of the Münchberg nappe pile, as well as in the unmetamorphosed nappes and in the parautochthon. The crystallization ages, as constrained by biostratigraphy or geochronology, range between approx. 510 and 450 Ma.

In the Münchberg nappe pile and its long-strike equivalents, the regular stacking order of the metamorphic and sedimentary nappes, with its complete inversion of stratigraphic ages and metamorphic grades, provides clear constraints for the reconstruction of the original relative positions of the individual units: the upper units are derived from more southeasterly positions (as shown in Fig. 4). Since the transport direction has changed from top to the SW towards top to the NW, Fig. 4 is a two-dimensional compromise. The features described in the previous chapters permit to suggest tentative geotectonic interpretations. From NW to SE, it is possible to distinguish the following units:

1. Thuringian facies Paleozoic rocks of the Frankenwald (very low grade), Fichtelgebirge (low to medium grade) and Erbendorf (very low grade): passive northwestern margin of the Saxothuringian rift basin.
2. Bavarian facies Paleozoic of the Münchberg a.o. nappes (very low grade): thinned continental crust (continental slope) at the northwestern margin of the Saxothuringian basin.
3. Greenschists of the Münchberg and similar nappes (low grade) and Erbendorf (medium grade, largely retrogressed): Proterozoic (?)island arc, belonging to the Cadomian basement, probably of the N margin of the Saxothuringian basin.
4. Liegend-Serie of the Münchberg and similar nappes (medium grade): Precambrian and/or Cambrian, with 500 Ma intrusions; Precambrian (Cadomian) basement and/or Cambrian cover, probably belonging to the extended northwestern margin of the Saxothuringian basin.
5. Eclogites and related rocks at the base of the high- to medium-pressure units (Münchberg a.o. nappes, Zones of Erbendorf-Vohenstrauß and of Tepla-Domazlice): Cambro-Ordovician ocean floor and overlying sediments of the Saxothuringian basin. The eclogites probably mark the suture. The boundary between the HS and LS in the Münchberg, and equivalent surfaces in the other klippen, would then represent the only outcrop of the true plate-boundary, though the whole nappe pile may be referred to as a suture zone.

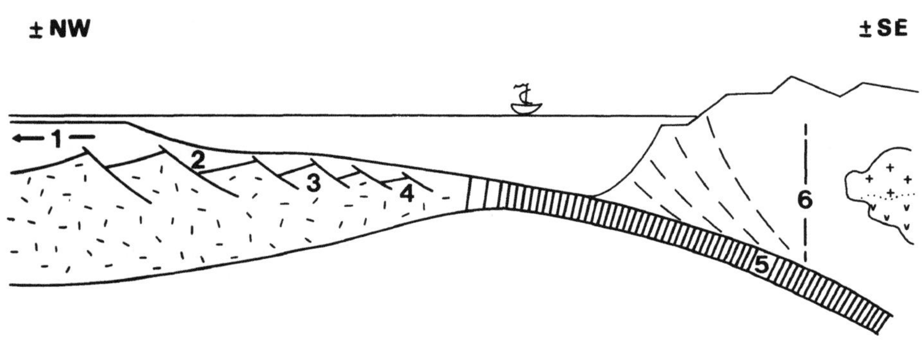

Fig. 4. Palinspastic reconstruction of the Münchberg nappe pile and neighboring units; *numbers* as in text

6. Higher parts of the high- to medium-pressure units (various metasediments and amphibolites; HS of the Münchberg nappe pile and equivalents in Wildenfels and Frankenberg, probably also the ZEV and ZTD with the exception of the high-pressure mafic rocks): At least partly: Proterozoic of the Cadomian basement, with Cambro-Ordovician intrusions. This basement is part of the Tepla-Barrandian (Bohemian) terrane, which is unconformably overlain, in the Barrandian Syncline near Prague, by unmetamorphosed and slightly folded Cambrian through Devonian rocks. The Tepla-Barrandian (Bohemian) terrane has bounded the Saxothuringian basin to the SE.

7.2 Tectonic Assemblage

After the opening of the Saxothuringian basin, in Cambrian through to Late Ordovician time, plate convergence started probably in the Latest Ordovician or Silurian (metamorphic age of the HP-rocks of the Winklarn area). Saxothuringian ocean floor was subducted under the Tepla-Barrandian (Bohemian) microplate and transformed into eclogite. The numerous indications of an early Devonian "event" around 380 Ma (U-Pb on zircons, K-Ar and Ar-Ar on hornblendes and micas) indicate rapid uplift and cooling, probably caused by collision of the Tepla-Barrandian with the northern margin of the Saxothuringian basin and subsequent tectonic underplating of continental crust. This is in accord with the composition of Famennian graywackes north of Erbendorf, at the southern flank of the Fichtelgebirge, which have been deposited on the passive margin sequence at the NW margin of the Saxothuringian basin: the graywackes contain detrital zircons, which exhibit the same U-Pb systematics as those from the ZEV (Dörr et al. 1991) together with chromite (Ludwig 1968), and are apparently derived from the Tepla-Barrandian margin to the SE. Hence, subduction of the Saxothuringian oceanic crust must have been completed before the Famennian. The Paleozoic sequence of the Prague Syncline in the center of the Bohemian block is topped by a clastic sequence of Givetian age (approx. 380–377 Ma after Harland et al. 1989). These rocks probably represent the sedimentary signal of continental collision and the resulting uplift.

During the collision, the high-pressure rocks now found at the NW margin of the Tepla-Barrandian (Bohemian) block were thrust up toward the SW. On its way to the surface, the hanging wall sampled the metamorphic levels of the foreland crust, which were sequentially accreted to the base of the growing tectonic pile, and, hence, appear in an inverted sequence. This process is highlighted by the K-Ar ages on muscovites, which are significantly younger in the lowest nappe (PPS phyllizes of Münchberg, Kreuzer et al. 1989b). At about the end of the Devonian (approx. 360 Ma), all metamorphic nappes were assembled, partly refolded and cooled through the closing temperature for argon in muscovite (350 ± 50 °C).

Since the Tournaisian flysch of the Bavarian facies contains already clasts of the pre-Carboniferous in Bavarian facies, the Bavarian facies nappes must have been accreted shortly after the assemblage of the crystalline nappes, and prior to 350 Ma (Tournaisian/Visean boundary after the time-scale of Harland et al. 1989). Thuringian facies pebbles in the youngest (Late Visean) flysch clastics indicate that, at a late stage of the tectonic development, accretion and uplift also incorporated parts of the Thuringian facies realm (Dörr et al. 1989).

As discussed in Chapter V.B.2, drag folds in the unmetamorphosed nappes and in the autochthon indicate transport of the metamorphic hanging wall grossly toward NW. This is corroborated by the orientation of the flysch basin, in which most of the sole markings show transport toward the SW or NE and only the more proximal deposits reflect direct input from the SE (dominant axial flow, Franke 1984b), thus constraining the orientation of the tectonic front. A linear arrangement with NW-ward transport is also suggested by large-scale comparisons between the Münchberg klippe and its eastern counterparts (Wildenfels, Frankenberg): all these nappe piles show essentially the same tectonic profile, with amphibolite facies rocks on top and sheared proximal flysch at the bottom (see also Chap. V.B.1). This lateral consistency over more than 150 km may be taken to indicate that the present-day alignment of the tectonic klippen (SW-NE) corresponds to the trend of the orogenic wedge during the Early Carboniferous thrusting.

Since the ductile mineral lineations in the metamorphic nappes of all the three klippen shows the same orientation (SW-NE), rotation of the individual klippen during their emplacement is unlikely. It appears rather that the rocks now contained in the Münchberg, Wildenfels, and Frankenberg klippen belonged to one coherent thrust unit, the major parts of which have been eroded away. In the area of the klippen, the maximum thickness of the nappe pile is constrained by the metamorphic grade of the underlying parautochthon, in which quartz has not been recrystallized. Maximum temperatures must have been less than approx. 300 °C, which suggests depths of burial less than 10 km. Towards the S, the thickness of the allochthon can be expected to increase according to the geometry of an orogenic wedge.

As described in section 2.4, directions of tectonic transport changed from top to the SW toward top to the NW during the period of accretion of the nappes. It is remarkable that this kinematic history and the conspicuous tectonic profiles with inverted metamorphic zonation are encountered in the Münchberg as well as in the related klippen of Wildenfels and Frankenberg. If one interprets the ductile foliation and SW-trending lineation in terms of nappe thrusting, this would imply that the tectonic sequence was consistent over more than 150 km (the distance Frankenberg-Münchberg) **in transport direction**. This appears unlikely, since nappe thrusts usually cut through the stratigraphic and metamorphic sequence at an angle, so that the tectonic sequence changes in transport direction (while it may remain relatively constant along the structural trend). The observed relationships would be easier to explain in a transpressional setting: in this case, the vertical sequence contained in the Münchberg complex would already have been available as a lateral sequence (Fig. 4), which would have been strongly sheared by dextral strike-slip. Of course, the inversion of the metamorphic grades which lead to the "obduction" of eclogites calls for some vertical component. Subsequent to the transpressional displacements, the NW-ward thrusting component became more important, and eventually lead to the emplacement of the pile on the Saxothuringian foreland.

Taken altogether, the existing data largely confirm the model of subduction and accretion, proposed first by Behr et al. (1982) and elaborated upon by Franke (1984a,b). Convergent plate motions, in the Saxothuringian belt, altogether span approx. 100 Ma, from the Ordovician/Silurian boundary through to the Late Visean. The only major discontinuities are the onset of collision at about 380 Ma, and the change in the tectonic transport direction from top SW (strike-slip with oblique obduction) towards top NW (normal thrusting), around 370–360 Ma.

The maximum preserved thickness of the Early Carboniferous flysch sediments is about 3 km, and the belt of outcrop of flysch sediments is today 80 km wide. Even if we allow for some erosion of flysch, and considerable tectonic shortening, it is difficult to explain the exhumation of the eclogites merely by uplift and erosion, especially so because the greater part of the uplift history (as indicated by the cooling ages) was accomplished already during the Devonian. We therefore conclude that ductile stretching under the force of gravity has contributed to the reduction of the overlying units. Gravitational spreading does not, however, serve to explain the reduction of the metamorphic profile **within** the Münchberg pile of nappes: in the case of gravitational spreading and thinning, the metamorphic grade of the mylonite zones separating the nappes should exhibit a normal sequence, with ductile faults at depth, and brittle faults at higher levels, instead of the inverted sequence that we observe. Since the allochthonous crystalline rocks do not reveal extensional features, the proposed normal faults should be sought for in northern parts of the Moldanubian realm (ZEV, ZTD).

Apart from the lack of extensional features, the most important uncertainty concerns the relationship of the exotic metamorphic nappes (ZEV, ZTD) with the underlying Moldanubian s.str. Though current investigations suggest SW-ward thrusting, the geotectonic significance of this event is uncertain. Collisional wedging (Vollbrecht et al. 1989; Weber and Vollbrecht 1989), or a general change of the tectonic polarity (Franke 1989a) have been considered, but remain at present entirely hypothetical. Answers to this question may only be expected from further studies in the eastern and southern parts of the Bohemian Massif, whose protolith ages and pressure-temperature-deformation-time paths are but poorly known.

References

Albat F, Grauert B, Hansen B (1989) Sr-Isotopenverteilung in einem Kleinbereichsprofil durch Gneise der Bohrung Püllersreuth (ZEV). KTB-Rep 89-3: 442

Albat F, Grauert B, Hansen B (1990) Sr-Isotopenverteilung in Quarzmineralisationen der Bohrung Püllersreuth (ZEV). KTB-Rep 90-4: 550

Behr HJ (1983) Intracrustal and subcrustal thrust-tectonics at the northern margin of the Bohemian Massif. In: Martin H, Eder FW (eds) Intracontinental fold belts. Springer, Berlin Heidelberg New York, pp 365–404

Behr HJ, Engel W, Franke W (1982) Variscan wildflysch and nappe tectonics in the Saxothuringian Zone (northeast Bavaria, West Germany). Am J Sci 282: 1438–1470

Behrmann JH, Lich S, Volp A, Heinisch H, Zadow A (1989) Scherkriterien im KTB: kontinuierliche Kernaufnahme duktiler Verformung von 480–2004 m Teufe. KTB-Rep 89-3: 418

Blümel P (1983) The western margin of the Bohemian Massif in Bavaria. Fortschr Mineral 61 (Beih 2): 171–195

Bowes DR, Aftalion M (1991) U-Pb zircon isotopic evidence for early Ordovician and late Proterozoic units in the Marianske Lazne complex, Central European Hercynides. Neues Jahrb Mineral Monatsh 1991: 315–326

Bosbach D, Stosch HG, Seidel E (1991) Magmatic and metamorphic evolution of metagabbros in the Münchberg Massif, N.E. Bavaria. Contrib Mineral Petrol 107: 112–123

Busch K (1970) Die Eklogitvorkommen des kristallinen Grundgebirges in NE-Bayern. – IV. Die Eklogite der Oberpfalz und ihr metamorpher Abbau. Neues Jahrb Mineral Abh 113: 138–178

Chappell BW, White AJR (1974) Two contrasting granite types. Pacific Geol 8: 173–174

Dörr W, Franke W, Kramm U (1989) U/Pb Daten von Zirkonen klastischer Sedimente – Beiträge zur Entwicklung der Böhmischen Masse. KTB-Rep 89-3: 351

Dörr W, Kramm U, Franke W, Gehlen K von (1991) U-Pb systematics of detrital zircons from the Saxothuringian Belt: constraints on the tectonic development. Terra Abstr 3/1: 206–207

Ducrot J, Lancelot JR, Marchand J (1983) Datation U-Pb sur zircons de l'eclogite de la Borie (Haut Allier, France) et conséquences sur l'évolution antehercynienne de l'Europe occidentale. Earth Planet Sci Lett 62: 385–394

Franke W (1984a) Late events in the tectonic history of the Saxothuringian Zone. In: Hutton DWH, Sanderson DJ (eds) Variscan tectonics of the North Atlantic Region. Blackwell, Oxford, pp 33–45

Franke W (1984b) Variszischer Deckenbau im Raume der Münchberger Gneismasse, abgeleitet aus der Fazies, Deformation und Metamorphose im umgebenden Paläozoikum. Geotekton Forsch 68: 1–253

Franke W (1989a) The geological framework of the KTB Drill Site, Oberpfalz. In: Emmermann R, Wohlenberg J (eds) The German Continental Deep Drilling Program (KTB). Site selection studies in the Oberpfalz and Schwarzwald. Springer, Berlin Heidelberg New York, pp 37–54

Franke W (1989b) Tectonostratigraphic units in the Variscan Belt of Central Europe. Geol Soc Am Spec Pap 230: 67–90

Franke W, Behrmann J, Moehrmann H (1992) Zur Deformationsgeschichte des Kristallins im Münchberger Deckenstapel. KTB-Rep 92-4: 225–240

Franz G, Thomas S, Smith DC (1986) High-pressure decomposition in the Weissenstein eclogite, Münchberg Gneiss Massif, Germany. Contrib Mineral Petrol 92: 71–85

Gebauer D (1994) Ein Modell der Entwicklungsgeschichte des ostbayerischen Grundgebirges auf der Basis plattentektonischer Vorstellungen und radiometrischer Datierungen. In: Erläuterungen zur geologischen Karte von Bayern 1: 25000, B1.6439 Tännesberg. Bayer Geol Landesamt (in press)

Gebauer D, Grünenfelder M (1979) U-Pb zircon and Rb-Sr mineral dating of eclogites and their country rocks. Example: Münchberger Gneiss Massif, northeast Bavaria. Earth Planet Sci Lett 42: 35–44

Gebauer D, Grünenfelder M (1982) Geological development of the Hercynian Belt of Europe based on age and origin of high-grade and high-pressure mafic and ultramafic rocks. Abstr 5th Int Conf on Geochronology, cosmochronology, isotope geology, Nikko Natl Park, Japan, pp 111–112

Gottstein O (1970) K-Ar ages of granitic and ectinitic rocks of the Teplá anticlinorium. Vestn Ustred Ustavo Geol 45: 201–205

Grauert B, Baumann A, Kleemann U (1990) Altersbestimmung einer Mylonitzone im Grenzbereich ZEV/Moldanubikum. KTB-Rep 90-4: 551

Grauert B, Hänny R, Soptrajanova G (1974) Geochronology of a polymetamorphic and anatectic gneiss region: the Moldanubian of the area Lam-Deggendorf, eastern Bavaria, Germany. Contrib Mineral Petrol 45: 37–63

Grüner F, Morteani G (1990) Hochdruckrelikte in goldführenden Metabasiten der Hangendserie der Münchberger Gneismasse, NE-Bayern. Beih Eur Mineral 2 (1): 84

Hammerschmidt K, Franz G (1988) Eklogitbildung und Deformation der Münchberger Gneismasse. Petrologische und isotopengeochemische Daten. Fortschr Miner 66 (Beih 1): 50

Hammerschmidt K, Franz G (1992) Retrograde evolution of eclogites – evidences from microstructures and Ar-Ar white mica dates, Münchberg Massif, northeastern Bavaria. Contrib Mineral Petrol 111: 113–125

Hansen BT, Teufel S, Ahrendt H (1989) Geochronology of the Moldanubian-Saxothuringian transition zone, northeast Bavaria. In: Emmermann R, Wohlenberg J (eds) The German Continental Deep Drilling Program (KTB) Springer, Berlin Heidelberg New York, pp 55–56

Harland WB, Armstrong RL, Cox AV, Craig LE, Smith AG, Smith DG (1989) A geologic time scale. University Press, Cambridge, 263 pp

Heinicke F (1987) Strukturgeologische Untersuchungen im Übergangsbereich zwischen der Zone Erbendorf-Vohenstrauß und dem Moldanubikum im Gebiet Vohenstrauß/Leuchtenberg, Oberpfalz. Diploma-thesis, Univ Göttingen, 84 pp

Hirschmann G, Stettner G, Weber K (1990) Stand und Probleme der geologischen Prognose für die KTB-Hauptbohrung. KTB-Report 90-4: 345

Hofmann B, von Drach V, Köhler H (1993) Nd model ages and Sr isotope systematics of drilled material from the KTB main bore hole. KTB Report 93-2: 389–390

Kleemann U, Stein E, Blümel P (1989) Mylonite im Grenzbereich zwischen ZEV und Moldanubikum. KTB-Report 89-3: 349

Klemd R, Matthes S, Okrusch M (1991) High-pressure relics in meta-sediments intercalated with the Weissenstein eclogite, Münchberg gneiss complex, Bavaria. Contrib Mineral Petrol 107: 328–342

Köhler H, Müller-Sohnius D, Cammann K (1974) Rb-Sr Altersbestimmunge an Mineral- u. Gesamtgesteinsproben des Leuchtenberger und Flossenbürger Granits, NE Bayern. Neues Jahrb Mineral Abh 123: 63–85

Kreuzer H, Seidel E (1989) Diskrete früh-devonische Ar-Ar-Alter der Hangend-Serie (Münchberger Masse). Ber DMG, Eur J Mineral 1 (Beih 1): 103

Kreuzer H, Okrusch M, Raschka H, Schüssler U, Seidel E, Vejnar Z (1989a) Ar-Ar-Datierungen am Westrand der Böhmischen Masse. KTB-Report 89-3: 427

Kreuzer H, Seidel E, Schüssler U, Okrusch M, Lenz K-L, Raschka H (1989b) K-Ar geochronology of different tectonic units at the northwestern margin of the Bohemian Massif. Tectonophysics 157: 149–178

Kreuzer H, Müller P, Carl C, Ebadi A, Höhndorf A, Patzak M (1990a) Mineral dating on core samples from 500 to 2000 m depth of the KTB drill hole. KTB-Report 90-4: 546

Kreuzer H, Müller P, Okrusch M, Patzak M, Schüssler U, Seidel E, Smejkal V, Vejnar Z (1990b) Ar-Ar confirmation for Cambrian, Early Devonian, and Mid-Carboniferous tectonic units at the western margin of the Bohemian Massif. Rundgespr Zeitmarken im europäischen Variszikum, 16–17 Nov 1990, Clausthal-Zellerfeld, Abstr

Kreuzer H, Vejnar Z, Schüssler U, Okrusch M, Seidel E (1992) K-Ar dating in the Teplá-Domazlice Zone at the western margin of the Bohemian Massif. In: Kukal Z (ed) Proc 1st Int Conf on the Bohemian Massif, Prague 1988, Czech Geol Survey Prague: 168–175

Lich S, Behrmann J, Heinisch H, Sprenger W, Volp A, Zadow A (1990) Evidenz für plastisches Fließen mit niedriger Drehzahl in den Gneisen der kontinentalen Tiefbohrung (KTB-Vorbohrung). Kurzfassungen und Poster, 3. Symp für Tektonik, Strukturgeologie, Kristallingeologie, Graz 1990, pp 131–132

Lister G, Price GP (1978) Fabric development in a quartz-feldspar-mylonite. Tectonophysics 49: 37–78

Lohoff B (1990) Petrographie und Geochemie von Metabasiten der Gabbroamphibolitmasse Neukirchen/Kdyne. Diploma thesis, Univ Würzburg, 87 pp

Ludwig V (1968) Zur Lithologie des "Kulms" bei Erbendorf/Oberpfalz (Bayern). Neues Jahrb Geol Palaeontol Monatsh 7: 407–417

Masch L, Koch N (1988) Deformationsstrukturen der Mylonite der Zone von Erbendorf-Vohenstrauß (ZEV) und Vergleich mit Basismyloniten der Silvretta-Decke (Graubünden). Erlanger Geol Abh 116: 81–100

Matte P, Maluski H, Rajlich P, Franke W (1990) Terrane boundaries in the Bohemian Massif: result of large-scale Variscan shearing. Tectonophysics 177: 151–170

Matthes S (1990) Die Erbendorfer Grünschieferzone. Ber DMG, Eur J Mineral 2 (Beih 2): 85–116

Matthes S, Richter P, Schmidt K (1975) Die Eklogitvorkommen des kristallinen Grundgebirges in NE-Bayern. IX Petrographie, Geochemie und Petrogenese der Eklogite des Münchberger Gneisgebietes. Neues Jahrb Mineral Abh 126: 45–86

Matthes S, Okrusch M, Röhr C, Schüssler U, Richter P, von Gehlen K (1995) Talc-chlorite-amphibole felses of the KTB pilot hole, Oberpfalz, Bavaria: protolith characteristics and phase relationships. Mineral and Petrol 52: 25–59

Miller H, Söllner F, Loske W (1990) U-Pb-Datierungen an Zirkonen aus Gneisen der KTB-Vorbohrung. KTB-Report 90-4: 544

Müller-Sohnius D, von Drach V, Horn P, Köhler H (1987) Altersbestimmungen an der Münchberger Gneismasse, Nordost-Bayern. Neues Jahrb Mineral Abh 156: 175–206

O'Brien PJ (1989) The petrology of retrograded eclogites of the Oberpfalz Forest, northeastern Bavaria, West Germany. Tectonophysics 157: 195–212

O'Brien PJ (1991) High pressure metamorphism in the NW Bohemian Massif: comparisons and contrasts between the Moldanubian Zone, Münchberg Massiv, ZEV, ZTT and Erzgebirge. KTB-Report 91-1: 1–12

O'Brien PJ, Carswell DA, Gebauer D (1990) Eclogite formation and distribution in the European Variscides. In: Carswell DA (ed) Eclogite facies rocks. Blackie, Glasgow, Chapman & Hall, New York, pp 204–224

O'Brien PJ, Röhr C, Okrusch M, Patzak M (1992) Eclogite facies relics and a multistage breakdown in metabasites of the KTB pilot hole, NE Bavaria: implications for the Variscan tectonometamorphic evolution of the NW Bohemian Massif. Contrib Mineral Petrol 112: 261–278

Okrusch M (1969) Die Gneishornfelse um Steinach in der Oberpfalz. Eine phasenpetrologische Analyse. Contrib Mineral Petrol 22: 32–72

Okrusch M (1971) Garnet-cordierite-biotite equilibria in the Steinach Aureole, Bavaria. Contrib Mineral Petrol 32: 1–23

Okrusch M, Seidel E, Schüssler U, Richter P (1989) Geochemical characteristics of metabasites in different tectonic units of the northeast Bavarian crystalline basement. In: Emmermann R, Wohlenberg J (eds) The German Continental Deep Drilling Program (KTB), site-selection studies in the Oberpfalz and Schwarzwald. Springer, Berlin Heidelberg New York, pp 67–79

Okrusch M, Matthes S, Schmidt K (1990a) Eklogite des Münchberger Gneisgebietes. Ber DMG, Eur J Mineral 2 (Beih 2): 55–84

Okrusch M, Schüssler U, Seidel E, Kreuzer H, Raschka H (1990b) Pre- to early Variscan magmatism in the Bohemian Massif. In: Franke W, Weber K (eds) Bohemian Massif. Conf on Paleozoic orogens in central Europe, Göttingen-Giessen, Aug–Sept 1990, Field Guide, pp 25–35

Okrusch M, Matthes S, Klemd R, O'Brien PJ, Schmidt K (1991) Eclogites at the northwestern margin of the Bohemian Massif: a review. Eur J Mineral 3: 707–730

Oppermann U (1985) Geochemie und Petrographie der Prasinite und Amphibolite am SW-Rand der Münchberger Gneismasse, NE-Bayern. Diploma thesis, Techn Univ Braunschweig, 140 pp

Patzak M (1991) Die Metabasite der KTB-Vorbohrung, Oberpfalz (NE-Bayern). Doctoral thesis, Univ Würzburg, 162 pp

Patzak M, Okrusch M, Röhr C (1991) Die Metabasite der KTB-Vorbohrung: Petrographie, Geochemie, Mineralchemie und Metamorphoseentwicklung. KTB-Report 91-1: 63–82

Pflug HD, Prössl KF (1990) Erste palynostratigraphische Ergebnisse aus der KTB-Vorbohrung. KTB-Report 90-4: 110–111

Pflug HD, Reitz E (1987) Palynology in metamorphic rocks: indication of early land plants. Naturwissenschaften 74: 386–387

Platt JP, Behrmann JH, Cunningham PC, Dewey JF, Helman M, Parish M, Shepley M, Wallis S, Weston PJ (1989) Kinematics of the Alpine arc and the motion history of Adria. Nature 337: 158–161

Pommerenke K (1985) Geochemie und Petrographie der Amphibolite im NE-Teil der Münchberger Gneismasse, NE-Bayern. Diploma thesis, Techn Univ Braunschweig, 135 pp

Puchelt H, Matthes S, Richter P (1978) Aussagemöglichkeiten der Seltenen Erden und anderer Spurenelemente für die Eklogite des Münchberger Gneisgebiets. Fortschr Mineral 56, Beih 1: 106–108

Reinhard J (1990) Metamorphism of the metapelites in the KTB drillcore: reaction sequence and thermobarometric constraints. KTB-Report 90-4: 535

Reitz E, Höll R (1988) Jungproterozoische Mikrofossilien aus der Habachformation in den mittleren Hohen Tauern und dem nordostbayerischen Grundgebirge. Jb Geol B-A Wien 131/2: 329–340

Röhr C, Kohl J, Hacker W, Keyssner S, Müller H, Sigmund J, Stroh A, Zulauf G (1990) German Continental Deep Drilling Programm (KTB) – geological survey of the pilot hole "KTB Oberpfalz VB". KTB-Report 90-8: B1–B55

Schalkwijk G (1991) Metabasites in the pilot hole of the German Continental Deep Drilling Project, Windischeschenbach, Eastern Bavaria. Structures and fabrics as documents of crustal evolution. Doctoral thesis Univ Bochum: 200 pp

Scheumann KH (1924a) Prävariskische Glieder des sächsisch-fichtelgebirgischen Kristallins. I. Die magmatisch-orogenetische Stellung der Frankenberger Gneisgesteine. Abh Math-Naturwiss Kl Sächs Akad Wiss 39 I: 1–16

Scheumann KH (1924b) Studien über die sächsisch-thüringischen Zwischengebirge I. Nachrichtenblatt Geol Paläontol Miner 1 (6): 10 pp

Scheumann KH (1924c) Studien über die sächsisch-thüringischen Zwischengebirge II. Die Gesteinsassoziationen im sächsischen Zwischengebirge. Nachrichtenblatt Geol Paläontol Miner 9 (10): 13 pp

Scheumann KH (1932) Über die petrogenetische Ableitung des roten Erzgebirgsgneises. Miner Petrogr Mitt 42: 413–454

Scheumann KH (1935) Die Rotgneise der Glimmerschieferdecke des sächsischen Granulitgebirges. Ber Math-Phys Kl Sächs Akad Wiss 87: 251–286

Scheumann KH (1938) Über die petrographische und chemische Substanzbestimmung der Gesteinsgruppe der Roten Gneise des sächsischen Erzgebirges und der angrenzenden Räume. Miner Petrogr Mitt 59: 391–440

Scheumann KH (1957) Über Gneise des Erzgebirges. Z Geol 6: 3–39

Schmädicke E (1991) Quartz pseudomorphs after coesite in eclogites from the Saxonian Erzgebirge. Eur J Mineral 3: 231–238

Schmädicke E, Okrusch M, Schmidt W (1992) Eclogite-facies rocks in the Saxonian Erzgebirge, Germany: high pressure metamorphism under contrasting P-T conditions. Contrib Mineral Petrol 110: 226–241

Schmidt W, Schmädicke E, Werner CD (1990) Eklogite des Erzgebirges. Ber DMG, Eur J Mineral 2 (Beih 1): 224

Schüller A (1954) Die kristalline Scholle von Wildenfels, ihr Stoffbestand und ihr tektonischer Bau im Rahmen des varistischen Gebirges. Z Geol 3: 707–749

Schüssler U (1990) Petrographie, Geochemie und Metamorphosealter von Metabasiten im KTB-Zielgebiet Oberpfalz. Geol Bavarica 95: 5–99

Schüssler U, Oppermann U, Kreuzer H, Seidel E, Okrusch M, Lenz K-L Raschka H (1986) Zur Altersstellung des ostbayerischen Kristallins. Ergebnisse neuer K-Ar-Datierungen. Geol Bavarica, 89: 21–47

Schüssler U, Richter P, Okrusch M (1989) Metabasites from the KTB Oberpfalz target area, Bavaria – geochemical characteristics and examples of mobile behaviour of "immobile" elements. Tectonophysics 157: 135–148

Schüssler U, Okrusch M, Patzak M, Müller P, Kreuzer H (1990) The polyphase thermal history of the Erbendorf-Vohenstrauß (ZEV) and Erbendorf Greenschist (EGZ) zones of NE Bavaria in the light of Ar-Ar spectra. Ber DMG, Eur J Mineral 2 (Beih 1): 231

Schüssler U, Vejnar Z, Okrusch M, Rose D, Seidel E (1992) Geochemistry of metabasites and gabbroic rocks from the Teplá-Domazlice Zone. In: Kukal Z (ed) Proc 1st Int Conf on the Bohemian Massif, Prague 1988, Czech Geol Survey Prague: 247–256

Schwan W (1974) Die Sächsischen Zwischengebirge und Vergleiche mit der Münchberger Gneismasse und anderen analogen Kristallinvorkommen im Saxothuringikum. Erlanger Geol Abh 99: 189 pp

Seibert J, Pauli Ch, Franke W, Behrmann J (1990) Zur Strukturentwicklung und Kinematik der Grenzzone Moldanubikum-Tepla/Barrandium im Bayerischen Wald. Kurzfassungen u. Poster, 3. Symp für Tektonik, Strukturgeologie, Kristallingeologie, Graz 1990, pp 208–210

Söllner F, Köhler H, Müller-Sohnius D (1981a) Rb/Sr-Altersbestimmungen an Gesteinen der Münchberger

Gneismasse (MM), NE-Bayern. Teil 1. Gesamtgesteinsdatierungen. Neues Jahrb Miner Abh 141: 90–112

Söllner F, Köhler H, Müller-Sohnius D (1981b) Rb/Sr-Altersbestimmungen an der Münchberger Gneismasse (MM), NE-Bayern. Teil 2. Mineraldatierungen. Neues Jahrb Miner Abh 142: 178–198

Steiger RH, Jäger E (1977) Subcommission on geochronology: convention on the use of decay constant in geo- and cosmochronology. Earth Planet Sci Lett 36: 359–362

Stein E (1988) Die strukturgeologische Entwicklung im Übergangsbereich Saxothuringikum/Moldanubikum in NE-Bayern. Geol Bavarica 92: 5–131

Stein E, Kleemann U (1992) Evidence for late Variscan emplacement of the ZEV. KTB-Report 92-4: 133–146

Stosch HG Lugmair GW (1987) Geochronology and geochemistry of eclogites from the Münchberg Massif, FRG. Terra Cognita 7: 163

Stosch HG, Lugmair GW (1990) Geochemistry and evolution of MORB-type eclogites from the Münchberg Massif, southern Germany. Earth Planet Sci Lett 99: 230–249

Teufel S (1988) Vergleichende U-Pb- und Rb-Sr-Altersbestimmungen an Gesteinen des Übergangsbereichs Saxothuringikum/Moldanubikum, NE-Bayern. Göttinger Arb Geol Paläontol 35: 1–87

Vejnar Z (1972) Regionally metamorphosed volcanic rocks from the West-Bohemian metabasite belt. Krystalinikum 9: 131–156

Vejnar Z (1977) The relationship between the metamorphic grade and composition of silicates in the West-Bohemian greenschists and amphibolites. Krystalinikum 13: 129–158

Vejnar Z (1986) The Kdyne Masif, South-West Bohemia – a tectonically modified basic layered intrusion. Sbor Geol Ved Geol 41: 9–67

Voll G (1960) Stoff, Bau und Alter der Grenzzone Moldanubikum/Saxothuringikum in Bayern unter besonderer Berücksichtigung gabbroider, amphibolitischer und kalksilikatführender Gesteine. Geol Jahrb Beih 42: 1–382

Vollbrecht A (1981) Tektogenetische Entwicklung der Münchberger Gneismasse. Göttinger Arb Geol Paläontol 24: 122 pp

Vollbrecht A, Weber K, Schmoll J (1989) Structural model for the Saxothuringian-Moldanubian suture in the Variscan basement of the Oberpfalz (northeastern Bavaria, F.R.G.) interpreted from geophysical data. Tectonophysics 157: 123–133

von Gehlen K, Schmitt GE (1989) Serpentinites from Winklarn (northeastern Bavaria, F.R.G.) and from Brezovica and Radusa (Yugoslavia): Relics of suboceanic mantle. Tectonophysics 157: 213–219

von Gehlen K, Matthes S, Okrusch M, Richter P, Röhr C, Schüssler U (1991) Ultramafische Einschaltungen in Metabasiten der KTB-Vorbohrung. KTB-Rep 91-1: 83–95

von Quadt A (1990) U-Pb-zircon and Sm-Nd analysis on metabasites from the KTB pilot bore hole. KTB-Report 90-4: 545

von Quadt, Gebauer D (1988) Sm/Nd, U-Pb-and Rb-Sr dating of high-pressure ultramafic to felsic rocks from the Moldanubian area of NE Bavaria (FRG) and the Saxonian Granulite Massif (GDR). Chem Geol 70: 15

von Quadt A, Gebauer D (1993) Sm-Nd and U-Pb dating of eclogites and granulites from the Oberpfalz, NE-Bavaria, Germany. Chem Geol 109: 217–339

Weber K (1984) Variscan events; Early Palaeozoic continental rift metamorphism and late Palaeozoic crustal shortening. In Hutton DHW, Sanderson DJ (eds) Variscan tectonics of the North Atlantic region. Geol Soc Spec Paper 14: 3–22

Weber K, Vollbrecht A (1989) The crustal structure at the KTB drill site, Oberpfalz. In: Emmermann R, Wohlenberg J (eds) The German Continental Deep drilling Program (KTB) Springer, Berlin Heidelberg New York, pp 5–36

Werner C-D (1989) Metamorphic mafic rocks of the Central Saxon Lineament zone. In: Variscan mafic rocks in the Saxothuringian zone. Guidebook of excursions in the German Democratic Republic May/June 1989. Academy of Sciences of the GDR, Central Institute for Physics of the Earth, Potsdam, pp 49–54

Note Added in Proof. Sm-Nd data of Schmädicke et al. (in press) have yielded 360 ± 7 Ma as the most reliable age of high-P metamorphism in the central Erzgebirge. Like in the Saxonian granulites, this age is at variance with the Late Devonian to Viséan sedimentation age of rocks which, today, structurally overlie the high-P rocks. This paradox requires either complex tectonic scenarios, or unconventional interpretations of the isotopic data.

Schmädicke E, Mezger K, Cosca MA, Okrusch M (in press) Variscan Sm-Nd and Ar-Ar ages of eclogite facies rocks from the Erzgebirge, Bohemian Massif. J metamorphic Geol 1995/13

V.C.2 Metamorphic Evolution

P. Blümel

1 Introduction

Within or at the periphery of the Saxothuringian Basin several isolated basement areas of medium or even high grades are found. These are the Münchberg Mass, the "Zwischengebirge" of Wildenfels and of Frankenberg, the Zone von Erbendorf Vohenstrauß and the Sächsische Granulitgebirge including its Schiefermantel (Fig. 1). The Erzgebirge Zone is incorporated into this chapter here since it may also be allochtonous. The last main metamorphism of most of the allochthonous units is of the MP type except for the Sächsische Granulitgebirge and its Schiefermantel, and some peripheral schists of the Erzgebirge Zone, which show an anomalously high geotherm.

2 The Münchberg Mass

The Münchberg Mass comprises five tectonic subunits which are piled in a synform with the vertical succession: Randschiefer-Serie, Phyllit-Grünschiefer-Serie, Rand-amphibolit-Serie, Liegend-Serie and Hangend-Serie (Fig. 1). This succession is derived from the geologic situation observed at the limbs of the Münchberg synform. Except for the Liegend-Serie all subunits were seen as equivalents of a (Cambro-) Ordovician volcanosedimentary association with metamorphic grades increasing from bottom to top.

Lenses of serpentinite, essentially of spinel-harzburgite type, are bound both to the low-grade and the high-grade subunits of the Münchberg Mass. In the low-grade parts it contains former dikes of olivine-gabbronoritic composition and thus resembles the basal part of ophiolites. Within the Liegend-Serie spinel-peridotite experienced HT-deformation and then recrystallized within the stability field of amphibole-peridotite (with pargasite) before serpentinization (Gayk and Seidel 1992).

The Liegend-Serie forms a reworked intrusive complex of gabbroids and granitoids emplaced within meta(psammo)-pelites and graphite-quartzite (Fig. 2). Observed mineral assemblages of metasediments are (+ plagioclase + quartz + rutile + ilmenite ± graphite): garnet + biotite + muscovite, garnet + kyanite + biotite + muscovite, staurolite + biotite + muscovite and kyanite + staurolite + biotite + muscovite. High-T mylonitization is widespread giving rise to neoblast aggregates of oligoclase, kyanite and biotite as well as dendritic garnet II (which rims garnet I or forms separate grains). The mica aggregates cause a flecky texture in hand specimens mistaken in early mapping as an indicator of contact metamorphism. Porphyroblastic, pre-mylonitic garnet I is homogeneous and rich in spessartine: $alm_{61.3}$ $pyp_{19.5}$ $gro_{3.1}$ $spe_{16.1}$. Post-mylonitic garnet II is inhomogeneous, relatively magnesian, but poor in manganese: $alm_{66.4}$ $pyp_{23.7}$ $gro_{6.4}$ $spe_{3.5}$ (averaged). PT estimates with GRAIL and GABI give 8.2 kb/610 °C. Garnet I cores often include sillimanite trails, kyanite being very rare, which indicates growth of first garnet at medium to high temperatures. More information on this early stage is lacking; it may have been anatexis or granulite facies (Blümel, KTB reports, unpubl.).

Metagabbros (e.g., Steinhügel) and meta-granodiorite (e.g., Weikenreuth) of the Liegend-serie, covering areas of less than 0.5 km^2, preserve undeformed rock portions with plagioclase crowded with needles of zoisite; adjacent amphibole and biotite are overgrown by tiny garnet grains of the composition $alm_{59.7}$ $pyp_{17.1}$ $gro_{19.6}$ $spe_{3.6}$. PT conditions are estimated at 11 kb/630 °C (garnet-plagioclase-biotite-muscovite-quartz thermobarometer, Hoisch; GABI) Blümel, KTB report, unpubl.; Bosbach et al. 1991).

Gabbro and granodiorite intruded around 500 Ma, thus pointing to a pre-Ordovician age of sedimentation for the paragneiss wallrock (if primary relations; Gebauer and Grünenfelder

Dallmeyer et al. (Eds.)
Pre-Permian Geology of Central and Eastern Europe

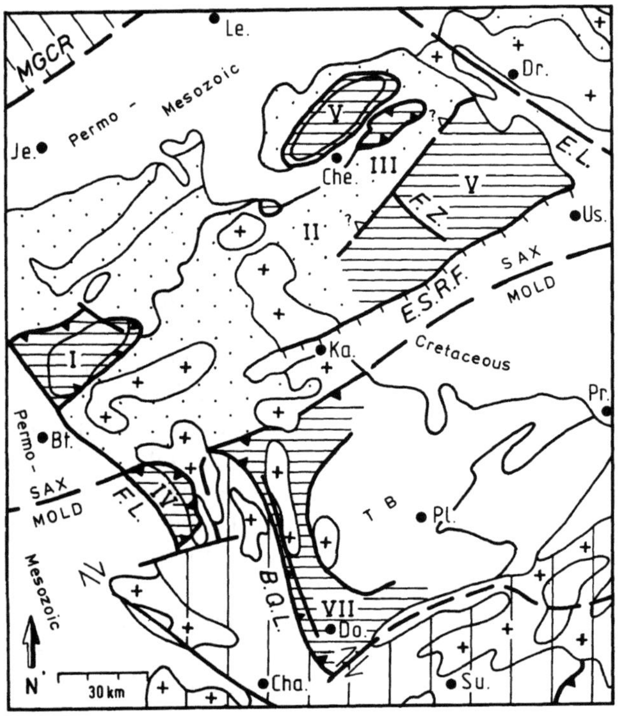

Fig. 1. Position of the exotic nappes (*horizontally ruled*) within the northwestern Bohemian Massif. *I* Münchberg Mass; *II* Wildenfelser Zwischengebirge; *III* Frankenberger Zwischengebirge; *IV* Zone von Erbendorf-Vohenstrauß; *V* Sächsisches Granulitgebirge incl. Schiefermantel; *VI* Erzgebirge Zone (gneiss region incl. schists envelope). *MGCR* Mid-German Crystalline Rise (*diagonal ruling*; mostly below post-Variscan cover); *MOLD* Moldanubian (*vertical ruling*); *SAX* Saxothuringian including exotic nappes *I–VI*; *TB* Teplá-Barrandean together with metamorphic Zone of Teplá-Domažlice (*VII*); *blank areas* show post-Variscan cover (incl. unmetamorphosed part of *TB*). *B.Q.L.* Bohemian Quartz Lode; *E.L.* Elbe Line; *E.S.R.F.* Erzgebirgesüdrand Fault; *F.L.* Fränkische Linie; *F.Z.* Flöha Zone. *Bt.* Bayreuth; *Cha.* Cham; *Che.* Chemnitz; *Do.* Domažlice; *Dr.* Dreden; *Je.* Jena; *Ka.* Karlovy Vary; *Le.* Leipzig; *Pl.* Plžen; *Pr.* Praha; *Su.* Susiče; *Us.* Ústí

1979; Stosch and Lugmair 1990). The cooling history of the Liegend-Serie was found to extend from 395 Ma (K-Ar/hornblende, Rb-Sr/muscovite) to 378 Ma (Rb-Sr/biotite); Rb-Sr whole rock model isochrons on paragneisses give Ordovician ages (Müller-Sohnius, et al. 1987).

The Hangend-Serie is well known for its wealth in eclogites, from which part of the associated amphibolites and hornblende gneisses may well be derived. Two variants of ecoglites occur. The light type contains omphacite + garnet + quartz + rutile + phengite + kyanite and/or zoisite and shows the composition of a high-Al basalt. Their "honeycomb"-structure points to a gabbro precursor. The mineral assemblage of the dark type is omphacite + garnet + quartz + rutile + hornblende or phengite and the rock composition corresponds to oceanic MORB. Primary omphacite contains 18–62 mol% jadeite without a systematic difference between both eclogite types. However, garnets of the light type are richer in pyrope (28–65 mol%) than the dark eclogite (15–35 mol%) with similar contents in grossular (9–28 mol% in the light versus 14–31 in the dark variant) and MnO (less than 2 wt%). The chemical zonation is prograde as indicated by the increase in pyrope at the expense of grossular and/or almandine and spessartine. This agrees well with the fact that phengite in the eclogite assemblage is richer in Si (6.95 atoms p.f.u.) as compared with phengite included within garnet (6.33–6.56 atoms p.f.u.) as well as with the presence of mineral inclusions of zoisite, kyanite and quartz. For detailed descriptions and data, the reader is referred to Okrusch et al. (1991).

For the dark eclogite of the Weißenstein the metamorphic history is reconstructed as follows (Franz et al. 1986; Okrusch et al. 1991; O'Brien 1993):

Eclogite stage : $P \geqslant 13\text{–}17$ kb
$T \geqslant 620 \pm 50\,°C$
Symplectite stage: $P = 8.5\text{–}12$ kb
$T \sim 620\,°C$
Amphibolite facies stage: at medium pressures.

The HP estimate is based on the reaction albite = jadeite + quartz (Holland calibration) and on the Si content of phengite (Massonne/Schreyer calibration). Barometry of first decomposition stages with stable clinopyroxene (less sodian) + plagioclase still yields high pressures of 14–15 kb, which are similar to the minimum pressures obtained from primary omphacites of maximal jadeite content (O'Brien 1993). Thus, pressures above 13–17 kb are possible – a confirmation of the so far doubted Münchberg origin of the coesite-bearing eclogite samples which are reported by Chesnokov and Popov (1965)? Actually, the symplectite stage integrates over a variety of differing PT stages, which are arranged to give a stepwise structured decompression path (O'Bien 1993).

Some eclogites probably equilibrated at $P^{H_2O} \approx P_{tot}$. In the absence of primary fluid inclusions in the HP assemblage, this condition is deduced from quartz veins within light eclogite which

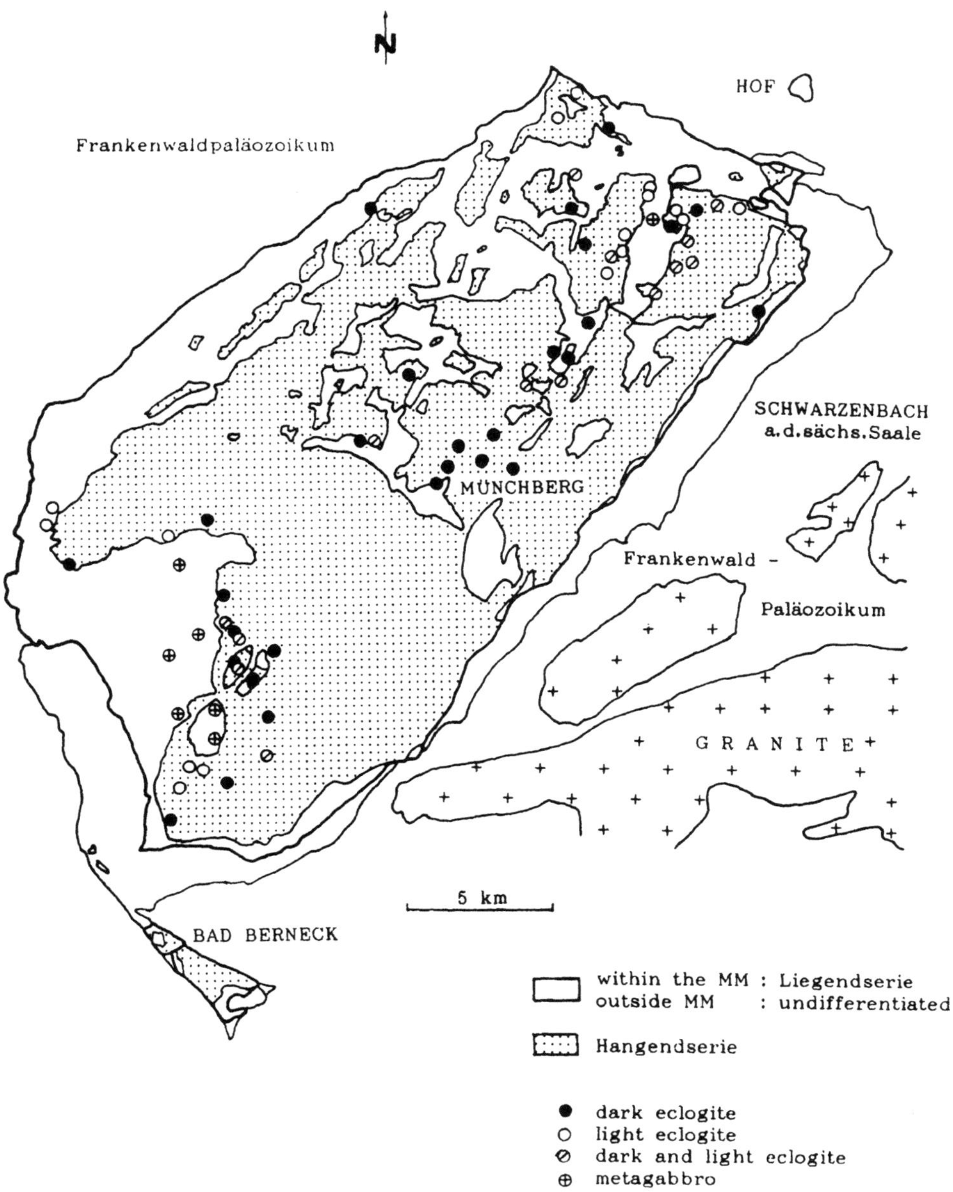

Fig. 2. Münchberg Mass with Hangend-Serie (*dotted area*); Liegend-Serie (*blank area* within *thick boundary*) and basal subunits of lower grades (undifferentiated). Small wedge on southern tip is the so-called Berneck Gneiskeil. Sudetic granites toward the E intrude Paleozoic schists of the Fichtelgebirge area

contain omphacite (jadeite$_{50}$), kyanite, zoisite, muscovite and occasionally graphite + Fe-sulfides; contacts in the eclogitic wallrock show grain-size enlargement without formation of hydrous minerals (Klemd 1989). The vein assemblage gives minimum pressures of 15 kb for $P^{H_2O} = p_{tot}$.

Thin calcareous metasediments intercalated within the Weißenstein eclogite surprisingly preserve relics of a former HP stage, which are: recrystallized hornblende-plagioclase aggregates, zoisite-rimmed kyanite and biotite-rimmed phengite. Late replacement of zoisite + kyanite by margarite occurred below 8–10 kb and 575 °C (Blümel 1986). In paragneiss garnet-phengite thermobarometry (with mica richest in Si) gives 620–650 °C (assuming 15 kb) and 640–680 °C (assuming 20 kb) respectively (Klemd et al. 1991).

Geochemistry and relic magmatic structures indicate that kyanite-bearing eclogites were probably former gabbros, which were modified by crustal contamination and crystal accumulation.

Conversely, the dark eclogites may have been extrusives, which are unmodified (Bosbach et al. 1991).

Eclogite of the Weißenstein gives crystallization ages between 525 + 40/−31 Ma (U-Pb/zircon/ upper discordia intercept; Gebauer and Grünenfelder 1979) and 480 ± 23 Ma (Sm-Nd reference isochron; Stosch and Lugmair 1990) followed by Middle-Upper Devonian HP metamorphism (380 + 14/−22 Ma, lower discordia intercept, Gebauer and Grünenfelder 1979; 395 ± 4 and 384 ± 3 Ma, Sm-Nd mineral isochron; Stosch and Lugmair 1990). Subsequent MP metamorphism of the Hangend-serie is approximated by hornblende (Ar-Ar and K-Ar) and muscovite (Rb-Sr) ages of 380 Ma (Kreuzer et al. 1989; Kreuzer et al. 1993). The Münchberg Mass did not reach in its present position before late Lower Carboniferous (ca. 325 Ma; Franke 1984). Shortly afterwards, LP regional metamorphism occurred in the LP-Moldanubian and adjacent Saxothuringian.

3 The Zone Erbendorf-Vohenstrauß (ZEV)

Being the host zone of the KTB well, the ZEV is one of the most thoroughly studied geological units of the Variscides. From the wealth of results, a condensed excerpt is given in this section.

Bordered toward the S and E by the Moldanubian, toward the N by the Erbendorfer Grünschiefer Zone and the "Erbendorfer Paleozoic", and cut toward the W by the Fränkische Linie, the Zone of Erbendorf-Vohenstrauß (ZEV) comprises metapsammopelites, orthogneiss, amphibolites with intercalated metagabbro and serpentinite, and rare eclogite-amphibolite (Fig. 3). On the basis of microfossils (?) – acritarchs and spores (?) – a Lower Devonian age was ascribed to some paragneisses of the KTB pilot drill hole (Pflug and Prössl 1989). This association is intruded near its eastern border by a plate-like, E-dipping granite pluton (Leuchtenberg granite).

In the metapsammopelites, which cover ca. three-fourths of the ZEV, three NNW-SSE trending main mineral zones are observed (Schulte and Blümel, internal DFG report 1993; from W to E): staurolite zone (±kyanite, ±sillimanite), garnet-aluminium silicate zone (±kyanite, ±sillimanite), and cordierite-sillimanite-andalusite zone. The latter evolved from garnet + kyanite + biotite + k-feldspar + quartz, predates HT

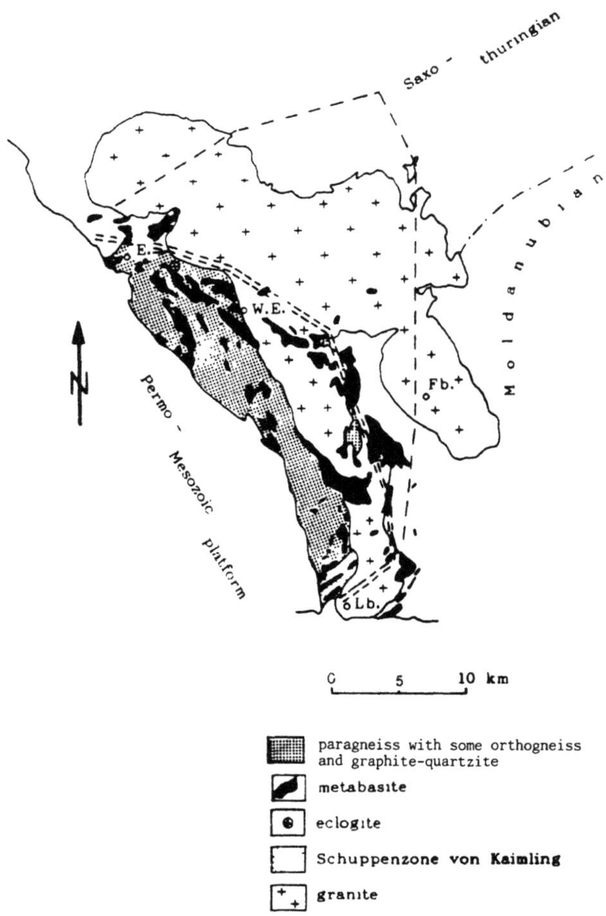

Fig. 3. Zone of Erbendorf-Vohenstrauß s.str. comprises paragneiss incl. few orthogneisses and graphite-quartzites (*shaded area*), and metabasites (*black*) incl. few eclogite-amphibolites and serpentintes; the Schuppenzone of Kaimling toward the S (*lightly dotted*) and the Erbendorfer Grünschiefer Zone (metabasites with associated phyllites N Erbendorf) are basal slices. Delimitation of ZEV by stratigraphic criteria (*dashed line*) or by tectonic criteria (*dashed double line*; Kleemann 1991). In the latter case, the Leuchtenberg granite (=southern limb of large composite pluton shown) would be an "interstitching" pluton intruded along the ZEV nappe thrust zone. A stratigraphically defined boundary would incorporate Saxothuringian low-grade metamorphic equivalents of the Moldanubian. *E.* Erbendorf; *Fb.* Flossenbürg; *Lb.* Leuchtenberg; *W.E.* Windisch-Eschenbach; ⊕ KTB well

mylonitization (→thrusting) and subsequent local static recrystallization at the Leuchtenberg granite. Similar rocks with a definite granulite precursor stage are part of the "Schuppenzone von Kaimling-Michldorf" (Kleemann 1991) at the southern edge of the ZEV.

In assemblages with Al_2SiO_5, garnets are chemically homogeneous, manganiferous (spessartine less than 10 mol%) and poor in calcium (grossular less than 3–6 mol%). In assemblages with staurolite, garnet displays bell-shaped Mn zonation profiles (spessartine less than 21 mol%) with thin retrograde rims. Coexisting muscovites are low in Si (less than 6.25 at. p.f.u.). PT conditions for garnet-staurolite gneiss are estimated at 7–7.5 kb/530–600 °C (GASP and GRIPS, locality Scherreuth) and for garnet-kyanite sillimanite gneiss with 6–7 kb/600–660 °C (GASP and GABI; Blümel unpubl. KTB report; Röhr and von Gehlen 1989; Kleemann 1991; Reinhard 1990; Schulte and Blümel, unpubl. KTB reports). Metamorphism of the paragneisses seems to ack any HP stage and to culminate in the MP amphibolite stage at 380 Ma (monazite, Teufel 1988). Rb-Sr mica ages are 373 ± 8 Ma for muscovite and 371 and 370 ± 5 Ma for biotite (Teufel 1988); the stratigraphic age remains just beyond these limits.

Associated amphibolites contain the assemblage hornblende (pargasitic-edenitic or magnesiohornblende) + plagioclase (oligoclase-labradorite) ± garnet (alm_{50-55} pyp_{12-17} spe_{1-4} gro_{25-35} adr_{0-3}) ± salite ± quartz (Schüßler 1990).

Garnet-amphibolites show reaction fabrics which point to an early HP stage (hornblende-plagioclase symplectites). Garnets may be rimmed by a plagioclase corona but themselves do not rim mineral relics. They probably formed by the prograde reaction: hornblende + plagioclase + ilmenite → garnet + clinopyroxene + quartz + rutile + H_2O. In the zonation profile, three stages are recorded, stage 1 + 2 being prograde (bell-shaped Mn profile) and stage 3 being retrograde (increase in Mn); transition from stage 1 to 2 is defined by the reversal from increasing to decreasing X^{Mg} values (grossular beginning to decrease) (Bühn and Okrusch 1993).

Garnet amphibolite and some of the ZEV amphibolites derive from metagabbro of predominantly normative MORB-type olivine tholeiite; lenses of metasomatized ultramafics may be associated in addition, containing assemblages with serpentines, amphibole, talc, chlorite and ores. Metagabbro displays complex reaction textures. Thus, igneous plagioclase decomposes in the presence of H_2O to zoisite + kyanite + quartz + less calcic plagioclase (MP amphibolite facies). Igneous augite exsolves rutile and transforms topotaxically to jadeitic clinopyroxene. Eclogite-facies garnet probably grows on alteration products of igneous olivine (or orthopyroxene) according to an allochemical reaction of the type: amphibole + kyanite + plagioclase + ilmenite = garnet + quartz + rutile + H_2O + alkali ions. Maximum P-conditions (in the presence of sodic plagioclase) are ca. 13 kb/580–650 °C (KTB pilot hole sample; Holland's calibration, among other). For eclogites at a subvertical distance of less than ca. 5 km (VSPE 1 and VBE 1 holes), HP_{min} conditions are obtained with 14–16 kb/760 °C and 13–14 kb/590 °C (O'Brien 1992).

The evolution of the metabasite complex (amphibolite-metagabbro-ultramafites etc.) starts (at latest) with the crystallization of the gabbro during the Lower Devonian (494 ± 3 Ma, idiomorphic zircons, upper discordia intercept) and continues to the HP and MP metamorphism in the (Silurian-)Devonian (398 Ma, Sm-Nd garnet-bulk rock isochron from metagabbro, von Quadt 1990; 371 Ma, averaged K-Ar and Ar-Ar ages of hornblende, Kreuzer et al. 1993).

At the southern edge of the ZEV, acid and basic granulites, orthogneiss, metagabbro, amphibolite, serpentinite and garnet pyroxenite are comprised within the Schuppenzone of Kaimling, which toward the S is in steep tectonic contact with prograde garnet micaschists (?Saxothuringian). The early granulite stage (391 Ma, zircon, upper discordia intercept. Teufel 1988) is estimated at 9–11 kb (GRAIL, GASP) and 800 °C (garnet-clinopyroxene, Ellis and Green 1979). Along its eastern border kyanite granulite is transformed into cordierite-sillimanite gneiss texturally differing from its Moldanubian analogue. Late cordierite – meanwhile also known from some ZEV paragneisses further toward the N (Schulte and Blümel 1992) – is formed by the incompatibility of garnet + Al_2SiO_5 (Kleemann 1991).

The ZEV is intruded by a late Variscan granite (Leuchtenberg granite; 326 ± 2 Ma, Siebel 1993) which produced a contact aureole with stable cordierite + sillimanite + garnet (X^{Mg} = 0.08–0.10; X^{Mn} = 0.06–0.11) + cordierite (X^{Mg} = 0.46–0.52) + biotite + k-feldspar (Steinach aureole, Okrusch 1971). Using Lonker's garnet-cordierite calibration, PT conditions near 4,5 kb (3.5 kb for X^{H_2O} = 0.5) and 620 °C are obtained.

The PT evolution of representative rocks of the ZEV, metagabbro, eclogite amphibolite and metapelite from the KTB pilot hole and adjacent drill holes is shown in Fig. 4. The early stage of the metabasite path connecting the shallow

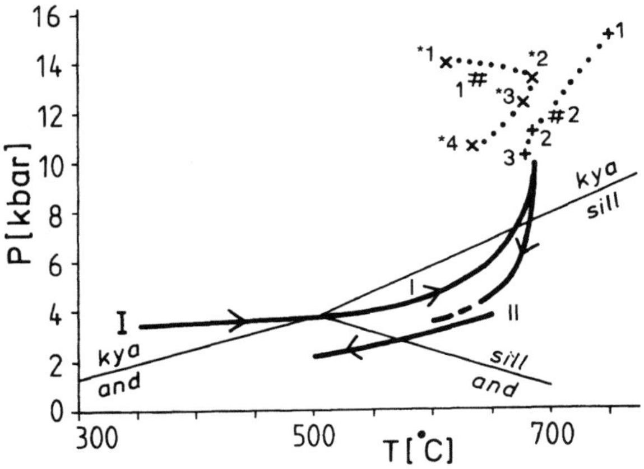

Fig. 4. PT evolution in the Zone of Erbendorf-Vohenstrauß, PT *path I* of paragneisses of the KTB well as inferred from garnet-biotite thermometry and paragenetic observation (Reinhardt 1990); eclogite VBE (KTB well 3875 m; *stages 1–4*); eclogite VSPE (drill hole close to KTB well, 48 m; *stages 1–3*); two metagabbros (KTB well 1268 m, 3620 m; all metabasite data from O'Brien et al. 1992). The paragneiss evolution culminates in the MP stage after isobaric LP heating followed by rapid down-thrusting; after juxtaposition with metabasites uplift along second part of *path I*. The uplifted ZEV association was emplaced onto the deforming Moldanubian (*path II*, Kleemann 1991)

emplacement level of gabbro with the HP stage is unknow. Yet for closely associated metapelite lacking indications of a HP stage, on the basis of the Al_2SiO_5 polymorph succession and GABI thermometry, an anticlockwise prograde section with strong isobaric heating (incipient LP stage?) followed by a clockwise ITD section was obtained. Metabasite and metapelite paths converge at 8–10 kb/ca. 675 °C. A subsequent joint PT evolution – though lacking thermobarometric criteria in metabasite – is deduced from the rock association.

The idea of an allochthonous nature of the ZEV overlying a quasi-autochthonous LP Moldanubian-Saxothuringian basement, was first raised because of the difference in the baric "type" of metamorphism and the existence of an intervening diaphthorite zone (Blümel 1983). Later attempts to compute the PT histories of both units into one family of PT loops (i.e., the postulate of an early ZEV cap) failed due to the impossibility to synchronize them at a reasonable thermal balance (Thompson-England models). Then the tectonic emplacement should post-date LP metamorphism, which integrated Saxothuringian sediments and "Proto"-Moldanubian gneisses into an LP unit (see Chap. VII.C.4). This timing is corroborated by the absence of LT equivalents of the LP unit adjacent to the ZEV, the PT trajectory of which is shown in Fig. 4. A maximum emplacement age is given by the crystallization age of the undeformed Leuchtenberg granite. This does not rule out the existence of locally younger mylonitization as dated by an Rb-Sr small-scale isochron with 305 Ma (Baumann et al. 1990).

According to geophysical investigations, the ZEV could extend toward the W, i.e., across the Franconian Line and below the Mesozoic platform, over a distance of ca. 10 km. Some 300 km further toward the WSW garnet-kyanite schists and -gneisses are found at the northern edge of the Variscan basement of the Black Forest (Wickert et al. 1990); a relation with the ZEV seems probable. Here, these MP rocks – tectonically justaposed with Permian sediments and volcanics – are taken to delineate the boundary between the Saxothuringian and Moldanubian.

Similarities in lithologic association and metamorphism of the ZEV with parts of the Münchberg Mass and the Zone of Teplá-Domažlice (ZTT = western portion of the Bohemicum) led to their interpretation as parts of a large MP unit in the western Bohemian Massif (Blümel 1984). However, the incorporation of the ZTT critically depends on the actual age of its well-known MP metamorphic zonation, which then was assumed to be Variscan – against the Cadomian view as based on a stratigraphic/tectonic discordance some 40 km out of area. New hints for a Cadomian age now comes from an undeformed granite intruded at the low-grade periphery of the ZTT, which is Cambrian (Stod granite, 518 Ma, Ar-Ar/biotite; Kreuzer et al. 1990). Nevertheless, Variscan events exist in the ZTT, Devonian as well as Carboniferous (e.g., Kreuzer et al. 1990), which, however, still need petrological and structural substantiation. The Carboniferous "rejuvenation" may well be ascribed to thrusting (similar to the ZEV). The eclogite-bearing Mariánské Lázně Complex shows Early Devonian HP metamorphism (379 ± 9 and 375 ± 5 Ma, Sm-Nd mineral isochron, Beard, et al. 1992) – similar to the Münchberg Mass – and thus is seen as a separate thrust sheet in the case of Cadomian MP metamorphism in the ZTT.

At the southwestern edge of the ZTT – beyond the layered igneous complex of Neukirchen-Kdyně – a zone of blastomylonitic gabbroamphibolites (concordant zircon fraction with 530 Ma, Gebauer and Grünenfelder 1982; K-Ar hornblende ages scattering between 463 and 329 Ma, Kreuzer et al. 1990) probably represents a Schuppenzone between the ZTT proper and the adjacent Moldanubian. Within the gabbroamphibolite (Folmava-Hoher Bogen-Rittsteig area) coronitic metagabbro, garnet-amphibolite (9 kb/570–600 °C, Bues 1993), garnet pyriclasite (6 kb/820–860 °C) (Büttner 1992) and cordierite-garnet-cummingtonite schist occur without lithologic contacts exposed. Slices of serpentinite (no Al phases), tremolite fels and talc schists as well as garnet-kyanite-sillimanite metatexite (between 6.5 kb/630 °C and 8.5 kb/690 °C, Büttner 1992) are localized within the gabbroamphibolite or close to the Moldanubian. The metamorphic grade of garnet- and gabbroamphibolite is probably similar at differing rock compositions; the granulite should represent an older stage.

Halfway between the ZEV and the ZTT – in the Moldanubian near Oberviechtach – serpentinite (with spinel), garnet pyroxenite (O'Brien 1991), eclogite-amphibolite and kyanite-garnet granulite are tectonically incorporated into cordierite-sillimanite gneiss. Eclogite-amphibolites passed through 15 kb/710 °C, 12 kb/800 °C, 8–10 kb/800 °C into the amphibolite facies; for garnet pyroxenite 28 kb are analyzed (O'Brien 1991). This association (Winkarner Serie, G. Fischer), which covers ca. 10 km^2, seems to have affinities with the Sächsische Granulitgebirge rather than the ZEV or ZTT.

4 The Saxonian Granulite Massif

The Saxonian Granulite Massif (SGM) consists of a core of partly LP-reworked former granulite facies rocks of acid, basic and sedimentary protoliths locally associated with garnet gneiss and cordierite gneiss (areal extension of SGM: $45 \times 20\,km^2$) and a metasedimentary envelope, the so-called Schiefermantel (width: 1–7 km), with an outwards decreasing grade of LP metamorphism (Fig. 5). Phyllites of the Schiefermantel have Cambrian to Lower Carboniferous deposition ages. Similarities, regarding rock association, mineral assemblages and reaction fabrics, to granulite terranes within the Moldanubian Zone exist. The exotic nature of the SGM so far was referred to thrusting (Pietsch 1963), diapirism (Watznauer et al. 1971; Behr 1978), diapiric folding (Weber and Behr 1983) and crustal extension (metamorphic core complex, Franke 1993). Granulite facies metamorphism was ascribed either to the Cadomian or the Variscan orogeny. Recent petrological, structural and geochronological investigations considerably increase our knowledge, though a unifying model of general acceptance is lacking.

Among the granulites, the acid variant containing the assemblage garnet (+kyanite) (+biotite) + perthite + oligoclase + quartz + rutile by far predominates. Kyanite and rutile may be overgrown by garnet and k-feldspar. Spinel (green)-anorthite coronas around garnet, sillimanite rimming kyanite, migmatization phenomena (including cordierite fleck-aploids) and late andalusite altogether portray stages of the post-granulite uplift. Another typical decompression reaction is: garnet + k-feldspar + rutile + $H_2O \rightarrow$ sillimanite + biotite + quartz + ilmenite; the product assemblage grows either in the mylonite schistosity or in stress shadows of garnet. Concomitantly, lattice-preferred orientation of quartz evolves from small-circle girdle via crossed girdle to two-point or single-point maxima (Behr 1978). Typical garnet compositions are $X^{Mg} = 0.18–0.35$, $X^{Ca} = 0.03–0.16$ and $X^{Mn} = 0.008–0.035$ (homogeneous core) and $Alm_{79-58}Pyp_{15-36}Gro_{2.5-4.3}Spe_{3-1}$ (retrogade rim) respectively; spinel is essentially a $hercynite_{60-73}$ – $spinel_{25-38}$ solid solution poor in gahnite (less than 3%; Rötzler 1992).

Locally, layers of pyroxene granulite or pyriclasite are interstratified leaving the impression of a "leptynite-amphibolite formation". Pyriclasite may contain diopside, hypersthene, garnet and secondary amphibole, biotite, magnetite and pyrrhotite. Garnet ($alm_{30-50}pyp_{35-60}gro_{6-18}adr_{2-4}spe_1$) is rimmed by orthopyroxene + anorthite. Of particular interest are layers with high-Mg rock variants containing jadeitic clinopyroxene ($jadeite_{20-40}$, Diethendorf quarry; Rötzler et al. 1992a).

Metaperidotite and layers or lenses of pyroxenite occur in three different associations (after Rötzler et al. 1992a):

1. Acid granulite-pyriclasite-garnet pyroxenite (the latter being layers within pyriclasite, which are thinly floored by serpentinite) (e.g., Diethensdorf quarry).
2. Granulite (retrogressed)-serpentinite (± garnet)-calc silicate gneiss (Waldheim area, eventually incl. Rubinberg quarry).

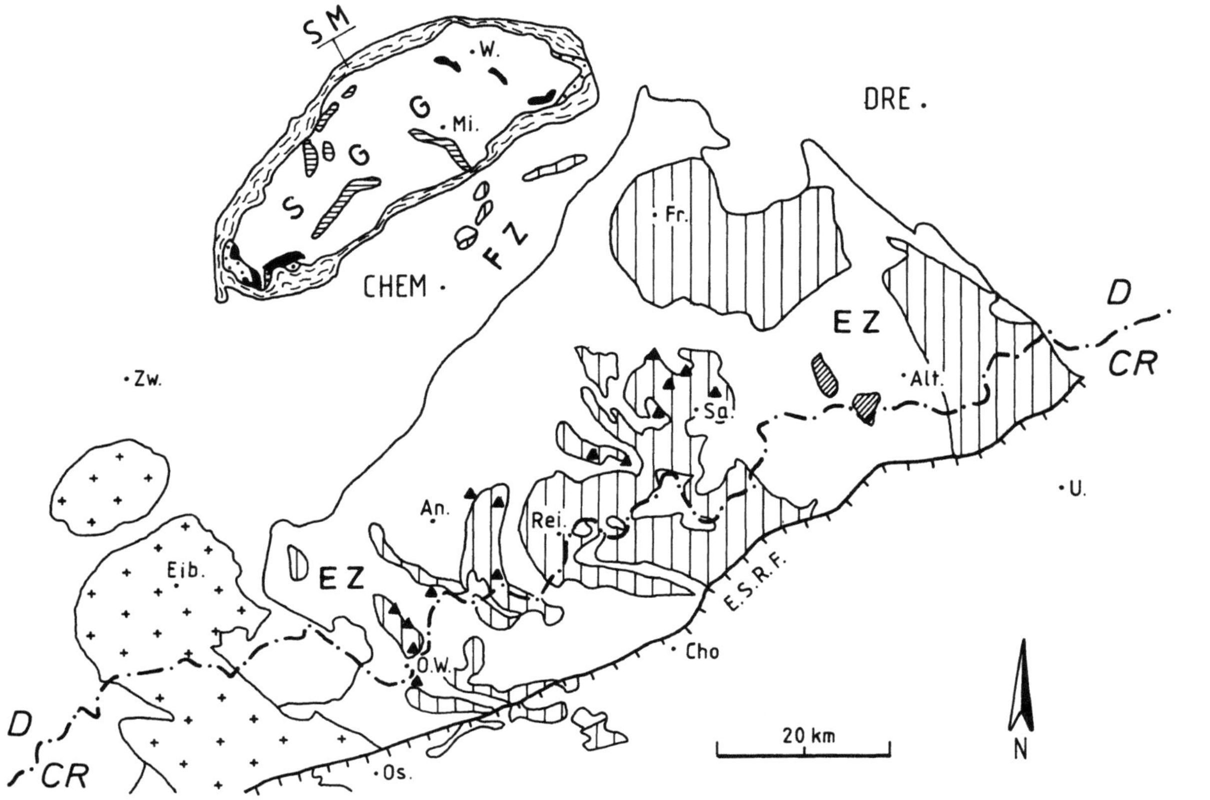

Fig. 5. Geology of the Sächsische Granulitgebirge (*SGG*) incl. Schiefermantel (*SM*), the basement portions of Frankenberger Zwischengebirge (*FZ*) and the Erzgebirge Zone (*EZ*). *SGG* Garnet and cordierite gneisses (hatched); serpentinite (*black*); diabase (*dotted*), and granulites incl. retrogressed derivatives (*blank*). *EZ* Gneissose or undeformed granitoids (*vertical ruling*); phyllites of Hermsdorf-Rehefeld (*diagonal ruling*); other gneisses or schists (undifferentiated; *blank*); and eclogite (amphibolite)s (*filled triangles*); Ohře valley basement (migmatites and granulites) beyond Erzgebirgssüdrand-Fault (*E.S.R.F.*) is probably part of the EZ. *Alt.* Altenberg; *An.* Annaberg; *CHEM.* Chemnitz; *Cho.* Chomutov; *DRE.* Dresden; *Eib.* Eibenstock; *Sa.* Sayda; *U.* Ústí; *Zw.* Zwickau

3. Noritic flaser gabbro-diabase-serpentinite squeezed into the boundary granulite core/ Schiefermantel (ophiolitic association? – eventually incl. Klatschmühle pyroxenite).

Garnet has the composition (Rubinberg quarry): $alm_{16.6}pyp_{71.4}gro_{3.0}adr_{1.1}spe_{1.3}uv_{6.6}$ in serpentinite and $alm_{35.0}pyp_{34.2}gro_{30.0}adr_{0.2}spe_{0.6}$ in pyroxenite. Clinopyroxene in websterite from Klatschmühle contains up to 27% jadeite.

Pyroxenite lenses within serpentinite in rock association 2 display megacrysts of clinopyroxene (less than 10 cm) with exsolved Ca-Fe garnet lamellae (Gilsberg quarry, near Reinsberg; Reiche and Bautsch 1985). In the case of an originally intermediate diopside-CaTs solid solution, an HP/T_{min} of ca. 17 kb/850 °C on a decompression path may be indicated (PT calibration, Carswell 1990).

The well-known kornerupine granulite of Waldheim, which is intercalated within a hercynite-bearing granulite (gneiss), is characterized by its enrichment in boron (0.45 oxide wt%), sodium (7.8 oxide wt%) and lithium. It probably derives from a sodium-rich zeolithic sediment into which boron was introduced either from evaporitic solutions or by hydrothermal exhalations. It contains kornerupine (X^{Mg} = 0.78, B_2O_3 = 3.4 wt%, Li_2O = 0.12 wt%), sapphirine (X^{Mg} = 0.75, Cr_2O_3 = 0.2 wt%, B_2O_3 = 0.81 wt%, and MgO/Al_2O_3 = 1.6–4.3), garnet (X^{Mg} = 0.61–0.77, MnO = 1.0–1.5 wt%, and CaO = 0.26–0.83 wt%), corundum, kyanite ± sillimanite ± andalusite, tourmaline and biotite all set in a matrix of oligoclase ($An_{12\text{-}13}Kf_{5\text{-}6}$). Reaction textures document a three-stage metamorphic history from stable garnet + kyanite + corundum through garnet + sillimanite + corundum +

sapphirine to a late stage with andalusite and dumortierite. The rock association near Waldheim also comprises a "calc-silicate-eclogite" with 50–70 vol% diopside ($Dio_{68}Hed_{21}Jad_3 CaTs_3 Fs_5$), 30–50 vol% garnet ($alm_8$ pyp_2 gro_{75} adr_{14} spe_1) and retrograde amphibole, biotite and plagioclase (Grew 1986; Bankwitz et al. 1990; Rötzler et al. 1992a).

Garnet gneiss and cordierite gneiss of a paleosome, metatexite or restite variant form isolated bodies (extension less than 12 km) within and at the periphery of the granulite core, with which both display mylonitized contacts. Both gneiss groups differ in bulk chemical composition, the former being rich in Al, Fe, Mg, Mn and the latter being rich in SiO_2 and K_2O at similar X^{Mg} (0.28 for garnet gneiss and 0.30 for cordierite gneiss). Garnet overgrows sillimanite, green spinel, plagioclase and quartz and is surrounded by a corona of ferroan cordierite (X^{Mg} = 0.38–0.56), cordierite + spinel ($hercynite_{ss}$ with 16% gahnite) or cordierite + k-feldspar. The late reactions yielding cordierite indicate a more or less isothermal decompression up to LP levels (Rötzler 1992).

Metapelites and -psammites of the Schiefermantel tectonically frame the granulite core toward the top to NW direction. LP metamorphism decreases from the sillimanite zone via staurolite-andalusite micaschists to garnet-biotite phyllites along a profile which is tectonically reduced (locally even to 2 km). The southwestern edge of the granulite core is tectonically overlain by an ophiolitic(?) serpentinite-flaser gabbro-diabase association (Reinhardt and Kleemann, unpubl.; Rötzler 1992; Rötzler et al. 1992a).

The PT evolution of the SGM (Fig. 6) is reconstructed from eclogite, pyroxene-, kornerupine-, and quartz-feldspar granulite, garnet gneiss and Schiefermantel rocks and their reaction textures using PT grids and divariant equilibria (GASP, GABI etc.). For *eclogite* (garnet + omphacite + quartz + rutile) peak conditions of 20.5 kb/960 °C were found, followed by a granulite stage (garnet rim + ferroan pargasite + labradorite + rutile/ilmenite) at 13.5 kb/870 °C/a^{H_2O} = 0.2 and by an amphibolite stage with ferroan pargasite + hercynitic spinel + plagioclase. Associated *pyroxene granulite* gives peak conditions (core compositions of garnet + clinopyroxene + andesine + ferroan pargasite + quartz) of 12 kb/820 °C/a^{H_2O} = 0.1 followed by a second granulite stage (hypersthene + Ti-rich biotite + labradorite symplectite

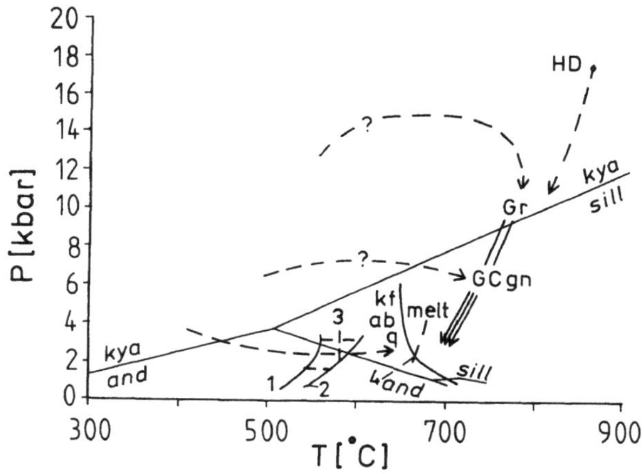

Fig. 6. PT evolution of rocks in the Sächsische Granulitgebirge and its Schiefermantel. *HD* PT-estimate inferred from assumed $diopside_{50}$ $CaTs_{50}$ solid solution (see text); *Gr* acid granulite (Rötzler 1992) together with kornerupine granulite (Grew 1986); *GCgn* garnet and cordierite gneisses (Rötzler 1992) *1* and *2* Lower T stability of cordierite + Al_2SiO_5 + biotite and upper T stability of staurolite + muscovite + quartz, respectively; *3* and *4* GABI temperatures for garnet-staurolite and garnet-sillimanite gneiss, respectively, of the Schiefermantel (Reinhardt and Kleemann 1994). Uplift paths of Ek, Gr and GCgn successively coverge and meet the Schiefermantel level near 3 kb

rimming garnet) at 8.5 kb/790 °C. For *kornerupine granulite* maximal PT conditions are represented by garnet + kyanite + corundum. Using the PT intersection of staurolite = garnet + Al_2SiO_5 + corundum with kyanite = sillimanite, minimum conditions of 11.5 kb/800 °C are obtained; if plagioclase already belonged to this assemblage, the actual peak pressure would not be much higher. *Garnet gneiss* equilibrated at 6.7 kb/720–780°C and some *Schiefermantel* rocks within the PT range of 2–3.5 kb/650–700 °C (Grew 1986; Reinhardt Kleemann 1994; Rötzler 1992; Rötzler and Kroner 1992).

PT paths of the SGM and Schiefermantel rocks (Fig. 6) are clearly simplistic – due to their 1–2 point pinning – and more complicated geometries seem to be plausible. Interestingly, for the Schiefermantel rocks, a simple heating path is inferred from the one-foliation microtexture, which shows syntectionic porphyroblasts of andalusite and staurolite with a concordant s_i/s_e relation (Reinhardt and Kleemann 1994). This path anticipates heat contributions during the tectonic emplacement (granites, shearing, fluids) in addition to contact metamorphism along the juxtaposed granulite body. However, it may well be that these three minor heat contributions are insufficient to produce LP/HT metamorphism; this the more so, since LP/HT rocks of regional extent are known throughout the Variscan. Nevertheless, the causes of such high geotherms (magmatic underplating, adiabatic uplift) are still in discussion.

Dating the polymetamorphic evolution of the SGM is far from being solved. The termination of the ductile regime is approximated by the zircon upper discordia intercept at 338 ± 5 Ma for synkinematic granite, which tectonically underlies the Schiefermantel (Berbersdorf granite, von Quadt 1993). From banded (?mylonitic) granulite, thin Rb-Sr slabs with ages of 342 (±8) and 321 (±30) Ma were obtained (Gorokov et al. 1987), which may be interpreted in terms of LP metamorphism and mylonitization (eventually including influx of fluids). The granulite stage proper seems to be poorly defined when looking at the spread of Sm-Nd mineral-whole rock ages from 346 (±2) to 308 (±9) Ma and zircon ages from 352 (±2) to 342 (±4) Ma (data from von Quadt 1993). Considering the various grades of post-granulitic re-equilibration, a minimum age of the granulite stage is inferred at ca. 350 Ma. Garnet pyroxenite (Klatschmühle) gave an Sm-Nd mineral isochron age of 380 ± 14 Ma, interpreted as HP stage, and a zircon upper discordia intercept at 338 ± 4 Ma (von Quadt 1993). An Sm-Nd whole rock isochron age of 471 ± 8 Ma (pooled over five samples) of mafic granulite is interpreted in terms of the crystallization age of the magmatic precursor (von Quadt 1993). It is interesting to note the similarity with the Rb-Sr whole rock age on acid granulites (425 ± 26 Ma, Jäger and Watznauer 1969). Thus, isotope geochronology opens a time span of metamorphic evolution from the Ordovician to the Lower Carboniferous, which is at variance with the stratigraphic record in the Saxothuringian Basin (see Chap. V.B.1). Obviously, we are confronted with two possible versions of the geologic evolution, a "one-orogeny" and a "two-orogenies" model.

5 The Erzgebirge Zone

In a schematic way, the Erzgebirge Zone (EZ) can be described as a gneiss region framed by Paleozoic micaschists and phyllites (the latter are metamorphosed members of the Thuringian Facies) – except for the southeastern delimitation which is given by the Eger (Ohře) graben of Tertiary age (Fig. 5). The Ohře valley basement within this graben is probably similar to the EZ. The gneiss region separates into three to four orthogneiss cores ("Rotgneise") which are surrounded by gneiss-mylonites of unknown (sedimentary or magmatic) origin. Idiomorphic zircons from three different granitoid core regions (Reitzenhain, Freiberg, Fürstenwalde-Lauenstein) gave 550–555 Ma ages (Kröner et al. 1991), which, however, do not contribute to settling the discussion on the stratigraphic age of the tectonically juxtaposed wallrocks.

The gneiss-mylonites locally contain lenses of (1) kyanite-granulite and its retrogressed equivalents (Zöblitz, Ohře valley, E Ostrov etc.), (2) garnet-serpentinite with folded layers of garnet-pyroxenite, and (3) metabasites with relics of kyanite-free eclogites, some of which with indications of a formerly stable coesite (Behr et al. 1965; Mathé 1990; Schmidt et al. 1990). Structurally, the EZ occupies the northeastern part of the Fichtelgebirge-Erzgebirge Antiform (with a SW-NE trend), which overprints W-E structures and is bisected by discordant "Sudetic" granite plutons (Erzgebirge-Kerngranit, Erzgebirge-Randgranit) leaving contact aureoles with sillimanite and/or andalusite in the Paleozoic wallrocks (Thuringian Facies).

For a granulite/garnet-peridotite assocziation in the Ohře valley basement, HP_{min} conditions of 15–22 kb/700–800 °C for kyanite-granulite, T = 900 °C for pyroxene-granulite, and 33 kb/860–1020 °C for garnet-peridotite are calculated. Peak conditions of granulite were followed first by (near) isobaric cooling and then by (near) isothermal decompression, both within the stability field of kyanite (Kotková 1993). For associated garnet-peridotite, two stages are reconstructed by zircon dating: Silurian partial melting (zircon core: 424 ± 4 Ma) and Lower Carboniferous HP metamorphism (zircon rim: 342 ± 5 Ma; Gebauer 1991).

Throughout the EZ several PT evolutions with differing baric precursor stages are recognized (K. Rötzler et al. 1992b, 1993; Willner et al. 1993; K. Rötzler, pers. comm. 1993; Schmädicke 1991; Schmädicke et al. 1991):

1. *HP/MT-HT unit*: PT_{max} in gneisses with 16–20 kb/750 °C and in coesite(?)-eclogite with ca. 28 kb/ca. 800 °C (Annaberg-Jöhrstadt and Sayda gneiss domes; Zöblitz muscovite

gneiss complex and western wallrocks of the Altenberg caldera).

2. *MP/MT unit*: gneiss domes with intercalated garnet-amphibolite giving a PT_{max} of 6–8 kb/580–650 °C (Reitzenhain, Freiberg and Fürstenwalde-Lauenstein).
3. *HP/LT unit*: micaschists (Cambrian?) with PT_{max} = 12–14 kb/520 °C and associated eclogite with P = ca. 20 kb (Oberwiesenthal-Schwarzenberg-Öderan-W Freiberg area as well as parts of the Hermsdorf-Rehefeld phyllites).
4. *MP/LT unit*: Ordovician phyllites with PT_{max} = 7–8 kb/450–480 °C (Johanngeorgenstadt-Fichtelberg area, Schwarzenberg-Öderan area, and part of the Hermsdorf-Rehefeld phyllites).
5. *LP/LT unit*: Ordovician phyllites with PT_{max} = 2.5 kb/400 °C (Schneeberg-Chemnitz-Süd area).

Obviously, the crustal segments of HP and MP provenance converged and tectonically recombined at a 8–6 kb level (T = ca. 600 °C) to give a new MP nappe unit, which in turn came into tectonic juxtaposition with the LP/LT unit. Further uplift results in the formation of a zonal structure of the EZ with HP and MP units of medium-high temperatures in core position followed successively toward the NW (and W) by HP, MP, and LP units of lower temperatures.

6 The "Zwischengebirge" of Frankenberg and Wildenfels

Located between the Sächsische Granulitgebirge and the Fichtelgebirge-Erzgebirge Anticline, both Zwischengebirge (Fig. 5) are small high-grade basement klippen which tectonically overlie Paleozoic slates of the Saxothuringian Basin. Together with the Münchberg Mass, they link up into the Zentralsächsische Lineament.

In the Frankenberg Massif (23 × 5 km^2) are associated: garnetiferous hornblende gneiss with aploid veins (±garnet), augen gneiss (±garnet) deriving from porphyric biotite granite, greenschists (probably deriving from lava and tuffs) and albite-orthoclase-biotite-magnetite schists (metarhyolite). Most of these rocks were transformed into diaphthorite or mylonite. Toward the NW and SE the massif is bordered by fault zones which dip below the massif (Werner 1990).

The *Wildenfels Massif* (2 × 1 km^2) rests on Ordovician and Lower Carboniferous slates and is transgressed by Rotliegend sediments. It contains garnetiferous two-mica gneisses, micaschists (partly with andalusite?), garnetiferous hornblende gneisses, and granitic gneisses with albite + clinozoisite. The rocks are widely transformed into diaphthorite or mylonite (Schueller 1954).

7 Attempts at Correlation

The LP/LT and MP/LT units of the EZ show similarities (incl. stratigraphy) with both border regions, Saxothuringicum/Moldanubicum and Cunean (Královsky hvozd) Micaschist/Bavarian Moldanubicum in NE and E Bavaria respectively (Chap. V.B.4). Paleozoic schists (Frauenbach-Sch.) halfway to NE Bavaria (near Nejdek/CR) contain a blueschist lens with crossite (Souček and Holub 1992). From this one would expect: (1) LP/HT gneisses below (or replaced by) the EZ gneiss region, and (2) a formerly complete LP basement (with an early HP stage) below the Paleozoic slates (Thur. F.). The EZ gneiss region possibly correlates with the Teplá portion of the ZTT, and the EZ schist envelope (units 3–5, from above) with the Slavskovský Les basement (Schlaggenwald-Kristallin) near Karlovy Vary (Karlsbad). The assumed relations may be blurred by further tectonic restructuring upon cooling.

References

Bankwitz P, Rötzler J, Neumann W (1990) Exkursionen zum DFG-Schwerpunkt "Stoffbestand, Struktur and Entwicklung der kontinentalen Unterkruste" in das Sächsische Granulitmassiv (Exkursionsführer)
Baumann A, Grauert B, Kleemann U (1990) Altersbestimmung einer Mylonitzone im Grenzbereich ZEV/Moldanubikum. KTB Rep 90-4: 551
Beard BL, Medaris LG, Johnson CM, Tonika J, Jelinek E (1992) Timing of eclogite and amphibolite metamorphism of the Márianské Lázně Complex, Czechoslovakia. AGU Spring Meet, EOS 73, No 14, Suppl, Washington
Behr HJ (1978) Subfluenz-Prozesse im Grundgebirgs-Stockwerk Mitteleuropas. Z Dtsch Geol Gesell 129: 283–318
Behr HJ, Fritsch E, Mansfeld L (1965) Die Granulite von Zöblitz im Erzgebirge als Beispiel für Granulitbildung in tiefreichenden Scherhorizonten. Krystalinikum 3: 7–29
Blümel P (1983) Intermediate and low-pressure facies series in the Variscan basement of southern Germany. Terra Cognita 3: 195

Blümel P (1984) Mitteldruck- und Niederdruckmetamorphose in den außeralpinen Varisziden Mitteleuropas. Fortschr Mineral 62 (Beih 1): 28–29
Blümel P (1986) Metamorphic processes in the Variscan crust of the central segment. In: Freeman R, Müller St, Giese P (eds) Proc 3rd Worksh on the European geotraverse (EGT) project: the central segment. Bad Honnef, 14–16 April 1986, pp 149–154
Bosbach D, Stosch HG, Seidel E (1991) Magmatic and metamorphic evolution in the Münchberg Massif, N.E. Bavaria. Contrib Mineral Petrol 107: 112–123
Bues CC (1993) Mineralchemie und Geothermobarometric der Jntrusiovgestein der Gabbroamphibolitmasse von Neukirchen b. Hl. Blut (Nordostbayern). Geol Bavarica 98: 119–157
Bühn B, Okrusch M (1993) Zur Petrologie und Interpretation der Granat-Amphibolite in der KTB Hauptbohrung. Z Geol Wiss 21: 37–47
Büttner St (1992) Druck- und Temperaturbedingungen der Regionalmetamorphose in der Grenzzone Gabbroamphibolitmasse von Neukirchen/Moldanubikum (E-Bayern). Diplomarbeit, Technische Hochschule, Darmstadt
Carswell T (ed) (1990) Eclogite facies rocks. Blackie, Glasgow
Chesnokov BV, Popov VA (1965) Increase in the volume of quartz grains in South Urals eclogite. Dokl Akad Nauk SSSR 162: 172–178
Ellis DJ, Green DH (1979) An experimental study of the effect of Ca upon garnet-clinopyvoxene Fe-Mg exchange equilibria. Contrib Mineral Petrol 71: 13–22
Franke W (1984) Variszischer Deckenbau im Raume der Münchberger Gneismasse – abgeleitet aus der Fazies, Deformation und Metamorphose im umgebenden Paläozoikum. Geotekt Forsch 68: 1–253
Franke W (1993) The Saxonian granulites: a metamorphic core complex? Geol Rundsch 82: 505–515
Franz G, Thomas S, Smith DC (1986) High pressure phengite decomposition in the Weissenstein eclogite, Münchberg Gneiss Massif, Germany. Contrib Mineral Petrol 92: 71–108
Gayk T, Seidel E (1992) Ultrabasite der Müchberger Gneismasse. 8. Rundge-spräch Geodynamik des europäischen Variszikums, Cheb/Westböhmen (Abstr)
Gebauer D (1991) Two Paleozoic high-pressure events in a garnet-peridotite of northern Bohemia, Czechoslovakia. Terra Abstr Suppl 6, Terra Nova 3: 5
Gebauer D, Grünenfelder M (1979) U-Pb Zircon and Rb-Sr mineral dating of eclogites and their country rocks. Example: Münchbeerg gneiss massif, northeast Bavaria. Earth Planet Sci Lett 42: 35–44
Gebauer D, Grünenfelder M (1982) Geological development of the Hercynian Belt of Europe based on age and origin of high-grade and high-pressure mafic and ultramafic rocks. 5th Int Conf on Geochronology, cosmochronology and isotope geology, Japan, Nikko, pp 111–112
Gorokhov IM, Anderson EB, Melnikov NN, Mitrofanov FP, Kutyavin EP, Bizunok MB, Rocheva ON (1987) Geochronology of metamorphic rocks of the Sächsische Granulitgebirge (GDR). In: Gerstenberger H (ed) Contributions to the geology of the Saxonian Granulite Massif (Sächsisches Granulitgebirge). ZFL Mitt 133: 43–62
Grauert B, Grosse-Westermann U, Albat F (1990) Interpretation von U-Pb-Monazitaltern moldanubischer Gneise. In: Emmermann R, Giese P (Hrsg.) KTB-Report 90-4, Beiträge zum 3. KTB-Kolloquium Gießen, 28.2.–2.3.1990, Niedersächsisches Landes-Amt für Bodenforschung, Hannover
Grew ES (1986) Petrogenesis of Kornerupine at Waldheim (Sachsen), German Democratic Republic. Z Geol Wiss 14: 525–558
Haase G, Gerstenberger H, Werner CD (1987) Rb-Sr geochronology of granites from the Sächsisches Granulitgebirge. Zentralinstitut für Isotopen- und Strahlenforschung-Mitteilungen 133, pp 89–97
Hofmann J, Mathé G, Werner CD, Stettner G, Shrbeny O, Ulrych J (1988) Saxothuringikum. In: Zoubek VE (ed) Precambrian in younger fold belts: the European Variscides, the Carpathians and the Balkanides. Wiley & Sons, 885 pp
Jäger E, Watznauer A (1969) Einige Rb/Sr-Datierungen an Granuliten des Sächsischen Granulitgebirges. Monatsber Dtsch Akad Wiss (5/6): 420–426
Kleemann U (1991) P-T-t-d Entwicklung im Grenzbereich zwischen der Zone von Erbendorf-Vohenstrauß (ZEV) und dem Moldanubikum in der Oberpfalz/NE-Bayern. Dissertation, Univ Bochum
Klemd R (1989) PT-evolution and fluid inclusion characteristics of retrograded eclogites, Münchberg Gneiss Complex, Germany, Contrib Mineral Petrol 102: 221–229
Klemd R, Matthes S, Okrusch M (1991) High-pressure relics in meta-sediments intercalated with Weissenstein eclogite, Münchberg gneiss complex, Bavaria. Contrib Mineral Petrol 107: 328–342
Köhler H, Müller-Sohnius D (1976) Ergänzende Rb/Sr-Altersbestimmungen an Mineral- und Gesamtgesteinsproben des Leuchtenberger und des Flossenbürger Granits (NE Bayern). Neues Jahrb Mineral Monatsh 8: 354–365
Kotková J (1993) Tectonometamorphic history of lower crust in the Bohemian Massif – example of north Bohemian granulites. Czech Geol Surv Spec Pap Prague
Kreuzer H, Seidel E (1989) Diskrete früh-devonische Ar/Ar Alter der Hangendserie (Münchberger Masse). Beih Eur J Mineral 1: 103
Kreuzer H, Seidel E, Schüßler U, Okrusch M, Lenz KL, Raschka H (1989) Ar-Ar geochronology of different tectonic units at the northwestern margin of the Bohemian Massif. Tectonophysics 157: 149–178
Kreuzer H, Müller P, Okrusch M, Patzak M, Schüßler U, Seidel E, Smejkal V, Vejnar Z (1990) Ar-Ar confirmation of Cambrian, Early Devonian, and Mid-Carboniferous events in tectonic units at the western margin of the Bohemian Massif. 6. Rundgespräch Geodynamik des europäischen Variszikums (extended Abstr)
Kreuzer H, Henjes-Kunst F, Seidel E, Bühn B (1993) Ar-Ar spectra on minerals from the KTB and related medium – pressure units. KTB Rep 93-2: 133–136

Kröner A, Frischbutter A, Bergner R, Hofmann J (1991) Zirkon-Evaporationsalter von granitoiden Gesteinen aus dem Erzgebirge und dem Rande der Lausitz und ihre geodynamische Bedeutung. 7. Rundgespräch Geodynamik des Variszikums Freiberg (Abstr)

Mathé G (1990) Zur Geologie der Serpentinitvorkommen im sächsischen Erzgebirge. Abh Staatl Mus Mineral Geol Dresden 37: 55–72

Müller-Sohnius D, von Drach V, Horn P, Köhler H (1987) Altersbestimmungen an der Münchberger Gneismasse Nordost-Bayerns. Neues Jahrb Mineral Abh 156: 175–206

O'Brien PJ (1991) High pressure metamorphism in the NW Bohemian Massif. In: Emmermann R, Luterjung J (eds) KTB-Rep 91-1: 1–12, Forschungsergebnisse des DFG-Schwerpunktoprogramms "KTB" 1986–1990, Niedersächsisches Landesamt für Bodenforschung, Hannover

O'Brien PJ (1993) Münchberg Mass eclogites: multiple metamorphic overprints and implications for P-T-d paths. Z Geol Wiss 21 (1/2): 217–224

O'Brien PJ, Röhr C, Okrusch M, Patzak M (1992) Eclogite facies relics and a multistage breakdown in metabasites of the KTB pilot hole, NE-Bavaria: implications for the Variscan tectonometamorphic evolution of the NW Bohemian Massif. Contrib Mineral Petrol 112: 261–278

Okrusch M (1971) Garnet-cordierite biotite equilibria in the Steinach-aureole, Bavaria. Contrib Mineral Petrol 32: 1–23

Okrusch M, Matthes S, Klemd R, O'Brien PJ, Schmidt K (1991) Eclogites of the northwestern margin of the Bohemian Massif. A review. Eur J Mineral 3: 707–730

Patzak M, Okrusch M, Röhr C (1991) Die Metabasite der KTB-Vorbohrung: Petrographie, Geochemie, Mineralchemie und Metamorphossentwicklung. In: Emmermann R, Lauterjung J (Hrsg) KTB-Rep 91-1: 63–82, Forschungsergebnisse des DFG-Schwerpunktprogramms "KTB" 1986–1990, Niedersächsisches Landes-Amt für Bodenforschung, Hannover

Pflug HD, Proessl KF (1989) Palynology in gneiss – results from the Continental Deep Driling Program. Naturwissenschaften 76: 565–567

Pietsch K (1963) Geologie von Sachsen. Deutscher Verlag der Wissenschaften, Berlin

Reiche M, Bautsch HJ (1985) Elecktron microscopical study of garnet exsolution in orthopyroxene. Phys Chem Minerals 12: 29–33

Reinhardt J (1990) Metamorphism of the metapelites in the KTB drill-core: reaction sequence thermobarometric constraints. In: Emmermann R, Giese P (eds) KTB-Rep 90-4: 535, Beiträge zum 3. KTB-Kolloquium Gießen, 28.2.–2.3.1990, Niedersächsisches Landes-Amt für Bodenforschung, Hannover

Reinhardt J, Kleemann U (1994) Extensional unroofing of granulitic lower crust and related low-pressure, high-temperature metamorphism in the Saxonicon granulite Massif, Germany. Tectonophysics 238: 71–94

Röhr C, von Gehlen K (1989) Geothermobarometrie eines Staurolith-führenden Paragneises aus der ZEV. KTB-Report 89-3: 428

Rötzler J (1992) Zur Petrogenese im Sächsischen Granulitgebirge – die pyroxenfreien Granulite und die Metapelite. Geotekt Forsch 77: 1–100

Rötzler J, Kroner U (1992) New petrological data on the Saxonian Granulite Massif and a correlation between metamorphic and deformative processes. 8. Rundgespräch Geodynamik des europäischen Variszikums, Cheb/Westböhmen (Abstr)

Rötzler J, Werner CD, Kroner U, Gerstenberger H (1992a) P-T-time path, deformation, history, and protoliths at the type locality of the granulites. IGCP. 304 Saxon Granulite Field Excursion/Field Guide Potsdam

Rötzler K, Schumacher R, Maresch WV (1992b) The metamorphic evolution in the western Erzgebirge. 8. Rundgespräch Geodynamik des europäischen Variszikums, Cheb/Westböhmen (Abstr)

Rötzler K, Schumacher R, Maresch WV (1993) PT-Entwicklung in Glimmerschiefern und Orthogneisen des Westerzgebirges. Eur J Mineral (Suppl) 5: 257

Scheumann KH, Huckenholz G (1961) Die Cordieritgneise und Granatgneise des sächsischen Granulitgebirges. Neues Jahrb Mineral Abh 96: 131–161

Schmädicke E (1991) Quartz pseudomorphs after coesite in eclogites from the Saxonian Erzgebirge. Eur J Mineral 3: 231–238

Schmädicke E, Schmidt W, Okrusch M (1991) Zur polymetamorphen Entwicklung der Eklogite des Erzgebirges. 7. Rundgespräch Geodynamik des europäischen Variszikums, Freiberg (Abstr)

Schmidt W, Schmädicke E, Werner CD (1990) Eklogite des Erzgebirges. Beih Eur J Mineral 2 (2): 125–169

Schüller A (1954) Die Kristalline Scholle von Wildenfels, ihr Stoffbestand und ihr tektonischer Bau im Rahmen des varistischen Gebirges. Geologie 3 (5–8): 707–749

Schulte B, Blümel P (1992) The staurolite isograd and late cordierite in metapelites of the ZEV (Zone von Erbendorf-Vohenstrauß). Zentralbl Geol Paläontol 7/8: 885–893

Schüßler U (1990) Petrographie, Geochemie und Metamorphosealter von Metabasiten im KTB-Zielgebiet Oberpfalz. Geol Bavarica 95: 5–99

Siebel W (1993) Geochronology of the Leuchtenberg Granite and the assoziated Redwitzites. KTB-Report 93-2: 411–415

Souček J, Holub FV (1992) Blueschist-greenschist metamorphism of metabasites in the western Krusne hory (Erzgebirge) Mts. 8. Rundgespräch Geodynamik des europäischen Variszikums, Cheb/Westböhmen (Abstr)

Stein E (1988) Die strukturgeologische Entwicklung im Übergangsbereich Saxothuringikum/Moldanubikum in NE-Bayern. Geol Bavarica 92: 5–131

Stosch HG, Lugmair GW (1990) Geochemistry and evolution of MORB-type eclogites from the Münchberg Massif, southern Germany. Earth Planet Sci Lett 99: 230–249

Teufel S (1988) Vergleichende U-Pb und Rb-Sr-Altersbestimmungen an Gesteinen des Übergangsbereiches

Saxothuringikum/Moldanubikum, NE-Bayern. Gött Arb Geol Paläontol 35: 87

Vinogradov VI, Pokrovsky BG (1987) 0-18 and Rb-Sr systematics in metamorphic rocks of the Saxonian Granulite Massif. Zentralinst Isotopen- Strahlenforsch Mitt 133: 73–87

von Quadt A (1990) U-Pb zircon and Sm-Nd analyses on metabasites from the KTB. In: Emmermann R, Giese P (eds) KTB-Rep 90-4: 545, Beiträge zum 3. KTB-Kolloquium Gießen, 28.2.–2.3.1990, Niedersächsisches Landes-Amt für Bodenforschung, Hannover

von Quadt A (1993) The Saxonian Granulite Massif. New aspects from geochronological studies. Geol Rundsch 82: 516–530

von Quadt A, Gebauer D (1990) Isotopic ages from the Saxonian Granulite Massif (GDR) and their geodynamic significance. In: Terranes in the circum-Atlantic Paleozoic orogens. Int Conf on Paleozoic orogens in central Europe, Göttingen-Gießen, Aug–Sept 1990, Abstr

Wagener-Lohse C, Blümel P (1984) Prograde Regionalmetamorphose vom Niederdruckp-Typ in der Grenzzone Saxothuringikum/Moldanubikum, E-Tirschenreuth/NE-Bayern. Fortschr Mineral 62 (Beih 1): 254–255

Watznauer A, Behr HJ, Mathé H (1971) Die Granulite Sachsens. Freiberger Forsch Hefte C 268: 91–111

Weber K, Behr HJ (1983) Geodynamic interpretation of the Mid-European Variszides. In: Martin H, Eder FW (eds) Intracontinental fold belts, Springer, Berlin Heidelberg New York

Werner CD (1990) Profil durch das Zwischengebirgskristallin von Frankenberg. Beih Eur J Mineral 2 (2): 152–269

Wickert F, Altherr R, Deutsch M (1990) Polyphase Variscan tectonics and metamorphism along a segment of the Saxothuringian – Moldanubian boundary: the Baden-Baden Zone, northern Schwarzwald (F.R.G.). Geol Rundsch 79: 627–647

Wienholz R, Hofmann J, Mathé HG (1979) Über Metamorphose, Tiefenbau und regionale Position des Erzgebirgskristallins. Z Geol Wiss 3: 385–395

Willner AP, Rötzler K, Maresch WV, Krohe A, Kopp J (1993) Hochdruckrelikte und Druck-Temperaturentwicklung in Orthogneisen des mittleren Erzgebirges. Beih Eur J Mineral 5: 256

VI Western Sudetes (Lugicum)

VI.A Introduction

A. Żelaźniewicz

Kossmat (1927) proposed a subdivision of Variscan Europe, in which he extended the Saxothuringicum east through Lusatia to the West Sudetes (Figs. 1, 2). He termed this region the Saxothuringian and Sudetian Zones. Suess (1926) referred to the region comprised of the Lusatia and the West Sudetes as Lugicum (after the province of Lausitz). Stille (1951) dismissed the possibility of linking Lugicum with Saxothuringicum and proposed the Elbe zone as a boundary between the two (Figs. 1, 2).

The Lugicum is bordered on the northeast by the Odra Fracture Zone (OFZ) and on the southwest by the Elbe Fracture Zone (EFZ). The southeastern border is defined by the Moravo-Silesian Thrust Zone (Moldanubian of Suess 1912), which juxtaposes Lugian rocks and the western margin of the Moravo-Silesicum (Fig. 1). The northwestern border of the Lugicum is marked by the Torgau-Doberlug-Göllnitz synclinorium, where Lugian rocks are hidden beneath Cambrian succession and covered by Upper Carboniferous-Triassic successsion (Fig. 2).

The Lugicum geologically represents the NE margin of the Bohemian Massif (Fig. 1). It may be geographically divided into the Lusatian Upland (west) and the Western Sudetes with Fore-Sudetic Upland (east). This region was formerly studied by Poles and Czechs (West Sudets) and East Germans (Lusatia). As a result there have been variations in approaching and interpreting geological problems of this region. This is exacerbated by poor exposure, as it is largely hidden beneath extensive Cenozoic deposits.

The Lugicum is an extremely complex mosaic region composed of upper Proterozoic and Paleozoic, mostly S-type granitoids. It also consists of a number of small, structural-stratigraphic domains (sometimes $<50\,km^2$), which are generally bounded by tectonic breaks (Fig. 2). These small domains may be grouped into eight larger terranes (Fig. 2): (1) the Lusatian Anticlinal Zone (LAZ) which consists of Upper Proterozoic graywackes which have been contact metamorphosed by emplacement of unfoliated Upper Proterozoic granodiorites and associated anatexites; (2) the Izera-Karkonosze Block (IKB) composed of Upper Proterozoic-Ordovician orthogneisses and low-to-mid-grade metasedimentary and metavolcanic rocks of Ordovician(?)-Early Carboniferous age along with the intervening area of unfoliated Variscan granite; (3) the Görlitz Syncline (GS) composed of an unmetamorphosed Cambrian-Lower Carboniferous succession; (4) the Góry Kaczawskie Block (GK) which consists of low-grade to very-low-grade metamorphosed succession of various sedimentary rocks and bimodal volcanites of Cambrian(?)-Early Carboniferous age; (5) the Góry Bardzkie Unit (GB) composed of unmetamorphosed autochthonous and allochthonous sedimentary rocks of Ordovician-Early Carboniferous age; (6) the Orlica-Śnieżnik Dome (OSD) with an orthogneiss core surrounded by variably metamorphosed sedimentary rocks and bimodal volcanic rocks of late Proterozoic-early Paleozoic age (with local inliers of eclogite and quartzo-feldspathic granulites); (7) the Góry Sowie Block (GSB) of high-grade metagraywackes of late Proterozoic-Early Cambrian age, with minor amphibolites of various provenance and mostly felsic granulites; and, (8) the Fore-Sudetic Block (FSB), located between the Sudetic Marginal Fault (SMF) and the Odra Fracture Zone (OFZ), and composed of high- to low-grade metamorphic rocks representing various sedimentary and igneous protoliths of late Proterozoic-Paleozoic age. All these domains host generally unfoliated Variscan granitoids (Fig. 2).

The Upper Paleozoic granitoids stitch and generally obscure most fault boundaries. Known ductile faults include the Main Intra-Sudetic Fault (MIF) which was interpreted by Don (1990) as a long-lived crustal fracture separating Caledonian and Hercynian elements throughout the Lugicum.

Dallmeyer et al. (Eds.)
Pre-Permian Geology of Central and Eastern Europe

Fig. 1. Location of the Lugicum (West Sudetes) (*stippled*) in central Europe. *H* Harz; *O* Odenwald; *S* Spessart; *Sd* Schwarzwald; *V* Vosges

In spite of marked differences in lithology, stratigraphy, and tectonism across the MIF, details of its character and kinematic evolution are poorly known. Also unclear is the nature of the border between the Lugicum and the East Sudetes (northern continuation of the Moravo-Silesian Zone). This coincides with a series of low-angle faults referred to as the Moravo-Silesian thrust zone (Misař et al. 1983), and commonly regarded as an extension of the Moldanubian thrust recognized by Suess (1912). The exact location of this boundary has been disputed. It has been defined as the Ramzova thrust (Bederke 1924), or the Nýznerov and Vacetín thrusts (Skacel 1989), where the Lugian rocks are in contact with Devonian metasedimentary rocks of the Moravo-Silesian Basin. It has also been located along exposures of tonalite or graphitic quartzite (Skácel 1977). Also important is the Marginal Sudetic Fault (MSF), which was active in the Neogene, but with significant Late Carboniferous-Permian movements. There is a 4–6-km difference in the Moho depth across this prominent fault (Guterch et al. 1975).

Stratigraphical and paleontological studies indicate that sedimentation in the West Sudetes had locally started in late Proterozoic, and continued to the Paleozoic, locally up to the early Late Carboniferous. On the other hand, the upper Proterozoic graywackes in the LAZ are otherwise correlated with flysch-like sediments of Proterozoic age exposed in the core of the NE-trending Schwarzburg Anticlinorium (along with the North-Saxonian graywacke). Thus, the Lusatian graywackes have been considered as part of the South Thuringian-Lusatian Anticlinal Zone, where sedimentation started at ca. 900–850 Ma and had ceased before the Paleozoic (Bankwitz 1988). The Lusatia therefore differs from the Western Sudetes, and direct linkages between the two areas are not warranted,and do not support recognition of the Lugicum as a geological entity in the sense of Suess (1926) and Stille (1951). The sedimentary and tectonothermal evolution of the Western Sudetes is definitely more akin to that of the Saxothuringicum with the Mid-German Grystalline Rise than to that of Lusatia. Thus, the suggestion of Kossmat (1927) to correlate the Western Sudetes with Saxothuringicum cannot be rejected (Franke et al. 1993).

The following chapters attempt to provide systematic description of the stratigraphic, structural, igneous and metamorphic characteristics of the Lugicum, with an outline of evolving ideas and major points of controversy.

Acknowledgments. Dave Dallmeyer kindly corrected the English.

Fig. 2. Geologic sketch of the Lugicum (West Sudetes). *Inset* shows location in the Bohemian Massif. *EFZ* Elbe Fracture Zone; *MIF* Main Intra-Sudetic Fault; *OFZ* Odra Fracture Zone; *SMF* Sudetic Marginal Fault. *GB* Görlitz Syncline; *GSB* Góry Sowie Block; *FSB* Fore-Sudetic Block; *IKB* Izera-Karkonosze Block; *ISB* Intra-Sudetic Basin; *Jp* Ještědské pohoři area; *KM* Kłodzko Metamorphic Unit; *LAZ* Lusatian Anticlinal Zone; *mom* Middle Odra metamorphic unit; *NSB* North-Sudetic Basin; *nk* Niemcza-Kamieniec metamorphic unit; *nm* Nové Město; *nk* Niemcza Zone; *OSD* Orlica-Śnieżnik Dome; *SD* Świebodźice Depression; *sm* Staré Město; *za* Zábřeh; *zb* Železný Brod

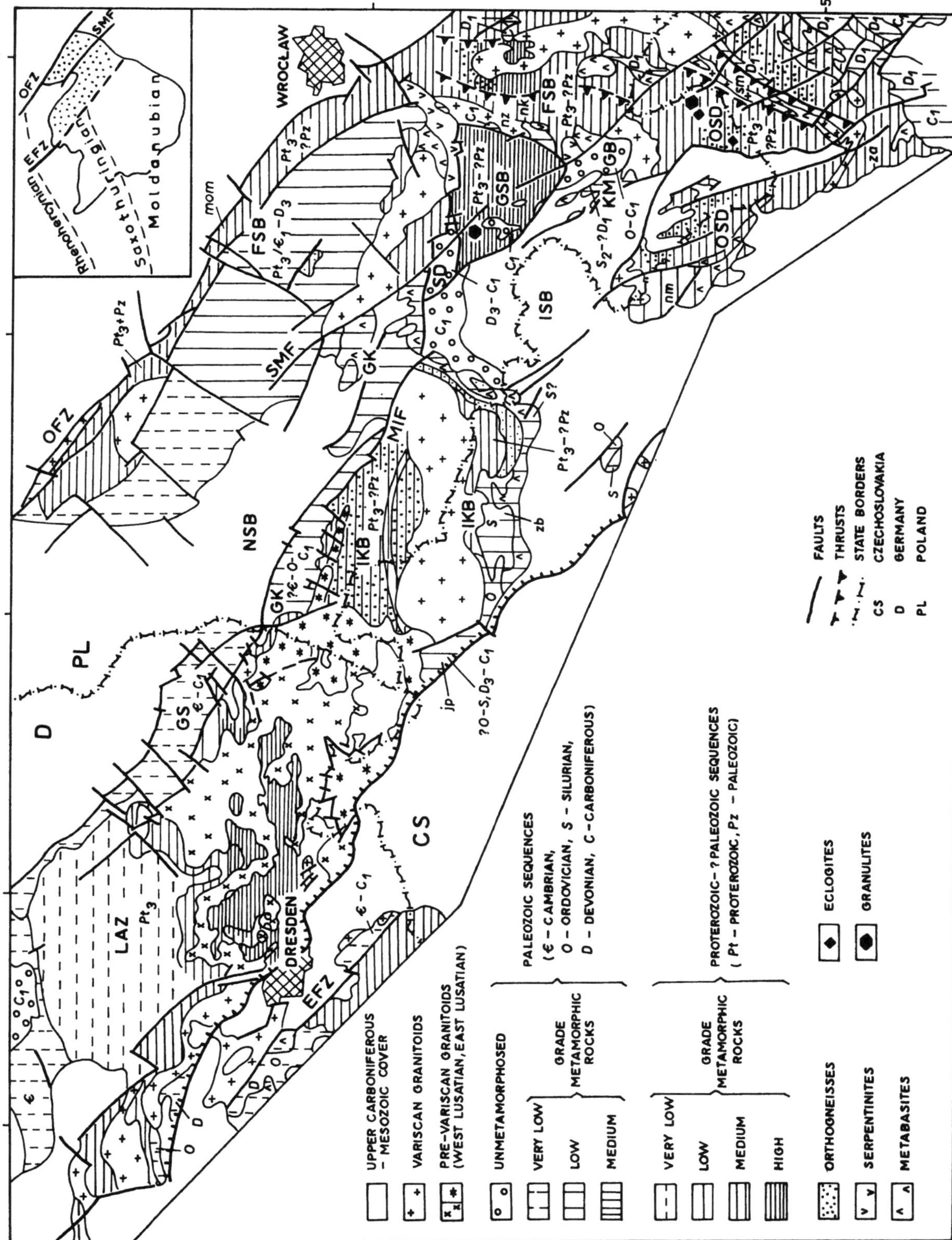
UPPER CARBONIFEROUS – MESOZOIC COVER
VARISCAN GRANITOIDS
PRE-VARISCAN GRANITOIDS (WEST LUSATIAN, EAST LUSATIAN)
UNMETAMORPHOSED
VERY LOW
LOW
MEDIUM
GRADE METAMORPHIC ROCKS
PALEOZOIC SEQUENCES
(Є – CAMBRIAN, O – ORDOVICIAN, S – SILURIAN, D – DEVONIAN, C – CARBONIFEROUS)
VERY LOW
LOW
MEDIUM
HIGH
GRADE METAMORPHIC ROCKS
PROTEROZOIC – ? PALEOZOIC SEQUENCES
(Pt – PROTEROZOIC, Pz – PALEOZOIC)
ORTHOGNEISSES
SERPENTINITES
METABASITES
ECLOGITES
GRANULITES
FAULTS
THRUSTS
STATE BORDERS
CS CZECHOSLOVAKIA
D GERMANY
PL POLAND
Rhenohercynian
Saxothuringian
Moldanubian
OFZ
EFZ
SMF
WROCLAW
DRESDEN
LAZ
GS
NSB
ISB
IKB
GSB
FSB
OSD
SD
GK
MIF
KM
GB
13°
14°
15°
16°
17°
50°
51°

References

Bankwitz P (1988) Schwarzburger Anticlinorium. In: Bankwitz E (ed) Klassische geologische Gebiete in Mitteleuropa, Fundament und Deckgebirge. Exursionführer, Potsdam Bederke E (1924) Das Devon in Schlesien und das Alter der Sudetenfaltung. Akademie der Wissenschaften der DDR, Zentralinstitut für Physik der Erde Fortschr Geol Paläontol 7: 1–5

Don J (1990) The differences in Paleozoic facies – structural evolution of the West Sudetes. Neues Jahrb Geol Paläontol Abh 179: 307–328

Franke W, Żelaźniewicz A, Porębski SJ, Wajsprych B (1993) Saxothuringian zone in Germany and Poland: differences end common features. Geol Rundsch 82: 583–599

Guterch A, Materzok R, Pajchel J, Perchuć E (1975) Seismic structure of Earth's crust and upper mantle along the international profile VII in the lights of studies by deep seismic soundings method. Przegl Geol 264: 153–163

Kossmat F (1927) Gliederung des varistischen Gebirgsbaues. Abh Sächs Geol Landesamts 1: 1–39

Misař Z, Dudek A, Havlena V (1983) Problemy regionalného geologického členeni Českého masivu. Věstn Ústřed Ústavu Geol 58: 305–313

Skácel J (1977) The character of contacts between the Śnieżnik gneisses and the Stare Miasto crystalline rocks. Acta Univ Wratislaviensis 378: 207–215

Skácel J (1989) On the Lugicum-Silesicum boundary. Acta Univ Wratislaviensis 1113: 45–55

Stille H (1951) Das mitteleuropäische varistische Grundgebirge im Bilde des gesamteuropäischen. Geolog Jahrb Beih 2: 138

Suess FE (1912) Die moravischen Fenster und ihre Beziehung zum Grundgebirge des Hohen Gesenkes. Denkschr Österr Akad Wiss Math – Naturwiss 88: 541–631

Suess FE (1926) Intrusiontektonik und Wandertektonik im varistischen Grundgebirge. Borntraeger, Leipzig

VI.B Stratigraphy

Z. Urbanek, A. Żelaźniewicz, H. Kemnitz, N. Hermsdorf, and U. Linnemann

1 Introduction

The geologic record of the late Proterozoic-Early Carboniferous has been represented in the Lugicum (West Sudetes) by a number of stratigraphic successions composed mostly of low- to high-grade metamorphic rocks, along with a wide range of igneous rocks and subordinate sedimentary rocks. The stratigraphic ranges of these successions are usually poorly documented, boundaries undefined, and their lateral spreads limited within small tectonic units (Fig. 2 in Chap. VI.A). This severely inhibits reliable chronocorrelations even between individual smaller units, the more so throughout the entire Lugicum. No wonder that such circumstances have resulted in many attempts to correlate individual units on various lithological bases, metamorphic grade inclusive, thus often bringing about dubious and/or arbitrary stratigraphic schemes. The following chapter therefore contains the actually known stratigraphic facts and mentions different, often contradictory, reconstructions of various local successions (Figs. 1, 2), without attempting at the moment an unwarranted generalization. The data presented below, though pointing to various flaws of current documentation, can be used to test both the models already proposed (e.g., Cwojdziński 1980; Michniewicz 1981; Don 1985; Aleksandrowski 1990; Matte et al. 1990), and those to come.

The quality of stratigraphic data currently in hand subdivides the chapter into two parts, one dealing with the Proterozoic and controversial Paleozoic successions (Fig. 1) and the other concerned with the proved, or most likely Paleozoic successions (Fig. 2).

2 Proterozoic-Paleozoic(?) Successions

2.1 The Lusatian Anticlinal Zone (LAZ)

The LAZ, comprising the area of 5000 km^2, is built almost exclusively of Proterozoic rocks (Fig. 1), represented by (1) graywackes and (2) so-called anatexites ("two-mica granodiorites" or "hybridic granites" of Ebert, 1935) and pre-Variscan S-type granodiorites (Eidam 1988; Tischendorf 1990), with the isotopically constrained Vendian to Cambrian emplacement age (Bielicki et al. 1990).

The group of anatexites (Stolpen Group, Fig. 1), although mineralogically and geochemically relatively homogenous, has come from two protolith sources, of which one is being identified as clayey-silty sediments and the other as sands and graywackes. Lithostratigraphy of these highly metamorphosed old crustal successions remains to a great extent hypothetical.

In the middle of the anatexite area (Fig. 2 in Chap. VI.A) occurs a W-trending, 9-km-wide belt of variegated rocks, rich in metabasites. Brause (1985) subdivided the Stolpen Group into a variegated key bed and a monotonous succession [bottom/roof (?) wall], assigning all to the Moldanubian main group (Middle Riphean). Hoth et al. (1985) proposed to correlate the Stolpen Group with the Medenec Group of the Erzgebirge (cf. Chap. V.B.1). Also it may be correlated with schist belts occurring within granite-gneisses of the IKB in the Western Sudetes.

The anatexites are overlain by the ca. 2-km-thick Lusatian graywackes (Lusatian Group), with contact thermally obscured by later granodiorites. The age of the Lusatian graywackes long remained unclear. Pietzsch (1962) supposed that part of the graywackes might be of Proterozoic and part of Early Carboniferous age. Hirschmann (1966) distinguished the Kamenz beds and the Görlitz beds, both of Proterozoic age, but a controversial

Dallmeyer et al. (Eds.)
Pre-Permian Geology of Central and Eastern Europe

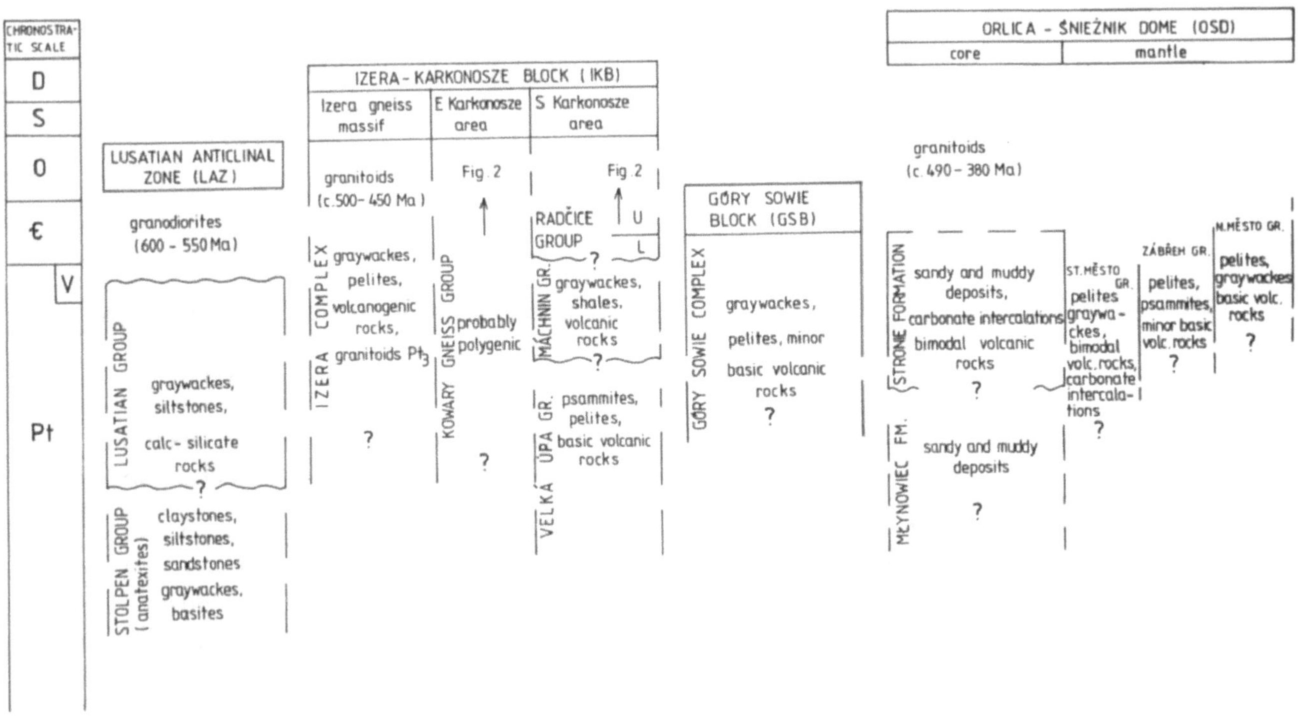

Fig. 1. Stratigraphic scheme of the Proterozoic-Paleozoic(?) successions in the Lugicum

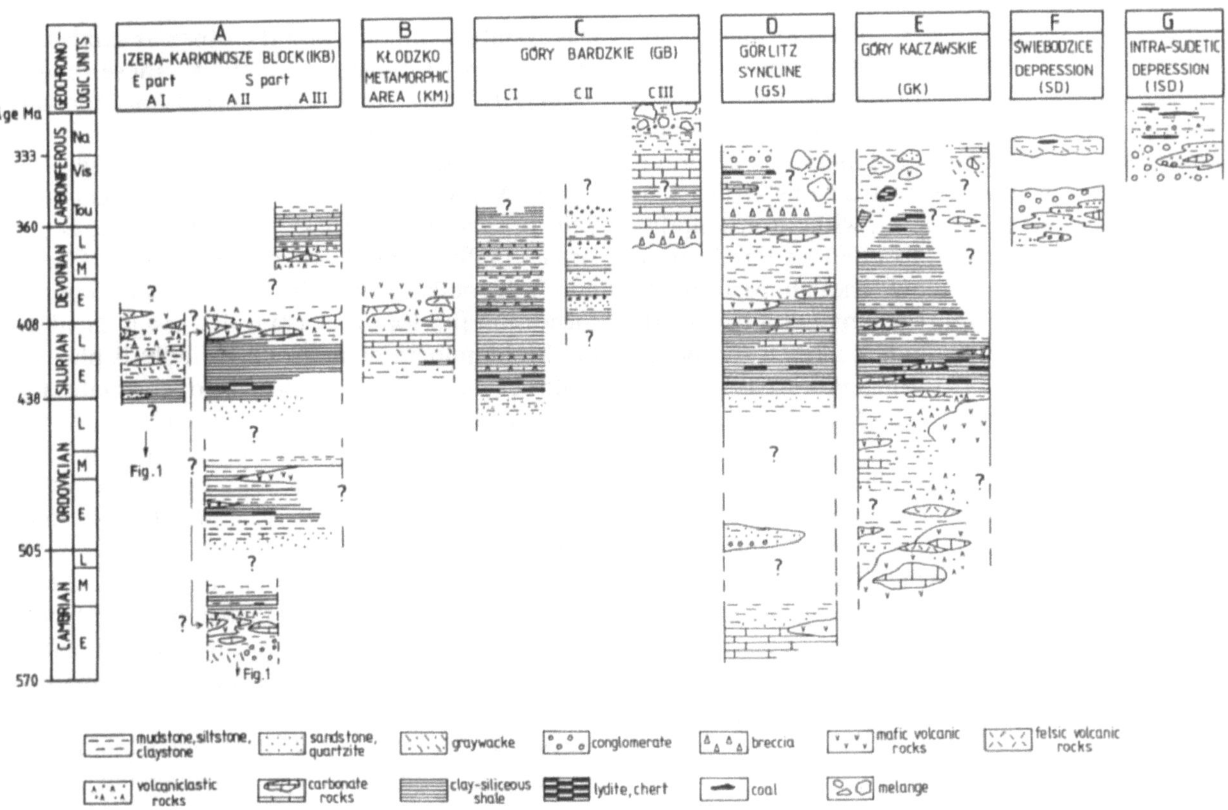

Fig. 2. Stratigraphic columns of the Paleozoic in the Lugicum. *AI* The East Karkonosze metamorphic area; *AII* the Železný Brod area; *AIII* the Ještědské pohoři; *CI* the Zdanów sequence; *CII* the Bardo-Młynów sequences; *CIII* the Kłodzko-Srebrna Góra sequences

boundary, nowhere exposed. Recent microfossil findings within the Kamenz beds (Weber et al. 1990) have confirmed the Vendian age, once determined by Burmann (1966), and pointed to a cold marine sedimentary environment.

Recently Kemnitz and Budzinski (1991) interpreted the Lusatian graywackes as pelagic turbidites developing under quite uniform facies conditions. The large amount of magmatogenic lithic fragments points to an orogenic source area of active shelf type, with only slight cratonic influence.

The graywackes now hidden to the NW under the Phanerozoic cover were most likely fed by metamorphic terrane (Hilmersdorfer sub-block of Brause 1969). The same could be true of the graywacke occurring east of the Westlusatian Fault (Frischbutter 1982).

On the SW flank of LAZ, NE of the low-grade, Paleozoic succession of the Elbe Zone (Fig. 2 in Chap. VI.A), there is a patchily exposed metagraywacke belt (Weesenstein Group = conglomerate, graywacke, quartzite, minor metabasite) intruded by ca. 600 Ma granodiorite (Dohna), which may either represent another source area for the Lusatian graywacke, or be part of it. Recently, Linnemann (1992) reconstructed there the proximal part of a submarine siliciclastic fan, with sedimentation controlled by a low stand of sea level, rift tectonism, and tholeiitic volcanism. The upper part of the Weesenstein Group contains quartzites defined as shelf sediments redeposited due to glacioeustatic sea-level changes most likely connected with the Early Vendian Varanger glaciation (global time mark 700–650 Ma ago).

2.2 The Izera-Karkonosze Block (IKB)

The IKB is a heterogenous complex, composed of the centrally located Upper Carboniferous Karkonosze granite and its low- and medium-grade metamorphic envelope representing various parts of the Proterozoic(?)-Carboniferous succession (Figs. 1, 2A). The IKB is usually divided into (1) the Izera gneiss massif (N of granite), (2) the East Karkonosze metamorphic area (E of granite), and (3) the South Karkonosze metamorphic area (Fig. 2 in Chap. VI.A). The bases of further subdivisions of the IKB into tectonostratigraphic units are unclear and are therefore omitted here, the more so that metamorphic grade has often been used as the main age criterion.

The Izera gneiss massif is built of the Izera complex (Fig. 1) and consists several gneiss variants, probably of different origin, accompanied by fine to coarse-grained granites, often with characteristic blue quartz grains (Rumburk granite). Less common are granodiorites, leucogranites, and mica schists with interbeds of calc-silicate rocks, quartzites, laminated paragneisses and amphibolites. The schistose rocks form three narrow, W-trending parallel belts (metamorphosed pelites, graywackes, and volcanogenic rocks), with the one adjacent to the Karkonosze granite being partly altered to cordierite-andalusite hornfelses and spotted schists (Kozłowski 1974; Żaba 1984; Oberc-Dziedzic 1988). The protoliths are assigned to the Proterozoic by taking into accunt regional relationships with the Lusatia (Chaloupský 1989).

According to Borkowska et al. (1980), the Izera gneisses and the Rumburk granite are alike from the petrological, geochemical, and geochronological points of view, representing one granitic complex of ca, 500–450 Ma in age (Rb-Sr whole rock isochrone ages of 501 ± 32 Ma and 462 ± 15 Ma for the Rumburk granite and Izera gneiss, respectively), which experienced post-Devonian regional metamorphism completed before 320–310 Ma. Oberc-Dziedzic (1988) assumes four developmental events, altogether bringing about (1) supracrustal series, (2) its metamorphosed derivatives, (3) augen gneisses, and (4) granites, and suggests that the Ordovician isotopic ages should rather be related to the intrusion of the Rumburk granite into the earlier Izera gneisses.

The metamorphic rocks of the East Karkonosze (= Rudawy Janowickie and Grzbiet Lasocki) consist of unfossiliferous gneisses and schists, with marbles, quartzites, bimodal metavolcanic rocks, and phyllites. Lithologic correlations with either the South Karkonosze or the Góry Kaczawskie resulted in different age estimations within the Proterozoic-early Paleozoic (Berg 1912; Oberc 1960; Szałamacha and Szałamacha 1967, 1968; Teisseyre JH 1971, 1973).

The metamorphic rocks of the East Karkonosze form two major lithostratigraphic units: the Kowary gneiss group (KGG) (Fig. 1) and the Rudawy Janowickie group (RJG) Fig. 2AI), differing in the degree of metamorphic reworking, but possibly jointly refolded (Teisseyre JH 1971, 1973).

The KGG is presumably polygenic, with part derived from graywackes, and equivalent to gneisses occurring in the South Karkonosze (Chaloupský 1989). Its age remains unknown, but it may have been Proterozoic or early Paleozoic, with a late Caledonian granitization overprint.

The RJG can be correlated with low-grade metasedimentary and metavolcanic rocks of Ordovician (?)-Silurian age cropping out in the region of Železný Brod (Fig. 2 in Chap. VI.A) (Teisseyre JH 1971). This group consists of two formations: (1) lower – the Czarnów schists fm. (originally pelites, psammites, graywackes, siliceous rocks, acidic and basic tuffs with carbonate intercalations), (2) upper – the Leszczyniec volcanic fm. (mostly hypabyssal and pyroclastic rocks compositionally matching a basalt-spilite keratophyre suite (Teisseyre JH 1971, 1973), possibly representing an island arc-trench environment (Narębski 1980).

In the South Karkonosze area, Chaloupský (1989) distinguished lithostratigraphically three groups (Figs. 1, 2AII) of which (1) the Velká Úpa Group (mica schists with quartzitic, graphitic, calcareous and basic intercalations) is to be of Middle(?) Proterozoic age, lithologically correlated with variegated group of the Moldanubicum, (2) the Machnin Group (ca. 300-m-thick monotonous metagraywackes and phyllites) of Late Proterozoic age, and (3) the Radčice Group (uppermost Proterozoic(?) to middle Cambrian) consisting in the lower part of phyllites and in the upper part of low-grade metamorphosed conglomerates and graywackes. The above subdivision is based entirely on lithological comparisons with other regions of the Bohemian Massif, and boundaries of the proposed units have not been stated clearly.

2.3 The Orlica-Śnieżnik Dome (OSD)

Fischer (1935) was the first to present a comprehensive stratigraphic scheme for the OSD rocks by distinguishing: (1) Archean complex consisting of the Młynowiec schists and paragneisses partly migmatized due to the injection of granitic magma and thus transformed into the fine-grained, migmatitic Gierałtów gneisses, (2) Algonkian-Cambrian complex of sedimentary-volcanogenic rocks deposited upon the Młynowiec-Gierałtów bedrock and metamorphosed to schists (Stronie schists) during Caledonian time, with the intervening intrusion of porphyritic granite subsequently changed dynamically to the Śnieżnik augen gneisses.

The Fischer lithostratigraphy has been adopted by many geologists, though with various modifications as to the order and origin of once proposed units.

The Proterozoic(?) age of the Młynowiec fm. is inferred from the relation to the approximately dated Stronie formation (Don et al. 1990).

The recent but controversial paleontologic data for marbles, quartzites, and paragneisses (Gunia 1990) point to the Late Proterozoic-Early Cambrian age of the Stronie formation.

The basal part of the ca. 4–5-km thick Stronie fm. protolith was dominated by sand-mud deposits, whereas carbonate-marl ones and associated volcanic rocks prevailed toward the top (Fischer 1935; Vangerow 1943). Such a rock suite is suggestive of an intense sedimentation accompanied by volcanic activity, possibly at a volcanic arc setting (Wojciechowska 1986).

The Śnieżnik gneisses are coarse or medium-grained rocks, with K-feldspar augens and typical fabrics of ductilely deformed granite, imparting several textural variants (Dumicz 1964; Żelaźniewicz 1984).

The Gierałtów gneisses are represented by light, fine, and even-grained, often thinly laminated two-mica rocks, locally accompanied by lensoid bodies of amphibolites, eclogites, and felsic granulites.

The contact of the two gneiss types has differently been interpreted as due to (1) metasomatism (Smulikowski K 1979), (2) migmatization (Don 1977), or (3) mylonitization and static recrystallization (Dumicz 1988).

The lithostratigraphic scheme of Fischer (1935), is remarkably consistent with recent isotopic data of Borkowska et al. (1990). Other opinions assumed that (1) all gneisses were of Proterozoic age and developed through granitization of the Stronie schists (Smulikowski K 1960; Teisseyre H 1968a); (2) an early Caledonian granite was gneissified (Śnieżnik type) prior to diapiric uprise of the migmatitic Gierałtów gneisses (Don 1964); (3) both types are themselves heterogenous genetically and temporally, with different but otherwise polyphase tectonometamorphic history during Variscan time (Dumicz 1988; Don et al. 1990).

The Młynowiec-Stronie group occurs in the core of the OSD, which is mantled by an epi- and mesozonally metamorphosed sedimentary-volcanogenic sequence of Proterozoic(?)-Silurian age, represented by the Kłodzko metamorphic area (KM) to the north, the Nové Město group (west), the Zabřeh group (south), and the Staré Město group (east) (Fig. 2 in Chap. VI.A; Fig. 1). The Nové Město phyllites and the underlying Zabřeh unit (paragneisses and pearl gneisses), traditionally linked with the Bohemicum (Kettner 1922), developed from pelitic and psammitic sedimentary rocks associated with abundant tholeiitic (MOR) rocks, supposedly of late Proterozoic(?) age (Opletal et al. 1980). The lower part of the Zabřeh group is considered to continue into the Staré Město group, both composed of lower amphibolite facies mica schists and paragneisses, with concordant metabasites (Pauk 1977).

The isotopic datings performed so far are not too consistent. The Śnieżnik type augen gneisses yielded a Sm-Nd age of 490 Ma (Liew and Hofmann 1988) and Rb-Sr whole rock isochrone age of 395 ± 35 Ma and 335 ± 5 Ma (mica and whole rock) interpreted as the intrusion and metamorphism age, respectively (Borkowska et al. 1990). However, van Bremeen et al. (1982) reported a Rb-Sr whole-rock regression age of 487 ± 11 Ma for the Śnieżnik granite emplacement. A Rb-Sr whole-rock age of 464 ± 35 Ma in the Gierałtów type gneisses is interpreted as an emplacement age probably coeval with the uprise of granulites (Borkowska et al. 1990). Eclogite bodies within granulites and Gierałtów gneisses show a Sm-Nd evidence of a HP event ca. 355–325 Ma ago (Brueckner et al. 1989). However, K-Ar biotite and phengite ages of ca. 380 Ma for the eclogites and Gierałtów gneisses (Bakun-Czubarow 1968) suggest earth movements, probably uplift, in the Devonian.

2.4 The Góry Sowie Block (GSB)

The GSB (Fig. 2 in Chap. VI.A) consists of at least four textural variants of biotite-oligoclase and two-mica + two-feldspar gneisses, with zonal distribution of sillimanite, and late cordierite overprint, complex orthoamphibolites, as well as dominantly felsic granulites in early tectonic slices, all cut by mostly N-trending pegmatite dykes. All these rocks form the Góry Sowie complex (Fig. 1) derived from mostly graywacke protolith, with scarcer semipelitic, psammitic, and minor volcanogenic intercalatons. This fairly lithologically monotonous complex is difficult for any more detailed stratigraphic subdivisions. However, structural and lithologic mapping seems to indicate more quartzofeldspathic sedimentary rocks at the base, passing up through graywacke to more argillaceous deposits, with flysch-like alternation of more psammitic and more pelitic beds (Żelaźniewicz 1987, 1990).

In view of the lithologic monotony and lack of fossils, the GSB complex has been equated with the Moldanubian Monotonous Group of the Bohemian massif core (Zoubek 1965; Chaloupský 1978), and assigned to the Proterozoic or even Archean. Quite recently, Gunia (1981a, 1984) retrieved some microfloristic assemblages from the metapelitic gneisses and concluded that the GSB complex has the Upper Riphean protolith. However, the first appearance of some taxa mentioned by him also took place in the Vendian and in the Early Cambrian; therefore one cannot exclude that the graywacke sedimenation had straddled the Precambrian-Cambrian boundary. A geotectonic setting of that sedimentation remains unknown, and so is the age of the earliest metamorphism of the protolith. The U-Pb and Rb-Sr isotopic study (van Bremeen et al. 1988) fingerprinted the waning stage of the peak metamorphic event (M_{2+3}) at 381 ± 2 Ma (monazite, U-Pb) followed by later pegmatite intrusions at 370 ± 4 Ma (Rb-Sr, book muscovite) under the epidote amphibolite facies conditions, altogether indicating rapid Devonian uplift and cooling, and otherwise consistent with stratigraphic evidence from the Frasnian-Tournaisian Świebodzice succession and from the Upper Devonian sediments in the Góry Bardzkie and Kłodzko region, which undoubtedly contain clastic fragments derived from the GSB (Sect. 3.3). The ca. 460–420 Ma K-Ar ages of the GSB gneisses obtained by Depciuch et al. (1980) are inconsistent with the U-Pb and Rb-Sr ages, and, being suggestive of Ordovician uplift, are in apparent conflict with the stratigraphic evidence.

2.5 The Fore-Sudetic Block (FSB)

The FSB (Fig. 2 in Chap. VI.A) is almost entirely covered with Cenozoic deposits and its low- and medium-grade metamorphic rocks of Proterozoic-Paleozoic age, emerging from beneath as isolated inliers, are known mostly from still rather scarce subsurface data. It consists of (1) Fore-Sudetic part of the Góry Sowie block along with adjacent basite-ultrabasite massifs, (2) epimetamorphic rocks being probably the eastern continuation of the Góry Kaczawskie, (3) the Middle Odra metamorphic rocks (MOM) passing to the south into (4) the Niemcza-Kamieniec schists (NK) and (5) rocks of the Niemcza zone (NZ). Element (1) is described in Section 2.2.4 and element (2) in Section 3.

East of the GSB there are tectonostratigraphic units built of amphibolite facies rocks, namely: (1) schist-quartzite sequence of the NZ, primarily composed of fine-grained graywackes, siltstones, and black graphitic sandstones, (2) schistose sequence of the Niemcza-Kamieniec region (NK) originally consisting of mudstones interbedded with limestones, sandstones, cherts, rhyolites and rhyolitic tuffs, and alkali basalts (Dziedzic 1985).

The age of these successions is still incompletely defined. The NZ sequence is at least partly Paleozoic, as evidenced by the presence of Carboniferous microflora (Visean-Namurian A) in metasiltstones (Dziedzic and Górecka 1965). The NK sequence has been assigned to Vendian and Lower Cambrian by virtue of the presence of acritarchs and problematic shelly fossils (Gunia 1981b).

The northeastern part of the FSB, totally hidden beneath the Cenozoic, is referred to as the Middle Odra metamorphic region (MOM) (Oberc 1968). The MOM is built of amphibolite facies metasedimentary and metavolcanic rocks, derived mostly from claystones and siltstones, graywackes, occasional carbonates, and basic volcanogenic rocks (Utzig 1975). In fact, they are represented by mica schists and thinly laminated paragneisses with alternations of marbles, quartzites, leptynites, and amphibolites. The series has been referred to as the Wrocław formation (Grocholski 1982). Locally, it contains younger granite-gneisses, petrographically and structurally equated with those of the IKB, and once believed to have developed through metasomatic granitization of a schistose series (Kozłowska-Koch 1959). Some elements of the Wrocław formation stick out from beneath the Cenozoic cover as small "islands" (e.g., nearby Wądroże Wielkie), or in metamorphic envelopes of the Strzegom-Sobótka and Strzelin granitoid massifs, then in the NK (Grocholski 1982). All these metamorphic rocks are of Proterozoic (Oberc 1972), possibly even early Cambrian age (Grocholski 1986). This notion, however, being based exclusively on lithological similarities to otherwise poorly documented Proterozoic complexes in the Western Sudetes, is not supported by paleontologic data, neither can any direct relationship be demonstrated. Also unclear are relations between low- and medium-grade metamorphic units throughout the FSB, which may represent either (1) primarily single-sequence but in parts having a different tectonometamorphic history (Grocholski 1982), or (2) two structural stages of different age (Oberc 1968).

Next to the amphibolite facies complex in central and western parts of the FSB (Fig. 2 in Chap. VI.A), there are low and very-low-grade metamorphic rocks assigned by Grocholski (1982) to two lithostratigraphic units referred to as the hypothetical Lusatian formation and the group of the Kaczawa formations. The latter unit will be featured along with the Góry Kaczawskie in Section 3. The Lusatian formation in the FSB has been folded in common with the Góry Kaczawskie formations, but their contact remains unknown. It is built of weakly metamorphosed flysch-like graywackes with subordinate conglomerate, mudstone, claystone, occasional calcareous interbeds, and probably volcanogenic rocks. The rich acritarch assemblage discovered in the NW part of the Lusatian formation area points to the late Proterozoic or early Cambrian (Jerzykiewicz and Grocholski 1975).

3 Paleozoic Successions

3.1 Cambrian and Ordovician

The best-documented Cambrian of the Lugicum is known from the Görlitz Syncline (Fig. 2 in Chap. VI.A; Fig. 2D), where a succession of carbonates, sandstones, and volcanic rocks is capped by argillites with the Lower Cambrian trilobites (Freyer 1977) and otherwise contains microfauna (Rosanov et al. 1969) and shelly fossils (O. Elicki pers. comm. 1991) of that age. There is no evidence for the Middle and Upper Cambrian.

The Ordovician is also incomplete. The 150 m thick Dubrau quartzite complex (conglomerates, quartzites, pelites) overlies the Lusatian graywackes. It contains only badly preserved brachiopods, hinting at the Tremadocian or even the upper Cambrian (Pietzsch 1962), and is covered by graywackes of supposed Late Ordovician age (Brause and Hirschmann 1969). In the GS itself, only evidence for the Lower Ordovician is provided by acritarchs (Burmann 1970). The "Eichberg" sandstone and underlying argillaceous shales may represent the uppermost Ordovician, because they occur directly below the Silurian graptolite-bearing shales.

Beyond the Görlitz Syncline, the Cambrian and Ordovician are poorly documented and difficult to distinguish. Possibly a thick volcanogenic suite with carbonates occurring at the base of faintly metamorphosed Paleozoic rocks in the Góry Kaczawskie (Fig. 2 in Chap. VI.A) is of Cambrian age (Fig. 2E). The traditional view (Dahlgrün 1934; Schwarzbach 1939; Teisseyre H 1956, 1980), assuming the Radzimowice slates as the oldest, Eocambrian element of the Kaczawa complex, is untenable in view of recent conodont findings proving that these rocks cannot be older that the Ordovician (Urbanek and Baranowski 1986). The Radzimowice slates are ca. 1000-m-thick turbidites with subordinate hemipelagites and pelagites laid down by cohesive gravity flows in a trench or trench-slope basin (Baranowski 1988).

The carbonates (= Wojcieszów limestones) of the Kaczawa complex are assigned to the Cambrian by correlation with the Görlitz Syncline (Schwarzbach 1933) and the presence of middle Cambrian(?) Tabulata(?) of the genus *Cambrotrypa* (Gunia 1967). They occur as up to 2000-m-long and 500-m-thick bodies set in volcanogenic rocks and were deposited in a shallow offshore marine basin with the floor extensively shaped by submarine volcanic activity (Lorenc 1983). The associated bimodal volcanogenic suite (pillow and massive lavas, keratophyres; see also Chap. VI.D) is to represent either initial rift or ocean island setting (Narębski et al. 1986; Furnes et al. 1989). Some part of this suite, passing continuously to the overlying Llandovery graptolite shales, can be of Ordovician age (Jerzmański 1965). Diabase sills occur within Ordovician(?) phyllites as well as Silurian lydites and graphite shales (Haydukiewicz and Urbanek 1986; Kryza and Muszyński 1988).

The main part of the Ordovician sequence (Fig. 2E) in the Góry Kaczawskie is represented by slates and metasandstones, ca. 1000 m thick, interpreted as distal turbidites probably derived from a continent (Baranowski 1975), with reworked and mixed Lower and Middle Ordovician conodont faunas (Baranowski and Urbanek 1972; Baranowski et al. 1990). The whole volcanogenic-sedimentary assemblage in the Kaczawa complex is probably both vertically and laterally replaced by mudstone-sandstone turbidites (Baranowski et al. 1990).

Another Paleozoic sequence in the Western Sudetes occurs in the Góry Bardzkie (Fig. 2 in Chap. VI.A; Fig. 2C), where ca. 20-m-thick pelagic clays and turbidites pass continuously upwards to the Lower Silurian graptolite shales (Wyżga 1987), which altogether form the oldest element of the Zdanów sequence (Fig. 2CI). This sequence is 200–300 m thick and has been reconstructed from various fragments actually found as olistoliths in the Carboniferous flysch (Fig. 2CII) of the Kłodzko-Srebrna Góra fm. (Wajsprych 1978, 1986).

The Cambrian and/or Ordovician probably occur also in the South Karkonosze (IKB), in the Železný Brod area (Fig. 2 in Chap. VI.A), but with no paleontologic constraint. Very diverse opinions about the lower Paleozoic rocks in this region make impossible to construct a reliable column (Fig. 2AII, AIII). More recently, Chaloupský (1989) proposed that the Lower and Middle Cambrian is represented by the upper part of the Radčice Group (Fig. 1), composed of metaconglomerates, metagraywackes, and abundant phyllites, with intercalations of carbonates and volcanogenic rocks of the Železný Brod complex. However, the latter (pillow and massive metabasites and metatuffs) has been assigned to the Silurian by Fediuk (1962). The Lower and Middle Ordovician is lacking in the recent Chaloupský scheme, although formerly most of rocks of the present Radčice Group were stated to be of early and middle Ordovician age, but with the Cambrian missing (e.g., Chaloupský and Teisseyre 1968). The Upper Ordovician(?) in the IKB (Fig. 2AI, AII) might be represented by ca. 80-m-thick quartzite, with conglomerate interbeds, which were declared to be the basis of another stratigraphic sequence (Poniklá Group) of latest Ordovician(?)-Silurian age (Chaloupský 1989). The quartzites overlie, with a marked unconformity, both the Proterozoic and the supposed Cambrian (Chaloupský 1989) or Ordovician (Chaloupský and Teisseyre 1968).

3.2 Silurian

The Silurian, represented mostly by black shales, is throughout the Lugicum always in a sedimentary contact with the Ordovician. The siliceous and alum shales of the GS (Fig. 2 in Chap. VI.A) are intercalated at the top with tuffs, quartzites, graywacke, and rare limestone (Fig. 2D). Except for the lack of the Ockerkalk bed, the whole sequence correlates well with that of the Saxothuringian zone and continues to the Devonian (Hirschmann 1966).

In the GB (Fig. 2 in Chap. VI.A; Fig. 2CI) the much condensed Silurian sequence is 50 m thich and consists of lydites, black and gray-green clay, and siliceous shales with tuffite intercalations (Porębska 1980), all interpreted as pelagites (Wyżga 1987). The graptolite faunas in the GK are less well known, but the sequence (Fig. 2E) here is very similar to that of the GB, being less than 100 m thick (Baranowski et al. 1990).

In the Ještědské pohoři (Fig. 2AIII) and Železný Brod (Fig. 2AII) areas of the IKB there are graptolite-bearing black shales and lydites of Wenclockian age (Chlupáč 1953; Horný 1964), then replaced by an assemblage of gray-greenish phyllites, crystalline limestones and dolomites and some volcaniclastic rocks (Svoboda 1955; Chaloupský 1989).

The weakly metamorphosed Silurian shales are also reported from the western part of the FSB (Fig. 2 in Chap. VI.A), with graptolite fauna of Early Ludlovian age (Jerzmański and Teller 1971), There are hints for the complete Silurian sequence to pass upwards to the Devonian and downwards to the Ordovician sandy and clayey-sandy deposits (Jerzmański 1970).

In the Kłodzko metamorphic area (KM; Fig. 2 in Chap. VI.A; Fig. 2B), on the northern periphery of the OSD, the Silurian sequence consists of phyllites with interbeds of lydite, metagraywacke, and crystalline limestone and bimodal volcanogenic suite, altogether 1500 m thick. Aleuritic-arenaceous flyschoid deposits passing gradually into carbonates and marls have been covered in turn by pyroclastic rocks with carbonate, silicic, and sandy interbeds, and capped by an effusive complex (Wojciechowska 1990). This bears MOR, WIP, and IAT signatures, and is interpreted as the initial rift sequence related to the so-called circum-Góry Sowie ophiolite (Narębski et al. 1988). Surprisingly, the supposed initial volcanic suite occurs at the top of the stratigraphic column established for the KM by Wojciechowska (1990).

3.3 Devonian and Lower Carboniferous

The pelagic sedimentation in the GK (Fig. 2 in Chap. VI.A; Fig. 2E) and GB (Fig. 2CI) areas continued to the Devonian as siliceous and clayey shales with Lochkovian (GK, GB) and Pragian (GB) graptolites (Zdanów sequence; Porębska 1984) and Middle and Upper Devonian conodonts (Haydukiewicz J 1979), with sporadic turbidite interbeds (Wyżga 1987). The Tournaisian black shales are included to the Zdanów sequence (Fig. 2CI) by Haydukiewicz J (1990), or treated by Wajsprych (1986) as a fragment of another tectonostratigraphic unit named the Kłodzko-Srebrna Góra sequences (Fig. 2CIII). They lie, at least partly, over the Góry Sowie gneisses and the low-grade metamorphic rocks of the Kłodzko and Nowa Ruda area. The Kłodzko-Srebrna Góra sequences (Upper Devonian-Lower/Upper(?) Carboniferous) are composed of several elements, all being deposited in one continuous Srebrna Góra basin (Wajsprych 1986). These are (Fig. 2CIII): (1) Upper Devonian (at least Famennian)-Lower Tournaisian breccia of gabbro, limestone and quartzofeldspathic fragments (gneiss, granite?) and pelagic carbonates, (2) Upper Tournaisian black shales, (3) Visean nodular limestone ("Kohlenkalk") with insert of allodapic limestone, (4) Upper(?) Carboniferous (starting in Visean) flysch and olistostromes (Srebrna Pass series) in the lower part and wildflysch (Ostróg series) in the upper part, with the fining-upward background sedimentation but the coarsening-upward allochtonous one; they were fed by both the Zdanów sequence (Ordovcian-Devonian) and Bardo-Młynów sequences (pelagites with up to 200-m-thick interbeds of proximal turbidites of Late Silurian/Early Devonian-Early Carboniferous age) (Fig. 2CI and CII).

In the Góry Kaczawskie (Fig. 2E) the ca. 100 m thick Devonian pelagic clay-siliceous shales contain graptolites the hercynicus-Zone (Jaeger 1964) and Emsian-Famennian conodonts (Urbanek 1978). The shales occur mostly as olistoliths in melanges and far less often within thrust sheets (Haydukiewicz A 1987a,b; see Chap. VI.C.3.4.). The sedimentary and tectonic melanges within shaly matrix contain blocky fragments (up to 1 km across) derived from the aforementioned Cambrian(?)-Devonian succession. They developed mostly during the early Carboniferous and possibly started in the late Devonian (Haydukiewicz A 1987a,b; Baranowski et al. 1990). It should be mentioned here that also the Radzimowice slates bear some features of a melange.

The paleontologically dated Carboniferous in the GK (Fig. 2E) is represented by detritic limestones occurring as blocks or thin layers amongst flysch deposits and containing the Upper Devonian-Lower Carboniferous (till Upper Visean) mixed conodont faunas (Chorowska 1978).

In the GS (Fig. 2 in Chap. VI.A; Fig. 2D) the graptolite-bearing Lochkovian (Jaeger 1964) and conodont-proved Devonian (Reichstein 1961; Freyer 1965, 1972) appears as monotonous series of alternating quartzites, pelites, and graywackes, with rare intercalations of limestones and basic volcanic rocks. The Lower Carboniferous begins with a reworked horizon and continues as typical flysch sediments, accompanied toward the top by chert, limestone, and significant conglomerate intercalations. Besides, fragments of Ordovician through Lower Carboniferous rocks occur in melange of Early Carboniferous age (Bankwitz 1988; Thomas 1990). The main folding in the GS took place in Late Visean-Westphalian A span (Hirschmann 1966).

The paleontologically proved, weakly metamorphosed, Middle Devonian-Lower Tournaisian rocks (phyllites, limestones, bimodal volcanogenic rocks) have also been found in the Ještědské pohoři (Fig. 2 in Chap. VI.A; Fig. 2AIII), the IKB (Chlupáč 1964; Zikmundová 1964; Galle and Chlupáč 1976; Chlupač and Hladil 1992), where, according to Chaloupský (1989), they overlie unconformably the previously folded Cambrian-Silurian and Proterozoic rocks.

In the FSB area (Fig. 2 in Chap. VI.A) the Devonian rocks have been drilled in several places, being invariably represented by weakly metamorphosed clay-sand and clay-diabase sequences with Givetian and Frasnian sporomorphs (Jerzmański 1970). Almost unmetamorphosed clay-siliceous shales with mudstone, carbonate and tuffite intercalations are proved to be of Frasnian and Famennian age (Chorowska 1982).

The Świebodzice depression (Fig. 2 in Chap. VI.A; Fig. 2F) contains the record of synorogenic Late Frasnian-Early Tournaisian sedimentation, where the crystalline basement (probably the GSB gneisses and unspecified Góry Kaczawskie diabases) is covered with a 4-km-thick siliciclastic Świebodzice succession, in turn unconformably overlain by coal-bearing deposits of Late Visean-Early Namurian age (Teisseyre H 1968b). The Świebodzice succession consists of four formations (Porębski 1981), where the polymictic conglomerates and coarse sandstones containing mudstone interbeds with Clymenia limestone (Gunia 1968) are followed by ca. 2-km-thick conglomerate apron with one facies derived almost exclusively from the Góry Sowie block (Książ Fm.) and the other being polymictic (Chwaliszów Fm.). The Świebodzice succession has recorded a progressive uplift of the nearby southern sourceland, and it is interpreted as having been deposited within a slope-type fan-delta system, generated by alluvial fans prograding directly onto a marine, structurally controlled slope (Porębski 1981).

The Świebodzice-type sedimentary clasts recycled into the Lower Carboniferous (pre-Late Visean) terrestrial conglomerates occur in the nearby Intra-Sudetic depression (Fig. 2 in Chap. VI.A; Fig. 2G), where the latter form so-called Older Kulm, 4 km thick, passing gradually to partly marine Upper Visean (Go α, Go β) deposits (Younger Kulm) encroaching even the GSB area (Żakowa 1963). The Older Kulm is composed of several megacyclothemes of polymictic conglomerates, wackes, and mudstones that developed in the easterly drainaged intramontane depression, bordered on the north, west, and south by active fault scarps (Teisseyre AK 1975). The same pattern of tectonic control was also exerted over the sedimentation of the Younger Kulm, which developed as the fan-delta complexes prograding into a low-energy marine embayment deepening and remaining open to the east (Mastalerz and Porębski 1987).

4 Conclusions

The geologic history of the Lugicum (West Sudetes) has been recorded in various, presently scattered, fault-bounded tectonostratigraphic units. Its consecutive stages may be approximately ordered into a succession of events which have resulted in the formation of the above-described sequences, now cropping out throughout the Lugicum. These were as follows: (1) the late

Proterozoic graywacke sedimentation upon the 900–800(?) Ma granitoidic(?) bed rock, then followed by granitoid plutonism (600–550 Ma); (2) the Riphean-early Cambrian graywacke and pelitic sedimentation on unknown basement; (3) the Cambrian-Ordovician thermal events, manifested by (3A) bimodal intraplate volcanism, (3B) granitoid plutonism (500–450 Ma) and (3C) low and medium grade metamorphism with HP elements; (4A) the Cambro-Ordovician through early Devonian pelitic-volcaniclastic sedimentation, followed by (4B) basic and/or bimodal volcanism of a late Silurian-early Devonian age, possibly continued to middle Devonian; (5) extensive early-middle Devonian MP/HT metamorphism and minor igneous activity. In regions escaping events 4B and 5 there was (6) the Cambrian-Ordovician through Late Visean sedimentation, with early Carboniferous flysch and melanges, terminated by a HP event followed by low to very-low-grade metamorphism. In areas affected by the early-middle Devonian metamorphism there was (7) the rapid orogenic uplift associated with Late Devonian through Late Visean coarse clastic sedimentation fed by the rising mid and high grade metamorphic complexes. A late to post-orogenic part of the history includes (8) the intense granitoid plutonism with 330–280 Ma cooling ages.

References

Aleksandrowski P (1990) Early Carboniferous strike-slip displacement at the northeast periphery of the Variscan belt in Central Europe. In: Int Conf on Paleozoic orogens in central Europe, Aug–Sept 1990, Göttingen-Giessen, JGSP 233

Bakun-Czubarow N (1968) Geochemical characateristic of the eclogites from environs of Nowa Wieś in the region of Śnieżnik Kłodzki. Archiw Mineral 28: 243–382

Bankwitz P (1988) Lausitzer Antiklinalzone und Görlitzer Synklinorium. In: Bankwitz E (ed) Klassische geologische Gebiete in Mitteleuropa. Exursionsführer, AdW DDR, Potsdam

Baranowski Z (1975) Metamorphosed flysch deposits in the northern Kaczawa Mts (Rzeszówek – Jakuszowa unit). Geol Sudet 10: 119–151

Baranowski Z (1988) Lithofacies characteristic of trench-fill metasediments in the Radzimowice slate (Paleozoic), Sudetes, SW Poland. Ann Soc Geol Pol 58: 325–383

Baranowski Z, Urbanek Z (1972) Ordovician conodonts from the epimetamorphic complex from Rzeszówek in the Kaczawa Mts. Bull Acad Pol Sci Sér Sci Terre 20: 211–216

Baranowski Z, Haydukiewicz A, Kryza R, Lorenc S, Muszyński A, Solecki A, Urbanek Z (1990) Outline of the geology of the Góry Kaczawskie (Sudetes, Poland). Neues Jahrb Geol Paläontol Abh 179: 223–257

Berg G (1912) Die kristallinen Schiefer des östlichen Riesengebirges. Abh Preuss Geol Landesanst 68: 1–188

Bielicki KH, Haase G, Eidam J, Kemnitz H, Schust F, Gerstenberger H, Hiller H, Habedank M (1990) Pb-Pb and Rb-Sr dating of granitoids from the Lusatian Block. In: 5th Working Meet Isotopes in nature. Proc, part I, 25–29 Sept 1989. Mitteilungen des Zentralinstitut für Isotopen- und Strahlenforschung, Leipzig, pp 23–45

Borkowska M, Hameurt J, Vidal P (1980) Origin and age of Izera gneisses and Rumburk granites in the Western Sudetes. Acta Geol Pol 30: 121–145

Borkowska M, Choukroune P, Hameurt J, Martineau F (1990) A geochemical investigation of the age, significance and structural evolution of the Caledonian-Variscan granite-gneisses of the Śnieżnik metamorphic area (Central Sudetes, Poland). Geol Sudet 25: 1–27

Brause H (1969) Das verdeckte Altpälaozoikum der Lausitz und seine regionale Stellung. Abh Dtsch Akad Wiss Berl 61: 1–143

Brause H (1975) Regionalbau Lausitz – Dolny Śląsk I. Anteil DDR Görlitzer Schiefergebirge. In: Exkursionsführer 22. Jahrestagung GGW Teil B, 5–24, 23–27 Sept 1975. Gesellschaft für Geologische Wissenschaften der DDR, Dresden-Görlitz

Brause H (1985) Granitgenese und Krustenbau in der Oberlausitz. Veröff Mus Westlausitz 9: 3–11

Brause H, Hirschmann G (1969) Überblick über die stratigraphische Entwicklung des Görlitzer Schiefergebirges. In: Exkursionsführer DGGW, Alt- und Vorpaläozoikum des Görlitzer Schiefergebirges und der westlichsten Westsudeten, 7–10 Sept 1969. Deutsche Gesellschaft für Geologische Wissenschaften, Berlin, pp 23–35

Brueckner HK, Medaris LG Jr, Bakun-Czubarow N (1989) Nd and Sm age of isotope patterns from Hercynian eclogites and garnet pyroxenites of the Bohemian Massif and the East Sudetes. Terra Abstr 1: 1

Burmann G (1966) Mikroreste aus der Lausitzer Grauwackenformation. Monatsber Dtsch Akad Wiss 8: 218–225

Burmann G (1970) Weitere organische Mikrofossilien aus dem Unteren Ordovizium. Palaontol Abh B 3/4: 289–332

Chaloupský J (1963) Die Konglomerate im Kristallin von Krkonoše (Riesengebirge). Sb Ústřed Ústavu Geol 28: 143–190

Chaloupský J (1978) The Precambrian tectogenesis in the Bohemian Massif. Geol Rdsch 67: 72–90

Chaloupský J (1989) Geological evolution of the Krkonoše-Jizerské hory crystalline complex in the Precambrian and Early Paleozoic. In: Chaloupský J (ed) Geologie Krkonoš a Jizerských hor. Ústřed Ústav Geol, Academia, Praha, pp 247–288

Chaloupský J, Teisseyre H (1968) Geology of the West Sudeten. In: Malkovský M (ed) Int Geol Congr

XXIII Session, Aug 1968. Excursion Guidebook, Úst Ústav Geol, Praha
Chlupáč I (1953) Nalez graptolitu v metamorfovanem siluru u Železnego Brodu v Podkrkonosi. Věstn Ústřed Ústavu Geol 28: 213–215
Chlupáč I (1964) Ein neuer Faunenfund im schwach metamorphierten Palaozoikum von Ještědské pohoři (Jeschkengebirge, Nordböhmen). Čas Mineral Geol 9: 27–35
Chlupáč I, Hladil J (1992) New Devonian occurrences in the Ještěd Mts., North Bohemia. Čas Miner Geol 37: 185–192
Chorowska M (1978) Visean limestones in the metamorphic complex of the Kaczawa Mts (Sudetes). Rocz Pol Tow Geol 48: 245–261
Chorowska M (1982) Stratigraphic investigations of metamorphic formations of the Sudetes. Biul Inst Geol 341: 117–139
Cwojdziński S (1980) Geologic evolution of South-Western Poland interpreted in terms of plate tectonics. Geol Sudet 15: 43–66
Dahlgrün F (1934) Zur Altersdeutung des Vordevons im westsudetischen Schiefergebirge. Z Dtsch Geol Ges 86: 385–393
Depciuch T, Lis J, Sylwestrzak H (1980) K-Ar ages of the Owl Mts gneiss raft. Acta Geol Pol 30: 507–517
Don J (1964) The Złote and Krowiarki Mts as structural elements of the Śnieżnik metamorphic massif. Geol Sudet 1: 79–117
Don J (1977) The new data on interrelations between the Śnieżnik and Gierałtów gneisses (Sudetes). Estud Geol 33: 287–292
Don J (1985) Model of expansional development of the Western Sudety Mts in the Paleozoic (a proposal). Przegl Geol 383: 109–116
Don J, Dowidar H (1988) Goszów quartzites and the problem of the Młynowiec formation (Śnieżnik metamorphic massif-Sudetes. Bull Pol Acad Sci Earth Sci 36: 239–252
Don J, Dumicz M, Wojciechowska I, Żelaźniewicz A (1990) Lithology and tectonics of the Orlica-Śnieżnik dome, Sudetes. Recent state of knowledge. Neues Jahrb Geol Paläontol Abh 179: 159–188
Dumicz M (1964) Geology of the crystalline massif of Bystrzyckie Mts. Geol Sudet 1: 169–208
Dumicz M (1988) Złoty Stok-Skrzynka structural element in the light of mesostructural analysis of Lądek-Śnieżnik. Geol Sudet 23 (2): 83–106
Dziedzic H (1985) Variscan rejuvenation of the Precambrian gneisses along the eastern margin of the Góry Sowie Massif, Fore-Sudetic block. Krystalinikum 18: 7–17
Dziedzic H, Górecka T (1965) On the occurrence of metamorphosed Carboniferous rocks in the Niemcza Zone (Sudetes). Bull Acad Pol Sci Sér Sci Géol Géogr 13: 161–165
Ebert H (1935) Hornfelsbildung und Anatexis im Lausitzer Massiv. Z Dtsch Geol Ges 87: 129–146
Eidam J (1988) Zur anatektischen und magmatischen Entwicklung der granitoiden Gesteine der Lausitzer Antiklinalzone. Z Geol Wiss 16: 393–406
Fediuk F (1962) Volcanic rocks of the Železný Brod metamorphic region. Rozpr Ústřed Ústavu Geol 29: 5–116
Fischer G (1935) Der Bau des Glatzer Schneegebirges. Jahrb Preuss Geol Landesanst 56: 712–732
Freyer G (1965) Conodontenfunde aus dem Paläozoikum der Lausitz. Abh Ber Naturkundemus Görlitz 40 (5): 1–12
Freyer G (1972) Eine unterdevonische Conodontenfauna von Litschen/Lausitz. Abh Ber Naturkundemus Görlitz 47 (10): 1–16
Freyer G (1977) Das Unterkambrium von Görlitz, seine geologische Erforschung und Fossilfuhrung – ein Überblick. Abh Ber Naturkundemus Görlitz 51 (7): 1–14
Frischbutter A (1982) Zur präkambrischen Entwicklung der Elbezone. Z Angew Geol 28: 359–366
Furnes H, Kryza R, Muszyński A (1989) Geology and geochemistry of Early Paleozoic volcanics of the Świerzawa Unit, Kaczawa Mts, W Sudetes, Poland. Neues Jahrb Geol Paläontol Monatsh 3: 136–154
Galle A, Chlupáč I (1976) Finds of corals in the metamorphic Devonian of the Ještědské pohoři Mountains. Věstn Ústřed Ústavu Geol 51: 123–128
Grocholski A (1982) Crystalline series of the Fore-Sudetic Block and the connected prospects of mineral resources. Biul Inst Geol 341: 97–116
Grocholski A (1986) Proterozoic and Palaeozoic of southwestern Poland in a light of new data. Biul Inst Geol 355: 7–29
Gunia T (1967) Cambrotrypa (Tabulata) from metamorphic rocks of the Western Sudetes. Rocz Pol Tow Geol 37: 417–427
Gunia T (1968) On the fauna, stratigraphy and conditions of sedimentation of the Upper Devonian in the Świebodzice Depression (Middle Sudetes). Geol Sudet 4: 115–220
Gunia T (1981a) Microflora from the paragneisses of Sowie Mts (Upper Pre-Cambrian, Sudetes). Geol Sudet 16 (2): 7–21
Gunia T (1981b) Microfossils from the metamorphic rocks series east of Niemcza, southwestern Poland. Geol Sudet 16 (2): 25–45
Gunia T (1984) Microflora from the fine-grained paragneisses of Jugowice vicinity (Sowie Mts-Sudetes). Geol Sudet 18 (1): 7–17
Gunia T (1990) Acritarchs and microproblematics of the crystalline limestones from the vicinity of Romanowo Górne (central Sudetes Mts, Krowiarki). Geol Sudet 24: 101–135
Haydukiewicz A (1987a) Sekwencja stratygraficzna w kompleksie kaczawskim. In: Oberc J (ed) Przewodnik 58 Zjazdu Polskiego Towarzystwa Geologicznego, 17–19 wrzesień 1987. Wałbrzych, Instytut Geologiczny & Uniwersytet Wrocławski, Kraków
Haydukiewicz A (1987b) Melanże Gór Kaczawskich. In: Przewodnik 58 Zjazdu Polskiego Towarzystwa Geologicznego, 17–19 wrzesień 1987. Wałbrzych, Instytut Geologiczny & Uniwersytet Wrocławski, Kraków
Haydukiewicz A, Urbanek Z (1986) The metamorphosed Devonian rocks in the Bolków unit (the Kaczawskie Mts, Sudetes). Geol Sudet 20: 185–196
Haydukiewicz J (1979) Stratigraphy of the Zdanów series in northern part of the Bardo unit on the basis of conodonts. Geol Sudet 14: 77–99

Haydukiewicz J (1990) Stratigraphy of Paleozoic rocks of the Góry Bardzkie and some remarks on their sedimentation (Poland). Neues Jahrb Geol Paläontol Abh 179: 275–284

Hirschmann G (1966) Assyntische und variszische Baueinheiten im Grundgebirge der Oberlausitz (unter spezieller Berücksichtigung der Geologie des östlichen Görlitzer Schiefergebirges. Freiberger Forschungsh C 212: 1–146

Horný R (1964) Noví graptoliti z metamorfovaného siluru Podkrkonoší. Čas Nár Muz Oddíl Přírodověd 83: 224

Hoth K, Brause H, Freyer G, Lorenz W, Palchen W, Wagner S (1985) Neue Ergebnisse zur Gliederung des Proterozoikum im Erzgebirge-Zapadne Sudety/Sudety Zachodnie – Antiklinorium sowie an seiner Nordflanke. Wiss Z Ernst-Moritz-Arndt-Univ Greifswald 34: 5–13

Jaeger H (1964) Monograptus hercynicus in den Westsudeten und das Alter der Westsudeten-Hauptfaltung. Geologie 13: 249–273

Jerzmański J (1965) Geology of north-eastern part of the Kaczawa Mts and their eastern extent. Biul Inst Geol 185: 109–193

Jerzmański J (1970) Neue Ergebnisse im westlichen Teil des Blok przedsudecki. Ber Dtsch Ges Geol Wiss A Geol Paläontol 15: 305–313

Jerzmański J, Teller L (1971) The Lower Ludlovian in borehole Biskupin IG-1 (western part of the Fore-Sudetic Block). Bull Acad Pol Sci Sér Sci Terre 19: 157–160

Jerzykiewicz J, Grocholski A (1975) Komunikat o występowaniu szczątków mikroflorystycznych w epimetamorficznych utworach bloku przedsudeckiego. Kwartalnik Geol 19: 923–924

Kemnitz H, Budzinski G (1991) Beitrag zur Lithostratigraphie und Genese der Lausitzer Grauwacken. Z Geol Wiss 19 (4): 433–441

Kettner R (1922) Algonkium na Morave. Cas Vlasten Mus Spolecnosti v Olomouci 33: 1–4

Kozłowska-Koch M (1959) On the granite-gneiss of Wądroże Wielkie. Arch Mineral 21: 383–408

Kozłowski K (1974) Crystalline schists and leucogranites of the Stara Kamienica-Świeradów Zdrój Belt (Western Sudetes). Geol Sudet 9: 7–100

Kryza R, Muszyński A (1988) Metamorphosed diabasic sill-sediment complex from Wojcieszów: mineralogical and geochemical evidence of Lower Paleozoic early-stage rifting in the Western Sudetes, SW Poland. Mineral Pol 19: 3–18

Liew DC, Hofmann AW (1988) Precambrian crustal components, plutonic associations, plate environment of the Hercynian Fold Belt of Central Europe: indications from a Nd and Sm isotopic study. Contrib Mineral Petrol 98: 129–138

Linnemann U (1992) Glazioeustatisch kontrollierte Sedimentationsprozesse im Oberen Proterozoikum der Elbezone (Weesensteiner Gruppe/Sachsen). Zentralbl Geol Paläontol I, 1991; 12: 2907–2934

Lorenc S (1983) Petrogenesis of the Wojcieszów crystalline limestones (Góry Kaczawskie, Sudetes). Geol Sudet 18: 61–119

Mastalerz K, Porębski SJ (1987) Fan-deltas complex in a Visean low-energy marine embayment, Intrasudetic Basin (southwestern Poland). In: Nemec W (ed) Int Symp Fan Deltas: sedimentology and tectonic settings, 11–12 June 1987, Bergen, Norway

Matte Ph, Maluski H, Rajlich P, Franke W (1990) Terrane boundaries in the Bohemian Massif: result of large-scale Variscan shearing. Tectonophysics 177: 151–170

Michniewicz M (1981) Early stages of the tectogenesis of Sudetes Mts as interpreted in the light of a derivative theory of deep diapirism and the concept of deep fractures. Geol Sudet 16 (2): 75–141

Narębski W (1980) Paleotectonic setting of the circum-Karkonosze Lower Paleozoic spilite-keratophyre suites based on geochemistry of iron group elements. Rocz Pol Tow Geol 50: 3–25

Narębski W, Dostal J, Dupuy C (1986) Geochemical characteristics of Lower Paleozoic spilite-keratophyre series in the Western Sudetes (Poland): petrogenetic and tectonic implications. Neues Jahrb Mineral Abh 155: 243–258

Narębski W, Wojciechowska I, Dostal J (1988) Initial rifting volanics in the Kłodzko Metamorphic Complex (Polish Middle Sudetes), evidenced by geochemical data. Bull Pol Acad Sci Earth Sci 36: 261–269

Oberc J (1960) Eastern Karkonosze tectonics and their position in the Sudeten Structures. Acta Geol Pol 10: 1–48

Oberc J (1968) Prekambr, Sudety. In: Budowa geologiczna Polski, stratygrafia część 1. Wydawnictwa Geol, Warszawa

Oberc J (1972) Budowa geologiczna Polski, Tektonika, Sudety i obszary przylegle. Wydawnictwa Geol, Warszawa

Oberc-Dziedzic T (1988) The development of gneisses and granites in eastern part of the Izera crystalline unit in the light of textural investigations. Acta Univ Wratislaviensis 997, Prace Geol Mineral 13: 1–181

Opletal M, Domečka K, Čech S (1980) Geology of the Orlické horý Mountains. Academia, Praha

Pauk F (1977) The nappe structure of the Orlické horý-Klodsko Dome. Práce a Studie – Přir, Pardubice 9: 7–32

Pietzsch K (1962) Geologie von Sachsen. VEB Dtsch Verlag Wiss, Berlin

Porębska E (1980) Stratigraphy, lithology and sedimentation of the Ordovician, Silurian and Lower Devonian in the Bardo Mts. In: Gunia T (ed) Development of the Bardo Structure (Sudetes) on the basis of new stratigraphical, sedimentological and tectonical investigations. Materiały Konf Terenowej, 20–21 Sept 1980, Srebrna Góra. Wrocław Univ Press

Porębska E (1984) Latest Silurian and Early Devonian graptolites from the Zdanów section, Bardo Mts (Sudetes). Ann Soc Geol Pol 52: 89–209

Porębski SJ (1981) Świebodzice succession (Upper Devonian-lowermost Carboniferous): a prograding, mass flow dominated fan-delta complex. Geol Sudet 16: 101–192

Porębski SJ (1990) Onset of coarse clastic sedimmentation in the Variscan realm of the Sudetes (SW

Poland): an example from the upper Devonian-lower Carboniferous Świebodzice succession. Neues Jahrb Geol Paläontol Abh 179: 259–274
Reichstein M (1961) Oberdevon-Kalkstein am Geiersberg bei Kunnensdorf, nordwestlich von Görlitz. Geologie 10: 246–247
Rosanov AJ, Missarshevsky WW, Volkova NA, Voronova LG, Krylov IN, Keller BM, Koroljuk IN, Lendzion K, Michniak R, Pychova NG, Sidorov AD (1969) Das Tommot und Probleme der unteren Abgrenzung des Kambriums. Trudy Geologiceskogo Instituta Akademii Nauk UdSSR, Moskau, pp 1–206
Schwarzbach M (1933) Das Cambrium der Oberlausitz. Abh Naturforsch Ges Görlitz 32 (2): 7–54
Schwarzbach M (1939) Die Tektonik der Bober-Katzbach-Gebirges. Jahrber Schles Ges Cult Naturwiss 193: 1–52
Smulikowski K (1960) Evolution of the granitegneiss in the Śnieżnik Mountains-East Sudetes. In: Sorgenfrei (ed) Rep 21st Int Geol Gongr, Det Berlingske Bogtrykkeri, Copenhagen 14: 120–130
Smulikowski K (1979) Polymetamorphic evolution of the crystalline complex of the Śnieżnik and Góry Złote in the Sudetes. Geol Sudet 14: 7–79
Svoboda J (1955) Vapence Krknoš a Jizerských hor Geotechnika 21
Szałamacha J, Szałamacha M (1967) The Niedamirów series in the light of a new structural subdivision of the eastern part of the Karkonosze Granite cover. Kwartalnik Geol 11: 243–257
Szałamacha J, Szałamacha M (1968) The metamorphic series of the Karkonosze-Izera Mountains Block. Biul Inst Geol 222: 33–76
Teisseyre AK (1975) Sedimentology and paleogeography of the kulm alluvial fans in the western Intrasudetic Basin (Central Sudetes, SW Poland). Geol Sudet 9: 5–125
Teisseyre H (1956) Świebodzice Depression as a geological unit. Biul Inst Geol 106: 5–60
Teisseyre H (1968a) On the stratigraphy and structural evolution of the metamorphic series in the Sudetes. Geol Sudet 4: 7–45
Teisseyre H (1968b) Stratigraphy and tectonics of the Świebodzice Depression. Biul Inst Geol 222: 77–106
Teisseyre H (1980) Precambrian in southwestern Poland. Geol Sudet 15: 7–40
Teisseyre JH (1971) On the age and sequence of beds in the metamorphic rocks of the Rudawy Janowickie Range and Lasocki Ridge. Geol Sudet 5: 165–210
Teisseyre JH (1973) Metamorphic rocks of the Rudawy Janowickie and Lasocki Grzbiet ranges. Geol Sudet 8: 7–118
Thomas U (1990) Unterkarbone Wildflysch-Ablagerungen im Südteil der DDR. Z Angew Geol 36 (5): 182–184
Tischendorf G (1990) Some comments on fractionation as one main procondition of formation of magmatogenic deposits. Z Geol Wiss 18: 749–762
Urbanek Z (1978) The significance of Devonian conodont faunas for the stratigraphy of epimetamorphic rocks of the northern part of the Góry Kaczawskie. Geol Sudet 13 (1): 7–30
Urbanek Z, Baranowski Z (1986) Revision of age of the Radzimowice slates from Góry Kaczawskie (Western Sudtes, Poland) based on conodonts. Ann Soc Geol Pol 56: 399–408
Utzig B (1975) Petrographic characteristics of crystalline basement of the Fore-Sudetic Block south of Wrocław, Kwartalnik Geol 19: 35–46
Van Breemen O, Aftalion M, Bowes DR, Dudek A, Misař Z, Povondra P, Vrána S (1982) Geochronological studies of the Bohemian massif, Czechoslovakia, and their significance in the evolution of Central Europe. Trans R Soc Edinb Earth Sci 73: 89–108
Van Breemen O, Bowes DR, Aftalion N, Żelaźniewicz A (1988) Devonian tectonothermal activity in the Góry Sowie gneissic block, Sudetes, southwestern Poland: evidence from Rb-Sr and U-Pb isotopic studies. Ann Soc Geol Pol 58: 3–19
Vangerow EF (1943) Das Normalprofil des Algonkium und Kambriums in den mittleren Sudeten. Geol Rdsch 34: 10–12
Wajsprych B (1978) Allochtonous Paleozoic rocks in the Visean of the Bardzkie Mts (Sudetes). Rocz Pol Tow Geol 48: 99–127
Wajsprych B (1986) Sedimentary record of tectonic activity on a Devonian-Carboniferous continental margin, Sudetes. In: Teisseyre AK (ed) IAS 7th Regional Meet, 23-25 May 1986. Excursion Guidebook, Kraków Ossolineum, Wrocław
Weber B, Kemnitz H, Paech K (1990) Zur Kenntnis der Mikrofossilien und Problematika aus der Lausitzer Grauwacke (Kamenzer Gruppe). Veröffentlichung des Museums der Westlausitz 14: Kamenz, 9–34
Wojciechowska I (1986) Metabasites in the NW part of Śnieżnik metamorphic unit (Kłodzko area, Sudetes, Poland). Geol Rdsch 73: 585–593
Wojciechowska I (1990) Geology of the Kłodzko metamorphic unit (Sudetes, Poland). Neues Jahrb Geol Paläontol Abh 179: 189–195
Wyżga B (1987) Lower Paleozoic of the Bardo Mountains (Sudetes): a sequence of deep-sea pelagic sediments. Geol Sudet 22: 119–145
Żaba J (1984) Genesis and metamorphic evolution of gneisses and granitoids of the Izera Stóg massif, Western Sudetes. Geol Sudet 19: 89–192
Żakowa H (1963) Stratigraphy and facial extents of the Lower Carboniferous in Sudetes. Kwartalnik Geol 7: 73–94
Żelaźniewicz A (1984) Synmetamorphic penetrative mylonitization in orthogneisses of the Bystrzyca Mts, Sudetes. Acta Geol Pol 34: 111–130
Żelaźniewicz A (1987) Tectonic and metamorphic evolution of the Góry Sowie, Sudetes Mts, SW Poland. Ann Soc Geol Pol 57: 203–348
Żelaźniewicz A (1988) Orthogneisses due to irrotational extension, a case from the Sudetes, Bohemian Massif. Geol Rdsch 77: 671–682
Żelaźniewicz A (1990) Deformation and metamorphism in the Góry Sowie gneiss complex, Sudetes, SW Poland. Neues Jahrb Geol Paläontol Abh 179: 129–157
Zikmundová J (1964) Finds of conodonts in the Devonian of the Ještědské pohoři Mountains. Věstn Ústřed Ústavu Geol 39: 455–457
Zoubek V (1965) Moldanubikum und seine Stellung im geologischen Bau Europas. Freiberger Forschungsh 190: 129–148

VI.C Structure

A. Żelaźniewicz, H. Kemnitz, and N. Hermsdorf

1 Introduction

The Lugicum consists of (1) Upper Proterozoic graywackes and granodiorites (in Lusatia) and (2) mosaic of tectonostratigraphic domains (in West Sudetes), some of which became folded and metamorphosed before the deposition of Upper Devonian clastics and pelagic limestones, and Visean clastics (Culm). This recognition once led Bederke (1929) to emphasize a major role of Caledonian orogeny in the West Sudetes, with minor effect of Variscan deformations. In view of much uncertain lithostratigraphy, some geologists doubted the presence of Caledonian events in the Sudetes and spoke about a continuous Caledono-Variscan cycle (Teisseyre H 1968a). The once vivid Barrandean syndrome of the Bohemian Massif geology assigned all metamorphosed rocks to the Precambrian until paleontologically proved otherwise, and resulted in ascribing most of the structure-forming events, prior to the Sudetic phase, to both early and late Cadomian (Assyntian) movements (Oberc 1972, 1977; Zoubek 1988), or to late Caledonian ones (Bederke 1924; Kodym and Svoboda 1948).

With the acquisition of more paleontologic and structural data, it has appeared that the West Sudetes contain sequences deformed and metamorphosed in various periods of the Paleozoic.

Recently, Don (1984, 1990) proposed subdividing this orogen into two portions, Caledonian and Variscan, situated on either side of the Main Intra-Sudetic Fault (Fig. 1 in Chap. VI.A), interpreted by him as a persistent, crustal-scale fracture zone, controlling, during the Paleozoic, both sedimentation and metamorphism in the adjacent areas.

Isotopic data (see Chap. VI.B.2) indicate, however, that Cadomian thermal events manifested by Late Proterozoic granitoid intrusions, and then Ordovician granites, spread throughout the Lugicum. In its Lusatian part, the Cadomian rocks had been exposed probably soon afterwards, whereas in the West Sudetes they may have become hidden beneath the deposits of various Paleozoic sedimentary basins, and subsequently subjected to Variscan deformation and metamorphism (Figs. 1, 2, 3). Therefore, a vital question is still open, whether (1) the Lusatian granodiorite-anatexite complex continues farther east to form a basement of the Late Proterozoic-Paleozoic successions in the West Sudetes, (2) the latter region has another type of basement rocks, or (3) Lugicum represents an amalgamate of Variscan terranes, with no common lower crust beneath.

Apart from speculative interpretations poorly rooted in important details of local geology (e.g., Quenardel et al. 1988; Matte et al. 1990; Aleksandrowski 1990), this question cannot at the moment be answered reliably, the more so that subsurface and isotopic data are still very scarce. Therefore any unbiased interpretation of the structure of the whole Lugicum can hardly be given at present. Here is given a brief review of literature data and a description of structural characteristics.

2 Precambrian Deformation

2.1 Lusatian Anticlinal Zone (LAZ)

The LAZ is a vast elevation of Proterozoic rocks flanked by very weakly metamorphosed Paleozoic sediments on the north (Görlitz Syncline) and west (Torgau-Doberlug-Göllnitz Syncline) and by strongly folded Paleozoic metasediments (Elbe Valley Syncline) on the south, where they are accompanied by Variscan granitoids (Meissen massif) and covered with Cretaceous deposits – now in tectonic contact through the Lusatian Thrust (the world's earliest recognized). On the east, the LAZ merges through the East Lusatian (Zawidów) granodiorite with metagranitoids of

Dallmeyer et al. (Eds.)
Pre-Permian Geology of Central and Eastern Europe

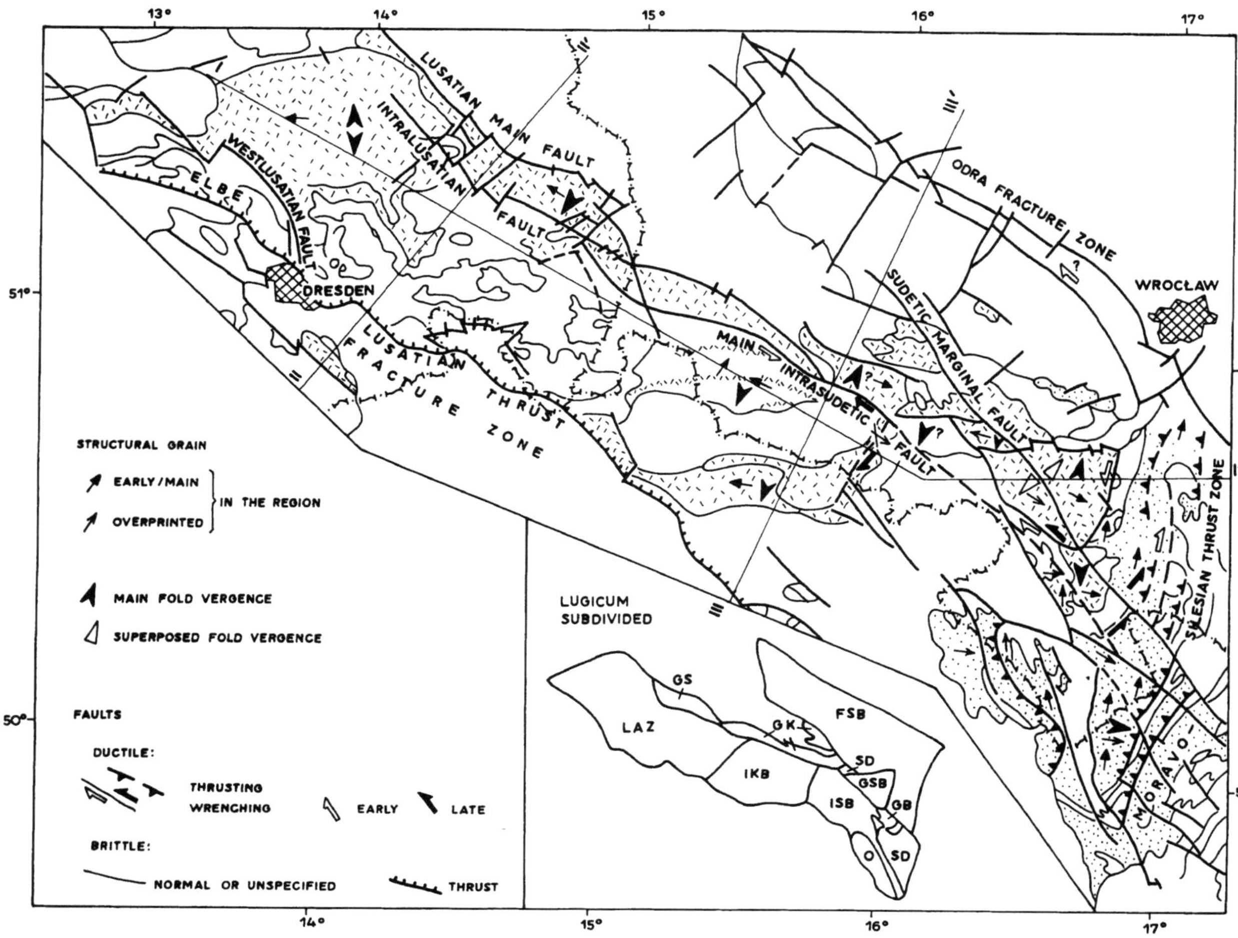

Fig. 1. Structural sketch of the Lugicum (Western Sudetes). See Fig. 1 in Chapter VI.A for names of the Lugian units. *Dashed* Domains with mainly W/NW-trending linear structures; *stippled* domains with mainly N/NE-trending linear structures; *II–II′* cross sections of Fig. 2

the IKB[1], with no obvious and clearly defined boundary (Fig. 1 in Chap. VI.A; Fig. 2).

The earliest tectonometamorphic event in the LAZ has been recorded by foliated rocks occurring as rafts and inclusions within anatexites. Neither timing (?Dalslandian) nor structural details about this event are available. Isotopic data (Rb-Sr, U-Pb; cf. Schust 1990) point clearly to late Cadomian (Panafrican) intrusion of granodiorite that thermally affected graywackes 600–570 Ma ago. The graywackes were involved in W-trending, roughly upright folds of 1B–1C class, with a width of some hundreds of meters, shortening of 0.6–0.7 (Brause 1975), and alternating northern and southern vergence (Fig. 1).

Very low strain intensity (d = 0.56; Bankwitz and Bankwitz 1988) increases occasionally only in zones of bedding-parallel shears and imbrications. Besides, three types of foliation can be discerned in the graywacke sequences (Kemnitz and Budzinski 1991), namely (1) S_1 – nonpenetrative crenulation cleavage and fracture cleavage associated with the late Cadomian folding, (2) S_2 – mimetic biotite growth in bedding planes due to contact metamorphism, and (3) S_3 – seen only in the East Lusatian rocks as very incipient schistosity, genetically related to the easterly increasing deformation of Paleozoic sediments and early Paleozoic granitoids through the S-C mylonitization in ductile strike-slip shear zones (?mostly sinistral and NW/WNW-trending), well evident farther west in the IKB.

The significance of those occasionally identifiable ductile shear zones is not fully recognized

[1] For full names of units and subdivision of the Lugicum (Western Sudetes), see Chapter VI.A.

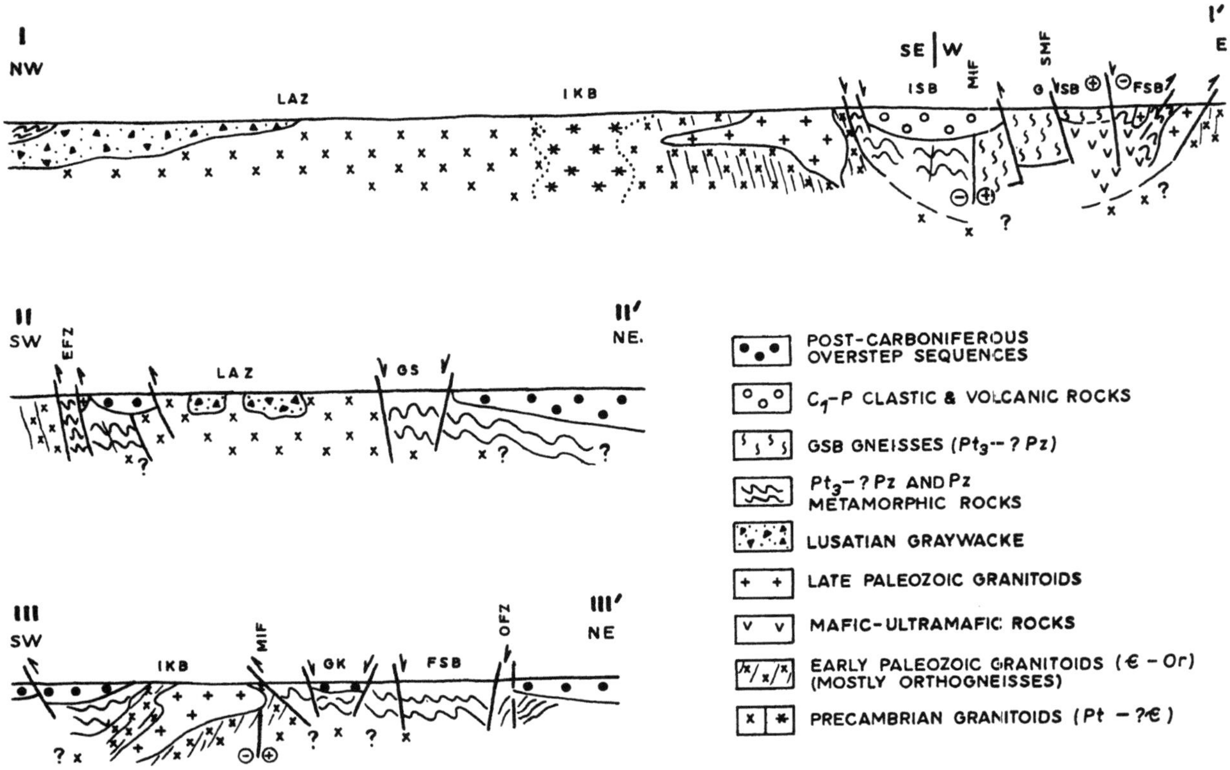

Fig. 2. Diagrammatic cross sections through the Lugicum. See Fig. 1 for locations

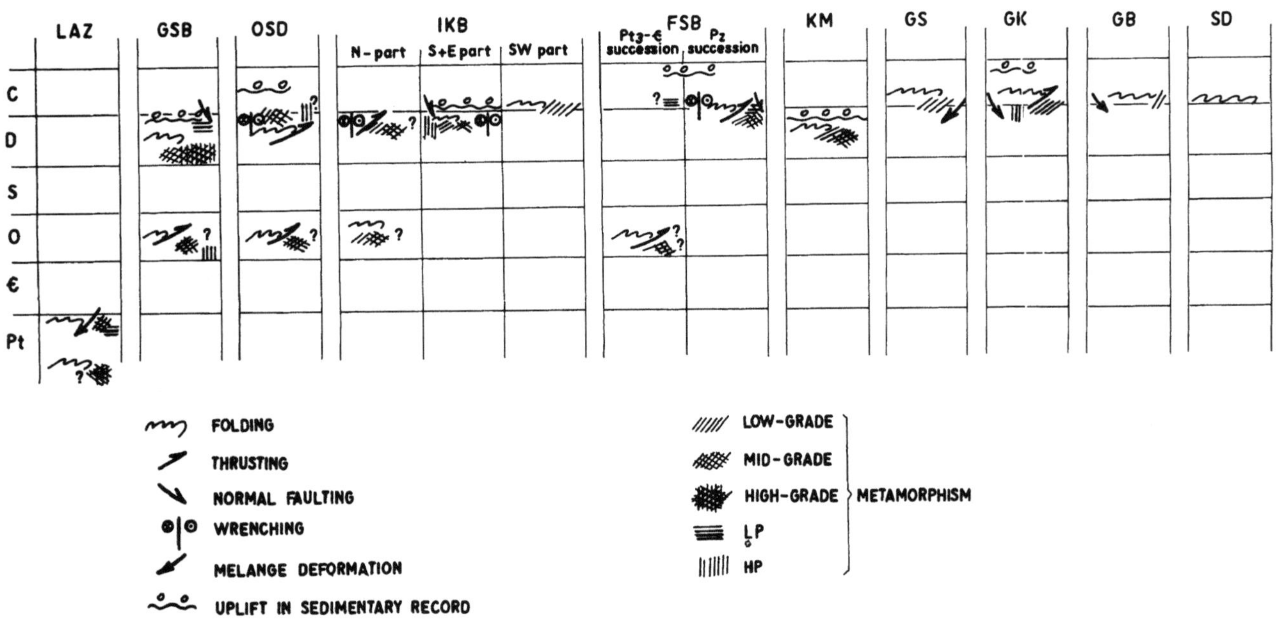

Fig. 3. Summary of tectonometamorphic history of the Lugicum (West Sudetes)

here. For instance, the one south of Löbau, with marked sinistral strike-slip component and strong mylonitic impact, was considered part of a boundary between LAZ and West Sudetes, as stated by Ebert (1943), Hirschmann (1962), and Möbus (1970). With its northwesterly trend it can perhaps be also related to the West-Lusatian Fault (Fig. 1) that had most likely been active since Late Proterozoic times and now defines the NE border of the Elbe Fracture (Lineament) Zone.

3 Paleozoic Deformation

3.1 Góry Sowie Block (GSB)

The GSB is in a brittle fault contact on the north with the Frasnian-?Tournaisian Świebodzice depression (Chap. VI.B.3), including a 2000-m-thick conglomerate derived almost exclusively from the GSB, and on the south again with the Frasnian-Upper Visean clastics (coming from GSB and nearby ophiolitic gabbroids) covered with the coal-bearing Namurian A. The central and northern part of the eastern margin of the GSB contains up to 1500-m-wide, N-trending zone (Figs. 1, 2) of ductilely sheared (sinistral wrench) gneisses at P-T conditions of the epidote amphibolite facies, then followed by cordierite blastesis. Also cordierite and andalusite (locally even sillimanite) occur in the adjacent Upper Visean metapelites and quartzites of the NZ (Niemcza zone), otherwise having but local evidence of ductile shearing (Dziedzic H 1985), which is, however, again more common farther east (Fig. 2).

The traditional assignment of the GSB rocks entirely to the Precambrian is no longer tenable (cf. Chap. VI.B.2.4), and they contain no legible record of any tectonometamorphic activity in the Precambrian.

The GSB is divided by the SMF (Fig. 1 in Chap. VI.A) into Sudetic and Fore-Sudetic parts. Both consist mostly of metagraywackes and metapelites developed as texturally varying gneisses, migmatized to various extent, and accompanied by minor metabasites with various protoliths, and dominantly felsic granulites (Kryza 1981; Żelaźniewicz 1987, 1990a). The Fore-Sudetic part of the GSB exposes levels of this complex a few kilometers deeper (Fig. 2), relatively richer in both amphibolites and retrograde granulites. The fairly monotonous gneissic lithology once prompted to correlate the GSB rocks with the Monotonous Group of the Moldanubicum (cf. Chap. VII), and eventually to interpret this high-grade inlier within low- and mid-grade rocks as a Moldanubian klippe being thrust northerly over the Saxothuringian domains, much in the same manner as has been suggested for the gneissic masses of Münchberg, Wildenfels, and Frankenberg (Kossmat 1927; Suess 1935; Behr et al. 1982).

Since nowadays practically only secondary, brittle fault boundaries of the GSB can be observed, it is rather difficult to verify the aforementioned interpretation (specifically in view of the lack of any pervasive ductile fabric in the neighboring and partly underlying basite-ultrabasite suite: Żelaźniewicz 1990b, interpreted as ophiolite: Pin et al. 1988), as also all other concepts concerned with the position of the GSB, ranging from a suspect terrane (Matte et al. 1990) and microcontinent (Cwojdziński 1980), through orogenic diapirism (Michniewicz 1981) to autochtonous intramontane mass (Cloos 1922; Bederke 1934; Oberc 1972).

At least four periods of synmetamorphic tectonism was reported by Grocholski (1967), the earliest N-trending folds being overprinted by at first the NW-SE and then the NE-SW sets. Having remapped the GSB area and restudied its structure and metamorphism, Żelaźniewicz (1987, 1990a) arrived at the conclusions as follows below (cf. Table 1 in Żelaźniewicz 1990a).

1. The D_1 event produced (a) generally N-verging, mostly recumbent folds (with occasional sheaths pointing to N/NE) and (b) zones of ductile shearing, in places merged to the N/NE-verging crustal thrusts that tectonically inserted the felsic granulites (800 °C, 15 kb; cf. Chap. VI.E) and peridotites (1000 °C, 20–30 kb; cf. Chap. VI.E) into the amphibolite facies gneisses (Żelaźniewicz 1985).

2. Most of the once dominantly horizontal D_1 (F_1, S_1) structures became steepened during superposed foldings (D_2 and D_3 events), bringing about well mappable interference structures. The upright-to-overturned macroscopic folds F_3, with either NE or SW vergence (Fig. 1), represent the main structures of the GSB and were probably due to the NE-SW to N-S regional shortening, thus oriented much like that of the D_1 event. The two events were separated by the temperature rise and mineral changes due to progressive metamorphism from the garnet + kyanite to sillimanite or locally even sillimanite + K-feldspar zone (Żelaźniewicz 1987, 1990a). The vaning stage of the metamorphism itself was fingerprinted by U-Pb monazite and zircon age of ca. 380 Ma (van Breemen et al. 1988).

3. The ensuing onset of uplift was due to the change in the overall tectonic regime, when

to the N-S compression a significant shear component was added (?postcollisional motion parallel to the orogen margin), giving rise to dextral transpression under the epidote amphibolite facies conditions. Then concurrently developed the N-trending, steep pegmatite veins (370 Ma), vertical folds, and possibly conjugate, NW-trending (dextral) and NNE-trending (sinistral) strike-slip shear zones, now partly exposed along the SW and E margins of the GSB (Fig. 1).

4. The last synmetamorphic deformational event was brought about by the vertical compression (due to uplift or extra burden from the top) giving rise to open, gravity collapse folds, with occasional brittle/ductile, directionally inconsistent shears along the subhorizonal axial planes (S_5), which often were taken up by the static, volatile assisted recrystallization and minor partial melting producing a granitoid-looking rock. The fabric reconstitution was induced by stress and ensuing decompression. Although this may hardly be told from the effects of earlier (D_3/M_3) nebulitic migmatization, it adds to an overall LP metamorphic impact on the GSB rocks, which is well compatible with the late LP metamorphic output characteristic of European Variscides.

3.2 The Orlica-Śnieżnik Dome (OSD)

The OSD boundaries are marked by (1) Upper Carboniferous-Cretaceous deposits of the Intra-Sudetic Depression (ISD) on the west, (2) Moravo-Silesian thrust zone on the east, and (3) brittle fault against the Paleozoic sequence of the GB. The OSD is naturally divided into two parts by the Upper Cretaceous graben – the eastern part is commonly referred to as the Śnieżnik metamorphic unit (SM), the western is known as the Orlické horý and Góry Bystrzyckie.

The OSD generally consists of the orthogneissic core, enveloped by a mottled series of mica schists and paragneisses with marbles and amphibolites, which is mantled in turn by phyllitic metapelites and metabasites (Fig. 1 in Chap. VI.A; Chap. VI.B.2.4).

The southern part of the SM is built of the N-trending anticlinoria (gneisess) and synclinoria (mica schists), which developed due to the generally upright refolding of the E-verging overturned and recumbent folds under amphibolite facies conditions (Teisseyre H 1968a, 1973). H. Teisseyre found mesoscopic evidence for five consecutively overprinted sets of fold and linear structures (paying less attention to planar ones), altogether matching the tectonic evolution of the SM. The earliest folds (F_1) with the penetrative axial planar foliation were represented by mostly N-trending and E-verging folds (Fig. 1), paralleled by composite lineation (mineral, intersection, corrugation + mullion, and boudin axes). These structures were overprinted by the NE-trending corrugations (F_2) and weak mineral lineation. Further deformation gave rise to the NW-trending, open to tight folds, accompanied locally by axial recrystallization, and still later to widespread kinking of the schistous units.

This mesostructural sequence has generally been confirmed, though with local modifications, in other parts of the OSD (Don 1964; Dumicz 1964; Kasza 1964; Cwojdziński 1976; Wojciechowska 1975; Żelaźniewicz 1978). However, the general accounts given by various authors differ widely (Don et al. 1990).

From the NW part of the OSD, Żelaźniewicz (1978, 1984) also reported the E-verging, overturned and recumbent folds making the earliest template of the OSD structure, with zonal yet penetrative axial shearing, which gave rise to the localized reorientation of linear structures in the S_2 foliation planes and within the E/SE-verging thrust zones.

The very characteristic feature of the OSD structure is common parallelism of various types of lineation (mineral, corrugation, intersection and mullions) with the axes of both large-scale, mappable, and small-scale folds of the F_1 set, and with the stretching lineation developed in the augen orthogneisses. Such a situation points to the extension of the former granite body, perpendicular to the regional W-E shortening (?early collision of Lugian and Moravo-Silesian blocks). The initial prolate geometry of regional strain (postcollisonal, orogen-parallel tectonic escape) was characteristic only of the early stage of the granite-gneiss transformation (Żelaźniewicz 1988), with ensuing static recrystallization under amphibolite facies conditions, which erased much of the early deformational fabric. Having overlooked this stage, one can arrive at a false impression of the only N-directed motion to have brought about the whole structure (Cymerman 1991). Then plane strain and localized flattening deformation under PT of the greenschist facies took place, obviously related to an overall northeasterly, late, dextral transpressive motion of the OSD complex. This most likely matches the late-stage Variscan transpression against the Moravo-Silesicum (Brunnovistulicum), reported from the SE margin of the Bohemian Massif, south of the Elbe Fracture Zone; however, with no legible sign of thin-skinned, northerly nappe emplacement recently recognized there (cf. Chap. VIII).

The easterly motion of rock units in the OSD was known long ago to Pauk (1953), who identified white schists as the mylonitized (ortho) gneiss and delineated two large nappes, with the mappable boundary between them located in the eastern part of the OSD. The easterly motion was also to cause a fan-like arrangement (virgation) of the OSD recumbent folds due to the push against the East Sudetes (Don 1964, 1990).

Don (1964, 1990) claimed that toward an end of the F_1 deformation (early Caledonian), the already metamorphosed mottled series (Stronie) had been intruded by a porphyritic granite (Śnieżnik) that became dynamically transformed at F_2 event (late Caledonian) into the augen orthogneiss during the easterly folding and napping. The ongoing amphibolite facies metamorphism and the increasing lack of a lateral room brought about the transversal (NW-SE), locally very intense warping (F_3), and resulted in the migmatization proceeding at the base of the entire pile. The migmatites were squeezed up diapirically and pierced through the overlying schist-orthogneiss complex in the form of domal masses of less deformed, fine-grained, laminated gneisses (Gierałtów type), locally burdened with some felsic granulites (800 °C, 20 kb; cf. Chap. VI.E) and eclogites that have been picked up from the lower crust (Don 1989) by the upgoing migmatites. The idea is partly corroborated with rather obscure evidence of the tectonic contact between these contrasting lithologies.

Dumicz (1979, 1988) drew attention to the role of intense Variscan transposition of the earlier flat-lying S-planes (S_1 and S_2 foliation in the recumbent folds) to the S_3 (S_{2+3}) foliation localized in ductile, more dynamically engaged zones (e.g., Skrzynka-Złoty Stok), where all the prior linear features became almost erased and the dynamometamorphism was followed by mostly static recrystallization, leading to local occurrences of migmatites (Gierałtów type). The subsequent folding (D_4) produced the actual synclinorial-anticlinorial outcrop pattern of the OSD.

The geochemistry-based model of Borkowska et al. (1990) assumes (1) the Gierałtów magma to have intruded the attenuated granulite crust ca. 465 Ma ago, (2) collisional eclogite metamorphism (not dated) with the production of Gierałtów (ortho)-gneisses at 420 Ma, (3) intrusion of the Śnieżnik magma during crustal thickening at 395 Ma, and (4) thermal overprint 335 Ma ago.

Fajst (1976) claimed that only the gneissic core along with the Stronie mica schists have had the N-trending lineation and folds (F_1) of ?Grenvillian age, which are absent from the mantle rocks (the Mové Město, Zabřeh and Staré Město series). The structural sequence in the latter started with the F_2 folds of Cadomian age, trending in the NNE direction in the E limb (SM) of the dome and in the NW/NNW direction in its W limb. Therefore he proposed a pre-Cadomian structural unconformity ("Orlica unconformity") to exist between the core and the mantle of the OSD.

In view of the stratigraphic ambiguities (cf. Chap. VI.B.2.4) Dumicz (1979, 1988) assumed the whole tectonothermal history of the OSD to have been accomplished during the early (Acadian phase) and late (Sudetic phase) Variscan orogenesis, whereas Don (1990) put it entirely to the Caledonian one. Still scarce and unclear isotopic data (e.g., scatter of Rb-Sr datings for the augen gneiss (≈ Śnieżnik) from 490 to 335 Ma and timing of eclogite at ca. 330 Ma (cf. Chaps. VI.B.2.4, VI.E) can support both Don's and Dumicz's interpretation, and yet they are not quite consistent with the account offered by Borkowska et al. (1990).

3.3 The Izera-Karkonosze Block (IKB)

The IKB contains the centrally situated Karkonosze (Krknoše in Czech) granitic massif (Fig. 1 in Chap. VI.A; Fig. 2), naturally dividing it into northern (Góry Izerskie), eastern (Rudawy Janowickie + Lasocki Grzbiet = East Karkonosze), and southern parts (Ješted + Železný Brod + Rýchory areas = South Karkonosze). On the west IKB merges, in a still unrevealed manner, with the East Lusatian (Zawidów) granodiorite, on the south and east it is in brittle fault contacts with Lower

Carboniferous-Upper Cretaceous sediments, while on the north there is a wide ductile fault zone, Main Intra-Sudetic Fault (MIF), at the border with the Góry Kaczawskie (cf. Don 1990). Although the latter is the best-exposed section of this fault, the kinematic story of MIF is still not fully known. Preliminary inspection to the MIF zone points to dextral wrenching, followed by generally S/SW-vergent thrusting, the two events being separated in time by the emplacement of the Karkonosze granite (Rb-Sr age of 328 Ma).

Most of the IKB consists of the Izera gneisses, having differently interpreted origin and evolution (cf. Chaps. VI.B, VI.D), but mainly being orthogneisses, with protolith emplaced between 500 and 450 Ma ago (Borkowska et al. 1980).

According to an earlier view (Oberc 1967) the Izera gneisses were, supposedly in Cadomian (Assyntian) times, involved in major NW-trending recumbent folds, with parallel mineral and corrugation lineation. Later deformation brought about both the S-verging folds (Fig. 1) and caused irregular reorientation of the earlier lineation.

Dziemiańczuk and Dziemiańczuk (1985) indicated that the overprint was manifested by a penetrative crenulation fabric (S_2), with associated development of mylonitic banding under retrograde greenschist facies conditions. The prominent intersection lineation (S_1/S_2) has shown a considerable scatter, while the S_1 foliation dips moderately to NE and often contains the down-dip-plunging extension lineation pointing to the N-S/NE-SW directed transport.

Oberc-Dziedzic (1988a) pointed out, however, that the significant stretching operated in the S_2 foliation planes, with mineral lineation plunging at various angles to NW and SE, and was accompanied by concurrent dynamic and static recrystallization (D_2). At the end of the extensional event this recrystallization added to the emplacement of fine-grained granite, and to the consecutive K-feldspar blastesis generally coarsening the granitic grainsize.

Accordingly, there is little doubt that the deformational history of the Izera gneisses is mainly that of a widespread ductile shearing, imposed on the granitoid protolith of 500–450 Ma age (Rb-Sr), though the kinematic and structural details of this process are still poorly recognized. The IKB rocks suffered mostly from the strike-slip, ductile mylonitization (?Hercynian) with a changeable sense of movement on the steep, WNW-trending foliation planes (generally at first sinistral and then dextral at lower, mostly greenschist temperatures). No direct evidence for the age of this shearing has been provided, yet it appears from the regional structural setting that the dextral wrench mylonitization was prior to the intrusion of the Karkonosze granite. Moreover, even rough inspection reveals a mylonitic status of the three narrow schistose belts set in the gneisses, of which at least two had most likely been derived from the same granitic protolith and then became the sheared channelways allowing abundant volatiles to migrate through and bring about extensive recrystallization as well as crystallization of new (also ore) minerals. The petrogenetic and structural history of these schist belts needs careful reexamination.

In the East Karkonosze it is difficult to tell the axial trend for the earliest folds, recognizable on both small and map-scale in mica schists and augen gneisses, but the NW/N direction seems to dominate (Teisseyre JH 1973). Their penetrative axial planar foliation (S_1) was effected with low-grade metamorphism followed by a static phase at garnet grade. This set of folds (F_1) is absent from the bimodal volcanogenic formation, which thus may have deposited, at least in part, after the earliest deformation. The F_2 folding, again on meso- and macroscale, with WNW-trending horizontal axes, and S/SW vergence, was accompanied by concurrent stilpnomelane and glaucophane blastesis. In augen gneisses, lying at the base of local succession, there is a conspicuous stretching lineation, pointing to a normal fault motion on the presently steeply ESE-dipping mylonitic foliation. However, the outcrop pattern shown on J.H. Teisseyre's (1973) maps point clearly to a dextral movement, with significant strike-slip component, on the steep, WNW-trending S_2 planes. This fits well the dextral transpression pattern recognizable in the MIF zone; thus one can suggest that the F_2 folds may have developed in a dextral transpressive regime.

The present steep attitude of the F_2 axes and mineral lineation was due to the rotation on the major NE-trending flexure (F_3), during updoming caused by the intruding Karkonosze granite (Teisseyre JH 1973). The foliation-parallel vertical motion on this flexure and the W-E extension are well matched by the steep normal faulting to ESE, toward the Dinantian Intra-Sudetic basin (Fig. 2), filled with up to 6500-m-thick early molasse clastics (Dziedzic K and Teisseyre AK 1990).

The F_3 small-scale folds developed by kinking, which was accompanied by retrograde metamorphism (chlorite grade), occurring in narrow zones of the S_3 axial shears (Teisseyre JH 1973). The shears were often connected with the NW/W-vergent reverse faulting and thrusting.

The East Karkonosze lithologic succession extends farther south to the Rýchory and Železný Brod areas. However, in the South Karkonosze, in view of many stratigraphic uncertainties, the timing of tectonic events is very controversial, and structural characteristics as described in the literature cannot be easily compared with those of the East Karkonosze. Chaloupský (1989) tried arbitrary correlations with other parts of the Bohemian Massif, and distinguished there fold structures which were to have developed during Grenvillian (NNE-trending), Cadomian (E-W trend), Caledonian (NW-SE trend), and Variscan (NE-SW trend) times, which resulted in the polyphase overprint relationships. Caledonian and Hercynian deformation under retrograde PT of greenschist facies were to erase the earlier structures. However, earlier Kodym and Svoboda (1948) interpreted the IKB in terms of two late Caledonian nappes differing by their stratigraphic contents and metamorphic grade. Máška (1954) also claimed Caledonian age of the deformation there, but rejected the nappe concept, since he found no difference in their small-scale structural inventory and pointed to the dominant E-W trend of S-verging folds, with numerous departures resulting from the competence contrast throughout the folded sequence. Other interpretations, much questioned by Chaloupský (1989), postulate the subduction-related, large-scale thrusting and concurrent HP metamorphism shown by Na-amphiboles in the Železný Brod metabasites (Cháb and Vrána 1979), or combined dextral strike-slip and thrust movements in ductile shear zones of Variscan age, with the dominant top-to-east motion, apparently bridging the South Karkonosze with the Erzgebirge (Rajlich and Synek 1987).

3.4 Góry Kaczawskie (GK)

The only exposed tectonic boundary of the GK is that on the south, defined by (1) MIF (cf. Chap. VI. C.3.3), (2) Visean and younger (even post-Permian) brittle faults against the Karkonosze granite and the Intra-Sudetic Depression, (3) post-Tournaisian, brittle, S-vergent thrusting over the SD clastics.

The tectonic interpretations of the GK structure have been changing with the acquisition of stratigraphic data (Chap. VI.B.3). Schwarzbach (1939) claimed a Caledonian orogenic fanning, with the N- and S-verging folds, respectively, on either side of the median line, now hidden beneath the Lower Permian graben. Quitzow (1939) supposed that the GK greenstones were transported from the south over the Ordovician-Silurian slates involved in the N-verging folds. However, in the southern GK, H. Teisseyre (1963) recognized the ESE-plunging anticlinal stack of four nappes (Caledonian), each being moved 10–15 km southwards, with the two upper units being later refolded with the Upper Devonian-Lower Carboniferous clastics of the Świebodzice Depression (Chap. VI.B.3).

More recent tectonic interpretation (Haydukiewicz 1987a,b), although contradicting the proposed age and geometry of these nappes, otherwise discloses thrusting as a prominent tectonic feature of the GK, defined by the common presence of (1) thrust sheets, a few hundreds up to 2000 m thick, composed of ?Cambrian-Silurian rocks (Devonian in one case only), subsequently involved in variously plunging folds (to NE, E, SE), and (2) polygenic melanges of Ordovician-Devonian bodies with strongly foliated Upper Devonian-Lower Carboniferous matrix. Such a tectonic style, combined with very badly exposed ground, greatly inhibits not only the identification of individual structures, but sense and even direction of early thrusting (different in different blocks), too. Nonetheless, at least two stages of Variscan horizontal tectonics seem detectable there (Baranowski et al. 1990), with distinct, S/SW-vergent, late (post-Visean) folding and thrusting (Fig. 2).

3.5 The Görlitz Syncline (GS)

The GK continues farther west, to the GS that has steep, brittle fault contacts with LAZ and Permo-Mesozoic rocks on the SW and NE, respectively (Fig. 1 in Chap. VI.A; Fig. 2).

Before the Devonian and Lower Carboniferous strata were proved, the GS was regarded as a region with strong Caledonian folding (Bederke 1924). Now it is obvious that the Paleozoic infilling of the GS was folded in Variscan times

(Sudetic phase) due to strong, NE-SW shortening, which brought about mostly the S-verging (Figs. 1, 2) isoclinal folds (Brause and Hirschmann 1969), with frequently met imbrications producing the inversion of stratification. The partly chaotic stratification was also explained as a result of olistolith (derived from Ordovician to Carboniferous rocks, with no Lower Cambrian ones) and olistostrome sedimentation (Bankwitz 1988), which is the situation well known from the GK and GB regions farther east. According to Thomas (1990), the gravitational mass movements occurred between Middle Visean and Dinantian III, with basic volcanism of Early/Late Devonian and ?Dinantian age (Hirschmann 1966).

3.6 The Świebodzice Depression (SD)

The SD has brittle boundaries on all sides: (1) steep, N-dipping normal fault with the GSB, (2) MSF, (3) S-vergent thrust of 6 km amplitude (Oberc 1972) bringing the GK over the SD, and (4) NW-trending reverse fault with the ISB, again with the GK rocks exposed along it (Teisseyre JH 1962).

The SD was distinguished from the Permo-Carboniferous Intra-Sudetic Depression (ISD) by virtue of different age, stratigraphy and tectonic style (Teisseyre H 1956). It is a complex Variscan structure which, besides the unmetamorphosed Frasnian-Tournaisian clastic succession (cf. Chap. VI.B.3), contains greenstones, slates, and quartzites with tectonic small-scale structures, altogether, resembling the GK rocks. The metamorphic sequence includes 200-m-thick cataclasites derived from a variety of bimodal volcanites. Since these rocks occur upon the Upper Devonian clastics, H. Teisseyre (1956, 1968b) assumed a gravity-driven, southerly sliding of the GK-derived elements toward the SD during the regional Bretonian compression. This caused both the updoming of the GK and the formation of the S-vergent, reclined folds in the SD sedimentary infilling. Those folds are characteristically en-echelon arranged relative to the NW-trending major faults (MIF, MSF), which might indicate the post-Tournaisian (?Visean) dextral transpression in the West Sudetes. Also a model of dextral strike-slip motion of crustal blocks with the resultant pull-apart opening has recently been proposed by Porębski (1990) to explain the origin and facies distribution in the Frasnian-Tournaisian sediments of the SD, and other synorogenic clastic deposits in the Sudetes.

3.7 The Góry Bardzkie (GB)

The GB boundaries are defined by (1) MSF, (2) complex, reverse, and normal, brittle fault zone with the GSB of mid-Visean (or older) age, (3) normal fault and Bardo thrust, at the border with the Kłodzko metamorphic unit, with Upper Devonian clastics cropped out alongside. Moreover, the GB is thermally affected (over 700°C) by the Kłodzko-Złoty Stok syenitoids and granitoids (Wojciechowska 1975; Bagiński 1990).

The internal tectonics of the GB is extremely complex as is its stratigraphy (cf. Chap. VI.B.3). Oberc (1972) distinguished over 50 NW/WNW-trending and S-verging folds (Fig. 1), developed along with thrusts during the Sudetic phase and then refolded by the W-verging folds during the Asturian event.

Oberc's account has tacitly assumed the GB complex in terms of one simple lithostratigraphic column. Wajsprych (1986), however, recognized there fragments of many rock complexes representing several different lithotectonic domains, altogether interpreted as ocean floor and trench deposits developing in the front of an accretionary wedge, with the SE subduction under, and NW obduction onto, the GSB, herein regarded as microcontinent shedding clasts to the Upper Devonian-Visean autochtonous apron in the GB.

3.8 The Fore-Sudetic Block (FSB)

The FSB is situated between the NW-trending major faults: MSF and OFZ (Fig. 1 in Chap. VI.A; Fig. 2). Being covered by Cenozoic deposits, it is rather poorly exposed, and thus known mostly from the subsurface data (Chap. VI.B.2.6). Due to manifold motion on the MSF during (Late) Carboniferous-Miocene time, the FSB now represents the crustal level by 4–6 km deeper than the West Sudetes Mts., although actually it is down-thrown and slightly tilted to the NE, abutting against the Odra Fracture Zone (OFZ). The latter appears as a few-km-wide vertical zone, "unseen" by seismic reflections, and thus liable to the interpretation as a major (?dextral, ?sinistral) wrench, compatible with the EFZ and possibly with the MIF.

The OFZ is consistent with the strike of the mid-grade metamorphic rocks of the MOM zone (cf. Chap. VI.B.2.6), commonly considered as the eastern continuation of the Mid-German Crystalline Rise. The low-grade metamorphic rocks, drilled SW of the MOM (Fig. 1 in Chap. VI.A; Fig. 2) are lithologically and stratigraphically closely akin to those of the Góry Kaczawskie, basic volcaniclastic sequence inclusive. They have most likely been involved in again SW-verging folds, occasionally including the ?Cadomian-?Ordovician metagranitoid (like that of IKB) basement, now spotted only as gneissic inliers, or perhaps even overthrust by the MOM (Oberc 1972).

Many evidence of ductile shears has been found in the N-trending, broad belt east of the Wrocław-Ząbkowice line. This belt is most likely an immediate continuation of the Moravo-Silesian thrust zone. Nonetheless, it is difficult to tell the kinematics from the unoriented drill-cores, otherwise recording highly changeable dip angles of the shear surfaces, with mylonitization accomplished mostly at low, retrograde P-T conditions (Cymerman and Jerzmański 1987).

The Moravo-Silesian thrust zone (Moldanubian; Nýznerov, Ramzova, and Vacetín thrusts), separating Lugicum and Moravo-Silesicum (Skácel 1989), had once been located immediately east of the GSB (Bederke 1929) in the Niemcza Zone (NZ). The NZ, however, consisting of fine-grained metapelites (some with Upper Visean spores; cf. Chap. VI.B) and graphictic quartzites along with metagraywackes, was also interpreted as (1) Silurian-Lower Carboniferous narrow graben (Finckh 1925), (2) intensely mylonitized GSB gneisses (Scheumann 1937), (3) dextral transform fault zone intruded by the differentiates of upper mantle magmas (Dziedzic H 1985), and carrying a late thermal overprint (andalusite, cordierite), but otherwise with only local evidence of ductile shearing and mylonitization (sinistral wrench). The GSB/NZ boundary was intruded by a late syntectonic hornblende granodiorite that remains mostly underformed, discounting the prominent, subhorizontal (NNE-SSW) igneous lineation. Thus Oberc (1966) draw the West/East Sudetes boundary farther east, along the Wzgórza Strzelińskie, where the ?Devonian quartzites have been folded together with the metagranitoidic Cadomian bedrock (Fig. 1 in Chap. VI.A; Fig. 2), matching the dated and well recognized situation in the Moravo-Silesian Zone, farther to the southeast (Jeseniký Mts).

No major structure has been recognized in the NZ metapelites, but they contain records of early vertical folds with biotite axial planar foliation involved in later open recumbent folds, and then upright kinking, altogether trending (Fig. 1) in the NNE-SSW direction (Dziedzic H 1985). The poorly exposed schists and gneisses between the NZ and the Moravo-Silesian thrust zone are otherwise structurally complex, having two to three sets of various (NE-, NW-trending) synmetamorphic (staurolite, almandine) folds (Oberc-Dziedzic 1988b) and zonally penetrative evidence for ductile shearing. There are indications for the top-to-N/NE oblique thrust and/or dextral wrench movements, and for complex kinematics of later, more brittle deformations, partly involving a low-angle extensional faulting.

4 Conclusions

1. It is at present very difficult to find a common denominator for the Lugian structure that has been studied with various methods at different times.

2. The Late Proterozoic deformation of the Lusatian graywacke brought about the W-trending open folds, verging to both S and N.

3. In the Pt -?Pz rock complexes (Caledonian of Don 1990) the early folds were recumbent, tight to isoclinal, and often accompanied by localized, yet penetrative thrusting. The folds became generally overprinted by the broadly coaxial, though not coplanar, roughly upright folds accompanied by a syn- to posttectonic metamorphic climax at P-T conditions of the amphibolite facies. Those early folds had WNW trends, but they apparently verged to the south in the IKB and to the north in the GSB, where the associated thrusts fetched up the lower crust granulite slivers. However, in the OSD, the eastern part of the FSB, and the eastern part of the IKB, the early folds were mostly E-verging. Yet, the superposed folds in these areas all trended generally in the WNW/NW direction. The reasons for such abrupt changes in the early structural trends are poorly recognized and understood, yet they most likely point to the presence of important tectonic breaks, being sealed or obscured at first in Late Devonian-Early Carboniferous and then once again in post-Visean

times, now locally manifesting in encountered fragments of large, ductile shear zones.

4. The Paleozoic rock complexes (Variscan of Don 1990) were apparently deformed in much more consistent manner, since they all have been folded around the W-E axes. These folds have commonly the southern vergence, with the exception for the GK, where the sense of an early tectonic motion remains unclear, but the N-S transport direction is a rule. The S/SW-vergent, post-Visean tectonic transport (often thrusting) is in common evidence. This south vergent structures are apparently in conflict with the generally N-driving Variscan orogenic fronts, found farther west in the Rhenohercynicum and Saxothuringicum.

5. The Lugicum, and in particular the West Sudetes, may be viewed as a Hercynian tectonic stack of Cadomian through Variscan elements. The Lugian structure, with its eastern part being most likely controlled by the presence of the Brunnovistulian (Moravian) indenter, developed due to complex response to the N-S overall compression repeatedly persistent throughout its tectonic history.

Acknowledgments. Professor J. Don is gratefully acknowledged for critically reading the manuscript.

References[2]

Aleksandrowski P (1990) Early Carboniferous strike-slip displacements at the northeast periphery of the Variscan belt in central Europe. In: Int Conf on Paleozoic orogens in central Europe, Aug–Sept 1990, Göttingen-Giessen, JGCP 233

Bagiński B (1990) Metamorfizm kontaktowy skał typu amfibolitów jako wynik termicznego oddziaływania intruzji kłodzko-złotostockiej. In: W kręgu zainteresowań Kazimierza Smulikowskiego, sesja 11–12 października 1990, Warszawa, Polskie Towarzystwo Mineralogiczne, Polska Akademia Nauk

Bankwitz P (1988) Lausitzer Antiklinalzone und Görlitzer Synklinorium. In: Bankwitz E (ed) Klassische Geologische Gebiete in Mitteleuropa, Exkursionfürher. AdW DDR, Potsdam

Bankwitz P, Bankwitz E (1988) Intensität der Deformation in der Saxothuringischen Zone, einschliesslich der Lausitz. Z Geol Wiss 16: 373–392

Baranowski Z, Haydukiewicz A, Kryza R, Lorenc S, Muszyński A, Solecki A, Urbanek Z (1990) Outline of the geology of the Góry Kaczawskie (Sudetes, Poland). Neues Jahrb Geol Paläontol Abh 179: 223–257

Bederke E (1924) Das Devon in Schlesien und das Alter der Sudetenfaltung. Fortschr Geol Paläontol 7: 1–50

Bederke E (1929) Die Grenze von Ost- und Westsudeten und ihre Bedeutung für die Einordnung der Sudeten in den Gebirgsbau Mitteleuropas. Geol Rdsch 20: 186–205

Bederke E (1934) Sudeten und Eulengneisproblem. Veröff Schles Ges Erdkd 21: 351–366

Behr HJ, Engel W, Franke W (1982) Variscan wildflysch and nappe tectonics in the Saxothuringian zone (north-east Bavaria, W Germany). Am J Sci 282: 1438–1470

Borkowska M, Hameurt J, Vidal P (1980) Origin and age of Izera gneisses and Rumburk granites in the Western Sudetes. Acta Geol Pol 30: 121–145

Borkowska M, Choukroune P, Hameurt J, Martineau F (1990) A geochemical investigation of the age, significance and structural evolution of the Caledonian-Variscan granite-gneisses of the Śnieżnik metamorphic area (Central Sudetes, Poland). Geol Sudet 25: 1–27

Brause H (1975) Regionalbau Lausitz-Dolny Śląsk I. Anteil DDR Görlitzer Schiefergebirge. In: Exkursionsführer 22. Jahrestagung GGW Teil B, 23–27 Sept 1975. Gesellschaft für Geologische Wissenschaften der DDR, Dresden-Görlitz, pp 5–24

Brause H, Hirschmann G (1969) Überblick über die stratigraphische Entwicklung des Görlitzer Schiefergebirges. In: Exkursionsführer DGGW, Alt- und Vorpaläozoikum des Görlitzer Schiefergebirges und der westlichsten Westsudeten, 7–10 Sept 1969. Deutsche Gesellschaft für Geologische Wissenschaften, Berlin, pp 23–35

Cháb J, Vrána S (1979) Crossite-actinolite, amphiboles of the Krkonoše-Jizera crystalline complex and their geological significance. Věstn Ústřed Ústavu Geol 54: 143–150

Chaloupský J (1989) Geology of the Krknoše and Jizerské hory Mts. Academia, Praha*

Cloos H (1922) Der Gebirgsbau Schlesiens und die Stellung seiner Bodenschätze. Borntraeger, Berlin

Cwojdziński S (1976) Meso- and macrostructural observations from the northern part of the Złoty Stok-Skrzynka tectonic zone. Kwartalnik Geol 20: 485–501*

Cwojdziński S (1980) Geologic evolution of South-Western Poland interpreted in terms of plate tectonics. Geol Sudet 15: 43–66*

Cymerman Z (1991) Ductile thrusting in the Śnieżnik terrain, the Sudetes; an axample of the Variscan orogeny. In: Moravian windows, Geol Worksh 26–30 April 1991. Moravský Krumlov

Cymerman Z, Jerzmański J (1987) Metamorphic rocks of eastern part of the Presudetic Block in the region of Niedźwiedź, near Ziębice. Kwartalnik Geol 31: 239–262*

Don J (1964) The Złote and Krowiarki Mts as structural elements of the Śnieżnik metamorphic massif. Geol Sudet 1: 79–117*

[2] Asterisk indicates paper with only abstract written in English.

Don J (1984) The Caledonides and Variscides in the Western Sudetes. Przegl Geol 382: 459–469*
Don J (1989) On the position of the eclogites within the gneisses of Międzygórze (Śnieżnik metamorphic massif, Sudetes). In: Żelaźniewicz A (ed) Indicators of structural evolution in polyphase deformed metamorphic complexes. Field Worksh in Poland, September 1989. Wrocław, Polska Akademia Nauk Instytut Nauk Geologiczorych
Don J (1990) The differences in Paleozoic facies-structural evolution of the West Sudetes. Neues Jahrb Geol Paläontol Abh 179: 307–328
Don J, Dumicz M, Wojciechowska I, Żelaźniewicz A (1990) Lithology and tectonics of the Orlica-Śnieżnik dome, Sudetes. Recent state of knowledge. Neues Jahrb Geol Paläontol Abh 179: 121–127
Dumicz M (1964) Geology of the crystalline massif of Bystrzyckie Mts. Geol Sudet 1: 169–208*
Dumicz M (1979) Tectogenesis of the metamorphosed series of the Kłodzko District: a tentative explanation. Geol Sudet 14 (2): 29–46*
Dumicz M (1988) Złoty Stok-Skrzynka structural element in the light of mesostructural analysis of Lądek-Śnieżnik. Geol Sudet 23 (2): 83–106*
Dziedzic H (1985) Variscan rejuvenation of the Precambrian gneisses along the eastern margin of the Góry Sowie Massif, Fore-Sudetic block. Krystalinikum 18: 7–17
Dziedzic K, Teisseyre AK (1990) The Hercynian molasse and younger deposits in the Intra-Sudetic Depression, SW Poland. Neues Jahrb Geol Paläontol Abh 179: 285–305
Dziemiańczuk E, Dziemiańczuk K (1985) On the origin of L_2 lineation in mica schists of western part of the Kamienica Belt (the Izera Block). Geol Sudet 20 (1): 227–273*
Ebert H (1943) Das granitische Grundgebirge der östlichen Lausitz. Preisschr Fürstl Jablonowsk Ges Leipzig 58
Fajst M (1976) A new unconformity in the Precambrian of the Bohemian Massif. Čas Mineral Geol 21: 257–275*
Finckh L (1925) Erläuterungen zur geologischen Karte von Preussen, Blatt Lauterbach. Preuss Geol Landesanstalt, Berlin
Grocholski W (1967) Structure of the Sowie Mts. Geol Sudet 3: 181–249*
Haydukiewicz A (1987a) Sekwencja stratygraficzna w kompleksie kaczawskim. In: Oberc J (ed) Przewodnik 58 Zjazdu Polskiego Towarzystwa Geologicznego, 17–19 wrzesień 1987. Wałbrzych, Instytut Geologiczny & Uniwersytet Wrocławski, Kraków
Haydukiewicz A (1987b) Melanże Gór Kaczawskich. In: Oberc J (ed) Przewodnik 58 Zjazdu Polskiego Towarzystwa Geologicznego, 17–19 wrzesień 1987. Wałbrzych, Instytut Geologiczny & Uniwersytet Wrocławski, Kraków
Hirschmann G (1962) Zur Frage des Seidenberger Granodiorits. Geologie 11: 620–621
Hirschmann G (1966) Assyntische und variszische Baueinheiten im Grundgebirge der Oberlausitz (unter spezieller Berücksichtigung der Geologie des östlichen Görlitzer Schiefergebirges. Freiberger Forschungs C 212: 1–146
Kasza L (1964) Geology of the upper basin of Biała Lądecka stream. Geol Sudet 1: 119–167*
Kemnitz H, Budzinski G (1991) Beitrag zur Lithostratigraphie und Genese der Lausitzer Grauwacken. Z Geol Wiss 19: 433–441
Kossmat F (1927) Gliedenrung des varistischen Gebirgsbaues. Abh Sächs Geol Landesamts 1: 1–39
Kodym O, Svoboda J (1948) The Caledonian nappe structure of Krkonoše and Jizerské Horý. Sb Štatniho Geol Ústavu ČSR 15: 109–180*
Kryza R (1981) Migmatization in gneisses of northern part of the Sowie Góry. Geol Sudet 16 (1): 7–100*
Máška M (1954) K tektonicke analyze krystalinika. Knihovna Ústřed Ústavu Geol 27: 1–258
Matte Ph, Maluski H, Rajlich P, Franke W (1990) Terrane boundaries in the Bohemian Massif: result of large-scale Variscan shearing. Tectonophysics 177: 151–170
Michniewicz M (1981) Early stages of the tectogenesis of Sudetes Mts as interpreted in the light of a derivative theory of deep diapirism and the concept of deep fractures. Geol Sudet 16 (2): 75–141*
Möbus G (1970) Anteile aus zwei Orogenen im homogenen Lausitzer Granodiorit. Ber Dtsch Ges Geol Wiss A 15: 289–304
Oberc J (1966) Geology of crystalline rocks of the Wzgórza Strzelińskie Hills, Lower Silesia. Stud Geol Pol 20: 1–187*
Oberc J (1967) Dispersion of B-lineation in the Izera Mts crystalline massif (Sudetes Mts). Rocz Pol Tow Geol 37: 373–383*
Oberc J (1972) Budowa geologiczna Polski, Tektonika, Sudety i obszary przyległe. Wydawnictwa Geol, Warszawa
Oberc J (1977) Besteht ein kaledonisches Tektogen in Südpolen? Neues Jahrb Geol Paläontol Monatsh 1: 56–63
Oberc-Dziedzic T (1988a) The development of gneisses and granites in eastern part of the Izera crystalline unit in the light of textural investigations. Acta Univ Wratislaviensis 997, Prace Geol Mineral 13: 1–181*
Oberc-Dziedzic T (1988b) Serie metamorficzne Wzgórz Strzelińskich i historia ich przeobrażeń. In: Oberc-Dziedzic T (ed) Budowa, rozwój i surowce skalne krystaliniku strzelińskiego, sesja 14–15 października 1988. Wrocław Instytut Nauk Geologicznych Uniwersytetn Wrocławskiego
Pauk F (1953) Poznamký ke geologii Orlických hor a Kralického Snežniku. Věstn Ústřed Ústavu Geol 28: 193–212*
Pin Ch, Majerowicz A, Wojciechowska I (1988) Upper Paleozoic oceanic crust in the Polish Sudetes: Nd-Sr isotope and trace element evidence. Lithos 21: 195–209
Porębski SJ (1990) Onset of coarse clastic sedimentation in the Variscan realm of the Sudetes (SW Poland): an example from the upper Devonian-lower Carboniferous Świebodzice succession. Neues Jahrb Geol Paläontol Abh 179: 259–274
Quenardel JM, Brochwicz-Lewiński W, Chorowska M, Cymerman Z, Grocholski A, Kossowska I, Piqué A,

Ploquin A, Santallier D (1988) The Polish Sudetes: a mosaic of Variscan terranes. Trabajos Geol Univ Oviedo 17: 139–144

Quitzow HW (1939) Der geologische Bau des nordöstlichen Bober-Katzbach-Gebirges und der anschliessenden Teile des Sudetenvorlandes. Jahrb Preuss Geol Landesanst 59: 558–586

Rajlich P, Synek J (1987) A cross section through the Moldanubian of the Bohemian Massif and the structural development in its ductile domains. Neues Jahrb Geol Paläontol Monatsh 11: 689–698

Scheumann KH (1937) Zur Frage nach den Vorkommen von Kulm in der Nimptscher Kristallinzone. Mineral Petrogr Mitt 49: 216–241

Schust F (1990) Radiometrische Altersdaten. In: Brause H (ed) Exkursionsführer Oberproterozoikum und Känozoikum der Lausitz. Gesellschaft für Geologische Wissenscheften, Berlin

Schwarzbach M (1989) Die Tektonik der Bober-Katzbach-Gebirges. Jahresber Schles Ges Cult Naturwiss 193: 1–52

Skácel J (1989) On the Lugicum-Silesicum boundary. Acta Univ Wratislaviensis 1113: 45–55*

Suess FE (1926) Intrusionstekonik und Wandertektonik im variszischen Grundgebirge. Borntraeger, Leipzig

Suess FE (1935) Der lugische Bau in seinem Verhältnis zur variszischen Orogene. Mitt Geol Ges 28: 1–36

Teisseyre H (1956) Świebodzice Depression as a geological unit. Biul Inst Geol 106: 5–60*

Teisseyre H (1963) The Bolków-Wojcieszów anticline – a representative Caledonian structure in the Western Sudetes. Prace Inst Geol 30: 279–300*

Teisseyre H (1968a) On the stratigraphy and structural evolution of the metamorphic series in the Sudetes. Geol Sudet 4: 7–45*

Teisseyre H (1968b) Stratigraphy and tectonics of the Świebodzice Depression. Biul Inst Geol 222: 77–106*

Teisseyre H (1973) Geology of the Śnieżnik Mountain Group. In: Smulikowski K (ed) Revue des problèmes géologiques des zones profondes de l'écorce terrestre en Basse Silésie. Session 6–16 September. Warszawa, Institut Des Sciences Géologiques De L'Academie Polonaise des Sciences

Teisseyre JH (1962) Structural features and development of the Struga element in the Middle Sudetes. Acta Geol Pol 12: 393–430*

Teisseyre JH (1973) Metamorphic rocks of the Rudawy Janowickie and Lasocki Grzbiet ranges. Geol Sudet 8: 7–118*

Thomas U (1990) Unterkarbone Wildflysch-Ablagerungen in Südteil der DDR. Z Angew Geol 36 (5): 182–184

Van Breemen O, Bowes DR, Aftalion N, Żelaźniewicz A (1988) Devonian tectonothermal activity in the Góry Sowie gneissic block, Sudetes, southwestern Poland: evidence from Rb-Sr and U-Pb isotopic studies. Ann Soc Geol Pol 58: 3–19

Wajsprych B (1986) Sedimentary record of tectonic activity on a Devonian-Carboniferous continental margin, Sudetes. In: Teisseyre AK (ed) IAS 7-th Regional Meet, 23–25 May 1986. Excursion Guidebook. Kraków, Ossolineum, Wrocław

Wojciechowska I (1975) Tectonics of the Kłodzko-Złoty Stok granitoid massif and its country rocks in the light of mesostructural investigations. Geol Sudet 10 (2): 61–121*

Żelaźniewicz A (1978) Major tectonic structures in the Polish part of the Góry Orlickie. Geol Sudet 13 (1): 67–86*

Żelaźniewicz A (1984) Synmetamorphic penetrative mylonitization in orthogneisses of the Bystrzyca Mts, Sudetes. Acta Geol Pol 34: 111–130

Żelaźniewicz A (1985) Granulitic inliers amidst a gneissic-magmatic complex of the Owl Mts, Sudetes. Acta Geol Pol 35: 157–171

Żelaźniewicz A (1987) Tectonic and metamorphic evolution of the Góry Sowie, Sudetes Mts, SW Poland. Ann Soc Geol Pol 57: 203–348*

Żelaźniewicz A (1988) Orthogneises due to irrotational extension, a case from the Sudetes, Bohemian Massif. Geol Rdsch 77: 671–682

Żelaźniewicz A (1990a) Deformation and metamorphism in the Góry Sowie gneiss complex, Sudetes, SW Poland. Neues Jahrb Geol Paläontol Abh 179: 129–157

Żelaźniewicz A (1990b) Gneisses associated with ophiolite during Late Devonian-Early Carboniferous uplife, Góry Sowie, SW Poland. In: Int Conf on Paleozoic orogens in central Europe, Aug-Sept 1990. Göttingen-Giessen, JGCP 233

Zoubek V (1988) Variscan belt. Conclusions, correlations. In: Zoubek V (ed) Precambrian in younger fold belts. Wiley, Chichester, pp 575–605

VI.D Igneous Activity

R. Kryza

1 Introduction

Pre-Variscan igneous activity in the Western Sudetes is recorded mainly as (1) mutually coexisting orthogneisses and granites within medium and high-grade metamorphic complexes; and (2) basic and bimodal volcanic series within poorly dated ?Proterozoic-Paleozoic metamorphic sequences, or within less metamorphosed and better documented Paleozoic successions (cf. Fig. 1 in Chap. VI.A). At present it is difficult to integrate information from different parts of the Sudetic mosaic into one model of the pre-Variscan magmatism for this region. Large-scale lateral displacements may have occurred, and so different elements of the mosaic have significantly different histories (Don 1990).

Variscan magmatism is represented by widespread granitic plutonism, and contemporaneous Carboniferous and Early Permian bimodal volcanism.

2 Pre-Variscan Igneous Rocks

2.1 Granitoids and Orthogneisses

2.1.1 Lusatian Anticlinal Zone (LAZ)

The oldest recognized two-mica granodiorite of the central and SW part of the LAZ (cf. Fig. 1 in Chap. VI.A) is intimately associated with ?Upper Riphean amphibolite facies rocks (e.g., Stolpen complex) and metatexites. It is interpreted as derived from graywacke-like metasediments by anatexis at relatively shallow crustal levels, and solidified in situ or intruded over small distances only (Förster et al. 1990).

A late Cadomian and pre-Late Ordovician age for the West-Lusatian biotite S-type granodiorite (b in Fig. 1) is constrained by thermal contacts with Lusatian graywackes of Vendian age, and no contact effects on adjacent lower Paleozoic sequences. This granodiorite has intruded the East-Lusatian (Zawidów) granodiorite (a in Fig. 1), which is locally strongly deformed, indicating the earlier emplacement of the latter. Radiometric data (Table 1) is consistent with an uppermost Proterozoic or Cambrian age for both granodiorites and also indicate their Variscan rejuvenation (Brause 1990).

2.1.2 Izera-Karkonosze Block (IKB)

The gneisses and granites of the Góry Izerskie area (c, d in Fig. 1), on the northern flank of the IKB, are texturally variable but display a simple mineral composition: quartz, plagioclase (An ~25), K-feldspar, chessboard albite, partly chloritized biotite, and white mica. Pinite pseudomorphs after cordierite, garnet, and sillimanite are locally found in the granites, including the coarse-grained Rumburk granite (Kozłowska-Koch 1965; Oberc-Dziedzic 1988a).

There was much controversy about an igneous versus metasomatic origin for the Izera granites and gneisses (cf. Oberc-Dziedzic 1988a). Borkowska et al. (1980) claim that the gneisses derived from an anorogenic intrusive granitic complex, related to continental rifting around 450–500 Ma ago (Table 1). However, Oberc-Dziedzic (1988a) argues that the dates are underestimated, and the complex represents a polygenetic assemblage metamorphosed mainly during the Cadomian orogeny.

The Kowary and Karkonosze gneisses in the southern part of IKB, of assumed pre-Late Ordovician age (Chaloupsky 1989), are very similar to the Izera gneisses.

2.1.3 Orlica-Śnieżnik Dome (OSD)

The internal part of the OSD is composed of the Śnieżnik gneisses (f in Fig. 1), the Gierałtów

Dallmeyer et al. (Eds.)
Pre-Permian Geology of Central and Eastern Europe

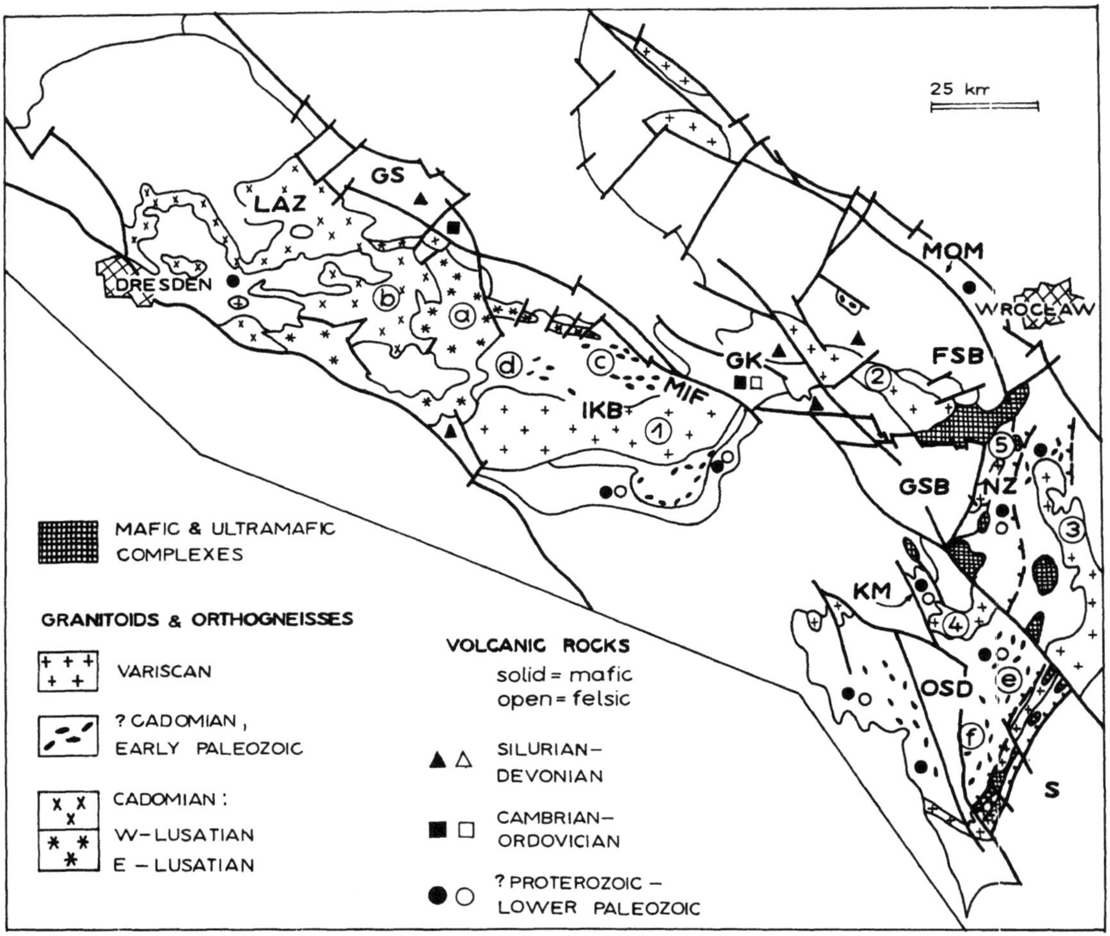

Fig. 1. Distribution of pre-Variscan and Variscan igneous and igneous-derived rocks in the Western Sudetes. *FSB* Fore-Sudetic Block; *GK* Góry Kaczawskie; *GS* Görlitzer Schiefergebirge; *GSB* Góry Sowie Block; *IKB* Izera-Karkonosze Block; *KM* Kłodzko Metamorphics; *LAZ* Lusatian Anticlinal Zone; *MIF* Main Intra-Sudetic Fault; *MOM* Middle Odra Metamorphics; *NZ* Niemcza Zone; *OSD* Orlica-Śnieżnik Dome; *S* Silesikum; *a–f* Cadomian and early Paleozoic granitoids and orthogneisses; *1–5* Variscan granitoids (see Tables 1, 2)

Table 1. Selected isotopic ages and $^{87}Sr/^{86}Sr$ initial ratios of pre-Variscan granites and gneisses of the Western Sudetes

Symbol in Fig. 1	Rocks	U-Pb and Pb-Pb ages	Rb-Sr age	Initial 87Sr/86Sr	Reference
a	E-Lusatian granodiorite	600 Zr	370 WR		1
		470–570 Fsp			1
b	W-Lusatian granodiorite	580 Zr	590 WR		1
		470–590 Fsp	530 WR		1
c	Izera gneiss		462 WR	0.7046	2
d	Rumburk granite		501 WR	0.7090	2
e	Gierałtów gneiss		464 WR	0.715	3
f	Śnieżnik gneiss		487 WR		4
			395 WR	0.713	3
			335 WR	0.7158	3

Fsp = feldspar, Mi = mica, WR = whole rock, Zr = zircon; References: 1 – from compilation by Schust, in Brause 1990; 2 – Borkowska et al. 1980; 3 – Borkowska et al. 1990; 4 – van Breemen et al. 1982.

migmatic gneisses (e in Fig. 1) with minor granulites and eclogites, and the supracrustal Młynowiec-Stronie group of assumed Proterozoic-Middle Cambrian age (cf. Chap. VI.B).

The Gierałtów gneisses are finely layered to massive rocks of aplitic appearance, composed of quartz, plagioclase (An 4-10), K-feldspar, two micas, and scarce garnet. The Śnieżnik gneisses are typically augen gneisses composed of perthitic K-feldspar, plagioclase (An 2-25), quartz, biotite, and muscovite, and rare spongy garnet (Smulikowski 1979).

Fisher (1935, vide Don et al. 1990) considered the Śnieżnik gneisses as orthogneisses derived from Caledonian porphyritic granites. In contrast, Smulikowski (1979) argued that both the Śnieżnik and Gierałtów gneisses represented different products of the metasomatic granitization of a metasedimentary sequence. Don (in Don et al. 1990), using structural criteria, suggests that the Gierałtów gneisses are products of migmatic mobilization postdating the presumably Ordovician Śnieżnik orthogneisses.

Borkowska et al. (1990) give petrological evidence that both gneisses are of igneous and continental-crustal derivation. The Gierałtów gneisses are alkaline, while those of Śnieżnik are of calc-alkaline affinity. Both are metaaluminous. The emplacement of the Gierałtów gneisses is thought to have been connected with a continental crustal attenuation episode, around 465 Ma ago. Accordingly, the Śnieżnik granite was emplaced around 395 Ma ago, in relation to crustal thickening, and the whole assemblage underwent Variscan metamorphism at about 335 Ma ago (Table 1). This model, however, is in disagreement with the Rb-Sr whole rock date of 487 Ma for the Śnieżnik gneiss by van Breemen et al. (1982).

2.1.4 Fore-Sudetic Block (FSB)

In the eastern part of the FSB, Oberc-Dziedzic (1988b) distinguished several varieties of gneisses within the metamorphic cover of the Strzelin granite (3 in Fig. 1). At least two of them, the porphyritic Strzelin gneiss and the Gościęcice augen gneiss, are of probable magmatic origin, as perhaps also the Doboszowice gneiss, east of the Niemcza Zone.

In the central part of the FSB, the coarse-grained porphyritic Wądroże Wielkie gneisses protrude a low-grade Paleozoic complex. They closely resemble the Izera gneisses and bear locally distinct mylonitic foliation (Kozłowska-Koch 1965).

2.2 Igneous Rocks in ?Proterozoic-Paleozoic Sequences

2.2.1 Izera-Karkonosze Block (IKB)

The Leszczyniec volcanic formation (LVF) in the eastern margin of the IKB consists of two main rock assemblages (Teisseyre 1973): (1) mafic lavas and shallow intrusives with minor felsic volcanic and sedimentary rocks; and (2) intermediate to felsic intrusives, clearly cross cutting the volcanic rocks, and ranging from diorites to albite-rich granites (Paczyn gneisses). The mafic rocks compare well geochemically with recent IAT. This chemical affinity, together with a record of HP metamorphism, were interpreted by Narębski (1990) as indicators of Silurian subduction. However, a Proterozoic age for the LVF has been argued (cf. refs. in Mierzejewski and Oberc-Dziedzic 1990).

The Zelezny Brod volcanic complex (ZBVC) on the southern edge of the IKB, of assumed Silurian-Early Devonian or Cambrian age (cf. Chap. VI.B), consists of low-grade metamorphic mafic lavas and volcaniclastics, minor diabase sills, and intercalations of acidic volcanic rocks. Trace element characteristics (Chaloupsky 1989) show inhomogenity within the complex. The basic sills in the lower part of the sequence are geochemically coherent and bear features of WPB. The remaining mafic rocks are more scattered – some of them are similar to the diabases, while others plot in the IAT field, like the metabasalts of LVF.

Mostly mafic volcanic rocks (diabases) of Late Devonian age are known from the low – grade metamorphic Jitrava Group in the Jested Mts, in the SW corner of the IKB (Chaloupski 1989).

2.2.2 Orlica-Śnieżnik Dome (OSD) and Kłodzko Metamorphics (KM)

The supracrustal Stronie formation of the OSD contains interlayers of variable laminated to massive amphibolites composed of hornblende, biotite, chlorite, epidote, zoisite, plagioclase, and locally clinopyroxene and garnet (Smulikowski 1979). Thin intercalations of two-feldspar schists are interpreted as acid metavolcanic rocks. Trace element and Nd-isotopic ratios in the basic metavolcanic rocks (Wojciechowska et al. 1989) indicate their considerable variability, from amphibolites resembling T-MORB, to metagabbros similar to rocks from suprasubduction or transform-fault regimes.

Low- to medium-grade complexes surround the core of the OSD at the SW, S, and SE (Nove Mesto, Zabreh, and Stare Mesto complexes, respectively). Basic and minor acidic metavolcanic rocks are widespread within all these sequences (Misar 1983). The metabasic rocks of the Nove Mesto complex, associated with psammitic-pelitic metasediments, have major and trace element characteristics comparable with recent OFT (Opletal et al. 1980).

The KM contains a several-hundred-meter-thick mafic effusive complex, with some metarhyolitic rocks and granitoids (mylonitic Ścinawka gneisses). Towards the SW, voluminous metagabbroic rocks occur. The rocks are thought to be Late Silurian or Early Devonian in age (Wojciechowska 1990).

The mafic rocks are geochemically diverse (Narębski et al. 1989). The metabasalts in the NE are of within-plate character and may represent an initial rift environment. The metagabbros are more variable and compare either with T-MORB or IAB. However, the considerable geochemical differences between the volcanic and gabbroic members make their genetic relations uncertain.

2.2.3 Fore-Sudetic Block (FSB)

Amphibolites and amphibole-biotite schists occur within mica schists of assumed Late Proterozoic age which, together with gneisses, mantle the Strzelin granites (3 in Fig. 1; Oberc-Dziedzic 1988b). A similar complex of paragneisses, mica schists, and minor marbles and quartzites, with intercalations of amphibolite (up to 35 m thick), was recognized in boreholes along the NE periphery of the FSB (Middle Odra Metamorphics sensu Oberc 1968).

Metabasic rocks with alkali-basaltic major element characteristics are known from the Niemcza-Kamieniec metamorphic suite, east of the Niemcza Zone. They are also associated with mica schists, minor quartzites and crystalline limestones of assumed Vendian-Cambrian age (cf. Chap. VI.B). Interlayers of felsic rocks of rhyolitic composition were controversially interpreted as metaarkoses, metavolcanic rocks, or mylonitized orthogneisses in the northern prolongation of the Ramzová overthrust (cf. Dziedzic 1985).

The large (ca. 3800-m-thick) amphibolite massif of Niedźwiedź, in the SE part of FSB, is formed of medium- to fine-grained, mostly streaky amphibolites composed of hornblende and plagioclase, and occasional garnet, epidote, quartz, and locally pyroxene. Amphibole schists and gneisses, quartzo-feldspatic rocks and crystalline limestones occur as minor intercalations. The age and position of the massif are problematic, though it was tentatively correlated with the Jesenik and Sobotin massifs of the Silesicum (Cymerman, Jerzmański 1988, vide Jerzmański 1989).

2.3 Igneous Rocks in Paleozoic Sequences

2.3.1 Görlitzer Schiefergebirge (GS) and Lusatian Anticlinal Zone (LAZ)

The older volcanic episode in the GS, documented as Early Cambrian (Hirschmann and Brause 1969), produced basic lavas (diabases) and volcaniclastic rocks, associated with clastic and carbonate sediments. The younger episode started in the Late Silurian, again with basic volcanic rocks, hosted by graptolitic slates and cherts. Basic rocks attain their maximum thickness (>100 m) in the Lower Devonian, and appear again in the Upper Devonian psammitic and pelitic sequence (Hirschimann and Brause 1969). The volcanic rocks display major element characteristics (Rösler and Werner 1979) typical of WPB or transitional between WPB and MORB.

In Upper Lusatia, numerous dykes, up to 100 m thick, of gabbro-norite, olivine gabbro and diorite, follow W-E- and N-S-trending deep extensional faults (Kramer 1989). Their trace element characteristics are inconclusive. They postdate the Lusatian granodiorite, but are older than late Variscan stock granites (Kramer 1989).

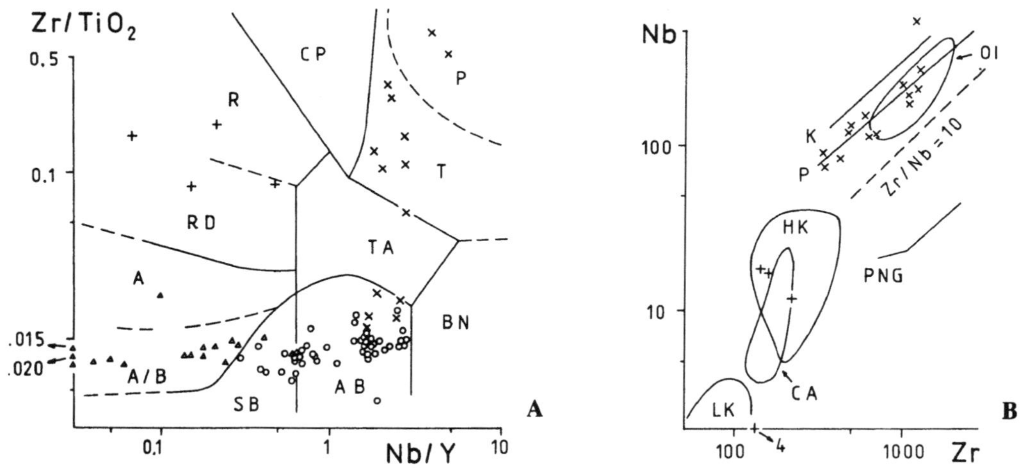

Fig. 2. A The Kaczawa Mts volcanics plotted on Zr/Ti – Nb/Y diagram. Lower part of the succession (?Cambrian-Ordovician): alkaline and transitional tholeiitic-alkaline WP basalts (*circles*), rhyodacitic volcanics (+) and alkaline bimodal lavas (×). Upper part of the succession (?Silurian): transitional WP-to-MOR, E-MOR and N-MOR basalts (*triangles*). *A* Andesite; *AB* alkali-basalt; *A/B* andesite/basalt; *BN* basanite, nephelinite; *CP* comendite, pantellerite; *PH* phonolite; *R* rhyolite; *RD* rhyodacite, dacite; *SB* subalk. basalt; *T* trachyte; *TA* trachyandesite. (After Furnes et al. 1990). **B** The Kaczawa Mts volcanics plotted on Nb-Zr diagram. *CA* Calc-alkaline; *HK* high-K; *LK* low-K, and *OI* ocean island lavas. *Continental rift volcanics*: *K* Kilombe, Kenya; *P* Pantelleria; subduction-related peralkaline rocks: *PNG* Papau New Guinea. The rhyodacitic volcanics (+) resemble CA lavas, while the alkaline bimodal rocks (×) follow recent continental rift volcanics. (After Furnes et al. 1990)

2.3.2 Góry Kaczawskie (GK) and Fore-Sudetic Block (FSB)

Volcanic activity is recorded mainly in the lower to middle part of the low-grade ?Cambrian to Lower Carboniferous Kaczawa succession. The ?Cambrian to Ordovician sequence is dominated by a composite volcanic assemblage comprising thick pillow lavas of transitional tholeiitic-alkaline within-plate affinity, bimodal alkaline lavas and shallow intrusions, and minor but significant felsic volcanogenic rocks of continental-crust derivation (Narębski et al. 1986; Furnes et al. 1990; Fig. 2). These volcanics are associated with gray slates and crystalline limestones.

The higher part of the complex is represented by a thick sequence of pillow lavas displaying E-MORB to N-MORB characteristics (Furnes et al. 1990), locally interbedded with Silurian black graptolitic slates and cherts.

The Kaczawa succession has been interpreted (Baranowski et al. 1990; Furnes et al. 1990) as representing a record of volcanism and sedimentation in a Paleozoic basin evolving from an initial rift regime (?Cambrian-Ordovician), through a mature basin with more evolved rifting, perhaps of MOR-type (Silurian-Devonian), to a collisional stage resulting in an accretionary prism (?Late Devonian-Early Carboniferous).

The western and central part of the FSB is composed of low grade volcanic-sedimentary rocks which are probably a continuation of the Kaczawa complex. Diabases and massive greenstones associated with grey slates crop out in the area north of the Strzegom granite (2 in Fig. 1). Similar lithologies were reported from boreholes, where their age was thought to be Early to Mid Devonian (cf. Chap. VI.B).

2.4 Paleozoic Ophiolites and Other Ultramafic Bodies

2.4.1 Circum Góry Sowie Block (GSB) Ophiolites

The GSB is surrounded on the NE, E, and S by ultramafic-mafic complexes locally displaying a nearly complete ophiolitic sequence (Majerowicz 1979, vide 1981, 1990). From gravimetric data (Znosko 1981; Jamrozik 1989), basic rocks are supposed to underlie the NE and E part of the GSB, and have also been reported from boreholes east of the Niemcza Zone (Jerzmański 1989).

According to Majerowicz (1990), the largest Ślęża ophiolite consists, from S to N, of the following members: (1) serpentinized peridotites (metamorphic peridotites) developed mostly out of harzburgites, and cut by gabbroic dykes; (2) ultramafic cumulates (pyroxenites and ?lherzolites) forming a narrow band of compositionally variable and partly serpentinized amphibole-pyroxene rocks; (3) mafic cumulates represented by coarse- to medium-grained, locally layered gabbros; and (4) a volcanic member of metadiabases and metabasalts of variable texture, from medium-grained to aphanitic. No pillow lava has yet been found.

The Nowa Ruda ophiolite, at the SSW side of GSB, from N to S, consists of: (1) metamorphic peridotites (poorly exposed); (2) ultramafic and mafic cumulates (Borkowska 1985); harzburgitic serpentinites, pyroxenites, troctolites, olivine and olivine-free gabbros, and plagioclasites; (3) diabases separated from the plutonic member by a steep shear zone, and with generally southward-fining textures; and (4) pillow lavas with well preserved volcanic features, which indicate the primary bedding dipping normally and moderately to the SW, in accord with the structural younding of the ophiolite.

According to Pin et al. (1988), the metabasites of the Mt. Ślęża and Nowa Ruda ophiolites show N-MORB trace element and Nd-isotopic characteristics and are considered to be products of normal oceanic spreading, or to have formed in a subduction-related marginal basin. The Sm-Nd whole rock isochron indicates an age of 353 ± 21 Ma for the Mt. Ślęża, and 351 ± 16 Ma for the Nowa Ruda ophiolites (Pin et al. 1988).

The other outcrops of mafic-ultramafic rocks southeast of the GSB (Braszowice-Brzeżnica, Szklary) are smaller and less continuous, but have also been interpreted as parts of this ophiolite association (Dziedzicowa 1989).

2.4.2 *Other Ultramafic Bodies*

Ultramafic rocks of upper mantle derivation (20–30 kb, 1100–1300 °C – Bakun-Czubarow 1981) occur as intercalations within the Góry Sowie felsic granulites. They are strongly serpentinized, but despite this, different varieties were classified as harzburgites, pyrope-bearing lherzolites, and minor wehrlites and websterites (Bakun-Czubarow 1981). Other small outcrops of ultramafic and mafic rocks (including hyperites), of diverse origin, are scattered all over GSB.

Ultramafic rocks are also known from several tiny outcrops along the southern and eastern margins of the IKB (Chaloupsky 1989, and references therein; Bakun-Czubarow and Smulikowski 1990), and in the NE part of the OSD (Bakun-Czubarow 1989). Most of them occur close to Variscan granitoids, and are supposed to have been emplaced along deep faults, from the upper mantle (Bakun-Czubarow and Smulikowski 1990).

The position of the mafic and ultramafic rocks of the Stare Mesto complex, situated west of the Myznerov thrust separating OSD from Silesicum, remains controversial. Misar (1983) consider them as part of the Moravo-Silesian ophiolite belt, while Narębski (1990) interprets the ultramafics as subduction-related intrusions of Alaskan type.

3 Variscan Granitoids

The Karkonosze granite (1 in Fig. 1) comprises two main facies: (1) porphyritic, rich in dark schlieren and containing scarce hornblende and relic pyroxene; (2) equigranular, more leucocratic, and of younger isotopic age (Table 2) and higher $^{87}Sr/^{86}Sr$ initial ratio (Duthou et al. 1990).

The Strzegom-Sobótka massif (2 in Fig. 1) is composed of four main facies (Majerowicz 1972): hornblende-biotite monzogranite and biotite monzogranite in the west, two-mica granite in the centre, and biotite granodiorite in the east. The two-mica granite is older (326 Ma) and has a weak foliation. The other facies are younger (~280 Ma). The Sr initial ratios are higher in the western part of the massif (Pin et al. 1988).

The Strzelin-Żulova massif (3 in Fig. 1) is not a single pluton but rather a metamorphic zone densely pierced by stocks and large granitic dykes (Morawski 1973, vide Oberc-Dziedzic 1988a). According to Lorenc (1988), two granitoid associations can be distinguished: (a) peraluminous granites and granodiorites of crustal affinity, and (b) a low-K group of diorites, tonalites, and granodiorites, derived from deeper sources.

The Kłodzko-Złoty Stok massif (4 in Fig. 1) is compositionally variable and consists mainly of monzogranodiorites, granodiorites and quartz monzodiorites. Monzodiorites, tonalites, and granites are minor facies. The hornblende

Table 2. Selected isotopic ages and $^{87}Sr/^{86}Sr$ initial ratios of Variscan Granitoids of the Western Sudetes

No. in Fig. 1	Massifs and rock types	K-Ar age	Rb-Sr age	Initial 87Sr/86Sr	Reference
1	Karkonosze (MG, GD)	302			1
	– porphyritic		328	0.7064	2
	– equigranular		310		3
2	Strzegom-Sobótka	266			4
	– MG (W part)		278	0.7098	5
	– 2-mica G (center)		326	0.705	5
	– GD (E part)		276	0.7058	5
3	Strzelin				
	– G and GD	278–288			6
	– T and D	257–272			6
4	Kłodzko-Złoty Stok (MG, GD, QtzMD)	298			7
5	Niemcza				
	GD	300			7
	QtzS, MD, V	272			7

D = diorite, G = granite, GD = granodiorite, MD = monzodiorite, MG = monzogranite, Qtz = quartz, S = syenite, SD = syenodiorite, T = tonalite, V = vaugnerite; References: 1 – Depciuch and Lis 1971; 2 – Pin et al. 1987; 3 – Duthou et al. 1990; 4 – Depciuch 1971; 5 – Pin et al. 1989; 6 – Depciuch 1972b; 7 – Depciuch 1972a.

contains relic clinopyroxene. The granitoids host numerous enclaves and xenoliths, and are of calc-alkaline metaaluminous character (Lorenc 1990).

The Niemcza granitoids (5 in Fig. 1) comprise mostly medium-grained granodiorites with a distinct magmatic foliation, and minor intrusions of finer-grained quartz syenite, monzodiorite, and vaugnerite. The last three lack any deformation structures. They are interpreted as having crystallized from mantle-derived gabbroic magmas, contaminated by crustal material (Puziewicz 1990).

Smaller granitic bodies of Variscan age are known from the LAZ (the Königshein and Stolpen monzogranites displaying K-Ar biotite ages of 330–350 and 230–290 Ma, respectively; cf. Brause 1990), and from the OSD (the Kudowa granites in the west which gave a ~320 Ma K-Ar biotite age; Przewłocki et al. 1962; and the Javornik and Bielice intrusions in the east). Relatively large granitic massifs have been recognized in boreholes in the NW part of the FSB (Fig. 1).

4 Conclusions

1. The Lusatian granodiorites, displaying the oldest isotopic ages, around 550–600 Ma, can be interpreted as a record of Cadomian plutonism. Younger ages of ca. 450–500 Ma from other orthogneisses (Izera granites, ?Śnieżnik, and some Gierałtów gneisses) may reflect a younger plutonic event or may result from Variscan rejuvenation.
2. Basic and bimodal volcanic suites of assumed Proterozoic and Lower to Mid Paleozoic sequences are conventionally classified into two groups. Group 1 comprises medium grade metamorphic rocks which often mantle gneissic domes (IKB, OSD, the eastern part of FSB). They are biostratigraphically poorly constrained, and their ages are variously assumed as Late-Proterozoic, Cambrian or, in some cases, up to Silurian and Early Devonian (IKB, KM). Their geochemical characteristics are diversified and often ambiguous, and they may represent suites of different age and provenance; if contemporaneous, they probably reflect complex geotectonic settings (Narębski 1990). Group 2 occurs within low-rade or unmetamorphosed and better paleontologically documented sequences of the GK, GS, and Jested Mts areas. The volcanic activity peaked in the Cambrian-Ordovician and continued through to Silurian and, in places, to Late Devonian times. In the GK, a geochemical trend in the volcanic rocks, possibly marking an evolution from early Paleozoic ensialic rifting, to a more evolved MOR-type regime during the Silurian, has been indicated.

3. The circum GSB ophiolite association has a thorough petrological documentation (Majerowicz 1990; Pin et al. 1988), though whether it represents a fragment of Late Devonian oceanic crust is still under discussion (Jamrozik 1989). Whether that ocean was a continuation of the basin recorded in the Kaczawa succession is also speculative. According to Narębski (1990), the ophiolite complexes along the eastern margin of the Western Sudetes represent fragments of oceanic crust of a narrow, Red Sea-type basin which formed by rifting and rapid spreading along the eastern periphery of the Bohemian massif.

4. The Variscan late and post-orogenic magmatism, in the form of intense granitic plutonism and contemporaneous bimodal volcanism, commenced probably 340–350 Ma ago, and continued until around 260 Ma, with at least two intense events around 325–330 and 280 Ma.

Acknowledgments. An earlier version of the manuscript was critically reviewed by J.P. Couturié, C. Pin, J. Zalasiewicz, and A. Żelaźniewicz.

References

Bakun-Czubarow N (1981) Ultrabasic rocks in granulites of the Góry Sowie Mts Block. In: Narębski W (ed) Ophiolites and initialites of northern border of the Bohemian Massif. Wissenschaftlichen Informationzentrums des Bergakademic Freiberg, Potsdam Freiberg, pp 79–86

Bakun-Czubarow N (1989) Ultramafic rocks in the eastern part of the Polish West Sudetes. In: Narębski W, Majerowicz A (eds) Lower and Upper Paleozoic metabasites and ophiolites of the Polish Sudetes. Inst Geol Sci, Wrocław University, Wrocław, May 1989, pp 223–237

Bakun-Czubarow N, Smulikowski W (1990) Mineralogiczno-petrograficzna charakterystyka pnia serpentynitowego na południe od Niedamirowa (Grzbiet Lasocki). In: Smulikowski K Symp, Inst Geol Sci, Pol Acad Sci, Warszawa, Abstr

Baranowski Z, Haydukiewicz A, Kryza R, Lorenc S, Muszyński A, Solecki A, Urbanek Z (1990) Outline of the geology of the Góry Kaczawskie (Sudetes, Poland). Neues Jahrb Geol Paläontol Abh 179: 223–257

Borkowska M (1985) Gabbroic rocks and their minerals from the Nowa Ruda massif. Geol Sudet 20: 3–35

Borkowska M, Hameurt J, Vidal P (1980) Origin and age of Izera gneisses and Rumburk granites in the Western Sudetes. Acta Geol Pol 30: 121–146

Borkowska M, Choukroune P, Hameurt J, Martineau F (1990) A geochemical investigation of the age, significance and structural evolution of the Caledonian-Variscan granite-gneisses of the Śnieżnik metamorphic area (Central Sudetes, Poland). Geol Sudet 25: 1–27

Brause H (ed) (1990) Oberproterozoikum und Känozoikum der Lausitz. Exkursionsführer, Tagung Hoyerswerda 14–17 Nov 1990. Deutsche Gesellschaft für Geologische Wissenschaften, Berlin, 50 pp

Chaloupsky J (1989) Geological evolution of the Krkonose-Jizerské hory crystalline complex in the Precambrian and Early Paleozoic. In: Chaloupsky J (ed) Geologie Krkonos a Jizerskych hor. Ustredni Ustaw Geol, Praha, 288 pp

Depciuch T (1971) Absolute age of the Strzegom granitoids determined by K-Ar method. Kwartalnik Geol 15: 862–869

Depciuch T (1972a) Absolute age (K-Ar) of granitoids from the Kłodzko-Złoty Stok area and the Niemcza zone. Kwartalnik Geol 16: 103–111

Depciuch T (1972b) Absolute age (K-Ar) of granitoids from Strzelin. Kwartalnik Geol 16: 95–102

Depciuch T, Lis J (1971) K-Ar absolute age of the Karkonosze massif granitoids. Kwartalnik Geol 15: 855–860

Don J (1990) The differences in Paleozoic facies-structural evolution of the West Sudetes. Neues Jahrb Geol Paläontol Abh 179: 307–328

Don J, Dumicz M, Wojciechowska I, Żelaźniewicz A (1990) Lithology and tectonics of the Orlica-Śnieżnick Dome, Sudetes – recent state of knowledge. Neues Jahrb Geol Paläontol Abh 179: 159–188

Duthou JL, Mierzejewski MP, Couturié JP, Pin C (1990) Postępy w datowaniu granitów Karkonoszy. In: Smulikowski K Symp Inst Geol Sci, Pol Acad Sci, Warszawa, Abstr

Dziedzic H (1985) Variscan rejuvenation of the Precambrian gneisses along the eastern margin of the Góry Sowie Massif, Fore-Sudetic Block, Krystalinikum 18: 7–27

Dziedzicowa H (1989) The Braszowice gabbro as a member of Variscan ophiolite complex. Acta Univ Wratislaviensis Prace Geol Mineral 17: 57–79

Förster H-J, Tischendorf G, Thomas R (1990) Geochemical signatures of barren and ore-generating granites, examplefied by magmatic rocks of the Lausitz and Erzgebirge. In: Paleozoic orogens in central Europe, IGCP Conf, Project 233, Terranes in the circum-Atlantic Paleozoic orogens, Göttingen-Giessen, Aug–Sept 1990, 4 pp

Furnes H, Kryza R, Muszyński A (1989) Geology and geochemistry of Early Paleozoic volcanics of the Świerzawa Unit, Kaczawa Mts, W Sudetes, Poland. Neues Jahrb Geol Paläontol Monatsh 3: 136–154

Furnes H, Kryza R, Muszyński A, Pin C, Garmann LB (1990) Trace element and Nd isotope geochemistry of the Kaczawa Mts (Polish Sudetes) volcanics: Evidence for Early Paleozoic continental break-up? In: Paleozoic orogens in central Europe, IGCP Conf, Project 233, Terranes in the circum-Atlantic Paleozoic orogens, Göttingen-Giessen, Aug–Sept 1990, 4 pp

Hirschmann G, Brause H (1969) Alt- und Vorpaläozoikum des Görlitzer Schiefergebirges. Exkursionsfürer, Görlitz 7–10 Sept 1969, Deutsche Gesellschaft für Geologische Wissenschaften Berlin, 115 pp

Jamrozik L (1989) Ofiolit Ślęży i jego pozycja tektoniczna. In: Niśkiewicz J, Jamrozik L (eds) Ślęża ophiolite and its ore mineralization. Symp Wrocław-Sobótka, 4-6 Sept 1989, Jnst Geol Sci, Wrocław University, pp 18-30

Jerzmański J (1989) Wyniki rozpoznania geologicznego niektórych anomalii geofizycznych w okolicy masywu Ślęży. Niśkiewicz J, Jamrozik L (eds) Ślęża ophiolite and its ore mineralization. Symp Wrocław-Sobótka, 4-6 Sept 1989, Jnst Geol Sci, Wrocław University, pp 48-52

Kozłowska-Koch M (1965) The granite-gneisses of the Izera Highlands. Arch Mineral 25: 123-259

Kramer W (1989) Early Hercynian gabbroic rocks of the Lusatian anticlinal zone. In: Guidebook of excursions in GDR, May/June 1989. Zentralinstitut für Physik der Erde, Potsdam, pp 44-48

Lorenc MW (1988) Granitoidy Wzgórz Strzelińskich. In: Budowa, rozwój i surowce Skalne krystaliniku strzelińskiego, Symp Wrocław, 14-15 Oct 1988, Inst Geol Sci, Wrocław University, pp 22-27

Lorenc MW (1990) Ewolucja magmowa masywu Kłodzko-Złotostockiego. In: Smulikowski K Symp, Inst Geol Sci, Pol Acad Sci, Warszawa, Abstr

Majerowicz A (1972) Strzegom-Sobótka granitic massif. Geol Sudet 6: 7-88

Majerowicz A (1981) Rock series of the Ślęża Mt group in the light of petrologic studies of ophiolite complexes. In: Narębski W (ed) Ophiolites and initialites of northern border of the Bohemian Massif. Wissenschaftlichen Informationzentrums des Bergakademic Freiberg, Potsdam Freiberg, pp 172-199

Majerowicz A (1990) Some petrological problems of the Ślęża ophiolite complex (Sudetes, Poland). In: Paleozoic orogens in central Europe, IGCP Conf, Project 233, Terranes in the circum-Atlantic Paleozoic orogens, Göttingen-Giessen, Aug-Sept 1990, 4 pp

Mierzejewski MP, Oberc-Dziedzic T (1990) The Izera-Karkonosze Block and its tectonic development (Sudetes, Poland). Neues Jahrb Geol Paläontol Abh 179: 197-222

Misar Z (1983) Lugicka oblast. In: Misar Z, Dudek A, Havlena V, Weiss (eds) Geologie CSRS I. Cesky masiv, Statni Pedagogické Nakladatelstvi, Praha, pp 173-198

Narębski W (1990) Petrotectonic affinities of Lower Silesian metabasites and the problem of origin of Paleozoic Sudetic ophiolites. In: Paleozoic orogens in central Europe, IGCP Conf, Project 233, Terranes in the circum-Atlantic Paleozoic orogens, Göttingen-Giessen, Aug-Sept 1990, 4 pp

Narębski W, Dostal J, Dupuy C (1986) Geochemical characteristics of Lower Paleozoic spilite-keratophyre series in the Western Sudetes (Poland): Petrogenetic and tectonic implications. Neues Jahrb Miner Abh 155: 243-258

Narębski W, Wojciechowska I, Dostal J, Pin C (1989) Geochemistry of metabasic rocks of the Kłodzko Metamorphic Unit: petrogenetic and paleotectonic implications. In: Narębski W, Majerowicz A (eds) Lower and Upper Paleozoic metabasites and ophiolites of the Polish Sudetes. Inst Geol Sci, Wrocław University, Wrocław May 1989, pp 157-174

Oberc J (1968) Blok przedsudecki na wschód od gnejsów sowiogórskich. In: Sokołowski S (ed) Budowa geologiczna Polski, 1, Stratygrafia. Wydawnictwa Geologiczne, Warszawa, pp 88-93

Oberc-Dziedzic T (1988a) The development of gneisses and granites in eastern part of the Izera crystalline unit in the light of textural investigations. Acta Univ Wratislaviensis Prace Geol Mineral 13: 184

Oberc-Dziedzic T (1988b) Serie metamorficzne Wzgórz Strzelińskich i historia ich przeobrażeń. In: Budowa, rozwój i surowce skalne krystaliniku strzelińskiego, Symp Wrocław, 14-15 Oct 1988. Inst Geol Sci, Wrocław University, pp 6-21

Opletal M, Domecka K, Cech S et al. (1980) Geology of the Orlicke hory Mountains. Ustredni Ustav Geol Praha: 202

Pin C, Mierzejewski M, Duthou JL (1987) Rb-Sr isochron age of the Karkonosze granite from the Szklarska Poręba Huta quarry and the determination of initial $^{87}Sr/^{86}Sr$ ratio in this granite. Przegl Geol 10: 512-517

Pin C, Majerowicz A, Wojciechowska I (1988) Upper Paleozoic oceanic crust in the Polish Sudetes: Nd-Sr isotope and trace element evidence. Lithos 21: 195-209

Pin C, Puziewicz J, Duthou JL (1989) Ages ond origins of a composite granitic massif in the Variscan belt: a Rb-Sr study of the Strzegom-Sobótka Massif, W Sudetes (Poland). Neues Jahrb Miner Abh 160: 71-82

Przewłocki K, Magda W, Thomas HH, Faul H (1962) Age of some granitic rocks in Poland. Geochim Cosmochim Acta 26: 1069-1075

Puziewicz J (1990) Post tectonic syenitic/dioritic rocks of the Niemcza zone (Sudetes, SW Poland). In: Paleozoic orogens in Central Europe, IGCP Conf, Project 233, Terranes in the circum-Atlantic Paleozoic orogens, Göttingen-Giessen, Aug-Sept 1990, 4 pp

Rösler HJ, Werner C-D (1979) Petrologie und Geochemie der variszischen Geosynklinalmagmatite Mitteleuropas. Freiberger Forchungsh C 336: 265

Smulikowski K (1979) Polymetamorphic evolution of the crystalline complex of Śnieżnik and Góry Złote Mts in the Sudetes. Geol Sudet 14: 7-76

Teisseyre JH (1973) Metamorphic rocks of the Rudawy Janowickie and Lasocki Grzbiet ranges. Geol Sudet 8: 7-120

Van Breemen O, Aftalion M, Bowes DR, Dudek A, Misar Z, Povondra P, Vrana S (1982) Geochronological studies of the Bohemian massif, Czechoslovakia, and their significance in the evolution of Central Europe. Trans R Soc Edinb Earth Sci 73: 89-108

Wojciechowska I (1990) Geology of the Kłodzko metamorphic unit (Sudetes, Poland). Neues Jahrb Geol Paläontol Abh 179: 189-195

Wojciechowska I, Narębski W, Dostal J, Pin C (1989) Paleotectonic setting of metabasites of NW part of the Śnieżnik Metamorphic Unit. In:

Narębski W, Majerowicz A (eds) Lower and Upper Paleozoic metabasites and ophiolites of the Polish Sudetes. Inst Geol Sci, Wrocław University, Wrocław May 1989, pp 175–189

Znosko J (1981) The problem of the oceanic crust and ophiolites in the Sudetes. Bull Acad Pol Sci Ser Sci Terre 29: 185–197

VI.E Metamorphic Evolution

R. Kryza

1 Introduction

The distribution of metamorphic complexes, their documented, or more often inferred, protolith ages and metamorphic grades are shown in Fig. 1 of Chapter VI.A. Here, a description of the complexes is given, generally according to their metamorphic grade. Particular attention is paid to eclogites and granulites which bear potentially important petrologic information.

2 Very Low- to Medium-Grade Complexes

2.1 Lusatian Anticlinal Zone (LAZ)

The southeastern part of the LAZ (cf. Fig. 1 in Chap. VI.A) is occupied by Cadomian granodiorites amalgamated with anatexites and amphibolite facies rocks (e.g., Stolpen complex), the latter interpreted as derived from a Proterozoic graywacke sequence intercalated with basic volcanics. Northwestwards, this crystalline basement is covered by Upper Proterozoic Lusatian graywackes which display only faint indications of very-low-grade metamorphism and deformation (crenulation cleavage; Kemnitz and Budzinski 1990). Both the West and East Lusatian granodiorites produced widespread contact aureoles within the graywackes, marked by biotite blastesis, grain coarsening, and migmatitic hornfelses with cordierite, sillimanite and andalusite (Kemnitz and Budzinski 1990).

Since Cadomian time, the LAZ has not been significantly buried, and only slightly thermally rejuvenated as shown by K-Ar mica ages from the Lusatian graywackes which are scattered between 560 and 360 Ma (data compiled by Schust in Brause, ed. 1990).

2.2 Góry Kaczawskie (GK) and Fore-Sudetic Block (FSB)

The ?Cambrian — L. Carboniferous volcanic-sedimentary Kaczawa complex was interpreted to represent the Barrowian greenschist facies (Maciejewski and Morawski 1979; Lorenc 1983). However, it has been found (Kryza et al. 1990) that the main phase of the greenschist facies metamorphism was preceded by a higher pressure episode, indicated by glaucophane overgrown successively by colourless to pale green amphiboles of actinolitic composition (Fig. 2A). Sodic amphiboles are found in metabasic rocks over the central and eastern part of the area (Kryza et al. 1990; Smulikowski W 1990; Fig. 1). This early HP episode has recently been confirmed by finding relics of jadeite in Qtz-normative metatrachytes in the southern GK (Kryza and Muszyński, unpubl. Fig. 2B).

The age of metamorphism has yet not been dated directly, but the latest greenschist facies event may have taken place as late as the Visean, since metasedimentary rocks of that age are part of the complex (cf. Chap VI.B). It is probable, but not yet proved, that the youngest part of the Kaczawa succession is less metamorphosed.

The central and northwestern area of the FSB is occupied by a Paleozoic greenschist facies sequence probably equivalent to part of the Kaczawa succession. The country rocks of the Strzegom-Sobótka granite (north of GSB in Fig. 1) represent the greenschist and epidote-amphibolite facies (phyllites, minor mica schists, calc-silicate and metabasic rocks with actinolitic amphibole). Sillimanite, andalusite, and altered cordierite are found only within the clearly zoned contact aureole of the granite (Majerowicz 1972).

The metamorphic grade in the GK and FSB becomes lower westwards (Maciejewski and Morawski 1979), and the rocks of the Görlitzer Schiefergebirge (GS) display hardly any metamorphism.

Dallmeyer et al. (Eds.)
Pre-Permian Geology of Central and Eastern Europe

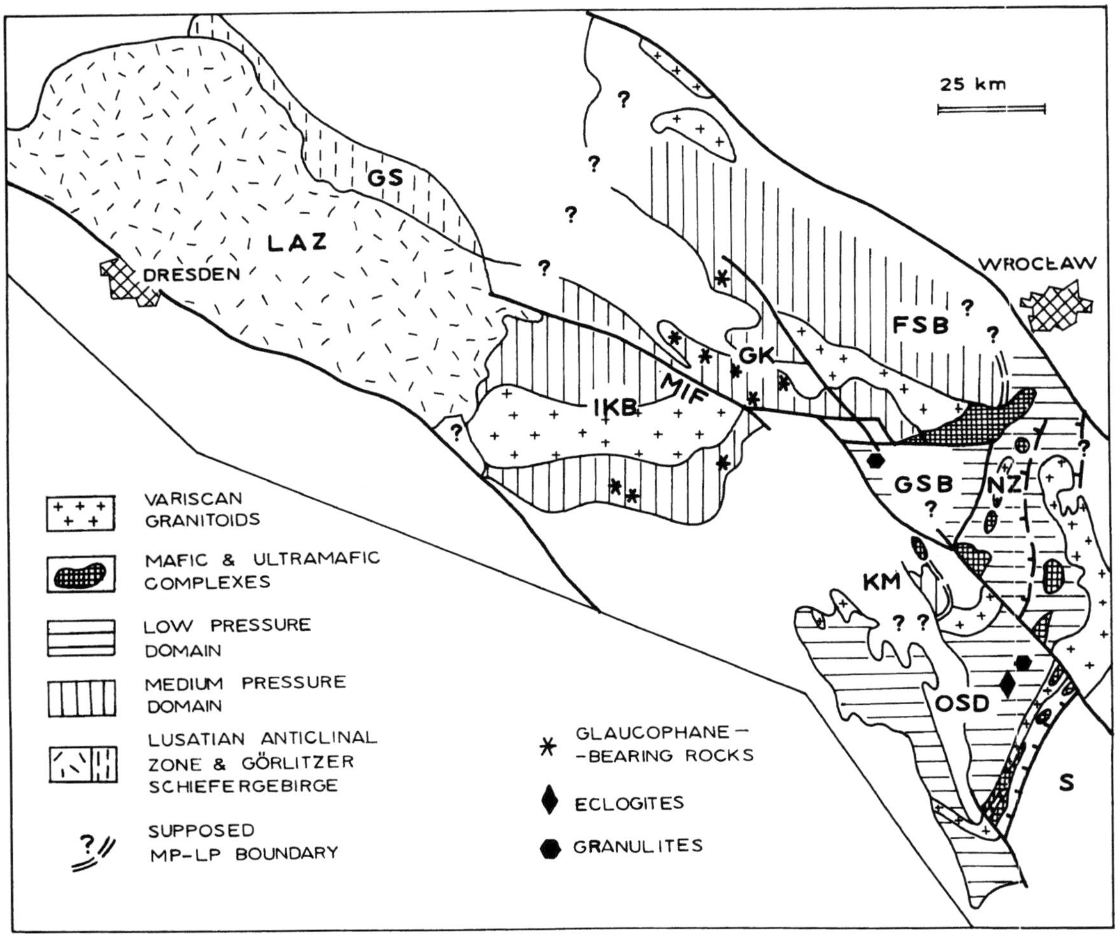

Fig. 1. Metamorphic domains in the Western Sudetes (a modification of Oberc-Dziedzic's 1989 model). *FSB* Fore-Sudetic Block; *GK* Góry Kaczawskie; *GS* Görlitzer Schiefergebirge; *GSB* Góry Sowie Block; *IKB* Izera-Karkonosze Block; *KM* Kłodzko Metamorphics; *LAZ* Lusatian Anticlinal Zone, *MIF* Main Intra-Sudetic Fault; *NZ* Niemcza Zone; *OSD* Orlica-Śnieżnik Dome; *S* Silesicum

2.3 Izera-Karkonosze Block (IKB)

In the Góry Izerskie area, north of the Karkonosze pluton, mica schists form three parallel aligned outcrops within gneisses and granites. The schists of the middle outcrop are composed of quartz, chlorite, biotite, and muscovite, and locally garnet, staurolite, and chloritoid (Żaba 1985). Kyanite, has been reported from the schists, but may represent a misidentification of chloritoid. Żaba (1985) interpreted the mineral parageneses as representing conditions transitional between the greenschist and the amphibolite MP facies (extreme P-T about 580–620 °C and 6 kb). The schists of the northern outcrop have rather simple mineral composition (mainly quartz, mica, and minor albite), while those in the southern belt are changed into hornfelses along the contact with the Karkonosze granite.

The para versus ortho origin of the Izera gneisses has caused controversy (cf. Oberc-Dziedzic 1988a). Some of the gneisses were found by Kozłowska-Koch (1965) to contain sillimanite, cordierite, andalusite, and corundum, and interpreted as remnants of the thermal influence of intrusive Izera granites. However, Oberc-Dziedzic (1988a) argues that andalusite and corundum occur only within the contact aureole of the Karkonosze granite, while cordierite and sillimanite are found in the Izera granites only, not in the gneisses. This may mean that the granites crystallized at higher T than the gneisses (Kozłowska-Koch 1965). A widespread, texturally late albite, which was interpreted as

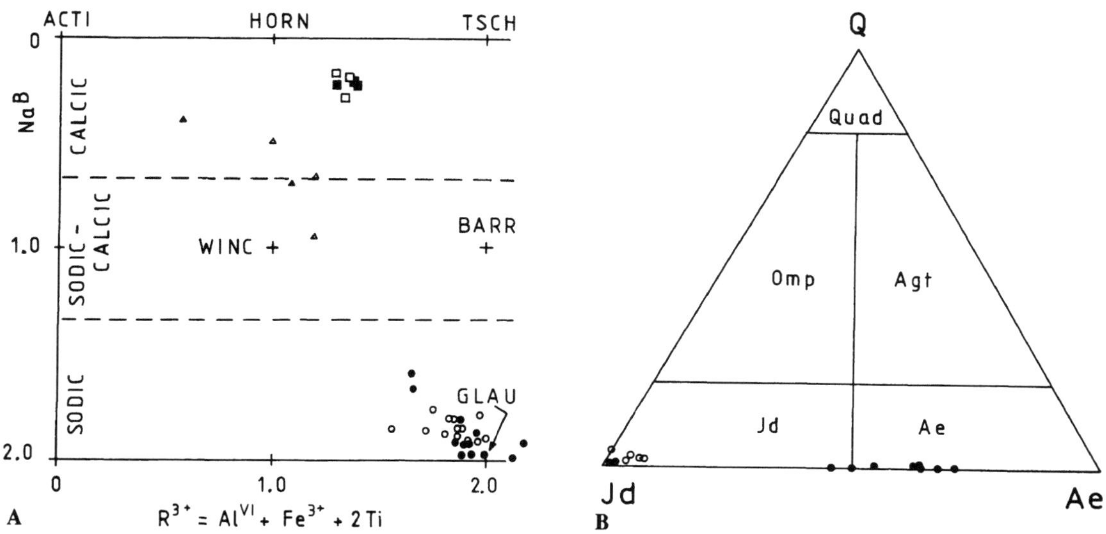

Fig. 2. A Compositions of amphiboles from two samples (*open* and *full symbols*) of metabasic rocks from the Góry Kaczawskie; *squares* igneous kaersutite; *circles* sodic amphiboles; *triangles* sodic-calcic and calcic amphiboles; based on total cations = 13 excluding Mn, Ca, and K. (Kryza et al. 1990). **B** Compositions of sodic pyroxenes from two samples (*open* and *full symbols*) of metatrachytes from the Góry kaczawskie; Fe3+ = Na+ − Al3+. (Kryza Muszyński, unpubl.)

evidence of metasomatism, may be a product of greenschist facies recrystallization (Maciejewski and Morawski 1979).

The internal part of the southern and eastern IKB (Velka Upa and Kowary Gneiss groups of assumed Proterozoic age; Chaloupsky 1989) is composed of gneisses and Bt-Grt mica schists. It is of amphibolite or epidote-amphibolits facies, partly retrograded during a subsequent greenschist facies event (Teisseyre 1973; Chaloupsky 1989). The external ?Upper Proterozoic to Silurian volcanic and sedimentary sequences of the IKB display south- and eastwards-decreasing metamorphic grades, down to lower greenschist facies. The tectonically bounded Jitrava Group in the SW corner of the IKB was subjected only to very low-grade metamorphism.

A significant feature of the southern and eastern margins of the IKB is the widespread occurrence of sodic amphibole (crossite) in metabasic rocks of the Leszczyniec volcanic formation and Zelezny Brod volcanic complex (cf. Chap VI.D; Wieser 1978; Chab and Vrana 1979; Guiraud and Burg 1984, and refs. therein). Actinolite or actinolitic hornblende and chlorite are apparently subsequent to the blue amphibole. The presence of Na amphiboles in the IKB has been interpreted variously as related to subduction (Wieser 1978; Chab and Vrana 1979; Narębski 1990) or as an effect of higher P conditions along a tectonic zone (Chaloupsky 1989). The age of metamorphism in the IKB is uncertain, in most part (excluding the Jitrava Group) probably pre-Late Devonian.

2.4 Orlica-Śnieżnik Dome (OSD) and Kłodzko Metamorphics (KM)

The OSD gneisses have traditionally been classified into two main groups: the Śnieżnik and Gierałtów gneisses, though actually they are texturally, and to a lesser extent also compositionally, variable. Their origin has also been disputed, particularly as to the ortho versus para derivation of the Śnieżnik gneisses (e.g., Smulikowski K 1979; Borkowska et al. 1990; Don et al. 1990; cf. Chap. VI.B).

Apart from gneisses, the northern and central parts of the OSD comprise the thick, medium-grade metamorphic supracrustal Młynowiec-Stronie group of inferred late Proterozoic-Cambrian age (Don et al. 1990). The dominant mica schists and paragneisses consist of two micas and quartz, with variable amount of plagioclase (An 2-27), and accessory garnet, staurolite, chlorite, and epidote. Locally, chloritoid, kyanite, and sillimanite were found (Ansilewski 1966; Smulikowski 1979). Textural relationships betweens the minerals indicate the superposition of a relatively higher T amphibolite facies on an earlier event with a higher P/T gradient (Smulikowski 1979). Evidence for a LP-HT metamorphic overprint (Cor, Sil, And) was also provided by Kozłowska-Koch (1973) from the mylonitic Haniak gneisses in the NE part of the OSD.

Within the Nove Mesto complex, which mantles the OSD to the SW, the metamorphism diminishes away from the dome, from the garnet zone in the NE, through Bt, to Chl zone in the SW (Opletal et al. 1980). This zonation is reflected also in the metabasic rocks, in which plagioclase changes from An 35 to An 10, and amphibole – from tschermakite to actinolite. The Zabreh complex, to the SE, represents the Grt zone, and occasional cordierite and sillimanite are interpreted as an effect of periplutonic metamorphism (Opletal et al. 1980). The Stare Mesto complex in the eastern part of the OSD comprises gneisses and mica schists of the staurolite zone (Misar 1983).

There are few radiometric constraints on the age of metamorphism in the OSD. However, the last event might have been around 335 ± 5 Ma, as indicated by a Rb-Sr whole rock-Bt-Ms isochron obtained from the Śnieżnik gneiss (Borkowska et al. 1990).

The mineral assemblages in the NE part of the Kłodzko Metamorphics are typical of the greenschist facies and, according to Wojciechowska (1990), the metamorphic grade increases south- and westwards, up to amphibolite facies. The metagabbros of the SW part often contain primary olive-green hornblende, and rarely also clinopyroxene. The hornblende is rimmed by bluish green barroisite-type amphibole. Plagioclase is strongly saussuritized, but its commonly high An content (30–40%) may be partly inherited from igneous feldspars. Minor components are epidote, clinozoisite, chlorite, and occasionally garnet, muscovite, and quartz (Kozłowska-Koch 1960).

An Early or Mid Devonian age of deformation and metamorphism in the KM is geologically well constrained (Wojciechowska 1990; cf. Chap. VI.B).

2.5 Circum Góry Sowie Block (GSB) Ophiolites

According to Majerowicz (in Majerowicz and Pin 1989), the mafic members of the Ślęża ophiolite have been metamorphosed. In the gabbros, the primary diallage is largely replaced by uralitic hornblende, while igneous plagioclase (up to 70% An) is partly changed into clinozoisite, epidote, and albite. In the metadiabases of the transitional plutonic/subvolcanic zone, secondary amphiboles are also common, ranging in composition from Mg-hornblende to actinolite, and the associated plagioclases vary from An 60 to albite. These mineral transformations may partly result from hydrothermal ocean floor metamorphism (Majerowicz and Pin 1989), which is also indicated by stable isotope data (Jędrysek 1989, vide Majerowicz and Pin 1989), and partly from subsequent regional metamorphism.

Similar mineral transformations were reported from the Nowa Ruda massif (Borkowska et al. 1989). On the contrary, the Braszowice gabbros are locally strongly deformed along shear zones, up to 1 m thick, into foliated amphibolites. Dziedzicowa (1989) interprets these rocks as an effect of "dynamic recrystallization" under amphibolite facies conditions. Internal structures in the flaser gabbros are concordant with those in the surrounding metamorphic rocks (Jamrozik 1979, vide Jamrozik 1989; Dziedzicowa 1989), suggesting that the mafic rocks were emplaced before the final deformation of the host rocks.

3 Medium- and High-Grade Complexes

3.1 Góry Sowie Block (GSB)

The GSB is composed of paragneisses and migmatites accompanied by minor mafic and ultramafic rocks, a few small granulite bodies, and tiny pods of marble and calc-silicate rocks. The paragneisses, traditionally believed to be Precambrian (?Moldanubian), are now considered to be Middle Riphean-Early Cambrian in age (cf. Chap. VI.B).

Most of the gneisses are composed of quartz, plagioclase (An 15-35), and biotite. Garnet and sillimanite are common, while K-feldspar and muscovite occur in considerable amounts only along the SW margin of the block. Trace amounts of relict kyanite in gneisses are widespread. Cordierite is limited to the NW corner and eastern edge of the block (Grocholski W 1967; Morawski 1973; Kryza 1981; Dziedzic 1985; Żelaźniewicz 1990).

The gneissic complex has undergone polymetamorphic evolution (Polański 1955; Kryza 1981; Dziedzic 1985; Żelaźniewicz 1990). The early stage of metamorphism was characterized by a relatively high P/T gradient, inferred from the widespread relict kyanite (Kryza 1981). The following metamorphic event took place under a moderate P/T gradient, within the high amphibolite facies, as indicated by sillimanite, K feldspar, and garnet. Subsequently, the apparently lowered pressure allowed locally intense anatectic melting producing nebulites and "in situ granites". A still high T during the latter event is indicated by texturally late sillimanite and subsequent cordierite in some of the nebulites (Kryża 1981).

According to Żelaźniewicz (1990), the gneissic complex experienced five tectonometamorphic events D_1–D_5, of which four (D_2–D_5) occurred close to one another during the Devonian (cf. Chap VI.C). The metamorphic climax and migmatization are ascribed to the D_2 and D_3 events, and their waning stage is interpreted to happen 380–370 Ma ago (U-Pb monazite and Rb-Sr biotite ages of van Breemen et al. 1988).

3.2 Fore-Sudetic Block (FSB)

The metamorphic cover of the Strzelin granites (east of NZ in Fig. 1) comprises two complexes (Oberc 1972; Oberc-Dziedzic 1988b): (1) an older complex, assumed Cadomian, composed of gneisses, mica schists, amphibolites, and calc-silicate rocks; and (2) a younger one, consisting of quartzites, schists, and metaconglomerates of inferred Early to Mid-Devonian age. According to Oberc-Dziedzic (1988b), the older complex underwent the first pre-Devonian metamorphic event under MP amphibolite facies conditions (Grt + Kya, local migmatization). The later, Variscan metamorphism occurred in both complexes, and was clearly zoned. In the sourth, the early phase was characterized by MP amphibolite facies conditions (Sil + Kfs + Grt) in both complexes, while in the north, the older complex was retrograded, and the younger rocks progressively metamorphosed under greenschist facies conditions. The later phase of the Variscan metamorphism was a LP-HT event which produced cordierite in both complexes all over the area. Afterwards, that event was overprinted by contact metamorphism near granite intrusions, as indicated by andalusite, sillimanite II, and corundum.

Clear evidence for polymetamorphism was also provided by Dziedzic (1985) from the areas east and west of the Niemcza zone (NZ). The general trend is from early MP series (Grt + Kya + Sil I in the west, and Grt + Kya + Sta in the east) to later LP-HT mineral assemblages (And, Sil II, Cor), the latter being interpreted as preceding the Variscan granitic plutonism (Dziedzic 1985). The distribution of metamorphic grades within the NZ itself is not yet clear.

The metamorphic complexes of the Strzelin and Niemcza Zone areas may continue north- and northwestwards, into the Middle Odra Metamorphics, where the pelitic lithologies often contain garnet and staurolite indicating MP amphibolite facies conditions (Utzig 1975, vide Grocholski A 1982).

4 Eclogites and Granulites

Predominantly felsic, Kya-bearing granulites occur in two areas: the GSB and OSD (Fig. 1). In the OSD, they are interlayered with minor omphacite granulites, and associated with small lensoid bodies of eclogite (Smulikowski K 1967). The eclogites have been studied in detail by K. Smulikowski (1967) and provided the basis for his general classification of eclogites (Smulikowski K 1972). All these rocks bear evidence to an original HP metamorphism (around 15–20 kb) and later decompression (Kryza et al. 1988; Pin and Vielzeuf 1988; Bakun-Czubarow 1991). However, their age and tectonic position remain disputable (Don 1989; Dumicz 1989).

4.1 Orlica-Śnieżnik Dome (OSD)

Eclogites occur as small lensoid intercalations in the Gierałtów gneisses (Rb-Sr whole rock age of 464 Ma; Borkowska et al. 1980), or as thin (up to ~10 cm) layers within the granulite complex of Stary Gierałtów.

Common eclogites (group C of Smulikowski 1967) consist of omphacite, garnet, and primary amphibole (carinthine), with occasional zoisite, phengite, quartz, rutile, kyanite, and dolomite. The so-called plagiclase eclogites (Kozłowski 1965) or omphacite granulites (Bakun-Czubarow 1991), contain primary oligoclase coexisting with omphacitic clinopyroxene. Larger bodies are mantled by retrogressive amphibolite facies rims, while smaller ones are completely retrograded.

According to Bakun-Czubarow (1990), the eclogites in different tectonic units of the OSD have different chemical compositions (tholeiitic-transitional to picrites, or calc-alkaline). They differ also in mineral chemistry and P-T conditions (Table 1).

Sm-Nd results from Grt-whole rock-Cpx define "ages" between 327 and 353 Ma, which have been interpreted as evidence for a Variscan HP metamorphic event (Brueckner et al. 1989).

The granulites appear in the core of the largest massif of the Gierałtów gneiss as a 0.5-km-wide and 9-km-long belt mantled by garnet-bearing gneisses (Kozłowski 1965; Pouba et al. 1985; Don et al. 1990). The dominant lithology is light granulite composed of quartz, perthitic K-feldspar, oligoclase, garnet, biotite, and minor kyanite, rutile, and locally sphene (Smulikowski 1979). The light granulites grade through "eclogite granulite", with minor poikiloblastic omphacite, to thin (up to 10 cm) layers of common eclogite composed of homogenous omphacite and garnet (Smulikowski 1979).

Pin and Vielzeuf (1988), using Grt-Kya-Pl-Qtz barometry, estimated the pressure of the granulite event at ca. 20 kb. Even higher pressures (20–25 kb) have recently been obtained from the OSD granulites by Kryza, Pin and Vielzeuf (unpubl.) from both Grt-Kya-Pl-Qtz and Cpx-Pl-Qtz barometers.

According to Don (1989), the Śnieżnik eclogites and the Gierałtów gneisses bear evidence only of the late D_3 tectonic event. The eclogite bodies are interpreted by that author as xenoliths derived from a deep seated basement, and probably transported upwards by Gierałtów migmatite masses, ascending after the D_2 event.

4.2 Góry Sowie Block (GSB)

Felsic granulites occur in three small outcrops in the northern part of the GSB. Orłowski (1983) described up to 150 m thick transitional zones in which the Bt content gradually increases at the expense of garnet and K-feldspar, and the rocks grade into gneisses. These boundary zones of the granulites were interpreted by Żelaźniewicz (1985) as blastomylonitic gneisses.

The granulites are fine-grained, generally weakly foliated rocks, composed of quartz, plagioclase, K-feldspar, garnet, biotite, and accessory kyanite, rutile, apatite, zircon, and opaques. Garnet porphyroblasts are distinctly zoned, with – from core to rim – an increase in Alm (from 50 to 59%) and Prp (31 to 38), a slight increase in Sps (1.1 to 1.7), and decrease in Grs (16.5 to 4.2%) contents (Kryza et al. 1988).

The temperature estimates in the aluminous granulites based on Bt-Grt thermometry are scattered between 600 and 900 °C, but a range of 750–850 °C is considered to be a reasonable estimate for the peak of the granulitic metamorphism. Pressures calculated from the Grt-Kya-Pl-Qtz barometer are 15–16 kb for the garnet cores, and 10 kb for the rims (at 800 °C). Lower pressures, between 8 and 12 kb, indicating decompression reactions, were also obtained from minor Cpx-bearing granulites, displaying Opx + Pl simplectite rims around garnet (Kryza et al. 1988).

The granulites host numerous lensoid bodies of upper mantle-derived ultramafic rocks, mostly harzburgites and lherzolites (with spinel or pyrope garnet; Bakun-Czubarow 1981). The P-T conditions of the primary equilibration of the ultramafic rocks were estimated at ~1300 °C and 20–30 kb (Bakun-Czubarow 1981). According to Żelaźniewicz (1985), the granulites and associated ultramafics represent tectonic slices upthrust along deep-seated ductile shear zones into their

Table 1. Selected mineral, PT, and Sm-Nd Grt-whole rock-Cpx age characteristics of the Śnieżnik area eclogites. (After Bakun-Czubarow 1990; Brueckner et al. 1989)

Eclogites from	Garnet		Ja in Px	T °C	P kb	Sm-Nd age
	Pyr	Gro				
Miedzygórze unit	36–55	18–27	30–39	620–740	13.5–16	337
Śnieżnik unit	29–34	27–31	39–45	650–750	16–17.5	330
Gierałtów complex				850	18	353

gneissic surroundings by the end of the pre-Late Devonian D_1 event.

5 Discussion and Conclusions

The original spatial metamorphic zonation in the Western Sudetes is obscured by the tectonic mosaic structure. Some tectonic units are homogeneous as to their metamorphic facies, some others display zonation (IKB, OSD). In some cases, adjacent units display different grades (e.g., IKB and GK), which may have come from the surposition of different grades in different times, or from subsequent important tectonic displacements.

Formerly, most of the metamorphic complexes were ascribed to the MP series and only two areas, i.e., GSB and Góry Izerskie, were considered as polymetamorphic units (Maciejewski and Morawski 1979). In contrast, Oberc-Dziedzic (1989) divided the Western Sudetes into two metamorphic zones of different P-T patterns of Variscan metamorphism. In the western zone, the early metamorphic stage was characterized by relatively high pressure, and was followed by an MP event. In the eastern zone, the later metamorphism was confined to several thermal cupolas, with a characteristic LP imprint on an ealier MP event (Dziedzic 1985; Oberc-Dziedzic 1989).

A simple sketch based on a modification of Oberc-Dziedzic's (1989) model is shown in Fig. 1. In respect of metamorphism, the Western Sudetes region can be divided into three main domains: (1) the LAZ in the NW, the basement of which was consolidated by medium- to high-grade metamorphism and anatexis during the Cadomian times; (2) the MP domain in the middle part of the region, comprising mainly low- to medium-grade complexes bearing evidence of an earlier, higher pressure episode followed by a MP event; and (3) the LP domain in the SE part of the Western Sudetes, characterized by mostly medium- to high-grade complexes with a clear LP-HT overprint on an earlier MP metamorphism.

The position of the GSB is intriguing; it is partly surrounded and probably also partly underlain by ophiolitic assemblages. They seem to coincide with the boundary zone between the MP and LP domains, which otherwise is difficult to define precisely. Also, the Main Intra-Sudetic Fault (Don 1990) and the Niemcza Zone may represent important regional structural boundaries, controlling the distribution of metamorphic phenomena.

Both the MP and LP domains are apparently polymetamorphic, but with little radiometric dating, it is usually uncertain whether we are dealing with two or more separate events, e.g., Cadomian, early and late Variscan, or with changing P-T conditions during individual metamorphic phases.

The metamorphism of high- and medium-grade gneissic massifs was traditionally referred to as Precambrian (Cadomian). However, recently obtained U-Pb and Rb-Sr ages from the GSB indicate a Devonian rather than Cadomian age for its main tectonothermal events and also a Devonian uplift (van Breemen et al. 1988). Yet more surprising are Variscan Sm-Nd ages obtained on the OSD eclogites, suggesting that rocks metamorphosed under P-T conditions typical of the lower crust, around 330–350 Ma age, were subsequently emplaced upwards during a late Variscan event (Brueckner et al. 1989).

The main metamorphic event in most parts of the area SW of the MIF is probably pre-Late Devonian (Don 1990), as geologically constrained in the southern part of IKB (Chaloupsky 1989), and in the KM (Wojciechowska 1990). Northwest of the MIF, the metamorphism seems to have been diachronous: the higher pressure event, indicated by relic glaucophane and jadeite, may have taken place in the Early Carboniferous (348 Ma ^{40}Ar-^{39}Ar age, H. Maluski, pers. comm. 1990), while greenschist facies conditions apparently persisted until the late Visean, since the youngest metamorphosed sediments are of that age (cf. Chap. VI.B). The late Variscan LP metamorphism was related to granitic plutonism, which displays two Rb-Sr age maxima around 330 and 280 Ma.

There seem to be some analogies between the MP and LP domains in Lugicum, and the MP and LP units distinguished by Blümel and Stein (1990) in the S and W parts of the Bohemian Massif. However, any detailed correlations would be premature.

Acknowledgments. Thanks are due to C. Nicollet, C. Pin, J. Zalasiewicz, and A. Żelaźniewicz for their comments on an earlier version of the manuscript.

References

Ansilewski J (1966) Petrography of the gneiss area of the Bialskie Mts. Geol Sudet 2: 121–260

Bakun-Czubarow N (1981) Ultrabasic rocks in granulites of the Góry Sowie Mts Block. In: Narębski W (ed) Ophiolites and initialites of northern border of the Bohemian Massif. Wissenschaftlichen Informationzentrums des Bergakademic Freiberg, Potsdam Freiberg, pp 79–86

Bakun-Czubarow N (1990) Eklogity Gór Śnieżnickich – 20 lat później. In: Smulikowski K Symp, Inst Geol Sci, Pol Acad Sci, Warszawa, Abstr

Bakun-Czubarow N (1991) HP omphacite granulites of the Złote Mts in the Sudetes – their geodynamic significance. Terra Abstr 3, EUG VI, Strasbourg

Blümel P, Stein E (1990) Metamorphic units within the Central-European Variscides. In: Paleozoic orogens in central Europe, IGCP Conf, Project 233, Terranes in the circum-Atlantic Paleozoic orogens. Göttingen-Giessen, Aug–Sept 1990, 4 pp

Borkowska M, Hameurt J, Vidal P (1980) Origin and age of Izera gneisses and Rumburk granites in the Western Sudetes. Acta Geol Pol 30: 121–146

Borkowska M, Marie FA, Narębski W, Pin C (1989) General mineralogical, geochemical and petrological characteristics of the Nowa Ruda gabbro-diabasic complex. In: Narębski W, Majerowicz A (eds) Lower and Upper Paleozoic metabasites and ophiolites of the Polish Sudetes. Inst Geol Sci, Wrocław University, Wrocław, pp 102–112

Borkowska M, Choukroune P, Hameurt J, Martineau F (1990) A geochemical investigation of the age, significance and structural evolution of the Caledonian-Variscan granite-gneisses of the Śnieżnik metamorphic area (Central Sudetes, Poland). Geol Sudet 25: 1–27

Brause H (ed) (1990) Oberproterozoikum und Känozoikum der Lausitz. Exkursionsführer, Tagung Hoyerswerda, 14–17 Nov 1990, Gesellschaft für Geologische Wissenschaften, Berlin, 50 pp

Brueckner HK, Medaris LG Jr, Bakun-Czubarow N (1989) First data on the Nd and Sr age and isotope patterns from Hercynian eclogites and garnet pyroxenite of the Śnieżnik metamorphic massif in the Sudetes and of the other occurrences within Moldanubian Zone of the Bohemian Massif. In: Żelaźniewicz A (ed) Indicators of structural evolution, Field Worksh in Poland, Pol Acad Sci, Inst Geol Sci, Wrocław University, Wrocław, Sept 1989, pp 72–74

Chab J, Vrana S (1979) Crossite-actinolite amphiboles of the Krkonose-Jizera crystalline complex and their geological significance. Vestn Ustred Ustavu Geol 54: 143–150

Chaloupsky J (1989) Geological evolution of the Krkonose-Jizerské hory crystalline complex in the Precambrian and Early Paleozoic. In: Chaloupsky J (ed) Geologie Krkonos a Jizerskych hor. Ustredni Ustaw Geol Praha: 288

Don J (1989) On the position of the eclogites within the gneisses of Miedzygórze (Śnieżnik metamorphic massif – Sudetes). In: Żelazniewicz A (ed) Indicators of structural evolution, Field Worksh in Poland, Pol Acad Sci, Inst Geol Sci, Wrocław University, Wrocław, Sept 1989, pp 44–58

Don J (1990) The differences in Paleozoic facies-structural evolution of the West Sudetes. Neues Jahrb Geol Paläontol Abh 179: 307–328

Don J, Dumicz M, Wojciechowska I, Żelaźniewicz A (1990) Lithology and tectonics of the Orlica-Śnieżnik Dome, Sudetes – recent state of knowledge. Neues Jahrb Geol Paläontol Abh 179: 159–188

Dumicz M (1989) The interrelationship of eclogites and gneiss series within the Międzygórze unit in the light of mesostructural study. In: Żelazniewicz A (ed) Indicators of structural evolution, Field Worksh in Poland, Pol Acad Sci, Inst Geol Sci, Wrocław University, Wrocław, Sept 1989, pp 31–43

Dziedzic H (1985) Variscan rejuvenation of the Precambrian gneisses along the eastern margin of the Góry Sowie Massif, Fore-Sudetic Block. Krystalinikum 18: 7–27

Dziedzicowa H (1989) The Braszowice gabbro as a member of Variscan ophiolite complex. Acta Univ Wratislaviensis Prace Geol Mineral 17: 57–79

Grocholski A (1982) Crystalline series of the Fore-Sudetic Block and the connected prospects for mineral resources. Biul Inst Geol 341: 97–116

Grocholski W (1967) Structure of the Sowie Mts. Geol Sudet 3: 181–249

Guiraud M, Burg JP (1984) Mineralogical and petrological study of a blueschist metatuff from the Zelezny Brod crystalline complex, Czechoslovakia. Neues Jahrb Miner Abh 149: 1–12

Jamrozik L (1989) Ofiolit Ślęży i jego pozycja tektoniczna. In: Niśkiewicz J, Jamrozik L (eds) Ślęża ophiolite and its ore mineralization, Symp Wrocław-Sobótka, 4–6 Sept 1989, Inst Geol Sci, Wrocław University, 18–30

Kemnitz H, Budzinski G (1990) Correlation and subdivision of monotonous series of the Lusatian greywackes, northern border of the Bohemian Massif. In: Paleozoic orogens in central Europe, IGCP Conf Project 233, Göttingen-Giessen, Aug–Sept 1990, 3 pp

Kozłowska-Koch M (1960) Granite-gneisses of Ścinawka and associated rocks of the metamorphic area of Kłodzko (Middle Sudeten, Poland), Arch Mineral 22: 349–400

Kozłowska-Koch M (1965) The granite-gneisses of the Izera Highlands. Arch Mineral 25: 123–259

Kozłowska-Koch M (1973) Polymetamorphites of the Złoty Stok – Skrzynka dislocation zone. Geol Sudet 8:121–160

Kozłowski K (1965) The granulitic complex of Stary Gierałtów – East Sudetes. Arch Mineral 25: 5–122

Kryza R (1981) Migmatization in gneisses of the northern part of the Sowie Góry, Sudetes. Geol Sudet 14: 7–100

Kryza R, Vielzeuf D, Pin C (1988) Evaluations des conditions P-T dans les granulites du massif de Sowie Góry, Sudetes. In: Lorenc S, Majerowicz A (eds) Petrologie et Geologie du Socle Varisque des Sudetes, Symp Wrocław, Sept 1988, Inst Geol Sci, Wrocław University, pp 48–60

Kryza R, Muszyński A, Vielzeuf D (1990) Glaucophane-bearing assemblage overprinted by greenschist-facies metamorphism in the Variscan Kaczawa complex, Sudetes, Poland. J Metamorph Geol 8: 345–355

Lorenc S (1983) Petrogenesis of the Wojcieszów crystalline limestones (Góry Kaczawskie. Sudetes). Geol Sudet 18: 61–119

Majerowicz A (1972) On the petrology of the granite massif of Strzegom-Sobótka. Geol Sudet 6: 7–96

Maciejewski S, Morawski T (1979) Metamorphism in Lower Silesia. Biul Inst Geol 318: 25–42

Majerowicz A, Pin C (1989) Recent progress in petrologic study of the Ślęża Mt ophiolite complex. In: Narębski W, Majerowicz A (eds) Lower and Upper Paleozoic metabasites and ophiolites of the Polish Sudetes. Inst Geol Sci, Wrocław University, Wrocław, pp 34–72

Misar Z (1983) Lugicka oblast. In: Misar Z, Dudek A, Havlena V, Weiss J (eds) Geologie CSRS I, Cesky masiv. Statni Pedagogické Nakladatelstvi Praha, pp 173–198

Morawski T (1973) The Sowie Góry area and its petrological problems. In: Smulikowski K (ed) XV Session de l'AZOPRO, Warszawa, 6–16 Sept 1973, Wydawnicłwa, Geologiczne, Warszawa, pp 44–58

Narębski W (1990) Petrotectonic affinities of Lower Silesian metabasites and the problem of origin of Paleozoic Sudetic ophiolites. In: Paleozoic orogens in central Europe, IGCP Conf, Project 233, Göttingen-Giessen, Aug–Sept 1990, 4 pp

Oberc J (1972) Budowa geologiczna Polski. IV, Tektonika, 2 – Sudety i obszary przyległe. Wydawnictwa Geologiczne, Warszawa, 307 pp

Oberc-Dziedzic T (1988a) The development of gneisses and granites in eastern part of the Izera crystalline unit in the light of textural investigations. Acta Univ Wratislaviensis Prace Geol Mineral 13: 184

Oberc-Dziedzic T (1988b) Serie metamorficzne Wzgórz Strzelińskich i historia ich przeobrażeń. In: Budowa, rozwój i surowce skalne krystaliniku strzelińskiego, Symp Wrocław, 14-15 Oct 1988; Inst Geol Sci, Wrocław University, pp 6–21

Oberc-Dziedzic T (1989) Variscan metamorphism of the Sudetes and Fore-Sudetic Block. Acta Univ Wratislaviensis Prace Geol Mineral 17: 115–138

Opletal M, Domecka K, Cech S et al. (1980) Geology of the Orlicke hory Mountains. Ustredni Ustav Geol Praha: 202

Orłowski R (1983) Granulity Gór Sowich – studium petrologiczne. Ph D Thesis, Inst Geol Sci, Pol Acad Sci, Warszawa, 97 pp

Pin C, Vielzeuf D (1988) Les granulites de haute-pression d'Europe moyenne témoins d'une subduction éo-hercynienne. Implications sur l'origine des groupes leptyno-amphibolitiques. Bull Soc Géol Fr 4/1: 13–20

Polański A (1955) On the metamorphism of the crystalline formations of the Sowie Góry (Middle Sudetes). Arch Mineral 18: 211–284

Pouba Z, Padera K, Fiala J (1985) Omphacite granulite from the NE marginal area of the Bohemian Massif (Rychleby Mts). Neues Jahrb Miner Abh 151: 29–52

Smulikowski K (1967) Eclogites of the Śnieżnik Mts in the Sudetes. Geol Sudet 3: 7–180

Smulikowski K (1972) Classification of eclogites and allied rocks. Krystalinikum 9: 107–130

Smulikowski K (1979) Polymetamorphic evolution of the crystalline complex of Śnieżnik and Góry Złote Mts in the Sudetes. Geol Sudet 14: 7–76

Smulikowski W (1990) Alkali amphiboles of the low grade metamorphic rocks of Kaczawa Mts, West Sudeten, Poland. Geol Sudet 25: 29–58

Teisseyre JH (1973) Metamorphic rocks of the Rudawy Janowickie and Lasocki Grzbiet ranges. Geol Sudet 8: 7–120

Van Breemen O, Bowes DR, Aftalion M, Żelaźniewicz A (1988) Devonian tectonothermal activity in the Góry Sowie gneissic block, Sudetes, southwestern Poland: evidence from Rb-Sr and U-Pb isotopic studies. Ann Soc Géol Pol 58: 3–19

Wieser T (1978) Glaucophane schists and associated rocks of Kopina Mt (Lasocki Range, Sudeten). Mineral Pol 9: 17–39

Wojciechowska I (1990) Geology of the Kłodzko metamorphic unit (Sudetes, Poland). Neues Jahrb Geol Paläontol Abh 179: 189–195

Żaba J (1985) Progressive regional metamorphism of the Izera Block, Western Sudetes (Poland). Acta Univ Carolinae Geol 1: 63–88

Żelaźniewicz A (1985) Granulitic inliers amidst a gneissic-migmatitic complex of the Owl Mts, Sudetes. Acta Geol Pol 35: 157–171

Żelaźniewicz A (1990) Deformation and metamorphism in the Góry Sowie gneiss complex, SW Poland. Neues Jahrb Geol Paläontol Abn 179: 129–157

VI.F Metallogenesis[1]

K. Mochnacka, M. Banaś, W. Kramer, and K. Pošmourný

1 Introduction

The Lugicum area is shared by the three countries: Czech Republic, Germany, and Poland.

From the metallogenic point of view, the whole Lugicum is characterized by the presence of a wide spectrum of ore deposits and occurrences (Fig. 1). Numbers in brackets which follow the names of specific deposits correspond to the numbers of localities shown in Fig. 1. In Fig. 1 also the elements Fe, Sn, W, U, Au, Ag, Cu, Pb, Zn, Hg, As, Bi, Ti, Co, Cr, and Ni stand for ore accumulations which represent metamorphic, magmatic, pegmatitic, hydrothermal, volcano-sedimentary, sedimentary, and residual types.

Rejuvenation of the mineralization and polycyclic nature of some orebodies has been shown. The presumed long-lasting metallogenic evolution of the Lugicum is confirmed by the estimated age of the source of lead – 1150 Ma (Vaněček et al. 1985).

Most of the localities are not of economic importance, others have already been worked out. The only economic-grade deposits currently being mined are Stanisławów and Boguszów (Ba-F and Ba, respectively) as well as Harrachov (Ba-F) in Bohemia.

The following classification of Lugicum ore deposits and occurrences is proposed:

Metamorphosed deposits related to submarine volcanism and sedimentation:

- magnetite deposits,
- hematite-magnetite deposits,
- hematite deposits,
- pyrite deposits,

Deposits related to pre-Variscan and Variscan magmatism:

a) Magmatic and pegmatite deposits:
   - Cr, Ti-Fe deposits and occurrences,
   - Cu-Ni occurrences,
   - U deposits and occurrences in granite,
   - Th deposit and occurrences in pegmatites,
b) skarn and other contact-metasomatic deposits,

Vein-type and stratabound hydrothermal deposits:

a) Vein-type:
   - U and polymetallic-U deposits and occurrences,
   - polymetallic (Cu, Pb, Zn, Ag, As, Au) deposits and occurrences,
   - Fe deposits,
   - barite-fluorite deposits,
b) stratabound Sn deposits,

Stratabound deposits of sedimentary affiliation:

- Cu-Ag deposits,
- U deposits and occurrences,
- Au, Ti alluvial deposits and occurrences,

Deposits and occurrences related to weathering:

- Ni-Mg-Fe deposits,
- Al deposits.

Fig. 1. Distribution of selected ore deposits and occurrences in the Lugicum (geology after Chap. IV.A; for region abbreviations see Sect. 1, this Chapter *a* Metamorphosed deposits related to submarine volcanism and sedimentation; *b* magmatic and pegmatite deposits; *c* skarn and other contact-metasomatic deposits; *d* vein-type deposits and occurrences; *e* stratabound Sn deposit; *f* stratabound ore deposits and occurrences of sedimentary affiliation; *g* alluvial placers; *h* deposits related to weathering

[1] Each author is responsible for data pertaining to his own country.

Dallmeyer et al. (Eds.)
Pre-Permian Geology of Central and Eastern Europe

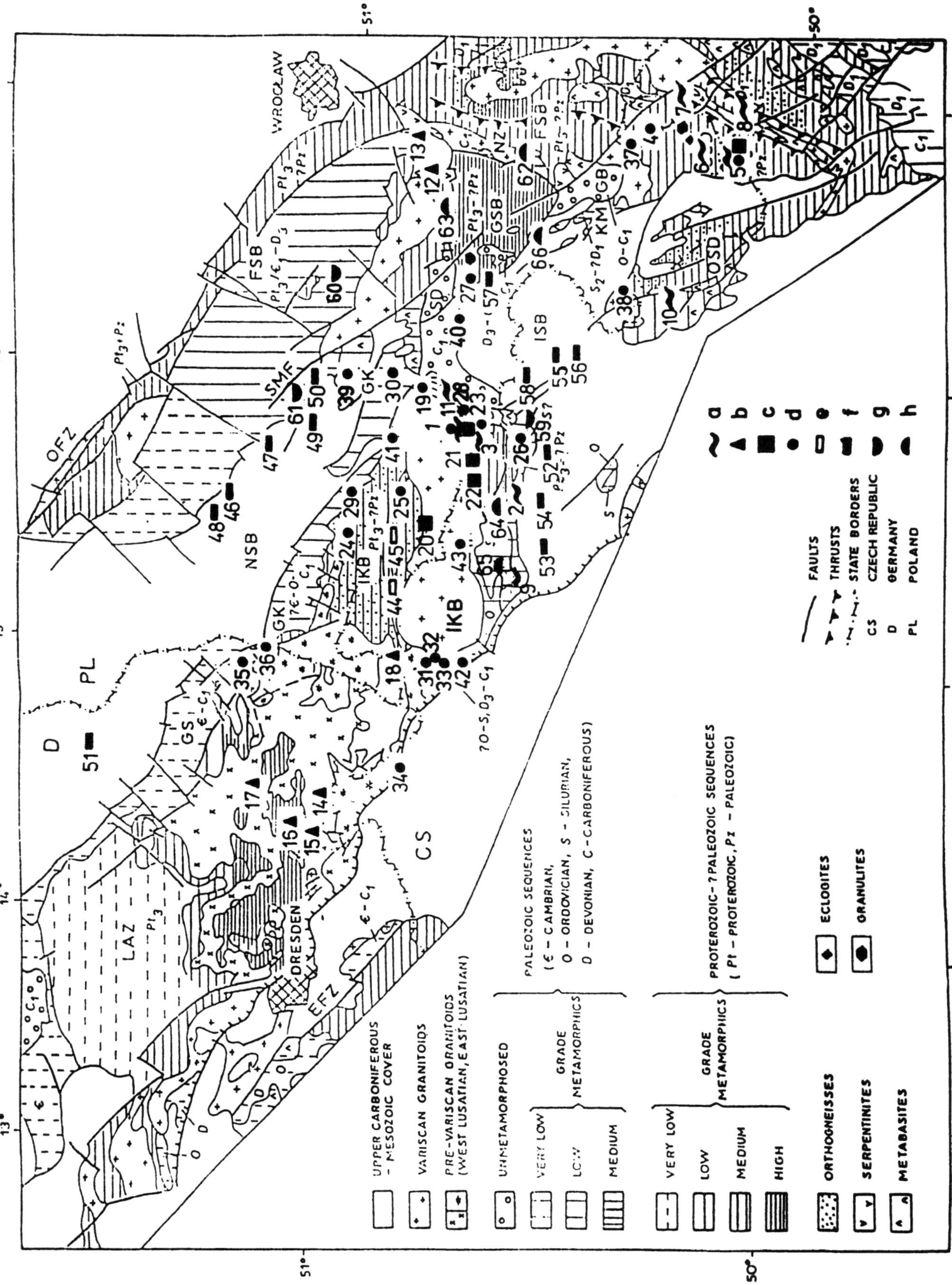
UPPER CARBONIFEROUS - MESOZOIC COVER
VARISCAN GRANITOIDS
PRE-VARISCAN GRANITOIDS (WEST LUSATIAN, EAST LUSATIAN)
UNMETAMORPHOSED
VERY LOW
LOW
MEDIUM
GRADE METAMORPHICS
PALEOZOIC SEQUENCES
(Є - CAMBRIAN, O - ORDOVICIAN, S - SILURIAN, D - DEVONIAN, C - CARBONIFEROUS)
VERY LOW
LOW
MEDIUM
HIGH
GRADE METAMORPHICS
PROTEROZOIC-?PALEOZOIC SEQUENCES
(Pt - PROTEROZOIC, Pz - PALEOZOIC)
ORTHOGNEISSES
SERPENTINITES
METABASITES
ECLOGITES
GRANULITES
FAULTS
THRUSTS
STATE BORDERS
CS CZECH REPUBLIC
D GERMANY
PL POLAND
WROCŁAW
DRESDEN
LAZ
EFZ
OFZ
SMF
FSB
NSB
ISB
IKB
GK
GS
OSD
KM
GB
GSB
NZ
SD
CS
D
PL
13°
14°
15°
16°
17°
50°
51°

2 Metamorphosed Deposits Related to Submarine Volcanism and Sedimentation

2.1 Magnetite Deposits

Stratiform iron deposits and occurrences are known from the Proterozoic? of both the Izera-Karkonosze Block (IKB in Polish, Jizerske hory-Krkonoše Block in Czech) and the Orlica-Śnieżnik Dome (Orlicke hory-Sněžnik Dome in Czech) (OSD).

Stratiform, lensoidal bodies of magnetite ores are often connected with limestones and amphibolites. In the Bohemian part of the IKB, the mafic rocks show a tholeiitic composition similar to Devonian tholeiites of the Jeseniky Mts. (Rambousek 1986) and are comparable with the recent island-arc basalts. The main ore mineral is magnetite, sometimes martitized and accompanied by maghemite. Gangue minerals are garnet (andradite), pyroxene (diopside, salite), epidote, clinozoisite, occasionally stilpnomelane, thuringite, and others. The presence of sulfides (pyrrhotite, chalcopyrite, pyrite, and others) is a characteristic feature.

The origin of these magnetite occurrences is controversial. Prevailing opinion refers the formation of these ores to the volcano-sedimentary processes, possibly combined with sedimentary ones (Pošmourný 1982). Alternatively, some authors regard them as products of skarn processes. For details of genetic discussion see Zimnoch (1961, 1967) and Pošmourný (1982).

A representative magnetite deposit is Kowary[1] (Poland), located in the Kowary Gneiss Group (KGG) (Teisseyre 1973). Magnetite forms flat lenses several centimeters to several meters thick. The ores are usually confined to the contacts between marbles and amphibolite schists, or are enclosed within the schists, or accompany the skarn bodies (Zimnoch 1961; Banaś and Mochnacka 1989). Orebodies consist of magnetite with martite and associated ilmenite, pyrite, arsenopyrite, and pyrrhotite. According to Zimnoch (1961), a ferruginous sediment formed with the participation of submarine effusive activity was regionally metamorphosed in the pre-Variscan epoch.

In the Bohemian part, two deposits are known: Herlikovice[2] and Malá Úpa-Smrčí[3], as well as several occurrences: Zálesi[4] in the Rychlebské hory Mts (OSD), and others in the IKB.

Iron deposits at Kletno[5] and Janowa Góra[6] (OSD) include magnetite layers and intercalations up to 2.5 m thick and/or disseminated structures hosted in mica schists and marbles (the Stronie Series, SS). The ores reveal compositions typical of skarns (hornblende, hedenbergite, ferrosalite, grossularite, andradite, and pistacite) (Banaś 1965a). Locally, sulfides are present. Iron skarns from Kletno are, for most part, the regionally metamorphosed sediments which contained primary ferruginous accumulations probably related to submarine volcanism, as suggested by the presence of amphibolites.

Magnetite deposits at Hraničná[7] in Rychlebské hory Mts and at Malé Vrbno[8] are located within the Staré Město crystalline complex. The ore forms two parallel layers similar to the banded magnetite ores of the Desná Dome (Sydvaranger type) described by Pouba (1970). The Hranična deposit is hosted mainly in gneisses, leptynites, amphibolites, crystalline limestones, and dolomites. In the whole OSD the predominantly basic, metamorphosed volcanics are associated with felsic metavolcanics. All the mafic rocks reveal a subalkaline, tholeiitic character, whereas the felsic rocks attain a calc-alkaline composition. Both rock types may have originated from the same subalkaline source magma, comparable with the basaltic magma of the recent mid-ocean ridges (Opletal et al. 1990).

Hematite-magnetite ores (itabirites) occur in the Zelezný Brod area (Vrát[9] deposit and others). These form conformable bodies within the epizonally metamorphosed Proterozoic or Early Paleozoic carbonates. Major ore components are hematite and quartz with subordinate magnetite and rare muscovite, and thuringite. Pošmourný (1982) related the origin of these iron occurrences to the volcano-sedimentary processes.

2.2 Hematite Deposits

Hematite ores are reported from the western part of the OSD (Nové Město Series), e.g., Kounov-Hluky.[10] The host rocks are epizonally metamorphosed graphitic phyllites, metabasics, and keratophyres of presumably Proterozoic age. The chemical composition of all the metavolcanics is identical with the metamorphosed volcanics from the Staré Město crystalline complex. The principal ore constituents are hematite and quartz accompanied by muscovite, chlorite (ripidolite), biotite, and albite. Close volcano-sedimentary affiliation has been suggested by Pošmourný (1982). The sedimentary basin of the ferruginous deposits of Nové Město Series was relatively deep-water, distant from the shore. It is supposed that considerable amounts of clays and pyroclastic material were mixed with rather scarce ore minerals.

2.3 Pyrite Deposit

In the eastern part of the metamorphic cover of the Variscan IKB Granite (Karkonosze Granite in Polish, Krkonose-Jizeské hory Granite in Czech) the Leszczyniec Volcanic Formation of the Rudawy Janowickie Group (RJG) (Teisseyre 1973) hosts the pyrite-bearing schists known from Wieściszowice[11] (Jaskólski 1964). The chlorite-sericite schists contain disseminated pyrite as a main sulfide. Rare chalcopyrite, galena, and sphalerite were also reported. Sporadically, carbonate veinlets with Fe, Cu, Pb, Zn sulfides and Sb-sulfosalts were observed (Piestrzyński and Salamon 1977). The presence of volcanics described as greenstones by Jaskólski (1964) is of genetic importance. The Wieściszowice deposit has been classified by Jaskólski (1964) as effusive-exhalative-sedimentary. A similar occurrence (Lišny near Zelezný Brod) is known from the southern part of the IKB.

3 Deposits Related to Pre-Variscan and Variscan Magmatism

3.1 Magmatic and Pegmatite Deposits

3.1.1 Cr, Ti-Fe Deposits and Occurrences

Chromite and Ti-magnetite accumulations are known from the Ślęża ophiolite (Majerowicz and Pin 1989), which borders the northernmost margin of the Góry Sowie Block (GSB). The ultrabasic unit hosts several chromite localities including the abandoned Tąpadła deposit[12] (Birecki 1962). The orebodies occur in serpentinites and contain chromite with minor magnetite and accessory pyrite, chalcopyrite, and pyrrhotite. The two latter sulfides presumably represent a later, superimposed mineralization phase, genetically unrelated to the ophiolite itself (Sałaciński 1979). This author relates the origin of ore minerals to magma crystallization, serpentinization of dunites and peridotites, and post-magmatic activity connected with both the Ślęża gabbro and the Strzegom granitoids.

Recent studies of the Ślęża ophiolite (Jamrozik et al. 1988) revealed the presence of ilmenite zone in the vicinity of the Tąpadła deposit (Strzegomiany[13]). The zone contains ilmenite accompanied by traces of magnetite and accessory rutile, and Fe and Cu sulfides.

3.1.2 Cu-Ni Occurrences

In its German part the early consolidated, block-like Lusatian Anticlinal Zone (LAZ) includes metallogenically barren, anatectic, and intrusive granodiorites of Precambrian to Lower Paleozoic age (Baumann and Tischendorf 1976). In the western part of Upper Lusatia, these magmatic rocks were intruded by pre-Upper Carboniferous gabbroic magmas. The latter show advanced fractionation which took place during magma rise in the intermediate chambers and, partly, within the small, dyke- and stock-shaped intrusions. The derivates are gabbro-norites, olivine gabbros and diorites which form single or composite bodies up to 100 m thick, distributed over the area of about 600 km^2. Brecciated, droplet-shaped (liquid-magmatic), and autometamorphic, per-crystalline sulfides (pyrrhotite, pentlandite, and Ag-Au-enriched chalcopyrite) (Rhode 1971; Kramer 1976; Kramer et al. 1991), are hosted especially in the orthopyroxene-rich rocks. The Sohland and Spree[14] orebody had been worked in the beginning of this century (Beck 1903). Its direct continuation has been found in the Bohemian part of the LAZ (Rožany). Accompanying Au and Pt have been quoted by Moravek et al. (1985, unpubl.). Small occurrences known from this area are Grenzland,[15] Dahrner Berg[16], and Bautzen[17] in Germany, as well as some others in Bohemia (Šluknov).

3.1.3 U Deposits and Occurrences in Granite

This type includes traces of disseminated bröggerite (Lis and Sylwestrzak 1977; Kucha et al. 1978) as well as veinlets of sooty pitchblende, autunite and marcasite, associated mostly with episyenitization (Lis and Sylwestrzak 1979), and hosted in the Karkonosze Granite.

3.1.4 Th Deposits and Occurrences in Pegmatites

The U and Th manifestations are reported from the northern contact zone of the Karkonosze Granite and its metamorphic cover (Bareja et al. 1982). The most interesting locality is Markocice near Bogatynia.[18] Here, the NNE-SSW-trending tectonic zone is accompanied by Th-bearing pegmatites (Banaś and Kucha 1984) called metasomatic syenitoids in the more recent publications (Jęczmyk and Juskowiakowa 1989). Diversified mineral assemblage includes Th-phosphates, monazite, thorite, thorogummite, Nb-rich phases, native phases and Fe, Cu, Zn and Pb sulfides (Marcinkowski 1985; Kucha 1989; Jęczmyk and Juskowiakowa 1989).

3.2 Skarn and Other Contact-Metasomatic Deposits

In the Lugicum area several ore deposits and occurrences are known to be genetically or only spatially related to the skarns (Banaś and Mochnacka 1988). In the Polish part of the Lugicum the skarns were reported from both the OSD and IKB.

In the OSD the early phase of skarn formation resulted from regional metamorphism and gave rise to the so-called regional skarns occurring in the Kletno deposit. The second phase of skarn formation developed at the contact of the Stronie Formation marbles with the Śnieżnik granites and produced disseminated, highly martitized magnetite and a sulfide assemblage with high-Fe sphalerite (up to 10% Fe), cosalite, and Cu-Fe sulfides (Banaś 1965a).

In the IKB area the contact-metasomatic aureole produced in the metamorphic cover of the Karkonosze Granite contains skarns genetically related to the granitic pluton. The examples are:

- Kowary[1] magnetites (see above) which were formed during the regional metamorphism but show a spatial relationship to the Kowary skarns,
- magnetites from the Miedzianka[19] deposit formed by the replacement of calcium silicate minerals (after Krusch, vide Zimnoch 1978). It seems that the influence of the Karkonosze Granite is genetically important,
- small magnetite deposit in Szklarska Poręba[20] with accompanying hematite and sulfides, located at the northern contact of the Karkonosze Granite with hornfelses of its metamorphic cover (Krajewski 1960).

In the Bohemian part of the Lugicum skarn and contact-metasomatic ores are represented by stratiform, polymetallic and tungsten localities hosted in the medium-grade and, partly, contact-metamorphosed Proterozoic series developed along the southeastern and eastern contact of the Karkonosze Granite. Lenticular polymetallic orebodies composed of pyrrhotite, arsenopyrite, chalcopyrite, and sphalerite are locally confined to the Ca-Al-silicate rocks (the term erlans is in local usage in Poland and in Bohemia) and marble layers. Tungsten mineralization is closely associated with the contact zone of the granite and its metamorphic cover. A typical example is the Obří Důl[21] deposit which contains scheelite mineralization. A high-temperature paragenesis reported from this locality includes cassiterite, stannite, malayaite, stockesite, and Sn-garnet.

The origin of this group of deposits is still a matter of discussion. Some authors advocate Variscan metallogenesis as having been involved as contact-metasomatic effects (skarns) of the IKB intrusion (Chrt 1959; Šrein, pers. comm.). On the other hand, Pertold (1978) believes that distinct metamorphic processes imprinted on these ores suggest their syngenetic origin.

Several small, isolated, polymetallic deposits containing mostly Cu, Pb and Zn ores are hosted in Ca-Al-silicate rocks (erlans) and marbles of various Proterozoic and Early Paleozoic Lugicum complexes. They differ from the group described above by the lower contents or absence of pyrrhotite, pyrite, and arsenopyrite and, sometimes, by the greater distance from the Karkonosze granite. Both the Rokytnice-nad-Jizerou[22] (IKB) and the Hraničná[7] (OSD) deposits are typical examples of such localities.

4 Vein-Type and Stratabound Hydrothermal Deposits

Numerous small deposits and occurrences of ore veins noticed in the Lugicum area (e.g., in the IKB, Mochnacka and Pošmourný 1981) are fault-controlled and contain the following main mineral parageneses:

- uranium and polymetallic-uranium,
- polymetallic (Pb, Zn, Cu, Ag, As, Au),

- iron,
- fluorite-barite,
- tin.

Typical veins (see Sect. 4.1, below) locally grade into or are accompanied by the metasomatic zones. Some occurrences can be related directly to the magmatic activity. A genetic link to the Variscan orogeny is suggested by structural and tectonic relationships as well as by the radiometric datings (see below). The latter also indicate the rejuvenation of mineralization processes during the later orogenic events.

Stratabound ores are represented by an only one type: tin deposits related to the specific zone of the schist complex (see Sect. 4.2).

4.1 Vein Type

4.1.1 Uranium and Polymetallic-Uranium Deposits and Occurrences

In the Kletno[5] area (OSD), orebodies are related to the tectonic zone and are hosted in the brecciated quartz mass formed in silicified marbles of the Stronie Series, close to the contact with the Śnieżnik ortogneisses. Nests and veins of black fluorite contain colloform pitchblende and its oxidation products (Banaś 1966, 1991). Accompanying minerals are: magnetite, martite, sulfides and sulfosalts of As, Cu, Pb, Ag, Bi, selenides of Pb, Cu, Ag, Bi, Hg, native Au, Ag, and Se and a weathering assemblage. Primary polymetallic mineralization form several parageneses (Banaś 1965b):

- quartz-fluorite-barite-Fe, Cu, Zn, Pb sulfides,
- quartz-fluorite,
- pitchblende-sulfides-selenides,
- uranium minerals-selenides.

In both the Orlické hory and the Rychlebské hory Mts (OSD), ore accumulations have been observed at the contact of Proterozoic mica schists with ortogneisses (e.g., Zalesi[4]), similar to the occurrences known from the IKB. Uranium mineralogy includes pitchblende and secondary uranium phases as well as Ni-Co-Bi-Se parageneses.

In the IKB area the classic example of vein-type deposit is Kowary.[1] Fissure veins and stockworks are hosted within marbles, skarns, erlans, schists, and magnetite bodies, close to the contact with the Karkonosze granite. Tectonic control is evident. Main gangue minerals are carbonates and quartz. Pitchblende and coffinite accompanied by a Co-Ni-As-Bi-Cu-Pb-Zn-Se assemblage form ore nests (Mochnacka 1967, 1991). Four stages of primary mineralization were distinguished, resulting in four paragenetic sequences: As-Co-Ni minerals, pyrite-pitchblende, sulfides-selenides, and carbonates. The Pb-U datings gave 265 and 70 Ma for pitchblende (Lis et al. 1971).

Other uranium occurrences known from the neigborhood of Kowary contain more uniform mineral assemblages accompanying pitchblende (Podgórze)[23] or are dominated with fluorite. Moreover, a high-temperature assemblage with brannerite was reported from a quartz vein cutting leucogranite in Wołowa Góra near Kowary (Lis et al. 1965).

Uranium-fluorite veins were found in the northern (Radoniów[24]) and central (Kopaniec-Mała Kamienica[25] and others) parts of the Isera Gneiss Massif (IKB). Orebodies are related to the tectonic zones and hosted in pink gneisses, leucogneisses, leucogranites, and schists (Jaskólski 1967; Mochnacka 1975). The Radoniów deposit contains pitchblende with pyrite, rare galena, and weathering uranium minerals. Occurrences from the central part of this area were dominated by secondary uranium minerals. Relationships to the alkaline metasomatosis have been suggested for these localities (Mochnacka 1975).

Uranium deposits occur also along the southern contact of the Karkonosze granite and its metamorphic cover as well as following the N-S-trending tectonic line which separates Proterozoic and Early Paleozoic groups of the IKB in the Labe River valley. Moreover, uranium occurrences were found to continue eastward, along the contact of Proterozoic mica schists with orthogneisses. An example is the Černý Důl[26] deposit, in which a Ni-Co-Bi paragenesis coexists with interesting Cu and Ag arsenides (see Hak et al. 1970).

Uranium-polymetallic occurrences have been reported from the contact of the Góry Sowie Gneisses with the Carboniferous sediments (Dziećmorowice,[27] ISB). Quartz-barite and quartz-calcite veins host Fe, Pb, Zn, Co, Cu, Bi, Ni, Sb sulfides and uranium minerals (Fedak and Lindner 1966). Finally, uranium minerals with minor sulfides accompany leucorhyolite veins cutting Carboniferous coal-bearing formation in the Lower Silesian Coal Basin (ISB) (Plewa 1968).

4.1.2 Polymetallic (Cu, Pb, Zn, Ag, As, Au, Co) Deposits and Occurrences

Localities of this type are known from the eastern metamorphic cover of the Karkonosze Granite. In Miedzianka[19] the so-called Czarnów Schists Formation (RJG, Teisseyre 1973) is cut by a multidirectional vein system dominated by Cu sulphides accompanied by Pb, Zn, As, and U minerals (Zimnoch 1978). The Czarnów[28] deposit located within the southern extension of the same schists series contains mostly arsenopyrite with pyrrhotite and sulfide assemblage (Zimnoch 1983) in a vein concordant with the enclosing schists. Hydrothermal origin is commonly suggested for this locality but, in the authors' opinion, contact-metasomatic features are evident.

Several quartz-sulfide veins are known from the western margin of the Góry Kaczawskie Block (GK) (e.g., Pilchowice[29]) (Domaszewska 1968; Paulo and Salamon 1973). Locally, pyrite-arsenopyrite lenses accompany the veins. Host rocks are low-grade metamorphics (graphitic and graywacke schists). Native Au is an interesting trace mineral.

The well-known Stara Góra[30] polymetallic deposit contains As, Zn, Pb, Sb, Cu, Au, and Ag minerals that form a vein system cutting the Early Paleozoic schists (Manecki 1965; Paulo and Salamon 1974). Recently, a Sn-W-Mo paragenesis has been reported from this locality (Sylwestrzak and Wołkowicz 1985). The vein system is fault-controlled. Relationship to the magmatic activity is discussed, but the sulfide veins are apparently younger than the adjacent porphyres (Paulo and Salamon 1974).

A group of small polymetallic deposits and manifestations is known from the northern and central parts of the Ještěd Crystalline Complex (Proterozoic-Lower Paleozoic) (IKB). Examples are Panenská Hůrka,[31] Andelská Hora[32], and Kryštofovo Udoli.[33] The presence of uranium and fluorite-barite accumulations with a Bi-Co-Ni assemblage is a spectacular feature of this ore district. Polymetallic ores are products of multistage formation processes (Pošmourný 1970).

Northwest from the Ještěd Mts. (Doubice[34]) Cu-Pb-Zn assemblage occurs in the upthrown blocks of Jurassic limestones and sandstones close to the Lužice Fault. Čadek and Malkovský (1966) suggested the origin of these ores in deep-circulating meteoric waters.

Small supergene deposits are known from the Görlitz Synclinorium. Co-Mn oxides are developed on hydrothermal quartz veins near Rengersdorf[35] and copper ores near Ludwigsdorf[36] (Richter 1959). The veins from Rengersdorf can be the differentiation products of the Königshain-Arnsdorf granite massif (Richter 1959) – one of the small, late-Variscan granitic intrusions of the LAZ.

The Złoty Stok[37] As-Au deposit (OSD) is located within the mica and amphibole schists and phyllites (the Stronie Schist Series), close to the contact with the Variscan Kłodzko-Złoty Stok intrusion. Its vein-type affiliation appears to be disputable, as the orebodies are mostly nests hosted within the dolomitic marbles, diopside skarns, and black serpentinites. Main ore minerals are arsenopyrite and löllingite with accessory magnetite, pyrrhotite, chalcopyrite, antimonite, hematite, and youngest, vein-type galena and sphalerite. Gold was traced mainly in löllingite, especially in its Co-rich variety (glaucopyrite) (Budzyńska 1971). Several genetic concepts have been developed for the Złoty Stok ores (for discussion see Kowalski 1967). A prevailing hydrothermal-metasomatic interpretation links the mineralization processes to the Variscan Kłodzko-Złoty Stok intrusion. In the authors' opinion, however, features typical of skarn deposits appear to be important. The average ore graded 3–5 ppm Au but up to 40 ppm were noticed in the richest parts (Krajewski 1960; Kowalski 1967). The Złoty Stok deposit (abandoned in 1961) appears to be largest primary gold accumulation in the Lugicum.

4.1.3 Fe deposits

Hematite veins were reported from the Orlické hory Mts (OSD) (Dańczów[38] and vicinity), where they fill the tectonic fissures in the contact zone of Variscan granites and metamorphics. Hematite shows collomorph structures and is accompanied by hydrohematite and gangue (quartz, barite) (Magda 1960).

Siderite and hematite veins were described from the south-eastern part of the Góry Kaczawskie (GK) (Wilcza) (Paulo 1973).

4.1.4 Barite-Fluorite Deposits

Both barite and fluorite are common gangue minerals in various vein-type deposits mentioned above. In some localities, however, barite (and/or fluorite) is the main commercial mineral and coexists with minor quartz and siderite. Most of the deposits and occurrences (including largest bodies) are spatially related to the large faults (Main Intra-sudetic Fault, Para-sudetic Fault, Jerzmański 1974). The final stage of the Variscan hydrothermal activity is proposed to be the source of mineralizing solutions, although younger ages cannot be excluded (for discussion see Jerzmański 1982).

The huge Stanisławów[39] barite-fluorite deposit is actually the largest exploited vein-type barite locality in Europe (primary reserves reached 10 mln tonnes $BaSO_4$). The 900-m-long, vertical vein is 1–5 m thick, in average (but locally swells to 10 m) and continues several hundreds of meters downdip. Wall-rocks are epimetamorphic greenschists and diabases. Barite is the main mineral with the accompanying quartz, siderite, Mn and Fe oxides and Zn and Pb sulfides (Paulo 1972, 1973). The fluorite content increases remarkably with depth.

The second exploited barite deposit – Boguszów[40] – is located in the Inner-Sudetic Basin. The composite vein is 1–3 m thick (locally up to 5 m) and cuts the Upper Carboniferous sediments and Variscan rhyolites. Barite is accompanied by quartz, minor fluorite, and Pb, Zn, Cu sulfides (including Ag-tetrahedrite) (Gruszczyk et al. 1970; Pawłowska 1970).

The remaining small barite deposits occur in the GK (e.g., Jeżów Sudecki[41]) and in the Góry Sowie (GSB). For details, see Jerzmański (1982) and Szałamacha (1976).

In the Bohemian part of the Lugicum, the most important barite-fluorite deposits are Křižany[42] and Harrachov[43] (IKB). The barite-fluorite veins are of multistage origin and specific mineralization periods are separated by distinct tectonic events (Reichmann 1975). Principal factors controlling the Ba-F mineralization are pre-Variscan and Variscan faults rejuvenated during the Saxonic geotectonic era.

Fluorite nests of hydrothermal origin hosted in marbles were exploited in the Kletno Mine[5] (OSD, see above). Some fluorite veins, however, can probably be connected also with the circulation of meteoric waters (e.g., Kopaniec,[25] IKB) (Mochnacka 1975).

4.2 Stratabound Tin Deposits

Tin-bearing zones are known from the western and central parts of the Precambrian schist belt (between Nové-Město-pod-Smrkem[44] in Bohemia through Gierczyn[45] and Przecznica in Poland) embedded within the Isera Gneiss Massif (IKB). Subeconomic cassiterite occurrences form flat, discontinuous, elongated lenses hosted by mica-chlorite-quartz schists. Cassiterite is usually of microscopic size and forms disseminated structures. Enclosing schists are enriched in chlorite, quartz, and garnets. Accompanying sulfides are pyrrhotite, chalcopyrite, pyrite, with minor arsenopyrite and sphalerite. Interesting are trace amounts of Bi-sulfosalts, native Bi (Jaskólski and Mochnacka 1958; Wiszniewska 1984; Kucha and Mochnacka 1986), scheelite and ferberite (Piestrzyński et al. 1990) and native Au (Piestrzyński et al. 1992). The origin of tin mineralization is a matter of discussion. Metamorphosed placer (Jaskólski 1962; Szałamacha and Szałamacha 1974), hydrothermal (Wiszniewska 1984; Mochnacka 1985, and others) or volcano-sedimentary (Lehmann and Schneider 1981) concepts have been proposed.

5 Stratabound Deposits of Sedimentary Affiliation

5.1.1 Cu-Ag Deposits

In the North-Sudetic Basin (NSB) copper ores were intensively exploited in Iwiny,[46] Grodziec,[47] Lubichów,[48] Nowy Kościół[49], and Wilków.[50] The host-rocks are clay-organic marls locally (Wilków) intercalated by limestones, all of Upper Permian (Zechstein) age. The thickness of Cu-marls reaches 2–3 m and copper contents vary from 0.1 to 2.3 wt%. Chalcocite, bornite, and chalcopyrite are the main commercial minerals (Konstantynowicz 1965; Banaś 1980). Copper is accompanied by trace elements, among which silver is of economic importance (Banaś 1960). Cu-marls are overlain by the Gervillia marls which contain galena. Sporadically noted calcite veinlets host a polymetallic paragenesis with Cu, Pb, Zn, Ni, Co sulfides, and U, for which a hydrothermal origin has been proposed (Lisiakiewicz 1969). The copper deposits from the NSB belong to the Kupferschiefer type genetically related to the Central European Zechstein basin of marine-salinary character. Primary ores have been affected by subsequent diagenetic processes.

Upper Permian stratabound copper deposits (Spremberg-Weisswasser,[51] Germany) occur in the northwestern extension of the North Sudetic Basin (Lusatian Basin). Sulfides (chalcocite, bornite, chalcopyrite) hosted in the organic-rich copper shale (Kupferschiefer) and in the underlying sandstones locally occur also in the overlying beds. Metal zonation: Zn-Pb, Pb, and Cu is distinct. The distribution pattern of ores follows the WNW-ESE-striking margin of the oxidized Kupfer-schiefer facies ("Röte Fäule") (Rentzsch and Langer 1963; Rentzsch and Knitzschke 1968).

The Kupferschiefer deposits in Central Europe were interpreted by Rentzsch (1981) as a uniform Central European Copper Belt, based on the regional tectonic control. Geochemical studies made by this author support the classic concept of the origin of metals from the Late Variscan magmatic, denudation, and sedimentation cycle. Recently, Hammer et al. (1990) reported trace elements, isotopes, and organic matter in the Kupferschiefer of the Sangerhausen basin in order to give an insight into the syngenetic to early diagenetic ore-forming processes, including the influx of meteoric saline waters.

In the Inner Sudetic Basin (ISB), the Lower Permian sequence includes Antracosia shales from which minor copper occurrences were reported (up to 0.5 wt%). Metal could be introduced due to the infiltration or hydrothermal activity connected with tectonic activation (Petrascheck, vide Osika 1987).

In the Bohemian part of the ISB, main deposits are Rudník,[52] Košťálov,[53] Horní Kalná,[54] Horní Vernéřovice[55], and others. Sulfides (chalcocite, bornite) prevail over the carbonates (malachite, azurite). Occasionally, uranium minerals have been noted (Chvaleč[56]). Sulfides are also commonly observed in relation to the coal-bearing formation. Copper occurrences are considered to be synsedimentary, diagenetically modified (Čadková 1971). The ores are commonly classified as red-bed type.

5.1.2 Uranium Deposits and Occurrences

Uranium minerals are hosted mainly within the Carboniferous sandstones (Grzmiąca deposit[57]) where metaautunite is accompanied by sooty pitchblende and minor sulfides (Kubisz and Smolarska 1960; Bareja 1981).

In the Okrzeszyn[58] deposit, uranium minerals are hosted in the Carboniferous coal seams in which 0.001–0.1 wt% U were assessed but no uranium minerals have been identified (Nielubowicz and Wróblewski 1963). Accompanying sulfides (Cu, Fe, Pb) were also reported. The origin of uranium accumulations in this area is related to the weathering and infiltration processes (Depciuch 1968). Uranium occurrences of similar character are also known from the Bohemian part of the ISB (Chvaleč[56], see above).

5.1.3 Au and Ti Alluvial Deposits and Occurrences

Alluvial gold placers appear to be common in the Lugicum area. In the Bohemian part, such occurrences accompany the primary gold deposits (with barite and cinnabar) hosted in a phyllite complex (Rýchory[59]). These localities are of historical importance and have not been verified recently. A relation to deluvial fossil placers known from the base of Carboniferous and Permian sequences is suggested. The Rýchory gold deposit is an outstanding locality because of high palladium contents (up to 10 wt% in native gold) and unique association of Pd minerals: potasite and stibiopalladinite (Malec and Veselovský 1985). Alluvial Au placers are also known from the vicinity of Staré Město (OSD).

Alluvial accumulations of Ti minerals with accompanied Nb ones, native Au and sapphires are known form the Jizerka deposit near Kořenov.

In the Polish part of the Lugicum, Au and cassiterite placers were noted in numerous localities in the northern metamorphic cover of the IKB (Jęczmyk 1979). Deposits, mostly of historical importance, are known from the Quaternary sandstones in the NSB (Legnickie Pole,[60] Złotoryja[61]) (Fedak and Lindner 1966; Grodzicki 1966; Domaszewska 1964; J. Banaś et al. 1985). Gold placers with zirconium and Ti minerals occur in similar geological conditions in the area of the Fore-Sudetic Block (FSB).

6 Deposits and Occurrences Related to Weathering

6.1.1 Ni, Mg, Fe Deposits

Mineral localities related to Miocene lateritization occur in the FSB. Weathering crusts have been developed on the basic and ultrabasic intrusions. Some of them contained increased amounts of Ni (pimelite, shuchardite with Fe-oxides and magnesite). In the largest, recently abandoned Szklary[62] deposit, the Ni-bearing zones were cut by silica veins containing high-quality chrysoprase (Ostrowicki 1965; Niśkiewicz 1967). Magnesite veins of similar origin are still being exploited at the Wiry[63] and Braszowice.

In Bohemia, residual limonite ores occur in the Zelezný Brod crystalline complex with the main deposits at Poniklá[64] and Ješenny.[65] Paleogeographic, paleoclimatic, and lithological controls are inferred. The deposits originated from Oligocene-Miocene weathering of Fe-rich metamorphics. Limonite concretions show partly increased contents of Mn, Ti and Zn.

6.1.2 Al Deposits

Lateritic bauxites are known from the Nové Město crystalline complex. These are located mainly at the bottom of the Cretaceous sequence and partly even in the Permian sediments, over the fossil weathering crusts developed on metabasics. Svoboda and Zoubek (1939) proposed a sedimentary origin for these accumulations, with the bauxites occupying several small, isolated basins. The most widespread mineral is boehmite. Neighboring alluvial sediments contain interesting occurrences of gold (Albrechtová 1972).

Similar bauxites are known from the ISB (Nowa Ruda[66] deposit). Gossan is here developed on the gabbros and diabases overlain by the Carboniferous coal-bearing formation (Górzyński 1966).

Acknowledgments. The authors are indebted to Dr. W. Mayer (University of Mining and Metallurgy, Kraków) for language corrections and improvement of the manuscript. Thanks are due to Prof. Z. Pouba (Charles University, Prague) for critical reading of the text.

References

Albrechtová E (1972) Occurrence of cassiterite, scheelite, wolframite and gold in the Orlické hory Mts. Sbor GPO 1: 141–143 (in Czech)

Banaś J, Grodzicki A, Salamon W (1985) Mineralogic-geochemical characterization of detrital native gold from the vicinity of Złotoryja and Wądroże Wielkie, Lower Silesia, SW Poland. Miner Pol 16: 97–108

Banaś M (1960) Silber in kupferhaltigen Zechstein Schiefern Niederschlesiens. Bull Acad Pol Sci Ser Geol Geogr 8: 197–201

Banaś M (1965a) Polimetamorphe Skarngesteine mit Eisen- und Zinkvererzung in den Sudeten. Freib Forschungsh C 186: 51–61

Banaś M (1965b) Signs of mineralization in the metamorphic complex of Snieżnik Kłodzki (Sudetes Mts). Geol Trans 27: 83 (in Polish)

Banaś M (1966) Products of oxidation of pitchblende from Kletno. Bull Acad Pol Sci Ser Geol Geogr 14: 23–26

Banaś M (1980) Zechstein copper deposits in Poland. In: Jankowieć S, Sillitoe RH (eds) European copper deposits. Dept of Econ Geol, Faculty of Mining and Geology, Belgrade, pp 136–145

Banaś M (1991) Pitchblende in Ag-Bi-Se paragenesis from Kletno Deposit, Sudety Mts, south-West Poland. In: Primary radioactive minerals. Theophrastus, Athens, pp 269–285

Banaś M, Kucha H (1984) Thorium and niobium mineralization connected with the Łużyce granitoid Massif in Poland. Proc 6th IAGOD Symp. Schweizerbart sche Verlagsbuchhandung, Stuttgart, pp 241–246

Banaś M, Mochnacka K (1988) Formation of skarns and other calc-silicate rocks from Sudetes. Ann Soc Geol Pol 58: 469–479

Banaś M, Mochnacka K (1989) Genetic types of uranium mineralization in Poland and their metallogenic relationships. In: Metallogenesis of uranium deposits. Panel Proc Ser IAEA Vienna, pp 187–200

Bareja E (1981) Uranium minerals in Upper Carboniferous rocks in the Nowa Ruda region. Kwart Geol 25: 287–292 (in Polish)

Bareja M, Jęczmyk M, Kanasiewicz J, Lis J, Miecznik JB, Sałdan M (1982) Radioactive elements in the Sudetes. Biul Inst Geol 341:259–289 (in Polish)

Baumann L, Tischendorf G (1976) Einführung in die Metallogenie-Minerogenie. Dtsch Verlag Grundstoffind, Leipzig, 458 pp

Beck R (1903) Die Nickelerzlagerstätte von Sohland a.d. Spree und ihre Gesteine. Z Dtsch Geol Ges 55: 296–330

Birecki T (1962) Occurrence of chromites at Tąpadła. Przegl Geol 3: 144–150 (in Polish)

Budzyńska H (1971) Mineralogy of arsenic deposits of the Złoty Stok. Arch Miner 29: 29–88 (in Polish)

Čadek J, Malkovský M (1966) Distribution and regularity of the exogenic ore deposits of the Bohemian Massif. Sbor Geol Věd LG 8: 91–101 (in Czech)

Čadkova Z (1971) Distribution of the metallic elements in the Permian sediments of the Intra-Sudetic basin. Sbor Geol Věd LG 14: 91–126
Chrt J (1959) Current results of geological research at the skarn-type deposits in the Krušne hory Mts. and the Krkonoše Mts. Geol Průzk 1: 3–4 (in Czech)
Depciuch T (1968) Geochemical properties and origin of uranium concentration in the Upper Carboniferous clastic deposits in the Intra-Sudetic basin. Biul Inst Geol 214: 75–180 (in Polish)
Domaszewska T (1964) Occurrence and exploitation of gold in Lower Silesia. Przegl Geol 4: 180–185 (in Polish)
Domaszewska T (1968) Mineralization in the region of Radomice and Klecza, north of Jelenia Góra. Przegl Geol 3: 123–126 (in Polish)
Fedak J, Lindner M (1966) Metallogenesis of the Sudety Mts. Prace Inst Geol 23: 5–315 (in Polish)
Górzyński Z (1966) The occurrence, properties and origin of bauxites at Nowa Ruda against the background of the geological structure of the area. Biul Inst Geol 196: 5–84 (in Polish)
Grodzicki A (1966) Auriferous-sand from Legnickie Pole – Mikołajowice – Wądroże Wielkie Region. Arch Miner 26: 473–499 (in Polish)
Grodzicki A (1971) Gold-bearing detrital deposits of the Karkonosze-Izera Mts Massif. Arch Miner 29: 305–343 (in Polish)
Gruszczyk H, Paulo A, Smolarska I (1970) The present knowledge on the barite deposit at Boguszów, Lower Silesia. Prace Inst Geol 59: 69–84 (in Polish)
Hak J, Johan Z, Skinner B (1970) Kutinaite: a new copper-silver arsenide mineral from Černy Důl, Czechoslovakia. Am Miner 55: 1083–1087
Hammer J, Junge F, Rosler HJ, Niese S, Gleisberg B, Stiehl G (1990) Element and isotope geochemical investigations of the Kupferschiefer in the vicinity of "Rote Faule", indicating copper mineralization (Sangerhausen basin, G.D.R.). Chem Geol 85: 345–360
Jamrozik L, Niśkiewicz J, Cholewicka-Meysner D, Farbisz J, Jodłowski S (1988) Discovery of Fe-Ti mineralized zone in gabbros of the Ślęza Massif (Fore Sudetic Block). Geol Sudet 23: 121–127 (in Polish)
Jaskólski S (1962) Erwägungen über die Genese der Zinnführende Schiefer im Isergebirge (Niederschlesien). Geol Trans 12: 33–53
Jaskólski S (1964) On the Origin of pyrite schists at Wieściszowice (Lower Silesia). Ann Soc Géol Pol 34: 29–63
Jaskólski S (1967) Polymetallic oxide and sulphide mineralization in gneissous granites of the Izera Mountains (Lower Silesia) and its origin. Geol Trans 43: 77 (in Polish)
Jaskólski S, Mochnacka K (1958) Tin deposits at Gierczyn in Isera Mountains, Lower Silesia; an attempt at elucidating their origin. Arch Miner 22: 17–106 (in Polish)
Jęczmyk M (1979) Heavy minerals in the deposits of streams in the Izera metamorphic area. Biul Inst Geol 319: 75–350 (in Polish)
Jęczmyk M, Juskowiakowa M (1989) Geology and geochemical characteristics of the crystalline rocks of the Bogatynia area (Western Sudetes Mts). Biul Inst Geol 360: 5–38 (in Polish)
Jerzmański J (1974) Principal sudetic faults and their significance for the metallogeny in the Lower Silesia area. Kwart Geol 18: 684–689 (in Polish)
Jerzmański J (1982) Barite and fluorite mineralization in the metallogenic evolution of the Sudetes. Biul Inst Geol 341: 319–327 (in Polish)
Konstantynowicz E (1965) Signs of mineralization in the Zechstein of the North-Sudetic Syncline. Geol Trans 28: 7–99 (in Polish)
Kowalski WM (1967) Metamorphic rocks from Złoty Stok (Lower Silesia). Geol Trans 42: 7–82 (in Polish)
Kowalski WM (1969) Ore minerals from Złoty Stok (Lower Silesia). Miner Trans 16: 23–41 (in Polish)
Krajewski R (1960) Iron ores. Biul Inst Geol: 22–38 (in Polish)
Kramer W (1976) Zur Petrologie und metallogenetischen Bedeutung der Dolerite (Lamprophyre) des Lausitzer Massivs. Z Geol Wiss 4: 975–994
Kramer W, Andrehs G, Heinrich C (1991) Gabbroic rocks and their nickel-copper sulphide mineralization in Central Europe. In: Symp on Sulphide deposits, Jinchang, Abstr, pp 68–72
Kubisz J, Smolarska I (1960) Meta-autunit from the Inner-Sudeten Trough. Bull Acad Pol Sci Ser geol Geogr 8: 31–35
Kucha H (1989) Native gold in thorium and rare earth (RE) ores of the Bogatynia region (Lower Silesia). Rudy Met Nieżel 34: 6–7 (in Polish)
Kucha H, Mochnacka K (1986) Preliminary report on bismuth minerals from the Gierczyn tin deposits, Lower Silesia, Poland. Miner Pol 17: 55–61
Kucha H, Lis J, Sylwestrzak H (1978) Thorium-bearing uraninite (bröggerite) from the Karkonosze Granite. Miner Pol 9: 105–110
Lehmann B, Schneider HJ (1981) Stratabound tin deposits. In: Wolf KH (ed) Handbook of stratabound and stratiform ore deposits, vol 9. Elsevier, Amsterdam, pp 743–771
Lis FJ, Kosztolanyi Ch, Coppens R (1971) Étude géochronologique du gîsement polymétallique de Kowary (Pologne). Miner Deposit 6: 95–102
Lis J, Sylwestrzak H (1977) On the occurrence of dispersed uraninite in the Karkonosze Granite. Przegl Geol 25: 297–300 (in Polish)
Lis J, Sylwestrzak H (1979) Episynites and perspectives of occurrence of intragranite uranium deposits in the Karkonosze Massif. Przegl Geol 27: 223–229 (in Polish)
Lis J, Stępniewski M, Sylwestrzak H (1965) Brannerite and minerals accompanying a quartz vein in Wołowa Góra near Kowary (Sudetes). Biul Inst Geol 193: 207–228
Lisiakiewicz S (1969) Geological structure and mineralogical analysis of copper deposit in the Grodziec Basin. Biul Inst Geol 217: 5–111 (in Polish)
Magda W (1960) Hematites of the Kudowa region. Biul Inst Geol: 38–44 (in Polish)
Majerowicz A, Pin C (1989) Recent progress in petrologic study of Ślęża Mt. In: Narębski W, Majerowicz A (eds) Lower and Upper Palaeozoic

metabasites and ophiolites of the Polish Sudetes. Guidebook of excursions in Poland. Wrocław, Polish Academy of Sciences, pp 34–72
Malec J, Veselovský F (1985) Gold mining in the neighbourhood of Svoboda nad Upou. Rozpr Nar Tech Muz 99: 149–160 (in Czech)
Manecki A (1965) Mineralogical and petrographical study of ore veins of the vicinity of Wojcieszów (Lower Silesia). Miner Trans 2: 70 (in Polish)
Marcinkowski B (1985) Traces of ore mineralization in the igneous-metamorphic complex in the vicinity of Bogatynia. Kwart Geol 29: 551–570 (in Polish)
Mochnacka K (1967) The geology of the polymetallic deposit at Kowary (Lower Silesia). Geol Trans 40: 73 (in Polish)
Mochnacka K (1975) Mineralization of metamorphic rocks of part of the Izera Mountains. Geol Trans 89: 86 (in Polish)
Mochnacka K (1985) Structures and textures of ores from the Gierczyn tin ore deposit (Sudetes, Poland) and their genetic interpretation. Miner Pol 16: 85–95
Mochnacka K (1991) Pitchblende-Coffinite Paragenesis in the Kowary deposit (the Sudety Mountains, Poland) In: Primary radioactive minerals. Theophrastus, Athens, pp 255–267
Mochnacka K, Pošmourný K (1981) Metallogenetic characteristics of the Palaeozoic and pre-Palaeozoic formations of the northern part of the Bohemian Massif (Krkonoše-Jizerské hory region). Časop Miner Geol 26: 29–43
Nielubowicz B, Wróblewski T (1963) A contribution to the recognition of the uranium mineralization in coals of the Radwanice Beds of Lower Silesia. Kwart Geol 7: 114–129 (in Polish)
Niśkiewicz J (1967) Geological structure of the Szklary Massif (Lower Silesia). Ann Soc Geol Pol 3: 387–416 (in Polish)
Opletal M, Jelínek E, Pecina V, Pošmourńy K, Poubova E (1990) Metavolcanites of the north-eastern part of Lugicum, their geochemistry and geotectonic interpretation. Sbor Geol Věd LG 45: 37–64 (in Czech)
Osika R (1987) Copper ores (Permo-Carboniferous formation of the Intra-Sudetic Basin). In: Osika R (ed) Geology of Poland, vol 6. Wydawnictwa Geologiczne, Warszawa, pp 315–316 (in Polish)
Ostrowicki B (1965) Nickel minerals of the weathering zone of serpentinites at Szklary (Lower Silesia). Miner Trans 1: 92 (in Polish)
Paulo A (1972) Mineralogy of the barite at Stanisławów (Lower Silesia). Miner Trans 29: 77 (in Polish)
Paulo A (1973) The barite deposit at Stanisławów against the background of metallogeny of Góry Kaczawskie. Geol Trans 76: 72 (in Polish)
Paulo A, Salamon W (1973) Native gold in ore veins of the western part of Góry Kaczawskie Mts (West Sudeten). Miner Pol 4: 85–91
Paulo A, Salamon W (1974) Contribution to the knowledge of a polymetallic deposit Stara Góra. Kwart Geol 18: 266–276 (in Polish)
Pawłowska J (1970) The barite mineralization in the northern part or the Intra-Sudetic trough. Prace Inst Geol 59: 97–123 (in Polish)
Pertold Z (1978) Prospects of the stratiform deposits of the Bohemian Massif. In: Kužwart M (ed) Theoretical basic prognoses of the raw materials in Czechoslovakia, pp 118–123 (in Czech)
Piestrzyński A, Salamon W (1977) New data on polymetallic mineralization of quartz veins in pyrite deposit from Wieściszowice (Lower Silesia). Kwart Geol 21: 27–35 (in Polish)
Piestrzyński A, Mochnacka K, Mayer W, Kucha H (1990) Scheelite and ferberite from the tin-bearing schists of the Kamienica range (the Sudety Mts., SW Poland). Miner Pol 21: 5–13
Piestrzyński A, Mochnacka K, Mayer W, Kucha H (1992) Native gold (electrum), Fe-Co-Ni arsenides and sulphoarsenides in the mica schists from Przecznica, the Kamienica Range, SW Poland. Miner Pol 23: 27–42
Plewa M (1968) Igneous rocks and signs of mineralization in the western and middle part of the Wałbrzych basin. Miner Trans 12: 65 (in Polish)
Pošmourńy K (1970) Polymetallic veins in the Ještědské pohoří Mts. Sbor Geol Věd LG 11: 7–38
Pošmourńy K (1982) Stratiform iron ores in the Crystalline of the north-eastern part of the Bohemian Massif (Krkonoše Mts, Zelezńy Brod region, the Orlické hory Mts). Sbor Geol Věd LG 23: 125–170 (in Czech)
Pouba Z (1970) Pre-Cambrian banded magnetite ores of the Desná Dome. Sbor Geol Věd LG 12: 7–64
Rambousek P (1986) Report of the geological investigation in Malá Úpa area in eastern part of the Krkonoše Mts. Zpr Geol Výzk 1984: 179–180 (in Czech)
Reichmann F (1975) Geology of the exploited fluorite deposits in the Bohemian Massif. Sbor Geol Věd LG 17: 39–58 (in Czech)
Rentzsch J (1981) Mineralogical-geochemical prospection methods in the Central-European Copper Belt. Erzmetall 34: 492–495
Rentzsch J, Knitzschke G (1968) Die Erzmineralparagenesen des Kupferschiefers und ihre regionale Verbreitung. Freib Forschungsh C 231: 189–211
Rentzsch J, Langer M (1963) Fazielle Probleme des Kupferschiefers von Spremberg-Weisswasser. Z Angew Geol 9: 507–513
Rhode G (1971) Über Pentlanditentmischungen in Pyrrhotinen aus Lausitzer Lamprophyren. Ber Dtsch Ges Geol Wiss B 16: 265–269
Richter P (1959) Bemerkungen über ein Vorkommen von Kobaltmanganerz bei Rengersdorf (Lausitz). Z Angew Geol 5: 482–484
Sałaciński R (1979) Chromite contents in serpentinites of the Tąpadła region, Lower Silesia. Arch Miner 35: 37–48 (in Polish)
Svoboda J, Zoubek V (1939) Contribution to the knowledge of the geological setting of the laterite occurrences at Rychnov nad Kněžnou. Věst Stát Geol Úst 15: 1–15 (in Czech)
Sylwestrzak H, Wołkowicz K (1985) A new assemblage of Sn-W-Mo minerals from Stara Góra (Lower Silesia) and its genetic significance. Przegl Geol 2: 73–75 (in Polish)
Szałamacha M, Szałamacha J (1974) Geological and petrographic characteristics of schists

mineralized with cassiterite on the basis of materials from the quarry at Krobica. Biul Inst Geol 279: 59–89 (in Polish)
Szałamacha M (1976) A barite-fluorite deposit in Jeżów Sudecki (Western Sudetes). Kwart Geol 20: 215–140 (in Polish)
Teisseyre J (1973) Metamorphic rocks of the Rudawy Janowickie and Lasocki Grzbiet ranges. Geol Sudet 8: 7–118 (in Polish)
Vaněček M, Patočka F, Pošmourný K, Rajlich P (1985) The use of isotopic composition of ore lead in metallogenic analysis of the Bohemian Massif. Rozpr Čs Akad Věd Ř Mat Přír Věd 95: 114
Wiszniewska J (1984) The genesis of ore mineralization of the Izera Schists in the Kamienickie Range (Sudetes). Arch Miner 40: 115–187 (in Polish)
Zimnoch E (1961) The magnetite series of Kowary. Biul Inst Geol 171: 7–91 (in Polish)
Zimnoch E (1967) Gîtes métamorphosés des minerals de fer dans les Sudètes comparés au point de vue de structure aux autres gîsements analogues. Geol Sudet 3: 251–296
Zimnoch E (1978) Ore mineralization of the Miedzianka deposit in the Sudetes. Biul Inst Geol 308: 91–134 (in Polish)
Zimnoch E (1983) Ore mineralization of the Czarnów deposit (Sudetes). Ann Soc Geol Pol 53: 289–306 (in Polish)

VII Moldanubian Region

VII.A Introduction

J. Fiala

The term Moldanubian region, as used here, follows the concept of Suess (1903), who defined Moldanubicum as the southern half of the Bohemian Massif, encompassing the rivers Moldau and Danube. This area includes two principal tectonostratigraphic units: a high-grade unit – the Moldanubian Zone (or the Moldanubicum s.s.) and a generally low-grade unit – the Teplá-Barrandian Zone (or Bohemicum as proposed by Malkovský 1979).

The Moldanubian Zone contains mainly high-grade gneisses and migmatites of supracrustal origin, orthogneisses, granulites, and numerous Variscan posttectonic granitoid intrusions. The Moldanubian supracrustals have traditionally been subdivided into the Monotonous Group with a variety of quartz-feldspathic gneisses, probably derived from a graywacke precursor, and the Varied Group consisting of a metagraywacke-metapelite assemblage with frequent intercalations of marble, calc-silicate rocks, graphitic schists, metabasites, and quartzites.

The Teplá-Barrandian Zone or, in the broader sense the Bohemicum (cf. Chap. VII.B.1), is largely composed of anchimetamorphic to greenschist-grade Precambrian sedimentary and volcano-sedimentary sequences overlain unconformably by Early Paleozoic strata (Lower Cambrian to Middle Devonian). The Precambrian sequences of the Bohemicum are supposed to be of Late Proterozoic age, as is shown by micropaleontologic correlations with Vendian and Riphean (Konzalová 1988) and sporadic geochronology (see Chap. VII.B.1). On the periphery of the Bohemicum there are several medium-grade crystalline units (e.g., Teplá and Domažlice complexes) which are, or seem to be, connected with the Bohemicum by metamorphic transitions. Variscan granitoid intrusions are located mostly on the periphery of the Bohemicum.

The mutual relations of the Moldanubicum and the Bohemicum have long been the subject of discussions. The contact of both units is either tectonic, obscured by Variscan intrusions, or covered by platform sediments.

Suess (1903) claimed continuity of both units and suggested that they differ only in the grade of Variscan metamorphic overprint. He considered the Moldanubian Monotonous Group as the stratigraphic equivalent of the Barrandian Proterozoic and the Varied Group as the equivalent of the Barrandian Paleozoic. Suess also regarded the Moldanubian granulites as metamorphic counterparts of the Barrandian Cambrian volcanics.

Máška and Zoubek (1960) abandoned the idea of stratigraphic correlations between the Barrandian and Moldanubicum and believed that the latter is a substantially older (Early Proterozoic to Archaean) complex forming also the basement of the Barrandian. According to Zoubek (1965, 1969) and Chaloupský (1978), high-grade metamorphism of the Moldanubicum is of Middle Proterozoic (Grenwillian) age. Later events should be of lower metamorphic grades except for thermal metamorphism related to the Variscan plutonic activity.

In contrast to Máška and Zoubek (1960), Losert (1967), Vejnar (1971), Stettner (1972) and Suk (1974) assume continuity between sedimentary basins of the Barrandian Upper Proterozoic and the Moldanubicum, the latter being affected more by the high-grade metamorphism at the Precambrian/Cambrian boundary (Assyntian orogeny). Ellenberger and Tamain (1980), however, regarded the Teplá-Barrandian as an integral part of the Saxo-Thuringian Zone with no direct relations to the Moldanubicum.

Tollmann (1982) interpreted the structure of the Moldanubian region in terms of large-scale Variscan nappe tectonics. According to this concept the nappe units, rooted within the Erbendorf line, were translated in general to the SE. Granulites and other high-grade metamorphic rocks are taken as remnants of the Gföhl nappe, the uppermost Moldanubian nappe unit. The Gföhl nappe was thrust over the

Dallmeyer et al. (Eds.)
Pre-Permian Geology of Central and Eastern Europe

Drosendorf nappe encompassing mostly the Moldanubian Varied and Monotonous Groups. The author extended the nappe concept also to the Bohemicum that is presumed to be thrust SE onto the Moldanubicum.

Weber and Behr (1983) emphasize the specific position of the Bohemicum and characterise this unit as an "intervening block" between the Saxo-Thuringicum and Moldanubicum that were, during Variscan orogeny, subjects of mutual collision along a suture-type discontinuity of Erbendorf line.

Rajlich (1987) interpreted the entire European Variscides as a major intracontinental shear belt and doubts the existence of any large nappe structure within the Bohemian Massif. This author defined two NW-SE-trending regional structural systems, the Bavarian and Elbe Shear Zones. He regarded the translations along this ductile shear zones as being the principal cause of Variscan orogeny in the Bohemian Massif. He also suggested that the overthrust structures in the Moldanubian Zone are of comparatively small extent and are results of local movements that compensate the main NW-SE-directed translations.

Matte et al. (1990) emphasized a number of analogies in the structure of the Bohemian Massif with that of the French sector of the Variscides. Their NW-SE section through the Bohemian Massif demonstrates a fan-like divergent form of the major structures with two high-grade metamorphic belts (the Saxo-Thuringian of NW vergence and the Moldanubian of E to SE vergence), separated by a less metamorphosed and much less eroded block – the Barrandian. From SE to NW, the Variscan belt is divided into six terranes: Moravian, Drosendorf, Gföhl, Barrandian, Münchberg-Teplá, and Saxo-Thuringian. All these terranes experienced different geological histories and are now separated by major thrusts or strike-slip faults. Major tectonothermal events are attributed to the Variscan orogeny (390–300 Ma), as are the terrane boundaries trending broadly SW-NE to SSW-NNE. Mentioned authors suppose this configuration to be a result of the complex Variscan history that includes an early phase of subduction and closure of an oceanic domain, followed by sporadic ocean floor obduction, continent-continent collision, subduction of continental crust and extensive strike-slip faulting.

Acknowledgments. The author is indebted to Prof. A. Kröner and Dr. F. Patočka for critical reading of the manuscript.

References

Chaloupský J (1978) The Precambrian tectogenesis in the Bohemian Massif. Geol Rdsch 67: 72–90

Ellenberger F, Tamain ALG (1980) Hercynian Europe. Episodes 1: 22–27

Konzalová M (1988) Bohemian Massif: Precambrian biostratigraphy. In: Zoubek V (ed) Precambrian in younger fold belts. Wiley, Chichester, pp 41–47

Losert J (1967) Contribution to the problem of the pre-Assyntian tectogenesis and metamorphism in the Moldanubicum of the Bohemian Massif. Krystalinikum 5: 61–84

Malkovský M (1979) Tektogenese der Plattformbedeckung des Böhmischen Massivs. Knihovna Ústř Úst Geol Praha 53: 176 pp

Máška M, Zoubek V (1960) Area of the Variscan intramontane block. In: Zoubek V (ed) Tectonic development of Czechoslovakia. NČSAV, Praha, pp 25–50

Matte P, Maluski H, Rajlich P, Franke W (1990) Terrane boundaries in the Bohemian Massif: Result of large-scale Variscan shearing. Tectonophysics 177: 151–170

Rajlich P (1987) Variszische duktile Tektonik im Böhmischen Massiv. Geol Rdsch 76: 755–786

Stettner G (1972) Zur geotektonischen Entwicklung im Westteil der Böhmischen Masse bei Berücksichtigung des Deformationsstils im orogenen Bewegungssystem. Z Dtsch Geol Ges 123: 291–326

Suess FE (1903) Bau und Bild der Böhmischen Masse. In: Diener C (ed) Bau und Bild Österreichs. F Temnky Wien, Leipzig, 322 pp

Suk M (1974) Lithology of Moldanubian metamorphics. Čas Mineral Geol 19: 373–379

Tollmann A (1982) Grossräumiger variszischer Deckenbau im Moldanubikum und neue Gedanken zum Variszikum Europas. Geotekton Forsch 64: 1–91

Vejnar Z (1971) Grundfragen des Moldanubikums und seine Stellung in der Böhmischen Masse. Geol Rdsch 60: 1455–1465

Weber K, Behr JH (1983) Geodynamic interpretation of the Mid-European Variscides. In: Martin H, Eder FW (eds) Intracontinental fold belts. Springer, Berlin Heidelberg New York, pp 427–469

Zoubek V (1965) Moldanubikum und seine Stellung im geologischen Bau Europas. Freiberger Forschungsh C 190: 129–148

Zoubek V (1969) Age relations in metamorphic terrains of the Bohemian Massif – some methods and results. Geol Assoc Can Spec Pap 5: 73–80

Zoubek V (1988) Moldanubian region: stratigraphic subdivision, main lithostratigraphic units. In: Zoubek V (ed) Precambrian in younger fold belts. Wiley, Chichester, pp 191–218

VII.B Teplá-Barrandian Zone (Bohemicum)

VII.B.1 Stratigraphy

J. Chaloupský, I. Chlupáč, J. Mašek, J. Waldhausrová, and J. Cháb

1 Introduction

The Bohemicum, an arch-like unit in the center of the Bohemian massif (Fig. 1), was established by Malkovský (1979) and accepted by some geologists (e.g., Chaloupský 1989). For the following description it is useful to subdivide it into two parts, the W one, represented by the continuous Teplá-Barrandian crustal segment, and the E one, composed of various units forming the basement of the Permo-Carboniferous and Cretaceous cover in the NE and E Bohemia and in an adjacent part of Moravia and rising to the surface round its margin.

2 The Teplá-Barrandian (TB) Crustal Segment

The TB segment crops out in the W and central Bohemia, in the NW part of the Železné hory Hills, and in the Central Bohemian Pluton as its roof pendants (Islet Zone); in the N it submerges under the Permo-Carboniferous and Cretaceous cover of the north-central Bohemia. It can be considered an independent tectonostratigraphic unit (Oczlon 1992).

Boundaries of the Bohemicum are tectonic: in the W, it is a shear-fault zone of the Bohemian Quartz Lode. In the S and SE, it is the Central Bohemian Suture, obliterated mostly by the Central Bohemian Pluton; the suture divides two seismically different crustal domains, the Moldanubian and the TB (Tomek and Vrána 1992). In the N, it is the Litoměřice thrust, the suture nature of which is indicated by the meta-ophiolite Mariánské Lázně Complex (MLC; Kastl and Tonika 1984) and a whole-crustal reflection zone dipping to the SE (Tomek and Vrána 1992). The NE boundary hidden under the young cover is usually called the Elbe line and its position, as well as its nature and existence, is an object of controversies. The easternmost boundary is obliterated by the Železné hory Hills Pluton.

The TB segment is composed of a Precambrian basement and its Lower Paleozoic cover. The Precambrian sequence was folded and anchimetamorphosed prior to the first Paleozoic transgression. The intensity of its metamorphism increases continuously toward the W, NW, N, and less impressively toward the NE and E. The age of this metamorphic imprint is uncertain (see Chap. VII.B.4). Granitic and mafic intrusions interpenetrate the Precambrian basement, especially in the W and NW of the segment.

2.1 Precambrian (Late Proterozoic)

The Precambrian age of the described sequence has been known since the third quarter of the 19th century, when the discordance between the Cambrian and its underground was found in the Barrandian area.

Figure 2 shows the chronostratigraphical position and the evolution of opinions on the lithostratigraphical subdivision of the TB Precambrian. The chronostratigraphical position is based on micropaleontological data (Konzalová 1981; Konzalová in Zoubek et al. 1988; Fatka 1994) and K-Ar and U-Pb radiometric ages (Dubanský 1984; Ordynec et al. 1984), which need modern revision. Recently, Dörr et al. (1992) determined the U-Pb Zrn age of a felsic volcanic pebble from a conglomerate of the ŠG – approximately 570 Ma B.P., which indicates that the described sequence could be partly the lowermost Cambrian.

The lithostratigraphy (Fig. 2) is based on rock assemblages and a several thousands of meters thick normal succession of beds in the area S of Prague. It should be noted that the BF represents most of the TB Precambrian (Fig. 1). The total thickness of the TB Precambrian probably exceeds 10 000 m.

The BF represents a shale-siltstone-graywacke flysch-type sequence with numerous mafic volcanic rocks concentrated in NE-SW-trending belts

Dallmeyer et al. (Eds.)
Pre-Permian Geology of Central and Eastern Europe

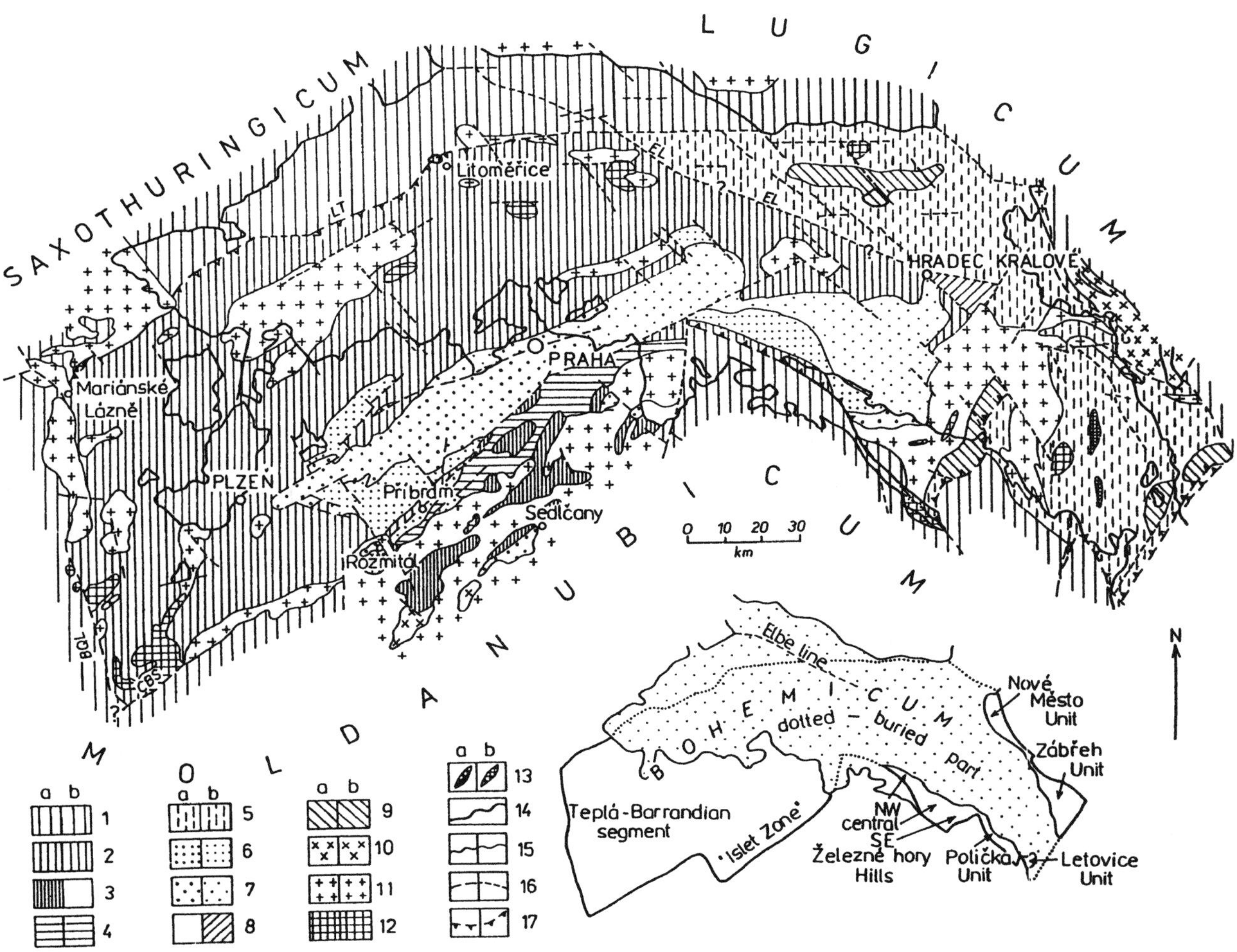

Fig. 1. Pre-Upper Carboniferous geology of the Bohemicum (J. Chaloupský, slightly modified). In the explanations *a* and *b* mean uncovered and covered areas (see the scheme below right). *1* Undistinguished low- to high-grade metamorphosed Precambrian and Lower Paleozoic rocks out of the Bohemicum; *2* TB Precambrian, Kralupy-Zbraslav Group, Blovice Formation, including the MLC; *3* TB Precambrian Kralupy-Zbraslav Group, Davle Formation; *4* TB Precambrian, Štěchovice Group; *5* undistinguished Precambrian and Paleozoic rocks of the NE to E part of the Bohemicum; *6* TB Cambrian; *7* TB Ordovician to Devonian; *8* Upper Devonian + Lower Carboniferous; *free square in a* means absence of the respective unit in uncovered areas; *9* pre-Upper Carboniferous Paleozoic rocks of the NE to E part of the Bohemicum; *10* metamorphosed granitic rocks (orthogneisses) in the roof pendants of the Central Bohemian Pluton and in the Orlice-Klodzko Crystalline unit out of the Bohemicum; *11* granitic intrusions; *12* mafic intrusions (diorites, gabbros); *13* ultramafic rocks; *14* boundary of the Upper Carboniferous and younger cover; *15* other nontectonic geological boundaries; *16* normal and strike-slipe faults; *17* thrust. *BQL* Bohemian Quartz Lode; *CBS* Central Bohemian Suture; *EL* Elbe Line; *LT* Litoměřice thrust

(Fig. 3). Sediments originated from dense currents, turbidites to gravitites, prevail absolutely. Large masses of the gravitites, mixed products of mudflows, sandflows, and slumps, accompany volcanites of the RK and NP belts. Numerous silicites, some of them with stromatolite-like structures (Pouba in Zoubek et al. 1988), frequently occur in the SE half of the outcropping area, beginning by the RK belt (Fig. 3). Black pyrite-enriched shales or slates accompany volcanites of the NP, RK, and SP belts and also occur as intercalations in flysch sediments between the RK and SP belts. Limestones or marbles form scarce, small intercalations in all volcanic belts.

Anchi- to high-grade metamorphosed volcanites of the belts can be characterized as follows (Fig. 4): SP belt – greenschists, amphibolites and allied rocks, scarce soapstone-related types, the chemical composition of which varies among picrites, tholeiitic basalts, and single trachytes (Fiala 1977).

Chronostratigraphy	Interregional lithostratigraphic unit	Local lithostratigraphic subdivision					Original lithology
		Kettner 1917	Röhlich 1965		Mašek and Zoubek jr. 1980, Mašek 1981		
Upper Riphean - Vendian	Brioverian	Postspilitic stage	Postspilitic series		Štěchovice Group (ŠG)		shale-siltstone-graywacke flysh, conglomerates, laminites
		Spilitic stage	Spilitic series	Lečice Beds	Kralupy-Zbraslav Group (KZG)	Lečice Member (LM)	dark siliceous shales, silicites
				Davle F.		Davle F. (DF)	shale-siltstone-graywacke flysh; CA volcanites
				F. with spilites		Blovice F. (BF)	shale-siltstone-graywacke flysh, gravitites; WP (rift-like) and MOR-type volcanites
		Prespilitic stage	Prespilitic series				

Fig. 2. The chronostratigraphical position and the survey of some lithostratigraphical schemes of the Precambrian of the TB segment

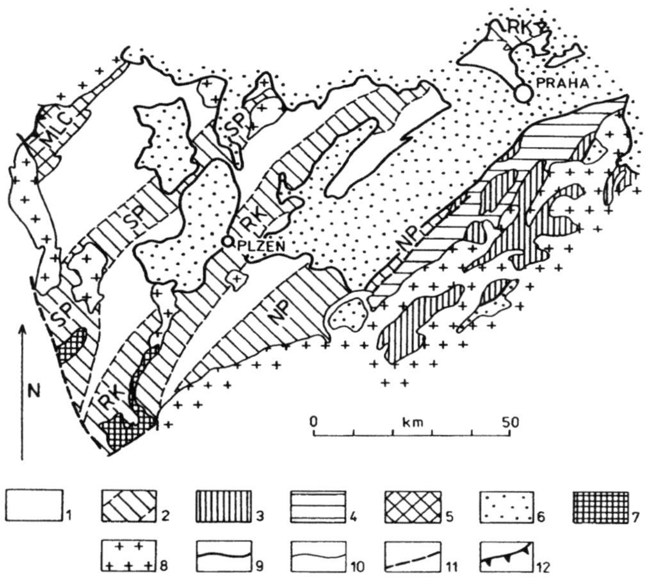

Fig. 3. Scheme of volcanic belts in the Precambrian Blovice Formation. *1* TB Precambrian, Kralupy-Zbraslav Group, Blovice Formation; *2* volcanic belts in the Blovice Formation: *SP* Stříbro-Plasy belt; *RK* Radnice-Kralupy belt; *NP* Nepomuk-Příbram belt; *3* TB Precambrian, Kralupy-Zbraslav Group, Davle Formation; *4* TB Precambrian, Štěchovice Group; *5*: *MLC* metaophiolite Mariánské Lázně Complex; *6* Lower Paleozoic and younger cover; *7* main mafic intrusions (diorites, gabbros); *8* main granitic intrusions; *9* boundary of the cover; *10* other nontectonic geological boundaries (or boundaries of an uncertain nature in the case of the SE edge of the MLC); *11* fault; *12* thrust

Most of the rocks resemble MORB, but ultramafic and alkaline members indicate a WP inter-arc (back-arc) extensional (rift-like) setting.

RK belt – very low- to high-grade metavolcanites. Tholeiitic basalts very similar to MORB predominate. Alkaline rocks are scarce with the exception of a several kilometers long body of alkaline basalts and andesites, which accompanies the RK belt along its NW side W of Prague. SSW of Plzeň, the metabasalts are connected with

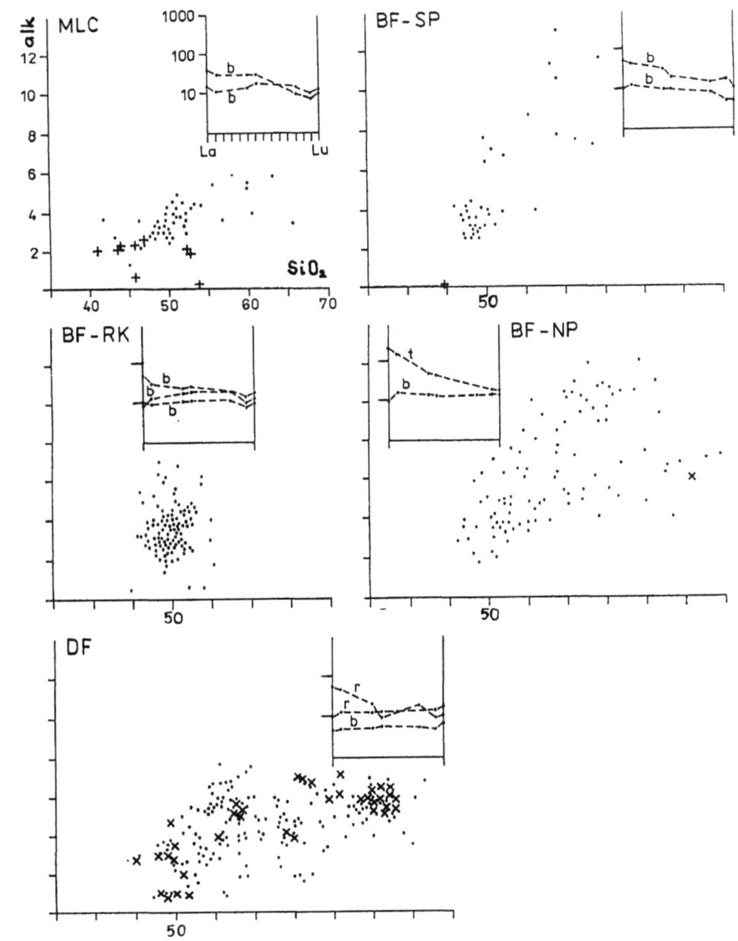

Fig. 4. Volcanites-metavolcanites of the TB Precambrian + Mariánské Lázně Complex (*MLC*), exemplified by the TAS diagrams accompanied by the chondrite-normalised REE distribution patterns of several typical samples (*b* metabasalt; *r* metarhyolite; *t* metatrachyte). *BF* Blovice Formation; *DF* Davle Formation; *SP* Stříbro-Plasy belt; *RK* Radnice-Kralupy belt; *NP* Nepomuk-Příbram belt. *Dots* Metavolcanites, including metagabbros in MLC; *crosses* ultramafic rocks; *oblique crosses* metamorphosed mafic dykes and metatonalite-metagranite suite. Sources of data: Čadková et al., unpubl.; Fiala 1958, 1976, 1977; Hejtman 1966; Kastl and Tonika 1984; Mašek 1984; Mašek et al. 1984; Mrázek, unpubl.; Palivcová 1956; Röhlich 1972; Tonika 1970; Vejnar 1972; Waldhausrová 1984; data of Waldhausrová, unpubl.; compiled by Waldhausrová and Cháb. H_2O- and CO_2-free basis, analyses with more than 4 wt% CO_2 were not used. MLC rocks with more than 55 wt% SiO_2 were described as migmatized in the respective source

metatuffs. The RK belt seems to indicate an ancient spreading center.

NP belt – very low- to low-grade metavolcanites, contact metamorphosed along the NW margin of the Central Bohemian pluton. This wide belt contains two different magmatic series, tholeiitic and alkaline ones. The former is composed of rocks similar to MORB, the latter is represented by the alkaline olivine basalt-trachyte assemblage, which – like the alkaline rocks of the SP belt – indicates a WP inter-arc (back-arc) extensional (rift-like) setting. In the E, in the Příbram Ore District, besides others, Si-rich, probably CA volcanites occur.

The thickness of the BF is estimated to 5000 to 10 000 m, its underground is unknown. The nature of the volcanites as well as the sediments (orogenic graywackes with a variable addition of a continental crustal material) indicates sedimentation on an oceanic or quasi-oceanic crust in back-arc or inter-arc settings (Cháb 1978, 1993; Jakeš et al. 1979). The MLC was considered by Cháb (1978) to be a relic of this crust including slices of serpentinized subcrustal peridotites. New radiometric data indicates, that its history was somewhat different (see Chap. VII.B.4). A new geochemical study (Okrusch et al. 1989) has shown that metamorphosed mafic igneous rocks of the MLC come from two different sources of magma, on the one hand from the primitive mantle source producing MOR-type basalts, on the other hand from the source producing more differentiated alkaline magmas, which indicate a WP setting.

The DF emerges from beneath the ŠG in cores of large brachyanticlines S of Prague and in roof pendants of the Central Bohemian Pluton. It consists of more than 2000 m thick sequence of anchi- to high-grade contact metamorphosed basalts, andesites, dacites, rhyolites, their tuffs, tuffites, and minor shales, siltstones, graywackes, and conglomerates; locally, the volcanites are accompanied by tonalite-granite intrusions. The sequence is topped by the 50- to 200-m-thick black siliceous shales and silicites (the Lečice Member, Röhlich 1965).

According to Waldhausrová (1984), the oldest volcanites and metavolcanites of the DF resemble MOR assemblage or LKT and related Si-enriched differentiates. Most of the volcanites of the DF represent K-poor CA association (Röhlich 1972; Fiala 1976; Waldhausrová 1984), indicating an IA setting (Fig. 4). The youngest tuff intercalations occur within the Lečice Member.

The most natural interpretation of an ancient tectonic setting and a mutual relation of both Formations of the KZG seems to be an island arc-back-arc (inter-arc) system, in which both units were formed contemporaneously, at least partially. Both formations indicate subduction orogeny and extensional processes above and behind a subduction (Wadati-Benioff) zone (Cháb 1993). An accretionary wedge is not preserved in the TB segment.

In the NW of the Železné hory Hills, a low-grade metamorphosed sequence of the BF- and DF-type crops out, which differs from that of the W and central Bohemia by the presence of metaconglomerates in the BF and metamanganolites connected with pyrite-rich black schists forming a local well-defined lithostratigraphic unit in a transitional part between the lower BF- and the upper DF-type portions of the sequence (Fiala and Svoboda 1956; Cháb et al. 1982).

The ŠG is restricted to the S vicinity of Prague. Laminites forming its first 200 to 400 m evolve continuously from the underlying LM. The laminites merge into a 3500- to 4000-m-thick flysch sequence composed of alternating shales, siltstones, and graywackes. There are irregularly distributed distal and proximal facies; the latter are most distinctly accumulated in the middle part of the sequence, where numerous conglomerate intercalations occur. The conglomerates contain besides dominant graywacke and siltstone pebbles even pebbles of volcanites and granitic rocks (Fiala 1948).

According to Kettner (1917), the beginning of the Postspilitic stage reflects an orogenic process; this idea has been broadly accepted up to the present. In contrast, Cháb (1993) assumes that the DF-ŠG transition reflects a cessation of the island arc and a coming of a hemipelagic sedimentation, which means the extinction of an orogenic activity with which the island arc was connected. A revival of a tectonic activity is recorded by the transition from the laminites to the flysch.

The assignment of the Sovolusky Group in the Železné hory Hills (the "Eocambrian" of Fiala and Svoboda 1956) to the ŠG is questionable; Fiala and Svoboda (1956) pointed out a possible discordance between the Sovolusky Group and its underground. The Group is composed of slightly metamorphosed shales, siltstones, conglomerates, and volcanites of an alkaline nature. The Cambrian age of this Group cannot be excluded.

2.2 Paleozoic

The main area of distribution of pre-Variscan Paleozoic rocks is the Barrandian, where Cambrian to Middle Devonian sequences show complex facies development and rich paleontological evidence, being famous as one of the world's classic areas since the works of Joachim Barrande. From the evolutionary and paleogeographical viewpoints we may distinguish: the Příbram-Jince basin filled with Cambrian deposits (Brdy Mts. etc.) and the Prague basin between Prague and Pilsen (Havlíček 1980, 1981) filled with Ordovician to Devonian sediments and volcanics (Figs. 1 and 7); recently, the development of this area has been summarized by Chlupáč et al. (1992).

Occurrences in the metamorphic mantle of the Central Bohemian Pluton belonging to the "Islet Zone" are only sporadically paleontologically documented; the sequence in the Sedlčany-Krásná Hora "Islet" is most complete (Chlupáč 1989). Cambrian to Devonian deposits bordered by apophyses of the Pluton are evidenced in the vicinity of Rožmitál SW of Příbram Ore District (Havlíček 1977).

The very low- to low-grade metamorphosed pre-Variscan Paleozoic of the NW part of the Železné hory Hills shows a close similarity to that of the Islet Zone. It continues toward the N under the Cretaceous cover. In contrast, non- to anchimetamorphosed Paleozoic rocks in the basement of the Bohemian Cretaceous Basin NE of Prague, which are known from boreholes, represent the NE to E continuation of the Barrandian.

All these occurrences are mutually similar and give evidence of a common development from Ordovician to Devonian.

In the E vicinity of Hradec Králové, an unmetamorphosed Upper Devonian to Lower Carboniferous sequence lying transgressively on low-grade schists of Ordovician or uncertain age was bored.

2.2.1 *Cambrian*

In the Barrandian, Cambrian deposits generally overlie the Proterozoic basement with a marked angular unconformity.

The main area of deposition was the Příbram-Jince basin in the SE flank of the Barrandian. In this intermontane depression an up to more than 2500-m-thick sequence of variegated conglomerates and other terrigenous continental deposits (Kukal 1971) was accumulated already during the Early Cambrian, showing only rare and episodic lagoonal influences (shale with *Kodymirus vagans*). The basal Cambrian conglomerate contains numerous granite pebbles, the age of one of which was determined 594 Ma B.P. by the U-Pb Zrn method (Dörr et al. 1992).

A marine transgression reached the Barrandian area in an early Middle Cambrian and fossiliferous deposits are known not only from the Příbram-Jince basin but also from the graben-like depression in the NW flank of the Barrandian (the Skryje-Týřovice area) and from the Železné hory Hills. The Middle Cambrian shales of the Jince Formation contain, especially at Jince and Skryje, a world-famous trilobite and other faunas, which show relationships to the Mediterranean, rather cool-water subprovince of the Prototethys realm (Havlíček 1971; Fatka 1994).

A marine regression falls during a late Middle Cambrian and no Upper Cambrian deposits are paleontologically evidenced (locally, continental sediments may, however, be present). The Late Cambrian epoch was distinguished by uplift and erosion, which reduced the extent of Cambrian deposits, and by strong volcanic activity.

The Cambrian Seník Formation in the Železné hory Hills is only poorly paleontologically documented, and its relation to the Sovolusky Formation is not known with certainty. No paleontologically evidenced Cambrian deposits are known from the Islet Zone and from the basement of the Bohemian Cretaceous Basin.

The volcanic activity is demonstrable during the whole Cambrian sedimentation; its peak falls in the Late Cambrian-early Tremadocian. Subaerial CA andesites, dacites, and rhyolites prevail absolutely. They exhibit higher K-concentrations than the volcanites of the Precambrian DF. Their chemical composition (Fig. 5) indicates a crustal source (Waldhausrová 1971), while the $^{87}Sr/^{86}Sr$ average ratio 0.704 (Vidal et al. 1975) resembles I-type granitic rocks. The volcanites indicate mature IA or ACM tectonic settings (Patočka et al. 1993).

2.2.2 *Ordovician*

The beginning of the Ordovician is marked by a stepwise marine transgression in Tremadocian to Llanvirnian times.

In the Barrandian (Prague basin) a sequence of marine siliciclastic and volcanic rocks accumulated in thickness frequently exceeding 2000 m. Diverse

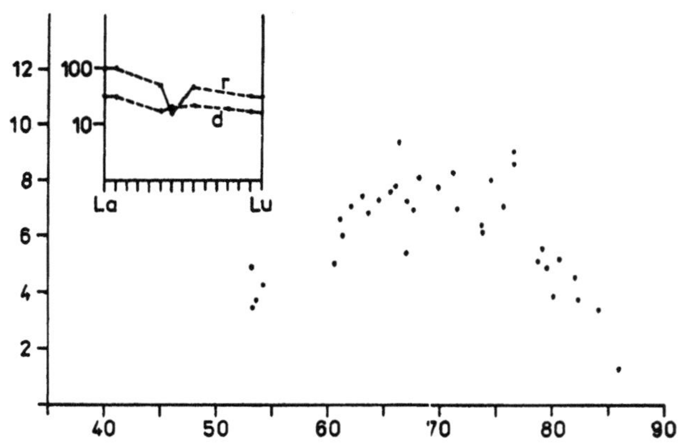

Fig. 5. Cambrian volcanites of the Barrandian in the TAS diagram with the example of chondrite-normalized REE distribution patterns in dacite (*d*) and rhyolite (*r*). Sources of data: Waldhausrová 1971; Šindelářová 1982

shallow-water sandstone and somewhat deeper-water mudstone facies contain locally rich faunas with dominant benthos (brachiopods, trilobites). These allow one to distinguish environmentally controlled life zones, changing influx of Mediterranean and North European elements, and turning points of event nature (Havlíček 1981, 1982; Chlupáč and Kukal 1988). The organic development was interrupted in the Kosovian (Hirnantian) by climatically controlled regressive events reflecting the cooling (Gondwana glaciation) documented even by diamictite layers in various parts of the Bohemian massif (Brenchley and Štorch 1989; Štorch 1990).

The Ordovician of the Islet Zone and the Železné hory Hills shows a similar facies development, although thicknesses are smaller and the paleontological evidence scarcer than in the Barrandian. The Ordovician near Rožmitál is developed as a monotonous shale sequence (Voltuš Formation), in which Arenigian and Caradocian ages are demonstrated by fossils (Havlíček 1977).

Ordovician sediments continue from the Barrandian and the Železné hory Hills into the basement of the Bohemian Cretaceous Basin, where anchi- to low-grade metamorphosed shales dominate. A lesser degree of the paleontological documentation and a similarity with Proterozoic rocks here caused difficulties in stratigraphic assignment.

The Cambro-Ordovician CA volcanism ceased during the early Ordovician; the last traces of this activity are known in Llanvirn. Contemporaneously, during Ashgill and Llanvirn, basaltic volcanism came and rapidly prevailed. This volcanic activity was concentrated in a NE-SW-trending belt, which is oblique to the present Variscan structures. Early to late Ordovician, predominantly basaltic volcanic rocks, reach a thickness up to 1000 m (Komárov Complex) in this belt (Fiala 1970, 1971). Volcanites are strongly altered; that is why their composition, as well as the composition of the Silurian volcanites, is exemplified by the MnO-TiO_2-P_2O_5 diagrams (Fig. 6) according to Mullen (1983). Patočka et al. (1994) have stressed their alkaline nature on the base of new detailed data. Their origin was probably influenced by a crustal extension behind the Cambro-Ordovician magmatic arc during the early Ordovician; in our opinion, they represent an evidence of a destruction of the N Gondwana margin.

In the Islet Zone and the Železné hory Hills, CA felsic metavolcanites occur, which are considered to be Ordovician in age by some authors; but, especially in the Železné hory Hills, their stratigraphic assignment is uncertain.

2.2.3 *Silurian*

A pronounced turning point – the onset of graptolitic shales – is generally characteristic for the beginning Silurian. The graptolitic shale facies persisted obviously during the whole Early Silurian and was replaced only locally, particularly close to the volcanic centers in the NW flank of the Barrandian, by shallow water carbonates and mixed volcano-sedimentary rocks with abundant benthic faunas (Fig. 7; Horný 1962; Havlíček and Štorch 1990; Kříž 1991).

The Upper Silurian strata are distinguished by increasing carbonate contents reflecting a warming-up climate and by rich development of diverse benthic assemblages known typically from the Barrandian (Havlíček and Štorch 1990).

Silurian in the Islet Zone and in the Železné hory Hills is paleontologically evidenced in its lower part (Llandoverian and Wenlockian); Upper Silurian carbonates are developed in some places, but determinable fossils are known from

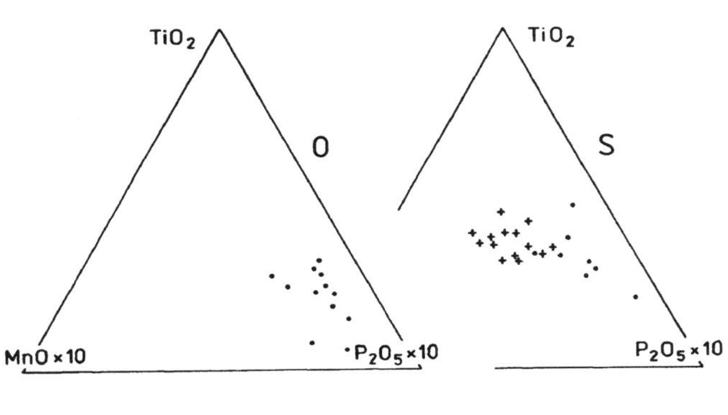

Fig. 6. Ordovician (*O*) and Silurian (*S*) mafic volcanites (*dots*) and subvolcanic mafic and ultramafic rocks (*crosses*) of the Barrandian (P. Štorch, unpubl.)

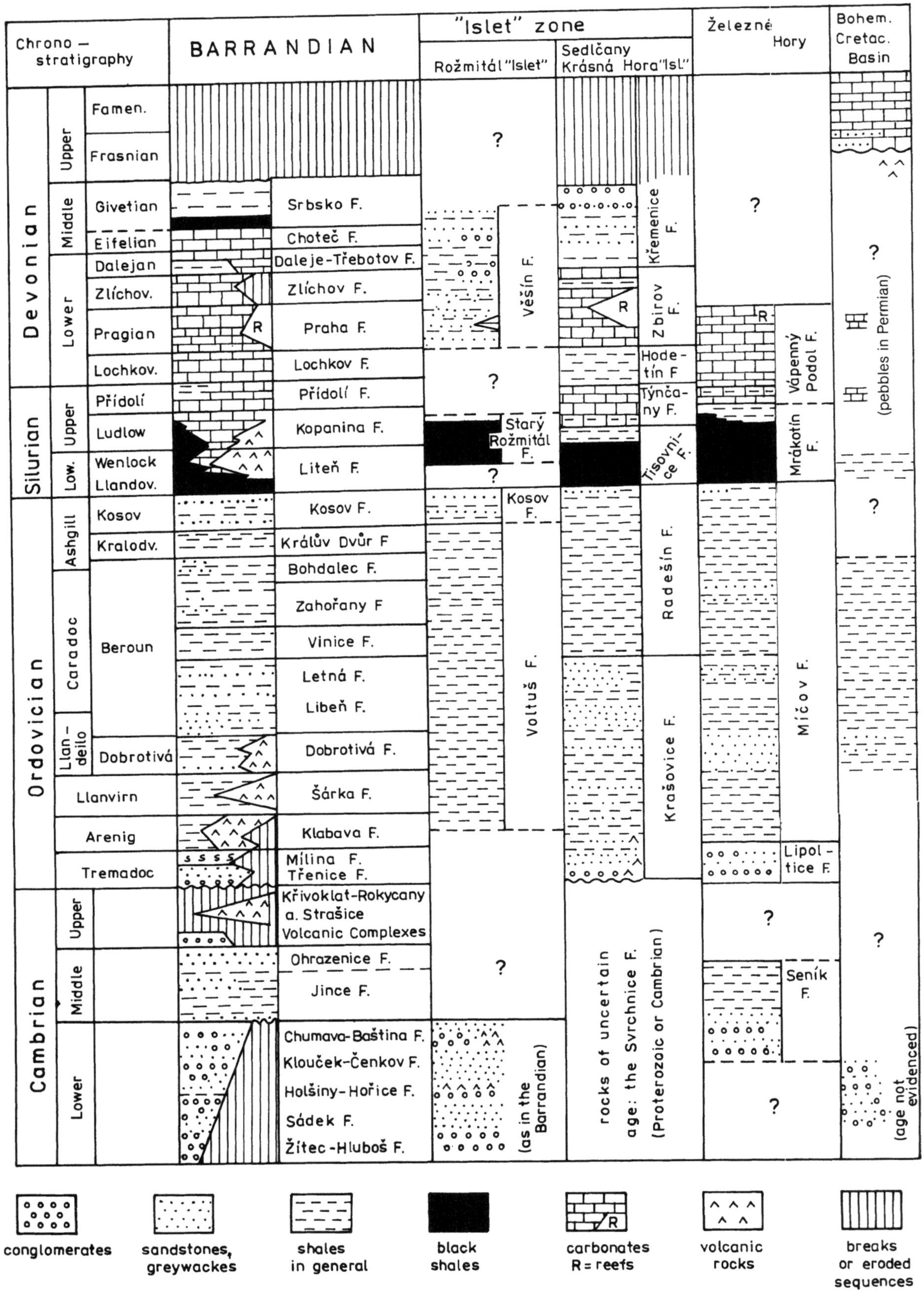

Fig. 7. Paleozoic stratigraphy in various parts of the Teplá-Barrandian segment of the Bohemicum

the Železné hory Hills only (dark limestones with *Scyphocrinites* in the lower part of the Vápenný Podol Formation).

No effects of late Caledonian orogenic events are demonstrable in the TB Paleozoic and particularly in the Barrandian the carbonate deposition and evolution of faunas continued from the Silurian into the Early Devonian without any marked breaks or lithological changes. Consequently, the Barrandian offers suitable conditions for defining the standard chronostratigraphic units of the uppermost Silurian (Přídolí) and Lower Devonian, which have here the internationally approved boundary stratotypes (Chlupáč et al. 1992; Kříž et al. 1986).

Silurian volcanic activity was concentrated in the NW flank of the Barrandian, where volcanic accumulations caused onset of shallow-water conditions starting with Wenlockian (Horný 1962; Kříž 1991); besides effusive rocks numerous mafic and ultramafic sills and dykes occur in the whole area of distribution of Silurian rocks and even in late Ordovician in the Barrandian. In comparison with the Ordovician volcanites, the Silurian volcanic and subvolcanic rocks show rather tholeiitic nature (Patočka et al. 1993) although alkaline rocks are present, too. The total association represents a WP rift-like type of the magmatic activity (Fiala 1978a,b), indicating progressive crustal extension (Patočka et al. 1993).

2.2.4 Devonian

The Lower and Middle Devonian of the Barrandian, preserved in the core of the brachysyncline SW of Praha, is characterized by complex carbonate development with repeating combinations of shallow-water high-energy bioclastic up to reef limestones and deeper-water low-energy micritic ones. Rich benthic and planctonic faunas allow detailed zonal subdivisions and evaluation of environmentally controlled changes and events of a great correlative value. Owing to this, the Barrandian became the type area of Lower Devonian standard stages Lochkovian and Pragian, and Zlíchovian and Dalejan as subdivisions of Emsian, which have here their stratotypes and reference sections (Chlupáč 1982, 1988; Chlupáč and Kukal 1988).

The carbonate development continues from the Silurian up to the Eifelian-Givetian boundary interval, where it is interrupted by the onset of black shales reflecting the Kačák event. Overlying coarser siliciclastics of the Roblín Member evidently reflect the onset of Variscan movements in the neighboring areas.

The metamorphic Lower Devonian, partly reef carbonates, is also distributed in the Islet Zone (the Zbirov Formation) and in the Železné hory Hills (the Vápenný Podol Formation, upper part). The single exception is represented by the occurrence near Rožmitál, where the Devonian is mostly siliciclastic (shales, siltstones, sandstones, and lenses of petromictic conglomerate in the uppermost part, Havlíček 1977). The youngest pre-Variscan unit of the Islet Zone (the Křemenice Formation, Chlupáč 1989) indicates coarse terrigenous input and probably reflects even here the early onset of Variscan movements.

Volcanic rocks are extremely scarce in the Devonian; several occurrences of basaltic rocks are known only in the Barrandian.

2.2.5 Upper Devonian-Lower Carboniferous

A transgressive Upper Devonian (Upper Frasnian, Famennian) clastic and carbonate sequence was found in boreholes in the NE part of the Bohemian Cretaceous Basin near Hradec Králové. It postdates the first effects of the Variscan orogeny in the central Bohemia. The lithology and fossils show a close similarity to some West Sudetes and Moravian occurrences (e.g., the Moravian Karst). The carbonate sedimentation continued conformably up to the Early Carboniferous and was followed by clastic Culm deposits (Chlupáč and Zikmundová 1976; Čech et al. 1989).

3 Units in the NE and E of the Bohemicum

According to Chaloupský (1989), the Precambrian of the TB type forms the whole Bohemicum and only in some places is covered by Early Paleozoic rocks. Assumed Precambrian (± Paleozoic) rocks emerge from beneath the Late Paleozoic and Cretaceous cover in the NE and E (Zvičina hills, Nové Město, and Zábřeh units), in the SE (Letovice and Polička units), and in the S of the SE part of the Železné hory Hills (Fig. 1).

Metamorphic rocks of the Zvičina Hills NW of Hradec Králové are considered to be Paleozoic (e.g., Cambrian, Havlíček 1980) without paleontological documentation. These metagraywackes, metaarkoses, quartzites, and phyllites seem to exhibit a close affinity to some Paleozoic occurrences in West Sudetes (Lugicum).

The Nové Město unit crops out in the Orlické hory Mts., where the Uhřínov ?thrust delimits it from the Orlice-Klodzko Crystalline Complex. Toward the SE, it merges into the Zábřeh unit. From the lithological point of view, the Nové Město unit is similar to the BF in the TB segment. It is composed of low- to medium-grade metamorphosed shales, siltstones, graywackes, and mafic volcanites, the chemical composition of which resembles that of the OFB (Opletal et al. 1980, 1990).

Similar rock assemblage also exists in the Zábřeh unit. Its knowledge is incomplete, however, and moreover, finds of Paleozoic rather than Precambrian microfossils cast doubts on its stratigraphic homogeneity or assignment (Koverdynský and Konzalová 1986).

The Letovice unit is composed of medium-grade schists, among which mafic (± ultramafic) rocks prevail. Especially these are considered to be dismembered and metamorphosed ophiolites (Jelínek et al. 1984). Similar rock assemblages continue to the N under the young cover; their relation to the Nové Město and Zábřeh units is unknown.

The Vítanov unit in the SE part of the Železné hory Hills and the Polička unit are composed of medium- to high-grade metasediments and metavolcanites.

A common feature of this heterogeneous group of the units is the absence of any kind of data concerning a high age of their protoliths. Their Precambrian (or Paleozoic) age is only hypothetical and in the case of the Zábřeh unit it is definitely uncertain; of course, the paleontological data need verification and the position of paleontologically productive rocks must also be revised.

Paleontologically well-documented Paleozoic occurrences in the E and NE of the Bohemicum are scarce, and their relationships and tectonic position remain questionable.

In the SE of the Železné hory Hills (Hlinsko Zone), the Lower Silurian graptolitic slates of the Llandoverian and Wenlockian age are evidenced by finds of graptolites (Horný 1956). The assignment of underlying quartz-rich rocks (supposed Ordovician) and the overlying Rychmburk graywackes, however, is doubtful, though micropaleontological finds in the latter exclude the Precambrian age (Konzalová and Vachtl 1976).

Close to the E edge of the Bohemicum, the only demonstrable Lower Paleozoic (Devonian) occurrence is near Městečko Trnávka, where calcareous siltstones yielded Givetian fossils (Chlupáč 1961). The nearly siliciclastic sequence of the "Mírov culm" syncline shows similarity to the Lower Carboniferous rocks of the Drahany Upland. Basal conglomerates of the Mírov culm rest on the rocks of the Zábřeh unit and merge upwards into alternating sandstones, siltstones, and shales. Sporadic finds of land plants are not sufficient for chronostratigraphic assignment.

4 Synopsis

The characteristic evolutionary features of the TB segment, the single part of the Bohemicum available for such summarization, can be briefly expressed as follows:

1. Predominant flysch type siliciclastic development of the Precambrian KZG and volcanism, indicating an IA and WP rift-like (back-arc or inter-arc) settings.
2. The KZG-ŠG (DF-ŠG) transition, interpretable as a "transgressive" event, a cessation of an IA and a transition to a hemipelagic environment, in which the laminites of the lowermost part of the ŠG originated.
3. The transition of the laminites into the flysch of the ŠG – a reactivation of tectonic processes.
4. An impressive angular unconformity between the Precambrian (+ lowermost Cambrian?) and the transgressive Cambrian or Ordovician. An important change of a plate-microplate boundary configuration along the N Gondwana margin.
5. Predominantly continental development of the Lower Cambrian, irregularly distributed in dependence on pretransgressive morphology and Cadomian to post-Cadomian tectonic regime.
6. The Middle Cambrian marine ingression followed by the late Middle Cambrian to Upper Cambrian regression. The Middle Cambrian faunas are relative to Mediterranean (N Gondwana margin) faunas.
7. Persistent Cambrian CA volcanism culminating during the late Cambrian-early Tremadocian time interval; the volcanism indicates a mature island arc or active continental margin settings.

8. A marked Lower Ordovician transgression and the siliciclastic development of the whole Ordovician, almost without carbonates. Faunistic relationships point to a rather cool peri-Gondwanian paleoclimatic zone.

9. The Ordovician basaltic volcanism showing alkaline features with the main maximum in the Arenig to Llandeilo, which probably indicate a beginning destruction of the N Gondwana margin. The extinction of the CA volcanism in the Llanvirnian.

10. The influence of the Gondwana glaciation in the late Ordovician, Kosovian (Hirnantian) sediments.

11. The beginning of the Silurian marked by the transgressive event, the onset of the graptolitic shales sedimentation; the beginning of a continental drift from cold to warm climatic zones.

12. Silurian sedimentation marked by widely distributed graptolitic shale facies in the lower and carbonate facies in the upper part of the sequence.

13. The alkaline and tholeiitic tendency of the Silurian volcanism, indicating a continuation of an extensional regime.

14. A demonstrable conformity between the Silurian and the Devonian evidenced by uninterrupted lineages of most fossil groups in terms of diverse carbonate facies of the Barrandian.

15. Richly differentiated, mostly carbonate, development of the Lower Devonian and Eifelian, with local reef facies; very sporadic volcanic activity.

16. The onset of the siliciclastic sedimentation near the Eifelian-Givetian boundary in the Barrandian, intepretable as a reflection of Variscan orogenic processes. The Givetian siliciclastic sediments, the last Lower Paleozoic rocks preserved in the Barrandian.

References

Brenchley PJ, Štorch P (1989) Environmental changes in the Hirnantian (Upper Ordovician) of the Prague Basin, Czechoslovakia. Geol J 25: 165–181

Čech S, Havlíček V, Zikmundová J (1989) Upper Devonian and Lower Carboniferous in north-eastern Bohemia. Věstn Ústřed Ústavu Geol 64: 65–75

Cháb J (1978) Návrh litostratigrafické a litologické terminologie pro svrchní proterozoikum tepelsko-barrandienské oblasti. Věstn Ústřed Ústavu Geol 53: 43–60

Cháb J (1993) General problems of the TB (Teplá-Barrandian) Precambrian, the Bohemian massif, Czechoslovakia. Věstn Česk Geol Ústavu 68: 1–6

Cháb J, Bouška V, Jelínek E, Pačesová M, Povondra P (1982) Petrology and geochemistry of the Upper Proterozoic FeMn-deposit Chvaletice (Bohemia, Czechoslovakia). Sb Geol Věd Ložisková Geol Mineral 23: 9–68

Chaloupský J (1989) Major tectonostratigraphic units of the Bohemian Massif. Geol Soc Am Spec Pap 230: 101–114

Chlupáč I (1961) Orientační výzkum některých menších výskytů devonu na Drahanské vysočině. Zprávy Geol Výzkumech Roce 1960: 89–95

Chlupáč I (with contributions by Jaeger H and Zikmundová J) (1972) The Silurian-Devonian boundary in the Barrandian. Bull Can Petrol Geol 20: 104–174

Chlupáč I (1982) The Bohemian Lower Devonian stages. Cour Forschungsinst Senckenb 55: 345–400

Chlupáč I (1988) The Devonian of Czechoslovakia and its stratigraphic significance. In: McMillan N, Embry AF, Glass DJ (eds) Devonian of the World, vol I: Regional syntheses. Can Soc Petrol Geol, Calgary, pp 481–497

Chlupáč I (1989) Stratigraphy of the Sedlčany – Krásná Hora Metamorphic "Islet" in Bohemia (Proterozoic? to Devonian). Čas Mineral Geol 34: 1–16

Chlupáč I, Kukal Z (1988) Possible global events and the stratigraphy of the Palaeozoic of the Barrandian (Cambrian-Devonian, Czechoslovakia). Sb Geol Věd Geol 43: 83–146

Chlupáč I, Zikmundová J (1976) The Devonian and Lower Carboniferous in the Nepasice bore in East Bohemia. Věstn Ústřed Ústavu Geol 51: 269–278

Chlupáč I, Havlíček V, Kříž J, Kukal Z, Štorch P (1992) Paleozoikum Barrandienu (kambrium-devon). Český Geol Ústav, Praha, 292 pp

Dörr W, Fišera M, Franke W (1992) Cadomian magmatic events in the Bohemian massif – U-Pb data from felsic magmatic pebbles. Abstr 7th Geol Worksh Styles of superposed Variscan nappe tectonics, Czechoslovakia. Czechoslovak Academy of Science, Praha, 3 pp

Dubanský A (1984) Stanovení radiogenního stáří kalium-argonovou metodou. (Geochronologická data z Českého masívu v oblasti ČSR). Sb Věd Pr Vys Šk Báňské Ostravě 30: 137–170

Fatka O (1994) Review of Precambrian and Cambrian microfossil assemblages in the Barrandian area (Czechoslovakia). Z Angew Geol (in press)

Fiala F (1948) Algonkické slepence ve středních Čechách. Sb Stát Geol Ústavu 15: 399–612

Fiala F (1958) Hlavní typy hornin v širším okolí Pramenů v Císařském lese. Geol Práce Zošit 50: 5–70

Fiala F (1970) Silurské a devonské diabasy Barrandienu. Sb Geol Věd Geol 17: 7–89

Fiala F (1971) Ordovický diabasový vulkanismus a biotitické lamprofyry Barrandienu. Sb Geol Věd Geol 19: 7–97

Fiala F (1976) Calc-alkaline trends of the Proterozoic volcanics of the Barrandian. Čas Mineral Geol 21: 31–54

Fiala (1977) Proterozoický vulkanismus Barrandienu a problematika spilitů. Sb Geol Věd Geol 30: 2–247
Fiala F (1978a) Proterozoic and Early Paleozoic volcanism of the Barrandian-Železné hory zone. Sb Geol Věd Geol 31: 71–90
Fiala F (1978b) The TiO_2-K_2O-P_2O_5 diagram and tectonomagmatic relations of the volcanics of the Barrandian area. Věstn Ústřed Ústavu Geol 53: 333–346
Fiala F, Svoboda J (1956) Problém subkambria a subkambrického zalednění v Železných horách. Sb Ústřed Ústavu Geol 22: 257–309
Havlíček V (1971) Stratigraphy of the Cambrian of Central Bohemia. Sb Geol Věd Geol 20: 7–52
Havlíček V (1977) The Paleozoic (Cambrian-Devonian) in the Rožmitál area. Věstn Ústřed Ústavu Geol 52: 81–94
Havlíček V (1980) Development of Paleozoic basins in the Bohemian Massif (Cambrian-Lower Carboniferous). Sb Geol Věd Geol 34: 31–65
Havlíček V (1981) Development of a linear sedimentary depression exemplified by the Prague Basin (Ordovician-Middle Devonian, Barrandian area, Central Bohemia). Sb Geol Věd Geol 35: 7–48
Havlíček V (1982) Ordovician in Bohemia: development of the Prague Basin and its benthic communities. Sb Geol Věd Geol 37: 103–136
Havlíček V, Štorch P (1990) Silurian brachiopods and benthic communities in the Prague Basin (Czechoslovakia). Rozpr Ústřed Ústavu Geol 48: 1–275
Hejtman B (1966) A contribution to the petrography and petrochemistry of the Jílové zone (Central Bohemia). In: Fediuk F, Fišera M (eds) Paleovolcanites of the Bohemian Massif. Charles University, Praha, pp 37–49
Horný R (1956) Nové poznatky o biostratigrafii skutečsko-hlinského siluru. Věstn Ústřed Ústavu Goel 31: 128–131
Horný R (1962) Das mittelböhmische Silur. Geologie 11: 873–916
Jakeš P, Zoubek J, Zoubková J, Franke W (1979) Graywackes and metagraywackes of the Teplá-Barrandian Proterozoic area. Sb Geol Věd Geol 33: 83–122
Jelínek E, Pačesová M, Mísař Z, Martinec P, Weiss Z (1984) Geochemistry of a dismembered metaophiolite complex, Letovice, Czechoslovakia. Trans R Soc Edinb Earth Sci 75: 37–48
Kastl E, Tonika J (1984) The Mariánské Lázně metaophiolite complex (West Bohemia). Krystalinikum 17: 59–76
Kettner R (1917) Versuch einer stratigraphischen Einteilung des böhmischen Algonkiums. Geol Rdsch 8: 169–182
Konzalová M (1981) Some Late Precambrian microfossils from the Bohemian massif and their correlation. Precambrian Res 15: 43–62
Konzalová M, Vachtl J (1976) On the age of the Rychmburk Greywacke. Věstn Ústřed Ústavu Geol 51: 129–138
Koverdynský B, Konzalová M (1986) Problematika stratigrafického zařazení zábřežské skupiny. Věstn Ústřed Ústavu Geol 61: 159–168
Kříž J (1991) Silurian of the Prague Basin (Bohemia) – tectonic, eustatic and volcanic controls on facies and faunal development. Spec Pap Palaeontol 44: 179–203
Kříž J, Jaeger H, Paris F, Schönlaub HP (1986) Přídolí – the fourth subdivision of the Silurian. Jahrb Geol Bundesanst 129: 291–360
Kukal Z (1971) Sedimentology of Cambrian deposits of the Barrandian area (Central Bohemia). Sb Geol Věd Geol 30: 53–100
Malkovský M (1979) Tektonogeneze platformního pokryvu Českého masívu. Knihovna Ústřed Ústavu Geol 53: 176
Mašek J (1981) Ke geologii proterozoika jv. křídla Barrandienu. Korelace proterozoických a paleozoických stratiformních ložisek 6: 14–28
Mašek J (1984) Svrchnoproterozoické vulkanity od Pňovic (u Rožmitálu pod Třemšínem). Věstn Ústřed Ústavu Geol 59: 77–83
Mašek J, Zoubek J (1980) Návrh vymezení a označování hlavních stratigrafických jednotek barrandienského proterozoika. Věstn Ústřed Ústavu Geol 55: 121–123
Mašek J, Havlíček V, Odehnal L, Straka J, Šalanský K, Štych J (1984) Vysvětlivky k základní geologické mapě ČSSR 1:25 000 12–423 Davle. Ústřed Ústav Geol, 67 pp
Mullen ED (1983) MnO-TiO_2-P_2O_5 a minor element discriminant for basaltic rocks of oceanic environments and its implications for petrogenesis. Earth Planet Sci Lett 62: 53–62
Oczlon MS (1992) Gondwana and Laurussia before and during the Variscan orogeny in Europe and related areas. Heidelb Geowiss Abh 53: 1–56
Okrusch M, Schüssler U, Seidel E, Kreuzer H, Rose D (1989) Metabasite der Zone Tepla-Taus (ZTT). Ber DFG-Projekt Ok/29-1. Geological Survey of Lower Saxony, Hannover
Opletal M, Domečka K, Čech S, Čuta M, Fajst M et al. (1980) Geologie Orlických hor. Oblastní Regionální Geol ČSR, Ústřed Ústav Geol, 202 pp
Opletal M, Jelínek E, Pecina V, Pošmourný K, Poubová E (1990) Metavulkanity jihovýchodní části lugika, jejich geochemie a geotektonická interpretace. Sb Geol Věd Geol 45: 37–64
Ordynec GJ, Žukova VJ, Habásko J (1984) Prevariská uranová mineralizace v proterozoiku Českého masívu. Čas Mineral Geol 29: 69–77
Palivcová M (1956) Geologicko-petrografické poměry v Kozohorském pásmu a jeho nejbližším okolí (Dobříšsko). Sb Ústřed Ústavu Geol Geol 22: 311–396
Patočka F, Vlašímský P, Blechová K (1993) Geochemistry of Early Paleozoic volcanics of the Barrandian Basin (Bohemian Massif, Czechoslovakia); implications for paleotectonic reconstructions. Jahrb Geol Bundesanst 136: 873–896
Röhlich P (1965) Geologische Probleme des mittelböhmischen Algonkiums. Geologie 14: 373–403
Röhlich P (1972) Petrografické poměry v severní části jílovského pásma. Sb Geol Věd Geol 22: 7–64
Šindelářová I (1982) Geochemie vulkanitů křivoklátsko-rokycanského pásma a strašického pásma. Thesis, Karlova Univ, Praha

Štorch P (1990) Upper Ordovician-Lower Silurian sequences of the Bohemian Massif, central Europe. Geol Mag 127: 225–239

Tomek Č, Vrána S (1992) Deep seismic reflection profiling in the South Bohemian Moldanubicum. Abstr 7th Geol Worksh Styles of superposed Variscan nappe tectonics, Czechoslovakia. Czechoslovak Academy of Science, Praha, 65 pp

Tonika J (1970) Mariánskolázeňský metabazitový komplex. Thesis, Charles Univ, Praha

Vejnar Z (1972) Regionally metamorphosed volcanic rocks from the West Bohemian metabasite belt. Krystalinikum 9: 131–156

Vidal Ph, Auvray B, Charlot R, Fediuk F, Hameurt J, Waldhausrová J (1975) Radiometric age of volcanics of the Cambrian "Krivoklat-Rokycany" complex (Bohemian Massif). Geol Rdsch 64: 563–570

Waldhausrová J (1971) The chemistry of the Cambrian volcanics in the Barrandian area. Krystalinikum 8: 45–75

Waldhausrová J (1984) Proterozoic volcanic and intrusive rocks of the Jílové zone in Central Bohemia. Krystalinikum 17: 77–97

Zoubek V, Cogné J, Kozoukharov D, Kräutner HG (eds) (1988) Precambrian in younger fold belts. Wiley, Chichester, 885 pp

VII.B.2 Structure

J. Holubec

The classic Barrandian region represents the type area of the Cadomian (Assyntian) orogeny of Central Europe (Kettner 1917; Svoboda et al. 1966; Holubec 1966, 1968). Transgression of Middle Cambrian and Ordovician rocks succeeded after strong erosion and almost complete beveling of the Cadomian structures. A well-known angular unconformity exists between Cadomian and Variscan structures in the Barrandian. Small-scale structural elements of two Cadomian deformation stages are also preserved in basal Cambrian pebbles.

The Proterozoic sequences are monotonous and consist of alternating graywackes, aleuropelites, pelites, and of types originated by their combination. Volcanic bodies occur in various levels. The primary structures make it possible to distinguish following lithofacies: (a) rocks with even bedding planes and planar structure. Rhythmic-graded bedding and internal lamination are characteristic; (b) rocks of irregular, lenticular, brecciated and fragmentary or olistromatic inner structures; (c) homogenous types. The discrimination and distribution of these types are decisive for the understanding of regional structures. In the Teplá-Barrandian area the following lithostratigraphic groups were distinguished in ascending order: the Rabštejn-Úslava groups A, B, the Zvíkovec group C and the Štěchovice group D (Fig. 1).

In 1990, pre-orogenic unconformity between Úslava and Zvíkovec groups was discovered by Holubec (Fig. 4). It divides the Proterozoic succession into two units differing in lithology, character of volcanism, and type of folding. There are two major cycles of volcanicity. The first is in the groups A, B, the second in the C group. The unconformity occurs in two places.

The first sharp unconformity exists at the basis of the Plzeň volcanic belt and is called the Balkovy unconformity. It cuts all the lithostratigraphic units of the Úslava group. The volcanic belt begins in the Hoher Bogen Mts. in Bavaria and extends in the NE to Plzeň and Mladá Boleslav. It is an intracontinental arc of basaltic rocks representing deep seated tectonic element of the

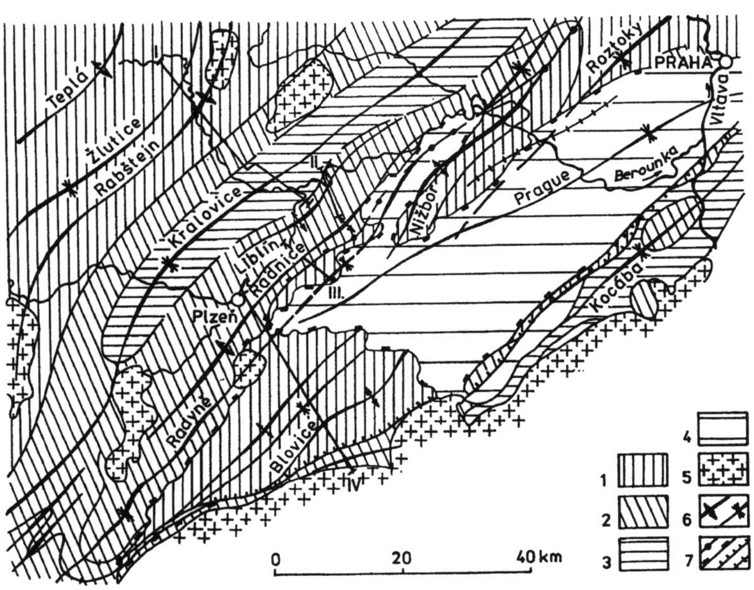

Fig. 1. Simplified geological map of the Teplá-Barrandian area. Proterozoic, groups folded during cadomian orogeny: *1* Rabštejn-Úslava; *2* Zvíkovec; *3* Štěchovice; *4* Lower Paleozoic rocks of Prague basin folded during Variscan orogeny; *5* granite; *6* anticline, syncline, *7* lithofacies boundary, unconformity, thrust fault. Location of cross sections *I–IV* given in Figs. 2 and 3.

Dallmeyer et al. (Eds.)
Pre-Permian Geology of Central and Eastern Europe

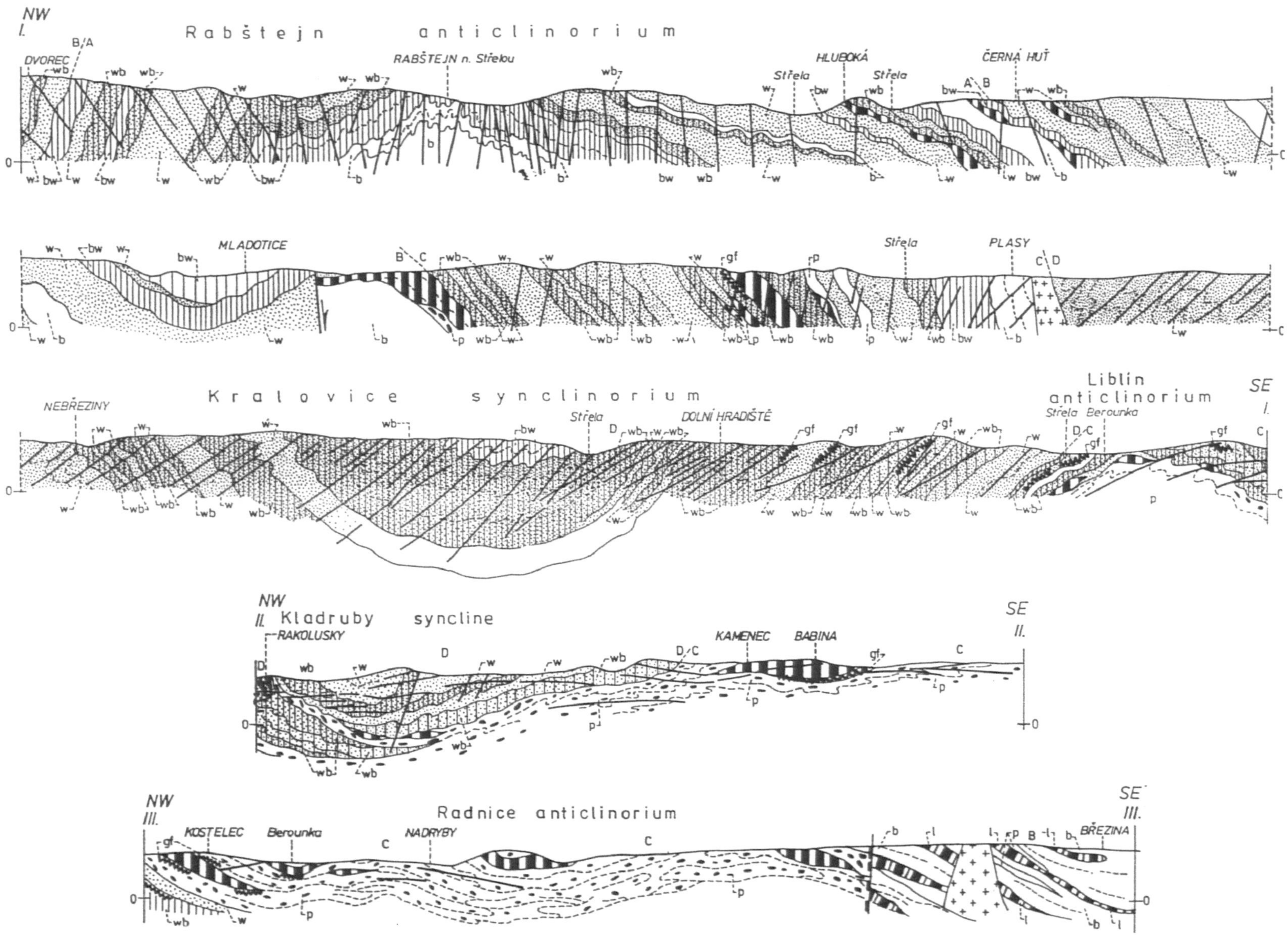

Fig. 2 I–III. Cross sections of the Střela river (I) of the Radnice brook (II) and across the central volcanic zone (III)

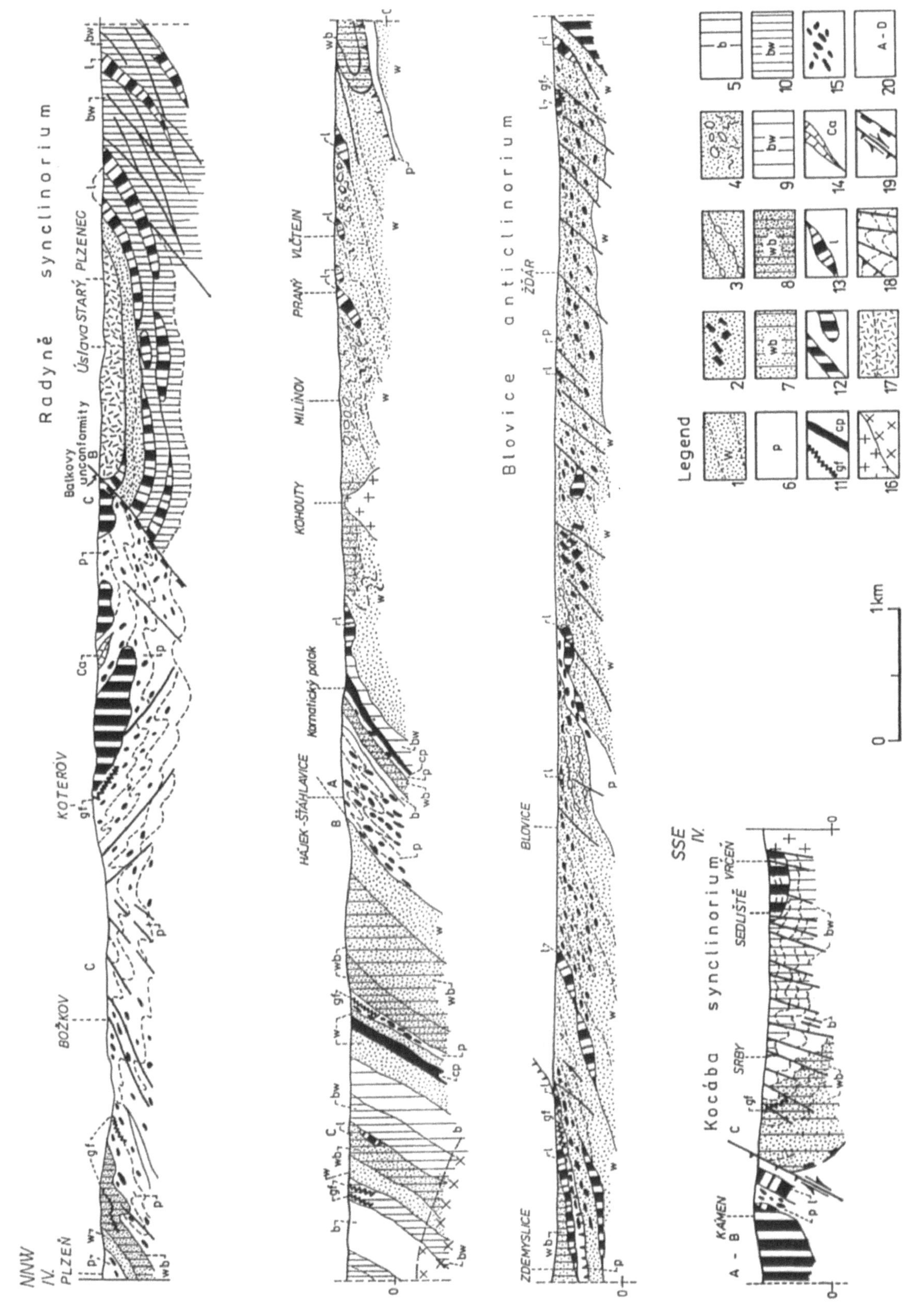
NNW
IV.
PLZEŇ
BOŽKOV
KOTEROV
Balkovy unconformity
ÚSLAVA STARÝ PLZENEC
Radyně synclinorium
HÁJEK-ŠŤÁHLAVICE
Kornatický potok
KOHOUTY
MILÍNOV
VLČTEJN
ZDEMYSLICE
BLOVICE
ŽDÁR
Blovice anticlinorium
KAMEN
SRBY
SEDLIŠTĚ
VRČEŇ
Kocába synclinorium
SSE
IV.
Legend
1km

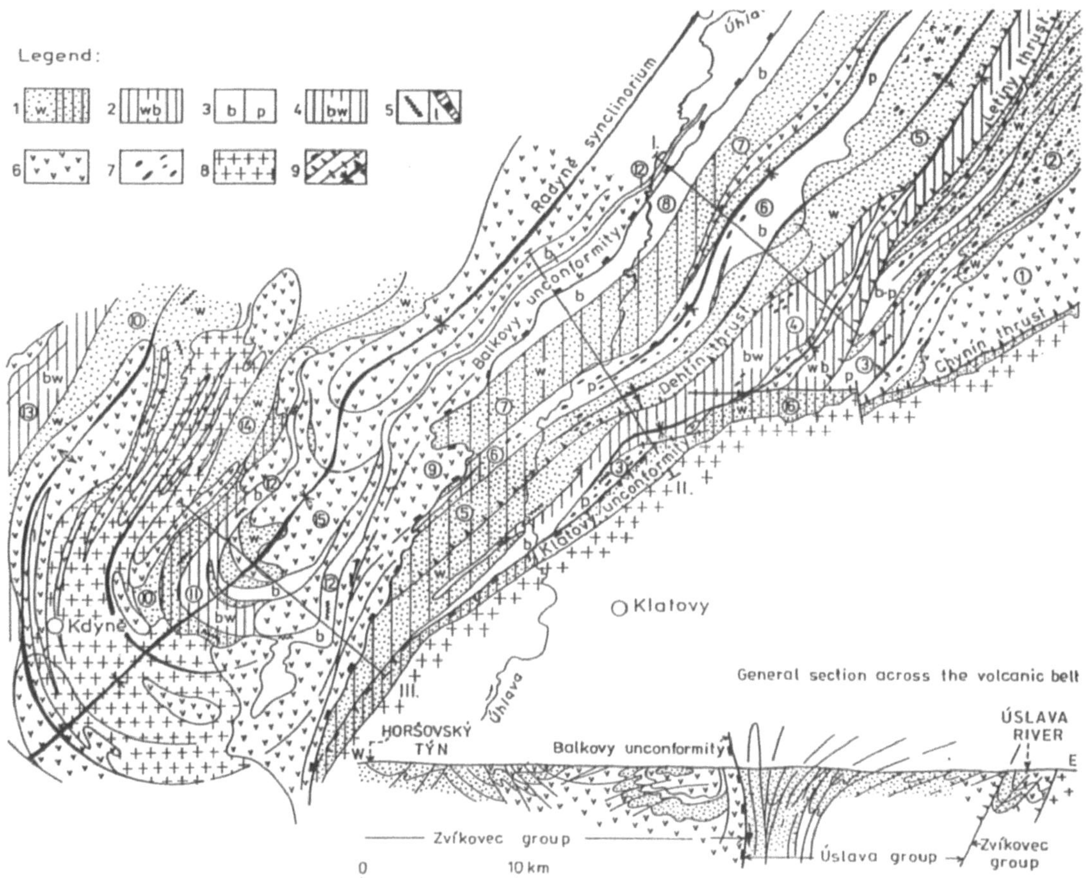

Fig. 4. Geological map through the central part of the Barrandian Proterozoic. Legend to Figs. 4 and 5: *1* massive and platy graywackes; *2* graywackes with aleuropelite layers; *3* aleuropelites, pelites; *4* aleuropelites with graywacke interbeds; *5* graphitic slates, lydite; *6* volcanics; *7* fragmentary olistostromatic facies; *8* plutonic rocks; *9* unconformity, thrust, fault, anticline, syncline. *Circled numbers* denote lithostratigraphic units in the Úslava groups: *1* volcanics of Buková Hill; *2* Blovice graywackes; *3* Měčín aleuropelites; *4* Kbelnice graded aleuropelites; *5* Vlčtejn graywackes; *6* Nezvěstice pelites; *7* Kozel graywackes; *8* Šťáhlavy aleuropelites. In stratigraphic order very likely 1, 2 + 5, 3 + 6, 4, 7, 8. In the Zvíkovec group from bottom to top: *9* Bítov volcanics: *10* Koloveč graywackes; *11* Modlín banded aleuropelites; *12* Černíkov black aleuropelites; *13* Zahořany banded graywackes; *14* Přetín graywackes; *15* Chudenice volcanics; *16* Žinkovy graywackes

Fig. 3. Cross section of the Úslava river. Summary captions to Figs. 2 and 3. *1* Graywackes; *2* primary brecciated graywackes; *3* lenticular graywackes; *4* graywackes with chaotic structures and conglomerates; *5* massive and laminated aleuropelites; *6* pelites; *7* graywackes with aleuropelite layers to 30%; *8* graywackes with aleuropelite layers above 30%; *9* aleuropelite with graywacke interbeds to 30%; *10* aleuropelite with graywacke layers above 30%; *11* graphitic slates, black shales; *12* volcanics; *13* lydite; *14* calcareous layer; *15* fragmentary facies; *16* granite, contact metamorphism; *17* Ordovician rocks; *18* bedding cut by S_3 surfaces and minor thrusts; *19* major thrust unconformity; *20* lithostratigraphic groups

Bohemian Massif. The second Klatovy unconformity lies at the basis of Žinkovy graywackes (Fig. 4, 16) which unconformably overlie various members of the Úslava groups. Basal conglomerate was not found.

An unconformity has been proposed, but not fully proved by geologists between the stratigraphically higher C and D groups.

All the minor and major fold structures and related metamorphic zonation in the Teplá-Barrandian domain are products of Precambrian Cadomian orogeny. It created shallow fold and thrust belts generally of semiplatform type. Some zones NW of Plazeň underwent intensive folding and, in contrast, others display a simple tectonic pattern, the rocks being virtually undeformed internally. Planar $S_{1\text{-}3}$ structures, thrusts, and thrust faults dominate in the Úslava group. The inner deformation of rocks around the community Blovice is the weakest in the whole Barrandian, and several volcanic bodies with primary circular caldera-like forms are well preserved. This complex is supposesd to lie on an older crystalline basement.

Cross sections across the whole area provide an insight into the lithostratigraphy and inner structures of the Proterozoic, and they are at the same time of wider importance for international correlation. The 80-km-long cross sections are shown in Figs. 1, 2, and 3. The principal structure of the Teplá Highland is the Rabštejn anticlinorium. It is a simple structure and in its limbs, bedding planes predominate. In the core, the beds became erected and cut by vertical S_3 surfaces. In the following Kralovice synclinorium the inner structures are complicated due to folding and dense shearing by S_3 surfaces.

The Liblín anticlinorium is situated at the lithofacies boundary and contains graphitic slates with pyrite, mined in the past. Here the flysch development of the Teplá Highland ends and a different irregular lithology of central parts of the basin begins. The structure is again very complicated due to thrusting. The axial S_3 cleavage is moderately inclined, and recumbent folds occur. The Kladruby syncline is simple. The NW limb is steep to overturned to the SE.

The Radnice anticlinorium is a flatly lying structure of the Plzeň volcanic belt. The rocks are in fragmentary dark pelitic facies. The $S_{1\text{-}3}$ surfaces and folds are subhorizontal.

The Radyně synclinorium (Fig. 3) is cut by the Balkovy unconformity. The pelitic rocks of fragmentary facies and volcanic bodies of the Zvíkovec C group occur in the NW, and banded aleuropelites with thick lydites of the Úslava B group in the SE. The SE limb of the Radyně synclinorium is steeply inclined and overturned more to the SW (Figs. 4, 5 I–III).

The Blovice anticlinorium in the Úslava river is composed mainly of coarse-grained graywackes of fragmentary facies with breccias and a horizon of conglomerates at the village Chlumánky. The NW limb is monoclinal, flat-lying without folds with local thrusting to the SE. The SE limb is steep and the beds form slices, being overturned and thrusted over the Kocába synclinorium. In the direction to the SW, the rocks are more intensively compressed, dipping steeply, and bedding planes are overturned. The same is valid

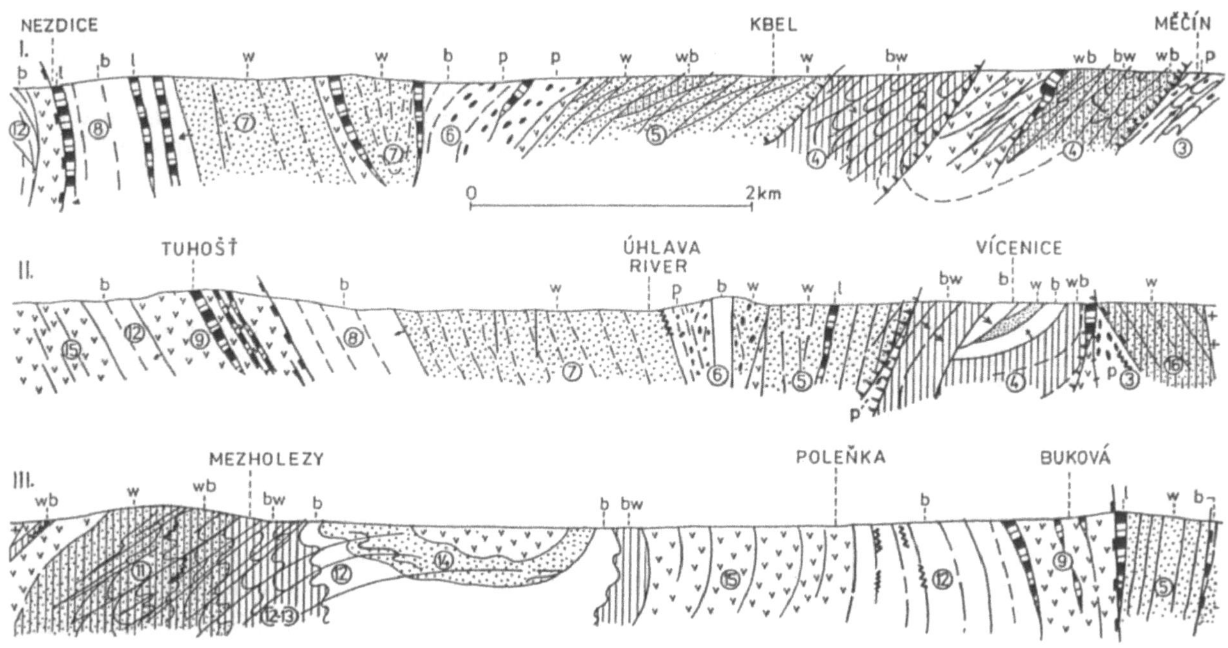

Fig. 5. Cross sections to the map in Fig. 4

for unconformity surfaces. A horst-like arrangement of graywacke and aleuropelitic zones is characteristic feature (Fig. 5 I–III) of the whole Úslava B group. The larger thrusts are designated Dehtín, Letiny, and Chynín. Many younger, mostly Hercynian, faults and thrusts along the granite pluton complicate the situation.

References

Holubec J (1966) Stratigraphy of the Upper Proterozoic in the core of the Bohemian Massif (The Teplá Barrandian Region). Rozpr Česk Akad Věd Řada mat přír Věd Praha 76 (S 4): 1–62

Holubec J (1968) Structural development of the geosynclinal Proterozoic and its relation to the deeper zones of the Earth's crust. Rozp Česk Akad Věd Řada mat přír Věd Praha 78(S 8): 1–77

Holubec J (1986) Structural map of the Bohemian Massif. Scale 1:500 000. Ústř Úst Geol Praha

Holubec J (1990) Struktura Českého masívu. Structure of the Bohemian Massif (Eng. res.). Ústř Úst Geol Praha: 1–214

Kettner R (1917) Versuch einer stratigraphischen Einteilung des böhmischen Algonkiums. Geol Rdsch Leipzig VIII (5/8): 169–188

Svoboda J (1966) Algonkian-Lower Carboniferous. In: Svoboda J et al. (eds) Regional geology of Czechoslovakia. Part I. The Bohemian Massif. Geol Surv Acad Prag: 281–340

VII.B.3 Igneous Activity

A. DUDEK

1 Introduction

One of the principal characteristics of the Bohemicum is the relative scarcity of plutonic rocks, which are so frequent in the Moldanubicum and Saxothuringicum. The overall positive gravity field of the whole area indicates that plutonites are missing also in deeper crustal levels. The bodies of small size are concentrated in the SW, near the boundary with Moldanubicum. The shapes and occurrences of these massifs are influenced by the NNW-SSE strike of the West Bohemian deep fault and by the NE-SW striking folds of the Upper Proterozoic rocks. The individual massifs may be divided into two groups: (a) basic massifs with some acidic differentiates, and (b) granitoid massifs. In both groups the presence of pre-Variscan (Cadomian) and Variscan members is taken for granted.

2 Basic Massifs

The basic massifs are represented mainly by gabbros and gabbronorites, diorites, gabbrodiorites, with subordinate ultramafic layers and syenitic and tonalitic differentiates. In marginal parts the bodies show frequently preferred orientation of the mineral constituents, but they are clearly discordant and postdating the regional folding and metamorphism of the Upper Proterozoic complexes. The phyllites to two-mica gneisses in their mantle are metamorphosed at the contact up to garnet-cordierite and orthopyroxene-cordierite hornfelses (Vejnar 1973, 1980). The main basic massifs are concentrated along the West Bohemian deep fault, whereas in the other parts of the Bohemicum only small bodies are found (Fig. 1).

The Kdyně (Neukirchen) massif is the largest body in the region and is situated on the crossing of the West and Middle Bohemian deep faults. According to Vejnar (1986) the massif has an imperfect layered structure, with a lower zone composed of olivine gabbros, olivine gabbronorites, leucogabbronorites, and olivine ferrodiorites, with rare layers of serpentinites, phlogopite peridotites, hortonolitites, and anorthosites. The middle zone is composed of diorites and the upper of quartz diorites. The Fe/Mg ratios of olivine in individual rock types vary from 0.22 to 0.91, the ratios of orthopyroxene from 0.20 to 0.83. In the massif two differentiation trends are apparent: (a) olivine gabbro-olivine gabbronorite-fayalite ferrodiorite, and (b) diorite-tonalite-trondhjemite. The age of this massif is not unequivocaly proved. It is supposed to be a Lower Cambrian intrusion, but Fischer et al. (1968) give a K/Ar hornblende date of 392 ± 12 Ma.

The Poběžovice massif is elongated in the NE-SW direction. It has a zoned structure with uralite gabbro in the center, gabbrodiorites nearer to the periphery, and diorites at the border. In places, layered structures as a result of fractionated crystallization are developed (Vejnar 1973). Quartz diorites and trondhjemites form a second generation of rocks, intruding mostly into the zones of brecciation and deformation. Sodium played a prominent role in the differentiation of the massif.

In the zone of the West Bohemian deep fault near Načetín, a circular boss composed mainly of gabbronorites of tholeiitic affinities is present. It displays a pronounced layered structure; the lower zone is of olivine gabbronorite and subordinate wehrlite, the main zone of gabbronorite and the upper zone of biotite norite (Vejnar 1980). In the norite, layers enriched with hercynite and corundum were found. At the margin the intrusion is accompanied by younger hornblende-biotite diorite to quartz diorite.

Peculiar Fe-rich rocks are present in a small hypabyssal boss by Mutěnín. This ring intrusion is composed of olivine and hypersthene diorites, amphibole diorite, and ferrosyenite, with typical

Dallmeyer et al. (Eds.)
Pre-Permian Geology of Central and Eastern Europe

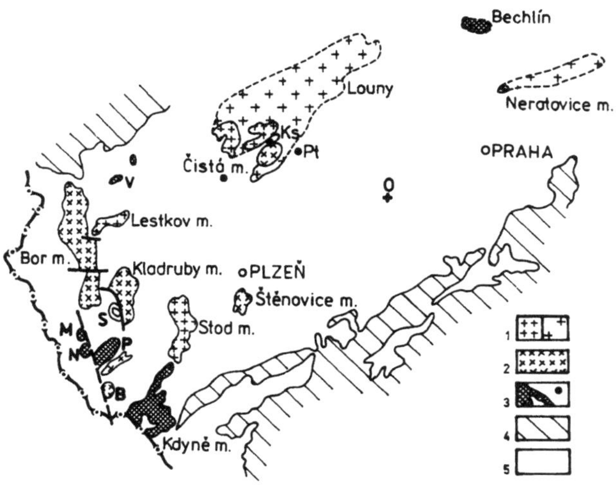

Fig. 1. Geological sketch of the plutonic bodies in the western part of Bohemicum. *1* Cadomian granitoids (outcrops, beneath sedimentary cover); *2* Variscan granitoids; *3* gabbroid massifs; *4* Variscan granitoid plutons in Saxothuringicum and Moldanubicum; *5* Upper Proterozoic volcanosedimentary complexes. *B* Babylon stock; *Ks* – Kosobody gabbro; *M* Mutěnín ring intrusion; *N* Načetín intrusion; *O* Otmíče diatreme with granitoid enclaves; *P* Poběžovice massif; *Pt* Petrovice melagabbro; *S* Sedmihoří ring intrusion; *V* Teplá and Výškovice gabbros

Fe-rich minerals – fayalite, eulite, ferrohastingsite, and lepidomelane (Vejnar 1975; Tonika 1979). The rocks have strong affinities to K-rich alkali basalts (Tonika 1979).

Small basic intrusions form also a part of the Teplá crystalline complex and of the Mariánské Lázně metabasite complex. These rocks are younger than the metabasites which were metamorphosed under HP-MT conditions, but were also influenced by younger metamorphic phases (development of garnet, Jelínek et al. 1994). They have tholeiitic affinities. The largest body composed of gabbros and peridotites is near Teplá, but detailed petrological data are missing. It lies in the center of a pronounced positive gravity anomaly, which may be followed farther to the NE for a distance of about 150 km and indicates the presence of basic magmatites in the depth. Radiometric U/Pb age was determined from the Výškovice body as $496 \binom{+106}{-4}$ and 494 ± 2 Ma, demonstrating its pre-Variscan origin (Bowes and Aftalion 1991).

Similar small bodies of gabbroid rocks may be found in volcanic zones of the Upper Proterozoic trending SW-NE. They are only some tens or hundreds meters long and their composition is variable. Well known is the Petrovice kaersutite melagabbro (Ulrych et al. 1976), which is characterized by high Ti and Na contents and has the chemistry of olivine alkali basalt. Another body near Kosobody (Fediuková and Fediuk 1980) is composed of hornblende-biotite gabbro with tholeiitic affinities. It is supposed that all this small bodies belong to the Variscan magmatic cycle, but are predating the intrusion of granitoid massifs. The hornblende K/Ar age of the Petrovice melagabbro was determined as 380 ± 15 Ma (Ulrych et al. 1976).

The above characteristics demonstrates that the basic plutonic rocks belong to different evolution series, beginning tholeiitic basalts over transition basalts to olivine alkali basalts. The tectonic and chronologic interpretation of this phenomenon is open; most probably the alkali types belong to a younger magmatic series (Fig. 2).

Contrary to these undoubtedly magmatic bodies, the basic rocks of the Neratovice complex (mostly covered by upper Cretaceous Sediments) are interpreted as products of metasomatic recrystallization of Proterozoic tholeiitic basalts (Fediuk et al. 1966). The age of this process was determined by K/Ar methods as Cadomian (535 to 573 Ma, Šmejkal and Melková 1969). Also of Proterozoic age (550 Ma) are pyroxene diorites

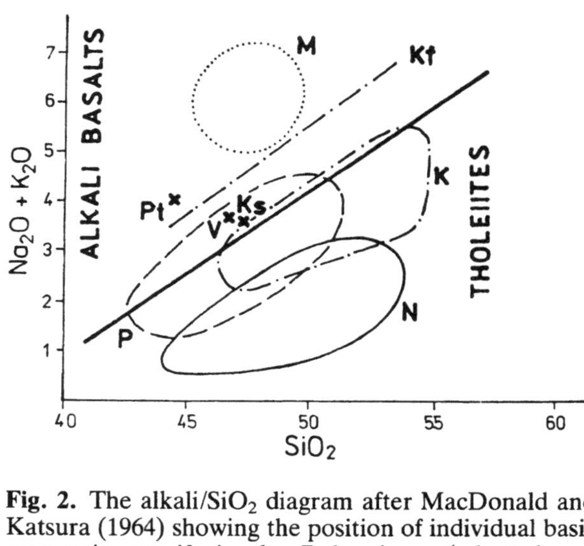

Fig. 2. The alkali/SiO_2 diagram after MacDonald and Katsura (1964) showing the position of individual basic magmatic massifs in the Bohemicum (adapted and completed after Vejnar 1980). *K* Kdyně massif; *Kf* ferrogabbro of the Kdyně massif; *Ks* Kosobody gabbro; *M* Mutěnín ferrodiorite ring intrusion; *N* Načetín intrusion; *P* Poběžovice massif; *Pt* Petrovice melagabbro; *V* Výškovice gabbro

from a massif found by drilling Bechlín in the basement of the North Bohemian cretaceous basin (Cháb 1975).

3 Granitoid Massifs

The Cadomian granitoid rocks are bound mainly on the units characterized by weaker metamorphism of the MP-LT type, absence of Variscan plutonism, weaker crust (Beránek et al. 1975) and positive gravity anomalies. The shape of the Cadomian granitoid massifs (Lestkov, Louny, Neratovice and partly Stod massif) is elongated SW-NE, in the strike of the Upper Proterozoic complexes.

The Stod massif is composed of alternating stripes of biotite hornhlende granodiorite to quartzdiorite and of younger biotite granite. The quartz diorite (K/Ar ages 510–530 Ma, Šmejkal and Vejnar 1967) may belong to the differentiation suit of the basic Kdyně massif, in the continuation of which the Stod massif lies (Tonika and Vejnar 1966). Some Cadomian hornblende biotite quartz diorites (K/Ar age 556 Ma) were also established in the mainly Variscan Bor massif. Some further massifs are attributed to this Cadomian granitoide suit, like the biotite granodiorite of Lestkov and the large Louny massif, mostly hidden below Upper Carboniferous and Upper Cretaceous sediments, but without geochronological evidence. The Tis granite of the Louny Massif is, according to its mineralogy and chemistry, comparable with the undoubtable pre-Variscan Rumburg granite of the Lusatian pluton (see Chap. VI.D). Its pre-Variscan age is supported also by its magnetic fabric (Chlupáčová et al. 1975) and by extremely low Th and U contents. The low Th and U contents in Cadomian granites is very distinctive in the Bohemian massif and differentiates them clearly from Variscan granites, where Th and U contents is several times higher than the clark values (Chlupáčová 1974).

Below Lower Paleozoic rocks of the Prague basin a hidden Precambrian granitic massif seems to be present as evidenced by a weak negative gravity anomaly (Buday et al. 1969) and by numerous inclusions of granitic rocks in Ordovician diabase diatreme in the vicinity of Otmíče, W. from Beroun. These biotite leucotonalites to granodiorites, thoroughly studied by Fiala (1983) are petrochemically similar to Proterozoic Cadomian granitoids of the Lusatian and Brno Plutons.

Very interesting is also a group of small bodies of aegirine-augite riebeckite syenites found S of Domažlice (Vejnar 1979), which are interpreted as a late fractionation product of alkali olivine magmas belonging to the Proterozoic volcanic series.

The massifs of Variscan age are bound either to the West Bohemian deep fault system on the boundary with Moldanubicum, or they form small stocks on some important faults or their crossing (Čistá and Štěnovice massifs). The large Bor massif is composed of old hornblende biotite quartz diorites, which were enclosed in Variscan porphyritic biotite granite (Vejnar et al. 1969). The Kladruby massif, located more to the East in a major anticline, is elongated in the N-S direction. This body is composed mainly of porphyritic biotite granites and granodiorites (Neužilová and Vejnar 1966; Vejnar 1967), which show petrochemical affinities to the Bor massif.

The minor intrusions – stocks – represent probably subvolcanic features of the Carboniferous calc-alkali volcanism. In the zone of the Bohemian Quartz Lode there is the small Babylon massif, composed of biotite and two-mica granites (Vejnar 1977). The most basic of these bodies is the hornblende biotite granodiorite of the Štěnovice massif, with some petrological and petrochemical affinities to the tonalitic group of the Middle Bohemian Pluton (Klomínský 1965).

The most striking ring structure has the Sedmihoří stock on the western side of the Kladruby massif (Dudek 1957; Vejnar 1967; Bartošek et al. 1969; Gnojek and Šťovíčková 1974). Its differentiation trends from porphyritic biotite granite to Na-rich granite of Sn-W-Li formation. The last minor intrusion with concentric structure is the Čistá stock, intruding into the Cadomian Louny Massif and composed of biotite granodiorites (Klomínský 1963b). On its W margin, an important zone of fenitization developed, where fenites and nephelinic syenites and cancrinite- nepheline dyke rocks may be found. These Variscan alkaline rocks are unique in the Bohemian massif, and their intricate genesis and history has been discussed by many authors (Klomínský 1963a; Kopecký et al. 1970; Chlupáčová et al. 1975; Kopecký 1987). In the latter paper they are interpreted as manifestations of a granitoid-alkali ultrabasic ring intrusion originated by subsurface cauldron subsidence.

The larger Železné hory pluton crops out in the eastern part of the Bohemicum, on the SE end of the Železné hory Mts. It is a complex plutonic body in which elements attributed to the Cadomian orogeny, as well as to the prevailing Variscan event, are present. The geochemical trend of the rocks also confirms that they do not constitute one continuous evolution series (Minařík et al. 1983). The southern part of the pluton is built by the Cadomian Všeradov complex, composed of albite granites, numerous albite granite dykes, and layers of quartz-keratophyres and keratophyres in the Upper Proterozoic sediments of the Vítanov group. This Na-rich volcano-plutonic complex is very similar to the acid volcanic and plutonic members of the Jílové zone in the Middle Bohemian Pluton (Vachtl 1977). The main part of the body is formed of hornblende biotite granodiorites, quartz diorites, and tonalites which include numerous xenolithes and small bodies of gabbros and diorites of various composition (K/Ar ages 332–336 Ma, Šmejkal 1964). These types build a strip running from W to E and then turning to the N, below the Upper Cretaceous sediments. There it was encountered by several drillings (Chaloupský 1974) and its shape ascertained by geophysical methods. The youngest member of the Železné hory pluton is the K-granite in the northern part, very variable in structural development (K/Ar dating 288 Ma, Šmejkal 1964).

A petrographic resemblance to the Všeradov complex is found in the small Cadomian Chvaletice massif, penetrating into marginal faults on the SW border of the Železné hory Mts. Pebbles similar to this biotite granites were found in the Upper Proterozoic conglomerates in the Železné hory Mts. (Fiala 1951).

References

Bartošek J, Chlupáčová M, Šťovíčková N (1969) Petrogenesis and structural position of small granitoid intrusions in the aspect of petrophysical data. Sb Geol Věd Řada Užitá Geofyzika 8: 37–68

Beránek B, Dudek A, Zounková M (1975) Rychlostní modely stavby zemské kůry v Českém masívu a Západních Karpatech. Sb Geol Věd Řada Užitá Geofyzika 13: 7–20

Bowes DR, Aftalion M (1991) U-Pb zircon isotopic evidence for early Ordovician and late Proterozoic units in the Mariánské Lázně Complex, Central European Hercynides. Neues Jahrb Miner Monatsh 1991: 315–326

Buday T, Dudek A, Ibrmajer J (1969) Některé výsledky interpretace gravimetrické mapy ČSSR v měř. 1:500000. Sb Geol Věd Řada Užitá Geofyzika 8: 7–35

Cháb J (1975) Intruzívní horniny strukturního vrtu Bechlín u Roudnice nad Labem. Sb Geol Věd Řada Geol 27: 55–82

Chaloupský J (1974) Geologie podloží křídy a permokarbonu. In: Malkovský M (ed) Geologie české křídové pánve a jejího podloží. Geofond, Praha, pp 40–71

Chlupáčová M (1974) Radioaktivita granitů západních a severozápadních Čech. Thesis, Charles Univ, Praha

Chlupáčová M, Hrouda F, Janák F, Rejl L (1975) The Fabric, Genesis and relative-age relations of the granitic rocks of the Čistá-Jesenice Massif (Czechoslovakia) as studied by Magnetic anisotropy. Garland Beitr Geophys 84: 487–500

Dudek A (1957) Sedmihoří – vypreparovaný složený peň. Sb Česk Společnosti Zeměpisné 62: 206–209

Fediuk F, Ichčinchorloo B, Ciniburk M (1966) The Neratovice complex – a product of metasomatic transformation of volcanites into rocks of plutonic appearance. Paleovolcanics of the Bohemian Massif, Karlova Univ, Praha, pp 51–60

Fediuková E, Fediuk F (1980) Gabbro od Kosobod na Rakovnicku. Acta Univ Carolinae, Geologica 1978: 365–392

Fiala F (1951) Příspěvek k poznání tzv. algonkických slepenců ze severozápadní části Železných hor. Sb Ústřed Ústavu Geol 18: 117–135

Fiala F (1983) Xenolity albitizovaných granitoidů v granulovaném diabasu od Otmíčů. Acta Univ Carolinae, Geologica, Hejtman Vol, pp 345–361

Fischer G, Schreyer W, Troll G, Voll G, Hart SR (1968) Hornblendealter aus dem ostbayerischen Grundgebirge. Neues Jahrb Mineral Monatsh 1968: 385–404

Gnojek I, Šťovíčková N (1974) The ring structure of the Sedmihoří granite stock. Sb Geol Věd Řada Užitá Geofyzika 12: 113–130

Jelínek E, Lang M, Tonika J, Ulrych J (1994) Metagabbro from Výškovice, Mariánské Lázně metabasite complex. Acta Univ Carolinae, Geologica (in press)

Klomínský J (1963a) Cancrinit-führende Alkalisyenite aus dem Čistá-Massiv, Westböhmen/Tschechoslovakei. Neues Jahrb Mineral Abh 99: 295–306

Klomínský J (1963b) Geologie čisteckého masívu. Sb Geol Věd Řada Geol 3: 7–29

Klomínský J (1965) Štěnovický granodioritový masiv. Sb Geol Věd Řada Geol 8: 75–99

Kopecký L (1987) The Čistá ring structure, Czechoslovakia. Proc 1st Sem on carbonatites and alkaline rocks of the Bohemian Massif. Geol Surv, Praha, pp 23–58

Kopecký L, Dobeš M, Fiala J, Šťovíčková N (1970) Fenites of the Bohemian Massif and the relations between fenitization, alkaline volcanism and deep fault tectonics. Sb Geol Věd Řada Geol 16: 51–112

MacDonald GA, Katsura T (1964) Chemical composition of Hawaiian lavas. J Petrol 5: 82–133

Minařík L, Vachtl J, Knotek M (1983) Geochemie Nasavrckého masívu. Stud Česk Akad Věd 7: 64

Neužilová M, Vejnar Z (1966) Geologie a petrografie hornin kladrubského masívu. Sb Geol Věd Řada Geol 11: 7–31

Šmejkal V (1964) Absolutní stáří některých vyvřelých a metamorfovaných hornin Českého masívu stanovené kalium-argonovou metodou (II.část). Sb Geol Věd Řada Geol 4: 121–136

Šmejkal V, Melková J (1969) Notes on some potassium-argon dates of magmatic and metamorphic rocks from the Bohemian Massif. Čas Mineral Geol 14: 331–338

Šmejkal V, Vejnar Z (1967) Zur Frage des praevaristischen Alters einiger Granitoide des Böhmischen Massivs. Sb Prací Vysoké Školy Báňské, Spec Vol, 1st Geochem Conf, 1965, Vysoká škola báňská, Ostrava, pp 123–128

Tonika J (1979) The Mutěnín ferrodiorite ring intrusion, West Bohemia. Krystalinikum 14: 195–208

Tonika J, Vejnar Z (1966) Geologie a petrografie stodského masívu. Čas Mineral Geol 11: 129–137

Ulrych J, Cimbálníková A, Fiala J, Kašpar P, Lang M, Minařík L, Palivcová M, Pivec E (1976) Petrology of the Petrovice melagabbro. Rozpr Česk Akad Věd Řada Mat Přír Věd 86 (9): 57

Vachtl J (1977) Comparison of two volcano-plutonic granite formations of the Nasavrky massif, Czechoslovakia. Acta Univ Carolinae, Geologica 1975: 199–220

Vejnar Z (1967) Petrogenetická korelace a metalogeneze některých západočeských granitoidních těles. Věstn Ústřed Ústavu Geol 42: 87–95

Vejnar Z (1973) Poběžovický masív a distribuce Mg, Fe v jeho minerálech. Sb Geol Věd Řada Geol 25: 85–143

Vejnar Z (1975) Highly ferrous silicates from the Mutěnín ferrodiorite ring intrusion, West Bohemia. Věstn Ústřed Ústavu Geol 50: 265–273

Vejnar Z (1977) The Babylone granite massif and its contact aureole, South-West Bohemia. Věstn Ústřed Ústavu Geol 52: 205–213

Vejnar Z (1979) The aegirine-augite – riebeckite syenite from the Domažlice crystalline area, West Bohemia. Věstn Ústřed Ústavu Geol 54: 199–205

Vejnar Z (1980) The spinel- and corundum-bearing basic intrusion of Drahotín, South-West Bohemia. Krystalinikum 15: 33–54

Vejnar Z (1986) The Kdyně massif, South-West Bohemia – a tectonically modified basic layered intrusion. Sb Geol Věd Řada Geol 41: 9–67

Vejnar Z, Neužilová M, Syka J (1969) Geologie a petrografie borského masívu. Věstn Ústřed Ústavu Geol 44: 247–256

All publications in Czech have an English summary.

VII.B.4 Metamorphic Evolution

J. Cháb, V. Suchý, and Z. Vejnar

1 Introduction

The most detailed data referring to metamorphism in the Bohemicum comes from the Teplá-Barrandian (TB) region to the W and from the Orlické hory Mts. (Nové Město unit) to the NE. Knowledge of other outcroppings along its E and SE border, the Železné hory Hills, the Polička, Letovice, and Zábřeh regions, is unsystematic. Only boreholes and isolated outcrops in small uplifted blocks provide information on most of the Bohemicum covered by Permo-Carboniferous and Cretaceous sediments.

2 The Teplá-Barrandian Region

The Cambrian to Devonian cover is unmetamorphosed or at most anchimetamorphosed, free of penetrative deformation, and thermally metamorphosed along the NW margin of the Central Bohemian Pluton. Within the pluton itself, in its roof pendants, rocks of undoubtedly Paleozoic (Ordovician to Devonian) age occur, which are weakly deformed, if at all; similar rock assemblages, deformed and regionally metamorphosed prior to thermal recrystallization, are also present. The grade of their last metamorphism is equal to the grade of metamorphism in the contact aureole mentioned above; Bt, And, Crd, as well as Grs, Ep, Ves, and Wo are the index minerals of metapelites and calc-silicate rocks, respectively.

The Precambrian basement underlying the Paleozoic cover is anchimetamorphosed (Jakeš et al. 1979; Suchý 1991). In an impressive shear zone near Příbram SW of Prague and possibly also N of Prague and in the E vicinity of Plzeň, the intensity of metamorphism reaches a transitional, "deep" part of the anchizone and even the chlorite zone (Dudek and Fediuk 1955; Mašek 1984; Suchý 1991). Within the anchizone, the outer boundary of which corresponds approximately with the last occurrences of Pu in mafic metavolcanites, there are several unimpressive, limited sectors, where weakly deformed rocks show only traces of a burial metamorphism (Suchý 1991). The white mica crystallinity between 0.28 to 0.22 $\Delta°2\theta$ characterizes most of the anchizone, corresponding to a region of isolated occurrences of Pe and to common occurrences of Pe + Pu in mafic metavolcanites in a marginal zone (Cháb and Bernardová 1974). The transition to the chlorite zone is indicated by the white mica crystallinity lower than 0.21 $\Delta°2\theta$; these values begin to dominate in the Pu + Act zone when its position is determinable.

To the S of the TB region, penetrative deformation increases toward the SE to S, from the S vicinity of Prague to the roof pendants of the Central Bohemian Pluton. In contrast, strongly anchimetamorphosed rocks ($\Delta°2\theta$ mostly below 0.21) also occur within the region of anchimetamorphism with no relation to deformations. The deformation is manifested above all by a regional cleavage (the Jílové cleavage, the main phenomenon of the Central Bohemian Shear Zone according to Rajlich et al. 1988). The Jílové cleavage and the regional metamorphism connected with it were considered to be Precambrian in age, but recent studies of the deformations cast doubts on this conclusion, although there is as yet no decisive proof. At present, it is not clear if any difference exists between the penetrative deformation of the Precambrian and Paleozoic rocks in the roof pendants of the Central Bohemian Pluton and along its NW edge. The original regional metamorphism reached the MP/MT amphibolite facies (Vrána and Cháb 1981).

To the W and NW of the TB region, the anchimetamorphosed Precambrian sequence becomes gradually more and more intensively metamorphosed. The anchimetamorphism increases into the transitional zone followed by the broad Chl zone. In the NW, the beginning of the Chl zone is indicated by the occurrence of Stp

Dallmeyer et al. (Eds.)
Pre-Permian Geology of Central and Eastern Europe

in metagraywackes (Cháb and Pelc 1973). The penetrative deformation is concentrated in narrow shear zones with an impressive cleavage in the "shallow" anchizone and acquires the regional distribution manifested by the metamorphic schistosity, an elongation of clastic grains forming an extended lineation, and a moderate preferred orientation of Qtz in the transitional zone (Holubec 1968; Franke 1975). The increase in metamorphism is also confirmed by the maturity of kerogen, indicating a gradual change from anthracite to metaanthracite in the anchizone to comparatively better crystallized subgraphitic varieties in the Chl and Bt zones (Jehlička et al. 1989; Rouzaud and Jehlička 1989; Suchý 1991).

Higher-grade metamorphic zones are clearly narrower and indicate a rapid increase in T and P to the W and NW edge of the TB region (Fig. 1). In the recent German literature, this region of more intensive metamorphism is often called Zone Tepla-Taus, shortened to ZTT. The nature of the metamorphic zoning is well known at present in its S, Domažlice (Taus) part (Vejnar 1966, 1972, 1977, 1982; Vejnar et al. 1984); Bt, Grt, St, Ky, and Sill zones follow after the Chl zone (Figs. 1 and 2). In an S sector, Sill precedes Ky. The shear (fault) zone of the Bohemian Quartz Lode in the W is accompanied by a steep cleavage and a retrograde metamorphism under Chl zone conditions and forms an arbitrary boundary to the TB region (see Chap. VII.B.1). This boundary is considered to be the real boundary between two different units by Chaloupský (1989), while Vejnar considers it to be a boundary

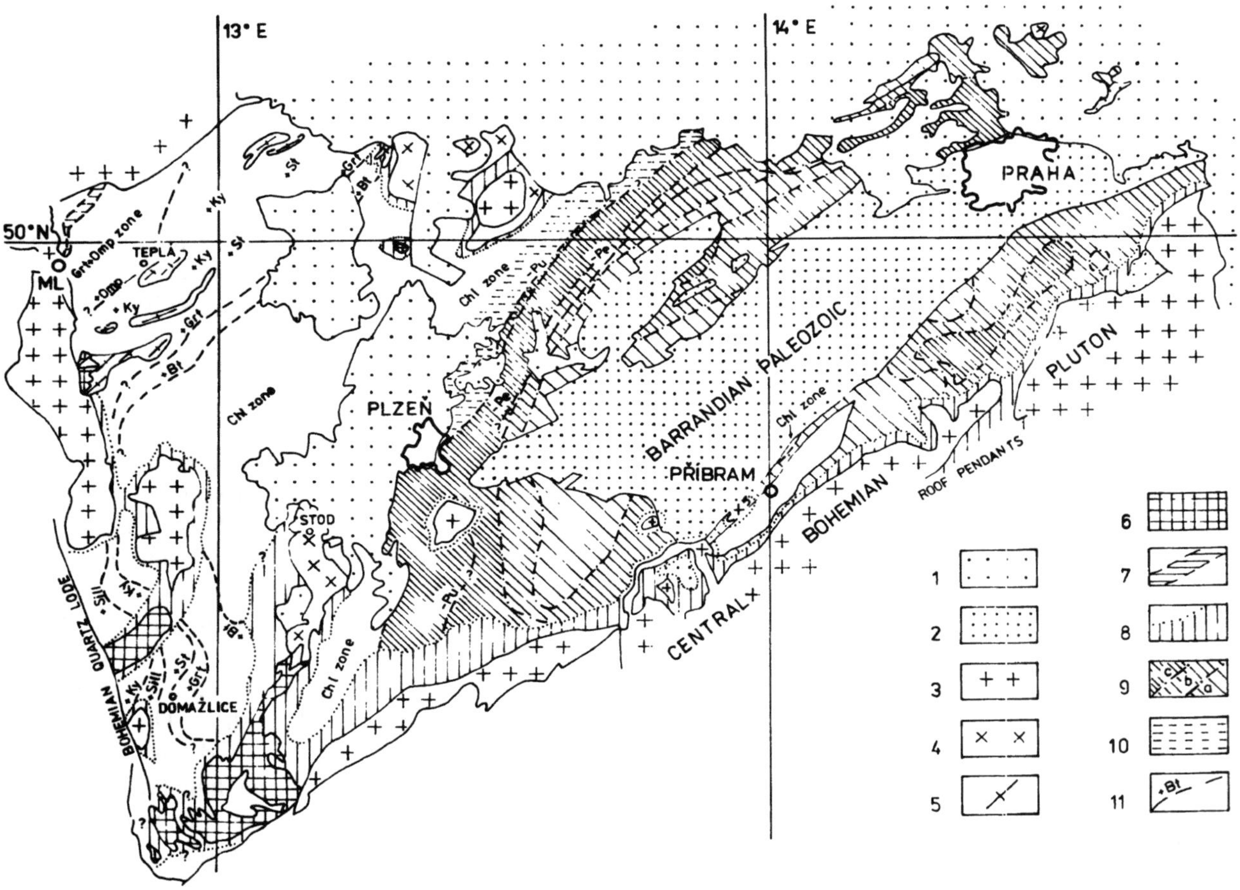

Fig. 1. Metamorphic zones in the TB region. *1* Permo-Carboniferous and younger cover; *2* Cambrian to Devonian (Barrandian sequences, unmetamorphosed, at most anchimetamorphosed locally); *3* Variscan granitic intrusions; *4* Pan-African granitic intrusions and granitic intrusions of an uncertain age; *5* orthogneisses; *6* mafic intrusions (Pan-African?-Variscan?); *7* ultramafic belt; *8* contact aureoles; *9* to *11* metamorphosed pre-Cambrian: *9* anchizones I (a), II (b), and III – transitional (c); *10* Stp in metagraywackes; *11* metamorphic isogrades. *ML* Mariánské Lázně

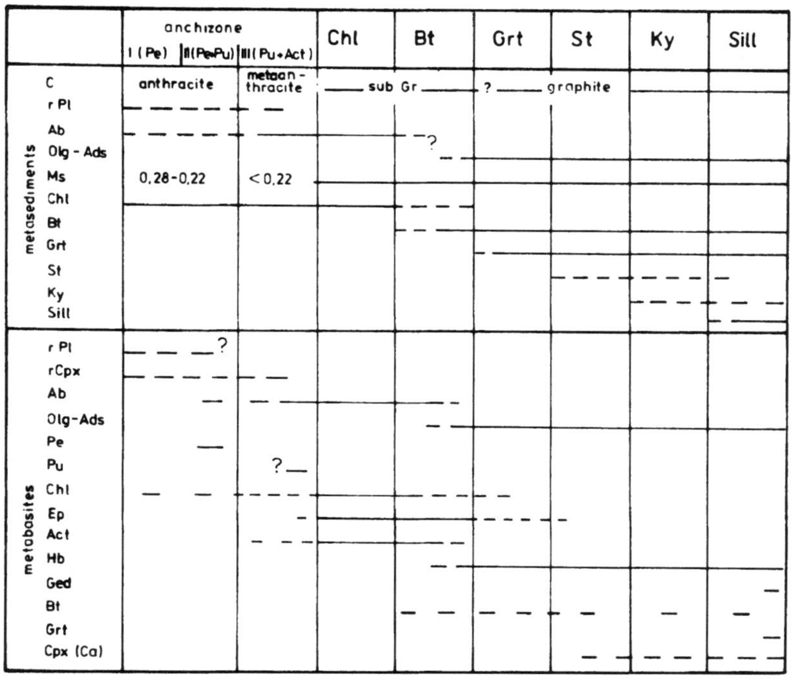

Fig. 2. Metamorphic zones in the Domažlice area (After Cháb and Suk 1977); *anchizones* are indicated by $\Delta°2\theta$ of white micas and by T °C of decomposition of kerogen (*r* relict minerals)

of local importance crossed by the Sill zone, which is followed by Sill + KFs and Crd zones originating from a later LP/HT process.

The regional metamorphic zones are superimposed by contact aureoles of mafic and granitic intrusions of Pan-African (Cadomian, see below) and Variscan ages largely in the Domažlice area (Vejnar 1981, 1990). The thermal metamorphism reached the pyroxene hornfels facies in aureoles of the mafic intrusions; Vejnar (1981) gave T between 600 and 800 °C. Around the granitic intrusions, the thermal metamorphism did not exceed the hornblende hornfels facies. Sometimes it is difficult to distinguish between the products of the thermal and regional metamorphisms.

In the Teplá (Tepla) area, the zoning (Figs. 1 and 3) differs from that of the Domažlice area (see Table 1).

The zoning of the Teplá part (including MLC) ends near Mariánské Lázně (Mariannebad) on a discontinuous belt of serpentinite bodies with small occurrences of Act, Tr, Chl + Tlc, and Chl schists. The belt passes across a large amphibolite body. Numerous relict eclogite pods occur in high-grade partly diablastic amphibolites south of the belt. Amphibolites to the N of the belt were metamorphosed under low-grade amphibolite facies conditions and were later thermally recrystallized in contact aureoles of Variscan granitic intrusions; these amphibolites probably do not belong to the TB region.

To the S of the MLC, the metamorphic evolution was complicated, as Medaris et al. (1991) and O'Brien (1991) show. An amphibolite facies metamorphism preceded the eclogite one (P > 14 kb, T = 680 °C) and then the development continued under the granulite (P − 12 kb, T = 750 °C) and again amphibolite facies conditions up to a closing low-grade metamorphic stage.

The Domažlice metamorphic zones series represents an MP/MT type, but some of its features put it nearer to the LP/HT series, especially the location of the transition between the greenschist and amphibolite facies and the partial precedence of Sill to Ky. In contrast, the Teplá series is the typical MP/MT type, turning into a transitional HP/MT type in its deeper part, if the MLC belongs to it. This pattern is somewhat disturbed by numerous meta-igneous rocks, the youngest of which, moderately deformed and metamorphosed granites and tonalites, have partially intrusive contacts accompanied by LP/HT metamorphism.

The metamorphic zones of the Teplá part continue toward the NE under the young cover, where they are cut off obliquely by contact with the high-grade crystalline unit of the Krušné hory Mts. Farthest to the N, the rocks of the TB region crop out in the Labe river valley near Litoměřice;

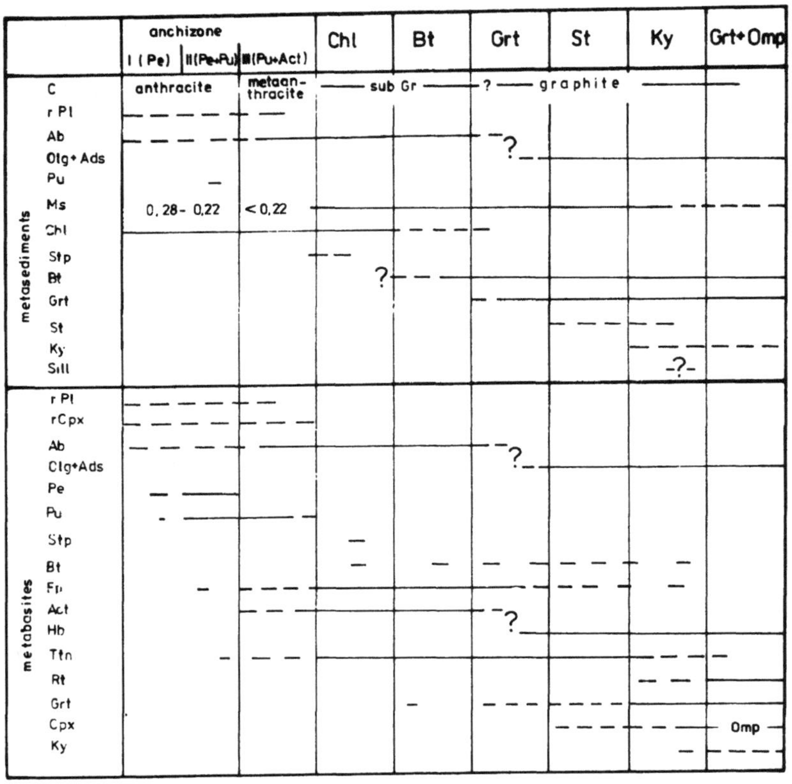

Fig. 3. Metamorphic zones in the Teplá area (same source and modification as in Fig. 2)

they were metamorphosed under low amphibolite facies conditions. To the N and NE of Prague, the Precambrian rocks in the underground of later sediments are metamorphosed in the Chl and Bt zones; occurrences of higher-grade rocks are exceptional (Chaloupský 1973, 1974).

The dating of the metamorphic zoning in the W and NW of the TB region is a subject of heated controversy. The anchimetamorphism, its transition to the Chl zone, and its increase in some shear zones plunging under the early Paleozoic cover are very probably Pan-African in age. The continuity of the metamorphic gradation from the Chl zone to the Sill (or Grt + Omp) zone has been explained by Czech geologists to be a result of Pan-African processes. In contrast, many German scientists have considered the metamorphism of the ZTT to be Variscan in age. Even recent radiometric data are contradictory. Gottstein's (1970) K/Ar data from the Teplá part are heterogeneous and do not enable a more precise interpretation. K/Ar analyses summarized by Šmejkal and Melková (1969) have shown that some granitic and mafic intrusions are Pan-African in age. Recently, one such older date (Šmejkal and Vejnar 1967) related to the Stod pluton in the Domažlice area has been confirmed by Ar/Ar dating (518 Ma, Kreuzer et al. 1989). For the Teplá area, new Ar/Ar data indicate exclusively Variscan events at 380 and 320 Ma (Kreuzer et al. 1992). U/Pb Zrn data from the MLC fall into the Vendian and the Early Ordovician (Bowes and Aftalion 1991) and are interpreted as protolith ages; the latter determination places the origin of the rock very close to the age of the Cambro-Ordovician CA volcanites of the Barrandian. Eclogite metamorphism developed at 433 and 379 Ma (Nd/Sm, Grt + Omp pairs, two localities, Beard et al. 1991). If further complications do not arise, and if recent interpretations of the radiometric data are correct, then we can infer that a very important mistake is hidden somewhere in the fundamental geological ideas about the structure and evolution of the ZTT (see footnote in Table 1).

3 The Železné hory Hills, NW Part

The Ordovician to Devonian cover is metamorphosed in the range of anchi- and Chl zones, not only in the Železné hory Hills but also farther to the N in the underground of younger sediments. Also, the Cambrian and so-called sub-

Table 1. The comparison of the series of the metamorphic zones in the Domažlice and Teplá sectors. (After Cháb and Suk 1977)[a]

Domažlice sector	Teplá sector[a]
1. Bt zone broader, more impressive	Bt zone narrow
In both sectors the location of the +Bt isograde is somewhat uncertain (Zoubek et al., in Suk et al.; Zoubek et al. 1974; Vejnar 1982)	
2. The transition between greenschist and amphibolite facies (the isogrades +Hb and +Olg) appears within the Bt zone	The transition between greenschist and amphibolite facies (the isogrades +Hb and +Olg) appears within the Grt zone
3. Ky zone is comparatively narrow	Ky zone is broad
4. Sill zone is present	Sill zone is absent; the importance of local Sill occurrences is not clear
5. Grt amphibolites are rare and known only from the Sill zone; they exhibit an abnormal chemical composition	Grt amphibolites are common in the Ky and Grt + Omp zones and exhibit a normal basaltic chemical composition
6. Ky is absent in amphibolites	Ky occurs in some Grt amphibolites of the Grt + Omp zone
7. Eclogites and related rocks are absent (= Grt + Omp zone is absent)	Eclogites and related rocks with Ky in some places are usual in the Grt + Omp zone; they contain its critical mineral assemblages
This comparison is impeded by varying degrees of knowledge of metamorphic zoning in the Domažlice and Teplá parts. Data on the Grt + Omp zone are taken from Tonika (1970).[a]	

[a] During the preparation of this chapter, the metamorphism in the Teplá sector was studied by Žáček and Cháb. This study verified the independent tectonometamorphic position of the MLC (Matte et al. 1990); this means that the Grt + Omp zone does not exist in the Teplá crystalline complex and that eclogites of its kyanite zone discovered by Tonika et al. (1992) represent heterogneous bodies exhibiting tectonometamorphic history different from the history of adjacent metapelites and metapsammites. These facts must be taken into account, because it was impossible to make precise corrections to the text before publication.

Cambrian exhibit similar very low-grade to low-grade metamorphism.

The Precambrian basement was metamorphosed under greenschist and lower amphibolite facies conditions. The Chl, Bt, and Grt zones and rare occurrences of St in the Grt zone were described by Losert (1967). The boundary between the greenschist and amphibolite facies is not well known; it is probably situated near the Grt isograde. This isograde itself is locally uncertain due to the presence of iron metamanganolites and pyrite-rich black schists enriched in Mn and P, in which Sps Grt occurs already in the chlorite zone. This horizon contains numerous, infrequent or rare minerals in rocks as well as in metamorphic segregations and postmetamorphic hydrothermal veins (tephroite, rhodonite, pyroxmangite, tirodite, Ba-feldspars, etc., Žák 1972a,b; Cháb et al. 1981). Moreover, the metamorphic gradation is complicated by an intricate tectonic segmentation along the SW edge of the Železné hory Hills, where moderately deformed and metamorphosed Chvaletice granite appears to form an independent tectonic slice. An approximately 1-km-wide mylonite zone separates this crystalline unit from the high-grade rocks of the Moldanubicum; the mylonitization developed under greenschist and lower amphibolite facies conditions (M. Fišera, pers. comm.).

4 Heterogeneous Parts of the E Sector of the Bohemicum

4.1 The Železné hory Hills, Middle Part

Most of the middle part of the Železné hory Hills is occupied by intrusive rocks of the Variscan Železné hory Hills Pluton, which is separated by a broad zone of low-temperature mylonites from the Precambrian and Lower Paleozoic rocks to the NW. High-grade Qtz-Fs schists and gneisses with rare intercalations of Bt schists, amphibolites, and calc-silicate rocks border the S edge of the pluton; according to Losert (1967), they were metamorphosed under high-grade Ky zone conditions, by which migmatite-producing anatectic processes were initiated.

4.2 The Železné hory Hills, SE Part

The influences of deformation + regional metamorphism and the following thermal metamorphism caused by the Železné hory Hills Pluton dominate in this segment, which is cut by the stripe of tectonic slices of paleontologically proved Silurian rocks. According to Sachsel's (1933) data, the intensity of metamorphism of these slices varies around the greenschist-amphibolite facies boundary; critical mineral assemblages contain Cld and/or St, and in one case even And and Grt in addition to Cld and St. Criteria enabling the distinction of products of the regional and thermal metamorphisms are, however, missing.

To the S of the Silurian stripe, rocks were metamorphosed under amphibolite facies conditions. To the N, the metamorphism reaches the greenschist facies grade and its intensity (together with the intensity of penetrative deformation) decreases rapidly; in the northernmost part of the segment, the deformation is negligible and products of the low-grade regional metamorphism are erased nearly completely by the later thermal metamorphism, which reached the hornblende hornfels facies. Crd hornfelses bordering the Železné hory Hills Pluton are its most conspicuous product. A narrow stripe of granitic mylonites separates the whole segment from the high-grade crystalline unit of the NE margin of the Moldanubicum.

The unit is metamorphosed in the range of the Chl to Grt zones and exhibits a later retrogres sion under the Chl zone conditions in its NW part (Opletal et al. 1980). The intensity of metamorphism increases toward the NE until the contact with the crystalline rocks of the Orlice-Klodzko unit and toward the SE in the direction of the Zábřeh unit. The position of the greenschist-amphibolite facies transition is somewhat uncertain; according to present data (Domečka and Opletal 1976; Opletal et al. 1980), it is situated in a higher-grade part of the Bt zone, probably close to the −Chl "isograde" in metasediments, which divides the Bt zone into the outer Bt + Chl + Ms and inner Bt + Ms subzones. In the vicinity of felsic as well as mafic intrusions, thermal aureoles exist, in which the thermal metamorphism does not overstep the hornblende hornfels facies.

The age of the progressive regional metamorphism is uncertain. K/Ar data, summarized by Opletal et al. (1980), indicate a late Variscan cooling (340 to 305 Ma, Bt concentrates from metasediments). The K/Ar age of one mafic intrusion (hornblende) falls into the same interval. On the other hand, K/Ar (Bt concentrates) ages of alkaline vein intrusions, which penetrated after the metamorphic and deformational processes, indicate the interval between 390 and 360 Ma (Opletal and Domečka 1976). That is why Opletal et al. (1980) consider the regional metamorphism to be pre-Variscan and regard the other data as a reflection of a Variscan rejuvenation. The retrograde metamorphism and granitic intrusions are classified as Variscan.

4.4 The Zábřeh Unit

A sedimentary cover compared usually with the culm facies of the Lower Carboniferous is weakly metamorphosed. However, more precise data are lacking, and only its E margin displays strong low-temperature mylonitic deformation.

The intensity of regional metamorphism of the basement increases from the middle part toward the S as well as to the N; the lowest-grade rocks occur probably to the E of the middle part, where their mineral assemblages can indicate the Bt zone. The metamorphic zoning is unknown. We can conclude from infrequent, inconsistent data that, toward the N, the metamorphism increases through the Grt and St zones to high-grade rocks, geneisses, and migmatites in a broader vicinity of tonalitic intrusions, from where occurrences of Sill and Crd were described. According to Opletal et al. (1980), the Zábřeh unit merges into the Nové Město unit on the NW; the transitional sector is characterized by a rapid decrease in metamorphic intensity over the Grt zone to the Bt one. The increase in metamorphic intensity toward the S, again to the amphibolite facies, is not well known.

4.5 The Letovice Unit

The metamorphism reaches an amphibolite facies grade, most probably the St zone grade, according to the critical mineral assemblage Grt + St in mica-schist. The conditions of the metamorphism were estimated roughly at 2 kb and 400 to 600 °C by Jelínek et al. (1984), which seems to be too low. Radiometric data (K/Ar and Ar/Ar ages, hornblendes, McIntyre et al. 1992) indicate metamorphism in the interval 350 to 335 Ma. The connection between the metamorphism and the polyphase deformation (Bowes et al. 1978, 1980) is not well known.

4.6 The Polička Unit

The metamorphism reaches a deep amphibolite facies grade as numerous occurrences of pearl gneisses (ophthalmitic migmatites) and migmatites show; but it is impossible to determine precisely either its nature or its zoning from scattered occurrences of Ky, Sill, and St. No metamorphic discordance was found between the Polička unit and the NE margin of the Moldanubicum.

References

Beard EL, Medaris LG Jr, Johnson CM, Mísař Z, Jelínek E (1991) Nd and Sr isotope geochemistry of Moldanubian eclogites and garnet peridotites, Bohemian massif, Czechoslovakia. 2nd Int Eclogite Conf, Terra Nova 3, Terra Abstr Suppl 6: 4

Bowes Dr, Aftalion M (1991) U-Pb zircon isotopic evidence for early Ordovician and late Proterozoic units in the Mariánské Lázně complex, Central European Hercynides. Neues Jahrb Mineral Monatsh 7: 315–326

Bowes DR, Hopgood AM, Mísař Z (1978) Polyphase deformation in amphibolites of the Letovice crystalline unit, Moravia. Věstn Ústřed Ústavu Geol 53: 273–280

Bowes DR, Hopgood AM, Mísař Z (1980) Characterization of tectonic regimes of the Letovice unit and Bíteš gneiss of the Moravicum using the structural imprint of polyphase deformation. Věstn Ústřed Ústavu Geol 55: 331–342

Cháb J, Bernardová E (1974) Prehnite and pumpellyite in the Upper Proterozoic basalts of the north-western part of the Barrandian region (Czechoslovakia). Krystalinikum 10: 53–66

Cháb J, Pelc Z (1973) Proterozoické droby severozápadní části Barrandienu. Sb Geol Věd Geol 25: 7–74

Cháb J, Suk M (1977) Regionální metamorfóza na území Čech a Moravy. Knihovna Ústřed Ústavu Geol 50: 156

Cháb J, Bouška V, Jelínek E, Pačesová M, Povondra P (1981) Petrology and geochemistry of the Upper Proterozoic FeMn-deposit Chvaletice. Sb Geol Věd Ložisková Geol Mineral 23: 9–68

Chaloupský J (1973) The basement of the Cretaceous and the Permo-Carboniferous of northern Bohemia. Geol Rdsch 62: 581–594

Chaloupský J (1974) Geologie podloží české křídy a permokarbonu. In: Malkovský M (ed) Geologie české křídové pánve a jejího podloží. Ústřední Ústrav Geologický, Praha, pp 40–71

Chaloupský J (1989) Major tectonostratigraphic units of the Bohemian Massif. Geol Soc Am Spec Pap 230: 101–144

Domečka K, Opletal M (1976) Stratigrafie, stavba a metamorfóza severní části novoměstské série. Sb Geol Věd Geol 28: 157–187

Dudek A, Fediuk F (1955) Zur Altersfrage der Metamorphose im barrandienischen Proterozoikum. Geologie 4: 397–403

Franke W (1975) K přednostní orientaci křemene v západní části středočeského proterozoika. Čas Mineral Geol 20: 159–173

Gottstein O (1970) K-Ar ages of granitic and ectinitic rocks of the Teplá anticlinorium. Věstn Ústřed Ústavu Geol 45: 201–205

Holubec J (1968) Structural development of the geosynclinal Proterozoic and its relations to the deeper zones of the Earth's crust. Rozpr Česk Akad Věd Řada Mat Přír 78 (8): 77

Jakeš P, Zoubek J, Zoubková J, Franke W (1979) Graywackes and metagraywackes of the Teplá-Barrandian proterozoic area. Sb Geol Věd Geol 33: 101–122

Jehlička J, Beny C, Rouzaud JN (1989) Graphitization of Precambrian kerogens: structural changes revealed by transmission electron microscopy and Raman micro-spectrometry. In: Kříbek B (ed) Metallogeny and anoxic sediments. Abstr Czech Working Group of IGCPP 254: 13–15

Jelínek E, Pačesová M, Martinec P, Mísař Z, Weiss Z (1984) Geochemistry of a dismembered metaophiolite complex, Letovice, Czechoslovakia. Trans R Soc Edinb Earth Sci 75: 37–48

Kreuzer H, Vejnar Z, Schüssler U, Okrusch M, Seidel E (1992) K-Ar dating in the Teplá-Domažlice zone at the western margin of the Bohemian massif. In: Kukal Z (ed) Proc 1st Int Conf on the Bohemian massif, Prague, Czechoslovakia, Sept 26–October 3 (1988), pp 168–175

Kreuzer H, Seidel E, Schüler U, Okrusch M, Lenz K-L, Raschka H (1989) K-Ar geochronology of different tectonic units at the northwestern margin of the Bohemian massif. Tectonophysics 157: 149–178

Losert J (1967) Contribution to the problem of the pre-assyntian tektogenesis and metamorphism in the Moldanubicum of the Bohemian Massif. Krystalinikum 5: 61–84

Mašek J (1984) Geologie proterozoika druhého břidličného pásma na Příbramsku a Dobříšsku. Sb Geol Věd Geol 39: 9–26

Matte P, Maluski H, Rajlich P, Franke W (1990) Terrane boundaries in the Bohemian massif: results of large-scale Variscan shearing. Tectonophysics 177: 151–170

McIntyre RM, Bowes DR, Hamidullah S, Onstott TC (1988) K-Ar and Ar-Ar isotopic study of amphiboles from metaophiolite complexes, eastern Bohemian Massif. Abstr Bohemian Massif Conf. Ústřední Ústav Geologický, Praha, p 55

Medaris LG, Beard BL, Mísař Z, Jelínek E (1991) Czech eclogites in the Moldanubian zone of the Bohemian Massif: Petrological characteristics and tectonic significance. 2nd Int Eclogite Conf, Terra Nova 3, Terra Abstr Suppl 6: 7

O'Brien PJ (1991) High pressure metamorphism in the NW Bohemian Massif: comparisons and contrasts between the Moldanubian zone, Münchberg Massif, ZEV, ZTT and Erzgebirge. KTB Rep 91-1: 1–12

Opletal M, Domečka K (1976) Przeglad wyników nowych badań geologicznych w Górach Orlickich. Prz Geol 7: 414–418

Opletal M, Domečka K, Čech S, Čuta M, Fajst M, Holub V, Kačura K, Líbalová J, Pošmourný K, Sekyra J, Střída M, Šalanský K, Šulcek Z, Tásler R, Valečka J (1980) Geologie Orlických hor. Ústřední Ústav Geologický, Praha, p 202

Pajlich P, Schulmann K, Synek J (1988) Strain analysis on conglomerates of the Central Bohemian Shear Zone. Krystalinikum 19: 119–134

Rouzaud JN, Jehlička J (1989) The presence of graphite in the Central Bohemian Upper Proterozoic shales and schists. Věstn Ústřed Ústavu Geol 64: 193–198

Sachsel E (1933) Beiträge zur Kenntnis der Geologie und Petrographie des Eisengebirges und der angrenzenden Gebiete. Mitt Österr Geol Ges 25: 195–245

Šmejkal V, Melková J (1969) Notes on some potassium argon dates of magmatic and metamorphic rocks from the Bohemian Massif. Čas Mineral Geol 14: 331–338

Šmejkal V, Vejnar Z (1967) Zur Frage der prävariszischen Alters einiger Granitoide des Böhmischen Massivs. In: Geochemie v Československu, Sborník prací 1. Geochem Konf, Ostrava 1965. VŠB, Ostrava, pp 123–128

Suchý V (1991) Anchimetamorfóza v sedimentech barrandienského svrchního proterozoika. Thesis, Charles Univ, Praha

Suk M et al. – Zoubek J et al. (1974) Nové poznatky z geologie prekambria Českého masívu. Výzkumné Práce Ústřed Ústavu Geol 7ú, 32 pp

Tonika J (1970) Mariánskolázeňský metabazitový komplex. Thesis, Charles Univ, Praha

Tonika J et al. (1992) Základní geologická mapa 1:25 000 list 11-412 Teplá + vysvětlující text. Unpubl map sheet and report, Czech Geological Survey, Praha

Vejnar Z (1966) The petrogenetic interpretation of kyanite, sillimanite, and andalusite in the south-western Bohemian crystalline complexes. Neues Jahrb Mineral Abh 104: 172–189

Vejnar Z (1972) Regionally metamorphosed volcanic rocks from the West-Bohemian metabasite belt. Krystalinikum 9: 131–156

Vejnar Z (1977) The relationship between the metamorphic grade and composition of silicates in the West-Bohemian greenschists and amphibolites. Krystalinikum 13: 129–158

Vejnar Z (1981) Cummingtonite-producing contact migmatization of hornfelses associated with the Poběžovice massif. Věstn Ústřed Ústavu Geol 56: 321–328

Vejnar Z (1982) Regionální metamorfóza peliticko-psamitických hornin domažlické oblasti. Sb Geol Věd Geol 37: 9–70

Vejnar Z (1990) Kontaktní dvur kdyňského masívu v jihozápadních Čechách. Sb Geol Věd Geol 45: 9–35

Vejnar Z, Doležal J, Hazdrová M, Kříž J, Mrňa F, Pokorný L, Rudolský J, Šefrna L, Tásler R, Tomášek M, Volšan V (1984) Geologie domažlické oblasti. Ústřední Ústav Geologický, 234 pp

Vrána S, Cháb J (1981) Metatonalite-metaconglomerate relation: the problem of the Upper Proterozoic sequence and its basement in the NE part of the Central Bohemian Pluton. Sb Geol Věd Geol 35: 145–187

Žák L (1972a) Genesis of the Chvaletice manganese-pyrite deposit. Čas Mineral Geol 17: 183–191

Žak L (1972b) Metamorphic paragenesis of the manganese-pyrite horizon in the Železné hory Mts./Bohemia/. Čas Mineral Geol 17: 345–356

VII.B.5 Metallogenesis

P. Mrázek and Z. Pouba

1 Introduction

Specific features of the geotectonic crustal development in the Late Proterozoic and in Early Paleozoic affected the course of the metallogenic process during which ore mineralization originated in the respective lithostratigraphical units of the Teplá-Barrandian Zone. Pouba (1986) assigned the Teplá-Barrandian Upper Proterozoic, including the isolated area of the Železné hory Mts. and the Barrandian Lower Paleozoic, to two metallogenic units. Geochemical and mineralogical differences between them are also apparent in the examples of ore mineralization conditioned by the later Variscan postplutonic mass mobilization (Fig. 1).

2 Late Proterozoic Metallogeny

Tectonomagmatic crustal development was the primary cause of mass differentiation leading to the metal accumulation in the Late Proterozoic. Cháb (1979) assumed that the Teplá-Barrandian Upper Proterozoic represents a combined petrotectonic association of rocks on the convergent boundary between the oceanic and continental plates with some divergent segments. The assemblage of tholeiitic, subalkalic, and alkalic metavoloanics is consequently considered as rocks of the island arcs, or of the active continental margin. On the other hand, Pouba and Zoubek (1986) consider the Teplá-Barrandian Upper Proterozoic as a part of the Proterozoic intracontinental rift in an advanced stage of development, during which mantle-derived magma ascended into the sedimentary basin. The authors admit two possibilities of movement in the final phase of the tectonic activity: with or without partial subduction of the rift margins beneath the plates in the Cadomian tectogenetic cycle. Mrázek et al. (1990) assume that relatively symmetrical arrangement of the occurrences of various types of volcanics and hydrothermalites, as well as of the gravity and magnetic anomalies coinciding with the occurrence of sulfides enriched in ^{34}S isotope, support the second variant presented by Pouba and Zoubek (1986).

Proterozoic ore mineralization can be denoted chiefly as stratiform or stratabound. Mrázek (1986) distinguished three types of metal accumulation connected genetically with the Proterozoic volcanic and postvolcanic thermal activity. They are considered to be submarine exhalative. The conditions of their origin were given by the distance from hydrothermal vents, the depth and temperature of deposition and other physical and chemical properties of the environment.

1. Pyrite and pyrrhotite mineralization with rare occurrence of chalcopyrite and sphalerite in low metamorphosed basic to intermediate metavolcanics overlain with silicified sediments and siliceous hydrothermalites enriched in Ag, As, Au, B, Ba, Cu, Hg, Mo, Pd, Sb, Sn, U, V, and Zr occur in the southeastern part of the Teplá-Barrandian Zone (Pouba and Mrázek 1981; Mrázek 1984; Morávek and Pouba 1987, 1990; Bernard 1991, and others). They are considered as proximal mineralization. Its structural and geochemical features and the geological setting of the representative localities (e.g., Struhadlo near Klatovy) are conformable to those of the Cyprus-type ore deposits (Mrázek 1986). High contents of gold coincident with stratiform Cu-Zn sulfidic mineralization in basic to acid metavolcanics and subvolcanic metaintrusives of the Jílové belt, which form a prominent geological boundary between the Bohemicum and the Moldanubicum, may also be assigned to the proximal type of ore mineralization. These indications are in some aspects similar to those of the greenstone belts of the Archean shield areas (Morávek and Pouba 1987, 1990).

Dallmeyer et al. (Eds.)
Pre-Permian Geology of Central and Eastern Europe

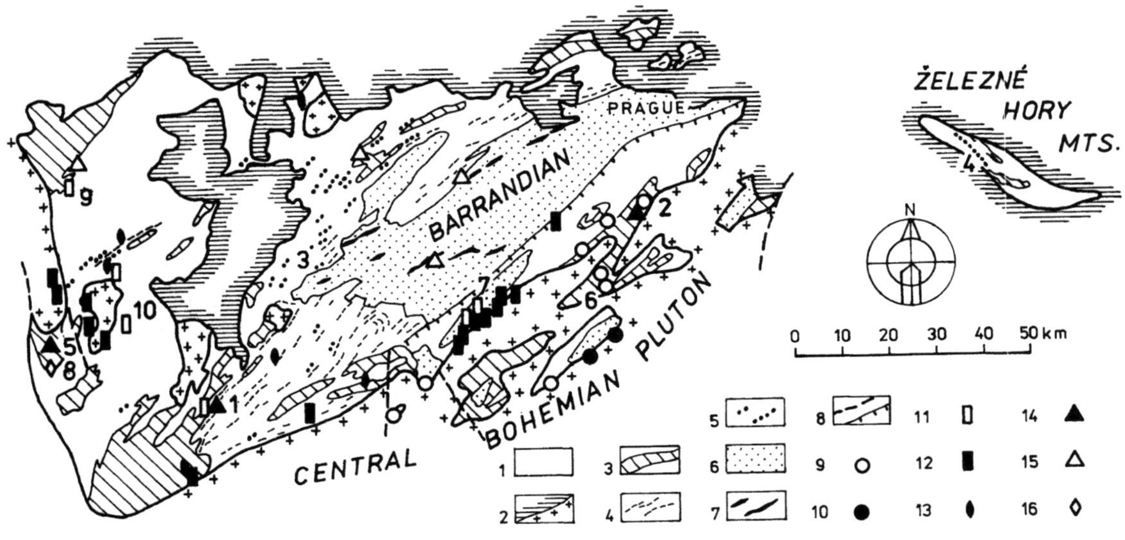

Fig. 1. Schematic map of the Teplá-Barrandian Zone with ore mineralizations. *1* Upper Proterozoic sediments; *2* Upper Proterozoic in contact with Variscan granites (*crosses*) and platform units (*horizontal hachure*); *3* Upper Proterozoic volcanic rocks; *4* silicites (*black cherts*); *5* black shales with pyrite (and manganese ores in Zelezné hory Mts.); *6* Lower Paleozoic rocks; *7* oolitic iron ores; *8* faults; *9* gold deposits and occurrences; *10* gold-stibnite deposits; *11* polymetallic ore veins; *12* uranium ore veins; *13* uranium-vanadium-iron oxide mineralizations; *14* stratiform polymetalic occurrences; *15* stibnite or mercury mineralizations; *16* quartz-chalcopyrite veins. Location of ore deposits and occurrences mentioned in text: *1* Struhadlo near Klatovy; *2* Jílové near Prague; *3* Hromnice and Kamenec; *4* Chvaletice; *5* Svržno near Horšovský Týn; *6* Čelina and Mokrsko; *7* Příbram; *8* Mutěnín; *9* Michalovy Hory; *10* Stříbro near Plzeň

2. The Fe^{III}-V-U mineralization in silicites, enriched locally also in all the chemical elements mentioned above, forms a specific geochemical formation in the southeastern and western parts of the Teplá-Barrandian Zone (Mrázek and Pouba 1975, 1976). It is considered to be a transitional type between the proximal and distal mineralization. Fe contents in the oxidic form reach up to 35%, contents of V and U, represented by V-garnets, roscoellite, coffinite, and uraninite disseminated in rocks, reach up to 0.2%. Metals probably precipitated from solutions on the geochemical barriers which had originated due to the production of oxygen around the bacteric-algal colonies.

3. Pyrite-bearing black shales, rarely enriched in Ni, Zn, and other base metals except Pb up to several tenths of a percent represent the distal mineralization. They occur along the main volcanic belt in the central part of the area (the Hromnice, Kamenec and other deposits) and in the Železné hory Mts. (the Chvaletice deposit). According to the isotopic composition of sulfur, some pyritic black shales of the Barrandian Proterozoic may have originated in an isolated (hypersaline?) basin. The basin may have formed during a temporary sea regression or in a petering-out of the Proterozoic sedimentary area (Mrázek et al. 1990). Pyritic black shales of the Železné hory Mts. associated with Mn-carbonates and accompanied with a graywacke-conglomerate sequence also deposited in shallow water environment, but presumably on the unstable shelf of the island arc or active continental margin (Cháb 1974). Contents of organic matter provide generally favorable conditions for the concentration of metals (Pouba and Kříbek 1986).

All the above-mentioned submarine-exhalative ores are very poor in Pb. Iron and base and precious metals probably had a single source – the volcanic rocks. The assemblage of newly generated minerals (chlorite, pumpellyite, actinolite, albite, leucoxene, quartz, carbonates, and sulfides), as well as the mass balance of the alteration of basaltic pillow lavas, indicate their subsea-floor metamorphism. Iron, together with other heavy metals, silica, and calcium was probably leached out of the volcanics during their interaction with sea water and accumulated in a favorable environment (Mrázek 1991). According to the isotopic study, sulfur of the rock-forming sulfides is of marine sulfate origin, which was reduced either bacterially at any place of the depositional basin or abiogenetically by the Fe^{2+} ions in areas of a high heat flow.

The last type of ore mineralization found in the area of the Teplá-Barrandian Upper Proterozoic is of problematic genesis. It mainly includes the occurrences of chalcopyrite, sphalerite, galena, pyrrhotite, pyrite, magnetite, and molybdenite in the volcano-sedimentary rock sequences metamorphosed under conditions of garnet to sillimanite zones (e.g., small ore deposite Svržno near Horšovský Týn). In places, they were acknowledged to be of metamorphic origin but there is no convincing evidence for it so far. The ores form banded, disseminated, or stringer structures.

3 Early Paleozoic Metallogeny

Metallogenetic processes of other types started in the Early Paleozoic of the Barrandian due to the changes of the facial conditions in the Central Bohemian sedimentary basin. Sediments were deposited in relatively shallow intracontinental sea, mostly in oxidic environment. Tectonic activity also did not reach the previous level.

Small deposits of Ordovician iron ores occur in many places of the Barrandian Lower Paleozoic. Lens-like bodies or layers of ore are interbedded with clay shales, sandstones, and altered basalt tuffs, above all on the lithological boundaries. The lower ore layers prevalently contain hematite, while Fe-leptochlorites, locally with magnetite, are more common in the upper part of the rock sequence. The main ore component in the whole geological profile is siderite. Oolitic structure is characteristic of all ore types.

The ore genesis is not clear so far. Most of the iron was probably imported to a depositional basin from the surrounding outcrops of weathered Proterozoic and Ordovician volcanic rocks. Ores were deposited mostly during local transgresions in shallow-water environment and then redeposited in a short distance (Skoček 1963). The mineralogical composition of ores is a product of diagenesis.

Small occurrences of layered siderite or siliceous Fe oxides are also known in Silurian rocks. Some of them are postulated to be metasomatic or submarine exhalative of the Lahn Dill type (Petránek 1976).

4 Variscan Metallogeny

The Variscan mass mobilization connected with several intrusions of anatectic plutonites comprises both the spatial redistribution of chemical elements including base and precious metals from the formerly mineralized rocks (protores), and the rising of chemical elements of subcrustal origin.

Along the Central Bohemian mobile zone bordering the Bohemicum to the SE, gold from the Proterozoic volcano-sedimentary units was presumably transferred during their granitization and contact metamorphism. Gold and scheelite concentrated in economically significant hydrothermal vein-type or stockwork ore deposits, e.g., Jílové near Prague, Čelina, and Mokrsko, etc. (Morávek and Pouba 1987, 1990). Similarly, during the later phase of the Variscan metallogeny, uranium together with a portion of organic carbon in variable oxidation stages was redistributed to hydrothermal veins along the western margin of the Central Bohemian Pluton such as, e.g., on the U-deposit Příbram (Mrázek and Pouba 1975; Pouba and Kříbek 1986). The enrichment of some Proterozoic and Paleozoic rocks in Hg, Sb, and Sn probably provided the possibility to form small accumulations of cinnabarite, stibnite, and stannite accompanied with barite in veinlets in the area SW of Prague.

Accumulation of sulfidic ores of Ag, Pb, Zn, and Cu in the Příbram ore district (the Central Bohemian mobile zone), small occurrences and deposits of chalcopyrite mainly in quartz veins scattered all over the Barrandian sequences (e.g., Mutěnín), as well as the Ni-Co-As and Pb-Zn-Cu ores in the ore district of Michalovy Hory near Mariánské Lázně (area of the Litoměřice deep fault bordering the Bohemicum to the NW) have a problematic geochemical relationship to the Proterozoic host rocks. A similar case could also be the Pb-Zn mineralization accompanied with barite veins in the vicinity of Stříbro near Plzeň. With respect to the isotopic composition of lead, subcrustal source was supposed there (Patočka et al. 1982).

References

Bernard JH (1991) Empirical types of ore mineralizations in the Bohemian Massif. Geol Surv Prague, pp 1–181

Cháb J (1974) Petrographical and geochemical investigations of the Chvaletice ore deposits: 1. Petrography (in Czech). Korelace proterozoických a paleozoických stratiformních lokižisek II. Fac Sci, Charles Univ, Prague, pp 42–56

Cháb J (1979) The Upper Proterozoic of the Bohemian Massif and its importance for development of the European continent (in Czech). Proc Sem K 60, Výročí Založení Ústřed Ústavu Geol Prague, pp 23–28

Morávek P, Pouba Z (1987) Precambrian and Phanerozoic history of gold mineralization in the Bohemian Massif. Econ Geol 82: 2098–2114

Morávek P, Pouba Z (1990) L'or dans la métallogénie du massif de Bohème. Mineral Deposit 25 (Suppl): 90–98

Mrázek P (1984) Trace elements in the Bohemian Upper Proterozoic rocks. Sbor Geol Věd Ložisk Geol Mineral Prague 26: 81–105 (in Czech)

Mrázek P (1986) Metallogeny of the West Bohemian Upper Proterozoic. Proc Conf Metallogeny of the Precambrian. IGCP Proj 91: 49–53 Geol Surv, Prague

Mrázek P (1991) Mass balance of metamorphism of the Proterozoic pillow lavas in western Bohemia (in Czech). Čas Mineral Geol Prague 36 (4): 219–234

Mrázek P, Pouba Z (1975) Relations between Fe-V-U mineralization and stromatolites (in Czech). Korelace proterozoických a paleozoických stratiformních ložisek III. Fac Sci, Charles Univ, Prague, pp 59–76

Mrázek P, Pouba Z (1976) The Fe-V-U formation in the Proterozoic rocks of the Svojšín and the Stříbro-Plasy volcanic belt (in Czech). Korelace proterozoických a paleozoických stratiformních ložisek IV. Fac Sci, Charles Univ, Prague, pp 53–71

Mrázek P, Hladíková J, Žák K (1990) Distribution of sulphur isotopes and tectonic development of the Barrandian Proterozoic (Bohemian Massif). Čas Mineral Geol Prague 35 (2): 135–146

Patočka F, Pošmourný K, Rajlich P, Richter A, Vaněček M (1982) Application of the Plumbotectonic model in the Bohemian Massif. Acta Univ Carol, Geol Pouba Prague 3: 217–229 (in Czech)

Petránek J (1976) Sedimentary iron ores of the Lahn Dill type: a new concept of their origin. Věstn Ústřed Úst Geol Prague 51: 203–207

Pouba Z (1986) A new metallogenetic division of sedimentary, volcano-sedimentary and weathering formations of the Bohemian Massif. In: Bernard JH, Pouba Z (eds) Ore deposits and metallogenesis of the Bohemian part of the Bohemian Massif. Geol Surv, Prague, pp 96–97 (in Czech)

Pouba Z, Kříbek B (1986) Organic matter and the concentration of metals in Precambrian stratiform deposits of the Bohemian Massif. Precambrian Res 33: 225–237

Pouba Z, Mrázek P (1981) Anomalous gold contents in the Proterozoic silicites of the Barrandian (in Czech). Korelace proterozoických a paleozoických stratiformních ložisek VI. Fac Sci, Charles Univ Prague, pp 81–89

Pouba Z, Zoubek V (1986) Upper Proterozoic metallogeny of the Bohemian Massif. Proc Conf Metallogeny of the Precambrian. Geol Surv, Prague, pp 23–33

Skoček V (1963) Petrographical composition and genesis of the iron ores from the area of Březina. Rozpr Česk Akad Věd Řada Mat Přír Věd Prague 73 (4): 1–108

VII.C Moldanubian Zone

VII.C.1 Stratigraphy

J. Fiala, G. Fuchs, and J.I. Wendt

1 General Characteristics of the Moldanubian Zone

J. Fiala

1.1 Boundaries

The Moldanubian Zone of the Bohemian Massif represents a major crystalline segment within the eastern part of the mid-Paleozoic Variscan foldbelt. It adjoins with the Saxo-Thuringicum to the NW, with the Bohemicum to the N, and with the Moravo-Silesian Zone to the E. The continuation of the Moldanubian Zone to the S and SW is covered by post-Variscan platform sediments.

The MO/ST boundary is conventionally identified with the boundary between mica-schists and paragneisses W of the town of Mariánské Lázně (e.g., Gaertner 1951). The lithological and metamorphic evolution near the MO/ST junction seem to indicate a transition between both units (Schreyer 1966). Weber and Behr (1983), however, supposed the MO/ST boundary to be a major suture-type discontinuity.

The MO/BO boundary in central Bohemia is obscured by Variscan intrusions of the Central Bohemian Pluton. However, the original character of a major vertical shear zone can be recognized (cf. Central Bohemian Shear Zone, Rajlich 1987).

In the West-Bohemian part of the NNW-SSE-trending MO/BO boundary void of the Variscan plutonic bodies, the high-grade Moldanubicum adjoins the medium-grade Domažlice complex of the Bohemicum. This local configuration led Vejnar (1971) to suggest the existence of metamorphic transitions between Bohemicum and Moldanubicum. The proper contact is, however, destroyed there by a post-Variscan quartz-vein filled wrench fault known as the Bohemian Quartz Pfahl.

At the northern MO/BO boundary, the Kutná Hora and Svratka Crystalline Complexes with medium- to HP-metamorphic grade are located. Both complexes are mostly taken as integral parts of Moldanubicum (e.g., Zoubek 1988). The boundary of these complexes with the Bohemicum, for the most part covered by the Bohemian Cretaceous Platform deposits, is either of tectonic nature (the Železné hory Mts. fault system) or is obscured by the Variscan intrusive body of the Železné hory Mts. Pluton. Losert (1967) suggests that a lithological continuity exists between the Bohemicum and the Kutná Hora Crystalline Complex as well as between the latter and the Moldanubicum. So, according this author, the MO/BO boundary would reflect there some intricately deformed metamorphic transition within a single Late Proterozoic volcano-sedimentary sequence.

The mostly NE-SW-striking Moldanubicum/Moravo-Silesicum boundary has the character of an overthrust (cf. Moldanubian Thrust, Suess 1912). This thrust manifests the late phase of a complex Variscan tectonic evolution. Structural analysis characterizes the MO/MS boundary mostly as a major ductile shear zone with a predominance of strike-slip movements (Schulmann 1990).

1.2 Subdivision

The problem of stratigraphic subdivision of high-grade metamorphic complexes such as the Moldanubian Zone is very difficult to solve as in other metamorphic terranes of the world.

Zoubek (1988) supposed four groups to exist within the Moldanubian lithostratigraphic sequence. The oldest one should be the Želiv Group (formerly the Monotonous Group) with predominating migmatised paragneisses, more or less void of intercalations. The younger, the Klet' Group, is built up mostly of felsic granulites (believed to be of a volcanogeneous origin) with intercalations of basic and ultrabasic rocks. The following Český Krumlov Group (formerly the

Dallmeyer et al. (Eds.)
Pre-Permian Geology of Central and Eastern Europe

Varied Group) consists of paragneisses with numerous calcareous, graphitic, and metabasite intercalations. The last of these, the Kaplice Group with muscovite-biotite paragneisses of quartz-pelitic lithology is taken as the youngest, flyschoid member of the Moldanubian sequence.

The appropriate lithostratigraphic (more exactly tectonostratigraphic) subdivision of the Moldanubian Zone needs more data on structural geology and geochronology. The best way for the present seems to be the subdivision into "terranes", i.e., units that experienced evidently different geological histories and are now separated by major thrusts or strike-slip faults (cf. Matte et al. 1990). According to the present level of knowledge, four terranes can be seriously characterized within the Moldanubian Zone:

The Gföhl terrane, firstly recognized in Austria as forming the Gföhl nappe (Fuchs and Matura 1976; Tollmann 1982), consists of a sheet of high-pressure granulites associated with garnet and spinel peridotites, pyroxenites, dunites, and minor eclogites as well as with high-grade (in part anatectic) ortho- and paragneisses, amphibolites, and metagabbros.

The Drosendorf terrane (Drosendorf nappe in Austria, see Tollman 1982), also largely identical with the previous Moldanubian Varied Group in Bohemia, consists essentially of a thick supracrustal sequences of graywacke-pelitic lithology with more or less frequent intercalations of marbles, calc-silicate gneisses, graphitic rocks, quartzites, leptynites, amphibolites, metadolerites, and metagabbros. Metamorphism of this sequences is medium- to high-grade without any HP granulite-grade indications. These probably Early Paleozoic sequences (see Sect. 2.4) were at least partly deposited on an old crystalline basement containing early Proterozoic I-type granitoid gneisses (Wendt et al. 1993).

According to Patočka (1991a,b), the amphibolites and plagioclase-biotite paragneisses of the Drosendorf terrane near Český Krumlov display the geochemistry pointing to premetamorphic origin in contrasting tectonic settings, i.e., in ocean-island and ocean-floor environments in case of the amphibolites and in active plate-margin environment in case of paragneisses. Thus, a tentative interpretation can be made, regarding the crystalline sequence of the Drosendorf terrane as a remnant of an accretionary wedge.

The Ostrong terrane (cf. Ostrong unit in Austria, Fuchs and Matura 1976) encompasses also the previous Moldanubian Monotonous and Kaplice Groups of Zoubek (1988) in South Bohemia. Both groups show mutual transitions with different intensity of HT-LP metamorphic overprints (Fiala et al. 1982). These monotonous sequences of quartz-pelitic and graywacke lithologies containing several orthogneiss bodies and rare amphibolite and calc-silicate intercalations are probably not older than Late Proterozoic (Kröner et al. 1988).

The Ostrong and Drosendorf terranes are separated by a major thrust zone (The South Bohemian Main Thrust), accompanied by retrograded eclogites and minor granulite occurrences (Fuchs and Scharbert 1979; Fiala et al. 1992).

The Bavarian terrane (the Bavaricum, Fuchs 1976) is largely composed of Variscan HT-LP-metamorphic anatexites having migmatite, pearl gneiss, or diatexite character. This area displays evidence of an evolution similar to that of the previously mentioned terranes before being almost completely obliterated by the HT-LP metamorphic event.

A tentative areal extent of terranes defined above is presented in Fig. 1.

2 Moldanubian Geochronology

J. Fiala and J.I. Wendt

2.1 Introduction

An unequivocal geological evidence for the age of deposition and metamorphism of the Moldanubian

Fig. 1. Tentative areal extent of terranes defined within the Moldanubian Zone (MZ). *1* Post-Permian platform cover; *2* Permian of Variscan intramontane furrows; *3* Late Variscan granitoid intrusions; *4* Drosendorf terrane; *5–8* Gföhl terrane: *5* extent of gneisses with disseminated exotics (lenses of eclogites, ultrabasics and granulites); *6* ultrabasics and eclogites (according to unpublished map of J. Machart, greatly exaggerated); *7* minor granulite occurrences; *8* major granulite occurrences (massifs); *9* Ostrong terrane; supposed overthrust boundaries to neighboring units; *10* Bavarian terrane; *11* MZ neighboring crystalline units (sometimes of conventional boudaries with MZ): *SXT* Saxothuringian Basin; *TC* Teplá Crystalline Complex; *DC* Domažlice Crystalline Complex; *KHC* Kutná Hora Crystalline Complex; *SC* Svratka Crystalline Complex; *MSZ* Moravo-Silesian Zone; *12* Barrandian-Železné hory Mts. Zone; *13* Conventional boundaries between units of mutual lithological transitions; *14* MZ overthrust boundaries with adjacent units; *15* Post-Variscan normal faults

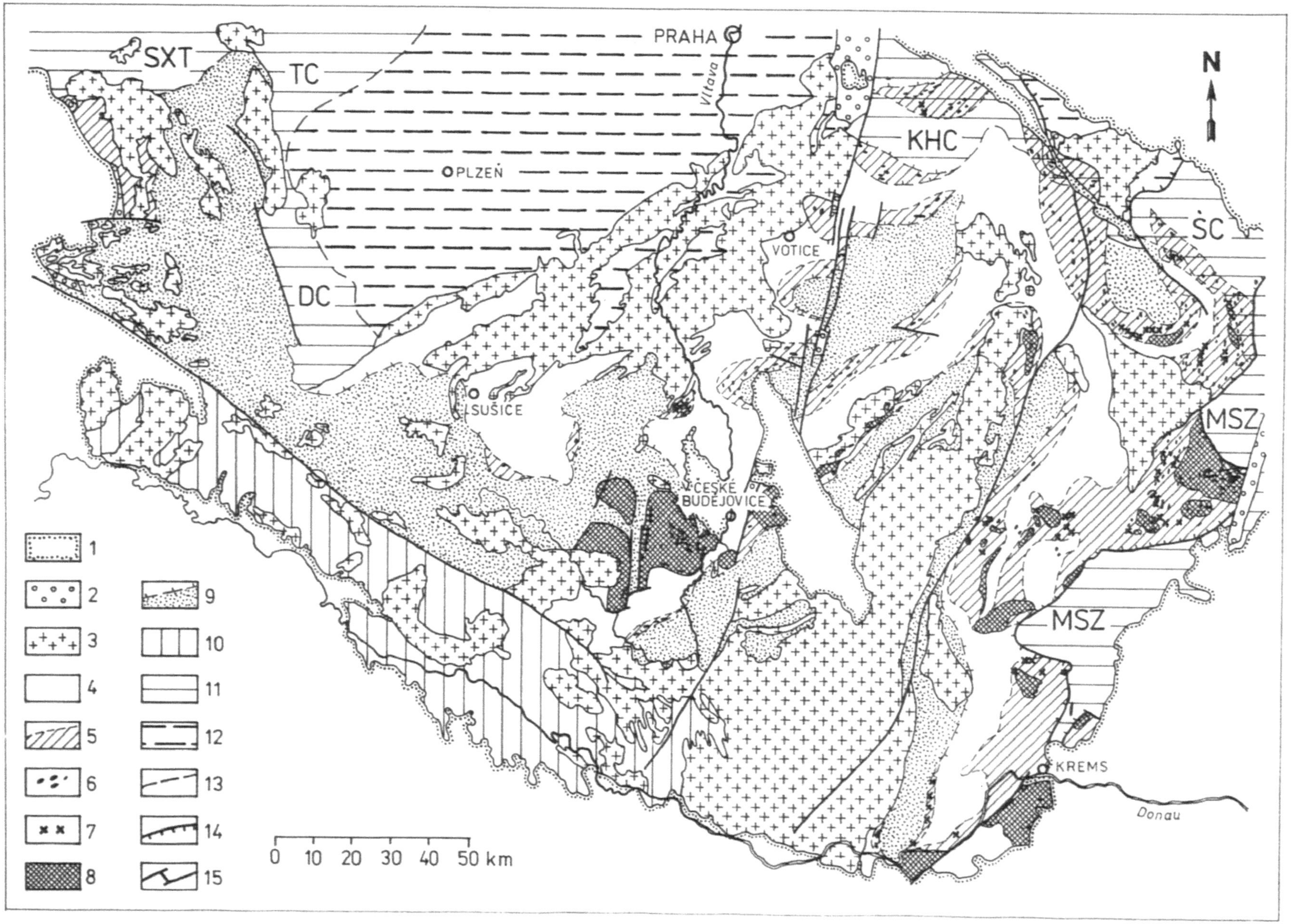

SXT
TC
PRAHA
Vltava
KHC
N
OPLZEŇ
VOTICE
ŠC
DC
SUŠICE
MSZ
ČESKÉ BUDĚJOVICE
MSZ
KREMS
Donau
1
2
3
4
5
6
7
8
9
10
11
12
13
14
15
0 10 20 30 40 50 km

metasediments does not exist. A minimum age for amphilbolite- and granulite-grade metamorphism is given by the occurrence of pebbles of Moldanubian paragneisses and granulites in early Visean conglomerates (Štelcl 1969). This fact provides evidence that high-grade metamorphism, at least in part of the Moldanubicum, is older than 333 Ma, using the time scale of Palmer (1983).

The age of Moldanubian metamorphism is discussed in Chapter VII.C.4. It seems to be useful at this point to present preliminary geochronological results on possible basement and deposition ages for Moldanubian supracrustals as well as on the evolution of the Moldanubian granulites.

2.2 The Problem of Early to Middle Proterozoic

Middle Proterozoic or still earlier ages of the Moldanubicum as a whole, proposed by Máška and Zoubek (1960) and Chaloupský (1978), are not clearly evidenced. Grauert et al. (1973), however, dated detrital zircons of the Moldanubian metasediments between values from 1800 to 2300 Ma. The grain of detrital zircon from a metasedimentary granulite of the Bory quarry (western Moravia) has yielded the oldest age – 2684 ± 14 Ma (Kröner et al. 1988). Early Archaean components, up to 3840 Ma in age, and Early to Middle Proterozoic metamorphic overprints were detected by ion microprobe dating of detrital zircons derived from various metasediments of the Moldanubicum in eastern Bavaria (Gebauer et al. 1989).

New U-Pb zircon ages on specific orthogneiss types in the Moldanubicum of South Bohemia (Wendt et al. 1993) provide first unequivocal evidences of some preserved remnants of an Early Proterozoic crust in the Bohemian Massif. These orthogneisses are of I-type granodiorite-tonalite-quartz diorite composition being located at the tectonic boundary between the Drosendorf and Ostrong terranes (Fig. 2). Conventional U-Pb zircon upper concordia intercepts, as well as ion microprobe and single grain evaporation ages of ca. 2060–2110 Ma, interpreted as reflecting the time of magmatic crystallization, together with an exceptionally high mean Nd-model crustal residence age of T_{DM} = 3.0 Ga, demonstrate a considerable antiquity of these rocks.

As seen from Fig. 2, a tectonostratigraphic position similar to that of dated Early Proterozoic orthogneisses is occupied by the Dobra orthogneiss in Austria (T_{DM} = 2.5 Ga, Liew and Hoffmann 1988). From this configuration a tectonic scheme can be constructed: the ancient orthogneisses that originally formed the basement of the Drosendorf supracrustals were first detached and then, involving their sedimentary envelope, thrust over the Ostrong terrane. This scenario agrees with statement of Fuchs and Matura (1980), characterizing the geological position of Dobra gneiss as follows: "The boundary with the underlying Ostrong unit is a deep-seated tectonic line accompanied by lenses of granulites and ultrabasites, whereas the Drosendorf unit seems to be transgressive onto the Dobra gneiss, with quartzites and arkosic gneisses at the base".

2.3 Age Limits for Deposition of Moldanubian metasediments

The youngest zircon within a detrital population may set an upper age-limit for the deposition of the metasediment in question.

Within the Monotonous Group of Ostrong terrane (locality Strážkovice, South Bohemia) the youngest detrital zircon age of 727 ± 106 Ma (Kröner et al. 1988, ion microprobe data) points to a Late Proterozoic maximum age for the original sediment deposition.

The Lhenice zone, South Bohemia, is usually taken as a lower part of the Varied Group (Drosendorf terrane) due to presence of a conspicuous K-feldspar-bearing quartzite horizon (see, e.g., Pletánek and Suk 1976). The study of a kinzigitic paragneiss from the Lhenice zone (locality Ktiš) by Wendt (1989) revealed two populations of detrital zircons. The rounded morphology of the older zircon population (somewhat heterogeneous in age – 1.6 to 2.0 Ga) indicates repeated sedimentary reworking after the erosion of source area of Middle to Early Proterozoic age. The younger prismatic, nearly euhedral zircons probably underwent only one sedimentary cycle; they show internal zonation and morphology typical for granitic rocks. Their age of 549 ± 5 Ma, therefore, probably corresponds to "Cadomian" intrusive activity in the source area. The value can also be interpreted as a maximum age of deposition of the studied Drosendorf metapelites that supports the idea of their Early Paleozoic derivation.

The knowledge of magmatic crystallization ages of suitable synsedimentary volcanic rocks

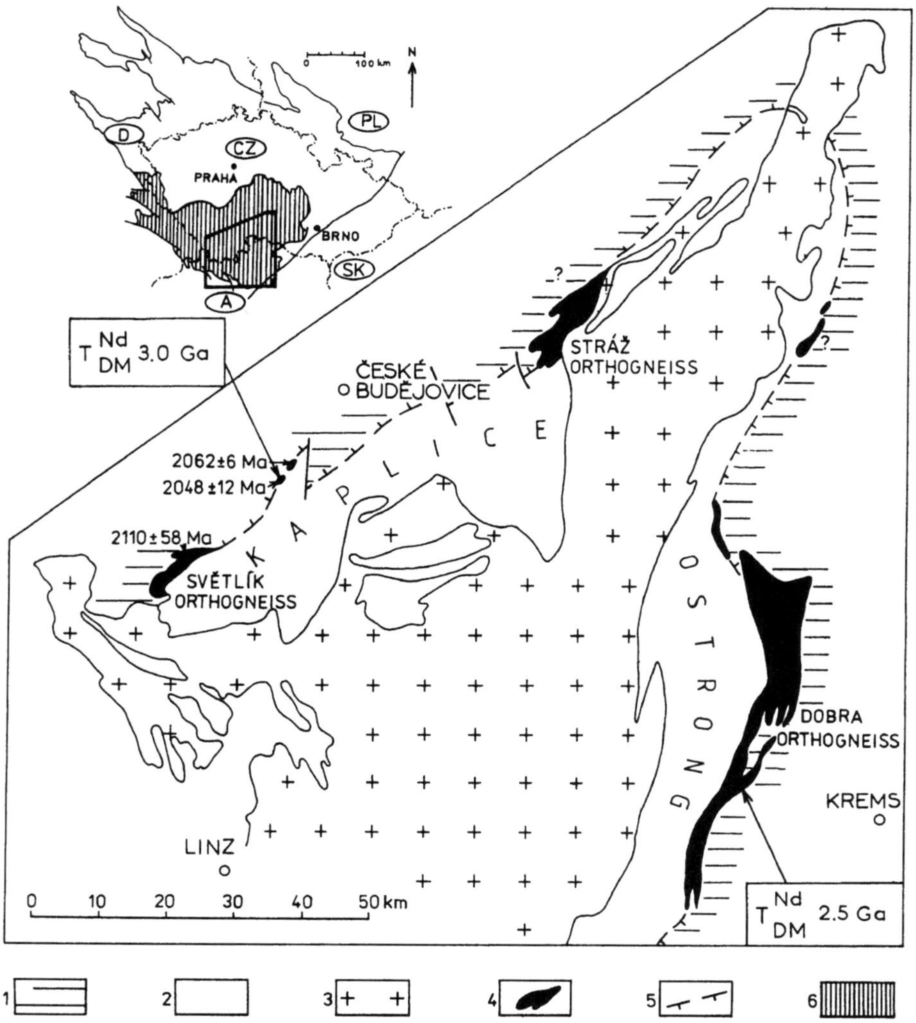

Fig. 2. Tectonostratigraphic position of Proterozoic orthogneisses in the southern part of the Moldanubian Zone. *1* Drosendorf and Gföhl Terranes (T_{DM}^{Nd} = 1.4–1.8 Ga); *2* Ostrong Terrane; *3* Central Moldanubian Pluton (Late Variscan); *4* orthogneisses (? – undated bodies); *5* supposed terrane boundary (South Bohemian Main Thrust) *6* Moldanubian Zone (within the *inset map* of the Bohemian Massif)

enables us to approximate deposition ages of volcano-sedimentary complexes in question.

Metamorphosed acid synsedimentary volcanic rocks (leptynites) were described by Fiala (1988) from the Moldanubicum of SW Bohemia. The volcano-sedimentary origin is documented by the shape of leptynite bodies that are up to 150 m thick and extend for tens of km along strike. The stable association with K-feldspar-bearing calc-silicate rocks and minor carbonate lenses supports this origin. The graywacke-leptynite-calcareous sequence in SW Bohemia directly underlies the typical metacarbonate sequence of the Sušice-Votice part of the Varied Group that could be ascribed to Drosendorf terrane.

The U-Pb zircon size fraction analyses of a leptynite (Chlum locality near the town of Kašperské Hory) by Wendt and Fiala (unpubl.) has identified two distinct zircon populations: one – clear, colorless, the other – dark brown and clouded. The latter is strongly metamict with subsequent new crystallization of small subindividuals. The fractions of the clear population plot near the concordia point of ca. 460 Ma which may be interpreted as the age of leptynite extrusion. The recrystallized metamict zircons indicate substantial Pb loss at ca. 315 Ma ago, probably related to the Variscan HT-LP metamorphism (cf. Teufel 1988). Regarding this overprint, the obtained leptynite age has to be taken as a minimum one.

The above data support independently the idea of Early Paleozoic deposition at least of some parts of the Drosendorf terrane.

2.4 Origin of the Moldanubian Granulite Complex

Within the Gföhl terrane, based on rock geochemistry (Fiala et al. 1987), three granulite precursors were presumed: (a) an aplogranite (or rhyolite) precursor of wet-granite minimum melt composition for the felsic granulites, (b) a metasedimentary precursor of meta-arenite to metapelite composition for the melanocratic (pyroxene-free) granulites, and (c) a calc-alkaline suite magmatic precursor for the mafic to intermediate pyroxene granulites.

The age of the Moldanubian HP-granulite grade metamorphism seems to be set well between ca. 340 and 350 Ma according to zircon U-Pb data (Var Breemen et al. 1982; Kröner et al. 1988; Aftalion et al. 1989). The new U-Pb zircon and Sm-Nd mineral-whole rock studies (Wendt et al. 1994) supply sufficient data to reconstruct the complex evolution of the Moldanubian granulite complex in southern Bohemia.

The felsic granulites were probably formed from a migmatitic to aplitic precursors resulting from an intracrustal anatectic event at ca. 370 Ma (minimum age), connected with extensional geotectonic settings (cf. Pin and Vielzeuf 1983). The upper concordia intercept ages for felsic granulites at ~1800 Ma, and the ages of detrital zircons from metasedimentary granulites point to crustal rocks having high content of a Proterozoic detrital component were subjects of the above-mentioned anatexis.

The nearly concordant U-Pb zircon age for pyroxene granulites, ca. 345 Ma, without striking indication for older components, may point to crystallization of the HP mineral assemblage from the Variscan calc-alkaline magmatic precursor intruding into the lower crust.

The Moldanubian granulites probably originated during Variscan continental collision, followed by extreme subduction of continental crust. This compressional event was preceded by some 25–40 Ma by a crustal extension connected, most probably with a supply of the mantle-derived heat.

2.5 Conclusions

From the preliminary geochronological data available for the Moldanubian Zone we can conclude: (a) The Drosendorf terrane consists most probably of Early Paleozoic volcano-sedimentary sequences, at least partly deposited on the crystalline basement containing Early Proterozoic granitoids of I-type characteristics. Derivation of the Drosendorf terrane as an Early Paleozoic active plate-margin complex cannot be excluded. (b) The Ostrong terrane consists of a monotonous quartz-pelitic sequence with passive margin affinity, probably not older than Late Proterozoic; a Cambro-Ordovician age cannot be excluded. (c) The granulites of the Gföhl terrane were formed by anatexis of evolved continental crust followed by HP granulite-grade metamorphism as a consequence of Variscan continental subduction. Origin from some Variscan accretionary-wedge complex seems to be less probable for the Gföhl terrane. (d) The concept of a major Variscan tectonothermal event affecting the entire Moldanubian Zone seems to be well supported by new geochronological data.

Acknowledgments. The authors are indebted to Prof. A. Kröner and Dr. F. Patočka for the critical reading of the manuscript.

3 The Austrian Part of the Moldanubicum

G. Fuchs

In the Waldviertel, the Moldanubicum builds up the region between the Variscan Moldanubian Pluton in the W and the Bíteš Gneiss of the Moravicum in the E (Fig. 3a,b). Thus the Moldanubicum of the Waldviertel is the direct continuation of that of Moravia and it is not surprising to find similar rock sequences there. There is also much resemblance between the Waldviertel and the South Bohemian Moldanubicum, particularly in the sequence of rock series. The Moldanubicum of South Bohemia, however, does not continue to the adjoining parts of Austria – the Mühlviertel. It was obliterated there during the Variscan orogenesis – except a few remains in the Böhmer Wald and Kropfmühl in the Passauer Wald. A new Variscan crystalline, the Bavaricum, was formed there (Fuchs 1976).

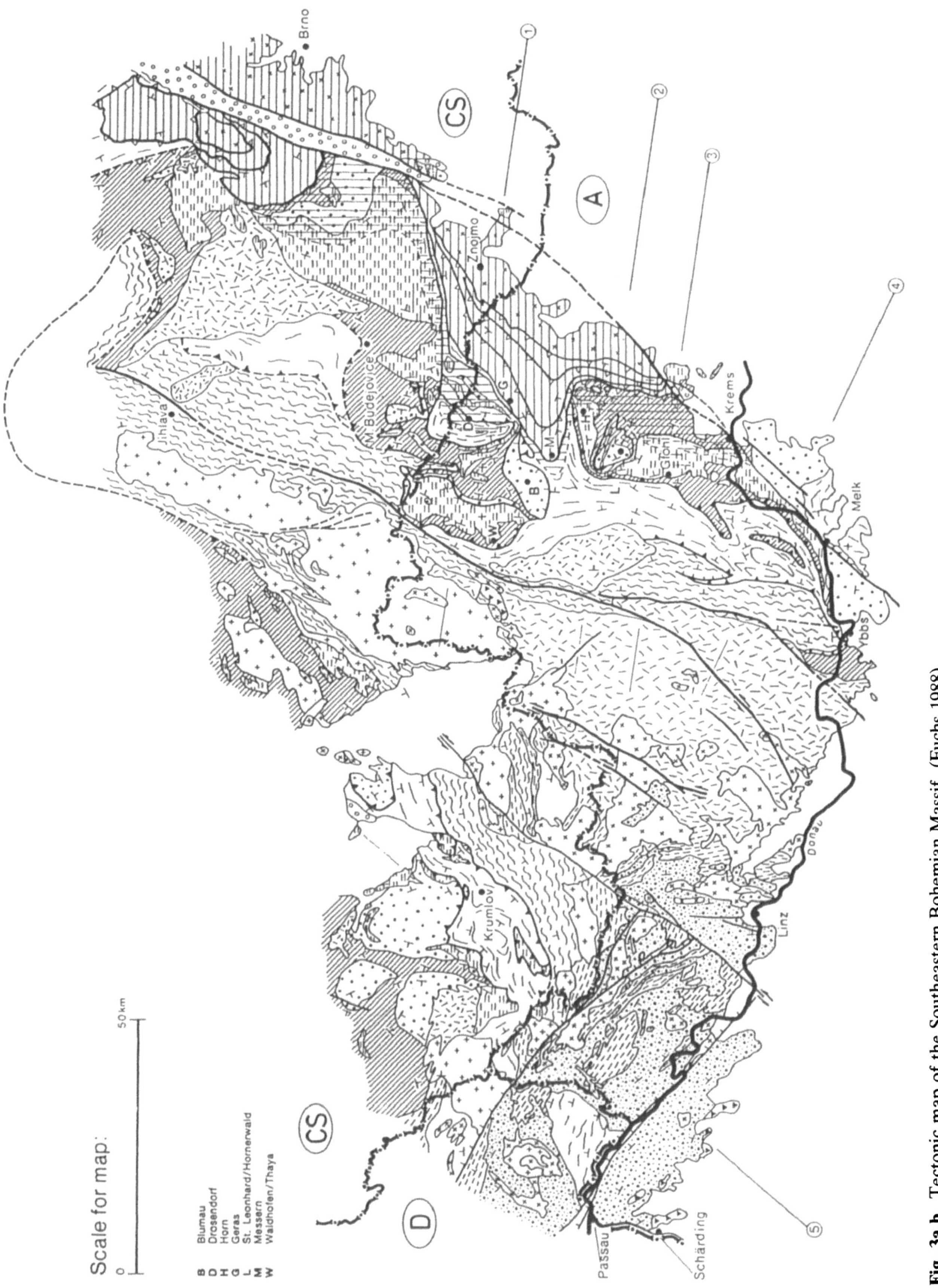

Fig. 3a,b. Tectonic map of the Southeastern Bohemian Massif. (Fuchs 1988)

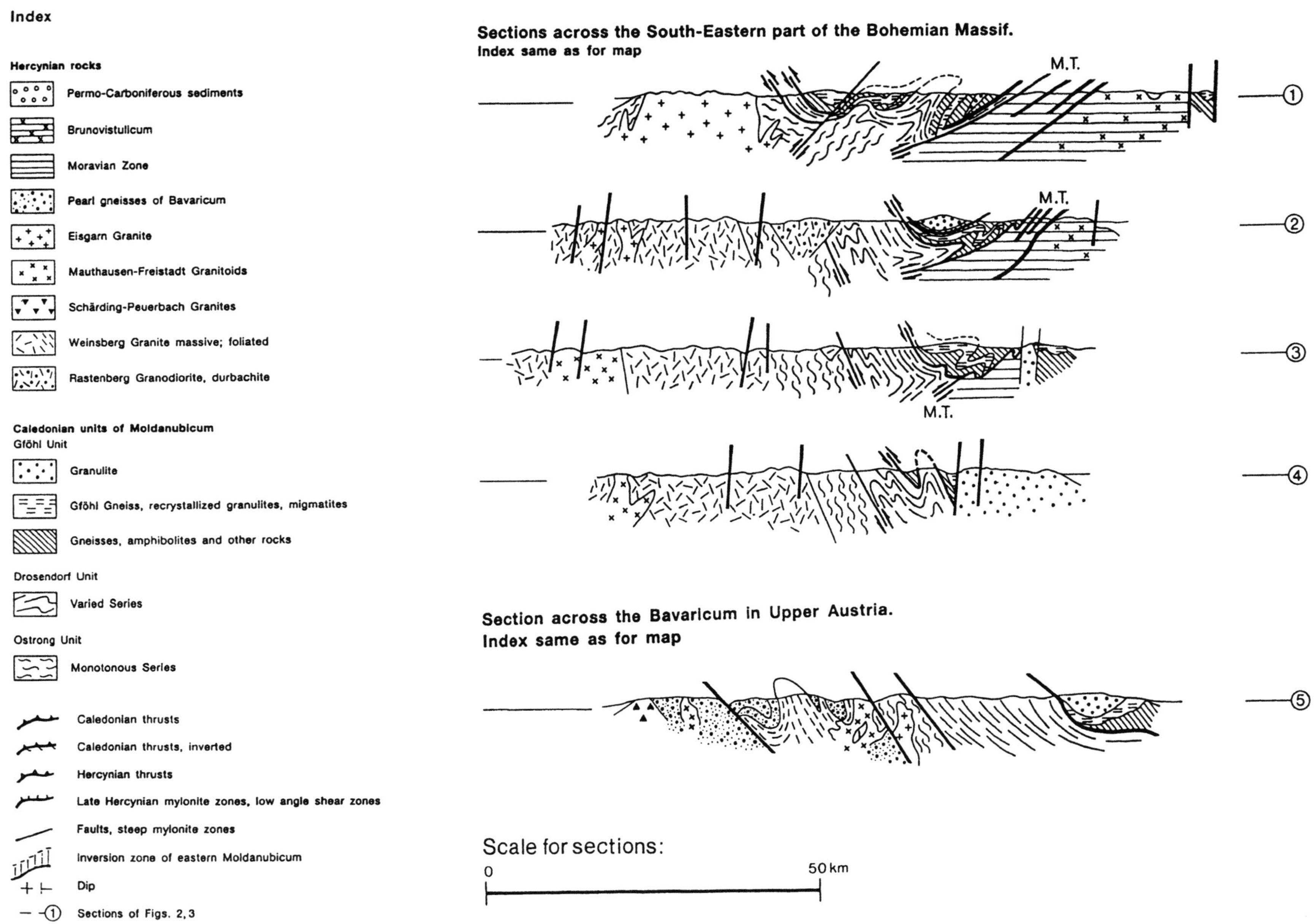
Index
Hercynian rocks
Permo-Carboniferous sediments
Brunovistulicum
Moravian Zone
Pearl gneisses of Bavaricum
Eisgarn Granite
Mauthausen-Freistadt Granitoids
Schärding-Peuerbach Granites
Weinsberg Granite massive; foliated
Rastenberg Granodiorite, durbachite
Caledonian units of Moldanubicum
Gföhl Unit
Granulite
Gföhl Gneiss, recrystallized granulites, migmatites
Gneisses, amphibolites and other rocks
Drosendorf Unit
Varied Series
Ostrong Unit
Monotonous Series
Caledonian thrusts
Caledonian thrusts, inverted
Hercynian thrusts
Late Hercynian mylonite zones, low angle shear zones
Faults, steep mylonite zones
Inversion zone of eastern Moldanubicum
Dip
Sections of Figs. 2, 3
Sections across the South-Eastern part of the Bohemian Massif.
Index same as for map
M.T.
1
M.T.
2
3
M.T.
4
Section across the Bavaricum in Upper Austria.
Index same as for map
5
Scale for sections:
0
50 km

Fig. 3b

The Waldviertel Moldanubicum strikes regionally NNE-SSW and dips towards the E. Thus the zoning Monotonous Group-Varied Group-Gföhl Gneiss-Granulite Complex represents an ascending sequence. This is, however, a tectonic and not a stratigraphic succession (Fig. 3a,b). The lowest unit, the **Ostrong Unit**, comprises the Monotonous Group. It is overthrust by the Dobra- and Spitz orthogneisses and the Varied Group, which build up the **Drosendorf Unit**. This, in turn, is overridden by the **Gföhl Unit**, consisting of characteristic amphibolites, migmatitic gneisses, the Gföhl Gneiss, Wolfshof Syenite-gneiss, granulites, and associated rocks. These tectonic units are distinct by certain rock assemblages and metamorphic characteristics; there are, however, no means to correlate these units and their geological age is disputable. In the following text the rock complexes are described in their tectonic order, which of course does not imply an age sequence.

The Monotonous Group (Ostrong Unit) consists of a thick uniform paragneiss complex-biotite-plagioclase gneisses with varying amounts of muscovite and potassium feldspar containing sillimanite, cordierite, and rare garnet. The lithological bending is poorly developed and the s-planes are predominantly uneven. Nebulitic textures indicate widespread anatexis. Nodules of calc-silicate rock are very common.

Leucocratic orthogneisses rich in alkali feldspar occur in the Ostrong area. They form a few major bodies or thin intercalations in the paragneiss. This suggests an acid volcanic origin.

Rare metaeclogites occur as small lenses. Obviously, the Monotonous Series was formed from nondiversified, thick, areno-argillaceous or graywacke series deposited in a subsiding basin along a continental margin. The main metamorphism was under low pressure/high temperature condition (630 °C/3 kb, Zaydan and Scharbert 1983).

The Monotonous Group is separated from the overlying Varied Group by a shear horizon. A lamella of granulite indicates that the deformation took place under high pressure/high temperature conditions.

The Drosendorf Unit starts with either the Dobra Gneiss of granitic composition or the Spitz Granodiorite-gneiss (Fuchs and Matura 1976). Apparently, they represent the basement of the Varied Group. In the following thick paragneiss sequence we find first a conspicuous band of calc-silicate rock or quartzite to arkosic gneiss respectively, and then the intercalations of marble, calc-silicate rocks, graphite schist, quartzite, and amphibolite overlie. This succession seems to reflect a stratigraphic order. The paragneisses, which form the matrix in this sequence, show well-defined bedding by alternation of more micaceous or quartz-feldspar rich varieties. They are biotite-plagioclase gneisses frequently containing muscovite, garnet, and sillimanite.

The source material of the Varied Group was an argillaceous-arenaceous-carbonate alternation with black shale and basic volcanic intercalations. This diversified series was deposited on a shelf, probably a passive continental margin. The volcanism may point to rifting.

As to the grade of metamorphism Zaydan and Scharbert (1983) suggested 670 °C/5 kb, whereas Högelsberger (1987) deduced ca. 700 °C/5, 5–7, 5 kb. From the composition of garnets, Petrakakis (1986) came to the conclusion that the Varied Group has undergone even 770 °C/7–9 kb conditions.

Regarding the age of the Varied Group there was much dispute after Andrusov and Čorná (1976) and Pacltová (1980, 1986) reported the finding of Siluro-Devonian microfossils. Obviously there occurred some contamination, because recent geochemical investigations by Frank et al. (1990) prove a Precambrian age of the carbonates. They studied pure massive marbles of the Varied Group and found consistently low Sr isotope ratio of 0.7060. This is interpreted as the nearly unaltered sea water ratio at the time of sedimentation. According to the knowledge of the Sr isotopic evolution of sea water during earth history, the results are a strong argument for Proterozoic age (minimum age 800–900 Ma).

Between the Varied Group of the Drosendorf Unit and the Gföhl Unit there is no well-defined boundary. A thrust contact, however, is expected, because the superposition of migmatites, magmatites, and metamorphics from the lower crust (granulites) and the Upper Mantle (ultramafitites) on a metasedimentary series is difficult to explain by a mechanism other than nappe tectonics.

The Gföhl Unit consists of typical rocks: the Rehberg Amphibolite is characterized by its banded alternation with orthogneiss and by the associated ultramafics. The paragneisses are frequently migmatitic and contain rare beds of graphite quartzite. The Gföhl Gneiss is mostly of granitic composition and exhibits a nebulitic texture. Small amounts of garnet, sillimanite, or kyanite are typical. Between the Gföhl Gneiss and its associated migmatites and the succeeding granulite series the Wolfshof Syenite-gneiss or leucogranites are found. Inclusions of the country rock and branching sills document their intrusive nature. Further, they show faint parallel structures and are predominantly massive. These magmatites form concordant intercalations in the regional structure, which suggests their intrusion during the nappe movements. Not yet in a solid state, they were sandwiched between the Gföhl Gneiss and the granulite series. The granulites form several large bodies. They are predominantly light, often banded rocks composed of perthitic orthoclase, oligoclase-andesine, garnet, and small amounts of biotite, kyanite, and sillimanite. There are also small bodies of pyroxene granulite and pyriclasite, which consist of andesine-labrador and varying amounts of ortho- and clinopyroxene, brown hornblende, garnet, and quartz. Pyrope peridotites or serpentinites, garnet-pyroxene amphibolites, and granulitic gneisses are associated with the granulites. The granulitic gneisses are rich in garnet, sillimanite, and kyanite and poor in biotite. They occur interbedded in the paragneisses or in the Gföhl Gneiss (Fuchs and Scharbert 1979).

Scharbert and Kurat (1974) conclude for the granulite series metamorphic conditions of 760 °C/ 11 kb and low p_{H_2O}. This suggests an origin from the lower crust, which is consistent with the observed association of ultramafics derived from the Upper Mantle.

For many years, the granulites were considered to be Precambrian (Scharbert 1964; Svoboda et al. 1966, and others). Radiometric age determinations gave a surprise: the granulites yielded Caledonian ages (Arnold and Scharbert 1973). Many other rocks in the Variscides of Europe gave such age too (see Zwart and Dornsiepen 1978; Dornsiepen 1979). Later, even Variscan age of the granulite metamorphism was suggested (Van Breemen et al. 1982). This implies that the high-grade metamorphism, the intra-Moldanubian nappe tectonics, the younger Moldanubian/ Moravian Thrust and the post-tectonic Hercynian plutonism occurred within only 15 Ma which from the heterogeneity of these events seems rather improbable (Fuchs 1986). Though there are open questions, we may expect Precambrian age for the educt of many Moldanubian rocks; their main metamorphism, however, was Palaeozoic (Caledonian, Variscan).

Finally, we have to deal with the **Mica-schist Zone**, which follows the Moldanubian/Moravian boundary. As the name says, the zone consists predominantly of mica-schists, which contain garnet, kyanite, and staurolite. In the mica-schists we find many other rocks intercalated. It is significant that the kind of these intercalations depends on the adjoining Moldanubian rock assemblage. In the case of the Varied Group, the mica-schists are interbedded with marble, graphite schist, quartzite, and amphibolite. If the Gföhl Unit is adjacent, we find as intercalations the Rehberg Amphibolite, ultramafics, graphite quartzite, Gföhl Gneiss, and granulites. Where the diverse rock series of the Moldanubicum strike obliquely to the Moravian boundary, it is possible to trace the mentioned rocks into the Mica-schist Zone. This is direct proof of the concept of Suess (1903, 1912), who accepted the Mica-schist Zone as a belt of retrogressive metamorphism. The paragneisses were altered to mica-schists, whereas the accompanying rocks retained their character better due to their composition. Thus the Mica-schist Zone does not represent an individual rock assemblage of the Moldanubicum or metasedimentary belt; it is just a zone of retrogressive metamorphism affecting different Moldanubian sequences.

References

Aftalion M, Bowes DR, Vrána S (1989) Early Carboniferous U-Pb zircon age for garnetiferous, perpotassic granulites, Blanský Les Massif, Czechoslovakia. Neues Jahrb Mineral Monatsh 1989: 145–152

Andrusov D, Čorná O (1976) Über das Alter des Moldanubikums nach mikrofloristischen Forschungen. Geol Práce Sb 85: 81–89

Arnold A, Scharbert HG (1973) Rb/Sr Altersbestimmungen an Granuliten der südlichen Böhmischen Masse in Österreich. Schweiz Mineral Petrogr Mitt 53: 61–78

Chaloupský J (1978) The Precambrian tectogenesis in the Bohemian Massif. Geol Rdsch 67: 72–90

Dornsiepen UF (1979) Rb/Sr whole rock ages within the European Hercynian. A Review. Kristalinikum 14: 33–49

Fiala J (1988) Tectonic evolution of crystalline complex in the Kašperské Hory ore district, Moldanubicum (in Czech). Acta Univ Carolinae Geol 1988: 427–436

Fiala J, Zeman J, Sobotková O (1982) Tektonischer Bau des Zentralteiles des Moldanubikums. Krystalinikum 16: 7–20

Fiala J, Matějovská O, Vaňková V (1987) Moldanubian granulites: Source material and petrogenetic considerations. Neues Jahrb Mineral Abh 157: 133–165

Fiala J, Fišera M, Jelínek E, Slabý J, Vrána S (1992) Guide to excursions. Int Worksh IGCP 304: High-pressure granulites – lower crustal metamorphism, Rohanov Czechoslovakia. Czech Geol Surv, Prague

Frank W, Hammer S, Popp F, Scharbert S, Thöni M (1990) Isotopengeologische Neuergebnisse zur Entwicklungsgeschichte der Böhmischen Masse. Proterozoische Gesteinsserien und variszische Hauptorogenese. Österr Beitr Meteorol Geophys H 3 1990: 185–228

Fuchs G (1976) Zur Entwicklung der Böhmischen Masse. Jahrb Geol Bundesanst 119: 45–61

Fuchs G (1986) Zur Diskussion um den Deckenbau der Böhmischen Masse. Jahrb Geol Bundesanst 129: 41–49

Fuchs G, Matura A (1976) Zur Geologie des Kristallins der südlichen Böhmischen Masse. Jahrb Geol Bundesanst 119: 1–43

Fuchs G, Matura A (1980) Die Böhmische Masse in Österreich. In: Oberhauser R (ed) Der geologische Aufbau Österreichs. Springer, Berlin Heidelberg New York, pp 121–128

Fuchs G, Scharbert HG (1979) Kleinere Granulitvorkommen im niederösterreichischen Moldanubikum und ihre Bedeutung für die Granulitgenese. Verh Geol Bundesanst 1979: 24–49

Gaertner HR (1951) Probleme des Saxothuringikums. Geol Jahrb 65: 409–450

Gebauer D, Williams IS, Compston W, Grünfelder M (1989) The development of Central European continental crust since the Early Archean based on conventional and ionmicroprobe dating of up to 3.84 b.y. old detrital zircons. Tectonophysics 157: 81–96

Grauert B, Hänny R, Soptrajanova G (1973) Age and origin of detrital zircons from pre-Permian basements of the Bohemian Massif and the Alps. Contrib Mineral Petrol 40: 105–130

Högelsberger H (1987) Petrologie der Marmore und Kalksilikatgesteine der Bunten Serie (Niederösterreichisches Moldanubikum). Dissertation, Formal- und Naturwiss Fak, Univ Wien

Kröner A, Wendt I, Liew TC, Compston W, Todt W, Fiala J, Vaňková V, Vaněk J (1988) U-Pb zircon and Sm-Nd model ages of high-grade Moldanubian metasediments, Bohemian Massif, Czechoslovakia. Contrib Mineral Petrol 99: 257–266

Liew TC, Hoffmann AW (1988) Precambrian crustal components, plutonic associations, plate environment of the Hercynian fold belt of Central Europe: Indications from a Nd and Sr isotopic study. Contrib Mineral Petrol 98: 129–138

Losert J (1967) Contribution to the problem of pre-Assyntian tectogenesis and metamorphism in the Moldanubicum of the Bohemian Massif. Krystalinikum 5: 61–84

Máška M, Zoubek V (1960) Area of the Variscan intramontane block. In: Zoubek V (ed) Tectonic development od Czechoslovakia. NČSAV Praha, pp 25–50

Matte P, Maluski H, Rajlich P, Franke W (1990) Terrane boundaries in the Bohemian Massif: result of large-scale Variscan shearing. Tectonophysics 177: 151–170

Pacltová B (1980) Further micropaleontological data for the Paleozoic age of the Moldanubian carbonate rocks. Čas Miner Geol 25: 275–279

Pacltová B (1986) Palynology of metamorphic rocks (Methodological Study). Rev Palaeobot Palyn 48: 347–356

Palmer AR (1983) The decade of North American geology geologic time scale. Geology 11: 503–504

Patočka F (1991a) Geochemistry and primary tectonic environment of the amphibolites from the Český Krumlov Varied Group (Bohemian Massif, Moldanubicum). Jahrb Geol B A 134: 117–133

Patočka F (1991b) The minor element geochemistry of paragneisses of the Český Krumlov Varied Group, the Šumava Mts. Moldanubicum (in Czech). Sb Jihočeskeho Muz Česk Budějovicích Přír Vědy 31: 118–125

Petrakakis K (1986) Metamorphism of high-grade gneisses from the Moldanubian zone, Austria, with particular reference to the garnets. Metamorphic Geol 4: 323–344

Pin C, Vielzeuf D (1983) Granulites and related rocks in Variscan median Europe: a dualistic interpretation. Tectonophysics 93: 47–74

Pletánek Z, Suk M (1976) Remarks to the stratigraphy and structure of the crystalline complex in South Bohemia. Vyzk Práce Ústřed Ústavu Geol 9: 1–28 (in Czech)

Rajlich P (1987) Variszische duktile Tektonik im Böhmischen Massiv. Geol Rdsch 76: 755–786

Scharbert HG (1963) Die Granulite des südlichen niederösterreichischen Moldanubikums (I. Teil). Neues Jahrb Mineral Abh 100: 59–86

Scharbert HG (1964) Die Granulite des südlichen niederösterreichischen Moldanubikums (II. und III. Teil). Neues Jahrb Mineral Abh 101: 27–66, 210–231

Scharbert HG, Kurat G (1974) Distribution of some elements between coexisting ferromagnesian minerals in Moldanubian granulite facies rocks, Lower Austria, Austria. Tschermaks Mineral Petrogr Mitt 21: 110–134

Schreyer W (1966) Metamorpher Übergang Saxothuringikum-Moldanubikum östlich Tirschenreut/Opf., nachgewiesen durch phasenpetrologische Analyse. Geol Rdsch 55: 491–509

Schulmann K (1990) Fabric and kinematic study of the Bíteš orthogneiss (southwestern Moravia): result of large-scale northeastward shearing parallel to the Moldanubian/Moravian boundary. Tectonophysics 177: 229–244

Suess FE (1903) Bau und Bild der Böhmischen Masse. In: Diener C (ed) Bau und Bild Österreichs F. Temnky Wien, Leipzig, 322 pp

Suess FE (1912) Die Moravischen Fenster und ihre Beziehung zum Grundgebirge des Hohen Gesenkes. Dankschr Österr Akad Wiss 88: 451–631

Svoboda J et al. (1966) Regional geology of Czechoslovakia, Part 1: The Bohemian Massif. Geol Surv Czech, 668 pp

Štelcl J (1969) Zur Entstehung und Entwicklung der paläozoischen Konglomerate auf dem Gebiete

Zentralmährens. Folia Fac Sci Net Univ Purkynianae Brun Geol 10: 1–89
Teufel S (1988) Vergleichende U-Pb- und Rb-Sr-Altersbestimmungen an Gesteinen des Übergangsbereiches Saxothuringikum/Moldanubikum, NE Bayern. Gött Arb Geol Paläontol 35: 1–87
Tollmann A (1982) Grossräumiger variszischer Deckenbau im Moldanubikum und neue Gedanken zum Variszikum Europas. Geotekton Forsch 64: 1–91
Van Breemen O, Aftalion M, Bowes DR, Dudek A, Mísař Z, Povondra P, Vrána S (1982) Geochronological studies of the Bohemian Massif, Czechoslovakia, and their significance in the evolution of Central Europe. Trans R Soc Edinb 75: 89–108
Vejnar Z (1971) Grundfragen des Moldanubikums und seine Stellung in der Böhmischen Masse. Geol Rdsch 60: 1455–1465
Weber K, Behr HJ (1983) Geodynamic interpretation of the Mid-European Variscides. In: Martin H, Eder FW (eds) Intracontinental fold belts. Springer, Berlin Heidelberg New York, pp 427–469
Wendt JI (1989) U-Pb Zirkondatierung und Sm/Nd Systematik sowie Petrologie und Geochemie an Granuliten und Gneisen aus dem Moldanubikum Südböhmens (ČSSR). Thesis, Johannes-Gutenberg-Univ, Mainz
Wendt JI, Kröner A, Fiala J, Todt W (1993) Evidence from zircon dating for existence of approximatelly 2.1 Ga old crystalline basement in southern Bohemia, Czech Republic. Geol Rdsch 82: 42–50
Wendt JI, Kröner A, Fiala J, Todt W (1994) U-Pb zircon and Sm-Nd dating of Moldanubian HP/HT granulites from South Bohemia, Czech Republic. J Geol Soc Lond 151: 83–90
Zaydan A, Scharbert HG (1983) Petrologie und Geochemie moldanubischer metamorpher Serien im Raume Persenbeug (südwestliches Waldviertel). Jahrb Geol Bundesanst 126: 181–199
Zoubek V (1988) Moldanubian Region: Stratigraphic subdivision, main lithostratigraphic units. In: Zoubek V (ed) Precambrian in younger fold belts. Wiley, Chichester, pp 191–218
Zwart HJ, Dornsiepen UF (1978) The tectonic framework of Central and Western Europe. Geol Mijnbouw 57: 627–654

VII.C.2 Structure

M. URBAN and J. SYNEK

1 Introduction

The Moldanubian Zone (MZ) was considered for many decades as an old nucleus (median mass) of the Bohemian Massif consolidated during the Cadomian or older orogenies, and surrounded by mobile Variscan belts (Stille 1951; Máška and Zoubek 1961; Zoubek et al. 1988; Chaloupský 1989). However, numrous investigations carried out in different parts of the MZ in Czechoslovakia, Austria, and Germany over the past 20 years have confirmed the early suggestions by Suess (1912, 1926), Kossmat (1927), and Kober (1938), who first interpreted the structural evolution of the MZ in terms of Variscan horizontal tectonics. There is a growing body of evidence indicating that the Moldanubian segment of the Bohemian Massif represents a mosaic of tectonic units with a distinct tectonometamorphic history which were finally assembled in the course of the Upper Paleozoic collision of Laurasia and Gondwana.

2 Lithotectonic Units and Their Boundaries

The Moldanubian Zone borders on three regional units (terranes) – the Saxothuringian (SZ), the Teplá-Barrandian (TBZ), and the Moravo-Silesian (MSZ) zones (Figs. 1, 2).

In the SW, the boundary between the Moldanubian and Saxothuringian zones is obscured by the final emplacement of a polymetamorphic nappe pile – the ZEV (Zone of Erbendorf-Vohenstrauß) nappe, which could once have been coherent with the ZTT (Zone of Teplá-Taus) of the Teplá-Barrandian Zone (Blümel 1983; Franke 1989; Weber 1990). The MZ was thrust towards the NW over the SZ; the actual character of the boundary can, however, be more complicated, as revealed by deep seismic profiles showing large crustal wedges of the MZ penetrating into the SZ (Vollbrecht et al. 1989).

The boundary of the MZ and the Teplá-Barrandian Zone s.s. (i.e., without the Železné hory Mts.) displays mostly strike-slip character. In the SW it corresponds to the sinistral Bohemian Pfahl and Mariánské Lázně shear zones. In the SE this boundary is represented by the dextral Central Bohemian Shear Zone (Rajlich et al. 1988), now largely masked by the intrusion of the Central Bohemian Pluton (331 ± 4 Ma Rb/Sr WR isochron, Van Breemen et al. 1982). The metamorphosed roof pendants of the Central Bohemian Pluton, which are mostly derived from the TBZ, suggest an originally hangingwall position of the TBZ with respect to the MZ prior to dextral shearing. Likewise, in the N and NE (in the Železné hory Mts.) the TBZ overlies the MZ represented here by the Kutná Hora-Svratka Complex.

The hangingwall position of the MZ with respect to the Moravo-Silesian Zone is evident along their whole mutual boundary.

Five more or less independent lithotectonic units can be distinguished in the MZ itself; these are the Monotonous (Ostrong), Variegated (Drosendorf), and Gföhl tectonic groups in the core of the MZ, and the Bavaricum (Fuchs 1976), and the Kutná Hora-Svratka Complex along the SW and NE margins of the MZ, respectively (Fig. 2).

The Monotonous and Variegated groups are sometimes collectively referred to as the Drosendorf unit or terrane (Fuchs 1976; Tollmann 1982; Matte et al. 1990). However, as the understanding of the Drosendorf unit is ambiguous (cf. Mísař, Chap. VIII.C.1) we propose that this term be avoided.

The Bavaricum represents an intensely reworked marginal part of the MZ in which remnants of the Monotonous, Variegated, and possibly the Gföhl groups can still be discerned (Fuchs 1976, 1990; Stettner 1988; Matte et al. 1990). The boundary of the Bavaricum with other Moldanubian units is a conventional one. It is

Dallmeyer et al. (Eds.)
Pre-Permian Geology of Central and Eastern Europe

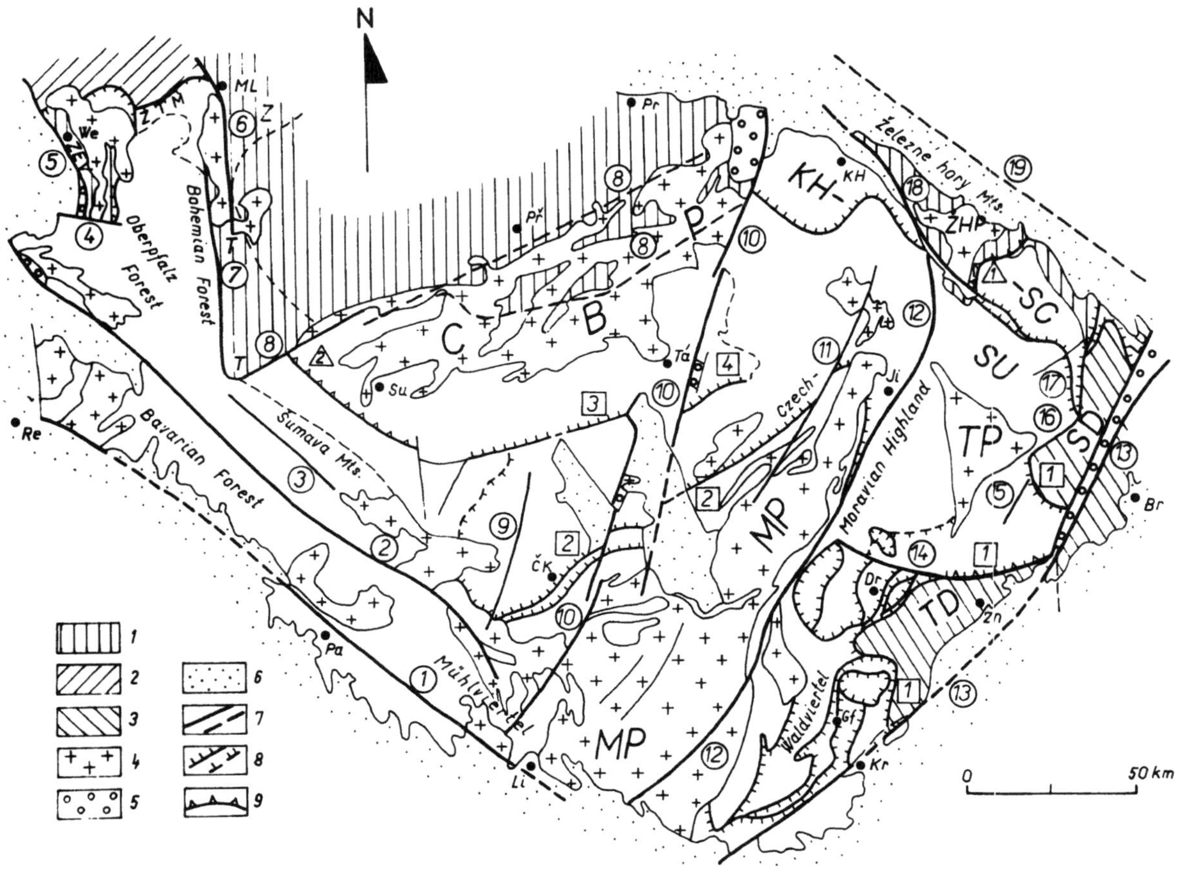

Fig. 1. Nomenclature of principal lithotectonic units and tectonic boundaries. *1* Teplá-Barrandian Zone; *2* Saxothuringian Zone; *3* Moravo-Silesian Zone; *4* Variscan granitoids; *5* Stephano-Autunian basins; *6* post-Lower-Permian sediments; *7* principal (mostly strike-slip) faults, *numbers in circles*: *1* Danube F.; *2* Bavarian Pfahl; *3* Runding F.; *4* Luhe F.; *5* Franconian F.; *6* Mariánské Lázně F.; *7* Bohemian Pfahl; *8* Central Bohemian Lineament (Shear Zone); *9* Lhenice F.; *10* Kouřim-Blanice-Rodl fault system; *11* Jindřichův Hradec F.; *12* Přibyslav-Vitis F.; *13* Boskovice-Diendorf F.; *14* Kravsko F.; *15* Mohelno F.; *16* Bíteš F.; *17* Vír F.; *18* Železné hory F.; *19* Elbe Lineament; *8* principal thrusts, *numbers in squares*: *1* Moldanubian T.; *2* South Bohemian T.; *3* Vodňany-Týn T.; *4* Choustník T.; *9* normal faults, *numbers in triangles*: *1* Hlinsko F.; *2* Sub-Šumava F. *Geologic units CBP* Central Bohemian Pluton; *MP* Moldanubian Pluton (South Bohemian Batholith); *ŽHP* Železné hory Pluton; *TP* Třebíč Pluton; *KHSC* Kutná Hora-Svratka Complex; *SU* Strážek Unit; *LU* Letovice Unit; *ZTM* Zone of Tirschenreuth-Mähring; *ZEV* Zone of Erbendorf-Vohenstrauß; *ZTT* Zone of Teplá-Domažlice (= Taus); *TD* Thaya Dome; *SD* Svratka Dome. *Towns Br* Brno; *ČK* Český Krumlov; *Dr* Drosendorf; *Gf* Gföhl; *Ji* Jihlava; *Kr* Krems; *KH* Kutná Hora; *Li* Linz; *ML* Mariánské Lázně; *Pa* Passau; *Pr* Prague; *Př* Příbram; *Re* Regensburg; *Su* Sušice; *Tá* Tábor; *We* Windischeschenbach; *Zn* Znojmo

marked by a gradual change in the foliation trend which is typically NW-SE in the Bavaricum and NE-SW in the Czech part of the MZ. This change corresponds with a northward decrease in intensity of dextral transcurrent movements along NW-SE-trending shear zones situated within and along the Bavaricum.

The significance of the Kutná Hora-Svratka Complex is still controversial as it is interpreted as (1) an integral part of the MZ (Suess 1926; Zoubek et al. 1988), sometimes correlated with the Gföhl Unit (Matte et al. 1990), (2) an independent unit or terrane (Koutek 1933; Mísař et al. 1983) or (c) a more metamorphosed part of the Železné hory Mts. Upper Proterozoic (e.g., Beneš 1964).

The Kutná Hora-Svratka Complex generally rests upon the Moldanubian s.s. units; an apparent structural inversion can be observed, however, in its easternmost part adjoining to the

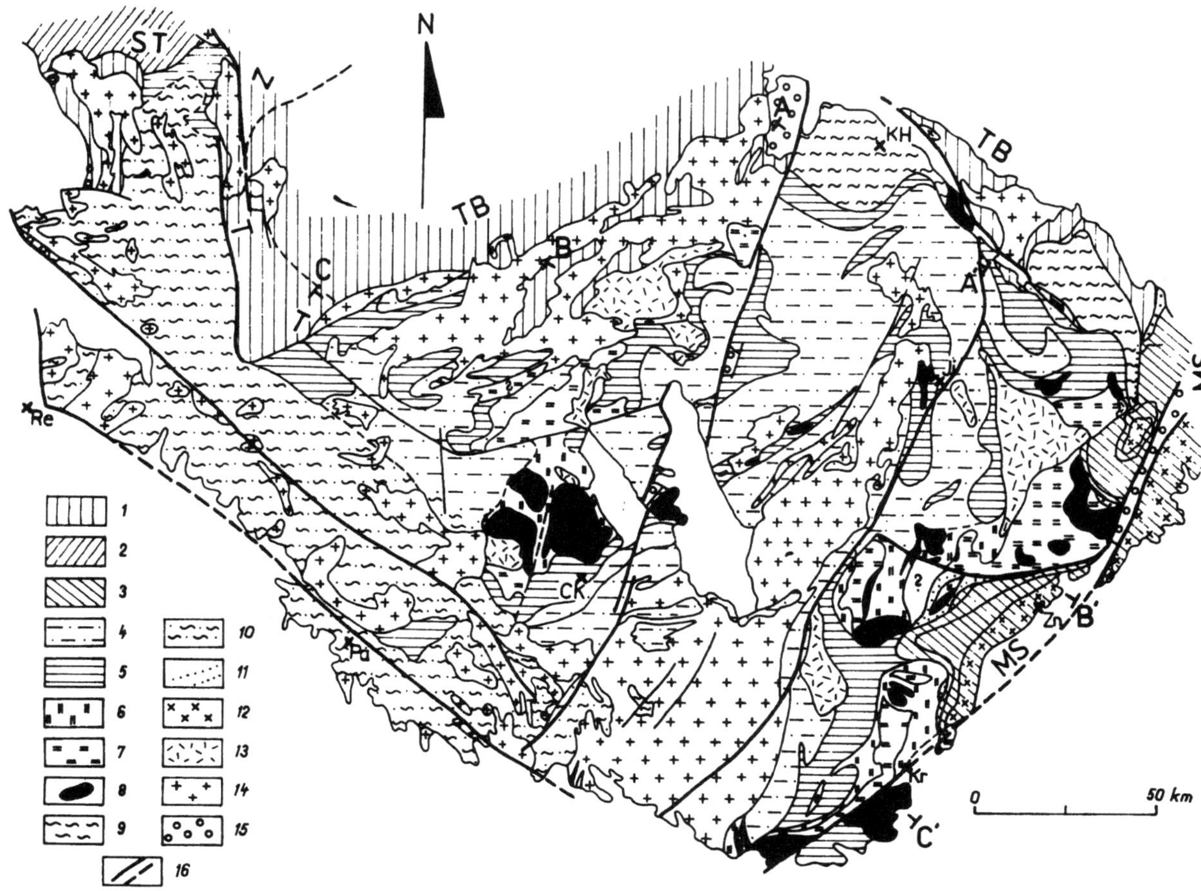

Fig. 2. Distribution of the main lithotectonic units. *1* Teplá-Barrandian Zone (including the Zone of Erbendorf-Vohenstrauß; *ZTT* Zone of Teplá-Domažlice (Taus); *2* Saxothuringian Zone (*ST*); *3* Moravo-Silesian Zone (*MS*); *4–10* Moldanubian Zone; *4* Monotonous Group; *5* Variegated Group; *6* Gföhl Group, undifferentiated (Raabs and Meißling Units); *7* Gföhl Gneiss; *8* granulite bodies; *9* Kutná Hora-Svratka Complex; *10* Bavaricum; *11* Moravian Micaschist Zone between the Moldanubian and Moravo-Silesian zones; *12* Cadomian granitoids; *13–14* Variscan granitoids; *13* durbachites; *14* other granitoids; *15* Permocarboniferous basins; *16* important faults: *ČK* Český Krumlov; *Ji* Jihlava; *KH* Kutná Hora; *Kr* Krems; *Pa* Passau; *Re* Regensburg; *Zn* Znojmo

Moravo-Silesian Zone (E of the Vir Fault, Figs. 1, 3).

3 Structural Evolution of the Moldanubian Zone

3.1 Geochronological Constraints

Recent radiometric data suggest that tectonic slices of old continental (ca. 2050 Ma, Wendt et al. 1988) and oceanic (ca. 1200 Ma, Quadt and Gebauer 1988) crust may be present within the predominantly metasedimentary sequences of the MZ. However, although the detrital zircons in Moldanubian metasediments were largely derived from Gondwana-type source terranes ranging in their age up to 3.8 Ga (Gebauer et al. 1989), the origin of basins filled with this material seems to be related to the breakup of Gondwana, which did not occur until latest Proterozoic to early Paleozoic time (Kröner et al. 1988; Gebauer et al. 1989). The post-Cadomian rifting was accompanied by intrusions of basic and acid magmas between ca. 600 and 490 Ma in both the Mz land the neighboring Zone of Erbendorf-Vohenstrauß (Gebauer and Grünenfelder 1982; Teufel 1988; Quadt 1990). Often disputed is the Caledonian event with a climax at ca. 470–460 Ma, which seems to have provoked thermal effects in various parts of the Mz (Gebauer

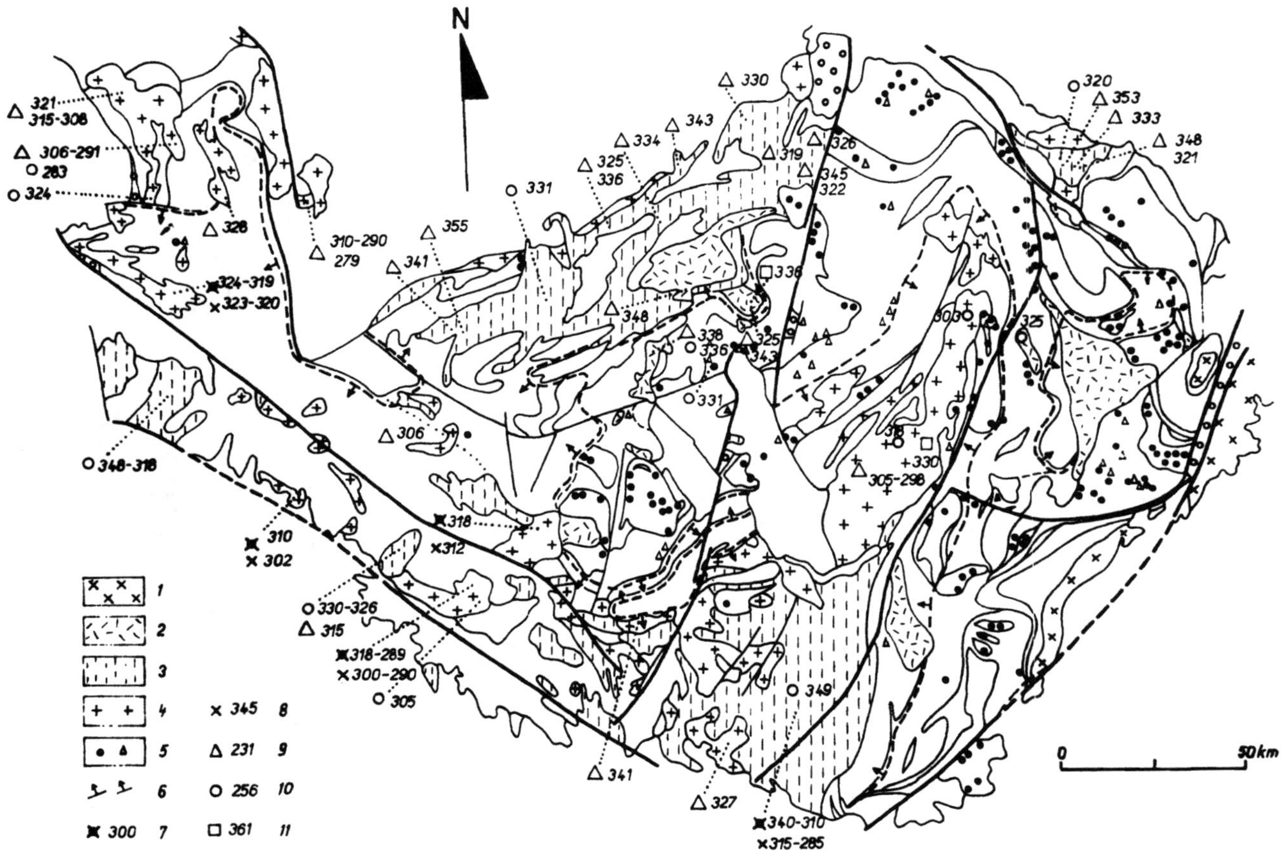

Fig. 3. Distribution of eclogites and peridotites, the age of granitoid intrusions in the MZ. *1* Cadomian granitoids of the Brno and Thaya Massifs in the Moravo-Silesian Zone; *2–4* Variscan granitoids: *2* durbachites; *3* older Moldanubian granitoids (Weinsberg granodiorite, Diorite I, Blatná granodiorite, etc.); *4* younger Moldanubian granites (Freistadt, Eisgarn, Leuchtenberg, Říčany, etc.); *5* distribution of peridotites (*black dots*) and eclogites (*triangles*) (mainly after Machart in Zoubek et al. 1988); *6* course of the cordierite-in line (after Dudek in Mísař et al. 1983); *7* muscovite K-Ar ages; *8* biotite K-Ar ages; *9* K-Ar ages undistinguished; *10* Rb-Sr WR ages; *11* Ar-Ar mineral ages

and Grünenfelder 1973; Arnold and Scharbert 1973; Grauert et al. 1974; Van Breemen et al. 1982; Gorokhov et al. 1983; Köhler and Müller-Sohnius 1985; Gebauer et al. 1989). Subsequent diachronous closure of previous basins is testified by several eclogite- and granulite-forming events during Early Silurian to Upper Devonian (or even Lower Carboniferous) period (Van Breemen et al. 1982; Kröner et al. 1988; Quadt and Gebauer 1988; Aftalion et al. 1989; Carswell and Jamtveit 1990; Beard et al. 1992).

The present architecture of the MZ is mainly due to the dominant MP/MT collisional event which led to stacking of all tectonic units at ca. 350–330 Ma (Van Breemen et al. 1982; Dallmeyer et al. 1990; Matte et al. 1990). This phase was succeeded by HT/LP extensional processes in the thickened crust and intrusions of numerous late- to postkinematic granitoids in several pulses between ca. 340 and 280 Ma (Van Breemen et al. 1982; Wendt et al. 1986; Kreuzer et al. 1989; Scharbert and Veselá 1990, and others).

The Variscan tectonic activity was terminated by important intracontinental strike-slip movements along two fault sets (NW-SE and NNE-SSW) vanning at around 290–260 Ma during the Permian (e.g., Brandmayr et al. 1990; Wallbrecher et al. 1991).

3.2 The Central Part of the MZ

Due to very strong Paleozoic tectonometamorphic imprints, the recognition of Cadomian or older events, and the separation of an older basement from younger metamorphosed cover sequences is very difficult. Radiometric investigations by Wendt et al. (1988) suggest that slices of Lower Proterozoic metagranitoids can be incorporated within the prevalently Upper Proterozoic to Lower Paleozoic volcanosedimentary sequence of the MZ while palynological studies (e.g., Pacltová 1980; Konzalová 1980) reveal that at least part of the metasediments (especially in the Variegated Group) could have developed from Silurian-Devonian protoliths. Likewise, Silurian spores were detected by Pflug and Reitz (1987) in micaschists of the Küni'sches Gebirge (Královský Hvozd Mts.) close to the boundary of the Moldanubian core with the Bavaricum. On the other hand, geochemical and geochronological data of Frank et al. (1991) indicate a pre-Cambrian age for the Variegated Unit of the Waldviertel.

Though relatively large portions of the MZ could have been subducted down to depths of 50 km or more (O'Brien et al. 1990), structures related to this early prograde evolution are mostly unknown. A Caledonian (Arnold and Scharbert 1973) or even higher age has been invoked for this HP phase in the past. However, recent radiometric studies (Van Breemen et al. 1982; Aftalion et al. 1989; Carswell and Jamtveit 1990; Beard et al. 1992) have not revealed ages higher than ca. 370–380 Ma. The HP parageneses of granulites, eclogites, and garnet peridotites) are best preserved in the Gföhl Unit (Scharbert and Kurat 1974; Dobretsov et al. 1984; Medaris and Carswell 1990; Becker and Altherr 1991; Carswell 1991), but HP relics are also present in other Moldanubian units (Fig. 3) including the Monotonous Group which is sometimes referred to as a LP unit (e.g., Blümel 1984). The low-pressure character of this unit resulted from a later HT/LP recrystallization under extensional conditions during the emplacement of Moldanubian plutons, as convincingly shown by the course of the cordierite-in line (Fig. 3). Typically, (except for the Strážek Unit in Moravia) the cordierite-in line does not cross the boundaryof the Gföhl Unit suggesting an already high position of the latter during the HT/LP recrystallization.

Most planar and linear structures observable in the core of the MZ can be ascribed to the main deformation phase related to Variscan collision tectonics between ca. 350 and 330 Ma.

Following the ideas of Suess (1912, 1926) and Kober (1938), and in agreement with more recent structural studies (e.g., Fuchs 1976, 1986; Matura 1976; Thiele 1976, 1984; Vrána 1979; Tollmann 1982; Matte et al. 1985, 1990; Rajlich et al. 1986; Rajlich 1987; Weber and Duyster 1990; Fritz 1991; Neubauer 1991; Schulmann et al. 1991; Urban, in press), two major phases of thrusting can be recognized in the core of the MZ: the intra-Moldanubian phase, and a relatively younger tectonic phase during which the already stacked Moldanubian nappes were emplaced over the Moravo-Silesian Zone (cf. Schulmann et al., Chap. VIII.C.2).

The dominant intra-Moldanubian phase, which was contemporaneous with the widespread MP/MT recrystallization and extensive anatexis, is responsible for the development of the regional composite foliation. The foliation trend, modified, however, by late regional folds, updoming in the vicinity of granitic plutons, strike-slip faulting and, possibly, oroclinal bending or northward indentation of the whole MZ, is shown in Fig. 4. The core of the Moldanubian Zone now represents a deeply eroded NNE-plunging megaantiform, the central part of which is occupied by the Moldanubian Pluton (South Bohemian Batholith).

The general trend of foliation and principal tectonic units (NE-SW in the W, and NNE-SSW in the E) is strikingly different from the younger NW-SE structural grain of the marginal parts of the MZ – the Bavaricum and the Kutná Hora-Svratka Complex.

The principal units in the center of the MZ are arranged more or less symmetrically around the axis of the Moldanubian Pluton (Fuchs 1976, 1990): the Monotonous Group occupies the innermost and lowermost position while the high-grade Gföhl Unit is situated in two synforms (in the Waldviertel and, less markedly, in southern Bohemia) on top of the sequence. The situation becomes more complicated in the Bohemian part of the MZ, where the occurrence of all three principal units seems to be doubled. Characteristically, vast areas of the MZ (especially in or close to the Variegated Group) exhibit low to medium foliation dips.

The hangingwall position of the Gföhl Unit with respect to generally lower grade (cf. Vrána et al., Chap. VII.C.4); Carswell 1991; Petrakakis and Richter 1991) Variegated and Monotonous Groups has long been considered a major argument in favor of the internal nappes within the MZ (Fuchs and Matura 1976). Moreover, as observed in the Waldviertel and Moravia, the boundaries of the Gföhl Unit often cut across the

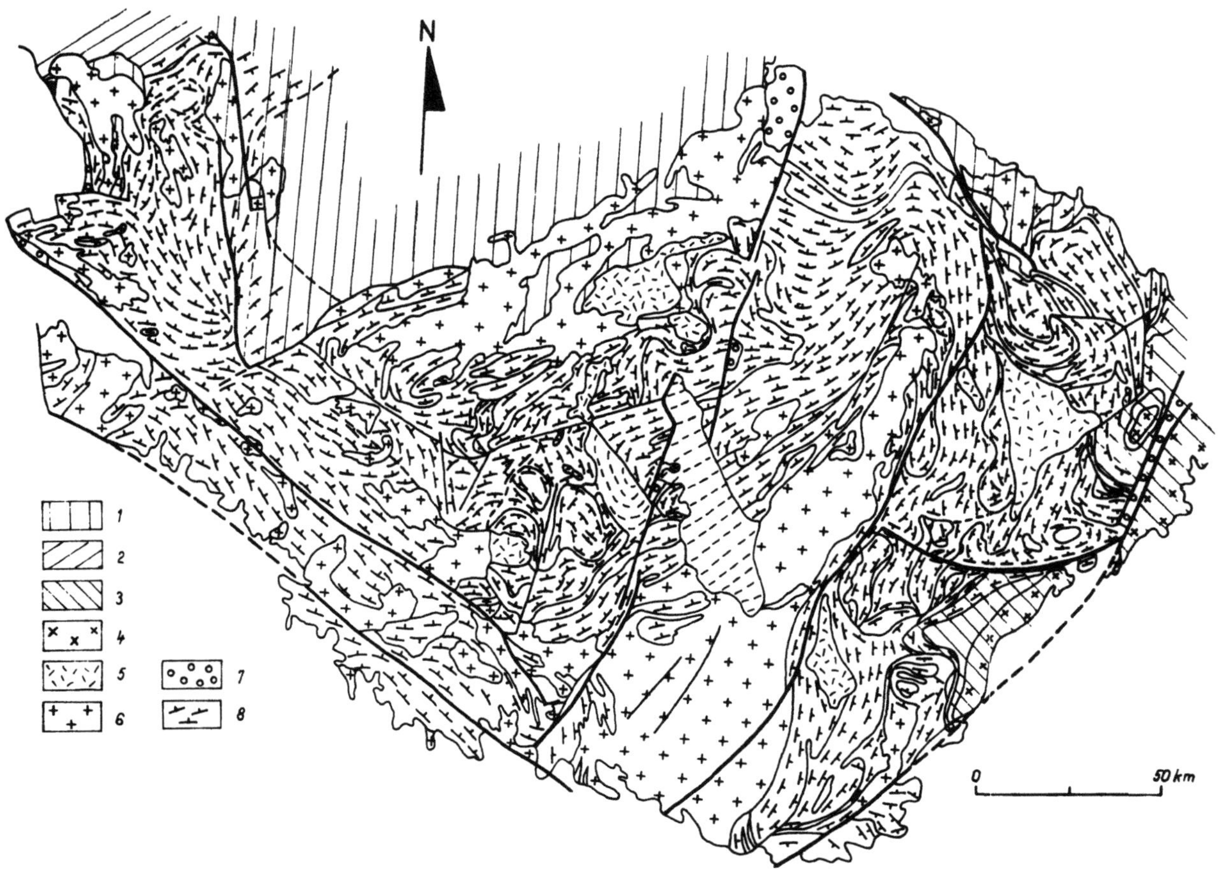

Fig. 4. Strike and sense of dip of the composite metamorphic foliation. *1* Teplá-Barrandian Zone; *2* Saxothuringian Zone; *3* Moravo-Silesian Zone; *4* Cadomian granitoids; *5–6* Variscan granitoids: *5* durbachites; *6* other granitoids; *7* Permocarboniferous basins; *8* strike and dip sense of the foliation

boundaries of the lower units. The Gföhl Unit itself can be divided into several separate thrust sheets with mylonites at their base; their recognition and spatial distribution, however, is not fully accepted (cf. Weber and Duyster 1990). The uppermost position in the Gföhl Unit can be occupied by a subhorizontal sheet of acid granulites (the St. Leonhard body in the Waldviertel, Fuchs and Matura 1976; Matte et al. 1985; the Jemnice body in W Moravia, Jenček 1984), more often, however, the position of granulites is not specific.

Another important thrust contact is often being placed at the boundary between the Monotonous and Variegated Groups. Despite frequent criticism of this concept (e.g., Thiele 1984), the tectonic contact is suggested by HT granulitic mylonites situated along this interface in the Waldviertel (Fuchs and Scharbert 1979; Petrakakis 1986) and farther N in Moravia, and by an inverted metamorphic gradient and occurrences of eclogites and Lower Proterozoic orthogneisses close to this boundary (South Bohemian Thrust) near Český Krumlov in S Bohemia (Vrána 1979, in press; Rajlich et al. 1986; Wendt et al. 1988).

The South Bohemian Thrust can be followed alongstrike from the Czechoslovak/Austrian frontier in a NE direction over a distance of ca. 70 km. This structure is composed of several individual thrusts, which facilitated the tectonic uplift of lower-crustal granulite complexes and slices of upper mantle ultramafics over the Variegated and Monotonous groups. Qualitative differences in P-T paths, and composition of individual granulite bodies suggest that the granulite body themselves are in a mutually allochthonous position (Vrána, in press). The prolongation of the South Bohemian Thrust further NE is marked by frequent scattered bodies of eclogites, peridotites, and granulites (Fiala et al. 1987); the width of the nappe sequence is, however, strongly reduced, partly due to sinistral strike-slip movements along the Jindřichův Hradec Shear Zone.

The prolonged activity of the intra-Moldanubian thrusting is reflected by late NW-dipping brittle-ductile shear zones (Kaplice Shear Zone, Vodňany-Týn Shear Zone), which cut older thrust contacts and/or provoke transposition of the dominant composite foliation (Vrána 1979).

The dominant linear structure of the MZ is a mineral or strectching lineation, which is often parallel to axes of intrafoliation folds. The trend of the stretching lineation is highly variable on both the mesoscopic and macroscopic scales. In general, two more or less perpendicular directions prevail – NW-SE (WNW-ESE) and NE-SW (NNE-SSW) – Fig. 5. The existence of these two dominant orientations of lineations is a source of a continuing discussion about the kinematics of the intra-Moldanubian thrusting.

In the classic paper by Stille (1951) as well as in more recent plate tectonic concepts (e.g., Behr et al. 1984; Lorenz and Nicholls 1984; Matte 1986) the MZ is considered as part of the SE-vergent flank of the bilateral Variscan orogen; the zone of structural parting, although not yet exactly defined, runs through the median part of the Teplá-Barrandian Zone. In fact, many structural models try to explain the tectonic evolution of the MZ in terms of general mono- or polyphase E- to SE-directed thrusting within the MZ and over the Moravo-Silesian Zone: some models suppose a far-traveled unique Gföhl nappe rooting either at the Erbendorf Line in the Oberpfalz (Tollmann 1982) or at the MZ/TBZ boundary (Matte et al. 1990) while others consider only a limited nappe transport (e.g., Thiele 1984; Rajlich and Synek 1987).

However, many recent structural investigations have shown that the tectonic evolution is more complex. Although kinematic indicators

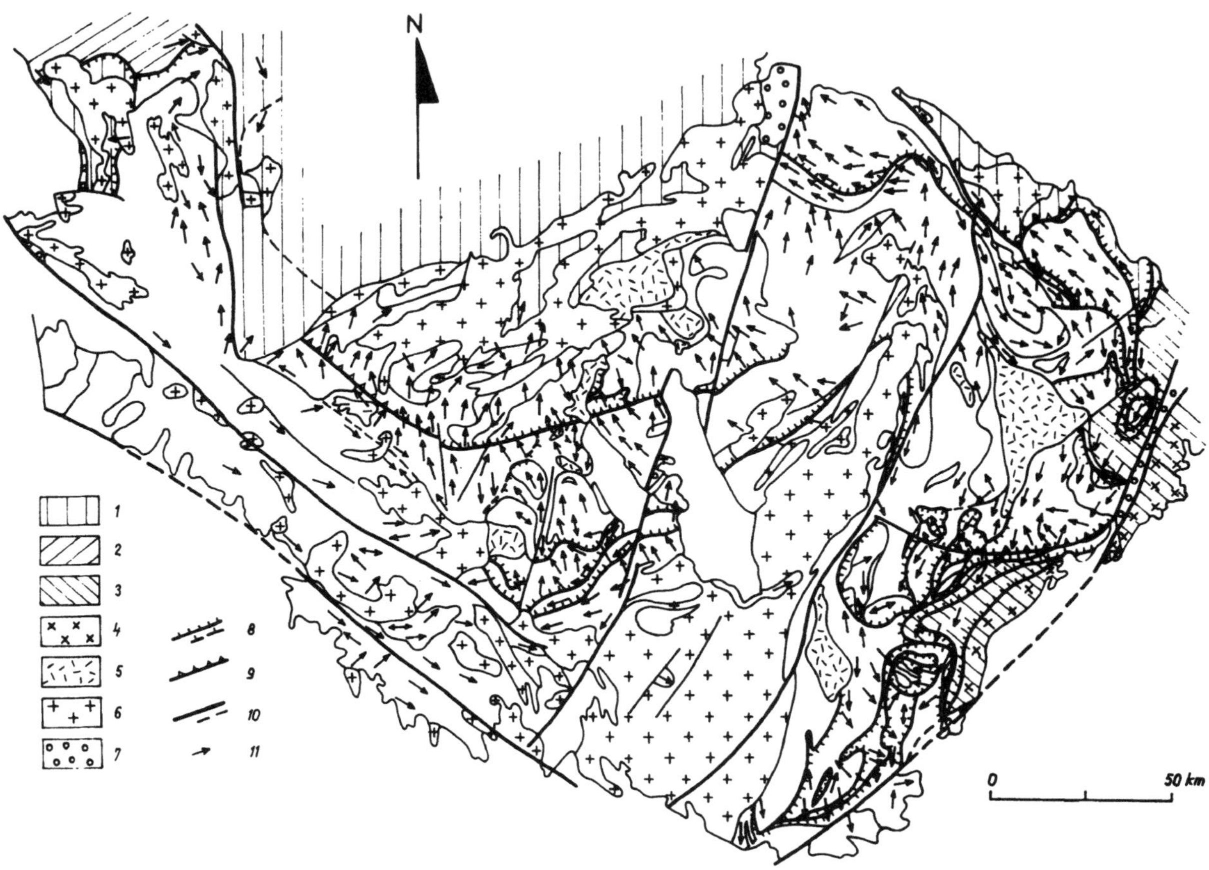

Fig. 5. Orientation of the stretching lineation. *1–6* See Fig. 4; *7* Permocarboniferous basins; *8* thrusts; *9* normal faults; *10* important (mostly strike-slip) faults; *11* azimuth and sense of dip of the stretching lineation (more generations)

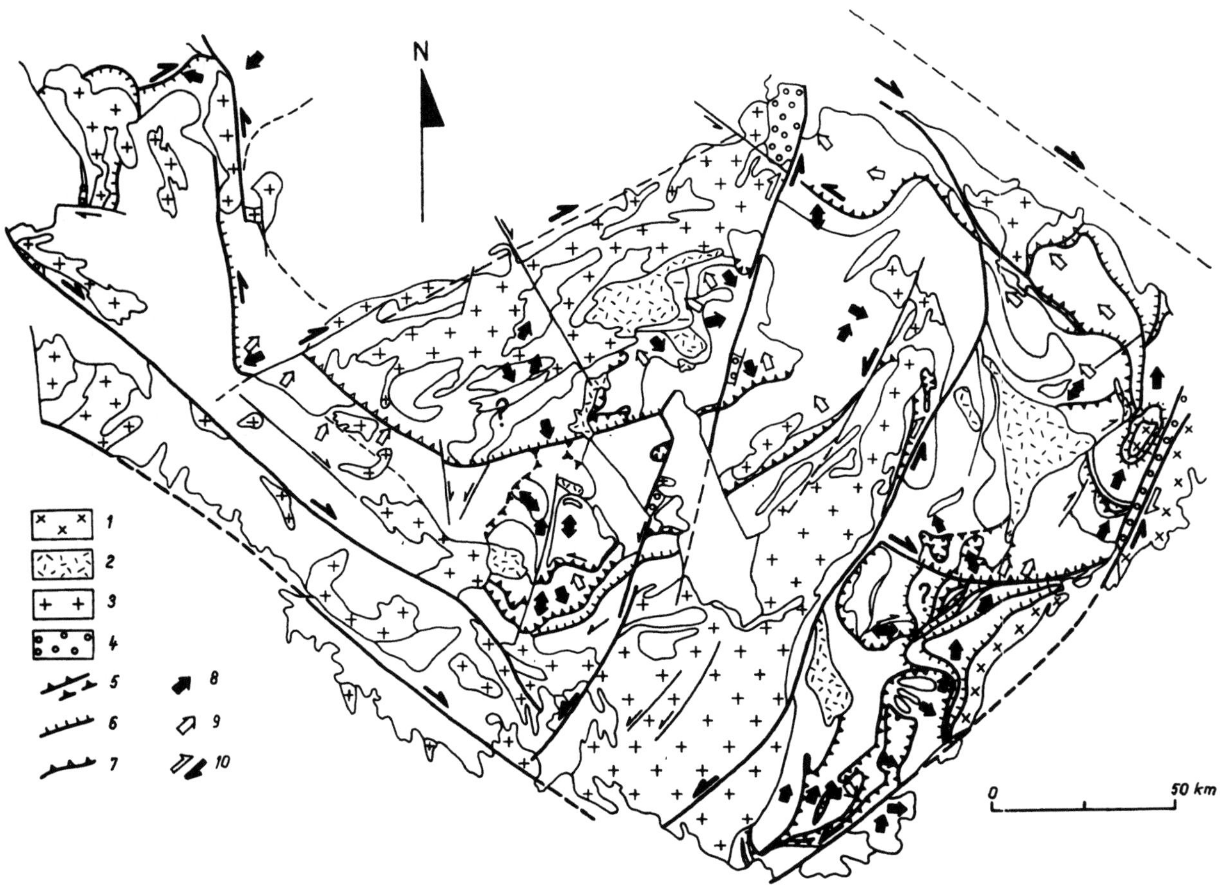

Fig. 6. Sense of tectonic transport. *1* Cadomian granitoids; *2–3* Variscan granitoids: *2* durbachites; *3* other granitoids; *4* Permocarboniferous basins; *5* intra-Moldanubian thrusts; *6* late intra-Moldanubian thrusts and thrusts between the MZ and surrounding units (or within the Moravo-Silesian Zone); *7* normal faults; *8* direction of thrusting; *9* direction of low-angle normal faulting; *10* sense of strike-slip movements (*empty arrows* indicate the sense of older movements if two senses are known)

often suggest SE-directed movements (e.g., Matte et al. 1985; Rajlich et al. 1986; Fig. 6), top-to-the-N- or NE-oriented thrusting (often combined with wrench processes) has been documented in many parts of the MZ (Fritz 1991; Neubauer 1991; Schulmann et al. 1991; Urban, in press). These generally N-directed movements can possibly be due to northward oroclinal bending of the entire MZ accompanied by lateral noncoaxial tectonic escape. Further investigations are also needed in order to explain the (north)westward nappe transport or backfolding considered (in a reverse time succession) by Fuchs (1976, 1986) and Weber and Duyster (1990) apart from the eastward one. As a whole, the kinematics of the intra-Moldanubian thrusting is far from being understood.

Moreover, it is necessary to bear in mind that vast areas of the MZ display mostly planar structures developed under conditions strongly deviating from plane strain geometry of deformation; kinematic interpretations based on the stretching direction and fabric asymmetry are then hard or impossible to perform.

On the other hand, well documentable is the late northward thin-skinned emplacement of the already stacked Moldanubian and Moravian nappes over the Brunovistulian Cadomian basement (Schulmann et al. 1991; Fritz 1991; Urban, in press).

The phases of crustal thickening (ca. 380–330 Ma) were followed by extensional processes under HT/LP conditions which slightly preceded or accompanied the emplacement of Moldanubian granitoids after ca. 330 Ma. (Note: an exceptional Rb-Sr WR age of 349 ± 4 Ma was reported, however, by Scharbert (1987a) from the late- to postkinematic Weinsberg Granite of the

Moldanubian Pluton. Similarly, high Rb-Sr ages obtained from the Eisgarn Granite further N in Czechoslovakia are considered by Scharbert and Veselá 1990 as geologically meaningless. Asymmetric structures such as shear bands coated with muscovite and fibrolitic sillimanite usually indicate top-to-the-NW normal movements in the Bohemian part of the MZ (e.g., Klečka et al. 1986; Kadounová et al. 1989) while conjugated SE-directed shearing has been observed especially in the Waldviertel (Neubauer 1991). Particularly well developed are the extensional structures in the Strážek Unit in Moravia, where the body of Bory granulites is cut by numerous extensional pegmatite veins extremely rich in cordierite: older structures of the surrounding anatectic sillimanite gneisses have been completely obliterated by the HT/LP recrystallization.

The Variscan tectonic activity in the MZ and the Bohemian Massif as a whole was terminated by a phase (or phases) of ductile to brittle strike-slip faulting, which, on a global scale, relates to dextral wrenching between Africa and Europe during the Carboniferous and Lower Permian (Arthaud and Matte 1977).

The most important strike-slip faults, often marked by zones of mylonitization, are arranged into a conjugate system of NW-SE dextral (Elbe Lineament, Bavarian Pfahl, Runding shear zone) and NNE-SSW sinistral (Blanice-Rodl shear zone, Přibyslav-Vitis and Boskovice-Diendorf faults) steep structures (Fig. 2). Their late Variscan age is constrained by frequent pull-apart and transtensional Stephano-Autunian basins situated along their course (Schröder 1987; Pašek and Urban 1991), and by 295 to 260 Ma muscovite $^{40}Ar/^{39}Ar$ plateau ages obtained from the Rodl shear zone mylonites (Brandmayr et al. 1990; Wallbrecher et al. 1991). Offsets along these faults may amount to several tens of kilometers,

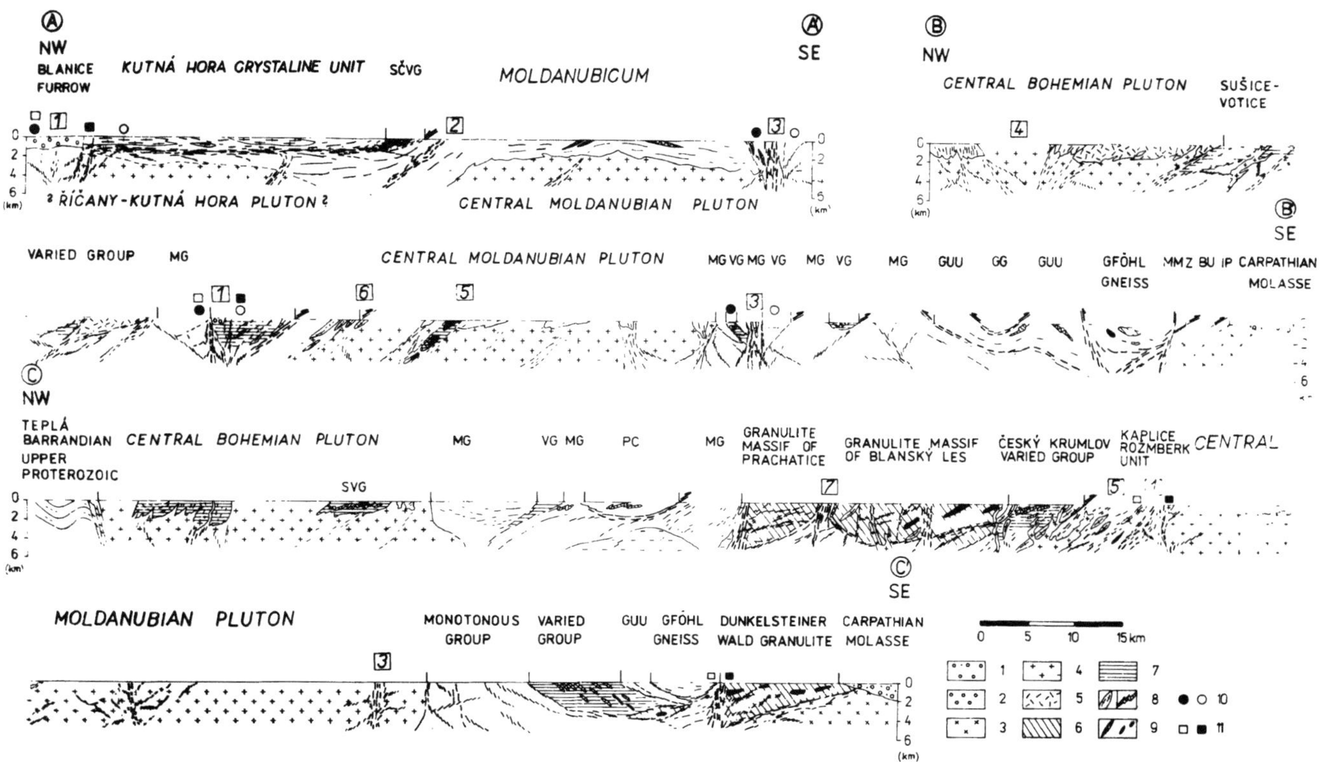

Fig. 7. NW-SE profiles accross the MZ (see Fig. 2 for localization). *1* Tertiary of the Carpathian foredeep; *2* Permocarboniferous sediments; *3* Cadomian granitoids (Brno Pluton); *4* Variscan granitoids; *5* durbachites; *6* granulites; *7* Variegated Group; *8* orthogneisses, marbles; *9* amphibolites, ultrabasites; *10* preplutonic dextral strike-slip shear zone; *11* postplutonic sinistral strike-slip shear zones. *Numbers in squares*: *1* Blanice-Rodl Zone; *2* Vlastějovice Z.; *3* Přibyslav Mylonite Z.; *4* Central Bohemian Shear Z.; *5* South Bohemian Thrust; *6* Choustník Thrust. *MG* Monotonous Group; *VG* Variegated Group; *SVG* Sušice Variegated Group; *ŠČVG* Šternberk-Čáslav Variegated Group; *GUU* Gföhl Unit undifferentiated; *GG* Gföhl Gneiss; *PC* Podolsko Complex (? an equivalent of the Gföhl Unit); *MMZ* Moravian Micaschist Zone; *BU* Bíteš Gneiss; *IP* Inner Phyllites

as indicated by the separation of the Pöchlarn and Dunkelsteiner Wald granulites along the Diendorf Fault in Austria by nearly 25 km (e.g., Matura 1976). Movements (sometimes with an inverted sense) along many strike-slip faults could have taken place even before the emplacement of the Moldanubian granitoids (Rajlich 1988; Urban and Mísař 1989; Figs. 6, 7).

Many previous strike-slip faults were reactivated during the Alpine collision (e.g., Bachmann et al. 1987; Malkovský 1987) when they controlled the development of Mesosoic and Tertiary basins such as those in S Bohemia.

3.3 Kutná Hora-Svratka Complex

The whole Kutná Hora-Svratka Complex (KHSC) is divided by the Lower Paleozoic of the Hlinsko Zone, and by the Přibyslav mylonite zone into two parts: the Kutná Hora Unit in the W and the Svratka Unit (or Svratka Antiform) in the E (Mísař et al. 1983). The entire complex is composed of a number of thrust units situated above the Moldanubian Zone s.s. Its lower units can be tentatively parallelized with the Moravian Micaschist Zone (a polymetamorphic unit lying between the Moravo-Silesian Zone and the Moldanubian Zone s.s.) and the Gföhl Unit, respectively, which are overridden by a sheet of peraluminous orthogneisses (Synek and Oliveriová, in prep.).

The oldest recognized deformation phase and related Barrovian metamorphism seem to be connected with the movements on the Elbe Zone, which was active in the central and northern parts of the Kutná Hora Unit, and along the today's transitional zone between the Svratka Unit and the Moldanubian Zone s.s. Contemporaneously, boudins of garnet peridotites, eclogites, and granulites were juxtaposed with the amphibolite facies rocks. The kinematics of this phase is masked by a very intense later overprint. However, the alignment of zones of ultrabasites, which are oblique to the present structural grain, and rare relic structures in leucocratic migmatites suggest that the KHSC could have been thrust (to the S or SE) over the neighboring Moldanubian Zone s.s. during a phase D1, which was roughly contemporaneous with the MP/MT phase of the intra-Moldanubian thrusting.

The younger deformation phase D2 is connected with prominent extensional movements along a low-angle normal fault, whose median plane is best preserved in planolinear orthogneisses ("Doubravčany pencil gneiss") and migmatites (Synek 1991) in the W part of the Kutná Hora Unit; in the E part of the KHSC the intensity of this deformation is generally lower. During D2 a strong subhorizontal NW-SE-trending stretching lineation was produced under amphibolite to greenschist facies conditions. Numerous kinematic indicators such as asymmetric feldspar clasts, rotated boudins, S-C structures, shear bands, and sheath folds document top-to-the-NW movements.

More complicated is the situation in the SE part of the KHSC, SE of the Vír Fault, where two mutually perpendicular (NW-SE and NE-SW) contemporaneous lineations are developed; the NW-SE lineation is connected with an oblate strain ellipsoid, while the NW-SE one rather corresponds to NW-directed constrictional flow (Melka et al., in press). Moreover, at the contact with the Moravo-Silesian Zone the relative succession of tectonic units known from the W part of the KHSC seems to be inverted: the Moldanubian s.s. sequence lies on the top, while the lower position is occupied by the KHSC (Micaschist Zone) with important relics of HP parageneses in micaschists (Johan et al. 1990; Lardeaux et al. 1991) at the base.

Limited sinistral ductile strike-slip movements could have occurred during D2 at the boundary of the KHSC and the Moldanubian Zone s.s., while dextral motions prevailed on NNE-SSW structures such as the Přibyslav and Blanice shear zones, which contributed to the deflection of the NW-SE-trending stretching lineation into a roughly N-S direction.

As a whole, the NW-vergent extensional movements in the KHSC are comparable to similar processes in the N part of the Moravian Svratka Dome (phase D2 of Schulmann et al. 1991 or D2L of Urban, in press) which postdate the main period of thrusting of the MZ over the Moravo-Silesian Zone. Their minimum age is constrained by a 325 Ma $^{39}Ar/^{40}Ar$ biotite plateau age obtained from orthogneisses in the Kutná Hora Unit (Matte et al. 1990). NW-directed normal faulting on the Hlinsko fault (Fig. 1) seems to be related to this extensional period.

During a still younger stage of evolution, dextral strike-slip movements took place along the Elbe Zone, which controlled the pull-apart intrusion of the Železné hory granitoid pluton dated at ca. 320 Ma (Scharbert 1987b), while conjugate sinistral movements took place along the Blanice and Přibyslav shear zones (Figs. 6, 7). The longevity of movements along the Elbe Lineament is documented by the formation of N-S ore veins in the Kutná Hora mining district at ca. 260 Ma, and by repeated movements leading to structural inversion along the Železné hory Fault during the Alpine period.

3.4 The SW Part of the Moldanubian Zone

The SW part of the Moldanubian Zone can be roughly divided into two parts comprising: (a) the antiformal wedge of the Oberpfalz Forest and the Bohemian Forest, and (b) the generally monoclinal area of the Bavarian Forest, Mühlviertel, Sauwald, and Šumava Mts., which mostly corresponds to the Bavaricum of Fuchs (1976).

The structural style of this part of the MZ has been recently studied, especially in the Oberpfalz area, close to the KTB drilling site. According to Weber and Vollbrecht (1986) and Vollbrecht et al. (1989), the Moldanubian/Saxothuringian boundary in the Oberpfalz is characterized by SE-dipping master décollements, some of which can be traced down to the Moho, and along which the MZ was thrust towards the NW and welded together with the SZ during the Middle Carboniferous HT/LP collisional event.

Numerous radiometric data from the marginal Moldanubian Zone of Tirschenreuth-Mähring (ZTM), as well as from other parts of the MZ in Bavaria and SW Bohemia (Gebauer and Grünenfelder 1973; Grauert et al. 1974; Köhler and Müller-Sohnius 1976, 1985; Wendt et al. 1986; Teufel 1988; Kreuzer et al. 1989) prove a rapid uplift and cooling at around 330 to 320 Ma, which was immediately followed by several pulses of posttectonic granite intrusions between 320 and 290 Ma.

The MZ/SZ boundary is represented by a HT-mylonitic belt along the NW rim of the ZTM, which is interpreted as a cryptic suture (Weber 1990). Steepening of the SW-NE-trending mylonitic foliation ascribed to the effects of dextral shearing is characteristic of the Bavarian part of the ZTM (Stein 1988; Weber 1990) while gentle E to SE dips can be observed on the Bohemian territory in both the Saxothuringian and Moldanubian zones.

Up to six deformation phases have been recognized in the Oberpfalz area (Stein 1988; Weber 1990) which can be divided into two groups: D1–D3, and D4 to D6. D1 seems to be related to a precollisional event which might be contemporaneous with the HP-MP metamorphism, the age of which, however, has not yet been sufficiently constrained. The HP event in eclogitic and granulitic rocks at Winklarn (Oberpfalz) has been dated at 427 ± 5 and 423 ± 8 Ma (Sm-Nd method, Quadt and Gebauer 1988). The Middle Devonian ages (390–370 Ma) typical of the MP event in the adjacent ZTT and ZEV (Kreuzer et al. 1989; Dallmeyer and Urban, in prep.) are scarce in the MZ (Teufel 1988). The D2 and D3 phases correlate with the peak of the HT/LP metamorphism. The D2 phase indicates a NW-thrusting and crustal wedging of the Moldanubian onto and into the Saxothuringian crust, while D3 refolds earlier structures around axes of regional NE-SW-trending open upright folds. The final emplacement of the ZEV nappe over the MZ/SZ boundary must have occurred immediately after D3.

The phases D4 to D6 are related to an intracontinental strike-slip regime. In the Oberpfalz this period has produced the postmetamorphic antiformal wedge (the D4 antiform of Stein 1988) between the dextral Bavarian Pfahl shear zone and the en-echelon array of sinistral shear zones in Bohemia (Bohemian Pfahl and Mariánské Lázně shear zones), which is followed by the course of post-tectonic granite intrusions.

Dextral shearing along WNW-ESE to NNW-SSE blastomylonitic to ultramylonitic shear zones is especially important in the Bavaricum, where it contributed to the development of various transpressional and transtensional flower structures (Reston and Meissner 1989) and SE-vergent asymmetric folds. The effects of transpression and transtension can be seen in late local SW-directed thrusting in the Mühlviertel and Sauwald (Fuchs 1976) and N- to NE-directed normal faulting in the Šumava Mts. The dextral shearing and the prominent WSW-ESE stretching lineation typical of the Bavaricum are clearly younger than the main period of thrusting in the core of the MZ as the older Moldanubian magmatites (Weinsberg Granite, Diorite I), which are mostly posttectonic in the Waldviertel, are strongly foliated in the Bavaricum (Fuchs 1976, 1990).

Acknowledgments. The authors are indebted to F. Neubauer for his careful and constructive review of the manuscript. H. Bryndová is thanked for drawing the figures.

References

Aftalion M, Bowes DR, Vrána S (1989) Early Carboniferous U-Pb zircon age for garnetiferous, perpotassic granulites, Blanský les Massif, Czechoslovakia. Neues Jahrb Mineral Monatsh 4: 145–152

Arnold A, Scharbert HG (1973) Rb-Sr Altersbestimmungen an Granuliten der südlichen Böhmischen Masse in Österreich. Schweiz Mineral Petrogr Mitt 53: 61–78

Arthaud F, Matte Ph (1977) Late Paleozoic strike-slip faulting in southern Europe and northern Africa: result of a right lateral shear zone between the Appalachians and the Urals. Geol Soc Am Bull 88: 1305–1320

Bachmann GH, Müller M, Weggen K (1987) Evolution of the Molasse Basin (Germany, Switzerland). Tectonophysics 137: 61–78

Beard BL, Medaris LG Jr, Johnson CM, Brueckner HK, Mísař Z (1992) Petrogenesis of Variscan high-temperature Group A eclogites from the Moldanubian Zone of the Bohemian Massif, Czechoslovakia. Contrib Mineral Petrol 111: 468–483

Becker H, Altherr R (1991) P-T evolution of granulites, garnet pyroxenites and garnet-bearing peridotites in the Moldanubian Gföhl unit in Lower Austria. Geol Worksh Moravian Windows, Moravský Krumlov, Prague, Abstr, p 34

Behr HJ, Engel W, Franke W, Giese P, Weber K (1984) The Variscan belt in Central Europe: main structures, geodynamic implications, open questions. Tectonophysics 109: 15–40

Beneš K (1964) Analýza vnitřní stavby moldanubicko-assyntské hraniční oblasti při sv. okraji moldanubického jádra. Rozpr Česk Akad Věd 74 (2): 1–80

Blümel P (1983) The western margin of the Bohemian massif in Bavaria. Fortschr Mineral 61: 171–195

Blümel P (1984) Mitteldruck- und Niederdruck-metamorphose in den außeralpinen Varisziden Mitteleuropas. Fortschr Mineral 62 (Beih 1): 28–29

Brandmayr M, Dallmeyer RD, Handler R, Wallbrecher E (1990) Age and kinematics of shear zones in the southern Bohemian Massif: evidence from the Rodl-Shear Zone. Proc IGCP 233 Conf Paleozoic Orogens in Central Europe, August–September 1990, Göttingen-Giessen

Carswell DA (1991) Variscan high P-T metamorphism and uplift history in the Moldanubian Zone of the Bohemian Massif in Lower Austria. Eur J Mineral 3: 323–342

Carswell DA, Jamtveit B (1990) Variscan Sm-Nd ages for the high pressure metamorphism in the Moldanubian Zone of the Bohemian Massif, Lower Austria. Neues Jahrb Mineral Abh 162: 69–78

Chaloupský J (1989) Major tectonostratigraphic units of the Bohemian Massif. Geol Soc Am Spec Pap 230: 101–114

Dallmeyer RD, Neubauer F, Höck V (1990) $^{40}Ar/^{39}Ar$ mineral age controls on the chronology of late Paleozoic tectonothermal activity in the southeastern Bohemian Massif, Austria (Moldanubian and Moravo-silesian zones). In: Franke W (ed) Bohemian Massif. Field guide, Conf IGCP 233, Paleozoic orogens in central Europe, Göttingen-Giessen 1990, pp 87–96

Dobretsov NL, Mísař Z, Popov EV (1984) The P-T conditions of equilibrium for some pyrope peridotites and their country rocks in the Moldanubian area at Mohelno (Eastern Moravia, Czechoslovakia). Mineral Slovaca 16: 87–95

Fiala J, Matějovská O, Vaňková V (1987) Moldanubian granulites and related rocks: petrology, geochemistry and radioactivity. Rozpr Česk Akad Věd 97 (1): 102

Franke W (1989) Tectonostratigraphic units in the Variscan belt of central Europe. Geol Soc Am Spec Pap 230: 67–90

Frank W, Scharbert S, Thöni M, Popp F, Hammer S (1991) Isotopengeologische Neuergebnisse zur Entwicklungsgeschichte der Böhmischen Masse. Proterozoische Gesteinsserien und variszische Hauptorogenese. Österr Beitr Meteorol Geophys 3: 185–228

Fritz H (1991) Structures and kinematics along the Moravian-Moldanubian boundary: preliminary results. Österr Beitr Meteorol Geophys 3: 77–96

Fuchs G (1976) Zur Entwicklung der Böhmischen Masse. Jahrb Geol Bundesanst 119: 1–43

Fuchs G (1986) Zur Diskussion um den Deckenbau der Böhmischen Masse. Jahrb Geol Bundesanst 129: 41–49

Fuchs G (1990) The Moldanubicum – an old nucleus in the Hercynian mountain ranges of Central Europe. In: Minaříková D, Lobitzer H (eds) Thirty years of geological cooperation between Austria and Czechoslovakia, Federal Geological Survey, Vienna, Geol Surv, Prague, pp 256–262

Fuchs G, Matura A (1976) Die Geologie des Kristallins der südlichen Böhmischen Masse. Jahrb Geol Bundesanst 119: 1–43

Fuchs G, Scharbert HG (1979) Kleinere Granulitvorkommen im niederösterreichischen Moldanubikum und ihre Bedeutung für die Granulitgenese. Verh Geol Bundesanst 1979: 29–49

Gebauer D, Grünenfelder M (1973) Vergleichende U/Pb-und Rb/Sr-Altersbestimmungen im bayerischen Teil des Moldanubikums. Fortschr Mineral Beih 50 (3)

Gebauer D, Grünenfelder M (1982) Geological development of the Hercynian Belt of Europe based on age and origin of the high-grade mafic and ultramafic rocks. Abstr 5th Int Conf on geochronology and isotope geology, Japan

Gebauer D, Williams IS, Compston W, Grünenfelder M (1989) The development of the Central European continental crust since the Early Archean based on conventional and ion-microprobe dating of up to 3.84-b.y.-old detrital zircons. Tectonophysics 157: 81–96

Gorokhov IM, Melnikov NN, Varsavskaja ES, Kutjavin EP (1983) Rb-Sr dating of magmatic and metamorphic events in the eastern part of the Bohemian Massif. Čas Mineral Geol 28: 349–361

Grauert B, Hänny R, Soptrajanova G (1974) Geochronology of a polymetamorphic and anatectic gneiss region: the Moldanubicum of the area Lam-Deggendorf; Eastern Bavaria, Germany. Contrib Mineral Petrol 45: 37–63

Jenček V (1984) Vysvětlivky k základní geologické mapě ČSSR 1:25000, 33-223 Vranov. Ústřed Ústav Geol Praha, 57 pp

Johan V, Autran A, Ledru P, Lardeaux JM, Melka R (1990) Discovery of relics of a high- pressure metamorphism at the base of the Moldanubian nappe complex. Proc Conf IGCP 233 Paleozoic orogens in central Europe, August–September 1990, Göttingen-Giessen

Kadounová Z, Rajlich P, Synek J, Šarbach M, Křibek B (1989) Shear kinematics on the contact between granulites and the Varied Group; Southern Bohemia, Moldanubian. Věstn Ústřed Ústavu Geol 64: 321–334

Klečka M, Rajlich P, Melka R (1986) Ductile shear zones and origin of orthogneisses in the thrust sheet of Choustník. Acta Montana 72: 36–62

Kober L (1938) Der geologische Aufbau Österreichs. Springer, Wien Berlin Heidelberg

Köhler H, Müller-Sohnius D (1976) Ergänzende Rb-Sr-Altersbestimmungen an Mineral- und Gesamtgesteinsproben des Leuchtenberger und des Flossenbürger Granits (NE-Bayern). Neues Jahrb Mineral Monatsh: 354–365

Köhler H, Müller-Sohnius D (1985) Rb-Sr Altersbestimmung und Sr-Isotopensystematik an Gesteinen des Regensburger Waldes (Moldanubikum NE Bayern), Teil 1: Par- agneisanatexite. Neues Jahrb Mineral Abh 151: 1–28

Konzalová M (1980) Zu der mikro- paläontologischen Erforschung graphitischer Gesteine in Südteil der Böhmischen Masse. Věstn Ústřed Ústavu Geol 55: 233–236

Kossmat F (1927) Gliederung des varistischen Gebirgsbaues. Abh Sächs Geol Landesamtes: 1–39

Koutek J (1933) Geologie posázavského krystalinika I. Věstn Stát Geol Ústavu 9: 319–333

Kreuzer H, Seidel E, Schüssler U, Okrusch M, Lenz KL, Raschka H (1989) K-Ar geo- chronology of different tectonic units at the northwestern margin of the Bohemian Massif. Tectonophysics 154: 149–178

Kröner A, Wendt I, Liew TC, Compston W, Todt W, Fiala J, Vaňková V, Vaněk J (1988) U-Pb zircon and Sm-Nd model ages of high- grade Moldanubian metasediments, Bohemian Massif, Czechoslovakia. Contrib Mineral Petrol 99: 257–266

Lardeaux JM, Johan V, Autran A, Ledru P (1991) Talcphengite-kyanite bearing meta- pelites: high-pressure metamorphism in the Moravian Zone and its tectonic significance. Abstr Geol Worksh Moravian Windows, Moravský Krumlov, Prague

Lorenz V, Nicholls IA (1984) Plate and intraplate processes of Hercynian Europe during the Late Paleozoic. Tectonophysics 107: 25–56

Malkovský M (1987) The Mesozoic and Tertiary basins of the Bohemian Massif and their evolution. Tectonophysics 137: 31–42

Máška M, Zoubek V (1961) Area of the Variscan intramontane block. The region of the Moldanubian elevation. In: Buday T, Kodymsan D, Mahel M, Máška M, Matěska A et al. (eds) Tectonic development of Czechoslovakia. Nakladatelství Česk Akad Věd Praha, pp 25–51

Matte Ph (1986) Tectonics and plate tec- tonics model for the Variscan belt of Europe. Tectonophysics 126: 329–374

Matte Ph, Maluski H, Echtler H (1985) Cisaillements ductiles varisques vers l'Est-Sud Est dans les nappes du Waldviertel (Sud Est du Massif de Bohème, Autriche). Données microtectoniques et radiométriques 39-Ar/ 40-Ar. C R Acad Sci Paris 301: 721–724

Matte Ph, Maluski H, Rajlich P, Franke W (1990) Terrane boundaries in the Bohemian Massif: result of large-scale Variscan shearing. In: Matte Ph (ed) Terranes in the Variscan Belt of Central Europe and Circum-Atlantic Paleozoic Orogens. Tectonophysics 177: 151–170

Matura A (1976) Hypothesen zum Bau und zur geologischen Geschichte des kristallinen Grundgebirges von Südwestmähren und des niederösterreichischen Waldviertels. Jahrb Geol Bundesanst 119: 63–74

Medaris LG Jr, Carswell DA (1990) The petr- ogenesis of Mg-Cr peridotites in European metamorphic belts. In: Carswell DA (ed) Eclogite facies rocks. Blackie, New York, pp 260–290

Melka R, Schulmann K, Schulmannová B, Hrouda F, Lobkowicz M, Figar Š (1994) The evolution of perpendicular linear fabrics during ductile thrusting in synkinematic granite (central Moravia, Bohemian Massif). J Struct Geol (in press)

Mísař Z, Dudek A, Havlena V, Weiss Z (1983) Geologie ČSSR I. Český masív. Státní Pedagogické Nakladatelství, Praha

Neubauer F (1991) Kinematics of the Moldanubian Zone in the southern Bohemian Massif: preliminary results from the Danube section. Österr Beitr Meteorol Geophys 3: 57–76

O'Brien PJ, Carswell DA, Gebauer D (1990) Eclogite formation and distribution in the Central European Variscides. In: Carswell DA (ed) Eclogite facies rocks. Blackie, New York, pp 204–220

Pacltová B (1980) Further micropaleontological data for the Paleozoic age of Moldanubian carbonate rocks. Čas Mineral Geol 25: 275–279

Pašek J, Urban M (1991) The tectonic evolution of the Plzeň Basin (Upper Carboniferous, West Bohemia): a review and reinterpretation. Folia Musei Rerum Naturalium Bohemiae Occidentalis, Geologica 32: 1–56

Petrakakis K (1986) Metamorphism of high grade gneisses from the Moldanubian zone, Austria, with particular reference to the garnets. J Metamorph Geol 4: 323–344

Petrakakis K, Richter W (1991) Metamorphose im niederösterreichischen Moldanubikum – eine Übersicht. Arbeitstag Geol Bundesanst 1991: 13–20

Pflug HD, Reitz E (1987) Palynology in metamorphic rocks: Indications of early land plants. Naturwissenschaften 74: 386–387

Quadt A (1990) U-Pb-zircon and Sm-Nd analyses on metabasites from the KTB pilot bore hole. KTB-Rep 90–4: 545

Quadt A, Gebauer D (1988) Sm/Nd-, U-Pb- and Rb-Sr dating of high-pressure ultramafic to felsic rocks from the Moldanubian area of NE Bavaria (FRG) and the Saxonian granulite massif (GDR). Abstr Conf Bohemian Massif, September 1988, Prague, p 71

Rajlich P (1987) Variszische duktile Tektonik im Böhmischen Massiv. Geol Rdsch 76: 755–786

Rajlich P (1988) Variscan shearing kinematics in the Bohemian Massif: Moldanubian and Moravo-Silesian junction. Acta Univ Carolinae Geol 4: 387–400

Rajlich P, Synek J (1987) A cross section through the Moldanubian of the Bohemian Massif and the structural development in its ductile domains. Neues Jahrb Geol Paläontol Monatsh 11: 689–698

Rajlich P, Synek J, Šarbach M, Schulmann K (1986) Hercynian-thrust related shear zones and deformation of the Varied Group on the contact of granulites (Southern Moldanubian, Bohemian Massif). Geol Rdsch 75: 665–683

Rajlich P, Schulmann K, Synek J (1988) Strain analysis of conglomerates from the Central Bohemian shear zone. Krystalinikum 19: 119–134

Reston T, Meissner R (1989) The three-dimensional structure of the Oberpfalz–an alternative interpretation of the DEKORP-KTB data. Tectonophysics 157: 1–11

Scharbert HG, Kurat G (1974) Distribution of some elements between coexisting ferromagnesian minerals in Moldanubian granulite facies rocks, Lower Austria, Austria. Tschermaks Mineral Petrogr Mitt TMPM 21: 110–134

Scharbert S (1987a) Rb-Sr Untersuchungen granitoider Gesteine des Moldanubikums in Österreich. Mitt Österr mineral Ges 132: 21–347

Scharbert S (1987b) Rb-Sr Analysen des Tonalits und Granits von der Lokalität Křižanovice (Železné hory). Čas Mineral Geol 32: 411–412

Scharbert S, Veselá M (1990) Rb-Sr systematics of intrusive rocks from the Moldanubicum around Jihlava. In: Minaříková D, Lobitzer H (eds) Thirty years of geological cooperation between Austria and Czechoslovakia. Fed Geol Surv Vienna Geol Surv Prague, pp 262–272

Schröder B (1987) Inversion tectonics along the western margin of the Bohemian Massif. Tectonophysics 137: 93–100

Schulmann K, Ledru P, Autran A, Melka R, Lardeaux JM, Urban M, Lobkowitz M (1991) Evolution of nappes in the eastern margin of the Bohemian Massif: a kinematic interpretation. Geol Rdsch 80: 73–92

Stein E (1988) Die strukturgeologische Entwicklung im Übergangsbereich Saxothuringikum/Moldanubikum in NE Bayern. Geol Bavarica 92: 5–131

Stettner G (1988) The Moldanubian region in the Bavarian segment of the Bohemian Massif (Federal Republic of Germany). In: Zoubek V, Cogné J, Kozhoukharov D, Krautner HG (eds) Precambrian in younger fold belts: European Variscides, the Carpathians and Balkans. Wiley, New York, pp 252–267

Stille H (1951) Das mitteleuropäische variszische Grundgebirge im Bilde des gesamteuropäischem. Geol Jahrb Beih 2: 1–138

Suess FE (1912) Die moravischen Fenster und ihre Beziehung zum Grundgebirge des Hohen Gesenkes. Österr Akad Wiss Denkschr 88: 541–631

Suess FE (1926) Intrusionstektonik und Wandertektonik im variszischen Grundgebirge. Bornträger, Berlin

Synek J (1991) Formation of sheath folds and reorientation of the stretching lineation in the western part of the Kutná Hora Crystalline Unit. Věstn Ústřed Ústavu Geol 66: 75–84

Teufel S (1988) Vergleichende U-Pb und Rb-Sr Altersbestimmungen and Gesteinen des Übergangbereiches Saxothuringikum/Moldanubikum, NE Bayern. Gött Arb Geol Paläontol 35: 1–87

Thiele O (1976) Ein westvergenter kaledonischer Deckenbau im niederösterreichischen Waldviertel? Jahrb Geol Bundesanst 119: 75–81

Thiele O (1984) Zum Deckenbau und Achsenplan des Moldanubikums der Südlichen Böhmischen Masse (Österreich). Jahrb Geol Bundesanst 126: 513–523

Tollmann A (1982) Großräumiger variszischer Deckenbau im Moldanubikum und neue Gedanken zum Variszikum Europas. Geotekton Forsch 64: 91

Urban M (1994) Kinematics of the Variscan thrusting in the E Moldanubicum (Bohemian Massif, Czechoslovakia): evidence from the Náměšť' granulite massif. Tectonophysics 201 (in press)

Urban M, Mísař Z (1989) Fold and shear structures of the southeastern Moldanubicum (Czechoslovakia). Krystalinikum 20: 111–130

Van Breemen O, Aftalion M, Bowes DR, Dudek A, Mísař Z, Povondra P, Vrána S (1982) Geochronological studies of the Bohemian Massif, Czechoslovakia, and their significance in the evolution of Central Europe. Trans R Soc Edinb Earth Sci 73: 89–108

Vollbrecht A, Weber K, Schmoll J (1989) Structural model for the Saxothuringian-Moldanubian suture in the Variscan basement of the Oberpfalz (Northeastern Bavaria, FRG) interpreted from geophysical data. Tectonophysics 157: 123–133

Vrána S (1979) Polyphase shear folding and thrusting in the Moldanubicum of southern Bohemia. Věstn Ústřed Ústavu Geol 54: 75–86

Vrána S (1994) The Moldanubian Zone in southern Bohemia: polyphase evolution of imbricated crustal and upper mantle segments. Proc Conf Bohemian Massif, September 1988, Prague (in press)

Wallbrecher E, Dallmeyer RD, Brandmayr M, Handler R, Maderbacher F, Platzer R (1991) Kinematik und Alter der Blattverschiebungszonen in der südlichen Böhmischen Masse. Arbeitstag Geol Bundesanst 1991: 35–48

Weber K (1990) Observations on the ductile deformation path of the paragneisses of the KTB pilot hole. KTB-Rep 90–8: J1–J19

Weber K, Duyster J (1990) Moldanubian Zone of the Waldviertel, Lower Austria. In: Franke W (ed) Bohemian Massif. Field guide, Conf IGCP 233

"Paleozoic orogens in central Europe", Göttingen-Giessen 1990, pp 87–96
Weber K, Vollbrecht A (1986) Kontinentales Tiefbohrprogramm der Bundesrepublik Deutschland, KTB-Ergebnisse der Vorerkundungsarbeiten, Lokation Oberpfalz. 2. KTB-Koll, September 1986, Seeheim/Odenwald
Wendt I, Kreuzer H, Müller P, Schmidt H (1986) Gesamtgesteins- und Mineraldatierungen des Falkenberger Granits. Geol Jahrb E 34: 5–60
Wendt I, Kröner A, Todt W, Fiala J, Rajlich P, Liew TC, Vaněk J (1988) U-Pb zircon ages and Nd whole-rock systematics for Moldanubian rocks of the Bohemian Massif, Czechoslovakia. Abstr Conf Bohemian Massif, Prague, 1988
Zoubek V, Cogné J, Kozhoukharov D, Krautner HG (1988) Precambrian in younger fold belts: European Variscides, the Carpathians and Balkans. Wiley, New York

VII.C.3 Igneous Activity

F.V. Holub, M. Klečka, and D. Matějka

1 Introduction

The Moldanubian zone is characterized by an abundance of Late Hercynian plutonic rocks of predominantly granitoid composition. These rocks are considered to be parts of one of the principal magmatic complexes – the Moldanubian (South Bohemian) Pluton and the Central Bohemian Pluton. Since both the large magmatic masses are of polygenetic batholithic character and are composed of several plutons, they are termed plutonic complexes in the following text. Some intrusive bodies exhibit rather independent characteristics and this is true especially for abundant dark ultrapotassic rocks and some dyke swarms.

2 The Moldanubian Plutonic Complex

The Moldanubian Plutonic Complex (MPC) is the largest Hercynian magmatic complex of the Bohemian Massif (about 6000 km^2). It is usually divided into the eastern (Bohemian) and the western (Bavarian) branches which are joined in the Waldviertel area in Austria (Fig. 1). The western branch begins in the environs of Regensburg (the more northerly granites belong to the West Bohemian Plutonic Complex) and extends to the SE as a number of separate granitoid bodies which probably cohere at some depth. The eastern branch (called the Central Massif) is more continuous and fills the core of a N-S anticlinal zone cutting obliquely across the older Hercynian metamorphic structures of the Moldanubicum.

The succession of the individual intrusive bodies has been interpreted in various ways by many authors cited in the synoptic papers by Waldmann (1951), Troll (1964), or Sattran and Klomínský (1970). Recently, this problem was solved using geochronological data by Finger and Höck (1986) and Liew et al. (1989) for the Austrian part and by Propach (1989) for the Bavarian part of the complex. The scheme presented here (Table 1) is based mainly on Finger and Höck's paper and supplemented by new geochronological data (Scharbert and Veselá 1990) and field evidence (Klečka and Matějka 1994) for the Bohemian part of the MPC.

The hitherto existing geochronological data from the MPC vary from 350 to 300 Ma (Köhler and Müller-Sohnius 1986; Scharbert 1987; Scharbert and Veselá 1990). The granitoid rocks can be divided into three successive groups corresponding to the evolutionary phases of the Hercynian orogen in this part of the Bohemian Massif: (1) an older synorogenic group (Early Carboniferous), (2) a younger late- to post-orogenic group (Upper Carboniferous) and (3) the youngest and least voluminous "anorogenic" group (probably Lower Permian). Durbachitic (ultrapotassic) rocks (Fig. 1) are not included in this scheme because of their petrogenetic independence and the lack of reliable geochronological data.

Both the two main granitoid groups consist of K_2O-rich granitoids with I-, transitional I/S- and typical S-character (Liew et al. 1989), where the latter are strongly peraluminous in nature. Selected chemical analyses are presented in Table 2.

The older granitoids (Group 1), which predominate in the Austrian part of the MPC, originated during the thermal climax of the Hercynian metamorphism (350–335 Ma according to Scharbert and Veselá 1990; Fiala et al., Chap. VII.C.1) and are associated with various types of migmatites (e.g., the pearl gneisses). The predominant granitoids of the Weinsberg type (together with the Kristallgranit I in Bavaria – Kraus 1962) are mostly coarse-grained biotite granites to granodiorites containing abundant K-feldspar megacrysts often up to 10 cm in size (Kurat 1965). The Schlierengranit type is biotite granite to amphibole-biotite granodiorite, often

Dallmeyer et al. (Eds.)
Pre-Permian Geology of Central and Eastern Europe

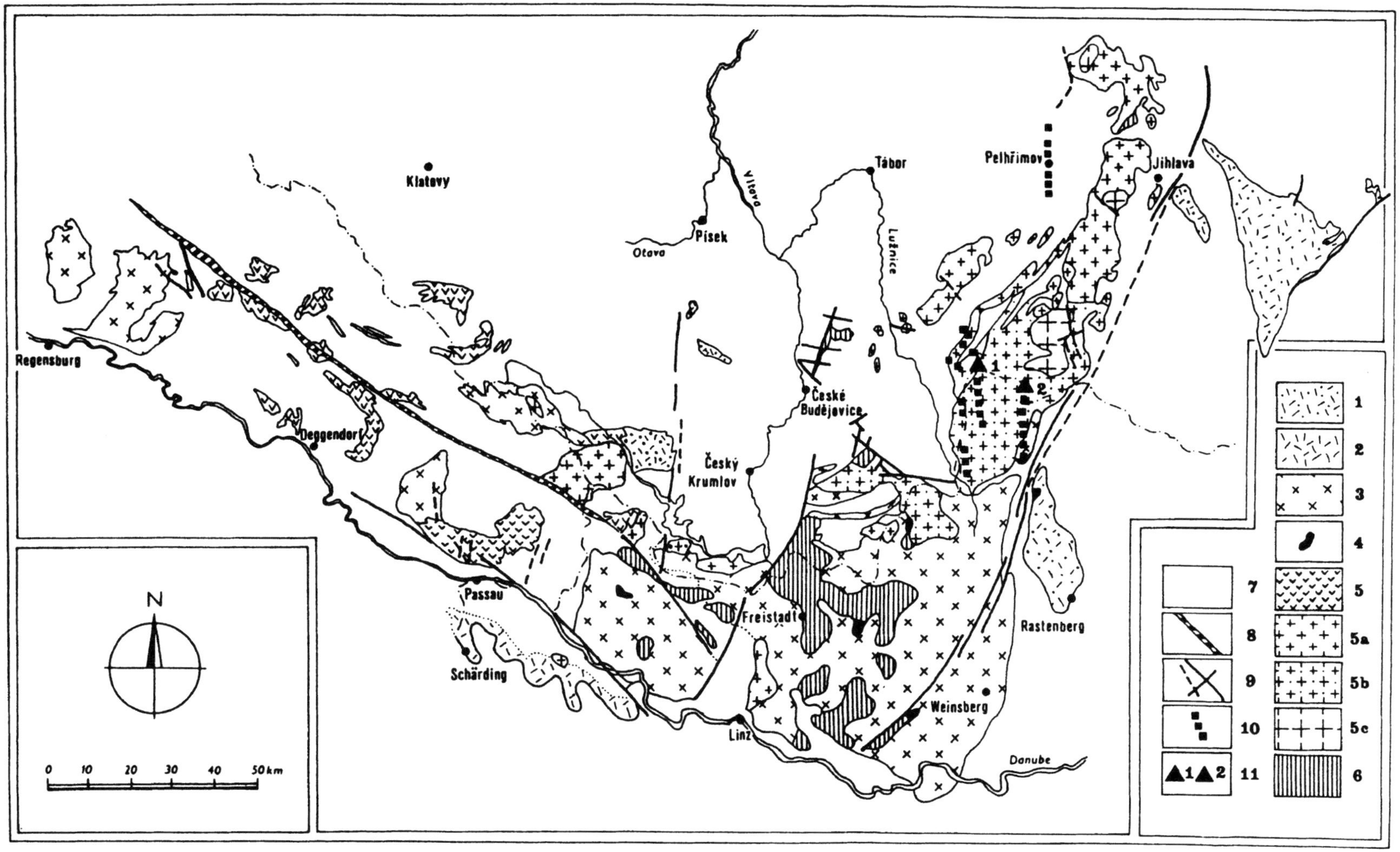

Fig. 1. Geological sketch map of the Moldanubian Plutonic Complex. *1* Ultrapotassic plutonites and the Rastenberg type; *2* Schärding and Peuerbach types (group 1A); *3* Weinsberg and Schlierengranit types (1B); *4* Quartz diorites and tonalites (1B); *5* Eisgarn type (2C, undistinguished) in the Bavarian branch; *5a* Eisgarn type (2C, undistinguished) and Lásenice type (2A) in the Central Massif; *5b* Eisgarn type – Číměř variety; *5c* Eisgarn type – Landštejn variety; *6* Freistadt, Mauthausen, Ševětín, and Pavlov types (2B); *7* Moldanubian metamorphics and younger sedimentary cover; *8* Quartz dykes; *9* faults; *10* Želiv-Lásenice volcanotectonic zone marked by dykes of felsitic subvolcanic rocks (group 3); *11* young granite stocks (group 3): *no. 1* Homolka; *no. 2* Kozí Hora

Table 1. Basic division scheme of granitoids in the Moldanubian Batholith. (After Finger and Höck 1986, supplemented and modified by Klečka and Matějka 1992)

GROUP 1	**Lower Carboniferous** synorogenic granitoids (variably deformed units)
SUBGROUP 1A Schärding type Peuerbach type	Old S-types
SUBGROUP 1B Weinsberg type ["Kristallgranit" type] "Schlierengranit" type Quartz diorites to tonalites	Old I- and I/S-(transitional) types 349 ± 4 Ma (Rb-Sr, whole rock)[a] [349 ± 11 Ma (Rb-Sr, whole rock)[b]]
GROUP 2	**Upper Carboniferous** late- to post-orogenic granitoids
SUBGROUP 2A Altenberg type [Lásenice type in the Czech part]	Locally slightly deformed S-types
SUBGROUP 2B Freistadt type Mauthausen type Ševětín type Pavlov type	Late I- and I/S-(transitional) types 329 ± 7 Ma (Rb-Sr, muscovite)[c] 324 ± 4 Ma (Rb-Sr, muscovite)[c]
SUBGROUP 2C Eisgarn type	Completely undeformed S-types 316 ± 7 Ma (Rb-Sr, whole rock)[a] 303 ± 6 Ma (Rb-Sr, whole rock)[d]
GROUP 3	**Upper Carboniferous to Lower Permian** anorogenic granitoids (extensional regime) connected with a N-S trending volcanotectonic zone
Subvolcanic felsic dykes *Small stocks:* Homolka type Kozí Hora type Šejby type	295 ± 5 Ma (Rb-Sr, whole rock)[e] 300 ± 41 Ma (Rb-Sr, whole rock)[f] 296 ± 31 Ma (Rb-Sr, whole rock)[g]

[a] data from the Austrian part (Scharbert 1987).
[b] data from the Bavarian branch (Köhler and Müller-Sohnius 1986).
[c] data from the Austrian part (Scharbert 1979).
[d] data from the Czech part (Scharbert and Veselá 1990).
[e] data from the Czech part (Klečka et al. 1994).
[f] data from the Czech part (Bendl et al. in print a).
[g] data from the Czech part (Bendl et al. in print b).

grading into migmatite rich in quartz-dioritic paleosomes. Far less common bodies of quartz diorite, quartz monzonite and tonalite composition often contain hypersthene, clinopyroxene, and biotite as their main mafic constituents (Koller and Niedermayr 1981). All these rocks are of I- or transitional I/S-character and may have originated by melting of an older calc-alkaline magmatic series or mixed rocks and, according to Liew et al. (1989), their isotopic composition does not provide any evidence for a major influx of juvenile mantle melts. Other granitoids of this older group (the Schärding and Peuerbach types) are of the S-type and were probably derived by melting of the high-grade paragneisses almost in situ.

Granitoid plutons of Group 2 intruded the uplifting and cooling metamorphic complex. Their contacts are mostly sharp and discordant. Many of these plutons are totally undeformed except for the proximity of some major shear zones, where foliation is sometimes developed (Altenberg and Lásenice types – subgroup 2A). The most common Eisgarn type (subgroup 2C) is a coarse-grained porphyritic two-mica granite which, together with the fine-grained Altenberg and medium-grained Lásenice type, belongs to the S-type granitoids. In Bohemia, the Eisgarn granite covers two main intrusive phases corresponding to the Číměř and Landštejn varieties (Klečka and Matějka 1994). The Mauthausen (Richter 1965), Freistadt (Klob 1971), Ševětín and Pavlov (Klečka and Matějka, 1992) types are mostly medium-grained rocks ranging in composition from biotite granite to biotite-pyroxene (± hornblende) quartz monzodiorite of the I/S- or I-type (subgroup 2B).

Group 3 is represented by anorogenic subvolcanic dykes and granite stocks, which are the youngest magmatic products associated with the extensional tectonic regime (Klečka and Vaňková 1988; Klečka and Matějka 1994; Matějka and Klečka 1992; Bendl et al. 1994). The dykes constitute a N-S-trending "volcanotectonic zone" (Želiv-Lásenice) about 85 km in length, with which some prominent circular, probably volcanotectonic structures are associated (Vrána 1990; Klečka 1994). The small granite intrusions are represented by topaz-bearing muscovite granite with Sn-(Nb-Ta) mineralization (the Homolka type) and greisenized biotite-muscovite granite with Mo-mineralization (the Kozí Hora type). These rocks may correspond either to extremely fractionated peraluminous S-granites or to highly fractionated granites of the A-type, associated with the Sn-W-(Mo) mineralization.

Table 2. Representative analyses of granitoids from the Moldanubian Plutonic Complex

Sample	1	2	3	4	5	6	7
Subgroup	1A	1B	1B	2C	2B	3	3
SiO_2	69.0	69.8	56.0	72.0	66.3	73.0	73.2
TiO_2	0.50	0.62	1.24	0.35	0.51	0.12	0.01
Al_2O_3	14.7	14.0	16.8	14.8	16.9	14.7	15.4
FeO tot.	2.75	3.45	7.81	1.89	3.20	2.48	0.49
MnO	0.05	0.06	0.13	0.02	0.05	0.05	0.08
MgO	1.01	0.71	4.85	0.53	1.59	0.07	0.05
CaO	1.38	1.58	6.04	0.91	3.77	0.67	0.45
Na_2O	3.02	2.69	3.35	2.73	3.83	4.05	4.21
K_2O	5.36	5.51	2.05	5.37	2.55	3.78	3.66
P_2O_5	0.30	0.19	0.20	0.25	0.19	0.44	0.90
mg	39.6	26.8	52.5	33.3	47.0	4.8	15.4
Rb	298	271	58	333	70	680	1397
Sr	178	136	394	75	75	22	33
Ba	649	678	471	452	660	40	30
Y	20	36	17	12	23	5	62
Zr	192	324	120	186	165	16	17
Nb	19	20	20	17	11	33	56
$^{87}Sr/^{86}Sr$	0.7085	0.7065	–	–	0.7053	0.71575	0.72439

Rock types: 1 – Schärding; 2 – Weinsberg; 3 – Quartz diorite; 4 – Eisgarn; 5 – Freistadt; 6 – Subvolcanic dyke; 7 – Homolka.
Analyses no. 1, 2 and 5 are from Liew et al. (1989). (Major oxides in weight percent, trace elements in ppm.)

3 The Central Bohemian Plutonic Complex

The Central Bohemian Plutonic Complex (CBPC) covers an area of about 3000 km^2 in central and south Bohemia (Fig. 2). The complex is elongated to a total length of about 145 km in the NE-SW direction along the Central Bohemian suture or shear zone which separates the Moldanubian zone from the Barrandian area. Its contacts are mostly sharp with strong effects of contact metamorphism on the neighboring Upper Proterozoic and Lower Paleozoic series at the NW margin and in numerous large relics of the roof (called the Islet Zone) including the Cadomian volcanics of the Jílové Belt. At the E and S contacts with the Moldanubian rocks, a zone of pearl gneisses and, in some places, of nebulitic to stromatitic migmatites is often developed.

On an average, the rocks of the CBPC are much less acidic and the mafic members are much more abundant than in the MPC. A great majority of the granitoid rocks correspond to the conventional I-type though some S-granitoids are also present.

Various rocks of the CBPC have been interpreted in terms of granitization (metamorphic recrystallization) of a volcanic or metasedimentary precursor (e.g., Palivcová 1965; Palivcová et al. 1989a,b). However, there is no special reason for such an interpretation, which contradicts the geological and petrographical evidence.

The rocks range in composition from scarce hornblendite through subordinate gabbro and diorite to predominant granodiorite and less frequent granite as well as from an extremely dark melasyenite to potassic melagranite. In addition to these plutonic rocks, abundant dykes occur. About 25 local petrographic types of granitoids have been defined and named after typical localities (for petrographic descriptions and petrochmical characteristics see, e.g., Kodym 1966; Steinocher 1969; Vejnar 1973, 1974). Some representative analyses of selected rock types are given in Table 3.

Using various criteria based on the rock chemistry, mineralogy, and field relations, the following principal groups or suites were found (Holub 1991), where the first three are arranged from the older to the younger ones.

1. A typical calc-alkaline plutonic suite of gabbro, diorite and biotite-hornblende tonalite to granodiorite (I-type) of the Sázava type, which

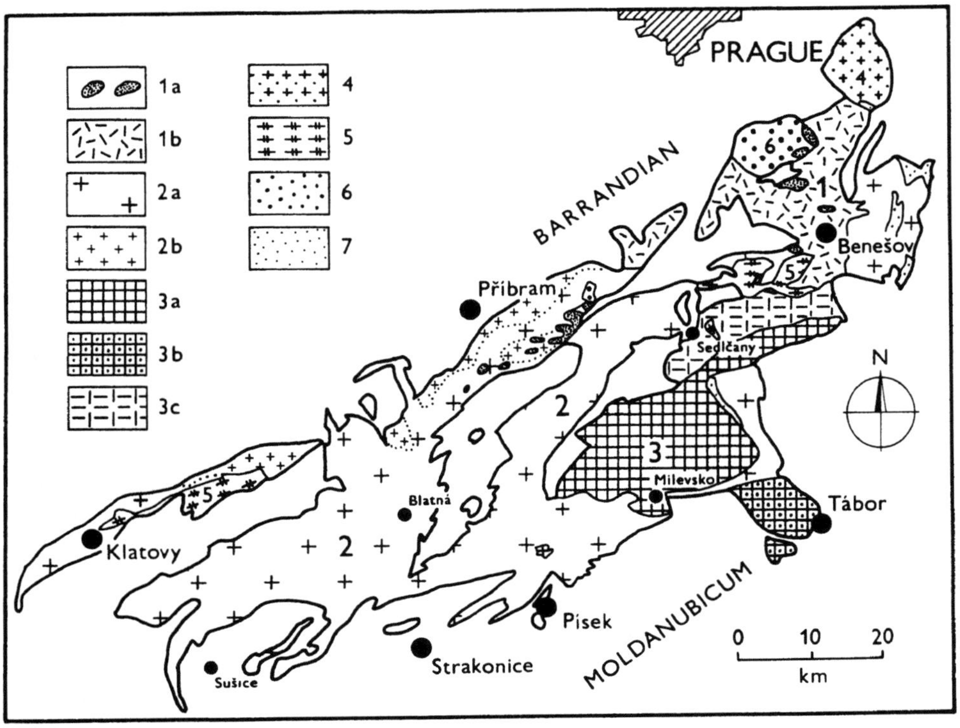

Fig. 2. Geological sketch map of the Central Bohemian Plutonic Complex. *1a* Gabbroic rocks; *1b* calc-alkaline I-granitoids (Sázava tonalite to granodiorite); *2a* high-K calc-alkaline granitoids (Kozárovice, Červená, Blatná, Těchnice, and Klatovy types); *2b* "marginal" granite; *3* ultrapotassic plutonites: *a* durbachite series (Čertovo břemeno type); *b* Tábor type; *c* Sedlčany granite; *4* Říčany granite; *5* S-type peraluminous granodiorites (Kozlovice, Maršovice, and Kosova Hora types); *6* Požáry type; *7* other granitoid rocks (mostly leucogranites)

Table 3. Representative analyses of selected rock types from the Central Bohemian Plutonic Complex

	1a	1b	2a	2b	3	4	5
SiO_2	51.9	65.9	63.8	67.5	66.5	70.2	65.28
TiO_2	1.15	0.46	0.68	0.51	0.52	0.31	0.76
Al_2O_3	17.8	15.3	15.8	15.2	15.1	14.9	15.9
Fe_2O_3	2.43	1.72	0.78	0.79	0.72	0.30	1.17
FeO	6.58	3.06	3.56	2.36	2.31	1.03	4.60
MnO	0.23	0.11	0.06	0.06	0.05	0.02	0.06
MgO	5.02	2.05	2.58	1.40	2.48	1.05	2.29
CaO	9.08	4.35	3.99	3.18	2.23	1.13	0.85
Na_2O	2.48	3.22	3.12	3.33	2.9	3.78	2.74
K_2O	1.46	2.94	3.80	4.02	5.76	5.68	3.80
P_2O_5	0.18	0.10	0.30	0.17	0.32	0.14	0.16
mg	50.5	44.2	51.9	44.8	59.9	59.0	41.9
Cr	70	15	55	45	125	40	90
Ni	25	10	–	15	30	15	–
Rb	40	90	165	190	330	310	145
Sr	460	330	405	350	355	320	145
Ba	550	1250	1300	1100	1240	1100	800
Zr	150	95	275	255	260	240	190
Th	5	13	20	20	31	34	11
U	5	4	7	10	18	7	7

1a – Quartz gabbro; 1b – Amphibole-biotite granodiorite (Sázava type); 2a – Amphibole-biotite granodiorite (Červená type); 2b – Biotite granodiorite to monzogranite (Blatná type); 3 – Amphibole-biotite granite (Sedlčany type); 4 – Biotite granite (Říčany type); 5 – Cordierite-biotite granodiorite (Maršovice type). (Major oxides in weight percent, trace elements in ppm).

probably represents the oldest part of the CBPC. Its geochemistry corresponds to granitoid plutons of the volcanic arc associations.

2. A high-K calc-alkaline suite involving some mafic members (biotite-hornblende melamonzonites, monzodiorites, quartz monzonites, e.g., Ledvinková 1985) of the shoshonitic affinity. The most voluminous rocks are often slightly porphyritic amphibole-biotite melagranodiorites and granodiorites up to granites of the Kozárovice, Červená, Blatná, and Klatovy types. The involvement of the lighter biotite granite of the "marginal" type in this group is advocated by Machart (1991) for petrochemical reasons. The rocks of this suite are more magnesian and contain more hygromagmatophile elements than the first group (Fig. 3). The Blatná granodiorite is the only rock type of the CBPC dated by the Rb-Sr method (van Breemen et al. 1982; Bendl and Vokurka 1989) which has yielded an age of 331 ± 4 Ma using whole-rock samples of the granodiorite and aplitic veins.

The rocks of these two suites are penetrated by numerous E-W-trending dykes of biotite lamprophyres and granodiorite or melasyenite to granite porphyries forming a conspicuous regional dyke swarm which extends southward out of the CBPC.

3. An ultrapotassic suite of highly magnesian melasyenite to melagranite of the Čertovo břemeno (i.e., the durbachite series) and Tábor types is of the same composition as the corresponding rocks outside the CBPC (see below.)

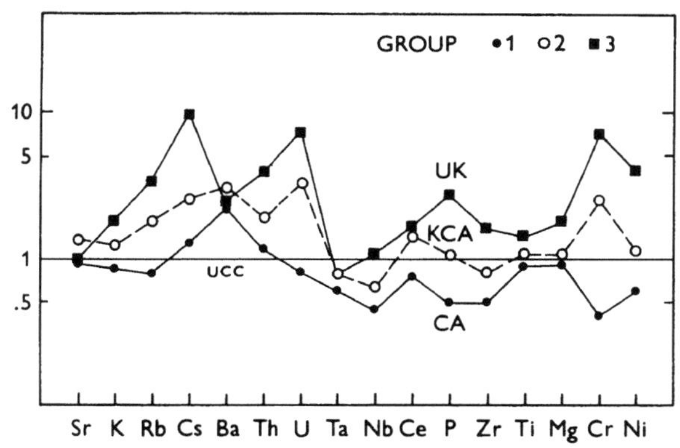

Fig. 3. Abundances of selected elements normalized to average composition of the upper continental crust (*UCC*) in representative analyses of (*1*) normal calc-alkaline, (*2*) high-K calc-alkaline, and (*3*) ultrapotassic granitoids (65% SiO_2) from the Central Bohemian Plutonic Complex. Normalizing values are from Taylor and McLellan (1985) except for P_2O_5 (= 0.2%)

The suite is younger than the dyke swarm (Kodym 1966; Holub and Žežulková 1978). A porphyritic amphibole-biotite granite of the Sedlčany type of somewhat less extreme composition may belong to this or to the next suite. Presence of numerous dark ultrapotassic inclusions is typical for these rocks.

The other plutonic suites may be mostly (except 6) contemporary with or somewhat younger than the ultrapotassic suite, but their interrelationships are poorly known.

4. Potassic CaO-poor granites containing various K-rich (up to ultrapotassic) microgranular inclusions, e.g., the Říčany type (biotite to muscovite-biotite granite) which is probably one of the youngest intrusions in the CBPC.

5. Strongly peraluminous, CaO-poor cordierite-biotite to muscovite-biotite granodiorite of definitely S-character (the Kozlovice, Maršovice, and Kosova Hora types).

6. CaO-rich and K_2O-poor acidic I-granitoids of biotite trondhjemite to granodiorite composition (the Požáry and Nečín types).

7. Small bodies and dykes of various leucogranites and aplitic rocks which probably belong to more than one group.

4 Ultrapotassic Rocks

The ultrapotassic rocks (the K-syenite series according to Sattran and Klomínský 1970) are characterized by their highly magnesian nature, accompanied by high concentrations of Cr (Ni) and of the hygromagmatophile elements like K, Rb, Cs, Th, U, etc. (Fig. 3; Table 4). They belong among rocks displaying the highest radioactivity in the Moldanubian Zone (Manová 1975, ms; Fiala et al. 1983). The most voluminous plutonic rocks of this group belong to the durbachite series (the Čertovo břemeno type) ranging in composition from extremely dark amphibole-biotite melasyenite (durbachite) to melagranite, and mineralogically as well as geochmically corresponding with the association durbachite – Granite des Cretes in the Vosges (see, e.g., Fluck 1980) and the Black Forest. Abundant dark microgranular inclusions are also of ultrapotassic melasyenitic composition (Holub 1977). The rocks of the durbachite series form a conspicuous zone of isolated large and small bodies between the E margin of the CBPC and the W branch of the MPC as well as the large (500 km²) Třebíč

Table 4. Representative analyses of ultrapotassic plutonic and dyke rocks from southern Bohemia (1, 2, 5, 6) and western Moravia (3, 4, 7)

	1	2	3	4	5	6	7
SiO_2	50.5	52.8	58.7	65.6	58.8	64.2	60.0
TiO_2	1.13	1.20	0.95	0.62	1.07	0.68	0.75
Al_2O_3	10.2	13.8	13.5	13.5	13.0	13.2	12.2
Fe_2O_3	1.32	1.07	0.59	0.70	1.00	0.70	1.84
FeO	5.66	5.65	4.63	2.75	4.82	2.97	2.36
MnO	0.13	0.11	0.08	0.05	0.10	0.06	0.08
MgO	14.4	8.35	6.13	3.77	7.23	4.02	5.65
CaO	5.74	4.57	3.88	2.68	3.54	2.51	4.07
Na_2O	1.15	2.14	1.97	2.7	1.77	2.70	3.3
K_2O	6.15	6.81	6.59	6.24	5.72	6.61	7.69
P_2O_5	1.34	1.13	0.87	0.43	0.65	0.53	0.97
mg	78.9	69.2	67.9	66.5	69.3	66.6	71.5
Cr	700	510	360	220	560	255	265
Ni	495	185	120	60	130	--	265
Rb	410	375	355	370	340	410	290
Sr	240	500	445	335	300	335	1320
Ba	1660	2360	1980	1250	1620	1300	5630
Zr	445	500	430	275	365	320	330
Th	26	48	43	39	43	46	60
U	10	19	18	21	13	29	12

1 – Extremely mafic amphibole-biotite plutonite; 2 – Durbachite; 3 – Durbachitic quartz melasyenite; 4 – Durbachitic melagranite (the Čertovo břemeno type); 5 – Olivine minette; 6 – Melagranite porphyry; 7 – Thuresite. (Major oxides in weight percent, trace elements in ppm).

pluton (Fig. 1). The second variety is represented by biotite-two-pyroxene melasyenite to melagranite of the Tábor and Jihlava types.

The chemical composition of these rocks is comparable to minette. Their origin can be best explained by partial melting of a highly anomalous upper mantle source which experienced depletion and consequently a strong enrichment in the hygromagmatophile elements probably by hydrous fluids released from a subducted slab (Holub 1990). Hybridization of the primitive ultrapotassic magma by acidic crustal melts during the ascent through the thickened anatectic continental crust played a crucial role in the petrogenesis of the more silicic members of the durbachite series (Holub 1988, 1990) as well as of the melagranite porphyries (Němec 1988). The hybridization (mixing) phenomena can explain the failure of Rb-Sr dating (see Scharbert and Veselá 1990).

Potassic to ultrapotassic dyke rocks are widespread and very variable. Minettes and melasyenite to melagranite porphyries prevail especially in the area of the CBPC (e.g., Němec 1974) but kersantites are more widespread in some parts of the Moldanubian crystalline complex (see Němec 1970). Although the majority of minettes are older than the ultrapotassic plutonic rocks, some dykes are younger. Sporadic dykes of ultrapotassic peralkaline microsyenite (thuresite), microgranite (karlsteinite) and similar rocks (e.g., Hackl and Waldmann 1935; Machart 1979) probably belong to the youngest magmatic rocks in the area.

5 Conclusions

The Hercynian syn- to postorogenic magmatic activity in the Moldanubian zone is of diverse character. Granitoids of the Moldanubian Plutonic Complex mostly represent the products of crustal anatexis of some older magmatic and sedimentary series. Rocks representing the fractionated and/or hybridized mantle melts are subordinate (the sporadic gabbroic and dioritic types, high-K mafic rocks). In contrast to the MPC, mafic magmas of several distinct types (calc-alkaline, shoshonitic, ultrapotassic) have played a much more important role in the petrogenesis of granitoids in the Central Bohemian Plutonic Complex where hybrid rocks (H-types according to Castro et al. 1991) predominate. The best example of hybridism and the extensive mantle-crust interaction is represented by the ultrapotassic rocks of the durbachite series. Though a close relationships of the magmatism to a contemporaneous or ancient subduction activity is highly probable, it is not possible to reconstruct the course and characteristics of the subduction zone (zones) without further research.

References

Bendl J, Vokurka K (1989) Strontium isotope model of formation of Blatná granodiorite. Geol Zb – Geol Carpathica 40: 655–664

Bendl J, Klečka M, Moneim M, Svobodová J (1994a) Rb-Sr dating of the topaz-bearing muscovite granite stock Homolka, Moldanubian Batholith. Mitt Österr Mineral Ges (in press)

Bendl J, Heřmánek R, Klečka M, Matějka D (1994b) Highly evolved granites (Šejby and Nakolice stocks) in the Nové Hrady Mts., Moldanubian Batholith: geochemistry and Rb-Sr dating. Mitt Österr Mineral Ges (in press)

Castro A, Moreno-Ventas I, Rosa JD de la (1991) H-type (hybrid) granitoids: a proposed revision of the granite type classification and nomenclature. Earth-Sci Rev 31: 237–253

Fiala J, Vaňková V, Wenzlová M (1983) Radioactivity of selected durbachites and syenites of the Bohemian Massif. Čas Mineral Geol (J Mineral Geol Prague) 28: 1–16

Finger F, Höck V (1986) Zur magmatischen Entwicklung des Moldanubikums in Oberösterreich. Jahrb Geol Bundesanst Wien 129: 641–642

Fluck P (1980) Métamorphisme et magmatisme dans les Vosges moyennes d'Alsace. Contribution à l'histoire de la chaîne varisque. Sci Géol Mém Strasbourg 62: 1–248

Hackl O, Waldmann L (1935) Ganggesteine der Kalireihe aus dem niederösterreichischen Waldviertel. Jahrb Geol Bundesanst Wien 85: 259–285

Holub FV (1977) Petrology of inclusions as a key to petrogenesis of the durbachitic rocks from Czechoslovakia. TMPM Tschermaks Mineral Petrogr Mitt 24: 133–150

Holub FV (1988) Potassic mafic magmatism of the Bohemian Massif. Abstr 1st Int Conf Bohemian Massif. Geol Surv, Prague, p 37

Holub FV (1990) Petrogenetic interpretation of chemical composition of potassic lamproid rocks in the European Hercynian Belt: an example from the central and southern parts of the Bohemian Massif. Thesis, Charles University Prague, 265 pp (in Czech)

Holub FV (1991) Contribution to petrochemistry of the Central Bohemian Plutonic Complex (English summary). In: Souček J (ed) Horniny ve vědách o Zemi. Charles Univ Prague, pp 117–140

Holub F, Žežulková V (1978) Relative ages of intrusives of the Central Bohemian Pluton near Zvíkov. Věstn Ústřed Ústavu Geol (Bull Geol Surv Prague) 53: 289–297 (English Summary)

Klečka M, Matějka D (1992) Modanubian Pluton as an example of the late Variscan crustal magmatism in the Moldanubian zone. In: 7th Geological Worksh Styles of superposed Variscan nappe tectonics. Kutná Hora, 24.-27.4.1992, Abstr, 13–14, Kutná Hora

Klečka M, Matějka D (1994) Moldanubian Batholith – an example of evolution of the late Paleozoic granitoid magmatism in the Moldanubian zone, Bohemian Massif (central Europe). In: Magmatism in relation to diverse tectonic settings (Prof. S.G. Karkare felicitation volume), Varanasi, India (in press)

Klečka M, Vaňková V (1988) Geochemistry of felsitic dykes from the vicinity of Lásenice near Jindřichův Hradec (South Bohemia) and their relation to Sn-W mineralization. Čas Mineral Geol (J Mineral Geol Prague) 33: 225–249

Klečka M, Bendl J, Matějka D (1994) Rb-Sr dating of acid subvolcanic dyke rocks – final magmatic products of the Moldanubian Batholith. Mitt Österr Mineral Ges (Wien) 139: 66–68

Klob H (1971) Der Freistädter Granodiorit im Österreichischen Moldanubikum. Verh Geol Bundesanst Wien 1: 98–142

Kodym O Jr (1966) Moldanubicum. In: Svoboda J (ed) Regional Geology of Czechoslovakia, Part I. The Bohemian Massif. Geol Surv, Prague, pp 40–99

Köhler H, Müller-Sohnius D (1986) Rb-Sr Altersbestimmungen und Sr-Isotopensystematik an Gesteinen des Regensburger Waldes (Moldanubikum NE Bayern), Teil 2: Intrusivgesteine. Neues Jahrb Mineral Abh 155: 141–219

Koller F, Niedermayr G (1981) Die Petrologie der Diorite im Nördlichen Waldviertel, Niederösterreich. TMPM Tschermaks Mineral Petrogr Mitt 28: 285–313

Kraus G (1962) Gefüge, Kristallgrössen und Genese des Kristallgranits I im Vorderen Bayerischen Wald. Neues Jahrb Mineral Abh 97: 357–434

Kurat G (1965) Der Weinsberger Granit im südlichen österreichischen Moldanubikum. TMPM Tschermaks Mineral Petrogr Mitt 9: 202–227

Ledvinková V (1985) Gabbroids in the Mirovice metamorphic islet. Sb Geol Věd Geol (J Geol Sci Geol Prague) 40: 35–61 (English Summary)

Liew TC, Finger F, Höck V (1989) The moldanubian granitoid plutons of Austria: chemical and isotopic studies bearing on their environmental setting. Chem Geol 76: 41–55

Machart J (1979) Alkali quartz microsyenite to microgranite (karlsteinite) from Jemnice near Strakonice (southern Bohemia). Čas Mineral Geol (J Mineral Geol Prague) 24: 165–173 (English summary)

Machart J (1991) Chemical types of granitoids in the southern part of the Central Bohemian Pluton. In: Souček J (ed) Horniny ve Vědách o Zemi. Charles Univ, Prague, pp 107–116 (English summary)

Matějka D, Klečka M (1992) Petrogenese und geodynamische Stellung des Moldanubischen Batholith,

Böhmische Masse. Zentralbl Geol Paläont T1 Stuttgart I, 7/8: 936–937
Němec D (1970) Lamprophyrische und lamproide Ganggesteine im Südteil des Böhmisch-Mährischen Anhöhe. TMPM (Tschermaks Mineralogische und Petrographische Mitteilungen) 14: 235–284
Němec D (1974) Petrochemistry of the dyke rocks of the Central Bohemian Pluton. Neues Jahrb Mineral Monatsh 1974: 193–209
Němec D (1988) Origin of syenite porphyries in the Central Bohemian Pluton by magma mixing. Neues Jahrb Mineral Abh 159: 59–71
Palivcová M (1965) The Central Bohemian Pluton – a petrographic review and an attempt at a new genetic interpretation. Krystalinikum (Prague) 3: 99–131
Palivcová M, Waldhausrová J, Ledvinková V (1989a) Granitization problem – once again. Geol Zb-Geol Carpathica (Bratislava) 40: 423–452
Palivcová M, Waldhausrová J, Ledvinková V (1989b) Precursors lithology and the origin of the Central Bohemian Pluton (Bohemian Massif). Geol Zb-Geol Carpathica (Bratislava) 40: 521–546
Propach G (1989) The origin of a comformable Variscan granite in Bavaria. Results of geochemical and geochronological investigation. In: Bonin B et al. (eds) Magma – crust interactions and evolution. Theophrastus, Athens, pp 193–210
Richter W (1965) Petrologische Untersuchungen am Mauthausener Granit im österreichischen Moldanubikum. TMPM Tschermaks Mineral Petrogr Mitt 10: 265–296
Sattran V, Klomínský J (1970) Petrometallogenetic series of igneous rocks and endogenous ore deposits in the Czechoslovak part of the Bohemian Massif. Sb Geol Věd Ložisková Geol Prague LG 12: 65–154
Scharbert S (1979) Geochronologische Datierungen. In: Schönlaub HP (ed) Das Paläozoikum in Österreich, Abh Geol Bundesanst Wien, pp 35–38
Scharbert S (1987) Rb-Sr-Untersuchungen granitoider Gesteine des Moldanubikums in Österreich. Mitt Österr Mineral Ges 132: 21–37
Scharbert S, Veselá M (1990) Rb-Sr systematics of intrusive rocks from the Moldanubicum around Jihlava. In: Minaříková D, Lobitzer H (eds) Thirty Years of geological cooperation between Austria and Czechoslovakia. Geol Surv Prague, pp 262–272
Steinocher V (1969) Die stoffliche Zusammensetzung, der provinziale Charakter und die petrologischen Verhältnisse des mittelböhmischen Plutons (German summary). Rozpr Česk Akad Věd Řada Mat Přír Věd Prague 79 (1): 1–99
Taylor SR, McLellan SM (1985) The continental crust: its composition and evolution. Blackwell, Oxford
Troll G (1964) Igneous and metamorphic rocks in the southern Bavarian Forest. Fortschr Mineral 52: 167–194
Van Breemen O, Aftalion M, Bowes DR, Dudek A, Mísař Z, Povondra P, Vrána S (1982) Geochronological studies of the Bohemian massif, Czechoslovakia, and their significance in the evolution of Central Europe. Trans R Soc Edinb Earth Sci 73: 89–108
Vejnar Z (1973) Petrochemistry of the Central Bohemian Pluton. Geochem Geochem Meth Data Prague 2: 5–116
Vejnar Z (1974) Trace elements in rocks of the Central Bohemian Pluton. Věstn Ústřed Ústavu Geol (Bull Geol Surv Prague) 49: 29–34
Vrána S (1990) The Pelhřimov volcanotectonic circular structure. Věstn Ústřed Ústavu Geol (Bull Geol Surv Prague) 65: 143–156
Waldmann L (1951) Das ausseralpine Grundgebirge Österreichs. In: Schaffer FX (ed) Geologie von Österreich. Franz Deuticke, Wien, pp 10–104

VII.C.4 Metamorphic Evolution

S. Vrána, P. Blümel, and K. Petrakakis

1 Introduction

The Moldanubian zone comprises a heterogenous assemblage of amphibolite and granulite facies crustal units intersliced with upper mantle rocks. The abundance of peridotite, of both spinel and pyrope types, eclogites, and garnet pyroxenites indicates that the region is an anomalous crustal domain and points to an efficient crust-mantle interaction. Certain granulites still retain the HP record (P = 12–15 kb), although later retrogression is widespread. Variscan granitoid intrusions were preceded and accompanied by regional LP periplutonic metamorphism and migmatization which, as a late process, results in a LP overprint dominating the regional pattern of metamorphic zoning (Fig. 1).

The essential features of the metamorphic zoning are shown in the metamorphic map of Czechoslovakia (Dudek et al. 1973; Cháb and Suk 1977) and other publications. There is no simple relationship between the regional pattern of metamorphic zonation and the main structural units. This is due mainly to a complex interplay of polymetamorphism and polyphase deformation. To a large extent, the metamorphic zoning shown in Fig. 1 can be taken as a picture of cumulative, superimposed effects. The record of older phases of metamorphism is often obliterated owing to high deformation rates of structures superimposed under amphibolite facies conditions within regional shear zones.

The main lithostratigraphic and tectonostratigraphic units, comprising the Gföhl unit and the Monotonous and Varied Groups, are characterized in the stratigraphy chapter.

While numerous petrological studies from the last three decades remain valuable, it should be noted that many conclusions must be modified in the light of the new geochronological, thermobarometric and tectonic data of the last few years:

1. Sm-Nd and U-Pb work plus reevaluation of the geology and petrology shows that the crustal rocks of the Moldanubian zone are heterogenous. Segments of older crust are characterized by magmatic zircon around 2050 Ma old; these include the Světlík orthogneiss and minor segments SW of České Budějovice, and probably the extensive Dobra gneiss in Austria (Wendt 1989; Kröner et al. 1988; Liew and Hofmann 1988; Fuchs and Matura 1976; Wendt et al. 1993). The assemblage Hbl + Bt + Pl + Qtz + Kfs + minor magnetite, allanite predominant in these orthogneisses has not been dated.
2. The age of MP-MT regional metamorphism of the Monotonous and Varied Groups is unresolved. It was generally interpreted as Cadomian because unmetamorphosed Lower Paleozoic rests transgressively on folded and phyllitized Upper Proterozoic in the Teplá-Barrandian area. Using Rb-Sr data, Fuchs and Matura (1976) interpreted the regional metamorphism of the major units in Austria as Caledonian. Geochronological information on the age of the granulite metamorphism and crystallization of upper mantle rocks combined with the relative chronology of deformation and recrystallization in gneisses point to an extensive Variscan regional metamorphism in Bohemia and Moravia (van Breemen et al. 1982 and others). The significance of Caledonian and Cadomian (Pan-African) metamorphism remains to be constrained by future geochronological work.

2 Early Metamorphic Evolution

High-grade metamorphism, migmatization, and polymetamorphism accompanying polyphase deformation (Vrána 1979) have minimized preservation of early mineral assemblages. Rare Mn-rich gneisses (~5 wt% MnO) bordering amphibolite near Tábor and SW of České Budějovice contain Grs-Sp-Alm garnet with a

Dallmeyer et al. (Eds.)
Pre-Permian Geology of Central and Eastern Europe

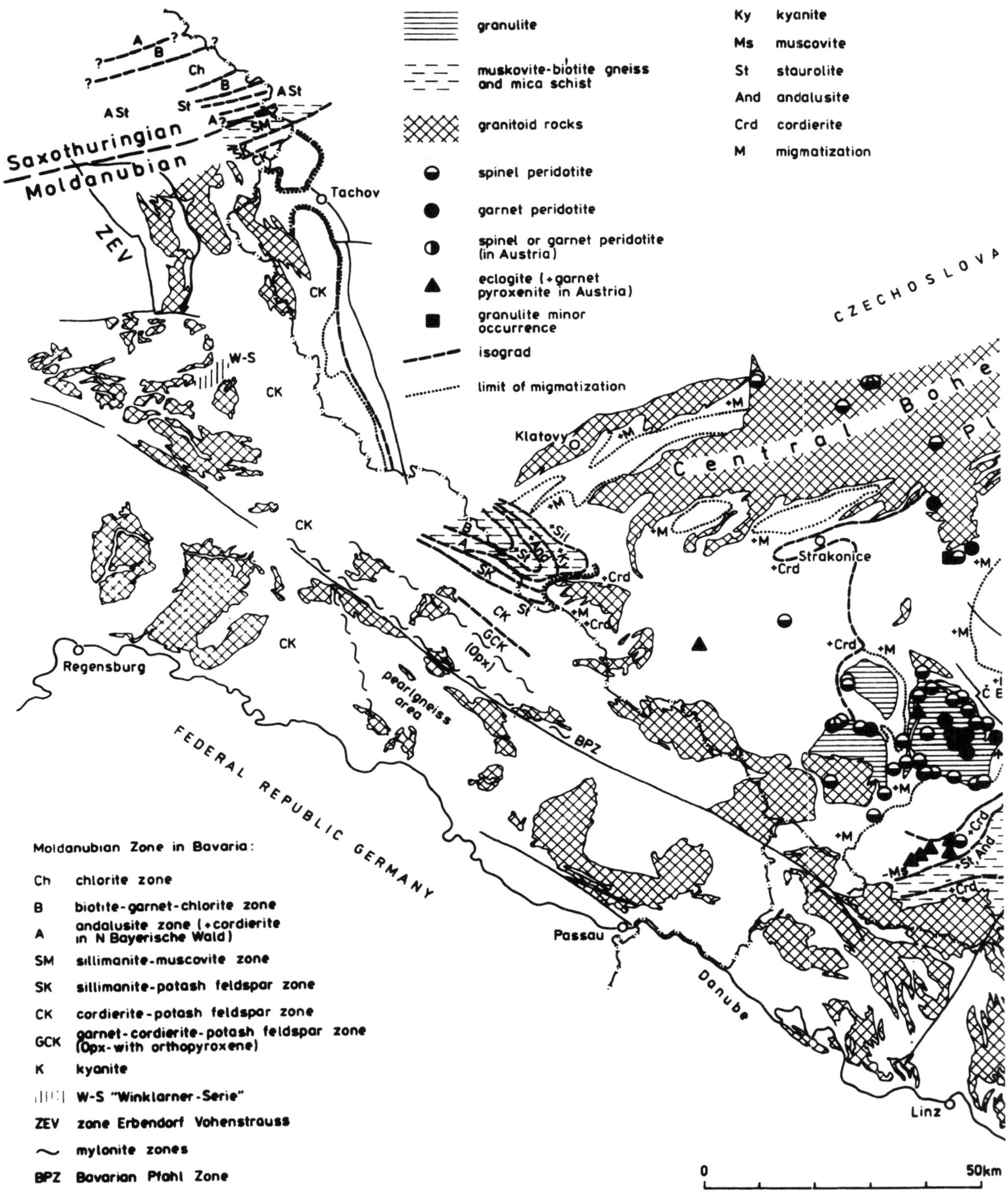

Fig. 1. Regional pattern of metamorphism in the Moldanubian Zone. After data from Dudek et al. (1973) the Geological map of Czechoslovakia 1 : 500 000, and Machart (1984, modified). The information for the Bavarian region of the Federal Republic Germany was compiled by P. Blümel

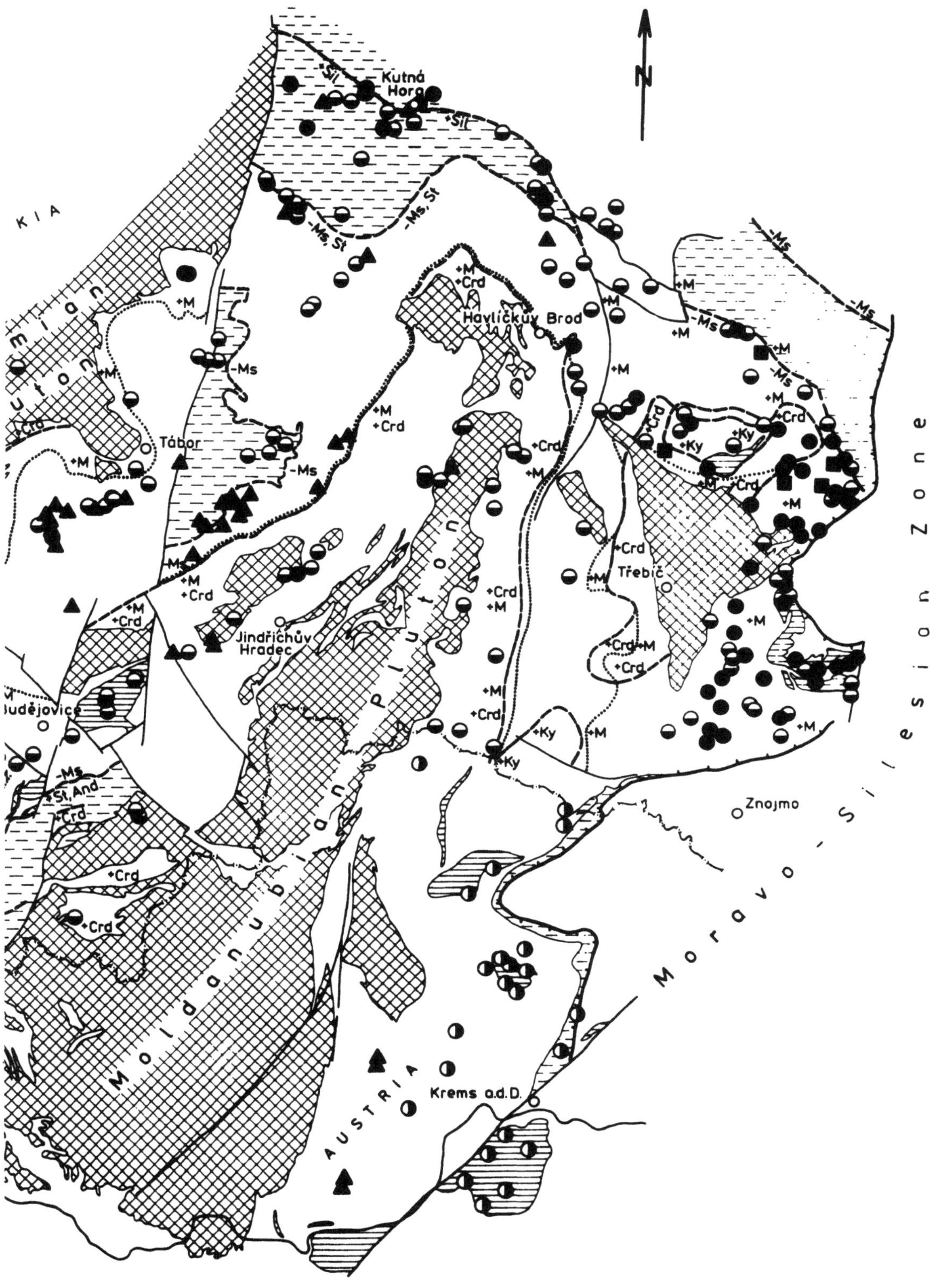
Kutná Hora
Havlíčkův Brod
Tábor
Jindřichův Hradec
Třebíč
Znojmo
Krems a.d.D.
AUSTRIA
Moldanubian Pluton
Moravo - Silesian Zone
+M
+Crd
-Ms
-Ms, St
+Sil
+Ky
St, And

delicate zoning pattern which probably survived from an early metamorphism. An early kyanite-zoisite assemblage in Al-rich bands in marble at Chýnov near Tábor points to $P_{min} = 6\,kb$. A later reaction to Pl + corundum took place.

Rare inclusions of chloritoid in garnet occur in muscovite-biotite gneiss SE of Klatovy (Blümel and Schreyer 1976; Baburek, unpubl.). Johan et al. (1990) report garnets carrying chloritoid /XMg = 0.30/+St I + paragonite + Ru + corundum and St + Ky + white mica + Bt + Pl + Qtz inclusions indicating 6.4 kb at 600 °C from micaschists at the base of the Moldanubian nappe along the Moldanubian/Moravian junction. In short, relics from low grade metamorphism are rare and localized. They suggest early MP/LT-MT conditions – a characteristic not necessarily valid for the whole Moldanubian zone.

Some orthogneisses can be interpreted as pre-Variscan intrusive granitoids. Relics of contact metamorphism accompanying intrusion, obscured by later regional metamorphism, may be present, but no convincing example has been found.

3 The Moldanubian Zone in the Czech Republic

S. Vrána

Metasedimentary gneisses are widespread and important in the delimitation of regional metamorphic zones:

1. muscovite-biotite gneiss (± Ky, Sil, And, St, Grt),
2. biotite and sillimanite-biotite gneiss,
3. cordierite-biotite gneiss and migmatite.

3.1 Muscovite-Biotite Gneisses

There are several belts of muscovite-biotite gneiss passing to sillimanite-biotite gneiss (Sil + Kfs zone). In the Chýnov-Mladá Vožice area E and NE of Tábor, muscovite-biotite gneisses (± Ky, Sil, St, Grt) show early generation fold structures. Sillimanite is the most common Al-rich accessory mineral (Cháb and Suk 1977). These rocks pass to sillimanite-biotite gneiss with minor Ms. The continuity of lithological layers and early deformation structures indicate that the low grade is a primary feature.

Muscovite-biotite gneisses (± Ky) in the Kutná Hora-Svratka area constitute the N and NE part of the Moldanubian zone (Losert 1967). The Ms + Bt + Ky assemblage occurs imprinted in metasedimentary gneisses, migmatites, and old granitoids transformed to orthogneisses. Ms-Sil zone gneisses show a transition to Sil + Bt gneisses. With reference to previous migmatization and granite intrusion, the assemblage Ms + Bt + Ky is a superimposed one. In the Svratka area (Němec 1968) the eastern minus Ms isograde marks the conventional border of the Moldanubian zone.

Muscovite-biotite gneisses in the Kaplice unit in S Bohemia extend 60 km alongstrike structurally above the Moldanubian Pluton. The NW-dipping monoclinal foliation is a late structure (S3). It overprints D1 structures along margins and elsewhere gneisses carry few relics of D2 structures. These relations indicate that Ms-Bt gneisses in this unit are retrograde rocks in a regional shear zone (Vrána 1979). Kyanite is preserved >2–3 km away of the Moldanubian Pluton and St, And, and Crd occur within 2 km of the granite contact. Early fibrolitic Sil I aggregates in well-aligned Ms reflect the retrogression history, and late fibrolitic Sil II replaces both Ky and And. The sequence of Al_2SiO_5 polymorphs is Sil I-Ky-And-Sil II-Sil III. The last type, Sil III, is prismatic and confined to the innermost contact. A superimposed crenulation cleavage S4 is accompanied by recrystallization of Ms + Bt + Chl + Pl + Qtz (M4) imprinted on the contact metamorphic products. Intercalations of quartzitic calc-silicate gneiss carry M3 actinolite ± clinozoisite, garnet, instead of pyroxene.

The muscovite-biotite gneiss area S of Klatovy on the Czech-FRG border shows a zoning sequence St-And-Ky-Sil on the NE side (Vejnar 1991) and a sequence from Bt zone to Crd + Kfs zone in the southern part (Blümel and Schreyer 1976).

3.2 Biotite and Sillimanite-Biotite Paragneisses

These rocks are widespread in large areas of Bohemia and Austria, in both the Monotonous and Varied Groups. The simple assemblage Sil + Bt + Pl + Qtz ± Grt, Kfs, Ms is typical. The minor amounts of Ms or Kfs suggest that both sides of the reaction Ms + Qtz = Sil + Kfs + H_2O may be represented; Ms in some occurrences is secondary. Rock compositions do not allow substantial Kfs. Varied Group gneisses with higher average Ca and Mn contents more frequently carry Grt. Paragneisses in central and S Bohemia exhibit two to three successive deformation-recrystallization events in amphibolite facies (Vrána 1979). This may indicate complexity in reequilibration and in the interpretation of element partitioning.

3.3 Cordierite-Biotite Gneiss and Migmatite

These rocks constitute a large part of the area, mainly around granitoid bodies of the Moldanubian Pluton (Fig. 1). Mineral assemblages witness to the decompression reaction Sil + Bt + Qtz = Crd + Kfs + H_2O or Crd + melt for which Holdaway and Lee (1977) obtained T ~650 °C and P = 3–4 kb. Besides some homogeneous gneisses which already contain this assemblage, migmatitic rocks are dominant. They show differentiation to stromatic-phlebitic up to nebulitic structures; Crd may occur in melanosomes and in many cases also in leucosomes, developed in part as subhedral-euhedral crystals. In some subareas, not indicated separately in Fig. 1, a somewhat higher T led to the reaction Sil + Bt + Qtz = Grt + Crd + melt. Some cordierite-biotite migmatites contain a disequilibrium assemblage of up to ten minerals (Zoubek 1927). Increased oxygen activity in the Bohemian-Moravian upland is indicated by minor magnetite.

A distinct metamorphic zoning shown by mineral assemblages in dolomitic marble intercalations in gneiss and migmatite of the NE part of the Moldanubian zone was defined (Novák 1989):

M1 (older metamorphism): Chl + Tr + Cal + Dol.

M2 (LP, Variscan): zone A: Phl + Tr + Cal + Dol, zone B: Fo + Chl I + Cal±Dol, zone C: Fo + Chl I + Spl + Cal±Chu±Dol, zone D: Fo + Spl + Cal±Chu±Dol. XF = 0.49 in clinohumite and mineral reactions indicate XCO_2 = 0.1–0.4. This zonality, more detailed than is available for gneisses, shows progression toward the Moldanubian Pluton. Low-T, retrogression M3 assemblages are of local occurrence.

3.4 Granulite-Gföhl Gneiss Units

The Gföhl nappe unit in Austria forms a nearly continuous NNE-trending belt in the upper part of a crustal stack. In S Bohemia and W Moravia, isolated, up to 20-km-long granulite massifs showing a complicated record of polyphase deformation (Kodym 1972; Urban and Mísař 1989) are typical. The case of the Saxonian granulite massif (Saxothuringikum) and a contrasting P-T-t record of some granulite bodies in S Bohemia (Vrána 1992) suggest significant autonomy of individual bodies. Slices of granulitic rocks tied to the Monotonous and Varied Groups interface (Fuchs and Scharbert 1979) also point to variation in tectonic position.

Felsic, often banded rocks composed of perthitic Kfs + Pl + Qtz + Ky + Grt ± Bt ± Sil predominate. By increase in Bt + Grt, the dark laminae in the banded sequence may pass to Bt + Grt – rich "granulitic gneiss" of apparent sedimentary derivation. The light-colored felsic portions (up to a few meters thick) in the banded sequence have a corundum-normative calc-alkaline leucogranite composition. There are also small pods and lenses of pyroxene granulite (granite B and granodiorite composition) and mafic granulite (pyriclasite) mostly corresponding to Qtz-poor granodiorite to monzodiorite which consists of Pl + Cpx + Opx ± Grt ± brown Hbl ± Ti-rich Bt. Pyrope and spinel peridotites, eclogites, and garnet pyroxenites in pod-like masses occur as inclusions of upper mantle derivation. The composition and petrology of granulites were described by Matějovská (1970), Kodym et al. (1978), Fediuková (1978), Vrána and Jakeš (1982), Fiala et al. (1982, 1987), Slabý (1983), Vrána (1989), and others.

High-rate deformation, strong flattening, and elongation dominate the tectonic pattern of the granulites. A traditional descriptive approach emphasized a laminated, tectonized granulite as a prototype, central to the solution of the granulite problem (Fiala et al. 1982; Zoubek in Zoubek et al. 1988). The recent search for unfoliated portions preserving early evolution stages produced finer-grained Grt + Ky hypersolvus leucogranite as a protolith to felsic granulite (Vrána and Jakeš 1982; Vrána 1989) and as a mobilisate component of felsic migmatites which evolved to banded granulites through advanced tectonic transposition. Grt + Ky hypersolvus leucogranite exhibits a stable assemblage of Ky +

mesoperthite, indicating the former coexistence of Ky + ternary feldspar ($\sim An_6Ab_{45}Or_{49}$) corresponding to T ~900 °C and P_{min} 12 kb. Such examples correspond to HP crystallization products from dry melts as explored by experimental studies on granite systems.

Garnetiferous perpotassic granulites (~12 wt% K_2O) from Plešovice, S Bohemia (Vrána 1989), and K-rich (5–7 wt% K_2O) varieties in the Saxonian granulite massif (Behr 1961) indicate the existence of HP calc-alkaline leucogranite-perpotassic alkali-feldspar syenite differentiation series as a consequence of noneutectic partial melting/crystallization. The high viscosity of dry HP melts probably hindered extensive occurrence of this differentiation.

Relatively abundant P-T estimates of granulite crystallization span a wide range, depending on the approach (Scharbert and Kurat 1974; Fiala et al. 1982; Carswell et al. 1989; Petrakakis 1991). The information on granulites in S Bohemia shows three prominent P-T regimes recorded on the general decompression/cooling P-T-t path (Table 1).

The origin of felsic granulites by dynamothermal metamorphism of thick volcanoclastic accumulations was long advocated (Zoubek et al. 1988). Recently, this interpretation became obsolete in the light of the data outlined below. Indications of HP/HT regimes early in granulite evolution make partial melting of nearly dry felsic rocks an obvious consequence. The search for ultimate protolith of felsic granulites leads to Grt + Ky hypersolvus leucogranite, not rhyolite. U-Pb isotopic studies and petrology point to the Variscan magmatic crystallization of zircon, particularly in perpotassic granulites. Zircon crystals up to 4 mm long in Kfs matrix and enclosed in garnets of granulitic composition plot on the concordia (Aftalion et al. 1989). Zircon grain-size fractions and a single crystal evaporation study by Wendt (1989) also point to granulite crystallization near 340–350 Ma and suggest igneous protolith rocks which preceded granulite crystallization only by ~10–20 Ma. A mineral and whole-rock Sm-Nd isochron on a mafic granulite suggest an age of 340 ± 20 Ma (Wendt 1989).

The Gföhl gneiss includes mainly migmatitic paragneiss, advanced migmatites to banded/folded leucocratic migmatites and rocks of granitic composition with nebulitic structure. Small amounts of Grt, Sil, or Ky besides Bt are typical; mesoperthite, Ru, Ilm, or late Ms may be present. The Gföhl gneiss, like granulites, often shows a variety of superimposed deformation fabrics. Amphibolite, graphite quartzite, and upper mantle rocks are also present. Dudek et al. (1974) interpreted the Gföhl gneiss in W Moravia as largely produced by recrystallization of felsic granulite. New data on the role of partial melting in evolution of felsic granulites suggest that Gföhl migmatites evolved probably in domains close to those of felsic granulites but with a limited separation of partial melts.

3.5 Peridotites and Eclogites

Peridotite and eclogite bodies ranging from <1 m to several-km-long lenticular masses are documented from more than 300 occurrences in Bohemia and W Moravia (Weiss 1966; Machart 1984). A significant number of localities is also known in Austria and few in Bavaria (Fig. 1). Eclogites occur as layers in peridotite, as isolated bodies in gneiss and migmatite loosely accompanied by peridotite, and as completely isolated lenses in gneiss. Deep boreholes (Sebířov, Holubov – Kodym et al. 1978) show that serpentinization of peridotite is a near surface process.

Table 1. Stages of metamorphism recorded in granulites

Facies	Assemblage	T °C	P, kb	Reference
Eclogite,	Grt + Ky + Pl + Qtz	~800	13–15	Pin and Vielzeuf (1988)
HP granulite	Ky + Grt + ternary Fs in felsic granulites	~900	>12	Vrána (1989)
Pyroxene granulite	Opx + Grt + Pl + Qtz in charnockitic granulites	~700 to 800	5.2	Unpubl. (Lišov massif)
Amphibolite facies retrogression	Pl + common Hbl, andalusite in felsic granulites	<600	<4	

Four types of occurrence of ultramafic rocks are distinguished in the Moldanubian zone (Machart 1988):

1. Garnet peridotites with eclogite layers and spinel peridotite, occurring in granulites and the Gföhl gneiss
2. Metaperidotite of spinel peridotite facies, sometimes accompanied by eclogite, in paragneiss and migmatite
3. Olivine-clinochlore-tremolite/actinolite schist as stratiform layers within large amphibolite bodies of the Varied Group. These rocks are probably cogenetic with mafic volcanites.
4. Ultramafic cumulates (pyroxenites, olivine pyroxenites to peridotites) within differentiated gabbro-peridotite and gabbro-norite intrusions.

Peridotites and metaperidotites of the first two groups correspond to Alpine-type peridotites of the mixed lherzolite-harzburgite type according to their tectonic fabric, residual composition, content of transition group elements, and composition of Al-Cr spinels (Machart 1988). The peridotites in granulites have preserved the primary assemblages in many places, sometimes readjusted to HP-HT granulite/eclogite facies conditions (Frýda and Rieder 1990). The phase transition of spinel to garnet was affected by whole-rock chemistry; some interlayered rocks such as pyropic garnetites and many eclogites contain garnet crystallized directly from melt. Dunites usually contain Cr-spinel only; both spinel and Grt and contained in lherzolites (Machart 1988). Peridotites in paragneiss, migmatite, and the Gföhl gneiss suffered amphibolite facies retrogression. Peridotites and eclogites of the Grt peridotite facies were originally equilibrated at T = 1000–1150 °C and P = 25–35 kb, spinel peridotites and associated eclogites at 800–950 °C and ~20 kb (Machart 1988; Fediuková 1989). Modern studies (Medaris et al. 1990) point to diversity in the P-T-t evolution and emplacement history of some contrasting Grt peridotite bodies in the granulite and Gföhl gneiss complex of W Moravia (Table 2).

Table 2. Thermobarometry and cooling rates for some Moldanubian garnet peridotites. (Medaris et al. 1990)

Locality	Host rock	Thermobarometry (mineral cores) (°C – kb)		Diffusion modeling cooling rates
Nové Dvory	Gföhl gneiss	1195	54.6	Initial 50–200 °C/Ma
		to 970	41.8	Later 10–50 °C/Ma
Mohelno	Felsic granulite	1265	27.9	Initial 1 °C/ year
		1070	23.4	Later 500 °C/ Ma

These authors suggest that the Nové Dvory body was emplaced into the crust with its garnetiferous assemblage intact and that the Mohelno body was emplaced into thickened lower crust as a hot slab of Spl peridotite in which Grt formed only along contacts as the body cooled in the Grt stability field. Some major peridotite bodies in the Blanský les granulite massif, S Bohemia, exhibit repetitive layering of Spl peridotite and Grt peridotite complicated by interlayers of eclogite, garnet clinopyroxenite, websterite, and garnetite with frozen drops of former Fe-Ti-O liquid (ilmenite + magnetite + spinel I).

Eclogites associated with peridotites and gneisses vary widely in composition from magnesian (strongly Prp garnet) to transitional (Alm-Prp garnet) and common (Grs – Prp – Alm garnet) (Dudek and Fediuková 1974). The prevailing common eclogites correspond to basalts and/or mafic cumulate, transitional and magnesian show ultramafic cumulate features. Compositions range from ne- to hy-normative and define a tholeiitic AFM trend. REE patterns are relatively flat to slightly LREE depleted at about 2–15 times chondrite abundance (Beard et al. 1989). Carnet pyroxenites (Scharbert and Carswell 1983) and websterites have low alkalis and high MgO giving rise to Prp-rich garnet and Jd-poor Cpx, while the common eclogites contain omphacite. The regional chemographic survey of Dudek and Fediuková (1974) shows a variation in $K_{D_{Fe-Mg}}^{Grt-Cpx}$: in the belt NW of the Moldanubian Pluton K_D = 3.48–5.63 with most values >4.0, while in the SE belt (W Moravia) the range is 2.01–4.15 with most values <3.0. The equilibration temperatures were thus higher in the SE belt (O'Brien et al. 1990). Sm-Nd geochronology on Grt-Cpx pairs from Bohemia yielded 324 and 338 Ma (Beard et al. 1989), 340–370 Ma from Austria (Carswell and Jamtveit 1990) and 374 Ma on Grt clinopyroxenite from Nihov, W Moravia (Brueckner et al. 1989). This indicates a Variscan age for the HP assemblage which evolved directly from upper mantle melt as cumulates with up to 30% trapped melt (Beard et al. 1989).

Besides the prevailing population of Variscan eclogites, there are some older bodies, such as eclogite in the allochtonous Winklarn unit near Oberviechtach in Bavaria (O'Brien 1989; von Quadt and Gebauer 1993). In the newly discovered Světlík eclogite belt in an upper structural level of the Monotonous Group in S Bohemia, prevailing eclogites of the ordinary Moldanubian types and Px-Grt hornblendites are accompanied by two occurrences of eclogite with a well-defined positive Eu anomaly, an absence of HREE enrichment, and an apparent crustal history as a protolith (O'Brien and Vrána, in press).

Peridotites and eclogites in the Moldanubian zone have been emplaced in a solid state as fragments of subcontinental upper mantle into crustal complexes by plastic flow and movements along tectonic zones. They are preferentially accumulated in the granulite and the Gföhl gneiss. In this environment, the evolution stage of felsic granulites as viscous dry melts under HP at the crust-mantle boundary facilitated an advanced dismembering of upper mantle rocks. High deformation rates following crystallization of felsic granulites also contributed to the highly tectonized mode of occurrence. Some Spl peridotites and minor granulite bodies have been emplaced along ductile shear zones. These processes result in a concentration of mantle rocks in a number of belts (Fig. 1; Machart 1984, 1988).

4 The Moldanubian Zone in Austria

K. Petrakakis

Scharbert (1963, 1964) carried out an extensive field study and petrographic examination of the Moldanubian granulites. He estimated the conditions of their equilibration at 525–710 °C and 8–10 kb. Scharbert and Kurat (1974) revised these estimates to 760 °C and 11 kb.

Zaydan and Scharbert (1983) compared gneisses and amphibolites from the southernmost parts of the Monotonous and Varied Groups. They defined two metamorphic zones, the Crd + Kfs zone, formed under 630 °C and 3 kb, and the Sil + Kfs zone formed under 670 °C and 5 kb, respectively (Zaydarn and Scharbert 1983, p. 182), in spite of the fact that the Grt + Cpx amphibolites of the Varied Group systematically yield higher metamorphic temperatures (~750 °C, p. 196).

Hödl (1985) studied the conditions of metamorphism of amphibolites and subordinate calc-silicate rocks at the base of the Gföhl unit (amphibolites of Rehberg). These rocks are commonly characterized by the assemblage Grt + Cpx + Pl + Hbl (1), Hbl + Pl + Cpx (2), Cpx + Scp + Pl + Cal (3), and Scp + Pl + Ep + Cpx + Hbl (4), suggesting a two-stage metamorphism. Temperature estimates of the older event spreading between 640 and 750 °C are based on the assemblage (1), commonly covering the upper T range, and (2). Lower temperature estimates of ~630 °C result from the overprinted assemblages (3) and (4).

These results may be compared with those of Kuschnig (1986), who described amphibolites typically found at the top of the Varied Group and immediately adjacent to the base of the Gföhl unit. Kuschnig distinguishes two metamorphic events mainly on the ground of a relict assemblage, such as (1) above, being partly or totally replaced by Act + Na–Pl. Temperatures suggested are 750 °C for the older and 530–590 °C for the younger metamorphic event. He also emphasizes the wide variation in the calculated T for the older event (650–800 °C).

Petrakakis (1986) studied metapelitic and quartzofeldspathic gneisses from the southern of part of the Varied Group. These rocks belong to the HT-facies type Grt + Sil + Bt + Kfs and show subordinate migmatitic features. They locally contain variable amounts of graphite. Ky is found as relict inclusions in Grt together with minute Ms, Qtz, and Kfs. Varying amounts of secondary Ms, forming neoblasts cross cutting the older textural elements and Chl are very common. The biotites are observed as inclusions in Grt, blasts adjacent to Grt, and as matrix crystals showing marked variation in size, composition, and alignment. These observations point to a late stage of the metamorphic evolution which is best characterized by cooling and partial rehydration.

The garnets are completely homogenized and show a typical cooling pattern at their rims (rapid decrease in XMg, steep increase in XMn within the outer 70–100 μm). This pattern is usually cross cut by resorption. Some garnets of quartzofeldspathic rocks exhibit relict, unusual Ca-rich cores which are enclosed in homogenized, Ca-poor garnet showing the cooling features described. The Ca-rich cores may represent a pre-HT growth pointing to an older stage of the metamorphic evolution.

The textural and paragenetic relationships (especially the inclusions of Ms, Qtz, Kfs, and Ky within Grt–Petrakakis 1986, Fig. 8) and geothermobarometric data suggest a prograde decompressive path of metamorphism attaining final conditions of probable equilibration as high as 700–770 °C and 7–9 kb at $a_{H_2O} << 1$. Preliminary calculations of a_{H_2O} under these conditions based on the equilibrium Ann + Sil + Qtz = Alm + Kfs + H_2O support the low a_{H_2O} values (Petrakakis 1988).

The rocks bound to the boundary between the Monotonous and Varied Groups show a typical granulitic appearance ("Granulite lamellae" of Fuchs and Scharbert 1979) and bear the same characteristic assemblage (Grt + Sil + Bt + Kfs) thus yielding P-T estimates consistent with other rocks of the Varied Group. Nevertheless, they show evidence of extensive deformation and partial recrystallization. On the other hand, scarce amphibolitic pods within the blastomylonitic gneisses show Opx + Pl symplectites partially replacing Grt. The same textural relations are observed in Grt amphibolites from the southernmost part of the Varied Group (Loja quarry). This feature suggests a rapid uplift and decompression. Both the blastomylonitic textures in the marginal granulitic gneisses and the Opx + Pl symplectites around Grt point to a major deformation event shortly after or during peak metamorphic conditions. Based on geothermobarometric, textural, and paragenetic data (especially the prevailing instability of Crd after uplift and decompression), Petrakakis (1986) suggests 530–630 °C and 2–4 kb as possible conditions of partial reequilibration and rehydration of the rocks of the Varied Group.

Högelsberger (1989) presents the results of investigation of marbles and calc-silicate rocks of the Varied Group which are intercalated with Grt + Cpx gneisses. The first two rock types are represented by varying mineral contents including Di, Tr, Phl, Scp, Pl, and Kfs. Högelsberger also describes samples bearing serpentinized Ol and Dol with Di, Tr, and Phl. He regards this as an invariant assemblage at constant P, suggesting HT conditions at some stage of the metamorphic evolution of the Varied Group. P-T estimates for this stage (700 °C and 7 kb) are mainly based on geothermobarometry of Grt + Cpx + Qtz + Pl + Kfs + Hbl + Bt-bearing gneisses and amphibolites from the Varied Group and the Gföhl Unit. Högelsberger emphasizes the equivalence of P-T estimates in both units and suggests T as high as 500 °C for the late stage of metamorphism mainly based on Cal-Dol geothermometry, fluid inclusion studies, and phase equilibria in the calc-silicate rocks and marbles of the Varied Group.

Petrakakis and Richter (1991) examined the paragenetic relations in the Px-free granulites and the Gföhl gneiss. Besides obvious differences in deformation grade, both rock types show the same characteristic assemblage as the gneisses of the Varied Group, i.e., Grt + Bt + Sil + Kfs. Similarly, Ky is metastable in respect to Sil and minute Ms, Kfs, and Qtz crystals are included in some garnets. These rocks show similar post-peak phenomena as described above.

Petrakakis (1991) used generalized consistent geothermobarometric methods in a Grt + Opx + Bt + Ru + Ilm gneiss and a Grt + Cpx + Ilm + Ru amphibolite, both belonging to the Gföhl Unit, and estimated the P-T-conditions of equilibration at 700–750 °C and 8.3 kb.

The ultramafic and ultrabasic rocks associated with the granulites and the Gföhl gneiss early drew the attention of geologists because of their significance in respect to the tectonometamorphic evolution of the Moldanubian crystalline complex in Lower Austria. Grt peridotite rocks are intensively serpentinized and intimately associated with Grt pyroxenites, which form layers of variable thickness of up to 40 cm or tectonically disrupted lenses and bodies of irregular shape. According to Scharbert and Carswell (1983) these rocks show Ol-tholeiitic to picritic compositions and now comprise Grt websterites (assemblages:Grt + Cpx + Opx, Grt + Cpx + Opx±Ol±Spl±Amp±Mg-rich Ilm) and Grt clinopyroxenites (assemblage: Grt + Cpx±Phl). The garnets attain X_{prp} values as high as 0.75, clinopyroxenes are al-augites or Cr-diopside carrying Opx exsolution lamellae. The rocks generally exhibit partial reequilibration under granulite facies conditions in accordance to the surrounding rocks, and are considered as crystallized products of originally solidified upper mantle partial melts. Carswell et al. (1989) observed Spr + Pl symplectites formed approximately at 1000 °C and P $\leqslant$ 20 kb by the breakdown of a deduced Ky-bearing eclogite originally formed under 1100–1150 °C and 30 kb. Carswell (1991) generalizes his petrologic and geochronologic data and proposes a six-stage P-T-t evolution of the Moldanubian Zone of Lower Austria.

The data presented above provide some insight into the metamorphic evolution of the Moldanubian complex of Lower Austria, but are still insufficient to answer many questions put especially by the recent geochronologic data. Most of the petrologic data point to a multiphase evolution of the Moldanubian Zone in Austria.

As regards the Monotonous Unit, which undoubtedly exhibits a distinct tectonic relation to the other units of the Moldanubian Zone, preliminary data (Linner, oral comm., Inst. Petrol. Univ. Vienna) most probably lead to a reevaluation of its conditions of metamorphism.

Despite earlier interpretations of these rocks (Zaydan and Scharbert 1983), Linner exphasizes the late formation of white mica through probable extensive rehydration on the ground of field and microscopic data. Thus the metamorphic conditions exceeded the stability field of Ms + Qtz and are characteristic of Crd + Bt + Sil + Kfs facies types. Relict Ky and primary Ms inclusions in Crd and Pl observed in some metapelitic samples are potential phases predating peak metamorphic conditions. This interpretation is consistent with the new promising observation of relict Ky and St in a Grt + Sil + Bt gneiss from the Monotonous Group.

The available data on the metamorphic conditions of the Varied Group clearly demonstrate T above ~700 °C and P of 7–9 kb. The apparent spread of the estimated values as a whole most probably lies within the uncertainities imposed by the analytical and geothermobarometric methods used. The sole exception are the significantly lower P-T values derived for gneisses by Zaydan and Scharbert (1983) which apparently are inconsistent with their values obtained from amphibolites. The current data on the Varied Group suggest a reduced a_{H_2O}, at least in rocks containing graphite. This is consistent with the small-scale migmatitic features shown by most of the rocks.

The pressures suggested by Scharbert and Kurat (1974) for the granulite formation are constrained by the tripple point of Althaus (1969) and are, therefore, about 2.5 kb higher than the value obtained from other rocks of the Gföhl Unit by means of generalized consistent geothermometry (Petrakakis 1991). The equilibration temperatures of 700–750 °C are in an excellent agreement. Carswell (1991, Fig. 10) suggests 750–800 °C and 12–15 kb for the granulite facies metamorphism (his stage IV), but, as he observes (pp. 323, 332), the mineral phases used might have failed to attain equilibrium. Preliminary investigations of fluid inclusions in quartz of a Grt + Opx + Bt granulite (Jawecki 1991, pers. comm., Inst. Petrol. Univ. Vienna) suggest two different types of inclusions trapped during different metamorphic stages: type I inclusions are composed of H_2O + $CO_2 \pm CH_4$ and show CO_2 homogenization T between −4.1 and + 28.6 °C. Type II inclusions are much more abundant and represent low salinity (7.5 wt% NaCl) aqueous fluids. Two inclusions of type I show exceptional densities higher than 0.95 g/cm^3 and yield isochores that pass near the P-T estimates suggested by Petrakakis (1991).

Striking is the approximate of the P-T estimates derived from rocks of both the Varied Group and the Gföhl Unit. This is consistent with the notable equivalence of the assemblage Grt + Sil + Bt + Kfs ± relict Ky. A distinction between both units on the ground of differing metamorphic conditions seems to be ambiguous. Thus a probable major boundary between both units either predates this metamorphic stage or postdates the latter, which would indicate only a minor displacement. In the last case this boundary may lie at any place above the lithologic boundary, most probably beneath the Gföhl gneiss-granulite complex (Frank et al. 1991).

Data referenced in the previous section point to late partial recrystallization and rehydration of the rocks in the Varied Group and the Gföhl Unit. Crude estimates spread between 450–650 °C at pressures not higher than 4 kb (see also stage VI of Carswell 1991).

5 The Moldanubian Zone in Bavaria

P. Blümel

In the Bayerische Wald region in Bavaria, the Moldanubian zone is dominated by relatively late, NW-SE-trending structures (Bavaricum). The gneiss region generally was metamorphosed under the P-T conditions of the Crd-Kfs zone (CK, Fig. 1) passing by decompression through the reaction boundary Sil + Bt + Qtz = Crd + Kfs + H_2O or Crd + melt. Using the calibration of Holdaway and Lee (1977), a confining P of 3–4 kb and a T near 650 °C were obtained. The

LP metamorphism is dated as Namurian (320 Ma, monazite, Grauert et al. 1990). Cooling ages vary from 310 to 290 Ma (Rb-Sr, biotite; Grauert et al. 1974). Mineral inclusions of Sil within the cores of cordierite grains suggest – in the presence of other reactant phases nearby – that this reaction did not go to completion, indicating a rapid uplift of the gneiss region. This agrees with the anomalous thermal gradient of LP regional metamorphism extending widely through the mid-European Variscan.

Pelitic paragneisses are differentiated into coarse-grained stromatitic or phlebitic leucosomes of tonalitic or granitoid composition with Bi and/or Crd and medium-grained melanosomes with Crd / + Sil/ + Bt at reduced amounts of feldspars and quartz. Progressive partial melting leads to the formation of Qtz-deficient melanosomes with Spl indicative of the reaction Sil + Bt = Crd + green Spl + melt. Nonmigmatitic paragneiss locally may contain rounded aggregates with fibrolitic Sil + Qtz (Sillimanit-flecken-gneiss, Oberpfalz). Hornblende gneisses may show tonalitic flecky leucosomes with Hbl + Bt + andesine + titanite + allanite (e.g., Zwiesel). Siliceous marble may contain wollastonite + diopside + garnet or forsterite (+ vesuvianite) (+ clinohumite) (Arber, Rauthmannsdorf, Donauleite). Folded and disrupted quartz veins are rimmed by serpentinized forsterite giving the impression of worm-like fossils.

Higher metamorphic grades appear in the Grt-Crd-Kfs zone (Cham-Regen; GCK, Fig. 1). This mineral zone results from the passage of the decompression path through the reaction boundary Sil + Bt + Qtz = Grt + Crd + melt. The host rock is a medium-grained garnet-cordierite-biotite diatexite. Using the P-T calibration of Lee and Holdaway (1977) for localities on the Grt-Crd-Kfs isograd, confining pressures of 4 kb and T of ~690 °C are obtained (composition of "invariant" cordierite: XMg = 0.356–0.517). Other garnet-cordierite gneisses consist of a granitoid leucosome with Grt + Crd + Sil and a restitic melanosome with Crd + Sil + Spl (green/ + Bt/ + Pl). The garnets within this melanosome are mantled by cordierite indicating a decompression reaction of the type Grt + Sil → Crd + Spl. An interesting "metasomatic" reaction is the replacement of melanosome-cordierite by both feldspars preserving the sillimanite inclusion trails of the reactant. This could be due to a granitoid partial melt which precipitated feldspar as a liquidus phase when dissolving cordierite. In ferroan and less aluminous rocks, Opx + Grt + Kfs assemblages are stable ("Opx", Fig. 1). Hypersthene (<3.66 wt% Al_2O_3) and garnet form large crystals (<1/2 cm) in the granitoid leucosome but are intimately intergrown in the melanosome. Less alkalic and more calcic gneisses contain two pyroxenes + plagioclase (+ biotite).

Transitions from the gneiss region toward lower grades are indicated by the sequence: sillimanite zone (SK, SM, Fig. 1) → andalusite-cordierite zone (A, Fig. 1) → biotite-garnet-chlorite zone (B, Fig. 1). Mineral reactions deduced from armored relics are (in addition to the cordierite-K-feldspar reaction):

1. St → Grt + Sil (sillimanite zone and cordierite-K-feldspar zone).
2. Grt + Ms → Sil + Bt + Qtz (giving the well-known biotite-fibrolite intergrowth; sillimanite zone).
3. And → Sil (precluding the direct PT passage from kyanite; sillimanite zone).
4. St + Chl + Ms → And + Bt (andalusite-cordierite zone).
5. St → Crd + And (cordierite rimming staurolite; andalusite-cordierite zone).
6. Chloritoid (XMg = 0.15) + Bt → Grt + Chl + Ms (chloritoid inclusions in garnet of the biotite-garnet-chlorite zone).

Zinc, though present in the educt rocks in traces (320–550 ppm) seems to stabilize Fe-staurolite and spinel in the Bt-Grt-Chl zone, where they occur in a few strongly fractionating grains (e.g., staurolite with 3.2 wt% ZnO).

Part of the gneiss region is reworked to an augenmylonite – a modification of Moldanubian Perlgneis – which contains "porphyroclasts" of feldspars and Crd, Sil, ±And, recrystallized Bt and plastically deformed Qtz (Fig. 1). For Crd pearlgneiss with two Al_2SiO_5 polymorphs and x^{Mg}_{Crd} = 0.425 maximum confining pressures of 3 kb and T_{max} 550 °C at strongly reduced X^{H_2O} (less than 0.5 using Holdaway and Lee 1977) are obtained. These subsolidus conditions are in accord with the absence of partial melting.

The LP gneiss in the NW part of the Moldanubian zone in Bavaria (E Oberpfalz region) is obviously polymetamorphic (kyanite relics) as are the interlayered metabasite (Ky-bearing eclogite-amphibolite, spinel meta-harzburgite, with secondary tremolite), and kyanite granulite, all of which are assembled in an allochtonous unit – the Winklarner Serie (G. Fischer) (Blümel 1986; De Nil 1989; O'Brien 1989; von Gehlen et al. 1991; Fig. 1). Elsewhere, eclogite-amphibolite (without Ky) occurs.

For the Winklarn eclogite a succession from prograde amphibolite facies through the HP stage (15 kb/710 °C), a HP granulite stage (12 kb/800 °C) and MP granulite stage (10–8 kb/800 °C) to the amphibolite facies (of unknown baric affiliation) is analyzed; associated garnet websterite equilibrated even at 28 kb (O'Brien 1991). The HP metamorphism is dated as Silurian (427 ± 5 Ma; von Quadt and Gebauer 1993).

Kyanite granulite, which is actually a former Grt-Ky restite, may have shared a HP stage (40 mol% Prp in Grt) prior to the LP stage (Crd corona with x^{Mg} = 0.73–0.80; Schmidt et al. 1990). The emplacement of the granulite was not later than the LP anatexis, since all petrographic transitions exist from potassium-free granulite to cordierite gneiss.

Most of the Winklarn "exotics" are foreign. Originating from deep lithospheric levels, they are brought by stepwise uplift into juxtaposition with supracrustal material which went down to at least MP levels.

Supracrustals consist of young sediments and Altkristallin; in the Moldanubian area, the former may be found in series with Devonian microfossils and the latter in series yielding Ordovician isochron ages. Further uplift into the LP region was probably countered by down-thrusting of the Paleozoic series of Thüringische Facies with subsequent reorganization yielding a LP zone. These metasediments are monometamorphic, since they lack Ky and pseudomorphs after Ky and instead preserve Sil pseudomorphs after And, as well as bell-shaped Mn zoning profiles in Grt.

In conclusion, it may be appropriate to emphasize the complexity and variability of the polymetamorphism patterns. Radiometric, petrological, and structural data supporting the Neo-Variscan crystallization of many HP granulites, eclogites, and some upper mantle rocks point to a significant crust/upper mantle interaction and stacking of crustal units during the Variscan. The prospect for an extensive correlation of the early and intermediate metamorphic and deformation events (early Variscan and pre-Variscan) is still remote. The recent discovery of intersliced units of a Lower Proterozoic crust and some pre-Variscan eclogites indicates additional complexities. It is mainly the late imprint of (periplutonic) LP Variscan metamorphism which contributes to a deceptive uniformity.

Low-pressure regional metamorphism commonly referred to reheating by delamination, underplating or extensive plutonism. For the Moldanubian zone this may not hold true on account of a crustal thickness of 30–35 km. Then the anomalous hot "cut-off" P-T array may be better explained by a rapid uplift as deduced from the microtexture in the cordierite gneiss region (see above). This event may correlate with the recently published Neo-Variscan eclogite ages and the extensive S-type magmatism.

References

Aftalion M, Bowes DR, Vrána S (1989) Early Carboniferous U-Pb zircon age for garnetiferous perpotassic granulites, Blanský les massif, Czechoslovakia. Neues Jahrb Mineral Monatsh 1989: 145–152

Althaus E (1969) Der Einfluss des Wassers auf metamorphe Mineralreaktionen. Neues Jahrb Mineral Monatsh 1968, 1969: 289–306

Beard BL, Medaris LG Jr, Valley JW, Brueckner HK, Mísař Z (1989) Mantle-derived Variscan eclogites from the Bohemian Massif, Czechoslovakia. 3rd Int Eclogite Conf, Würzburg. Blackwell, Oxford, pp 2–3

Behr HJ (1961) Beiträge zur petrographischen und tektonischen Analyse des Sächsischen Granulitgebirges. Freiberger Forschungsh C 119: 5–64

Blümel P (1986) Metamorphic processes in the Variscan crust of the central segment. Proc 3rd EGT Worksh Bad Honnef, Eur Sci Found, Strasbourg, pp 149–155

Blümel P, Schreyer W (1976) Progressive regional low-pressure metamorphism in Moldanibian metapelites of the northern Bavarian Forest, Germany. Krystalinikum 12: 7–30

Brueckner HK, Medaris LG Jr, Bakun-Czubarow N (1989) Nd and Sr age and isotope patterns from Hercynian eclogites and garnet pyroxenites of the Bohemian Massif and the East Sudetes. Int Eclogite Conf, Würzburg. Blackwell, Oxford, 4 pp

Carswell DA (1991) Variscan high P-T metamorphism and uplift history in the Moldanubian Zone of the Bohemian Massif in Lower Austria. Eur J Mineral 3: 323–342

Carswell DA, Jamtveit B (1990) Variscan Sm-Nd ages for the high-pressure metamorphism in the Moldanubian zone of the Bohemian Massif, Lower Austria. Neues Jahrb Mineral, Abh 162: 69–78

Carswell DA, Möller Ch, O'Brien PJ (1989) Origin of sapphirine-plagioclase symplectites in metabasites from Mitterbachgraben, Dunkelsteinerwald granulite complex, Lower Austria. Eur J Mineral 1: 455–466

Cháb J, Suk M (1977) Regionální metamorfóza na území Čech a Moravy. Ústřed Ústav Geol, Prague, 156 pp

De Nil J (1989) Geologie und Petrogaphie der Eklogitamphibolit-Granulit-Metaperidotit-Anatexit-Serie im Moldanubikum der mittleren Oberpfalz bei Oberviechtach. Dipl Thesis, Ruhr-Universität Bochum, 88 pp

Dudek A, Fediuková E (1974) Eclogites of the Bohemian Moldanubicum. Neues Jahrb Mineral Abh 121: 127–159

Dudek A, Cháb J, Chaloupský J, Kamenický J, Krist E, Mísař Z, Suk M (1973) Mapa metamorfní stavby 1:1 000 000. Ústřed Ústav Geol, Prague

Dudek A, Matějovská O, Suk M (1974) Gföhl orthogneiss in the Moldanubicum of Bohemia and Moravia. Krystalinikum 10: 67–78

Fediuková E (1978) Mafic minerals from granulites of the borehole Holubov (South Bohemian Moldanubicum). J Geol Sci Ser Econ Geol-Mineral Prague 19: 169–198

Fediuková E (1989) Eclogites in Czechoslovakia. Krystalinikum 20: 27–48

Fiala J, Lang M, Obrda J, Pivec E, Ulrych J (1982) Petrology of some garnet-kyanite-K-feldspar leptynites in the Czech Moldanubicum, Czechoslovakia. Rozpr Česk Akad Věd Mat-Nat Sci Ser 97: 102 pp

Fiala J, Matějovská O, Vaňková V (1987) Moldanubian granulites and related rocks: petrology, geochemistry and radioactivity. Rozpr Česk Akad Věd Math-Nat Sci Ser 97: 102

Frank W, Scharbert S, Thöni M, Popp F, Hammer S (1991) Isotopengeologiche Neuergebnisse zur Entwicklungsgeschichte der Böhmischen Masse. Österr Beitr Meteorol Geophys 3: 185–228

Frýda J, Rieder M (1990) Crystallization history of a pyroxenite xenolith in a granulite inferred from chemical and single-crystal X-ray data. In: Spencer RJ, I-Ming Chon (eds) Fluid-mineral interactions: a tribute to H.P. Eugster Geochem Soc Spec Publ 2: 85–102

Fuchs G, Matura A (1976) Zur Geologie des Kristallins der Südlichen Böhmischen Masse. Jahrb Geol Bundesanst 119: 1–43

Fuchs G, Scharbert HG (1979) Kleinere Granulitvorkommen im niederösterreichischen Moldanubikum und ihre Bedeutung für die Granulitgenese. Verh Geol Bundesanst 2: 29–49

Grauert B, Hänny R, Soptrajanova G (1974) Geochronology of a polymetamorphic and anatectic gneiss region: The Moldanubicum of the Area Lam-Deggendorf, Eastern Bavaria, Germany. Contrib Mineral Petrol 45: 37–63

Grauert B, Grosse-Westermann U, Albat F (1990) Interpretation von U-Pb-Monazit-Altern moldanubischer Gesteine. KTB Rep 90-4: 549

Hödl M (1985) Petrologie und Geochemie des Rehberger Amphibolites im Niederösterreichischen Moldanubikum. PhD Thesis, Univ Vienna

Högelsberger H (1989) Die Marmore und Kalksilikatgesteine der Bunten Serie - Petrologische Untersuchungen und geologische Konsequenzen. Jahrb Geol Bundesanst 132: 213–230

Holdaway MJ, Lee SM (1977) Fe-Mg cordierite stability in high-grade pelitic rocks based on experimental, theoretical and natural observation. Contrib Mineral Petrol 63: 175–198

Johan V, Autran A, Ledru P, Lardeaux JM, Melka F (1990) Discovery of relics of a high-pressure metaorphism at the base of the Moldanubian nappe complex. Terranes in the circum-Atlantic Paleozoic orogens. Univ Göttingen, Göttingen, p 2

Kodym O (1972) Multiphase deformation in the Blanský les granulite massif (southern Bohemia). Krystalinikum 9: 91–105

Kodym O, Jakeš P, Schovánek P (1978) Granulite and ultramafische Gesteine aus der Strukturbohrung Holubov. J Geol Sci Geol Ser 32: 7–47

Kröner A, Wendt I, Liew TC, Compston W, Todt W, Fiala J, Vaňková V, Vaněk J (1988) U-Pb zircon and Sm-Nd model ages of high-grade Moldanubian metasediments, Bohemian Massif, Czechoslovakia. Contrib Mineral Petrol 99: 257–266

Kuschnig K (1986) Geochemie und Petrologie der Amphibolite der Bunten Serie. PhD Thesis, Institute of Petrology, Univ Vienna

Lee SM, Holdaway MJ (1977) Significance of Fe-Mg cordierite stability relations on temperature pressure and water pressure in cordierite granulites. AGU Monogr 20 (The Earth's Crust): 79–94

Liew TC, Hofmann AW (1988) Precambrian crustal components, plutonic associations, plate environment of the Hercynian Fold Belt of central Europe: indications from a Nd and Sr isotopic study. Contrib Mineral Petrol 98: 129–138

Losert J (1967) Contribution to the problem of the pre-Assyntian tectogenesis and metamorphism in the Moldanubicum of the Bohemian Massif. Krystalinikum 5: 61–84

Machart J (1984) Ultramafic rocks in the Bohemian part of the Moldanubicum and Central Bohemian islet zone (Bohemian Massif). Krystalinikum 17: 13–32

Machart J (1988) Petrology and position of ultramafic rocks in the Moldanubian region of the Bohemian Massif. In: Zoubek V, Cogné J, Kozhoukharov D, Kräutner HG (eds) Precambrian in younger fold belts. John Wiley, Chichester, pp 233–238

Matějovská O (1970) Composition of coexisting garnet and biotite from some granulites of Moldanubicum, Czechoslovakia. Neues Jahrb Mineral Monatsh 1970: 249–263

Medaris LG Jr, Wang HF, Mísař Z, Jelínek E (1990) Thermobarometry, diffusion modelling and cooling rates of crustal garnet peridotites: two examples from the Moldanubian zone of the Bohemian Massif. Lithos 25: 189–202

Němec D (1968) Die Metamorphose des NE-Randes des Kernes der Böhmischen Masse. Verh Geol Bundesanst Wien 1968: 187–203

Novák M (1989) Metamorphism of dolomitic rocks at the north-eastern margin of the Moldanubicum, western Moravia, Czechoslovakia. Acta Mus Moraviae Sci Nat 74: 7–51

O'Brien PJ (1989) The petrology of retrograded eclogites of the Oberpfalz Forest, NE Bavaria, West Germany. Tectonophysics 157: 195–212

O'Brien PJ (1991) High Pressure Metamorphism in the NW Bohemian Massif: Comparisons and Contrasts between the Moldanubian Zone, Münchberg Massif, ZEV, ZTT and Erzgebirge. KTB Rep 91-1: 1–12

O'Brien PJ, Vrána S. The petrology and geochemistry of eclogites and related rocks from the Světlík area, Moldanubian Zone, Czech Republic. Geol Rundsch (in press)

O'Brien PJ, Carswell DA, Gebauer D (1990) Eclogite formation and distribution in the European Variscides. In: Carswell DA (ed) Eclogite facies rocks. Blackie, Glasgow, pp 204–224

Petrakakis K (1986) Metamorphism of high grade gneisses from the Moldanubian zone, Austria, with particular reference to the garnets. J Metamorph Geol 4: 323–344

Petrakakis K (1988) Hydrobarometrie an metamorphen Gneisen aus der Bunten Serie. Fortschr Mineral 66 (Beih 1): 122 (Abstr)

Petrakakis K (1991) Metamorphosebedingungen in der Gföhler Einheit. Österr Beitr Meteorol Geophys 3: 137–148

Petrakakis K, Richter W (1991) Gföhler Gneise und Granulite aus dem mittleren und südlichen NÖ-Moldanubikum: eine paragenetische und texturelle Gegenüberstellung. Österr Beitr Meteorol Geophys 3: 121–135

Pin C, Vielzeuf D (1988) Les granulites de haute-pression d'Europe moyenne témoins d'une subducion éo-hercynienne. Implications sur l'origine des groupes leptyno-amphiboliques. Bull Soc Géol Fr 1988 t IV: 13–20

Scharbert HG (1963) Die Granulite des südlichen niederösterreichischen Moldanubikums, I. Neues Jahrb Mineral Abh 100: 59–86

Scharbert HG (1964) Die Granulite des südlichen niederösterreichischen Moldanubikums, II. Neues Jahrb Mineral Abh 101: 27–66

Scharbert HG, Carswell DA (1983) Petrology of garnet-clinopyroxene rocks in a granulite facies environment, Bohemian massif of Lower Austria, Austria. Bull Minéral 106: 761–774

Scharbert HG, Kurat G (1974) Distribution of some elements between coexisting ferromagnesian minarals in Moldanubian Granulite Facies Rocks, Lower Austria, Austria. Tschermaks Mineral-Petrogr Mitt 21: 110–134

Schmidt I, Blümel P, Müller W (1990) Phasenanalyse von Granat/Orthopyroxen/Plagioklas/Cordierit-Reaktionsgefügen in Disthengranulit (O. Viechtach). KTB Rep 90-4: 542

Slabý J (1983) Modální složení a petrochemie granulítů lišovského masívu a masívu Blanského lesa v jižnícn Čechách. Čas Mineral Geol 28: 41–60

Urban M, Mísař Z (1989) Fold and shear structure of the southeastern Moldanubicum (Czechoslovakia). Krystalinikum 20: 111–130

Van Breemen O, Aftalion M, Bowes DR, Dudek A, Mísař Z, Povondra P, Vrána S (1982) Geochronological studies of the Bohemian massif, Czechoslovakia, and their significance in the evolution of Central Europe. Trans R Soc Edinb Earth Sci 73: 89–108

Vejnar Z (1991) The metamorphic zonal pattern in the Moldanubicum of the NW part of the Šumava Mts, Královsky hvozd unit. Bull Geol Surv Prague 66: 129–140

Von Gehlen K, Schmitt GE, Spies O (1991) Zur Genese der Serpentinite der Oberpfalz. KTB Rep 91-1: 96–105

von Quadt A, Gebauer D (1993) Sm-Nd and U-Pb dating of eclogites and granulites from the Oberpfalz, NE Bavaria, Germany. Isotope Geosci (Chem Geol) 108 pp (in press)

Vrána S (1979) Polyphase shear folding and thrusting in the Moldanubicum of southern Bohemia. Bull Geol Surv Prague 54: 75–86

Vrana S (1989) Perpotassic granulites from southern Bohemia – a new rock-type derived from partial melting of crustal rocks under upper mantle conditions. Contrib Mineral Petrol 103: 510–522

Vrána S (1992) The Moldanubian zone in southern Bohemia: polyphase evolution of imbricated crustal and upper mantle segments. Conf Bohemian Massif, Czech Geol Survey, Prague, pp 331–336

Vrána S, Jakeš P (1982) Orthopyroxene and two-pyroxene granulites from a segment of charnockitic crust in southern Bohemia. Bull Geol Surv Prague 57: 129–143

Weiss J (1966) Ultrabasic rocks of the west Moravian crystalline complex. Krystalinikum 4: 171–181

Wendt JI (1989) U-Pb Zirkondatierung und Sm-Nd Systematik sowie Petrologie und Geochemie an Granuliten und Gneisen aus dem Moldanubikum Südböhmens (ČSSR). Thesis, Univ Mainz

Wendt JI, Kröner A, Fiala J, Todt W (1993) Evidence from zircon dating for existence of approximately 2.1 Ga old crystalline basement in southern Bohemia, Czech Republic. Geol Rundsch 82: 42–50

Zaydan A, Scharbert HG (1983) Petrologie und Geochemie moldanubischer Serien im Raume Persenbeug (südwestliches Waldviertel). Janrb Geol Bundesanst 126: 181–199

Zoubek V (1927) O injekčním a kontaktním metamorfismu okolí Pelhřimova. Sb Stát Geol Ústavu 7: 263–413

Zoubek V, Fiala J, Vaňková V, Machart J, Stettner G (1988) Moldanubian region. In: Zoubek V, Cogné J, Kozhoukharov D, Kräutner HG (eds) Precambrian in younger fold belts. John Wiley, Chichester, pp 183–267

VIII Moravo-Silesian Zone

VIII.A Introduction

Z. Mísař and M. Urban

Forming the easternmost part of the Bohemian Massif, the Moravo-Silesian Zone (MSZ) can be followed from Krems in Austria in a NNE direction to Moravia (Czech Republic) and farther N probably up to the Odra Fault in Poland. It includes the autochthonous Cadomian basement (Brunovistulicum; Dudek 1980) with its Devonian to Carboniferous sedimentary cover, and the classic allochthonous Variscan units – the Moravicum and Silesicum (Suess 1912). All these units (except for the puzzling Drosendorf Window; see below) are situated E of the Moldanubian and Ramzová (or Nýznerov) thrusts (Suess 1912) – Fig. 1, or the Moravo-Silesian Fault Zone (Mísař 1967), a major Variscan tectonic interface separating the MSZ from the overriding Moldanubian and Lusatian nappes.

The immediate hangingwall to the MSZ is in many places constituted by micaschist units (Staré Město Micaschist Zone, Letovice Unit, Moravian Micaschist Zone – Fig. 1) which are traditionally treated as parts of the Lusatian, Bohemian (Teplá-Barrandian) and Moldanubian Zones, respectively, but all contain similar HP relics in both metaigneous and metapelitic (Johan et al. 1990) lithologies and/or dismembered metaophiolites (Jelínek et al. 1984). Following the original proposal by Suess (1912), these micaschist units are regarded by some authors as zones of diaphtorites at the base of the Moldanubian Thrust (e.g., Fuchs 1976).

In the E, the MSZ submerges beneath the Carpathian foredeep extending farther E as far as to the deep boundary of the Bohemian Massif – the Peripieninic Lineament (Máška and Zoubek 1960) or even below the Central Carpathians, where it may reappear as a basement reactivated during the Alpine orogeny (Grecula and Roth 1978).

The northern termination of the MSZ (and the Central European Variscides) adjacent to the Tornquist Line or Hamburg-Dobruja Lineament is still a controversial issue. The arcuate structure of the Variscan zones as expressed by the Rhenohercynian and Moravo-Silesian zones as well as the more internal ones (Kossmat 1927; Stille 1951; Ellenberger and Tamain 1980; Engel et al. 1983) has been strongly modified by NW-SE transverse faults parallel to the mobile margin of the East European Platform (Dvořák and Paproth 1969, 1988; Franke 1989) and is also controlled by the stable block of the Brunovistulicum (Dudek 1980).

From the geotectonic and paleofacial point of view, the MSZ is characterized by the transgression of Lower to Middle Devonian clastics onto pre-Devonian (Upper Proterozoic to Silurian) basement of the Brunovistulicum and Silesicum, and by synorogenic flysch sedimentation starting diachronously in the uppermost Devonian or Lower Carboniferous, and passing upwards into the parallic and continental coal-bearing molasse. The uppermost Visean conglomerates already contain pebbles of metamorphic rocks derived from the Moravicum as well as high-grade Moldanubian gneisses and granulites (Štelcl 1960; Dvořák 1973). Correspondingly, the Rb-Sr and Ar-Ar age determinations of the Bíteš gneiss (Morauf and Jäger 1982; Dallmeyer et al. 1990) indicate that the Moravicum must have attained high crustal levels during the Lower Carboniferous.

Many units of the MSZ can be well correlated with the principal zones of the European Variscides (Fig. 2). The coal-bearing molasse of the Upper Silesian Basin corresponds to the Sub-Variscan Zone of W Europe, and the sedimentary and volcanic record of the Moravo-Silesian Devonian and Culm is in many aspects similar to that of the Rhenohercynian Zone (Dvořák and Paproth 1969; Engel et al. 1983). Moreover, it is tempting to correlate the basement to the transgressive Devonian in the E part of the Silesicum with the Mid-German Crystalline Rise (Mísař et al. 1983). On the other hand, the paleotectonic position of the Brunovistulicum with its Cadomian granitoids (Scharbert and Batík 1981; Van Breemen et al. 1982) and the transgressive Devonian is not

Dallmeyer et al. (Eds.)
Pre-Permian Geology of Central and Eastern Europe

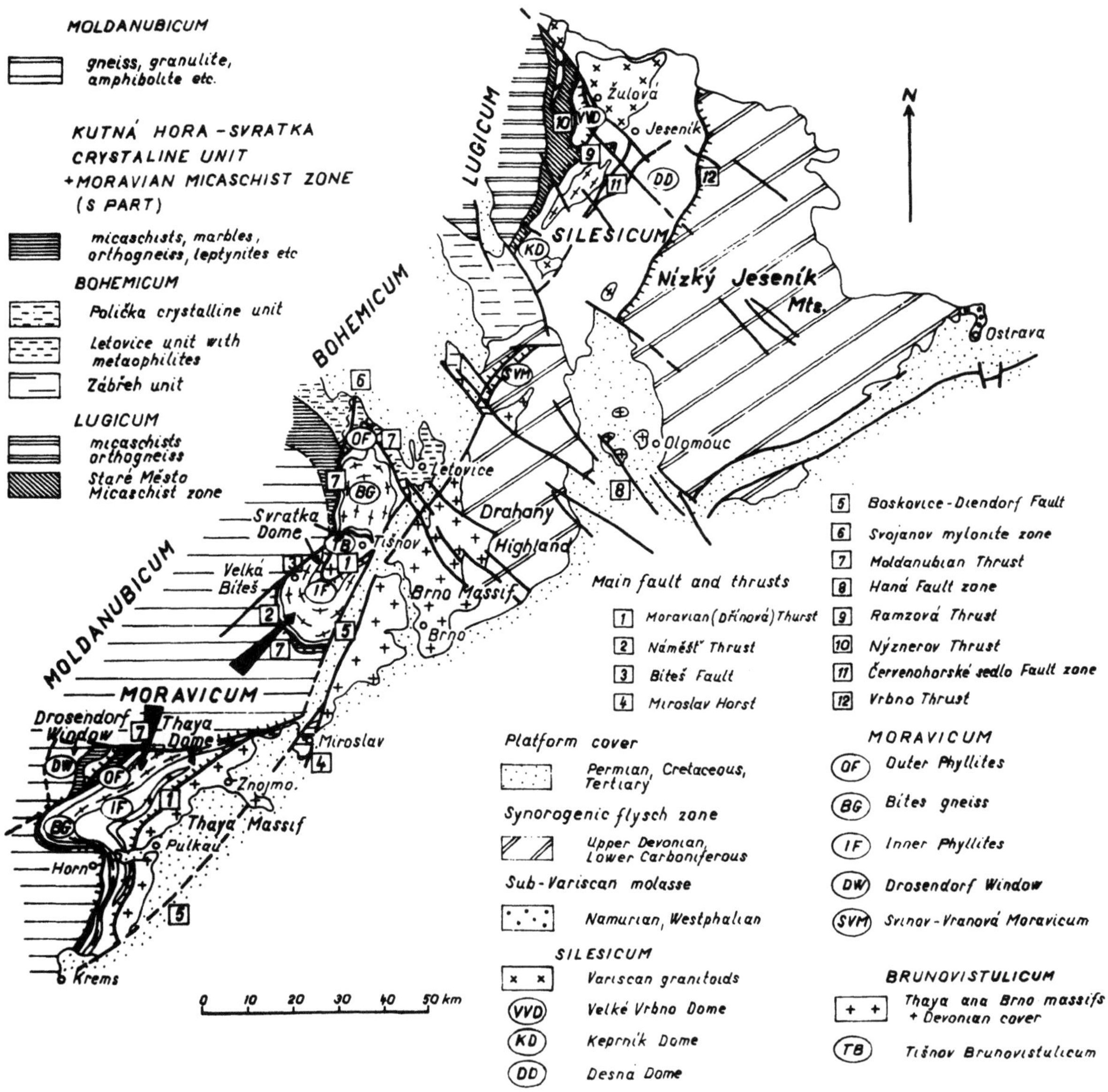

Fig. 1. Geological scheme of the Moravo-Silesian Zone. (Compiled by Z. Mísař 1991)

perfectly understood. It may either represent part of the south, Gondwana-derived continent (Tollmann 1982; Matte et al. 1990) or a prong of the East European Platform (Adam et al. 1979; Weiss 1980; Grygar et al. 1991).

As a whole, the MSZ can be divided into two parts – the northern and the southern segments. The boundary between these two segments corresponds to a wide, repeatedly rejuvenated NW-SE transverse fault zone (the Haná Fault Zone, Mísař 1963) representing part of the Elbe Lineament. In both the northern and southern segments it is possible to discern autochthonous and allochthonous domains. The intensity of Variscan deformation generally decreases eastwards: the western part of the MSZ, close to the boundary with the overriding Moldanubian and Lusatian zones, is composed of a number of allochthonous thrust sheets including strongly reactivated Cadomian basement and its Devonian cover (Jaroš and Mísař 1974; Cháb and Opletal 1984; Fritz 1991; Crygar et al. 1991; Schulmann et al. 1991), while the E part (Nízký Jeseník Mts., E of the Stephano-Autunian Boskovice Furrow) displays mostly autochthonous position.

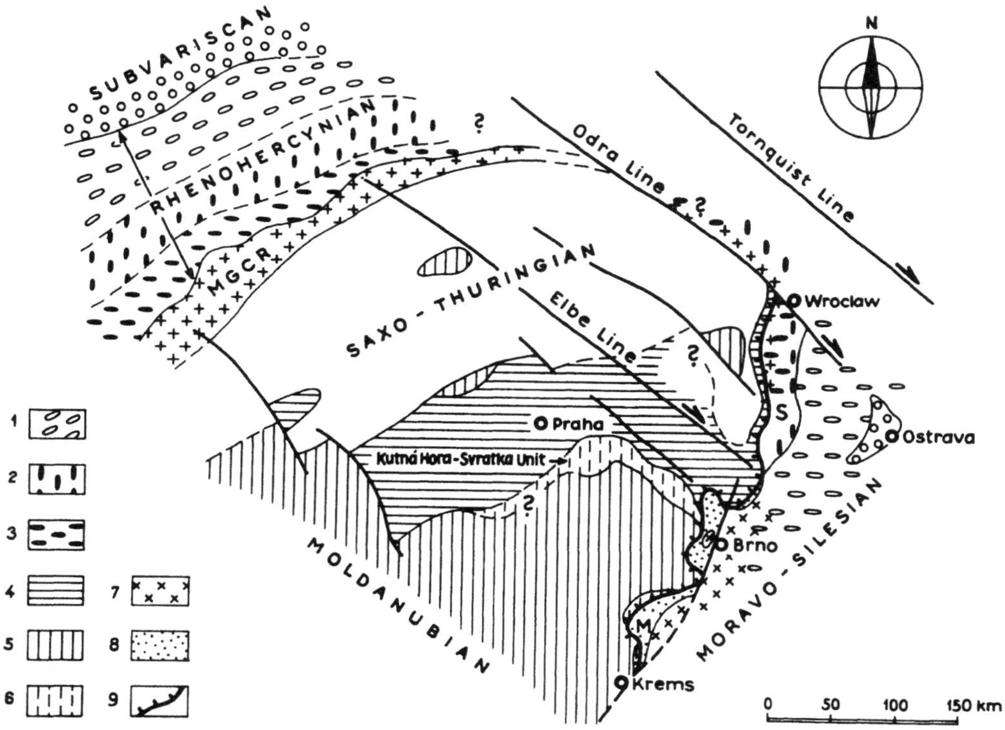

Fig. 2. Correlation scheme of the Central European Variscides. *1–3* Rhenohercynian Zone; *1* flysch; *2* Giessen-Harz Z.; *3* Northern Phyllite Z; *4–6* Moldanubian Zone; *4* Teplá-Barrandian Z.; *5* Moldanubian Z. s.s.; *6* Kutná Hora-Svratka Z.; *7–8* Moravo-Silesian Zone; *7* Brunovistulicum (out-cropping portions); *8* Moravicum; *9* western boundary of the MSZ (Moldanubian and Ramzová (or Nýznerov) thrusts); *MGCR* Mid-German Crystalline Rise; *S* Silesicum

The principal units of the MSZ are then as follows (Fig. 1):

A. The northern segment (from W to E)
1. The Velké Vrbno Dome – an allochthonous unit equivalent to the Moravicum in the S segment.
2. The Keprník Dome – allochthonous unit composed of slices of pre-Devonian metamorphics and problematic Lower Paleozoic, in many aspects comparable with the Moravicum.
3. The Desná Dome – allochthonous unit composed of slices of pre-Devonian metamorphics (?equivalent to the Mid-German Crystalline Rise) and metamorphosed Devonian.
4. Synorogenic flysch (Culm) of the Nízký Jeseník Mts. and Oderské vrchy Hills – autochthonous or paraautochthonous (cf. Tomek 1988; Synek et al. 1990) unit overlying the Devonian sediments and the Brunovistulian basement.
5. Parallic and continental molasse of the Upper Silesian Basin

B. The southern segment
1. Moravicum – clearly allochthonous unit emerging in two tectonic windows, the Svratka and Thaya Domes, from the Moldanubian nappe. It further appears as small tectonic slices E of the Boskovice Furrow at Krhovice and Miroslav. Close to the boundary with the N segment (between Svinov and Vranová), it forms a tectonic wedge surrounded by Culm sediments.
2. Drosendorf Window – appearing within the Moldanubian Zone, this allochthonous segment can be, with respect to its lithological and metamorphic character, divided into two tectonic units which can be correlated with the Vranov-Olešnice Group of the Moravicum and the Micaschist Zone, respectively (Jenček and Dudek 1971) rather than with the Moldanubian Variegated Group (Fuchs 1976; Matte et al. 1990).
3. The Cadomian Brno and Thaya granitoid massifs, (part of the Brunovistulicum) partly

split by the sinistral Boskovice-Diendorf Fault (e.g., Jaroš and Mísař 1967; Matura 1976). Along with their metamorphic host rocks and the transgressive Devonian and Carboniferous sediments, the granitoids are in a prevailingly autochthonous position. Only along their western margin were they involved in a NE-directed Variscan dextral transpression and overthrust by the nappes of the Moravicum and Moldanubicum (Matte et al. 1990; Fritz 1991; Schulmann et al. 1991). Beneath these nappes, the Brunovistulicum may extend as far as to the Býteš fault or even the Přibyslav-Vitis fault (Weiss 1980; Dudek 1980).

Acknowledgments. The authors wish to thank H. Fritz for helpful comments on the manuscripts, and H. Bryndová for drawing the figures.

References

Adam Z, Beránek B, Weiss J (1979) Contribution of deep geophysical investigations to the problem: Solution of the contact between the Carpathians and the Bohemian Massif. Proc Conf Czechoslovak geology and global tectonics, 19–21 May 1976, Czechoslovak Academy of Science Smolenice, pp 257–268

Cháb J, Opletal M (1984) Příkrovová stavba východního okraje skupiny Červenohorského sedla v Hrubém Jeseníku. Věstn Ústřed Ústavu Geol 59: 11–21

Dallmeyer RD, Neubauer F, Höck V (1990) $^{40}Ar/^{39}Ar$ mineral age controls on the chronology of late Paleozoic tectonothermal activity in the southeastern Bohemian Massif, Austria (Moldanubian and Moravo-silesian zones). In: Franke W (ed) Bohemian Massif. Field Guide, Conf IGCP 233, Paleozoic orogens in central Europe, Göttingen-Giessen 1990, pp 87–96

Dudek A (1980) The crystalline basement block of the Outer Carpathians in Moravia: Bruno-Vistulicum. Rozpr Česk Akad Věd 90 (8): 1–85

Dvořák J (1973) Synsedimentary tectonics of the Paleozoic of the Drahany Upland (Sudeticum, Moravia, Czechoslovakia). Tectonophysics 17: 35–39

Dvořák J, Paproth E (1969) Über die Position und die Tektogenese des Rhenohercynikums und des Sudetikums in den mitteleuropäischen Varisziden. Neues Jahrb Geol paläontol Monatsh 18: 65–88

Dvořák J, Paproth E (1988) Trends of the Variscan development near the SW border of the East European Platform. Z Angew Geol 34: 353–359

Ellenberger F, Tamain ALG (1980) Hercynian Europe. Episodes 22 (7): 7–12

Engel W, Franke W, Langenstrassen F (1983) Paleozoic sedimentation in the northern branch of the Mid-European Variscides; essay of an interpretation. In: Martin H, Eder FW (eds) Intracontinental fold belts. Springer, Berlin Heidelberg New York, pp 9–41

Franke W (1989) Variscan plate tectonics in the Variscan belt of Europe – current ideas and open questions. Tectonophysics 169: 221–228

Fritz H (1991) Structures and kinematics along the Moravian-Moldanubian boundary: preliminary results. Österr Beitr Meteorol Geophys 3: 77–96

Fuchs G (1976) Zur Entwicklung der Böhmischen Masse. Jahrb Geol Bundesanst 119: 45–61

Grecula P, Roth Z (1978) Kinematický model Západních Karpat v souborném řezu. Zb Geol Věd 32: 49–73

Grygar R, Adamusová M, Doležalová J, Kalendová J, Kreml P, Vavro M (1991) Strukturně-kinematická analýza sv. okraje Českého masívu. Vys Šk Báňská, Ostrava, 92 pp

Jaroš J, Mísař Z (1967) Problém hlubinného zlomu boskovické brázdy. Sb Geol Věd Geol 12: 131–147

Jaroš J, Mísař Z (1974) Deckenbau der Svratka Kuppel und seine Bedeutung für das geodynamische Modell der Böhmischen Masse. Sb Geol Věd Geol 26: 69–82

Jelínek K, Pačesová M, Martinec P, Mísař Z, Weiss Z (1984) Geochemistry of a dismembered metaophiolite complex, Letovice, Czechoslovakia. Trans R Soc Edinb Earth Sci 75: 37–48

Jenček V, Dudek A (1971) Beziehungen zwischen dem Moravikum und Moldanubikum am Westrand der Thaya-Kuppel. Věstn Ústřed Ústavu Geol 46: 331–338

Johan V, Autran A, Ledru P, Lardeaux JM, Melka R (1990) Discovery of relics of a high-pressure metamorphism at the base of the Moldanubian nappe complex. Proc Conf IGCP 233 Paleozoic Orogens in Central Europe, August–September 1990, Göttingen-Giessen

Kossmat F (1927) Gliederung des varistischen Gebirgsbaues. Abh Sächs Geol Landesamtes 1: 1–39

Máška M, Zoubek V (1960) The principal division of the West Carpathians and their pre-Neoidic basement. In: Buday T, Kodym O, Mahel M et al. (eds) Tectonic development of Czechoslovakia. Nakladatelství Česk Akad Věd Praha, pp 139–151

Matte Ph, Maluski H, Rajlich P, Franke W (1990) Terrane boundaries in the Bohemian Massif: Result of large-scale Variscan shearing. Tectonophysics 177: 151–170

Matura A (1976) Hypothesen zum Bau und zur geologischen Geschichte des kristallinen Grundgebirges von Südwestmähren und dem niederösterreichischen Waldviertel. Jahrb Geol Bundesanst 119: 63–74

Mísař Z (1963) Pre-Devonian geological development of the northeastern margin of the Bohemian Massif. Rozpr Česk Akad Věd 73 (17): 1–60

Mísař Z (1967) Die Entwicklung und Gliederung der Mährisch-schlesischen Zone der Böhmischen Masse. Acta Univ Carolianae, Geologica 1: 11–27

Mísař Z, Dudek A, Havlena V, Weiss J (1983) Geologie ČSSR I, Český masív. Státní pedagogické nakladatelství, Praha, 333 pp

Morauf W, Jäger E (1982) Rb-Sr whole rock ages for the Bites-gneiss, Moravicum, Austria. Schweiz Mineral Petrogr Mitt 62: 327–334

Scharbert S, Batík B (1981) The age of the Thaya (Dyje) Pluton. Verh Geol Bundesanst 3: 325–331

Schulmann K, Ledru P, Autran A, Lardeaux JM, Urban M, Lobkowitz M (1991) Evolution of nappes

in the eastern margin of the Bohemian Massif: a kinematic interpretation. Geol Rdsch 80: 73–92
Stille H (1951) Das mitteleuropäische variszische Grundgebirge im Bilde des gesamteuropäischem. Geol Jahrb Beih 2: 1–138
Štelcl J (1960) Petrografie kulmských slepenců j. části Drahanské vysočiny. Folia Fac Sci Nat Univ Brno I/1, pp 1–47
Suess FE (1912) Die moravischen Fenster und ihre Beziehung zum Grundgebirge des Hohen Gesenkes. Österr Akad Wiss Denkschr 88: 541–631
Synek J, Rajlich P, Urban M (1990) Transpression and transtension in the Devonian and Culm of the Jeseníky Mts. Acta Univ Carolinae, Geologica 2: 209–234
Tollmann A (1982) Großräumiger Variszischer Deckenbau im Moldanubikum und neue Gedanken zum Variszikum Europas. Geotekton Forsch 64: 1–64
Tomek Č (1988) Eastern boundary of European Variscides: large-scale oblique continental collision zone. 28th Int Geol Congr, Washington 1988, Abstr 3: 343
Van Breemen O, Aftalion M, Bowes DR, Dudek A, Mísař Z, Povondra P, Vrána S (1982) Geochronological studies of the Bohemian Massif, Czechoslovakia, and their significance in the evolution of Central Europe. Trans R Soc Edinb Earth Sci 73: 89–108
Weiss J (1980) Geologischer Bau des Mährischen Block. Z Geol Wiss 8: 557–565

VIII.B Autochthon

VIII.B.1 Stratigraphy

J. Dvořák

1 Introduction

The Rhenoherzynikum and the Moravian Paleozoic Basin are mirror symmetrical external basins bordering the Variscan foreland (Fig. 1). In Moravia, this foreland is formed by the Pre-Carpathian platform to the southeast (Dvořák and Paproth 1988). Two sub-basins are distinguished in the Moravian basin: the Jeseníky sub-basin (Jeseníky Mountains) to the north, and the Drahany sub-basin (Drahany Upland) to the south. The more mobile basement of the Juseníky sub-basin consists mainly of gneiss. A stable granitoid mass of Proterozoic age forms the southern closure of this sub-basin (Dvořák 1985a).

2 Drahany Upland

Due to its stable Proterozoic granitoid basement, the Drahany sub-basin is a transversely segmented structure with a brachysynclinal closure to the south of Brno, where the Lower Devonian marine deposits pass into purple Old Red arkoses and quartz conglomerates with a maximum thickness of 1500 m (Fig. 2).

The Drahany Upland consists of several transverse blocks which are separated by deep-seated faults (Figs. 2, 3; Dvořák 1973). Differences in the succession and thickness of the sediments and volcanic deposits on the individual blocks reflect the differential mobility of the basement. From north to south, both the polarity and the geosynclinal character of the sedimentation on the individual transverse blocks decrease and are replaced by platform-like facies on the southernmost blocks (Fig. 3).

The oldest Paleozoic sediments (Llandoverian to Ludlovian graptolite-bearing shales, and Pragian and Emsian deposits) occur in a transverse depression along the Nectava-Konice Fault. The connection between the Barrandian sea and the Moravian sea was situated along the western prolongation of this fault (the Elbe lineament). Thin basal sandstones in the central part of the Moravian basin are overlain by dark gray to black Emsian to Upper Devonian shales of the Stínava-Chabičov Formation. Their deposition was accompanied by spilite volcanism. From the Eifelian onwards, the marine transgression extended to the east and south beyond the margin of this transverse depression. The up to 1000-m-thick reef limestones of the Macocha Formation (Eifelian to Frasnian) reflect a warm climate, continuous subsidence of the basement, a low relief because of the absence of clastic material in the deposits, and open-marine, well-aerated conditions (Dvořák 1980). The lower part of the carbonate sequences is predominantly dark-coloured and dolomitic. The presence of early diagenetic dolomite suggests evaporation in shallow marine embayments. The light-gray upper portion of the reef limestones contains a highly diverse reef-building fauna. These limestones accumulated in a basin with a minimum influx of freshwater from the continent. This produced a vertical seawater circulation due to the intensive evaporation which resulted in an increasing salt content of the surficial water. This heavy, salt-loaded water sank to the bottom whereas deep, light water rose to the surface. The high oxygen content of the salt-loaded water completely destroyed the organic material of the substrate. This mechanism created a very vulnerable ecosystem which was highly sensitive to external influences and characteristic of transgressive conditions (Dvořák 1990a). The reef development came to an end between the Middle Frasnian and the Lower Famennian, presumably due to the tectonic relaxation during that timespan. However, the gradual change from transgressive into regressive conditions was a diachronic process as can be deduced from the local differences in the timing of the reef extinction. It is related to the tilting towards the basin center found in many of the eastern and southern foreland blocks (Fig. 4; Dvořák et al. 1976).

Dallmeyer et al. (Eds.)
Pre-Permian Geology of Central and Eastern Europe

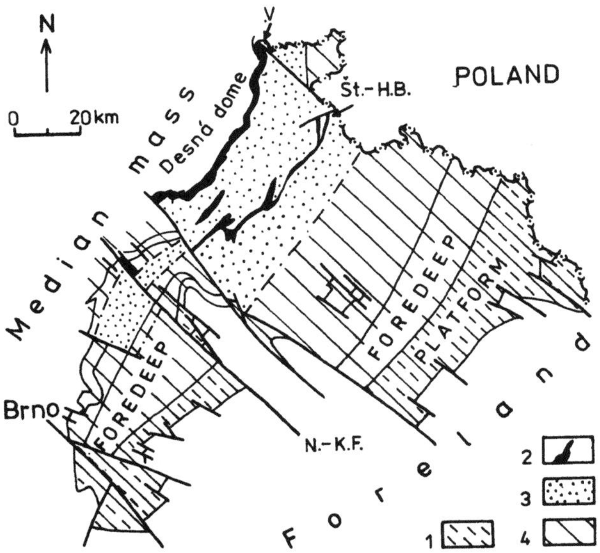

Fig. 1. Schematic geologic map of Moravo-Silesian Palaeozoic. *1* Unfolded rocks on the platform; *2* volcano-sedimentary formations on the surface (*Št-H.B.* Šternberk Horní Benešov belt; *V* Vrbno belt); *3* volcanosedimentary formations below the flysch; *4* carbonate formations on the surface or below the flysch and molasse; *N.-K.F.* Nectava-Konice Fault

Thin sequences of nodular, micritic (Křtiny) limestone line the margin of these blocks. When purple, lateritic clay (Al_2O_3/Na_2O – 80 to 130; Dvořák 1986a, 1990a) was washed down from the slowly rising ridges of the Proterozoic crystalline rocks into the basin, a red nodular limestone was formed in a shallow oxygen-rich environment. Limestone breccias and slumps occur locally. Condensed, variegated nodular limestones interfinger with dark-grey shales with limestone nodules. The dark-coloured sediments reflect anoxic, slightly deeper environments during the Famennian and Lower Tournaisian.

The subsidence of some of the eastern platform blocks started during the Upper Famennian with the deposition of light gray, sugar-like massive pelbiosparites of the 400-m-thick Dražovice Formation in a shallow-marine, oxygen-rich environment on carbonate flats and subaquaceous bars. Dolomite intercalations and the abundance of authigenic quartz crystals suggest evaporitic conditions (Stoček 1979).

Some small blocks in the more mobile zones became downwarped again from the Middle Famennian onward (Annulata event: Dvořák et al. 1988). This is at least indicated by the thickness of the dark gray, biodetritic (Hády-Říčka) limestones, which is about ten times that of the nodular limestones on the blocks near the shoreline. Intensive micritization of bioclasts in residual shell beds is common. In the southeastern part of the Moravian Karst, this limestone accumulation came to an end during the Middle Visean. The northward shift of the subsidence of small blocks took place in the southern part of the Moravian Karst between the Upper Famennian and the Lower Visean. It was accompanied by gravitational folding and the movement of large olistolites with a northern vergency (Dvořák et al. 1987).

Black siliceous shales of the Famenno-Tournaisian Ponikev Formation were deposited in the central and northern parts of the Drahany Upland. The Lower and Middle Visean are characterized by red and green trilobite-bearing shales of the Březina Formation.

The onset of the flysch deposits is diachronic (Fig. 5). The flysch facies, which replaced the limestone facies, started in the northwest near the transverse axial depression along the Nectava-Konice Fault around the Devono-Carboniferous boundary. It expanded on the western part of the Drahany transverse block during the Tournaisian. Rapid warping of this area during the Lower Visean was matched by the accumulation of the more than 2000 m thick greywackes of the Protivanov Formation. The coeval strata on the Moravian Karst block to the south have a total thickness of only some tens of metres. The clastic material was largely derived from the Proterozoic crystalline rocks of the Svratka Anticline to the northwest. During the Middle Visean, the zone of maximum subsidence shifted eastwards and included the main part of the Drahany Upland. The depocenter of the about 1500-m-thick alternation of graywacke, siltstone, and shale of the Rozstání Formation shifted to the south, parallel to the strike of the basin axis, during the lowermost Upper Visean. The Rozstání Formation wedges to the east and to the south.

The up to 3000-m-thick Myslejovice Formation (Upper Visean, Go α–γ zones) forms the final molasse unit of this sedimentary succession. An impressive fan of coarse conglomerates was laid down in the southern part of the Drahany Upland. The largest (rounded) boulders have a diameter of up to 2 m.

Towards the north-north-east, these conglomerates grade into graywackes which, further on, pass into laminated shales. The relative amount of pebbles derived from the Moldanubian area

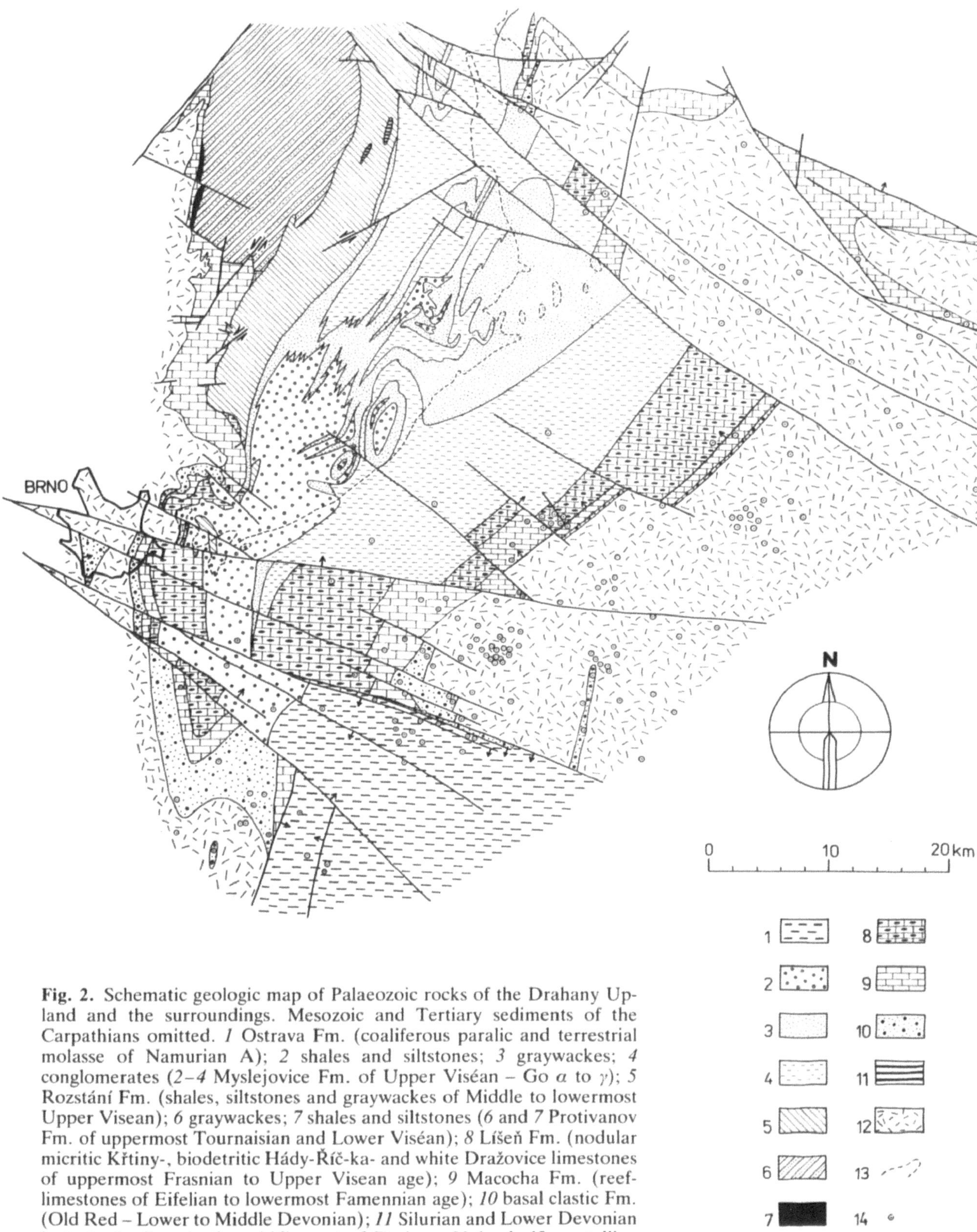

Fig. 2. Schematic geologic map of Palaeozoic rocks of the Drahany Upland and the surroundings. Mesozoic and Tertiary sediments of the Carpathians omitted. *1* Ostrava Fm. (coaliferous paralic and terrestrial molasse of Namurian A); *2* shales and siltstones; *3* graywackes; *4* conglomerates (*2–4* Myslejovice Fm. of Upper Viséan – Go α to γ); *5* Rozstání Fm. (shales, siltstones and graywackes of Middle to lowermost Upper Visean); *6* graywackes; *7* shales and siltstones (*6* and *7* Protivanov Fm. of uppermost Tournaisian and Lower Viséan); *8* Líšeň Fm. (nodular micritic Křtiny-, biodetritic Hády-Říč-ka- and white Dražovice limestones of uppermost Frasnian to Upper Visean age); *9* Macocha Fm. (reef-limestones of Eifelian to lowermost Famennian age); *10* basal clastic Fm. (Old Red – Lower to Middle Devonian); *11* Silurian and Lower Devonian in tectonic slices in the middle part of Drahany Upland; *12* crystalline rocks of Proterozoic age (mainly granitoids); *13* eastern boundary of the Bohemian Massif on the surface

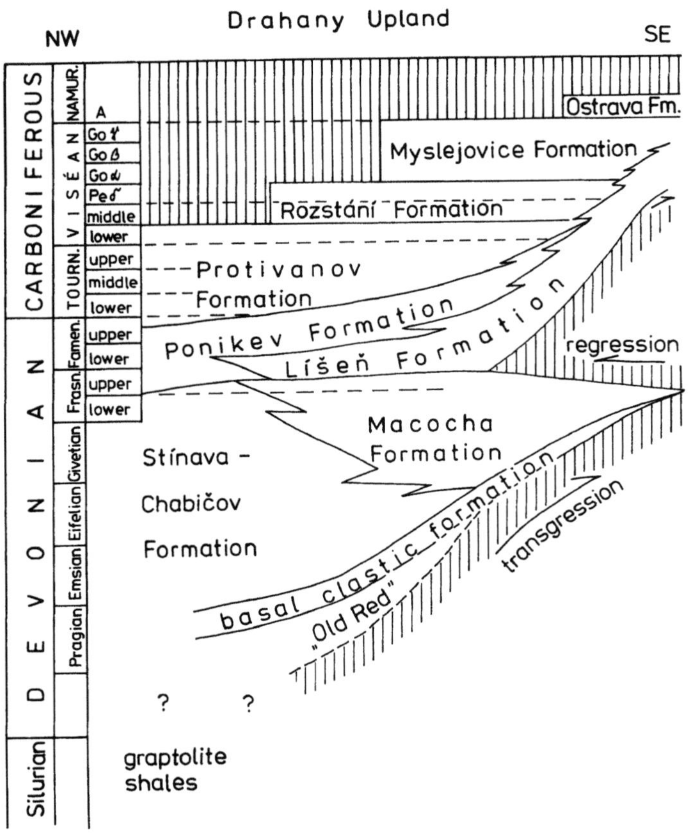

Fig. 3. Schematic stratigraphic table of Drahany area

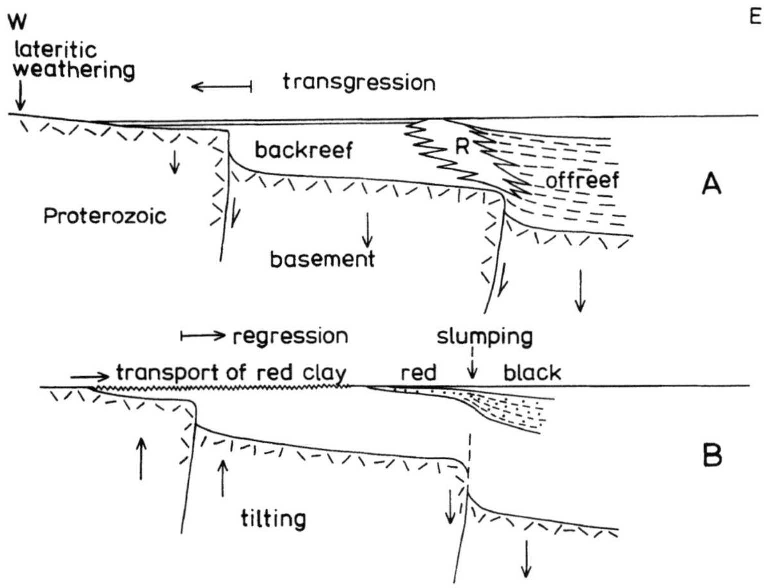

Fig. 4A,B. Schematic sections through the Moravian Karst, showing facies and thickness of Givetian and Frasnian (**A**) and Famennian (**B**) deposits controlled by synsedimentary tectonics of Proterozoic basement (reef extinction)

increases from being fairly common at the base to abundant in the upper portion of the Myslejovice Formation. Pebbles of Variscan granites have not been found, but pebbles of acid volcanic rocks and tuffs are common (Dvořák 1973).

The conglomerate deposition shifted to the south into the western part of the Nesvačilka block during Go β time. This 2500-m-thick mass of conglomerates laterally grades into graywackes and black shales deposited during Go β–γ time.

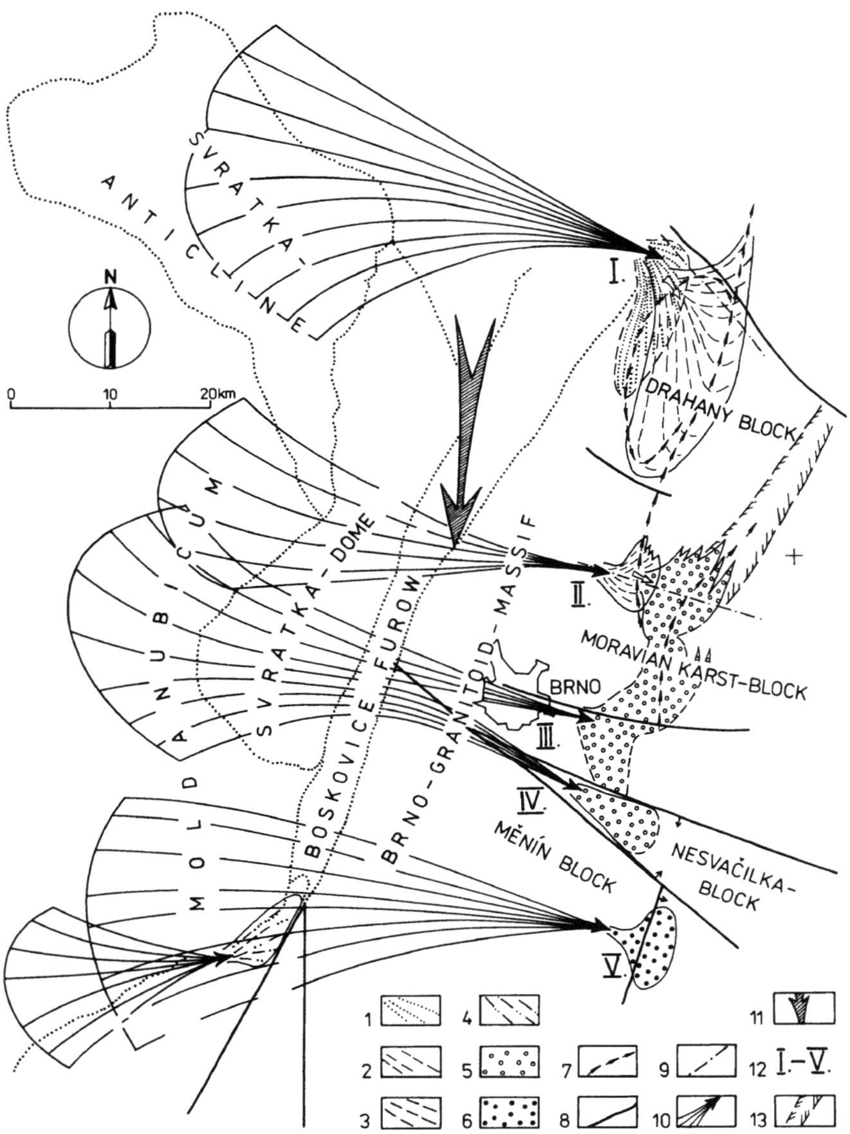

Fig. 5. Schematic map illustrating the interrelationships between uplifting source regions, transport of clastic material, transverse blocks, mouth of streams and distribution of sediments within the basin of the Drahany Upland in the time of deposition of flysch and molasse. This scheme clearly indicates migration of movements from N to S, but also the migration of the clastic wedges in relation to the long axis of the basin of the Drahany Upland. *1* Graywackes and shales (Upper Tournaisian); *2* graywackes (Lower Visean); *3* graywackes (lowermost part of Upper Visean); *4* graywackes and conglomerates (Upper Visean); *5* conglomerates (Upper Visean); *6* conglomerates (Namurian A); *7* current directions in the basin; *8* deep-seated faults; *9* transversal axial depressions; *10* directions of streams, transporting clastic material and situation of the mouths; *11* direction indicating regional succession of general movements (uplift and subsidence); *12* stages of development: *I* Tournaisian and Lower Visean; *II* Middle and Upper Visean (Go α mainly); *III* Upper Visean (Go α and β); *IV* Upper Visean (Go β and γ); *V* Namurian A; *13* the zone of maximum subsidence during the deposition of the upper part of Myslejovice Formation (upper part of Go α)

More than 1000-m-thick, nonmarine coal-bearing Namurian A sediments were laid down in a marginal, transverse depression in the southeastern part of the Měnín block. Several hundred meters of coarse conglomerates occur at the western border of this area. Moldanubian granulite pebbles prevail (Dvořák 1989). The transport of clastic material from the northwest to the southeast suggests that the uplift of the median mass was contemporaneous with the downwarp of the

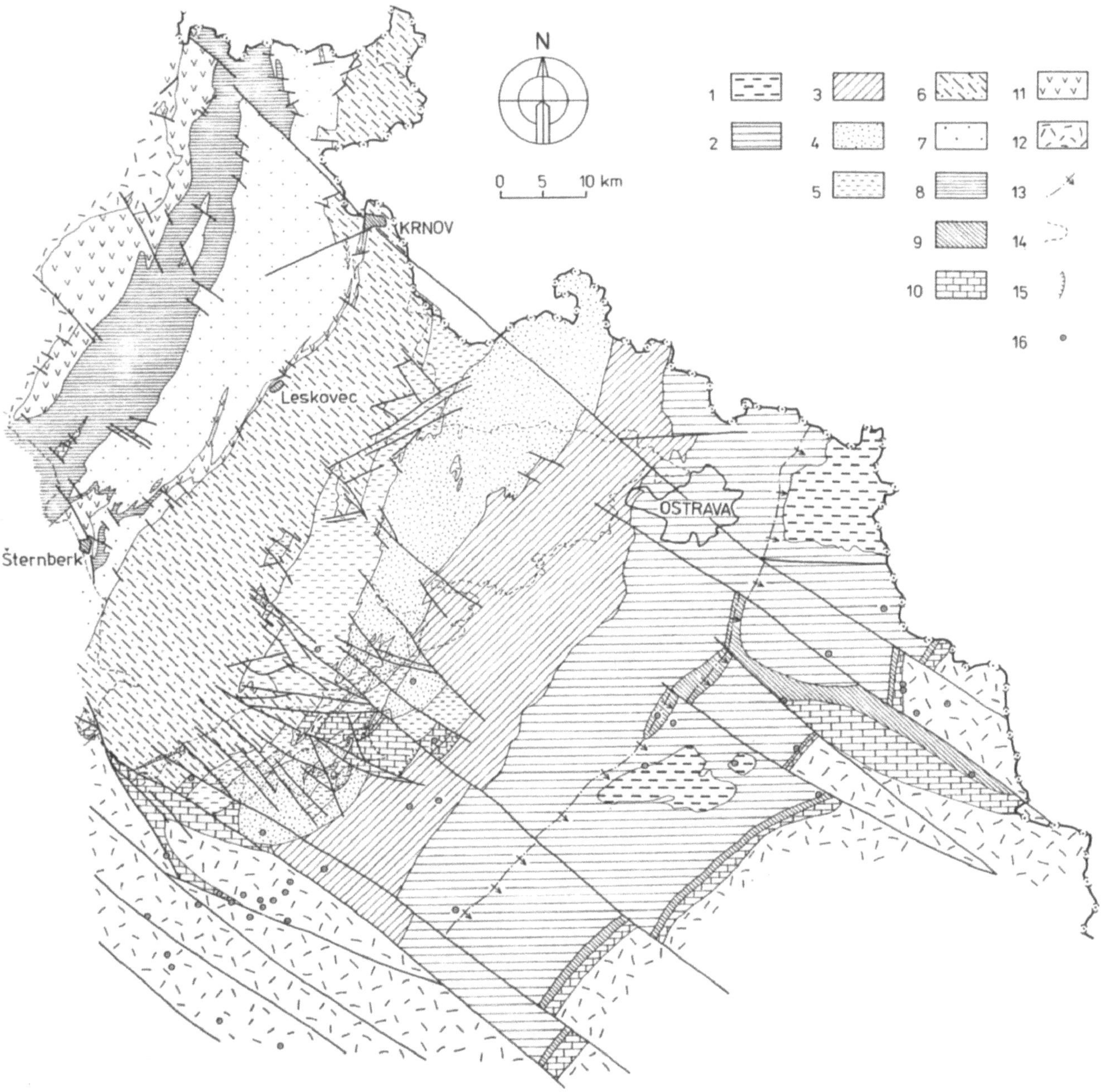

Fig. 6. Schematic geological map of the Paleozoic formations of Jeseníky Mts. and the continuation below the Carpathians. *1* Karviná Formation (terrestrial coaliferous molasse – Upper Namurian and Westphalian A); *2* Ostrava Fm. (paralic coaliferous molasse – lower part of Namurian A); *3* alternation of shales, siltstones and graywackes; *4* graywackes, locally with intercalations of conglomerates near the base (*3* and *4* Hradec-Kyjovice Fm.-Upper Visean, Go γ zone and base of Namurian A); *5* shales and siltstones with intercalations of graywackes (upper part of Moravice Fm.-Upper Visean, Go β zone); *6* shales and siltstones locally with intercalations of graywackes (lower part of Moravice Fm.-Upper Visean, Go α zone); *7* graywackes, locally alternation of shales, siltstones and graywackes (Horní Benešov Fm.-Lower and Middle Visean); *8* alternation of shales, silt-stones and graywackes locally with pebbly mudstones (Andělská Hora Fm.-Famennian and Tournaisian); *9* shales, in upper part with intercalations of sandstones (Upper Visean, upper part of Go β zone and Go γ zone, base of Namurian A); *10* limestones (Macocha- and Líšeň Fm.-Upper Eifelian to Upper Visean); *11* volcano-sedimentary development of Vrbno and Šternberk-Horní Benešov Zones (basal quartzites, shales, spilites and keratophyres, limestones, siliceous shales, quartz-conglomerates and sandstones – Pragian to Upper Tournaisian); *12* undifferentiated basement of Proterozoic age; *13* Orlová fault with vergency. It is the eastern border of the folded foredeep. To the east horizontally lying formations on the foreland platform; *14* eastern border of the Bohemian massif on the surface; *15* some owerthrusts; *16* boreholes in the Carpathians

basin. The boundary between the more mobile blocks to the north and the less mobile ones to the south shifted gradually southward during the deposition of flysch and molasse.

3 Jeseníky Mountains

The pre-flysch stage in the western part of the Jeseníky Mountains is characterized by intense spilite volcanism along three parallel NNE-SSW running deep-seated fault zones (Fig. 6). The relative amount of rare earths in the basic rocks (diabase and spilite) is similar to that in alkaline basalt. The intermediate keratophyres are comparable with alkali-trachyte or trachy-andesite (Přichystal 1990). The chemical character of the volcanism remained unchanged from the Lower Devonian until the Lower Viséan. The narrow volcanic zones are separated by gneiss domes without any volcanism (Fig. 8). The volcanic activity (marked by up to 1000-m-thick rocks)

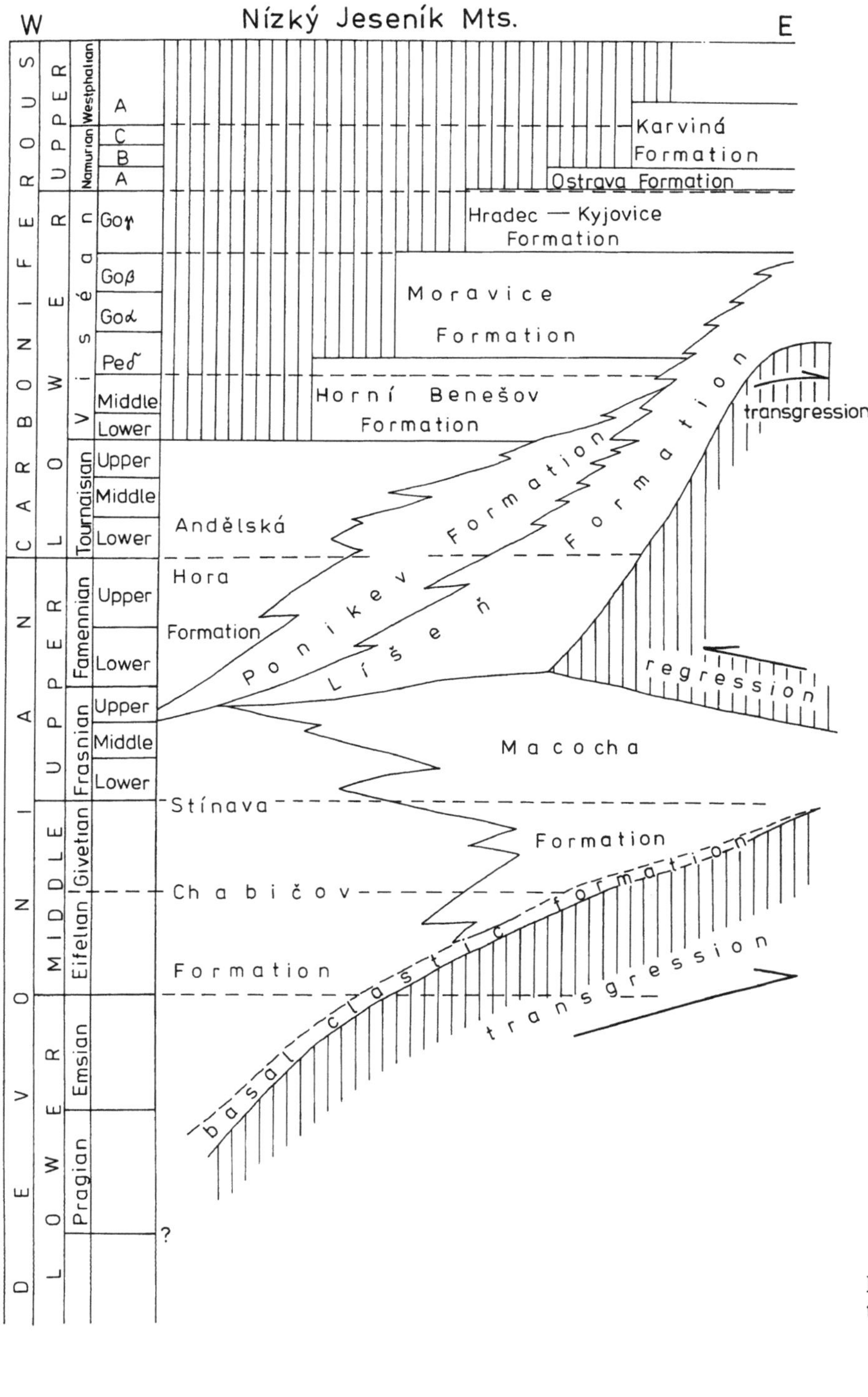

Fig. 7. Schematic stratigraphic table of the Jeseníky area

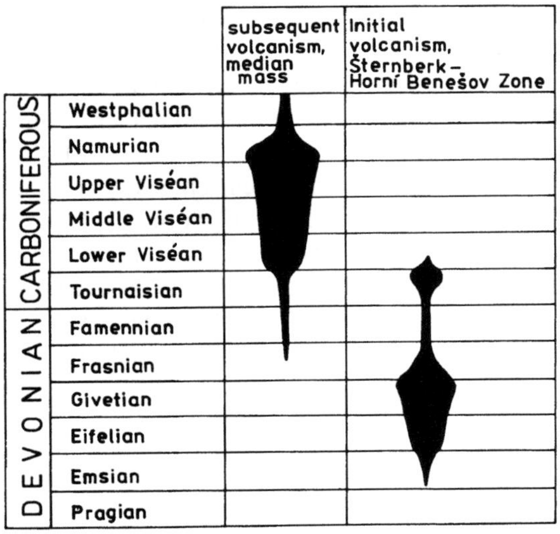

Fig. 8. Stratigraphic chart of the time course of volcanism in the Jeseníky Mts. and in the median mass (time coincidence of inicial and subsequent volcanism)

occurred reached its acme during the Late Emsian and Middle Devonian in each of the three fault zones. The volcanism ceased in the Vrbno zone during the Famennian, and in the Šternberk-Horní Benešov zone during the Lower Viséan. The youngest volcanic extrusions interfinger with typical flysch deposits (Fig. 8).

Sedimentation in the volcanic mobile zone presumably started before the Siegenian and perhaps already during the Silurian. In Pragian (Siegenian) time, the sea flooded the margins of the Proterozoic domes. Quartzites with a rich Rhenish-type fauna (Chlupáč 1981) were deposited. They pass into dark argillaceous rocks of the Stínava-Chabičov Formation in the central part of the mobile zones. The summits of the linear volcanic belts along the fault zones possibly rose above sea level during the peak of volcanic activity and the herewith associated maximum rate of subsidence during the Emsian and Middle Devonian. They formed the base of thin reef limestones of Eifelian to Famennian age (Dvořák 1994). Black, green and red siliceous shales with silicites of the Ponikev Formation accumulated along the margins of these volcanic belts during the Upper Frasnian to Lower Visean. A condensed sequence of these shales was deposited in quiet, shallow water.

Before the onset of the flysch deposition, the crystalline basement ridges were slightly uplifted as suggested by the presence of sandy limestone breccias, quartz conglomerates and coarse-grained sandstones (more than 1 km^3) in narrow north-south trending depressions between these volcanic ridges (nonmarine Upper Tournaisian sediments capping black, radiolite-bearing siliceous shales; Dvořák 1994).

Deposition in the eastern part of the Jeseníky Mountains started only during the Middle Devonian. The almost 800-m-thick reef limestones of the Macocha Formation accumulated on the slightly tilted border of the foreland during the Givetian and Lower Frasnian. A very slow partial regression from the Upper Frasnian onwards is indicated by the condensed succession of the nearshore, nodular Křtiny Limestone. The up to 150-m-thick, dark gray biodetrital Hády-Řícka Limestone at the easternmost border of this basin characterizes the shallow, high-energy environments in an more rapidly downwarping area during the Upper Famennian into Upper Viséan (Go β zone). Lenticular limestone breccias with phosphate nodules line the flanks of the uplifted former reefs. Many local sedimentary gaps of various ages occur in this lithofacies, whereas a more continnous depositional record exists for the depressions in between.

The shortening of the basin by means of block tilting (without intensive plastic fold deformation) was simultaneous with the thrust movements. This is emphasized by the thrust of Lower Frasnian reef limestones (with important sedimentary gaps in their upper portion) over the depressions with uninterrupted sedimentation (Dvořák 1992).

The flysch facies differs geochemically from the pre-flysch deposits (Fig. 9). Intensively chemically altered clastic material had been deposited during the pre-flysch period; but this situation changed abruptly at the onset of the flysch sedimentation. A fresh, chemically unaltered detritus was washed down from the rapidly rising central massif, wherein mechanical disintegration of the rocks prevailed over chemical weathering in a tropical climate. Elsewhere in the basin, however, the deposition of intensively chemically altered sediments (Al_2O_3/Na_2O ratio up to 100) continued during the flysch development. Presumably, the rising mountains caused a locally abundant precipitation, which in turn produced the rapid transport of fresh detritus into the basin (Dvořák 1990a).

The luxurious vegetation in coastal swamps reflects the influence of a tropical climate associated with nonmarine volcanism (CO_2 production). Plant remains become more frequent in the

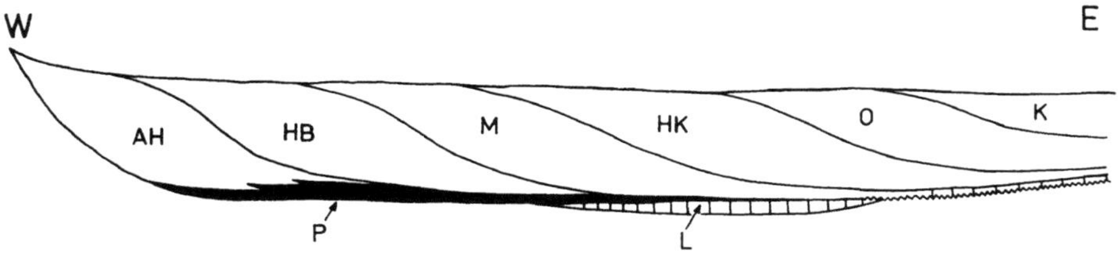

Fig. 9. Oversimplified section through the Jeseníky area. Flysch: *AH* Andělská Hora Fm. (Famennian and Tournaisian); *HB* Horní Benešov Fm. (Lower and Middle Visean); *M* Moravice Fm. (Upper Visean Go α and β); *HK* Hradec-Kyjovice Fm. (Upper Visean-Go γ and lowermost Namurian A). Paralic molasse: *O* Ostrava Fm. (lower part of Namurian A). Terrestrial molasse: *K* Karviná Fm. (upper part of Namurian and Westphalian A); *P* Ponikev Fm. (siliceous shales); *L* Líšeň Fm. (biodetritic limestones). Gradual thinning away of the very thick flysch formations and transition in nonflysch facies. Migration of the zones of maximal thickness from W to E (not to scale)

younger flysch deposits, whereas coal seams formed in a paralic molasse regime. The oldest flysch sediments lack the characteristic goniatites and pelecypods, which would gradually adapt themselves to the newly developing ecosystem from the Upper Viséan onwards.

The clastic detritus from the rising central massif to the west was transported by rivers into the NNE-SSW-oriented basin (Fig. 10). The sediments of the clastic fans at the river mouths were subsequently redistributed by suspension currents within the SSW-NNE-directed zone of maximum subsidence. Kumpera (1983) emphasized that the flute casts in these deposits are parallel to the earliest fold axes. This proves that the individual flysch troughs developed as synclinal depressions in a compressional regime. The sea reached its maximum depth in the axial portion of these depressions, whereas an eastward shallowing of the basin is perpendicular to that direction as indicated by the presence of the siliceous shales of the Ponikev Formation and the simultaneous absence of more coarse-grained siliciclastics.

Thus, the thickness of the individual layers of clastic sediment increases eastwards and then wedges again when different lithofacies develop. Coarse-grained clastics occur in the axis of the zone of maximum downwarp, whereas the western shore remained rather in the shade of the depositional environment as suggested by the presence of more fine-grained, clayey to silty clastics. The margins of the clastic fans are locally characterized by the presence of lenses of shallow-marine limestones with an abundant fauna.

Such clastic fans are poorly developed in the oldest Andělská Hora Formation, but they abound in the younger Moravice and Hradec-Kyjovice formations (Dvořák 1994a). The deposition of the clastic material must have been rapid although not continuous as suggested by thin intercalations

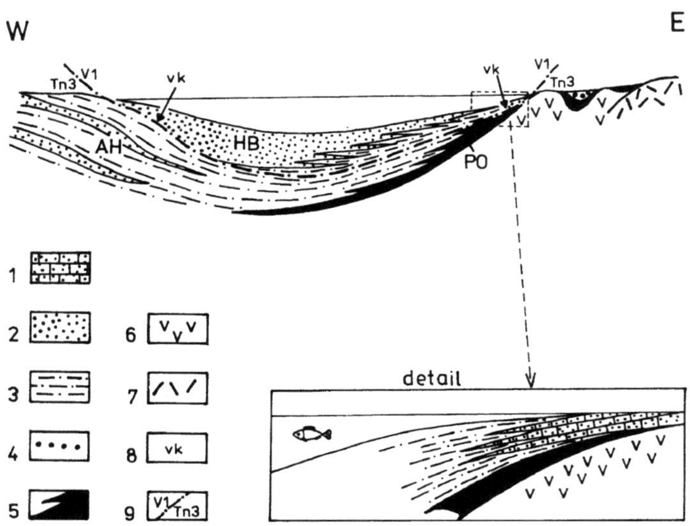

Fig. 10. Schematic facial section through the western part of the Jeseníky area. *1* Calcareous sandstones; *2* graywackes; *3* Shales and siltstones; *4* quartz conglomerates; *5* siliceous shales; *6* basic volcanic rocks; *7* crystalline rocks of Proterozoic age; *8* volcanic quartz; *9* biostratigraphic boundary Tournaisian Visean, *AH* Andělská Hora Fm.; *HB* Horní Benešov Fm.; *PO* Ponikev Fm. (not to scale)

of black tuffaceous shales (Dvořák 1986b). The well-developed graded bedding of the oldest flysch deposits gradually disappears in the younger ones.

The considerable spatial reduction of the original basins, as well as the denudation of their western margin, suggest that the maximum width of the area with flysch deposition cannot have exceeded 50 km during any of the individual depositional stages. The western margin of the flysch facies always showed regressive tendencies (advancing clastic wedges), whereas a transgressive tendency marked the eastern margin (Fig. 11; Dvořák 1989). The fill of the zone of maximum subsidence exhibits an imbricated structure. The maximum thickness of the sediments in the various segments varies between 1500 and 2000 m. The rate of migration of the zone of maximum subsidence from the median mass, into the foreland increased when the compressional forces became more intensive (Dvořák 1975). Also the deposition of clastic material increased in response to the higher rate of uplift of the progressively more extensive source area. This is emphasized by the increasing diameter of the largest pebbles in the conglomerate intercalations, even when these were derived from more distant sources.

The upwarp of some of the blocks of the median mass if was linked to the intrusion of enormous Variscan granitoids at depth and to the herewith associated subsequent acid volcanism in the newly developing mountain range (Dvořák and Paproth, 1988). This volcanic activity started during the Upper Frasnian with the deposition of thin tuffite intercalations. The onset of the subsequent nonmarine volcanism was coeval with the earliest flysch deposits.

Apart from the tuff and tuffite intercalations, the existence of this volcanism can be deduced from the presence of pebbles and minute clasts of acid volcanites and their tuffs in the conglomerates and greywackes of the flysch facies. Intensive eruptions of nonmarine volcanoes accompanied the onset of the deposition of the Horní Benešov Formation and the eastward shift of the zone of maximum smbsidence. The increased rate of deposition of coarser-grained material emphasizes the increased rate of upwarp of the source area where, apart from granodiorites and phyllites, also metamorphites of a higher metamorphic grade occurred. The number of pebbles of acid volcanites and their biotite-bearing tuffs increased during that period (Kukal 1980).

The Desná Dome, as well as the formerly uplifted blocks of the Proterozoic basement between the Vrbno zone and the Šternberk-Horní Benešov Belt, were downwarped again during the deposition of the Andělská Hora and Horní Benešov formations, whereas the blocks to the east of the Šternberk-Horní Benešov belt became uplifted (Fig. 12). The folds which developed

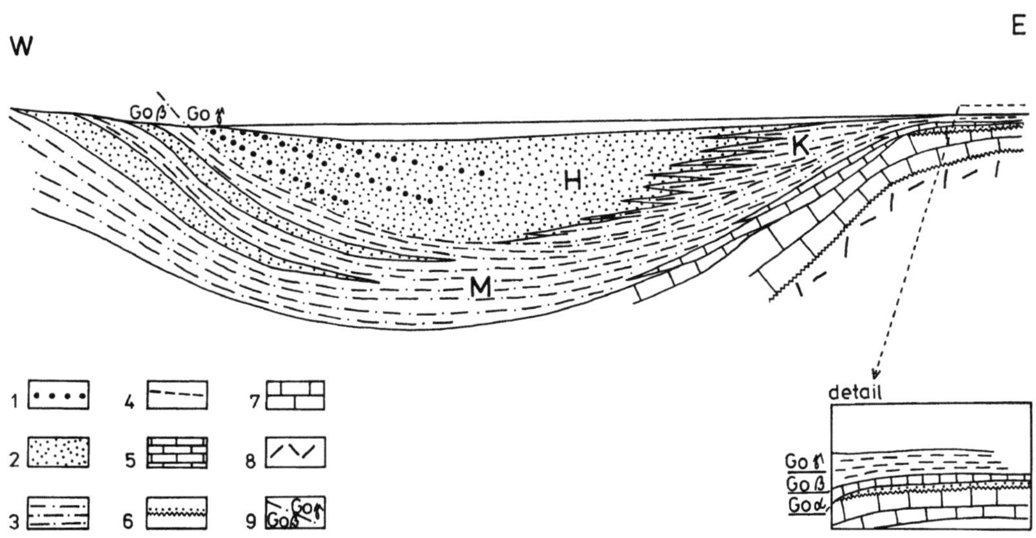

Fig. 11. Schematic facial section through the eastern part of the Jeseníky area. *1* Polymict conglomerates; *2* graywackes; *3* shales and siltstones; *4* fine clay shales; *5* biodetritic limestones (Tournaisian and Visean); *6* quartz sandstones (Upper Visean) overlying reef-limestones of Frasnian age; *7* reef-li-mestones (Givetian and Frasnian); *8* Proterozoic basement; *9* biostratigraphic boundary between the Go β and GO γ zones, *H* Hradec – graywackes; *K* Kyjovice – shales (Hradec-Kyjovice Fm.); *M* Moravice Fm. (not to scale)

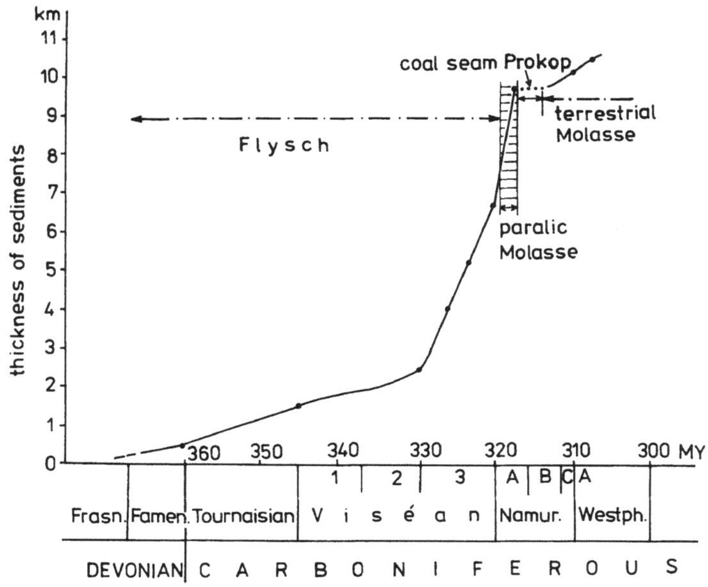

Fig. 12. Rates of deposition in the time of flysch and molasse development in Jeseníky area

during this timespan were marked by a westward vergency. The synclinorium to the west of the Šternberk-Horní Benešov belt rapidly subsided during the Lower and Middle Visean and became filled by relatively coarse-grained clastics, which showed the effect of intense upward heath flows, folding and cleavage. The basic initial volcanism continued in the downwarped area at the foot of the Šternberk-Horní Benešov belt until the Middle Visean.

Devonian volcanites and sediments of the Šternberk-Horní Benešov belt were listrically pushed upwards and integrated in a complex fan-shaped anticlinorium together with the rocks of the Horní Benešov Formation during the Lower and Middle Visean. The height of this upward-pushed anticlinorium exceeds 3000 m (Dvořák 1994b). This is documented by the low-grade thermal metamorphism of organic matter in the Lower Devonian to Lower Carboniferous sediments in the apical portion of the anticlinorium of the Šternberk-Horní Benešov belt as compared with the adjacent deep synclinorial structure to the west with its infill of Horní Benešov deposits. The Lower Viséan rocks at the bottom of the Dětřichov-1 borehold at 2400 m depth display an epizonal metamorphism (Dvořák 1994a, b).

The deep synclinorium to the west of the Šternberk-Horní Benešov belt is the result of a synsedimentary collision between two blocks of the Proterozoic basement which reduced the original sedimentary space. The zone of maximum subsidence shifted to the east of the Šternberk-Horní Benešov belt during the Upper Visean. This subsidence was at maximum 1500 m. The western margin of the basin shifted into the position of the actual Šternberk-Horní Benešov belt during the deposition of the lower portion of the Moravice Formation. This is documented by fragments of calcareous sandstone and limestone of Middle to Upper Visean age, which were reworked in slump conglomerates after a short-distance transport (Dvořák 1994a).

The zone of maximum subsidence shifted further eastwards during the deposition of the uppermost Viséan Hradec-Kyjovice Formation, when the area to the west of the Andělská Hora Formation had already become a source for siliciclastics. The actual foredeep of the Variscan tectogen (the western part of the Upper Silesian Basin) formed the easternmost depocenter, where some 1300-m-thick sediments were deposited around the Visean-Namurian boundary before the onset of sedimentation of the paralic Ostrava Formation (Dvořák 1990b). The marine foredeep deposits prior to the paralic sedimentation show no flysch characteristics.

The transition from marine to paralic sedimentation started during the period of maximum subsidence in the foredeep. Nevertheless, the rate of sedimentation exceeded that of the subsidence, so that the basin was completely filled (Fig. 13). The Ostrava Formation was still gently folded (Havlena 1982). It is capped by the 5-m-thick Prokop Coal Seam which accumulated during the upper half of the Namurian A and the lower half

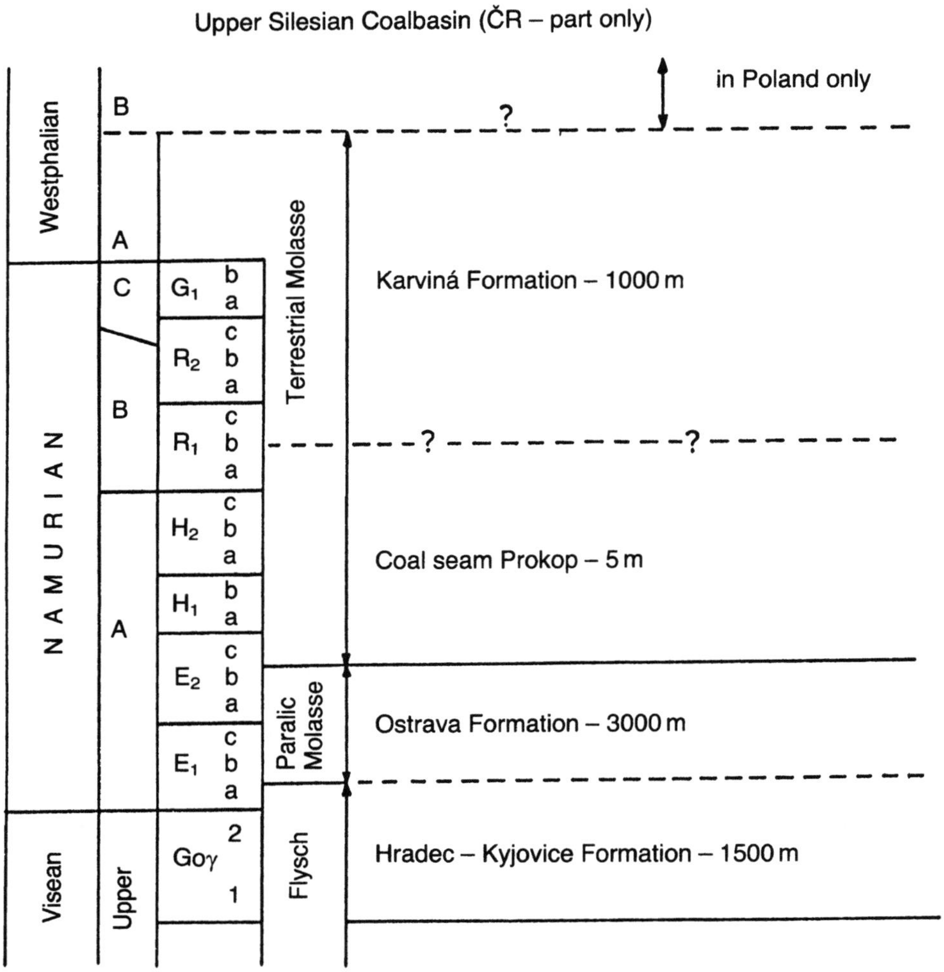

Fig. 13. Stratigraphic table of the Upper Silesian Basin

of the Namurian B, a period about twice as long as the time needed for the deposition of the 3000-m-thick Ostrava Formation at the end of the main stage of the Variscan deformation.

The area of deposition shifted further eastwards onto the platform margin of the Variscan foreland where undeformed nonmarine Upper Namurian and Westphalian deposits display minor thicknesses.

The Lower Namurian A peak of the tectonic activity matches the maximum subsidence in the foredeep and the progressive rise of the folded tectogen. Intensive volcanic and tectonic activity is documented by thick tuffite intercalations in the Ostrava Formation.

The tectogenesis of the European Variscides was the result of the collision between Gondwana and Laurasia in the circumequatorial zone (Dvořák 1985b). The forelands played an active role, since they were thrust under the tectogen. The underthrusted blocks of the continental crust of the forelands were melted because of the steep temperature gradient. The melt migrated below the median mass where it intruded as low-density granitoids. The flysch stage reflects the peak of this very rapid process.

Whereas the deposition of the Andělská Hora Formation at the base of the succession needed approximately 20 Ma and that of the Horní Benešov Formation about 15 Ma, the sedimentation of the Moravice Formation took place in only 3 Ma and that of the Hradec-Kyjovice Formation (including the more than 1300-m-thick infill of the youngest zone of maximum subsidence underlying the Ostrava Formation) perhaps 2 Ma. The flysch facies reflects the rapid rise of the central massif due to the intrusion of granitoids and the subsequent volcanism.

Whereas the foreland blocks in Moravia and Silesia had been underthrusted (subclucted) from the ESE to the WNW, the zones of maximum subsidence shifted in the opposite direction from the median mass onto the foreland, i.e., from the WNW to the ESE. The individual flysch troughs (the zones of maximum subsidence) mark the place of underthrusting or subduction of the various blocks. This process resulted in the significant spatial reduction of the original sedimentation area. However, no increased thickness of the Earth's crust is observed today because of the melting in the deepest part of the crust. This mechanism is deduced from the extensive positive gravity anomaly as well as from the preserved reduced thickness of the crust of less than 30 km (Blížkovský et al. 1977).

References

Blížkovský M, Dvořák J, Kadlec E, Novotný A (1977) Interpretation of derived gravity maps from the Jeseníky Mts. and adiacent areas. Sb Geol Věd Užitá Geofyz 14: 7–21

Chlupáč I (1981) Homalonotid trilobites from the metamorphic Devonian of the Hrubý Jeseník Mts., Czechoslovakia. Čas Mineral Geol 26: 361–370

Dvořák J (1973) Synsedimentary tectonics of the palaeozoic of the Drahany Upland (Sudeticum, Moravia, Czechoslovakia). Tectonophysics 17: 359–391

Dvořák J (1975) Interrelationship between the sedimentation rate and the subsidence during the flysch and molasse stage of the Variscan geosyncline in Moravia (Sudeticum). Neues Jb Geol Paläontol Monatsh 6: 339–342

Dvořák J (1980) Geotectonic conditions of the forming and extinction of the reef complex, notably in the Devonia of Moravia. Věst Ústř Úst Geol 55: 203–208

Dvořák J (1985a) Tektogeneze devonu a spodního karbonu Horno-slezské pánve. Zem Plyn Nafta 30: 1–14

Dvořák J (1985b) The development of Variscan Orogene in Europe. Congr Int Strat Geol Carbonifere 10, Madrid, 1983, 3: 549–551

Dvořák J (1986a) The Famennian of Moravia (ČSSR): the relation between tectonics and sedimentary facies. Ann Soc Géol Belg Spec Vol Aachen 1986: 131–136

Dvořák J (1986b) Autochtonous basinal sediments of the Variscan Flysch in the Nízký Jeseník Mountains (Moravia, ČSSR). Neues Jb Geol Paläontol Monatsh 1986: 80–84

Dvořák J (1989) Beziehungen zwischen Tektonik und Paläogeographie im mährischen Karbon. Geol Jb Hessen 117: 37–51

Dvořák J (1990a) Paleofacial and paleoclimatic aspects of Devonian and Carboniferous sedimentation in Rhenohercynikum and Moravia. Věst Ústřed Ústavu Geol 65: 65–74

Dvořák J (1990b) Geology of the Palaeozoic sediments of the deep borehole Jablůnka 1 (Beskydy Mts, NE Moravia) – comparison with the deep borehole Münsterland 1. Sbor Geol Věd Geol 45: 65–90

Dvořák J (1992) Geology of the Carbonate Evolution of the Devonian and the Lower Carboniferous near Grygov, Přerov, Sobíšky and Hranice. Scr Univ Purkyn Brun Geol 21: 37–62

Dvořák J (1994a) Variský flyšový vývoj v Nízkém Jeseníku na Moravě a ve Slezsku. Czech Geol Surv Spec Pap 3: 77 pp

Dvořák J (1994b) Geology of the Palaeozoic rocks of the Šternberk-Horní Benešov zone (Nízký Jeseník Mts, northern Moravia). Sbor Geol Věd Geol 48: 50–101

Dvořák J, Paproth E (1988) Trends of the Variscan development near the SW border of the East European Platform. Z Angew Geol 34: 353–359

Dvořák J, Friáková O, Lang L (1976) Block structure of the old basement as indicated by the facies development of the Devonian and the Carboniferous in the Moravian Karst (Sudeticum, Moravia, ČSSR). Geol Palaeontol 10: 153–160

Dvořák J, Friáková O, Hladil J, Kalvoda J, Kukal Z (1987) Geology of Paleozoic in the vicinity of quarries of Mokrá Cement Factory (Moravian Karst). Sbor Geol Věd Geol 42: 41–88

Dvořák J, Friáková O, Kullman J (1988) Influence of Volcanism on Upper Devonian Black Limestone and shale Deposition, Czechoslovakia. Devonian of the World, II. Can Soc Petrol Geol Mem 14: 393–397

Havlena V (1982) The Namurian deposits of the Upper Silesian Coal Basin. Rozpr Česk Akad Věd Řada Mat Přír Věd 92 (7): 80

Kukal Z (1980) The sedimentology of Devonian and Lower Carboniferous deposits in the western part of the Nízký Jeseník Mountains, Czechoslovakia. Sbor Geol Věd Geol 34: 131–207

Kumpera O (1983) Geologie spodního karbonu jesenického bloku. Knih Ústřed Ústavu Geol 59: 172

Přichystal A (1990) Hlavní výsledky studia paleozoického vulkanismu ve šternbersko-hornobenešovském pruhu (Nízký Jeseník). Sb Geol Věd Ložisk Geol mineral 29: 41–66

Skoček V (1979) Litologie paleozoika v karpatské předhlubni (úsek střed). Sb Geol Věd Geol 33: 7–46

VIII.B.2 Structure

H. FRITZ and F. NEUBAUER

1 Introduction

The Moravo-Silesian Parautochthon comprises foreland-related units along the eastern margins of the Bohemian Massif (Fig. 1). It represents an assemblage of units which escaped extensive tectonic transport during Variscan orogeny. This assemblage exhibits a variety of units including a Cadomian basement and Silurian to Carboniferous cover sequences. Their structural evolution involved late Proterozoic to Cambrian consolidation, Silurian-Devonian extension, Carboniferous plate convergence, post-Variscan extension, and probable effects of Alpine collision. The basement units are exposed in two tectonic windows; the Svratka window in the north and the Thaya window in the south (Figs. 1, 2a). Cover sequences occur mainly east of the Svratka window (Fig. 1).

2 Tectonostratigraphic Units

The basement units include: (1) the late Proterozoic to early Cambrian composite Thaya and Brno batholiths; and, (2) a Cadomian pre-intrusive complex. This complex consists of the pre-intrusive metamorphic Deblin Fm. and probable equivalents in Austria (Therasburg Fm.; Fig. 1). Post-Cadomian cover sequences comprise: (1) A Silurian to lower Carboniferous (Tournaisian) pre-flysch sequence including platform and continental slope sequences with quartzite, shale, limestones and mafic volcanics; (2) Lower and Middle Visean flysch which grades partly to Late Visean molasse-type sediments (Culm); and, (3) Namurian to Westfalian coal-bearing molasse (Ostrava Fm.: Dvorak 1973, 1982, 1989). Post-Variscan sediments of late Westfalian to Permian age occur in the NNE-trending Boskovice graben.

3 Metamorphism

Proterozoic metamorphism (up to amphibolite facies) is evident in the Deblin Fm. and in some xenoliths within the Brno batholith (Jaros and Misar 1976). In the Therasburg Fm. the age of initial metamorphism is uncertain because of an intense Variscan overprint which systematically increases from north to south. Paleozoic sedimentary bodies exposed in the CR and the late Proterozoic Brno batholith record a very low grade metamorphic overprint. However, low grade metamorphism has been reported from the Thaya batholith in Austria (Höck 1975, VIII.C.3). In addition, new $^{40}Ar/^{39}Ar$ mineral ages of white mica from the Brno granitoid body gave late Proterozoic plateau ages without hints of post-intrusive reheating. This is contrasted by heating experiments on pegmatitic white mica from the Thaya batholith which indicate diffusive argon loss in low temperature heating increments (Dallmeyer, unpubl. data).

4 Variscan Collisional Structures

The regional distribution of basement units largely reflects kinematics of Variscan tectonics. A more detailed picture may be obtained by investigation of map-scale and small-scale structures, which suggest a two-step Variscan collisional history, and reveal systematic differences in structural associations from north to south and between basement and cover sequences.

Basement units exposed along western margins of the Moravo-Silesian zone form two dome-like structures; the Thaya and Svratka windows (Figs. 1, 2a). Both windows are interpreted as result of basement duplexes during NE-directed thrusting in an oblique collision regime (Fritz and Neubauer 1993). Kinematics associated with nappe stacking reflect noncoaxial flow towards the NE and produced penetrative

Dallmeyer et al. (Eds.)
Pre-Permian Geology of Central and Eastern Europe

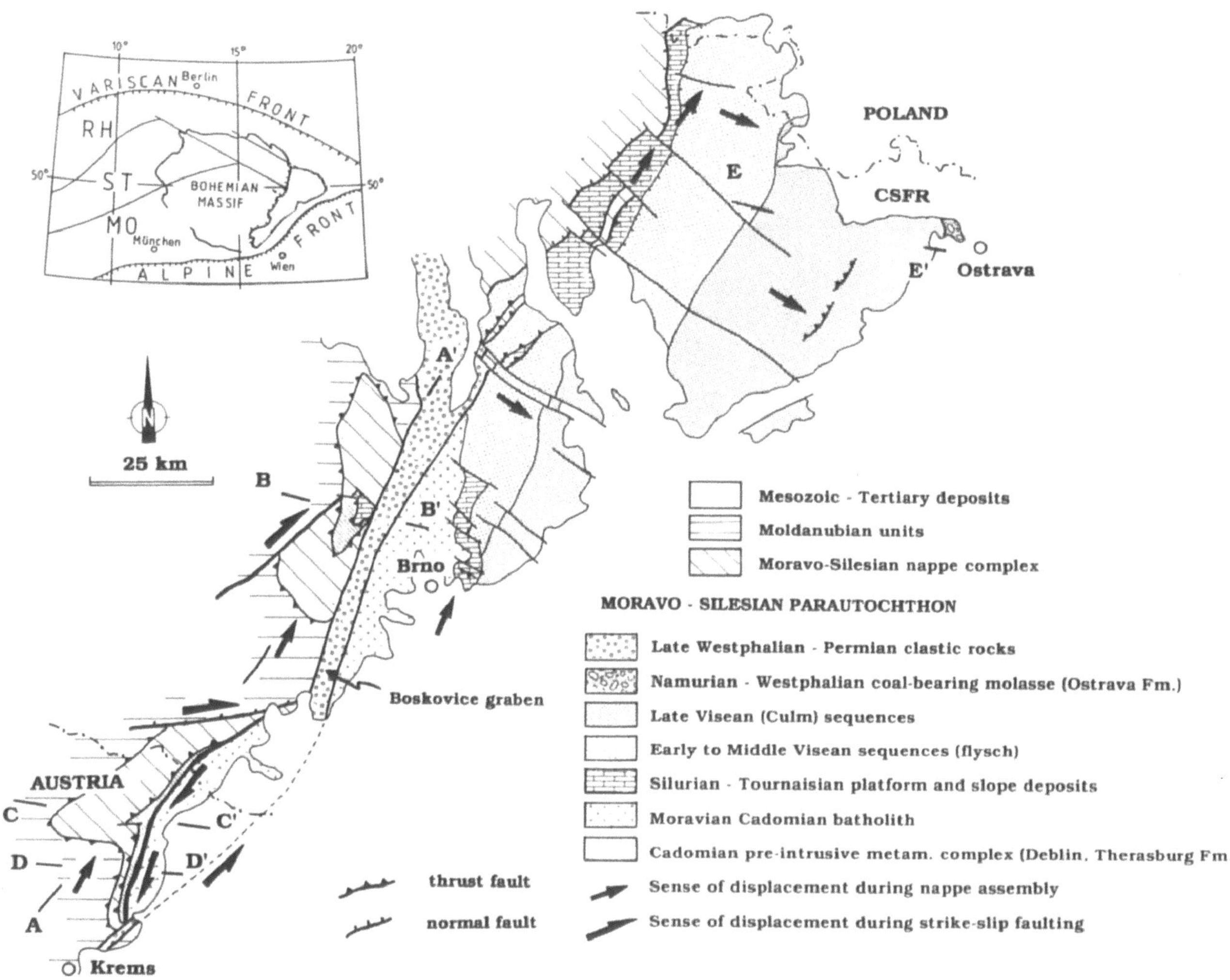

Fig. 1. Simplified tectonic map of the Moravo-Silesian unit. *Insert*: *MO* Moldanubian unit; *ST* Saxothuringian unit; *RH* Rhenoherzynian unit; *stippled* Moravo-Silesian unit. Location of cross sections are indicated: *A-A'* *B-B'* (Svratka dome), *C-C'* and *D-D'* (both Thaya dome) and *E-E'* (Culm basin)

structures in the Moravian nappes, the Deblin Fm., the Therasburg Fm., and in upper portions of the Moravo-Silesian batholith (Schulmann et al. 1991; Fritz and Neubauer 1993). Nearly exclusively NE-displacement directions are found in southern parts of the crystalline basement whereas northeastern parts are more frequently characterized by E-directed flow (Rajlich 1990; Schulmann 1990; Schulmann et al. 1991; Fritz and Neubauer 1993). A Variscan age of thrusting is proved by incorporation of fossil-bearing Devonian limestones in the Svratka dome (Schulmann et al. 1991).

A pericollisional flysch basin developed NE of the crystalline basement. Maximum subsidence occurred in the Visean (approximately coeval with nappe stacking in the hinterland: Dvorak 1973, 1982, 1989; Dallmeyer et al. 1992). The basin axis trends approximatey NE-SW and is parallel to the overall strike of the mountain chain (Dvorak 1973; Cizek and Tomek 1991).

Structural association in the Devonian platform sedimentary units support oblique collision in a dextral transpressional regime. Ductile strain estimated from Devonian fossils along northwestern margins of the Moravo-Silesian Parautochthon indicates NE-SW stretching (Rajlich 1989) which correlates with thrusting. Associated en echelon-oriented folds trend approximately NW. Thrust planes with partial overthrust of Brno granitoid over Devonian sedimentary rocks and associated folds occur in the southern part of the basin

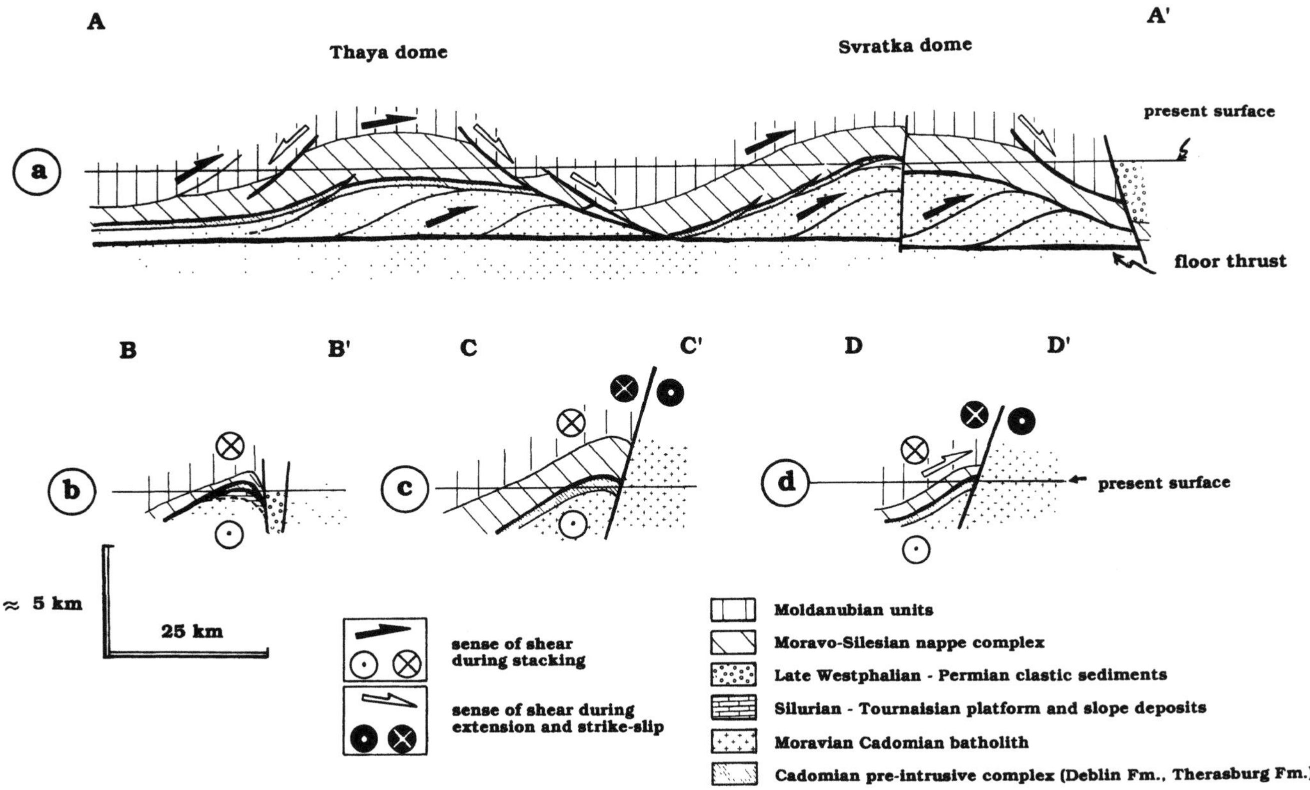

Fig. 2. Generalized tectonic cross sections through the southeastern Bohemian Massif. *Arrows* indicate displacement directions during Variscan orogeny. Locations indicated in Fig. 1

and generally trend WNW (Schulmann et al. 1991).

Northeastward stacking of basement units was progressively followed by W-E shortening and N-S extension in an overall strike slip regime (Figs. 1, 2b, c, d). Related structures include a major dextral fault zone which developed on top (west) of the Thaya and Brno batholiths and a set of dextral, second order faults. These have acute angles to the master fault and are interpreted as macro-scale Riedl shears (Fig. 1). E-vergent folds with E-directed displacement along associated faults occur apart from the strike slip zones and accommodated W-E shortening. N-S extension was as achieved by low-angle, normal faulting north and south of the domes (Fig. 2a). N-S extension affected mainly the Moravian nappes and only minor sectors of the Moravo-Silesian Parautochthon. N-trending antiformal structures of the basement (Figs. 1, 2b, c, d) and the uplift of the domes correlate with the described structures.

Southeastward propagation of a flysch front reflected the changed W-E compression and E-directed nappe stacking in the hinterland. As an effect of the eastward thrust propagation basinal sequences where increasingly incorporated in nappe stacking and suffered E-directed imbrication (Figs. 1, 3). A hinterland-dipping imbricate fan developed out of a master fault. The basal decollement is situated within Devonian limestones, close to the crystalline basement (Cizek and Tomek 1991). NW-directed back thrusts may be observed in northwestern parts of the basin (Fig. 3). Fault-related folds with NNE-trending axes frequently occur in northwestern sectors of the foredeep and die out towards the southeast. Stretching directions progressively change towards southeast (Rajlich 1989), and strain intensity decreases from NW to SE.

Synsedimentary WNW-ESE-oriented faults split the basin in blocks with contrasting sedimentary evolutions (Dvorak 1973). These faults record major horizontal displacement and may be interpreted as tear faults associated with SE-directed thrusting.

The collisional environment was terminated by deposition of coal-bearing molasse-type sedi-

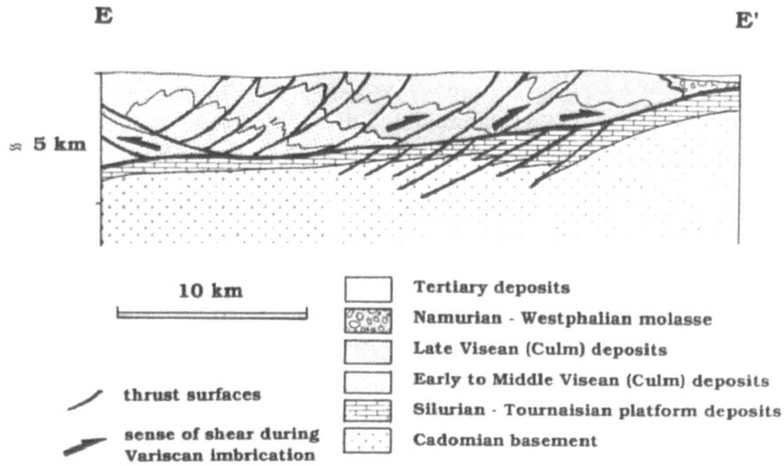

Fig. 3. Interpreted seismic section through the Moravo-Silesian basin. (After Cizek and Tomek 1991)

ments of the Ostrava Fm. along eastern margins of the pericollisional basin.

5 Postcollisional Strike-Slip Displacement

Late Westfalian to Permian sedimentary rocks occur locally within synsedimentary graben fillings (Boskovice graben) NW of Brno (CR) and near Krems (Austria) (Fig. 1). Sedimentation was controlled by a sinistral transtensional fault zone with an horizontal offset of up to 30 km. This fault is part of a conjugate set of WNW striking dextral, and NNE striking sinistral faults which accommodated bulk N-S shortening (Handler et al. 1991; Brandmayr et al., in press). The faults penetrated the already consolidated Bohemian Massif tectonic unit which include Moldanubian and Moravo-Silesian units.

6 Alpine Reactivation

The Alpine/Carpathian front is located within a few kilometers of the Bohemian Massif along its southern and southeastern margins (Fig. 1, insert). Therefore, reactivation of late Variscan faults during Alpine collision (Apulian microplate) with the Bohemian Massif was likely. Arguments for Alpine reactivation of the conjugate set of faults arise from 40Ar/39Ar spectra of white mica from mylonites of bouth sets of the Moldanubian shear zones (Brandmayr et al., in press). Low-temperature heating increments record apparent ages of approximately 170–200 Ma, and are interpreted to record the resetting of the intracrystalline argon isotopic system during Alpine deformation. In a similar way, reactivation of synsedimentary faults within the pericollisional Culm basin is likely due to convergence of the Carpathian belt.

7 Final Remarks

Structural evolution of the Moravo-Silesian Parautochthon reflects Variscan collision in a dextral transpressive regime (e.g., Matte et al. 1990) as a result of indentation of Moldanubian units with the Moravo-Silesian block (Fritz and Neubauer 1991). Inner parts of the orogen (metamorphic basement units and parts of the Devonian platform sedimentary rocks) were incorporated in a regional wrench corridor and record NE-directed thrusting, progressively followed by E-displacement. Southern parts of the Moravo-Silesian Parautochthon are interpreted as a root zone during Variscan collision. Here structures related to stacking are preferedly preserved. In external parts, a pericollisional basin formed due to lithospheric bending (Dallmeyer et al. 1992). In the pericollisional flysch basin a second deformation increment involving SE-directed imbrication developed. We assign progressive clockwise change of displacement directions as recorded in the Moravo-Silesian units to be typical for dextral transpressional orogens (Fritz and Neubauer, 1993).

References

Brandmayr M, Dallmeyer RD, Handler R, Wallbrecher E (1994) Conjugate shear zones in the southern Bohemian Massif: implications for Variscan vs. Alpine tectonothermal activity. Tectonophysics (in press)

Cizek P, Tomek C (1991) Large-scale thin-skinned tectonics in the eastern boundary of the Bohemian Massif. Tectonics 10: 273–286

Dallmeyer RD, Neubauer F, Hoeck V (1992) 40Ar/39Ar mineral age controls on the chronology of late Paleozoic tectonothermal activity in the southeastern Bohemian Massif, Austria (Moldanubian and Moravo-Silesian zones). Tectonophysics 202: 135–153

Dvorak J (1973) Synsedimentary tectonics of the Palaeozoic of the Drahany Upland (Sudeticum, Moravia, Czechoslovakia). Tectonophysics 17: 359–391

Dvorak J (1982) The Devonian and Lower Carboniferous in the basement of the Carpathians south and southeast of Ostrava (Upper Silesian Coal Basin, Moravia, Czechoslovakia). Z Dtsch Geol Ges 133: 551–570

Dvorak J (1989) Beziehungen zwischen Tektonik und Paläogeographie im mährischen Karbon. Geol Jahrb Hessen 117: 37–51

Fritz H, Neubauer F (1991) Tectonics of the SE Bohemian Massif as a result of Early Carboniferous indentation. Terra Abstr 3: 206

Fritz H, Neubauer F (1993) Kinematics of crustal stacking and dispersion in the south-eastern Bohemian Massif. Geol Rundsch 82: 556–565

Handler R, Brandmayer M, Wallbrecher E (1991) The Rodl Shear Zone in the southern Bohemian Massif. Zentralbl Geol Paläont 1: 69–86

Höck V (1975) Mineralzonen in Metapeliten und Metapsammiten der Moravischen Zone in Niederösterreich. Mitt Geol Ges 67: 49–60

Jaros J, Misar Z (1976) Nomenclature of the tectonic and lithostratigraphic units in the Moravian Svratka Dome (Czechoslovakia). Vestnik UUG 51: 113–122

Matte P, Maluski H, Rajlich P, Franke W (1990) Terrane boundaries in the Bohemian Massif: Result of large-scale Variscan shearing. Tectonophysics 177: 151–170

Rajlich P (1989) Strain anlysis of Devonian fossils from the NE part (Rhenohercynian zone) of the Bohemian Massif. Ann Tecton 3: 44–51

Rajlich P (1990) Strain and tectonic styles related to Variscan transpression and transtension in the Moravo-Silesian basin, Bohemian Massif, Czechoslokia. Tectonophysics 174: 351–367

Schulmann K (1990) Fabric and kinematic study of the Bites orthogneiss (southwestern Moravia): result of large-scale northeastward shearing parallel to the Moldanubian/Moravian boundary. Tectonophysics 177: 229–244

Schulmann K, Ledru P, Autran A, Melka R, Lardeaux JM, Urban M, Lobkowics M (1991) Evolution of nappes in the eastern margin of the Bohemian Massif: a kinematic interpretation. Geol Rdsch 90: 73–92

VIII.B.3 Igneous Activity (Cadomian Plutonism in the Moravo-Silesian Basement)

F. Finger, G. Frasl, A. Dudek, E. Jelínek, and M. Thöni

1 Introduction

The extraordinary abundance of plutonic basement rocks of Cadomian age constitutes certainly one of the most outstanding lithological features of the Moravo-Silesian unit.

Although only to a minor part directly exposed at the surface, the plutonites are known from drillings and geophysical reseach to occupy at least one third of the entire basement. Most plutonites are found in the southwestern sector of the Moravo-Silesian unit, where they form a large coherent batholitic terrain. It is uncertain, how far this batholitic terrane actually continues in northern direction below the thick Devonian and Carboniferous Silesian sediments, because in this area only few drillings reached the crystalline basement (Figs. 1, 3). Conversely, the northeastern half of the Moravo-Silesian basement is already known to host only few Cadomian plutons and to consist mainly of a pre-batholitic monotonous, metamorphic series of volcano-sedimentary derivation.

At the surface, the Cadomian plutonites can only be studied in the outermost west of Moravia and Silesia, where basement rocks crop out in the **Thaya Dome**, the **Svratka Dome**, the **Brno Massif** and the **Keprnik and Desna Domes** (Fig. 1).

Due to the increasing Variscan deformation of the Moravo-Silesian basement towards the Moldanubian boundary (Schulmann 1990), there are some uncertainties about the actual western limitation of the Cadomian plutonism. Most regional geologists agree that the very characteristic, strongly sheared granitoid pencil gneisses of western Moravia and Silesia (Bittesch gneiss, Keprnik gneiss – Fig. 1) still represent Cadomian plutons, although the bodies are probably more or less dislocated from the large autochthonous Moravo-Silesian basement block, which has been outlined and described as Brunovistulicum by Dudek (1980).

Some authors argue that even some bodies of granitoid gneisses deep within the classic Moldanubian unit (Dobra gneiss, Spitz gneiss) might represent a Variscan-reworked Moravo-Silesian plutonic basement (e.g., Finger and Frasl 1990). This concept is, however, still under debate, and reliable age data are missing. We will, therefore, preliminarily not consider these possible Cadomian plutons inside of the Moldanubian realm.

2 Regional Distribution and Petrographic Outlines of the Plutonic Rock Types

2.1 Synopsis

The large batholitic terrain in the southwestern half of the Moravo-Silesian basement consists of two separate domains, the smaller **Thaya (Dyje) Dome** and the larger **Brno Batholite** (Figs. 1–3). Both these domains are today dissected from each other by a major post-Variscan sinistral fault system (Diendorf fault, Boskovice furrow), but were formerly connected in such a manner that the eastern margin of the Thaya Dome (the Thaya Batholite, see below) joined the Brno Batholite in the west (see Schermann 1965). The size of the original coherent batholitic complex (Thaya Batholite + Brno Batholite) can be estimated at least at 10 000 km^2; the all-over Cadomian age is proven by several radiometric data (Dudek and Melkova 1975; Scharbert and Batik 1980; Van Breemen et al. 1982).

To the northeast of the Brno Batholite, some separate plutonic bodies have been found (**massifs of Olomouc, Vlkoš, Rusava and Jablunkov**), the age of which is also regarded as Cadomian (Dudek 1980).

Furthermore, there are the allochthonous to parautochthonous bodies of **granitoid gneisses along the western margin of the Moravo-Silesian unit** (western Thaya Dome, Svratka Dome,

Dallmeyer et al. (Eds.)
Pre-Permian Geology of Central and Eastern Europe

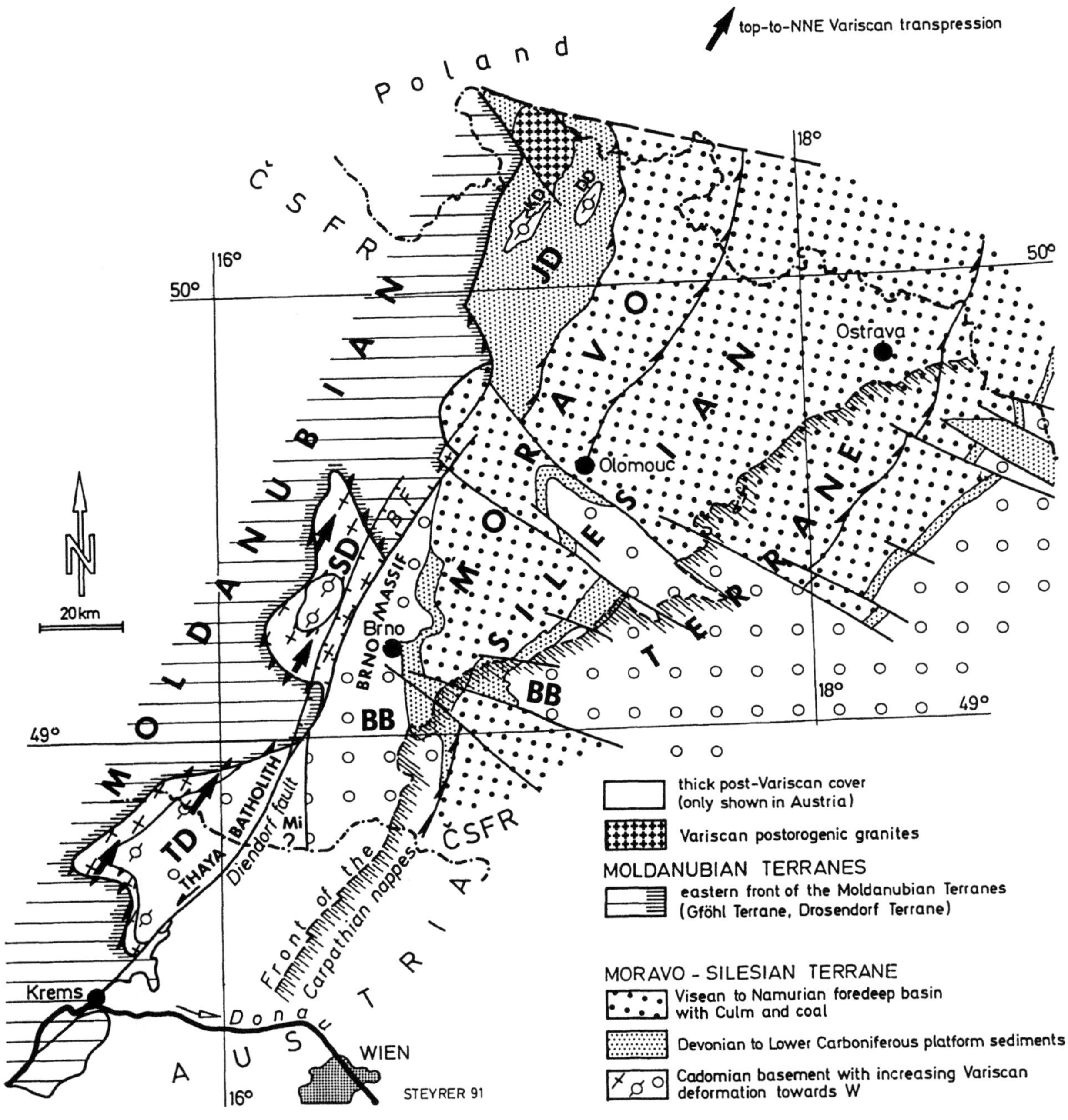

Fig. 1. Sketch map of the Variscan units at the southeastern side of the Bohemian Massif, where the Variscides descend below the Neogene Molasse Basin and the front of the Carpathian Flysch Nappes of Moravia (Czech Republic). Geological subdivisions mainly after Matte et al. (1990). *BB* Brno Batholite; *DD* Desná Dome; *JD* Jesenik Dome; *KD* Keprnik Dome; *TD* Thaya Dome; *SD* Svratka Dome

Keprnik Dome), which should be also considered as Cadomian plutons (see age determinations of Scharbert 1977; Van Breemen et al. 1982; Morauf and Jäger 1982). The primary spatial relations of these bodies relative to each other and to the Brno and the Thaya Batholite are not exactly known, probably they were located further to the southwest prior to the Variscan strike-slip event that affected the Moldanubian/Moravo-Silesian boundary (Schulmann 1990).

2.2 The Thaya Dome

In the Thaya Dome (Fig. 2), three N-S-elongated and generally west to northwest-dipping tectonic

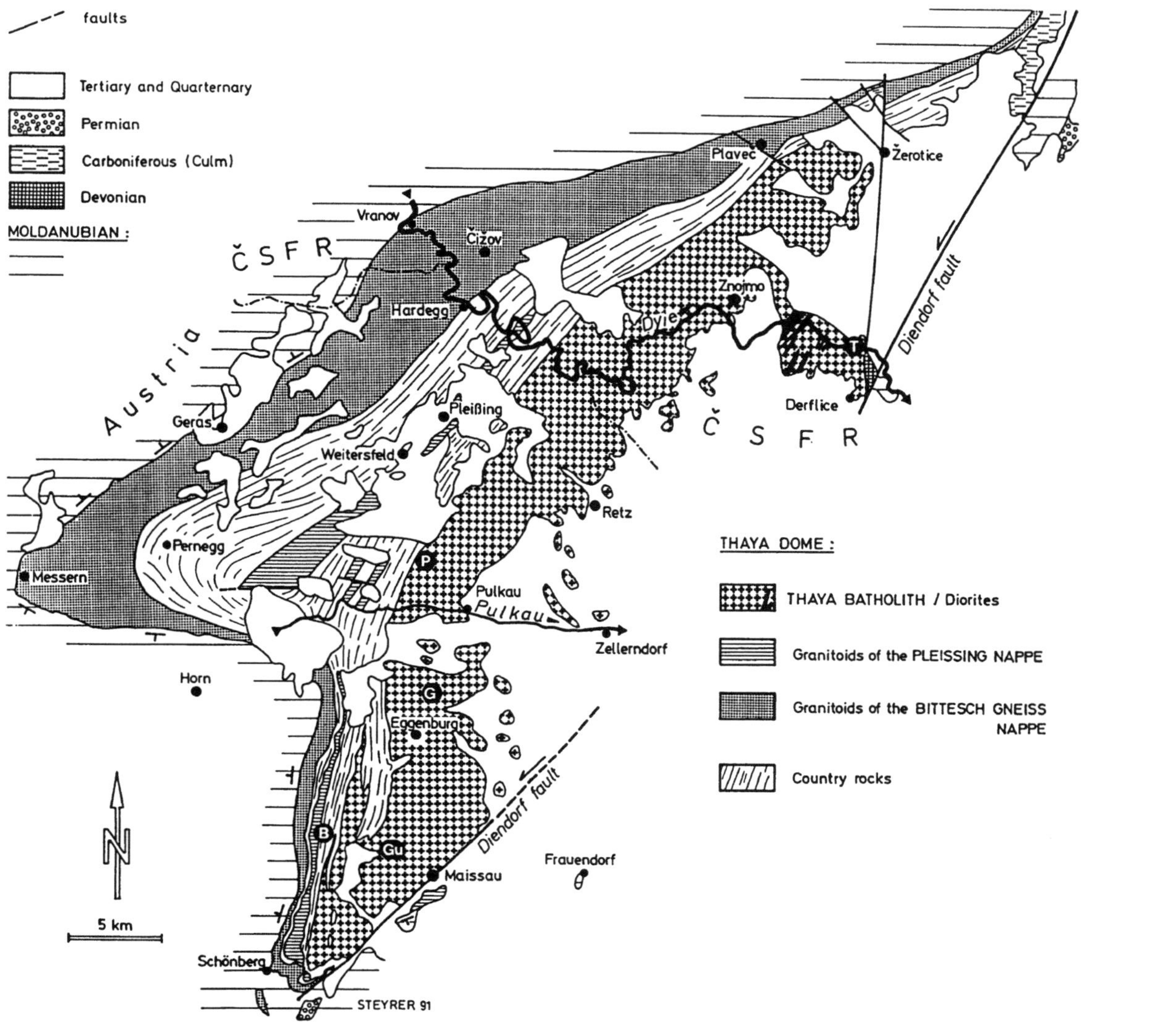

Fig. 2. Geological sketch map showing the distribution of granitoids in the Thaya Dome (data source: Fuchs and Matura (1976); Batik (1984); unpubl. mapping results of Frasl (see Finger et al. 1989). *B* Buttendorf; *G* Gauderndorf; *Gu* Gumping; *P* Passendorf; *T* Tasovice

units are usually distinguished (Preclik 1926; Frasl 1968), each consisting of a granitoid core and a schist cover. These units are from east to west the autochthonous Thaya Batholite, the allochthonous (parautochthonous?) Pleissing Nappe and the allochthonous Bittesch Gneiss Nappe.

2.2.1 The Thaya Batholite

In the Thaya Batholite, the plutonic basement is much better preserved than in the two higher, more or less allochthonous tectonic units, where most rocks became penetratively converted into intensely sheared gneisses.

Five main plutonic rock types can be distinguished *in the Austrian part of the Thaya Batholite* (mainly according to Finger et al. 1989):

The granitoids of the type Eggenburg/Maissau are medium to coarse grained rocks of granitic composition (Fig. 4a) and make up most of the southern half of the batholite. The light-gray to pink granites contain only some few percent of biotite (4–6%); in some rare cases amphibole occurs in small amounts as additional mafic phase. At the southern termination of the Thaya Batholite, the Eggenburg/Maissau granites grade into an aplitic, turmaline-bearing facies.

The Gumping type granitoids occur also in the southern half of the Thaya Batholite (note village of Gumping in Fig. 2) and form there a NNE-SSW-striking body, which is at least 8 km long and 2 km wide. The rocks contain blocky K-feldspar phenocrysts, 2–3 cm in length, are relatively rich in biotite (10–20%) and have granodioritic to quartz-monzodioritic composition (Fig. 4a).

The granitoids of the type Gauderndorf are fine grained biotite-granites to granodiorites (biotite contents 6–10%) and occur as small stocks mainly in the area north of Eggenburg (note village of Gauderndorf in Fig. 2).

The granitoids of the Pulkau/Zellerndorf type are medium-grained light granodiorites to granites (adamellites) with low biotite contents (4–7%). The rocks build up most of the northern half of the Thaya Batholite in Austria. They are macroscopically rather similar to the Eggenburg/Maissau granites in the south, both were formerly summarized as "Hauptgranit" by Preclik (1937) and Finger et al. (1989). However, the Pulkau/Zellerndorf granites and granodiorites must be distinguished from the Eggenburg/Maissau granites from geochemical grounds (see Sect. 3).

The granitoids of the type Passendorf are fine to medium grained biotite-tonalites, which occur along the northwestern margin of the Thaya Batholite between the Pulkau valley and the Thaya (Dyje) valley (Fig. 2). The biotite contents of these tonalitic rocks vary between 10 and 40%, amphibole contents are usually low, even in the melanocratic endmembers.

The ČSFR end of the Thaya Batholite was investigated by Batik (1984) and consists predominantly of biotite granites to granodiorites, which probably correspond largely to the Pulkau/Zellerndorf type of Austria. Approximately 5 km E of Znojmo, Batik (1984) mapped several small NNE striking bodys of fine grained biotite-hornblende-quartz-diorites, and, near the Diendorf fault (see Fig. 2), a distinct type of a fine grained pink granodiorite (*type Tasovice*). No pendantes of these both rock types are as yet recognized in the Austrian part of the Thaya Batholite.

Seen as a whole, felsic granodiorites and granites make up by far most of the Thaya Batholite (ca. 90%), intermediate rocks like tonalites or diorites do not exceed 10%.

2.2.2 The Pleissing Nappe

The Pleissing Nappe hosts at least two separate (north)westward dipping granitoid layers, both are strongly deformed. The higher one in the west consists of aplitic and granitic gneisses (Weitersfeld type). The lower layer in the southeast contains mainly less felsic granodioritic and tonalitic gneisses (Buttendorf type), being sometimes considerably rich in amphibole.

2.2.3 The Bittesch Gneiss Nappe

The Bittesch Gneiss Nappe comprises several variants of mostly severely sheared granites and granodiorites with up to 5% secondary muscovite and 2–7% biotite (no amphibole). The gneisses are generally leucocratic, less felsic metaplutonic rocks (e.g., tonalites, diorites) are absent in the Bittesch Gneiss Nappe.

Three different Rb-Sr WR isochrons have been published for metagranitoids of the Bittesch Gneiss Nappe. A first one by Scharbert (1977) yielded an age of 796 ± 11 Ma, a second one by Morauf and Jäger (1982) an age of 570 ± 44 Ma (both results have been interpreted as primary magmatic ages by the respective authors). Conversely, Van Breemen et al. (1982) obtained an isochron of 480 ± 50 Ma for the Bittesch Gneiss of the Svratka Dome and interpreted this isochron as being influenced by metamorphic resetting of the Rb-Sr system.

2.3 The Brno Batholite

The Brno Batholite has been investigated by Štelcl and Weiss (1986) in its small outcropping part, the Brno massif, which is about 800 km^2 in size, and by Dudek (1980) in the much larger covered part (Fig. 3), where much information is available from boreholes and from geophysics.

Generally, it was found that the Brno Batholite consists predominantly of rocks of granodioritic mode (ca. 50%). For the macroscopically somewhat variable subtypes of granodiorites, quite a row of local names is in use (see Štelcl and Weiss 1986; Dudek 1980).

The granodiorites of the Brno Batholite can be described as fine- to medium-grained, rarely coarse-grained rocks, often containing amphibole as an additional mafic phase besides biotite. A distinct body of a fine- to medium-grained leucocratic granodiorite has been mapped as *Stupava type* granodiorite by Dudek (1980) (Fig. 3).

Rocks of granitic composition are clearly less frequent than in the Thaya Batholite and

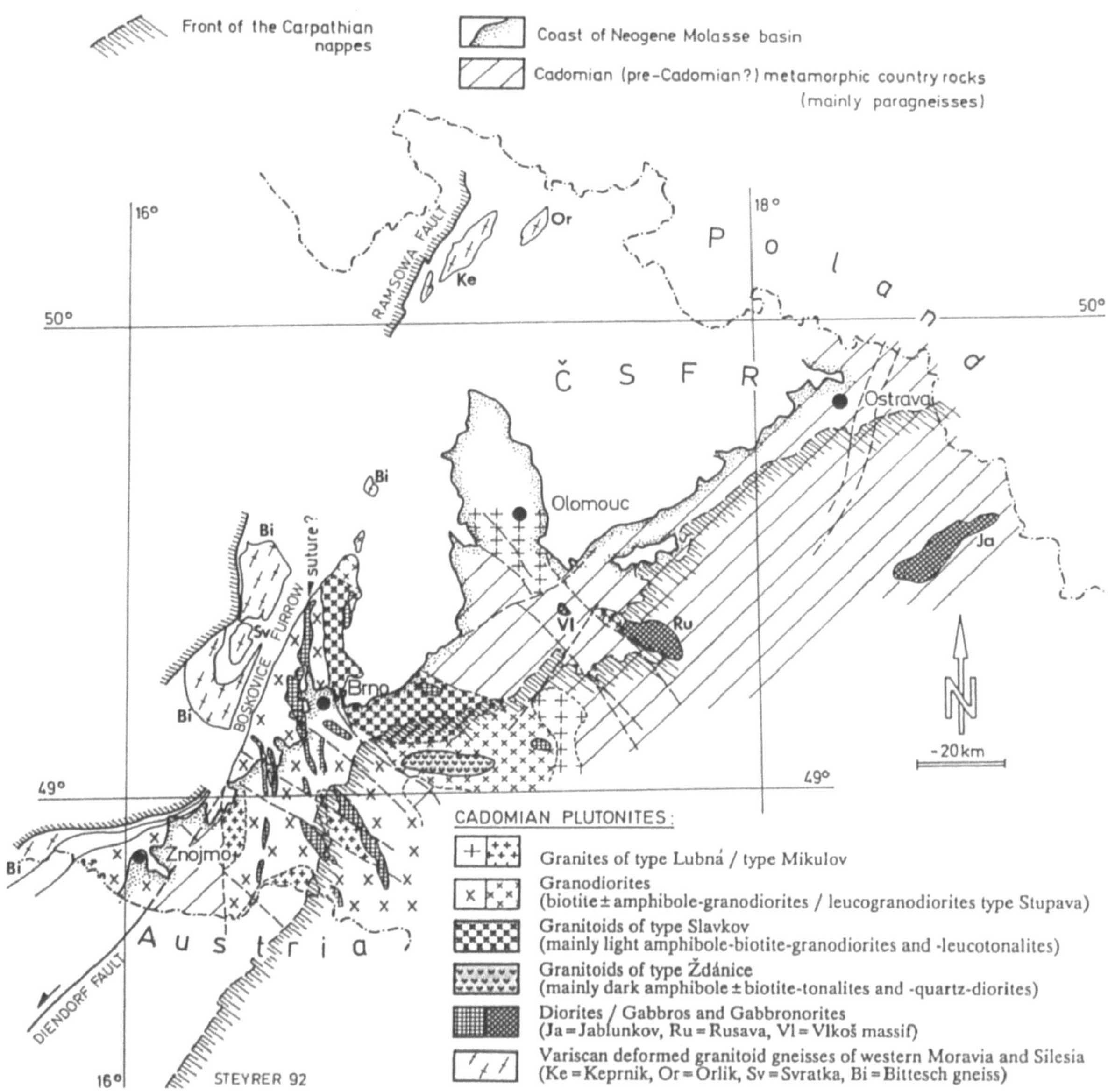

Fig. 3. Geological sketch map showing the distribution of Cadomian granitoids in the mostly covered Czech part of the Moravo-Silesian basement, as far as known from outcrops, drillings, and geophysics. Map compiled mainly according to data of Dudek (1980, and unpubl.) and official maps of the CSFR

do not exceed 20%. Two different subtypes of granite have been distinguished: a medium-grained biotite-granite in the southwestern part of the batholite (**Mikulov type**), which grades into the surrounding granodiorite, and a medium to coarse grained pinkish leukogranite (**Lubná type**), which forms an independent intrusion at the eastern margin of the batholite.

Granites similar to those of the Lubná type build up the large plutonic body in the area of Olomouc (Fig. 3), which appears to be somewhat separated from the Brno Batholite.

Unlike the Thaya Batholite, the Brno Batholite hosts also a considerable number of plutonic rocks of intermediate to basic composition. A large body of a distinct medium grained, leucocratic hornblende-biotite-tonalite to granodiorite (**Slavkov-type**), has been found to make up much of the basement east of Brno. A dark hornblende-biotite-quartz-diorite to -tonalite (**Ždánice type**) forms an almost 100 km^2 large body approximately 30 km ESE of Brno (Fig. 3).

In the median part of the Brno Batholite, a north-south-trending zone with several elongated bodies of diorite has been recognized. These dioritic rocks are usually fine- to medium-grained, have hornblende as major mafic constituent (with or without biotite), and, in part, contain enclaves of gabbros and ultrabasic rocks. One diorite in the southwestern Brno massif was dated with 584 ± 5 Ma (U-Pb zircon age) by Van Breemen et al. (1982).

In the exposed Brno Massif, the large diorite body NNW of Brno is accompanied at its eastern margin by metabasalts. Several CSFR workers argue that this formation might represent a part of a dismembered ophiolite suite (e.g., Hrouda in Štelcl and Weiss 1986), and, therefore, could indicate a Cadomian or pre-Cadomian suture or terrane boundary, respectively (see Fig. 3).

2.4 The Basic Massifs of Vlkoš, Rusava, and Jablunkov

Not directly connected with the Brno Batholite are the three basic massifs in the northeastern half of the Moravo-Silesian basement (Fig. 3), which comprise mainly gabbros and gabbronorites with layered cummulate textures.

The small Vlkoš massif, which contains also some serpentinized dunites, is penetrated by various dyke rocks, which appear to be similar in composition to intermediate granitoids of the Brno Batholite. It is therefore reasonable to assume that the basic massifs were the earliest members of the Moravo-Silesian Cadomian plutonism. The same sequence of intrusion is also suggested by the observation that gabbros locally occur as enclaves in the diorites and tonalites of the Brno Batholite.

2.5 The Allochthonous to Parautochtonous Granitoid Gneisses of Western Moravia and Silesia

The largest and most important of these gneissic bodies is the **Bittesch gneiss**, which forms an elongated unit of relatively small thickness, stretching for ca. 120 km along the Moravo-Silesian/Moldanubian boundary. It is a major constituent of the Thaya Dome and the Svratka Dome as well. Most of the currently available "granitological" data of the Bittesch gneiss come from the Austrian part of the Thaya Dome (see Sects. 2.1 and 3). Significant differences between the Bittesch gneiss of the Thaya and the Svratka Dome have, however, never been reported.

In a deeper structural level is the **Svratka gneiss** (Fig. 3), which crops out within a geological window in the core of the Svratka Dome. The rock is commonly interpreted as parautochthonous continuation of the Brno Batholite.

The **Keprnik gneiss** occurring in the centre of the Jesenik-Desná Dome (Figs. 1, 3) has been dated by Van Breemen et al. (1982) at ca. 550 Ma and is apparently also a Cadomian pluton. The rock resembles the Bittesch gneiss with reference to its structure and granitic composition (Suess 1926).

A probable equivalent of the Keprnik gneiss is the Orlik gneiss about 50 km to the east. The Orlik gneiss was recently investigated by Fišera and Patočka (1989) and interpreted as blastomylonitic granite.

3 Geochemistry

The chemistry of the Cadomian plutonism of the Moravo-Silesian basement has been studied by Dudek (1980) and Jelinek and Dudek (1993) in the Czech Republic, and by Finger et al. (1989) in the Austrian Thaya Dome. In both areas it was found that the plutonic rocks are very similar in composition to those of modern destructive plate margins.

Especially the Czech Republic sector (Table 1) provides all the features commonly regarded as typical of subduction-derived (Cordilleran) I-type plutonism (e.g., Pitcher 1983; Pearce et al. 1984), for example:

- Extremely wide compositional spectrum from gabbros over diorites, tonalites to granodiorites and granites (Fig. 4).
- Calc-alkaline alkali-lime index in the sense of Peacock (1931).
- Generally metaluminous to weakly peraluminous characteristics with a frequent occurrence of hornblende.
- Significant predominance of sodium over potassium (except for a few, more fractionated variants with high SiO_2).
- Low Sr-initial ratios of ca. 0.704 (Table 1, Fig. 7).
- Low Rb, Y, Nb, Yb, Ta and high Sr, Ba (Table 1, Fig. 5).

Table 1. Typical chemical analyses of the main plutonic rock types of the Brno Batholite (data source: Jelinek and Dudek 1993 and unpubl.; except for $^{87}Sr/^{86}Sr_i$ values, which are model initial ratios calculated with T = 550 Ma according to unpublished analyses of M. Thöni, Vienna)

No.[a]	188	99	104	73	164	203	192
SiO_2	72.01	71.83	68.36	62.46	54.69	53.49	51.78
TiO_2	0.40	0.19	0.26	0.65	1.76	1.07	0.37
Al_2O_3	13.67	14.68	15.90	17.53	16.95	16.71	16.92
Fe_2O_3	1.32	0.37	0.92	2.10	6.80	3.56	2.02
FeO	1.13	1.64	1.62	2.18	2.30	4.85	4.66
MnO	0.04	0.06	0.07	0.07	0.21	0.13	
MgO	0.50	0.76	0.78	1.90	3.94	5.02	9.41
CaO	0.99	0.88	2.44	4.46	6.23	7.04	8.75
Na_2O	4.36	3.89	5.04	4.56	3.71	3.34	1.95
K_2O	4.05	4.42	2.41	2.11	1.54	1.14	0.55
P_2O_5	0.43	0.05	0.12	0.20	0.24	0.40	0.32
CO_2	0.75	0.01	0.09		0.47	0.35	0.69
H_2O+	0.69	1.48	1.21	1.36	0.67	2.91	1.80
H_2O-		0.12	0.10	0.10	0.22	0.28	0.23
S	0.12			0.02			0.16
Total	100.46	100.38	99.32	99.71	99.79	100.29	99.61
Rb	93	115	48	24	34	13	10
Sr	190	226	734	1054	717	710	160
Ba	1095	558	972	1112	513	351	51
Nb	16	19	11	11	8	6	2
Ga	17	20	18	21	18	18	14
Y	21	18	11	15	21	19	10
Zr	153	110	141	174	106	93	15
V	22	14	2.5	68	155	147	114
Co	3.6	1.6	2.0	3.0	23.5	29.0	40.0
Cr	29	120	128	73	165	101	422
Ni	17	2	2	6	17	24	128
Cu	192		5	8	154	42	73
Pb	43	17	15	7	1	1	0
Zn	366	28	38	65	86	87	45
As	30	2	4	8	10	7	5
Sc	3.6	2.4	3.3	7.5	22.3	24.2	29.2
La	29.00	12.30	18.50	29.00	13.70	15.60	2.30
Ce	57.0	32.0	38.0	65.0	34.0	44.0	5.5
Nd	20.0	13.0	17.5	34.0	17.0	23.0	3.0
Sm	3.40	2.50	2.60	5.80	5.60	5.50	0.98
Eu	0.68	0.49	0.69	1.35	1.42	1.46	0.45
Gd	2.6	2.4	2.5	3.9	4.4	4.3	0.7
Tb	0.40	0.32	0.27	0.45	0.83	0.66	0.28
Tm	0.30	0.13	0.10	0.25	0.30	0.33	0.33
Yb	1.60	0.75	0.80	1.10	2.10	2.20	0.96
Lu	0.18	0.13	0.14	0.14	0.31	0.48	0.19
Hf	5.2	3.3	3.6	4.8	3.0	3.2	0.7
Ta	1.22	0.77	0.45	0.70	0.44	0.42	0.10
Cs	2.80	1.10	0.80	0.70	1.10	0.40	1.10
Th	18.00	5.20	2.70	3.45	3.10	1.80	0.40
U	3.6	0.5	1.5		1.5	2.0	0.2
$^{87}Sr/^{86}Sr_i$			0.704	0.704	0.703		

[a] Sample locations: 188: Lubna type granite, drilling Lubná 8, 1856.0 m
99: Mikulov type granite, drilling H. Dunajovice 1, 1809.5 m
104: Stupava type leucogranodiorite, drilling Ždánice 5, 1020.1 m
73: Slavlkov type hornblende-biotite-leucoquartzdiorite, drilling Slavkov 2, 1340 m
164: Ždánice type biotite-hornblende-quartzdiorite, drilling Unřice 5, 1976.3 m
203: biotite-hornblende-diorite, drilling Ježov 3, 2087.7 m
192: gabbronorite of Rusava massif, drilling Rusava 9, 822.3 m.

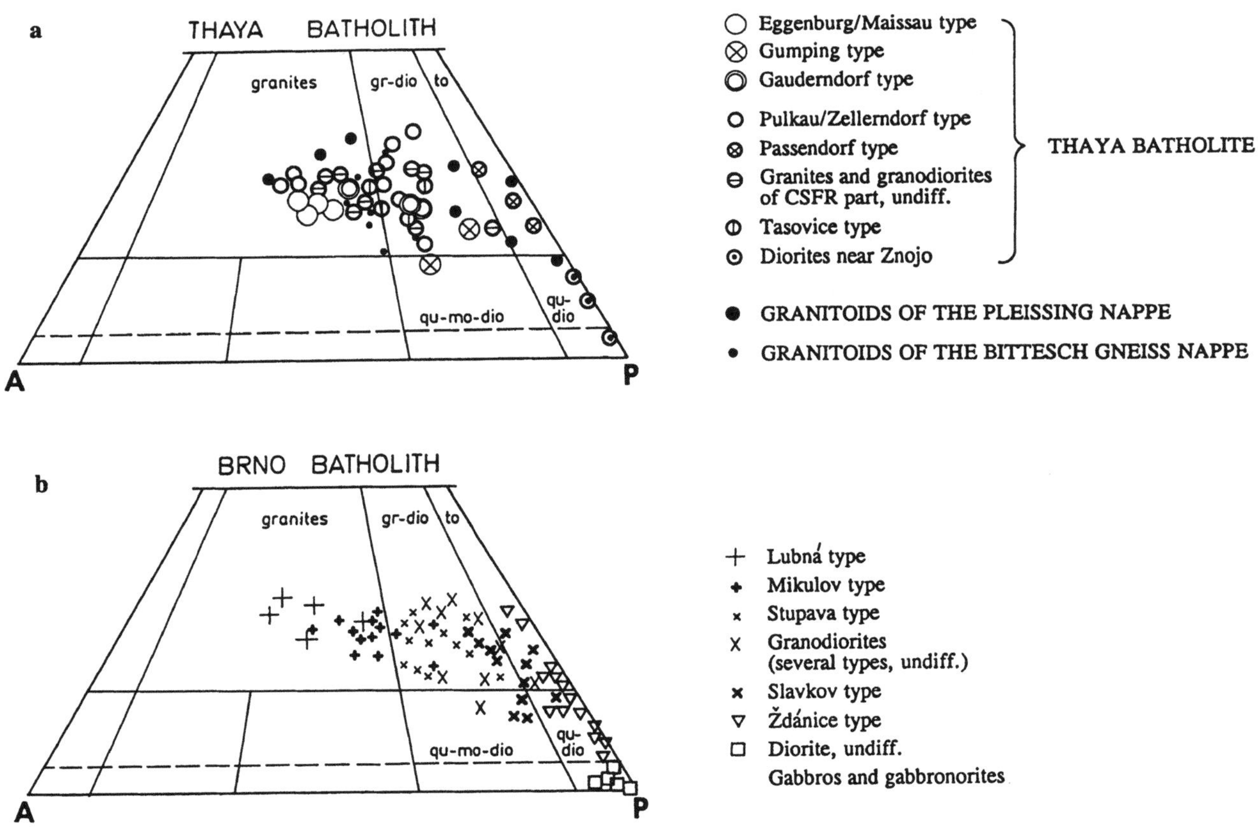

Fig. 4a,b. APQ-diagrams according to Streckeisen (1967) with plots of Cadomian plutonic rock types of the Thaya Dome (**a**) and the Brno Batholite (**b**)

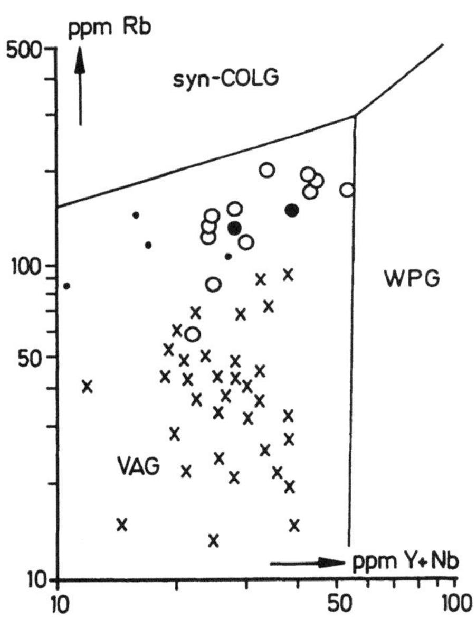

Fig. 5. Rb vs. Y + Nb diagram for the tectonic interpretation of granitic rocks (Pearce et al. 1984). *VAG* volcanic arc granites; *WPG* within-plate granites; *syn-COLG* syn-collision granites. Shown are plots of plutonic rocks of the Brno Batholite (*crosses*) and the Thaya Dome: *open circles* Thaya Batholite; *big full circles* Pleissing Nappe; *small dots* Bittesch gneiss

In the AFM plot (Fig. 6), the rocks of the Brno Batholite follow a Fe-enriched, tholeitic trend, as it is often observed in the case of island arcs and immature continental arcs (see e.g., Brown 1982). The usually very low Rb and K_2O contents and Rb/K ratios of the granitoids, as well as the primitive Sr isotope characteristics (Table 1), are also suggestive of an island arc setting rather than of a continental arc.

On the other hand, it can be shown by means of Harker diagrams that CaO is quite low in the Brno Batholite, whereas Na_2O is rather high at a given SiO_2, when compared to typical modern island arc plutonic suites of the Pacific (e.g., Whalen 1985). According to the work of Plank and Langmuir (1988), low CaO/Na_2O ratios appear to be indicative of arcs built upon thick crust. If the Brno Batholite was, in fact, emplaced within a considerably thick crustal pile, we would expect this crust, however, to be compositionally rather immature and an island arc type crust rather than a cratonic crust, because no signs of mature crust are recognizable in the chemistry and isotopic systematics of the plutonic rocks.

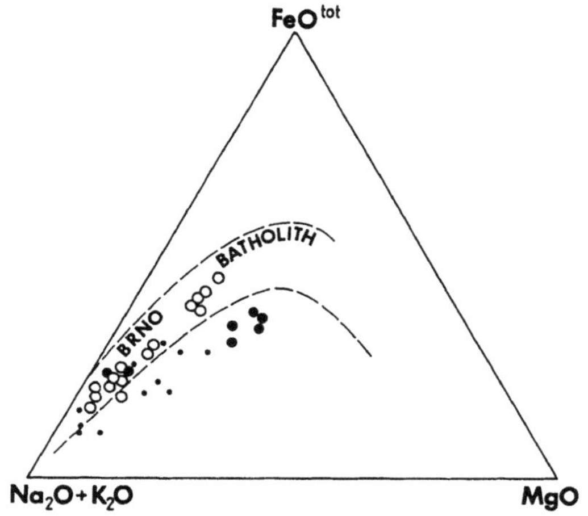

Fig. 6. AFM diagram showing the Fe-enriched tholeitic trend of the Brno Batholite and plots of granitoids of the Thaya Dome (symbols as in Fig. 5). Note the different, calc-alkaline trend of the granitoids of the Pleissing and Bittesch Gneiss Nappes (*full circles* and *dots*)

A modern lithological and chemical investigation of the metamorphic country rocks of the Brno Batholite is presently missing, but would certainly shed more light on the primary batholitic environment.

In the **Austrian Thaya Dome**, the SiO_2-variation of the Cadomian plutonism is not so extended and, on average, much more felsic (ca. 60–75% SiO_2 – see Table 2); however, the rocks can be also generally outlined as I-type granitoids on the basis of their high Na_2O contents and low mol $Al_2O_3/CaO + Na_2O + K_2O$ ratios (Chappell and White 1974). Like the rocks of the Brno Batholite, they display a trace element characteristics typical of volcanic arcs (Pearce et al. 1984 – see Fig. 5). Nevertheless, it is noticeable that Rb, e.g., is on average clearly higher in the Thaya Dome than in the Brno Batholite (Fig. 5).

Unlike the Brno Batholite, for which a common magma source and a classical differ-

Table 2. Typical chemical analyses of the main plutonic rock types of the Thaya Dome (data source: Finger et al. 1989, except for $^{87}Sr/^{86}Sr_i$ values, which are model initial rations calculated with T = 550 Ma according to analyses of Scharbert and Batik 1980)

	Thaya Batholith					Pleissing Nappe		B. G.
Sample No.[a]	Fi-37	Fi-38	Fi-28	Fi-35	Fi-32	Fi-23	Fi-64	Fi-27
SiO_2	62.69	63.20	71.06	72.46	73.03	61.73	69.60	74.00
TiO_2	0.59	0.91	0.29	0.14	0.19	0.65	0.38	0.06
Al_2O_3	17.76	15.70	14.95	13.98	13.54	14.97	14.90	14.37
FeO_{tot}	4.59	5.05	2.16	2.06	2.00	5.14	2.48	1.06
MnO	0.10	0.10	0.01	0.01	0.03	0.06	0.04	0.01
MgO	1.60	1.53	0.64	0.23	0.22	4.02	0.88	0.30
CaO	5.02	3.47	2.03	1.03	1.12	4.62	0.99	1.31
Na_2O	3.94	4.07	4.43	4.30	4.29	2.63	4.07	3.93
K_2O	2.01	3.64	3.12	4.54	4.35	3.91	4.92	3.55
P_2O_5	0.25	0.28	0.07	0.03	0.05	0.38	0.27	0.06
LOI	1.24	1.14	1.19	0.84	1.10	1.00	1.43	1.28
Total	99.79	99.09	99.95	99.62	99.96	99.11	99.96	99.93
Rb	58	142	123	142	181	146	207	145
Sr	441	245	222	316	76	706	116	313
Ba	556	831	771	543	608	1291	860	618
Nb	9	12	7	12	12	16	6	8
Y	13	25	17	13	33	23	7	8
Zr	131	295	167	107	224	173	196	70
$^{87}Sr/^{86}Sr_i$				0.710	0.709			0.713

[a] Sample locations: Fi-37: Passendorf type tonalite, roadcut near Passendorf
Fi-38: Gumping type granitoid, near village of Gumping
Fi-28: Gauderndorf type granitoid, quarry at Gauderndorf
Fi-35: granodiorite of Pulkau/Zellerndorf type, roadcut N Pulkau
Fi-32: granite of Eggenburg/Maissau type, quarry at Feldberg
Fi-23: Buttendorf type metatonalite, Buttendorf
Fi-64: Weitersfeld type granite gneiss, quarry at Weitersfeld
Fi-27: Bittesch gneiss, Klein Meiseldorf.

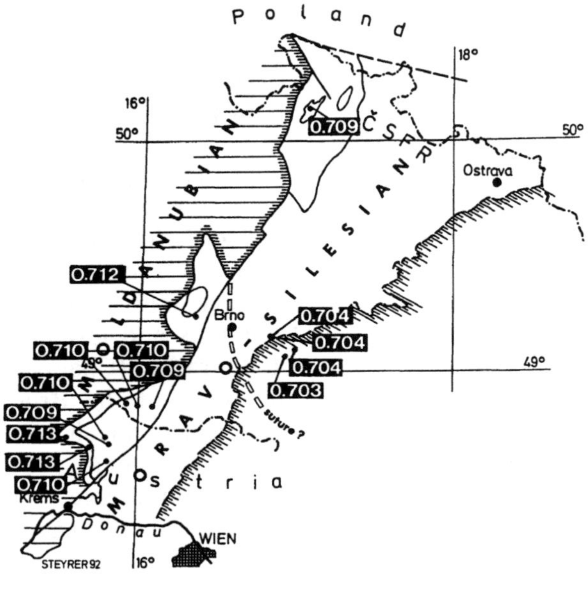

Fig. 7. Sketch map of the Moravo-Silesian basement according to Fig. 1 with $^{87}Sr/^{86}Sr$ model initial ratios (T = 550) of Cadomian plutonic rocks (data source: Thaya, Svratka and Keprnik Dome: Morauf and Jäger 1982; Scharbert and Batik 1980; Van Breemen et al. 1982. Brno Batholite: four unpublished analyses by M. Thöni, Vienna)

entiation mechanism by fractional crystallization can be inferred from field evidence and from the regular element variations in chemical diagrams (Jelinek and Dudek 1993), the Austrian Thaya Dome is more composite and it hosts at least three separate plutonic suites.

A first suite is constituted by the diorite-tonalite-granodiorite-granite association of the northern Thaya Batholite (diorites of Znojmo; Passendorf and Pulkau/Zellerndorf type granitoids – see Sect. 2.2.1). This northern suite most closely matches the plutonic suite of the Brno Batholite chemically, in that it is also high in Ba and Sr, rather low in Rb at low SiO_2, tholeitic in the AFM diagram (Fig. 6), and calc-alkaline in terms of the alkali-lime index. From this point of view, the northern Thaya Batholite could be derived from the same or a similar parental magma as the Brno Batholite. However, Scharbert and Batik (1980) reported Rb-Sr isotope data for granites and granodiorites of the northern Thaya Batholite, which indicate high Sr model initial ratios of ca. 0.710 (Fig. 7). These high values contrast with the generally low $^{87}Sr/^{86}Sr$ ratios of the Brno Batholite and imply a considerable role of old crustal components in the magmatic evolution of the northern Thaya Batholite. Further isotopic and chemical work (involving also the more primitive lithologies) will be, however, necessary to decide whether the magma source of the northern Thaya Batholite was basically different from that of the Brno Batholite, or whether the differences in isotopic composition are simply due to a combined fractionation/crustal contamination mechanism involving the same primary melt.

The **second plutonic suite** of the Thaya Dome is defined by the Gumping and Eggenburg/Maissau type granitoids of the southern Thaya Batholite (see Sect. 2.2.1). This suite is also tholeitic in the AFM diagram (Fig. 6), but alkali-calcic in the sense of Peacock (1931), and it follows a K-feldspar enriched quartz-monzonitic to granitic trend in the Streckeisen diagram (Fig. 4a). Furthermore, the assembly is high in Rb and distinct HFS elements (Zr, Y) and low in Sr (Table 2), and, by that, displays some affinities to A-type granite suites (Whalen et al. 1987). Locally, even transitions into peralkaline compositions have been found.

It is important to know that rocks of this southern second suite mainly define the 551 ± 6 Ma Rb-Sr isochron published by Scharbert and Batik (1980) for the Thaya Batholite. We therefore consider the age of 551 Ma to be close to the formation age of the I/A-transitional-type granitoid suite of the southern Thaya Batholite but not necessarily valid for the entire batholite.

In view of the fact that a zircon age of 584 ± 5 Ma has been detected for a diorite of the Brno area (Van Breemen et al. 1982), it appears possible that the plutonic suite of the Brno Batholite and the chemically rather similar rocks of the northern Thaya Batholite are somewhat older than those of the southern Thaya Batholite.

The **third group of plutonic rocks** of the Thaya Dome is defined by a calc-alkaline trend in the AFM diagram. Most granitoids of the Pleissing Nappe and the Bittesch Gneiss Nappe apparently follow such a calc-alkaline trend and not the Fe-enriched tholeitic trend of the Brno and the Thaya Batholite (Fig. 6). The plutonic rocks of both nappes are, furthermore, mostly distinctly enriched in Ba and remarkably low in Y (Table 2).

However, a particularly low Rb/Sr ratio allows subdivision of the generally felsic rocks of the Bittesch Gneiss Nappe from most of the felsic plutonic end members of the Pleissing Nappe. It is noteworthy that similarly low Rb/Sr ratios have also been reported for the Bittesch gneiss of the Svratka Dome and the Keprnik gneiss of Silesia by Van Breemen et al. (1982), and correspondingly low Rb concentrations were measured by Fišera and Patočka (1989) in the case of the Orlik gneiss of Silesia, implying a close genetic relationship of all these gneissic bodies of the western margin of Moravo-Silesia. Perhaps these allochthonous gneisses have to be considered as an independent suite of Cadomian granitoids.

4 Evidence for a Chemical East-West Asymmetry of Plutonism

Seen as a whole, there seems to be a chemical east-west asymmetry over the Moravo-Silesian plutonic terrain. The high abundance of granites (s.s.) in the Thaya Dome and the clear predominance of granodiorites and tonalites in the Brno Batholite are a first important indicator of such a polarity. Even if we neglect the possibly younger I/A-transitional types of the southern Thaya Batholite, there remains a clear regional difference between the primitive quartzdiorite-tonalite-granodiorite lithology of the Brno Batholite and the more evolved granodiorite-granite lithology of the northern Thaya Batholite, which joined the Brno Batholite originally westwards (see Streckeisen diagrams in Fig. 4). A parallel trend can be observed on much smaller scale in the exposed Brno massif, where the hornblende contents diminish in the granodioritic variants from east to west, and the granodiorites become generally more evolved from east to west with reference to their K-feldspar/plagioclase/quartz ratios (Štelcl and Weiss 1986).

Another independent argument for an east-west compositional change of the plutonism is that larger bodies of basic end members (gabbros, gabbronorites and ultrabasics) have been found only in the (north)east of the batholitic terrain.

A conventional interpretation of all these chemical east-west asymmetries would be to propose a westerly inclined Cadomian subduction zone beneath the Moravo-Silesian basement block, an ocean and a trench in the east, and a continental hinterland in the west.

Such a tectonic model is also consistent with the Sr isotopic data available to date, and the calculated Sr model initial ratios of the plutonic rocks. The data clearly suggest an increased input of old continental crust to the plutonism from E to W (Fig. 7). In this context it is also interesting to note that the Bittesch gneiss of the Austrian Thaya Dome yielded a strikingly high Nd crustal residence age of more than 2.5 Ga (Liew and Hoffmann 1988), and that both the Bittesch gneiss and the Keprnik gneiss contain a clear pre-Cadomian, probably mid-Proterozoic memory in the U/Pb system of their zircons (Van Breemen et al. 1982). Note that the dated diorite of the Brno area lacks such a memory.

All this indicates that old continental crust became an increasingly important source of the Cadomian plutonism towards the west. The presence of an ancient continental basement could also explain why the westernmost gneisses in the Thaya Dome tend to follow the typical calc-alkaline AFM trend of modern continental arcs (Brown 1982) instead of a tholeitic trend.

The presently available data are, however, insufficient to decide whether the compositional east-west changes of the Cadomian plutonism of Moravo-Silesia are:

1. more or less gradually progressive, and, for example, triggered by an increasing thickness of a continental wedge from east to west, or
2. rather abrupt along a medial line.

The latter case, which seems presently more likely to us, could indicate an ancient terrane boundary within the pre-batholitic crust. For example, an accreted oceanic island arc terrane could have existed in the east, and a continental terrane in the west, a possible suture could have been the suspect ophiolite zone of the Brno Batholite (see Sect. 2.3).

5 Possible Plate Tectonic and Paleogeographic Implications

There are two contrasting possibilities how to place the Moravo-Silesian basement rocks paleogeographically in Cadomian times:

1. They could have been positioned near or at the margin of the Fennosarmatian craton.
2. They could have been situated in the foreland of Gondwana being part of the Pan-African orogenic belt.

While the first interpretation is the classic one, which has been mainly put foreward by CSFR workers (Suk 1986; cf. also Finger et al. 1989), the alternative "Gondwana model" was raised and increasingly supported during the past years by various experts in the Variscan event. These people argued, mainly on the basis of the tectonic subdivision and evolution of the central European Variscides, that the Moravo-Silesian unit was a part of the southern rather than of the northern flank of the mainly Gondwana-derived Variscan fold belt (Matte et al. 1990).

From the viewpoint of plutonism, a Gondwana derivation of the Moravo-Silesian unit appears well possible, because the Pan-African of the Arabian shield and NE Africa, is, in fact, characterized by a marked abundance of granitic rocks. These rocks are, moreover, often of the "subduction I-type", and very primitive in Sr-isotopic composition (Engel et al. 1980), just like, e.g., the granitoids of the Brno Batholite.

Many of the subduction-related plutonites of the Pan-African are, however, older than 600 Ma (Gass 1982). An important exception is Saudi Arabia, where subduction activity, accompanied by calc-alkaline plutonism, is thought to proceed to the end of the Pan-African orogenic cycle, i.e., almost to 500 Ma (Nasseef and Gass 1977). If the Moravo-Silesian basement is, indeed, Africa-derived, the Pan-African of Saudi-Arabia could be, therefore, a suitable point of contact.

A very characteristic signature of all Pan-African plutonic terranes is that the subduction-related calc-alkaline magmatic activity is usually postdated by subalkaline or alkaline granite intrusions with A-type affinities (Gass 1982) and, therewith, interesting parallels arise again to the Cadomian plutonism of the Moravo-Silesian unit (see Sect. 3).

On the other hand, from the viewpoint of plutonism, it is not so unproblematic to assume a derivance of the Moravo-Silesian basement from Fennosarmatia, mainly because there has been until now no real evidence for a Cadomian active plate margin situation, with the appropriate magma production, along the Fennosarmatian continent elsewhere in Europe or Asia.

Furthermore, one could argue that a westerly inclination of a fossile subduction zone, as strongly suggested by the regional polarity of the Moravo-Silesian plutonism, would not fit such a model proper without assuming dissection and strong rotation of the whole Moravo-Silesian block relative to the Fennosarmatian mainland in post Cadomian times. However, such rotations are not fully impossible and could have, for example, occurred during the Variscan orogeny (cf. Brochwicz-Lewinsky et al. 1985).

Acknowledgments. We would like to thank P. Blümel (Darmstadt) for reviewing this chapter, and H.P. Steyrer (Salzburg) for drafting the figures (manuscript submitted and accepted in 1991).

References

Batik P (1984) Geologicka stavba moravika mezi biteŝskou rulou a dyjskym masivem. Vĕstn Ustřed Ustavu Geol 59: 322–330

Brochwicz-Lewinsky W, Pozaryski W, Tomczyk H (1985) Ruchy Przesuwcze w Poludniowej Polsce w Paleozoiku, Prz Geol 12 (386): 651–658

Brown GC (1982) Calc-alkaline intrusive rocks: their diversity, evolution and relation to volcanic arcs. In: Thorpe RS (ed) Andesites. Wiley, New York, pp 437–460

Chappell BW, White AJR (1974) Two contrasting granite types. Pac Geol 8: 173–174

Dudek A (1980) The crystalline basement block of the Outer Carpathians in Moravia: Bruno-Vistulicum. Rozpr Česk Akad Věd Řada Mat Přir Ved 90/8: 1–85

Dudek A, Melkova J (1975) Radiometric age determination in the crystalline basement of the Carpathian Foredeep and of the Moravian Flysch. Vĕstn Ustřed Ustavu Geol 50: 257–264

Engel AEJ, Dixon TH, Stern RJ (1980) Late Precambrian evolution of Afro-Arabian crust from ocean arc to craton. Geol Soc Am Bull 91: 699–706

Finger F, Frasl G (1990) Pre-, syn- and postorogenic granitoids in the Austrian sector of the Variscan orogen. Abstr IGCP 233, Göttingen, pp 36–40

Finger F, Frasl G, Höck V, Steyrer HP (1989) The Granitoids of the Moravian Zone of Northeast Austria: products of a Cadomian Active Continental Margin? Precambrian Res 45: 234–245

Fišera M, Patočka F (1989) Geochemistry of Variscan blastomylonites of the Vidly pod Pradĕdem locality, the Hruby Jesenik Mts.: paleotectonic implications. Vĕstn Ustřed Ustavu Geol 64: 301–312

Franke W (1989) Tectonostratigraphic units in the Variscan belt of central Europe. Geol Soc Am Spec Pap 230: 67–89

Frasl G (1968) The Bohemian Massif in Austria, Moravian Zone. Guide to Exkursions 32 C, Int Geol Congr XXIII Sess, Prague, pp 13–24

Fuchs G, Matura A (1976) Zur Geolopie des Kristallins der südliches Böhmischer Masse. Jahrb Geol Bundesanst 119: 1–43

Gass IG (1982) Upper Proterozoic (Pan-African) calcalkaline magmatism in northeastern Africa and Arabia. In: Thorpe RS (ed) Andesites. Whiley, New York, pp 591–609

Jelinek E, Dudek A (1993) Geochemistry of Precambrian plutonites of the covered part of the

Brunovistulicum (Variscides, Czechoslovakia). Precambrian Res 62: 103–125
Liew TC, Hoffmann AW (1988) Precambrian crustal components, plutonic associations, plate environment of the Hercynian Fold Belt of central Europe: Indications from a Nd and Sr isotopic study. Contrib Mineral Petrol 98: 129–138
Matte Ph, Maluski H, Rajlich P, Franke W (1990) Terrane boundaries in the Bohemian Massif: result of large-scale Variscan shearing. Tectonophysics 177: 151–170
Morauf W, Jäger E (1982) Rb-Sr whole-rock ages for the Bites Gneiss, Moravicum, Austria. Schweiz Mineral Petrogr Mitt 62: 327–334
Nasseef AO, Gass IG (1977) Granitic and metamorphic rocks of the Taif area, western Saudi Arabia. Geol Soc Am Bull 88: 1721–1730
Peacock MA (1931) Classification of igneous rock series. J Geol 39: 65–70
Pearce JA, Harris NBW, Tindle AG (1984) Trace element discrimination diagrams for the tectonic interpretation of granitic rocks. J Petrol 25 (4): 956–983
Pitcher WS (1983) Granite type and tectonic environment. In: Hsü K (ed) Mountain building processes. Academic Press, London, pp 19–40
Plank T, Langmuir ChT (1988) An evaluation of the global variations in the major element chemistry of arc basalts. Earth Planet Sci Lett 90: 349–370
Preclik K (1926) Die Moravische Phyllitzone im Thayatale. Sb Statniho Geol Ustavu Czech Repub 6: 180–192
Preclik K (1937) Das Nordende des Thayabatholiten. Věstn Statniho Geol Ustavu Czech Repub 12: 34–61
Scharbert S (1977) Neue Ergebnisse radiometrischer Altersdatierungen an Gesteinen des Waldviertels. Arbeitstag Geol Bundesanst Wien Waldviertel 1997: 11–15
Scharbert S, Batik P (1980) The age of the Thaya (Dyje) Pluton. Verh Geol Bundesanst (Austria) 1980: 325–331
Schermann O (1965) Über Horizontalseitenverschiebungen am Ostrand der Böhmischen Masse. Mitt Ges Geol Bergbaustud 16: 89–103
Schulmann K (1990) Fabric and kinematic study of the Biteš orthogneiss (southwestern Moravia): result of large-scale northeastward shearing parallel to the Moldanubian/Moravian boundary. In: Matte Ph (ed) Terranes in the Variscan Belt of Europe and Circum-Atlantic Paleozoic Orogens. Tectonophysics 177: 229–244
Štelcl J, Weiss J (1986) Brněnsky Masiv. Univ JE Purkyně, Brno 254 pp
Streckeisen AL (1967) Classification and nomenclature of igneous rocks. Neues Jalub Min Abh 107: 140–240
Suess FE (1926) Intrusionstektonik und Wandertektonik im Variszischen Grundgebirge. Bornträger, Berlin, pp 268
Suk M (1986) Das Präkambrium der Böhmischen Masse und des Brunovistulikums in der Tschechoslowakei. Fortsch Mineral 64 (2): 227–234
Van Breemen O, Aftalion M, Bowes DR, Dudek A, Misar Z, Povondra P, Vrana S (1982) Geochronological studies of the Bohemian Massif, Gzechoslovakia, and their significance in the evolution of Central Europe. Trans R Soc Edinb Earth Sci 73: 89–108
Whalen JB (1985) Geochemistry of an Island-Arc Plutonic Suite: the Uasilau-Yau Yau Intrusive Complex, New Britain, P.N.G. J Petrol 26: 603–632
Whalen JB, Currie KL, Chappell BW (1987) A-type granites: geochemical characteristics, discrimination and petrogenesis. Contrib Mineral Petrol 95: 407–419
White AJR, Chappell BW (1983) Granitoid types and their distribution in the Lachlan Fold Belt, southeastern Australia. In: Roddick JA (ed) Circum-Pacific Plutonic Terranes. Geol Soc Am Mem 159: 21–34

VIII.B.4 Metamorphic Evolution

A. Dudek

1 Introduction

The basement and the foreland of the Moravian nappes at the eastern margin of the Bohemian Massif is composed of the Brunovistulicum, a block consolidated during the Cadomian orogeny. Complexes attributed to the Brunovistulicum crop out mainly in the Brno and Dyje (Thaya) granitoid massifs, both the intrusive rocks and their metamorphic roof are covered by the sediments of the Carpathian Foredeep and Flysch nappes, and are mostly known from numerous oil drillings. Metamorphic rocks build up the entire basement north of the line Vyškov-Zlín, and smaller relics in the mantle of the Brno and Dyje massifs south of this line (Fig. 3, Chap. VIII.B.3). To the Brunovistulicum are also attributed metamorphic rocks in the cores of the Svratka and Dyje Domes of the Moravicum, and in the Desná Dome of the Silesicum (Dudek 1988; Mísař and Urban, Chap. VIII.A).

The Brunovistulian rocks were affected by several metamorphic events. The oldest one gave rise to schists and gneisses which were later intruded by the large Cadomian postkinematic Brno pluton and smaller gabbroid massifs towards the end of this cycle. The minimum age of this metamorphic event is constrained by 514–660 Ma K/Ar biotite and muscovite mineral datings, and 550 Ma hornblende datings in amphibolites (Dudek and Melková 1975; Dudek 1980).

The Cadomian metamorphic and magmatic rocks were affected by younger tectonometamorphic events connected with the Variscan orogeny and the formation of an intricate nappe structure along the eastern margin of the Bohemian Massif. During this time, the gneisses in the cores of the Svratka and Desná Domes were intensively retrogressed. This low temperature-middle pressure metamorphism caused mylonitization and new mineral growth, and its influence may be traced farther to the east where reworking was less penetrative and the neomineralization limited only to zones of more intense shearing. This metamorphic event may be recorded by the K/Ar biotite ages obtained from gneisses of the Chorynĕ-9 drilling (340 and 480 Ma, Dudek and Melková 1975). Still younger metamorphic phenomena can be discerned in the zone of the Boskovice Furrow on the W side of the Brunovistulicum. Mylonitization and shearing took place most probably at the same time as in the Blanice Furrow farther to the W, where they were dated at 280 Ma (Brandmayr et al. 1990).

2 The Cadomian Metamorphism

The metamorphic basement complex is divided by the Haná fault zone into two different parts. North of this zone, in the North Moravian Block (Fig. 1), the protolith of the metamorphic rocks was a sequence of clastic rocks, with alternating aleuropelitic, psammitic (graywackes) and subordinate pelitic layers. Several systems of lineation originated during penetrative deformation accompanying medium- to high-grade regional metamorphism (connected with intensive folding). The rocks were converted to various types of schists and gneisses showing the following typical mineral associations:

biotite-muscovite-quartz-albite-oligoclase ± garnet
quartz-oligoclase-muscovite-biotie (± staurolite)
quartz-oligoclase-andesine-biotite-muscovite-garnet (± sillimanite).

Small amounts of potash feldspar are rarely present. Low Al_2O_3 contents are typical for the whole complex, so that the Al-rich minerals characteristic of individual metamorphic zones are only sporadically encountered. The only specimen with staurolite found in the southern part of the North Moravian Block is indicative of the metamorphism in the staurolite-quartz sub-

Dallmeyer et al. (Eds.)
Pre-Permian Geology of Central and Eastern Europe

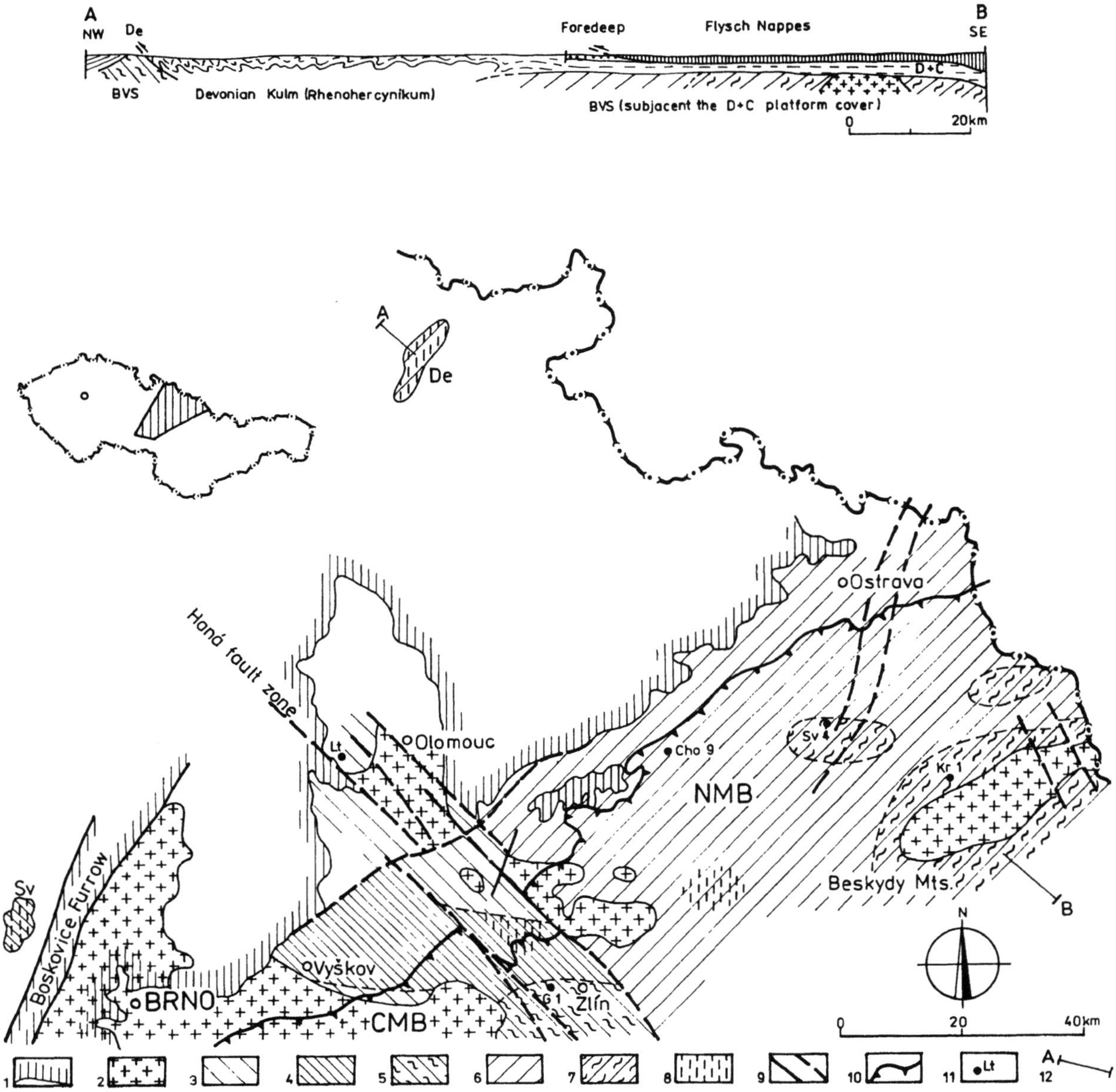

Fig. 1. Metamorphic complexes of the northern part of the Brunovistulicum (covered by Carpathian Foredeep and Flysch nappes). *1* The exposed parts of the Bohemian Massif (mostly Devonian and Carboniferous sediments); *2* intrusive rocks of the Brno pluton and small gabbroid massivs; *3–5* metamorphosed volcano-sedimentary sequence with frequent metabasites (3 chlorite-sericite phyllites, *4* muscovite-biotite schists and biotite gneisses, *5* migmatized garnet- and sillimanite-bearing muscovite-biotite paragneisses); *6–7* metamorphosed clastic sequence (*6* fine-grained muscovite-biotite to biotite paragneisses, *7* migmatized garnet-bearing muscovite-biotite paragneisses), *8* strongly retrograded rocks; *9* important faults; *10* front of the Carpathian nappes; *11* drillings mentioned in the text (*Sv* Kozlovice SV 4, *Kr* Krásná 1, *Lt* Lutín, *Cho* Choryně 9, *G* Gottwaldov 1); *12* line of the section.
Sv, *De* Brunovistulicum in the cores of the Svratka and Desná domes as basement of the moravosilesian nappes; *BVS* Brunovistulicum (in the section); *NMB* North Moravian Block; *CMB* Central Moravian Block

facies of the amphibolite facies. The intensity of metamorphism increases toward the E and ENE, where (beneath the Beskydy Mts.) some migmatized gneisses penetrated by granodiorite and diorite dykes were encountered (in the drillholes Krásná-1 and Kozlovice SV-4). The migmatization is connected with recrystallization and grain size coarsening, and mobilization of the felsic constituents. K-feldspar was found only as rare individual grains in a few specimens. The chemical composition of migmatites lies within the variation field of gneisses, and no material import took place during the migmatitization (Dudek 1980, 1988).

South of the Haná fault zone, in the Central and South Moravian Blocks, the metamorphic rocks are represented by volcanosedimentary complexes, which, in addition to phyllites and schists, contain frequent metabasites. In the belt between the marginal faults of the Haná fault zone the metamorphism increases rapidly toward the ESE, from greenschists facies chlorite-sericite and biotite (± garnet) phyllites to micaschists and weakly migmatized paragneisses showing the association

oligoclase-quartz-biotite-muscovite-garnet-sillimanite.

The metabasites in the respective zones are represented by mineral associations such as:

chlorite-actinolite-albite, chlorite-albite-epidote or actinolite-epidote-albite

in the greenschist facies,

actinolitic hornblende-oligoclase-andesine-quartz ± epidote

in the amphibolite facies, and

hornblende-labradorite ± quartz

in rocks associated with the migmatitic paragneisses (drilling Gottwaldov 1). The metabasites originated from tholeiitic basalts and tuffs. It has not been possible to determine more exact P-T conditions of metamorphism to the present.

As observed in many drillings in rocks which had been intensely folded and sheared, recrystallization outlasted the tectonic movements, as evidenced by granoblastic structures of felsic minerals, transverse biotites, growth of quartz and biotite porphyroblasts in tectonic shadows and in the fold cores. All this suggests that the rocks, having undergone dynamic recrystallization, were additionally subject to elevated temperature during the decompression period of the metamorphic cycle (England and Thompson 1984), or, in some places, to contact metamorphism. Contact effects associated with the intrusion of the Brno pluton may be assumed mainly in the phyllitic rocks in the Haná fault zone (drilling Lutín), where large porphyroblasts of andalusite developed in layers of suitable chemical composition.

The signs of low-temperature hydrothermal retrogressive changes in the eastern parts of the Brunovistulicum, where metamorphic recrystallization of the Devonian platform sediments could not be determined, suggest that a low-temperature retrogression connected with the final stages of the Cadomian tectonic activity may have also taken place, prior to the younger Variscan processes.

3 The Variscan Metamorphism

All Cadomian rocks, both metamorphic and plutonic, were affected by younger retrograde metamorphism, which mainly reworked the rocks at the western margin of Brunovistulicum – in the tectonic windows of the Desná, Svratka, and Dyje Domes of the Moravicum and Silesicum. Metamorphism also affected the Devonian rocks of the sedimentary cover. In the eastern part of the Brunovistulicum, the retrogression is mainly confined to zones of intense tectonic fracturing and limited shearing. The retrograde metamorphism in this region seems to have been fostered by low-temperature hydrothermal activity, which resulted in albitization of plagioclase, chloritization of biotite, carbonatization, sericitization, and epidotization of plagioclase. These associations correspond to greenschist facies conditions (quartz-albite-muscovite-chlorite subfacies).

A strong Variscan reworking of the basement is visible especially in those parts of the Brunovistulicum which were overridden by the Moravian and Silesian nappes. In general, the intensity of reworking increases toward the NNE. The external parts of the Dyje massif were strongly sheared and metamorphosed to medium-grade assemblages, resulting in the development of a pronounced stretching lineation and biotite growth (Frasl 1991; Lobkowicz et al. 1991). The internal parts of the Dyje massif show only low-grade metamorphism, which is confined to zones of shearing and mylonitization, containing newly crystallized epidote, zoisite, albite-oligoclase and, rarely, garnet (Frasl 1991).

In the core of the Svratka Dome, the Cadomian basement (Deblín Group) consists of migmatized biotite gneisses with metabasite bodies, micaschists and phyllites and intrusions of late Cadomian granitic rocks (Fajst 1969; Jaroš and Mísař 1974). This basement and its Devonian sedimentary cover were later affected by Variscan low-grade greenschist facies metamorphism (Schulmann et al. 1991; Kratochvíl 1981). Farther east, the Cadomian granitic rocks of the Brno pluton were much less affected by the Variscan events.

The basement rocks in the core of the Desná Dome, represented by garnet-bearing biotite augengneisses, muscovite-biotite gneisses, and granodiorite bodies, are covered by transgressive Devonian rocks. These complexes were folded, sheared, and metamorphosed during Variscan tectonism and form complicated nappe structures. Three deformation events and three metamorphic phases, of low- to medium-grade (greenschist and amphibolite facies) have been detected in the Variscan tectonometamorphic cycle. The basement rocks were strongly reworked and retrogressed to gneissic mylonites, blastomylonites, and relic metagranitoids. Their mineral associations are newly formed and only some garnets I are relics of the pre-Variscan metamorphic event (Fediuková et al. 1988). The new mineral associations in individual Variscan metamorphic phases are:

quartz-albite-chlorite-muscovite, or quartz-albite-epidote-biotite,
quartz-muscovite-biotite-plagioclase-garnet II-staurolite ± chloritoid
quartz-plagioclase-muscovite-chlorite, and quartz-muscovite-biotite-plagioclase ± garnet, staurolite.

The P-T conditions of this Variscan metamorphism (progressive, in the basement retrogressive) were estimated as 350–450 °C and 0.3 GPa for the first phase, and 350–550 °C and 0.2–0.5 GPa for the second phase (Cháb et al. 1984). The sedimentary envelope was progressively metamorphosed to various types of schists with garnet, staurolite, and chloritoid.

All these metamorphic processes are probably followed by a still younger, weak metamorphic event, the "Jungmoravische Phase" of Frasl (1968). This event is characterized by chloritization, sericitization, and sometimes by the growth of stilpnomelane in the zones of mylonitization.

References

Brandmayr M, Dallmeyer RD, Handler R, Wallbrecher E (1990) Age and kinematics of shear zones in the southern Bohemian Massif: evidence from the Rodl-shear zone. In: Terranes in the Circum-Atlantic Paleozoic orogens. Abstr Int Conf Paleozoic orogens in central Europe, Göttingen-Giesen, Aug–Sept 1990

Cháb J, Fišera M, Fediuková E, Novotný P, Opletal M, Skácelová D (1984) Problémy tektonického a metamorfního vývoje východní části Hrubého Jeseníku. Sb Geol Věd Geol 39: 27–72 (English summary)

Dallmeyer RD, Neubauer F, Höck V (1900) $^{40}Ar/^{39}Ar$ mineral age controls on geochronology of Late Paleozoic tectonothermal activity in the southeastern Bohemian Massif, Austria (Moldanubian and Moravo-Silesian zones). In: Terranes in the Circum-Atlantic Paleozoic orogens. Abstr Int conf Paleozoic orogens in Central Europe, Göttingen–Giesen, Aug – Sept 1990

Dudek A (1980) The crystalline basement block of the Outer Carpathians in Moravia: Bruno-Vistulicum. Rozpr Česk Akad Věd Řada Mat Přír Věd 90/8: 1–85

Dudek A (1988) Brunovistulicum. In: Zoubek V, Cogné J, Kozhoukharov D, Krautner HG (eds) Precambrian in younger fold belts. Wiley, Chichester, pp 287–295

Dudek A, Melková J (1975) Radiometric age determination in the crystalline basement of the Carpathian Foredeep and of the Moravian Flysch. Věstn Ústřed Ústavu Geol 50: 257–264

England PC, Thompson AB (1984) Pressure-temperature-time paths of regional metamorphism. I. Heat transfer during the evolution of thickened continental crust. J Petrol 25: 894–928

Fajst M (1969) Granitisation und Mylonitisation im NW-Teile des Svratka Komplexes. Acta Univ Carol Geol 1968: 309–328

Fediuková E, Fišera M, Cháb J, Kopečný V, Opletal M, Rybka R (1988) Garnets of the pre-Devonian rocks in the eastern part of the Hrubý Jeseník Mts. (North Moravia, Czechoslovakia). Acta Univ Carolinae, Geologica 1985: 197–234

Frasl G (1968) The Bohemian Massif in Austria, Moravian Zone. Guide to Excursion 32 C, Int Geol Congr XXIII Session, Prague, pp 13–24

Frasl G (1991) Das Moravikum der Thaya-Kuppel als Teil der variszisch deformierten Randzone des Bruno-Vistulikum – eine Einführung. Arbeitstag Geol Bundesanst Wien, September 1991, pp 49–62

Jaroš J, Mísař Z (1974) Deckenbau der Svratka-Kuppel und seine Bedeutung für das geodynamische Modell der Böhmischen Masse. Sb Geol Věd Geol 26: 69–82

Kratochvíl M (1981) Analýza mikrostavby vápenců v tektonických stylech devonu Českého masívu rentgenovou difrakcí. PhD Thesis, Charles Univ, Praha

Lobkowicz M, Schulmann K, Melka R (1991) Variscan reactivation of the Cadomian basement in the cores of the Moravian windows. Geol Worksh Moravian Windows, Moravský Krumlov 26.-30.4 1991, Abstr, p 18

Schulmann K, Ledru P, Autran A, Melka R, Lardeaux JM, Urban M, Lobkowicz M (1991) Evolution of nappes in the eastern margin of the Bohemian Massif: a kinematic interpretation. Geol Rdsch 80: 73–92

VIII.B.5 Metallogenesis

J. Aichler, B. Fojt, and M. Vaněček

1 Introduction

The Moravo-Silesian zone represents an example of intensive Variscan metallogenic development evidently connected, especially in its northern part, with the Devonian crustal extension and subsequently affected by Variscan compressional events (Sawkins and Burke 1980). The northern part of the Moravo-Silesian Zone (the Silesicum) belongs to the so-called East-Sudetic subprovince and in the southern part of this zone, the Brunnia and the West Moravian metallogenic regions can be distinguished (Fig. 1; Vaněček et al. 1985).

2 Silesicum

The East-Sudetic metallogenic subprovince is characterized by the presence of abundant iron, base metal and gold ores, and by the absence of uranium mineralization.

To the mineralization syngenetically connected with the pre-Devonian crystalline complexes belong banded magnetite ores locally graded into an Al-Mn rich horizon in the Desná Dome gneisses and ranged to the Sydvaranger type (Pouba 1970). The metamorphosed iron deposit Hraničná with impregnation and a thin layer of massive polymetallic mineralization is bound to dolomites of the variegated part of the Proterozoic Staré Město Group (Litochleb 1975).

The origin of ore accumulations during the Variscan metallogenic process in the Silesicum took place from the Early Devonian to the Late Permian and was related to: (1) submarine mafic and felsic volcanism (iron ores of the Lahn Dill type and stratabound Fe-Pb-Zn ± Cu ± Au ores respectively), (2) remobilization and redeposition of ore matter within sedimentary and volcanic sequences by the Variscan metamorphism and deformation (vein and "stockwork" gold-bearing and base metal mineralizations), and (3) post-orogenic plutonism (molybdenite occurrences and scheelite-bearing scarns) (Fig. 2; Skácel 1987).

The first stage of the Variscan metallogenic process probably started with the formation of small iron ore deposits of the Lahn-Dill type confined mostly to the submarine mafic volcanics and less to volcano-sedimentary and sedimentary deposits. The iron ore deposits and occurrences consist of lenticular and often folded ore bodies formed mainly by magnetite, hematite, lepto-chlorites in various amounts, and rare siderite (Skácel 1966). Tens to hundreds of iron deposits of this type were mined in both the Vrbno Group and in the Šternberk-Horní Benešov belt and gave rise to the local metallurgical industry, having culminated in the 18th and especially in the 19th century.

The stratabound Fe-Pb-Zn ± Cu ± Au sulfide deposits in the Silesicum area, analogously to Rammelsberg, Meggen, and sulfide deposits of Iberian Pyrite Belt, seem to have been formed in a similar geotectonic environment with dominance of crustal extension (Sawkins and Burke 1980). The total tonage of the Silesicum polymetallic stratabound deposits (Zlaté Hory, Horní Město, Oskava, Horní Benešov) was estimated at about 100 Mt of mostly low grade ore. These stratabound base-metal deposits occur within and/or in the close vicinity of the Devonian felsic volcanics.

The Zlaté Hory ore district has been ranked among the most important base-metal and copper ore sources in the Czech Republic, having estimated reserves of 203 kt Cu, 714 kt Zn, 174 kt Pb, and 20 t Au (Vaněček 1988). The district is bound to the huge overturned anticline of volcano-sedimentary rock sequence metamorphosed in greenschist facies in the northern part of the Devonian Vrbno Group (Fig. 3). The ore bodies, often developed in chloritic quartzites or quartz keratophyres, are formed predominantly by disseminated, less by banded ores, rarely by massive iron sulfides. They are accompanied by chalcopyrite, sphalerite, gold, and minor galena. Primary zoning was recognized in the ore district,

Dallmeyer et al. (Eds.)
Pre-Permian Geology of Central and Eastern Europe

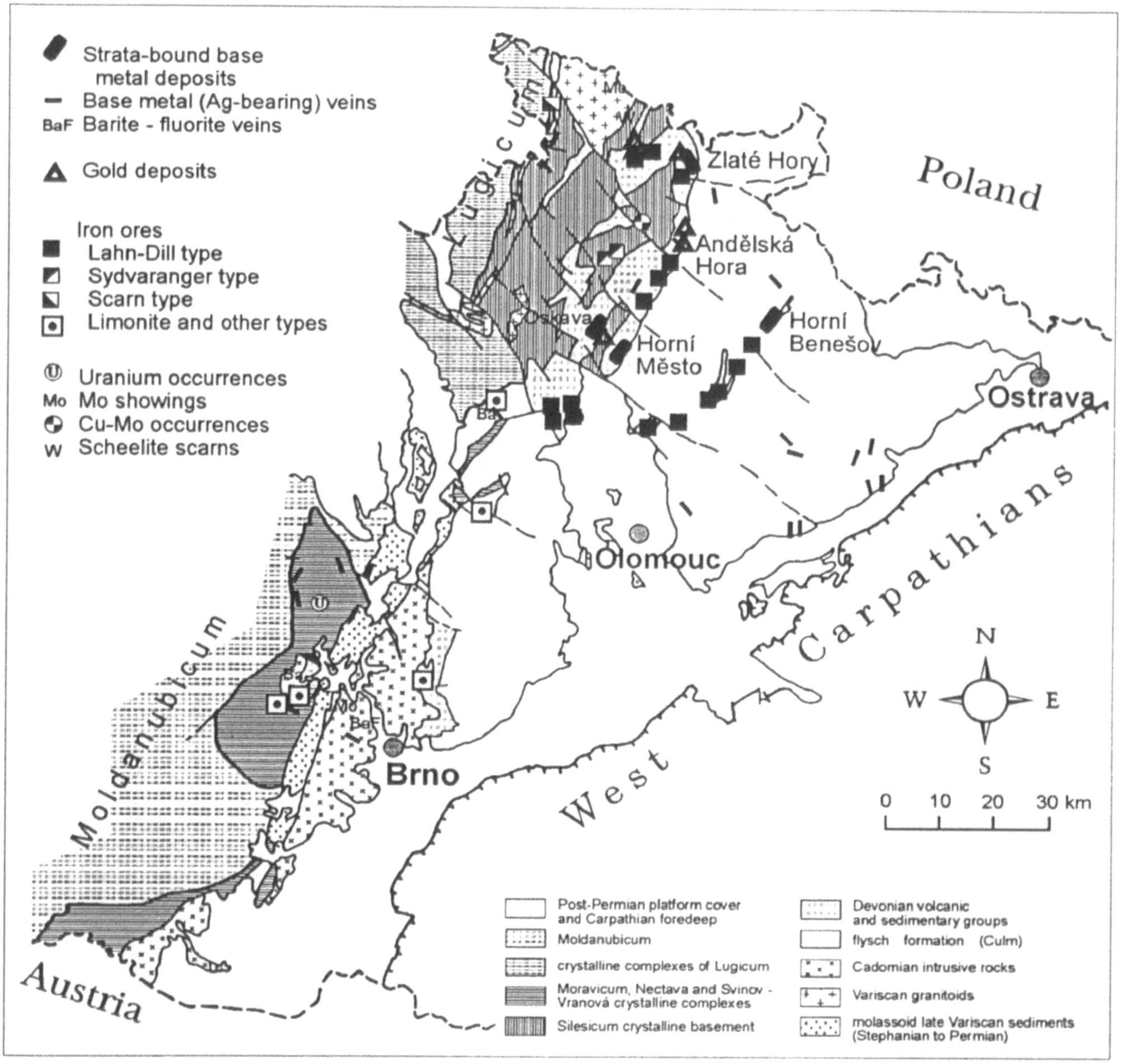

Fig. 1. Ore deposits and main occurrences of the Moravo – Silesian zone. (After Ilavský and Sattran 1968, 1981)

since copper ores are found in central and deeper parts of elevation, whereas zinc-lead ores are developed at the periphery (Hettler et al. 1977). In the western part of the district, gold is present in some enriched tectonic zones of ENE-WSW strike. The origin of the mineralized structures may be caused by the development of ductile shear zones (Kalenda 1986). A comparison between the Zlaté Hory deposits and other volcanogenic deposits indicated the close resemblance of this deposit to the Rosebery type, and individual features point to certain original similarity with the Kuroko-type deposits (Patočka and Vrba 1989).

The Horní Město ore district, now abandoned, situated in the southern part of the Devonian Vrbno Group, is formed by the row of lenticular to E-declining ore bodies bound to tectonized sericitic schists (felsic metatuffs) and quartz keratophyres. Pyrite, sphalerite, and galena prevail in disseminated ore (Fojt 1962). The same character of mineralization has been ascertained

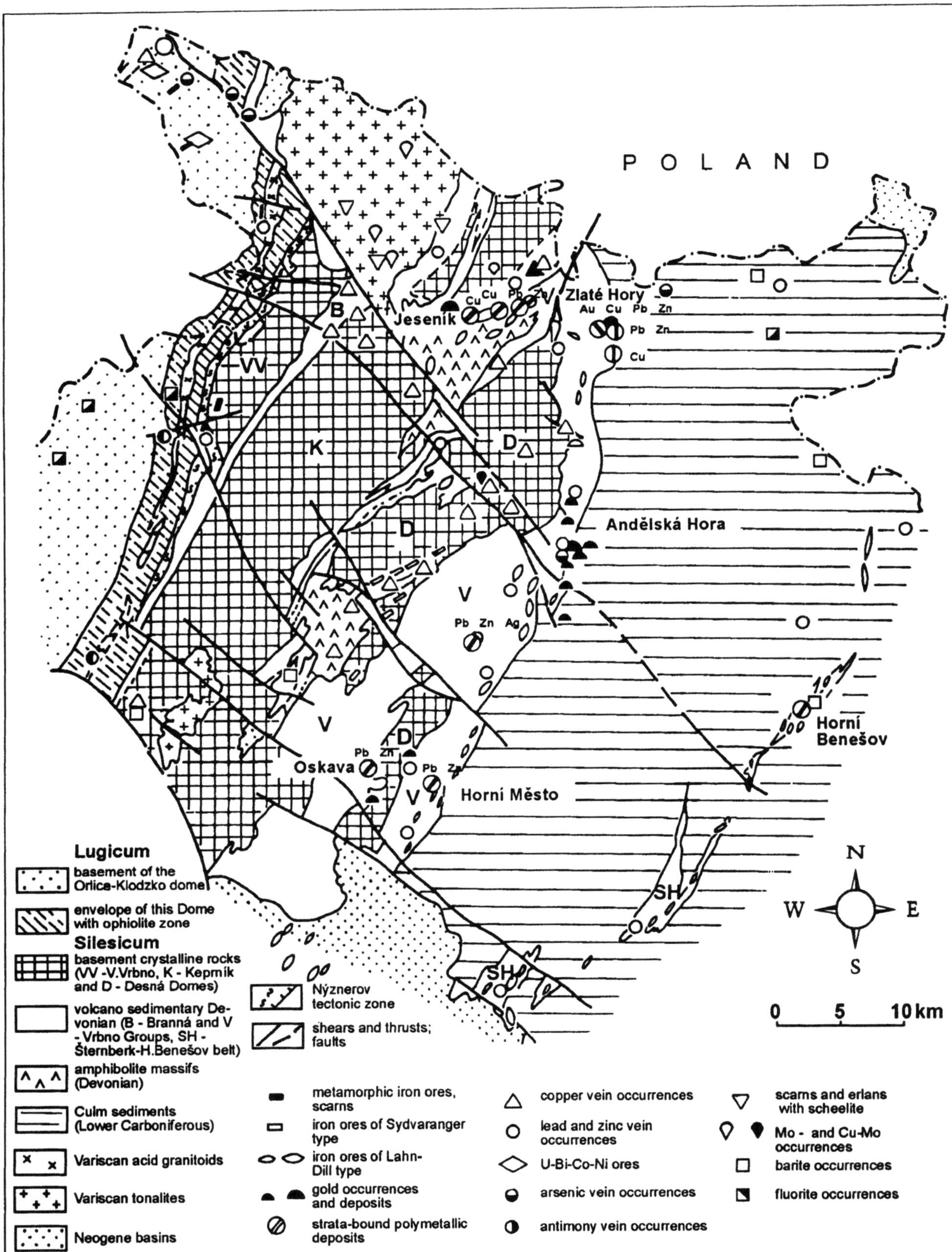

Fig. 2. Generalized geology, main ore occurrences, and deposits of the Silesicum and adjacent part of the Lugicum. (After Skácel 1987)

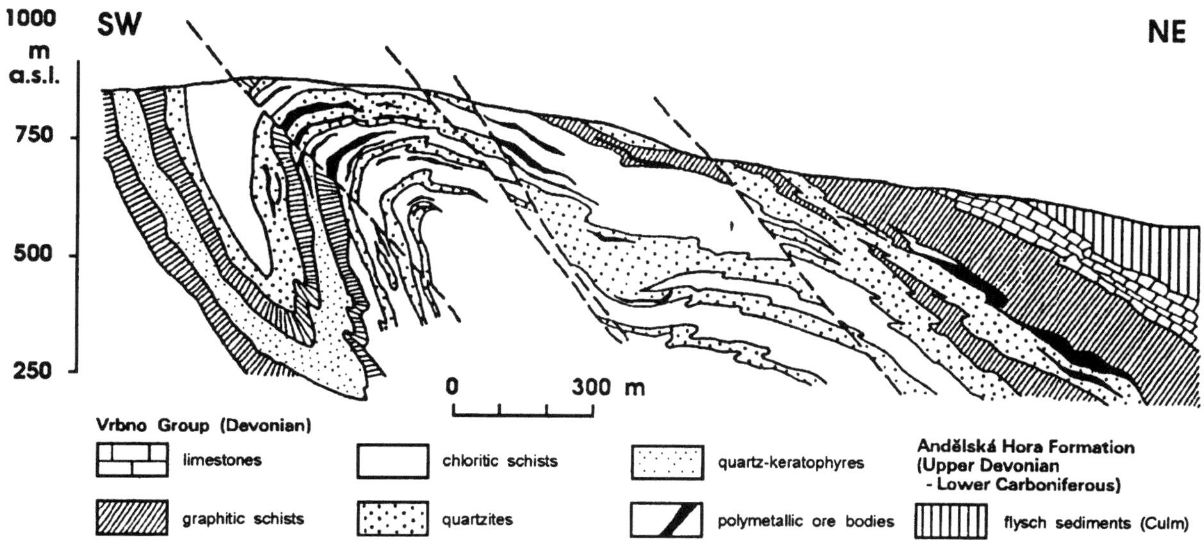

Fig. 3. Cross section through the central part of the Zlaté Hory ore district. (Čabla et al. 1979)

at the newly discovered Oskava deposit, situated to the W from Horní Město (Koverdynský et al. 1986).

Another important polymetallic ore deposit, Horní Benešov, is located in a tectonically thrusted and sheared, slightly metamorphosed Devonian slice surrounded by flysh rocks of Lower Carboniferous. Its ore bodies, located en echelon, are presented by disseminated nest and stockwork, and less breccia and atoll-like ores in massive quartzose rocks and silicified marbles. In sericitic schists in their surroundings, the streaky and banded ores are located. Pyrite, sphalerite, and galena are the main ore minerals, barite is common in upper levels of the deposits (Urbánek 1988).

Most of the primary textural and structural features of the mentioned stratabound deposits have been affected by metamorphism and most of the ores have been completely recrystallized. The polyphase Variscan metamorphism released a huge amount of water and a portion of ore and gangue matter from volcano-sedimentary sequences. It caused their migration and gave origin to mostly steeply inclined discordant vein bodies of both sulfides and gold-bearing quartz ores in Zlaté Hory and in the other districts.

The Variscan metamorphism and deformation had an insignificant influence on isotopic composition of sulfur. The $\delta^{34}S$ values of barites from the Silesian stratabound deposits are consistent with the $\delta^{34}S$ values of barites from the Meggen and Rammelsberg deposits (Hladíková et al. 1992). Whereas the $\delta^{34}S$ values of base metal sulfides from Rammelsberg, Meggen, and other similar deposits are positive (Ohmoto 1986), most of the $\delta^{34}S$ values of base-metal sulfides from the Silesicum deposits lie between 0 and $-12‰$, with a maximum in the range of -2 to $-8‰$. Both sulfur derived from the marine sulfate and sulfur mobilized by hydrothermal solutions from the surrounding sedimentary and volcanic rocks played a major role during the formation of this stratabound deposits (Hladíková et al. 1990). An important influence of upper crustal rocks as the source of metals during formation of these deposits and the majority of other types of deposits is indicated by high radiogenic lead of these ores (Vaněček et al. 1985).

Most authors advocate the concept of volcanogenic submarine hydrothermal and/or exhalative origin of the Silesicum base-metal deposits (Havelka et al. 1963; Čabla et al. 1979; Orel 1986). Submarine hydrothermal and exhalative activity is indicated in the Silesicum, besides numerous iron deposits of the Lahn-Dill type, by different types of rock alterations, namely K-metasomatism, silicification, chloritization, carbonatization, possibly tourmalinization, and biotitization (Mixa and Orel 1990). The primary ore-forming solution was similar to seawater. Convective geothermal systems accompanying crustal extension in the Devonian, comparable to modern systems at spreading centers, are assumed to have existed in this region. The sulfide accumulations formed probably mostly in nonlithified sediments and pyroclastics (Čabla et al. 1979).

Besides the gold of base-metal ores, bound mostly to pyrite and chalcopyrite, native gold occurs in quartz in mesozonally metamorphosed rocks (e.g., Zlatý chlum deposit) and low sulfide quartz-carbonate veins and disseminations in carbonaceous slates (Andělská Hora ore district). The fineness of free gold is mostly very high in anomalously segregated quartz parts of metamorphites, while in the stratabound sulfide deposits it probably depends on topomineral influences of coexisting mineral phases. The ratio of pure gold and gold fixed in the structure of sulfides (mainly pyrite) is believed to depend on the degree of metamophic reworking of the ores (Fojt et al. 1987).

The origin of gold-bearing vein mineralizations in the Silesicum is attributed by many authors to the Variscan regional metamorphism. Flow of the released fluids was controlled by the existing P-T gradient during the Variscan tectonic processes.

Low sulfide Au-quartz vein mineralization of the Andělská Hora ore district is hosted by carbonaceous slates, altered tuffs and dolerites of the Devonian Vrbno Group. Carbonate-quartz veins and veinlets with low content of sulfides (pyrite, pyrrhotite, chalcopyrite, less arsenopyrite, galena, sphalerite, sulfoarsenides) are located in different planes of foliation, cleavage, and joints connected with superposed fold deformation and polyphase metamorphism of greenschist facies. Sulfidic sulfur originated mainly from bacterial reduction of marine sulfate (Hladíková et al. 1990). Gold with fineness 830–940 is mostly free, forming grains at the margins of veins/veinlets bound to intergranular space among gangue minerals and/or to cracks in sulfides (Malec 1987; Aichler et al. 1990). The polyphase-folded, thrusted, and metamorphosed Fe-rich carbonaceous chlorite-muscovite slates enriched in Fe-sulfides and Fe-Mg carbonates appear to have been the favorable screen for reducing of the ascending metamorphic – hydrothermal fluids. Massive influx of the H_2O-CO_2 fluids (Ďurišová 1990), derived from devolatilization of the metamorphic pile, caused extensive berezite-listvaenite metasomatic alterations of rocks. Precipitation of gold might be ascribed to changing P-T conditions, decreasing both pH and H_2S activity at low fo_2 (Aichler 1992).

Mineralizations connected with post-orogenic plutonism occur in the Žulová massif and its mantle. The granitic pluton seems to have been the source (at least of heat) for the latter hydrothermal polymetallic vein occurrences, and by showings of molybdenite and scheelite-bearing scarns (Skácel 1987).

Indications of Late Variscan veins in the East-Sudetic subprovince are bound mainly to the NW-SE fault zones. Chalcopyrite predominated in the inner zone of the Silesicum (the Hrubý Jeseník Mts) while the Pb-Zn sulfides prevail in the veins within the Lower Carboniferous Culm facies in the Nízký Jeseník and Oderské vrchy Uplands (Skácel 1987). Losert (1957) placed the genesis of the Pb-Zn veins in the Asturian phase of the Variscan orogeny. The isotopic composition of radiogenic leads suggests that the source of the ore veins was a homogenized reservoir – either the Culm complex or the thick pile of underlying Devonian limestones (Vaněček et al. 1985).

3 Brunnia Metallogenic Region

The Brunnia metallogenic region (Vaněček et al. 1985) is of little importance. The quartz-sericite-molybdenite mineralization was discovered in tectonic zones in vicinity of Černá Hora (Kolek 1985) within the altered and mylonitized granitoids of the Brno Massif (584 ± 5 Ma, van Breemen et al. 1982) accompanied by rare chalcopyrite and pyrite.

4 Moravicum

The polymetallic zone of the Moravicum (Ilavský and Sattran 1980) forms the eastern part of the West Moravian metallogenic region (Vaněček et al. 1985). This zone is characterized by the presence of hydrothermal vein and metasomatic occurrences (in marbles, Češková and Orel 1971) of Pb-Zn-Cu sulfides in the Svratka Dome. The upper content of copper (chalcopyrite, bournonite) is typical for the veins in surroundings of Štěpánov and Borovec. The polymetallic veins are partly accompanied by fluorite – barite mineralization, e.g., to the W of Tišnov (Ilavský and Sattran 1980).

References

Aichler J (1992) Origin of gold-bearing mineralization in the Andělská Hora ore district in the Jeseníky Mts. PhD Thesis, Czech Geol Surv, Prague

Aichler J, Fojt B, Cháb J, Kvaček M, Novotný P, Orel P, Vaněček M (1990) Gold-bearing mineralization of the locality Suchá Rudná in the Jeseníky Mts. Sb Geol Věd Ložisk Geol Mineral Praha 29: 79–10 (in Czech)

Čabla V, Hettler J, Tomšík J (1979) Ore deposits of the Zlaté Hory ore district from the viewpoint of plate tctonics theory. Sb GPO Ostrava 20 (12): 5–69 (in Czech)

Češková L, Orel P (1971) Polymetallic mineralization of the vein-metasomatic type at Heroltice. Sb Geol Věd Ložisk Geol Praha 14: 41–63 (in Czech)

Ďurišová J (1990) Fluid inclusion study in the Jeseníky ore district (Czechoslovakia). Sb Geol Věd Ložisk Geol Mineral Praha 29: 167–186 (in Czech)

Fojt B (1962) Petrographic characteristics of the sulfide and base-metal deposits of the Horní Město district. Pr Brněnské Základny Česk Akad Věd Praha, 34, 9: 385–444 (in Czech)

Fojt B, Hauk J, Vodová E (1987) Gold in Jeseníky stratiform deposits. Gold in Western Carpathians, pp 35–39, Geol Surv Dionýz Štúr, Bratislava (in Czech)

Havelka J, Palas M, Scharm B (1963) New metallogenic concept of the nonferrous metal ore deposits in the Jeseníky Mts province and its consequences in prospecting. Geol Průzk 5: 225–227 (in Czech)

Hettler J, Skácel J, Tomšík J (1977) Zlaté Hory ore district. Sb GPO Ostrava 13 III: 3–149

Hladíková J, Šmejkal V, Fojt B, Pertold Z, Aichler J (1990) Isotopic composition of sulfur, carbon and oxygen from selected localities of the Jeseníky Mts. Sb Geol Věd Ložisk Geol Mineral Praha 29: 141–165 (in Czech)

Hladíková J, Aichler J, Mixa P (1992) Source of sulfur of base metal sulphide deposits at the NE margin of the Bohemian Massif (Czechoslovakia). Abstr 29th IGC Kyoto, Kyoto, vol 3, 757

Ilavský J, Sattran V (eds) (1968) Metallogenic map of Czechoslovakia, 1:1 000 000. Ústřed Ústav Geol Praha

Ilavský J, Sattran V (1980) Explanations to metallogenic map of Czechoslovakia 1:500 000. Geol Ústav Dionýza Štúra Bratislava

Ilavský J, Sattran V (eds) (1981) Metallogenic map of Czechoslovakia, 1:500 000. Ústřed Ústav Geol Praha (in Czech)

Kalenda F (1986) New information on prognosis of ore structures at the Zlaté Hory – west deposits. In: Sb 25. Symp Hornická Příbram Vědě a technice 1986, sekce G, Příbram, pp 238–257 (in Czech)

Kolek P (1985) The assessment of the lithogeochemical Mo-prospecting in the Brno Massif. In: Štelcl J (ed) Final report. The relationship of intracrustal and supracrustal pattern of the Moravian and the Silesian blocks. University of Jan Evangelista Purkyně, Brno, pp 30–38 (in Czech)

Koverdynský B, Daňko J, Fojt B et al. (1986) Results of prognosis research of the new occurrence of polymetallic ores near Oskava in Jeseníky Mountains. In: Sb 25. Symp Hornická Příbram Vědě a Technice 1986, sekce Ložisková geologie, Příbram, pp 258–263 (in Czech)

Litochleb J (1975) Magnetite and sulphide mineralization at the deposit Hraničná. Čas Slez Muz Opava Sér A 24: 149–167

Losert J (1957) The deposits and occurrences of the lead-zinc ores in the North Moravian Culm (the Odra Uplands surroundings of Hrabůvka). Rozpr Česk Akad Věd Praha 67 (4): 61 (in Czech)

Malec J (1987) Chemical composition of gold from occurrences in the Jeseníky Mts area. Manuscript, Instit Raw Materials, Kutná Hora (in Czech)

Mixa P, Orel P (1990) Evidence of submarine geothermal Paleozoic activity in the eastern part of Middle European Variscides (Czechoslovakia). 8th IAGOD Symp, Program with Abstr, A95, Geological Survey of Canada, Ottawa

Ohmoto H (1986) Stable isotope geochemistry of ore deposits. In: Valley JW, Taylor HP (eds) Stable isotopes in high temperature geological processes. Reviews in Mineralogy, vol 16. Mineralogical Society of America, Chelsea, MI

Orel P (ed) (1986) Prospective – economic geological assessment of the Šternberk – Chabičov structure. Manuscript, Czech Geol Surv, Brno (in Czech)

Patočka F, Vrba J (1989) The comparison of stratabound massive sulfide deposits using the fuzzy-linguistic diagnosis of the Zlaté Hory deposits, Czechoslovakia, as and example. Mineral Deposit 4: 192–198

Pouba Z (1970) Pre-Cambrian banded magnetite ores of the Desná Dome. Sb Geol Věd Prague LG 12: 7–64 (in Czech)

Sawkins FJ, Burke K (1980) Extensional tectonics and Mid-Paleozoic massive sulfide occurrences in Europe. Geol Rdsch 69: 349–360

Skácel J (1966) Iron ore deposits of the Moravo-Silesian Devonian. Rozpr Česk Akad Věd Řada Mat Přír Věd Praha 76 II: 1–59 (in Czech)

Skácel J (1987) The Silesicum and adjacent parts of the Lugicum: ore deposits and metallogeny. Excursion guide for Inaugural Meet of IGCP Proj 254: 15–23 Geol Survey, Prague

Urbánek J (1988) Remarks on the checking exploration of the deposit Horní Benešov. Geol Průzkum Praha 30 (7): 193–198 (in Czech)

Van Breemen O, Aftalion M, Bowes DR et al. (1982) Geochronological studies of the Bohemian Massif, Czechoslovakia, and their significance in the evolution of Central Europe. Trans R Soc Edinb Earth Sci Edinb 73: 89–108

Vaněček M (1988) Metallogeny of the Bohemian Massif. Conf Bohemian Massif, Geol Surv, Prague

Vaněček M, Patočka F, Pošmourný K, Rajlich P (1985) The use of the isotopic composition of ore lead in metallogenic analysis of the Bohemian Massif. Rozpr ČSAV, Řada Mat Přír Věd Praha 95: 1–114

VIII.C Allochthonous Units

VIII.C.1 Stratigraphy

Z. Mísař

1 Introduction

The allochthonous units of the Moravo-Silesian Zone (MSZ) represent an independent crustal segment surrounded by the Moldanubicum, Bohemicum, and Lugicum in the W, and the Brunovistulicum in the E, the latter being considered as an autochtonous part of the MSZ (Dudek 1980). Both the allochthonous and autochthonous parts are divided by the NW-SE transverse Haná Fault Zone (Mísař 1967) into two, segments, northern and southern. The main constituents of the allochthonous part of the MSZ are the Moravicum (in the S) and the Silesicum (in the N) (Fig. 1 in Mísař and Urban, Chap. VIII.A)

In our definition, the Moravicum consists of different lithological groups occurring in the Thaya and Svratka domes (the Moravian Windows of Suess 1912), which are limited by the Brunovistulicum in the footwall, and the Moldanubicum, Kutná Hora-Svratka Unit, and Letovice Unit in the hangingwall. Some elements of the Moravicum can also be found as tectonic relics W and E of the Thaya Dome (Drosendorf Window and Miroslav Horst, respectively), and NE of the Svratka Dome (Svinov-Vranová Crystalline Unit).

There is no paleontological evidence for the age of lithological groups in the Thaya and Svratka domes. The only radiometric determinations were performed on the Bíteš gneiss, giving a broad spectrum of Rb-Sr WR ages (790 Ma – Scharbert 1977; 560 Ma – Morauf and Jäger 1982; 480 Ma – Van Breemen et al. 1982). The hornblende $^{39}Ar/^{40}Ar$ plateau ages of 346 ± 7 Ma and 328.7 ± 3.3 Ma obtained from the Bíteš gneiss by Matte et al. (1990) and Dallmeyer et al. (1990), respectively, do not help much to decipher the age of the protolith.

In the following paragraphs the succession of lithological units in the Thaya and Svratka domes (from top to bottom) will be described without taking into account different views on their internal tectonic subdivision into individual nappes (e.g., Suess 1912; Matura 1980; Tollmann 1982; Finger et al. 1989; Frasl 1991; Schulmann et al. 1991) or the idea of a tectonic uniformity within a single Moravian Nappe (Jaroš and Mísař 1974). The lithological character of individual units is also summarized in Tables 1–7, added to which are the descriptions of hangingwall and footwall regional units. As occurrences of Moravicum rocks such as those in the Drosendorf Window and Miroslav Horst are also mentioned, the nature of their mostly tectonic contacts will also be briefly outlined.

2 The Lithostratigraphy of the Moravicum

There exists a puzzling geological unit – the Micaschist Zone (sometimes called the Moravian Micaschist Zone in the geographic sense) bordering the western margin of the Moravicum. F.E. Suess (1912) interpreted the Micaschist Zone as a zone of strongly retrogressed Moldanubian rocks. A similar interpretation is advocated by Matura (1980), Frasl (1991), and others. On the other hand, according to Jenček and Dudek (1971) and Jenček (1984), the Micaschist Zone W of the Thaya Dome (i.e., their Šafov Group or the Podhradí Group in the Drosendorf Window; see below) represents part of the Moravicum. Mísař (1963) argues that the Micaschist Zone is an independent lithological unit containing micaschists, marbles, amphibolites, graphite quartzites, orthogneisses, and ultrabasic rocks. Along the northern closure of the Svratka Dome, the northern part of the Micaschist Zone joins the Letovice crystalline complex with dismembered metaophiolites (Jelínek et al. 1984). Recently, relics of HP mineral parageneses were discovered in micaschists (Johan et al. 1990).

Dallmeyer et al. (Eds.)
Pre-Permian Geology of Central and Eastern Europe

The Thaya Dome (and likewise the Svratka Dome in the N) is, according to the classic terminology of Suess (1912), composed of three tectonic subunits, namely the Outer Phyllites, the Bíteš Gneiss, and the Inner Phyllites. In a recent terminology (Tollmann 1985; Frasl 1991), the main tectonic units are the Bíteš Gneiss Nappe on the top, the intermediate Pleissing Nappe, and the Thaya Batholith forming together with its old metamorphic roof the autochthonous or parautochthonous basement. The Olbersdorf Group with basal quartzites, pelites, and spilite-keratophyres is transgressive on the Thaya Batholith, and is very probably Devonian in age (Frasl 1991); it is tectonically overlain, however, by gneisses of the Pleissing Nappe.

Beneath the Micaschist Zone and the Moldanubian Gföhl Nappe, the first lithological unit attributed to the Moravicum is the Vranov Group (Table 1, Fig. 1). Being predominantly composed of metapelites, this unit contains very frequent intercalations of marbles, calcitic shales, sandstones, and basic volcanic rocks. The lithological character of the Vranov Group perfectly corresponds to that of the Olešnice Group in the Svratka Dome. Therefore, the term Vranov-Olešnice Group has been proposed by Mísař (1963) for this neritic variegated facies of the Moravicum.

The Bíteš Group mostly consists of two-mica orthogneisses with rare intercalations of micaschists within the orthogneiss body but with relatively frequent layers of basic metavolcanites and metatuffites (some basic dykes may also be present) near the contact with the overlying Vranov Group.

Table 1. Lithological units of the Moravicum in the Thaya Dome (+ overlying and underlying units)

Unit	Characteristics
Micaschist zone (1)[a] Šafov Group (3); Jenček and Dudek (1971)	Two-mica garnet (± kyanite) micaschists, amphibolites, peridotites, two-mica orthogneisses
.... Moldanubian Thrust	
Moravicum (1)	
Outer Phyllites (2) (Suess 1912) or Vranov Group (3) (Jenček and Dudek 1971) or Vranov-Olešnice Group (3) (Mísař 1963)	Massive or banded fine-grained (garnet) paragneisses and micaschists with interlayers of calc-silicate schists, tremolite marbles, graphite schists, amphibolites, and quartzites
....	
Bíteš Group (3) (Bíteš orthogneiss)	Intercalations of the Bíteš orthogneiss with the lower part of the Vranov Group, a sheet-like body interpreted as a metagranitoid (e.g., Frasl 1954, 1970) or as a syntectonic intrusion (Dudek 1960; Mísař 1961), metamorphically heterogenous, prevailing type: two-mica augen-gneiss with numerous blastomylonitic layers
.... Intrusive contact	
Inner Phyllites (2) (Suess 1912) or Lukov Group (3) (Batík 1984) or Pernegg Formation (3) (Höck and Vetters 1975, Frasl et al. 1990)	Garnet (± staurolite) micaschists with interlayers of marbles, quartzites and amphibolites and with a marking horizon of the Fugnitz calc-sillicate schist on the top
.... Tectonic contact	
Pleissing Tectonic Zone (Pleissing Nappe) (2)	
Weitersfeld Pencilgneiss Pleissing orthogneiss Buttendorf orthogneiss	Mylonitized granitoids comagmatic with the Thaya granitoids, their metamorphic roof (Therasburg Formation) Slices of the Lukov-Pernegg Group
....	
Therasburg Formation (3) (Höck and Vetters 1975; Höck et al. 1990)	An equivalent of the Deblín Group of the Svratka Dome
.... Moravian Thrust	
Brunovistulicum (1) (Thaya Pluton and its metamorphic roof – Therasburg Formation)	Mylonitized granitoids, biotite gneisses and micaschists, amphibolites, calc-silicate rocks

[a] Numbers after unit names in Table 1 (and following tables) denote: (1) regional unit; (2) mostly tectonic unit; (3) lithological unit.

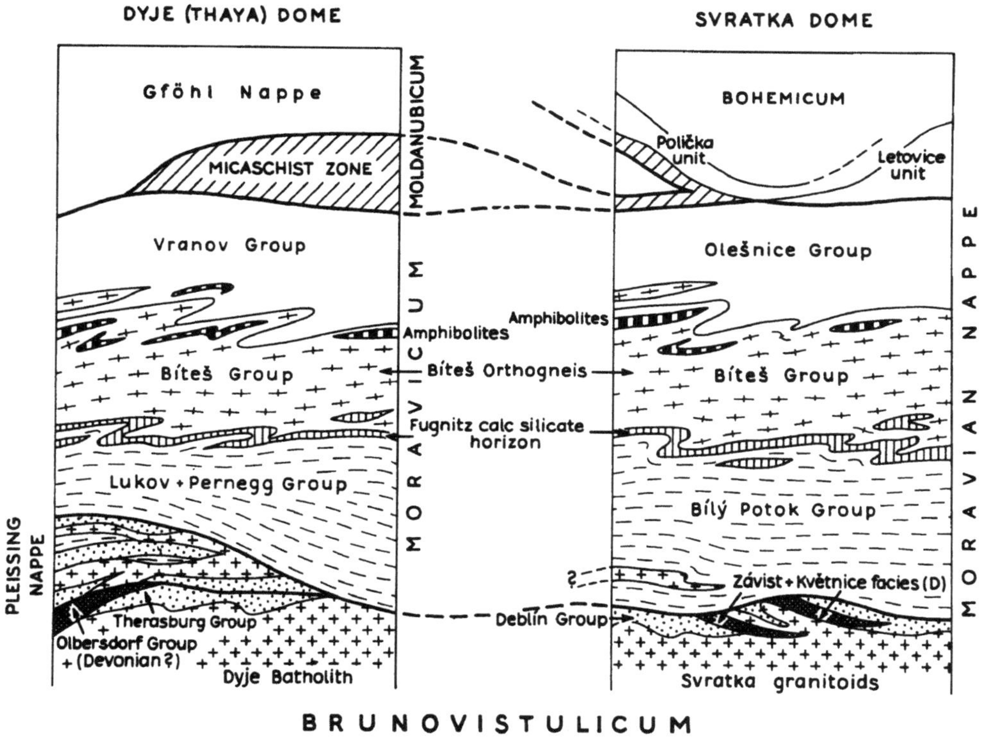

Fig. 1. Correlation scheme of lithological units in the Thaya (Dyje) and Svratka domes of the Moravicum

The Lukov Group (on the Czechoslovak territory) and the Pernegg Group (in Austria) resemble the neritic facies of the Vranov Group. They contain a very distinct marker horizon of calcitic shales and impure limestones situated almost continuously at the contact between the Bíteš Orthogneiss and the Lukov-Pernegg Group – the so-called Fugnitz calcsilicate schist.

A very complex Pleissing Tectonic Zone or the Pleissing Nappe (Tollmann 1985; Frasl 1991, and others) separates the Lukov-Pernegg Group from the autochthonous or paraautochthonous Thaya Batholith (Thaya granite) and its old metamorphic roof). In the N of the Thaya Dome, the Pleissing Tectonic Zone is tectonically reduced, and, consequently, the Lukov Group is in direct contact with the Thaya Batholith. The internal structure of the Pleissing Tectonic Zone (Nappe) is not yet perfectly understood. On the basis of lithology, individual laminae comparable to the Therasburg and Lukov-Pernegg groups, and laminae of magmatic rocks (the Pleissing, Weitersfeld or Buttendorf orthogneisses) can be discerned within this unit.

The old roof of the Thaya Batholith (the Therasburg Group; Höck and Vetters 1975; Höck et al. 1990) is composed of metamorphosed pelitic and arenitic sediments with some intercalations of limestones and calcitic shales. In places it was injected by apophyses of the Thaya Batholith.

The Drosendorf Window situated W of the Thaya Dome is tectonically separated from the surrounding Moldanubian Gföhl Nappe. It is interpreted either as a tectonic window of the Moldanubian Variegated Group emerging from the Gföhl Nappe (e.g., Fuchs 1976; Matte et al. 1990) or as a tectonic segment of the Micaschist Zone and the Moravicum (Jenček and Dudek 1971). Jenček and Dudek (op. cit.) distinguished the Podhradí Group corresponding by its rather monotonous lithology to the rock assemblage of the Micaschist Zone, and the Vratěnín Group which is a typical variegated facies with graphitic rocks, marbles, quartzites, and metabasic rocks similar to the Vranov(-Olešnice) Group (Table 2).

Because of great uncertainties about the position and interpretation of the Drosendorf Window, the use of the term Drosendorf Terrane (Matte et al. 1990) or the Drosendorf Variegated Group (Rajlich 1987) for one of the principal Moldanubian units should be avoided.

The lithological character of the Moravicum in the Miroslav (Misslitz) Horst in the vicinity

Table 2. Lithological units of the Drosendorf Window

Unit	Characteristics
Moldanubian zone (1) – Gföhl Nappe (2)	Granulites, amphibolites, migmatites
.... Tectonic contact	
Podhradí Group (3) (Jenček and Dudek 1971) = Micaschist zone (1)	Garnet micaschists and two-mica gneisses, orthogneisses, amphibolites
.... ?	
Moravicum (1) Vratěnín Group (3) (Jenček and Dudek 1971) = an equivalent of Vranov Group	Garnet micaschists, graphite schists, marbles, quartzites, metabasites
.... Tectonic contact	
Moldanubian zone (1) – Gföhl Nappe (2)	Granulites, amphibolites, migmatites

Table 3. Lithological units of the the Moravicum in the Svratka Dome

Unit	Characteristics
Micaschist zone (1) part of the Letovice Crystalline Unit (1) in the N, part of the Kutná Hora-Svratka Crystalline Unit (2) in the S (Mísař 1963)	Two-mica micaschists, amphibolites, graphite quartzites, white schists (Schulmann et al. 1991) Two-mica micaschists, tourmaline migmatites and orthogneisses, marbles, amphibolites, leptynites
.... Moldanubian Thrust	
Moravicum (1) Outer Phyllites (2) (Suess 1912) Olešnice Group (3) (= Vranov-Olešnice Group) Mísař (1963), Jaroš and Mísař (1976)	Amphibolites and metagabbros Garnet gneisses Garnet micaschists with amphibolites Marbles and dolomitic marbles Quartzites Metatuffs and metatuffites Graphite micaschists and phyllites Graphite and tremolite marbles Quartzites Garnet and staurolite micaschists Amphibolites, hornblende schists, marbles
.... Intrusive contact	
Bíteš Group (3)	Garnet orthogneiss Biotite-muscovite (ortho) gneiss Muscovite or Sericite orthogneiss In the upper part alternation of the Bíteš Gneiss with the lower part of the Olešnice Group
....	
Inner Phyllites (2) (Suess 1912) Bílý Potok Group (3) (Jaroš and Mísař 1976)	Calc-silicate horizon (an equivalent of the Fugnitz calc-silicate schist in the Thaya Dome) Muscovite-sericite and biotite (± garnet) micaschists Quartzites, quartzitic phyllites Graphite garnet phyllites Banded phyllites Sericite-chlorite phyllites Metabasites (greenschists)

of Krhovice N of Znojmo and at Maissau in Austria (Fig. 1) is equivalent to that of the Thaya and Svratka domes (micaschists, graphite schists, marbles, quartzites). The occurrence of Moldanubian rocks above the Moravicum in the Miroslav Horst (Dudek 1962) was already considered by Suess (1912) as an argument for the Moldanubian overthrust.

The classic sequence of the Outer Phyllites, Bíteš Geniss and the Inner Phyllites (Suess 1912) is particularly well developed in the *Svratka Dome*. The overlying tectonic units belong either to the Moldanubicum or to the Micaschist Zone and Letovice Crystalline Unit (Fig. 1). All lithological units of the Svratka Dome form together a single Moravian Nappe in the sense of Jaroš and Mísař (1974) which was thrust on the autochthonous and parautochthonous Brunovistulian basement (the Svratka granite and the Deblín Group). The succession of lithological units forming the Moravicum of the Svratka Dome (i.e., from top to bottom, the Olešnice Group, the Bíteš gneiss Group, and the Bílý potok Group) is shown in Table 3 and Fig. 1.

The Olešnice Group is a neritic sedimentary facies very rich in intercalations of graphite schists (with graphite horizons), marbles (in part dolomitic), quartzites, and basic volcanic rocks. The host rocks, which are mostly pelitic in the lower part of the Olešnice Group, become more

Table 3. (*Contd.*)

Unit	Characteristics
.... Moravian (Dřínová) Thrust	
Autochtonous or parautochthonous Brunovistulicum (1)	
Deblín Group (3) (= roof of the Brno (= Svraeka granite) (Jaroš and Mísař 1976)	Biotite phyllites with interlayers of amphibolites and calcsilicates, biotite gneisses, migmatites, aplites and pegmatites
.... Intrusive contact	
Brno (= Svratka) granite	Mylonitized granitoids, Aplites, dykes of metabasites

Table 4. Lithostratigraphy of the Devonian in the core of the Svratka Dome. (Jaroš and Mísař 1974, 1976)

Lateral facies	Characteristics
Závist facies (Givetian)	Laminated limestones with cherts, dolomites Sandy dolomitic limestones Sericite quartzites and siltstones Sericite monomict conglomerates Sericite polymict conglomerates
Květnice facies (Givetian)	Dark laminated graphite limestones light, in part dolomitic, massive or laminated limestones Laminated limestones with cherts Sericite quartzites

arenitic towards the top of this unit. Basic volcanic rocks and tuffs are distributed irregularly within the sedimentary sequence, while basalts and gabbros dominate on its top.

The Bíteš Group consists predominantly of a biotite-muscovite (sericite) orthogneiss. However, local interlayers of micaschists, amphibolites, and hornblende-biotite gneisses are also present.

The Bílý potok Group is composed of flysch-like sediments with a marker horizon of marbles and calcitic shales (the Fugnitz calcsilicate schist) on the top. Quartzitic layers and basic volcanic rocks occur close to the base of the sequence.

Based on the not yet successfully determined intrusive age of the Bíteš orthogneiss (see above), the volcanosedimentary sequence of the Svratka Dome (and the whole Moravicum) can be considered as being Upper Proterozoic to Lower Paleozoic in age.

The Brunovistulian basement below the Moravian Nappe shows quite a different lithology compared with the Moravicum. The Svratka granite (an equivalent of the Brno granite farther E) and accompanying aplites are intrusive into the Deblín Group, a metamorphic sequence of sedimentary and volcanic rocks. Both members of the basement are transgressively overlain by basal Devonian sandstones and conglomerates of the Závist and Květnice facies. The Květnice facies becomes richer in limestones towards the top while in the Závist facies the clastic sedimentation predominates in its whole extent (Table 4). Rocks belonging to both facial developments occur now as separate duplexes alternating with duplexes of the pre-Devonian basement; they document the minimum paleogeographical extension of the Devonian basin towards the W. The stratigraphy of the Devonian metasediments is based on relatively well preserved fossils (Brachiopoda, Stromatoporoidea, Crinoidea) of Lower to Middle Givetian age (Jaroš and Mísař 1968, 1974, 1976; Bosák 1983; Koverdynský and Hladil 1985).

Equialents of the Moravicum rocks represented by graphite phyllites, micaschists, graphite marbles, and rare quartzites and amphibolites occur near Svinov and Vranová in the Drahany Highland. Along with the Brunovistulian cataclastic granites, they form a tectonic slice wedged into the Devonian and Lower Carboniferous (Culm) sediments.

3 The Lithostratigraphy of the Silesicum

In the Silesicum the nappe tectonics affected both the Devonian cover and its metamorphic basement (Kettner 1950; Cháb and Opletal 1984; Cháb et al. 1990). The whole unit is divided by the Červenohorské Sedlo Tectonic Zone into the western *Keprník Dome*, and the eastern *Desná Dome*. Individual nappes are easier to follow in the Desná Dome due to generally lower metamorphic grade (greenschist to amphibolite facies) and the presence of fossiliferous Devonian basal metasediments which were tectonically incorporated into the pre-Devonian basement. In the deeper-eroded Keprník Dome the presumed Devonian age of sediments have not been paleontologically verified. The metamorphic grade of prevailing in the Keprník Dome orthogneisses and migmatites goes up to the amphibolite facies.

In our definition, the Silesicum of the Vysoký Jeseník Mts. is tectonically bounded on both the W and E sides. The western boundary – against the Staré Město Micaschist Zone – is defined by the Ramzová and Nýznerov overthrusts, while in the E the Vrbno Shear Zone (Synek et al. 1990) represents a distinct boundary of the Silesicum and the Flysch unit of the Nízký Jeseník Mts.

As already mentioned by Suess (1912), a variegated graphite-bearing formation emerges from beneath the Staré Město Micaschist Zone in the Lusatian terrane W of the Ramzová Thrust. The lithologic content of this unit – currently known as the Velké Vrbno Dome (Květoň 1951; Mísař et al. 1983) – is frequently correlated with that of the Moravicum.

The lithological units of the *Velké Vrbno Dome*, which, unfortunately, contain no fossils, have been thoroughly studied by Květoň (1951) and Skácel (1979); their superposition is summarized in Table 5 and Fig. 2. From the three main subunits distinguished by Květoň (1951), the Graphite Formation shows the strongest links to the Moravicum. On the contrary, the Upper Clastic and Lower Clastic formations of the Velké Vrbno Dome have a specific lithology with greater abundances of basic volcanic and intrusive rocks. Judging just from their lithological character, Květoň (1951) suggested an Ordovician to Silurian age for these three subunits, while Mísař (1967) regards them as Upper Proterozoic.

Also the stratigraphy of the *Keprník Dome* is hampered by the lack of paleontological data. The core of the dome is formed of the Keprník gneiss and migmatites (Mísař 1960) which have yielded a U/Pb zircon lower intercept age of 546 ± 6 Ma (van Breemen et al. 1982). A similar age can also be anticipated for the metamorphic rocks (the Keprník Monotonous Group) in the roof of the Keprník Orthogneiss (Table 6). The age of the overlying Keprník Variegated Group (Mísař et al. 1983) with graphite schists, marbles, quartzites, and amphibolites is uncertain, as is the case with the Graphite Formation of the Velké Vrbno Dome.

Quartzites, conglomerates, marbles, and metapelites of the Branná Group (Table 6) along the western flank of the Keprník Dome, and quartzitic rocks, calc-silicate rocks, amphibolites, and micaschists in the Červenohorské Sedlo Tectonic Zone are presumed to be of Devonian age, but without direct proof.

Table 5. Lithological units of the Velké Vrbno Dome. (After Květoň 1951)

Unit	Characteristics
Staré Město Micaschist Zone (1)	Two-mica gneisses and micaschists metagabbros, serpentinites, amphibolites, marbles (a body of diorite-tonalite)
.... Nýznerov Thrust in the W	
.... Ramzová Thrust in the E	
Velké Vrbno Dome (1)	
Upper Clastic Formation (3)	Plagioclase amphibolites Garnet amphibolites, metagabbros Hybrid migmatitic gneisses Two-mica gneisses and micaschists
? equivalent of the Bíteš Gneiss	Two-mica and hornblende ortho-gneisses, pegmatites
Graphite Formation (3) (= ?equivalent of the Vranov-Olešnice Group)	Quartzitic phyllites, hornfels gneisses, metabasites, marbles Graphite and pyrite schists with quartzites Dolomitic marbles and dolomites graphite marbles and graphite beds Tremolite marbles Graphite metacherts Quartzites and graphite schists
Lower Clastic Formation (3)	Micaschists (quartzitic) and phyllites Metabasite horizon (metatuffs and metatuffites) Banded garnet amphibolites Small layers of orthogneisses Garnet micaschists Quartzitic phyllites Banded micaschists, fine-grained gneisses
.... Basement unknown	

The *Desná Dome*, limited by the Červenohorské Sedlo Tectonic Zone in the W and the Vrbno Shear Zone in the E, displays a complicated imbricated or nappe structure (Pouba et al. 1962; Cháb et al. 1990) consisting of slices of pre-Devonian metamorphic basement and metamorphosed (up to amphibolite facies)

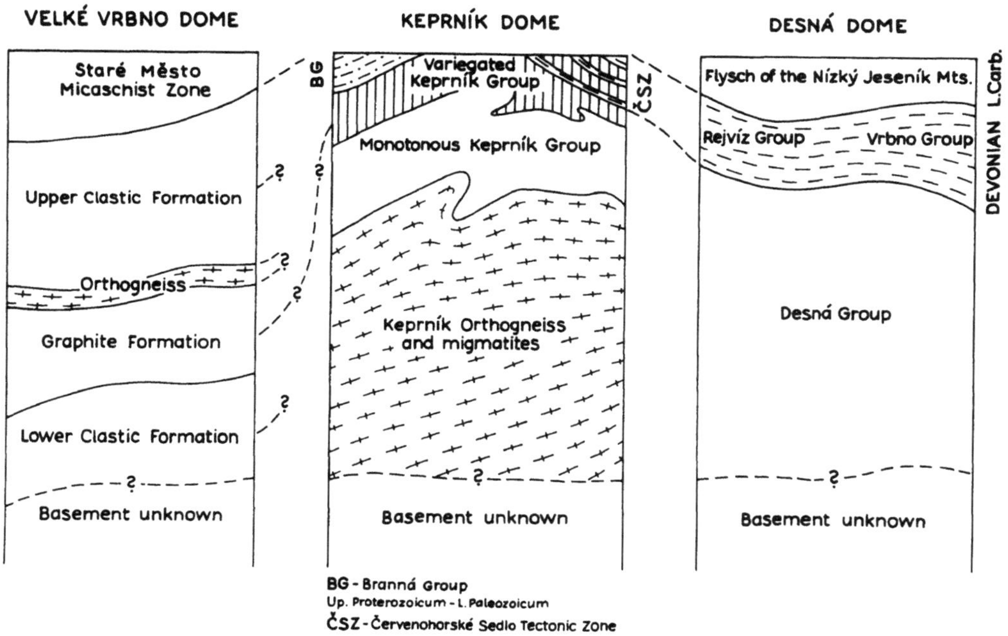

Fig. 2. Correlation scheme of lithological units in the Velké Vrbno, Keprník and Desná domes of the Silesicum. *BG.* Branná Group (Upper Proterozoic-Lower Paleozoic?); *ČSZ.* Červenohorské sedlo tectonic zone

Table 6. Lithological units of the Keprník Dome. (Mísař et al. 1983)

Unit	Characteristics	Unit	Characteristics
Staré Město Micaschist Zone (1) and Velké Vrbno Dome (1)	Gneisses, micaschists, quartzites Amphibolites, serpentinites Calc-silicates, tonalites Ophiolite sequence (Mísař 1970)	.... tectonic contact	
		Keprník Dome (1) Variegated Keprník Group (3)	Micaschists, calc-silicates, quartzitic gneisses, graphite schists
.... Ramzová Thrust		Monotonous Keprník Group (3)	Staurolite micaschists, gneisses, migmatites, pearl gneisses
Branná Group (3)	Light marbles with organic matter (possibly Devonian) Banded phyllites, impure marbles metabasites, graphite quartzites quartzites and polymict conglomerates (base of the Devonian ?) banded and quartzitic phyllites, impure marbles, metabasites	Keprník orthogneiss and migmatites	Two-mica orthogneiss with zones of granitization (Mísař 1960)
		.... Basement unknown	

Devonian sediments and volcanic rocks. The certainly pre-Devonian metagraywackes and the Formation of Banded iron ores (Pouba 1961) of the Desná Group (Table 7) are thought to be Upper Proterozoic in age; however, no radiometric or paleontologic data are available. The paleontologically evidenced (Pragian) transgressive littoral clastic sediments (e.g., Chlupáč 1989) and Upper Devonian pelites and marbles in the Vrbno and Rejvíz groups (Table 7) show a close spatial relation to Devonian basic volcanic and intrusive rocks with an island arc

Table 7. Lithological units of the Desná Dome. (Mísař et al. 1983)

Unit	Characteristics
Lugodanubian nappe (1) (Suess 1912)	
Hypothetic nappe of Cháb et al. (1990)	
.... Thrust	
Desná Dome (1)	
Devonian rocks in tectonic slices and thrust sheets of the Červenohorské Sedlo Tectonic Zone (2) (Pouba et al. 1962; Cháb and Opletal 1984)	Garnet-staurolite micaschists, chloritoide micaschists, calcsilicates, quartzites Sobotín and Jeseník metabasite complexes
Rejvíz Group (Devonian) (3) (Chlupáč 1989)	Marbles, graphite schists, garnet micaschists, metabasites, Greenshists Jeseník metabasite complex
Vrbno Group (Devonian) (3) (Chlupáč 1989)	Quartzites Graphite phyllites Metabasites Marbles Sericite phyllites
.... Primary transgressive contact strongly tectonically modified	
Nappes of the Desná Group (3)	Porphyroids, fine-grained biotite gneisses, amphibolites, talc schists, orthogneisses and migmatites, banded iron ores
.... ?...?...?...?	
Autochthonous and parautochthonous Brunovistulicum or ? equivalent of Mid-German Crystalline Rise	

affinity (Jakeš and Patočka 1982; Mísař et al. 1984).

Acknowledgments. W. Frisch is thanked for constructive criticism of the first version of the manuscript.

References

Batík P (1984) Geologická stavba moravika mezi bítešskou rulou a dyjským masívem. Věstn Ústřed Ústavu Geol 59: 321–330

Bosák P (1983) Palinspastická rekonstrukce sedimentačního prostoru tišnovských brunid. Věstn Ústřed Ústavu Geol 58: 341–347

Cháb J, Opletal M (1984) Příkrovová stavba východního okraje skupiny Červenohorského sedla v Hrubém Jeseníku. Věstn Ústřed Ústavu Geol 59: 11–21

Cháb J, Fediuková E, Fišera M, Novotný P, Opletal M (1990) Variská orogeneze v silesiku. Sborník geologických Věd, Ložiskova Geol Mineral 29: 9–40

Chlupáč I (1989) Fossil communities in the metamorphic Lower Devonian of the Hrubý Jeseník Mts., Czechoslovakia. Neues Jahrb Geol Paläontol Abh 17: 367–392

Dallmeyer RD, Neubauer F, Höck V (1990) $^{40}Ar/^{39}Ar$ mineral age controls on the chronology of late Paleozoic tectonothermal activity in the southeastern Bohemian Massif, Austria (Moldanubian and Moravo-silesian zones). In: Franke W (ed) Bohemian Massif. Field guide, Conf IGCP 233, Paleozoic orogens in central Europe, Göttingen-Giessen 1990, pp 87–96

Dudek A (1960) K otázce moldanubického nasunutí v severní části dyjské klenby. Věstn Ústřed Ústavu Geol 35: 229–232

Dudek A (1962) Zum Problem der Moldanubischen Überschiebung im Nordteile der Thayakuppel. Geologie 11: 757–791

Dudek A (1980) The crystalline basement block of the Outer Carpathians in Moravia: Bruno-Vistulicum. Rozpr Česk Akad Věd Řada Mat Přír Věd 90 (8): 1–85

Finger F, Frasl G, Höck V, Steyrer HP (1989) The granitoids of the Moravian Zone of Northeast Austria: products of a Cadomian active continental margin. Precambrian Res 45: 235–245

Frasl G (1954) Anzeichen schmelzflüssigen und hochtemperierten Wachstums und den grossen Kalifeldspaten einiger Porphyrgranite, Porphyrgranitgneise und Augengneise Österreichs. Sonderabdruck, Jahrb Geol Bundesanst 47 (1): 71–131

Frasl G (1970) Zur Metamorphose und Abgrenzung der Moravischen Zone im niederösterreichischen Waldviertel. Nachr Dtsch Geol Ges 2: 55–61

Frasl G, Höck V, Finger F (1990) The Moravian Zone in Austria. Field guide, Conf IGCP 233, Paleozoic Orogens in Central Europe, Göttingen-Giessen 1990, pp 127–142

Frasl G (1991) Das Moravikum der Thaya Kuppel als Teil der variszisch deformierten Randzone des Bruno-Vistulikums – eine Einführung. Arbeitstag Geol Bundesanst 1991: 49–62

Fuchs G (1976) Zur Entwicklung der Böhmischen Masse. Jahrb Geol Bundesanst 119: 45–61

Höck V, Vetters W (1975) Bericht 1974 über geologischen Aufnahmen auf Blatt Horn (21). Verh Geol Bundesanst 1975: A22–A26

Höck V, Marschallinger R, Topa D (1990) Granat-Biotit-Geothermometrie in Metapeliten der Moravischen Zone Österreich. Österr Beitr Meteorol Geophys 3: 149–167

Jakeš P, Patočka F (1982) Compositional variations of the Devonian volcanic rocks of the Jeseníky Mts. Věstn Ústřed Ústavu Geol 57: 193–204

Jaroš J, Mísař Z (1968) Stratigrafické postavení vápenců na Tišnovsku. Věstn Ústřed Ústavu Geol 43: 9–13

Jarŏs J, Mísař Z (1974) Deckenbau der Svratka Kuppel und seine Bedeutung für das geodynamische Modell der Böhmischen Masse. Sb Geol Věd Geol 26: 69–82

Jaroš J, Mísař Z (1976) Nomenclature of the tectonic and lithostratigraphic units in the Moravian Svratka dome. Věstn Ústřed Ústavu Geol 51: 113–122

Jelínek K, Pačesová M, Martinec P, Mísař Z, Weiss Z (1984) Geochemistry of a dismembered metaophiolite complex, Letovice, Czechoslovakia. Trans R Soc Edinb Earth Sci 75: 37–48

Jenček V (1984) Vysvětlivky k základní geologické mapě ČSSR 1:25 000, 33–223 Vranov. Ústřed Ústav Geol Praha 8: 1–36

Jenček V, Dudek A (1971) Beziehungen zwischen dem Moravikum und Moldanubikum am Westrand der Thaya-Kuppel. Věstn Ústřed Ústavu Geol 46: 331–338

Johan V, Autran A, Ledru P, Lardeaux JM, Melka R (1990) Discovery of relics of a high-pressure metamorphism at the base of the Moldanubian nappe complex. Proc Conf IGCP 233, Paleozoic Orogens in Central Europe, August-September 1990, Göttingen-Giessen

Kettner R (1950) Geologické profily devonem mezi Heřmanovicemi a Vrbnem ve Slezsku. Věstn Česk Spol Nauk 10: 1–28

Koverdynský B, Hladil J (1985) Stáří vitošovských vápenců (devon, severozápadní Morava). Věstn Ústřed Ústavu Geol 60: 1–7

Květoň P (1951) Stratigrafie krystalinických sérií v okolí severomoravských grafitových ložisek. Sb Ústřed Ústavu Geol 18: 1–60

Matte Ph, Maluski H, Echtler H (1985) Cisaillements ductiles varisques vers l'Est-Sud Est dans les nappes du Waldviertel (Sud-Est du Massif de Boheme, Autriche). Données microtectoniques et radiométriques $^{39}Ar/^{40}Ar$. C R Acad Sci Paris 301 (II): 721–726

Matte Ph, Maluski H, Rajlich P, Franke W (1990) Terrane boundaries in the Bohemian Massif: result of large-scale Variscan shearing. Tectonophysics 177: 151–170

Matura A (1980) Die Moravische Zone. In: Oberhauser R (ed) Geologischer Aufbau von Österreich, Geol Bundesanst, Vienna

Mísař Z (1960) Metasomatic granitization and its zonality in the Keprník Dome in the Hrubý Jeseník Mts. Rozpr Česk Akad Věd 70 (9): 1–52

Mísař Z (1961) Geologické postavení bítešské ortoruly. Čas Mineral Geol 6: 289–296

Mísař Z (1963) Pre-Devonian geological development of the northeastern margin of the Bohemian Massif. Rozpr Česk Akad Věd 73 (17): 1–60

Mísař Z (1967) Die Entwicklung und Gliederung der Mährisch-schlesischen Zone der Böhmischen Masse. Acta Univ Carolianae, Geologica 1: 11–27

Mísař Z (1970) Texturen ultrabasischer Gesteine der Staré Město Glimmerschieferzone. Acta Univ Carolinae, Geologica 2: 69–93

Mísař Z, Dudek A, Havlena V, Weiss J (1983) Geologie ČSSR I, Český masív. Státní Pedag Nakl, Praha, 333 pp

Mísař Z, Jelínek E, Souček J, Tonika J (1984) The correlation of gabbro-peridotite complexes of the Bohemian Massif. Krystalinikum 17: 99–113

Morauf W, Jäger E (1982) Rb-Sr whole rock ages for the Bites-gneiss, Moravicum, Austria. Schweiz Mineral Petrogr Mitt 62: 327–334

Pouba Z (1961) Magnetic ores of Sydvaranger type in Czechoslovakia. 21st Int Geol Congr, Copenhagen, 16: 164–169

Pouba Z (ed) (1962) Vysvětlivky k přehledné geologické mapě ČSSR 1:200 000, M-33-XVIII Jeseník. Geol Surv Praha: 1–86

Rajlich P (1987) Variszische duktile Tektonik im Böhmischen Masse. Geol Rdsch 76: 755–786

Scharbert S (1977) Neue Ergebnisse radiometrischen Altersdatierungen an Gesteinen des Waldviertels. Arbeitstag Geol Bundesanst 21: 11–15

Schulmann K, Ledru P, Autran A, Lardeaux JM, Urban M, Lobkowitz M (1991) Evolution of nappes in the eastern margin of the Bohemian Massif: a kinematic interpretation. Geol Rdsch 80: 73–92

Skácel J (1979) Tektonické plochy na styku východních a středních Sudet. Acta Univ Palackianae, Geologica-Geographica 17: 97–106

Suess FE (1912) Die moravischen Fenster und ihre Beziehung zum Grundgebirage des Hohen Gesenkes. Österr Akad Wiss Denkschr 88: 541–631

Synek J, Rajlich P, Urban M (1990) Transpression and transtension in the Devonian and Culm of the Jeseníky Mts. Acta Univ Carolinae, Geologica 2: 209–234

Tollmann A (1982) Großräumiger variszischer Deckenbau im Moldanubikum und neue Gedanken zum Variszikum Europas. Geotekton Forsch 64: 91

Tollmann A (1985) Geologie von Österreich, Bd II. Deuticke, Vienna

Van Breemen O, Aftalion M, Bowes DR, Dudek A, Mísař Z, Povondra P, Vrána S (1982) Geochronological studies of the Bohemian Massif, Czechoslovakia, and their significance in the evolution of Central Europe. Trans R Soc Edinb Earth Sci 73: 89–108

VIII.C.2 Structure

K. Schulmann, M. Lobkowicz, R. Melka, and H. Fritz

1 Introduction

The eastern margin of the Bohemian massif is a classical area where the Variscan nappe tectonics has been known since the early work of Suess. According to the concept of Suess (1912, 1926), the Moldanubian unit, consisting of high-grade metamorphic rocks including HP relics, overthrusted from west the Moravian unit, characterized by medium- to low-grade metamorphism. The high-grade schists and gneisses at the base of the Moldanubian nappe were strongly retrogressed during eastward overthrusting, thus giving rise to micaschists.

The Moravian zone appears recently in the form of two large-scale tectonic windows. It was affected by strong deformation and mesozonal metamorphism as a result of overthrusting of hot Moldanubian plate. The intensity of deformation and metamorphism decreases simultaneously downwards, so that undeformed and unmetamorphosed rocks remain preserved in the lowermost autochthonous granitic cores.

2 Principal Structural Zonation of the Moravian Zone

The Moravian zone displays similar litho-tectonic zonation in both the Svratka and Dyje windows (Fig. 1). The superposition of particular domains and nappes from the bottom to the top is:

1. **The autochthonous and parautochtonous Brunovistulian domains** (Dudek 1980) comprise pre-Variscan basement and its Devonian cover.
2. **The lower Moravian unit** comprises mostly metasedimentary varied group. A slice of strongly mylonitized Weitersfeld orthogneiss occurs at its base in the Dyje dome.
3. **The upper Moravian unit** consists of the Bíteš orthogneiss and its metasedimentary host rocks.

The boundary of the Moravian zone with the overlying Moldanubicum is transitional zone. It is represented by **the Moravian micaschist zone** comprising monotonous sedimentary sequence including HP relictual assemblages.

3 General Tectonic Aspect of the Moravian Zone; Definition of Major Interfaces

It is generally accepted that Moldanubian and Moravian zones are separated by a large-scale synmetamorphic shear zone (Suess 1912, 1926) represented by the micaschist zone. In agreement with the original terminology, this zone is termed **the main Moldanubian thrust zone** (Fig. 1). Critical high-pressure, low-temperature metamorphism assemblages (Johan et al. 1990) within the micaschist zone indicate that subduction-related relics were involved in later synmetamorphic (Barrovian) nappe stacking.

Three major thrust zones have been distinguished in Moravian zone (Fig. 1). **The intermediate Moravian thrust** separates two synmetamorphic internal nappes which are generally distinguished in the Moravian zone. The major interface between the lower Moravian unit and the metasedimentary host rocks of the Dyje dome is marked by synmetamorphic thrust zone – **the lower Moravian thrust**. The whole Moravian nappe pile is separated from the autochthonous domain in the Svratka dome by a low-grade thrust zone (the Dřínová overthrust of Jaroš and Misař 1974). This late thrust is correlated with a low-grade décollement zone separating the metasedimentary host rock from the Dyje granitic massif. Both shear zones are grouped together as **the boundary Moravian thrust zone**.

The geometry and P-T conditions of at least the last stages of nappe stacking differ in both the Svratka and Dyje windows.

Dallmeyer et al. (Eds.)
Pre-Permian Geology of Central and Eastern Europe

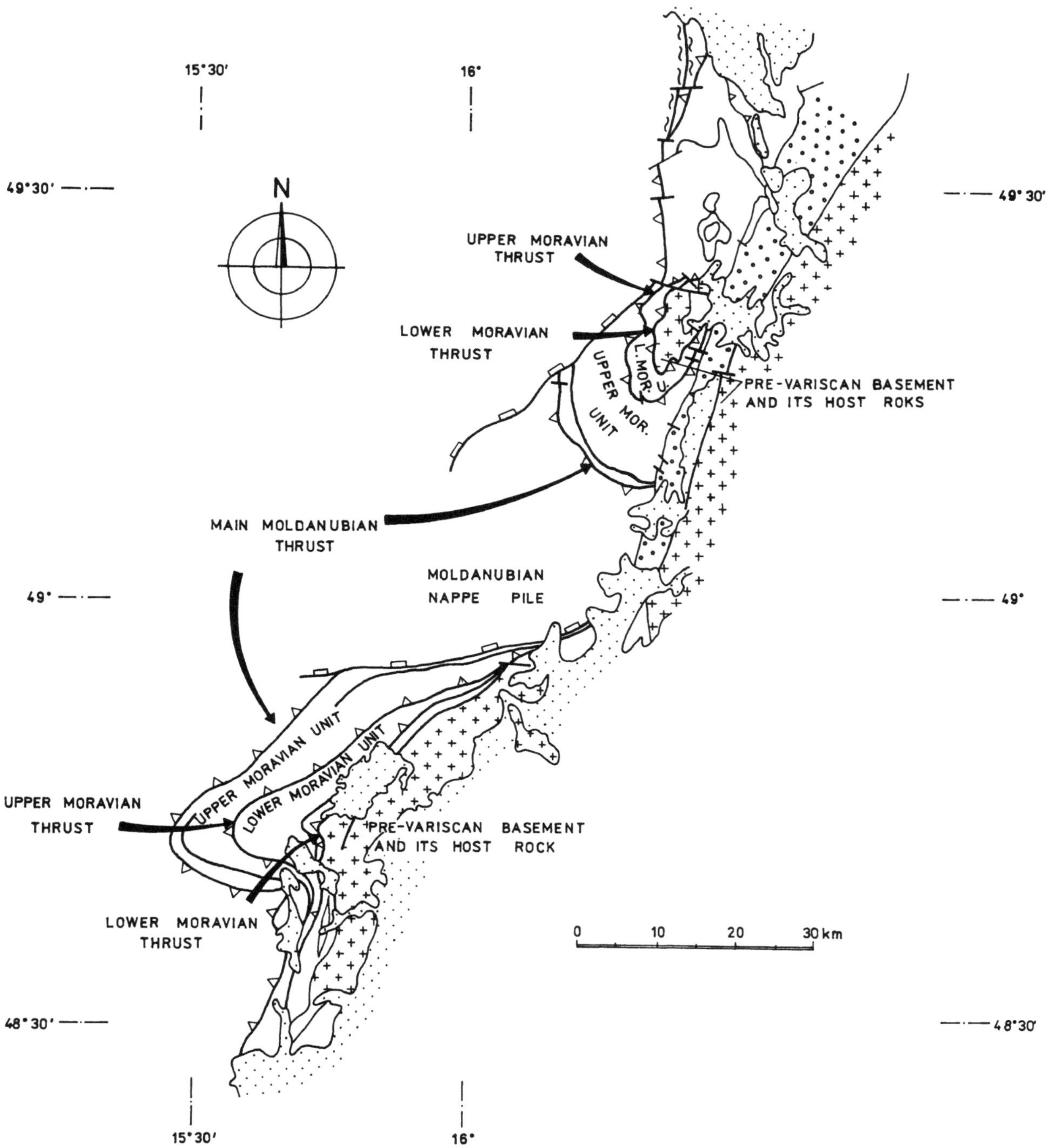

Fig. 1. Simplified tectonic sketch map showing the major thrust surfaces and tectonic units

4 Succession of Main Deformation Events: Building of the Nappe Pile

Two major phases of ductile deformation (D_1 and D_2) have been recognized in the Moravian zone, both characterized by flat-lying foliations and subhorizontal stretching lineations. The D_2 late deformation phase is represented by postfolial folding of the stacked nappe pile and brittle-ductile shear zones within the basement.

4.1 D_1 Deformation Phase: Stacking of the Moravian Nappes

D_1 deformation results from overthrusting of Moldanubian hot plate over the cold Moravian basement along a major crustal shear zone – the main Moldanubian thrust.

This deformation phase is responsible for subsequent formation of the highly ductile Moravian nappes that are separated from each other by the intermediate Moravian thrust. This deformation event is contemporaneous with the Barrovian metamorphism that affected the upper and lower Moravian nappes. In the southern Dyje dome, the Brunovistulian basement was reactivated under amphibolite facies conditions (Lobkowicz et al. 1994) along the lower Moravian thrust zone. This MT reworking of the basement is not present in northern Svratka dome.

4.1.1 D_1 Deformation in Moravian Nappes

The D_1 deformation is characterized by composite S_1 foliation, L_1 lineation and F_1 synschistose isoclinal folds and sheath folds.

S1 Metamorphic Foliation. The S_1 metamorphic foliation in metapelites of the upper Moravian unit is defined by the presence of characteristic mineral assemblages and textures: flattened quartz aggregates and parallel arrangement of micas. Synkinematical garnet, staurolite, and kyanite, being partly transformed to synfolial sillimanite, grow in the S_1 foliation.

On the other hand, mineral assemblages that form the S_1 metamorphic foliation in the lower Moravian unit differ in the Dyje and Svratka domes:

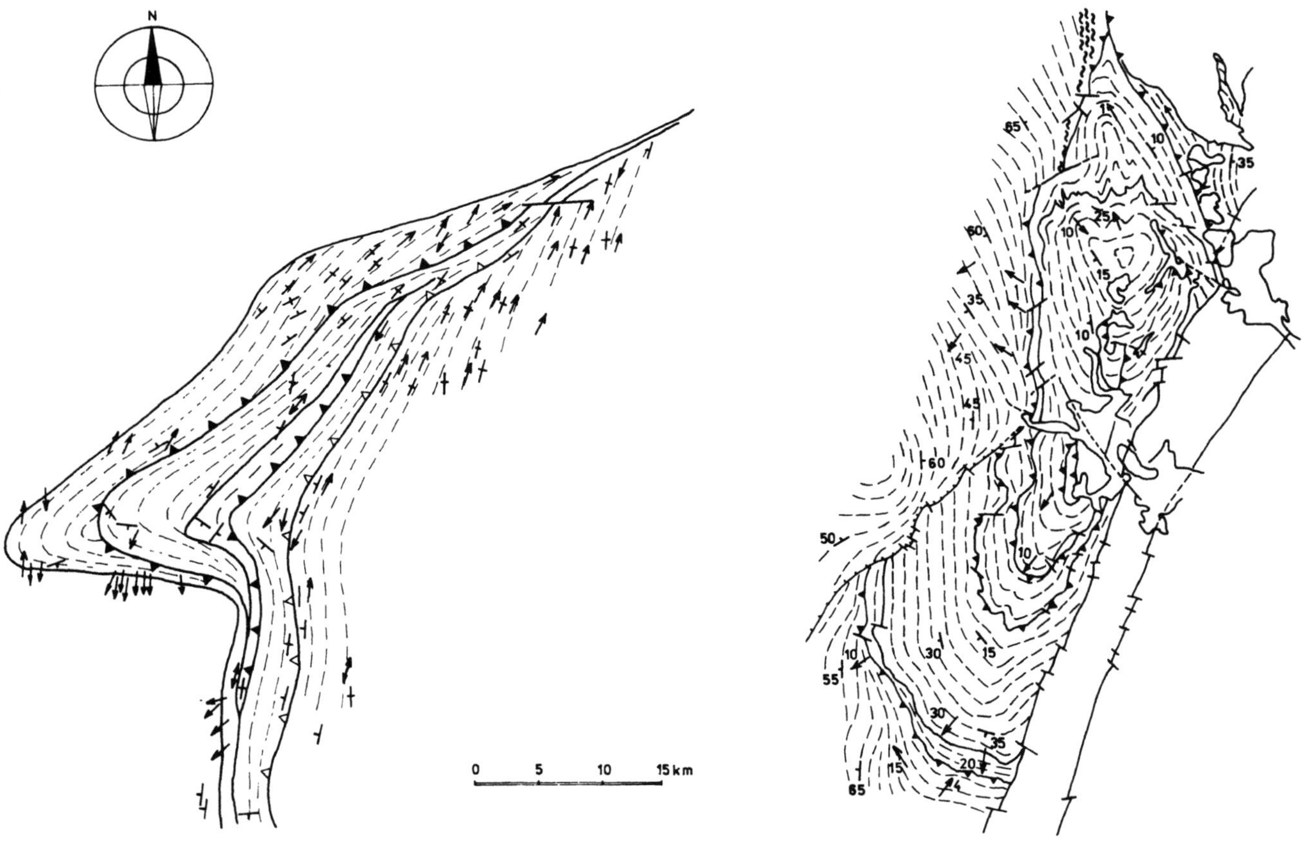

Fig. 2. Foliation and lineation map of the Moravian windows showing typical onion-like foliation distribution and transversal character of regional stretching independent of the strike of foliation planes. Note the change of stretching direction in the northern part of the Svratka dome

1. S_1 metamorphic foliation in metapelites of the lower Moravian unit of the Dyje dome is defined by biotite, muscovite and contains synkinematical garnet and syn- to post-kinematical staurolite.
2. S_1 foliation in metapelites of the Svratka dome is marked by syntectonic growth of chlorite, muscovite, and biotite. In several sites, the original lithological S_0 surfaces may even be observed.

In the Bíteš and the Weitersfeld orthogneiss, the S_1 is marked by flattened quartz ribbons, separated from bands of predominantly feldspar matrix by layers of recrystallized micas.

The composite S_1 foliation is perfectly concordant with the lithological boundaries of individual units, the latter forming the classical onion structures (Suess 1926) of both the Moravian windows (Fig. 2).

4.2 L1 Stretching Lineation, D_1 Kinematics and Finite Strain

The L1 stretching lineation is marked by preferred orientation of Barrovian minerals in metapelites and by elongation of feldspar clasts and quartz ribbons in orthogneisses. The subhorizontal L1 lineation trends NE-SW in the whole Dyje window and in the southern part of the Svratka window, but changes its trend to NW-SE in the northern part of the Svratka window. The L1 stretching lineation is effectively transversal (Fig. 2) and independent of the orientation of lithological limits and foliation directions (Suess 1926; Schulmann 1990).

Rotated garnets in metapelites of the lower and upper Moravian nappes represent an important criterion indicating noncoaxial strain path during the Barrovian metamorphic stage. In the Bíteš orthogneiss the asymmetrical sigmoidal K-

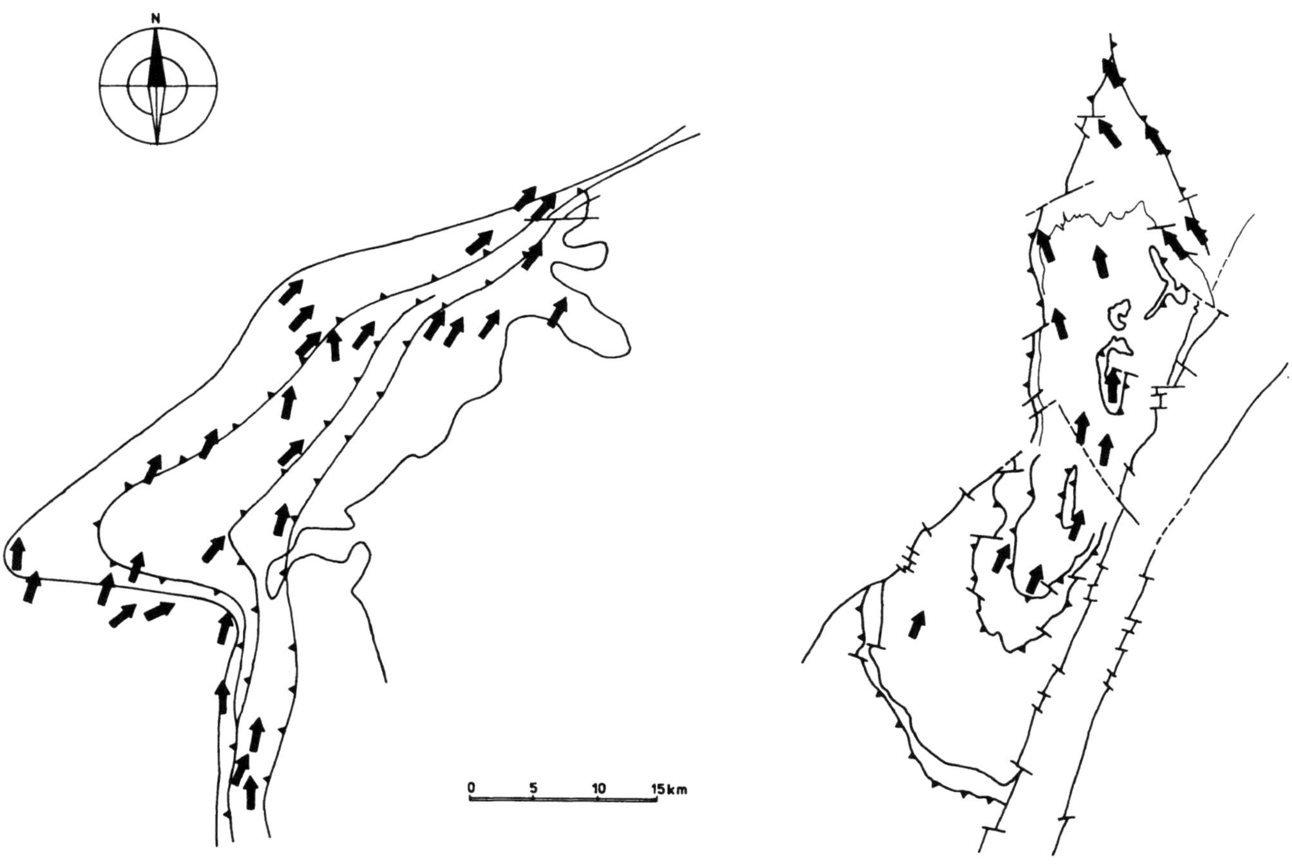

Fig. 3. Map of consistently northvergent kinematics in both Moravian windows. Both D_1 and D_2 sense of shear indicators are used. See text for discussion

feldspar porphyroclasts with recrystallized feldspar/quartz tails indicate HT conditions of noncoaxial flow. The spiral inclusions observed in the XZ sections of rotated garnets and sigmoidal feldspars and S-C relationships in deformed orthogneisses all indicate in north-eastward sense of displacement of the hangingwall segments (Fig. 3).

Detailed strain analyses performed in the Bites orthogneiss of the Dyje window (Schulmann 1990; Fritz 1990) revealed two zones exhibiting different strain symmetries and intensities.

The E-W-striking dip-slip domain of the Messern Arc and the northernmost part of the Dyje dome display oblate low-intense fabrics. Similar strain patterns have also been found in the core of the Bites orthogneiss body. These fabrics are associated with HT well-annealed granoblastic microstructures. Quartz c-axis-preferred orientations associated with oblate strains and low strain intensities can be approximated to small circles centered around the Z-axis of finite strain. The opening angle of 40° yields according to calibration of Hara and Nishimira (1977) a temperature of 550–600 °C (Fig. 4).

Plane-strain to constrictional fabrics are typical for the NW-dipping strike-slip domain occurring at the western margin of the Dyje dome; textures associated with plane strain fabrics show later reworking (Schulmann 1990). The intensity of deformation increases in conjunction with prolateness of strain from the core of the Bíteš orthogneiss body towards its boundaries; the foot-wall boundary represented by the inter-

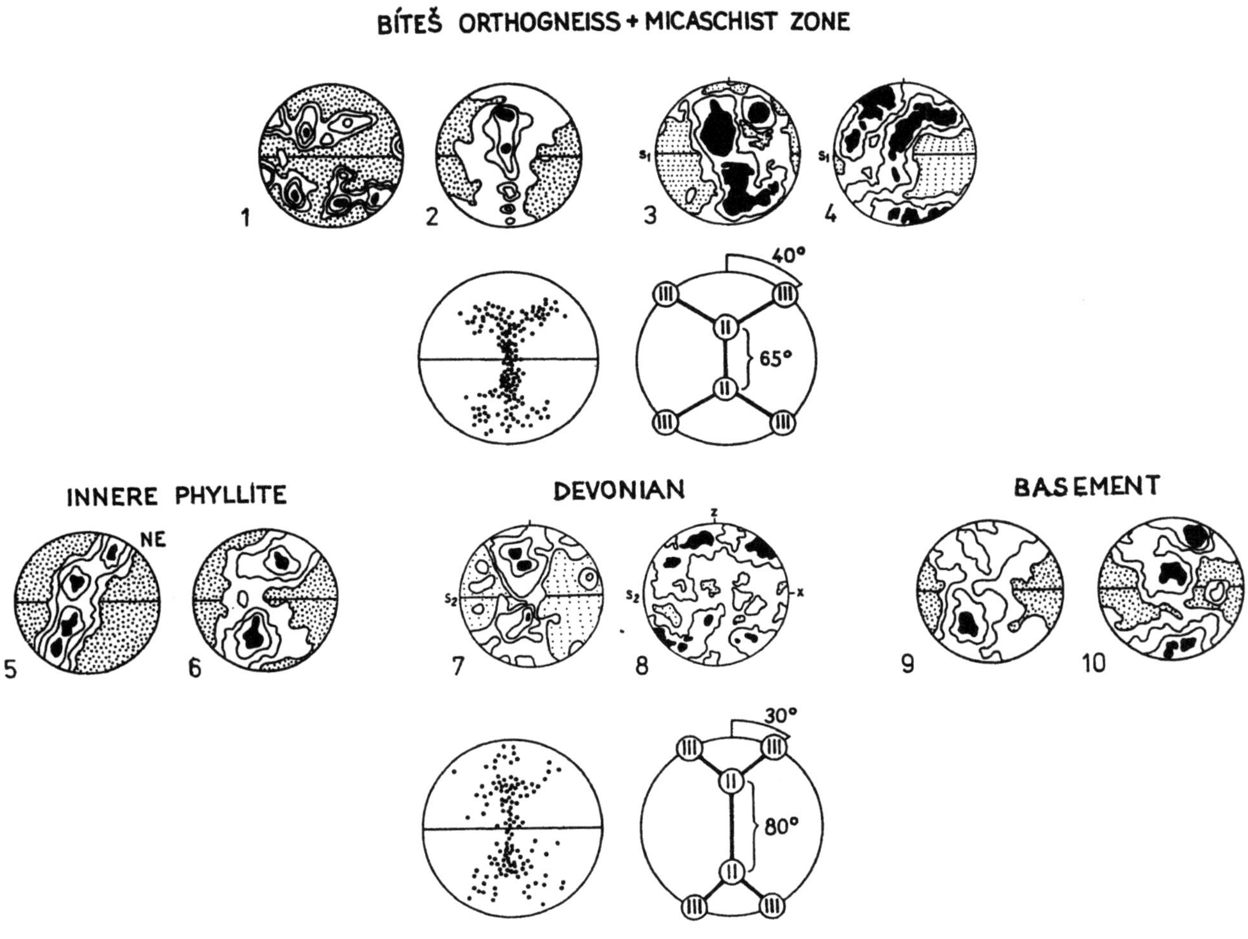

Fig. 4. The most typical quartz c-axis patterns from individual units of the Moravian zone. Note the skeletal outlines defined by *points* representing centers of density maxima from the whole Moravian zone. Higher opening angles and lower distances between maxima II in the upper Moravian unit are in agreement with higher metamorphic conditions of the shearing. Lower hemisphere equal area projections. *Stippled areas* represent densities less than one times uniform distribution

mediate Moravian thrust shows lower strain intensity (Preclik 1926). An asymmetrical Lister type I crossed girdle quartz c-axis pattern characterizes this deformation (Fig. 4:1, 2, 3, 4).

4.3 F1 Folds

Two types of synschistose folds have been recognized in Moravian rocks (Bowes et al. 1980; Plášil 1980; Kratochvil and Schulmann 1984; Schulmann 1990). (1) Tight to isoclinal recumbent intrafolial F1 folds, similar in geometry (Ramsay 1967), with axes parallel to L1 lineation and (2) mildly to strongly noncylindrical sheath folds with noses parallel to the regional stretching. This type of folding is restricted to the areas with strong lithological anisotropy, e.g., at the upper intrusive contacts of the Bíteš and Weitersfeld orthogneisses with its host rock. The finite strain symmetry in these areas is of a plane-strain to constrictional type.

4.3.1 D_1 Deformation of the Basement Rocks

The Cadomian basement in the Dyje dome was strongly reworked during Variscan nappe tectonics, which led to the formation of highly foliated mylonite belt some 2–3 km thick at the western part of the Dyje massif (Lobkowicz et al. 1994).

The north-northeast-trending mylonitic foliation S_1 dips steeply (50–70°) to the west at the southwestern part of the Dyje dome. At the northern termination of the Dyje dome, the angle of foliation dip decreases to 20–30° (Fig. 2). The foliation in the mylonite zone is marked by flattened recrystallized feldspar aggregates, quartz ribbons, and biotite clusters.

The very strong north-northeast-striking stretching lineation L2 is characterized by elongation of biotite clusters, elongated dioritic inclusions, and by recrystallized tails of quartz and micas around larger feldspar porphyroclasts. Numerous kinematic markers such as S-C fabrics and asymmetrical feldspar clasts indicate top to north-northeast shearing (Fig. 3).

Prolate fabrics predominate at the westernmost part of the deformed granite body, while plane strain or even oblate fabrics occur in the core of the mylonite belt. Quartz appears in recrystallized ribbons exhibiting strong c-axis-preferred orientation of Lister type I (Fig. 4:9, 10).

Further to the east, the deformation becomes inhomogeneous and large volumes of isotropic granite are preserved within the sheared rock. Zones of homogeneously foliated orthogneiss anostomose around lens-shaped pods of undeformed granite elongated in NE-SW direction.

4.4 D_2 Deformation Phase: Reactivation of the Nappe Pile

The Moravian boundary thrust, separating the Moravian nappe pile from the intensely deformed basement, is a major result of the D_2 deformation phase.

The D_2 deformation is retromorphic and generally reaches the greenschist facies conditions. The late, D_2, retromorphic thrusts separate individual tectonic units, reactivate earlier synmetamorphis nappe interfaces and are subparallel to the lithologic boundaries. The deformation is related to a strong noncoaxial NE- or N-oriented shearing, that is connected with the formation of brittle-ductile and ductile shear zones and asymmetrical A-type folds.

4.4.1 D_2 Deformation in the Moravian Nappe Pile

In the Moravian nappes, the D_2 deformation is characterized by (1) formation of retromorphic shear zones either developed by reactivating of earlier tectonic boundaries (base of the Bíteš and the Pleissing nappes) or penetrating the inner parts of individual lithological units (2) intense refolding of primary intrusive roofs of the Bíteš and the Weitersfeld orthogneisses. These asymmetrical NW-facing folds were firstly discussed by Dudek (1962).

The D_2 mylonite shear zones reactivate the early MT foliations connected with a destabilization of the Barrovian mineral assemblages. The D_2 results in intense constrictional fabrics in the Bíteš orthogneiss nappes (Schulmann 1990). Marbles, calc-silicate rocks, and metapelites of the upper part of the lower Moravian unit were imbricated, together with the basal parts of the Bites orthogneiss body during the D_2 phase.

Stable muscovite and chlorite characterize the D_2 mineral assemblages, while biotite and oligoclase were destabilized during the D_2 (Bernroider 1986). The S_2 foliation planes are often covered by large amount of sericite. L_2

lineation is marked by corrugation of S_2 planes or by arrangement of smeared-out chloritized biotite and linear arrangement of sericite. Distinct NNE-dipping shear bands are indicative of the general NNE-ward direction of the displacement. Quartz in quartzites of the lower Moravian unit in the Svratka dome and in detached metasedimentry roof of the Dyje granite occurs in form of recrystallized mosaic. Typical oblique girdles or asymmetrical clef girdles indicate a combination of basal ⟨a⟩ and prism ⟨a⟩ glide during northeastward noncoaxial shearing (Fig. 4: 5, 6).

The roofs of the Bíteš and the Weitersfeld orthogneiss bodies are rich in leucocratic sheets, that cross cut the overlying metasedimentary formations. They represent regions that are strongly anisotropic from the point of view of the mechanical behavior. Therefore, this part of the Moravian nappes was intensely refolded as a result of the D_2 deformation. North- to north-northeast-trending recumbent folds F_2, one to several meters in size, with subhorizontal axial planes, predominate in the Dyje dome and in the southern part of the Svratka dome. On the other hand, northwest-trending folds prevail in the northern part of the Svratka dome.

4.4.2 D_2 Deformation in the Autochthonous Domain

The autochthonous and parautochthonous basement in the cores of the Dyje and the Svratka tectonic windows display different structural patterns, with respect to the late D_2 stacking of the Moldanubain/Moravian nappe pile. Two major differences have been observed (Figs. 5, 6).

1. The Moravian boundary thrust in the Svratka window separates the lower Moravian unit from the basement rocks and its Devonian cover. The same boundary thrust is represented in the Dyje dome as a décollement phyllonitic zone separating the granitic Dyje core from its metamorphic original host rock (Figs. 5, 6).

2. The D_2 deformation in the Svratka dome is penetrative and is characterized by mylonitization of basement rocks and its Devonian cover within basement and cover duplexes. In the Dyje dome, the D_2 deformation is strongly heteregeneous and is marked by the presence of brittle-ductile shear zones and arrays of slickenside surfaces.

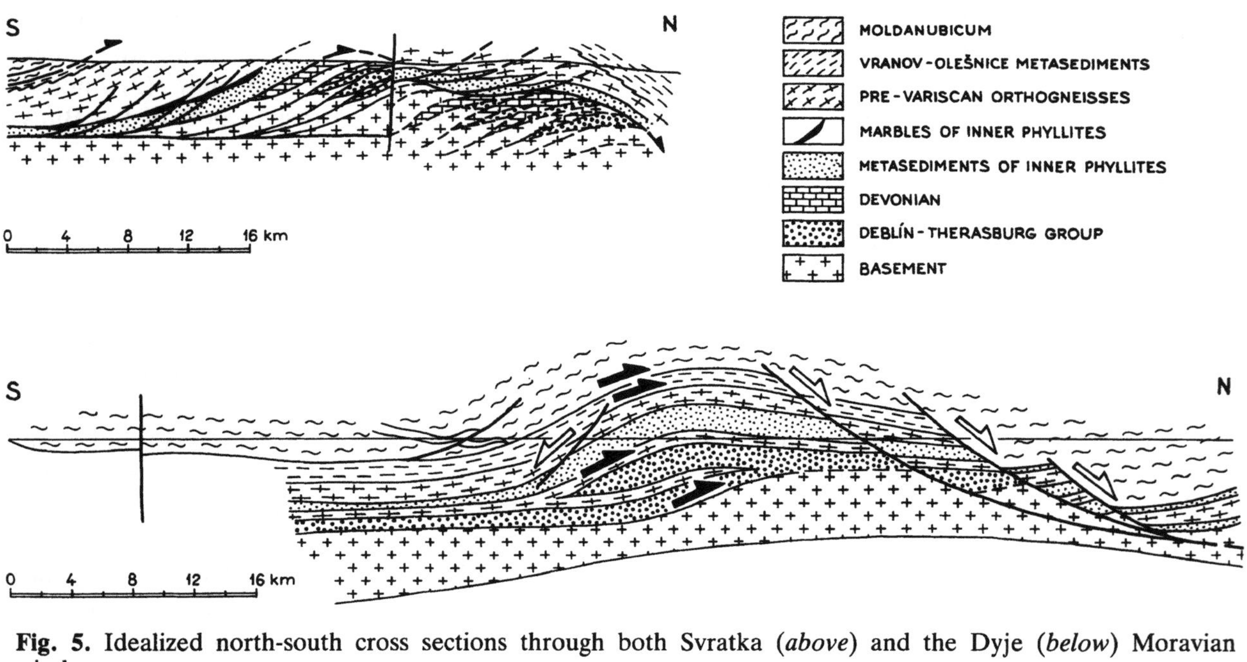

Fig. 5. Idealized north-south cross sections through both Svratka (*above*) and the Dyje (*below*) Moravian windows

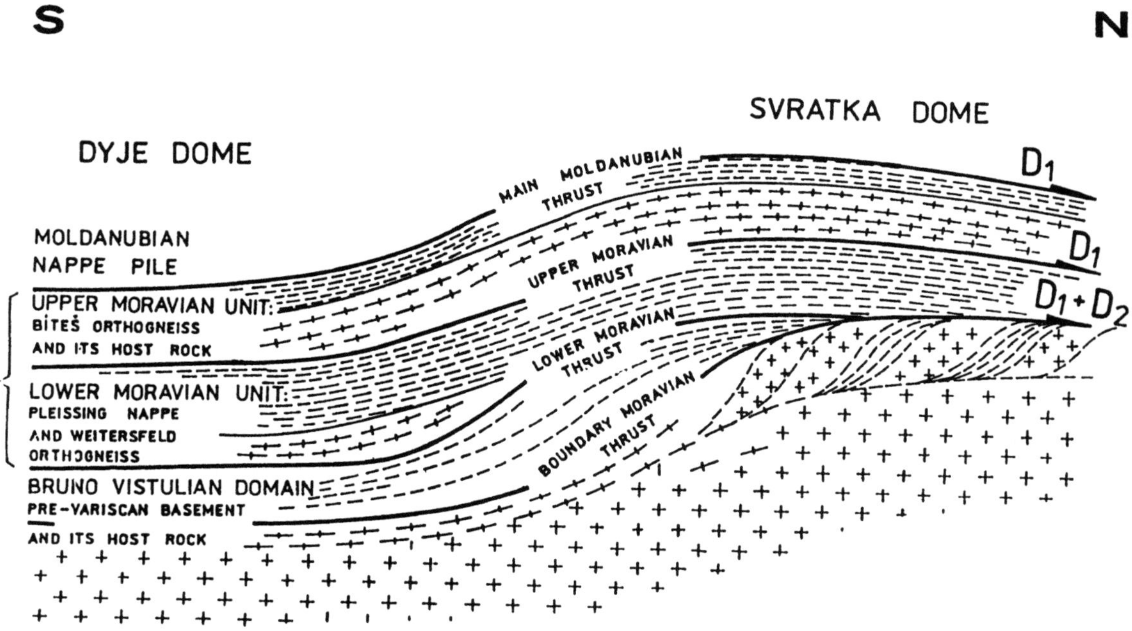

Fig. 6. Idealized sketch explaining the used terminology of Variscan structural pattern of Moravian windows. Note the difference of degree of involvement of Cadomian crust in Variscan tectonics in south and north. See text for discussion

4.4.3 D_2 Deformation in the Autochthonous Domain in the Dyje Dome

Metasediments of the mantle of the Dyje massif reveal a high degree of D_2 mylonitization. The zone of homogeneously retrogressed and phyllonitized micaschists locally reaches a thickness of 1 km. This huge phyllonite zone is associated with a decollement of most of the mantle rocks from the Thaya granite.

Phyllonitization is characterized by the occurrence of fine-grained chlorite-sericite-albite phyllites with a great amount of quartz exudates, occurring mainly in the neck zones of the foliation boudinage. The alternation of chlorite-sericite phyllites with biotite schists has been interpreted as a result of imbrications of the mantle of the Dyje granite below the Pleissing nappe (Preclik 1926).

Ubiquitous NNE-dipping C_2 surfaces are consistent with the NNE sense of shearing, that is parallel with the L_2 lineation. Intersection of the C_2 surfaces with S1 foliation produces almond-shaped structures. The inclusion spirals in rotated garnets are often disrupted by late shear surfaces at the boundaries with fine-grained matrix. Asymmetric pressure shadows around garnets are filled by chlorite and quartz. All these criteria are indicative of a pretectonic origin of the garnets with respect to D_2.

The D_2 deformation in the autochthnous granite is manifested by the occurrence of subvertical NNE-SSW-trending brittle-ductile shear zones producing fine-grained ultramylonites and phyllonites (Lobkowicz et al. 1994). These zones either cut the S1 mylonitic foliation of previously deformed granite in the marginal mylonitic zone of the batholith or evenly affect the isotropic granitic rocks in the core of the massif.

There is only one dominant set of shear zones corresponding to the northeastward sense of shear. The conjugated set is not present. According to Gapais et al. (1987), this shear zone pattern may be interpreted as a result of noncoaxial deformation during dextral wrenching.

Sets of striated microfaults associated with D_2 shear zones have been used for paleostress determinations (Lobkowicz et al. 1994). The data obtained are fully compatible with the orientations of extension and compression directions linked to brittle-ductile subvertical deformation zones. The brittle deformation and brittle-ductile shear zones are thus linked to the same dextral wrenching regime, which prevailed during D_2 deformation of the Dyje granite.

4.4.4 D_2 Tectonics in the Autochthonous Domain in the Svratka Dome

Granitoids, high-grade metamorphic rocks, and Devonian sedimentary cover in the Tišnov window were involved in thin-skinned north-vergent thrust tectonics. Generally, flat-lying mylonite zones with a typical duplex geometry separate different lithologic facies. Two large-scale tectonic slices, those at Závist and Květnice, containing the basement rocks as well as the Devonian cover, have been identified by Zapletal (1932) and Jaroš and Mísař (1986). They differ essentially in the content of Devonian clastics and limestones and therefore in the degree of involvement of cover sediments into hinterland dipping duplexes. The mylonite foliation in sheared granitoids and migmatites of the basement rocks (Deblín group, Jaroš and Mísař 1974) is characterized by flattened quartz blebs and micaceous layers composed of the product of destabilization of feldspars (clinozoisite, micas, quartz) and relict biotite (Fajst 1968; Kachlík 1989). S2 foliation in Devonian clastics is marked by XY planes of finite strain of deformed pebbles (Plášil 1977) and in deformed limestones by flattening of calcite grains (Kratochvíl 1981).

North-northeastward-trending L2 stretching and aggregate lineation is marked by pulled-apart distorted fragments of elongated feldspar clasts, alignment of relictual biotite, chlorite, and epidote in mylonitized basement rocks and by strong elongation of deformed quartz pebbles in Devonian clastics.

Different types of shear sense criteria, e.g., S-C relationships in deformed granitoids, shear bands, and assymetrical sigmoidal pebbles (Fig. 3.) demonstrate the noncoaxial character of D_2 deformation.

A finite strain ellipsoid with plane strain to slightly constrictional symmetry has been obtained by Plášil (1977) from deformed quartz pebbles. Strain data show an increasing strain magnitude from bottom to top and from east to west. The increase in strain intensity is connected with the increasing prolateness of the strain ellipsoid.

Relictual porphyroclastic textures are locally preserved in quartzitic deformed pebbles and in granitoids, which can be attributed to the low-temperature plastic deformation (Bouchez and Pécher 1981). Syntectonic recrystallization is connected with plastic deformation of quartz grains mostly in deformed conglomerates. Quartz c-axis microfabrics in deformed basement rocks and Devonian quartz pebbles show either Lister type-I crossed girdle patterns (Lister 1977) or unique girdle oblique with respect to the XY finite strain axes (Fig. 4: 7, 8). The asymmetry of quartz c-axis diagramms is in agreement with the imposed northward shear. This type of quartz microfabric has been called Moravian type of by Němec (1961, 1979, 1980).

Open to close, upright to recumbent, generally east-vergent folds with north-northeastward oriented fold axes deform the S2 mylonitic foliation and lithological boundaries in the basement of the Svratka dome (Plášil 1980; Kratochvíl 1984; Kachlík 1989). These subhorizontaly striking F2 folds vary from cm to km in amplitude (Plášil 1980), and have very different shapes and degrees of flattening.

5 Discussion

The recent tectonic pattern of the Moravian zone is a result of a long-lasting N-S-directed convergence of the Moldanubian and Brunovistulian units. Collision of both plates produced prograde inverse syntectonic metamorphism, intense imbrication of the crust, development of large-scale basement and cover duplexes, and the recent antiformal structure of the Moravian zone.

The allochthonous Moldanubian and autochthonous Brunovistulian domains are separated by stacked Moravian nappe pile, derived from the southern part of the Brunovistulian microplate.

The main structural feature of the Moravian zone is inverted polyphase metamorphism and stable kinematics developed in the highly ductile Moravian nappes. The downward migration of thrust boundaries of subsequently piled nappes, as well as large-scale strain partitioning, are typical features of the nappe stacking. The temperature also decreased from the south to the north, so that the metamorphism and style of deformation differ in individual units at the Svratka and the Dyje dome. This pattern may be best explained by using the classical model of overthrusting of hot Moldanubian plate over cold Brunovistulian basement. The roots of the Moldanubian nappe were probably situated in the south, which is indicated by constant north-northeastward kinematics and systematically decreasing degree of metamorphism from the south to the north.

During the initial stage of Moldanubian overthrusting, a large portion of pre-Variscan basement rocks (including the roof of Cadomian granite and its host rocks) were cut from the main basement body. The transported huge slice of basement rocks was transported with the Moldanubian nappe, producing first metamorphic foliation, stretching lineation, and sheath folds. The whole nappe behaved as an extremely ductile body, for which large-scale train partitioning is typical. This process led to the formation of the upper Moravian nappe.

In the following stage, the pile of the Upper Moravian unit and Moldanubian nappe cut another northward-lying portion of the basement rocks producing the lower Moravian nappe. Although the upper Moravian unit consumed agreat amount of conducted heat, the behavior of the lower Moravian nappe was still very ductile. The presence of syntectonic Barrovian assemblages in metamorphic foliation, intense northeastward stretching lineation, sheath folds, and large-scale strain partitioning are characteristic of this stage.

The heat acquired by the lower Moravian unit during under thrusting is responsible for the metamorphism in garnet-biotite zone in metapelites and for amplibolite facies mylonitization of granitoids of the basement rocks.

This complete evolution of nappe stacking is developed only in the southern Dyje dome. At the northern Svratka dome, only the upper Moravian nappe is fully developed, comprising a slice of granitic rocks and its host rocks, while the lower Moravian nappe lacks the granitic base. This may be explained by insufficent thermal and mechanical energy of the overlying nappes, which were unable to cut the basement rocks together with part of the granitic core. Instead, only the metasedimentary roof is detached.

Such contrasting patterns in the Dyje and Svratka domes (Fig. 6) may be explained by time migration of deformation in vertical direction downwards in the nappe pile and also by migration of the thrust boundaries horizontally northwards. Hence, the dropped thermomechanical energy at the northen termination of the Moravian nappe pile produced only typical thinn-skinned tectonics below the boundary Moravian thrust in the Svratka dome.

References

Bernroider M (1986) Zur Geologie und Petrographie Moravischer Gesteine im Gebiet NW Weitersfeld, Niederösterreich. Dissertation Univ Salzburg

Bouchez JL, Pecher A (1981) The Himalayan Main Central Thrust pile and its quartz-rich tectonites in Central Nepal. Tectonophysics 78: 23–50

Bowes DR, Hopgood AM, Misař Z (1980) Characterization of tectonic regimes of the Letovice unit and Bíteš gneiss of the Moravicum using the structural imprint of polyphase deformation. Věstn ÚÚG 56: 331–342

Dudek A (1962) Zum Problem der moldanubischen Uberschiebung im Nordteil der Thayakuppel. Geologie 11: 757–792

Dudek A (1980) The crystalline basement block of the Outer Carpathians in Moravia-Brunovistulicum. Rozpr Cesk Akad Věd Rada Mat Prir 90 (8): 85

Fajst M (1968) Granitisation und Mylonitisation im NW Teile des Svratka-Komplexes. Acta Univ Carol Geol 4: 309–328

Fritz H (1990) Structures and kinematics along the Moravian-Moldanubian boundary: preliminary results. Öterr Beitr Meteoroe Geophys 3: 77–96

Gapais D, Bale P, Choukroune P, Cobbold PR, Mahjoub Y, Marquer D (1987) Bulk kinematics from shear zone patterns: some field examples. J Struct Geol 9: 419–427

Hara I, Nishimira Y (1977) Boundary between subbasal I and subbasal II fields of quartz deformation in geological conditions. Tectonophysics 39: 273–286

Jaroš J, Misař Z (1974) Der Deckenbau der Svratka Kuppel and seine Bedeutung fur das geodynamische Modell der Böhmischen Masse. Sb Geol Ved Geol 26: 69–82

Jaroš J, Mísař Z (1986) Geologicko-strukturní charakteristika moravika svratecké klenby a přilehlé části moravské svorové zony. Rep Charles Univ, Prague (unpubl)

Johan V, Autran A, Ledru P, Lardeaux J-M, Melka R (1990) Discovery of relics of high pressure metamorphism at the base of the Moldanubian nappe complex. IGCP 233, Terranes in the Circum-Atlantic Paleozoic orogens, 24–27 Aug, Göttingen, Conf Abstr

Kachlik V (1989) A contribution to the tectonomagmatic history of the Moravian unit in the core of the Svratka dome. Krystalinikum 20: 49–64

Kratochvíl M (1981) Analýza mikrostavby vápenců v tektonických stylech devonu Českého masívu rentgenovou difrakcí. PhD Thesis, Charles Univ, Prague (unpubl)

Kratochvíl M (1984) Profil údolí Nedvědičky mezi Jabloňovem a Nedvědicemi. Rep Charles Univ, Prague, (unpubl)

Kratochvíl M, Schulmann K (1984) Correlation of the Moravicum and Moldanubicum in the Dyje valley on the basis of structures of polyphase deformation. Cas Mineral Geol 29: 337–352

Lobkowicz M, Schulmann K, Melka R (1994) Variscan reactivation of the Cadomian basement in the

Dyje tectonic window (Bohemian Massif). Acta geodinamica (submitted)
Lister GS (1977) Crossed girdle c-axis fabrics in quartzites plastically deformed by plane strain and progressive simple shear. Tectonophysics 39: 51–54
Nemec D (1961) Mikrostavba kremene v moraviku a prilehlem moldanubiku zapadni Moravy. Knihovna UUG, 197 pp
Nemec D (1979) Der Blockbau der Mahrischen Scholle vom Standpunkt des Quartzgefüges der Gesteine. Geol Rdsch 68: 775–792
Nemec D (1980) Beziehungen zwischen dem Metamorphosegrad und den Quartzgefüge-Gürtelbildern in der Böhmischen Masse. Z Geol Wiss Berlin 8: 1169–1180
Preclik K (1926) Das Nordende der Thayakuppel. Sb SGU 6: 373–398
Plášil M (1977) Deformační analýza křemenných valounů devonských konglomerátů, v jižní části závistské jednotky v jádře svratecké klenby. Diploma work, Charles Univ, Prague, 114 pp
Plášil M (1980) Analýza polyfázové vrásové deformace v tělese bítešských rul svratecké klenby moravika. PhD Thesis, Charles Univ, Prague
Ramsay JG (1967) Folding and fracturing of rocks. McGraw- Hill, New York, 568 pp
Schulmann K (1990) Fabric and kinematic study of the Bites orthogneiss (southwestern Moravia): result of large-scale northeastward shearing parallel to the Moldanubian/Moravian boundary. Tectonophysics 177: 229–244
Suess FE (1912) Die moravischen Fenster und ihre Beziehung zum Grundgebirge des Hohen Gesenkes. Denkschr Österr Akad Wiss Mat Naturwiss 88: 541–631
Suess FE (1926) Intrusionstektonik und Wandertektonik im variszichem Grundgebirge. Bornträger, Berlin, 212 pp
Zapletal K (1932) Geologie a petrografie země moravskoslezské. Vlastivědné Publikace Moravskoslezské, pp 1–280

VIII.C.3 Metamorphic Evolution

V. Höck

1 Introduction

When F.E. Suess (1903) divided the eastern part of the Bohemian Massif into the structural deeper Moravian unit and the upper Moldanubian zone, one of his most important criteria was the changing degree of metamorphism between both units: The high-grade metamorphic Moldanubian was thrusted over the low grade Moravicum forming the retrogressive Glimmerschieferzone in between. Later on in 1912, Suess subdivided the Moravian unit into the Thaya window or dome (Dyje dome) in the S and the Schwarzawa window or dome (Svratka dome) in the N (Fig. 1). Both units show a comparable lithology and metamorphism, but with a certain variability. He also realized the decreasing metamorphism in both domes from the outer, structurally higher units to the inner, lower parts. These findings have later been confirmed by other authors, e.g., Preclik (1927, 1937); Frasl (1968, 1970, 1983); Höck (1975, 1983); Höck et al. (1991) for the Thaya dome and Jaroš and Mísař (1976); Jaroš (1992); Schulmann et al. (1991); Cháb and Suk (1977) for the Svratka dome.

Despite these early achievements, only little was accomplished in the following half century. It was only recently, after 1970, that more detailed investigations on the metamorphism in the Moravian zone were carried out. This has been done mainly for the Thaya dome, therefore the stage of knowledge is unequal between the Thaya dome and the Svratka dome. As a consequence, both units will be treated separately in respect to their metamorphic evolution after a short introduction into their general geology and lithology.

2 Lithology and Geological Setting of the Moravian Zone

2.1 The Thaya Dome (Fig. 1)

The Thaya batholith forming the deepest structural unit is a metamorphosed and deformed granitoid complex. According to Scharbert and Batík (1980) it is of Cadomian age (550 Ma based on a Rb-Sr whole-rock isochrone). In the west, the batholith is overlain by its primary roof, in the east it is covered by Tertiary sediments, at its southern end it is directly cut by the Diendorf fault. The Brno massif is believed to be the continuation of the Thaya batholith separated by the Diendorf fault. Four different lithologies can be distinguished in Austria:

1. The Hauptgranit, is a medium grained, light granite to granodiorite occupying more than two thirds of the surface of the batholith (cf. Preclik 1937).
2. The Gumping type comprises gneissic biotite-rich granodiorites and quartz-monzodiorites with blocky K-feldspar phenocrysts and amphibole altered to biotite.
3. The Passendorf type, essentially a fine- to medium-grained tonalite and metatonalite occurs mainly at the NW side of the batholith.
4. The Gauderndorf type is a fine-grained granitic to granodioritic rock. Dikes of this type locally cross cut the Hauptgranit type.
5. In the CR a fifth granite type has to be added, the granodiorite of Tasovice (Preclik 1937; Dudek 1960; Batík 1984).

Aplites and pegmatites occur in all parts and are frequent in some areas. Most of them show the same metamorphic overprint as their plutonic host rocks. Rarely, spessartites occur (Batík 1984).

East of Znojmo near the village of Tasovice, a small section of the autochtonous cover of the Thaya batholith is preserved, consisting of

Dallmeyer et al. (Eds.)
Pre-Permian Geology of Central and Eastern Europe

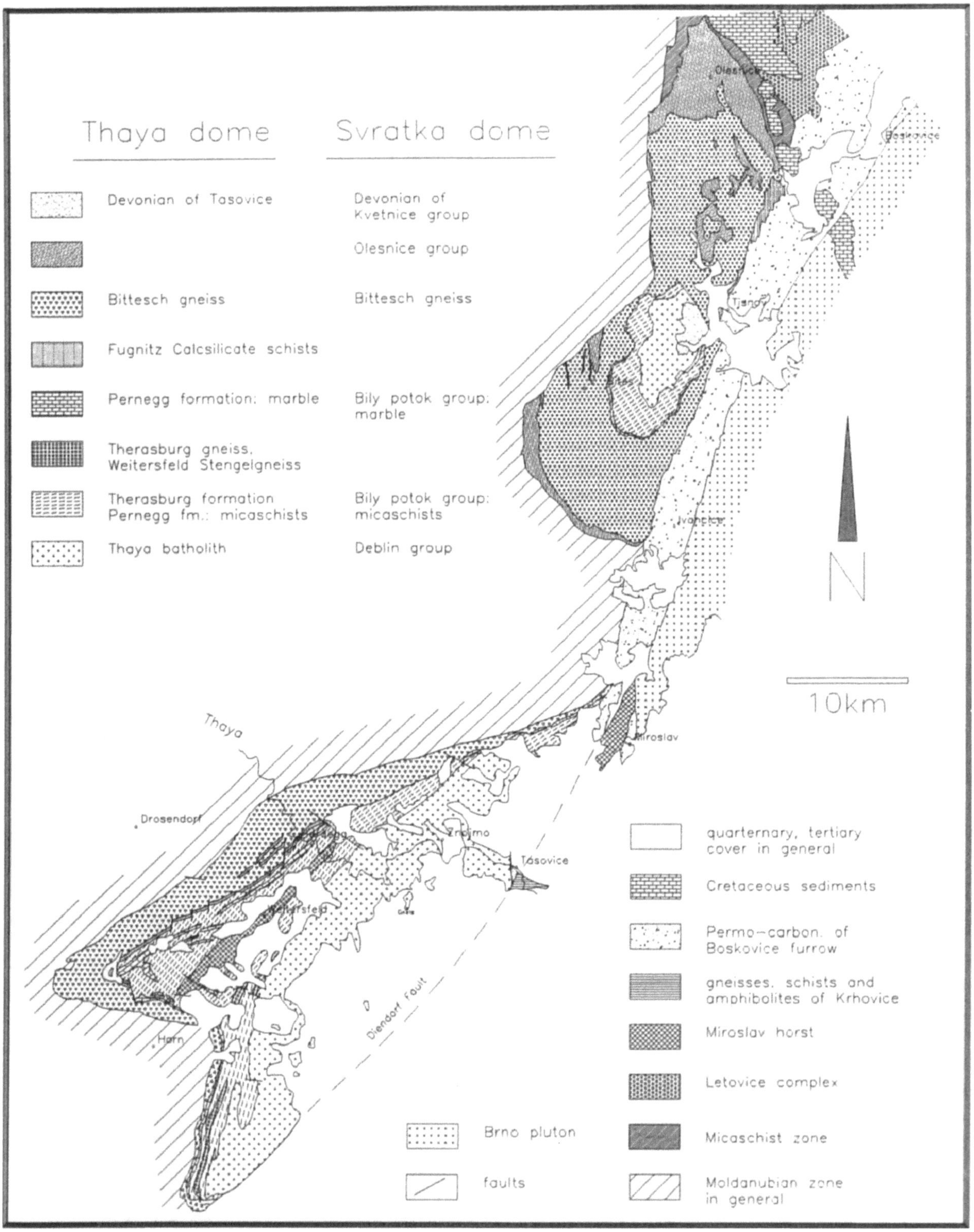

Fig. 1. Geological sketch map of the Moravian zone with the Thaya dome in the south and the Svratka dome in the north

Devonian conglomerates, quartzites, arcoses and sandstones (Preclik 1926; Dudek 1960). A presumed continuation of this cover at the very southern end of the Thaya dome might be the series of Olbersdorf(?) with quartzites, black phyllites, and marbles (Frasl 1977).

Towards the west, the gneisses of the Thaya batholith are overlain by a metamorphosed pelitic to psammitic sequence called the Therasburg formation (Höck et al. 1991). It consists of micaschists partly with a considerable amount of albite and/or oligoclase eventually leading to fine-grained gneisses. Shales and graywackes may be considered as educt of this rocks. The more feldspar-rich varieties form fine-grained strongly foliated gneisses (Therasburg gneisses). Quartzites are intercalated with the gneisses and micaschists. The large amount of magnetite and ilmenite in these rocks is interpreted as coming from Precambrian black sands (Libowitzky 1990).

A distinct gneiss body with granitic composition, the Stengelgneiss of Weitersfeld, separates the Therasburg formation from the tectonically higher sequence of the Pernegg formation (Höck et al. 1991). Based on field and petrographic evidence, it has to be separated from the Therasburg gneisses and is restricted to the central part of the Moravian zone.

The Pernegg formation comprises micaschists, calcschists, and pure marbles grading into each other. The marbles prevail in the upper part of the sequences as coherent layers. The protolith is a shale-limestone sequence. The uppermost part of the marbles is formed by a very distinct horizon of calosilicate schists, the Fugnitzer Kalksilikatschiefer. It is an only several-meters-thick layer consisting of quartz, plagioclase, K-feldspar, amphibole, partly clinopyroxene, calcite, epidote, and sphene. Sometimes the calc-silicate schists are found as small layers and lenses within the above-lying Bittesch gneiss.

The distinction between the Therasburg and the Pernegg formation becomes difficult where the separating layer of the Stengelgneiss of Weitersfeld is missing, because of the similarities of the micaschists. Therefore, in the CR part both units are termed Lukov formation (Batík 1984; Batík and Fediuková 1992) with an upper and a lower member being roughly identical with the Pernegg and Therasburg formation respectively.

The assumed stratigraphic position of the Therasburg formation is inferred from some preserved intrusive contacts and migmatites of the Cadomian Thaya batholith as pre-Cadomian. Therefore the whole sequence must be older than 550 to 600 Ma. The age of the Pernegg formation is still questionable, but a recent indication comes from Sr-isotopes study in the marbles (Frank et al. 1992). $^{87}Sr/^{86}Sr$ ratios of around 0.707 in mica-poor carbonates argue for an Upper Proterozoic age for the marbles according to the Sr-isotope variation curve in the seawater (Faure 1982).

The Bittesch gneiss is the uppermost unit of the Moravian zone and has been considered its most typical member. It is a highly deformed orthogneiss with a well-developed Augen structure. Dark amphibolite layers up to 50 cm thick are restricted to the uppermost 20 to 30 m. These layers, parallel to the regional s-planes, are repeated many times in a relatively small section, and were taken as indication of a volcanic origin at least for the upper part of the Bittesch gneiss, with interlayers of rhyolitic and basic lava flows (Frasl 1970). Recently, Frasl (1989) is more inclined to interpret these amphibolites as highly deformed dikes in a former granitic body. The discussion on the age of the Bittesch gneiss centers around three conflicting age determinations:

790 Ma (S. Scharbert 1977)	Rb/Sr whole rock
560 Ma (Morauf and Jäger 1982)	Rb/Sr whole rock
480 Ma (van Breemen et al. 1982)	Rb/Sr whole rock

It is widely accepted that the Thaya dome consists of three tectonic units (Waldmann 1930; Frasl 1970; Tollmann 1985): an autochtonous basement comprising the Thaya pluton and its cover, the Therasburg formation. The so-called Pleissing nappe (Stengelgneiss of Weitersfeld and the Pernegg formation) and the highest unit, the Bittesch gneiss nappe, form the allochthonous bodies which are overthrusted in turn by the Glimmerschiefer zone, the lowest part of the Moldanubian. The direction of the thrusting was believed generally to be with top to the east (Tollmann 1985; Fuchs 1976; Thiele 1976; Matte et al. 1985), but recently Fritz (1991), Schulmann (1990), and Schulmann et al. (1991) showed that a dextral transpression with a top to the NNE movement between the Moldanubian and the Moravian and within the Moravian nappes was an important thrusting event.

2.2 The Svratka Dome (Fig. 1)

According to Jaroš and Mísař (1976) and Jaroš (1992), the deepest unit in the Svratka dome are the Tišnov Brunnides. They consist of a crystalline basement, the Deblín group with the Svratka granitoids, and the Devonian Závist-Květnice group.

The former is built up of migmatites, gneisses, some metabasites, migmatized micaschists, and rare calc-silicate rocks. The gneisses, migmatites, and micaschists are intruded by granites, granodiorites and aplites, the Svratka granitoids which are now cataclastic. The Závist-Květnice group contains conglomerates, quartzites, siltstones, limestones, dolomitic limestones, and dolomites mainly of Devonian age. It defines the original sedimentary cover of the Deblín group and the Svratka granitoids. The whole sequence is believed (Dudek 1980) to be part of the Brno unit, and is separated from the Moravian zone proper by the Dřínova overthrust.

The Bílý potok unit, according to Jaroš and Mísař (1976) and Jaroš (1992), forms the lowest nappe unit of the Moravian zone. It is divided into a lower division with cataclastic granites, quartzites, muscovite-biotite phyllites, and some metabasic rocks. The upper division in composed of graphitic phyllites, some quartzites, garnet-bearing micaschists, limestones, and calc-silicate rocks. These divisions correspond approximately to the Therasburg formation and the Pernegg formation in the Thaya dome.

The Bítesch gneiss, overlying the Bílý potok group, has the same appearance as in the Thaya dome including the amphibolite layers in the uppermost section. The Olešnice group is the highest member of the Moravian nappes and is considered to be the primary cover of the Bittesch gneiss. It consists of garnet-staurolite micaschists, garnet-gneisses, phyllites with limestones and quartzites, amphibolites, and some types of metagabbros. Whether an equivalent group exists around the Thaya dome is still debated among geologists from Austria and the CR. The latter are in favor of such a zone (Vranov-Olešnice group), the former are mainly advocating the concept of the micaschist zone (Suess 1912) which would be not a part of the Moravicum. For this reason it is omitted around the Thaya dome in Fig. 1.

The tectonic situation in the Svratka dome is similar to that in the Thaya window. The Deblín group, together with the Svratka granitoid, forms the (par)autochthonous basement overthrusted by the Bílý potok unit (Dřínova overthrust according to Jaroš and Míšar 1976). The Bittesch gneiss and its primary (?) cover, the Outer Phyllites form the highest Moravian nappes (Jaroš and Mísař 1976; Jaroš 1992; Schulmann et al. 1991). Also in the Svratka window the sense of movement is seen differently. Jaroš and Mísař (1976) and Jaroš (1992) argue for an eastward thrusting, Schulmann et al. (1991) favor a top to the NNE movement.

The Moravian Zone as a whole can be regarded as the former western margin zone of the so-called Brunovistulian block sensu Dudek (1980), which is an old triangle-shaped Cadomian consolidated microcontinent at the eastern part of the Bohemian Massif consisting of Cadomian granitoids and migmatized paragneisses. During the Variscan orogeny, the western marginal parts of the Brunovistulian block were overthusted by a hot nappe pile of the Moldanubian mobile belt. Thrusting occurred in connection with a strong dextral transpression between the Moldanubicum and the western flank of the Brunovistulicum (Schulmann et al. 1991; Fritz 1991). The transpressional movements were followed by local updoming (Thaya dome, Svratka dome).

3 Metamorphism in the Thaya Dome

Frasl (1968 and 1970) carried on the results and ideas of Suess (1912), Waldmann (1930), and Preclik (1927) regarding the metamorphic evolution and presented finally a three-stage concept for the metamorphism:

1. the old Moravian phase connected with the intrusion of the Thaya batholith,
2. the middle Moravian phase causing the regional metamorphism, and
3. the young Moravian phase of a retrograde stage.

The first stage is mainly restricted to the immediate vicinity of the Thaya batholith, where the intrusion of the granitoids formed a migmatitic contact aureole which was later overprinted by the middle Moravian metamorphism and deformation event. The area with the best relicts is found in the northernmost part in Austria east of the ruin Kaja, close to the Thaya valley, where migmatitic structures such as stromatites and

A 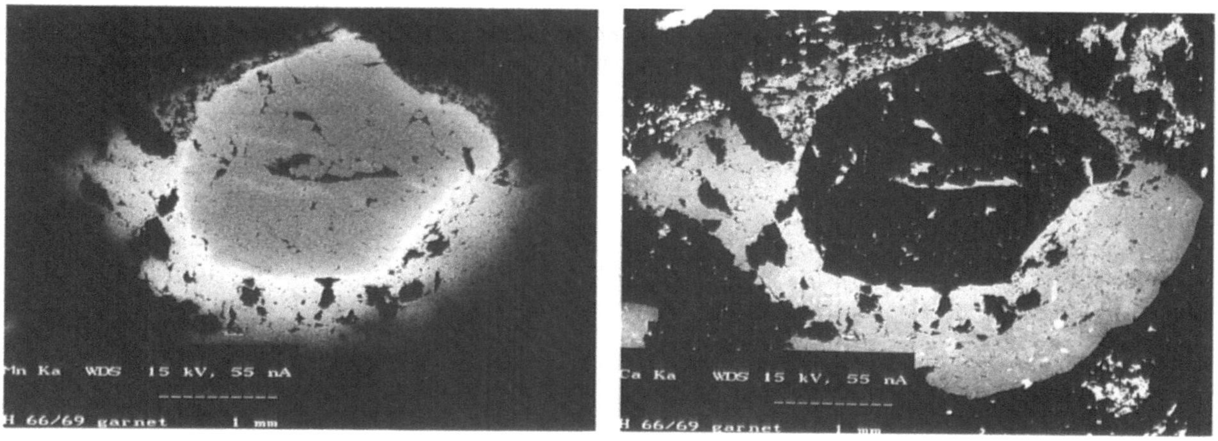 B

Fig. 2A,B. X-ray maps showing the distribution of Mn (**A**) and Ca (**B**) in garnet 66/69 from paragneisses surrounding the Thaya batholith in the west, with a strong influence of the old Moravian metamorphism. Both elements are homogenously distributed in the core, but increase dramatically at the core/rim boundary. Towards the rim they decrease systematically. While Ca shows a sudden break between core and rim, Mn has a steep but distinct gradient towards the core, probably due to diffusion of Mn into the core during the growth of the rim

cross cutting dikes of leucogranites and aplites are still preserved. In other parts, only dike-like intrusions into the overlying metasediments of the Therasburg formation are observed.

Mineralogical relicts are possible pseudomorphs after cordierite (Frasl 1968) and almandine-rich cores of garnets with a distinct two-phase growth pattern (Höck et al. 1991). The cores are fairly homogenous with a very weak zoning in respect to Fe, Mg, and Mn, with Fe and Mg very slightly decreasing towards the outer part of the core (Figs. 2, 3). Almandine is around 85 mol%, pyrope ~9%, spessartine and grossular are around 1–2% (Table 1: 66/69). The rim has formed later with an element distribution coinciding exactly with the zoning pattern of the garnets formed in the Pernegg formation during the middle Moravian event (Table 1: 66/69 c/r and 34/85 c). The core-rim boundary is of special interest (Figs. 2, 3). There is a sharp break in the Ca concentration but a more continuous increase in Mn from the core to the rim. This is probably due to the diffusion of Mn into the core during the formation of the Mn-rich rim.

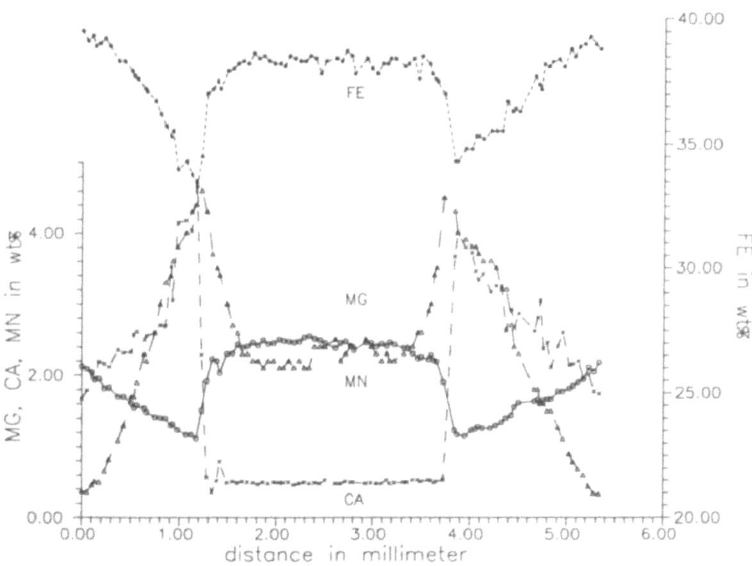

Fig. 3. Element profiles of Fe, Mn, Mg and Ca across the garnet 66/69. Fe and Mg are high in the core, low at the inner rim increasing towards the outer rim. The same features as in Fig. 2 can be recognized

Table 1. Microprobe analyses of garnet, Fe^{3+} has been calculated based on an ideal structural formula (12 O). The end members are given in addition. *c* core; *r* rim; *66/69 micaschist* Therasburg formation close to the Thaya valley; *34/85 micaschist* Pernegg formation, central part of the Thaya dome; *129/90* garnet-bearing phyllite, southern end of the Thaya dome

Sample	66/69			34/85		129/90	
	c	c/r	r	c	r	c	r
SiO_2	36.14	35.39	35.32	36.48	37.01	36.02	36.30
Al_2O_3	20.73	20.76	20.62	21.35	21.38	20.39	20.29
FeO	38.94	34.18	38.18	32.25	35.59	25.71	27.34
MnO	0.38	4.42	2.63	4.47	0.82	15.70	12.31
MgO	2.18	1.23	2.36	1.82	2.26	0.77	0.88
CaO	1.70	3.65	0.51	3.52	3.41	1.53	2.66
Total	100.07	99.63	99.62	99.89	100.47	100.12	99.78
Si	2.942	2.907	2.904	2.957	2.974	2.957	2.977
Al^{IV}	0.058	0.093	0.096	0.043	0.026	0.043	0.023
Al^{VI}	1.932	1.917	1.903	1.997	1.998	1.930	1.938
Fe^{3+}	0.082	0.088	0.102	0.013	0.008	0.080	0.068
Fe^{2+}	2.569	2.260	2.523	2.173	2.384	1.685	1.807
Mn	0.026	0.308	0.183	0.307	0.056	1.092	0.855
Mg	0.265	0.151	0.289	0.220	0.271	0.094	0.108
Ca	0.148	0.321	0.045	0.306	0.294	0.135	0.234
alm	0.854	0.744	0.830	0.723	0.794	0.561	0.602
sps	0.009	0.101	0.060	0.102	0.019	0.363	0.285
pyr	0.088	0.050	0.095	0.073	0.090	0.031	0.036
grs	0.008	0.062	0.000	0.095	0.094	0.005	0.044
anr	0.041	0.043	0.014	0.006	0.004	0.040	0.034

The question whether the Fugnitz calc-silicate schists are originally formed by a contact metamorphism (thus showing also remnants of an old Moravian event), caused by the intrusion of the Bittesch gneiss magma, an interpretation favored by earlier workers and more recently advocated by Frasl (1983) and Bernroider (1989), remains an open problem. The same is true for the question of the age of this contact metamorphism, because the age dating of the Bittesch gneiss gave no clear results (cf. Sect. 2). However, the mineral assemblage now observed originated from the middle Moravian event.

The regional metamorphism of the middle Moravian phase has taken place during the Hercynian orogeny. Recent $^{40}Ar/^{39}Ar$ plateau ages on amphiboles from amphibolites in the Bittesch gneiss give 328.7 ± 3.3, muscovites from the Bittesch gneiss 328.7 ± 0.3 and the Stengelgneiss of Weitersfeld yielded 328.5 ± 0.7 Ma (Dallmeyer et al. 1990, 1992). Already Suess (1912) and later Preclik (1937) and Frasl (1970) pointed out that the metamorphic grade is highest in the western part and decreases toward the south, the east, and the north. Therefore, the metamorphism is inverse in respect to the internal Moravian nappe structure with the structurally highest units exhibiting the high-grade part of the metamorphic zonation. Höck (1975) introduced four mineral zones with different assemblages in the metapelitic rocks (Pernegg and Therasburg formations) striking somewhat oblique (SSW-NNE) to the regional structures (S-N to SW-NE) especially in the southern and northern part of the Moravicum (Fig. 4). The first and southernmost zone (1) shows assemblages with only phengite and chlorite. With increasing temperature, biotite appears in addition (zone 2) and later on garnet (zone 3). In the westernmost zone (4), staurolite + garnet + biotite is the typical assemblage, while chlorite disappears as a stable phase.

Some modifications have to be made on this mineral zonation, especially at the southern low grade part of the Moravian zone (Fig. 4). It is doubtful whether phengite and chlorite, the characteristic assemblages of zone 1 in the metapelites of the Therasburg and Pernegg formations respectively, are actually part of the middle

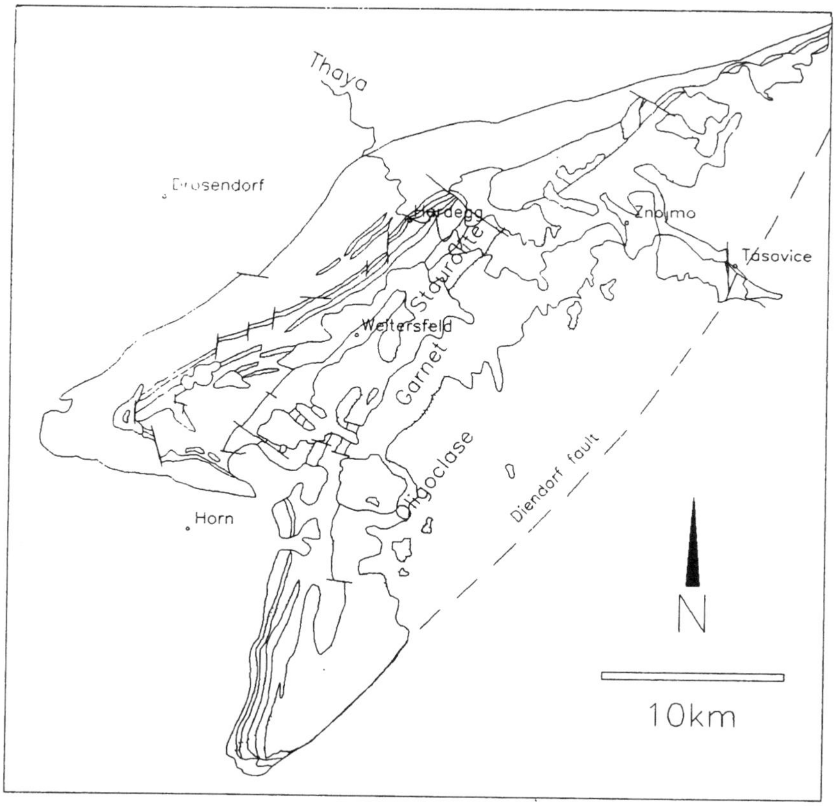

Fig. 4. Geological contour map of the Thaya dome with the distribution of the mineral zones. The oligoclase, almandine-rich garnet, and staurolite zone cut obliquely through the geological boundaries in the northern and southern end of the Thaya dome

Moravian phase. It might be equally possible that they formed at the expence of biotite during the younger Moravian phase (see below). Some tiny relicts of a Mn-rich garnet and acidic oligoclase together with albite are occasionally found. The garnet has a slight zoning pattern with alm 60, spess 28, gross 4.5, pyp 3.5 mol% at the rim, and alm 56, spess 36, gross 0.5, pyp 3 mol% in the core (Table 1: 129/90), the oligoclase has An 15 (Table 2: 129/90). Where the younger overprint is weak, biotite appears, but due to the very rare occurrence of the spessartine-rich garnet, the assemblage biotite-garnet has not been observed so far.

Towards the north, a more almandine-rich garnet appears with a composition of alm 65–70, spess 12–15, gross 10–14, and pyp 5–7 coexisting with biotite. Quartz, muscovite, chlorite, and oligoclase (~An 20) occur as additional phases. Further northwest, approximately where the Pernegg formation and the Bittesch gneiss turn from south-north striking into an east-west striking (Fig. 4), staurolite comes in as a critical phase forming an assemblage with garnet, muscovite, biotite, plagioclase (An ~ 22), and quartz. The composition of these minerals is given in Tables 1–4; sample 34/85.

The growth of garnets in this area is of particular interest. A two-phase growth is widespread with an inclusion-free or inclusion-poor core and a distinct rim, rich in inclusions at its inner part. The outermost zone is again poor in inclusions (quartz and some ilmenite). In some cases, radial growth patterns in the core are developed with some also radial-orientated quartz inclusions. The cores of the garnet often show relicts of older fabrics, often oblique to the external s-planes, sometimes synkinematically rotated into an s-shaped s_i. In several cases, the outer rims of the garnets with the higher FeO and MgO contents show a synkinematic growth in respect to the main foliation.

Chemically the garnets are strongly zoned with Ca- and Mn-rich cores and Fe- and Mg-rich rims (Table 1: 34/85). In a profile, the pattern is typically bell-shaped (Fig. 5), but when mapped, the distribution of the chemical isolines are often irregular, following the radial inclusion patterns (Fig. 6). How these patterns form is still unclear. One explanation could be some deformation

Table 2. Microprobe analyses of plagioclases. Structural formulas based on 8 O. Samples 129/90, 34/85 as in Table 1. *54/87* Bittesch gneiss: southern end of the Thaya dome; *2/88* granite gneiss: Thaya batholith near Thaya valley; *51/87* Fugnitz calc-silicate schist: southern end of the Thaya dome; *82/73* Fugnitz calc-silicate schist: central part of the Thaya dome

Sample	129/90	34/85	54/87	2/88	51/87	82/73
SiO_2	64.05	62.09	64.66	62.31	62.47	57.71
Al_2O_3	22.20	23.71	21.91	23.41	23.72	26.42
FeO	0.24	0.25	0.00	0.00	0.17	0.07
CaO	3.14	4.68	3.04	4.76	5.00	8.74
Na_2O	9.68	9.16	10.07	8.62	8.55	6.45
K_2O	0.07	0.08	0.11	0.19	0.21	0.23
Total	99.38	99.98	99.79	99.29	100.12	99.62
Si	2.841	2.756	2.856	2.776	2.764	2.594
Al	1.161	1.240	1.140	1.229	1.237	1.400
Fe	0.009	0.009	0.000	0.000	0.006	0.003
Ca	0.149	0.223	0.144	0.227	0.237	0.421
Na	0.833	0.788	0.862	0.745	0.734	0.562
K	0.004	0.005	0.006	0.011	0.012	0.013
an	0.151	0.219	0.142	0.231	0.241	0.423
ab	0.845	0.776	0.852	0.758	0.747	0.564
or	0.004	0.005	0.006	0.011	0.012	0.013

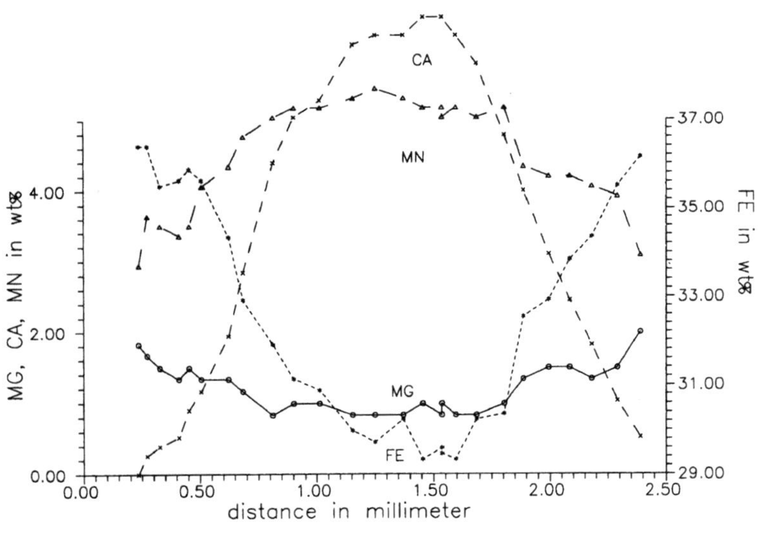

Fig. 5. Element profiles of Fe, Mn, Mg, and Ca across garnet 34/85 showing the typical bell-shaped element distribution with Ca and Mn decreasing from core to the rim, while Fe and Mg are increasing

partitioning between garnet and matrix in the sense of Bell (1985) and Bell and Johnson (1989) with quickly changing directions of shear in the early stages of garnet growth.

All these findings justify the delineation of some mineral zones in the metapelitic rocks with (1) a spessartine-rich garnet and(?)/or biotite in the south, (2) the incoming of an almandine-rich garnet, and (3) finally staurolite. When mapped (Höck 1975, 1983), these zones run oblique to the regional strike of the metapelites, cutting the lithological and tectonical boundaries in some places. If this is true, increasing metamorphism should also be found in other lithologies along strike. Frasl 1983 introduced an albite/oligoclase transition boundary within the different gneiss bodies (Bittesch gneiss, Thaya batholith), showing an appearance similar to the mineral zones in the metapelites. Even within the oligoclase zone, an increase in metamorphism can be demonstrated

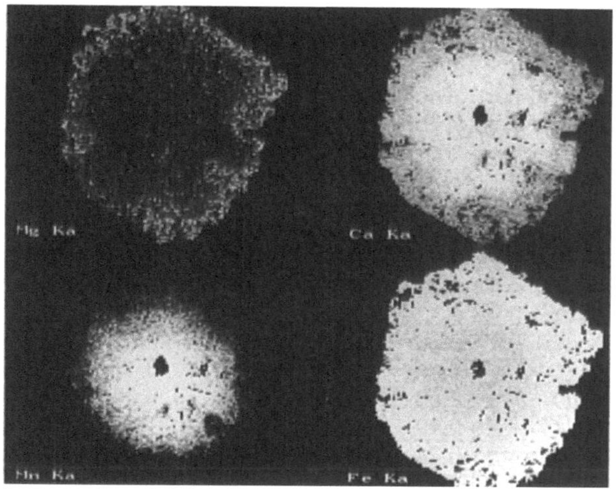

Fig. 6. X-ray maps showing the distribution of Fe, Mn Mg, and Ca in garnet 34/85 from the Pernegg formation which formed during the middle Moravian metamorphism. The radial growth pattern can still be recognized. The core is poor in Fe and Mg but rich in Mn and Ca

along the strike of orthogneisses. The An-content, e.g., increases from An 14% in the S to An 23% in the central part (Table 2: 54/87 and 2/88). The same increase can be observed with the Na content in K-feldspar coexisting with plagioclase.

Comparable observations can be made in the Fugnitz calc-silicate schists. The An content in their plagioclase (rim composition) increases from 25% (S) to 45% in the garnet-biotite-staurolite zone (Table 2: 51/87 and 82/73). Mg-rich, Al-poor amphibole is found only in the southernmost parts, while the occurrence of diopsidic clinopyroxene is confined to the zone of staurolite in the pelitic rocks.

The zonation of the metamorphism is fairly well established in the Austrian part of the Thaya dome, in the CR parts systematic investigations are still missing; but qualitative descriptions by Preclik (1926, 1937), Batík (1984), Batík and Fediuková (1992) and Gregerová and Koch (1988) indicate a continuous extension of these mineral zones into the Czech parts of the Thaya dome. Again, they seem to cut through the lithological and tectonic boundaries (Fig. 4).

This hitherto qualitative approach of the spatial distribution of metamorphism can be quantified by using some geothermobarometers. Coexisting pairs of garnet and biotite (Tables 1 and 4) from samples without any retrogressive indications were used for thermometric calculations. The most Mg-rich compositions of garnet (Fig. 5) with adjacent biotites yielded temperatures, according to the model of Hodges and Spear (1982), around 580 °C for the garnet-biotite zone, and temperatures between 580°–600 °C for the staurolite zone (Figs. 4, 7). These temperatures are based on seven samples with at least five independent measurements for each sample. The error caused by analytical uncertainty is estimated between 10 and 25 °C (Höck et al. 1991). Small amounts of ferric iron in biotite could lower the temperature by several degrees. According to the exclusive occurrence of ilmenite as ironoxide phase not to much Fe^{3+} is expected to be incorporated into biotite. Also the calculated andradite component in garnet is very small (cf. Table 1).

The transition from the garnet + biotite + chlorite zone and the garnet + biotite + staurolite zone is due to the discontionous reaction:

$$Mu + Chl + Cr = Staur + Bio + H_2O$$

According to Thompson (1976), the equilibrium temperature of this reaction should be at 580 °C at 5 kb P_{H2O}, which fits well the temperatures deduced from garnet-biotite geothermometry.

Bernroider (1989) reported also temperature calculations from garnet-biotite pairs from the

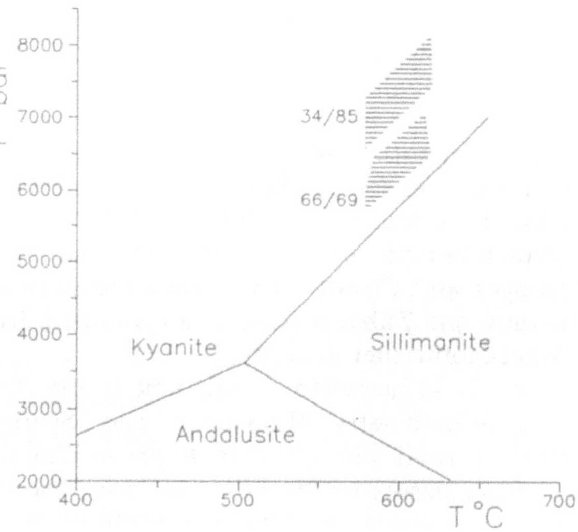

Fig. 7. P-T diagram depicting the metamorphic conditions for the staurolite zone in the metapelites exemplified for two samples 34/85 and 66/69. Temperatures were calculated from coexisting garnet(rim)-biotite pairs according to Hodges and Spear (1982), the pressures from the garnet-muscovite-plagioclase-biotite geobarometer (Hodges and Crowley 1985). The *arrow* indicates the increasing metamorphic conditons from the southern part of the Thaya dome to the staurolite zone

staurolite zone in the range of 530–590 °C according to the model by Ferry and Spear (1978). Recalculations of his data with the Hodges and Spear model (1982) gave temperatures consistent with those above. Temperature values obtained by Bernroider from coexisting minerals in calc-silicate rocks or marbles such as clinopryroxene + amphibol + calcite + clinozoisite + plagioclase + quartz range between 520 to 540 °C at very low X_{CO2}. These temperatures are based on the situation of univariant reaction curves and invariant points in a TX_{CO2} section for a total pressure of 4 kbar only. The more likely pressures of 6–7 kbar (see below) would raise the temperatures again to values in accordance with those obtained by the garnet-biotite geothermometry.

The temperatures in the southernmost area are more difficult to assess. Calculations of the two feldspar geothermoneter in gneisses with coexisting K-feldspar and plagioclase have not been carried out yet, but the occurrence of oligoclase in orthogneisses indicates temperatures around 450 to 500 °C, compared to the incoming of oligoclase in the Hercynian granite gneisses of the Hohe Tauern (e.g., Morteani and Raase 1974; Höck 1980).

Al_2SiO_5 phases are extremely rare; findings of kyanite are reported only a few times from the high-grade rocks (Batík 1984; Preclik 1937) and from the low-grade southern part (Frasl 1977, 1991); but in all cases the relation of kyanite to the rest of the mineral assemblages in unclear. Even though it cannot be used in geobarometry, it gives at least some indication that the metamorphism took place in the kyanite field. Indeed, calculation of the gar-mus-plag-bio geobarometer (Hodges and Crowley 1985) gave pressures between 6 and 7 kbar for temperatures taken from garnet-biotite thermometry.

This is in accordance with first results from phengite-barometry (Massonne and Schreyer 1987), carried out on K-feldspar and biotite-bearing gneisses of the Moravian zone. Si contents in phengites of 3.20 to 3.27 pfu (Table 3: 31/72 and 10/82) result in 6 to 8 kbar pressure at temperatures between 580 and 600 °C (Fig. 8). It is interesting to note that muscovites from gneisses in the southernmost parts (H 54/87 in Table 3 and Fig 8) have a lower excess of Si. This results in pressures around 4–5 kbar at temperatures of 450 to 500 °C estimated for the southern end of the Moravian zone. The arrow in Fig. 7 indicates the increasing path of metamor-

Table 3. Microprobe analyses of muscovites. Structural formula based on 22 O. Samples 34/85, 2/88, 54/87 as in Tables 1 and 2. *31/73* Stengelgneiss of Weitersfeld: central part of the Thaya dome; *10/82* Bittesch gneiss: central part of the Thaya dome

Sample	31/72	10/82	2/88	34/85	54/87
SiO_2	47.97	48.04	47.07	45.27	45.74
TiO_2	0.32	0.49	0.46	0.26	0.54
Al_2O_3	30.00	30.20	31.33	37.21	31.66
FeO	2.82	3.40	3.17	0.82	4.16
MnO	0.06	0.00	0.05	0.00	0.05
MgO	1.66	1.84	1.39	0.41	1.18
CaO	0.00	0.00	0.00	0.00	0.08
Na_2O	0.19	0.32	0.19	1.77	0.26
K_2O	10.74	10.65	10.93	9.09	10.17
Total	93.76	94.94	94.59	94.83	93.84
Si	6.540	6.488	6.387	6.017	6.275
Al^{IV}	1.460	1.512	1.613	1.983	1.725
Al^{VI}	3.360	3.295	3.397	3.846	3.394
Ti	0.033	0.050	0.047	0.026	0.056
Fe	0.322	0.384	0.360	0.091	0.477
Mn	0.007	0.000	0.006	0.000	0.006
Mg	0.337	0.370	0.281	0.081	0.241
Ca	0.000	0.000	0.000	0.000	0.012
Na	0.050	0.084	0.050	0.456	0.069
K	1.868	1.835	1.892	1.541	1.780

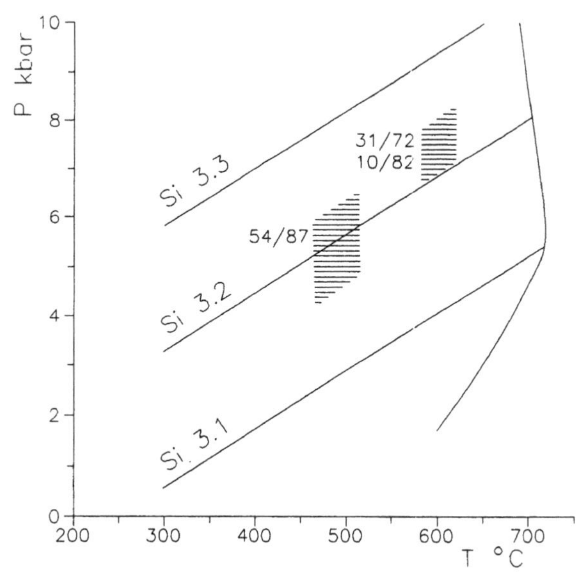

Fig. 8. PT diagram depicting the phengite geobarometer (Massonne and Schreyer 1987). Phengite-bearing gneisses from the staurolite zone (31/72, 10/82) show pressures around 5 to 7 kbar at temperatures of 580–600 °C taken from the metapelites (Fig. 7). The phengite from one sample in the southern Thaya dome (54/87) indicates lower pressures around 4 to 5 kbar at a given T of approximately 450 to 500 °C

Table 4. Microprobe analyses of biotite. Structural formula based on 22 O. Samples as in Table 1

Sample	66/69	34/85
SiO_2	34.71	34.77
TiO_2	1.49	2.40
Al_2O_3	18.96	19.35
FeO	22.05	21.36
MgO	8.01	8.19
Na_2O	0.28	0.21
K_2O	8.83	8.76
Total	94.33	95.04
Si	5.398	5.341
Al^{IV}	2.602	2.659
Al^{VI}	0.873	0.844
Ti	0.174	0.277
Fe	2.868	2.744
Mg	1.857	1.875
Na	0.084	0.063
K	1.752	1.717

phism from the south to the central part of the Thaya dome.

The retrograde, young Moravian phase is not very well known. Both the Therasburg and the Pernegg formations are affected by this late event. The older (middle Moravian) minerals such as staurolite, garnet, and biotite are completely changed to chlorite and muscovite; but often only retrograde chlorite is found as reaction rims around staurolite and garnet. In areas where the chloritization is not so penetrative, extremly small rims enriched in Mn around garnets can be observed. Sometimes biotite has quantitatively disappeared, leaving only some garnets and staurolite relicts in a chlorite + muscovite + quartz matrix.

4 Metamorphism in the Svratka Dome

In contrast to the Thaya dome, where modern investigations on metamorphism have been carried out (Höck 1975; Höck et al. 1991; Bernroider 1989) no such efforts have been made in the Svratka dome. Modern structural studies (e.g., Schulmann et al. 1991) rely also on the poor knowledge of the metamorphism.

A general overview was given by Cháb and Suk (1977). They adopted the three-stage model proposed by Frasl (1968, 1970) of the old, middle, and young Moravian phase also for the Svratka dome. Based on the original studies by Suess (1912) and later Preclik (1937), there is a general agreement that the Variscan metamorphism increases from structural deeper units, i.e., the Tišnov Brunnides sensu Jaroš and Mísař (1976), Jaroš (1992) towards the upper units, i.e., the Bittesch gneiss and the Olešnice group. In these respects, the metamorphic evolution has some common features in both domes. The old Moravian phase is restricted to the pre-Devonian basement of Deblín group in the Tišnov Brunnides. It is expressed by the formation of gneisses and micaschists, marbles, and calc-silicate rocks, as well as their metamorphism. If the Svratka granites intruding these formation are of the same age as the Brno pluton (584 ± 5 Ma U/Pb ages, van Breemen et al. 1982), this metamorphism is at least of Cadomian age.

The middle Moravian phase is developed as low-grade overprint with chloritization and a weak metamorphism on the Devonian sediments of the Závist-Květnice group. The relevant mineral assemblages are muscovite and chlorite.

In the Bílý potok group, typical greenschist facies assemblages prevail: biotite-bearing micaschists with some rare garnets (especially in the lower section, Schulmann et al. 1991), marbles contain muscovite and biotite. The mineralogy of some calc-silicate schists at the basis of the Bittesch gneiss is not known in detail. In the tectonic highest portion of the Bittesch gneiss and the Olešnice group, the mineral assemblages are characterized by albite (oligoclase ?), muscovite, biotite, garnet and staurolite – a typical amphibolite facies association. According to Schulmann et al. (1991), biotite-sillimanite occurs in some orthogneisses. High-grade pelitic mineral assemblages such as garnet-staurolite-kyanite (sillimanite) are confined to the micaschist zone W of the Svratka dome.

At least these minerals and their distribution describe an increasing metamorphism towards the higher unit, but at the present stage, it remains open whether the individual tectonic units show each some defined metamorphic assemblages or whether the metamorphic zonation cuts the tectonic structure obliquely as in the Thaya dome.

5 Conclusions

It is agreed by all authors so far – as demonstrated above – that in the Moravian zone the metamorphism decreases from the W to the E, i.e., from the higher to the lower units. To explain this fact two models remain to be discussed:

1. Each tectonic subunit (e.g., the Deblín group, the Bílý potok group) has a separate metamorphic evolution and the metamorphic assemblages are superposed by nappe movements in such a way that the lowest unit has a weak overprint, while the highest nappes exhibit an amphibolite grade metamorphism. The main thermal event would have taken place before the main nappe movements.
2. The metamorphic zones cut obliquely to the lithological and tectonic boundaries, causing an inverse metamorphic zonation. The metamorphism in this case took place after, or at least together with, the nappe movements.

The first model is preferred by Schulmann et al. (1991) for the evolution of the Svratka dome, postulating a first synmetamorphic (?) nappe emplacement causing the superposition of high grade units over lower grade units. This metamorphic event should correspond to the middle Moravian phase. A second stage of deformation would be northwards oriented under low temperature conditions (younger Moravian phase?) transporting the Moravian nappes over the autochthonous basement of the Delin group.

The second model is favored by Höck (1975) and Höck et al. (1991) for the Thaya dome, where it could be demonstrated that there is an increase in metamorphism along strike at least from the southern end to its central part. This is not in full agreement with the ideas of Fritz (1991), who believes, based on structural investigations in the Bittesch gneiss, that the Moravian nappes have formed together with the overthrust of the Moldanubian block over the Moravian. The geometry of the mineral zones indicates that the zonation is compatible with an NNE-directed movement of the Moldanubian over the Moravian unit, but the internal Moravian nappes should have formed prior to the maximum stage of the metamorphic evolution. The thickness of the overriding Moldanubian plate must have thinned considerably towards the south, the east, and the north, as suggested by the lower temperatures and probably pressures as well as the almost unmetamorphic original Devonian sedimentary cover at the northeastern end of the Thaya batholith.

Many problems in the metamorphic evolution in both domes remain unsolved yet, especially regarding the relation between the P-T paths and the deformation and nappe tectonics. These problems can only be resolved by a careful systematic study and comparison of the mineral evolution and the deformation-crystallization relation in both domes, a project currently being carried out by a joint research group from Prague and Salzburg.

Acknowledgments. The research has been supported by a grant of the BMWF and a grant of the FWF (Project No. S4704). The author likes to thank K. Schulmann, A. Dudek, and J. Leichmann (CR) for numerous discussions and the improvement of earlier versions of the manuscript. D. Topa kindly carried out the microprobe analyses, R. Marschallinger and Ch. Stejskal helped with the computer graphics. The help and support of all individuals and institutions is gratefully acknowledged.

References

Batík P (1984) Geologická stavba moravika mezi bíteškou rulou a dyjským masívem. Věst UUG 596: 321–330
Batík P, Fediuková E (1992) Garnet chemistry as an indicator of metamorphic development and deformations of the Moravicum in the Dyje dome. Vestn Cesk Geol Ust 67/1: 1–24
Bell TH (1985) Deformation partioning and porphyroblast rotation in metamorphic rocks: a radical reinterpretation. J Metamorph Geol 3: 109–118
Bell TH, Johnson SE (1989) Porphyroblast inclusion trails: the key to orogenesis. J Metamorph Geol 7: 279–310
Bernroider M (1989) Zur Petrogenese präkambrischer Metasedimente und cadomischer Magmatite im Moravikum. Jahrb Geol B-A 132/2: 349–373
Cháb J, Suk M (1977) Regionalní metamorfóza na území Čech a Moravy. Knihovna UUG Praha 50: 156
Dallmeyer RD, Neubauer F, Höck V (1990) $^{40}Ar/^{39}Ar$ mineral age controls on the chronology of Late Paleozoic tectonothermal activity in the southeastern Bohemian Massif, Austria (Moldanubian and Moravo-Silesian zones). In: Franke W (ed) IGCP 233 Field guide, Int Conf Göttingen/Giessen, pp 87–96
Dallmeyer RD, Neubauer F, Höck V (1992) Chronology of late Paleozoic tectonothermal activity in the southeastern Bohemian massif, Austria (Moldaubian and Moravo-Silesian zones) – $^{40}Ar/^{39}Ar$ mineral age controls. Tectonophysics 210: 135–153
Dudek A (1960) Kristalline Schiefer und Devon östlich von Znojmo. Sb UUG 26/1959 Odd Geol 1: 101–141
Dudek A (1980) The crystalline basement block of the Outer Carpathians in Moravia, Bruno-Vistulicum. Rozpr Česk Akad Věd Řada Mat Přír Věd 90 (8): 185

Faure G (1982) Principles of isotope geology. Wiley, New York
Ferry J, Spear FS (1978) Experimental calibration of the partioning of Fe and Mg between biotite and garnet. Contrib Mineral Petrol 66: 113–117
Frank W, Scharbert S, Höck V (1992) Rb/Sr Untersuchungen an Karbonatgesteinen der Böhmischen Masse. Schwerpunktprogramm S47 GEO: Präalpidische Kruste in Österreich. Ber 1991: 9–10
Frasl G (1968) The Bohemian Massif in Austria, Moravian Zone. Guide to Excursions 32C, Int Geol Congr, XXII Sess Prague, pp 3–24
Frasl G (1970) Zur Metamorphose und Abgrenzung der Moravischen Zone im niederösterreichischen Waldviertel. Nachr Dtsch Geol Ges 2: 55–60
Frasl G (1977) Contributions on the excursion guide Waldviertel of the Geol B-A Wien 1977
Frasl G (1983) Zur Geologie des Kristallins und Tertiärs der weiteren Umgebung von Eggenburg. In: Höck V, Frasl G, Steininger F, Vetters W (eds) Exkursiousführer. Österr Geol Ges 1983, No 1, 4–18
Frasl G (1989) Zur Entstehung der Moravischen Zone. Abstr ÖGG-Meet (Working group on metamorphism), Vienna, 2 pp
Fritz H (1991) Structures and kinematics along the Moravian-Moldanubian border: preliminary results. Österr Beitr Meteorol Geophys H3 1990: 77–96
Fuchs G (1976) Zur Entwicklung der Böhmischen Masse. Jahrb Geol B-A 119: 45–61
Gregerová M, Koch Z (1988) Příspěvek k poznání povahy granátu z rul pláště dyjského masívu. Scr Fac Sci Nat Univ Purk Brun 18/7: 251–260
Höck V (1975) Mineralzonen in Metapeliten und Metapsammiten der Moravischen Zone in Niederösterreich. Mitt Geol Ges 66–67: 49–60
Höck V (1980) Distribution maps of minerals of the Alpine metamorphism in the Penninic Tauern window, Austria. Mitt Österr Geol Ges 71/72: 119–27
Höck V (1983) Zur Geologie des Kristallins und Tertiärs der weiteren Umgebung von Eggenburg. In: Höck V, Frasl G, Steininger F, Vetters W (eds) Exkursionsführer. Österr Geol Ges 1983, No 1, 42–48
Höck V, Marschallinger R, Topa D (1991) Granat-Biotit-Geothermometrie in Metapeliten der Moravischen Zone in Österreich. Österr Beitr Meteorol Geophys H3 1990: 149–167
Hodges KV, Crowley PD (1985) Error estimation and empirical geothermobarometry for pelitic systems. Am Mineral 70: 702–709
Hodges KV, Spear FS (1982) Geothermometry, geobarometry and the Al_2SiO_5 triple point at Mt. Moosilauke, New Hampshire. Am Mineral 67: 1118–1134
Jarős J (1992) The nappe structure in the Svratka dome. In: Kukal Z (ed) Proc 1st Int Conf Bohemian Massif, Prague 1988: 137–140
Jaroš J, Mísař Z (1976) Nomenclature of the tectonic and lithostratrigraphic units in the Moravian Svratka dome. Vest UUG 51/2: 113–123
Libowitzky E (1990) Precambrian blacksands as precursors of magnetite and ilmenite bearing chlorite micaschists, Bohemian Massif, Austria. Mineral Petrol 43: 147–160
Massonne HJ, Schreyer W (1987) Phengite geobarometry based on the limiting assemblage with Kfeldspar, phlogopite, and quartz. Contrib Mineral Petrol 96: 212–224
Matte P, Maluski H, Echtler H (1985) Cisailements ductiles varisques vers l'Est-Sud-Est dans les nappes du Waldviertel (Sud-Est du Massif de Boheme, Autriche). Données microtectoniques et radiométric $^{39}Ar/^{40}Ar$. C R Acad Sci Paris 301 Ser II (10): 721–726
Morauf W, Jäger E (1982) Rb-Sr whole rock ages for the Bites gneiss, Moravicum, Austria. Schweiz Mineral Petrogr Mitt 62: 327–334
Morteani G, Raase P (1974) Metamorphic plagioclase crystallisation and zones of equal anorthite content in epidote bearing, amphibolefree rocks of the western Tauernfenster, eastern Alps. Lithos 2/74: 101–111
Preclik K (1926) Das Nordende der Thayakuppel. Sbor SGU 6: 373–399
Preclik K (1927) Zur Tektonik und Metamorphose der moravischen Aufwölbung am Ostrand der Böhmischen Masse. Geol Rdsch 18: 81–103
Preclik K (1937) Das Nordende des Thayabatholithen. Věst SGU 13: 34–61
Scharbert S (1977) Neue Ergebnisse radiometrischer Alterdatierungen an Gesteinen des Waldviertels. Arbeitstag Geol B A 1977, Waldviertel: 11–15
Scharbert S, Batík P (1980) The age of the Thaya (Dyje) Pluton. Verh Geol B-A 1980/3: 325-331
Schulmann K (1990) Fabric and kinematic study of the Býteš orthogeniss (southwestern Moravia). Result of large-scale northeastward shearing parallel to the Moldanubian/Moravian boundary. Tectonophysics 177: 229–244
Schulmann K, Ledru P, Autran A, Melka R, Lardeaux JM, Urban M, Lobkowicz M (1991) Evolution of nappes in the eastern margin of the Bohemian Massif: a kinemtic interpretation. Geol Rdsch 80: 73–92
Suess FE (1903) Bau und Bild der Böhmischen Masse. Wien 1903, Tempsky and Freytag, Leipzig
Suess FE (1912) Die moravischen Fenster und ihre Beziehung zum Grundgebirge des Hohen Gesenke. Denkschr Akad Wiss Mat Naturwiss Kl 83: 541–631
Thiele O (1976) Ein westvergenter Deckenbau im niederösterreichischen Waldviertel? Jahrb Geol BA 119: 75–81
Thompson AB (1976) Mineral reactions in pelitic rocks: I. Prediction of PTX (Fe-Mg) phase relations. Am J Sci 276: 401–424
Tollmann A (1985) Geologie von Österreich, Bd 2. Deuticke, Wien
Van Breemen O, Aftalion M, Bowes DR, Dudek A, Mísař Z, Povondra P, Vrána S (1982) Geochronological studies of the Bohemian Massif, Czechoslovakia, and their significance in the evolution of Central Europe. Trans R Soc Edinb Earth Sci 73: 89–108
Waldmann L (1930) Zum geologischen Bau der Thayakuppel und ihre Metamorphose. Mitt Geol Ges Wien 21: 133–152

IX The North Variscan Foreland

D. Franke

1 Introduction

The analysis of an orogen remains incomplete if there is no discussion about the development of its foreland. With regard to the mid-European Variscides, the situation is considerably complicated because the corresponding Paleozoic rocks of the foreland are overlain by the thick Young Paleozoic to Cenozoic platform cover of the Central European Basin. This is why our present knowledge about the lithological character, the stratigraphic succession, the thickness conditions, as well as the tectonic structure of the foreland formations is still very restricted. Only a few boreholes have reached the pre-Permian in the subsurface, mostly at the border zones. Geophysical surveys give some information, but the geological interpretation is often controversial. Therefore in order to construct a comprehensive picture of the paleogeographical and tectonical development, models must be used. This picture, of course, is only hypothetical. New drilling results and improved geophysical methods may lead to corrections and better refinement.

As regards the northern Variscan foreland, in this account only the central part is considered. In the west it is delimited by the Caledonian Brabant Massif, in the east by the Polish section of the Tornquist Zone. The southern border is marked by the Variscan front, the northern border by the southern edge of Baltica (Fig. 1). Concerning the duration of geologic time, the description is confined to the interval of the Variscan epoch proper including the preceding Old Paleozoic (Caledonian) period.

2 Stratigraphy

2.1 Pre-Devonian

The oldest series encountered so far in the foreland area are unfossiliferous siliciclastic rocks of possibly Young Proterozoic (Vendian) age. They are proved by boreholes in Schleswig-Holstein (northern Germany) and Jutland (southern Denmark) as well as in Mecklenburg-Vorpommern (NE Germany). These are folded and cleaved phyllitic slates, quartzitic slates, and fine- to medium-grained quartzites. Radiometric age determinations carried out on several samples (Frost et al. 1981; Ziegler 1982) show a Caledonian metamorphic overprint (400–450 Ma).

At the present time, only Ordovician and Silurian formations are known from the Lower Paleozoic. The Ordovician, drilled on Rügen island, in the southern Baltic Sea, and in Pomerania (NW Poland), consists of more than 2000 m of Caledonian folded and thrust graptolite-bearing shales and siltstones (Llanvirn, Caradoc, Ashgill) and a more coarse-grained clastic complex with sandstones, quartzites, and graywackes (Llandeilo). The stratigraphically oldest parts are represented by gray, greenish and (secondary) reddish shales and siltstones of presumed pre-Llanvirnian age. The Silurian, just evident from northern Poland near the Teisseyre-Tornquist Zone, is made up of tectonically deformed gray to dark gray graptolite shales and siltstones, reaching a thickness of at least 1500 m (Tomczyk 1987). From the genetic point of view these "geosynclinal" Lower Paleozoic complex is divergently interpreted as Gondwana-derived terrane (Pożaryski 1990; Franke 1990c), as Baltica-related marginal thrust belt (Berthelsen 1992b) or as back-arc basin behind a Caledonian North German volcanic arc area (Krauss 1993).

The pre-Devonian in the more southern parts of the foreland, that is beneath the internal zone of the Central European Basin, is anticipated at depths of about 10–13 km, but boreholes have not reached these horizons thus far. According to present ideas founded on regional comparisons with the adjacent areas, an assemblage of Precambrian to Lower Paleozoic terranes is presumed amalgamated to an East Avalonian

Dallmeyer et al. (Eds.)
Pre-Permian Geology of Central and Eastern Europe

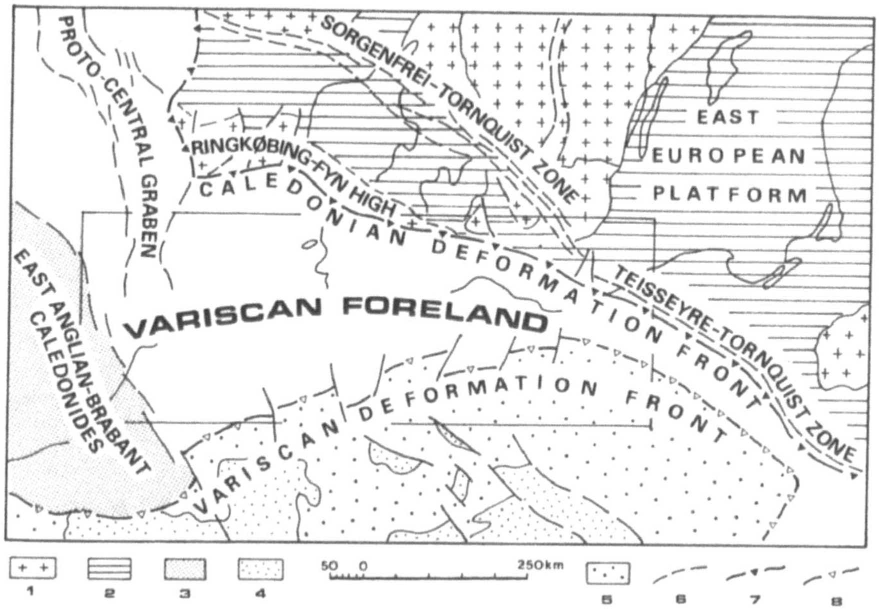

Fig. 1. Position of the investigation area within the structural framework of northern Central Europe. *1* Pre-Cadomian basement of the East European Platform; *2* unfolded Cambro-Silurian of the East European Platform; *3* folded Upper Proterozoic to Cambro-Silurian of the East Anglian-Brabant Caledonides; *4* folded pre-Permian of the mid-European Variscides, outcropping; *5* folded pre-Permian of the mid-European Variscides, covered by younger formations; *6* important faults; *7* northern front of Caledonian deformation; *8* northern front of Variscan deformation

superterrane (North German Massif after Franke et al. 1989) during their Caledonian accretion against Baltica (Fig. 2).

2.2 Devonian

At the transition between the Silurian and the Devonian, the area which later became the northern Variscan foreland was finally consolidated during the Young Caledonian diastrophism. As far as can be concluded from the situation at the northern border region of the foreland, proved by boreholes, this is well documented by the intense folding of the Lower Paleozoic series in this region, by their general uplift and associated deep erosion, by the formation of a north adjoining Upper Silurian foreland basin, and by a sharp angular unconformity between the Caledonian deformed complexes and the overlying post-Silurian sediments. This comprehensive change in the tectonic regime after collision of the North German part of East Avalonia with Baltica is also reflected in the sedimentary record. The "geosynclinal" character of the Caledonian folded Cambro-Silurian was replaced by a Devonian-Carboniferous platform sedimentation.

In the area concerned, the Devonian begins with the Caledonian molasse, the Old Red Sandstone formation. In NE Germany (Rügen island), this Old Red Sandstone is characterized by up to 1700 m of alternating reddish to greenish and gray continental (mostly fluviatile) sandstones, siltstones, and claystones of Emsian to Lower Givetian age. Faunal elements in some marly and limey intercalations point to short marine incursions (K. Zagora 1993). Towards the east in the adjoining Polish territories (Pomerania), a similar development is evident. There also the section consists of clastic continental deposits; but, in contrast to Rügen, the Old Red Sandstone type sedimentation occurred locally – interrupted by Middle Givetian marine carbonates, claystones, and siltstones – up to the Upper Givetian (Miłaczewski 1987). In the western foreland area, a Lower to Middle Devonian Old Red Sandstone is also proved by boreholes in Schleswig-Holstein, the southern North Sea, as well as the Campine Basin on the eastern flank of the Brabant Massif.

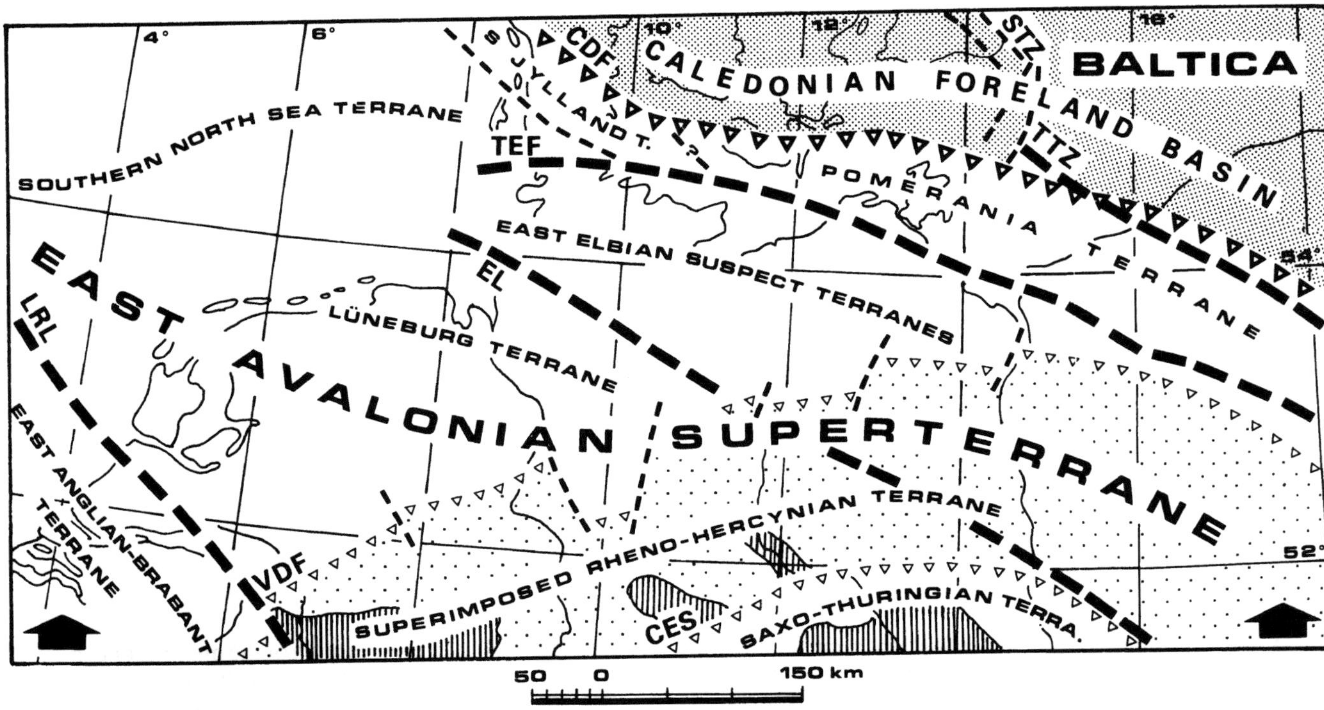

Fig. 2. Tentative model of the Lower Palaeozoic structural plan in northern Central Europe (according to the terrane model). *CDF* Caledonian deformation front; *STZ* Sorgenfrei-Tornquist Zone (in Lower Paleozoic times only of subordinated importance); *TTZ* Teisseyre-Tornquist Zone; *TEF* Trans-European Fault (strike direction still indeterminately); *CES* Central European Suture (Variscan Rheno-Herzynian/Saxo-Thuringian boundary); *VDF* Variscan deformation front; *EL* Elbe lineament; *LRL* Lower Rhine lineament; *arrows* general direction of Caledonian accretion

In the Upper Givetian, at the latest in Frasnian times, a general marine transgression took place. The continental formations gave way to marine ones. On Rügen island this marine sequence begins with gray-colored silty-sandy rocks of the Upper Givetian (350 m). The Frasnian (up to 1000 m) is mainly composed of an alternation of dark claystones and marlstones. Occasionally there are also intercalations of sandstones, dolomites, and coral-stromatoporoid limestones. In the Famennian (max. 170 m) the marine lower part is similar to the Frasnian development, whereas in the brackish to littoral influenced upper part also reddish sediments and anhydrites occur (K. Zagora 1993).

Immediately comparable to the Rügen section are the corresponding sequences in the southern Baltic Sea and in Pomerania. Alternations of marine claystones, marlstones, and coral limestones predominate. Siltstones, sandstones, and dolomites are intercalated. The whole Upper Devonian thickness amounts to 1000 m (Miłaczewski 1987).

Knowledge of the higher sections of the Devonian in the western part of the foreland is very restricted. Nevertheless, the few results from deep boreholes both at the northern edge (Schleswig-Holstein, North Sea) and the southern transition zone to the Variscan region (Campine Basin, Münsterland) make it possible to imagine that in the intervening unexplored area also a continuous marine sedimentation took place (cf. Fig. 3), either on a more or less homogeneous carbonate platform with biostromal and lagoonal limestones, or in a more differentiated shelf area with pelitic sediments in the troughs and calcareous ones on the swells (Krebs 1975; Franke et al. 1977; Ziegler 1982).

2.3 Lower Carboniferous

The Lower Carboniferous development is characterized by the sharp contrast between the Variscan Culm facies and the Carboniferous Limestone (Kohlenkalk) facies of the foreland

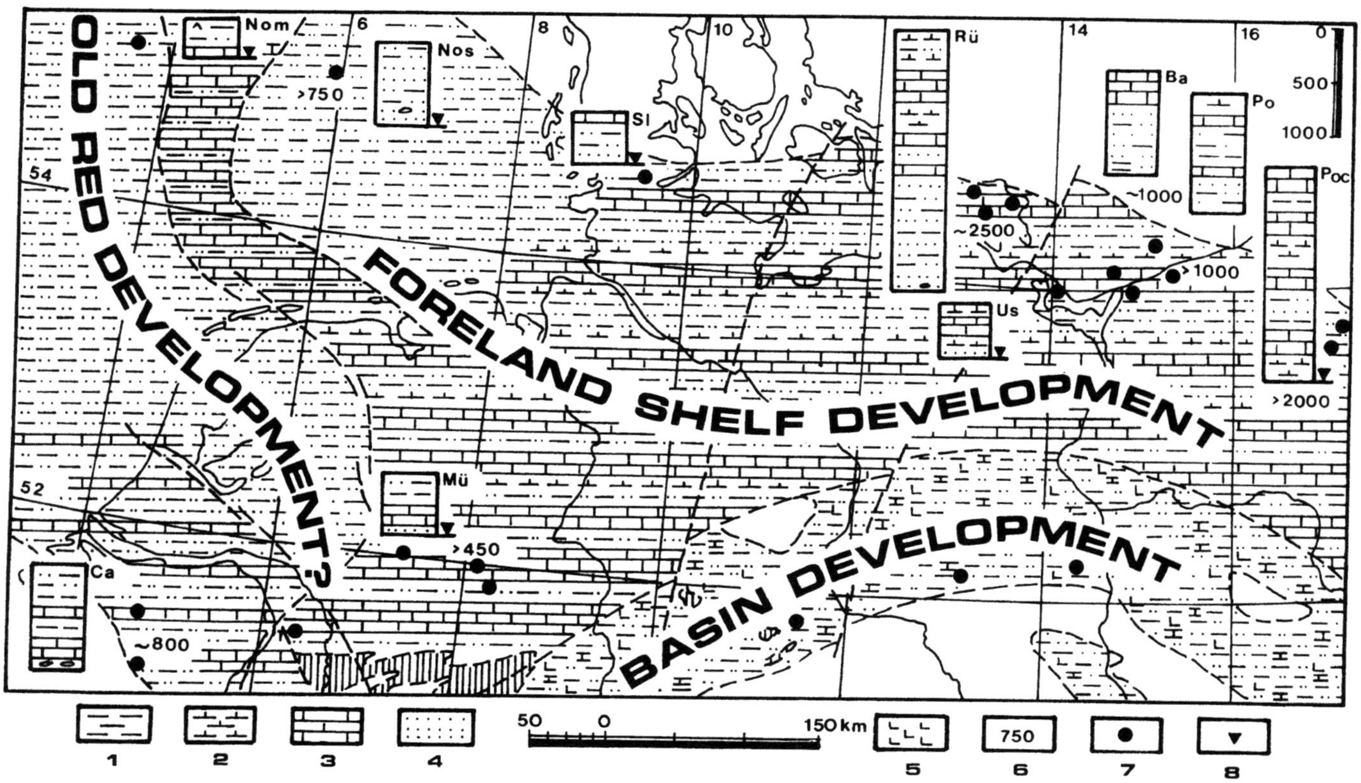

Fig. 3. Tentative model of the present distribution, tectonic position and possible lithofacial development of the Devonian in northern Central Europe. *1* Pelites; *2* marly rocks; *3* calcareous rocks; *4* sandstones, quartzites; *5* effusive rocks; *6* regional thickness development; *7* important borehole; *8* not penetrated profile. *Vertical lines* Outcropping Devonian; *vertical scale* scale of profiles (m); *regional position of profiles*: *Poc* Central Pomerania; *Po* Northwest Pomerania; *Ba* Southern Baltic; *Us* Usedom island; *Rü* Rügen island; *Sl* Schleswig; *Nos* SE North Sea; *Nom* Middle North Sea; *Mü* Münsterland area; *Ca* Campine Basin

(Fig. 4). Type profiles of the latter are known from Rügen, Pomerania, and the Campine Basin.

The lower part of the Rügen sequence is composed of a 100–200 m thick alternation of clayish and limey marlstones (Lower and Middle Tournaisian) which is followed by 100–300 m claystones and clayish marlstones with single limestone layers (Upper Tournaisian). Higher up, the sequence consists of 400–850 m alternating limestones, marlstones, and claystones (Lower Visean). The maximum of limestone sedimentation (500–750 m) was reached in the Middle Visean and lower Upper Visean. Marlstones of the middle Upper Visean (>300 m) form the youngest member of the Carboniferous Limestone here (Hoffmann et al. 1975).

Not far to the west, on Hiddensee island, a somewhat different development is known. There the limestone content decreases considerably. Clayish and silty lithologies predominate. South of Rügen island, on the NE German mainland (Vorpommern), a further changed Lower Carboniferous (Visean) sequence has been encountered in superdeep boreholes (I. Zagora 1990). Typical are limestones, marlstones, and brecciated carbonate rocks (approx. 200 m), a volcanic-pyroclastic series (800 m) and a "Culm-like" sequence with silty claystones and limey fine sandstones (400 m).

The Lower Carboniferous of Pomerania shows not only analogies but also certain differences compared with the NE German occurrences. This is based mainly on its near-shore position. Claystones, siltstones, several kinds of marlstones and limestones, plus arkosic and quartz sandstones have been found. Peculiarities are tuffitic rocks as well as gypsiferous and anhydritic carbonates. Thicknesses vary considerably from a few hundreds metres up to more than 1000 m. Stratigraphically, a nearly complete sequence from the Lower Tournaisian to the Upper Visean is present (Żelichowski 1987).

A similar development to that at Pomerania has been proved on Falster island (Denmark).

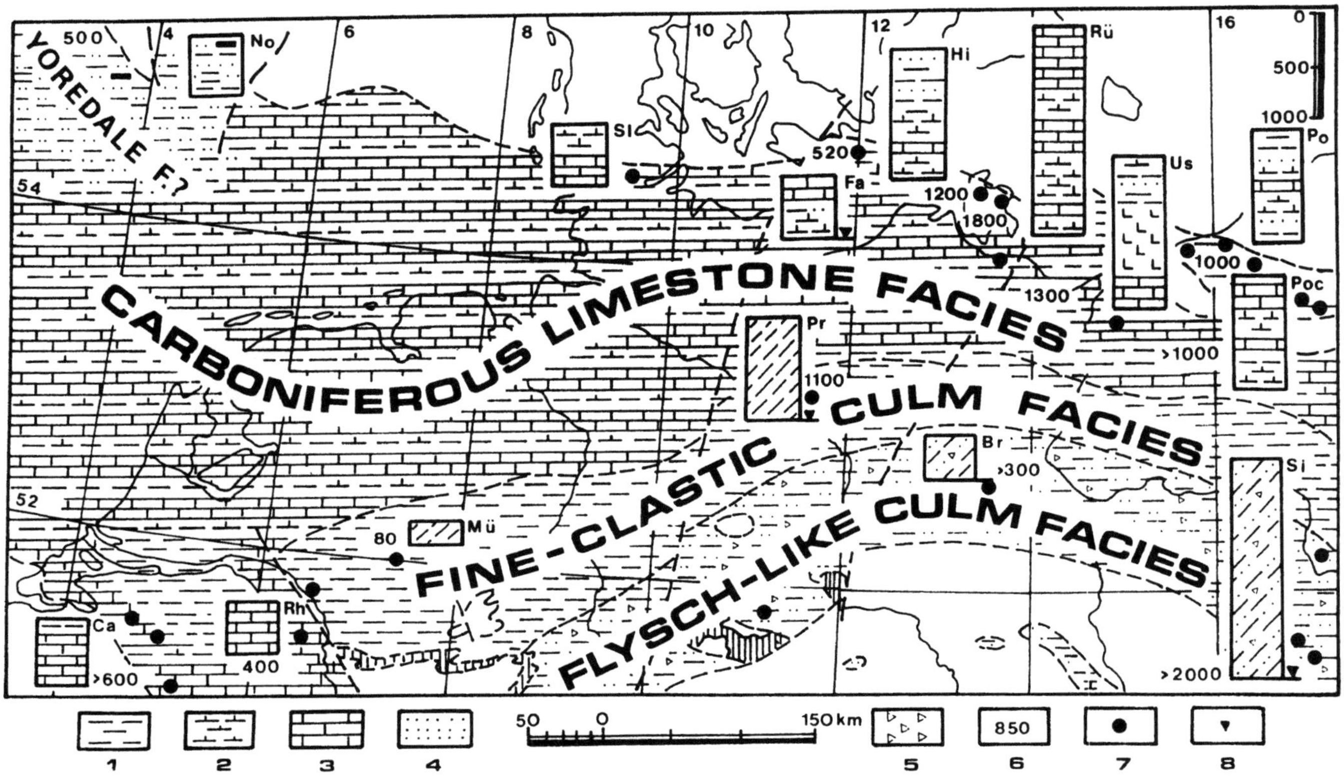

Fig. 4. Tentative model of the present distribution, tectonic position and possible lithofacial development of the Lower Carboniferous in northern Central Europe. *1* Pelites; *2* marly rocks; *3* calcareous rocks; *4* sandstones, quartzites; *5* graywackes; *6* regional thickness development; *7* important borehole; *8* not penetrated profile. *Vertical lines* Outcropping Lower Carboniferous; *vertical scale* scale of profiles (m); *regional position of profiles*: *Poc* Central Pomerania; *Po* Northwest Pomerania; *Us* Usedom island; *Rü* Rügen island; *Fa* Falster island; *Sl* Schleswig; *No* Central North Sea; *Ca* Campine Basin; *Rh* Lower Rhine region; *Mü* Münsterland area; *Pr* Prignitz area; *Br* Brandenburg; *Si* Lower Silesia

There, more than 500 m of Lower to Middle Visean were penetrated consisting of carbonatic-pelitic rocks. An adjacent shore line is indicated by sandy intercalations, plant-bearing beds and signs of coal (Michelsen 1971).

The boundary between this carbonatic foreland facies in the north and the clastic Culm facies of the Variscan region in the south can only be traced with difficulty. From the area around the Brabant Massif typical Carboniferous Limestone is known (Bless et al. 1976). It is involved in the Variscan tectogene (Dinant Basin) and part of its foreland (Campine Basin). On the other hand, to the east several deep boreholes in the Münsterland area, in SW Mecklenburg, and in Middle Poland encountered the fine clastic series of the Culm facies, which documents the extreme northwards occurrence of Variscan influences. There the southern edge of the carbonate platform of the foreland must be situated still farther to the north, at the present time concealed by the greater than 8–10 km thick sedimentary cover of the Central European Basin (Franke 1990c).

2.4 Upper Carboniferous

During the Upper Carboniferous, the marked differences between the Variscan development and that of its foreland decreased more and more. The upfolding of the orogen, which had already begun in Lower Carboniferous times and continued into Upper Westphalian times, resulted in the formation of a marginal deep in front of it and extending from the British Isles in the west to Poland in the east (Fig. 5). This marginal deep developed continuously out of the Variscan flysch troughs and the north adjoining argillaceous-siliceous distal Culm, respectively. Typical is the forelandward displacement of the main depression axes. Accordingly the Namurian reached its greatest thickness immediately north of the

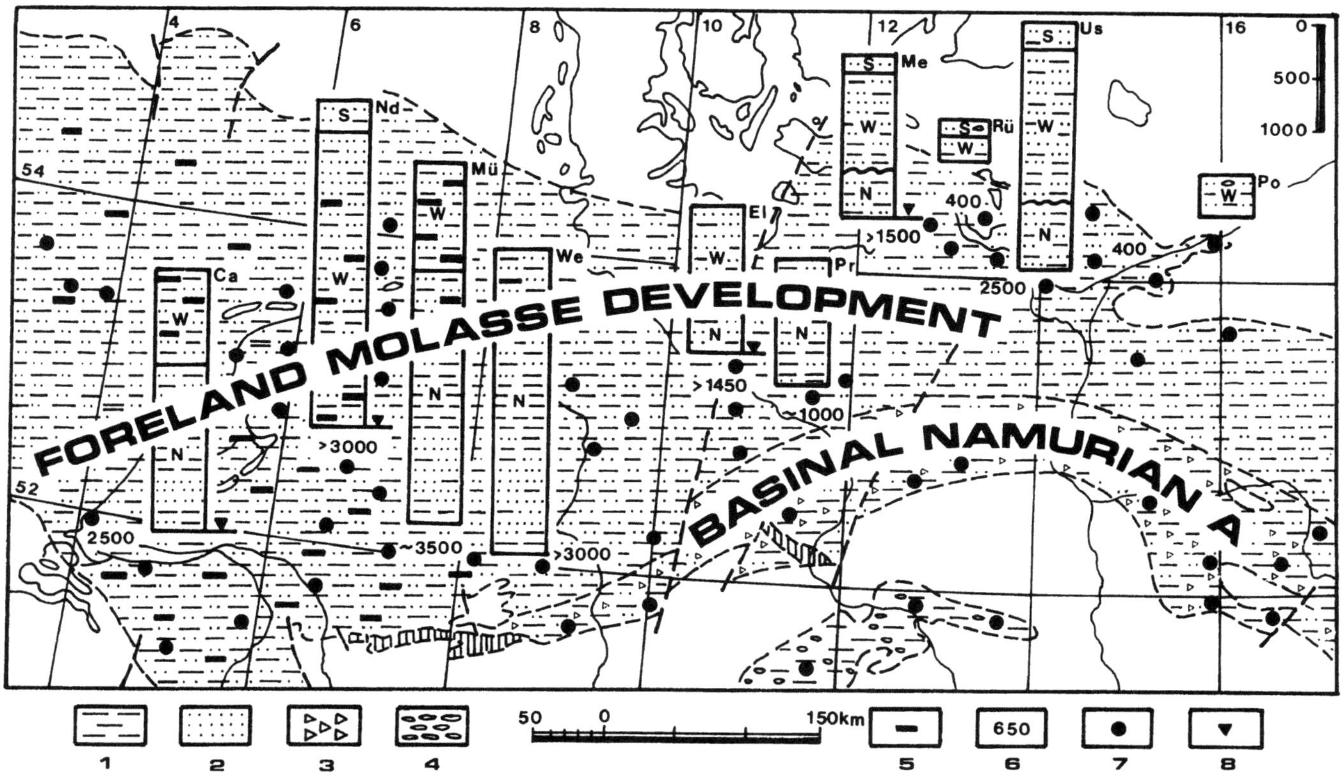

Fig. 5. Tentative model of the present distribution, tectonic position and possible lithofacial development of the Upper Carboniferous in northern Central Europe. *1* Pelites; *2* sandstones, quartzites; *3* graywackes; *4* conglomerates; *5* coal; *6* regional thickness development; *7* important borehole; *8* not penetrated profile. *Vertical lines* Outcropping Upper Carboniferous; *vertical scale* scale of profiles (m); *regional position of profiles: Po* Pomerania; *Us* Usedom island; *Rü* Rügen island; *Me* Mecklenburg; *Pr* Prignitz area; *El* Lower Elbe region; *We* SE Westfalia; *Mü* Münsterland area; *Nd* N Netherlands/NW Germany; *Ca* northern Campine Basin; *stratigraphy of profiles: s* Stephanian; *w* Westphalian; *n* Namurian

Lower Carboniferous Culm depocenters while the thickest Westphalian is found north of the maximum of Namurian subsidence (Katzung and Krull 1984; Paproth 1989).

Lithologically, the Upper Carboniferous sequences are relatively monotonous. Only the Namurian A still shows the orogen-foreland differentiation, in the south (southern Niedersachsen, Altmark, Brandenburg, SW Poland) with an alternation of claystones, sandstones, and graywackes of the marine flyschoid Culm (up to 2000 m), farther to the north (Ruhr district, Münsterland, SW Mecklenburg) with dark sapropelitic claystones (only up to 200 m) and at the northern border of the basin (Vorpommern) with a clastic series (<100 m) containing some continental features. In the Namurian B/C the molasse stage in the whole foredeep was reached. For the first time, a paralic environment predominated, i.e., a frequent change between marine and continental conditions. As well as claystones, siltstones, and sandstones, sporadically thin coal seams occur; marine horizons are still common. The maximum thickness amounts to more than 3000 m (Nordrhein-Westfalen, southern Niedersachsen).

During the Westphalian a gradual change from humid-paralic into a limnic-fluviatile up to an arid-continental facies took place. Characteristic is a cyclic alternation of claystones, siltstones, and sandstones with intercalations of coal seams. The latter reach their greatest distribution in the Westphalian A/B. The youngest seams are proved at the Westphalian C/D boundary. In the Upper Westphalian D only rootlet beds are developed. Also the thickness of the marine horizons decreases up the sequence. On the other hand, beds with nonmarine fauna and plant remains become more abundant, while the sand content increases from the older to the younger Westphalian stages (Hedemann and Teichmüller 1971; Hedemann et al. 1984).

It must be emphasized that in the western part of the foreland basin both lithology and thicknesses are of considerable continuity over a long distance of about 300 km from its southern edge up to the North Sea region. Contrary to this, and in an easterly direction, pronounced variations are evident. These are peculiarities in the lithological composition and stratigraphic succession (with only few coals), a different range of marine horizons, as well as a temporally changed tectonic development (Żelichowski 1987; Hoth et al. 1990). This implies that nearly N-S-striking elements separating distinct sedimentation areas played an essential role in the palaeogeographical development of the region (Franke 1990c).

In the Stephanian, the formerly wide foreland basin was very much restricted. The lithofacies development is characterized by a regionally and temporally uniform position within the arid-continental environment, resulting in a continuous deposition of red beds. Above a weak erosional basal disconformity, the sequence mostly begins with a base sandstone followed by alternating reddish brown claystones, siltstones, and sandstones. Sometimes thin marine-influenced limestone or claystone horizons are present. The thicknesses range between 200 m and 600 m (Hedemann et al. 1984).

Several peaks of the orogenic collisional movements within the external Variscan zone are reflected in the foreland area by breaks in the sedimentary record, denudation processes, erosional discordances, and weak angular unconformities. Important are regionally changing Upper Namurian (Erzgebirgian) and Upper Westphalian (Asturian) events. At the end of the Carboniferous and at Lower Permian times, Late Variscan extension resulted in the formation of deep-reaching NW-SE and N-S trending transverse fault zones. The associated post-collisional intense magmatism consists of tholeiitic microgabbro-basalt series in the northern border zone and more than 2000 m thick rhyolite-andesite series in the central parts of the foreland (Kramer 1988; Korich 1990).

3 Structure

The results of deep boreholes, geophysical measurements and regional geological compilations suggest a conspicuous heterogeneity of crustal structure in the area of the mid-European Variscan foreland. However, the few data available do not enable unambiguous statements about the composition, age, and structural configuration of this region. Therefore, several models have been developed which reflect the divergence of opinion.

3.1 The Central Segment

The central segment is that part of the foreland with the thickest Young Paleozoic to Cenozoic overburden. For this reason, our knowledge of the deep subsurface is much more restricted than in the southern or northern border regions. Only geophysical data give a broad overview, although interpretations are often contradictory.

The multiform gravimetric and magnetic anomalies, the inhomogeneities proved by deep seismic soundings and other geophysical features in the area concerned hint at a basement structure built up by a mosaic of varied crustal blocks (Dohr 1989; Wever et al. 1990). Starting on the supposition that these blocks were drifting from high southern latitudes to their present positions during Early Paleozoic times, it is presumed that the putative Tornquist Sea situated between Baltica in the north and peri-Gondwana (East Avalonian) fragments in the south gradually was closed. In the course of this process the different Gondwana-derived terranes, as already mentioned before, have probably been amalgamated to a substantially and structurally heterogenous superterrane accreted to the Laurussian continent during the Young Caledonian (Scandian) tectogenesis at the latest. However, it is still being vehemently discussed where the subsequent suture zone of the Tornquist Sea is located.

In the absence of real proof it is impossible to distinguish pre-Cadomian, Cadomian and Caledonian parts of the superterrane. Often the positive gravimetric/magnetic anomalies reflecting the higher crustal densities and/or magnetizations are regarded as blocks with stronger metamorphism and greater age (e.g., the well-known East Elbian Massif). However, the anomalies are sometimes explained as expressions of a varying Moho depth or of younger (Permo-Carboniferous, Cretaceous, or Tertiary) magmatism (cf. Bachmann and Grosse 1989). Possibly a mixture of these effects might explain the present picture.

Apart from the interpretation of the central segment as a huge Cadomian/Caledonian superterrane or massif, it is possible to observe an internal differentiation into specific crustal domains. The results of deep seismic measurements show that the lower crust can be subdivided in a southern one with good lamination and a northern one with low reflectivity. In the same sense, the lower crustal velocities differ from 6.4 km/s in the south to 6.9 km/s in the north. The border line between both areas is formed by the Elbe Lineament. This obviously very important structure also separates regions of variable Moho topography with constant depths of 30 km south and a shallowing to 26 km north of it (Prodehl and Giese 1990; Aichroth et al. 1992). Moreover, the crustal densities and the magnetization of the crust increases north of the lineament (Grosse et al. 1992; Bosum and Wonik 1991). Even electrical resistivity models derived from magnetotelluric and magneto-variational investigations underline its exciting significance (ERCEUGT Group 1992).

Surprising is the divergency between the E-W-oriented strike direction of the superterrane on the whole and the NW-SE striking of the Elbe Lineament situated within it. The latter coincide with other essential elements in the structural plan of northern Central Europe, that is the Tornquist Zone, the suspect Trans-European Fault and smaller Mesozoic and Cenozoic re-activated fractures. Like these, a Late Precambrian (Cadomian) to Early Paleozoic (Caledonian) primary genesis is very probably.

According to this basement structure, it is conceivable that "exotic" crustal blocks of different age and origin abut each other, eventually with narrow "home-made" Caledonian fold belts in between. Strike-slip movements along the faults cannot be excluded. They would more easily explain the present juxtaposition of various elements of the crystalline crust and would support the terrane model for the region investigated. On the other hand, a Caledonian orogen model with a Caledonian arc and accretionary slate-schist belt north of the Elbe Lineament and a Gondwana-derived Cadomian massif ("Lüneburg Massif") south of it (Berthelsen 1992b) could also explain the marked crustal inhomogeneities.

In addition to the NW-SE direction, meridional fractures are also of great importance for the structural division of the superterrane. In the west the Lower Rhine Lineament perhaps abuts against the Brabant Massif. Farther to the east, the Ems, Hessen, and Rheinsberg Lineaments are of significance not only for the crystalline crust, but also for the overlying sedimentary cover (Fig. 6). Left-lateral as well as right-lateral strike-slip movements of the order of 20–60 km along these faults are derived from regional displacements of various Phanerozoic complexes (Franke 1990b).

Of course, most of these considerations are still more or less hypothetical. Further research activities should lead to a better understanding of existing interpretations.

3.2 The Southern Border

The southern border of the Variscan foreland can be drawn in two different ways: firstly an original (primary) one lying anywhere beneath the external Variscan zone with temporally changed position; and secondly a present-day (secondary) one identical with the northern Variscan deformation front.

For the original southern foreland border, it is hard to place a definite line because from Devonian times onwards a continuous shifting of the Variscan domain relative to its northern foreland took place. In such a way, areas belonging to the foreland shelf, for instance in the Middle Devonian, have been involved in the Variscan regime during the Lower Carboniferous due to the southwards subduction and pre-collisional shortening of the external orogenic zone (W. Franke 1992).

These paleogeographically determined boundaries do not necessarily coincide with true crustal boundaries. The latter could possibly be represented by the so-called Norddeutsche Linie (Bachmann and Grosse 1989). At this nearly E-W-striking element, significant changes of the crust's physical parameters occur. The gravity, in particular, shows a remarkable sinking in a southerly direction along the anomalies of Bramsche-Vlotho, Hannover, Magdeburg, and Dessau. In the same direction, seismic soundings indicate a deepening of the Conrad discontinuity. Also magnetotelluric data point to an old crustal boundary. Precambrian origin with a Variscan and younger overprinting (inclusively intense magmatic activity) is likely. Conformably with that, the Norddeutsche Linie could be interpreted

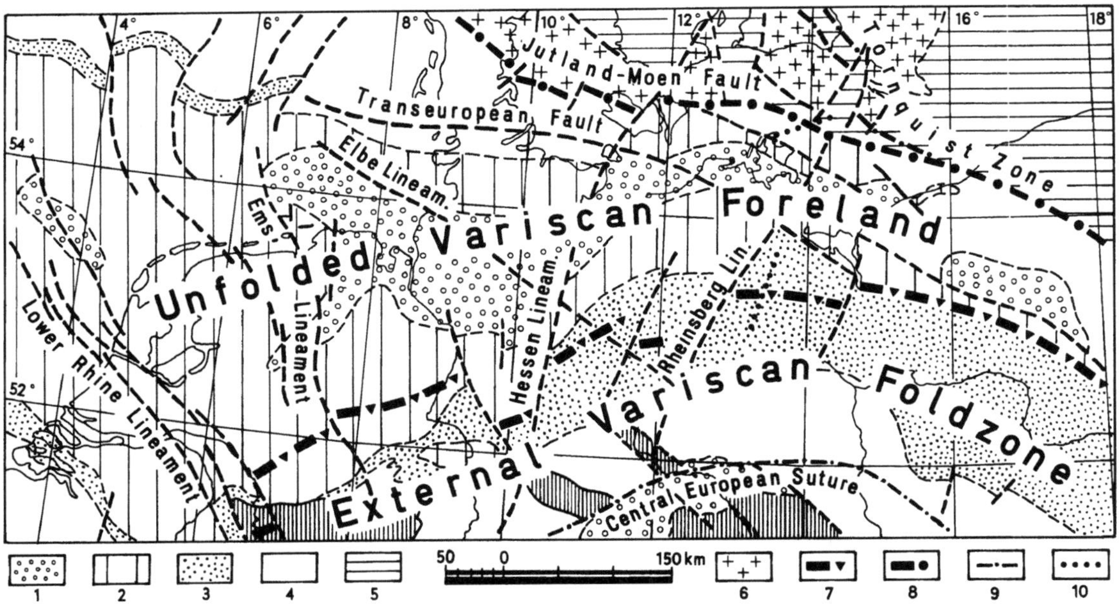

Fig. 6. Pre-Permian subcrop in the area of the northern Variscan foreland. *1* Stephanian; *2* Westphalian; *3* Namurian; *4* pre-Upper Carboniferous, undivided; *5* unfolded Cambro-Silurian of the East European Platform; *6* pre-Cadomian basement of Baltica; *7* Variscan deformation front; *8* Caledonian deformation front; *9* Central European Suture; *10* profiles of Figs. 7 and 8

as the boundary between heavy (Cadomian) crust of the foreland and light (Paleozoic) crust of the mid-European Variscides.

The DEKORP 2N deep seismic profile reveals another significant crustal boundary situated beneath the northern Rhenish slate mountains south of the Ebbe and Venn anticlines. There the reflective Variscan lower crust contacts under the transparent lower crust of the foreland (W. Franke et al. 1990). From this and additional data (Eckergneis of the Harz mountains, high-grade metamorphic xenoliths of the Hessen depression) it could be concluded that the most part of the Rheno-Hercynian zone, perhaps up to its southern contact with the Saxo-Thuringian zone, is underlain by Cadomian (and even pre-Cadomian?) East Avalonian crust (Berthelsen 1992a).

The present (secondary) border between the Variscan orogen and its foreland is characterized by the Variscan deformation front, separating folded Paleozoic complexes in the south from an unfolded Devonian-Carboniferous platform development in the north. Although there exists no general consensus of opinion about its possible course, the few data available from boreholes and some geophysical criteria permit a regional delineation as shown on Fig. 6.

In contrast with the high quality results reached on this matter in Western Europe (BIRPS-ECORS profiles), in the mid-European part a clear geophysical distinction between orogen and foreland cannot be made, either by the DEKORP or by the EGT deep seismic soundings. The reason may be the different tectonic character of the foreland and the associated variability of the transition zone. Presumably, long-reaching thrust zones (like the Midi overthrust) with uplifted rigid foreland blocks (like the Brabant Massif) are largely missing here. On the other hand, the overlying thick platform cover complicates the delineation of deep-seated structures.

For the structure of the deformation front, often a gradually decreasing deformation intensity towards the foreland is assumed. This has been deduced from the results of exploration activities particularly in Nordrhein-Westfalen north of the Ruhr district, as well as from deep seismic sounding along the DEKORP 2N profile (W. Franke et al. 1990). Additionally, new geophysical indications gained in the surroundings of the East Elbian Massif (Altmark area and northern Brandenburg) also hint at thrust-like elements (Fig. 7). There perhaps a forelandward-directed nappe tectonics exists. Information from deep boreholes seems to support such a supposition

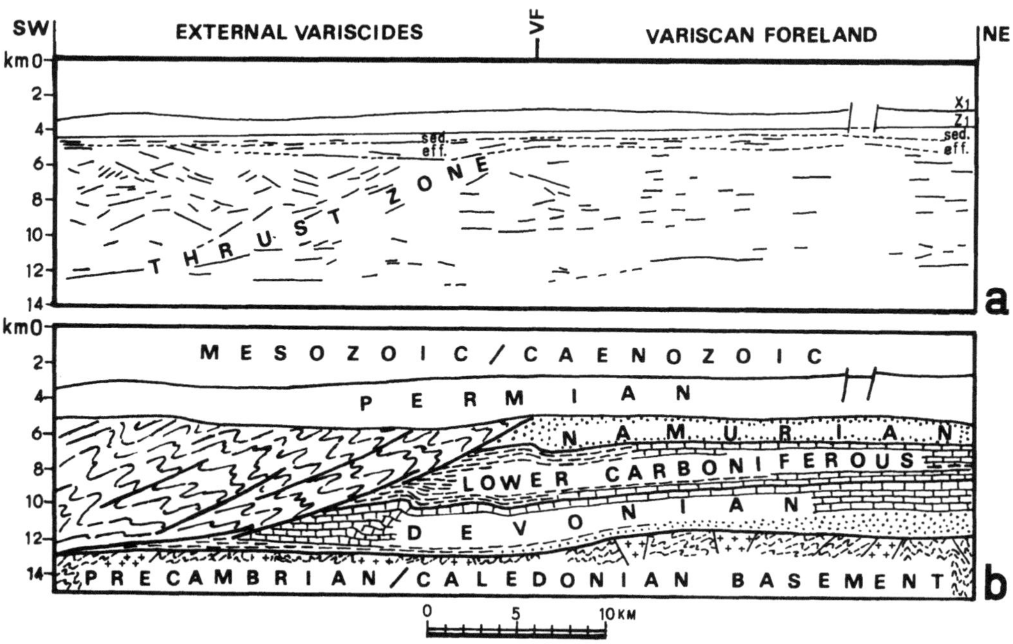

Fig. 7a,b. Line drawing (**a**) of a deep seismic profile across the Variscan deformation front and its geological interpretation (**b**). X_1 Reflector base Mesozoic; Z_1 reflector base Zechstein; *sed.* sedimentary Rotliegendes; *eff.* effusive Rotliegendes; *VF* Variscan front. Position of profile, see Fig. 6

(Franke et al. 1989). Finally, a fault-controlled delimitation of the Variscan folding is worth discussing. It may be imagined that, especially in Middle Poland, deep-reaching NW-SE fractures parallel to the Teisseyre-Tornquist Zone influenced Variscan tectogenesis.

A northward-directed subduction of Variscan crust beneath the foreland, as it is considered at times, cannot be confirmed paleogeographically or paleotectonically. On the contrary, a southward subduction of external Variscan (Rhenohercynian) crust beneath internal Variscan (Saxothuringian) crust is probable (Behr et al. 1984).

3.3 The Northern Border

The northern border of the mid-European Variscan foreland is drawn here coincident with the southern edge of Baltica. This boundary between the pre-Cadomian Fennosarmatian area in the north and the presumed Cadomian/Caledonian regions south of it has been a matter of dispute in the last few years. New drilling activities and recent geophysical investigations have led to progress in the definition of the regional position and tectonic character of this important crustal boundary.

For many years, the Tornquist Zone was considered the southwestern border of the East European Platform including its elevated northwestern part, the Baltic Shield. Today, we know that only its SE section (Teisseyre-Tornquist Zone) represents a true platform boundary, whereas the NW part (Sorgenfrei-Tornquist Zone) is actually a significant intra-plate structure within the Precambrian platform (Eugeno-S Working Group 1988; Franke 1990a). In the area of the southern Baltic Sea, the zone loses its fundamental geotectonic meaning. From there the proper platform boundary diverges in a more E-W direction (cf. Fig. 1). However, up to now, no general agreement has been reached about its real position. At least two different versions are possible (Franke 1993).

According to the first version, a remarkable fault zone bordering the southern flank of the Danish Ringkoebing-Fyn High and its eastern prolongation (North Arkona Block) could be interpreted as the platform boundary. This dislocation zone, called the Jutland-Moen Fault, separates two paleogeographically und tectonically different units (Fig. 8). In the northern part, true "Baltic" platform conditions are evident. This is indicated by geophysical data as well as by drilling results.

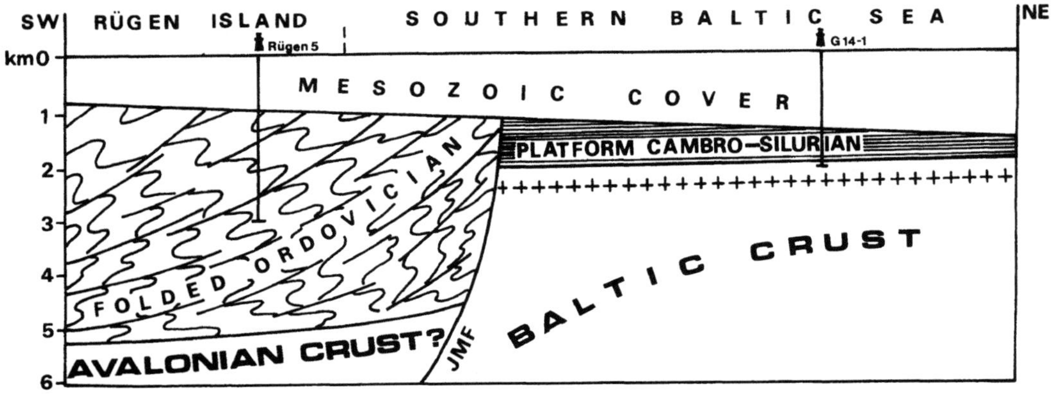

Fig. 8. Schematic geological section across the boundary between the Caledonian folded complexes of the Variscan foreland and the Cambro-Silurian platform development of Baltica (not to scale). *JMF* Jutland-Moen Fault; wildcat well Rügen 5; G14-1 off-shore well North Arkona Block 1 (G14-1); distance between both wells about 30 km. Position of section, see Fig. 6

The latter have shown that the Lower Paleozoic platform development known from Bornholm or Scania reaches farther to the south and southeast than was formerly supposed. On the North Arkona Block in the southern Baltic Sea north of Rügen island, an offshore well encountered a nearly complete undisturbed Cambro-Silurian standard profile down to the Precambrian crystalline basement, which proved its unequivocal association with Baltica (Piske and Neumann 1990; Franke 1990a).

In contrast to this, south of the Jutland-Moen Fault, quite another situation can be observed. There the Lower Paleozoic not only differs in lithology, thickness, and faunal content, but was also intensely deformed by Caledonian movements. These significant heterogeneities on both sides of the fault have been the motive for its interpretation as a terrane boundary approximately coincident with the Caledonian deformation front. In this sense, south of Baltica a Caledonian terrane (Pomerania Terrane) adjoins that is accreted against and partly overthrust onto the Precambrian crust (Pożaryski 1990; Franke 1990c).

The second version for a southern border of Baltica is founded on the results of deep seismic soundings carried out as part of the European Geotraverse Project (Eugeno-S Working Group 1988). These measurements show a distinct change in seismic basement structure about 50 km south of the suspected Caledonian deformation front in the area of Schleswig-Holstein. Here a shield-like crust in the north is in immediate contact with an extremly thin (12–15 km) crystalline crust in the south. Between the two different crustal types the southern boundary of Baltica can be anticipated. The Caledonian complexes proved by boreholes north of this boundary (called the Trans-European Fault) are interpreted as shieldward-directed nappe structures of a marginal belt overthrust above "Baltic" crust (Berthelsen 1992b).

From these different interpretations of the boundary between the Variscan foreland and Precambrian Baltica, two basic concepts for the general structural plan result. In accordance with the terrane model, the foreland basement consists of several Gondwana-derived pre-Cadomian, Cadomian, and Caledonian terranes fitted together during Early Paleozoic times and accreted to Baltica during the Young Caledonian tectogenesis at the latest. According to the marginal belt model, it is built up by a North German-Polish Caledonian mountain chain composed of Caledonian-deformed accretionary belts that locally surround older (Cadomian or pre-Cadomian) basement massifs. The reliability of both these concepts still remains problematic.

4 Conclusions

The northern foreland of the mid-European Variscides shows, in both its lithostratigraphical and structural development, special peculiarities compared with the adjoining areas. After a mobile Lower Paleozoic (Caledonian) epoch, the platform stage started in Lower Devonian times with molassoid Old Red Sandstone formations.

Upper Devonian and Lower Carboniferous sequences are characterized by predominantly pelitic-carbonatic shelf deposits. During the Upper Carboniferous, the marine foreland basin became a paralic to continental marginal deep in front of the uprising Variscan orogen.

Interpretation of the geotectonic situation has given rise to different opinions. A pre-Cadomian as well as a Cadomian/Caledonian basement age are discussed. Various indications point to a Young Caledonian age for final consolidation. The crustal heterogeneity of this basement is proved by geophysical measurements. According to the data available, it can be interpreted either as an internal zone of a North German-Polish Caledonian mountain chain or as a Cadomian/Caledonian terrane complex composed of several microblocks.

The southern border of the foreland basement against the Variscan domain proper presumedly is marked by a significant E-W-striking crustal boundary (Norddeutsche Linie) or a more southerly located deep fracture proved beneath the Rhenish slate mountains. From there, the old crust is overridden by the external Variscan upper crustal sedimentary complexes bounded northwards by the only uncertain Variscan deformation front. In the north, the suspect Cadomian/Caledonian basement is delimited by the southern edge of Baltica the real position of which (Jutland-Moen Fault, Trans-European Fault) is still problematic.

It is suggested that the Variscan development immediately influenced the Devonian-Carboniferous sedimentary and tectonic evolution of the foreland. On the other hand, the composition and structural plan of the foreland had an inportant effect on the geologic history of the Variscides, especially in the northern external zones. However, the lack of sufficient information, particularly on the deeply subsided central parts of the foreland, permits only the construction of more or less hypothetical models which may be revised with the acquisition of new results from improved geophysical and drilling activities.

Acknowledgments. The author wishes to express his gratitude to Dr. Alfred Whittaker, British Geological Survey, Nottingham, who kindly revised the English text.

References

Aichroth B, Prodehl C, Thybo H (1992) Crustal structure along the Central Segment of the EGT from seismic-refraction studies. Tectonophysics 207: 43–64

Bachmann GH, Grosse S (1989) Struktur und Entstehung des Norddeutschen Beckens – geologische und geophysikalische Interpretation einer verbesserten Bouguer-Schwerekarte. Niedersächs Akad Geowiss Veröff 2: 23–47

Behr HJ, Engel W, Franke W, Giese P, Weber K (1984) The Variscan belt in Central Europe: main structures, geodynamic implications, open questions. Tectonophysics 109: 15–40

Berthelsen A (1992a) Mobile Europe. In: Blundell D, Freeman R, Mueller St (eds) A continent revealed. The European Geotraverse. Cambridge Univ Press, pp 11–32

Berthelsen A (1992b) From Precambrian to Variscan Europe. In: Blundell D, Freeman R, Mueller St (eds) A continent revealed. The European Geotraverse. Cambridge Univ Press, pp 153–164

Bless MJM, Bouckaert J, Boozet P, Conil R, Cornet P, Fairon-Demaret M, Groessens E, Longerstaey PJ, Meesen JPMT, Paproth E, Pirlet H, Streel M, Van Amerom HWJ, Wolf M (1976) Dinantian rocks in the subsurface North of the Brabant and Ardenno-Rhenish massifs in Belgium, the Netherlands and the Federal Republic of Germany. Meded Rijks Geol Dienst Nieuwe Ser 27: 81–195

Bosum W, Wonik TH (1991) Magnetic anomaly pattern of Central Europe. Tectonophysics 195: 253–259

Dohr G (1989) Ergebnisse geophysikalischer Arbeiten zur Untersuchung des tieferen Untergrundes in Norddeutschland. Niedersächs Akad Geowiss Veröff 2: 4–22

ERCEUGT-Group (1992) An electrical resistivity crustal section from the Alps to the Baltic Sea (central segment of EGT). Tectonophysics 207: 123–139

Eugeno-S Working Group (1988) Crustal structure and tectonic evolution of the transition between the Baltic Shield and the North German Caledonides. Tectonophysics 150: 253–348

Franke D (1990a) The north-west part of the Tornquist-Teisseyre zone – platform margin or intraplate structure? Z Angew Geol 36: 45–48

Franke D (1990b) Submeridionale Transcurrent-Störungen im präpermischen Untergrund der Mitteleuropäischen Senke. Z Angew Geol 36: 201–207

Franke D (1990c) Der präpermische Untergrund der Mitteleuropäischen Senke – Fakten und Hypothesen. Niedersächs Akad Geowiss Veröff 4: 19–75

Franke D (1993) The southern border of Baltica – a review of the present state of knowledge. Precambrian Res 64: 419–430

Franke D, Krull P, Pfeiffer H (1977) Palaeogeographic and tectonic development of the external zones of the Central European Variscides and their Northern Foreland. In: La chaîne varisque d'Europe moyenne et occidentale. Coll Int CNRS, Rennes, No 243, pp 515–529

Franke D, Hoffmann N, Kamps J (1989) Alter und struktureller Bau des Grundgebirges im Nordteil der DDR. Z Angew Geol 35: 289–296
Franke W (1992) Phanerozoic structures and events in Central Europe. In: Blundell D, Freeman R, Mueller St (eds) A continent revealed. The European Geotraverse. Cambridge Univ Press, pp 164–180
Franke W, Bortfeld RK, Brix M, Drozdzewski G, Dürbaum HJ, Giese P, Janoth W, Jödicke H, Reichelt CH, Scherp A, Schondl J, Thomas R, Thünker M, Weber K, Wiesner MG, Wong HK (1990) Crustal structure of the Rhenish Massif: results of deep seismic reflection lines DEKORP 2-North and 2-North-Q. Geol Rdsch 79: 523–566
Frost RTC, Fitch FJ, Miller JA (1981) The age and nature of the crystalline basement of the North Sea Basin. In: Illing LV, Hobson GD (eds) Petroleum geology of the Continental Shelf of North-West Europe. Inst Petrol Heyden, London, pp 43–57
Grosse S, Behr HJ, Edel JB, Heinrichs T (1992) The gravity field along the Central Segment of the EGT. Tectonophysics 207: 97–121
Hedemann HA, Teichmüller R (1971) Die paläogeographische Entwicklung des Oberkarbons. Fortschr Geol Nordrhein-Westf 19: 129–142
Hedemann HA, Schuster A, Stancu-Kristoff G, Lösch J (1984) Die Verbreitung der Kohlenflöze des Oberkarbons in Nordwestdeutschland und ihre stratigraphische Einstufung. Fortschr Geol Nordrhein-Westf 32: 39–88
Hoffmann N, Lindert W, Weyer D, Illers KH (1975) Zum Unterkarbonvorkommen auf den Inseln Rügen und Hiddensee. Z Geol Wiss 3: 851–873
Hoth K, Weyer D, Kahlert E, Döring H (1990) Das Standardprofil des Oberkarbons von Vorpommern. Z Angew Geol 36: 361–368
Katzung G, Krull P (1984) Zur tektonischen Entwicklung Mittel-und Nordwesteuropas während des Jungpaläozoikums. Z Angew Geol 30: 163–173
Korich D (1990) Zu Stoffbestand und Genese der sauren bis intermediären Vulkanite des Darß-Uckermark-Eruptivkomplexes in Mecklenburg-Vorpommern. Z Angew Geol 20: 337–350
Kramer W (1988) Magmengenetische Aspekte der Lithosphärenentwicklung. Schriftenr Geol Wiss 26: 1–136
Krauss M (1993) Zur stofflichen und strukturellen Entwicklung und Situation im Untergrund von Rügen. Exkursionsführer Rügen-Bornholm. Ernst-Moritz-Arndt-Univ Greifswald, pp 15–32
Krebs W (1975) Geologische Aspekte der Tiefen-exploration im Paläozoikum Norddeutschlands und der südlichen Nordsee. Erdöl-Erdgas-Z 91: 277–284
Michelsen O (1971) Lower Carboniferous foraminiferal faunas of the boring Ørslev No. 1, island of Falster, Denmark. Danmarks Geol Undersøgelse II R (98): 1–86
Miłaczewski L (1987) Dewon. In: Raczyńska A (ed) Budowa geologiczna wału pomorskiego i jego podłoża. Prace Inst Geol 119: 16–26
Modliński Z (1987) Ordowik. In: Raczyńska A (ed) Budowa geologiczna wału pomorskiego i jego podłoża. Prace Inst Geol 119: 10–12
Paproth E (1989) Die paläogeographische Entwicklung Mittel-Europas im Karbon. Geol Jahrb Hessen 117: 53–68
Piske J, Neumann E (1990) Zur Ausbildung des Altpaläozoikums im Gebiet der südwestlichen Ostsee. Z Angew Geol 36: 342–346
Pożaryski W (1990) Kaledonidy środkowej europy – orogenem przesuwczyn złożonym terranów. Prz Geol 38: 1–8
Prodehl C, Giese P (1990) Seismic investigations around the EGT in Central Europe. In: Freeman R, Giese P, Mueller ST (eds) The European Geotraverse: Integrative Studies. Eur Sci Found, pp 77–97
Tomczyk H (1987) Sylur. In: Raczyńska A (ed) Budowa geologiczna wału pomorskiego i jego podłoża. Prace Inst Geol 119: 12–16
Wever T, Meissner R, Sadowiak P, Dekorp Group (1990) Deep reflection seismic data along the central part of the European Geotraverse in Germany: a review. Tectonophysics 176: 87–101
Zagora I (1990) Zur Foraminiferenfauna des Kohlenkalkes im Untergrund der Ostseeküste der DDR. FRG Subcomm for Stratigraphy of the Carboniferous, Saalfeld
Zagora K (1993) Zur Litho- und Biofazies des Devons von Rügen. Internationale Exursions- und Vortragstagung Rügen-Bornholm, Kurzfassungen: 44, Greifswald
Żelichowski AM (1987) Development of the Carboniferous of the SW margin of the East European platform in Poland. Prz Geol 5: 230–237
Ziegler PA (1982) Geological atlas of Western and Central Europe. Elsevier, Amsterdam

X Paleomagnetic Constraints on the Paleogeographic Evolution of Europe During the Paleozoic

V. Bachtadse, T.H. Torsvik, J.A. Tait, and H.C. Soffel

1 Introduction

It is now generally accepted that the Caledonian and Hercynian orogenies are the result of collision between Laurentia (comprising N America, Greenland, W Spitsbergen, Rockall, and Northern Britain), Baltica (consisting of Scandinavia, the Russian platform, and E Spitsbergen) and the southern continents of Gondwana. Within this framework, several minor continental fragments have been identified (e.g., Avalonia, Armorica), which have been moving independently for various lengths of time and which are now incorporated into the mosaic of the European Paleozoic Massifs. In this chapter we present a series of paleogeographic reconstructions for key periods in the early and mid-Paleozoic.

2 Apparent Polar Wander Paths

The apparent polar wander (APW) paths for Laurentia (Van der Voo 1993), Baltica (Torsvik et al. 1992), Armorica (as defined by Torsvik et al. 1990a), and Eastern Avalonia (Trench and Torsvik 1991a) are portrayed in Fig 1. For lower Paleozoic times, the path for Armorica has been interpolated between the Ordovician and late Devonian/Carboniferous poles using a spherical line fitting procedure (Jupp and Kent 1987). The individual data have been discussed at length by Kent and Van der Voo (1990), Van der Voo (1988), Torsvik et al. (1990a), Livermore et al. (1985), and Bachtadse and Briden (1990), with the exception of more recent paleomagnetic data especially from Baltica (Torsvik et al. 1990b; Torsvik and Trench 1991; Trench and Torsvik 1991b; Perroud et al. 1992) and the western edge of the Bohemian Massif (Tait et al. 1994a,b).

The APW paths for Laurentia, Eastern Avalonia, and Baltica all converge during mid-Silurian to mid-Devonian times towards the so-called Siluro-Devonian corner. This convergence of the APW paths marks the formation of the Old Red Continent. The precise timing of amalgamation of Armorica to the Old Red Continent is uncertain. Originally, Van der Voo (1979) postulated that Armorica and Laurentia/Baltica collided during the early Silurian, causing the Taconic orogenic event. However, this scenario has subsequently been revised, and collision between Armorica and Laurentia/Baltica during the early-mid-Devonian was proposed by Perroud et al. (1984). This issue is still unresolved, mainly due to the lack of reliable Siluro-Devonian data for Armorica. Comparing the APW path for Armorica to those for Laurentia and Eastern Avalonia (Fig. 1), however, suggests that these three lithospheric elements were at similar paleolatitudes from the early Devonian onward.

The APW path for Gondwana is less straightforward (Fig. 2). Two basic paths have been postulated (termed X and Y by Morel and Irving, 1978; Fig. 2A,B), each of which has slight variations, as discussed by Bachtadse and Briden (1990, 1991). The differences between these two basic paths focus on the question as to whether or not the well-defined Ordovician pole position in NW Africa can be directly connected with the equally well-defined cluster of Carboniferous pole positions to the southeast of southern Africa (Fig. 2A). This was originally proposed by McElhinny et al. (1968), and was subsequently identified by Morel and Irving as path X. Alternatively, the African APW path may be more complex (Fig. 2B), involving at least one pronounced loop during Silurian times (path Y, Morel and Irving 1978). This more complex solution has been controversial, and until recently was based on paleomagnetic data from the Eastern Australian Lachlan Fold Belt. These data are controversial with regards to the APW path for Gondwana, as their interpretation depends on the geodynamic history of the Lachlan Fold Belt, which is generally considered to comprise allochthonous terranes.

Dallmeyer et al. (Eds.)
Pre-Permian Geology of Central and Eastern Europe

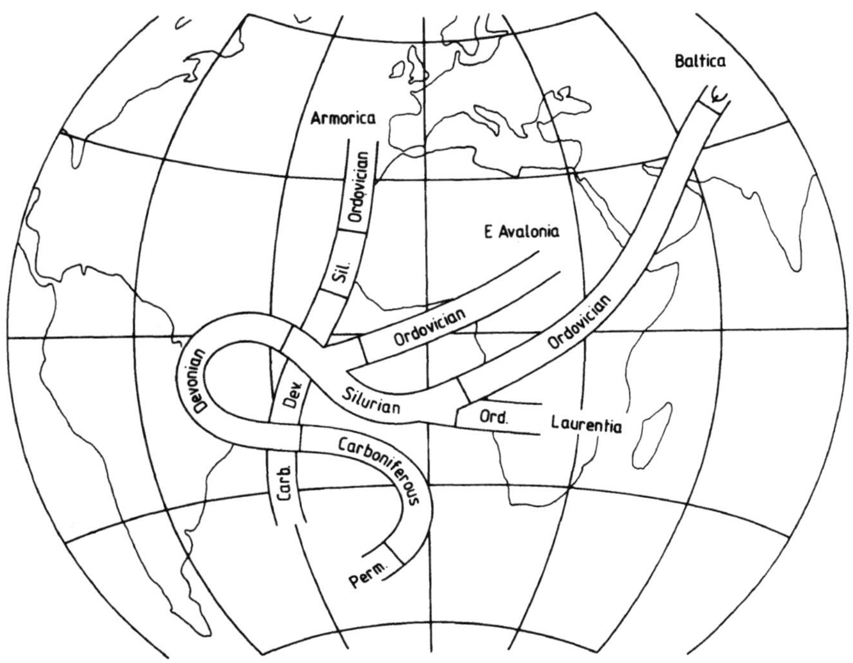

Fig. 1. Apparent polar wander paths for Baltica (Torsvik et al. 1992), Laurentia (Van der Voo 1993), Eastern Avalonia (Trench and Torsvik 1991a), and Armorica (Torsvik et al. 1990a) for Ordovician to Permian times. Note that the APW path for Laurentia has been rotated into European coordinates. (Bullard et al. 1965)

More recently, however, this more complex path has received additional support from a result published by Hargraves et al. (1987), who reported paleomagnetic data from the Silurian age Aïr igneous ring complexes in Niger, and from Li et al. (1990, 1991) following a re-study of sedimentary sequences from the cratonic Australian Amadeus basin.

Further constraints for the shape of the APW path during Devonian times have also been obtained by paleomagnetic studies from Australia (Hurley and Van der Voo 1987; Li et al. 1988; Thrupp et al. 1991), northern Africa (Aïfa et al. 1990; Bachtadse and Briden 1991) and southern Africa (Bachtadse et al. 1987). These studies yield groups of well-defined and robust paleo-south poles situated in northern and central Africa. If the positions of these paleopoles are accepted, they require the paleo-south pole to have moved from the northwestern margin of Gondwana in the Ordovician, to the west of Chile (in conventional Gondwana reconstruction) in the Silurian, then back into central Africa by mid-Devonian times, and finally to the east of southern Africa by Carboniferous times (Fig. 2B).

An alternative to this rather complex model has been proposed by Bachtadse and Briden (1991). They suggest a more direct path (Fig. 2C), which is reminiscent to that originally proposed by McElhinny et al. (1968). This alternative solution is based on the identification of secondary magnetizations in Sudanese rocks of Devonian age (Bachtadse and Briden 1991). These remagnetizations yield paleopoles very similar to those reported by Hargraves et al. (1987) from Silurian age rocks, and thus prompted Bachtadse and Briden (1991) to suggest that they may also be remagnetized.

However, the problem as to which path should be adopted as representative for Gondwana remains unsolved, and will be discussed further in Section 3.2.1.

Fig. 2A–C. Alternative apparent polar wander paths for Gondwana during the Paleozoic. **A** McElhinny et al. (1968); **B** Bachtadse and Briden (1990); **C** Bachtadse and Briden (1991). Note the similarity between **A** and **C**, which are both variations of the X-type path, whereas **B** is reminiscent of the Y-type path proposed by Morel and Irving (1978). The shape of the apparent polar wander path in **C** is constrained by the Devonian age paleo-south pole in Northern Africa (Bachtadse and Briden 1991) and their reinterpretation of the magnetization age of the results of Hargraves et al. (1987) from the Air Complex (see text for details)

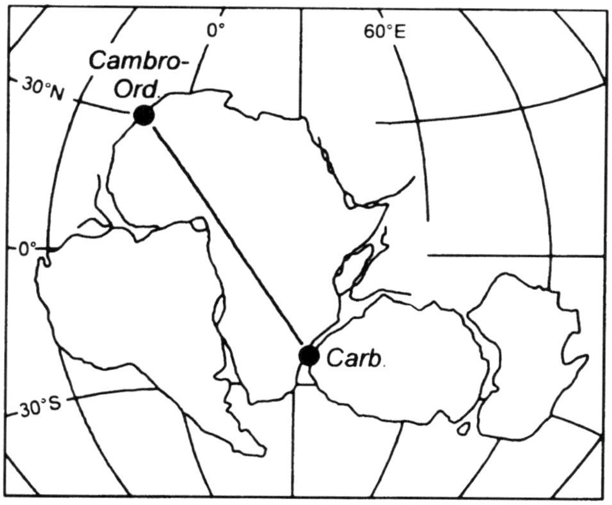

A

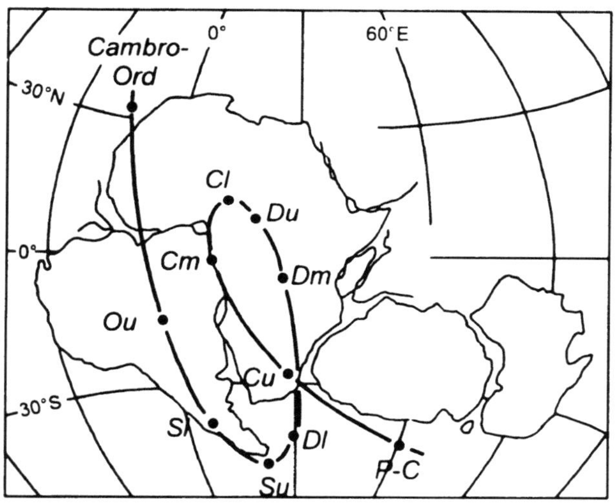

B

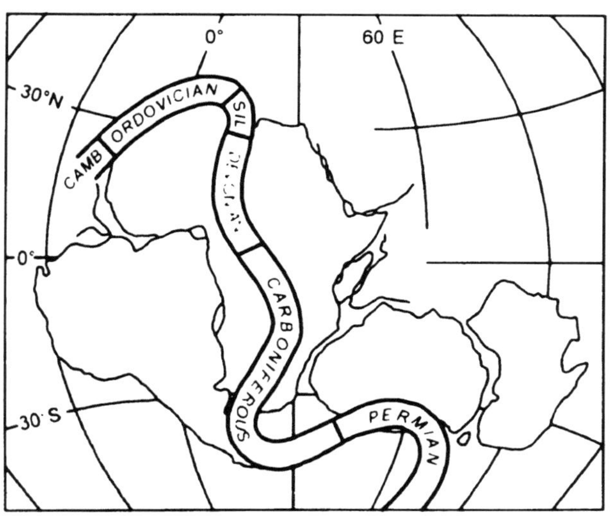

C

3 Paleogeographic Reconstructions

3.1 Ordovician

3.1.1 Gondwana

There is little ambiguity as to the position of Gondwana during Ordovician times. The paleomagnetic data set (see review by Bachtadse and Briden 1990) identifies a well-defined paleo-south pole in N Africa. This is supported by both paleobiogeographical evidence (Boucot 1990), and lithofacies indicators, such as the occurrence of glacial deposits in N Africa/Morocco (Boeuf et al. 1971) and the predominance of cold water faunas (the Mediterranean Facies) in N Africa during the Ordovician (Spjeldnaes 1961). Combination of all these lines of evidence indicates that the northern edge of Gondwana lay at high (>70°), peri-polar southern latitudes during this period, as shown in Fig. 3A.

3.1.2 Baltica

The Ordovician age sediments of Baltica comprise predominantly platform sediments (nodular carbonates and shales), and the faunas are indicative of temperate to warm water environments (Webby 1984). This is in marked contrast to the cold water faunas of Gondwana, and the faunal facies are markedly different to those of Laurentia (Whittington and Hughes 1972). These differences can most plausibly be explained by the existence of the Iapetus Ocean between Baltica/Gondwana and Laurentia, and the Tornquist Ocean between Baltica and Gondwana (Cocks and Fortey 1982; Scotese and McKerrow 1990). However, the width and geodynamic evolution of the Tornquist Ocean has been a matter of some debate over the last decade. One of the main arguments is whether the warm water lithofacies of present day southern Scandinavia and the 'Baltic Faunal' province indicate tropical or more temperate paleolatitudes (Cocks and Fortey 1982).

Recently published paleomagnetic data from Scandinavia (Perroud et al. 1992; Torsvik et al. 1990b; Trench and Torsvik 1991b) imply that Baltica occupied temperate southerly paleolatitudes in late Cambrian times (Torsvik and Trench 1991) and then moved northwards during the Ordovician whilst undergoing an anti-clockwise rotation of approximately 170°. If this scenario is valid, then it implies that Baltica was inverted with respect to its present-day position, i.e., its present-day southern margin pointed towards the northwest during the Ordovician (Fig. 3A). This, therefore, places the warm water carbonates of present-day southern Scandinavia at lower paleolatitudes (40°–50°) than the present-day northeastern margin, which was close to 70°S and faced Gondwana in early Ordovician times. Since the northern margin of Gondwana itself was up to 30° further south than this (Sect. 3.1.1), this reconstruction allows for a separation of 2000–3000 km between Baltica and Gondwana (Fig. 3A). By late Ordovician times the separation between Baltica and Gondwana and increased to approximately 6000 km (Fig. 3B) due to the northward drift of Baltica.

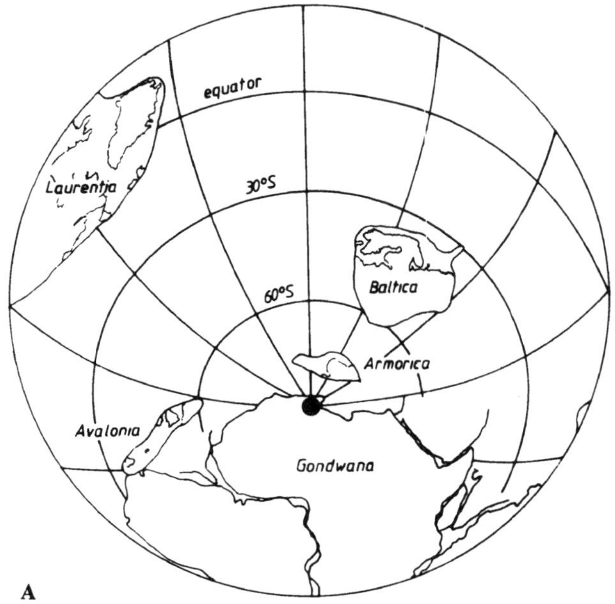

A

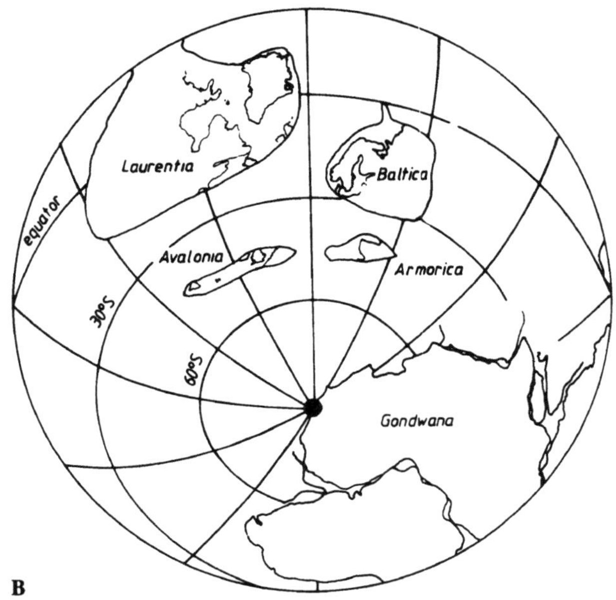

B

Fig. 3A,B. Early **A** and late Ordovician **B** paleogeographic reconstructions of the Atlantic-Bordering Continents

3.1.3 Armorica

Paleomagnetic evidence, mostly from the Armorican Massif (Duff 1979; Hagstrum et al. 1979; Perigo et al. 1983; Perroud and Van der Voo 1985; Perroud et al. 1986; Ruffet et al. 1990), is in good agreement with paleontological and lithological indicators that Armorica occupied high latitudes during the Ordovician, and probably formed part of Gondwana (Fig. 3A). Structural evidence, such as the pronounced "Cadomian unconformity" in Armorica, which is thought to be coeval to the latest Pan-African unconformity identified in northern Africa, lends further support to this reconstruction. Paleomagnetic data from the eastern continuation(?) of Armorica (the Bohemian Massif) are still sparse and ambiguous. Studies of Ordovician sediments and volcanic rocks from Bohemia (Krs et al. 1986) yielded substantially shallower paleolatitudes for Bohemia than would be predicted on the basis of the Armorican data-set. They are also in apparent contrast the numerous lithological and paleontological indicators which suggest high peri-polar latitudes (Storch 1990). Paleomagnetic data from the western margin of the Bohemian Massif (Böhm, 1989) and the Barrandian Basin (Tait et al. 1994a), however, yield steep inclinations in excess of 70°, and, therefore, confirm the suspicion that the earlier data from Bohemia reflect secondary palaeomagnetic overprints rather than primary, Ordovician magnetizations.

3.1.4 Eastern Avalonia

During early Ordovician times (Fig. 3A), when, from faunal evidence, the Iapetus Ocean is believed to have reached its maximum width (McKerrow and Cocks 1976, 1986), lithofacies indicators in Eastern Avalonian and Armorican sediments suggest that there was a link between these two units (Noblet and Lefort 1990). This has been substantiated by recent paleomagnetic data by Torsvik and Trench (1992), who report paleolatitudes in excess of 60°S for early Ordovician rocks in South Wales. Comparing this with results from coeval rocks in Scotland (part of Laurentia), a separation of some 5000 km between Laurentia and Eastern Avalonia during early Ordovician times is indicated (Torsvik and Trench 1992). Paleomagnetic data from mid-Ordovician volcanic rocks of Wales position Eastern Avalonia at paleolatitudes of 51°S (McCabe and Channell 1990). In contrast to this, Trench and Torsvik (1991a) report paleomagnetic directions for similar aged rocks from Wales which correspond to significantly lower paleolatitudes of around 40°S.

However, paleomagnetic data from upper Ordovician age rocks place Eastern Avalonia at 35–40°S (Trench and Torsvik 1991a), thus narrowing both the Iapetus and Tornquist Oceans to approximately 1500 km (Fig. 3B).

3.2 Silurian

3.2.1 Gondwana

As mentioned above, there still exists some debate as to the latitudinal movement of Gondwana between Ordovician and Carboniferous times. If the complex motion of the paleo-south pole over Gondwana (Fig. 2B) is interpreted in its geodynamic context, then this implies the closure of a wide ocean between Baltica/Laurentia and Gondwana by Silurian times (Fig. 4A) with subsequent re-opening of this ocean during the early Devonian (Fig. 5A). From the late Devonian onward, this ocean then closed again and the classic Pangea assemblage was established in the Carboniferous.

This model demands high drift rates for Gondwana, in excess of 25 cm/a during the Silurian (Bachtadse and Briden 1991). It is obvious, however, that such a scenario depends on (a) the geodynamic evolution of the Australian Lachlan Fold Belt, from where all the Earliest Devonian data supporting the loop have been derived and (b) the exact age and timing of magnetization of the Aïr pole (Hargraves et al. 1987).

The alternative path for Gondwana as proposed by Bachtadse and Briden (1991) and shown in Fig. 2C suggests steady northward movement of Gondwana until the Carboniferous. The minor loop that they then propose is consistent with anti-clockwise rotation of Africa during the Carboniferous, a motion pattern in good agreement with geologic evidence (Arthaud and Matte 1977). These two alternative paleogeographic reconstructions for Silurian times (Fig. 4A,B) illustrate the main differences in the APW paths for Gondwana shown in Figs. 2B and C. It has to be pointed out, however, that much more defini-

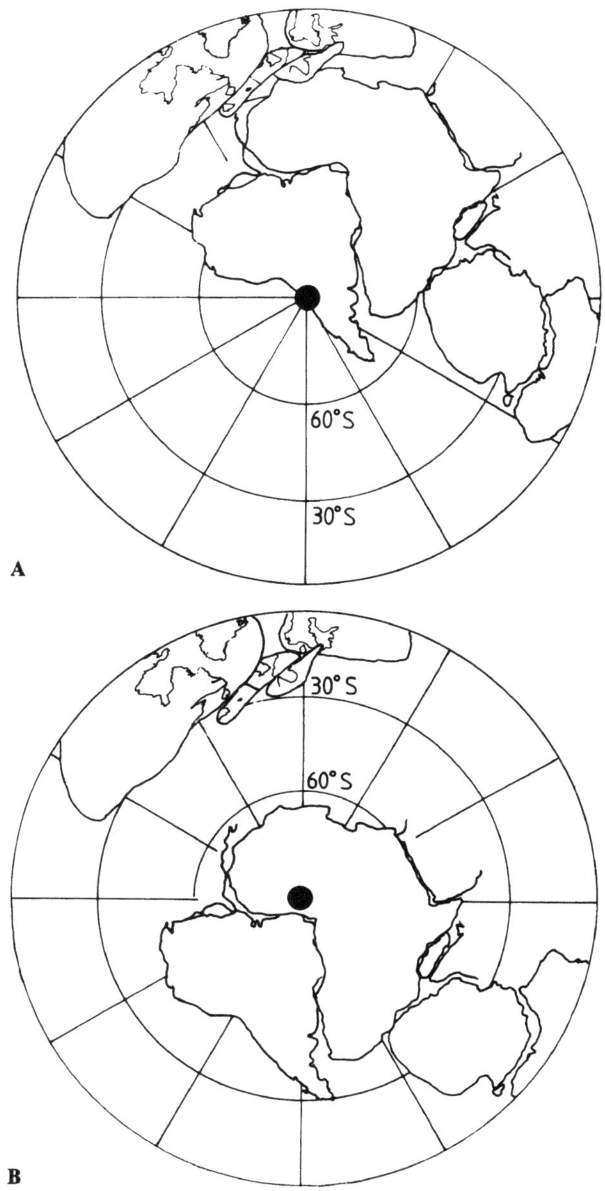

Fig. 4A,B. Paleogeographic reconstructions of the Atlantic Bordering Continents in mid-Silurian times using both the Y-type **A** and X-type **B** paths for Gondwana tive high-quality data from all Gondwana continents are needed, before a robust APW path for Gondwana, and hence a reliable paleogeographic reconstruction for Silurian times can be presented.

3.2.2 Baltica

By late Ordovician times, warm water carbonates, including reefs (Webby 1984) document Baltica's approach to temperate climates, and by Silurian times paleomagnetic poles indicate an equatorial position for Baltica (Fig. 4). Mid-Silurian paleopoles from Baltica and Laurentia are comparable, hence indicating the convergence and collision of these two continents (Scandian Orogeny) at that time (Torsvik et al. 1990a; 1992).

3.2.3 Eastern Avalonia

Only a few paleopoles from S Britain are available to constrain the movement of Eastern Avalonia during the Silurian. Thus, the APW path (Fig. 1) is essentially constructed by interpolation between mid-Ordovician poles and "Siluro-Devonian corner" poles (Trench and Torsvik 1991a). The Somerset and Gloucester pole of Piper (1975) is the only reported mid-Silurian paleopole, and its position is in agreement with the typical "Siluro-Devonian corner" poles identified elsewhere in Eastern Avalonia. However, Torsvik et al. (1993) have recently questioned this pole position. They have obtained new paleomagnetic data from the Mendip Hills in SW England which suggest that Eastern Avalonia was comparable paleolatitudes to Baltica in mid-Silurian times, as portrayed in Fig. 4. This indicates, therefore, that the final convergence between Eastern Avalonia and Baltica occurred in the late Ordovician (ca. 450–440 Ma), i.e., prior to the collision between Baltica and Laurentia (the Scandian Orogeny, ca. 425 Ma). A more detailed description and discussion of the Silurian paleogeography is given in Torsvik et al. (1993).

3.2.4 Armorica

As for Gondwana, the paleogeographic location of Armorica during Silurian times is unclear. The only data available for the Silurian of Hercynian Europe are two pole positions from the Iberian peninsula. These results from the San Pedro red beds (Perroud and Bonhommet 1984) and the Almaden volcanic rocks (Perroud et al. 1991), both yielding rather shallow inclinations. If the Gondwana path Y is adopted, then these results imply coherence of Armorica with Gondwana (Fig. 4A), as postulated by Robardet et al. (1990) based on paleontological grounds. If on the other hand, Gondwana's path X is adopted, then the results indicate that Armorica had rifted away from the northern margin of Gondwana and had migrated northwards as an independent tectonic unit (Fig. 4B).

However, the Almaden pole is rather similar to results from late Devonian rocks from the center of the Almaden syncline as well as to underlying volcanic rocks of Ordovician age. Thus, although the results pass a fold test, their reliability as a primary magnetization can be argued (Parés and Van der Voo 1992). The San Pedro results also pass a fold test and, therefore, predate the late-Carboniferous folding. However, a pre- or syn-folding remagnetization of both the Almaden volcanic rocks as well as of the San Pedro red beds cannot be ruled out (Parés and Van der Voo 1992; Van der Voo 1993).

Recent results from carbonate and volcanic rocks sampled at the Siluro-Devonian boundary in the Berounka Valley, (Barrandian Basin, Czech Republic) yield northeasterly and intermediate paleomagnetic directions (Tait et al. 1994b). These pass both a fold and a reversal test, and are thus interpreted as having been acquired during or shortly after deposition. These directions describe a paleoposition for the Bohemian Massif at about 24°S by Siluro-Devonian times and are used for the palaeogeographic reconstructions shown in Fig. 4A and B. However, whether the new Siluro-Devonian data from the Bohemian Massif are representative for the whole of Armorica (sensu stricto), or whether we have to modify the Armorica concept and imply the existence of several smaller lithospheric units, is still unclear.

3.3 Devonian

3.3.1 Gondwana

The central African location of the late Devonian paleo-south pole which was based on the results of Hailwood (1974) has been questioned repeatedly, mainly on the basis of paleoclimatological (Wendt 1985; Scotese and Barrett 1990) and radiometric evidence (Salmon et al. 1986). However, more recently, high-quality paleomagnetic data for the late Devonian of cratonic Australia (Hurley and Van der Voo 1987), northern (Bachtadse and Briden 1991) and southern Africa (Bachtadse et al. 1987) are in strong support of the central African pole position. When the paleolatitudes for the northern coast of Africa, based on this pole, are compared with the paleolatitudes for the southern margin of the Old Red Continent (see Sect. 3.3.2), an ocean of up to 4500 km in width is indicated for late Devonian times (Fig. 5A).

3.3.2 Baltica

Paleomagnetic data for the Devonian of Baltica are still rather sparse. However, early Devonian poles from the Ukraine (Smethurst and Khramov 1992) and Spitsbergen are comparable to the "Siluro-Devonian corner" poles (Torsvik et al. 1992). For mid- and late Devonian times there are very few reliable paleomagnetic poles for Baltica, as most results have been shown to comprise overprints (Torsvik et al. 1990b). However, as discussed in Section 3.2.2, there is sufficient evidence that Baltica and collided with Laurentia by the Silurian (Torsvik et al. 1990a), thus closing the intervening Iapetus Ocean and forming the Old Red Continent (Fig. 5A).

3.3.3 Eastern Avalonia

Comparing the Devonian paleomagnetic data for Eastern Avalonia and the resulting APW path (see review by Torsvik et al. 1990a) to the one for Laurentia (Van der Voo 1993) shows that the two paths converge during Silurian times (Fig. 1), thus closing the British sector of the Iapetus Ocean, as shown in Fig. 5A.

3.3.4 Armorica

Unfortunately, the paleomagnetic data-set for Armorica during the Devonian is equally as sparse as the one for the Silurian. So far, only two pole positions have been published for late Devonian rocks from the Central European Harz Mountains and the Franconian Forest (Bachtadse et al. 1983).

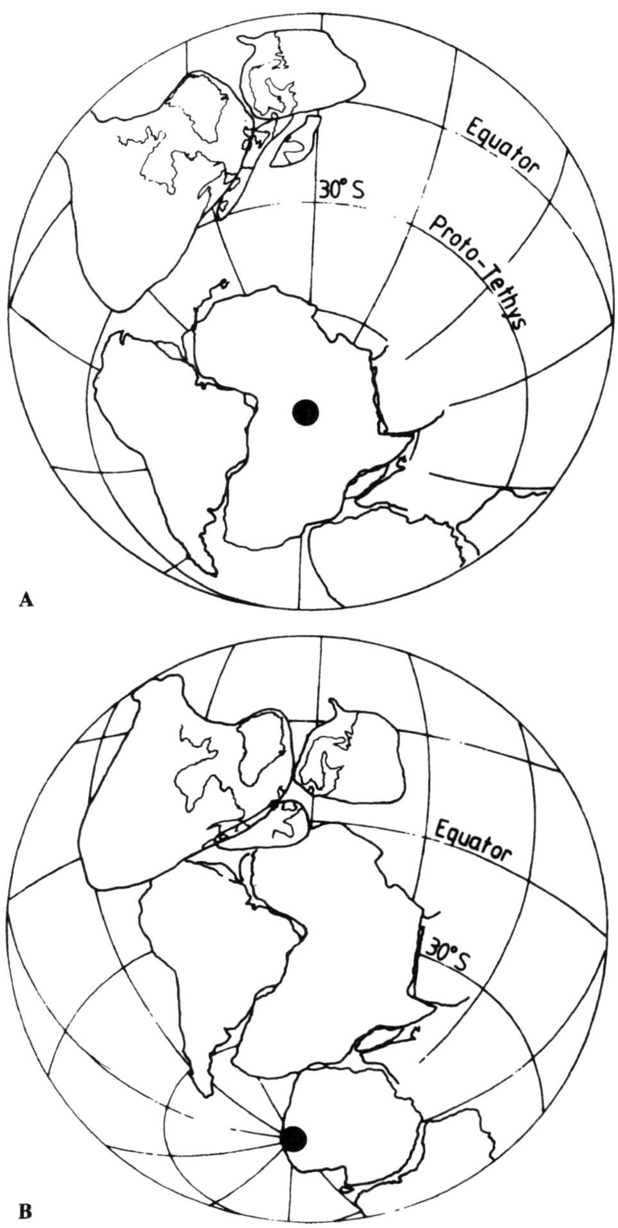

Fig. 5A,B. Late Devonian **A** and Permo-Carboniferous **B** paleogeographic reconstructions of the Atlantic Bordering Continents

These pass fold tests and are believed to represent primary magnetizations. The resulting paleolatitudes of 13°S indicate the proximity of Armorica to Baltica (Fig. 5A) and the closure of any intervening ocean. This is in agreement with data from the Devonian-age Montmartin red beds in Britanny (Jones et al. 1979), which indicate paleolatitudes for Armorica similar to the ones from the Harz and Franconian Forest. However, these results have since been shown to fail the fold test

(Perroud et al. 1985) and, therefore, postdate Carboniferous folding. In contrast to this, data from the Massif Central yield significantly steeper inclinations, implying still some 2000 km of separation between Armorica and the Old Red Continent (Edel 1987). However, in the absence of unambiguous structural control, the paleogeographic significance of the Massif Central results remain debatable and the reconstruction in Fig. 5A is based on the data of Bachtadse et al. (1983).

3.4 Carboniferous

In marked contrast to the sparse paleomagnetic data-set for Devonian and Silurian times, large numbers of Carboniferous paleomagnetic pole positions are available for all the major continents as well as from the microplates, such as Armorica and Eastern Avalonia. Although most of the Carboniferous data from Armorica and Eastern Avalonia are based on directions ascribed to Hercynian remagnetization (Bachtadse and Van der Voo 1986; Perroud 1986), the similarity between Armorican, Baltic, and Laurentian pole positions is evident (Fig. 1). This indicates, therefore, the proximity of Armorica to the Old Red Continent at this time. The same is also true for Gondwana; the paleolatitude of its northern margin was about 20°S, thus reducing the width of the intervening ocean to less then 1000 km by Carboniferous times (Fig. 5B).

Plotting the European Carboniferous and Devonian declination data on the map, a systematic pattern of declination deviation becomes apparent, which is most pronounced around the Bay of Biscay, the Ibero-Armorican arc (Fig. 6). These deviations have been identified as being a direct function of the change in general strike along the northern margin of the Hercynian mountain belt, as discussed by Perroud (1986), Bachtadse and Van der Voo (1986), and Bachtadse (1990). One possible interpretation of this declination pattern is that it reflects large-scale thrust rotations (especially around the Ibero-Armorican Arc, the Rhenish Massif and the Ardennes) and/or pervasive strike-slip tectonics in the more internal parts of the orogen (Harz Mountains, Franconian Forest, Massif Central). Indentation of the irregularly shaped northern margin of Gondwana

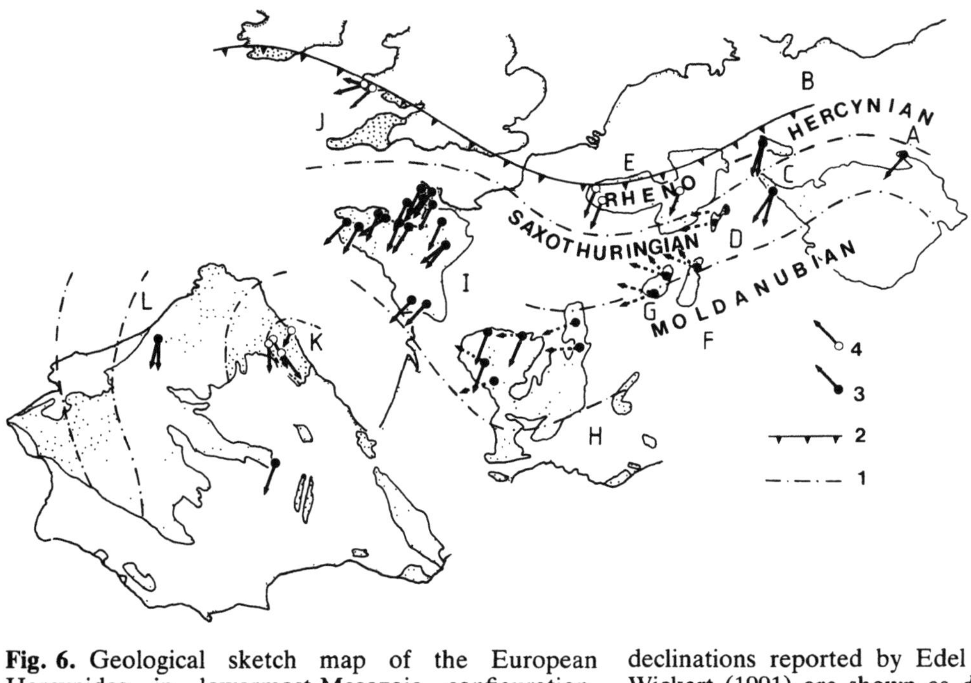

Fig. 6. Geological sketch map of the European Hercynides in lowermost-Mesozoic configuration. Shown are the structural trend *1*, the northern overthrust *2*, and the Devonian and Carboniferous paleomagnetic declinations for autochthonous *3* and allochthonous *4* regions of the orogen. Paleomagnetic declinations reported by Edel (1987) and Edel and Wickert (1991) are shown as *dashed arrows*. *A* Holy Cross Mts; *B* Harz Mts; *C* Franconian Forest; *D* Odenwald/Spessart; *E* Rhenish Massif/Ardennes; *F* Black Forest; *G* Vosges; *H* Massif Central; *I* Armorican Massif; *J* South Wales; *K* Cantabria; *L* Galicia-Castilla

into Europe during the final stages of suturing between Gondwana and the northern continents is one possible explanation for this deformation pattern (Bachtadse and Van der Voo 1986).

A different interpretation of this late Paleozoic declination pattern has been brought forward by Edel (1987) and Edel and Wickert (1991). They reported northwesterly directions of magnetization (termed the C-direction by Edel 1987) in the Massif Central, the Vosges, the Ardennes, the Odenwald, and the Spessart in various rock types of earliest Carboniferous age. These directions differ significantly from the ones expected for Europe based on the Armorican APW path as depicted in Fig. 1 by up to 50° in declination. In order to explain these magnetic directions, Edel (1987) and Edel and Wickert (1991) postulated rather complex rotational movements [90° anti(?)-clockwise, followed by 40° clockwise] of Hercynian Europe during the Carboniferous. The fact, however, that the C-directions display inclination values between +40 and −40° as in the Massif Central (Edel 1987) or are only poorly defined as in the Odenwald (see Fig. 2 in Edel and Wickert 1991) leads us to doubt the validity of Edel and Wickert's (1991) conclusion until more high quality paleomagnetic data for the internal regions of the Hercynian belt become available.

4 Conclusion

Despite the fact that some details of the paleogeographic evolution of the Atlantic Bordering Continents are still controversial, such as the precise drift history of Gondwana during the Silurian, a rather clear picture of the overall paleogeographic, and by inference, paleotectonic evolution during Paleozoic times has emerged during the past few years. All the paleomagnetic data discussed in the previous sections are summarized in Fig. 7. This figure shows the paleolatitudes for each unit versus time and clearly illustrates their interaction during the Paleozoic.

There is very good paleomagnetic evidence to position the northern margin of Gondwana at very high latitudes during the early Ordovician and also to infer that Western Avalonia, Eastern Avalonia, and Armorica (including the Bohemian Massif) were also close to, or attached to, Gondwana. Sometime during the Ordovician, Baltica commenced its northward drift accompanied by an anti-clockwise rotational motion. Eastern Avalonia seems to have rifted off Gondwana during the Ordovician, whereas the timing of breakoff for Armorica is not yet known in detail. Laurentia remained more or less fixed in an equatorial position throughout Paleozoic times (Kent and Van der Voo 1990).

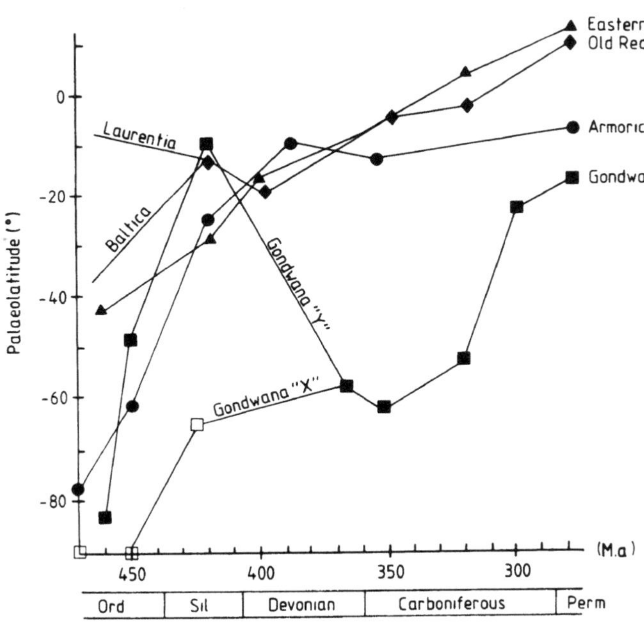

Fig. 7. Paleolatitude versus time for given localities in central England (Eastern Avalonia), central Europe (Armorica), northern Newfoundland (Laurentia), and northern Africa (Gondwana). Note that the paleolatitudes for Gondwana have been calculated using the apparent polar wander paths shown in Fig. 2B and C

The geodynamic scenario gains some complexity during the Silurian, mainly because of the different interpretations of the Gondwana APW path. If one accepts the cluster of Silurian and earliest Devonian pole positions to the northwest of Chile, then an almost equatorial position of the northern margin of Gondwana during Silurian times has to be implied. This position also implies collision with Laurentia during the Silurian, giving rise to the Acadian orogeny in North America (Kent and Van der Voo 1990). Alternatively, Gondwana did not collide with the northern continents until the Carboniferous (Bachtadse and Briden 1991).

New data from the Bohemian Massif suggest that this part of Europe had already traveled across the Tornquist Ocean by the end of the Silurian and had reached close proximity to Baltica. Thus there is good evidence that by Siluro-Devonian times Armorica and Eastern Avalonia were in close proximity, and the Iapetus ocean between Baltica and North America had already closed. This is similar to the model proposed for the Armorican microplate by Perroud et al. (1984), but more reliable data from the various European Hercynian Massifs are needed to elucidate details of their drift history, and thus to conclusively demonstrate the coherence of Armorica during the Paleozoic.

The final suturing between the consolidated northern continents and Gondwana, and the closure of the Rheic Ocean did not occur until the end of the Carboniferous.

Acknowledgments. The German Research Council (DFG) provided financial support to VB, JAT and HCS (SFB69, Fr668/6-2). THT acknowledges support through the Norwegian Research Council for the Humanities and Sciences (NAVF) and the Norwegian Geological Survey (NGU). This chapter benefited greatly from thorough reviews by Rob Van der Voo and Jean Bernard Edel, which helped to improve the manuscript and to clarify several points. F. Maier read an earlier version of this manuscript and his valuable comments are gratefully acknowledged.

References

Aïfa T, Feinberg H, Pozzi JP (1990) Devonian/Carboniferous palaeopoles for Africa. Consequences for Hercynian geodynamics. Tectonophysics 179: 288–304

Arthaud F, Matte P (1977) late Paleozoic strike slip faulting in southern Europe and northern Africa: result of a right-lateral shear zone between the Appalachians and the Urals. Geol Soc Am Bull 88: 1305–1320

Bachtadse V (1990) Hercynian palaeomagnetism of Europe: arguments for large scale complex thin-skinned and thick-skinned rotations. Cuad Geol Iberica 12: 147–162

Bachtadse V, Briden JC (1990) Paleomagnetic constraints on the position of Gondwana during Ordovician to Devonian times. In: McKerrow WS, Scotese CR (eds) Palaeozoic palaeogeography and biogeography. Geol Soc Mem 12: 43–48

Bachtadse V, Briden JC (1991) Palaeomagnetism of Devonian Ring Complexes from the Bayuda Desert, Sudan – new constraints on Gondwanas apparent polar wander path. Geophys J Int 104: 635–646

Bachtadse V, Van Der Voo R (1986) Paleomagnetic evidence for crustal and thin-skinned rotations in the European Hercynides. Geophys Res Lett 13: 161–164

Bachtadse V, Heller F, Kröner A (1983) Palaeomagnetic investigations in the Hercynian mountain belt of Central Europe. Tectonophysics 91: 285–299

Bachtadse V, Van der Voo R, Haelbich IW (1987) Paleomagnetism of the Western Cape Fold belt, South Africa, and its bearing on the Paleozoic apparent polar wander path for Gondwana. Earth Planet Sci Lett 84: 487–499

Böhm V (1989) Vergleichende paläomagnetische Untersuchungen an verschiedenen Gesteinen des mitteleuropäischen Variszikums. PhD Thesis, Ludwig-Maximilians-Universität, München

Boeuf S, Biju-Duval B, De Charpal O, Rognon P, Gariel O, Bennacef A (1971) Les grès du Paleozoïque inférieur au Sahara. Sedimentation et discontinuité. Évolution structurale d'un craton. Technip, Paris

Boucot AJ (1990) Silurian biogeography. In: McKerrow WS, Scotese CR (eds) Palaeozoic palaeogeography and biogeography. Geol Soc Mem 12: 191–196

Bullard EC, Everett J, Smith AG (1965) The fit of the continents around the Atlantic. Phil Trans R Soc Lond A258: 41–45

Cocks LRM, Fortey RA (1982) Faunal evidence for oceanic separations in the Palaeozoic of Britain. J Geol Soc Lond 139: 465–478

Duff BA (1979) The palaeomagnetism of Cambro-Ordovician redbeds, the Erquy Spilite Series, and the Trégastel-Ploumanac'h granite complex, Armorican Massif (France and the Channel Islands). Geophys J R Astron Soc 59: 345–365

Edel JB (1987) Palaeopositions of the western Europe Hercynides during the late Carboniferous deduced from palaeomagnetic data: consequences for "stable" Europe. Tectonophysics 139: 31–42

Edel JB, Wickert F (1991) Palaeopositions of the Saxothuringian (Northern Vosges, Pfalz, Odenwald, Spessart) in Variscan times: palaeomagnetic investigations. Earth Planet Sci Lett 103: 10–26

Hagstrum JT, Van der Voo R, Auvray B, Bonhommet N (1979) Eocambrian-Cambrian palaeomagnetism of the Armorican Massif, France. Geophys J R Astron Soc 61: 489–517

Hailwood EA (1974) Palaeomagnetism of the Msissi Norite (Morocco) and the Palaeozoic reconstruction

of Gondwanaland. Earth Planet Sci Lett 23: 376–386
Hargraves RB, Dawson EM, Van Houten FB (1987) Palaeomagnetism and age of mid Palaeozoic ring complexes in Niger, West Africa, and tectonic implications. Geophys J R Astron Soc 90: 705–729
Hurley NF, Van der Voo R (1987) Paleomagnetism of Upper Devonian reefal limestones, Canning Basin, Western Australia. Geol Soc Am Bull 98: 138–146
Jones M, Van der Voo R, Bonhommet N (1979) late Devonian to early Carboniferous palaeomagnetic poles from the Armorican Massif, France. Geophys J R Astron Soc 58: 287–308
Jupp PE, Kent JT (1987) Fitting smooth paths to spherical data. Appl Stat 36: 34–36
Kent DV, Van der Voo R (1990) Palaeozoic palaeogeography from palaeomagnetism of the Atlantic bordering continents. In: McKerrow WS, Scotese CR (eds) Palaeozoic Palaeogeography and Biogeography. Geol Soc Mem 12: 49–56
Krs M, Krsova M, Pruner P, Havlicek V (1986) Paleomagnetism, paleogeography and multicomponent analysis of magnetization of Ordovician rocks from the Barrandian area of the Bohemian Massif. Sb Geol Věd 20: 9–45
Li ZX, Schmidt PW, Embleton BJJ (1988) Palaeomagnetism of the Hervey Group, central New South Wales, and its tectonic implications. Tectonics 7: 351–367
Li ZX, Powell CMcA, Thrupp GA (1990) Australian Palaeozoic palaeomagnetism and tectonics – II. A revised apparent wander path and palaeogeography. J Struct Geol 12: 567–575
Li ZX, Powell CMcA, Embleton BJJ, Schmidt PW (1991) New palaeomagnetic results from the Amadeus Basin and their implications for stratigraphy and tectonics. In: Korsch RJ, Kennard JM (eds) Geological and geophysical studies in the Amadeus Basin, Central Australia. Bur Miner Resour Aust, pp 349–360
Livermore RA, Smith AG, Briden JC (1985) Palaeomagnetic constraints on the distribution of continents in the late Silurian and early Devonian. Phil Trans R Soc Lond B 309: 29–56
McCabe C, Channell JET (1990) Paleomagnetic results from volcanic rocks of the Shelve inlier, Wales: evidence for a wide late Ordovician Iapetus Ocean in Britain. Earth Planet Sci Lett 96: 458–469
McElhinny MW, Briden JC, Jones DL, Brock A (1968) Geological and geophysical implications of palaeomagnetic results from Africa. Rev Geophys 6: 201–238
McKerrow WS, Cocks LRM (1976) Progressive faunal migration across the Iapetus Ocean. Nature 143: 237–239
McKerrow WS, Cocks LRM (1986) Ocean, island arcs and olistostromes: the use of fossils in distinguishing sutures, terranes and environments around the Iapetus Ocean. J Geol Soc Lond 143: 185–191
Morel P, Irving E (1978) Tentative paleocontinental maps for the early Phanerozoic and Proterozoic. J Geol 86: 535–561
Noblet Ch, Lefort JP (1990) Sedimentological evidence for a limited separation between Armorica and Gondwana during the early Ordovician. Geology 18: 303–306
Parés JM, Van der Voo R (1992) Paleozoic paleomagnetism of Almaden, Spain: a cautionary note. J Geophys Res 97: 9353–9356
Perigo R, Van der Voo R, Auvray B, Bonhommet N (1983) Paleomagnetism of late Precambrian-Cambrian volcanics and intrusions from the Armorican Massif, France. Geophys J R Astron Soc 75: 235–260
Perroud H (1986) Paleomagnetic evidence for tectonic rotations in the Variscan Mountain Belt. Tectonics 5: 205–214
Perroud H, Bonhommet N (1984) A Devonian pole for Armorica. Geophys J R Astron Soc 77: 839–845
Perroud H, Van der Voo R (1985) Palaeomagnetism of the late Ordovician Thouars Massif, Vendée province, France. J Geophys Res 90: 4611–4625
Perroud H, Van der Voo R, Bonhommet N (1984) Palaeozoic geodynamic evolution of the Armorican plate on the basis of paleomagnetic data. Geology 12: 579–582
Perroud H, Robardet M, Van der Voo R, Bonhommet N, Paris F (1985) Revision of the age of magnetization of the Montmartin red beds, Normandy, France. Geophys J R Astron Soc 80: 541–549
Perroud H, Bonhommet N, Thebault JP (1986) Palaeomagnetism of the Ordovician Moulin de Chateaupanne formation, Vendée, western France. Geophys J R Astron Soc 85: 573–582
Perroud H, Calza F, Khattach (1991) Palaeomagnetism of the Silurian volcanism at Almaden, southern Spain. J Geophys Res 96: 1949–1962
Perroud H, Robardet M, Bruton DL (1992) Palaeomagnetic constraints upon the palaeogeographic position of the Baltic Shield in the Ordovician. Tectonophysics 201: 97–120
Piper JDA (1975) Palaeomagnetism of Silurian lavas from Somerset and Gloucestershire, England. Earth Planet Sci Lett 25: 355–360
Robardet M, Paris F, Racheboeuf PR (1990) Palaeogeographic evolution of southwestern Europe during early Palaeozoic times. In: McKerrow WS, Scotese CR (eds) Palaeozoic Palaeogeography and Biogeography. Geol Soc Mem 12: 411–420
Ruffet G, Perroud H, Féraud H (1990) $^{40}Ar/^{32}9Ar$ dating of a late Proterozoic palaeomagnetic pole for the Armorican Massif (France). Geophys J Int 102: 397–409
Salmon E, Montigny R, Edel JB, Pique A, Thuizat R, Westphal M (1986) The Msissi Norite revisited: K/Ar dating, Petrography and Paleomagnetism. Geophys Res Lett 13: 741–743
Scotese CR, Barrett SF (1990) Gondwana's movement over the South Pole during the Paleozoic: Evidence from lithologic indicators of climate. In: McKerrow S, Scotese CR (eds) Palaeozoic palaeogeography and biogeography. Geol Soc Mem 12: 75–86
Scotese CR, McKerrow WS (1990) Revised World maps and introduction. In: McKerrow WS, Scotese

CR (eds) Palaeozoic palaeogeography and biogeography. Geol Soc Mem 12: 1–21

Smethurst MA, Khramov AN (1992) A new Devonian palaeomagnetic pole for the Russian platform and Baltica, and related apparent polar wander. Geophys J Int 108: 179–192

Spjeldnaes N (1961) Ordovician climatic zones. Nor Geol Tidsskr 41: 45–77

Storch A (1990) Upper Ordovician-lower Silurian sequences of the Bohemian Massif, central Europe. Geol Mag 127: 225–239

Tait J, Bachtadse V, Soffel H (1994a) New palaeomagnetic constraints on the position of central Bohemia during Early Ordovician times. Geophys J Int 116: 131–140

Tait J, Bachtadse V, Soffel H (1994b) Silurian paleogography of Armorica: new paleomagnetic data from central Bohemia. J Geophys Res 99: 2897–2907

Thrupp GA, Kent DV, Schmidt PW, Powell CMcA (1991) Palaeomagnetism of red beds of the late Devonian Worangie Point Formation, SE Australia. Geophys J Int 104: 179–201

Torsvik TH, Trench A (1991) The lower-middle Ordovician palaeofield of Scandinavia: southern Sweden "revisited". Phys Earth Planet Interiors 65: 283–291

Torsvik TH, Smethurst MA, Briden JC, Sturt BA (1990a) A review of Palaeozoic palaeomagnetic data from Europe and their palaeogeographical implications. In: McKerrow WS, Scotese CR (eds) Palaeozoic palaeogeography and biogeography. Geol Soc Mem 12: 25–41

Torsvik TH, Oleson O, Ryan PD, Trench A (1990b) On the palaeogeography of Baltica during the Palaeozoic: new palaeomagnetic data from the Scandinavian Caledonides. Geophys J Int 103: 261–279

Torsvik TH, Smethurst MA, Van der Voo R, Trench A, Abrahamsen N, Halvorsen E (1992) BALTICA – a synopsis of Vendian-Permian palaeomagnetic data and their palaeotectonic implications. Earth Sci Rev 33: 133–152

Torsvik TH, Trench A, Svensson I, Walderhaug HJ (1993) Palaeogeographic significance of mid Silurian palaeomagnetic results from southern Britain – major revision of the Apparent Polar Wander Path for eastern Avalonia. Geophys J Int 113: 651–668

Trench A, Torsvik TH (1991a) A revised Palaeozoic apparent polar wander path for Southern Britain (Eastern Avalonia). Geophys J Int 104: 227–233

Trench A, Torsvik TH (1991b) The lower Palaeozoic apparent polar wander path for Baltica: palaeomagnetic data from Silurian limestones of Gotland, Sweden. Geophys J Int 107: 373–379

Van der Voo R (1979) Paleozoic assembly of Pangea: a new plate tectonic model for the Taconic, Caledonian and Hercynian orogenies. EOS Trans Am Geophys Union 60: 241

Van der Voo R (1988) Paleozoic paleogeography of North America, Gondwana and intervening displaced terranes: comparison of palaeomagnetism with palaeoclimatology and biogeographical patterns. Bull Geol Soc Am 100: 311–324

Van der Voo R (1990) Palaeozoic palaeogeography from palaeomagnetism of the Atlantic-bordering continents. In: McKerrow WS, Scotese CR (eds) Palaeozoic palaeogeography and biogeography. Geol Soc Mem 12: 49–56

Van der Voo R (1993) Paleomagnetism of the Atlantic, Tethys and Iapetus Oceans. Cambridge Univ Press, New York, 411 pages

Van der Voo R, Peinado J, Scotese CR (1984) A paleomagnetic evaluation of Pangea reconstructions. In: Van der Voo R, Scotese CR, Bonhommet N (eds) Plate reconstructions from Palaeozoic palaeomagnetism. Geodyn Ser, AGU, Washington, pp 11–26

Webby BD (1984) Ordovician reefs and climates: a review. In: Bruton DL (ed) Aspects of the Ordovician system. Palaeontol Contrib Univ Oslo, pp 89–100

Wendt J (1985) Disintegration of the continental margin of northwestern Gondwana: late Devonian of the eastern Anti-Atlas (Morocco). Geology 13: 815–818

Whittington HB, Hughes CP (1972) Ordovician geography and faunal provinces deduced from Trilobite distribution. Phil Trans R Soc Lond 263: 235–278

XI Geodynamic Evolution

W. FRANKE, R.D. DALLMEYER, and K. WEBER

1 Introduction

Geodynamic models may be formulated by assessing large-scale plate motions through paleomagnetic, biogeographic, and climatological records. Although these methods help constrain large-scale relationships, they are fairly inaccurate in detail. Consideration of field relationships allows more precise location of individual tectonic boundaries, and helps to resolve crustal thickness, direction of subduction, chronology of collision, and the overall kinematic evolution of collisional stages. This approach is inherently biased by enhanced preservation of later orogenic events (collisional stage), and therefore fails to constrain the actual amounts of plate tectonic transport. These contrasting approaches reflect different scales of observation, and are seldom discussed in a synoptical evaluation. Summarizing the foregoing chapters of this book offers a unique opportunity for a thorough assessment of various geodynamic models for the central European orogens.

2 Tectonostratigraphic Units Defined by Field Relationships

Essentially all of Variscan Europe is composed of contrasting, fault-bounded tectonic units which initially experienced a late Proterozoic (Cadomian or Panafrican) tectonothermal development. This indicates that Gondwana initially extended northwards of the Tornquist-Teysseyre Zone and the lapetus suture of the British Isles (Fig. 1) In spite of differences in the pre-Cadomian record, about which little is known, the Variscan structural units are characterized by tectonic development between the Cambrian and the Carboniferous. It is possible to distinguish several contrasting lithotectonic elements (terranes ?). These will be characterized in order of their inferred palinspastic sequence from north to south.

2.1 East Avalonia

East Avalonia comprises SE England, N France, Belgium, and N Germany (i.e., the autochthonous parts of the Rhenohercynian Zone and the Subvariscan Foredeep: Kossmat 1927). East Avalonia has previously been variably referred to as the northern part of the "mid-European microplate" (Franke 1989a,b), the "Lüneburg Massif" (Berthelsen 1992), or the "London-Brabant Massif" (Ziegler 1990). East Avalonia is largely composed of Cadomian crust which has been detected in eastern England, along southern margins of the Rhenish Massif, and in the Harz Mts. East Avalonia accreted with Laurentia/ Baltica in the late Ordovician/early Silurian, thereby completing assembly of the Old Red Continent.

West of the Rhine river (in the Ardennes, Brabant massif, and SE England), Cadomian basement is unconformably overlain by Cambrian through Silurian sequences. These locally record Silurian (Caledonian) deformation and low- to very low-grade metamorphism. This probably was associated with collision of Avalonia and Baltica. East of the Rhine, Caledonian deformation is absent, with the exception of areas close to the Trans-European-Tornquist Fault Zone (Fig. 2). Throughout the Rhenish Massif, and especially in areas East of the Rhine, pronounced crustal attenuation occurred in the Devonian, with detritus supplied from northern Caledonian sources.

2.2 Rheic and Rhenohercynian Oceans

Crustal extension in the early Devonian led to formation of oceanic crust, which has been pre-

Dallmeyer et al. (Eds.)
Pre-Permian Geology of Central and Eastern Europe

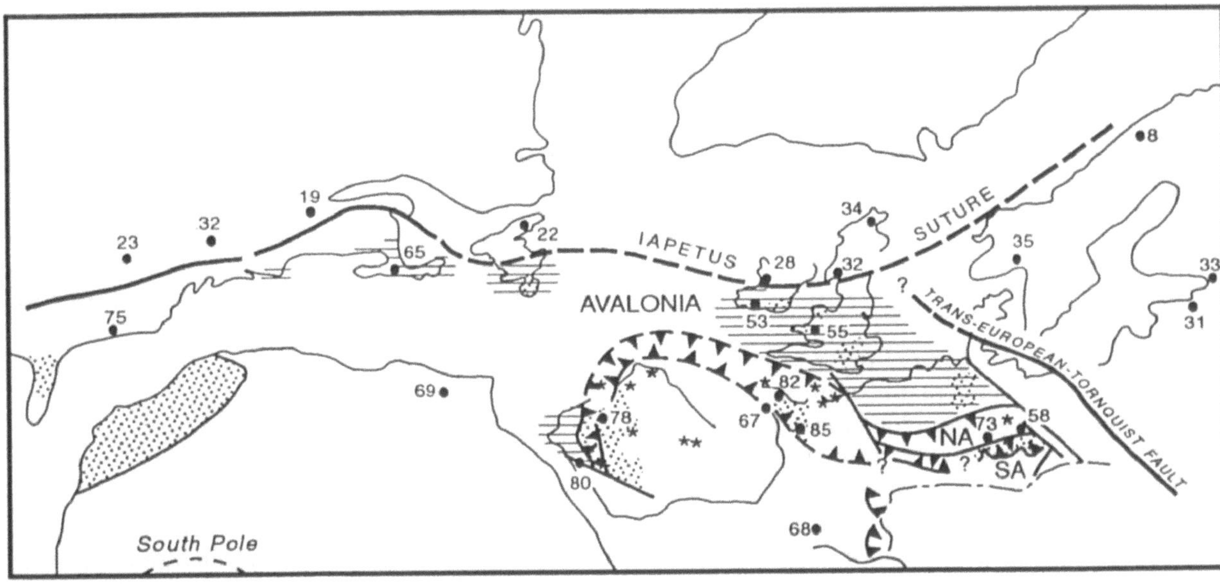

Fig. 1. Permian continental assembly, with the main tectonic subdivision. Kinship of Variscan terranes is highlighted by exposures of Cadomian basement, Ordovician paleomagnetic inclinations, and the traces of the latest Ordovician "Saharan" glaciation. (Franke 1992)

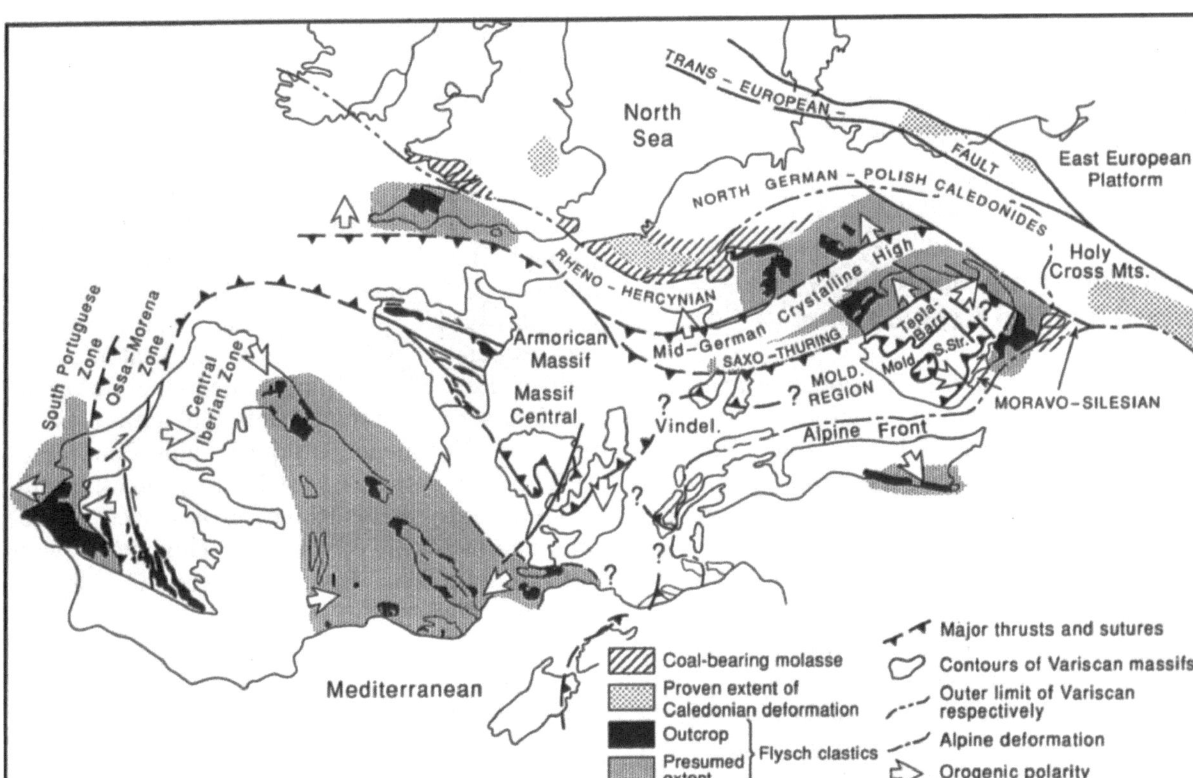

Fig. 2. Structural map of the Variscan Belt in Central and Eastern Europe. (Franke 1992)

served in the Giessen/Harz nappe and in southeasternmost parts of the Northern Phyllite Zone. The suture resulting from the closure of the Devonian (Rhenohercynian) basin is located between the Northern Phyllite Zone and the Mid-German Crystalline High. The Rhenohercynian belt can be traced westwards into SW England and correlates with the South- Portuguese Zone; eastwards, it probably continues into the Moravo-Silesian belt (Fig. 2).

Paleomagnetic evidence and geochemical data from the Taunus segment of the Northern Phyllite Zone suggest that the Devonian ocean most likely had an Ordovician precursor. On the basis of the age of the relevant rocks, and their relative position within the Variscan orogen, the Ordovician ocean probably corresponds to the "Rheic ocean" of Cocks and Fortey (1982, 1990). It is possible that the Rheic ocean closed more or less completely before onset of the Rhenohercynian rifting episode. In this scenario, late Ordovician/Silurian calc-alkaline rocks preserved in the Taunus segment of the Northern Phyllite Zone might represent an island arc which formed during plate convergence between East Avalonia and Armorica. Due to penetrative early Carboniferous tectonic reworking in the early Carboniferous, boundaries of this island arc are difficult to locate. The Northern Phyllite Zone is a tectonic collage, which probably comprises (1) southernmost parts of East Avalonia, (2) remnants of the island arc, and (3) northernmost parts of Armorica. This assembly was tectonically juxtaposed in the early Carboniferous.

2.3 "Armorica"

Tectonic units described below (Mid-German Crystalline High, Saxothuringian basin, Tepla-Barrandian), comprise a plate-tectonic unit which has been referred to as Armorica in recent paleomagnetic reconstructions. However, field evidence reveals that Armorica consisted of at least two different terranes, represented by the Mid-German Crystalline High and Saxothuringian autochthon and the Tepla-Barrandian. The latter two appear to have been separated by a relatively narrow Saxothuringian ocean.

2.3.1 Mid-German Crystalline High

The Mid-German Crystalline High represents the northern margin of Armorica. Present tectonometamorphic characteristics reflect the effects of late Devonian/early Carboniferous convergence, during which it represented a northwest-driving active plate margin (i.e., the southeastern margin of the closing Rhenohercynian basin). The Crystalline High comprises subduction-related plutonic rocks of late Devonian/early Carboniferous age, and acted as a source for synorogenic clastic sediments within the Rhenohercynian basin. Metamorphic events have been documented in the early Carboniferous (Spessart) and in the Devonian to early Carboniferous (Odenwald).

The pre-Devonian history is poorly constrained. Protolith ages are only known from the Spessart segment [Silurian quartzites, orthogneiss (Rotgneis) with a late Ordovician Rb/Sr-whole rock crystallization age]. The "post-collisional" character of Rotgneis meta-granitoids might record subduction of an Ordovician Rheic ocean prior to Devonian rifting. The chemical composition of metamagmatic rocks exposed in the eastern Odenwald (Böllstein) is compatible with this scenario (Altenberger et al. 1990).

A western segment of the Mid-German Crystalline High (basement underlying the Permo-Carboniferous Saar/Nahe basin) records a markedly different development. Middle Devonian-Permian sedimentary rocks unconformably overly crystalline basement (?400 Ma granitoid). Alhough this area has conventionally been included with the Mid-German Crystalline High, it may represent an exotic element emplaced by strike-slip translation.

2.3.2 Saxothuringian Basin

The Saxothuringian basin represents a Cambro-Ordovician rift basin, which probably developed on Cadomian crust. Crustal extension is documented by thick neritic clastic sequences and bimodal volcanic rocks. A deeper-water Bavarian facies has been thrust northwestward over sequences of the neritic/hemipelagic Thuringian realm. Infill of the Saxothuringian basin can be interpreted to have occurred along the southern, passive margin of an unknown tectonic element which was situated northward (basement of the Mid-German Crystalline High?).

Eclogites developed from early Paleozoic (ca. 500 Ma) MORB-type basalts occur within a separate nappe sheet now situated between the Saxothuringian passive margin sequences and the Tepla-Barrandian terrane. They probably represent fragments of Saxothuringian ocean floor.

2.3.3 Tepla-Barrandian (Bohemian)

Folded and unmetamorphosed Cambrian-Middle Devonian sequences unconformably overlie a weakly metamorphosed Cadomian basement. The stratigraphic and tectonic records of the Tepla-Barrandian, as well as its tectonic position in the orogen, are similar to that of the Central Armorican Zone in France.

Deeper structural levels of the Tepla-Barrandian unit are exposed along its northwestern and southwestern margins. These deeper levels record an early Variscan (ca. 380–400 Ma) metamorphism and penetrative cleavage. Bimodal Cambro-Ordovician intrusions are exposed in southwestern sectors of the Tepla-Barrandian (Neukirchen-Kdyne). This probably related to early Paleozoic rifting documented in the Barrandian synform.

It is uncertain whether the Barrandian Paleozoic rocks represent part of a separate basin, or belong to epicratonic, marginal parts of one of the adjacent basins (Saxothuringian or Moldanubian).

2.4 Moldanubian s.str.

2.4.1 Gföhl Unit of the Moldanubian s.str.

The Gföhl unit (sensu Tollmann 1982) consists of various metaigneous and metasedimentary rocks, which include granulites, eclogites, and peridotites. Protolith ages are not well constrained. However, Proterozoic mafic rocks were involved. Gföhl rocks occur in nappe units which overlie lower-grade rocks of the Drosendorf unit. Because nappe units involving high-pressure and mantle rocks usually develop from passive margins, we infer the existence of another major basin (Moldanubian basin), which probably correlated with the French Massif Central Ocean (Tollmann 1982; Matte 1986; Matte et al. 1990).

Wendt et al. (1993) reported U/Pb zircon ages of 2–2.1 Ga from metagranodiorites at the structural base of the Gföhl unit. These are the oldest protoliths described from the Bohemian Massif, and could represent basement to the Gföhl and/or the Drosendorf terrane.

2.4.2 Drosendorf Unit of the Moldanubian s.str.

The Drosendorf unit (sensu Tollmann 1982) consists essentially of fairly monotonous metasedimentary sequences which include quartzites, graphitic quartzites, and carbonates. Paleozoic microfloras have been identified (including one Silurian locality). The presence of pre-Variscan basement rocks has not been proven.

2.5 Moravo-Silesian

The Moldanubian region is bounded on the southeast by the Moldanubian overthrust. This is actually a zone of important dextral strike slip with transtensional and transpressional increments (Schulmann et al. 1991). Thrusting emplaced previously assembled Drosendorf and Gföhl rocks over the Moravian crystalline rocks. The latter consist of a variety of para- and orthogneisses of unknown protolith ages. Lithological comparisons suggest that the Moravian crystalline rocks may represent equivalents of less metamorphosed, autochthonous Silesian sequences exposed to the east.

Silesian sectors of the Moravo-Silesian terrane consist of Cadomian basement (granitoids) overlain by neritic clastic sediments (early Paleozoic through Devonian), carbonates (Devonian), and early Carboniferous flysch. External, southeastern segments are partly concealed beneath Carpathian nappes. Here, the Silesian belt includes an early Carboniferous carbonate platform which was time-equivalent with flysch clastics deposited further west, and a late Carboniferous coal-bearing molasse basin. The Silesian belt also contains Devonian and early Carboniferous mafic volcanic rocks. The depositional evolution of the Silesian belt is comparable with the Rhenohercynian belt in Germany, and suggests a correlation of these areas in an arc structure around eastern parts of the Bohemian Massif (for an alternative view, see Chap. VIII.B.1).

3 Origin and Width of the Major Basins

Relative palinspastic relationships of the major Variscan lithotectonic elements are best constrained for the Cambrian-Ordovician. Paleomagnetic data suggest that all basement units (from Avalonian crust underlying the Rhenish Massif, southwards to Morocco) were spatially related to northern sectors of Gondwana. An important phase of rifting initiated in the Cambrian. This is reflected in the sedimentary record within Barrandian and Saxothuringian sequences. This extensional episode could also have affected the Moldanubian basin, but protolith ages of the Moldanubian metasedimentary rocks are largely unknown. Rifting attained oceanic dimensions. However, all of the basins developed between the northern margin of the Saxothuringian basin and main parts of Paleoafrica must have been relatively narrow, because glacial sedimentary rocks relating to the late Ordovician (Saharan) glaciation extend northwards to the Saxothuringian autochthon in North Bavaria and Thuringia. The oldest subduction-related metamorphic rocks exposed in the Massif Central and in the Bohemian Massif record ca. 430 Ma ages (early Silurian according to the time scale of Harland et al. 1989). Therefore, it is not likely that the Saxothuringian and more southerly basins could have extended at a later time. These basins, therefore, most likely represented evolved rifts or narrow oceans.

The Rhenohercynian basin is a relatively young tectonic feature. Accumulation of thick Devonian shelf and hemipelagic deposits, associated with Emsian to early Carboniferous bimodal volcanic rocks, clearly indicate considerable crustal extension. In addition, the widespread occurrence of MORB-type mafic rocks (Lizard-Giessen-Harz nappes) suggests formation of Rhenohercynian ocean floor in the early Devonian.

There was initially no substantial evidence of pre-Devonian separation between the Rhenohercynian basement and areas further south. When it became clear that the Northern Phyllite Zone contained Silurian metavolcanic rocks with arc-like geochemical signatures, Franke (1989a,b) and Franke and Oncken (1990) still preferred to regard these magmatic events as earliest indications of later (essentially Devonian) rifting (Fig. 3A). However, there is increasing evidence suggesting that the Silurian arc-type rocks record closure of an Ordovician (Rheic) precursor to the Devonian basin:

1. Paleomagnetic data suggest that "Avalonia" drifted northward more rapidly than Armorica and Gondwana. As a result, by the late Ordovician, there may have been as much as 33° latitudinal difference between northwestern margins of the Saxothuringian basin and Wales. This would have represented a latitudinal distance of ca. 3500 km. Because ca. 1000 km of continental crust presently exists between these areas, and error circles of nearly 10° allow deviations of ±1000 km, the latitudinal distance between these sites could actually have been as little as 1500 km. This distance must be reduced by conservative estimates of shortening in the intervening continental crust (at least several 100 km: Behrmann et al. 1991). The remaining 1000 km represent a minimum estimate for the latitudinal width of the Ordovician (Rheic) ocean. The actual separation may have been significantly less if part of the latitudinal changes were accomplished by strike-slip motions. On the other hand, the separation suggested by the Ordovician paleopoles appears to be quite realistic because, at a rate of separation of 5 cm/a and a time span from Arenig to Ashgill (i.e., ca. 50 Ma), 2500 km of ocean could have been generated.

2. A number of exotic rock units exposed in southern parts of the Rhenohercynian belt display sedimentological and/or zoogeographic affinities with Armorica. Such rocks occur as Ordovician-early Devonian clasts in mass-flow deposits underlying the Lizard nappe in south Cornwall, in the Gießen nappe in Germany (Paleozoic rocks of the Lindener Mark near Gießen), and in parts of the Northern Phyllite Zone in the Harz Mts. (Wippra Zone). All of these areas were situated north of the suture marking closure of the Devonian basin (i.e., now adjacent to the Rhenohercynian parautochthon: Avalonia). However, they contain sediments and faunas with Armorican affinities. These include Ordovician quartzites (Armorican quartzite), trilobites of the *Neseuretus*-province in the Ordovician quartzites of Cornwall (Sadler 1974), and a similar fauna in the Andreasteich-Quarzit of the Lindener Mark. Silurian and Pragian (= mid-early Devonian) limestones of the Lindener Mark contain benthic ostracodes closely related to the Tepla-Barrandian of the Bohemian Massif. These affinities suggest that the depositional sites of these Ordovician-Pragian exotic rocks, were separated from the Rhenohercynian autochthon (Avalonia).

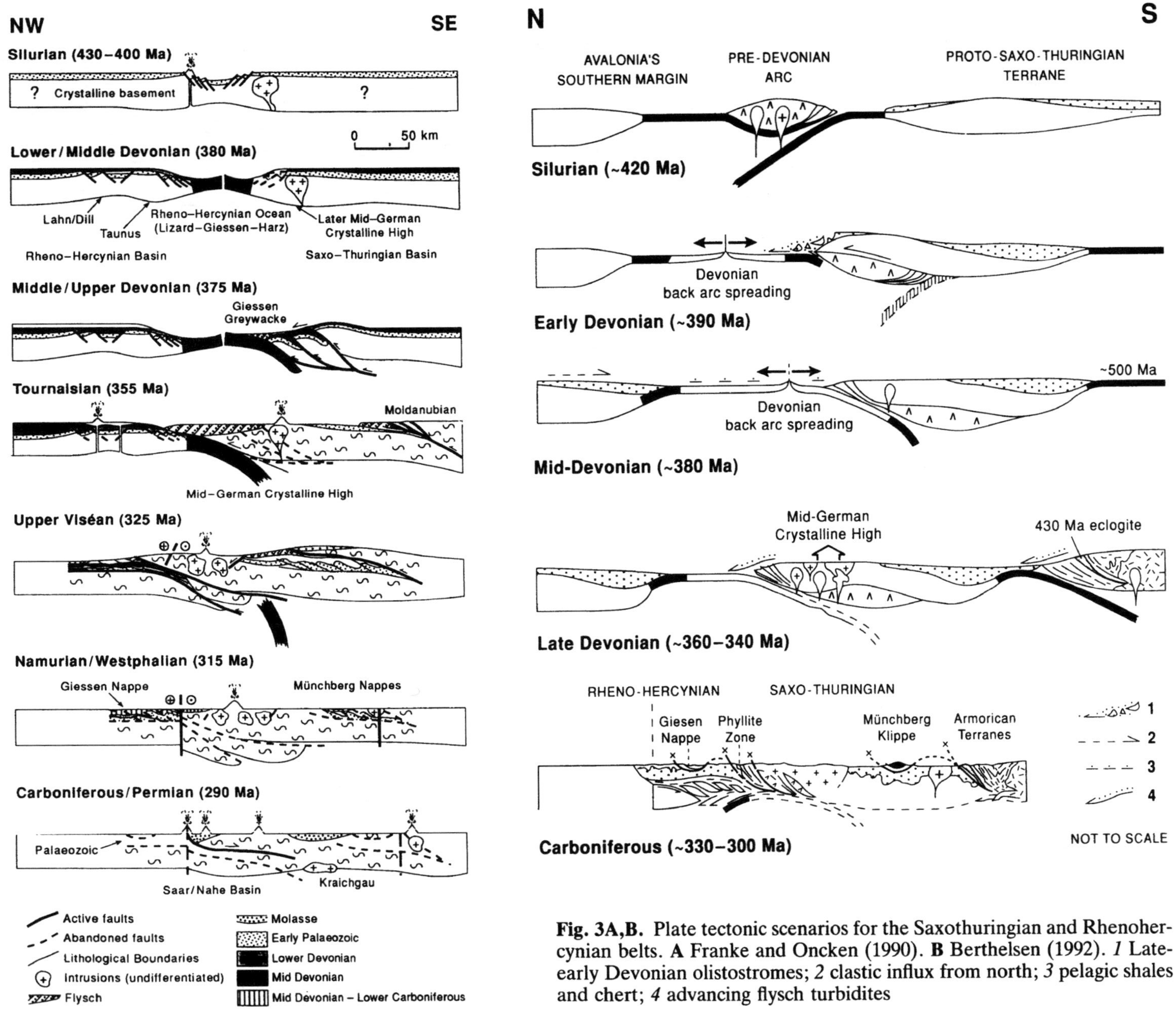

Fig. 3A,B. Plate tectonic scenarios for the Saxothuringian and Rhenohercynian belts. **A** Franke and Oncken (1990). **B** Berthelsen (1992). *1* Late-early Devonian olistostromes; *2* clastic influx from north; *3* pelagic shales and chert; *4* advancing flysch turbidites

The available evidence suggests the following evolution (Fig. 3B):

1. Separation of Avalonia from Armorica in the early Ordovician, with creation of the Rheic ocean.
2. Convergence of Avalonia/Armorica with formation of a magmatic arc (calc-alkaline rocks of the S-Taunus and the Mid-German Crystalline High). Subsequently, this arc may either have accreted to Avalonia, or else accreted/subducted under northern margin of Armorica (this latter option is shown in Fig. 3B); ultimate closure of the Rheic ocean occurred by the end of the Pragian (mid-early Devonian).
3. Renewed extension onward from the early Emsian; parts of north Armorica (olistoliths in south Corwall and Lindener Mark, parts of the Wippra Zone) were stranded along northern margins of the opening Rhenohercynian ocean. Spreading from the Emsian to the early Frasnian (i.e., > ca. 10 Ma) may have produced a latitudinal width of 500 km. The Devonian (Rhenohercynian) basin therefore is undetectable by available paleomagnetic data.
4. Plate convergence onward from the Frasnian.

Behr et al. (1984) noted that Devonian-early Carboniferous extension in the narrow Rhenohercynian basin is difficult to reconcile with compression recorded in sourrounding areas. The area in question was too small to allow comparisons with modern oceans, where spreading along the ridge and subduction under the active margin are systematically associated. It has therefore been proposed, that a mid-European microplate (Armorica plus Avalonia) overrode the spreading center of the lapetus ocean or Tornquist's Sea, and was divided by ridge subduction (Matthews 1978; Franke 1989a). This is compatible with geochemical characteristics of the Rhenohercynian volcanic rocks.

Several workers have suggested back-arc extension to explain the sedimentary and volcanic record of the Rhenohercynian belt (e.g., Ziegler 1990). This hypothesis tacitly assumes the existence of a northward-descending lithospheric slab in the Devonian to early Carboniferous. However, the only potential active margin volcanics presently observed are Silurian (Northern Phyllite Zone). There is no evidence of Devonian ocean crust in southern areas which could have been subducted northwards to create the Rhenohercynian basin.

Devonian back-arc spreading appears to have been possible within the Saxothuringian belt (Fig. 4), where Frasnian volcanism and the rise of metamorphic domes (Saxonian Granulite) record considerable extension. This activity might have related to southeastward subduction of Rhenohercynian ocean crust. As in the Rhenohercynian belt, Devonian extension was time-equivalent with convergence at the southern margin (Tepla-Barrandian), documented in cooling ages and synorogenic clastic sediments.

Some paleomagnetic data (Chap. X) may be taken to suggest that a Devonian ocean existed between Armorica and Africa. However, the APW for the Gondwana mainland is poorly constrained, and the presence of Devonian reef carbonates and identical associations of land

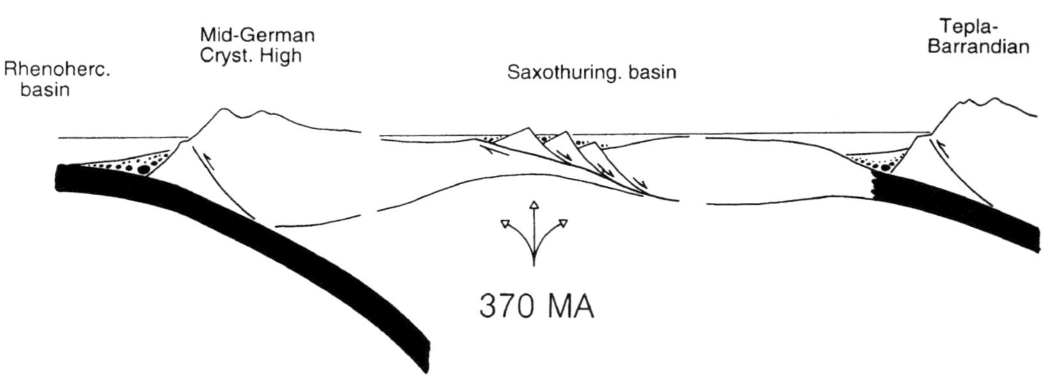

Fig. 4. Plate-tectonic scenario for the Rhenohercynian and Saxothuringian basins in the early upper Devonian (ca. 370 Ma)

plants in northern Africa and Europe argue against a significant separation (see, e.g., McKerrow and Scotese 1990).

4 Onset and Early Stages of Plate Convergence

Early convergence in the Rhenohercynian belt is documented by deposition of late Devonian flysch graywackes on radiolarites and MORB-type basalts. The earliest flysch of this type has largely the same age from Southwest England to the Harz Mountains (Frasnian in the Lizard complex, Frasnian in the S Hunsrück, Frasnian in the Giessen nappe, late Frasnian/Famennian in the Werra uplift, and Frasnian in the Harz Mts). Biostratigraphic uncertainty does not permit documentation of any heterochroneity along the active margin. A cylindroidal facies pattern was maintained into the early Carboniferous, where the front of synorogenic clastic sedimentation consistently parallels the tectonic trends. The timing of foreland basin sedimentation was the same even in distant along-strike equivalents of the Rhenohercynian basin (South Portugal and Moravo-Silesian belt, Franke and Engel 1986).

Subduction-related magmatism occurred within the Mid-German Crystalline High (Odenwald segment), whose 360-Ma calc-alkaline Frankenstein gabbro intruded upper crustal levels of pre-existing continental crust. If metamorphism in the country rocks of the pluton was "early" Variscan (suggested by hornblende cooling ages), uplift may have related to continental underplating during collision. In the Tournaisian, a few turbidity currents already deposited graywacke beds along the northern, passive margin of the Rhenohercynian basin (Chap. III.B.1). If Devonian subsidence in the Saar/Nahe segment of the Mid-German Crystalline High was controlled by extension related to continental escape, the onset of deposition in the middle Devonian may also constrain collisional chronology.

Because the oldest unequivocal indications of a Rhenohercynian ocean are early Devonian, and convergence appears to have commenced in the Frasnian, subduction along the tectonic front was apparently short-lived. This also indicates a rather narrow Rhenohercynian ocean. Northwestward migration of flysch facies and northwestward tectonic vergence in the Rhenohercynian belt indicate a southward-directed subduction.

S of the Saxothuringian basin (Bavaria and eastern Germany), there is no evidence for subduction-related magmatism. In southwestern Germany (northern Vosges), early Carboniferous calc-alkaline plutons postdated the closure of the basin. It is possible that the Saxothuringian basin was too narrow to result in a subduction-related magmatic arc.

Ultimate closure of the Saxothuringian basin is recorded by Famennian graywacke turbidites (Thuringian facies sequence of Erbendorf) developed in southernmost parts of the northwestern, passive basin margin. This is consistent with a tectonic record of transpressional strike-slip in continental basement rocks of juxtaposed units beginning at ca. 380 Ma. Further evidence of collision may be derived from mineral ages in metamorphic rocks of the Tepla-Barrandian terrane. These range between ca. 390 (zircon) and 365 (white mica), and indicate rapid uplift and cooling, probably due to continental underplating and isostatic uplift.

In the Tepla-Barrandian and Moldanubian (s.str.), the existence and position of Paleozoic basins is uncertain. However, mid-late Devonian plutonic rocks exposed south of the Tepla-Barrandian (Kosler et al. 1993) may record northward subduction of the Moldanubian basin floor. Upper time limits for closure of any basin in the Moldanubian (s.str), are provided by mineral ages within the high-grade metamorphic rocks in the Gföhl allochthon, and in shear zones associated with their emplacement. These include U/Pb-ages (zircons) in granulites (ca. 360–340 Ma), and Ar/Ar data from biotites in the shear zone (323 ± 7 Ma, see the compilation in Franke 1989b).

Sedimentary, tectonic, and metamorphic records are mutually compatible, and indicate that any oceanic crust within the Rhenohercynian and Saxothuringian basins must have been subducted by the end of the Devonian. This is also true for the inferred basin within the Moldanubian (s.str.). Clear indications of subduction-related magmatism are only present in the Mid-German Crystalline High.

5 Kinematics During "Main Stage" of Collision

The collisional stage in the European Varicides commenced in the late Devonian or early Carboniferous, and continued into the Early Permian (i.e., > ca. 80 Ma). The tectonometamorphic record of this collisional stage is very complex, and, as yet, poorly resolved. However, two features may be distinguished: (1) older thrusting and strike-slip displacements along the terrane boundaries; (2) younger strike-slip faults cutting across the sutures (Figs. 5, 6).

5.1 Displacements at Terrane Boundaries

Evidence for collisional include the tectonic and sedimentary record of the less metamorphosed external zones and the geological relationships of crystalline, central parts of the Variscan orogen (ductile faulting and thrusting, metamorphism and plutonism). An integrative approach of these contrasting processes requires precise geochronological constraints of all tectonic episodes.

Sedimentary and paleomagnetic records clearly indicate that the Variscan belt represents a terrane collage. Several tectonic fronts developed during plate convergence and collision. The finite state is a fan-shaped array of largely time-equivalent thrust belts, with the Rhenohercynian and Saxothuringian belts verging northwest, and the Moldanubian belt verging southeast (Fig. 5). This bilateral symmetry has been central to all modern tectonic models (starting with Bard et al. 1980), although the relative importance and detailed kinematics of the different belts have remained controversial.

5.1.1 Rhenohercynian Belt and Mid-German Crystalline High

From the late Devonian, the Mid-German Crystalline High acted as an active margin, with Rhenohercynian basin floor subducted southward. As suggested by Wunderlich (1964), a northwestward propagating "orogenic wave" migrated across the Rhenohercynian foreland from the Visean to the Westphalian. This is documented in records of synorogenic clastic sedimentation and isotopic age of cleavage formation (Ahrendt et al. 1977, 1983; Reuter 1987).

The development of the active margin (Mid-German Crystalline High) was complex; segments west of the Rhine river (western Odenwald and eastern Odenwald plus Spessart) appear to have very different tectonometamorphic records. Significant changes occur across the Rhine area, and it is possible that a Devonian transform fault ("Loke Shear" of Berthelsen 1992) remained active in the collisional stage: in more extended crust east of the shear, tectonic shortening could have been more important, and the fracture itself

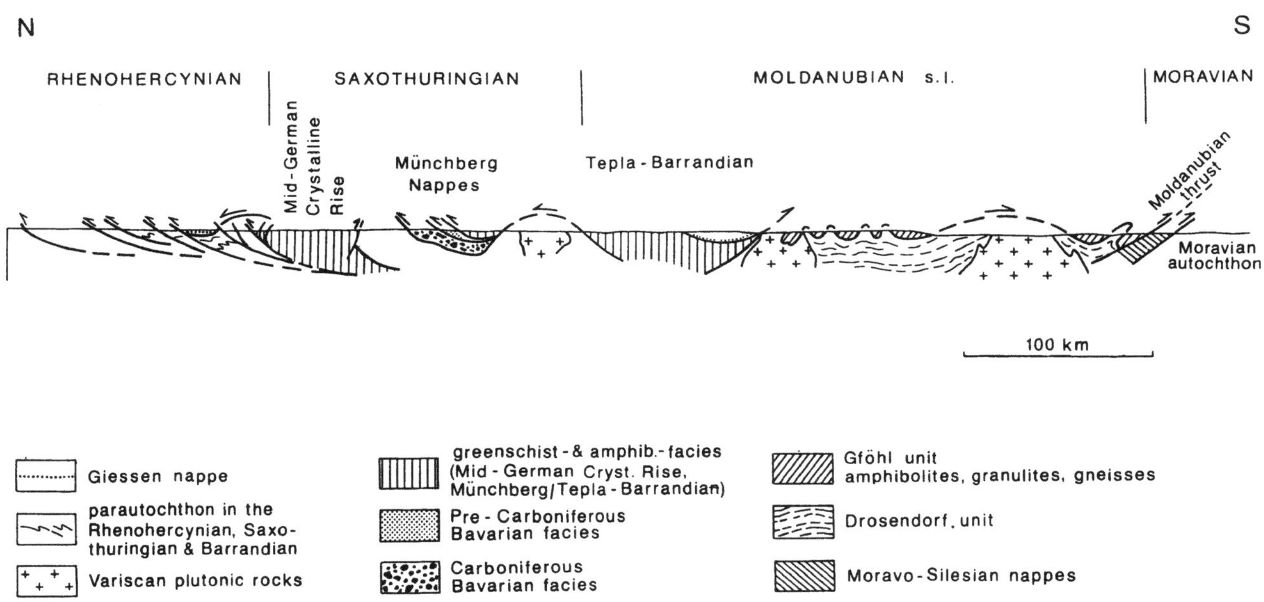

Fig. 5. Structural cross section of the Variscan Belt along a traverse from the Rhenohercynian to the Moravo-Silesian belt. (Behr et al. 1984)

could have been reactivated into a sinistral shear belt (Otzberg fault, Chap. IV.C.4).

The Saar/Nahe area subsided in the middle Devonian to Carboniferous, while the Odenwald was metamorphosed and exhumed. The Saar/Nahe segment of the Mid-German Crystalline High is therefore interpreted as a strike-slip-related feature, either as a transtensional basin controlled by a sinistral strike-slip fault located behind the subduction zone, or as an area of collisional escape.

Southern margins of the Rhenohercynian belt must have undergone important strike-slip displacements. This is documented by the tectonic record as well as by paleogeographic "misfits": In SW England, Devonian rocks north of the Culm Synclinorium consist of thick clastic sediments (Old Red Sandstone facies), whereas hemipelagic sequences exposed south of the Culm Synclinorium are entirely devoid of silt- and sandstones. Likewise, late Devonian flysch graywackes in the S Hunsrück (west of the Rine River) contrast with coeval marine carbonates in the Saar/Nahe segment of the Mid-German Crystalline High.

Strike-slip displacements along southern margins of the Rhenohercynian belt may have been time-equivalent with the northwestward tectonic transport observed to the northwest: both effects could have been reconciled by strain-partitioning in a transpressional setting. However, at least one component of dextral strike-slip has been shown to postdate northwest-directed transport (Oncken 1988).

Important transpressional faulting has also been resolved for the boundary between the Mid-German Crystalline High and adjacent sequences of the Saxothuringian basin. The latter do not record any indications of Silurian arc magmatism. In addition, the Mid-German Crystalline High was an important source for Frasnian to Early Namurian synorogenic graywackes deposited in the Rhenohercynian basin. Some clastic input should also be detected in the Saxothuringian basin. However, the Thuringian facies of the Saxothuringian basin received only minor Devonian quartzite beds of unknown provenance (Franke 1984). More voluminous and proximal clastic fans might have been excised by important, yet unidentified strike-slip displacements.

5.1.2 Saxothuringian Belt and Moldanubian Regions

The Saxothuringian was characterized by development of an early Carboniferous flysch basin, in which tectonic structure and facies patterns reveal an accretionary prism and a basin axis both parallel with the later tectonic trend (southwest/northeast, with northwest-facing tectonic and sedimentary transport). An important component of strike-slip translation occurred along the active margin. Devonian ductile shear, documented in the metamorphic nappes, is best explained in terms of dextral transpression. A later change to northwestward transport directions (Chap. V.C.1) might reflect a change in plate kinematics (from strike-slip to more orthogonal compression), but could likewise be related to strain partitioning in a regime of dextral transpression. Segments originally rooted in the main shear zone could have gradually migrated across the foreland, accreted parts of the footwall, and, as moving away from the main fault, record an increasing component of thrusting normal to the fault zone. In the main fault zone, dextral strike-slip remained dominant (Lapp et al. 1992; Zulauf 1992).

Strike-slip movements are also reflected in the regional distribution of metamorphic grades in the Tepla/Barrandian (Bohemian) terrane. In the center of an extensive synformal structure, Cambrian to Middle Devonian sedimentary rocks of very low metamorphic grade are preserved. This precludes involvement of the entire Tepla-Barrandian terrane in large-scale crustal stacking. Thrusting was apparently limited to margins of the Tepla-Barrandian terrane: southwestern parts were emplaced by southwestward thrusting over the Moldanubian s.str. Along northwestern margins of the Tepla-Barrandian, progressively deeper structural levels are exposed towards the Saxothuringian/Moldanubian suture. This reflects northwestward thrusting over Saxothuringian rocks. Central parts of the Tepla-Barrandian were not underplated and uplifted; instead, they subsided in the Devonian, and received clastic sediments, while high-grade metamorphic rocks now preserved along northwestern margins were uplifted and exhumed. This kinematic development (though not the general setting) is similar to the "steady-state exhumation model" of Jamieson (1991) for the Alpine Fault in New Zealand, where high-grade rocks have been uplifted in a narrow belt at the front of a transpressional zone, whereas the hinterland was affected by relative normal displacement.

Strike-slip movements may also be deduced for southeastern margins of the Tepla-Barrandian terrane, where low-grade metasedimentary rocks are juxtaposed, along the Central Bohemian shear zone, against the high-grade rocks of the Moldanubian s.str. (Gföhl and Drosendorf units). The boundary has been affected by dextral shearing (Matte et al. 1990), with normal faulting components (Seibert et al. 1990). Dextral strike slip is also recorded in the Moldanubian overthrust, which was active in the early Carboniferous (e.g., hornblende and muscovite cooling ages between 341 and 329 Ma, or 329–326 Ma, Dallmeyer et al. 1992).

Prior to strike-slip displacements, the Moldanubian (s.str.) must have undergone important southeastward thrusting (Gföhl over Drosendorf). The probable root zone of the Gföhl nappe(s) is now concealed by the Central Bohemian Pluton (Tollmann 1982; Matte et al. 1990). Similar developments have been recorded from Southwest Germany (Black Forest, Vosges: Flöttmann 1988) and from the French Massif Central.

In summary, it appears that earlier stages of tectonic development were characterized by a bilateral subduction and thrusting (e.g., Flöttmann and Kleinschmidt 1989; Wickert and Eisbacher 1988; Matte et al. 1990; Dallmeyer et al. 1992). Later, dextral strike-slip displacements became increasingly important. Continued underthrusting from both sides under central parts of the orogen must have created an increasingly acute space problem under the central zone. As a result, a kinematic change towards strikeslip occurred. Continued bilateral underthrusting of continental crust is ruled out, where the belts of opposed vergence are directly juxtaposed, without an intervening buffer. This is the case in southwest Germany. The divergent thrust zones (Lalaye/Lubine-Baden-Baden on the north, ligne des klippes-Badenweiler/Lenzkirch on the south) include, between them, a "pop-up" only 50–110 km wide which exposes high-grade Moldanubian rocks (Eisbacher et al. 1989). Both faults have been shown to represent important zones of dextral transpression (Eisbacher et al. 1989; Wickert and Eisbacher 1988; Wickert et al. 1990). This arrangement may be characterized as a crustal-scale flower structure.

5.2 Cross-Cutting Strike-Slip Zones

A sinistral, NNE-trending shear zone occurs between the Vosges and the Black Forest (SW Germany). It is now largely masked by the Rhine Graben, but appears to have displaced the Saxothuringian/Moldanubian boundary (Edel and Fluck 1989). All other cross cutting shear zones in central Europe appear to be dextral. One example is the Thuringian Forest, where early Permian (Rotliegend) basins formed as pull-parts along NW-trending faults. The same explanation may probably be applied to isolated Permocarboniferous basins developed along southwestern margins of the Bohemian Massif (Franconian line, Franke 1989b; Vollbrecht et al. 1989). Together with sinistral, NE-trending faults within and at the SE margin of the Bohemian massif, NW-trending dextral faults probably represent a shear couple recording NS-shortening (Fig. 6; Handler et al. 1991).

A regional-scale dextral wrench zone beneath the Mesozoic foreland SW of the Bohemian Massif has been inferred, by Vollbrecht et al. (1989) and Edel and Weber (in press). Along this fault zone, concealed equivalents of the Tepla-Barrandian in the subsurface of south Germany are supposed to be in contact with the Saxothuringian outcrop of north Bavaria. This implies dextral offset of more than 100 km. However, there is no unequivocal evidence for such large-scale displacements. Figure 6 represents, therefore, the conventional, cylindroidal correlation of the Variscan basement outcrops.

NW-trending fault zones become more important as one approaches the southwestern margin of the East European Platform. The first element of this kind is the Elbe Zone, which is apparent, on geological maps, as a Cretaceous basin associated with a northwest-trending transpressional fault related to the Alpine collision. However, this fault reactivated the late Variscan Mid-Saxonian fault zone, which borders the early Permian Döhlen basin. This fault probably initiated as a dextral shear zone, and was later reactivated as a top-to-SW reverse fault. Dextral offset is apparent from the map patterns: the Rhenohercynian Hörre/Gommern quarzite is displaced for ca. 30 km between the Harz Mts. and the Flechtingen Hills. Correlation of the Krkonosce/Izera dome with the antiform of the Eastern Erzgebirge also implies ca. 50 km dextral offset.

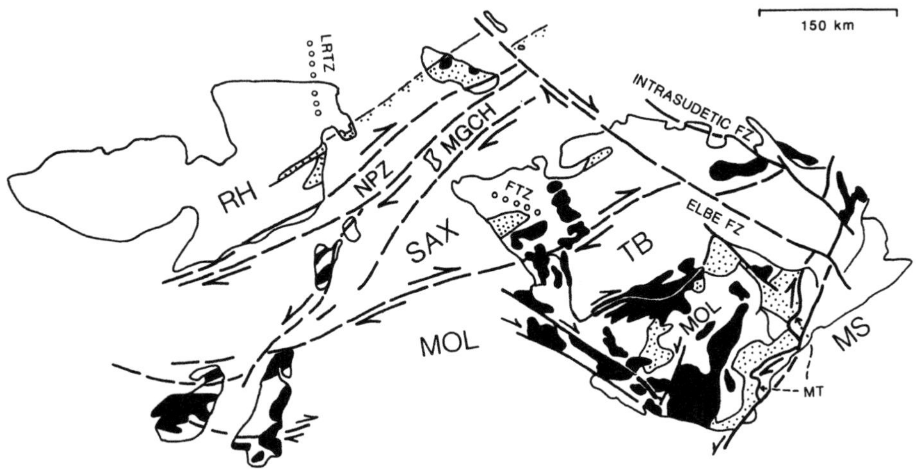

Fig. 6. Simplified structural map, showing the location and sense of shear of main collisional and postcollisional shear zones. *Black* Late- and post-tectonic granites; *stippled* allochthons; *FTZ* Frankenwald transverse zone; *LRTZ* Lippstadt-Ramsbeck transverse zone; *MGCH* Mid-German Crystalline High; *MOL* Moldanubian s.str.; *MS* Moravo-Silesian Zone; *MT* Moldanubian "thrust" (dextral transpression); *NPZ* Northern Phyllite Zone; *RH* Rhenohercynian Zone; *SAX* Saxothuringian basin; *TB* Tepla-Barrandian Zone

The northeastern margin of the Lausitz/Sudetes block is marked by the Intrasudetic Main Fault. With the exception of the Góry Sowie gneiss, areas northeast of the fault are of relatively low metamorphic grade and underwent early Carboniferous deformation and metamorphism. By contrast, parts of the higher-grade terrains to the southwest were affected by early Variscan events (often referred to as Caledonian).

A number of elongate basins with thick marine and continental clastic infill occur along the Intrasudetic Main Fault. These include the Swiebodzice basin (Famennian to Viséan), the Bardo basin (starting in the Famennian, but with strong subsidence only in the Viséan), the Intrasudetic Basin (Tournaisian to early Permian) and the Swierzawa Graben (early Permian). These basins probably originated as pull-apart structures along the Intrasudetic fault zone, which was apparently active at least from the Famennian onwards.

Early stages in the development of the Sudetes resemble that of the Saxothuringian west of the Elbe Zone: Cambro-Ordovician bimodal magmatism (Izera gneiss and Góry Kaczawskie), Devonian bedded cherts (Góry Kaczawskié and Góry Bardzkié), and early Carboniferous wildflysch (Góry Bardzkié), are equivalent to the Bavarian facies. The ca. 380-Ma zircons from the Góry Sowié Gneiss recall medium-pressure rocks of the same age in the Teplá-Barrandian and related terranes. Within the Devonian, development of the Sudetes began to diverge from that of the Saxothuringian: the initial collisional collage into a variably dispersed strike-slip collage. This change possibly reflects the final oblique collision of the Variscan terranes in Central Europe with the East-European Platform. Later, shortening in Central Europe was apparently accommodated by dextral strike-slip along the platform margin.

The importance of dextral strike-slip displacements along northwest/southeast-trending faults has been highlighted by Arthaud and Matte (1977), Le Pichon et al. (1977), Lorenz and Nicholls (1984), Lorenz (1976), and Matte (1986). This kinematic pattern was probably controlled by plate motions of Africa, which first collided with northern blocks, and was then displaced westwards along the east- to eastnortheast-trending Variscan sutures. As a result, old sutures where transformed into strike-slip faults. Dextral displacements at northwest/southeast-trending fault systems (relating to the platform margin) might partly correspond to Riedel shears (Le Pichon et al. 1977). However, the timing and kinematics of the late collisional strike-slip processes has not been resolved in detail.

Another problem exists in correlation of Moravo-Silesian rocks with the Rhenohercynian belt. If this is valid, it might be explained by oroclinal bending around the core region (Tepla-Barrandian) of the Bohemian Massif. Any such plate rotation remains to be proven by paleomagnetic studies in relevant areas.

6 Late- and Post-Tectonic Events

Exhumation of high grade rocks requires an important amount of extensional faulting. These effects are documented in Southwest Germany (Wickert et al. 1990; Eisbacher et al. 1989), and possibly also in the Odenwald and Spessart segments of the Mid-German Crystalline High. In the Bohemian Massif, metamorphic domes of the Saxonian Granulites and the Erzgebirge represent extensional core complexes, but it is quite possible that the very high-grade rocks belong to a pre-Variscan basement. This would greatly reduce the amount of Variscan extension required for exhumation. Extensional faults responsible for exhumation of the Münchberg, Wildenfels, and Frankenberg crystalline nappes remain to be detected. They can be expected to occur in northern parts of the Tepla-Barrandian terrane. Likewise, extensional faults relating to the exhumation of the Moldanubian high-grade nappes have to be postulated: they may be masked by the central Bohemian pluton.

Collisional deformation was accompanied by widespread and important thermal effects. A pervasive low-pressure/high-temperature metamorphism affected all central sectors of the orogen (Moldanubian s.l.) and also parts of the Mid-German Crystalline High. From published P-T paths, it appears that this metamorphism was not only produced by underthrusting, but was significantly aided by rising isograds associated with a high thermal gradient. This thermal regime was maintained throughout the Carboniferous and early Permian.

Apart from minor late Devonian-earliest Carboniferous, subduction-related plutons exposed in the Mid-German Crystalline High, most granitoids were intruded during the early-late Carboniferous. Some were emplaced still in the Earliest Permian. Most granite intrusions post-date ductile deformation of the country rocks. In the Black Forest and Vosges, late lower Carboniferous to Namurian granites were intruded into a crust, which underwent gravitational collapse (subsequent to early Carboniferous convergence) and dextral strike-slip movements (Wickert et al. 1990; Eisbacher et al. 1989). The Central Bohemian Pluton, which probably crystallized in the late Lower Carboniferous (Rb/Sr whole rock, 331 ± 4 Ma, van Breemen et al. 1982), was intruded along the Central Bohemian shear zone (dextral transtensional shear).

Other plutons are oriented northwest/southeast (e.g., the Eibenstock and Leuchtenberg-Falkenberg granites in the Saxothuringian/Moldanubian boundary region). They cut across the suture and are largely undeformed. These cross-cutting plutons possibly reflect extension under northwest-southeast-directed main horizontal stress. The same interpretation may apply to posttectonic magmatic rocks in the Rhenohercynian belt (Brocken and Ramberg plutons in the Harz Mts. and related, northnorthwest-trending dike swarms).

A WNW-trending string of very small posttectonic granite bodies in the western part of the Saxothuringian belt (Frankenwald transverse zone, FTZ in Fig. 6) is also recognizable as a metamorphic anomaly, in which syntectonic minerals indicate high anchizone to greenschist grade (Franke 1984). A similar, NNW-trending syntectonic anomaly occurs between Lippstadt and Ramsbeck in the NE part of the Rhenish Massif (LRTZ in Fig. 6; see Weber 1972; Hoyer et al. 1974; Teichmüller et al. 1979). These syntectonic high-temperature zones apparently relate to convective heating in fracture zones cutting across the Variscan structural grain.

Emplacement of late- and post-tectonic Variscan granites largely postdated LP/HT metamorphism and concomitant tectonic activities. However, they could well represent later expressions of the same thermal event. The tectonic regime appears to have changed, during the late Devonian and Carboniferous, from subduction/collision, to dextral strike-slip and gravitational extension. In the Permian, continued extension accompanied by bimodal volcanism indicates mantle involvement. The thermal processes observed in the late collisional stage could have developed by delamination of continental lithosphere, following subduction and transpression. Delamination would have exposed a previously thickened crust to heating, magmatic intrusion, and renewed extension.

References

Ahrendt H, Hunziker JC, Weber K (1977) K/Ar-Altersbestimmungen an schwach-metamorphen Gesteinen des Rheinischen Schiefergebirges. Z Dtsch Geol Ges 129: 229–247

Ahrendt H, Clauer N, Hunziker JC, Weber K (1983) Migration of folding and metamorphism in the Rheinsiche Schiefergebirge deduced from K/Ar and

Rb/Sr age determinations. In: Martin H, Eder FW (eds) Intracontinental fold belts. Springer, Berlin Heidelberg New York, pp 323–338

Altenberger H, Berch T, Mocek B, Zaipeng Y, Yong S (1990) Geochemie und Geodynamik des Böllsteiner Odenwaldes. Mainzer Geowiss Mitt 19: 183–200

Arthaud F, Matte Ph (1977) Late Palaeozoic strike-slip faulting in southern Europe and northern Africa; result of a right-lateral shear zone between the Appalachians and the Urals. Geol Soc Am Bull 88: 1305–1320

Bard JP, Burg JP, Matte Ph, Ribeiro A (1980) La chaîne hercynienne d'Europe occidentale en termes de tectonique de plaques. In: Cogné J, Jianshy (eds) 26th Int Geol Congr, Paris, C6. Soc Géol du Nord (Villeneuve d'Ascq), BRGM (Orléans), pp 285–299

Behr H-J, Engel W, Franke W, Giese P, Weber K (1984) The Variscan belt in central Europe; main structures, geodynamic implications, open questions. Tectonophysics 109: 15–40

Behrmann J, Drozdzewski G, Heinrichs T, Huch M, Meyer W, Oncken O (1991) Crustal-scale balanced cross-sections through the Variscan fold belt, Germany – the central EGT segment. Tectonophysics 196: 1–21

Berthelsen A (1992) Mobile Europe. In: Blundell D, Freeman R, Mueller St (eds) A continent revealed: the European Geotraverse. Cambridge Univ Press, Cambridge, pp 11–32

Besang C, Harre W, Kreuzer H, Lenz H, Müller P, Wendt I (1976) Radiometrische Datierung, geochemische und petrographische Untersuchungen der Fichtelgebirgsgranite. Geol Jahrb E 8: 3–71

Carl C, Wendt I, Wendt JI (1989) U/Pb whole rock and mineral dating of the Falkenberg granite in northeast Bavaria. Earth Planet Sci Lett 94: 236–244

Cocks LRM, Fortey RA (1982) Faunal evidence for oceanic separations in the Palaeozoic of Britain. J Geol Soc Lond 139: 465–478

Cocks LRM, Fortey RA (1990) Biogeography of Ordovician and Silurian faunas. In: McKerrow WS, Scotese CR (eds) Palaeozoic palaeogeography and biogeography. Geol Soc Lond Mem 12: 97–104

Dallmeyer RD, Neubauer F, Höck V (1992) Chronology of late Paleozoic tectonothermal activity in the southeastern Bohemian Massif, Austria (Moldanubian and Moravo-Silesian zones): $^{40}Ar/^{39}Ar$ mineral age controls. Tectonophysics 210: 135–153

Edel JB, Fluck P (1989) The upper Rhenish Shield basement (Vosges, Upper Rhinegraben, Schwarzwald): main structural features deduced from magnetic, gravimetric and geological data. Tectonophysics 169: 303–316

Edel JB, Weber K (1994) Cadomian terranes, wrench-faulting and thrusting in the Central European Variscides – geophysical and geological evidence. Geol Rundsch (in press)

Eisbacher G, Lüschen E, Wickert F (1989) Crustal-scale thrusting and extension in the Hercynian Schwarzwald and Vosges, Central Europe. Tectonics 8/1: 1–21

Flöttmann T (1988) Strukturentwicklung, P-T-Pfade und Deformationsprozesse im Zentralschwarzwälder Gneiskomplex. Frankfurter Geowiss Arb, A, 6: 1–206

Flöttmann T, Kleinschmidt G (1989) Structural and basement evolution in the Central Schwarzwald Gneiss Complex. In: Emmermann R, Wohlenberg J (eds) The German Continental Deep Drilling Program (KTB). Springer, Berlin Heidelberg New York, pp 265–275

Franke W (1984) Variszischer Deckenbau im Raume der Münchberger Gneismasse, abgeleitet aus der Fazies, Deformation und Metamorphose im umgebenden Paläozoikum. Geotekton Forsch 68: 253 pp

Franke W (1989a) Tectonostratigraphic units in the Variscan Belt of Central Europe. Geol Soc Am Spec Pap 230: 67–90

Franke W (1989b) The geological framework of the KTB drill site. In: Emmermann R, Wohlenberg J (eds) The German Continental Deep Drilling Program (KTB). Springer, Berlin Heidelberg New York: 37–54

Franke W (1992) Phanerozoic structures and events in Central Europe. In: Blundell D, Freeman R, Mueller St (eds) A continent revealed: the European Geotraverse. Cambridge Univ Press, Cambridge: 164–179

Franke W, Engel W (1986) Synorogenic sedimentation in the Variscan Belt of Europe. Bull Soc Géol Fr 1: 25–33

Franke W, Oncken O (1990) Geodynamic evolution of the North-Central Variscides – a comic strip. In: Freeman R, Giese P, Mueller St (eds) The European Geotraverse: integrative studies. Eur Sci Found, pp 187–194

Handler R, Brandmayr M, Wallbrecher E (1991) The Rodl shear zone in the southern Bohemian Massif. Zentralbl Geol Paläontol I (1): 69–86

Hoyer P, Clausen CD, Leuteritz K, Teichmüller R, Thome KN (1974) Ein Inkohlungsprofil zwischen dem Gelsenkirchener Sattel des Ruhrkohlebeckens und dem Ostsauerländer Hauptsattel des Rheinischen Schiefergebirges. Fortschr Geol Rheinland Westf 14: 161–172

Jamieson RA (1991) P-T-t paths of collisional orogens. Geol Rdsch 80/2: 321–332

Hambrey MJ, Harland WB (eds) (1981) Earth's pre-Pleistocene glacial record. Cambridge Univ Press, Combridge, 1004 pp

Harland WB, Armstrong RL, Cox AV, Craig LE, Smith AG, Smith DG (1989) A geologic time scale. Cambridge Univ Press, Cambridge, 263 pp

Köhler H, Müller-Sohnius D, Cammann K (1974) Rb-Sr Altersbestimmungen an Mineral- und Gesamtgesteinsproben des Leuchtenberger und Flossenbürger Granits, NE Bayern. Neues Jahrb Mineral Abh 123 (1): 63–85

Kosler J, Aftalion M, Bowes DR (1993) Mid-late Devonian plutonic activity in the Bohemian Massif: U-Pb zircon isotopic evidence from the Stare Sedlo and Mirotice gneiss complexes, Czech Republic. Neues Jahrb Mineral Abh 9: 41–431

Kossmat F (1927) Gliederung des varistischen Gebirgsbaues. Abh Sächs Geol Landesamt 1: 1–39

Lapp M, Oncken O, Weber K (1992) Strukturelle Entwicklung im Bereich der Tillengebirgszone (TZ) im südlichen Saxothuringikum. Frankf Geowiss Arb Ser A 11: 118–119

Le Pichon X, Sibuet JC, Francheteau J (1977) The fit of the continents around the North Atlantic ocean. Tectonophysics 38: 169–209

Lorenz V (1976) Formation of Hercynian subplates; possible causes and consequences. Nature 262: 374–377

Lorenz V, Nicholls IA (1984) Plate and intraplate processes of Hercynian Europe during the Late Palaeozoic. Tectonophysics 107: 25–56

Matte Ph (1986) Tectonic and plate tectonics model for the Variscan Belt of Europe. Tectonophysics 126: 329–374

Matte Ph, Maluski H, Rajlich P, Franke W (1990) Terrane boundaries in the Bohemian Massif: Results of large-scale Variscan shearing. Tectonophysics 177: 151–170

Matthews SC (1978) Caledonian connexions of Variscan tectonism. Z Dtsch Geol Ges 129: 423–428

McKerrow WS, Scotese CR (eds) (1990) Palaeozoic Palaeogeography and Biogeography. Geol Soc Lond Mem 12: 435 pp

Oncken O (1988) Geometrie und Kinematik der Taunuskammüberschiebung – Beitrag zur Diskussion des Deckenproblems im südlichen Schiefergebirge. Geol Rdsch 77: 551–575

Reuter A (1987) Implications of K-Ar ages of whole-rock and grain-size fractions of metapelites and intercalated metatuffs within an anchizonal terrane. Contrib Mineral Petrol 97: 105–115

Robardet M, Doré F (1988) The late Ordovician diamictite formations from southwestern Europe: North-Gondwana glaciomarine deposits. Palaeogeogr Palaeoclimatol Palaeoecol 66: 19–31

Sadler P (1974) Trilobites from the Gorran Quartzites, Ordovician of S Cornwall. Palaeontology 17: 71–93

Seibert J, Pauli C, Franke W, Behrmann J (1990) Zur Strukturentwicklung und Kinematik der Grenzzone Moldanubikum – Tepla-Barrandium im Bayerischen Wald. Abstr TSK III, Graz 1990, pp 208–210

Schulmann K, Ledru P, Autran A, Melka R, Lardeaux JM, Urban M, Lobkowicz M (1991) Evolution of nappes in the eastern margin of the Bohemian Massif: a kinematic interpretation. Geol Rdsch 80: 73–92

Teichmüller M, Teichmüller R, Weber K (1979) Inkohlung und Illit-Kristallinität. Vergleichende Untersuchungen im Mesozoikum und Paläozoikum von Westfalen. Fortschr Geol Rheinland Westf 27: 201–276

Tollmann A (1982) Großräumiger variszischer Deckenbau im Moldanubikum und neue Gedanken zum Variszikum Europas. Geotekton Forsch 64: 91

Van Breemen O, Aftalion M, Bowes DR, Dudek A, Misar Z, Povondra P, Vrana S (1982) Geochronological studies of the Bohemian Massif, Czechoslovakia, and their significance in the evolution of central Europe. Earth Sci 73: 89–108

Vollbrecht A, Weber K, Schmoll J (1989) Structural model for the Saxothuringian-Moldanubian suture in the Variscan basement of the Oberpfalz (Northeastern Bavaria, F.R.G.) interpreted from geophysical data. Tectonophysics 157: 123–133

Weber K (1972) Kristallinität des Illits in Tonschiefern und andere Kriterien schwacher Metamorphose im Rheinischen Schiefergebirge. Neues Jahrb Geol Paläontol Abh 141: 333–363

Weber K, Vollbrecht A (1989) The crustal structure at the KTB Drill Site, Oberpfalz. In: Emmermann R, Wohlenberg J (eds) The German Continental Deep Drilling Program (KTB). Springer, Berlin Heidelberg New York, pp 5–36

Wendt JI, Kröner A, Fiala J, Todt W (1993) Evidence from zircon dating for existence of approximately 2.1 Ga old crystalline basement in southern Bohemia, Czech Republic. Geol Rundsch (1993) 82: 42–50

Wickert F, Eisbacher GH (1988) Two-sided Variscan thrust tectonics in the Vosges Mountains, northeastern France. Geodin Acta (Paris) 2 (3): 101–120

Wickert F, Altherr R, Deutsch M (1990) Polyphase Variscan tectonics and metamorphism along a segment of the Saxothuringian-Moldanubian boundary: the Baden-Baden Zone, northern Schwarzwald (FRG). Geol Rdsch 79/3: 627–647

Wunderlich HG (1964) Maß, Ablauf und Ursachen der orogenen Einengung am Beispiel des Rheinischen Schiefergebirges, Ruhrkarbons und Harzes. Geol Rdsch 54: 561–582

Ziegler PA (1990) Geological Atlas of Western and Central Europe (2nd and completely revised edition). Shell Int Petroleum Maatschappij – Geol Soc Publ House, 239 pp 56 plates (separate volume)

Zulauf G (1992) Kinematik und tektonometamorphe Entwicklung im Grenzbereich Saxothuringikum/Moldanubikum am Westrand der Böhmischen Masse. Frank Geowiss Arb Ser A 11: 186–189

Subject Index

Nomenclature, Abbreviations

The Nomina Anatomica 1972 term is in italics

1. sulci (s.) and cisterns (c)
(smaller types only are used)

ac	acoustic sulcus
amb c.	ambient or circumpeduncular cistern, between *cisterna interpeduncularis* and quadrigeminal cistern (cisterna venae magnae Galeni or cisterna laminae tecti)
calc	*s. calcarinus,* calcarine sulcus
ca-po	common trunk of calcarine and parieto-occipital sulci (s. antecalcarinus Cunningham), tronc commun des scissures calcarine et pariéto-occipitale
cc	*s. corporis callosi,* pericallosal sulcus, sillon du corps calleux
cent	*s. centralis,* rolandic fissure, sillon de Rolando
chor	*fissura choroidea*
chor c.	choroid cistern. This seems to be the most adequate common name for different "cisterns" described on pneumoencephalographic pictures, such as "wings of ambient cistern", "retropulvinarian cistern", "cisterna veli interpositi", "citerne soustrigonale" etc. designating respectively the lower, posterior and superior parts of a characteristic C shaped air filling in the lateral view (see also the figures and text on choroid and posterior cerebral arteries in Ch. IV-F)
ci	*s. cinguli,* cingulate sulcus, callosomarginal sulcus
ci′	intralimbic sulcus
coll	*s. collateralis*
cu	*s. cunei*
d	sulcus diagonalis Eberstaller, on F_3 (p), between precentral sulcus and anterior vertical branch of sylvian fissure
fi	*s. frontalis inferior*
fm	s. frontalis medius, a more or less well-developed sagittal sulcus often fragmented in several parts, dividing the middle frontal gyrus, especially its anterior part. If well developed, it gives rise to a four gyrus pattern of the frontal lobe (Eberstaller) instead of the classic three convolutions
fma	s. frontomarginalis Wernicke
fs	*s. frontalis superior*
fs′	s. frontalis intermedius on upper surface of superior frontal gyrus
h	*s. hippocampi*
ic	incisura capitis, branch of inferior frontal sulcus, cutting in F_3 (c) from above
ifop	incisure of frontal operculum (Déjerine), s. subcentralis anterior (Eberstaller)
ins	*sulcus circularis insulae,* marginal sulcus of insula
ins (a)	anterior marginal sulcus of insula
ins (i)	inferior marginal sulcus of insula
ins (s)	superior marginal sulcus of insula
ip	*s. intraparietalis*
ip c.	*cisterna interpeduncularis*
ipo	*incisura preoccipitalis,* preoccipital notch
ipop	incisure of parietal operculum (Déjerine) s. subcentralis post. (Eberstaller)
marg	marginal ramus of cingulate sulcus, branche marginale du sillon callosomarginal
olf	*s. olfactorius*
orb	*s. orbitales*
orb (t)	transverse orbital sulcus, central part of Broca's incisure en H, anteriorly convex, roughly parallel to the first transversal part of the sylvian fissure. This anterior boundary of F_3 (o) could be considered as a continuation of the inferior frontal sulcus on orbital lobe
parc	s. paracentralis, anterior boundary of paracentral lobule
pj	s. parietalis intermedius of Jensen, separating supramarginal and angular gyri. Usually, a branch of the intraparietal sulcus
po	*s. parietooccipitalis,* scissure perpendiculaire interne
poc	*s. postcentralis,* postrolandic sulcus
poc (i)	inferior postcentral sulcus
poc (s)	superior postcentral sulcus
prc	*s. precentralis,* prerolandic sulcus
prc (i)	inferior precentral sulcus
prc (s)	superior precentral sulcus
qu c.	quadrigeminal cistern, cisterna venae magnae Galeni
rh	*s. rhinalis*
ro	s. rostralis (Eberstaller), incisure sus-orbitaire (Broca)
sp	*s. subparietalis*
st tr	*s. temporalis transversus.* The deep opercular sulcus posterior to the great or anterior transverse gyrus of Heschl, anterior limit of the planum temporale
sy	*s. lateralis,* sylvian fissure, scissure de Sylvius
sy (f)	*fossa lateralis cerebri* designates the cavity between the orbito-fronto-parieto-temporal opercula and the insula
sy (h)	*ramus anterior,* anterior horizontal branch of sylvian fissure between triangular (F_3c) and orbital

(F$_3$o) parts of inferior frontal gyrus

sy (p) — *ramus posterior* of sylvian fissure, junction of frontoparietal and temporal opercula

sy (p asc) — ascending terminal part of ramus posterior, cutting in deeply the inferior parietal lobule (supramarginal gyrus)

sy (t) — trunk of sylvian fissure

sy (v) — *ramus ascendens*, anterior vertical branch of sylvian fissure between triangular (F$_3$c) and opercular (F$_3$p) parts of inferior frontal gyrus

ti — third of inferior temporal sulcus (N.A.: *sulcus occipitotemporalis*) separating the inferior temporal gyrus from the fusiform gyrus

tm — middle or second temporal sulcus. The latest Nomina Anatomica rebaptized this furrow *sulcus temporalis inferior*. To avoid confusion, however, we continue to use the classical term, found in most textbooks of neuroanatomy. The middle temporal gyrus is thus bounded at its lower limit by a fragmented furrow, the general direction of which is parallel to the superior temporal sulcus. Its terminal posterior part follows similarly an ascending direction, somewhat slanting towards the occipital pole (=anterior occipital sulcus of Wernicke)

ts — *s. temporalis superior* or parallel sulcus

ts (asc) — ascending continuation of ts, sometimes also called angular sulcus, cutting in deeply the inferior parietal lobule

2. Gyri (g) landmarks etc.
(abbreviations begin with capital letters).

Ang — *g. angularis*, pli courbe

Cc — *corpus callosum*

Cc (g) — *genu* } corporis callosi

Cc (spl) — *splenium* } corporis callosi

Cer — cerebellum, cervelet

Ci — *gyrus cinguli* or upper limbic gyrus

Coll — *eminentia collateralis*

Cu — *cuneus*

Dent — *gyrus dentatus* or fascia dentata, corps godronné

F — *fornix*, trigone

F_1 — *g. frontalis superior*, première circonvolution frontale

F_1m — *g. frontalis medialis*, mesial aspect of superior frontal gyrus

F_1 (p) — pied de F_1, posterior end of F_1, anastomosing with precentral gyrus

F_2 — *g. frontalis medius*, deuxième circonvolution frontale

F_2 (p) — pied de F_2, posterior end of F_2, anastomosing with precentral gyrus

F_3 — *g. frontalis inferior*, troisième circonvolution frontale

F_3 (c) — *pars triangularis*, cap de la troisième circonvolution frontale

F_3 (o) — *pars orbitalis*, of g. frontalis inferior, partie orbitaire de la troisième circonvolution frontale

F_3 (p) — *pars opercularis* of g. frontalis inferior, pied de la troisième circonvolution frontale (Broca). (Actually, all three parts of the inferior frontal gyrus are "opercular" insofar as they cover the corresponding part of the insula)

Falx — *falx cerebri*

FM — *foramen interventriculare* Monro

Fus — *g. fusiformis*, fourth temporal gyrus, T_4. Also *called gyrus occipitotemporalis medialis* (NA)

H — *hippocampus*, Ammon's horn, corne d'Ammon

Ins — *insula*

L — *g. lingualis*

Mes — *mesencephalon*

Op (a) — *operculum frontale*, anterior operculum; covering the anterior aspect of insula, it comprizes the orbital and the triangular parts of gyrus frontalis inferior F_3(o) and F_3(c)

Op (s) — *operculum frontoparietale*, superior operculum. Covers the superior aspect of the insula. Extends between anterior and posterior ascending branches of sylvian fissure [sy (v)–sy (p asc)]

Op (i) — aspect of the insula. This is the superior or supratemporal surface of T_1

Op (t) or — *operculum temporale*, inferior operculum covering the inferior

P_1 — *lobulus parietalis superior*

P_2 — *lobulus parietalis inferior*

Parc — *lobulus paracentralis*

Pcu — *precuneus*, lobule quadrilatère

Ped — *pedunculus cerebri*

Ph — *g. parahippocampalis*

Pl t or Pl temp — planum temporale: terminal part of the floor of the sylvian fissure, situated behind the great transverse gyrus of Heschl (from sulcus temporalis transversus principalis). This posterior part of the supratemporal (opercular) surface of the gyrus temporalis superior can show secondary transverse gyri and sulci (generally more pronounced on the left side, as compared to the flatter right planum)

Poc — *g. postcentralis*, posterior central or postrolandic gyrus. Circonvolution pariétale ascendante

Pons — *pons* Varoli, protubérance annulaire

Prc — *g. precentralis*, anterior central or prerolandic gyrus. Circonvolution frontale ascendante.

Pulv — *pulvinar thalami*

Re — *g. rectus*

Sm — *g. supramarginalis*

St — supratemporal–opercular–surface of T_1

St pol — planum polare: concave anterior part of supratemporal surface, anterior to Heschl's gyrus

Stt p — g. temporalis transversus primus: great transversal temporal gyrus of Heschl. Circonvolution temporale profonde antérieure

T_1 — *g. temporalis superior*

T_2 — *g. temporalis medius*

T_3 — *g. temporalis inferior*

Tc or Tent — *tentorium cerebelli*, tente du cervelet

Tci — *incisura tentorii*

Th — *thalamus*

U	*uncus* gyri parahippocampalis. Crochet de l'hippocampe
V_3	*ventriculus tertius*
V_4	*ventriculus quartus*
Va	atrium or trigonum of lateral ventricle, carrefour ventriculaire in the French terminology, where frontal, temporal and occipital horns join
Vf	*cornu anterius ventriculi lateralis,* frontal horn, corne frontale
V lat	*ventriculus lateralis*
Vo	*cornu posterius ventriculi lateralis* occipital horn, corne occipitale
Vt	*cornu inferius ventriculi lateralis,* temporal horn, corne temporale
Pl chor	*plexus choroideus,* choroid plexus

Orientation lines and planes

Ac	*commissura anterior*
Pc	*commissura posterior*
ACPC	bicommissural line of Talairach, joining the upper border of anterior commissure to the lower border of posterior commissure
ACPC-plane	bicommissural plane perpendicular to interhemispheric plane
ACPC //	axial plane parallel to the preceeding
ACPC ⊥	frontal plane perpendicular to both ACPC and interhemispheric planes
CENT //	central or rolandic plane, parallel to the axis of the central sulcus and perpendicular to interhemispheric plane
SY //	plane parallel to the direction of posterior branch of sylvian fissure (not taking into account the ascending terminal part) perpendicular to the interhemispheric plane
SY ⊥	plane perpendicular to SY //
VAC	vertical orientation line, perpendicular to ACPC drawn at posterior border of anterior commissure
VPC	vertical orientation line, perpendicular to ACPC drawn at anterior border of posterior commissure
VAC plane, VPC plane	frontal planes perpendicular to both ACPC and midline

3. Arteries (a) and Veins (v)

To designate cerebral blood vessels, the Nomina Anatomica terms only will be used in this book. In radiologic and in some anatomic descriptions a detailed nomenclature of individual cortical arteries is frequently used. With minor modifications, this nomenclature is derived from the descriptions of Foix and collaborators. For several arterial branches, the names of sulci or gyri are borrowed, as for instance "rolandic artery", "angular gyrus artery", "calcarine artery". Such names, however, imply a specific relationship between the arterial branching pattern and different cortical structures. Perusal of the original descriptions as well as our own material shows the lack of such specific relationships. We therefore prefer to avoid these terms which might be misleading for the reader.

In the light of our studies, angiographic localization of cortical structures is inscribed in the three-dimensional course of the cerebral vascular tree, in the composite deep sulcal vascular laminae and *not* in the highly variable overall course of inconstant individual "arteries".

For the purpose of localization of cortical structures at least, the simple basic division of the Nomina Anatomica seems therefore quite satisfactory.

If, for other purposes, a more detailed nomenclature of cortical blood vessels should be elaborated and convened upon internationally, the use of less specific, regional designations would probably help to avoid misunderstanding. "Central branches of the sylvian artery", seems less equivocal than "rolandic artery", "superomedial branch of the posterior cerebral artery" seems more accurate than "calcarine and parietooccipital arteries" etc.

a. Bas.	*arteria basilaris*
a. CI.	*a. carotis interna*
a. comm. a.	*a. communicans anterior*
a. chor. a.	*a. choroidea anterior*
a. comm. p.	*a. communicans posterior*
a. CA	*a. cerebri anterior*
a. CA (o)	*rami orbitales* arteriae cerebri anterioris. Orbital branches of the anterior cerebral artery ("orbito-frontal artery" in radiologic literature)
a. CA (f)	*rami frontales* arteriae cerebri anterioris. Frontal branches of anterior cerebral artery. (Frontopolar, anterior-internal, middle internal, posterior internal, paracentral lobule arteries in the literature)
a. CA (p)	*rami parietales* arteriae cerebri anterioris. Parietal branches of anterior cerebral artery ("superior internal parietal and inferior internal parietal arteries")
a. CM	a. cerebri media. Middle cerebral or sylvian artery
a. CM (o)	rami orbitales a. cerebri mediae: they leave the sylvian fissure crossing the orbital operculum (F_3o), anterior horizontal branch Sy (h) included ("orbito-frontal artery")
a. CM (f)	*rami frontales* a. cerebri mediae leaving the sylvian fissure in the anterior vertical branch Sy (v) or crossing the frontal operculum ("prefrontal and precentral arteries")
a. CM (p)	*rami parietales* a. cerebri mediae. Parietal branches of sylvian artery. Branches leaving the sylvian fissure through the parietal operculum and its notches, if present, including the posterior ascending branch [sy (p asc)]. (Central, anterior, and posterior parietal, angular gyrus arteries in the literature)
a. CM (t)	*rami temporales* a. cerebri mediae. Temporal branches of sylvian artery, they leave the fissure crossing its temporal lip, from the temporal pole to the end of the posterior limb. ("temporooccipital, posterior, middle, anterior temporal and temporopolar" arteries)
r. str.	*rami striati* a. C.M.
a. CP	*a. cerebri posterior,* posterior cerebral artery
a. CP (t)	*rami temporales,* temporal branches of a. cerebri posterior ("temporooccipital branches")
a. CP (o)	*rami occipitales.* Occipital branches of a. cerebri posterior ("posterior temporo-occipital and calcarine arteries" in the usual descriptions)

a. CP (po) *ramus parietooccipitalis.* Actually, the plurial rami would be more adequate, as there generally are two branches penetrating in the parieto-occipital fissure ("parieto-occipital artery" in the literature)

r. chor. p. *rami choroidei posteriores,* posterior choroid arteries. Several branches of the posterior cerebral artery enter the choroid sulcus

r. chor. p. (1) It is customary to describe a posterolateral and a posteromedial

r. chor. p. (m) choroidal artery, the former going to the plexus of the lateral, the latter to the plexus of the third ventricle

v. bas. *vena basalis,* basal vein of Rosenthal

v.G. *vena cerebri magna* great cerebral vein of Galenus

v.c.i. *vena cerebri interna,* internal cerebral vein

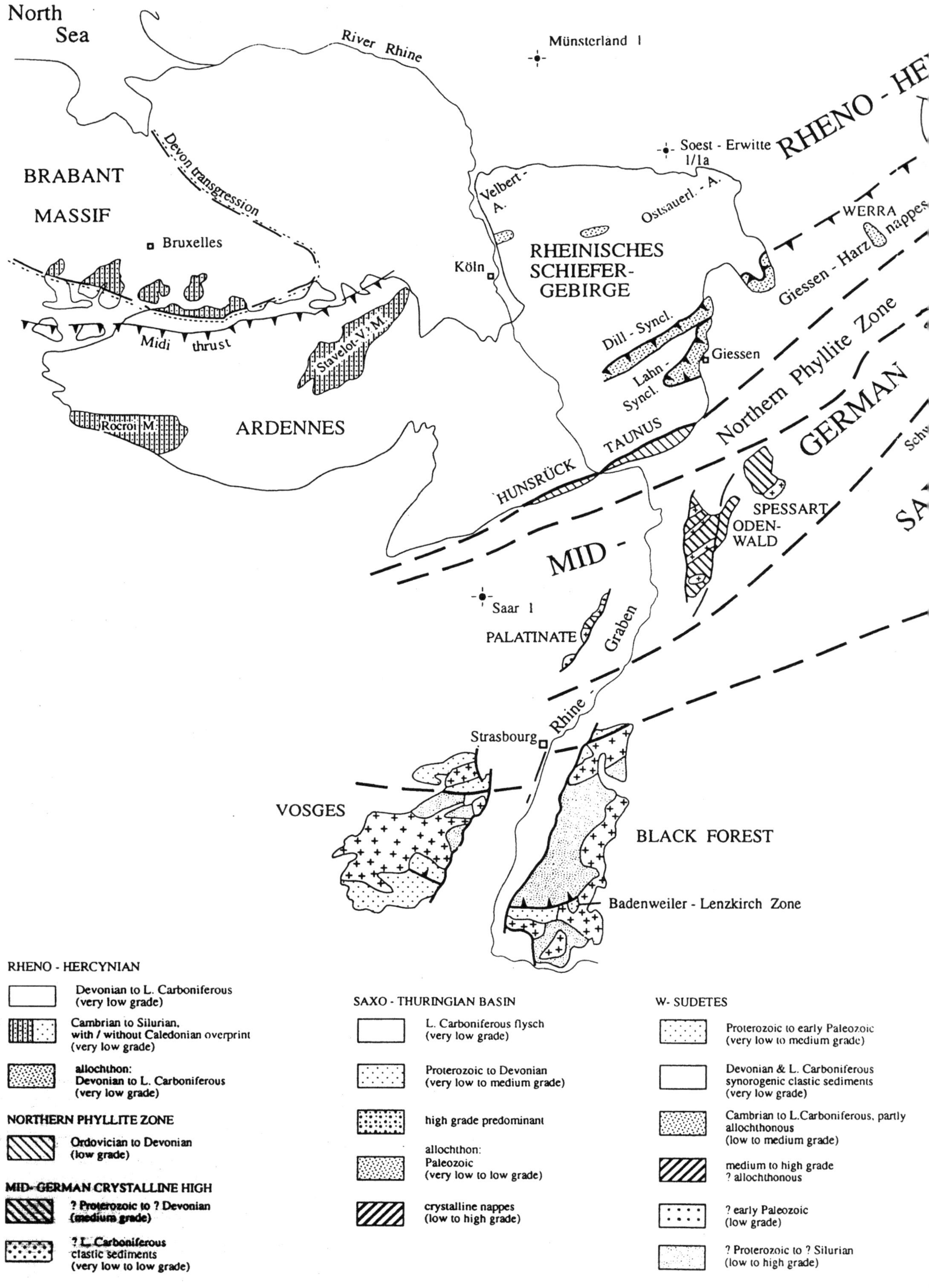

North Sea
River Rhine
Münsterland 1
RHENO - HE
Soest - Erwitte 1/1a
BRABANT MASSIF
Devon transgression
Velbert - A.
Ostsauerl. - A.
WERRA
nappes
Giessen - Harz
Bruxelles
RHEINISCHES SCHIEFER-GEBIRGE
Köln
Dill - Syncl.
Giessen
Lahn - Syncl.
Northern Phyllite Zone
GERMAN
Midi thrust
Stavelot-V. M.
Rocroi M.
ARDENNES
TAUNUS
HUNSRÜCK
SPESSART
ODEN-WALD
SA
MID
Saar 1
PALATINATE
Graben
Rhine
Strasbourg
VOSGES
BLACK FOREST
Badenweiler - Lenzkirch Zone
RHENO - HERCYNIAN
Devonian to L. Carboniferous (very low grade)
Cambrian to Silurian, with / without Caledonian overprint (very low grade)
allochthon: Devonian to L. Carboniferous (very low grade)
NORTHERN PHYLLITE ZONE
Ordovician to Devonian (low grade)
MID-GERMAN CRYSTALLINE HIGH
? Proterozoic to ? Devonian (medium grade)
? L. Carboniferous clastic sediments (very low to low grade)
SAXO - THURINGIAN BASIN
L. Carboniferous flysch (very low grade)
Proterozoic to Devonian (very low to medium grade)
high grade predominant
allochthon: Paleozoic (very low to low grade)
crystalline nappes (low to high grade)
W- SUDETES
Proterozoic to early Paleozoic (very low to medium grade)
Devonian & L. Carboniferous synorogenic clastic sediments (very low grade)
Cambrian to L.Carboniferous, partly allochthonous (low to medium grade)
medium to high grade ? allochthonous
? early Paleozoic (low grade)
? Proterozoic to ? Silurian (low to high grade)

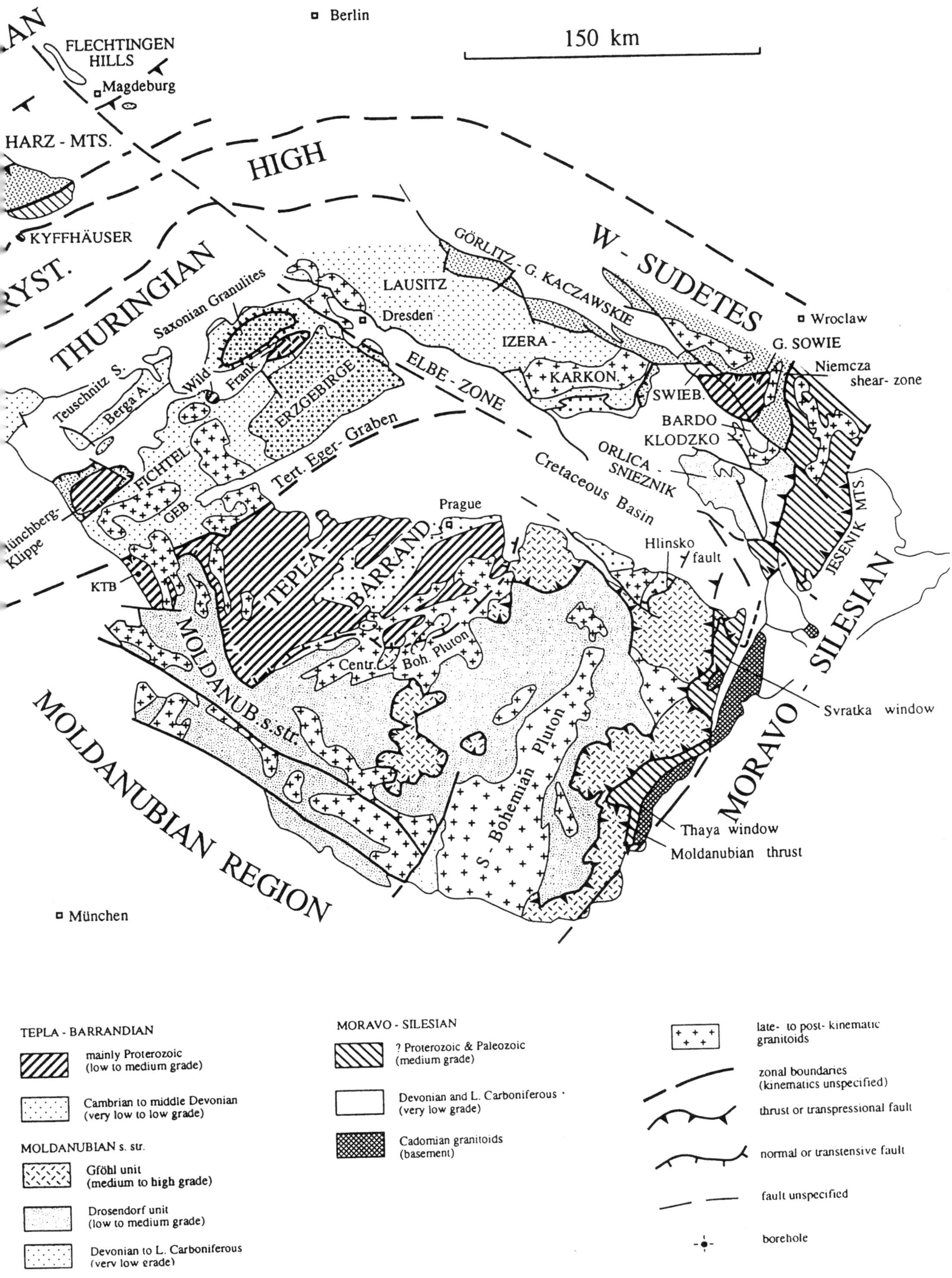

Berlin
150 km
FLECHTINGEN HILLS
Magdeburg
HARZ - MTS.
KYFFHÄUSER
HIGH
W - SUDETES
THURINGIAN
Saxonian Granulites
LAUSITZ
Dresden
GÖRLITZ - G. KACZAWSKIE
Wroclaw
G. SOWIE
IZERA -
KARKON.
SWIEB.
Niemcza shear- zone
Teuschnitz S.
Berga A.
Wild.
Frank.
ERZGEBIRGE
ELBE - ZONE
BARDO
KLODZKO
ORLICA - SNIEZNIK
Tert. Eger- Graben
FICHTEL GEB.
Cretaceous Basin
Münchberg- Klippe
Prague
TEPLA -
BARRAND
Hlinsko fault
JESENIK MTS.
KTB
MOLDANUB.s.str.
Centr. Boh. Pluton
S - Bohemian Pluton
SILESIAN
MORAVO -
Svratka window
MOLDANUBIAN REGION
Thaya window
Moldanubian thrust
München
TEPLA - BARRANDIAN
mainly Proterozoic (low to medium grade)
Cambrian to middle Devonian (very low to low grade)
MOLDANUBIAN s. str.
Gföhl unit (medium to high grade)
Drosendorf unit (low to medium grade)
Devonian to L. Carboniferous (very low grade)
MORAVO - SILESIAN
? Proterozoic & Paleozoic (medium grade)
Devonian and L. Carboniferous (very low grade)
Cadomian granitoids (basement)
late- to post- kinematic granitoids
zonal boundaries (kinematics unspecified)
thrust or transpressional fault
normal or transtensive fault
fault unspecified
borehole

MIX
Papier aus verantwortungsvollen Quellen
Paper from responsible sources
FSC® C105338

If you have any concerns about our products,
you can contact us on
ProductSafety@springernature.com

In case Publisher is established outside the EU,
the EU authorized representative is:
Springer Nature Customer Service Center GmbH
Europaplatz 3, 69115 Heidelberg, Germany

Printed by Libri Plureos GmbH
in Hamburg, Germany